Many-Body Theory Exposed!

Propagator Description of Quantum Mechanics
in Many-Body Systems

2nd Edition

Many-Body Theory Exposed!

Propagator Description of Quantum Mechanics
in Many-Body Systems

2nd Edition

Willem H Dickhoff

Washington University in St Louis, USA

Dimitri Van Neck

Ghent University, Belgium

World Scientific

NEW JERSEY · LONDON · SINGAPORE · BEIJING · SHANGHAI · HONG KONG · TAIPEI · CHENNAI

Published by

World Scientific Publishing Co. Pte. Ltd.

5 Toh Tuck Link, Singapore 596224

USA office: 27 Warren Street, Suite 401-402, Hackensack, NJ 07601

UK office: 57 Shelton Street, Covent Garden, London WC2H 9HE

British Library Cataloguing-in-Publication Data
A catalogue record for this book is available from the British Library.

MANY-BODY THEORY EXPOSED!
Propagator Description of Quantum Mechanics in Many-Body Systems
2nd Edition

ISBN-13 978-981-281-379-4
ISBN-10 981-281-379-9
ISBN-13 978-981-281-380-0 (pbk)
ISBN-10 981-281-380-2 (pbk)

to J

to Lut, Ida, and Cor

Preface

We are thrilled that a second enlarged edition of our book is now available. Positive feedback of colleagues who have used it as a textbook has been very stimulating, and helped motivate us to increase the scope of the material. Naturally, we have used the book with great pleasure in teaching our own quantum many-particle courses. In addition to eliminating typos and mistakes, two chapters have been added. We now include a chapter on finite-temperature Green's functions that contains some recent applications. This canonical material was omitted in the first edition simply because we ran out of steam. The development of the dispersive optical model in nuclear physics has also been included in a new chapter. Such an analysis utilizes the *framework* of the Green's function method to correlate elastic scattering data with complex single-particle potentials that can be related to the self-energy. We hope that the new material will make the book more comprehensive.

In the first edition, we expressed our thanks to numerous teachers, colleagues, and students in an anonymous way. We will take the risk of omission when we express our gratitude now specifically to the following people: Klaas Allaart, Smain Amari, Paul Ayers, Carlo Barbieri, Egbert Boeker, Sigfrido Boffi, Mario Brand, Mark Burnett, Bob Charity, John Clark, Piotr Czerski, Yves Dewulf, Peter Domitrovich, Amand Faessler, Chris Gearhart, Wouter Geurts, Carlotta Giusti, Henry Glyde, Pim Hengeveld, Willem Hesselink, Kris Heyde, Louk Lapikás, Jürgen Meyer-ter-Vehn, Jonathan Morris, Jon Mueller, Herbert Müther, Artur Polls, Marco Radici, Angels Ramos, Güstl Rijsdijk, Avraham Rinat, Arnau Rios, Neil Robertson, Stefan Rombouts, Libby Roth-Stoddard, Jan Ryckebusch, Demetrios Sarantites, Lee Sobotka, Veronique Van Speybroeck, Gerard van der Steenhoven, Manfred Trefz, Brian Vonderfecht, Seth Waldecker, Michel Waroquier, Shi-shu

Wu and Jie Yuan for their invaluable contributions to our understanding of the subject of Green's functions in many-particle physics.

We anticipate unavoidable corrections also to the second edition of the text, but hopefully less than for the first edition. Readers can track these at http://wuphys.wustl.edu/~wimd.

Willem H. Dickhoff, St. Louis
wimd@wuphys.wustl.edu

Dimitri Van Neck, Ghent
Dimitri.VanNeck@UGent.be

January 2008

Preface to the first edition

Surveying the available textbooks that deal with the quantum mechanics of many-particle systems, one might easily arrive at the incorrect conclusion that few new developments have taken place in the last couple of decades. We only mention the recent discovery of Bose–Einstein condensation of dilute vapors of atoms at low temperature to make the point that this is not the case. In addition, coincidence experiments involving electron beams have clarified in wonderful detail the properties of electrons in atoms and protons in nuclei, since the majority of textbooks have been written. Also, most of them do not provide a satisfactory transition from the typical single-particle treatment of quantum mechanics to the more advanced material. Our experience suggests that exposure to the properties and intricacies of many-body systems outside the narrow scope of one's own research can be tremendously beneficial for practitioners as well as students, as does a unified presentation. It usually takes quite some time before a student of this material masters the subject sufficiently so that new research can be initiated. Any reduction of that time facilitated by a student-friendly textbook, therefore appears welcome. For these reasons, we have made an attempt at a systematic development of the quantum mechanics of nonrelativistic many-boson and many-fermion systems.

Some material originated as notes that were made available to students taking an advanced graduate course on this subject. These students typically take a one-year course in graduate quantum mechanics without actually seeing many of the topics that deal with the many-body problem. We note that motivated undergraduate students with one semester of upper-level quantum mechanics are also able to absorb the material, if they are willing to fill some small gaps in their knowledge.

As indicated above, an important goal of the presentation is to provide a unified perspective on different fields of physics. Although details differ greatly when one studies atoms, molecules, electrons in solids, quantum liquids, nuclei, nuclear/neutron matter, Bose–Einstein or fermion condensates, it is helpful to use the same theoretical framework to develop physically relevant approximation schemes. We therefore emphasize the Green's function or propagator method from quantum field theory, which provides this flexibility, and in addition, is formulated in terms of quantities that can often be studied experimentally. Indeed, from the comparison of the calculation of these quantities with data, it is often possible to identify missing ingredients of the applied approximation, suggesting further improvements.

The propagator method is applied to rederive essential features of one- and two-particle quantum mechanics, including eigenvalue equations (discrete spectrum) and results relevant to scattering problems (continuum problem). Employing the occupation number representation (second quantization), the propagator method is then developed for the many-body system. We use the language of Feynman diagrams, but also present the equation of motion method. The important concept of self-consistency is emphasized, which treats all the particles in the system on an equal footing, even though the self-energy and the Dyson equation singles out one of the particles. Atomic systems, the electron gas, strongly correlated liquids including nuclear matter, neutron matter, and helium systems, as well as finite nuclei illustrate various levels of sophistication needed in the description of these systems. We introduce the mean-field (Hartree–Fock) method, random phase approximation (ring diagram summation), summation of ladder diagrams, and further extensions. A detailed presentation of the many-boson problem is provided, containing a discussion of the Gross–Pitaevskii equation relevant to Bose–Einstein condensation of atomic gases. Spectacular features of many-particle quantum mechanics in the form of Bose–Einstein condensation, superfluidity, and superconductivity are also discussed.

Results of these methods are, where possible, confronted with experimental data in the form of excitation spectra and transition probabilities or cross sections. Examples of actual theoretical calculations that rely on numerical calculations are included to illustrate some of the recent applications of the propagator method. We have relied, in some cases, on our own research to present this material for the sole reason that we are familiar with it. References to different approaches to the many-body problem are sometimes included, but are certainly not comprehensive.

The book offers several options for use as an advanced course in quantum mechanics. The first six chapters contain introductory material and can be omitted when it was covered in the standard sequence on quantum mechanics. Starting from Ch. 7, canonical material is developed and supplemented by topics that have not been treated in other textbooks. It is possible to tailor the material to the specific needs of the instructor by emphasizing or omitting sections related to Bose–Einstein condensation, atoms, nuclei, nuclear matter, electron gas, *etc.* In addition to standard problems, we also introduce a few computer exercises to pursue interesting and illustrative calculations. We have attempted a more or less self-contained presentation, but include a sizable list of references for further study. By providing detailed steps, we have tried to reduce the level of frustration many students encounter when first confronting this challenging material. We hope that the book will also be useful to researchers in different fields.

As usual with a text of this kind, it is impossible to cover all available material. We have refrained from discussing important topics in solid state physics, confident that these are more than adequately covered in appropriate textbooks. We have also omitted the finite-temperature formalism of many-body perturbation theory, since it is well-documented in other texts.

It is a pleasure to thank the many colleagues, students, and others who have contributed to the material in this book, in particular those who have collaborated on the research reported here and those from the Department of Subatomic and Radiation Physics at the University of Ghent. Without their scholarship and interest, we would not have been motivated to complete this lengthy project. A special thanks goes to our colleagues who have provided us with data and information that allowed us to construct many of the figures in the text.

We anticipate unavoidable corrections to the text. Readers can track these at `http://wuphys.wustl.edu/~wimd`.

Willem H. Dickhoff, St. Louis
`wimd@wuphys.wustl.edu`

Dimitri Van Neck, Ghent
`Dimitri.VanNeck@UGent.be`

Contents

Chapter 1

Identical particles

In this chapter, we develop some basic concepts associated with identical particles. Section 1.1 presents simple estimates that help to identify under what conditions quantum phenomena associated with identical particles occur. Section 1.2 discusses the theoretical and experimental background which suggest that only certain many-particle states are realized in nature. We briefly review the notation pertaining to one-particle quantum mechanics, and continue with the case of two identical particles in Sec. 1.3. In Sec. 1.4, some illustrative examples are presented that clarify some experimental consequences related to identical particles. Finally, in Sec. 1.5 we construct states with N identical fermions or bosons and discuss their properties.

1.1 Some simple considerations

In a quantum many-body system, particles of the same species are completely indistinguishable. Moreover, even in the absence of mutual interactions, they still have a profound influence on each other, as the number of ways in which the same quantum state can be occupied by two or more particles is severely restricted. This is a consequence of the so-called spin-statistics theorem, further discussed in the next section. One may expect that such effects do not play a role when the number of possible quantum states is much larger than the number of particles, since it is unlikely that two particles would then occupy the same quantum state. This argument provides a rough-and-ready estimate when quantum phenomena, related to identical particles, become important.

Consider the energy levels for a particle of mass m enclosed in a box with volume $V = L^3$,

$$\varepsilon_{n_x,n_y,n_z} = \frac{h^2}{8mL^2}(n_x^2 + n_y^2 + n_z^2), \tag{1.1}$$

where h is Planck's constant and the n_i can be any nonzero positive integer. The number of states $\Omega(E)$ below an energy E is given by

$$\Omega(E) = \frac{\pi}{6}\left(\frac{8mL^2E}{h^2}\right)^{3/2} = \frac{\pi}{6}\left(\frac{8mE}{h^2}\right)^{3/2}V, \tag{1.2}$$

assuming E to be large enough so that $\Omega(E)$ is essentially a continuous function of energy (see *e.g.* [McQuarrie (1976)]). If we take the average energy of a particle to be $E = \frac{3}{2}k_BT$, where k_B is Boltzmann's constant and T the temperature in Kelvin, one can check that in a box with $L = 10$ cm and at $T = 300$ K, the number of states Ω for an atom with mass $m = 10^{-25}$ kg is about 10^{30}. This is much larger than the number of atoms N in the box under normal conditions of temperature and pressure. Generalizing this argument, while requiring $N \ll \Omega$, quantum indistinguishability effects will not play a role when

$$1 \gg Q \equiv \frac{N}{\Omega} = \frac{6}{\pi}\rho\left(\frac{h^2}{12mk_BT}\right)^{3/2}, \tag{1.3}$$

where $\rho = N/V$ is the particle density and Eq. (1.2) was used with E replaced by $\frac{3}{2}k_BT$. Large particle mass, high temperature, and low density favor this condition. Small mass, low temperature, and high density on the other hand favor the appearance of quantum effects associated with identical particles.

The dimensionless quantity Q is listed in Table 1.1 for a number of many-body systems. Even for the lightest atoms and molecules, quantum effects are only expected at low temperatures. However, for electrons in metals, the condition (1.3) is already dramatically violated at 273 K. In a white dwarf star, the temperature is much higher, but a quantum treatment of the electrons is still mandatory because of the extreme density. For the protons and neutrons in nuclei, at a typical nuclear energy scale of about 1 MeV or 10^{10} K, the condition (1.3) is also severely violated. The same holds true for the neutrons in a neutron star at $T = 10^8$ K, which is rather cool according to nuclear standards. A dilute vapor of alkali atoms (rubidium), exhibits a spectacular quantum effect when cooled down to extremely low temperatures: the formation of the so-called Bose–Einstein condensate, recently achieved experimentally [Wieman and Cornell (1995)].

Table 1.1 Q-parameter for different systems.

System	T (K)	Density (m^{-3})	Mass (u)	Q
He (l)	4.2	1.9×10^{28}	4.0	1.1
He (g)	4.2	2.5×10^{27}	4.0	1.4×10^{-1}
He (g)	273	2.7×10^{25}	4.0	2.9×10^{-6}
Ne (l)	27.1	3.6×10^{28}	20.2	1.1×10^{-2}
Ne (g)	273	2.7×10^{25}	20.2	2.5×10^{-7}
e$^-$ Na metal	273	2.5×10^{28}	5.5×10^{-4}	1.7×10^3
e$^-$ Al metal	273	1.8×10^{29}	5.5×10^{-4}	1.2×10^4
e$^-$ white dwarfs	10^7	10^{36}	5.5×10^{-4}	8.5×10^3
p,n nuclear matter	10^{10}	1.7×10^{44}	1.0	6.5×10^2
n neutron star	10^8	4.0×10^{44}	1.0	1.5×10^6
^{87}Rb condensate	10^{-7}	10^{19}	87	1.5

The dimensionless quantity Q, given in Eq. (1.3), for a number of many-body systems, using representative values of densities and temperatures. The mass of the particles is given in atomic mass units (u). Helium and neon are considered at atmospheric pressure, with the liquid phase at boiling point. Electrons in the metals sodium and aluminum can be compared to electrons in white dwarf stars. Protons and neutrons at saturation density of nuclear matter (the density observed in the interior of heavy nuclei) are considered as well as neutrons in the interior of neutron stars. The last entry is the Bose–Einstein condensate of a dilute vapor of ^{87}Rb atoms, magnetically trapped and cooled to ca. 100 nK.

Similar estimates for the importance of quantum effects are obtained by considering the thermal wavelength of a particle given by

$$\lambda_T = \left[\frac{h^2}{2\pi m k_B T} \right]^{1/2} \qquad (1.4)$$

for a particle with mass m and energy $k_B T$. When λ_T^3 becomes comparable to the volume per particle (V/N), one expects the identity of particles to play a significant role.

1.2 Bosons and fermions

Spin and statistics are related at the level of quantum field theory [Streater and Wightman (2000)]. The Dirac equation for a spin-$\frac{1}{2}$ fermion cannot be quantized without insisting that the field operators obey anticommutation relations. These, in turn, lead to Fermi–Dirac statistics and the Pauli exclusion principle for fermions. The latter comprise all fundamental particles with half-integer intrinsic spin. Similarly, the quantization of Maxwell's equations without sources and currents, is only possible when commutation relations between the field operators are imposed, leading to Bose–Einstein statistics. Bosons can be identified by integer intrinsic spin appropriate for fundamental particles like photons and gluons. A wonderful historical perspective on the development of quantum statistics can be found in [Pais (1986)].

Several important many-particle systems contain fermions as their basic constituents. Without recourse to quantum field theory one can treat the consequences of the identity of spin-$\frac{1}{2}$ particles as a result that is based on experiment. Indeed, this is how Pauli came to formulate his famous principle [Pauli (1925)]. By analyzing experimental Zeeman spectra of atoms, he concluded that electrons in the atom could not occupy the same single-particle (sp) quantum state. To incorporate this observation, it is necessary to postulate that quantum states which describe N identical fermions must be antisymmetrical upon interchange of any two of these particles. A similar postulate, requiring symmetric states upon interchange, applies to quantum states of N identical bosons. Here too, experimental evidence can be invoked to insist on symmetric states to account for Planck's radiation law [Pais (1986)]. It appears that only symmetric or antisymmetric many-particle states are encountered in nature.

1.3   Antisymmetric and symmetric two-particle states

To implement these postulates and study their consequences, it is useful to repeat a few simple relations of sp quantum mechanics that also play an important role in a many-particle setting. Texts on Quantum Mechanics where this background material can be found are [Sakurai (1994)], [Messiah (1999)], and [Gottfried and Yan (2004)]. A sp state is denoted in Dirac notation by a ket $|\alpha\rangle$, where α denotes a complete set of sp quantum numbers. For a fermion, α can represent the position quantum numbers, r, its total spin s (which is usually omitted), and m_s the component of its spin along the z-axis. For a spinless boson the position quantum numbers, r, may be appropriate. Many other possible complete sets of quantum numbers can be considered, depending on the specific problem which holds true in a many-particle setting as well. This choice will be further discussed when the independent-particle model is introduced in Ch. 3. To keep the presentation general, the notation $|\alpha\rangle$ will be used. When discussing specific examples, an appropriate selection of sp quantum numbers will be employed.

The sp states form a complete set with respect to some complete set of commuting observables like the position operator, the total spin, and its third component. They are normalized such that

$$\langle \alpha | \beta \rangle = \delta_{\alpha, \beta} \tag{1.5}$$

where the Kronecker symbol is used to include the possibility of δ-function normalization for continuous quantum numbers. For eigenstates of the position operator one has for example

$$\langle \boldsymbol{r}, m_s | \boldsymbol{r}', m_s' \rangle = \delta(\boldsymbol{r} - \boldsymbol{r}') \delta_{m_s, m_s'} \tag{1.6}$$

for a spin-$\frac{1}{2}$ fermion. For a spinless boson

$$\langle \boldsymbol{r} | \boldsymbol{r}' \rangle = \delta(\boldsymbol{r} - \boldsymbol{r}') \tag{1.7}$$

is appropriate in this representation. The completeness of the sp states makes it possible to write the unit operator as

$$\sum_{\alpha} | \alpha \rangle \langle \alpha | = 1. \tag{1.8}$$

In the case of continuous quantum numbers, an integration instead of a summation is required, or a combination of both in the case of a mixed spectrum.

The complex vector space, relevant for N particles, can be constructed as the direct product space of the corresponding sp spaces [Messiah (1999)]. Complete sets of states for N particles are obtained by forming the appropriate product states. The essential ideas can already be elucidated by considering two particles. In this case the notation (note the rounded bracket in the ket)

$$| \alpha_1 \alpha_2) = | \alpha_1 \rangle | \alpha_2 \rangle \tag{1.9}$$

is introduced. The first ket on the right-hand side of this equation refers to particle 1 and the second to particle 2. Such product states obey the following normalization condition

$$(\alpha_1 \alpha_2 | \alpha_1' \alpha_2') = \delta_{\alpha_1, \alpha_1'} \delta_{\alpha_2, \alpha_2'} \tag{1.10}$$

and completeness relation

$$\sum_{\alpha_1 \alpha_2} | \alpha_1 \alpha_2)(\alpha_1 \alpha_2 | = 1. \tag{1.11}$$

While these product states are sufficient for two nonidentical particles, they do not incorporate the correct symmetry required to describe identical bosons or fermions. Indeed, for $\alpha_1 \neq \alpha_2$ we note that

$$|\alpha_2\alpha_1) \neq |\alpha_1\alpha_2), \qquad (1.12)$$

representing a difficulty when interpreting a measurement on the system if the two particles are identical. If α_1 is obtained for one particle and α_2 for the other, it is unclear which of the states in Eq. (1.12) represents the two particles. In fact, the two particles could as well be described by

$$c_1|\alpha_1\alpha_2) + c_2|\alpha_2\alpha_1) \qquad (1.13)$$

which leads to an identical set of eigenvalues when a measurement is performed. This exchange degeneracy presents a difficulty because a specification of the eigenvalues of a complete set of observables does not uniquely determine the state, as expected on the basis of general postulates of quantum mechanics [Dirac (1958)].

We will employ permutation operators to illustrate how the antisymmetrization and symmetrization postulates avoid this difficulty. The permutation operator P_{12} is defined by

$$P_{12}|\alpha_1\alpha_2) = |\alpha_2\alpha_1). \qquad (1.14)$$

While introduced as interchanging the quantum numbers of the particles, this operator can also be viewed as effectively interchanging them. Clearly,

$$P_{12} = P_{21} \text{ and } P_{12}^2 = 1. \qquad (1.15)$$

Consider the Hamiltonian of two identical particles:

$$H = \frac{\boldsymbol{p}_1^2}{2m} + \frac{\boldsymbol{p}_2^2}{2m} + V(|\boldsymbol{r}_1 - \boldsymbol{r}_2|). \qquad (1.16)$$

The observables, like position and momentum, must appear symmetrically in the Hamiltonian, as in the classical case. To study the action of P_{12}, consider an operator A_1 acting on particle 1

$$A_1|\alpha_1\alpha_2) = a_1|\alpha_1\alpha_2) \qquad (1.17)$$

where a_1 is an eigenvalue of A_1 contained in the set of quantum numbers α_1. Similarly, an identical operator A_2 acting on particle 2 will yield

$$A_2|\alpha_1\alpha_2) = a_2|\alpha_1\alpha_2). \qquad (1.18)$$

We note that

$$P_{12}A_1|\alpha_1\alpha_2) = a_1 P_{12}|\alpha_1\alpha_2) = a_1|\alpha_2\alpha_1) = A_2|\alpha_2\alpha_1) \tag{1.19}$$

and

$$P_{12}A_1|\alpha_1\alpha_2) = P_{12}A_1P_{12}^{-1}P_{12}|\alpha_1\alpha_2) = P_{12}A_1P_{12}^{-1}|\alpha_2\alpha_1). \tag{1.20}$$

From these two results, one deduces that

$$P_{12}A_1P_{12}^{-1} = A_2, \tag{1.21}$$

since Eqs. (1.19) and (1.20) hold for any state $|\alpha_1\alpha_2)$. It follows that

$$P_{12}HP_{12}^{-1} = H \tag{1.22}$$

or

$$[P_{12}, H] = 0, \tag{1.23}$$

implying that both operators can be diagonal simultaneously. In the case that $\alpha_1 \neq \alpha_2$, the normalized eigenkets of P_{12} are:

$$|\alpha_1\alpha_2)_+ = \frac{1}{\sqrt{2}}\{|\alpha_1\alpha_2) + |\alpha_2\alpha_1)\} \tag{1.24}$$

and

$$|\alpha_1\alpha_2)_- = \frac{1}{\sqrt{2}}\{|\alpha_1\alpha_2) - |\alpha_2\alpha_1)\}, \tag{1.25}$$

with eigenvalues $+1$ and -1, respectively. While these states normally do not yet correspond to eigenstates of the two-particle Hamiltonian given in Eq. (1.16), they now have the correct symmetry, so that eigenstates of H will be linear combinations of these symmetric or antisymmetric two-particle states, depending on the identity of the particles involved.

We define the symmetrizer

$$\mathcal{S}_{12} = \frac{1}{2}(1 + P_{12}) \tag{1.26}$$

and the antisymmetrizer

$$\mathcal{A}_{12} = \frac{1}{2}(1 - P_{12}). \tag{1.27}$$

When applied to any linear combination of $|\alpha_1\alpha_2)$ and $|\alpha_2\alpha_1)$, these operators will automatically generate the symmetric or antisymmetric state. For

identical fermions, the Pauli exclusion principle results from the requirement that an N-particle state must be antisymmetrical upon interchange
of any two particles. In the case of two particles this implies that the
relevant state is the antisymmetrical one (dropping the $-$ subscript):

$$|\alpha_1\alpha_2\rangle = \frac{1}{\sqrt{2}}\{|\alpha_1\alpha_2\rangle - |\alpha_2\alpha_1\rangle\}. \tag{1.28}$$

This state vanishes when $\alpha_1 = \alpha_2$, thus incorporating Pauli's principle. The
symmetric state for two bosons [Eq. (1.24)] is not yet properly normalized
when $\alpha_1 = \alpha_2$, while demonstrating the possibility that bosons can occupy
the same sp quantum state. The properly normalized two-boson state is
given by

$$|\alpha_1\alpha_2\rangle_S = \left[\frac{1}{2n_\alpha!n_{\alpha'}!...}\right]^{1/2}\{|\alpha_1\alpha_2\rangle + |\alpha_2\alpha_1\rangle\}, \tag{1.29}$$

where n_α denotes the number of particles in sp state α and the notation
$n_\alpha!n_{\alpha'}!...$ represents $\prod_\alpha n_\alpha!$, containing the occupation n_α for each sp state
α. Obviously

$$\sum_\alpha n_\alpha = 2 \tag{1.30}$$

here. From now on, the states for more than one particle which have angular
brackets will denote the antisymmetric or symmetric states. It should also
be noted that as required for fermions

$$|\alpha_2\alpha_1\rangle = -|\alpha_1\alpha_2\rangle \tag{1.31}$$

and both kets therefore represent the same physical state. Only one of
these should be counted when the completeness relation for two identical
fermions is considered. In practice this can be accomplished by ordering the
sp quantum numbers. Suppose one has a set of sp states labeled by discrete
quantum numbers $|1\rangle, |2\rangle, |3\rangle, ...$ etc. For two particles the completeness
relation in terms of antisymmetric states then reads *e.g.*

$$\sum_{i<j} |ij\rangle\langle ij| = 1. \tag{1.32}$$

It is also possible to use an unrestricted sum, if one corrects for the number
of equivalent states

$$\frac{1}{2!}\sum_{ij} |ij\rangle\langle ij| = 1. \tag{1.33}$$

For bosons the completeness relation for ordered sp quantum numbers is also expressed by Eq. (1.32), whereas

$$\sum_{ij} \frac{n_1! n_2! \dots}{2!} |ij\rangle \langle ij| = 1 \tag{1.34}$$

applies for unrestricted sums.

1.4 Some experimental consequences related to identical particles

Scattering experiments represent an ideal tool to illustrate the consequences of dealing with identical particles. Consider two particles that have identical mass and charge, but can be distinguished in some other way, say their color being red or blue. A scattering experiment analyzed in the center of mass of these particles can have two separate outcomes for the same scattering angle. Suppose detectors able to distinguish red and blue, are located in the direction θ (detector D_1) and $\pi - \theta$ (detector D_2) with the z-axis. If the red particle approaches in the z-direction, the (quantummechanical) cross section for the red particle in D_1 and the blue particle in D_2 reads

$$\frac{d\sigma}{d\Omega}(red\ D_1, blue\ D_2) = |f(\theta)|^2, \tag{1.35}$$

where $f(\theta)$ is the scattering amplitude. The cross section for the red particle in D_2 and the blue particle in D_1 is given by

$$\frac{d\sigma}{d\Omega}(red\ D_2, blue\ D_1) = |f(\pi - \theta)|^2. \tag{1.36}$$

If the detectors are colorblind, one cannot distinguish between these processes and the cross section for a count in D_1 becomes the sum of the two probabilities

$$\frac{d\sigma}{d\Omega}(particle\ in\ D_1) = |f(\theta)|^2 + |f(\pi - \theta)|^2. \tag{1.37}$$

With identical bosons both processes cannot, even in principle, be distinguished. This implies that the probability amplitudes must be added before squaring, to obtain the cross section which therefore reads

$$\frac{d\sigma}{d\Omega}(bosons) = |f(\theta) + f(\pi - \theta)|^2, \tag{1.38}$$

Many-body theory exposed!

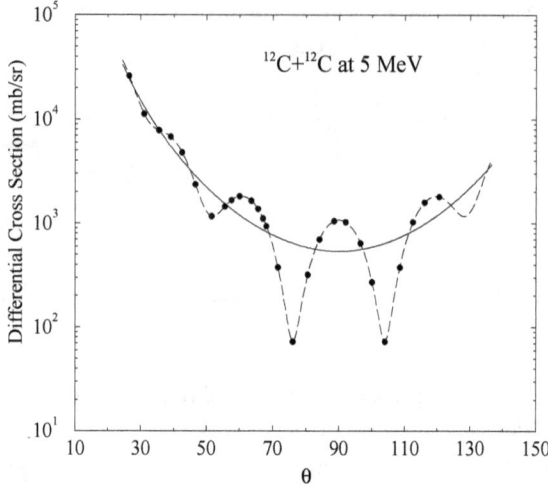

Fig. 1.1   Differential cross section of ^{12}C on ^{12}C scattering at 5.0 MeV center-of-mass energy as a function of the scattering angle θ also in the center of mass. The full curve is obtained from the Coulomb scattering amplitude by using Eq. (1.37) while the dashed line employs the correct expression (1.38) for identical bosons. The experimental data are taken from [Bromley *et al.* (1961)].

and now includes an interference term. The result of the interference is that at $\theta = \pi/2$ the cross section for bosons is twice that for distinguishable particles (but colorblind detectors). This prediction is confirmed by experiment as shown in Fig. 1.1. Here the differential cross section for the scattering of two identical ^{12}C nuclei at low energy (5 MeV) is plotted as a function of the center-of-mass scattering angle. The full line employs the Coulomb scattering amplitude [Sakurai (1994)] according to Eq. (1.37) whereas the dashed line employs Eq. (1.38). The comparison with the data [Bromley *et al.* (1961)] unambiguously points to the identical boson–boson cross section as the correct description. In the case of identical fermions one only observes the interference when both particles have identical spin quantum numbers. The resulting cross section is given by

$$\frac{d\sigma}{d\Omega}(fermions) = |f(\theta) - f(\pi - \theta)|^2 . \qquad (1.39)$$

Consequently no particles will be detected at all at $\theta = \pi/2$! This type of experiment does, however, require the beam and target spins to be polarized in the same direction.

Observe that in Fig. 1.1 the differential cross section for the scattering of composite particles is shown. This example demonstrates that it is necessary to consider these carbon nuclei as identical bosons, at least at those energies where no internal excitation of one or both nuclei can occur. The critical ingredient deciding the identity of this nucleus is the total number of fermions present, 12 here. For an even number of constituent fermions the composite particle behaves as a boson, while it acts as a fermion when this number is odd. Examples for atoms are the ^{3}He and ^{4}He isotopes. In each case the number of electrons and protons is 2. Since ^{3}He has only one neutron, its total number of fermions is odd and a collection of such atoms will behave as identical fermions. The two neutrons in the ^{4}He nucleus are responsible for its boson character. While chemically identical, these He liquids exhibit spectacularly different quantum effects. The same reasoning demonstrates that the ^{87}Rb atom, referred to in Tab. 1.1, represents a boson.

1.5 Antisymmetric and symmetric many-particle states

In dealing with N particles one can proceed in a similar way as in Sec. 1.3 for two particles. Product states are denoted by

$$|\alpha_1\alpha_2...\alpha_N) = |\alpha_1\rangle\,|\alpha_2\rangle\,...\,|\alpha_N\rangle \tag{1.40}$$

with orthogonality in the form

$$(\alpha_1\alpha_2...\alpha_N|\alpha_1'\alpha_2'...\alpha_N') = \langle\alpha_1|\alpha_1'\rangle\langle\alpha_2|\alpha_2'\rangle...\langle\alpha_N|\alpha_N'\rangle$$
$$= \delta_{\alpha_1,\alpha_1'}\delta_{\alpha_2,\alpha_2'}...\delta_{\alpha_N,\alpha_N'}. \tag{1.41}$$

The completeness relation reads

$$\sum_{\alpha_1\alpha_2...\alpha_N} |\alpha_1\alpha_2...\alpha_N)(\alpha_1\alpha_2...\alpha_N| = 1. \tag{1.42}$$

Again, these product states do not incorporate the correct symmetry. A projection onto symmetric or antisymmetric states is therefore required. To accomplished this for fermions we apply the antisymmetrizer for N-particle states

$$\mathcal{A} = \frac{1}{N!}\sum_{p}(-1)^p P, \tag{1.43}$$

where the sum is over all $N!$ permutations. P is a permutation operator for N particles and the sign indicates whether the corresponding permutation is even or odd.[1] A symmetrizer must be used for N identical bosons

$$S = \frac{1}{N!} \sum_p P. \tag{1.44}$$

Normalized antisymmetrical states are then given by

$$|\alpha_1\alpha_2...\alpha_N\rangle = \sqrt{N!}\, \mathcal{A}\, |\alpha_1\alpha_2...\alpha_N), \tag{1.45}$$

while for bosons one obtains

$$|\alpha_1\alpha_2...\alpha_N\rangle = \left[\frac{N!}{n_\alpha! n_{\alpha'}!...}\right]^{1/2} \mathcal{S}\, |\alpha_1\alpha_2...\alpha_N) \tag{1.46}$$

with $\sum_\alpha n_\alpha = N$.

A consequence of this explicit construction of antisymmetric states for N fermions is that no sp state can be occupied by two particles, *i.e.* the quantum numbers represented *e.g.* by α_1 cannot occur twice in any antisymmetric N-particle state. Pauli's exclusion principle is therefore incorporated. For any antisymmetric N-particle state there are $N!$ physically equivalent states generated by a permutation of the sp quantum numbers. Only one physical state corresponds to these $N!$ states. Some details are presented on the next page for three particles.

By using a standard ordering of the sp quantum numbers one can write the completeness relation for N particles as

$$\sum_{\alpha_1\alpha_2...\alpha_N}^{ordered} |\alpha_1\alpha_2...\alpha_N\rangle \langle\alpha_1\alpha_2...\alpha_N| = 1. \tag{1.47}$$

In the case of a 1-dimensional harmonic oscillator this ordering procedure is obvious but in other instances no ambiguity need arise. If no ordering is employed, completeness can be written as

$$\frac{1}{N!} \sum_{\alpha_1\alpha_2...\alpha_N} |\alpha_1\alpha_2...\alpha_N\rangle \langle\alpha_1\alpha_2...\alpha_N| = 1. \tag{1.48}$$

---

[1]The N-particle permutation operator can be written as a product of two-particle permutation operators. The number of the latter terms decides the even or odd character of the N-particle operator.

Example: For three fermions for example, one has

$$|\alpha_1\alpha_2\alpha_3\rangle = \frac{1}{\sqrt{6}}\{|\alpha_1\alpha_2\alpha_3\rangle - |\alpha_2\alpha_1\alpha_3\rangle + |\alpha_2\alpha_3\alpha_1\rangle$$
$$- |\alpha_3\alpha_2\alpha_1\rangle + |\alpha_3\alpha_1\alpha_2\rangle - |\alpha_1\alpha_3\alpha_2\rangle\}.$$

It should be clear now that antisymmetry upon interchange of any two particles is included since

$$|\alpha_1\alpha_2\alpha_3\rangle = -|\alpha_2\alpha_1\alpha_3\rangle$$

and so on. For three bosons a possible state reads

$$|\alpha_1\alpha_1\alpha_2\rangle = \frac{1}{\sqrt{3!2!}}\{|\alpha_1\alpha_1\alpha_2\rangle + |\alpha_1\alpha_1\alpha_2\rangle + |\alpha_1\alpha_2\alpha_1\rangle$$
$$+ |\alpha_2\alpha_1\alpha_1\rangle + |\alpha_2\alpha_1\alpha_1\rangle + |\alpha_1\alpha_2\alpha_1\rangle\}$$
$$= \frac{1}{\sqrt{3}}\{|\alpha_1\alpha_1\alpha_2\rangle + |\alpha_1\alpha_2\alpha_1\rangle + |\alpha_2\alpha_1\alpha_1\rangle\}.$$

Symmetry upon interchange of any two particles is again incorporated since

$$|\alpha_1\alpha_1\alpha_2\rangle = +|\alpha_1\alpha_2\alpha_1\rangle$$

and so on.

Normalization for states with ordered sp quantum numbers has the form

$$\langle\alpha_1\alpha_2...\alpha_N|\alpha_1'\alpha_2'...\alpha_N'\rangle = \langle\alpha_1|\alpha_1'\rangle\langle\alpha_2|\alpha_2'\rangle...\langle\alpha_N|\alpha_N'\rangle$$
$$= \delta_{\alpha_1,\alpha_1'}\delta_{\alpha_2,\alpha_2'}...\delta_{\alpha_N,\alpha_N'}, \tag{1.49}$$

whereas, if the sp states are not ordered, the result is obtained in the form of a determinant

$$\langle\alpha_1\alpha_2...\alpha_N|\alpha_1'\alpha_2'...\alpha_N'\rangle = \begin{vmatrix} \langle\alpha_1|\alpha_1'\rangle & \langle\alpha_1|\alpha_2'\rangle & ... & \langle\alpha_1|\alpha_N'\rangle \\ \langle\alpha_2|\alpha_1'\rangle & \langle\alpha_2|\alpha_2'\rangle & ... & \langle\alpha_2|\alpha_N'\rangle \\ \vdots & \vdots & \ddots & \vdots \\ \langle\alpha_N|\alpha_1'\rangle & \langle\alpha_N|\alpha_2'\rangle & ... & \langle\alpha_N|\alpha_N'\rangle \end{vmatrix}. \tag{1.50}$$

The normalized N-particle wave function of an antisymmetric state is given by

$$\psi_{\alpha_1\alpha_2...\alpha_N}(x_1x_2...x_N) = (x_1x_2...x_N|\alpha_1\alpha_2...\alpha_N\rangle, \tag{1.51}$$

where

$$(x_1 x_2 ... x_N| = \langle x_1 | \langle x_2 | ... \langle x_N | \tag{1.52}$$

and $x_1 = \{r_1, m_{s_1}\}$. Often this wave function is written in determinantal form

$$\psi_{\alpha_1 \alpha_2 ... \alpha_N}(x_1 x_2 ... x_N) = \frac{1}{\sqrt{N!}} \begin{vmatrix} \langle x_1 | \alpha_1 \rangle & ... & \langle x_N | \alpha_1 \rangle \\ \langle x_1 | \alpha_2 \rangle & ... & \langle x_N | \alpha_2 \rangle \\ \vdots & \ddots & \vdots \\ \langle x_1 | \alpha_N \rangle & ... & \langle x_N | \alpha_N \rangle \end{vmatrix}. \tag{1.53}$$

This wave function is commonly called a Slater determinant [Slater (1929)]. Exchange of rows and columns in such a determinant does not change their practical use, so both conventions are found in the literature. In practice it is very cumbersome to work with Slater determinants to calculate matrix elements of operators between many-particle states. For this reason a more practical method is introduced in the next chapter.

There is no restriction on the occupation of sp states for N-boson states. In fact, *all* particles can occupy the same sp state! For a given symmetric N-particle state there are $N!$ physically equivalent states, resulting from a permutation of the sp quantum numbers. In addition, one can have multiple occupation of a sp state. Such states should only be counted once in the completeness relation. In an unrestricted sum over quantum numbers for $N = 3$ all states

$$|\alpha_1 \alpha_1 \alpha_2\rangle = |\alpha_1 \alpha_2 \alpha_1\rangle = |\alpha_2 \alpha_1 \alpha_1\rangle \tag{1.54}$$

occur. The appropriate weighting of these states is obtained by including factorial factors $n_\alpha!$ in the completeness relation as follows

$$\sum_{\alpha_1 \alpha_2 ... \alpha_N} \frac{n_\alpha! n_{\alpha'}! ...}{N!} |\alpha_1 \alpha_2 ... \alpha_N\rangle \langle \alpha_1 \alpha_2 ... \alpha_N| = 1. \tag{1.55}$$

When ordering of the sp states is introduced, no such factors need be included

$$\sum_{\alpha_1 \alpha_2 ... \alpha_N}^{ordered} |\alpha_1 \alpha_2 ... \alpha_N\rangle \langle \alpha_1 \alpha_2 ... \alpha_N| = 1 \tag{1.56}$$

as for fermions. Normalization for states with ordered sp quantum numbers has the form

$$\langle \alpha_1 \alpha_2...\alpha_N | \alpha_1' \alpha_2'...\alpha_N' \rangle = \langle \alpha_1 | \alpha_1' \rangle \langle \alpha_2 | \alpha_2' \rangle ... \langle \alpha_N | \alpha_N' \rangle$$
$$= \delta_{\alpha_1,\alpha_1'} \delta_{\alpha_2,\alpha_2'} ... \delta_{\alpha_N,\alpha_N'}, \tag{1.57}$$

whereas if the sp states are not ordered, we find

$$\langle \alpha_1 \alpha_2..\alpha_N | \alpha_1' \alpha_2'..\alpha_N' \rangle = \frac{1}{[n_\alpha!..n_{\alpha'}!..]^{1/2}} \sum_P \langle \alpha_1 | \alpha_{p_1}' \rangle \langle \alpha_2 | \alpha_{p_2}' \rangle .. \langle \alpha_N | \alpha_{p_N}' \rangle. \tag{1.58}$$

The sum on the right-hand side is called a permanent. The normalized N-particle wave function of a symmetric state becomes

$$\psi_{\alpha_1 \alpha_2...\alpha_N}(x_1 x_2...x_N) = (x_1 x_2...x_N | \alpha_1 \alpha_2...\alpha_N). \tag{1.59}$$

1.6 Exercises

(1) Determine the expectation value of the kinetic energy for N particles, $T_N = \sum_{i=1}^{N} \frac{p_i^2}{2m}$, in terms of the relevant single-particle matrix elements, by employing the Slater determinant given in Eq. (1.53).
(2) Suppose that the single-particle Hilbert space has finite dimension D and is spanned by an orthonormal basis set $\{|\alpha\rangle\}$, $\alpha = 1, \ldots, D$. What is the dimension of the N-fermion space? Comment on the result that the same dimension is found for $D - N$ fermions.

Chapter 2

Second quantization

In this chapter, we present a method that greatly facilitates working with many-fermion or many-boson states. For this purpose, the fermion addition operator is defined in Sec. 2.1 and the Fock space is introduced. After determining the action of the adjoint of the particle addition operator, we proceed to derive the important anticommutation relations among these operators. Many-particle states with the correct symmetry properties can quite easily be constructed by acting with these operators on the state without particles, the so-called vacuum state. Similar results for bosons are presented in Sec. 2.2. The form of one- and two-body operators in terms of particle addition and removal operators is discussed in Secs. 2.3 and 2.4, respectively. Some basic applications follow in Sec. 2.5.

2.1 Fermion addition and removal operators

Dealing with symmetric or antisymmetric many-particle states is simplified considerably by the use of the occupation number representation (second quantization). In this section the relevant concepts for fermions are presented. A key point is not to work in the space of a fixed number of particles. Instead, the vector space is employed which is the direct sum of the vacuum state with no particles $|0\rangle$, the complete set of sp states $\{|\alpha\rangle\}$, the complete set of antisymmetric two-particle states $\{|\alpha_1\alpha_2\rangle\}$, and so on until infinite particle number. This space is referred to as Fock space. Completeness of the states in this space, using ordered sp quantum numbers, is expressed by

$$\sum_{N=0}^{\infty} \sum_{\alpha_1\alpha_2..\alpha_N}^{ordered} |\alpha_1\alpha_2...\alpha_N\rangle \langle\alpha_1\alpha_2...\alpha_N| = 1. \qquad (2.1)$$

States with different particle number are automatically orthogonal.

   An important quantity is the fermion addition operator, often called a creation operator. It is defined by

$$a_\alpha^\dagger \, |\alpha_1\alpha_2...\alpha_N\rangle \equiv |\alpha\alpha_1\alpha_2...\alpha_N\rangle \qquad (2.2)$$

and adds a particle with quantum numbers α to an antisymmetric state in which N particles occupy sp levels $\{\alpha_1, \alpha_2, ..., \alpha_N\}$. An antisymmetric $N+1$-particle state is the result. Note that if the level characterized by α is already occupied, the outcome is zero. Observe also that the $N+1$-particle state containing α may not yet be ordered and the ordering of α among the α_i could consequently result in an extra minus sign.

   The adjoint of $a_\alpha^\dagger$ is called a particle removal (destruction) operator based on the following result

$$a_\alpha \, |\alpha_1\alpha_2...\alpha_N\rangle = \sum_{M=0}^{\infty} \overset{ordered}{\sum_{\alpha_1'\alpha_2'..\alpha_M'}} |\alpha_1'\alpha_2'..\alpha_M'\rangle \, \langle\alpha_1'\alpha_2'..\alpha_M'| \, a_\alpha \, |\alpha_1\alpha_2..\alpha_N\rangle$$

$$= \sum_{M=0}^{\infty} \overset{ordered}{\sum_{\alpha_1'\alpha_2'..\alpha_M'}} |\alpha_1'\alpha_2'..\alpha_M'\rangle \, \langle\alpha_1\alpha_2..\alpha_N| \, a_\alpha^\dagger \, |\alpha_1'\alpha_2'..\alpha_M'\rangle^*$$

$$= \sum_{M=0}^{\infty} \overset{ordered}{\sum_{\alpha_1'\alpha_2'..\alpha_M'}} |\alpha_1'\alpha_2'..\alpha_M'\rangle \, \langle\alpha_1\alpha_2..\alpha_N|\alpha\alpha_1'\alpha_2'..\alpha_M'\rangle^*. \qquad (2.3)$$

We arrive at the last line by using the definition of the particle addition operator given in Eq. (2.2). In addition, $M = N-1$, since states containing different particle number are orthogonal. As discussed in Sec. 1.5, the normalization of antisymmetric states can be given in terms of real numbers. The complex conjugation sign in Eq. (2.3) can thus be omitted. It is also clear that once α has been ordered among the α' states, Eq. (1.49) can be applied. Suppose α must be placed before α_i'. If $i = 1$, no sign change will occur, ordering therefore leads to the phase $(-1)^{i-1}$. Equation (1.49) then gives

$$\langle\alpha_1\alpha_2...\alpha_N|\alpha_1'\alpha_2'...\alpha\alpha_i'...\alpha_M'\rangle = \delta_{\alpha_1,\alpha_1'}\delta_{\alpha_2,\alpha_2'}...\delta_{\alpha_i,\alpha}\delta_{\alpha_{i+1},\alpha_i'}...\delta_{\alpha_N,\alpha_{N-1}'}. \qquad (2.4)$$

As a result, we find

$$a_\alpha \, |\alpha_1\alpha_2...\alpha_N\rangle = (-1)^{i-1} \, |\alpha_1\alpha_2...\alpha_{i-1}\alpha_{i+1}...\alpha_N\rangle \qquad \text{if } \alpha = \alpha_i, \qquad (2.5)$$

and

$$a_\alpha \, |\alpha_1\alpha_2...\alpha_N\rangle = 0 \qquad \text{if } \alpha \neq \alpha_i, i = 1, ..., N. \qquad (2.6)$$

As a consequence, one also has

$$a_\alpha \, |0\rangle = 0. \qquad (2.7)$$

The operator a_α therefore has the property that its action upon an anti-symmetric N-particle state produces an *antisymmetric* $N-1$-particle state, provided the sp state α is occupied (otherwise the result is zero).

The fermion addition and removal operators obey the following, extremely important, operator relations (sometimes called fundamental anti-commutation relations):

$$\{a_\alpha, a_\beta^\dagger\} = a_\alpha a_\beta^\dagger + a_\beta^\dagger a_\alpha = \delta_{\alpha,\beta}, \qquad (2.8)$$

$$\{a_\alpha, a_\beta\} = \{a_\alpha^\dagger, a_\beta^\dagger\} = 0. \qquad (2.9)$$

We now present a typical analysis to obtain one of these results. For an N-particle ket in which α is not occupied, we have

$$a_\alpha a_\alpha^\dagger \, |\alpha_1\alpha_2...\alpha_N\rangle = a_\alpha \, |\alpha\alpha_1\alpha_2...\alpha_N\rangle = |\alpha_1\alpha_2...\alpha_N\rangle . \qquad (2.10)$$

In addition

$$a_\alpha^\dagger a_\alpha \, |\alpha_1\alpha_2...\alpha_N\rangle = 0. \qquad (2.11)$$

These two results combined, show that

$$\{a_\alpha, a_\alpha^\dagger\} \, |\alpha_1\alpha_2...\alpha_N\rangle = |\alpha_1\alpha_2...\alpha_N\rangle . \qquad (2.12)$$

When the N-particle ket does contain the sp state α, we can assume without loss of generality that $\alpha_1 = \alpha$. Then

$$a_\alpha a_\alpha^\dagger \, |\alpha\alpha_2...\alpha_N\rangle = 0 \qquad (2.13)$$

and

$$a_\alpha^\dagger a_\alpha \, |\alpha\alpha_2...\alpha_N\rangle = a_\alpha^\dagger \, |\alpha_2...\alpha_N\rangle = |\alpha\alpha_2...\alpha_N\rangle \qquad (2.14)$$

which yield

$$\{a_\alpha, a_\alpha^\dagger\} \, |\alpha\alpha_2...\alpha_N\rangle = |\alpha\alpha_2...\alpha_N\rangle . \qquad (2.15)$$

Since this procedure can be applied for any N and, as shown, for fixed N for any state, Eq. (2.8) holds for $\alpha = \beta$. A similar strategy works for the proof of the other identities.

Antisymmetric N-particle states can now be generated by repeated application of particle addition operators to the vacuum state

$$
|\alpha_1\alpha_2\alpha_3...\alpha_N\rangle = a^\dagger_{\alpha_1} |\alpha_2\alpha_3...\alpha_N\rangle = a^\dagger_{\alpha_1} a^\dagger_{\alpha_2} |\alpha_3...\alpha_N\rangle = ...
$$
$$
= a^\dagger_{\alpha_1} a^\dagger_{\alpha_2}...a^\dagger_{\alpha_N} |0\rangle = \prod_i a^\dagger_{\alpha_i} |0\rangle . \tag{2.16}
$$

Note that Eq. (2.9) automatically ensures that the Pauli principle is incorporated in the above construction. Indeed, one can write for example

$$
|\alpha_1\alpha_2...\alpha_N\rangle = a^\dagger_{\alpha_1} a^\dagger_{\alpha_2}...a^\dagger_{\alpha_N} |0\rangle = -a^\dagger_{\alpha_2} a^\dagger_{\alpha_1}...a^\dagger_{\alpha_N} |0\rangle
$$
$$
= -|\alpha_2\alpha_1...\alpha_N\rangle , \tag{2.17}
$$

which shows that the state with $\alpha_1 = \alpha_2$ does not exist. Another useful notation for the antisymmetric states in Fock space identifies which sp states are present in the ket. The corresponding occupation numbers can be zero or one for fermions as in the following example

$$
|n_{\alpha_1} = 1, n_{\alpha_2} = 0, n_{\alpha_3} = 1, 0, ..., 0, ...\rangle = |\alpha_1\alpha_3\rangle . \tag{2.18}
$$

This notation can be used for any state in Fock space that corresponds to an antisymmetrized direct product state, and illustrates that antisymmetric states form the basis in the occupation number representation for identical fermions.

2.2 Boson addition and removal operators

In dealing with boson addition and removal operators, it is convenient to use the notation that characterizes the occupation of each sp state

$$
|\alpha_1\alpha_2...\alpha_N\rangle = \left[\frac{N!}{n_\alpha! n_{\alpha'}!...} \right]^{1/2} \mathcal{S} |\alpha_1\alpha_2...\alpha_N) \equiv |n_\alpha n_{\alpha'}...\rangle . \tag{2.19}
$$

The n_α again correspond to the number of particles that occupy the sp level α, *etc.* It is customary to include only those n_α in Eq. (2.19) that are different from zero. There is, however, no limit on these occupation numbers like there is for fermions. Addition and removal operators may be introduced as for fermions. For sp states, one has

$$|\alpha\rangle = a_\alpha^\dagger |0\rangle . \qquad (2.20)$$

For two-particle states

$$|\alpha\beta\rangle = a_\alpha^\dagger a_\beta^\dagger |0\rangle \qquad (2.21)$$

when $\alpha \neq \beta$. If $\alpha = \beta$, we must include an extra normalization factor

$$|\alpha\alpha\rangle = |n_\alpha = 2\rangle = \frac{1}{\sqrt{2}} a_\alpha^\dagger a_\alpha^\dagger |0\rangle . \qquad (2.22)$$

In the general case

$$|n_\alpha n_\beta ... n_\omega\rangle = \frac{1}{[n_\alpha! n_\beta! ... n_\omega!]^{1/2}} \left(a_\alpha^\dagger\right)^{n_\alpha} \left(a_\beta^\dagger\right)^{n_\beta} ... \left(a_\omega^\dagger\right)^{n_\omega} |0\rangle . \qquad (2.23)$$

The boson addition and removal operators obey the following fundamental commutation relations:

$$[a_\alpha, a_\beta^\dagger] = a_\alpha a_\beta^\dagger - a_\beta^\dagger a_\alpha = \delta_{\alpha,\beta}, \qquad (2.24)$$

$$[a_\alpha, a_\beta] = [a_\alpha^\dagger, a_\beta^\dagger] = 0. \qquad (2.25)$$

These results can be generated in a similar way as for fermion operators and are related to the requirement that symmetric states are obtained after the action of an addition or removal operator of a boson sp state. The commutation relations for a given sp state are identical to those for harmonic oscillator quanta. It is then not surprising that the following relations hold

$$a_\alpha^\dagger |n_\alpha n_\beta ... n_\omega\rangle = \sqrt{n_\alpha + 1} |n_\alpha + 1 \, n_\beta ... n_\omega\rangle , \qquad (2.26)$$

$$a_\alpha |n_\alpha n_\beta ... n_\omega\rangle = \sqrt{n_\alpha} |n_\alpha - 1 \, n_\beta ... n_\omega\rangle , \qquad (2.27)$$

and likewise for operators involving other sp quantum numbers. The results of Eqs. (2.26) and (2.27) can be verified by using Eq. (2.23) and the commutation relations.

2.3   One-body operators in Fock space

Relevant operators in many-particle systems involve only the coordinates of
one or two (and in unusual cases three) particles. It is therefore important
to translate the action of such operators into the language of particle addition and removal operators. We will follow a similar strategy as in [Blaizot
and Ripka (1986)]. Consider first an operator which acts only on one particle. Such a one-body operator, F, acting in a sp space can be written
as

$$F = \sum_\alpha \sum_\beta |\alpha\rangle \langle\alpha| F |\beta\rangle \langle\beta| \qquad (2.28)$$

and is completely determined by all its matrix elements $\langle\alpha| F |\beta\rangle$ in a chosen
sp basis. In an N-particle space, the corresponding extension is simply

$$F_N = F(1) + F(2) + ... + F(N) = \sum_{i=1}^N F(i), \qquad (2.29)$$

where each $F(i)$ acts only on particle i. Using Eq. (2.28) the action of $F(i)$
on a product state (note the round bracket) is given by

$$F(i)|\alpha_1\alpha_2\alpha_3...\alpha_N) = |\alpha_1\rangle |\alpha_2\rangle ... |\alpha_{i-1}\rangle \left\{ \sum_{\beta_i} |\beta_i\rangle \langle\beta_i| F |\alpha_i\rangle \right\} |\alpha_{i+1}\rangle ... |\alpha_N\rangle$$

$$= \sum_{\beta_i} \langle\beta_i| F |\alpha_i\rangle |\alpha_1...\alpha_{i-1}\beta_i\alpha_{i+1}...\alpha_N). \qquad (2.30)$$

The matrix elements of F do not depend on which particle is considered.
The number $\langle\beta_i| F |\alpha_i\rangle$ in the above expression will consequently be the
same for any particle. Calculation of this matrix element for another particle will in fact only involve a change in dummy variables. For the operator
F_N we can write

$$F_N|\alpha_1\alpha_2\alpha_3...\alpha_N) = F(1) |\alpha_1\rangle |\alpha_2\rangle ... |\alpha_N\rangle + ... + |\alpha_1\rangle |\alpha_2\rangle ...F(N)|\alpha_N\rangle$$

$$= \sum_{\beta_1} \langle\beta_1| F |\alpha_1\rangle |\beta_1\alpha_2...\alpha_N) + ... + \sum_{\beta_N} \langle\beta_N| F |\alpha_N\rangle |\alpha_1\alpha_2...\beta_N)$$

$$= \sum_{i=1}^N \sum_{\beta_i} \langle\beta_i| F |\alpha_i\rangle |\alpha_1\alpha_2...\alpha_{i-1}\beta_i\alpha_{i+1}...\alpha_N). \qquad (2.31)$$

To obtain the action of F_N on an antisymmetric or symmetric N-particle
state, we observe that F_N is symmetric and therefore commutes with the

antisymmetrizer $\mathcal{A}$ or the symmetrizer $\mathcal{S}$ (remember the example of two particles in which H commutes with P_{12} discussed in Sec. 1.3). As a result

$$F_N \,|\alpha_1\alpha_2\alpha_3...\alpha_N\rangle = \sum_{i=1}^{N}\sum_{\beta_i} \langle\beta_i|\, F\,|\alpha_i\rangle \,|\alpha_1\alpha_2...\alpha_{i-1}\beta_i\alpha_{i+1}...\alpha_N\rangle. \quad (2.32)$$

We can now show that the Fock-space operator (note the "ˆ" notation for such an operator)

$$\hat{F} = \sum_{\alpha\beta} \langle\alpha|\, F\,|\beta\rangle \, a^\dagger_\alpha a_\beta \quad (2.33)$$

gives the same result for any N when acting on Eq. (2.16) for fermions and Eq. (2.23) for bosons. Consider the following commutator in the case of fermions

$$[\hat{F}, a^\dagger_{\alpha_i}] = \sum_{\alpha\beta} \langle\alpha|\, F\,|\beta\rangle \,[a^\dagger_\alpha a_\beta, a^\dagger_{\alpha_i}] = \sum_{\alpha\beta} \langle\alpha|\, F\,|\beta\rangle \,(a^\dagger_\alpha a_\beta a^\dagger_{\alpha_i} - a^\dagger_{\alpha_i} a^\dagger_\alpha a_\beta)$$

$$= \sum_{\alpha\beta} \langle\alpha|\, F\,|\beta\rangle \, a^\dagger_\alpha (a_\beta a^\dagger_{\alpha_i} + a^\dagger_{\alpha_i} a_\beta) = \sum_{\alpha\beta} \langle\alpha|\, F\,|\beta\rangle \, a^\dagger_\alpha \delta_{\beta,\alpha_i}$$

$$= \sum_{\alpha} \langle\alpha|\, F\,|\alpha_i\rangle \, a^\dagger_\alpha = \sum_{\beta_i} \langle\beta_i|\, F\,|\alpha_i\rangle \, a^\dagger_{\beta_i}, \quad (2.34)$$

where the fundamental anticommutation relation (2.8) has been used. This result can be employed in the following manipulation

$$\hat{F}\,|\alpha_1\alpha_2\alpha_3...\alpha_N\rangle = \hat{F}a^\dagger_{\alpha_1}a^\dagger_{\alpha_2}...a^\dagger_{\alpha_N}\,|0\rangle$$
$$= [\hat{F}, a^\dagger_{\alpha_1}]a^\dagger_{\alpha_2}...a^\dagger_{\alpha_N}\,|0\rangle + a^\dagger_{\alpha_1}\hat{F}a^\dagger_{\alpha_2}...a^\dagger_{\alpha_N}\,|0\rangle$$
$$= [\hat{F}, a^\dagger_{\alpha_1}]a^\dagger_{\alpha_2}...a^\dagger_{\alpha_N}\,|0\rangle + a^\dagger_{\alpha_1}[\hat{F}, a^\dagger_{\alpha_2}]...a^\dagger_{\alpha_N}\,|0\rangle + ... + a^\dagger_{\alpha_1}a^\dagger_{\alpha_2}...[\hat{F}, a^\dagger_{\alpha_N}]\,|0\rangle$$
$$= \sum_{i=1}^{N}\sum_{\beta_i} \langle\beta_i|\, F\,|\alpha_i\rangle \, a^\dagger_{\alpha_1}...a^\dagger_{\alpha_{i-1}}a^\dagger_{\beta_i}a^\dagger_{\alpha_{i+1}}...a^\dagger_{\alpha_N}\,|0\rangle$$
$$= \sum_{i=1}^{N}\sum_{\beta_i} \langle\beta_i|\, F\,|\alpha_i\rangle \,|\alpha_1...\alpha_{i-1}\beta_i\alpha_{i+1}...\alpha_N\rangle, \quad (2.35)$$

which proves the equivalence for fermions of Eqs. (2.32) and (2.35) for a given N. Since this outcome can be generated for any N, we conclude that Eq. (2.33) has the required form of a one-body operator in Fock space. A similar procedure works for bosons to establish Eq. (2.33).

2.4 Two-body operators in Fock space

The two-body interaction V provides an example of an operator involving the coordinates of two particles. In order to establish the corresponding Fock-space operator, consider first the two-body operator V acting on states in the two-particle space of product states

$$V = \sum_{\alpha\beta}\sum_{\gamma\delta}|\alpha\beta)(\alpha\beta|V|\gamma\delta)(\gamma\delta|. \tag{2.36}$$

In an N-particle space the extension of this operator is given by

$$V_N = \begin{cases} V(1,2)+ V(1,3)+ V(1,4)+ \ldots + V(1,N)+ \\ \qquad\quad V(2,3)+ V(2,4)+ \ldots + V(2,N)+ \\ \qquad\qquad\qquad V(3,4)+ \ldots + V(3,N)+ \\ \qquad\qquad\qquad\qquad\qquad \ddots \qquad \vdots \\ \qquad\qquad\qquad\qquad\qquad\qquad V(N-1,N) \end{cases}$$

$$= \sum_{i<j=1}^{N} V(i,j) = \frac{1}{2}\sum_{i\neq j}^{N} V(i,j), \tag{2.37}$$

where each operator $V(i,j)$ acts only on particles i and j. The action of $V(i,j)$ on a product state of N particles yields

$$V(i,j)|\alpha_1..\alpha_i..\alpha_j..\alpha_N) = \sum_{\beta_i\beta_j}(\beta_i\beta_j|V|\alpha_i\alpha_j)$$

$$\times |\alpha_1..\alpha_{i-1}\beta_i\alpha_{i+1}..\alpha_{j-1}\beta_j\alpha_{j+1}..\alpha_N). \tag{2.38}$$

The matrix elements of V do not depend on the selected pair of particles. The numbers $(\beta_i\beta_j|V|\alpha_i\alpha_j)$ in the above expression will thus be identical for any pair of particles as long as the same quantum numbers are involved. The action of the operator V_N therefore yields

$$V_N|\alpha_1\alpha_2\alpha_3...\alpha_N) = \sum_{i<j=1}^{N}\sum_{\beta_i\beta_j}(\beta_i\beta_j|V|\alpha_i\alpha_j)|\alpha_1...\beta_i...\beta_j...\alpha_N). \tag{2.39}$$

When V_N acts on an antisymmetric or symmetric N-particle state, we note that V_N is symmetric, and consequently commutes with the antisymmetrizer $\mathcal{A}$ or symmetrizer $\mathcal{S}$. As a result,

$$V_N\,|\alpha_1\alpha_2\alpha_3...\alpha_N\rangle = \sum_{i<j=1}^{N}\sum_{\beta_i\beta_j}(\beta_i\beta_j|V|\alpha_i\alpha_j)\,|\alpha_1...\beta_i...\beta_j...\alpha_N\rangle\,. \tag{2.40}$$

The following Fock space operator (again with the " ^ " notation)

$$\hat{V} = \frac{1}{2} \sum_{\alpha\beta\gamma\delta} (\alpha\beta|V|\gamma\delta) a_\alpha^\dagger a_\beta^\dagger a_\delta a_\gamma \tag{2.41}$$

generates the same result for any N when acting on Eq. (2.16) for fermions and Eq. (2.23) for bosons. For fermions, consider the following commutator

$$[\hat{V}, a_{\alpha_i}^\dagger] = \sum_{\beta_i \beta_j \alpha_{i'}} (\beta_i \beta_j|V|\alpha_i \alpha_{i'}) a_{\beta_i}^\dagger a_{\beta_j}^\dagger a_{\alpha_{i'}}. \tag{2.42}$$

It is obtained in a similar way as Eq. (2.34) by making use of the symmetry $V(i,j) = V(j,i)$ which implies

$$(\alpha\beta|V|\gamma\delta) = (\beta\alpha|V|\delta\gamma). \tag{2.43}$$

We employ this result to demonstrate that

$$\hat{V}|\alpha_1\alpha_2\alpha_3...\alpha_N\rangle = \sum_{i=1}^{N}\sum_{j>i}^{N}\sum_{\beta_i\beta_j} (\beta_i\beta_j|V|\alpha_i\alpha_j) a_{\alpha_1}^\dagger ... a_{\beta_i}^\dagger ... a_{\beta_j}^\dagger ... a_{\alpha_N}^\dagger |0\rangle \tag{2.44}$$

which proves the equivalence. In this derivation, it is necessary to use

$$\sum_{\beta_j\alpha_{i'}} f(\beta_j, \alpha_{i'})[a_{\beta_j}^\dagger a_{\alpha_{i'}}, a_{\alpha_j}^\dagger] = \sum_{\beta_j} f(\beta_j, \alpha_j) a_{\beta_j}^\dagger \tag{2.45}$$

for each $j > i$. Equation (2.44) is equivalent to Eq. (2.39) and holds for any N. Equation (2.41) therefore represents the extension of the two-particle operator V_N in Fock space. For bosons, one proceeds in a similar fashion. The two-body interaction in Fock space is also represented by Eq. (2.41). An alternative form for $\hat{V}$ in the case of fermions can be written as

$$\hat{V} = \frac{1}{4} \sum_{\alpha\beta\gamma\delta} \langle\alpha\beta|V|\gamma\delta\rangle a_\alpha^\dagger a_\beta^\dagger a_\delta a_\gamma, \tag{2.46}$$

where

$$\langle\alpha\beta|V|\gamma\delta\rangle \equiv (\alpha\beta|V|\gamma\delta) - (\alpha\beta|V|\delta\gamma) = \langle\alpha\beta|\hat{V}|\gamma\delta\rangle. \tag{2.47}$$

The last term in Eq. (2.47) contains the second quantized two-body operator $\hat{V}$. Note that in the expressions for $\hat{V}$ in Eqs. (2.41) and (2.46) the order of the quantum numbers γ and δ in the matrix element is different from the ordering of the corresponding particle removal operators. Depending on the nature of the interaction V, it can be useful to choose the unsymmetrized version of $\hat{V}$ given in Eq. (2.41) or the symmetrized version in

Eq. (2.46). Using this Fock space formulation of one- and two-body operators, the Hamiltonian of a many-particle system in second quantized form
can be written as

$$\hat{H} = \hat{T} + \hat{V}$$
$$= \sum_{\alpha\beta} \langle\alpha|\,T\,|\beta\rangle\, a_\alpha^\dagger a_\beta + \frac{1}{2} \sum_{\alpha\beta\gamma\delta} (\alpha\beta|V|\gamma\delta) a_\alpha^\dagger a_\beta^\dagger a_\delta a_\gamma \qquad (2.48)$$

for both fermions and bosons. The alternative result for the two-body
interaction given in Eq. (2.46), is also useful in the case of bosons but includes a plus sign. For a given choice of sp basis and two-body interaction
V, it is possible to calculate the relevant one- and two-body matrix elements which appear in $\hat{H}$ (although this can be very tedious and computer
time consuming). The calculation of matrix elements of $\hat{H}$ between many-
particle states is therefore reduced to manipulating particle addition and
removal operators using their (anti)commutation relations. This procedure
represents a substantial practical advantage over other methods that deal
with the calculation of matrix elements of operators between symmetric or
antisymmetric many-particle states.

When an explicit three-body interaction (symmetric in the coordinates
of the particles) needs to be included, one can use the first quantized version
in the N-particle space

$$W_N = \sum_{i<j<k=1}^{N} W(i,j,k) \qquad (2.49)$$

and the Fock space operator

$$\hat{W} = \frac{1}{6} \sum_{\alpha\beta\gamma} \sum_{\alpha'\beta'\gamma'} (\alpha\beta\gamma|W|\alpha'\beta'\gamma') \, a_\alpha^\dagger a_\beta^\dagger a_\gamma^\dagger a_{\gamma'} a_{\beta'} a_{\alpha'}. \qquad (2.50)$$

2.5   Examples

As an example of a second quantized one-body operator consider

$$\hat{N} = \sum_\alpha a_\alpha^\dagger a_\alpha. \qquad (2.51)$$

From Eqs. (2.34) and (2.35) it follows that

$$\hat{N}\,|\alpha_1...\alpha_N\rangle = N\,|\alpha_1...\alpha_N\rangle \qquad (2.52)$$

for any N and any set α_i. The operator $\hat{N}$ can therefore be called the number operator since it simply counts the number of particles in the state on which it acts. If the state has a fixed number of particles, it is an eigenstate of $\hat{N}$.

To explain the name second quantization, it is useful to mention the confusing convention to denote the addition operators for particles with quantum numbers r, m_s by

$$\psi^\dagger_{m_s}(r) \equiv a^\dagger_{rm_s} \tag{2.53}$$

and similarly for the removal operators. In this basis the kinetic energy matrix element becomes

$$\langle rm_s | T | r'm_s' \rangle = \frac{-\hbar^2}{2m} \delta(r - r') \nabla'^2 \delta_{m_s, m_s'}. \tag{2.54}$$

A conventional spin-independent local two-particle interaction yields

$$(r_1 m_{s_1} \; r_2 m_{s_2} | V(r, r') | r_3 m_{s_3} \; r_4 m_{s_4}) = \delta(r_1 - r_3) \delta(r_2 - r_4) \tag{2.55}$$
$$\times \, \delta_{m_{s_1}, m_{s_3}} \delta_{m_{s_2}, m_{s_4}} V(|r_3 - r_4|).$$

The Hamiltonian can now be rewritten as

$$\hat{H} = \sum_{m_s} \int d^3r \; \psi^\dagger_{m_s}(r) \{ \frac{-\hbar^2}{2m} \nabla^2 \} \psi_{m_s}(r) \tag{2.56}$$
$$+ \frac{1}{2} \sum_{m_s m_s'} \int d^3r \int d^3r' \; \psi^\dagger_{m_s}(r) \psi^\dagger_{m_s'}(r') V(|r - r'|) \psi_{m_s'}(r') \psi_{m_s}(r).$$

The above expression can easily lead to the wrong interpretation when one mistakenly thinks of ψ as a wave function. In order to avoid this pitfall the notation $a^\dagger_{rm_s}$ will be used to denote an operator which adds a particle with sp quantum numbers $\{rm_s\}$ to a many-particle state (and a_{rm_s} for the removal operator).

A change of sp basis in sp space can be written in the following way

$$a^\dagger_\alpha |0\rangle = |\alpha\rangle = \sum_\lambda |\lambda\rangle \langle\lambda|\alpha\rangle = \sum_\lambda a^\dagger_\lambda |0\rangle \langle\lambda|\alpha\rangle. \tag{2.57}$$

This procedure can be repeated for $a^\dagger_\alpha$ (a_α) acting on any state in Fock space and leads to the operator identities

$$a^\dagger_\alpha = \sum_\lambda \langle\lambda|\alpha\rangle a^\dagger_\lambda \tag{2.58}$$

and

$$a_\alpha = \sum_\lambda \langle \alpha | \lambda \rangle a_\lambda. \tag{2.59}$$

Such transformations are unitary since they correspond to a basis transformation.

2.6   Exercises

(1) Perform the analysis to generate the remaining anticommution relations in Eqs. (2.8) and (2.9).
(2) Obtain the commutation relations for boson addition and removal operators Eqs. (2.24) and (2.25).
(3) Check Eqs. (2.26) and (2.27).
(4) Calculate the commutator in Eq. (2.42) for both the fermion and the boson case and check the relation (2.44).
(5) Determine the second-quantized form of

- the density operator

$$\rho_N(\boldsymbol{r}) = \sum_{i=1}^{N} \delta(\boldsymbol{r} - \boldsymbol{r}_i)$$

- the electrical current density operator

$$j_N(\boldsymbol{r}) = \frac{1}{2} \sum_{i=1}^{N} \left\{ \frac{\boldsymbol{p}_i}{m} \delta(\boldsymbol{r} - \boldsymbol{r}_i) + \delta(\boldsymbol{r} - \boldsymbol{r}_i) \frac{\boldsymbol{p}_i}{m} \right\}$$

- the three components of the spin density operator

$$s_N(\boldsymbol{r}) = \sum_{i=1}^{N} s_i \delta(\boldsymbol{r} - \boldsymbol{r}_i)$$

in the case of fermions and using the $\{\boldsymbol{r}, m_s\}$ basis.
(6) Determine the second-quantized form of the two-body Coulomb interaction for fermions in the $\{\boldsymbol{r}, m_s\}$ basis.
(7) Show that for any n-body operator $\hat{F}^{(n)}$ the relation

$$\sum_\alpha a_\alpha^\dagger [a_\alpha, \hat{F}^{(n)}] = n \hat{F}^{(n)}$$

holds, for both fermions and bosons.

(8) Use the result from the previous problem to show that, in case of a Hamiltonian $\hat{H} = \hat{T} + \hat{V}$ consisting of a one-body operator $\hat{T}$ and a two-body operator $\hat{V}$, the energy expectation value $E_0^N = \langle \Psi_0^N | \hat{H} | \Psi_0^N \rangle$ in an arbitrary N-body state $|\Psi_0^N\rangle$ can be written as

$$E_0^N = \frac{1}{2} \left\{ \langle \Psi_0^N | \hat{T} | \Psi_0^N \rangle \right.$$

$$\left. + \sum_\alpha \sum_{\nu(N-1)} [E_0^N - E_\nu^{N-1}] \, | \langle \Psi_\nu^{N-1} | a_\alpha | \Psi_0^N \rangle |^2 \right\},$$

where the $|\Psi_\nu^{N-1}\rangle$ form the complete set of eigenstates of $\hat{H}$ in the $(N-1)$-particle space.

Chapter 3

Independent-particle model for fermions in finite systems

We can obtain useful descriptions for finite systems comprised of fermions by identifying a one-body potential that already generates some of the physics associated with the interaction between the particles. This one-body potential combined with the kinetic energy and pertinent external potentials, like the Coulomb attraction to the nucleus for electrons in atoms, forms a good baseline to discuss the physics of these systems. In particular, it leads to a shell-model picture that is appropriate for atoms, nuclei, and other localized many-fermion systems. In addition, such a Hamiltonian may be used as a starting point of the perturbation expansion of relevant physical quantities, to be presented in Ch. 8. The general discussion of this independent-particle description, where fermions do not interact but are aware of each other's identity, will be given in Sec. 3.1. The application of the independent-particle model to electrons in atoms is presented in Sec. 3.2, and for nucleons in nuclei in Sec. 3.3. An important hypothetical system entitled "nuclear matter" is introduced in Sec. 3.3.1. It describes an idealized way to represent and study the global properties of the interior of nuclei which exhibit strikingly similar features for nuclei heavier than ^{16}O. In Sec. 3.4, the method of second quantization is applied to the concept of isospin, as it is relevant for the description of nuclei.

3.1 General results and the independent-particle model

We obtain a solvable many-particle problem by considering the following decomposition of the original Hamiltonian

$$\hat{H} = \hat{T} + \hat{V} = \hat{H}_0 + \hat{H}_1, \tag{3.1}$$

where

$$\hat{H}_0 = \hat{T} + \hat{U} \tag{3.2}$$

and

$$\hat{H}_1 = \hat{V} - \hat{U}, \tag{3.3}$$

with $\hat{U}$ being a suitably chosen one-body operator. When only $\hat{H}_0$ is considered, the corresponding many-particle problem can be solved straightforwardly. Note that a one-body external field $\hat{U}_{ext}$ can also be included in $\hat{H}_0$, if appropriate.

There are various situations in which the choice of $\hat{U}$ is very important: it can be used for example to include the average effect of the two-body interaction $\hat{V}$. The remaining effects of $\hat{V}$ may then be small, as for atoms. If the actual ground state of the system breaks a symmetry that the Hamiltonian respects, the choice of $\hat{U}$ is critical. Systems with spontaneous magnetization provide an example. It is advantageous to include in the Hamiltonian $\hat{H}_0$ a term that includes the symmetry-breaking effect, and yields a noninteracting ground state which displays this behavior. Such a starting point in perturbation theory suggests better convergence properties than departing from a noninteracting state with the "wrong" symmetry. A possible feature of $\hat{U}$ is therefore that it can speed up the convergence of the perturbation expansion. In the case of nuclei, we can use $\hat{U}$ to localize the nucleons in a potential well. The same holds for the corresponding many-particle states that are eigenstates of $\hat{H}_0$. Without localization, plane-wave many-particle states must be employed. These are the eigenstates of $\hat{T}$ thus complicating the description of the nucleus substantially.

In the case of spherical symmetry, the sp problem can be solved straightforwardly, although a numerical solution might be required. We denote the relevant eigenstates and energies of H_0 by

$$H_0 \, |\lambda\rangle = (T + U) \, |\lambda\rangle = \varepsilon_\lambda \, |\lambda\rangle . \tag{3.4}$$

The second-quantized Hamiltonian $\hat{H}_0$ using this $\{|\lambda\rangle\}$ basis can be written as

$$\hat{H}_0 = \sum_{\lambda\lambda'} \langle\lambda| \, (T + U) \, |\lambda'\rangle \, a_\lambda^\dagger a_{\lambda'}$$
$$= \sum_{\lambda\lambda'} \varepsilon_{\lambda'} \, \delta_{\lambda,\lambda'} \, a_\lambda^\dagger a_{\lambda'} = \sum_\lambda \varepsilon_\lambda a_\lambda^\dagger a_\lambda \tag{3.5}$$

using Eq. (3.4). All the many-particle eigenkets of $\hat{H}_0$ for N particles are of the form

$$\left|\Phi_n^N\right\rangle = |\lambda_1\lambda_2...\lambda_N\rangle = a_{\lambda_1}^\dagger a_{\lambda_2}^\dagger ... a_{\lambda_N}^\dagger |0\rangle \qquad (3.6)$$

with eigenvalue

$$E_n^N = \sum_{i=1}^N \varepsilon_{\lambda_i}. \qquad (3.7)$$

The above result can be generated by employing Eqs. (2.34) and (2.35). In the present example, Eq. (2.34) leads to

$$\left[\hat{H}_0, a_{\lambda_i}^\dagger\right] = \varepsilon_{\lambda_i} a_{\lambda_i}^\dagger. \qquad (3.8)$$

For obvious reasons, the states $\left|\Phi_n^N\right\rangle$ are called independent-particle states, since the only correlation they include pertains to the Pauli principle preventing the particles to occupy the same sp state. The state with the lowest energy for N particles corresponds to filling the lowest sp levels in accord with the Pauli principle. This state is nondegenerate when N coincides with a shell closure (a situation with complete occupation according to the degeneracies of the occupied levels). It may be written as

$$\left|\Phi_0^N\right\rangle = \prod_{\lambda_i \leq F} a_{\lambda_i}^\dagger |0\rangle, \qquad (3.9)$$

where F characterizes the energy level above which all levels are empty, and below which all levels are completely occupied. The ket $\left|\Phi_0^N\right\rangle$ is sometimes referred to as the Fermi sea. The solution of the many-fermion problem associated with $\hat{H}_0$ is now complete, since all the eigenstates and eigenenergies for any particle number N have been determined. Only the solution of the relevant sp problem given by Eq. (3.4) is required, plus the proper inclusion of the Pauli principle which is facilitated by the use of second quantization.

3.2 Electrons in atoms

When dealing with electronic problems it is convenient to use the system of *atomic units* (a.u.), in which the electron mass m_e and elementary charge e are taken as units of mass and charge. In addition, the choice for the units of length and time is such that the numerical values of $\hbar$ and $4\pi\epsilon_0$ (where

ϵ_0 is the permittivity in free space) are unity as well. The atomic unit of length then corresponds to the Bohr radius a_0,

$$\text{a.u. (length)} = a_0 = \frac{4\pi\epsilon_0\hbar^2}{e^2 m_e} \approx 5.29177 \times 10^{-11} \text{ m.} \qquad (3.10)$$

The atomic unit of time is

$$\text{a.u. (time)} = \frac{a_0}{\alpha c} \approx 2.41888 \times 10^{-17} \text{ s,} \qquad (3.11)$$

where

$$\alpha = \frac{e^2}{4\pi\epsilon_0\hbar c} \approx \frac{1}{137.036} \qquad (3.12)$$

is the fine-structure constant. The atomic unit for energy is the Hartree,

$$E_H = \frac{\hbar^2}{m_e a_0^2} \approx 27.2114 \text{ eV,} \qquad (3.13)$$

which is twice the ground-state energy of the Bohr hydrogen atom. The atomic unit of magnetic dipole moment,

$$\text{a.u. (magnetic moment)} = \frac{\hbar e}{m_e} = 2\mu_B \qquad (3.14)$$

is twice the Bohr magneton, $\mu_B \approx 5.78838 \, 10^{-5} \text{eVT}^{-1}$. Up to date values of these quantities can be found at the website of the National Institute of Standards and Technology (NIST).

In the description of atoms, most aspects of the physics can be understood on the basis of the following Hamiltonian [Lindgren and Morrison (1982)]

$$H_N = \sum_{i=1}^N \frac{p_i^2}{2} - \sum_{i=1}^N \frac{Z}{|r_i|} + \frac{1}{2}\sum_{i\neq j}^N \frac{1}{|r_i - r_j|} + V_{mag}. \qquad (3.15)$$

The various parts of this Hamiltonian for N electrons correspond, in this order, to: the kinetic energy of the electrons, the attraction to the nucleus of charge Z which is assumed to be infinitely heavy, the Coulomb repulsion between the electrons, and finally, the magnetic interactions. The latter comprise the spin-own-orbit interaction, the spin-other-orbit interaction, and the less important spin-spin and orbit-orbit interactions. The actual form of the magnetic interactions can be obtained systematically for a single-electron system by using a nonrelativistic reduction of the Dirac equation [Cohen-Tannoudji *et al.* (1992)]. For a many-electron system this procedure cannot be applied, and is still a subject of study. Since these magnetic interactions are rather small perturbations compared to the other three contributions to the Hamiltonian, it is conventional to lump them together in an effective one-body spin-orbit interaction. The latter simulates most of the magnetic interaction effects:

$$V_{mag} \Rightarrow V_{so}^{eff} = \sum_i \zeta_i \, \boldsymbol{\ell}_i \cdot \boldsymbol{s}_i, \tag{3.16}$$

where the sum runs only over electrons in the open shells, and the ζ_i describe the strength of this effective sp spin-orbit interaction.

The Hamiltonian for the electrons in an atom is theoretically well-founded. Nevertheless, it should be clear that nonrelativistic calculations must at some stage be complemented by explicitly including relativity. This becomes more urgent for heavier atoms since the hydrogen-like binding increases with Z^2, becoming 10% of the electron's rest mass for $Z = 60$. The increased localization of the $1s$ wave function also leads to high-momentum components.

Sensible atomic many-body calculations can be performed by neglecting the magnetic interactions altogether, as will be done below. A characteristic feature of electron shell structure is exemplified by the ionization energies for neutral atoms shown in Fig. 3.1. These exhibit marked jumps at the noble gases. A simple starting point to describe shell closures for atoms is provided by the choice

$$H_0^N = \sum_{i=1}^N H_0(i) \tag{3.17}$$

with

$$H_0(i) = \frac{\boldsymbol{p}_i^2}{2} - \frac{Z}{r_i} + U(\boldsymbol{r}_i). \tag{3.18}$$

The auxiliary sp potential U must contain a large portion of the effect of the Coulomb repulsion between the electrons. Already for $U = 0$, the simple atomic shell model with the appropriate sp levels is generated by the hydrogen-like Hamiltonian with nuclear charge Z. Clearly, the results of Sec. 3.1 identify the ground state of the noninteracting many-electron system associated with $\hat{H}_0$ immediately: electrons will occupy the lowest sp energy states in accordance with the Pauli principle. For each energy level there is a $(2\ell + 1) * (2s + 1)$-fold degeneracy, which stems from the

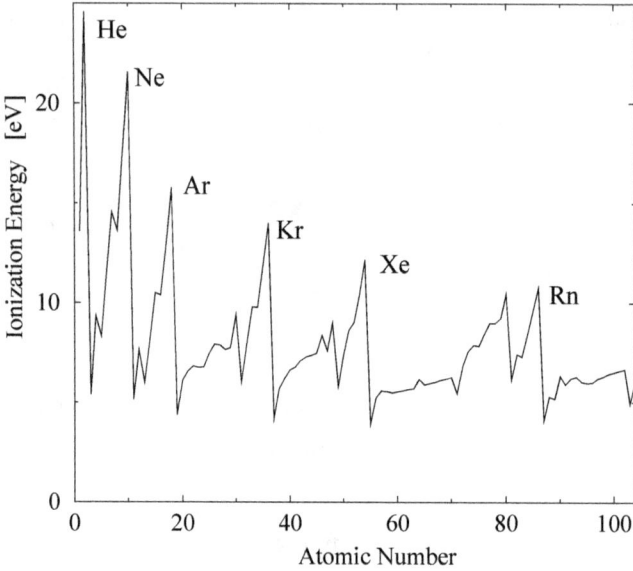

Fig. 3.1   Ionization energies for neutral atoms. At the positions of the noble gases, indicated in the figure, large jumps occur, illustrating shell closures. Other, less pronounced maxima identify subshell closures, *e.g.* at $Z = 80$ corresponding to mercury. Experimental data were taken from [Martin *et al.* (2002)].

rotational invariance and spin-independence of the Hamiltonian. An additional, accidental, degeneracy exists for this problem, yielding sp energies that are only determined by the principal quantum number n and given by $-Z^2/2n^2$ in atomic units.

The degeneracy is lifted when the effect of the closed shells is approximately included in U. This effect is an essential ingredient to explain the observed ionization energies illustrated in Fig. 3.1. A simple example, including U, is provided by considering the alkali atoms. These atoms have one electron outside a closed shell. The presence of the closed electron shell(s) screens the nuclear charge for the last electron leading to an attraction of only one unit of charge at large distances. Very close to the nucleus, the electron will experience the full attraction of the nuclear charge Z. Both features are illustrated in Fig. 3.2. The presence of closed shells is expected to generate a spherically symmetric field, due to the filling of all m_ℓ and m_s substates. A smooth interpolation between the two extreme cases therefore provides a reasonable picture of U, as shown for Na in Fig. 3.2. The effective potential lifts the accidental degeneracy of the hydrogen-like Hamiltonian

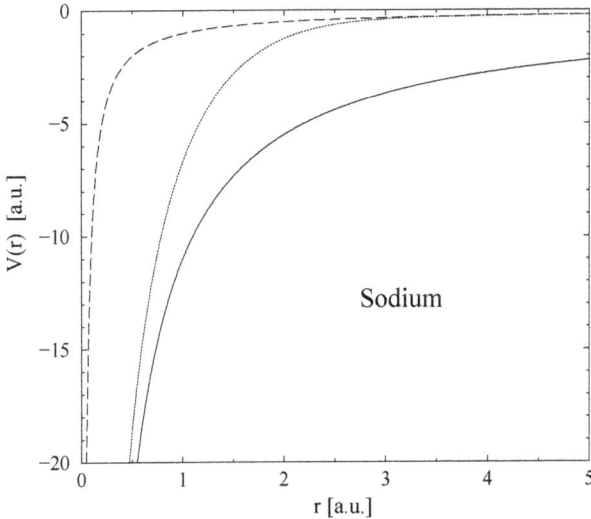

Fig. 3.2 Illustration of the effective Coulomb potential for the last electron in Na (dotted line). It is given in atomic units (1 H = 27.2 eV) as a function of the distance in units of the Bohr radius $a_0 = 5.29 \times 10^{-11}$ m. Close to the nucleus the potential approaches the full attraction of the nucleus ($-Z/r$) shown by the full line, whereas at larger distances the screening of the other electrons reduces this attraction to $-1/r$ shown by the dashed line.

in such a way, that the level with the highest probability to be close to the nucleus will profit most in energy. The s state is thus lowered with respect to the p state, and so on, for a given shell with principal quantum number "n". The latter becomes a quantum number distinguishing the states with the same ℓ value. For the sodium atom this leads to the filling of the corresponding $1s, 2s$, and $2p$ shells with the notation: $(1s)^2(2s)^2(2p)^6$. The last electron will therefore occupy the $3s$ as illustrated in Fig. 3.3.

Figure 3.3 also identifies the excited states of sodium that can be obtained by removing the last electron from the $3s$ and placing it in one of the other empty levels. Suppose the states $|n\ell m_\ell m_s\rangle$ are eigenkets of H_0 in Eq. (3.18) with an appropriate auxiliary potential

$$H_0 |n\ell m_\ell m_s\rangle = \varepsilon_{n\ell} |n\ell m_\ell m_s\rangle , \qquad (3.19)$$

where n no longer refers to the hydrogen-like quantum number, but still characterizes the radial behavior of the corresponding wave functions. The ground state representing Na in this approximation can now be written in

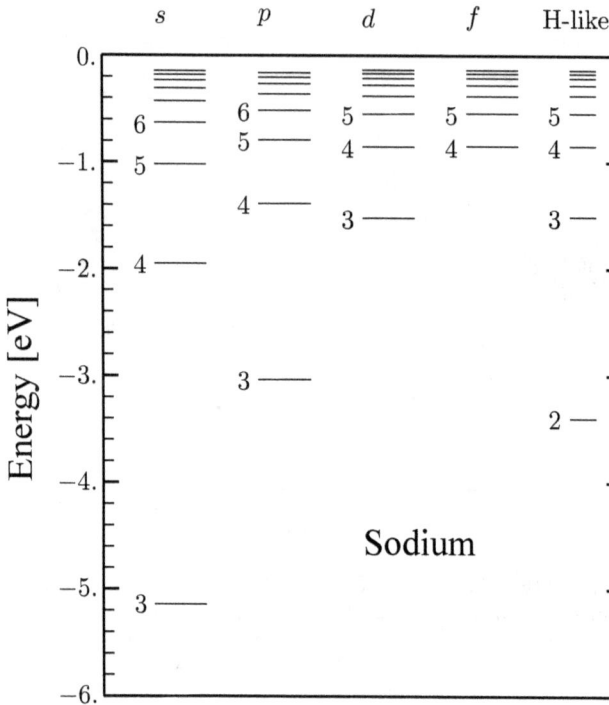

Fig. 3.3 Energy levels of the Na atom illustrating the tendency of levels with increasing "n" and ℓ to be more hydrogen-like. The latter energies are displayed in the last column. The first column corresponds to the available s states for the last electron, the second column to available p states, *etc.* Experimental data were taken from [Martin *et al.* (2002)].

the following way

$$|300m_s, 211\tfrac{1}{2}, 211 -\tfrac{1}{2}, ..., 100\tfrac{1}{2}, 100 -\tfrac{1}{2}\rangle$$
$$= a^\dagger_{300m_s} a^\dagger_{211\frac{1}{2}} a^\dagger_{211-\frac{1}{2}} ... a^\dagger_{100\frac{1}{2}} a^\dagger_{100-\frac{1}{2}} |0\rangle , \qquad (3.20)$$

where for each occupied state the four quantum numbers $n\ell m_\ell m_s$ are indicated. A similar interpretation can be given for the spectra of the other alkali atoms. The possibility of a simple understanding of these atoms based on such straightforward considerations, indicates that it must indeed be possible to represent the effect of the mutual interaction of the electrons by an average sp potential. The determination of this average sp potential from the electron–electron interaction requires the Hartree–Fock procedure

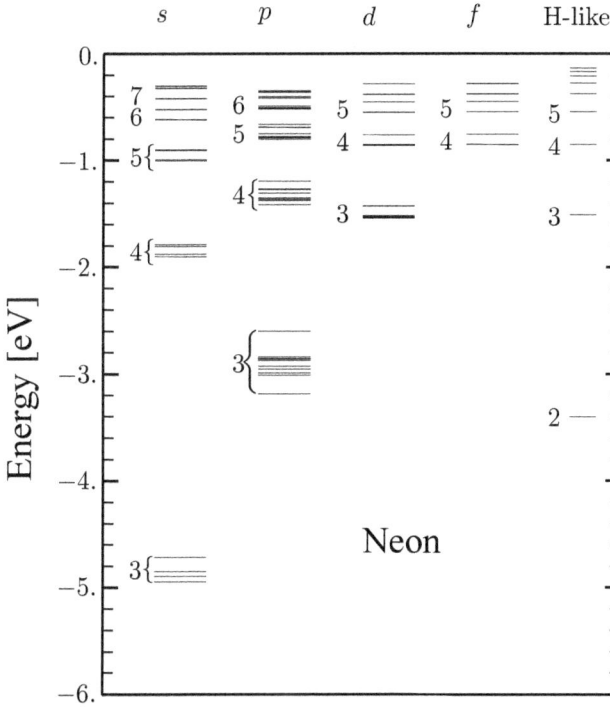

Fig. 3.4 Energy levels of the Ne atom that can be interpreted as the promotion of a $2p$ electron to the available empty orbitals, starting with $3s$ *etc.* The energies of the states become more hydrogen-like with increasing "n" and ℓ also for this atom. Experimental data were taken from [Martin *et al.* (2002)].

as discussed in Ch. 10.

Another simple confirmation of the atomic shell-model picture is provided by the excited states of the neon atom shown in Fig. 3.4. This atom has the $(1s)^2(2s)^2(2p)^6$ configuration occupied in the ground state and corresponds to a closed-shell system. All the excited levels can be understood in terms of the promotion of the last occupied $2p$-electron to an unoccupied orbital, starting with the $3s$, $3p$, $3d$, $4s$, $4p$ and so on. In terms of particle addition and removal operators, these states can be obtained from the closed-shell ground state, $\left|\Phi_0^N\right\rangle$ as follows

$$a_{n\ell}^\dagger a_{2p}\left|\Phi_0^N\right\rangle. \tag{3.21}$$

The excitation energy of such a state is then given by $\varepsilon_{n\ell} - \varepsilon_{2p}$, reflecting the results of Eqs. (3.6) and (3.7). The presence of more than one energy level at the position of a state like Eq. (3.21), is due to the splitting that results from the inclusion of the magnetic interactions and the two-body Coulomb interaction. We return to some of the subtleties of this spectrum in Sec. 13.5.

3.3 Nucleons in nuclei

The shell structure observed in atoms is also found in nuclei. However, its origin is quite different. In addition, it is not as easy to understand as in the atomic case. In atoms it can be demonstrated by considering the ionization energy as in Fig. 3.1. Shell closures at 2, 10, 18, 36, 54, and 86, which signal the position of the noble gas atoms, are then clearly visible. For nuclei, a similar quantity, called separation energy, exhibits equivalent features. For neutrons it is defined by

$$S_n(N, Z) = B(N, Z) - B(N - 1, Z) \qquad (3.22)$$

and for protons by

$$S_p(N, Z) = B(N, Z) - B(N, Z - 1), \qquad (3.23)$$

where B describes the nuclear binding energy for the nucleus as a function of N, the number of neutrons, and Z, the number of protons. B is defined by decomposing the total mass of the nucleus as follows

$$M(N, Z) = E(N, Z)/c^2 = N\, m_n\ +\ Z\, m_p\ -\ B(N, Z)/c^2. \qquad (3.24)$$

The separation energy exhibits shell closures as illustrated for $N = 126$ in Fig. 3.5. A shell closure appears for fixed values of the difference $N - Z$ as a function of N, but does display an odd-even staggering that can be interpreted in terms of additional stability of systems with an even number of neutrons. Staggering can be eliminated by considering the separation energy only for odd N with Z even, as a function of the number of neutrons (see Fig. 3.5), and likewise for protons. From this analysis shell closures emerge at $N= 50$, 82, and 126 and at $Z= 50$ and 82. Less clear, but deduced from other data like spectra and magnetic moments, are shell closures for both N and Z corresponding to 2, 8, 20, and 28 [Bohr and Mottelson (1998)]. Historically, it was more difficult to relate these "magic" numbers to shell closures than in the atomic case.

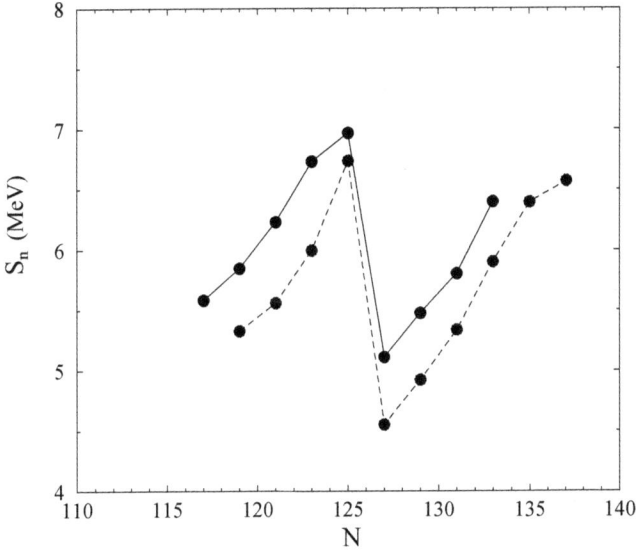

Fig. 3.5   Neutron separation energies for nuclei in the Pb region. Only even Z nuclei are used in this plot. The full line corresponds to $N-Z = 41$, the dashed line to $N-Z = 43$. Data were taken from [Audi *et al.* (2003)].

The sp potential that is responsible for this type of shell structure is generated by the nucleons themselves, since there is no center of attraction as for electrons in atoms. Ultimately, the sp potential must be related to the interactions between the nucleons, which are responsible for the binding of the system. For now, it is useful to introduce an empirical sp potential that provides an adequate description of nuclear shell structure. It can be written as [Bohr and Mottelson (1998)]

$$U = V f(r) + V_{\ell s} \left(\frac{\boldsymbol{\ell} \cdot \boldsymbol{s}}{\hbar^2} \right) r_0^2 \frac{1}{r} \frac{d}{dr} f(r) \tag{3.25}$$

with

$$f(r) = \left[1 + \exp \left(\frac{r - R}{a} \right) \right]^{-1}. \tag{3.26}$$

This form is referred to as a Woods–Saxon shape. The depth of the potential is given by

$$V = \left[-51 \pm 33 \left(\frac{N - Z}{A} \right) \right] \text{ MeV}, \tag{3.27}$$

where the plus sign is for neutrons, the minus sign for protons, and $A = N + Z$ the total number of nucleons. The radius parameter is expressed as

$$R = r_0 \, A^{1/3}, \tag{3.28}$$

with $r_0 = 1.27$ fm, the diffuseness parameter $a = 0.67$ fm, and the strength of the spin-orbit interaction $V_{\ell s} = -0.44V$. For protons one must also include the effect of their mutual Coulomb repulsion, which can be reasonably represented by the potential of a homogeneous sphere with charge Z and radius $R_C = R$. Such a parametrization is successful in describing some of the low-energy properties of nuclei with one more or less particle, with respect to a doubly magic nucleus like ^{208}Pb. A comparison with experimental data is shown in Fig 3.6. In the independent-particle model, the ground state of this nucleus is generated by filling the relevant proton and neutron shells. The lowest-energy states for the $A = 209$ system, are reached by adding a proton or neutron in the corresponding, lowest available, empty shells. The observed and calculated positions of these levels show a good correspondence. The same holds for odd nuclei neighboring ^{16}O, ^{40}Ca, ^{48}Ca, and ^{56}Ni that have closed shells for both protons and neutrons. The energy of an additional proton or neutron for an $A = 209$ system in this simple model is given by

$$\hat{H}_0 \, a_\alpha^\dagger \, \big|^{208}\mathrm{Pb}_{g.s.}\big\rangle = \big[\varepsilon_\alpha + E(^{208}\mathrm{Pb}_{g.s.})\big] \, a_\alpha^\dagger \, \big|^{208}\mathrm{Pb}_{g.s.}\big\rangle, \tag{3.29}$$

where α represents the quantum numbers of an unoccupied proton or neutron state. The experimental information is obtained by subtracting the ground-state energy of ^{208}Pb from the one of ^{209}Bi or ^{209}Pb, yielding the position of the first "empty" level for an extra proton or neutron, respectively. The position of the other experimental levels can then be established by adding their excitation energy to the sp energy corresponding to the ground state of the appropriate $A = 209$ system. Note that this procedure allows a comparison with the sp levels calculated from the Woods–Saxon potential, although the comparison presupposes that the independent-particle model is appropriate. For the states in the $A = 207$ systems, the independent-particle description suggests

$$\hat{H}_0 \, a_\alpha \, \big|^{208}\mathrm{Pb}_{g.s.}\big\rangle = \big[E(^{208}\mathrm{Pb}_{g.s.}) - \varepsilon_\alpha\big] \, a_\alpha \, \big|^{208}\mathrm{Pb}_{g.s.}\big\rangle. \tag{3.30}$$

The calculated position of the levels again corresponds to ε_α and can be compared with experiment for the last occupied "sp" state by subtracting the ground-state energy of the relevant $A = 207$ nucleus from $E^{(0)}(^{208}\mathrm{Pb})$.

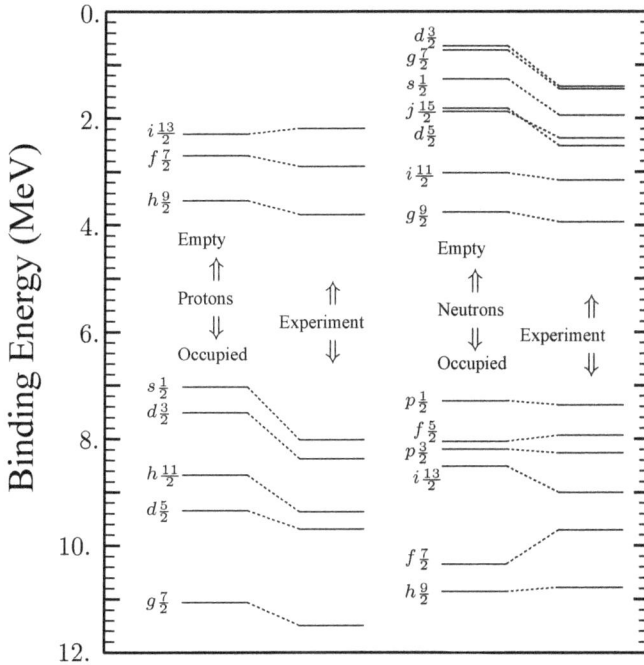

Fig. 3.6 Energy levels of particles and holes with respect to ^{208}Pb. The results for the empirical potential are shown in the first column for protons and the third column for neutrons. These levels are compared with the corresponding experimental data for protons in the second and for neutrons in the fourth column.

Higher excited states then occur lower in energy (are more deeply bound) as shown in Fig 3.6. The position of the sp levels compares favorably with the experimental data, although there are clearly some details missing. The present interpretation of the experimental data remains relevant for a correlated system, as shown in Sec. 21.5.

A Woods–Saxon potential has a finite depth with a finite number of bound states associated with the exponential fall-off at large r, which implies that all bound states are well localized. In contrast, the hydrogen-like potential has an infinite number of bound states due to the r^{-1} behavior. The latter also generates orbits with very large radii for weakly bound states. The central part of the Woods–Saxon potential can be reasonably

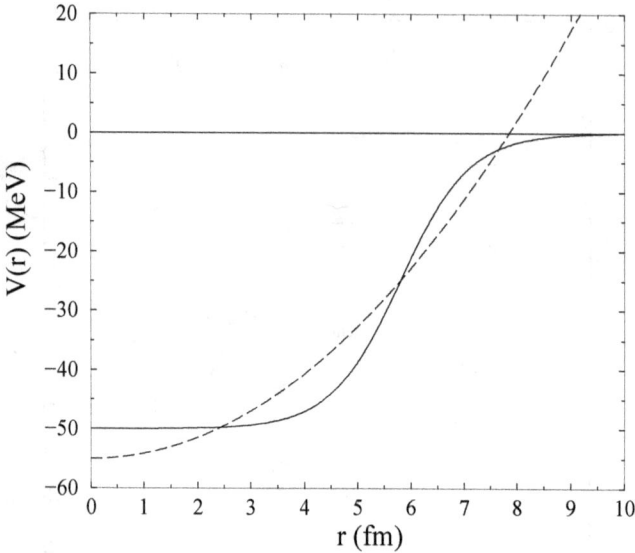

Fig. 3.7 Comparison of the central part of the Woods–Saxon potential given in Eq. (3.25) with an adjusted HO potential for $A = 100$.

approximated by a three-dimensional harmonic oscillator (HO) potential

$$U_{HO}(r) = \frac{1}{2}m\omega^2 r^2 - V_0. \tag{3.31}$$

The oscillator frequency ω and constant shift V_0 can be adjusted to resemble the Woods–Saxon well as shown in Fig. 3.7. The HO potential has only discrete eigenstates. Positive energy states, which correspond to scattering states for the Woods–Saxon well, therefore require special attention. The eigenvalues of the HO potential read

$$H_{HO}|n\ell m_\ell m_s\rangle = \left(\hbar\omega(2n + \ell + \tfrac{3}{2}) - V_0\right)|n\ell m_\ell m_s\rangle \tag{3.32}$$

with

$$n = 0, 1, 2, \ldots$$
$$\ell = 0, 1, 2, \ldots$$
$$-\ell < m_\ell < \ell. \tag{3.33}$$

The total number of oscillator quanta is given by

$$N = 2n + \ell, \tag{3.34}$$

which implies, according to Eq. (3.32), that for each oscillator energy only states with the same parity are degenerate. The HO potential leads to the magic numbers 2, 8, 20, 40, 70, 112, and 168, as can be easily verified. Although the first three shell closures correspond to experimental observations, the others show little resemblance to experiment.

A Nobel prize winning suggestion was made by [Goeppert-Mayer (1949)] and [Jensen *et al.* (1949)] who, independently, introduced a strong one-body spin-orbit potential, similar to the one given in Eq. (3.25). The effect of this potential is mostly experienced at the surface of the nucleus, since the derivative of $f(r)$ peaks there [see Eq. (3.26)]. The presence of the $\boldsymbol{\ell} \cdot \boldsymbol{s}$ operator requires a change of sp basis to states with good total angular momentum

$$|n(\ell s)jm_j\rangle = \sum_{m_\ell m_s} |n\ell m_\ell m_s\rangle \left(\ell \ m_\ell \ s \ m_s| \ j \ m_j\right). \tag{3.35}$$

The transformation bracket is usually referred to as a Clebsch–Gordan coefficient [Sakurai (1994)] (see also App. B). Using the operator identity

$$\boldsymbol{\ell} \cdot \boldsymbol{s} = \tfrac{1}{2}(\boldsymbol{j}^2 - \boldsymbol{\ell}^2 - \boldsymbol{s}^2), \tag{3.36}$$

one finds

$$\frac{\boldsymbol{\ell} \cdot \boldsymbol{s}}{\hbar^2} |n(\ell s)jm_j\rangle = \tfrac{1}{2}\left(j(j+1) - \ell(\ell+1) - \tfrac{1}{2}(\tfrac{1}{2}+1)\right) |n(\ell s)jm_j\rangle. \tag{3.37}$$

Obviously for an *s*-wave, the spin-orbit potential does not contribute. For other ℓ-values one obtains for $j = \ell + \tfrac{1}{2}$ the eigenvalue $\tfrac{1}{2}\ell$, whereas for $j = \ell - \tfrac{1}{2}$ one has $-\tfrac{1}{2}(\ell+1)$ from Eq. (3.37). Combining this result with the sign of $V_{\ell s}$ and that of the derivative of $f(r)$, shows that the spin-orbit interaction can substantially lower the energy of the subshell with the largest orbital angular momentum and $j = \ell + \tfrac{1}{2}$. This is confirmed by experiment in light nuclei, but the shifts are not so large as to alter the main shell closures at 2, 8, and 20. The first deviation occurs for the $0f\tfrac{7}{2}$ shell which becomes a major shell of its own, leading to the observed shell closure at 28. In higher shells the lowering of this $\ell_{max} + \tfrac{1}{2}$ orbit is so large that it comes to reside among the different-parity orbitals of the $N-1$ major shell. This occurs for the $0g\tfrac{9}{2}$ shell, leading to the shell closure at 50, the $0h\tfrac{11}{2}$ with a resulting closure at 82, and finally with the $0i\tfrac{13}{2}$ yielding a shell closure at 126. These features are schematically indicated in Fig. 3.8, where the first column indicates the energy quantum number (not to scale) of the major shells of the HO together with their parity π. The corresponding

$N = 6, \pi+ \quad - \; 0i, 1g, 2d, 3s$ $\cdots\cdots\cdots\cdots\cdots\cdots\cdots$ [126]

$0i\frac{13}{2}$ ——— [14]

$2p\frac{1}{2}$ — [2]

$2p\frac{3}{2}$ ——— [4]

$1f\frac{5}{2}$ ——— [6]

$N = 5, \pi- \quad - \; 0h, 1f, 2p$ $1f\frac{7}{2}$ ——— [8]

$0h\frac{9}{2}$ ——— [10]

$\cdots\cdots\cdots$ [82]

$0h\frac{11}{2}$ — [12]

$2s\frac{1}{2}$ ——— [2]

$1d\frac{3}{2}$ ——— [4]

$N = 4, \pi+ \quad - \; 0g, 1d, 2s$ $1d\frac{5}{2}$ ——— [6]

$0g\frac{7}{2}$ ——— [8]

$\cdots\cdots\cdots$ [50]

$0g\frac{9}{2}$ ——— [10]

$1p\frac{1}{2}$ ——— [2]

$0f\frac{5}{2}$ ——— [6]

$N = 3, \pi- \quad - \; 0f, 1p$ $1p\frac{3}{2}$ ——— [4]

$0f\frac{7}{2}$ ——— [8]

$\cdots\cdots\cdots$ [28]

$\cdots\cdots\cdots$ [20]

$0d\frac{3}{2}$ ——— [4]

$N = 2, \pi+ \quad - \; 0d, 1s$ $1s\frac{1}{2}$ ——— [2]

$0d\frac{5}{2}$ ——— [6]

$\cdots\cdots\cdots$ [8]

$0p\frac{1}{2}$ ——— [2]

$N = 1, \pi- \quad - \; 0p$ $0p\frac{3}{2}$ ——— [4]

$\cdots\cdots\cdots$ [2]

$N = 0, \pi+ \quad - \; 0s$ $0s\frac{1}{2}$ ——— [2]

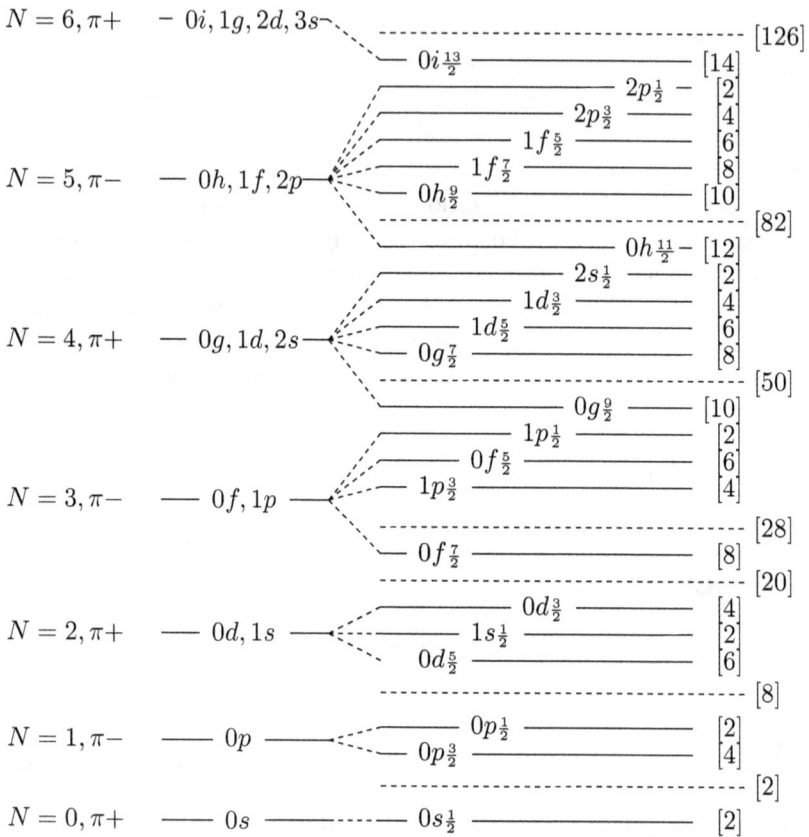

Fig. 3.8 Schematic energy levels for nucleons with inclusion of a substantial spin-orbit potential. The left column indicates the major shells together with their parity π. Appropriate shell closures are listed in the rightmost column.

quantum numbers are given in the next column. Additional splitting of the HO degeneracy, with the more realistic Woods–Saxon well, will also favor the higher ℓ orbitals. This feature, together with the aforementioned spin-orbit effects, is incorporated in the schematic splitting of the sp levels shown with appropriate sp quantum numbers. The latter are given together with the number of nucleons they can contain.

The interaction between nucleons is not completely understood theoretically. In principle, one would like to derive the interaction between nucleons from the perspective of quantum chromodynamics (QCD). This feat is not expected in the near future. Protons and neutrons will remain relevant degrees of freedom to describe nuclei, on account of the energy scales involved for nuclear excitations. We therefore proceed in a practical way, and employ experimental data that characterize the two-nucleon system to construct interactions that represent these data accurately. Different interactions describe the experimental details, but they differ at distances not probed by energies corresponding to elastic nucleon-nucleon (NN) scattering. Additional features of these interactions are discussed in Ch. 4. Dealing with nucleons implies the use of nonrelativistic quantum mechanics and the corresponding Schrödinger or Lippmann–Schwinger equations. At 140 MeV excitation energy in the two-nucleon system, it becomes possible to create an additional pion. Such mesonic degrees of freedom are usually not included explicitly when considering nuclear excitations below this threshold. Consequently, one often studies the nuclear many-body problem with a Hamiltonian of the form given by Eq. (2.48), with a two-body interaction V that accounts for all low-energy two-nucleon data in a nonrelativistic framework. The long-range part of the interaction is accurately represented by the exchange of a virtual pion. The pion is virtual since not enough energy is available for its actual production. In commonly used language, such experimentally constrained interactions are characterized as "realistic". The nuclear many-body problem is thus defined, with obvious restrictions: excitations at energies higher than 140 MeV cannot be accounted for explicitly. Nevertheless it will be possible to understand much of the many-particle aspects of the nucleus, by considering such a Hamiltonian. This is hardly surprising since the coupling to the physical states above 140 MeV is, albeit indirectly, experimentally constrained. Medium modifications of the interaction and the properties of nucleons are most sensitive to an energy scale which is related to shells near the Fermi energy, *i.e.* a scale associated with nucleons moving from occupied to empty levels.

3.3.1 *Empirical mass formula and nuclear matter*

Important qualitative aspects of nuclei, that require an explanation from many-body theory, are revealed by the systematics of nuclear binding as a function of N and Z. The experimental observation that the density in the interior of nuclei is constant and about the same for nuclei heavier than ^{16}O, is also significant [Frois and Papanicolas (1987)]. The systematics of nuclear binding is shown in Fig. 3.9, where the binding energy [Eq. (3.39)] per nucleon is plotted as a function of $A = N + Z$. For each value of A the most stable nucleus was used for the experimental point. A smooth curve through these experimental data is given by the so-called semi-empirical

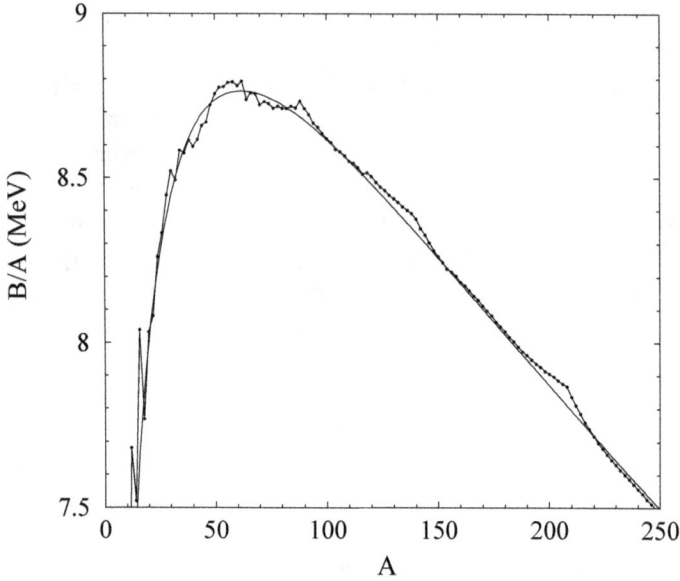

Fig. 3.9 Binding energy per nucleon according to the empiciral mass formula given in Eq. (3.38), compared with the experimental binding for the most stable nucleus at a given A as a function of A. Data were obtained from [Audi *et al.* (2003)].

mass formula [Weizsäcker (1935); Bethe and Bacher (1936)]

$$B = b_{vol}A - b_{surf}A^{2/3} - \frac{1}{2}b_{sym}\frac{(N-Z)^2}{A} - \frac{3}{5}\frac{Z^2 e^2}{R_c}. \qquad (3.38)$$

A relevant set of values of the parameters includes [Bohr and Mottelson (1998)]: $b_{vol} = 15.56$ MeV, $b_{surf} = 17.23$ MeV, $b_{sym} = 46.57$ MeV, and $R_c = 1.24A^{1/3}$ fm. Most nuclei have a binding energy of about 8 MeV per particle, which is rather small compared to the rest mass of the nucleon which is about 939 MeV. A plot of the binding energy per nucleon, as a function of A, according to Eq. (3.38) (evaluated for a given A at the most stable value of N), is compared with the corresponding experimental data in Fig. 3.9. The term proportional to the number of particles in the mass formula, is called the volume term. The second term represents the loss of attraction due to the presence of the surface, and reflects the lack of neighbors to interact with in this region. These first two terms suggest a saturation of the nuclear interaction, implying that nucleons, on the average, experience attraction from other nucleons only at rather short-range. The third term incorporates the tendency of the nuclear force to favor nuclei with $N = Z$ and is called the symmetry energy. The last term represents the energy of a uniformly charged sphere of radius R_c.

Theoretically switching off the Coulomb interaction between the protons leaves only the volume term for $N = Z$ in the limit of infinite volume and constant density. This limit is extremely relevant since, as already mentioned, the central density in the interior of nuclei is found to be constant. Elastic electron scattering experiments reveal the properties of the charge density at the center of a nucleus; multiplying this number with A/Z results in the aforementioned constant central density of 0.16 nucleons per fm^3. Such a hypothetical system is referred to as "nuclear matter". It embodies essential global properties of nuclei. The value of the saturation density of nuclear matter, and the corresponding binding energy per particle of about 16 MeV, requires a theoretical explanation that starts from a realistic two-body interaction. Indeed, the issue remains a subject of study to this day. The goal of the nuclear matter problem is then to explain why a minimum in the energy per particle of -16 MeV occurs at a density of 0.16 nucleons per fm^3. In addition, one must quantitatively reproduce these numbers applying many-particle methods. A typical failure has been that when the correct energy at the minimum of the energy per particle was obtained, the corresponding density is about a factor of two too high, or when the correct saturation density was calculated, the energy is only about half of what is required. This is discussed in more detail in Sec. 16.3.4.

3.4 Second quantization and isospin

Neutrons and protons display the same magic numbers, and therefore follow the same shell structure. This cannot be accidental. In effect, the mass difference of the neutron and the proton is only about one part in a thousand of the average of the proton and neutron mass. According to quantum mechanics, this degeneracy must reflect a symmetry of the Hamiltonian describing the strong interaction [Heisenberg (1932)]. In other words, there is an observable which commutes with the Hamiltonian H_S governing the strong interaction. As a consequence, simultaneous eigenstates of this observable and H_S can be found. Assuming that the strong force is independent of particle type, one can further neglect the weak and electromagnetic interactions that do distinguish between the proton and the neutron, and eliminate the small mass difference.

In the following discussion [Georgi (1982)] an explicit distinction is made between particle addition and removal operators for protons and neutrons. For example, the operator $p_\alpha^\dagger$ adds a proton with quantum numbers α to any state in Fock space, and $n_\alpha^\dagger$ does the same for a neutron. These operators also obey anticommutation relations, yielding

$$\{p_\alpha^\dagger, p_\beta\} = \delta_{\alpha,\beta} \quad \text{and} \quad \{n_\alpha^\dagger, n_\beta\} = \delta_{\alpha,\beta}, \tag{3.39}$$

with all other anticommutators equal to zero, including those involving proton and neutron operators. A state with Z protons and N neutrons is then given by

$$|\alpha_1\alpha_2...\alpha_Z; \beta_1\beta_2...\beta_N\rangle = p_{\alpha_1}^\dagger p_{\alpha_2}^\dagger...p_{\alpha_Z}^\dagger n_{\beta_1}^\dagger n_{\beta_2}^\dagger...n_{\beta_N}^\dagger |0\rangle. \tag{3.40}$$

Observing that at the sp level the interchange of a proton for a neutron with otherwise identical quantum numbers does not change the energy, one can postulate that this should also be valid for any collection of protons and neutrons. The corresponding operators are easily written down in second quantization. The operator which changes neutron into proton states, while leaving all other quantum numbers unchanged, is represented by

$$\hat{T}^+ = \sum_\alpha p_\alpha^\dagger n_\alpha. \tag{3.41}$$

The operator

$$\hat{T}^- = \sum_\alpha n_\alpha^\dagger p_\alpha \tag{3.42}$$

accomplishes the opposite. Based on the degeneracy of the neutron and proton energy, the assumption is that

$$[\hat{H}_S, \hat{T}^\pm] = 0. \tag{3.43}$$

Consider now the following commutator of $\hat{T}^+$ and $\hat{T}^-$ denoted by $\hat{T}_3$

$$\hat{T}_3 = \frac{1}{2}[\hat{T}^+, \hat{T}^-] = \frac{1}{2}\sum_{\alpha\beta}(p_\alpha^\dagger n_\alpha n_\beta^\dagger p_\beta - n_\beta^\dagger p_\beta p_\alpha^\dagger n_\alpha)$$

$$= \frac{1}{2}\sum_{\alpha\beta}(p_\alpha^\dagger p_\beta \delta_{\alpha,\beta} - n_\beta^\dagger n_\alpha \delta_{\alpha,\beta}) = \frac{1}{2}\sum_\alpha(p_\alpha^\dagger p_\alpha - n_\alpha^\dagger n_\alpha). \tag{3.44}$$

This operator merely counts the number of protons and subtracts the number of neutrons (multiplied by $\frac{1}{2}$). On physical grounds, it is clear that

$$[\hat{H}_S, \hat{T}_3] = 0. \tag{3.45}$$

It is also possible to show that

$$[\hat{T}_3, \hat{T}^{\pm}] = \pm \hat{T}^{\pm}. \tag{3.46}$$

These operators therefore satisfy the same algebra as the angular momentum operators. Indeed, defining

$$\hat{T}_1 = \frac{1}{2}(\hat{T}^+ + \hat{T}^-) \tag{3.47}$$

and

$$\hat{T}_2 = \frac{1}{2i}(\hat{T}^+ - \hat{T}^-), \tag{3.48}$$

there is a one-to-one correspondence between the triplet $(T_1, T_2, T_3)^1$ and the triplet of angular momentum operators (J_x, J_y, J_z), including identical commutation relations. The spectrum of the angular momentum operators $\boldsymbol{J}^2$ and J_z is solely determined by the commutation relations between $J_x, J_y,$ and J_z. Thus we can simply relabel all these results in terms of new quantum numbers related to the operators $\boldsymbol{T}^2$ and T_3, that are referred to as the total isospin (squared) and its 3-projection. Isospin invariance of the strong interaction implies that its Hamiltonian is unchanged under isospin rotations that are generated by the operators $T_1, T_2,$ and T_3, in complete analogy with the angular momentum case. Rotations in "iso"-space about a direction $\hat{\boldsymbol{n}}$ can be written as

$$R(\hat{\boldsymbol{n}}) = \exp\{-i\hat{\boldsymbol{n}} \cdot \boldsymbol{T}\}. \tag{3.49}$$

Physical states are labeled with isospin quantum numbers T and M_T, for total isospin and its third component, respectively. Although only states with T_3 as a good quantum number are observed, the full apparatus of angular momentum algebra, summarized in App. B, can be applied in making use of the isospin symmetry of the strong interaction. Clebsch–Gordan coefficients can then also be employed to couple states to good total isospin. The proton and the neutron represent a doublet with total isospin $t = \frac{1}{2}$. The proton is, arbitrarily, assigned isospin 3-projection $m_t = \frac{1}{2}$ and the neutron $m_t = -\frac{1}{2}$. Historically, this assignment was opposite in nuclear physics to make T_3 positive for nuclei with neutron excess, which represent the vast majority of stable nuclides.

---

[1]Although the kinetic energy is also represented by this symbol, no ambiguity should arise.

Instead of dealing separately with neutrons and protons, the isospin formalism can be used. The complete set of nucleon quantum numbers must then include isospin. Hence, a proton at $\boldsymbol{r}$ with spin projection m_s can be denoted by

$$|\boldsymbol{r}m_s\rangle_p = |\boldsymbol{r}m_s m_t = \tfrac{1}{2}\rangle\,, \qquad (3.50)$$

where the total isospin $t = \tfrac{1}{2}$ has been suppressed just like the total spin quantum number $s = \tfrac{1}{2}$. The neutron is then assigned $m_t = -\tfrac{1}{2}$. The eigenvalue equations for this doublet read

$$\boldsymbol{T}^2|\boldsymbol{r}m_s m_t\rangle = \tfrac{1}{2}(\tfrac{1}{2}+1)|\boldsymbol{r}m_s m_t\rangle \qquad (3.51)$$

and

$$T_3|\boldsymbol{r}m_s m_t\rangle = m_t|\boldsymbol{r}m_s m_t\rangle\,. \qquad (3.52)$$

Particles in an isospin multiplet are looked upon as identical particles, with T_3 as just another quantum number. For this reason the proton addition and removal operators anticommute with those for neutrons. Examples of the utility of the isospin concept abound in nuclear physics.

It is useful to determine first the angular momentum of a closed-shell system before its isospin. Consider the third component of the total angular momentum operator in second quantization

$$\hat{J}_z = \sum_{n\ell jm}\sum_{n'\ell'j'm'}\langle n\ell jm|j_z|n'\ell'j'm'\rangle a^\dagger_{n\ell jm}a_{n'\ell'j'm'}$$
$$= \sum_{n\ell jm}\hbar m\, a^\dagger_{n\ell jm}a_{n\ell jm}. \qquad (3.53)$$

Without loss of generality, we can let this operator act on one full shell where all the particles have quantum numbers n,ℓ,j and all components of j_z are occupied

$$\hat{J}_z|n\ell j; m = -j, -j+1, ..., j\rangle$$
$$= \sum_m \hbar m\, a^\dagger_{n\ell jm}a_{n\ell jm}|n\ell j; m = -j, -j+1, ..., = j\rangle$$
$$= \left\{\sum_{m=-j}^{j}\hbar m\right\}|n\ell j; m = -j, -j+1, ..., j\rangle$$
$$= 0 \times |n\ell j; m = -j, -j+1, ..., j\rangle\,. \qquad (3.54)$$

We conclude that the z-component of the total angular momentum vanishes. By applying the raising and lowering operator to the closed shell in a similar way, more vanishing results are generated, demonstrating that the total angular momentum of this state is zero. The same analysis holds for the total isospin. Note that for a given shell, both proton and neutron states must be completely filled to yield a total isospin of zero.

3.5 Exercises

(1) Consider the magnetic moment of Z electrons in an atom with nuclear charge Z (atomic units are used)

$$\mu = -\frac{1}{2} \sum_{i=1}^{Z} (\boldsymbol{\ell}_i + 2\boldsymbol{s}_i).$$

  a) Determine the second quantized form of the z-component of this operator using the sp basis formed by the eigenstates of the sp Hamiltonian which treats the other electrons in central-field approximation (see Eq. (3.18)). Be sure to evaluate the relevant sp matrix elements in this basis.
  b) Consider an atom with one electron outside a set of closed shells, occupying the lowest sp state that is not filled. Denote the quantum numbers of this state by n, ℓ, m_ℓ, m_s. Assume that for the last electron $m_\ell = \ell$ and $m_s = \frac{1}{2}$. Determine the magnetic moment of this atomic state (with one electron outside the closed shells) by calculating the corresponding expectation value of the operator obtained in part a) (this is the actual definition of the magnetic moment). When evaluating the magnetic moment, you should carefully consider all possible contributions, including the one from the closed shells (if any). Compare your result with experimental data for alkali atoms.

(2) Calculate the magnetic moment of ^{15}O, ^{15}N, ^{17}O, and ^{17}F in the independent-particle approximation. It is defined by

$$\mu = \langle JM_J = J | \hat{\mu}_z | JM_J = J \rangle,$$

where for A nucleons the magnetic moment operator (in first quantization) is given by

$$\mu^A = \sum_{i=1}^{A} \{g_\ell(i)\ell_i + g_s(i)s_i\}.$$

The factor g_ℓ is 1 in units of nuclear magnetons for protons, and 0 for neutrons. The notation assumes that the orbital and spin angular momentum are in units of $\hbar$. A nuclear magneton is given by $e\hbar/2m_pc$. In these units the spin factor g_s is 5.58 for protons and -3.82 for neutrons. Assume that the above nuclei correspond to either a missing $p\frac{1}{2}$ proton or neutron in the double closed-shell ^{16}O, or a $d\frac{5}{2}$ proton, or neutron added to it. Employ the second-quantized operator for the magnetic moment and calculate the sp matrix elements. Use the projection theorem [see Eq.(3.10.40) in [Sakurai (1994)]]. Produce an argument to demonstrate that closed shells don't contribute. Express your results in nuclear magnetons and compare with experiment.

(3) Write down the charge density operator for the nucleus in first quantization including isospin. Construct the corresponding second-quantized operator. Show that this operator may be written as

$$\hat{\rho}_C(r) = \frac{1}{4\pi}e \sum_{\ell jm_jm_t} (\tfrac{1}{2} + m_t)a^\dagger_{r\ell jm_jm_t} a_{r\ell jm_jm_t}$$

in the appropriate basis $\{|r(ls)jm_jm_t\rangle\}$. This operator is appropriate for a closed-shell system which has no angular momentum. In that case the nuclear charge distribution depends only on r. One may therefore divide the charge density operator by 4π, and integrate over all angles to obtain the above expression. Evaluate the expectation value of this operator for the ground state of a doubly closed-shell nucleus.

(4) Use the hydrogen-like Hamiltonian for the He atom to approximate the ground state in the independent-particle model. Calculate the energy of the ground state with the inclusion of the expectation values of the two-body electron–electron interaction (first-order perturbation theory) and compare with experiment.

(5) Show that the convention where proton operators commute with those of neutrons (*i.e.* $[p^\dagger_\alpha, n_\beta] = [p^\dagger_\alpha, n^\dagger_\beta] = 0$, *etc.*) is equivalent to the choice of anticommuting operators made in Sec. 3.4, and leads to the same results for the expectation value of operators that conserve the number of protons and neutrons.

Chapter 4

Two-particle states and interactions

The present chapter deals with simple symmetry considerations for two-particle states that clarify the consequences of the Pauli principle, when such states are coupled to good total angular momentum or isospin. In Sec. 4.1, we study free particles and the transformation to total and relative momentum. By first reviewing antisymmetric two-particle states for nucleons, it is possible to simplify to other systems with smaller degeneracies, like two electrons or two ^{3}He atoms. Possible states for two free identical bosons are obtained in the same way. In each case important contraints on the possible angular momentum states are encountered. These results can be generalized to two particles (holes) outside closed shells, and several examples are discussed in Sec. 4.2, leading to restrictions on the possible angular momentum and isospin states for particles in the same orbit. In Sec. 4.3, some general observations are presented on the subject of two-body interactions. Examples of relevant interactions for various systems are described in Sec. 4.4.

4.1 Symmetry considerations for two-particle states

It is important to develop the consequences of the symmetry of two-particle states beyond the obvious ones discussed so far. It is frequently practical, or even necessary, in dealing with two particles to employ the angular momentum basis. This is important for short-range interactions which only influence a limited number of angular momentum states. The relevant transformation proceeds from plane waves to spherical waves, and is usually treated at the sp level. The consequences of symmetry are substantial however, and will be illustrated first for two nucleons in free space. These results can then be simplified for other identical fermion systems.

4.1.1 *Free-particle states*

We start by briefly discussing the plane-wave states associated with the kinetic energy operator

$$T = \frac{\boldsymbol{p}^2}{2m}. \tag{4.1}$$

Suppressing possible discrete quantum numbers, the momentum eigenstates of Eq. (4.1)

$$\frac{\boldsymbol{p}^2}{2m}\,|\boldsymbol{p}'\rangle = \frac{\boldsymbol{p}'^2}{2m}\,|\boldsymbol{p}'\rangle \tag{4.2}$$

describe particles that are not influenced by any interaction. The wave function associated with these states, the plane wave, can be written as

$$\langle \boldsymbol{r}|\boldsymbol{p}\rangle = \frac{1}{(2\pi\hbar)^{3/2}}e^{\frac{i}{\hbar}\boldsymbol{p}\cdot\boldsymbol{r}}. \tag{4.3}$$

The present choice leads to the normalization condition

$$\langle \boldsymbol{p}'|\boldsymbol{p}\rangle = \frac{1}{(2\pi\hbar)^3}\int d\boldsymbol{r}\; e^{\frac{i}{\hbar}(\boldsymbol{p}-\boldsymbol{p}')\cdot\boldsymbol{r}} = \delta(\boldsymbol{p}'-\boldsymbol{p}). \tag{4.4}$$

It is often practical to use wave vectors with $\boldsymbol{k} = \boldsymbol{p}/\hbar$. We introduce the convention here that momenta are denoted by $\boldsymbol{p},\boldsymbol{p}',\boldsymbol{P}$, etc. The notation reserved for wave vectors is $\boldsymbol{k},\boldsymbol{K},\boldsymbol{q}$, etc. The wave function for the state with wave vector $\boldsymbol{k}$ becomes

$$\langle \boldsymbol{r}|\boldsymbol{k}\rangle = \frac{1}{(2\pi)^{3/2}}e^{i\boldsymbol{k}\cdot\boldsymbol{r}} \tag{4.5}$$

with normalization

$$\langle \boldsymbol{k}'|\boldsymbol{k}\rangle = \delta(\boldsymbol{k}'-\boldsymbol{k}). \tag{4.6}$$

For some applications, it is more convenient to use the so-called box normalization. The particle is then confined to a cubic box with sides L and volume $V = L^3$, yielding the wave function

$$\langle \boldsymbol{r}|\boldsymbol{p}\rangle = \frac{1}{\sqrt{V}}e^{\frac{i}{\hbar}\boldsymbol{p}\cdot\boldsymbol{r}}. \tag{4.7}$$

Boundary conditions imposed at the edges of the box, allow only discrete values of the momentum $\boldsymbol{p}$. The normalization then reads

$$\langle \boldsymbol{p}'|\boldsymbol{p}\rangle = \delta_{\boldsymbol{p}',\boldsymbol{p}}, \tag{4.8}$$

which can be obtained from

$$\langle \boldsymbol{p}'|\boldsymbol{p}\rangle = \int_{box} d\boldsymbol{r} \ \langle \boldsymbol{p}'|\boldsymbol{r}\rangle\langle \boldsymbol{r}|\boldsymbol{p}\rangle = \frac{1}{V}\int_{box} d\boldsymbol{r} \ e^{\frac{i}{\hbar}(\boldsymbol{p}-\boldsymbol{p}')\cdot\boldsymbol{r}} = \delta_{\boldsymbol{p}',\boldsymbol{p}}. \quad (4.9)$$

The inclusion of spin and other discrete quantum numbers like isospin is straightforward.

4.1.2 Pauli principle for two-particle states

A free nucleon can be described by quantum numbers corresponding to momentum, spin $(\frac{1}{2})$, spin projection, isospin $(\frac{1}{2})$, and isospin projection

$$|\boldsymbol{p} \ s = \tfrac{1}{2} \ m_s \ t = \tfrac{1}{2} \ m_t\rangle \equiv |\boldsymbol{p}m_s m_t\rangle. \quad (4.10)$$

These states can, for example, be normalized in the box as

$$\langle \boldsymbol{p}'m_s'm_t'|\boldsymbol{p}m_s m_t\rangle = \delta_{\boldsymbol{p}',\boldsymbol{p}}\delta_{m_s,m_s'}\delta_{m_t,m_t'}. \quad (4.11)$$

Antisymmetry for two nucleons requires the two-body state to be constructed as follows

$$
\begin{aligned}
&|\boldsymbol{p}_1 m_{s_1} m_{t_1}; \boldsymbol{p}_2 m_{s_2} m_{t_2}\rangle \\
&= \frac{1}{\sqrt{2}}\{|\boldsymbol{p}_1 m_{s_1} m_{t_1}\rangle |\boldsymbol{p}_2 m_{s_2} m_{t_2}\rangle - |\boldsymbol{p}_2 m_{s_2} m_{t_2}\rangle |\boldsymbol{p}_1 m_{s_1} m_{t_1}\rangle\} \\
&= \frac{1}{\sqrt{2}}\sum_{SM_S}\sum_{TM_T}\{(\tfrac{1}{2} \ m_{s_1} \ \tfrac{1}{2} \ m_{s_2} \ |S \ M_S)(\tfrac{1}{2} \ m_{t_1} \ \tfrac{1}{2} \ m_{t_2} \ |T \ M_T) \\
&\quad\times |\boldsymbol{p}_1 \ \boldsymbol{p}_2 \ S \ M_S \ T \ M_T) - (\tfrac{1}{2} \ m_{s_2} \ \tfrac{1}{2} \ m_{s_1} \ |S \ M_S)(\tfrac{1}{2} \ m_{t_2} \ \tfrac{1}{2} \ m_{t_1} \ |T \ M_T) \\
&\quad\times |\boldsymbol{p}_2 \ \boldsymbol{p}_1 \ S \ M_S \ T \ M_T)\}, \quad (4.12)
\end{aligned}
$$

where the individual spins and isospins have been coupled to total spin and isospin in the second equality. Since the dynamics is related to the relative motion of the particles, it is appropriate to switch to a basis involving the center of mass (total) and relative momentum

$$\boldsymbol{P} = \boldsymbol{p}_1 + \boldsymbol{p}_2 \quad (4.13)$$
$$\boldsymbol{p} = \tfrac{1}{2}(\boldsymbol{p}_1 - \boldsymbol{p}_2). \quad (4.14)$$

Consequently, the states in the last line of Eq. (4.12) both have the same total momentum but opposite relative momentum, $\boldsymbol{p}$ and $-\boldsymbol{p}$, respectively. The transformation of the relative momentum quantum number to the basis with its magnitude, orbital angular momentum, and its projection for these two cases is given by

$$|\boldsymbol{p}\rangle = \sum_{LM_L} |pLM_L\rangle \langle LM_L|\hat{\boldsymbol{p}}\rangle = \sum_{LM_L} |pLM_L\rangle Y^*_{LM_L}(\hat{\boldsymbol{p}}) \qquad (4.15)$$

and

$$|-\boldsymbol{p}\rangle = \sum_{LM_L} |pLM_L\rangle \langle LM_L|\widehat{-\boldsymbol{p}}\rangle = \sum_{LM_L} |pLM_L\rangle (-1)^L Y^*_{LM_L}(\hat{\boldsymbol{p}}). \qquad (4.16)$$

The following property of the spherical harmonics has been used in the last equation

$$Y^*_{LM_L}(\widehat{-\boldsymbol{p}}) = Y^*_{LM_L}(\pi - \theta_p, \phi_p + \pi) = (-1)^L Y^*_{LM_L}(\hat{\boldsymbol{p}}). \qquad (4.17)$$

The symmetry property of the Clebsch–Gordan coefficients (see App. B) for spin

$$(\tfrac{1}{2}\, m_{s_2}\, \tfrac{1}{2}\, m_{s_1}\, |S\, M_S) = (-1)^{\frac{1}{2}+\frac{1}{2}-S}\, (\tfrac{1}{2}\, m_{s_1}\, \tfrac{1}{2}\, m_{s_2}\, |S\, M_S) \qquad (4.18)$$

and isospin

$$(\tfrac{1}{2}\, m_{t_2}\, \tfrac{1}{2}\, m_{t_1}\, |T\, M_T) = (-1)^{\frac{1}{2}+\frac{1}{2}-T}\, (\tfrac{1}{2}\, m_{t_1}\, \tfrac{1}{2}\, m_{t_2}\, |T\, M_T), \qquad (4.19)$$

are used to write Eq. (4.12) as

$$|\boldsymbol{p}_1 m_{s_1} m_{t_1}; \boldsymbol{p}_2 m_{s_2} m_{t_2}\rangle \qquad (4.20)$$

$$= \frac{1}{\sqrt{2}} \sum_{SM_STM_TLM_L} (\tfrac{1}{2}\, m_{s_1}\, \tfrac{1}{2}\, m_{s_2}\, |S\, M_S)\,(\tfrac{1}{2}\, m_{t_1}\, \tfrac{1}{2}\, m_{t_2}\, |T\, M_T)\, Y^*_{LM_L}(\hat{\boldsymbol{p}})$$

$$\times \left[1 - (-1)^{L+S+T}\right] |\boldsymbol{P}\, p\, LM_LSM_S\, TM_T\rangle$$

$$= \frac{1}{\sqrt{2}} \sum_{SM_STM_TLM_LJM_J} (\tfrac{1}{2}\, m_{s_1}\, \tfrac{1}{2}\, m_{s_2}\, |S\, M_S)\,(\tfrac{1}{2}\, m_{t_1}\, \tfrac{1}{2}\, m_{t_2}\, |T\, M_T)\, Y^*_{LM_L}(\hat{\boldsymbol{p}})$$

$$\times (L\, M_L\, S\, M_S\, |J\, M_J) \left[1 - (-1)^{L+S+T}\right] |\boldsymbol{P}\, p\, (LS)JM_J\, TM_T\rangle.$$

The coupling to total angular momentum has been performed after the last equality sign. The main point of Eq. (4.20) is the appearance of the factor $\left[1 - (-1)^{L+S+T}\right]$. Its presence demonstrates that only when $L + S + T$ is odd, a physical antisymmetric state can occur. In the case of an S-wave interaction ($L = 0$) only two possibilities exist: either $S = 0$ and $T = 1$ or $S = 1$ and $T = 0$. The two other combinations of spin and isospin are excluded. The spectroscopic notation for the different channels is given by $^{2S+1}L_J$ where the actual values of S and J are inserted, and the letter notation for L is used ($L = 0$ corresponds to S, $L = 1$ to P, $L = 2$ to D *etc.*). The two S-wave channels for nucleons are thus denoted by $^1S_0$ and $^3S_1$. The strong interaction conserves parity and is a scalar with respect to rotations generated by $\boldsymbol{J}$ and $\boldsymbol{T}$; the total angular momentum and total isospin. This implies that the coupling between different channels must conserve J and T and can change the L-value by $\Delta L = 0, \pm 2$. It follows that the $^1S_0$ two-proton channel is uncoupled, whereas the proton–neutron channel allows a coupling between the $^3S_1$ and $^3D_1$ channels. The latter coupling is realized in nature, due to the presence of the so-called tensor force, which is instrumental in binding the deuteron and giving it its quadrupole moment. For this reason the coupling to total angular momentum in Eq. (4.20) is necessary for nucleons.

If we now turn to antisymmetric two-particle states for electrons or ^{3}He atoms which have spin $\frac{1}{2}$, one can start from Eq. (4.12) and simply remove all reference to isospin. The corresponding factor that decides which partial wave channels are physically allowed then becomes $\left[1 + (-1)^{L+S}\right]$. This shows that an S-wave interaction implies a total spin of zero, whereas a P-wave requires a total spin of one, *etc.* Since there is no need to consider tensor or spin-orbit forces for these systems, L and S are separately conserved and the coupling to total angular momentum states can be omitted. If only one spin projection of the species is present, the consequence of the Pauli principle is even more dramatic. The Pauli factor now becomes $\left[1 - (-1)^L\right]$, demonstrating that there can be no S-wave interaction. Recent efforts to cool fermionic atoms in magnetic and optical traps to temperatures substantially below the Fermi temperature, have to deal with this lack of S-wave interaction. For spinless bosons, the above analysis shows that only states with even L survive, consistent with the experimental observations discussed in Sec. 1.4.

4.2 Two particles outside closed shells

In finite systems with spherical symmetry, the coupling of angular momentum states for two particles must be considered. The consequences of the Pauli principle are also striking here. We will illustrate this by using second quantization. Consider two particles added to a closed-shell nucleus such as discussed in Sec. 3.3. We assume that these two particles are either two protons or two neutrons in the same sp shell, characterized by total angular momentum j. Such a state can be written as

$$|\Phi_{jm,jm'}\rangle = a^\dagger_{jm} a^\dagger_{jm'} |\Phi_0\rangle . \qquad (4.21)$$

It is immediately clear that a total angular momentum of $J = 2j$ is not allowed, since this would require both particles to have the same maximal projection of the angular momentum. The possible total angular momentum states can be obtained by coupling the sp angular momenta using Clebsch–Gordan coefficients. Employing their symmetry properties and the anticommutation relation of the particle addition operators, one finds

$$|\Phi_{jj,JM}\rangle = \sum_{mm'} (j\ m\ j\ m'\ |J\ M) |\Phi_{jm,jm'}\rangle = (-1)^J\ |\Phi_{jj,JM}\rangle , \qquad (4.22)$$

where a change of dummy indices has been employed. The factor $(-1)^J$ ensures that only even values of J yield physical states. An example of this situation is illustrated in Fig. 4.1 for two protons or two neutrons added to the closed-shell system ^{40}Ca. In the independent-particle model of Ch. 3, these nucleons will occupy the $0f\frac{7}{2}$ sp state. The allowed values of the total angular momentum for two such nucleons are therefore 0, 2, 4, and 6. The corresponding levels in ^{42}Ca and ^{42}Ti are indicated by solid lines. Other levels with a more complex interpretation are indicated by dashed lines. For the ^{42}Sc nucleus all values between 0 and 7 are possible for the total angular momentum when proton and neutron $0f\frac{7}{2}$ are involved. Some of these states are present at low energy in this nucleus. It is instructive to analyze these same spectra using the isospin formalism. The similarity of the spectra of ^{42}Ca and ^{42}Ti certainly confirms the importance of the isospin description.

To include isospin we proceed in a similar fashion. Denoting the uncoupled states by

$$|\Phi_{jmm_t,jm'm'_t}\rangle = a^\dagger_{jmm_t} a^\dagger_{jm'm'_t} |\Phi_0\rangle , \qquad (4.23)$$

one finds

$$|\Phi_{jj,JM,TM_T}\rangle = \sum_{mm'm_tm'_t}(j\ m\ j\ m'\ |J\ M)(\tfrac{1}{2}\ m_t\ \tfrac{1}{2}\ m'_t\ |T\ M_T)|\Phi_{jmm_t,jm'm'_t}\rangle$$
$$= (-1)^{J+T+1} |\Phi_{jj,JM,TM_T}\rangle . \qquad (4.24)$$

Consequently $J+T$ must be odd. This result is consistent with Eq. (4.22), since for two protons or two neutrons the total isospin must be one. The spectrum of ^{42}Sc in Fig. 4.1 thus contains both $T = 0$ and 1 states. The

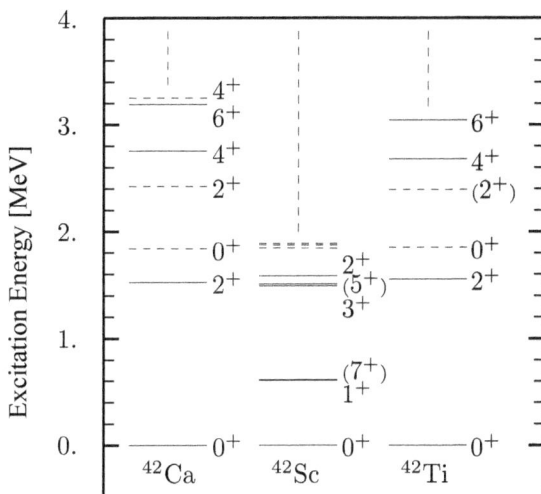

Fig. 4.1 Lowest energy levels for the $A = 42$ system illustrating the effect of the Pauli principle. The relevant sp level for the nucleons is the $0f\frac{7}{2}$. Those levels that can be interpreted as $0f\frac{7}{2}^2$ are indicated by solid lines. Total angular momentum and parity of the levels are indicated where appropriate. When this assignment is not certain round brackets are used. The three columns corresponds to the nuclei that can be reached from ^{40}Ca by adding two nucleons. The levels displayed for ^{40}Ca and ^{42}Ti all have isospin 1 and correspond closely to each other. The 0^+ and 2^+ levels in ^{42}Sc are likewise $T = 1$ states. The other levels shown, correspond to $T = 0$ configurations. The presence of additional levels at higher energy is indicated by the vertical dashed lines. Data were taken from [Sing and Cameron (2001)].

even J states correspond to similar states in the other nuclei with about the same excitation energies. For the odd J states, with $T = 0$, there is no counterpart.

For electrons in the same atomic orbital we can apply an identical strategy. The uncoupled states can be denoted by

$$|\Phi_{\ell m_\ell m_s,\ell m'_\ell m'_s}\rangle = a^\dagger_{\ell m_\ell m_s} a^\dagger_{\ell m'_\ell m'_s} |\Phi_0\rangle, \tag{4.25}$$

and one obtains

$$|\Phi_{\ell,LM_L,SM_S}\rangle$$
$$= \sum_{m_\ell m'_\ell m_s m'_s} (\ell\, m_\ell\, \ell\, m'_\ell\, |L\, M_L)\, (\tfrac{1}{2}\, m_s\, \tfrac{1}{2}\, m'_s\, |S\, M_S)\, |\Phi_{\ell m_\ell m_s,\ell m'_\ell m'_s}\rangle$$
$$= (-1)^{L+S} |\Phi_{\ell\ell,LM_L,SM_S}\rangle. \tag{4.26}$$

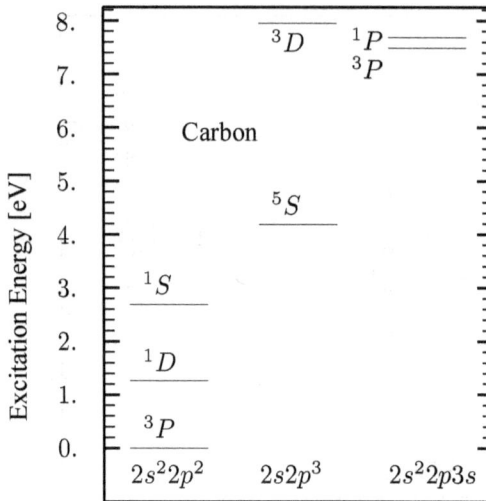

Fig. 4.2 Lowest energy levels of the carbon atom illustrating the effect of the Pauli principle. The three columns are labeled at the bottom by their main electronic configuration. The first column corresponds to two electrons in the same $2p$ orbit which leads to the restriction $L + S$ even, implied by Eq. (4.26). This restriction does not apply to the other configurations. Spectroscopic notation, ^{2S+1}L, is employed to identify the L and S values of the different levels. In the case of $S = 1$, the additional splitting, corresponding to the total angular momentum J, is too small to be visible. Data were obtained from [Martin *et al.* (2002)].

We conclude that $L + S$ must be even, which is illustrated in Fig. 4.2 for the three lowest states in the carbon atom. These states have $2p$ electrons outside the $1s^2 2s^2$ configuration. With the coupling scheme of Eq. (4.26) these last two electrons can couple to $S = 0$ and 1. The possible L-values are given by 0, 1, and 2, denoted by S, P, and D, respectively. The ground state has $S = 1$ and then must have $L = 1$ on account of the Pauli principle. This configuration is denoted by ^3P. The even L-values necessarily have $S = 0$ and are labeled by ^1S and ^1D, respectively. These levels can be found in the first column of Fig. 4.2. For $S = 1$, additional splittings occur since different values of the total angular momentum are possible. They are too small to be distinguished in the figure. It is instructive to count the number of states of the $2p^2$ system, including all the possible degeneracies, as suggested in one of the exercises of this chapter. The coupling to individual total angular momentum j, as for nucleons, and the subsequent coupling to total angular momentum should of course lead to the same number of states, when the Pauli principle is taken into account. A few other low-energy configurations are also listed in the second and third column of Fig. 4.2. The Pauli principle does not restrict the values of L and S for these states.

4.3   General discussion of two-body interactions

The main issue in the many-body problem is an appropriate treatment of the interaction between the constituent particles. Before giving some explicit examples of relevant interactions, it is useful to put this discussion in a wider perspective. All interactions between the spin-$\frac{1}{2}$ fermions of the "standard model" of the electroweak and strong interactions, take place by the exchange of spin-1 bosons. These fermions include all the quarks which come in three colors and six flavors, and all the leptons which include the electron, the muon, and the tau together with their corresponding neutrino's. The exchanged bosons include the photon of quantum electrodynamics (QED), the gluons of quantum chromodynamics (QCD), and the $W^{\pm}$ and Z bosons complementing the electroweak interaction.

In general, interactions between particles in **any** setting can be pictured in terms of a generalized exchange mechanism. Depending on the circumstances, the "particle" that is exchanged between the constituent fermions may be a low-energy bosonic excitation (of any integer spin) of the medium. For example, electrons in a solid can exchange the lattice vibrations (phonons) of the core atoms. It will be useful to keep this exchange mechanism in mind, even though it is not always transparently at work. An illustration is provided by the instantaneous Coulomb repulsion between two electrons which does originate from the one-photon exchange mechanism [Sakurai (1967)].

Apart from their obvious thermodynamic relevance, the importance of the low-energy excitations of a many-particle system cannot be overemphasized. Their excitation energy provides a new energy scale which is not present in the vacuum. This can be discussed at several levels for nuclei. The nucleons making up the nucleus, are themselves composite objects made up of quarks in such a way that no explicit color is present. A nucleon can thus be considered as the lowest bound state of three quarks with total angular momentum and isospin $\frac{1}{2}$. Experimentally, it has been impossible to isolate quarks up to now. The energy scale associated with generating individual quarks is therefore infinite for all practical purposes. This is true even though QCD in the high-energy domain describes a weakly interacting system of massless quarks, predominantly interacting by the exchange of single massless gluons. In the low-energy domain, where nonperturbative effects and confinement dominate, the lowest excited states of QCD are found at excitation energies of the order of hundreds of MeV. A particularly important example is the Δ-isobar at 1232 MeV, which has spin and isospin $\frac{3}{2}$. The energy difference between the nucleon and the Δ provides a new energy scale. In addition, there are bosonic excitation modes of QCD which can be interpreted in terms of quark–antiquark states. The lowest-energy state is the pion with angular momentum 0 and isospin 1, which also has opposite parity from the nucleon and the Δ. The energy of the pion is about 140 MeV. Due to its low energy and strong coupling to the nucleon, the exchange mechanism, discussed above, is relevant. Consequently, part of the interaction between two nucleons is represented by the exchange of individual pions. Since they are the lowest-mass mesons, their exchange generates the long-range part of the interaction, illustrating the connection between the mass (energy) of the exchanged particle and the range of the interaction. The idea of a meson-exchange mechanism to describe the strong interaction, dates back to [Yukawa (1935)]. Exchange mechanisms of higher-energy mesons with other quantum numbers, can be used to describe the interaction between two nucleons at shorter range. The energy scale, which involves the explicit excitation of other QCD states, therefore starts at 140 MeV with an additional important state (the Δ) at 300 MeV.

This discussion demonstrates that the elementary excitations of the QCD field theory are subject to a different energy scale, compared to the noninteracting free field theory in which the quarks and gluons have no mass. In the interacting theory, the colorless bound states dominate at low energy, and explicit single quark and gluon degrees of freedom with color are effectively at infinite excitation energy. Depending on the objectives one must choose the relevant degrees of freedom that most efficiently describe the properties of the system. In the case of one nucleon, one may attempt a solution of QCD on the lattice. For many nucleons, it is more fruitful to start from physical nucleons that interact by means of the meson-exchange mechanism, using input from experimental data. This approach certainly makes sense on account of the overwhelming experimental evidence that nucleons retain much of their identity when they are brought together with other nucleons in nuclei.

In general, one can say that the low-lying excitations of a system play an important role in understanding the physics of the many-particle system.

Sometimes these states are referred to as elementary modes of excitation or quasiparticles. To understand the physics, it may be important to treat the interaction between these modes. In liquid ^{4}He we find bosonic collective modes like phonons and rotons. One level of understanding is achieved by describing the liquid in terms of atoms interacting with each other, another level is achieved by working with phonons and rotons [Nozières and Pines (1990)]. Whereas the former mode of description has certainly not yet been completely successful in bringing about a microscopic understanding of rotons, the latter description mode is restricted by its inherent phenomenological character. Indeed, even if one can numerically calculate certain properties of the system microscopically, this does not imply a deep understanding of the physics at the same time.

For a nucleus it is wise to maintain the description in terms of nucleons until QCD can be solved with an accuracy better than the lowest energy scale relevant for the system under study. In a nucleus the lowest excitation modes have energies of the order of MeV's in light nuclei, but in heavy nuclei the lowest excited state (of boson character) may be at about 50 keV. Again, one is dealing with a new energy scale that is introduced because nucleons are localized. Nucleons experience an overall mutual attraction forming a self-bound system, the nucleus. Its size is dictated by the interactions between the nucleons in the nuclear medium. From a sp point of view, the nucleons find themselves bound in the average attractive field of the other nucleons. Such a potential well has a range corresponding to the size of the nucleus and consequently, introduces a new energy scale related to its sp energies. An empirical potential of this kind was presented in Sec. 3.3. Associated with the new energy scale, one encounters collective behavior of nuclei that are best understood in terms of nucleons moving coherently between these sp levels. Such bosonic excitations may be successfully employed to interpret other excited states of the many-particle system.

In a similar vein, one is not interested in understanding the helium liquids in terms of a Hamiltonian at the level of the Coulomb interactions between the electrons, the alpha particles, and the electrons and the alpha particles (or ^{3}He nuclei). Instead, many-particle theory attempts to explain the properties associated with the relevant energy scales of the whole liquid. This scale in kelvin (energy/k_B) is associated with the many-particle nature of the system. The composite helium atoms experience a mutual interaction characterizing their collisions in free space. It reflects the polarization effect of the electron cloud of one atom on that of the other, representing the long-range part of the interaction. At short distances, the effect of the Pauli principle between the electrons in the different atoms, leads to a strongly repulsive component. The resulting interaction between pointlike helium atoms, is thus used to simulate effects associated with degrees of freedom, important at higher energy scales. Facilitating this approach is the neat separation of electronic excitation energies that exceed those of the liquid by four to five orders of magnitude.

A similar simplification must be made for electrons in a molecule or solid. The original Hamiltonian, describing the Coulomb interaction between nuclei, nuclei and electrons, and electrons and electrons, is too general to provide a realistic starting point for the description of the solid state [Anderson (1963); Anderson (1984)]. Instead, the nuclei are assumed to be localized in a lattice with most of the atomic electrons still tightly bound to them. Only the electrons that become delocalized, form the relevant electronic degrees of freedom that are subject to a periodic potential leading to the observed band structure.

4.4   Examples of relevant two-body interactions

After this general perspective we present some examples of interactions used in many-particle calculations. We will also evaluate some of the relevant two-body matrix elements that appear in the Fock-space two-body operator in Eq. (2.41). Often, an interaction is grounded in theory, but in some cases a certain amount of phenomenology, constrained by experiment, is necessary. Such constraints may involve an accurate description of the corresponding fermion–fermion scattering in free space. In many systems the basic interaction only depends on the relative distance between the particles, and no explicit spin (or isospin) dependence needs to be considered. Spherical symmetry reduces the dependence further to the magnitude of the relative distance. For such an interaction two-body matrix elements in coordinate space yield (suppressing discrete quantum numbers)

$$
(r_1 r_2 | V | r_3 r_4) = (Rr | V | R'r')
$$
$$
= \delta(R - R') \langle r | V | r' \rangle = \delta(R - R')\delta(r - r')V(r). \qquad (4.27)
$$

To obtain this result, a transformation to center-of-mass and relative coordinates has been employed

$$
R = \tfrac{1}{2}(r_1 + r_2) \qquad\qquad (4.28)
$$

$$r = r_1 - r_2, \tag{4.29}$$

and similarly for the primed coordinates. An interaction with such matrix elements in coordinate space is called a local interaction, since it is diagonal in the relative coordinate. An important example is the Coulomb interaction between charges $q_1 e$ and $q_2 e$

$$V_C(r) = \frac{q_1 q_2 e^2}{r}. \tag{4.30}$$

Also useful is the so-called Yukawa interaction given by

$$V_Y(r) = V_0 \frac{e^{-\mu r}}{\mu r}. \tag{4.31}$$

In the case of nucleons, a considerable number of operators is required to describe the interaction accurately. The simplest of these involves the spin-spin interaction which is usually written as

$$V_{spin} = V_\sigma(r) \boldsymbol{\sigma}_1 \cdot \boldsymbol{\sigma}_2, \tag{4.32}$$

where the dot product involves the Pauli spin matrices of the two particles. Since $\boldsymbol{\sigma}_1 \cdot \boldsymbol{\sigma}_2$ corresponds to $4 \boldsymbol{s}_1 \cdot \boldsymbol{s}_2 / \hbar^2$, two-particle states coupled to good total spin are needed to construct the eigenstates of this operator. Using the identity

$$2 \boldsymbol{s}_1 \cdot \boldsymbol{s}_2 = \boldsymbol{S}^2 - \boldsymbol{s}_1^2 - \boldsymbol{s}_2^2, \tag{4.33}$$

where $\boldsymbol{S} = \boldsymbol{s}_1 + \boldsymbol{s}_2$, one obtains

$$\langle S'M_S' | \boldsymbol{\sigma}_1 \cdot \boldsymbol{\sigma}_2 | S M_S \rangle = (2S(S+1) - 3) \, \delta_{S,S'} \delta_{M_S,M_S'}, \tag{4.34}$$

which yields -3 for $S = 0$ and 1 for $S = 1$. Spin–spin interactions are not present in the basic interaction between electrons or ^3He atoms, but do appear in an effective form when the interaction between these fermions is considered inside the medium. Nuclear interactions in addition carry an explicit isospin dependence which leads to

$$V_{isospin} = V_\tau(r) \boldsymbol{\tau}_1 \cdot \boldsymbol{\tau}_2, \tag{4.35}$$

where the $\boldsymbol{\tau}$ matrices are the isospin equivalent of the Pauli spin matrices. The result, corresponding to Eq. (4.34), reads in this case

$$\langle T'M_T' | \boldsymbol{\tau}_1 \cdot \boldsymbol{\tau}_2 | T M_T \rangle = (2T(T+1) - 3) \, \delta_{T,T'} \delta_{M_T,M_T'}, \tag{4.36}$$

showing that states with good total isospin are necessary. It is possible to give an accurate account of the scattering of two nucleons up to the threshold of pion production by employing an interaction of the following form [Wiringa *et al.* (1984)]:

$$v^{14}(1,2) = \sum_{p=1,14} [v_\pi^p(r) + v_I^p(r) + v_S^p(r)] \, O_{12}^p. \qquad (4.37)$$

The local radial dependence ($r = |\boldsymbol{r}_1 - \boldsymbol{r}_2|$) is governed by a long-range pion-exchange term, v_π^p, an intermediate-range part v_I^p, and a short-range contribution, v_S^p. The radial dependence of the pion-exchange interaction contains the Yukawa form of Eq. (4.30). The charged π^+ and π^- have a mass of $\mu_\pi \hbar c = m_\pi c^2 = 139.6$ MeV, whereas the π^0 corresponds to 135 MeV, together usually treated as an isospin triplet. Fourteen operators O_{12}^p need to be considered

$$
\begin{array}{cccc}
1 & \boldsymbol{\tau}_1 \cdot \boldsymbol{\tau}_2 & \boldsymbol{\sigma}_1 \cdot \boldsymbol{\sigma}_2 & \boldsymbol{\sigma}_1 \cdot \boldsymbol{\sigma}_2 \, \boldsymbol{\tau}_1 \cdot \boldsymbol{\tau}_2 \\
S_{12} & S_{12} \, \boldsymbol{\tau}_1 \cdot \boldsymbol{\tau}_2 & \boldsymbol{L} \cdot \boldsymbol{S} & \boldsymbol{L} \cdot \boldsymbol{S} \, \boldsymbol{\tau}_1 \cdot \boldsymbol{\tau}_2 \\
\boldsymbol{L}^2 & \boldsymbol{L}^2 \, \boldsymbol{\tau}_1 \cdot \boldsymbol{\tau}_2 & \boldsymbol{L}^2 \, \boldsymbol{\sigma}_1 \cdot \boldsymbol{\sigma}_2 & \boldsymbol{L}^2 \, \boldsymbol{\sigma}_1 \cdot \boldsymbol{\sigma}_2 \, \boldsymbol{\tau}_1 \cdot \boldsymbol{\tau}_2 \\
(\boldsymbol{L} \cdot \boldsymbol{S})^2 & (\boldsymbol{L} \cdot \boldsymbol{S})^2 \, \boldsymbol{\tau}_1 \cdot \boldsymbol{\tau}_2 & &
\end{array}
\qquad (4.38)
$$

to account for all the details exhibited by the data. This set of operators contains the usual Pauli spin and isospin matrices, the tensor operator

$$S_{12}(\hat{\boldsymbol{r}}) = 3 \, (\boldsymbol{\sigma}_1 \cdot \hat{\boldsymbol{r}}) \, (\boldsymbol{\sigma}_2 \cdot \hat{\boldsymbol{r}}) - \boldsymbol{\sigma}_1 \cdot \boldsymbol{\sigma}_2, \qquad (4.39)$$

the relative orbital angular momentum $\boldsymbol{L}$, and the total spin $\boldsymbol{S}$ of the pair. Employing a partial-wave basis, one can use standard angular momentum techniques to determine the matrix elements of these operators [Sakurai (1994); Messiah (1999)], as summarized in App. B. Clearly, it is necessary to include a coupling to total angular momentum to keep such calculations manageable. This coupling was outlined in Sec. 4.1. An example of the radial dependence of the nucleon–nucleon interaction in the two possible *S*-wave channels is shown in Fig. 4.3. The interaction was taken from [Reid (1968)] and is referred to as the Reid soft-core (RSC) potential. It can be Fourier transformed to momentum space, unlike a hard-core potential. It is noteworthy that the channel which binds the deuteron, displays less attraction than the $T = 1$ interaction which is attractive but doesn't bind the two-nucleon system. The explanation of this feature is the importance of the role of the tensor force which acts in the coupled $^3S_1$-$^3D_1$ channel, as discussed in detail in Ch. 15.

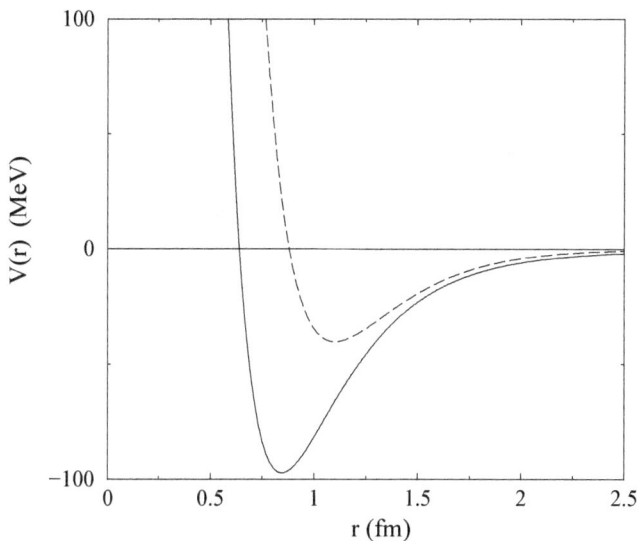

Fig. 4.3 The radial dependence of the $^1S_0$ nucleon-nucleon interaction is shown as the full line, while the central part of the $^3S_1$-$^3D_1$ channels corresponds to the dashed line [Reid (1968)]. The long-range part of these interactions is described by the pion-exchange interaction.

Matrix elements of interactions in momentum space are needed when we consider scattering problems or large homogeneous systems. Using the transformation to total and relative momenta, presented in Eqs. (4.13) and (4.14), one obtains for a central spin and isospin independent interaction (suppressing these discrete quantum numbers)

$$(\boldsymbol{p_1 p_2}|V|\boldsymbol{p_3 p_4}) = (\boldsymbol{Pp}|V|\boldsymbol{P'p'}) = \delta_{\boldsymbol{P,P'}} \langle \boldsymbol{p}|\, V\, |\boldsymbol{p'} \rangle . \qquad (4.40)$$

Wave vectors can also be employed for the description of the relative motion

$$\langle \boldsymbol{k}|\, V\, |\boldsymbol{k'} \rangle = \frac{1}{V} \int d^3 r\, \exp\{i(\boldsymbol{k'} - \boldsymbol{k}) \cdot \boldsymbol{r}\} V(r). \qquad (4.41)$$

The matrix element can be manipulated further by using the standard expansion

$$\exp\{i\boldsymbol{q} \cdot \boldsymbol{r}\} = 4\pi \sum_{\ell m} i^\ell Y^*_{\ell m}(\hat{\boldsymbol{r}}) Y_{\ell m}(\hat{\boldsymbol{q}}) \mathrm{j}_\ell(qr), \qquad (4.42)$$

where j_ℓ is the spherical Bessel function. Inserting this result in Eq. (4.41) and performing the angular integration, one obtains

$$\langle \boldsymbol{k} | \, V \, | \boldsymbol{k'} \rangle = \frac{4\pi}{V} \int dr \; r^2 \; \mathrm{j}_0(qr) V(r), \tag{4.43}$$

with $q = |\boldsymbol{k} - \boldsymbol{k'}|$. For the Yukawa interaction of Eq. (4.31) the last integral can be performed analytically (see *e.g.* [Gradshtein and Ryzhik (1980)]) yielding

$$\langle \boldsymbol{k} | \, V_Y \, | \boldsymbol{k'} \rangle = \frac{4\pi}{V} \frac{V_0}{\mu} \frac{1}{\mu^2 + (\boldsymbol{k'} - \boldsymbol{k})^2}. \tag{4.44}$$

This result can be employed to generate the matrix element of the Coulomb interaction

$$\langle \boldsymbol{k} | \, V_C \, | \boldsymbol{k'} \rangle = \frac{4\pi}{V} \frac{q_1 q_2 e^2}{(\boldsymbol{k'} - \boldsymbol{k})^2}, \tag{4.45}$$

where the case $\boldsymbol{k} = \boldsymbol{k'}$ requires special care but can usually be omitted on account of cancellations, as for the homogeneous electron gas [Fetter and Walecka (1971); Mahan (1990); Gross *et al.* (1991); Mattuck (1992)], discussed in Sec. 5.2.

Another type of interaction that can be encountered is of the following form

$$V(r) = A \, e^{-\alpha r}. \tag{4.46}$$

Such a term can be used to describe the short-range part of the atom–atom repulsion [Aziz *et al.* (1979)] allowing Fourier transformation unlike r^{-12}-type interactions. The Fourier transform of the interaction in Eq. (4.46) yields

$$\langle \boldsymbol{k} | \, V \, | \boldsymbol{k'} \rangle = V(q) = -\frac{4\pi A}{V} \frac{d}{d\alpha} \frac{1}{\alpha^2 + q^2}, \tag{4.47}$$

with $q = |\boldsymbol{q}| = |\boldsymbol{k'} - \boldsymbol{k}|$ the magnitude of the transferred wave vector. This form is also useful for obtaining matrix elements in a partial-wave basis. The long-range part of the atom–atom interaction contains the attractive r^{-6} van der Waals tail, but for accurate results is supplemented by r^{-8} and r^{-10} contributions [Aziz *et al.* (1979)]. The corresponding interaction is shown in Fig. 4.4.

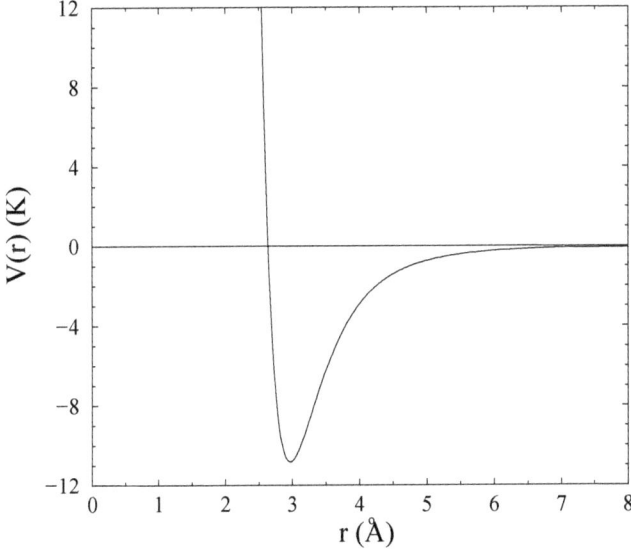

Fig. 4.4 Helium–helium interaction from [Aziz *et al.* (1979)]. The qualitative similarity with the nuclear interactions of Fig. 4.3 is evident.

In a partial-wave basis, matrix elements are required of the form

$$\langle kLM_L| V |k'L'M'_L\rangle = \int d\hat{\boldsymbol{k}} \, \langle LM_L|\hat{\boldsymbol{k}}\rangle \int d\hat{\boldsymbol{k}}' \, \langle \hat{\boldsymbol{k}}'|L'M'_L\rangle \, \langle \boldsymbol{k}| V(r) |\boldsymbol{k}'\rangle. \tag{4.48}$$

In the case of a Yukawa interaction, Eq. (4.44) can be written as

$$\langle \boldsymbol{k}| V_Y(r) |\boldsymbol{k}'\rangle = \frac{4\pi}{V} \frac{V_0}{\mu} \frac{1}{2kk'} \frac{1}{\frac{\mu^2+k^2+k'^2}{2kk'} - \cos\theta_{kk'}}. \tag{4.49}$$

This last fraction can be expanded using the following relation between Legendre functions Q_ℓ and Legendre polynomials P_ℓ

$$\frac{1}{\frac{\mu^2+k^2+k'^2}{2kk'} - \cos\theta_{kk'}} = \sum_{\ell=0}^{\infty} (2\ell+1) \, Q_\ell\left(\frac{\mu^2+k^2+k'^2}{2kk'}\right) P_\ell(\cos\theta_{kk'})$$

$$= \sum_{\ell=0}^{\infty} \sum_{m=-\ell}^{\ell} 4\pi \, Q_\ell\left(\frac{\mu^2+k^2+k'^2}{2kk'}\right) Y^*_{\ell m}(\hat{\boldsymbol{k}}) Y_{\ell m}(\hat{\boldsymbol{k}}'). \tag{4.50}$$

In the last equality the addition theorem for spherical harmonics was used. Note that the argument of the Legendre function must be larger than 1, while the argument of the Legendre polynomial must be less than 1. Inserting these results, one obtains after performing the angular integrations

$$\langle kLM_L| V |k'L'M'_L\rangle = \delta_{L,L'} \,\delta_{M_L,M'_L} \frac{(4\pi)^2 V_0}{V\mu 2kk'} Q_L\left(\frac{\mu^2 + k^2 + k'^2}{2kk'}\right). \quad (4.51)$$

The first three Legendre functions are given by

$$Q_0(z) = \frac{1}{2}\ln\left(\frac{z+1}{z-1}\right)$$

$$Q_1(z) = \frac{z}{2}\ln\left(\frac{z+1}{z-1}\right) - 1$$

$$Q_2(z) = \frac{3z^2-1}{4}\ln\left(\frac{z+1}{z-1}\right) - \frac{3}{2}z. \quad (4.52)$$

4.5   Exercises

(1) Calculate the normalization of the states given in Eqs. (4.22), (4.24), and (4.26).

(2) Determine the possible states allowed by the Pauli principle when two nucleons are added in the $0d\frac{5}{2}$ shell to ^{16}O. Include isospin in your analysis. Which states are allowed when two particles are removed in the $0p\frac{1}{2}$ state? Compare your results with the experimental spectra for these nuclei.

(3) Consider two nucleons that can occupy the $0d\frac{5}{2}, 1s\frac{1}{2}$, or $0d\frac{3}{2}$ orbit. Evaluate the number of states allowed by the Pauli principle for each possible value of J and T.

(4) Determine all the possible states for the $2p^2$ electron configuration in carbon when a basis with good total sp angular momentum j is employed. Compare your number with the L-S scheme.

(5) Evaluate the matrix elements of the fourteen operators in Eq. (4.37) in the partial wave basis with the total angular momentum J and isospin T as good quantum numbers.

(6) Calculate the matrix elements of Eqs. (4.45) and (4.47) in a partial wave basis.

(7) The $^1S_0$ component of the Reid soft-core interaction is given by

$$V^{^1S_0}(r) = -h\frac{e^{-\mu_\pi r}}{\mu_\pi r} - 1650.6\frac{e^{-4\mu_\pi r}}{\mu_\pi r} + 6484.2\frac{e^{-7\mu_\pi r}}{\mu_\pi r}. \quad (4.53)$$

Calculate the matrix elements in momentum space ($h = 10.463$ MeV).

Chapter 5

Noninteracting bosons and fermions

In Ch. 3, we discussed the consequences of the Pauli principle for an assembly of noninteracting fermions localized in space. For atoms and nuclei, the resulting shell model, or independent-particle model, provides a useful starting point for further study. When dealing with a large homogeneous system, it is practical to take advantage of translational invariance in choosing a sp basis. The special role of the momentum or wave vector basis is therefore clear. The corresponding "shell model" of such an infinite system is referred to as the Fermi gas. Details are presented in Sec. 5.1. An important idealization of a system of electrons in a metal, the electron gas, is introduced in Sec. 5.2. Fermi-gas considerations apply to several other infinite systems that are briefly reviewed in Sec. 5.3 for nuclear and neutron matter, and in Sec. 5.4 for the ^{3}He liquid.

After reviewing some statistical mechanics in Sec. 5.5, the occupation number representation is employed to derive some standard results for noninteracting bosons and fermions at finite temperature. The phenomenon of Bose–Einstein condensation is discussed in Sec. 5.6. Bosons in an infinite homogeneous system are considered in Sec. 5.6.1. A preliminary presentation of Bose–Einstein condensation in traps is given in Sec. 5.6.2 with attention to the thermodynamic limit in Sec. 5.6.3. Fermions at finite temperature are briefly dealt with in Sec. 5.7.

5.1 The Fermi gas at zero temperature

The bulk properties of homogeneous systems of interacting fermions at a certain density ρ is of great interest. For such a system, the Fermi gas, where the interparticle interactions are neglected, provides a good starting point. It is instructive to study it first at zero temperature. Applications involving fermions at finite T will be presented in Sec. 5.7. In the Fermi gas each particle only contributes its kinetic energy

$$H_0 = T = \frac{\boldsymbol{p}^2}{2m}.$$ (5.1)

For the momentum eigenstates of Eq. (5.1), we will apply the box normalization discussed in Sec. 4.1.1. It is convenient to introduce periodic boundary conditions as suggested by translational invariance. In the x-direction, for example, one requires

$$e^{ik_x x} = e^{ik_x(x+L)} = e^{ik_x x} e^{ik_x L},$$ (5.2)

where $p_x = \hbar k_x$. This result implies that

$$\cos(k_x(x+L)) + i\sin(k_x(x+L)) = 1,$$ (5.3)

which is fulfilled when

$$k_x = n_x \frac{2\pi}{L} \qquad \text{where} \qquad n_x = 0, \pm 1, \pm 2, \ldots$$ (5.4)

and similarly for k_y and k_z. It follows that each allowed triple $\{k_x, k_y, k_z\}$ corresponds to a triple of integers $\{n_x, n_y, n_z\}$. The Pauli principle allows only a fixed number of fermions in each sp momentum eigenstate, depending on the spin and/or isospin degeneracy. The ground state is then obtained by filling the momentum states up to a maximum value; the Fermi momentum $p_F = \hbar k_F$. The maximum wave vector k_F can be determined by calculating the expectation value of the number operator [see Eq. (2.51)] in the ground state

$$|\Phi_0\rangle = \prod_{|\boldsymbol{k}| < k_F, \mu} a^\dagger_{\boldsymbol{k}\mu} |0\rangle ,$$ (5.5)

where μ labels spin/isospin quantum numbers. We will consider the limit when both $N \to \infty$ and $V \to \infty$ such that their ratio, the density $\rho = N/V$, remains constant. This procedure is referred to as taking the "thermodynamic limit." Summations can be replaced by integrations over continuous quantum numbers like wave vectors, as follows

$$\sum_{\boldsymbol{k}\mu} f(\boldsymbol{k}, \mu) = \sum_{n_x n_y n_z} \sum_\mu f\left(\frac{2\pi \boldsymbol{n}}{L}, \mu\right)$$

$$L \to \infty \quad \Rightarrow \quad \int d\boldsymbol{n} \sum_\mu f(\frac{2\pi \boldsymbol{n}}{L}, \mu) = \frac{V}{(2\pi)^3} \int d\boldsymbol{k} \sum_\mu f(\boldsymbol{k}, \mu)$$ (5.6)

for any function f. The transition from discrete triples $\{n_x, n_y, n_z\}$ to continuous variables can be made in the case of large L, since any physical quantity described by f will change slowly when one of the discrete variables changes by one unit. To obtain the Fermi wave vector, we consider

$$N = \langle \Phi_0 | \hat{N} | \Phi_0 \rangle = \sum_{\boldsymbol{k}\mu} \langle \Phi_0 | a_{\boldsymbol{k}\mu}^\dagger a_{\boldsymbol{k}\mu} | \Phi_0 \rangle = \sum_{\boldsymbol{k}\mu} \theta(k_F - k)$$

$$= \frac{V}{(2\pi)^3} \sum_\mu \int d^3k \, \theta(k_F - k) = \frac{\nu V}{6\pi^2} k_F^3, \tag{5.7}$$

where ν represents the spin/isospin degeneracy and θ denotes the step function. The relation between the Fermi wave vector and the density therefore becomes

$$k_F = \left\{ \frac{6\pi^2 N}{\nu V} \right\}^{1/3} = \left\{ \frac{9\pi}{2\nu} \right\}^{1/3} \frac{1}{r_0}. \tag{5.8}$$

In the last equality, r_0 has been introduced, representing the radius of a sphere containing on average one particle

$$\frac{V}{N} = \frac{1}{\rho} = \frac{4}{3}\pi r_0^3. \tag{5.9}$$

Clearly, r_0 also serves as a measure of the interparticle spacing. Conversely, one can write the density as

$$\rho = \frac{N}{V} = \nu \frac{k_F^3}{6\pi^2}. \tag{5.10}$$

Equations (5.8) and (5.10) show that for a fixed density, a smaller Fermi wave vector is found when the degeneracy factor ν is larger.

The energy of the ground state of the Fermi gas is obtained by employing the kinetic energy operator

$$\hat{T} = \sum_{\boldsymbol{k}\mu} \sum_{\boldsymbol{k}'\mu'} \langle \boldsymbol{k}\mu | \frac{\hbar^2 \boldsymbol{k}^2}{2m} | \boldsymbol{k}'\mu' \rangle a_{\boldsymbol{k}\mu}^\dagger a_{\boldsymbol{k}'\mu'} = \sum_{\boldsymbol{k}'\mu'} \frac{\hbar^2 \boldsymbol{k}'^2}{2m} a_{\boldsymbol{k}'\mu'}^\dagger a_{\boldsymbol{k}'\mu'}, \tag{5.11}$$

with the result

$$\hat{T} | \Phi_0 \rangle = \left(\sum_{\boldsymbol{k}'\mu'} \frac{\hbar^2 \boldsymbol{k}'^2}{2m} a_{\boldsymbol{k}'\mu'}^\dagger a_{\boldsymbol{k}'\mu'} \right) \prod_{|\boldsymbol{k}| < k_F \mu} a_{\boldsymbol{k}\mu}^\dagger | 0 \rangle. \tag{5.12}$$

Using Eq. (2.34), we obtain a kinetic energy contribution from each sp state that is occupied in $|\Phi_0\rangle$

$$\hat{T}|\Phi_0\rangle = \left(\sum_{|\boldsymbol{k}|<k_F,\mu} \frac{\hbar^2 \boldsymbol{k}^2}{2m} \right) |\Phi_0\rangle. \tag{5.13}$$

The energy of the ground state is then generated by taking the appropriate continuum limit discussed above

$$E_0 = \sum_{|\boldsymbol{k}|<k_F,\mu} \frac{\hbar^2 \boldsymbol{k}^2}{2m} = \frac{V}{(2\pi)^3} \sum_{\mu} \int d^3k \frac{\hbar^2 k^2}{2m} \theta(k_F - k)$$

$$= V \frac{\nu}{(2\pi)^3} \, 4\pi \, \frac{\hbar^2}{2m} \frac{1}{5} k_F^5. \tag{5.14}$$

In the thermodynamic limit, the energy per particle is thus given by

$$\frac{E_0}{N} = \frac{V}{N} \frac{\nu}{2\pi^2} \frac{\hbar^2 k_F^5}{10m} = \frac{3}{5} \frac{\hbar^2 k_F^2}{2m} = \frac{3}{5} \varepsilon_F = \frac{3}{5} k_B T_F. \tag{5.15}$$

Equation (5.10) has been used here while the free Fermi energy ε_F and Fermi temperature T_F have also been introduced. Boltzmann's constant has been denoted by k_B.

5.2   Electron gas

Various systems qualify to be considered in terms of this simple Fermi gas model. An example is provided by the homogeneous electron gas, which provides a first approximation to a metal or a plasma. The positive ions are represented by a static, uniform background. This yields a combined system that is electrically neutral. One neglects therefore the motion of the ions which, at least in first approximation, is permissible due to their much larger mass. The uniform background assumption is less appropriate, but the system is nevertheless of great importance. We note that the spin degeneracy factor $\nu = 2$. It is also customary to use the Bohr radius $(a_0 = \hbar^2/me^2 = 5.29 \ 10^{-11} \text{ m} = 0.529\text{Å})$ to introduce a dimensionless parameter for this system written as

$$r_s = \frac{r_0}{a_0}. \tag{5.16}$$

A metallic element contains N_A (Avogadro's number) atoms per mole and ρ_m/A moles per cm^3, where ρ_m is the mass density (in grams per cm^3), and A is the atomic mass. If one assumes that each atom contributes Z electrons to conduction, the number of free electrons per cubic centimeter is

$$\rho = \frac{N}{V} = 6.02205 \times 10^{23} \frac{Z \rho_m}{A}. \tag{5.17}$$

Taking $Z = 1$ for alkali metals yields densities (at a temperature of 5K) of 0.91 for Cs, 1.15 for Rb, 1.40 for K, and 2.65 for Na times $10^{22}/\text{cm}^3$. Corresponding values of r_s are 5.62 for Cs, 5.20 for Rb, 4.86 for K, and 3.93 for Na, respectively. This translates into Fermi wave vectors ranging from 0.65 Å^{-1} for Cs to 0.92 Å^{-1} for Na. The corresponding range of Fermi energies and temperatures is given by 1.59 eV and 1.84 $\times 10^4$ K for Cs, and 3.24 eV and 3.77 $\times 10^4$ K for Na. In its relativistic version, the electron gas model can also be used to model white dwarf stars. One typically assumes that the star contains the nuclei of He atoms (α particles) and electrons [Landau and Lifshitz (1980); Huang (1987)]. Note that for realistic conditions it is necessary to treat the electrons relativistically, due to the high density of such a system.

Perhaps surprisingly, properties of the interacting (not free) electron gas are also used extensively in quantum chemistry, within the framework of modern density functional theory (DFT). In DFT, the energy of any electronic system (also strongly inhomogeneous ones like atoms or molecules), is expressed as a universal functional of the local electron density in the system. While the structure of this functional is unknown, it should become equal to the electron gas result in the limit of slowly varying electron densities. This feature is used as an important constraint in the construction of phenomenological density functionals [Dreizler and Gross (1990)].

We now consider the electron-gas Hamiltonian in more detail, using atomic units (see Sec. 3.2). The electron gas consists of a homogeneous distribution of electrons at density ρ, interacting through their mutual Coulomb repulsion. Electrical neutrality is restored, by adding an inert background distribution of positive charge with the same density. The electrostatic energy of the positive background is simply

$$E_b = \frac{1}{2} \int d\mathbf{r}_1 \int d\mathbf{r}_2 \, \frac{\rho^2}{|\mathbf{r}_1 - \mathbf{r}_2|} = \frac{1}{2} \rho^2 V \int d\mathbf{r} \frac{1}{r}, \tag{5.18}$$

which shows that the energy per particle, (E_b/N), diverges like $V^{2/3}$ in the thermodynamic limit because of the long-range nature of the Coulomb potential. As the system is globally charge-neutral, we expect a finite result if we add E_b to similar contributions of the electron–electron and electron–background interaction. To study this cancellation in a controlled manner, it is convenient to momentarily replace the Coulomb inverse power law $1/r$ with the Yukawa-type function $e^{-\mu r}/r$. The energy of the background now reads

$$E_b = \tfrac{1}{2}\rho^2 V \int dr \frac{e^{-\mu r}}{r} = \tfrac{1}{2}\rho^2 V \frac{4\pi}{\mu^2}. \qquad (5.19)$$

The interaction between the electrons and the background charge distribution gives rise to the following one-body potential for the electrons

$$U(r) = -\rho \int dr' \frac{e^{-\mu|r-r'|}}{|r-r'|} = -\rho \int dr' \frac{e^{-\mu r'}}{r'} = -\rho \frac{4\pi}{\mu^2}. \qquad (5.20)$$

This potential is just a constant, as a consequence of translational invariance. The contribution to the total energy of the N electrons is therefore

$$E_{e-b} = N\left(-\rho \frac{4\pi}{\mu^2}\right) = -\rho^2 V \frac{4\pi}{\mu^2}. \qquad (5.21)$$

The matrix elements of the Yukawa interaction are given in Eq. (4.44). Including the spin projection $m_s = \pm\frac{1}{2}$, we have

$$(\boldsymbol{p}_1 m_1, \boldsymbol{p}_2 m_2| V |\boldsymbol{p}_3 m_3, \boldsymbol{p}_4 m_4) = \qquad (5.22)$$

$$\delta_{m_1,m_3}\delta_{m_2,m_4}\delta_{\boldsymbol{P},\boldsymbol{P}'} \frac{4\pi}{V} \frac{1}{(\boldsymbol{p}-\boldsymbol{p}')^2 + \mu^2}.$$

We introduced the center-of-mass and relative momenta by

$$\boldsymbol{P} = \boldsymbol{p}_1 + \boldsymbol{p}_2, \quad \boldsymbol{p} = \tfrac{1}{2}(\boldsymbol{p}_1 - \boldsymbol{p}_2)$$
$$\boldsymbol{P}' = \boldsymbol{p}_3 + \boldsymbol{p}_4, \quad \boldsymbol{p}' = \tfrac{1}{2}(\boldsymbol{p}_3 - \boldsymbol{p}_4). \qquad (5.23)$$

For the second-quantized form of the electron–electron interaction we then find

$$\hat{V} = \frac{1}{2}\sum_{\boldsymbol{p}_i m_i} (\boldsymbol{p}_1 m_1, \boldsymbol{p}_2 m_2| V |\boldsymbol{p}_3 m_3, \boldsymbol{p}_4 m_4)\, a_{\boldsymbol{p}_1 m_1}^\dagger a_{\boldsymbol{p}_2 m_2}^\dagger a_{\boldsymbol{p}_4 m_4} a_{\boldsymbol{p}_3 m_3} \quad (5.24)$$

$$= \frac{1}{2}\sum_{\substack{\boldsymbol{P}\boldsymbol{p}\boldsymbol{p}' \\ m_1 m_2}} \frac{4\pi}{V} \frac{1}{(\boldsymbol{p}-\boldsymbol{p}')^2 + \mu^2} a_{\boldsymbol{P}/2+\boldsymbol{p}m_1}^\dagger a_{\boldsymbol{P}/2-\boldsymbol{p}m_2}^\dagger a_{\boldsymbol{P}/2-\boldsymbol{p}'m_2} a_{\boldsymbol{P}/2+\boldsymbol{p}'m_1}.$$

Let us now isolate the part of $\hat{V}$ where no relative momentum is transferred, *i.e.* we split $\hat{V} = \hat{V}_d + \hat{V}'$ where $\hat{V}_d$ contains the diagonal contributions in Eq. (5.24) having $\boldsymbol{p} = \boldsymbol{p}'$, and $\hat{V}'$ contains the remainder. Since the interaction matrix element in $\hat{V}_d$ is a constant, the result can be expressed in terms of the number operator,

$$\hat{V}_d = \tfrac{1}{2} \frac{4\pi}{V\mu^2} \sum_{\substack{p_1 p_2 \\ m_1 m_2}} a^\dagger_{p_1 m_1} a^\dagger_{p_2 m_2} a_{p_2 m_2} a_{p_1 m_1} = \tfrac{1}{2} \frac{4\pi}{V\mu^2} \hat{N}(\hat{N} - 1). \quad (5.25)$$

The contribution of $\hat{V}_d$ to the total energy then becomes [note that in the thermodynamic limit $N(N-1) \to N^2$],

$$E_d = \tfrac{1}{2} \frac{4\pi}{V\mu^2} N^2 = \tfrac{1}{2} \frac{4\pi}{\mu^2} \rho^2 V. \quad (5.26)$$

This is the classical electrostatic repulsion of the electron charge density, which nicely cancels with the corresponding terms E_b and E_{e-b} in Eqs. (5.19) and (5.21), arising from the positive background.

The final electron-gas Hamiltonian then becomes

$$\hat{H} = \hat{T} + \hat{V}' + E_d + E_{e-b} + E_b = \hat{T} + \hat{V}', \quad (5.27)$$

where $\hat{V}'$ is given by

$$\hat{V}' = \tfrac{1}{2} \sum_{\substack{P, p \neq p' \\ m_1 m_2}} \frac{4\pi}{V} \frac{1}{(p - p')^2} a^\dagger_{P/2+p\,m_1} a^\dagger_{P/2-p\,m_2} a_{P/2-p'\,m_2} a_{P/2+p'\,m_1}. \quad (5.28)$$

Since the summation is restricted to $p \neq p'$, there is now no danger in setting $\mu = 0$ and going back to the genuine Coulomb force. It is useful to evaluate the expectation value of this interaction in the noninteracting ground state [see Exercise (1) of this chapter]. We will take up the discussion of the electron gas again in later chapters.

5.3 Nuclear and neutron matter

The hypothetical infinite system with $N = Z$ and no Coulomb interaction between protons, is called nuclear matter and was introduced in Sec. 3.3.1. The system should reflect two essential numbers in nuclear physics that characterize global properties of nuclei. The first number is associated with the observed density in the interior of nuclei, equal to 0.16 nucleons per fm^3. According to Eq. (5.8), the corresponding wave vector, using a degeneracy factor $\nu = 4$ for nuclear matter, is therefore $k_F = 1.33$ fm^{-1}.

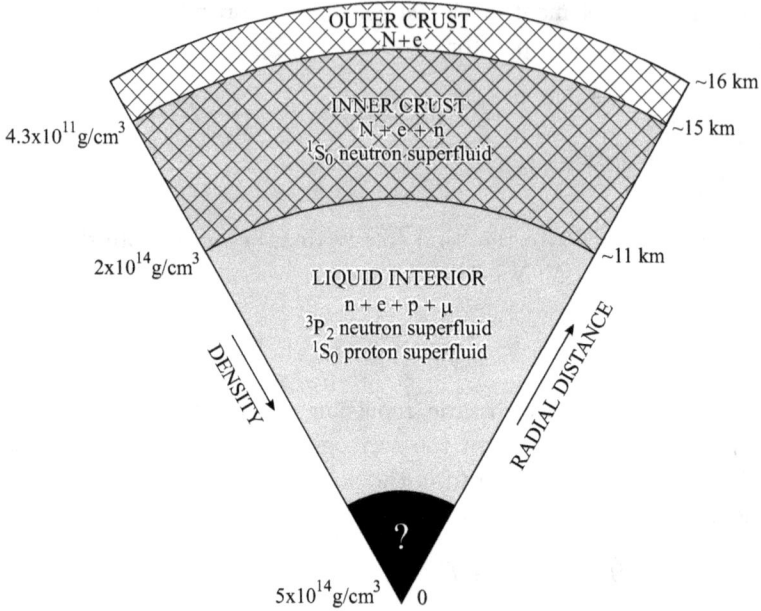

Fig. 5.1   The structure of a neutron star of about 1.4 solar masses. Values of radial distances and central density are only indicative because of uncertainties in the neutron matter equation of state.

The interparticle spacing becomes $r_0 = 1.14$ fm, comparable to the minimum in the interaction in the $T = 0$ channel (see Fig. 4.3). This density is referred to as the saturation, or normal density of nuclear matter. Since only the volume term of the empirical mass formula applies, the binding at saturation density is expected to be about 16 MeV per particle. Nuclear matter calculations, starting from realistic interactions, must explain these saturation properties. A more complete discussion will be presented in Ch. 16.

Another important application of infinite nuclear systems involves the study of the interior of neutron stars, which is schematically illustrated in Fig. 5.1. The physics of neutron stars is intricate and interesting [Baym and Pethick (1975); Baym and Pethick (1979)]. The low-density exterior corresponds to a solid crust of a Coulomb lattice of neutron-rich nuclei, immersed in a degenerate electron fluid. With increasing density, it is energetically favorable to "remove" electrons by the process of inverse β-decay. This β-decay turns the protons in nuclei to neutrons, until the nuclei with large neutron excess begin to "drip" neutrons because they can no longer bind them. The matter density for this "neutron drip" corresponds to 4.3 $\times 10^{11}$ g cm^{-3}. The neutrons form a superfluid of condensed $^1S_0$ pairs. At higher density, electrons, nuclei, and free neutrons coexist and determine the state of lowest energy. With increasing density the neutron proton ratio increases further, and the neutron fluid essentially determines the properties of the system above 4×10^{12} g cm^{-3}, where the neutrons start to provide more pressure than the electrons. The pressure exerted by neutrons therefore supports the neutron star against gravitational collapse. At about 2 $\times 10^{14}$ g cm^{-3} the nuclei have disappeared and the liquid is composed of neutrons, while a small fraction of protons, electrons, and muons is present. This neutron fluid with degeneracy factor $\nu = 2$, can be considered a huge nucleus with a lower density than normal nuclear matter. At higher densities the interactions between neutrons is important, and must be included in determining the equation of state. In this density regime, neutrons may also exhibit superfluidity in the coupled $^3P_2$-$^3F_2$ channel, and protons may become superconducting with $^1S_0$ quantum numbers. The highest density region has an uncertain composition. Strange particles, pion condensation, or quarks degrees of freedom may have to be considered. More details of neutron superfluidity are discussed in Ch. 22.

5.4 Helium liquids

Helium atoms come in two isotopic varieties, ^{3}He and ^{4}He. The lighter isotope is a fermion, the heavier one a boson. Systems of both types of particles exhibit spectacular quantum features. At zero pressure both systems remain liquid, when the temperature approaches 0 K. Both exhibit superfluidity and other remarkable properties, when the temperature is lowered. The binding forces between the atoms that form the liquid, are van der Waals type forces. They originate from the polarization induced in the electron clouds when the atoms approach each other (for a simple discussion see [Sakurai (1994)]). In addition to this attractive component, which is quite weak for He atoms, there is an effective short-range repulsion between the atoms which is almost hard-core like. This is due to the Pauli principle when the electron clouds start to overlap. Because of their light mass there is a delicate balance between the kinetic and potential energy. At zero pressure both systems remain liquid when the temperature approaches 0 K. For ^{3}He the observed density at zero pressure is equal to 36.84 cm^3/mole. This translates into 0.0163 atoms/Å^3 which in turn yields $k_F = 0.784$ Å^{-1}, using a degeneracy factor $\nu = 2$ since the total spin of the atom is $\frac{1}{2}$. At saturation the binding energy per atom is 2.52 K, a tiny amount compared to the atomic energy scales. It makes sense to discuss the system from the perspective of atoms interacting through (a dominant) two-body interaction. The latter can be studied from a more microscopic starting point and constrained by information at higher temperature. There, quantum effects no longer play a role and the kinetic and potential energy separate. Attempts have been underway for a long time to form fully spin-polarized ^{3}He. Such a system would have a degeneracy $\nu = 1$, with interesting consequences for the way in which the interaction is sampled (see Sec. 4.1). The boson counterpart ^{4}He forms a system that is more bound. This is hardly surprising, since there is no Pauli principle that leads to a substantial kinetic energy contribution. At a density of 0.0218 atoms/Å^3 the binding energy per atom is 7.14 K.

5.5 Some statistical mechanics

A microscopic description of a system at finite temperature can be obtained by evaluating the ensemble average of the statistical operator. We will consider the statistical operator in the grand canonical ensemble

$$\hat{\rho}_G = \frac{e^{-\beta(\hat{H}-\mu\hat{N})}}{Z_G}, \tag{5.29}$$

where $\beta = (k_B T)^{-1}$, μ is the chemical potential, and the grand partition function is given by

$$Z_G = \text{Tr}\left(e^{-\beta(\hat{H}-\mu\hat{N})}\right)$$
$$= \sum_N \sum_n \langle \Psi_n^N | e^{-\beta(\hat{H}-\mu\hat{N})} | \Psi_n^N \rangle = \sum_N \sum_n e^{-\beta(E_n^N - \mu N)}. \tag{5.30}$$

In the evaluation of the trace, the basis with energy and particle number eigenstates was used. A standard result from statistical mechanics [Landau and Lifshitz (1980)] yields the thermodynamic potential

$$\Omega(T, V, \mu) = -k_B T \ln Z_G, \tag{5.31}$$

so that the statistical operator can be written as

$$\hat{\rho}_G = e^{\beta(\Omega - \hat{H} + \mu\hat{N})}. \tag{5.32}$$

The ensemble average $\langle \hat{O} \rangle$ of any operator $\hat{O}$ is obtained by evaluating

$$\langle \hat{O} \rangle = \text{Tr}\left(\hat{\rho}_G \hat{O}\right) = \frac{\text{Tr}\left(e^{-\beta(\hat{H}-\mu\hat{N})}\hat{O}\right)}{\text{Tr}\left(e^{-\beta(\hat{H}-\mu\hat{N})}\right)}. \tag{5.33}$$

Detailed results can be derived by studying the noninteracting many-fermion or many-boson systems. By replacing $\hat{H}$ by $\hat{H}_0$, the complete set of states of the independent-particle model can be used in the occupation number representation to evaluate the grand partition function. We employ the notation of Eqs. (2.18) and (2.19), introduced for fermions and bosons, respectively. Summing over the complete set of states in Fock space, can also be accomplished by summing over all possible occupations of the sp states. The operator $\hat{H}_0$ and the number operator, can then be replaced by their eigenvalues for a given state as follows

$$\hat{H}_0 \left| n_1...n_\infty \right\rangle = \sum_i n_i \varepsilon_i \left| n_1...n_\infty \right\rangle \tag{5.34}$$

$$\hat{N} \left| n_1...n_\infty \right\rangle = \sum_i n_i \left| n_1...n_\infty \right\rangle. \tag{5.35}$$

The grand partition function now reads

$$\begin{aligned}
Z_0 &= \sum_{n_1...n_\infty} \exp\left\{\beta(\mu n_1 - \varepsilon_1 n_1)\right\} ... \exp\left\{\beta(\mu n_\infty - \varepsilon_\infty n_\infty)\right\} \\
&= \prod_{i=1}^{\infty} \text{Tr}\left(\exp\left\{-\beta(\varepsilon_i - \mu)n_i\right\}\right), \tag{5.36}
\end{aligned}$$

where the trace includes a summation over the possible occupation numbers for sp state i. All occupation numbers must be included for bosons, yielding

$$Z_0^B = \prod_{i=1}^{\infty} \sum_{n=0}^{\infty} \left[\exp\left\{\beta(\mu - \varepsilon_i)\right\}\right]^n = \prod_{i=1}^{\infty} \left[1 - \exp\left\{\beta(\mu - \varepsilon_i)\right\}\right]^{-1}. \tag{5.37}$$

The thermodynamic potential for noninteracting bosons then reads

$$\begin{aligned}
\Omega_0^B(T,V,\mu) &= -k_B T \ln \prod_{i=1}^{\infty} \left[1 - \exp\left\{\beta(\mu - \varepsilon_i)\right\}\right]^{-1} \\
&= k_B T \sum_{i=1}^{\infty} \ln\left[1 - \exp\left\{\beta(\mu - \varepsilon_i)\right\}\right]. \tag{5.38}
\end{aligned}$$

The average number of particles is obtained by differentiating the thermo-dynamic potential with respect to the chemical potential, keeping T and V (meaning ε_i) fixed [Landau and Lifshitz (1980)]

$$\langle N \rangle \equiv \sum_{i=1}^{\infty} n_i^0 = \sum_{i=1}^{\infty} \frac{1}{\exp\{\beta(\varepsilon_i - \mu)\} - 1}. \tag{5.39}$$

The mean occupation number of a sp state i is denoted by n_i^0.

For fermions the restriction of the occupation number to 0 or 1 yields, according to Eq. (5.36),

$$Z_0^F = \prod_{i=1}^{\infty} \sum_{n=0}^{1} [\exp\{\beta(\mu - \varepsilon_i)\}]^n = \prod_{i=1}^{\infty} [1 + \exp\{\beta(\mu - \varepsilon_i)\}]. \tag{5.40}$$

As a result, the thermodynamic potential for noninteracting fermions reads

$$\Omega_0^F(T, V, \mu) = -k_B T \sum_{i=1}^{\infty} \ln[1 + \exp\{\beta(\mu - \varepsilon_i)\}]. \tag{5.41}$$

The number of particles is given by

$$\langle N \rangle \equiv \sum_{i=1}^{\infty} n_i^0 = \sum_{i=1}^{\infty} \frac{1}{\exp\{\beta(\varepsilon_i - \mu)\} + 1}. \tag{5.42}$$

We note that the average occupation numbers of a sp state i, denoted by n_i^0, has the familiar form for bosons in Eq. (5.39) and for fermions in Eq. (5.42).

5.6   Bosons at finite T

5.6.1   *Bose–Einstein condensation in infinite systems*

The ground state of a noninteracting system of bosons is obviously the one in which all the bosons occupy the lowest sp level. This state is the limit that one approaches when the temperature is lowered towards $T = 0$. First we will consider a collection of bosons in a box with the usual periodic boundary conditions. The sp energy becomes

$$\varepsilon(k) = \frac{\hbar^2 k^2}{2m}. \tag{5.43}$$

Summation over sp states can be replaced by an integral over wave vectors, as discussed in Sec. 5.1,

$$\sum_i \rightarrow \frac{\nu V}{(2\pi)^3} \int d\boldsymbol{k}, \tag{5.44}$$

where ν is the degeneracy factor associated with discrete quantum numbers such as spin. It is common to transform the integral over k to one over the energy, as determined by Eq. (5.43),

$$\frac{\nu V}{(2\pi)^3} 4\pi k^2 \, dk = \frac{\nu V}{2\pi^2} \left(\frac{2m}{\hbar^2}\right)^{3/2} \frac{\varepsilon d\varepsilon}{2\varepsilon^{1/2}} = \frac{\nu V}{4\pi^2} \left(\frac{2m}{\hbar^2}\right)^{3/2} \varepsilon^{1/2} d\varepsilon. \tag{5.45}$$

The thermodynamic potential for an ideal Bose gas can now be written as

$$\Omega_0^B = k_B T \frac{\nu V}{4\pi^2} \left(\frac{2m}{\hbar^2}\right)^{3/2} \int_0^\infty d\varepsilon \, \varepsilon^{1/2} \ln\left[1 - \exp\left\{\beta(\mu - \varepsilon)\right\}\right]$$

$$= -\frac{\nu V}{4\pi^2} \left(\frac{2m}{\hbar^2}\right)^{3/2} \frac{2}{3} \int_0^\infty d\varepsilon \, \frac{\varepsilon^{3/2}}{\exp\left\{\beta(\varepsilon - \mu)\right\} - 1}, \tag{5.46}$$

where in the last line, a partial integration has been performed. Similarly, we obtain for the average energy

$$E = \sum_i n_i^0 \varepsilon_i = \frac{\nu V}{4\pi^2} \left(\frac{2m}{\hbar^2}\right)^{3/2} \int_0^\infty d\varepsilon \, \frac{\varepsilon^{3/2}}{\exp\left\{\beta(\varepsilon - \mu)\right\} - 1}, \tag{5.47}$$

and average particle number

$$N = \sum_i n_i^0 = \frac{\nu V}{4\pi^2} \left(\frac{2m}{\hbar^2}\right)^{3/2} \int_0^\infty d\varepsilon \, \frac{\varepsilon^{1/2}}{\exp\left\{\beta(\varepsilon - \mu)\right\} - 1}. \tag{5.48}$$

Using the thermodynamic relation $\Omega = -PV$, one recovers from Eqs. (5.46) and (5.47) the standard expression for an ideal gas, $PV = \frac{2}{3}E$. The denominator in the integrals above reflects the occupation probability associated with energy ε. As a result, this denominator may not become negative. For this reason μ must always be such that for a given sp Hamiltonian with a corresponding spectrum $\varepsilon - \mu \geq 0$ (for any ε). In the present case ε may be zero so $\mu \leq 0$. If the density in Eq. (5.48) is kept fixed while the temperature is lowered, the absolute value of the chemical potential is expected to decrease, to allow the density to remain constant. The limit $\mu = 0$ is then reached for a temperature T_0, for which Eq. (5.48) yields

$$
\begin{aligned}
N &= \frac{\nu V}{4\pi^2} \left(\frac{2m}{\hbar^2}\right)^{3/2} \int_0^\infty d\varepsilon \, \frac{\varepsilon^{1/2}}{\exp\{\varepsilon/k_B T_0\} - 1} \\
&= \frac{\nu V}{4\pi^2} \left(\frac{2mk_B T_0}{\hbar^2}\right)^{3/2} \int_0^\infty dx \, \frac{x^{1/2}}{\exp(x) - 1} \\
&= \frac{\nu V}{4\pi^2} \left(\frac{2mk_B T_0}{\hbar^2}\right)^{3/2} \zeta(\tfrac{3}{2}) \tfrac{1}{2}\sqrt{\pi},
\end{aligned}
\tag{5.49}
$$

where $\zeta(\tfrac{3}{2}) = 2.612$ (Riemann ζ-function). The result can be rewritten as

$$
T_0 = \frac{3.31}{\nu^{2/3}} \frac{\hbar^2}{mk_B} \left(\frac{N}{V}\right)^{2/3}.
\tag{5.50}
$$

For temperatures below T_0 it is clear that the integral will give only the number of particles with $\varepsilon > 0$, while the occupation of the lowest state becomes macroscopically large. The latter state does not contribute to the integral, due to $\sqrt{\varepsilon}$ weighting which occurs when the summation is replaced by an integration. Therefore, it is necessary to track the number of particles in the lowest sp state. Equation (5.49) with T_0 replaced by T, continues to represent the number of particles with $\varepsilon > 0$, since μ remains zero. This implies that for $T < T_0$

$$
N_{\varepsilon>0}(T) = N \left(\frac{T}{T_0}\right)^{3/2}.
\tag{5.51}
$$

The remaining particles must then be in the lowest-energy sp state with $\varepsilon = 0$ according to

$$
N_{\varepsilon=0}(T) = N \left[1 - \left(\frac{T}{T_0}\right)^{3/2}\right].
\tag{5.52}
$$

The macroscopic occupation (proportional to N) of a single quantum state is referred to as Bose–Einstein condensation. At T_0 a discontinuity in the slope of the specific heat at constant volume is present [Landau and Lifshitz (1980); Fetter and Walecka (1971)]. The density of the ^{4}He liquid, $\rho = 0.145$ g cm^{-3}, yields $T_0 = 3.14$ K for the transition temperature. The actual experimental behavior is somewhat reminiscent of this result, but the specific heat has the shape of a λ, and is different from the ideal gas prediction. The experimental transition takes place at 2.2 K, below which ^{4}He exhibits superfluid properties. While superfluidity cannot be explained by the ideal gas description, it is nevertheless presumed to be related to Bose–Einstein condensation (see Sec. 19.1).

5.6.2 *Bose–Einstein condensation in traps*

The original papers by Bose and Einstein are discussed in context by [Pais (1986)]. Reviews from just before 1995 are collected in [Griffin *et al.* (1995)] while the more recent developments are discussed in [Dalfovo *et al.* (1999); Ketterle (1999); Burnett *et al.* (1999)]. New books on the subject have also been written recently [Pethick and Smith (2002); Pitaevskii and Stringari (2003)]. In 1995 Bose–Einstein condensation was first observed in experiments on rubidium (^{87}Rb) [Anderson *et al.* (1995)] and sodium (^{23}Na) atoms [Davis *et al.* (1995)], leading to the 2001 Nobel prize in physics, awarded to Wieman, Cornell, and Ketterle. Both types of atoms have an odd number of protons and an even number of neutrons in the nucleus. Together with the odd number of electrons, the total number of fermion constituents corresponds to an even integer, which makes these atoms bosons (for energies small compared to their internal excitation energies).

Historically, the superfluidity of liquid ^{4}He has been linked to Bose–Einstein condensation. Backed by theory, evidence has been gathered experimentally from neutron scattering measurements at high momentum transfer (see Sec. 19.2.1), that the occupation of the lowest sp state is close to 10% [Sokol (1995)] towards 0 K. This represents a macroscopic occupation of the lowest sp state; the zero momentum state. In the case of ^{4}He it is however, hard to find an unambiguous signature of Bose–Einstein condensation. Only recently [Wyatt (1998)], has tentative evidence for the condensate in ^{4}He become available by using quantum evaporation. This technique involves the scattering of phonons (excitations generated in the liquid) with atoms, somewhat analogous to the photoelectric effect. When such a phonon is incident on the free surface of the liquid, it is absorbed by an atom which is released into the vacuum above the liquid with conservation of momentum parallel to the surface. Measuring the momentum of the phonon and the evaporated atom allows determination of the momentum of the atom in the liquid, before it absorbed the phonon. The measurements of [Wyatt (1998)] at 100 mK indicate a macroscopic occupation, of the zero momentum state, but further refinements are necessary to pin down the actual fraction of condensed atoms.

The first efforts to achieve Bose–Einstein condensation with atomic gases were focused on experiments with hydrogen atoms, but it was not until recently [Fried *et al.* (1998)] that Bose–Einstein condensation has been observed for spin-polarized hydrogen. Critical to these new developments for atomic gases have been laser-based methods, such as laser cooling and magneto-optical trapping [Chu (1998); Cohen-Tannoudji (1998); Phillips (1998)]. After trapping, further lowering of the temperature can be accomplished by evaporative cooling in which the depth of the trap is reduced, allowing the most energetic atoms to escape. The remaining atoms rethermalize at lower temperature. It should be noted that these systems must be studied in a metastable gas phase, since the equilibrium configuration is the solid phase, except for hydrogen. Typical temperatures achieved to study Bose–Einstein condensation range from 500 nK to 2μK, with densities between 10^{14} and 10^{15} atoms per cm^3. The largest condensates in sodium have 20 million atoms, and in hydrogen 1 billion. Condensates have also been obtained for ^{7}Li atoms [Bradley *et al.* (1995)]. Other vapors of cesium, potassium, and helium are under study. The shape of the condensates depends on the magnetic trap and can be round with a diameter of 10 to 50 μm, or cigar-shaped with a diameter of about 15 μm and a length of 300 μm.

The confinement in traps implies that these systems are highly inhomogeneous, reflecting a substantial density variation in a finite region of space. This implies that Bose–Einstein condensation can be observed in coordinate space as well, since the relevant sp wave functions are localized. The magnetic traps used for alkali atoms, yield confining potentials that can be very well approximated by a quadratic form

$$V_{ext}(\boldsymbol{r}) = \frac{1}{2}\, m \left(\omega_x^2 x^2 + \omega_y^2 y^2 + \omega_z^2 z^2 \right). \tag{5.53}$$

The actual analysis of the experimental data requires the inclusion of the interaction between the atoms, as discussed in Ch. 12. Nevertheless, some useful results can be generated by assuming the atoms to be identical, pointlike, noninteracting particles in a harmonic potential according to Eq. (5.53). The eigenvalues of the corresponding three-dimensional harmonic oscillator potential are given by

$$\varepsilon_{n_x n_y n_z} = (n_x + \tfrac{1}{2})\,\hbar\omega_x + (n_y + \tfrac{1}{2})\,\hbar\omega_y + (n_z + \tfrac{1}{2})\,\hbar\omega_z, \tag{5.54}$$

where $\{n_x, n_y, n_z\}$ are nonnegative integers. The ground state of the noninteracting system with N atoms is obtained by putting all atoms in the lowest sp state with $n_x = n_y = n_z = 0$. The corresponding sp wave function is simply the product of three ground state oscillator wave functions:

one for each of the three dimensions considered. It has the form

$$\phi_{000}(\boldsymbol{r}) = \left(\frac{m\omega_{HO}}{\pi\hbar}\right)^{3/4} \exp\left\{-\frac{m}{2\hbar}(\omega_x x^2 + \omega_y y^2 + \omega_z z^2)\right\}, \qquad (5.55)$$

where the normalization factor can be written in terms of the geometric average of the three oscillator frequencies

$$\omega_{HO} = (\omega_x \omega_y \omega_z)^{1/3}. \qquad (5.56)$$

The density distribution for the ground state with N such bosons

$$|\Phi_0^N\rangle = \frac{1}{\sqrt{N!}}(a_{000}^\dagger)^N |0\rangle, \qquad (5.57)$$

then becomes

$$\rho(\boldsymbol{r}) = N |\phi_{000}(\boldsymbol{r})|^2. \qquad (5.58)$$

The density grows with N, but the size of the cloud is independent of N. It is determined by the trap potential, yielding the characteristic harmonic oscillator length

$$a_{HO} = \left(\frac{\hbar}{m\omega_{HO}}\right)^{1/2}. \qquad (5.59)$$

This length corresponds to the average width of the Gaussian, represented by Eq. (5.55). In the available experiments it is typically of the order $a_{HO} \approx 1 \,\mu\text{m}$. At finite temperature the atoms will also occupy excited states of the oscillator potential. The radius of this cloud will be larger than a_{HO}. The effect can be estimated by using the classical Boltzmann distribution for the density corresponding to a spherical potential $V_{ext}(r)$ [Landau and Lifshitz (1980)]

$$\rho_{cl}(r) \propto \exp\left\{-\frac{V_{ext}(r)}{k_B T}\right\} \qquad (5.60)$$

and assuming $k_B T \gg \hbar\omega_{HO}$ so that a significant fraction of the particles populates the higher oscillator levels. If

$$V_{ext}(r) = \frac{1}{2}m\omega_{HO}^2 r^2, \qquad (5.61)$$

the width of the classical density distribution is given by

$$R_T = a_{HO}\left(\frac{k_B T}{\hbar\omega_{HO}}\right)^{1/2}, \qquad (5.62)$$

and therefore larger than a_{HO}. It illustrates that Bose–Einstein condensation in harmonic traps shows up in the form of a sharp peak in the central region of the density distribution. By taking the Fourier transform of the ground state wave function, the momentum distribution of the atoms in the condensate is obtained. Since the wave function in momentum space is also Gaussian, the distribution is centered at zero momentum with a width proportional to a_{HO}^{-1}. As a result, the condensate appears both in coordinate and momentum space as a narrow peak. This is quite different from the case of a uniform gas, where the particles go to the zero momentum state, but there is no signature in coordinate space, since condensed and noncondensed particles fill the same volume. The experimental signal of Bose–Einstein condensation has been detected as the occurrence of a sharp peak over a broader distribution in both the velocity and spatial distributions. By switching off the trap, one lets the condensate expand and measures the density of the expanding cloud by light absorption. If there is no interaction between the atoms, the expansion is ballistic and the imaged distribution can be related to the initial velocity (momentum) distribution. For the spatial distribution one measures the density of the atoms in the trap directly by dispersive light scattering. The effect of the interaction between the atoms modifies the results substantially, as discussed in Ch. 12. The symmetry of the confining potential leads to important signatures at the noninteracting level. These features persist when interactions are taken into account. The first experiments were carried out with axial symmetry for which the ground state wave function can be written as

$$\phi_{000}(\boldsymbol{r}) = \frac{\lambda^{1/4}}{\pi^{3/4} a_\perp^{3/2}} \exp\left\{-\frac{1}{2a_\perp^2}(r_\perp^2 + \lambda z^2)\right\}, \qquad (5.63)$$

where $a_\perp = (\hbar/m\omega_\perp)^{1/2}$ is the oscillator length in the xy-plane. Since $\omega_\perp = \lambda^{-1/3}\omega_{HO}$, it follows that $a_\perp = \lambda^{1/6} a_{HO}$. The momentum space wave function has a corresponding asymmetry governed by the parameter λ. If condensation has taken place, the shape of the expanding cloud is an ellipse with an aspect ratio of $\sqrt{\lambda}$, whereas one expects spherical symmetry for a thermal distribution. The actual values of this ratio are strongly influenced by the interaction between the atoms, but still show this important anisotropy which is used to identify the condensate.

5.6.3 Trapped bosons at finite temperature: thermodynamic considerations

At temperature T, the number of particles follows from Eq. (5.39), which yields

$$N = \sum_{n_x n_y n_z} \frac{1}{\exp\left\{\beta(\varepsilon_{n_x n_y n_z} - \mu)\right\} - 1}, \qquad (5.64)$$

while the total energy is given by

$$E = \sum_{n_x n_y n_z} \frac{\varepsilon_{n_x n_y n_z}}{\exp\left\{\beta(\varepsilon_{n_x n_y n_z} - \mu)\right\} - 1}. \qquad (5.65)$$

Statistical mechanics is complicated by the fact that the usual thermodynamic limit is not appropriate for these gases. Indeed, due to the inhomogeneity of the system, it is not possible to take N and V to infinity, while keeping their ratio constant. As in the uniform case however, one may separate the lowest energy ε_{000} from the sum in Eq. (5.64) and denote by N_0 the number of particles in this state. N_0 can be of order N, when the chemical potential approaches the energy of the lowest state

$$\mu \to \mu_c = \frac{3}{2}\hbar\bar\omega, \qquad (5.66)$$

where $\bar\omega = (\omega_x + \omega_y + \omega_z)/3$ is the average frequency. The limit is reached for a critical temperature $T = T_c$. Equation (5.64) can be written as

$$N - N_0 = \sum_{n_x \neq 0, n_y \neq 0, n_z \neq 0} \frac{1}{\exp\left\{\beta\hbar(\omega_x n_x + \omega_y n_y + \omega_z n_z)\right\} - 1}. \qquad (5.67)$$

The sum can be evaluated numerically for finite N, but for $N \to \infty$ it can be replaced by an integral

$$N - N_0 = \int_0^\infty dn_x dn_y dn_z \frac{1}{\exp\left\{\beta\hbar(\omega_x n_x + \omega_y n_y + \omega_z n_z)\right\} - 1}. \qquad (5.68)$$

This approximation is referred to as a semiclassical description of the excited states. It implies that the relevant excitation energies are much larger than the level spacing which is fixed by the oscillator frequencies. Valid for large N and $k_B T \gg \hbar\omega_{HO}$, it can always be checked numerically using Eq. (5.67). The integral (5.68) can be evaluated with the following result

$$N - N_0 = \zeta(3) \left(\frac{k_B T}{\hbar\omega_{HO}}\right)^3, \qquad (5.69)$$

where the Riemann ζ function has a value given by $\zeta(3) \approx 1.202$. It leads to the transition temperature for Bose–Einstein condensation. Imposing that $N_0 \to 0$ at T_c, one obtains

$$k_B T_c = \hbar \omega_{HO} \left(\frac{N}{\zeta(3)} \right)^{1/3} = 0.94 \, \hbar \omega_{HO} \, N^{1/3}. \qquad (5.70)$$

Note that the physically relevant "thermodynamic limit" of this harmonically trapped system would consist of $N \to \infty$, while keeping $N\omega_{HO}^3$ fixed. Substituting Eq. (5.70) in (5.69), the T dependence of the condensate fraction for $T < T_c$ becomes

$$\frac{N_0}{N} = 1 - \left(\frac{T}{T_c} \right)^3 . \qquad (5.71)$$

By evaluating the energy in a similar way, all thermodynamic quantities can be generated. A further useful quantity is the density of thermal particles out of the condensate, $\rho_T(\mathbf{r})$. The sum of $\rho_T(\mathbf{r})$ and the condensate density, $\rho_0(\mathbf{r}) = N_0 |\phi_{000}(\mathbf{r})|^2$, generate the total density distribution. The classical expression

$$\rho_T(\mathbf{r}) = \int \frac{d\mathbf{p}}{(2\pi\hbar)^3} \left[\exp\{\beta\varepsilon(\mathbf{p},\mathbf{r})\} - 1 \right]^{-1}, \qquad (5.72)$$

can be used in this case, where

$$\varepsilon(\mathbf{p},\mathbf{r}) = \frac{\mathbf{p}^2}{2m} + V_{ext}(\mathbf{r}) \qquad (5.73)$$

is the semiclassical energy in phase space. Equation (5.72) then yields

$$\rho_T(\mathbf{r}) = \frac{g_{3/2}(e^{-\beta V_{ext}(\mathbf{r})})}{\lambda_T^3}, \qquad (5.74)$$

where the function $g_{3/2}$ is a Bose function $g_\alpha(z) = \sum_{n=1}^{\infty} \frac{z^n}{n^\alpha}$, discussed in [Huang (1987)]. Important finite size corrections and crucial modifications of these results occur when the interaction between the atoms is taken into account [Dalfovo et al. (1999)] (see also Ch. 12). Nevertheless, the above results, involving noninteracting bosons, already yield some useful notions, relevant for the understanding of the recent experiments on Bose–Einstein condensation.

5.7 Fermions at finite T

5.7.1 *Noninteracting fermion systems*

The noninteracting Fermi gas at finite temperature yields thermodynamic quantities that exhibit characteristic differences with their bosonic counterparts. Using fermion occupation numbers, the thermodynamic potential of the homogeneous Fermi gas becomes

$$\Omega_0^F = -\frac{\nu V}{4\pi^2} \left(\frac{2m}{\hbar^2}\right)^{3/2} \frac{2}{3} \int_0^\infty d\varepsilon \, \frac{\varepsilon^{3/2}}{\exp\{\beta(\varepsilon - \mu)\} + 1}, \tag{5.75}$$

after performing similar steps as for the noninteracting Bose gas. Likewise, we obtain for the energy

$$E = \sum_i n_i^0 \varepsilon_i = \frac{\nu V}{4\pi^2} \left(\frac{2m}{\hbar^2}\right)^{3/2} \int_0^\infty d\varepsilon \, \frac{\varepsilon^{3/2}}{\exp\{\beta(\varepsilon - \mu)\} + 1}, \tag{5.76}$$

and particle number

$$N = \sum_i n_i^0 = \frac{\nu V}{4\pi^2} \left(\frac{2m}{\hbar^2}\right)^{1/2} \int_0^\infty d\varepsilon \, \frac{\varepsilon^{1/2}}{\exp\{\beta(\varepsilon - \mu)\} + 1}. \tag{5.77}$$

Using $\Omega = -PV$, one recovers from Eqs. (5.75) and (5.76) also for fermions the standard result for an ideal gas; $PV = \frac{2}{3}E$. The denominator in the integrals above, reflects the occupation probability associated with energy ε. It is clear that this occupation number is always less or equal to one, for all values of μ and T. It is useful to check the zero-temperature limit which recovers the result from Sec. 5.1 with $\mu = \varepsilon_F$ at $T = 0$, resulting in full occupation of states with $k < k_F$, the Fermi sea. Relevant thermodynamic quantities are straightforward to work out [Fetter and Walecka (1971); Landau and Lifshitz (1980)].

5.7.2 *Fermion atoms in traps*

While the transition to a Bose–Einstein condensate for atomic gases is abrupt, the crossover to quantum degeneracy for fermionic atoms is gradual. It was recently demonstrated for the first time for ^{40}K atoms [DeMarco and Jin (1999)]. The potassium atom has an odd number of protons, neutrons, and electrons, thereby making it a fermion. The electrons combine to atomic spin $F = \frac{9}{2}$. It is more difficult to cool fermions by using the evaporation technique, mentioned earlier. In this process the highest energy atoms are removed and the remaining gas equilibrates by elastic collisions at a lower temperature. The elastic collision rate drops sharply for fermions when the temperature is lowered, because the interaction is dominated by the S-wave. As discussed in Ch. 4, S-wave scattering is prohibited for single species fermions. This problem was overcome by using a mixture of two spin states for which such collisions are allowed. For the trap used in the experiment, one expects a Fermi temperature of 0.6 μK. The actual experiment cooled about 8×10^5 atoms to approximately $T = 0.5T_F$. At that point, one of the spin states was removed and a single-component gas remained.

We proceed in a similar fashion as for bosons, to describe the sp Hamiltonian for the fermions

$$H = \frac{p_x^2 + p_y^2 + p_z^2}{2m} + \tfrac{1}{2}m\omega_r^2 \left(x^2 + y^2 + \lambda^2 z^2 \right), \qquad (5.78)$$

where ω_r and $\omega_z = \lambda\omega_r$ are the trap frequencies in the radial and axial directions, respectively. The sp energy is given by Eq. (5.54), with appropriate changes in notation for the oscillator frequencies. In the actual experiments the thermal energies far exceed the level spacing ($k_B T \gg \hbar\omega_r$), like for the boson systems discussed above. It is therefore allowed to replace the discrete sp spectrum by a continuous one, employing the appropriate density of states. The density of states at ε can be expressed in general according to [Economou (1983)]

$$\rho(\varepsilon) = \sum_n \delta(\varepsilon - \varepsilon_n), \qquad (5.79)$$

where ε_n is given by the spectrum under consideration. For the spectrum of the Hamiltonian given by Eq. (5.78), one calculates

$$\rho(\varepsilon) = \frac{\varepsilon^2}{2\lambda(\hbar\omega_r)^3}. \qquad (5.80)$$

A practical way to obtain this density is to first determine the number of states with energy below $\varepsilon + d\varepsilon$, then subtract the number below ε, and divide by $d\varepsilon$ (for $d\varepsilon \to 0$). The fermion equivalent of Eq. (5.64) can now be used to study the chemical potential [Butts and Rokhsar (1997)]

$$N = \sum_{n_x n_y n_z} \left[\exp\left\{ \beta(\varepsilon_{n_x n_y n_z} - \mu) \right\} + 1 \right]^{-1} = \int d\varepsilon \, \frac{\rho(\varepsilon)}{e^{\beta(\varepsilon-\mu)} + 1}. \qquad (5.81)$$

At zero temperature, all levels below the Fermi energy are occupied, the ones above empty. Integrating Eq. (5.81) yields

$$\varepsilon_F \equiv \mu(T = 0, N) = \hbar\omega_r \left[6\lambda N\right]^{1/3}, \tag{5.82}$$

which yields the energy scale for the atomic cloud. A characteristic length scale is given by the extent of the orbit of a classical particle with energy ε_F in the trap potential

$$R_F \equiv \left[\frac{2\varepsilon_F}{m\omega_r^2}\right]^{1/2} = (48\lambda N)^{1/6} a_r, \tag{5.83}$$

where a_r is the radial oscillator parameter. A characteristic wave number is obtained from the momentum of a particle with energy ε_F

$$k_F \equiv \left[\frac{2m\varepsilon_F}{\hbar^2}\right]^{1/2} = (48\lambda N)^{1/6}\frac{1}{a_r} = \frac{(48\lambda N)^{1/3}}{R_F}, \tag{5.84}$$

which shows that k_F is about equal to the inverse of the interparticle spacing in the gas. For a general temperature, the chemical potential can be determined numerically from Eq. (5.81). Analytic results are available at low temperature ($k_B T \ll \varepsilon_F$) from Sommerfeld's expansion [Landau and Lifshitz (1980)]

$$\mu(T, N) = \varepsilon_F \left[1 - \frac{\pi^2}{3}\left(\frac{k_B T}{\varepsilon_F}\right)^2\right]. \tag{5.85}$$

In the classical limit, $k_B T \gg \varepsilon_F$, one finds

$$\mu(T, N) = -k_B T \ln\left[6\left(\frac{k_B T}{\varepsilon_F}\right)^3\right]. \tag{5.86}$$

A similar procedure yields the energy and, subsequently, the specific heat of the trapped gas from

$$C_N \equiv \frac{1}{N}\frac{\partial E}{\partial T}\bigg|_N, \tag{5.87}$$

where $E(T, N)$ is the total energy. The experimental data from [DeMarco and Jin (1999)] clearly confirm the expectations for the deviation of the energy from the classical result as a function of temperature, as well as the behavior of the specific heat given by Eq. (5.87). Further experiments are geared towards studying situations in which the interaction between atoms near the Fermi energy is effectively attractive. The expectation is that new insights into pairing may be gleaned from such systems.

5.8   Exercises

(1) Evaluate the ground state energy of the electron gas in first-order perturbation theory. Make a plot of the energy per particle as a function of r_s and determine the minimum.

(2) Assume that nucleons interact by means of a two-body interaction given by

$$V = V_0(r) + V_\tau(r)\boldsymbol{\tau}_1 \cdot \boldsymbol{\tau}_2 + V_\sigma(r)\boldsymbol{\sigma}_1 \cdot \boldsymbol{\sigma}_2 + V_{\sigma\tau}(r)\boldsymbol{\sigma}_1 \cdot \boldsymbol{\sigma}_2\boldsymbol{\tau}_1 \cdot \boldsymbol{\tau}_2,$$

where each radial dependence is governed by a Yukawa form with different masses and constants. Evaluate the ground state energy of nuclear matter in first-order perturbation theory, using this interaction.

(3) Calculate the energy and corresponding specific heat at constant volume of the noninteracting Bose gas, below and above T_0.

(4) Calculate the energy of a trapped Bose gas and its specific heat. Compare these quantities with the corresponding ones for the uniform Bose gas.

(5) Provide the details for the calculation of the density of states given in Eq. (5.80). Use Sommerfeld's expansion to determine the energy of the Fermi gas in the trap at low temperature, and calculate the specific heat according to Eq. (5.87). Compare your results with [DeMarco and Jin (1999)].

Chapter 6

Propagators in one-particle quantum mechanics

In order to master the concept of a sp propagator in a many-particle system, it is instructive to pose the problem for one particle in this language. In Sec. 6.1, the time evolution of a quantum state, as generated by the Hamiltonian of the system, is reviewed. The relation between a state at an earlier and a later time, suggests the definition of the propagator and relates it to the intuitive notion of Huygens' principle. The expansion of the propagator in terms of a known, or unperturbed propagator is introduced in Sec. 6.2, with special emphasis on its diagrammatic representation. A solution method for bound state problems is illustrated in Sec. 6.3, whereas a discussion of scattering in the propagator language is presented in Sec. 6.4.

6.1 Time evolution and propagators

Time evolution in physics is determined by the Hamiltonian of the physical system. In quantum mechanics, the state of a particle with quantum numbers α at time t_0 can be denoted by $|\alpha, t_0\rangle$. At t later than t_0, one obtains the state $|\alpha, t_0; t\rangle$, which has evolved from the initial one, according to

$$|\alpha, t_0; t\rangle = e^{-\frac{i}{\hbar}H(t-t_0)} |\alpha, t_0\rangle, \tag{6.1}$$

for a Hamiltonian that does not depend on time. The correctness of Eq. (6.1) can be checked by substituting this expression in the Schrödinger equation

$$i\hbar \frac{\partial}{\partial t} |\alpha, t_0; t\rangle = H |\alpha, t_0; t\rangle. \tag{6.2}$$

Equation (6.1) for $|\alpha, t_0; t\rangle$ can be written in terms of the wave function of the particle at time t as follows

$$\begin{aligned}
\psi(\boldsymbol{r}, t) &= \langle \boldsymbol{r}|\alpha, t_0; t\rangle = \langle \boldsymbol{r}| e^{-\frac{i}{\hbar}H(t-t_0)} |\alpha, t_0\rangle \\
&= \int d\boldsymbol{r}' \, \langle \boldsymbol{r}| e^{-\frac{i}{\hbar}H(t-t_0)} |\boldsymbol{r}'\rangle \langle \boldsymbol{r}'|\alpha, t_0\rangle \\
&= i\hbar \int d\boldsymbol{r}' \, G(\boldsymbol{r}, \boldsymbol{r}'; t-t_0)\psi(\boldsymbol{r}', t_0),
\end{aligned} \tag{6.3}$$

where G is referred to as the propagator or Green's function

$$G(\boldsymbol{r}, \boldsymbol{r}'; t-t_0) = -\frac{i}{\hbar} \langle \boldsymbol{r}| e^{-\frac{i}{\hbar}H(t-t_0)} |\boldsymbol{r}'\rangle . \tag{6.4}$$

Its physical meaning can be illustrated by recalling Huygens' principle. Indeed, Eq. (6.3) demonstrates that the wave function at $\boldsymbol{r}$ and t is determined by the one at the original time t_0, receiving contributions from all $\boldsymbol{r}'$ which are weighted by the amplitude G. Note that knowledge of the initial wave function and the propagator G thus allows the construction of ψ at any time $t > t_0$.

Several alternative ways of writing the propagator may be obtained by using

$$H |n\rangle = \varepsilon_n |n\rangle \tag{6.5}$$

for the exact eigenstates of H. Assuming a discrete spectrum to simplify the notation, these alternative ways of writing the propagator include

$$\begin{aligned}
G(\boldsymbol{r}, \boldsymbol{r}'; t-t_0) &= -\frac{i}{\hbar} \langle \boldsymbol{r}| e^{-\frac{i}{\hbar}H(t-t_0)} |\boldsymbol{r}'\rangle = -\frac{i}{\hbar} \langle 0| a_{\boldsymbol{r}} e^{-\frac{i}{\hbar}H(t-t_0)} a_{\boldsymbol{r}'}^\dagger |0\rangle \\
&= -\frac{i}{\hbar} \sum_n \langle 0| a_{\boldsymbol{r}} |n\rangle \langle n| a_{\boldsymbol{r}'}^\dagger |0\rangle \, e^{-\frac{i}{\hbar}\varepsilon_n(t-t_0)} \\
&= -\frac{i}{\hbar} \sum_n u_n(\boldsymbol{r}) u_n^*(\boldsymbol{r}') e^{-\frac{i}{\hbar}\varepsilon_n(t-t_0)},
\end{aligned} \tag{6.6}$$

employing standard notation for energy eigenfunctions. To incorporate the causality condition $t > t_0$ explicitly, it is convenient to include the step function $\theta(t - t_0)$ in this expression. For practical calculations, it is essential to consider the Fourier transform (FT) of the propagator to arrive at nonperturbative (all-order) solution methods. To work out this FT, the following representation of the step function can be employed

$$\theta(t - t_0) = -\int \frac{dE'}{2\pi i} \frac{e^{-iE'(t-t_0)/\hbar}}{E' + i\eta}. \tag{6.7}$$

Note that $\eta \downarrow 0$ is implied. For $t > t_0$, the integration path can be closed in the lower half plane and the contribution of the enclosed pole yields a result equal to 1. For $t < t_0$, one can close the contour in the upper half plane which yields a vanishing outcome, since no pole is enclosed. At $t = t_0$, the step function jumps from 0 to 1. Its derivative is given by

$$\frac{d}{dt}\theta(t - t_0) = \delta(t - t_0). \tag{6.8}$$

The FT of the propagator then reads in various alternative forms

$$G(\boldsymbol{r}, \boldsymbol{r}'; E) = -\frac{i}{\hbar} \int_{-\infty}^{\infty} d(t - t_0)\, e^{\frac{i}{\hbar}E(t-t_0)}$$

$$\times \left\{ \theta(t - t_0) \sum_n u_n(\boldsymbol{r})u_n^*(\boldsymbol{r}')e^{-\frac{i}{\hbar}\varepsilon_n(t-t_0)} \right\}$$

$$= \sum_n \frac{u_n(\boldsymbol{r})u_n^*(\boldsymbol{r}')}{E - \varepsilon_n + i\eta} = \sum_n \frac{\langle 0|\, a_{\boldsymbol{r}}\, |n\rangle\, \langle n|\, a_{\boldsymbol{r}'}^\dagger\, |0\rangle}{E - \varepsilon_n + i\eta}$$

$$= \langle 0|\, a_{\boldsymbol{r}} \frac{1}{E - H + i\eta} a_{\boldsymbol{r}'}^\dagger\, |0\rangle = \langle \boldsymbol{r}|\, \frac{1}{E - H + i\eta}\, |\boldsymbol{r}'\rangle. \tag{6.9}$$

The presence of the $i\eta$ term in the denominator originates from the inclusion of condition $t > t_0$ (time going forward). The formulation in Eq. (6.9) assumes a spinless boson. The inclusion of spin quantum numbers for a fermion is straightforward. Some of the expressions for G will have their counterpart for the sp propagator in a many-particle system. It is important to realize that one can study the propagator in any sp basis

$$G(\alpha, \beta; E) = \langle 0|\, a_\alpha \frac{1}{E - H + i\eta} a_\beta^\dagger\, |0\rangle, \tag{6.10}$$

where α represents an appropriate set of sp quantum numbers to identify a possible state of the particle.

6.2 Expansion of the propagator and diagram rules

The exact propagator can be related to an approximate one by using a decomposition of the Hamiltonian

$$H = H_0 + V, \tag{6.11}$$

where H_0 is referred to as the unperturbed Hamiltonian for which the corresponding propagator $G^{(0)}$ is readily available. The following operator identity

$$\frac{1}{A - B} = \frac{1}{A} + \frac{1}{A} B \frac{1}{A - B}, \tag{6.12}$$

with $A = E - H_0 + i\eta$ and $B = V$ may then be employed. This operator equation relates G, involving H,

$$G = \frac{1}{E - H + i\eta} \tag{6.13}$$

to the corresponding operator $G^{(0)}$, involving H_0, and the potential V

$$\begin{aligned} G &= G^{(0)} + G^{(0)} \, V \, G \\ &= G^{(0)} + G^{(0)} \, V \, G^{(0)} + G^{(0)} \, V \, G^{(0)} \, V \, G^{(0)} + ... \end{aligned} \tag{6.14}$$

The unperturbed propagator, which is given by

$$G^{(0)}(\alpha, \beta; E) = \langle 0| \, a_\alpha \frac{1}{E - H_0 + i\eta} a_\beta^\dagger \, |0\rangle, \tag{6.15}$$

can then be used to obtain

$$\langle \alpha| \, \frac{1}{E - H + i\eta} \, |\beta\rangle = \langle \alpha| \, \frac{1}{E - H_0 + i\eta} \, |\beta\rangle \tag{6.16}$$

$$+ \sum_{\gamma\delta} \langle \alpha| \, \frac{1}{E - H_0 + i\eta} \, |\gamma\rangle \, \langle \gamma| \, V \, |\delta\rangle \, \langle \delta| \, \frac{1}{E - H + i\eta} \, |\beta\rangle$$

or

$$G(\alpha, \beta; E) = G^{(0)}(\alpha, \beta; E) + \sum_{\gamma, \delta} G^{(0)}(\alpha, \gamma; E) \, \langle \gamma| \, V \, |\delta\rangle \, G(\delta, \beta; E). \tag{6.17}$$

6.2.1 *Diagram rules for the single-particle propagator*

It is possible to generate a series of diagrams that represent the contributions to the sp propagator in a perturbation expansion in the potential V. These terms can be derived algebraically by iterating the equation for the sp propagator. It is convenient to choose $\{|\alpha\rangle\}$ to be eigenstates of H_0 with eigenvalues $\{\varepsilon_\alpha\}$. One then obtains

$$G^{(0)}(\alpha, \beta; E) = \frac{\delta_{\alpha,\beta}}{E - \varepsilon_\alpha + i\eta}. \tag{6.18}$$

For a contribution of kth order in V we find:

Rule 1 Draw a directed line with k zigzag (horizontal) interaction lines V and $k+1$ directed unperturbed propagators $G^{(0)}$

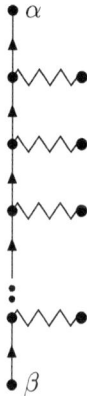

Rule 2 Label external points (α and β)
Label each V

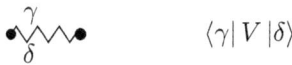 $\langle \gamma | V | \delta \rangle$

For each full line with arrow write

 $G^{(0)}(\mu, \nu; E)$

Rule 3 Sum (integrate) over all internal quantum numbers

Examples of diagrams in the single-particle problem

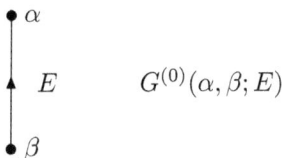 $G^{(0)}(\alpha, \beta; E)$

$$\sum_{\gamma,\delta} G^{(0)}(\alpha,\gamma;E) \langle \gamma| V |\delta\rangle G^{(0)}(\delta,\beta;E)$$

$$\sum_{\gamma,\delta} G^{(0)}(\alpha,\gamma;E)$$
$$\times \left\{ \sum_{\epsilon\theta} \langle \gamma| V |\epsilon\rangle G^{(0)}(\epsilon,\theta;E) \langle \theta| V |\delta\rangle \right\} G^{(0)}(\delta,\beta;E)$$

An extension to third and higher order in V is straightforward. The representation of the perturbation series given in Eq. (6.14), may appear superfluous at this point. Similar diagrams will be generated when the sp propagator is considered in a many-body system however, and it is useful to illustrate some of the possible resummations of the diagrams. To avoid cluttering notation, the operator form [Eq. (6.14)] can be employed to rearrange the series in several ways

$$\begin{aligned} G &= G^{(0)} + G^{(0)} \, V \, G^{(0)} + G^{(0)} \, V \, G^{(0)} \, V \, G^{(0)} + ... \quad (6.19) \\ &= G^{(0)} + G^{(0)} \, V \, \{G^{(0)} + G^{(0)} \, V \, G^{(0)} + ...\} = G^{(0)} + G^{(0)} \, V \, G \\ &= G^{(0)} + \{G^{(0)} + G^{(0)} \, V \, G^{(0)} + ...\} \, V \, G^{(0)} = G^{(0)} + G \, V \, G^{(0)} \\ &= G^{(0)} + G^{(0)} \{V + V \, G^{(0)} \, V + ...\} \, G^{(0)} = G^{(0)} + G^{(0)} \, \mathcal{T} \, G^{(0)}, \end{aligned}$$

where

$$\begin{aligned} \mathcal{T} &= V + V \, G^{(0)} \, V + V \, G^{(0)} \, V \, G^{(0)} \, V + ... \\ &= V + V \, G^{(0)} \, \{V + V \, G^{(0)} \, V + ...\} \\ &= V + V \, G^{(0)} \, \mathcal{T} = V + \mathcal{T} \, G^{(0)} \, V = V + V \, G \, V. \quad (6.20) \end{aligned}$$

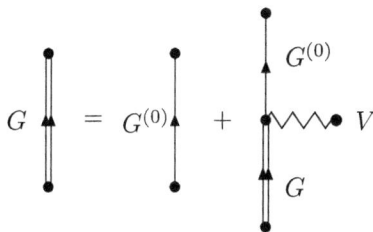

Fig. 6.1 Diagrammatic representation of the summation in Eq. (6.19) for sp propagator G in terms of $G^{(0)}$ and V.

A compact way to draw $G = G^{(0)} + G^{(0)}VG$ diagrammatically is given in Fig. 6.1. The double line depicts G. Similar diagrammatic results can be generated for the other forms of Eq. (6.19). $\mathcal{T}$ is simply the $\mathcal{T}$-matrix familiar from the calculation of the scattering amplitude. The corresponding equations in Eq. (6.20) illustrate possible forms for the Lippmann–Schwinger equation. The diagrammatic representation of Eq. (6.20) is illustrated in Fig. 6.2. This formulation in terms of $\mathcal{T}$-matrices, each diagrammatically represented by a double zigzag line, can be quite practical in the case of continuum solutions. Considering again the problem in which V corresponds to a localized potential and $H_0 = T$, one can choose the sp basis $\{|\alpha\rangle = |\boldsymbol{p}\rangle\}$ to obtain

$$\langle \boldsymbol{p}_1 | \mathcal{T}(E) | \boldsymbol{p}_2 \rangle = \langle \boldsymbol{p}_1 | V | \boldsymbol{p}_2 \rangle$$
$$+ \int d\boldsymbol{p} \, \langle \boldsymbol{p}_1 | V | \boldsymbol{p} \rangle \, \frac{1}{E - p^2/2m + i\eta} \, \langle \boldsymbol{p} | \mathcal{T}(E) | \boldsymbol{p}_2 \rangle \quad (6.21)$$

for a particle without spin. There should be no confusion between the kinetic energy T and the symbol for the scattering quantity $\mathcal{T}(E)$.

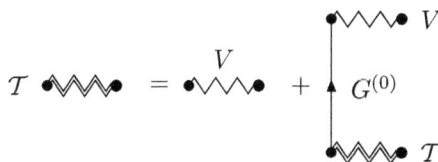

Fig. 6.2 A possible diagrammatic representation of Eq. (6.20) for $\mathcal{T}$.

6.3   Solution for discrete states

A general procedure is available to generate the possible discrete states of H from the propagator equation for G. It can also be applied to solve the propagator equation in the medium with both an energy-independent, as well as an energy-dependent potential. In the present case familiar results will be obtained, but it is useful to illustrate the procedure, since it can easily be adapted in more complicated situations. The exact eigenkets and energies of H are denoted by

$$H \, |m\rangle = \varepsilon_m \, |m\rangle \tag{6.22}$$

for possible discrete states ($\varepsilon_m < 0$) and by

$$H \, |\mu\rangle = \varepsilon_\mu \, |\mu\rangle \tag{6.23}$$

for continuum states ($\varepsilon_\mu > 0$). The completeness relation for the exact eigenstates of H

$$1 = \sum_m |m\rangle \, \langle m| + \int d\mu \, |\mu\rangle \, \langle \mu| \tag{6.24}$$

can be used to rewrite Eq. (6.10) as

$$G(\alpha, \beta; E) = \sum_m \frac{\langle \alpha|m\rangle \langle m|\beta\rangle}{E - \varepsilon_m + i\eta} + \int d\mu \, \frac{\langle \alpha|\mu\rangle \langle \mu|\beta\rangle}{E - \varepsilon_\mu + i\eta}. \tag{6.25}$$

Assume that H_0 is given by the kinetic energy T. For V an energy-independent, localized, but not necessarily local potential, may be considered. We choose to work with $\{|\alpha\rangle\} = \{|\boldsymbol{p}\rangle\}$ representing the eigenstates of T. To generate the equation for the bound state energies from the propagator equation, the following practical recipe can be employed. Calculate:

$$\lim_{E \to \varepsilon_n} (E - \varepsilon_n)\{G = G^{(0)} + G^{(0)} \, V \, G\}. \tag{6.26}$$

The three limits for each of the terms in this equation will be inspected one at a time. The limit of the left-hand side of Eq. (6.26) yields

$$\lim_{E \to \varepsilon_n} (E - \varepsilon_n)\left\{ \sum_m \frac{\langle \alpha|m\rangle \langle m|\beta\rangle}{E - \varepsilon_m + i\eta} + \ldots \right\} = \langle \alpha|n\rangle \langle n|\beta\rangle$$

$$\Rightarrow \langle \boldsymbol{p}|n\rangle \langle n|\boldsymbol{p}'\rangle. \tag{6.27}$$

For the first term on the right side of Eq. (6.26) we find

$$\lim_{E \to \varepsilon_n} (E - \varepsilon_n) \langle \alpha | \frac{1}{E - T + i\eta} | \beta \rangle \Rightarrow \lim_{E \to \varepsilon_n} (E - \varepsilon_n) \frac{\delta(\boldsymbol{p} - \boldsymbol{p}')}{E - \frac{\boldsymbol{p}^2}{2m} + i\eta}$$

$$= 0, \tag{6.28}$$

and for the last term

$$\lim_{E \to \varepsilon_n} (E - \varepsilon_n) \times \sum_{\gamma\delta} \langle \alpha | \frac{1}{E - T + i\eta} | \gamma \rangle \langle \gamma | V | \delta \rangle \left\{ \sum_m \frac{\langle \delta | m \rangle \langle m | \beta \rangle}{E - \varepsilon_m + i\eta} + \ldots \right\}$$

$$= \sum_{\gamma\delta} \langle \alpha | \frac{1}{\varepsilon_n - T} | \gamma \rangle \langle \gamma | V | \delta \rangle \langle \delta | n \rangle \langle n | \beta \rangle$$

$$\Rightarrow \int d\boldsymbol{p}'' \frac{1}{\varepsilon_n - \frac{\boldsymbol{p}^2}{2m}} \langle \boldsymbol{p} | V | \boldsymbol{p}'' \rangle \langle \boldsymbol{p}'' | n \rangle \langle n | \boldsymbol{p}' \rangle, \tag{6.29}$$

respectively. Collecting the two remaining terms, one finds

$$\langle \boldsymbol{p} | n \rangle = \frac{1}{\varepsilon_n - \frac{\boldsymbol{p}^2}{2m}} \int d\boldsymbol{p}'' \langle \boldsymbol{p} | V | \boldsymbol{p}'' \rangle \langle \boldsymbol{p}'' | n \rangle, \tag{6.30}$$

or, rearranging slightly and noting that $\langle \boldsymbol{p} | n \rangle = \phi_n(\boldsymbol{p})$, the momentum space wave function, one obtains

$$\frac{\boldsymbol{p}^2}{2m} \phi_n(\boldsymbol{p}) + \int d\boldsymbol{p}'' \langle \boldsymbol{p} | V | \boldsymbol{p}'' \rangle \phi_n(\boldsymbol{p}'') = \varepsilon_n \phi_n(\boldsymbol{p}). \tag{6.31}$$

This corresponds to the Schrödinger equation in momentum space that yields the bound-state energies and corresponding eigenfunctions.

Instead of the momentum representation, we can maintain the general notation and collect the limits of Eqs. (6.27)–(6.29) with the result

$$\langle \alpha | n \rangle = \sum_{\gamma\delta} \langle \alpha | \frac{1}{\varepsilon_n - H_0} | \gamma \rangle \langle \gamma | V | \delta \rangle \langle \delta | n \rangle. \tag{6.32}$$

By multiplying this equation with $\langle \beta | (\varepsilon_n - H_0) | \alpha \rangle$ and summing over α, it reduces to

$$\sum_{\alpha} \langle \beta | (\varepsilon_n - H_0) | \alpha \rangle \langle \alpha | n \rangle = \sum_{\delta} \langle \beta | V | \delta \rangle \langle \delta | n \rangle. \tag{6.33}$$

The matrix form of the Schrödinger eigenvalue equation in the basis $\{|\alpha\rangle\}$ can now easily be recognized

$$\varepsilon_n \langle \beta | n \rangle = \sum_{\alpha} \{ \langle \beta | H_0 | \alpha \rangle + \langle \beta | V | \alpha \rangle \} \langle \alpha | n \rangle. \tag{6.34}$$

Another useful exercise in anticipation of later many-body applications, is to consider the issue of normalization. This is not a particularly illuminating problem for the usual situation, when V does not depend on the energy. In the many-body problem we will, however, encounter similar equations in which the quantity corresponding to V is energy dependent. It is then still possible to obtain the appropriate eigenvalue equation, provided one assumes that V is well-behaved near ε_n. In the resulting eigenvalue equation, V appears at the eigenvalue ε_n as follows

$$\langle \alpha | n \rangle = \sum_{\gamma\delta} \langle \alpha | \frac{1}{\varepsilon_n - H_0} | \gamma \rangle \langle \gamma | V(\varepsilon_n) | \delta \rangle \langle \delta | n \rangle. \tag{6.35}$$

Near the discrete eigenvalue ε_n the propagator can be written as

$$G(\alpha, \beta; E \to \varepsilon_n) \Rightarrow \frac{\langle \alpha | n \rangle \langle n | \beta \rangle}{E - \varepsilon_n} + f_{\alpha\beta}(E), \tag{6.36}$$

where f is well-behaved near ε_n. The smooth behavior of $G^{(0)}$ and V near ε_n implies

$$\frac{G^{(0)}(E)V(E)}{E - \varepsilon_n} = \frac{G^{(0)}(\varepsilon_n)V(\varepsilon_n)}{E - \varepsilon_n} + \left. \frac{\partial G^{(0)}V}{\partial E} \right|_{\varepsilon_n}. \tag{6.37}$$

Inserting Eq. (6.36) in the propagator equation (6.17), using (6.37) and the smoothness of f, yields

$$\frac{\langle \alpha | n \rangle \langle n | \beta \rangle}{E - \varepsilon_n} + f_{\alpha\beta}(\varepsilon_n) = G^{(0)}(\alpha, \beta; \varepsilon_n)$$

$$+ \sum_{\gamma\delta} G^{(0)}(\alpha, \gamma; E) \langle \gamma | V(E) | \delta \rangle \left\{ \frac{\langle \delta | n \rangle \langle n | \beta \rangle}{E - \varepsilon_n} + f_{\delta\beta}(\varepsilon_n) \right\}$$

$$= G^{(0)}(\alpha, \beta; \varepsilon_n) + \sum_{\gamma\delta} G^{(0)}(\alpha, \gamma; \varepsilon_n) \langle \gamma | V(\varepsilon_n) | \delta \rangle \frac{\langle \delta | n \rangle \langle n | \beta \rangle}{E - \varepsilon_n}$$

$$+ \sum_{\gamma\delta} G^{(0)}(\alpha, \gamma; \varepsilon_n) \langle \gamma | V(\varepsilon_n) | \delta \rangle f_{\delta\beta}(\varepsilon_n)$$

$$+ \sum_{\gamma\delta} \left. \frac{\partial G^{(0)}(\alpha, \gamma; E) \langle \gamma | V(E) | \delta \rangle}{\partial E} \right|_{\varepsilon_n} \langle \delta | n \rangle \langle n | \beta \rangle. \tag{6.38}$$

The first term on the left and the second on the right just represent the original eigenvalue equation. Together these terms represent the singular contributions in the propagator equation and therefore cancel each other here. It is useful to multiply the remaining terms by $\langle n | (\varepsilon_n - H_0) | \alpha \rangle$

while summing over α. The terms containing f are then equal, provided the alternative form of the eigenvalue equation $G = G^{(0)} + G\,V\,G^{(0)}$ [see Eq. (6.19)] is used. Eliminating these contributions, the remainder can be further manipulated to generate the normalization condition in the form

$$\sum_{\alpha}\langle n|\alpha\rangle\langle\alpha|n\rangle - \sum_{\alpha\beta}\langle n|\alpha\rangle\langle\alpha|\left.\frac{\partial V(E)}{\partial E}\right|_{\varepsilon_n}|\beta\rangle\langle\beta|n\rangle = 1. \qquad (6.39)$$

For an energy-independent potential one finds

$$1 = \sum_{\alpha}|\langle\alpha|n\rangle|^2, \qquad (6.40)$$

which can of course also be directly stated by invoking the normalization of $|n\rangle$ and the completeness of the states $|\alpha\rangle$. Filling in the steps outlined here for the more general energy-dependent case, is recommended as an exercise. The normalization to 1 occurs due the presence of $G^{(0)}$ in the propagator equation when V does not depend on the energy. If not, additional terms contribute to the normalization as in Eq. (6.39).

6.4 Scattering theory using propagators

The elastic scattering process in free space is completely determined by one particular matrix element of the $\mathcal{T}$-matrix. It is instructive to work out this result explicitly using the propagator method. To this end one starts from Eq. (6.19) and uses its various incarnations to perform the required analysis. Employing the δ-function normalization (see Sec. 4.1.1), we choose the wave vector basis associated with the unperturbed Hamiltonian, $H_0 = T$. Assuming a spinless particle, the noninteracting propagator reads

$$G^{(0)}(\boldsymbol{k},\boldsymbol{k}';E) = \delta(\boldsymbol{k}-\boldsymbol{k}')\frac{1}{E-\hbar^2k^2/2m+i\eta}. \qquad (6.41)$$

Inserting this result in Eq. (6.19), yields

$$\begin{aligned}
G(\boldsymbol{k},\boldsymbol{k}';E) &= G^{(0)}(\boldsymbol{k},\boldsymbol{k}';E) + G^{(0)}(\boldsymbol{k};E)\int d\boldsymbol{q}\,\langle\boldsymbol{k}|V|\boldsymbol{q}\rangle G(\boldsymbol{q},\boldsymbol{k}';E)\\
&= G^{(0)}(\boldsymbol{k},\boldsymbol{k}';E) + G^{(0)}(\boldsymbol{k};E)\langle\boldsymbol{k}|\mathcal{T}(E)|\boldsymbol{k}'\rangle G^{(0)}(\boldsymbol{k}';E), \quad (6.42)
\end{aligned}$$

where

$$G^{(0)}(\boldsymbol{k},\boldsymbol{k}';E) = \delta(\boldsymbol{k}-\boldsymbol{k}')G^{(0)}(\boldsymbol{k};E) \qquad (6.43)$$

is the noninteracting propagator. The second equality in Eq. (6.42) is particularly useful for the asymptotic analysis, to be explored below. The usual results from scattering theory are derived in the coordinate representation. The required double FT of the propagator is given by

$$G(\boldsymbol{r}, \boldsymbol{r}'; E) = \int \frac{d\boldsymbol{k}}{(2\pi)^{3/2}} \int \frac{d\boldsymbol{k}'}{(2\pi)^{3/2}} e^{i\boldsymbol{k}\cdot\boldsymbol{r}} G(\boldsymbol{k}, \boldsymbol{k}'; E) e^{-i\boldsymbol{k}'\cdot\boldsymbol{r}'}. \qquad (6.44)$$

The transform of the noninteracting propagator only involves one integration due to the presence of the δ-function in Eq. (6.43)

$$\begin{aligned}
G^{(0)}(\boldsymbol{r}, \boldsymbol{r}'; E) &= \int \frac{d\boldsymbol{k}}{(2\pi)^{3/2}} \int \frac{d\boldsymbol{k}'}{(2\pi)^{3/2}} \, e^{i\boldsymbol{k}\cdot\boldsymbol{r}} G^{(0)}(\boldsymbol{k}, \boldsymbol{k}'; E) e^{-i\boldsymbol{k}'\cdot\boldsymbol{r}'} \\
&= \int \frac{d\boldsymbol{k}}{(2\pi)^3} \, e^{i\boldsymbol{k}\cdot(\boldsymbol{r}-\boldsymbol{r}')} G^{(0)}(\boldsymbol{k}; E). \qquad (6.45)
\end{aligned}$$

Equation (6.42) can now be transformed to yield

$$\begin{aligned}
G(\boldsymbol{r}, \boldsymbol{r}'; E) &= G^{(0)}(\boldsymbol{r}, \boldsymbol{r}'; E) \qquad\qquad\qquad\qquad\qquad\qquad (6.46)\\
&\quad + \int d\boldsymbol{r}_1 \int d\boldsymbol{r}_2 \, G^{(0)}(\boldsymbol{r}, \boldsymbol{r}_1; E)\langle \boldsymbol{r}_1|V|\boldsymbol{r}_2\rangle G(\boldsymbol{r}_2, \boldsymbol{r}'; E) \\
&= G^{(0)}(\boldsymbol{r}, \boldsymbol{r}'; E) + \int d\boldsymbol{r}_1 \int d\boldsymbol{r}_2 \, G^{(0)}(\boldsymbol{r}, \boldsymbol{r}_1; E)\langle \boldsymbol{r}_1|T(E)|\boldsymbol{r}_2\rangle G^{(0)}(\boldsymbol{r}_2, \boldsymbol{r}'; E).
\end{aligned}$$

With this result, an asymptotic analysis can be developed, leading to the cross section.

We first need to calculate $G^{(0)}(\boldsymbol{r}, \boldsymbol{r}'; E)$ by means of the FT of Eq. (6.41), as given in Eq. (6.45). The on-shell wave vector k_0 is defined by

$$E \equiv \frac{\hbar^2 k_0^2}{2m}. \qquad (6.47)$$

Performing the angular integrals and extending the integration limit in k to $-\infty$, leads to (replacing E by k_0)

$$\begin{aligned}
G^{(0)}(\boldsymbol{r}, \boldsymbol{r}'; E) &= \frac{2m}{\hbar^2} \frac{1}{i|\boldsymbol{r}-\boldsymbol{r}'|} \frac{1}{8\pi^2} \int_{-\infty}^{\infty} dk \, k \frac{e^{ik|\boldsymbol{r}-\boldsymbol{r}'|} - e^{-ik|\boldsymbol{r}-\boldsymbol{r}'|}}{(k_0 - k + i\eta)(k_0 + k + i\eta)} \\
&= \frac{2m}{\hbar^2} \frac{-1}{4\pi|\boldsymbol{r}-\boldsymbol{r}'|} e^{ik_0|\boldsymbol{r}-\boldsymbol{r}'|}. \qquad (6.48)
\end{aligned}$$

Contour integration in the complex wave-vector plane has been used for the last equality. When $r' \gg r$, we find

$$k_0|\boldsymbol{r} - \boldsymbol{r}'| = k_0 r' \sqrt{1 + \left(\frac{r}{r'}\right)^2 - \frac{2\boldsymbol{r}\cdot\boldsymbol{r}'}{r'^2}} \approx k_0 r' - k_0 \hat{\boldsymbol{r}}' \cdot \boldsymbol{r}. \qquad (6.49)$$

Equation (6.48) then becomes

$$G^{(0)}(\mathbf{r}, \mathbf{r}'; E) \rightarrow -\frac{m}{2\pi\hbar^2} \frac{e^{ik_0 r'}}{r'} e^{-ik_0 \hat{\mathbf{r}}' \cdot \mathbf{r}}. \qquad (6.50)$$

Substituting the outcome in the second part of Eq. (6.46) for both $r' \gg r$ and $r' \gg r_2$, and using the finite range of V, demonstrates that G is separable and can be written as

$$G(\mathbf{r}, \mathbf{r}'; E) = -\frac{m}{2\pi\hbar^2} \frac{e^{ik_0 r'}}{r'} \psi_{k_0}(\mathbf{r}) \qquad (6.51)$$

in the asymptotic domain. In turn, Eq. (6.51) can be inserted in Eq. (6.46), to arrive at the standard integral equation for the wave function. In addition, the appropriate formulation for the asymptotic wave function is obtained, providing the link with the scattering amplitude

$$\psi_{k_0}(\mathbf{r}) = e^{-ik_0 \hat{\mathbf{r}}' \cdot \mathbf{r}} + \int d\mathbf{r}_1 \int d\mathbf{r}_2 \, G^{(0)}(\mathbf{r}, \mathbf{r}_1; E) \langle \mathbf{r}_1 | V | \mathbf{r}_2 \rangle \psi_{k_0}(\mathbf{r}_2)$$

$$= e^{-ik_0 \hat{\mathbf{r}}' \cdot \mathbf{r}} + \int d\mathbf{r}_1 \int d\mathbf{r}_2 \, G^{(0)}(\mathbf{r}, \mathbf{r}_1; E) \langle \mathbf{r}_1 | T(E) | \mathbf{r}_2 \rangle e^{-ik_0 \hat{\mathbf{r}}' \cdot \mathbf{r}_2}. \qquad (6.52)$$

We identify the origin of the motion $\hat{\mathbf{r}}'$, in the direction of the negative z-axis, so that $\mathbf{k} \equiv -k_0 \hat{\mathbf{r}}'$ points into the positive z-direction. Consider r values, much larger than the range of the potential, and therefore much larger than any contributing value of r_1. Equation (6.50) can now be used again in the second part of Eq. (6.52), to identify the coefficient multiplying the outgoing spherical wave $e^{ik_0 r}/r$ as the scattering amplitude. A double FT of the T-matrix element back to wave vector space, finally yields the scattering amplitude

$$f_{k_0}(\theta, \phi) = -\frac{4m\pi^2}{\hbar^2} \langle \mathbf{k}' | T(E) | \mathbf{k} \rangle. \qquad (6.53)$$

The angles θ and ϕ are associated with the direction of $\hat{\mathbf{r}}$ and $\mathbf{k}' \equiv k_0 \hat{\mathbf{r}}$, the latter corresponding to the wave vector of the detected particle. It has the same magnitude k_0 as the initial state. The differential cross section for the direction (θ, ϕ) is then simply the absolute square of Eq. (6.53)

$$\frac{d\sigma}{d\Omega} = |f_{k_0}(\theta, \phi)|^2. \qquad (6.54)$$

The present formulation is closely tailored to the conventional experimental situation, where a collimated beam, characterized by a given energy or momentum, propagates along the z-axis toward a target situated at the origin. Detection takes place in a direction pointing away from the origin specified by the angles θ and ϕ. A similar derivation can be presented for a particle with spin, to obtain the elastic scattering amplitude ($|\boldsymbol{p}_f| = |\boldsymbol{p}_i|$) in momentum space

$$f_{m_f,m_i}(\theta,\phi) = -\frac{m\hbar}{2\pi}(2\pi)^3 \langle \boldsymbol{p}'m_f| T(E = \frac{\boldsymbol{p}^2}{2m}) |\boldsymbol{p}m_i\rangle. \qquad (6.55)$$

The direction of $\boldsymbol{p}'$ coincides with the wave vector $\boldsymbol{k}'$ above and $\boldsymbol{p}$ points into the z-direction. The polarized cross section then reads

$$\frac{d\sigma^{m_f,m_i}}{d\Omega} = |f_{m_f,m_i}(\theta,\phi)|^2. \qquad (6.56)$$

6.4.1   *Partial waves and phase shifts*

Often the interaction is of short range and spherically symmetric. When this is the case, it is invariably useful to analyze the scattering process in an angular momentum basis since only a limited set of ℓ-values will contribute. The transformation from states with wave vectors to those with orbital angular momentum must be employed

$$|\boldsymbol{k}\rangle = \sum_{\ell m_\ell} |k\ell m_\ell\rangle \langle \ell m_\ell|\hat{\boldsymbol{k}}\rangle = \sum_{\ell m_\ell} |k\ell m_\ell\rangle Y_{\ell m_\ell}^*(\hat{\boldsymbol{k}}). \qquad (6.57)$$

In the latter basis, the noninteracting propagator becomes

$$
\begin{aligned}
G^{(0)}(k\ell m_\ell, k'\ell' m_{\ell'}; E) &= \frac{\delta(k-k')}{k^2}\delta_{\ell,\ell'}\delta_{m_\ell,m_{\ell'}} \frac{1}{E - \hbar^2 k^2/2m + i\eta} \\
&= \frac{\delta(k-k')}{k^2}\delta_{\ell,\ell'}\delta_{m_\ell,m_{\ell'}} G^{(0)}(k; E).
\end{aligned} \qquad (6.58)
$$

Since the energy has no angular dependence, the energy denominator is the same as in the wave-vector basis. Expressing Eq. (6.19) in the angular momentum basis and assuming that the interaction is rotationally invariant, we find

$$
\begin{aligned}
G_\ell(k,k';E) &= \frac{\delta(k-k')}{k^2}G^{(0)}(k;E) + G^{(0)}(k;E)\int_0^\infty dq q^2 \langle k|V^\ell|q\rangle G_\ell(q,k';E) \\
&= \frac{\delta(k-k')}{k^2}G^{(0)}(k;E) + G^{(0)}(k;E)\langle k|T^\ell(E)|k'\rangle G^{(0)}(k';E).
\end{aligned} \qquad (6.59)
$$

The equation for the iterated interaction, or $\mathcal{T}$-matrix, can then be written as

$$\langle k|T^\ell(E)|k'\rangle = \langle k|V^\ell|k'\rangle + \int_0^\infty dq\ q^2\langle k|V^\ell|q\rangle G^{(0)}(q;E)\langle q|T^\ell(E)|k'\rangle. \quad (6.60)$$

The coordinate space version of Eq. (6.59) is obtained by a double Fourier–Bessel transform

$$G_\ell(r,r';E) = \frac{2}{\pi}\int_0^\infty dk\ k^2\int_0^\infty dk'\ k'^2\ \mathrm{j}_\ell(kr)\mathrm{j}_\ell(k'r')G_\ell(k,k';E). \quad (6.61)$$

The transformation proceeds from angular momentum states with wave vector to those with position, employing the spherical Bessel function

$$\langle k\ell m_\ell|r\ell'm_{\ell'}\rangle = \delta_{\ell,\ell'}\delta_{m_\ell,m_{\ell'}}\sqrt{\frac{2}{\pi}}\mathrm{j}_\ell(kr). \quad (6.62)$$

The result for the noninteracting part of the propagator, represented by the first term in Eq. (6.59), reduces to

$$G_\ell^{(0)}(r,r';E) = \frac{2}{\pi}\int_0^\infty dk\ k^2\ \mathrm{j}_\ell(kr)\mathrm{j}_\ell(kr')G^{(0)}(k;E). \quad (6.63)$$

The Fourier–Bessel transform of Eq. (6.59) has the following form

$$\begin{aligned}
G_\ell(r,r';E) &= G_\ell^{(0)}(r,r';E) + \int_0^\infty dr_1 r_1^2\int_0^\infty dr_2 r_2^2 \\
&\quad \times G_\ell^{(0)}(r,r_1;E)\langle r_1|V^\ell|r_2\rangle G_\ell(r_2,r';E) \\
&= G_\ell^{(0)}(r,r';E) + \int_0^\infty dr_1 r_1^2\int_0^\infty dr_2 r_2^2 \\
&\quad \times G_\ell^{(0)}(r,r_1;E)\langle r_1|T^\ell(E)|r_2\rangle G_\ell^{(0)}(r_2,r';E).
\end{aligned} \quad (6.64)$$

If the interaction V is local in coordinate space, only one integral in the first equality remains. The second equality can be used to study the asymptotic behavior of the propagator outside the range of the interaction. The integral in Eq. (6.63) can be performed analytically by employing contour integration in the complex wave-vector plane, as discussed in [Gottfried and Yan (2004)]. The spherical Bessel functions are well-behaved so all the singularities are contained in the denominator of Eq. (6.63). For $r' > r$ the following equality can be used

$$\mathrm{j}_\ell(kr') = \frac{1}{2}\left[\mathrm{h}_\ell(kr') + \mathrm{h}_\ell^*(kr')\right], \quad (6.65)$$

involving the spherical Hankel functions. The term with $j_\ell h_\ell$ decreases exponentially in the upper half k-plane, allowing the contour to be closed in the upper half plane. The other term requires a contour in the lower half plane. The final result combines to

$$G_\ell^{(0)}(r, r'; E) = -ik_0 \frac{2m}{\hbar^2} j_\ell(k_0 r_<) h_\ell(k_0 r_>). \qquad (6.66)$$

The coordinate argument in the spherical Hankel function must be the larger of r and r' and is denoted by $r_>$, while the argument of the spherical Bessel function is the smaller, and denoted by $r_<$. For simplicity, it will be assumed in the current analysis that the interaction has a finite range, $\langle r|V^\ell|r'\rangle = 0$ for r, r' larger than some r_0. Substituting Eq. (6.66) in the second part of Eq. (6.64) for $r' > r$ and $r' > r_0$ yields

$$G_\ell(r, r'; E) = -ik_0 \frac{2m}{\hbar^2} \Big\{ j_\ell(k_0 r) h_\ell(k_0 r')$$

$$+ \int_0^\infty dr_1 r_1^2 \int_0^\infty dr_2 r_2^2 \, G_\ell^{(0)}(r, r_1; E)\langle r_1|T^\ell(E)|r_2\rangle j_\ell(k_0 r_2) h_\ell(k_0 r') \Big\}$$

$$= -ik_0 \frac{2m}{\hbar^2} \psi_{\ell k_0}(r) h_\ell(k_0 r'), \qquad (6.67)$$

where

$$\psi_{\ell k_0}(r) = j_\ell(k_0 r) \qquad (6.68)$$

$$+ \int_0^\infty dr_1 r_1^2 \int_0^\infty dr_2 r_2^2 \, G_\ell^{(0)}(r, r_1; E)\langle r_1|T^\ell(E)|r_2\rangle j_\ell(k_0 r_2).$$

The propagator therefore separates as a product of a function of r and a different one of r'. Substituting the last line of Eq. (6.67) into the first part of Eq. (6.64) generates the integral equation for the wave function ψ (under the condition that $r' > r_0$)

$$\psi_{\ell k_0}(r) = j_\ell(k_0 r) \qquad (6.69)$$

$$+ \int_0^\infty dr_1 r_1^2 \int_0^\infty dr_2 r_2^2 \, G_\ell^{(0)}(r, r_1; E)\langle r_1|V^\ell|r_2\rangle \psi_{\ell k_0}(r_2).$$

The version of this integral equation with a local interaction V can be found in standard textbooks (see *e.g.* [Gottfried and Yan (2004)]). Equation (6.69) is derived here, to demonstrate the relation between the propagator and the wave function for a scattering problem.

The asymptotic analysis of the propagator can be performed by using Eq. (6.66) in Eq. (6.64), under the assumption that the propagator is considered for $r < r'$, with both larger than the range of the interaction, r_0. Values of r_1 and r_2 in Eq. (6.64) larger than r_0, yield no contributions to the integral. As a result, the effective interaction, $\mathcal{T}$, has a range similar to V. Using the relation between spherical Bessel and Hankel functions (6.65), the asymptotic behavior of the propagator is obtained from the second part of Eq. (6.64)

$$
G_\ell(r, r'; E) \to -i \left(\frac{m}{\hbar^2}\right) k_0 \mathrm{h}_\ell(k_0 r') \left\{ \mathrm{h}_\ell^*(k_0 r) + \mathrm{h}_\ell(k_0 r) \left[1 - 4i\frac{m}{\hbar^2} k_0 \right.\right.
$$
$$
\left.\left. \times \int_0^\infty dr_1 r_1^2 \int_0^\infty dr_2 r_2^2 \, \langle r_1 | \mathcal{T}^\ell(E) | r_2 \rangle \mathrm{j}_\ell(k_0 r_1) \mathrm{j}_\ell(k_0 r_2) \right] \right\} \qquad (6.70)
$$
$$
= -i\frac{m}{\hbar^2} k_0 \mathrm{h}_\ell(k_0 r') \left\{ \mathrm{h}_\ell^*(k_0 r) + \mathrm{h}_\ell(k_0 r) \left[1 - 2\pi i \left(\frac{m k_0}{\hbar^2}\right) \langle k_0 | \mathcal{T}^\ell(E) | k_0 \rangle \right] \right\}.
$$

In the last step of Eq. (6.70), one can return to the on-shell matrix element of the $\mathcal{T}$-matrix in wave-vector space, which completely determines the outcome of the scattering process. The term in square brackets corresponds to the $\mathcal{S}$-matrix element that defines the phase shift

$$
\langle k_0 | \mathcal{S}^\ell(E) | k_0 \rangle = \left[1 - 2\pi i \left(\frac{m k_0}{\hbar^2}\right) \langle k_0 | \mathcal{T}^\ell(E) | k_0 \rangle \right] \equiv e^{2i\delta_\ell}. \qquad (6.71)
$$

This result can be represented by

$$
\tan \delta_\ell = \frac{\mathrm{Im}\, \langle k_0 | \mathcal{T}^\ell(E) | k_0 \rangle}{\mathrm{Re}\, \langle k_0 | \mathcal{T}^\ell(E) | k_0 \rangle}, \qquad (6.72)
$$

which explicitly shows that a nonzero imaginary part of the iterated interaction is required to obtain a nonvanishing phase shift. In turn, the imaginary part of the interaction only appears for energies where the noninteracting propagator has a nonvanishing imaginary part. For scattering, this pertains to all positive energies. By substituting the explicit form of the spherical Hankel functions for $\ell = 0$ in Eq. (6.70), the asymptotic propagator for the s-wave channel reads

$$
G_{\ell=0}(r, r'; E) \to -\frac{2m}{k_0 \hbar^2} \frac{1}{rr'} e^{i(k_0 r' + \delta_0)} \sin(k_0 r + \delta_0). \qquad (6.73)
$$

A standard result is contained in this equation: the imaginary part of Eq. (6.73) is simply the product of the asymptotic wave functions, as a function of r and r', respectively. Finally, the scattering amplitude can be written in terms of the on-shell T-matrix elements or phase shifts

$$f(\theta, \phi) = \sum_l \frac{2l+1}{k_0} \left\{ \frac{-mk_0\pi}{\hbar^2} \right\} \langle k_0|T^\ell(E)|k_0\rangle P_\ell(\cos\theta)$$

$$= \sum_\ell \frac{2\ell+1}{k_0} e^{i\delta_\ell} \sin\delta_\ell P_\ell(\cos\theta), \qquad (6.74)$$

leading to the differential cross section in Eq. (6.54). For the total cross section we find the usual result

$$\sigma_{tot} = \frac{4\pi}{k_0^2} \sum_\ell (2\ell+1) \sin^2\delta_\ell. \qquad (6.75)$$

6.5   Exercises

(1) Consider the second-order diagram contributing to G. Perform the inverse FT to obtain this contribution as a function of the time difference $t - t'$.
(2) Construct all diagrammatic representations of Eqs. (6.19) and (6.20).
(3) Work out the details of the normalization of the state $|n\rangle$ in the case of an energy-dependent potential $V(E)$. Follow the outline given below Eq. (6.38) to obtain Eq. (6.39).
(4) Perform all the operations that lead to Eq. (6.66).
(5) Extend the formalism discussed in Sec. 6.4.1 to the case of a spin-$\frac{1}{2}$ fermion, which also experiences a spin-orbit potential like the one introduced in Sec. 3.3.

Chapter 7

Single-particle propagator in the many-body system

In analogy to the case of the sp problem, it is possible to consider a propagator in the many-body system. Its definition for fermions is presented in Sec. 7.1 employing the Heisenberg picture [Sakurai (1994); Messiah (1999)]. For completeness, some results related to various pictures in quantum mechanics are collected in App. A. The propagator is defined in terms of either adding, or removing, a particle from the correlated N-particle ground state. The latter (removal) process is a new feature, not present in the sp problem of Ch. 6, which corresponds to $N = 0$. Complementary information is contained in the addition and removal amplitudes. Both are accessible experimentally: the addition process in terms of elastic scattering from the correlated ground state, and the removal process in coincidence experiments of the form $(e, 2e)$ for atoms and $(e, e'p)$ for nuclei. Various incarnations of the sp propagator in the many-body system are discussed. A particularly important one, the Lehmann representation is studied in Sec. 7.2 and requires a FT to the energy formulation.

An important connection with experimental data can be made for the imaginary part of the sp propagator. The relevant quantities are the spectral functions describing the removal and addition probability density of particles, with specified quantum numbers, to and from the correlated ground state. They are introduced in Sec. 7.3. Section 7.4 demonstrates that expectation values of one-body operators in the ground state can be calculated from the one-body density matrix, which is related to the "hole" part of the sp propagator. In addition to all one-body expectation values, it is possible to determine the energy of the ground state from the hole part of the sp propagator, when the Hamiltonian contains at most two-body operators. Basic examples for the sp propagator in a noninteracting system are discussed in Sec. 7.5. A simple presentation of knockout reactions and their relation to spectral functions is given in Sec. 7.6. Experimental data from the $(e, 2e)$ reaction on atoms are discussed in Sec. 7.7. Corresponding results for the $(e, e'p)$ reaction on nuclei are reviewed in Sec. 7.8. From comparison with experiment, a clear "single-particle" picture of the atom and the nucleus arises, that further motivates the development of the propagator description of the many-fermion system.

7.1 Fermion single-particle propagator

The sp propagator in a many-particle system is defined as

$$G(\alpha, \beta; t, t') = -\frac{i}{\hbar} \left\langle \Psi_0^N \right| \mathcal{T}[a_{\alpha_H}(t) a_{\beta_H}^\dagger(t')] \left| \Psi_0^N \right\rangle. \qquad (7.1)$$

The expectation value, with respect to the exact ground state of the system of N particles, samples an operator that represents both particle as well as hole propagation. The latter term is naturally absent in the one-particle problem. The state $\left| \Psi_0^N \right\rangle$ is the normalized Heisenberg ground state for the N-particle system and E_0^N the corresponding eigenvalue

$$\hat{H} \left| \Psi_0^N \right\rangle = E_0^N \left| \Psi_0^N \right\rangle. \qquad (7.2)$$

The Heisenberg picture is briefly reviewed in App. A. The particle addition and removal operators in the definition of the sp propagator are given in the Heisenberg picture by

$$a_{\alpha_H}(t) = e^{\frac{i}{\hbar}\hat{H}t} a_\alpha e^{-\frac{i}{\hbar}\hat{H}t} \qquad (7.3)$$

and

$$a_{\alpha_H}^\dagger(t) = e^{\frac{i}{\hbar}\hat{H}t} a_\alpha^\dagger e^{-\frac{i}{\hbar}\hat{H}t}, \qquad (7.4)$$

respectively. The time-ordering operation[1] $\mathcal{T}$, appearing in Eq. (7.1), is defined here to include a sign change when two fermion operators are interchanged and can be written, using step functions, as

$$\mathcal{T}[a_{\alpha_H}(t) a_{\beta_H}^\dagger(t')] = \theta(t - t') a_{\alpha_H}(t) a_{\beta_H}^\dagger(t') - \theta(t' - t) a_{\beta_H}^\dagger(t') a_{\alpha_H}(t). \quad (7.5)$$

The procedure puts operators with the later time to the left of earlier ones and includes a sign when a change of order is required. The reason for defining the propagator as a time-ordered product is related to the availability of a perturbative expansion, as will become clear in Ch. 8.

_____

[1] No confusion should arise with the scattering quantity.

Invoking the definition of the Heisenberg picture operators and the time-ordering operation for fermion operators, one obtains

$$
\begin{aligned}
G(\alpha, \beta; t - t') = -\frac{i}{\hbar} \Big\{ & \theta(t - t') e^{\frac{i}{\hbar} E_0^N (t-t')} \left\langle \Psi_0^N \middle| a_\alpha e^{-\frac{i}{\hbar} \hat{H}(t-t')} a_\beta^\dagger \middle| \Psi_0^N \right\rangle \\
& - \theta(t' - t) e^{\frac{i}{\hbar} E_0^N (t'-t)} \left\langle \Psi_0^N \middle| a_\beta^\dagger e^{-\frac{i}{\hbar} \hat{H}(t'-t)} a_\alpha \middle| \Psi_0^N \right\rangle \Big\} \quad (7.6)
\end{aligned}
$$

$$
\begin{aligned}
= -\frac{i}{\hbar} \Big\{ & \theta(t - t') \sum_m e^{\frac{i}{\hbar}(E_0^N - E_m^{N+1})(t-t')} \left\langle \Psi_0^N \middle| a_\alpha \middle| \Psi_m^{N+1} \right\rangle \left\langle \Psi_m^{N+1} \middle| a_\beta^\dagger \middle| \Psi_0^N \right\rangle \\
& - \theta(t' - t) \sum_n e^{\frac{i}{\hbar}(E_0^N - E_n^{N-1})(t'-t)} \left\langle \Psi_0^N \middle| a_\beta^\dagger \middle| \Psi_n^{N-1} \right\rangle \left\langle \Psi_n^{N-1} \middle| a_\alpha \middle| \Psi_0^N \right\rangle \Big\}.
\end{aligned}
$$

As expected, the propagator depends only on the time difference $t-t'$. Note that the completeness of the exact eigenstates of $\hat{H}$ for both the $N+1$ as well as the $N-1$ system, has been used together with

$$
\hat{H} \middle| \Psi_m^{N+1} \rangle = E_m^{N+1} \middle| \Psi_m^{N+1} \rangle \qquad (7.7)
$$

and

$$
\hat{H} \middle| \Psi_n^{N-1} \rangle = E_n^{N-1} \middle| \Psi_n^{N-1} \rangle. \qquad (7.8)
$$

The first term in Eq. (7.6) is called the addition part of the propagator, or alternatively, the "particle" or forward propagating part. The second term is likewise referred to as the removal, the "hole", or the backward propagating part. While the similarity with Eq. (6.6) is evident (apart from the presence of the hole part), it is also clear that Eq. (7.6) contains relevant information about the many-body system.

7.2 Lehmann representation

As in the sp problem, one can introduce the FT of the sp propagator which is more convenient for practical calculations, but also brings out the information that is contained in the sp propagator more clearly

$$
G(\alpha, \beta; E) = \int_{-\infty}^{\infty} d(t - t') \, e^{\frac{i}{\hbar} E(t-t')} \, G(\alpha, \beta; t - t'). \qquad (7.9)
$$

As before, it is recommended to use the integral representation of the step function as given in Eq. (6.7). The result of this FT can be expressed in various equivalent ways. The FT of the last version of $G(\alpha, \beta; t - t')$ in Eq. (7.6) yields

$$
\begin{aligned}
G(\alpha, \beta; E) =& \sum_m \frac{\langle \Psi_0^N | a_\alpha | \Psi_m^{N+1} \rangle \langle \Psi_m^{N+1} | a_\beta^\dagger | \Psi_0^N \rangle}{E - (E_m^{N+1} - E_0^N) + i\eta} \\
&+ \sum_n \frac{\langle \Psi_0^N | a_\beta^\dagger | \Psi_n^{N-1} \rangle \langle \Psi_n^{N-1} | a_\alpha | \Psi_0^N \rangle}{E - (E_0^N - E_n^{N-1}) - i\eta} \\
=& \langle \Psi_0^N | a_\alpha \frac{1}{E - (\hat{H} - E_0^N) + i\eta} a_\beta^\dagger | \Psi_0^N \rangle \\
&+ \langle \Psi_0^N | a_\beta^\dagger \frac{1}{E - (E_0^N - \hat{H}) - i\eta} a_\alpha | \Psi_0^N \rangle. \quad (7.10)
\end{aligned}
$$

The first equality is known as the Lehmann representation [Lehmann (1954)] of the sp propagator. The last line is obtained by removing the complete set of exact $N + 1$-eigenstates in the first term and the complete set of $N - 1$-eigenstates in the second one, after replacing the eigenvalues E_m^{N+1} and E_n^{N-1} by $\hat{H}$. Note that **any** sp basis can be used in this formulation of the propagator. Many texts choose to specialize either to coordinate space, or momentum representation. However, this choice is not always appropriate, especially when dealing with finite systems where comparisons with experimental data are possible. It is also instructive to compare this form of the sp propagator with the corresponding one for the sp problem [see Eq. (6.9)]. Apart from the hole term, which is naturally absent in the sp case, there is a clear similarity between the two results. Indeed, the matrix elements involving the addition and removal operators, obey Schrödinger-like equations, as will be discussed in Sec. 9.5.

7.3   Spectral functions

For finite systems one can relate essentially all the information contained in the sp propagator to experimental data. Consider first the information in the denominator of the first equality in Eq. (7.10). The positions of the poles signal the location of the excited states in the $N + 1$ or $N - 1$-particle systems with respect to the ground state of the N-particle system. Note that it should be possible to reach those states by the addition (or removal) of a particle with sp quantum numbers α to (or from) the ground state of the system. It is useful to visualize this as a physical process that can be realized experimentally. Second, the numerator determines the distribution of the corresponding transition strength from the ground state of the N-particle system to these states in the $N \pm 1$ systems. This information is a crucial measure of the strength of the correlations in the system, as they induce behavior which deviates from the independent-particle model. A good tool to develop intuition for the effect of correlations on sp properties is provided by the spectral functions. At energy E, its hole part is the combined probability density for removing a particle with quantum numbers α from the ground state, while leaving the remaining $N - 1$-system at an energy $E_n^{N-1} = E_0^N - E$. The relation to the imaginary part of the diagonal element of the sp propagator reads

$$S_h(\alpha; E) = \frac{1}{\pi} \operatorname{Im} G(\alpha, \alpha; E) \qquad\qquad E \le \varepsilon_F^-$$

$$= \sum_n \left| \langle \Psi_n^{N-1} | a_\alpha | \Psi_0^N \rangle \right|^2 \delta(E - (E_0^N - E_n^{N-1})). \qquad (7.11)$$

A detailed presentation of available experimental information about this quantity in atoms and nuclei will be discussed in Secs. 7.7 and 7.8, respectively. The probability density for the addition of a particle with quantum numbers α, leaving the $N + 1$-system at energy $E_m^{N+1} = E_0^N + E$ similarly reads

$$S_p(\alpha; E) = -\frac{1}{\pi} \operatorname{Im} G(\alpha, \alpha; E) \qquad\qquad E \ge \varepsilon_F^+$$

$$= \sum_m \left| \langle \Psi_m^{N+1} | a_\alpha^\dagger | \Psi_0^N \rangle \right|^2 \delta(E - (E_m^{N+1} - E_0^N)). \qquad (7.12)$$

Equation (7.12) defines the particle spectral function. The Fermi energies introduced in Eqs. (7.11) and (7.12) are given by

$$\varepsilon_F^- = E_0^N - E_0^{N-1} \qquad\qquad (7.13)$$

and

$$\varepsilon_F^+ = E_0^{N+1} - E_0^N. \qquad\qquad (7.14)$$

In obtaining the imaginary part of the propagator, the very practical identity

$$\frac{1}{E \pm i\eta} = \mathcal{P} \frac{1}{E} \mp i\pi\delta(E) \qquad\qquad (7.15)$$

has been employed, where the symbol $\mathcal{P}$ denotes the principal value. The above expressions for the spectral functions are particularly useful for

analyzing finite systems where discrete bound states exist and for certain
problems involving band structure, localization or external magnetic fields
in condensed matter systems. In finite systems, like nuclei, there can be a
considerable difference between ε_F^- and ε_F^+. In infinite systems which are
not superfluids or superconductors this difference vanishes in the thermo-
dynamic limit.

The occupation number of a sp state α can be generated from the hole
part of the spectral function by evaluating

$$n(\alpha) = \left\langle \Psi_0^N \middle| a_\alpha^\dagger a_\alpha \middle| \Psi_0^N \right\rangle = \sum_n \left| \left\langle \Psi_n^{N-1} \middle| a_\alpha \middle| \Psi_0^N \right\rangle \right|^2$$

$$= \int_{-\infty}^{\varepsilon_F^-} dE \sum_n \left| \left\langle \Psi_n^{N-1} \middle| a_\alpha \middle| \Psi_0^N \right\rangle \right|^2 \delta(E - (E_0^N - E_n^{N-1}))$$

$$= \int_{-\infty}^{\varepsilon_F^-} dE \, S_h(\alpha; E). \tag{7.16}$$

The depletion number is determined by the particle part of the spectral
function

$$d(\alpha) = \left\langle \Psi_0^N \middle| a_\alpha a_\alpha^\dagger \middle| \Psi_0^N \right\rangle = \sum_m \left| \left\langle \Psi_m^{N+1} \middle| a_\alpha^\dagger \middle| \Psi_0^N \right\rangle \right|^2$$

$$= \int_{\varepsilon_F^+}^{\infty} dE \sum_m \left| \left\langle \Psi_m^{N+1} \middle| a_\alpha^\dagger \middle| \Psi_0^N \right\rangle \right|^2 \delta(E - (E_m^{N+1} - E_0^N))$$

$$= \int_{\varepsilon_F^+}^{\infty} dE \, S_p(\alpha; E). \tag{7.17}$$

An important sum rule exists for $n(\alpha)$ and $d(\alpha)$ which can be deduced by
employing the anticommutation relation for a_α and $a_\alpha^\dagger$

$$n(\alpha) + d(\alpha) = \left\langle \Psi_0^N \middle| a_\alpha^\dagger a_\alpha \middle| \Psi_0^N \right\rangle + \left\langle \Psi_0^N \middle| a_\alpha a_\alpha^\dagger \middle| \Psi_0^N \right\rangle = \left\langle \Psi_0^N \middle| \Psi_0^N \right\rangle = 1. \tag{7.18}$$

This distribution between occupation and emptiness of a sp orbital in the
correlated ground state is a sensitive measure of the strength of correlations,
provided a suitable sp basis is chosen, to be discussed in Secs. 7.7 and 7.8.

7.4   Expectation values of operators in the correlated ground state

The sp propagator will also provide the expectation value of any one-body operator in the ground state

$$\left\langle \Psi_0^N \middle| \hat{O} \middle| \Psi_0^N \right\rangle = \sum_{\alpha,\beta} \langle \alpha| O |\beta\rangle \left\langle \Psi_0^N \middle| a_\alpha^\dagger a_\beta \middle| \Psi_0^N \right\rangle = \sum_{\alpha,\beta} \langle \alpha| O |\beta\rangle \, n_{\alpha\beta}. \quad (7.19)$$

Here, $n_{\alpha\beta}$ is the one-body density matrix element that can be obtained from the sp propagator using the Lehmann representation

$$
\begin{aligned}
n_{\beta\alpha} &= \int \frac{dE}{2\pi i} \, e^{iE\eta} \, G(\alpha,\beta;E) \quad\quad\quad\quad\quad\quad\quad\quad (7.20)\\
&= \int \frac{dE}{2\pi i} \, e^{iE\eta} \sum_n \frac{\left\langle \Psi_0^N \middle| a_\beta^\dagger \middle| \Psi_n^{N-1} \right\rangle \left\langle \Psi_n^{N-1} \middle| a_\alpha \middle| \Psi_0^N \right\rangle}{E - (E_0^N - E_n^{N-1}) - i\eta}\\
&= \sum_n \left\langle \Psi_0^N \middle| a_\beta^\dagger \middle| \Psi_n^{N-1} \right\rangle \left\langle \Psi_n^{N-1} \middle| a_\alpha \middle| \Psi_0^N \right\rangle = \left\langle \Psi_0^N \middle| a_\beta^\dagger a_\alpha \middle| \Psi_0^N \right\rangle.
\end{aligned}
$$

Note the convergence factor in the integral with an infinitesimal (positive) η which requires closing the contour in the upper half of the complex E-plane. Consequently, only the removal or hole part of the spectral amplitude contributes. Directly using the imaginary part of the propagator yields

$$n_{\beta\alpha} = \frac{1}{\pi} \int_{-\infty}^{\varepsilon_F^-} dE \,\, \mathrm{Im}\, G(\alpha,\beta;E) = \left\langle \Psi_0^N \middle| a_\beta^\dagger a_\alpha \middle| \Psi_0^N \right\rangle. \quad (7.21)$$

Knowledge of the sp propagator G therefore allows the calculation of the expectation value of any one-body operator in the correlated ground state, according to Eq. (7.19).

Surprisingly, the energy of the ground state can also be determined from the sp propagator provided that, as has been assumed up to now, there are only two-body interactions between the particles. Two-body forces usually dominate in most systems, but for light nuclei (and thus for nuclear systems in general), the consideration of at least three-body forces is necessary to account for all experimental details. Most discussions in this book will not require the explicit inclusion of three-body forces. The energy sum rule for two-body interactions was first clarified by [Galitskii and Migdal (1958)] and later applied to finite systems by [Koltun (1972); Koltun (1974)]. It requires only the hole part of the propagator. Consider the following integral

$$
\begin{aligned}
I_\alpha &= \frac{1}{\pi} \int_{-\infty}^{\varepsilon_F^-} dE \; E \;\; \mathrm{Im}\; G(\alpha, \alpha; E) = \int_{-\infty}^{\varepsilon_F^-} dE \; E \; S_h(\alpha; E) \\
&= \sum_m (E_0^N - E_m^{N-1}) \langle \Psi_0^N | a_\alpha^\dagger | \Psi_m^{N-1} \rangle \langle \Psi_m^{N-1} | a_\alpha | \Psi_0^N \rangle \\
&= \langle \Psi_0^N | a_\alpha^\dagger a_\alpha \hat{H} | \Psi_0^N \rangle - \sum_m \langle \Psi_0^N | a_\alpha^\dagger E_m^{N-1} | \Psi_m^{N-1} \rangle \langle \Psi_m^{N-1} | a_\alpha | \Psi_0^N \rangle \\
&= \langle \Psi_0^N | a_\alpha^\dagger a_\alpha \hat{H} | \Psi_0^N \rangle - \langle \Psi_0^N | a_\alpha^\dagger \hat{H} a_\alpha | \Psi_0^N \rangle = \langle \Psi_0^N | a_\alpha^\dagger [a_\alpha, \hat{H}] | \Psi_0^N \rangle .
\end{aligned}
\tag{7.22}
$$

Using Eqs. (2.34) and (2.42), the commutator in Eq. (7.22) reads

$$
[a_\alpha, \hat{H}] = \sum_\beta \langle \alpha | T | \beta \rangle a_\beta + \sum_{\beta\gamma\delta} (\alpha\beta | V | \gamma\delta) a_\beta^\dagger a_\delta a_\gamma .
\tag{7.23}
$$

Inserting Eq. (7.23) into (7.22) finally yields

$$
I_\alpha = \sum_\beta \langle \alpha | T | \beta \rangle \langle \Psi_0^N | a_\alpha^\dagger a_\beta | \Psi_0^N \rangle + \sum_{\beta\gamma\delta} (\alpha\beta | V | \gamma\delta) \langle \Psi_0^N | a_\alpha^\dagger a_\beta^\dagger a_\delta a_\gamma | \Psi_0^N \rangle .
\tag{7.24}
$$

Summing this expression over α (see also Exercises (7) and (8) in Ch. 2), we find

$$
\sum_\alpha I_\alpha = \langle \Psi_0^N | \hat{T} | \Psi_0^N \rangle + 2 \langle \Psi_0^N | \hat{V} | \Psi_0^N \rangle .
\tag{7.25}
$$

Combining Eq. (7.25) with the expectation value for the kinetic energy and applying Eq. (7.19), the desired result can be expressed as

$$
\begin{aligned}
E_0^N &= \langle \Psi_0^N | \hat{H} | \Psi_0^N \rangle \\
&= \frac{1}{2\pi} \int_{-\infty}^{\varepsilon_F^-} dE \sum_{\alpha,\beta} \{ \langle \alpha | T | \beta \rangle + E\, \delta_{\alpha,\beta} \}\; \mathrm{Im}\; G(\beta, \alpha; E) \\
&= \frac{1}{2} \left(\sum_{\alpha,\beta} \langle \alpha | T | \beta \rangle n_{\alpha\beta} + \sum_\alpha \int_{-\infty}^{\varepsilon_F^-} dE \; E \; S_h(\alpha; E) \right) ,
\end{aligned}
\tag{7.26}
$$

where the validity of the last equality can be checked by inserting the definition of $S_h(\alpha; E)$ given in Eq. (7.11).

7.5 Propagator for noninteracting systems

When the many-particle problem, for example the one with the Hamiltonian $\hat{H}_0$, does not contain a two-body interaction, the sp propagator becomes

$$G^{(0)}(\alpha, \beta; t - t') = -\frac{i}{\hbar} \langle \Phi_0^N | T[a_{\alpha_I}(t)a_{\beta_I}^\dagger(t')] | \Phi_0^N \rangle, \qquad (7.27)$$

where $|\Phi_0^N\rangle$ is the nondegenerate ground state of $\hat{H}_0$ for N particles with eigenvalue

$$E_{\Phi_0^N} = \sum_{\alpha < F} \varepsilon_\alpha \qquad (7.28)$$

as in Eq. (3.9) and therefore

$$\hat{H}_0 |\Phi_0^N\rangle = E_{\Phi_0^N} |\Phi_0^N\rangle. \qquad (7.29)$$

For the present considerations, the state $|\Phi_0^N\rangle$ can correspond to the Slater determinant of an infinite Fermi system, a closed-shell atom, or a closed-shell nucleus. The perturbation expansion of propagators for open-shell systems (having a degenerate ground state) is nontrivial and not well-developed. It will not be further discussed here. The particle addition and removal operators in the definition (7.27) of the so-called unperturbed sp propagator, are given by the equivalent of Eqs. (7.3) and (7.4) with the replacement $\hat{H} \to \hat{H}_0$. This substitution yields these operators in the so-called interaction picture, summarized in App. A. They are given by

$$a_{\alpha_I}(t) = e^{\frac{i}{\hbar}\hat{H}_0 t} a_\alpha e^{-\frac{i}{\hbar}\hat{H}_0 t} \qquad (7.30)$$

and

$$a_{\alpha_I}^\dagger(t) = e^{\frac{i}{\hbar}\hat{H}_0 t} a_\alpha^\dagger e^{-\frac{i}{\hbar}\hat{H}_0 t}, \qquad (7.31)$$

respectively. Assuming that H_0 is diagonal in the sp basis $\{|\alpha\rangle\}$ and using the corresponding interaction picture operators given by Eqs. (A.15) and (A.16) together with the time-ordering operation for fermion operators, one obtains

$$G^{(0)}(\alpha, \beta; t - t') = G_+^{(0)}(\alpha, \beta; t - t') + G_-^{(0)}(\alpha, \beta; t - t') \qquad (7.32)$$

$$= -\frac{i}{\hbar}\delta_{\alpha\beta}\{\theta(t - t')\theta(\alpha - F)e^{-\frac{i}{\hbar}\varepsilon_\alpha(t-t')} - \theta(t' - t)\theta(F - \alpha)e^{\frac{i}{\hbar}\varepsilon_\alpha(t'-t)}\}.$$

Equation (7.32) represents the propagation of a particle or a hole on top of the noninteracting ground state. As in Ch. 3, we observe that, in the sp basis $\{|\alpha\rangle\}$ associated with H_0, the states with one particle added or removed, are eigenstates of $\hat{H}_0$ according to

$$\hat{H}_0 \, a_\alpha^\dagger \, |\Phi_0^N\rangle = (E_{\Phi_0^N} + \varepsilon_\alpha) \, a_\alpha^\dagger \, |\Phi_0^N\rangle \qquad \alpha > F \qquad (7.33)$$

and

$$\hat{H}_0 \, a_\alpha \, |\Phi_0^N\rangle = (E_{\Phi_0^N} - \varepsilon_\alpha) \, a_\alpha \, |\Phi_0^N\rangle \qquad \alpha < F. \qquad (7.34)$$

Choosing another sp basis leads to a slightly more involved result that can be related to the present one by a double basis transformation both for the particle addition as well as the particle removal operator, according to Eqs. (2.58) and (2.59).

Again it is useful to consider the FT of the unperturbed sp propagator which generates

$$G^{(0)}(\alpha, \beta; E) = \delta_{\alpha,\beta} \left\{ \frac{\theta(\alpha - F)}{E - \varepsilon_\alpha + i\eta} + \frac{\theta(F - \alpha)}{E - \varepsilon_\alpha - i\eta} \right\}. \qquad (7.35)$$

One can also arrive at Eq. (7.35) by replacing $\hat{H}$ by $\hat{H}_0$ and $|\Psi_0^N\rangle$ by $|\Phi_0^N\rangle$ in Eq. (7.11) for the exact propagator

$$G^{(0)}(\alpha, \beta; E) = \langle \Phi_0^N | \, a_\alpha \frac{1}{E - (\hat{H}_0 - E_{\Phi_0^N}) + i\eta} a_\beta^\dagger \, |\Phi_0^N\rangle$$

$$+ \langle \Phi_0^N | \, a_\beta^\dagger \frac{1}{E - (E_{\Phi_0^N} - \hat{H}_0) - i\eta} a_\alpha \, |\Phi_0^N\rangle. \qquad (7.36)$$

The spectral functions for the noninteracting system are particularly simple. Using again the sp basis $\{|\alpha\rangle\}$ which diagonalizes H_0, one has the following hole spectral function

$$S_h^{(0)}(\alpha; E) = \frac{1}{\pi} \mathrm{Im} \, G^{(0)}(\alpha, \alpha; E) \qquad E < \varepsilon_F^{(0)-}$$

$$= \delta(E - \varepsilon_\alpha) \, \theta(F - \alpha). \qquad (7.37)$$

Its particle counterpart yields

$$S_p^{(0)}(\alpha; E) = -\frac{1}{\pi} \mathrm{Im} \, G^{(0)}(\alpha, \alpha; E) \qquad E > \varepsilon_F^{(0)+}$$

$$= \delta(E - \varepsilon_\alpha) \, \theta(\alpha - F). \qquad (7.38)$$

The transition strength in the unperturbed spectral function is therefore located at the sp energies that correspond to the eigenvalues of the sp Hamiltonian H_0. The sp states that are occupied contribute to the hole spectral function, those that are empty to the particle part. In the independent-particle model description for an atom, a nucleus, or a Fermi gas, the hole spectral function therefore displays δ-function peaks with strength equal to 1. This corresponds to the certainty that it is possible to remove a particle from such occupied orbitals. The same holds for the particle spectral function, where this certainty relates to the possibility of adding a particle to an empty orbit. The simplicity of these results is related to the choice of the sp basis. In another sp basis the numerators change while the position of the poles in the sp propagator does not. As an example, consider the sp propagator in the $\{|rm_s\rangle\}$ representation

$$G^{(0)}(rm_s, r'm_s'; E) = \left\langle \Phi_0^N \left| a_{rm_s} \frac{1}{E - (\hat{H}_0 - E_{\Phi_0^N}) + i\eta} a_{r'm_s'}^\dagger \right| \Phi_0^N \right\rangle$$

$$+ \left\langle \Phi_0^N \left| a_{r'm_s'}^\dagger \frac{1}{E - (E_{\Phi_0^N} - \hat{H}_0) - i\eta} a_{rm_s} \right| \Phi_0^N \right\rangle \quad (7.39)$$

$$= \sum_\alpha \left\{ \frac{\langle rm_s|\alpha\rangle\langle\alpha|r'm_s'\rangle\theta(\alpha - F)}{E - \varepsilon_\alpha + i\eta} + \frac{\langle rm_s|\alpha\rangle\langle\alpha|r'm_s'\rangle\theta(F - \alpha)}{E - \varepsilon_\alpha - i\eta} \right\}.$$

Note that the numerators in Eq. (7.39) contain again the relevant sp wave functions which in this simple example describe the transition matrix elements of the particle addition and removal operators in the coordinate representation. Occupation numbers are most easily evaluated in the $\{|\alpha\rangle\}$ basis, yielding, not surprisingly,

$$n^{(0)}(\alpha) = \int_{-\infty}^{\varepsilon_F^{(0)^-}} dE\, \delta(E - \varepsilon_\alpha)\, \theta(F - \alpha) = \theta(F - \alpha). \quad (7.40)$$

7.6 Direct knockout reactions

The hole spectral function introduced in Sec. 7.3 can be experimentally observed in so-called knockout reactions. The general idea is to transfer a large amount of momentum and energy to a particle of a bound N-particle system (*e.g.* an electron in an atom or molecule, or a nucleon in a nucleus). The particle is subsequently ejected from the system, and one ends up with a fast-moving particle and a bound $N - 1$-particle system. By observing the momentum of the knocked-out particle, it is possible to reconstruct the spectral function of the system, provided that the interaction between the ejected particle and the remainder is sufficiently weak.

Let's assume that the N particle system is initially in its ground state,

$$|\Psi_i\rangle = |\Psi_0^N\rangle, \tag{7.41}$$

and makes a transition to a final N-particle eigenstate

$$|\Psi_f\rangle = a_{\boldsymbol{p}}^\dagger |\Psi_n^{N-1}\rangle, \tag{7.42}$$

composed of a bound $N-1$-particle eigenstate, $|\Psi_n^{N-1}\rangle$, and a particle with momentum $\boldsymbol{p}$.

For simplicity we consider the transition matrix elements for a scalar external probe $\rho(\boldsymbol{q}) = \sum_{j=1}^N \exp(i\boldsymbol{q} \cdot \boldsymbol{r}_j)$, which transfers momentum $\hbar\boldsymbol{q}$ to a particle. Suppressing other possible sp quantum numbers, like *e.g.* spin, the second-quantized form of this operator is given by

$$\hat\rho(\boldsymbol{q}) = \sum_{\boldsymbol{p},\boldsymbol{p}'} \langle \boldsymbol{p}| \exp(i\boldsymbol{q} \cdot \boldsymbol{r}) |\boldsymbol{p}'\rangle \, a_{\boldsymbol{p}}^\dagger a_{\boldsymbol{p}'} = \sum_{\boldsymbol{p}} a_{\boldsymbol{p}}^\dagger a_{\boldsymbol{p}-\hbar\boldsymbol{q}}. \tag{7.43}$$

The transition matrix element now becomes

$$\begin{aligned}
\langle \Psi_f | \hat\rho(\boldsymbol{q}) |\Psi_i\rangle &= \sum_{\boldsymbol{p}'} \langle \Psi_n^{N-1}| a_{\boldsymbol{p}} a_{\boldsymbol{p}'}^\dagger a_{\boldsymbol{p}'-\hbar\boldsymbol{q}} |\Psi_0^N\rangle \\
&= \sum_{\boldsymbol{p}'} \langle \Psi_n^{N-1}| \delta_{\boldsymbol{p}',\boldsymbol{p}} a_{\boldsymbol{p}'-\hbar\boldsymbol{q}} + a_{\boldsymbol{p}'}^\dagger a_{\boldsymbol{p}'-\hbar\boldsymbol{q}} a_{\boldsymbol{p}} |\Psi_0^N\rangle \\
&\approx \langle \Psi_n^{N-1}| a_{\boldsymbol{p}-\hbar\boldsymbol{q}} |\Psi_0^N\rangle.
\end{aligned} \tag{7.44}$$

The last line is obtained in the so-called *Impulse Approximation*, where it is assumed that the ejected particle is the one that has absorbed the momentum from the external field. It is a good approximation whenever the momentum $\boldsymbol{p}$ of the ejectile is much larger than typical momenta for the particles in the bound states; the neglected term in Eq. (7.44) is then very small, as it involves the removal of a particle with momentum $\boldsymbol{p}$ from $|\Psi_0^N\rangle$.

There is one other assumption in the derivation: the fact that the final eigenstate of the N-particle system was written in the form of Eq. (7.42), *i.e.* a plane-wave state for the ejectile on top of an $N-1$-particle eigenstate. Again, this is a good approximation, if the ejectile momentum is large enough, as can be understood by rewriting the Hamiltonian in the N-particle system as

$$H_N = \sum_{i=1}^N \frac{\boldsymbol{p}_i^2}{2m} + \sum_{i<j=1}^N V(i,j) = H_{N-1} + \frac{\boldsymbol{p}_N^2}{2m} + \sum_{i=1}^{N-1} V(i,N). \tag{7.45}$$

The last term in Eq. (7.45) represents the *Final-State Interaction*, or the interaction between the ejected particle N and the other particles $1..N - 1$. If the relative momentum between particle N and the others is large enough their mutual interaction can be neglected, and $H_N \approx H_{N-1} + p_N^2/2m$. The result (7.44) is called the *Plane Wave Impulse Approximation* or PWIA knockout amplitude, for obvious reasons. It is precisely a removal amplitude (in the momentum representation) appearing in the Lehmann representation of the sp propagator [see Eq. (7.10)].

The cross section of the knockout reaction, where the momentum and energy of the ejected particle and the probe are either measured or known, is according to Fermi's golden rule proportional to

$$d\sigma \sim \sum_n \delta(\hbar\omega + E_i - E_f)|\langle \Psi_f| \hat{\rho}(\boldsymbol{q}) |\Psi_i\rangle|^2. \qquad (7.46)$$

The energy-conserving δ-function contains the energy transfer $\hbar\omega$ of the probe, and the initial and final energies of the system are $E_i = E_0^N$ and $E_f = E_n^{N-1} + p^2/2m$, respectively. Note that the internal state of the residual $N - 1$ system is not measured, hence the summation over n in Eq. (7.46).

Defining the missing momentum $\boldsymbol{p}_{miss}$ and missing energy E_{miss} of the knockout reaction as[2]

$$\boldsymbol{p}_{miss} = \boldsymbol{p} - \hbar\boldsymbol{q} \qquad (7.47)$$

and

$$E_{miss} = \boldsymbol{p}^2/2m - \hbar\omega = E_0^N - E_n^{N-1}, \qquad (7.48)$$

the PWIA knockout cross section can be rewritten as

$$d\sigma \sim \sum_n \delta(E_{miss} - E_0^N + E_n^{N-1})|\langle \Psi_n^{N-1}| a_{\boldsymbol{p}_{miss}} |\Psi_0^N\rangle|^2$$
$$= S_h(\boldsymbol{p}_{miss}; E_{miss}). \qquad (7.49)$$

Equation (7.49) is therefore exactly proportional to the hole spectral function defined in Eq. (7.11). This is of course only true in the PWIA, but when the deviations of the impulse approximation and the effects of the final-state interaction are small, as for atoms, or well under control as for

---

[2] We will neglect here the recoil energy of the residual $N - 1$ system, *i.e.* we assume the mass of the N and $N - 1$ system to be much heavier than the mass m of the ejected particle.

nuclei (see also Sec. 21.5), it is possible to obtain precise experimental information on the hole spectral function of the system.

In many instances the N-particle target system is probed by its interaction with an external electromagnetic radiation field. This obviously occurs when the target is placed in a real photon beam, but it is also true when a beam of electrons is incident on the target, *e.g.* in nuclear $(e, e'p)$ reactions, or $(e, 2e)$ reactions on atoms, molecules and solids. The scattering process of the electron off a particle in the target can be described on a deeper level as the exchange of a virtual photon between the electron and a target particle. In general a (real or virtual) photon is characterized by its 4-momentum $(\hbar\omega/c, \hbar\boldsymbol{q})$ and its polarization 4-vector $(\epsilon_0, \boldsymbol{\epsilon})$. The interaction Hamiltonian, describing the coupling of the electromagnetic field with the charges and currents in the target, is proportional to

$$\hat{H}_{int} \sim \int d\boldsymbol{r} e^{i\boldsymbol{q}\cdot\boldsymbol{r}} \left(\epsilon_0 \hat{\rho}(\boldsymbol{r}) - \boldsymbol{\epsilon} \cdot \hat{\boldsymbol{J}}(\boldsymbol{r}) \right),$$

$$= \left(\epsilon_0 \hat{\rho}(\boldsymbol{q}) - \boldsymbol{\epsilon} \cdot \hat{\boldsymbol{J}}(\boldsymbol{q}) \right), \quad (7.50)$$

where $\hat{\rho}$ and $\hat{\boldsymbol{J}}$ are the charge density operator and current density operator of the N-particle system. The charge density corresponds to the scalar probe that was discussed above. The vector nature of the current density somewhat complicates the discussion, but it can be shown that also in this case the important proportionality in Eq. (7.49) holds [Frullani and Mougey (1984)].

7.7 Discussion of $(e, 2e)$ data for atoms

The presentation of $(e, 2e)$ data must begin with the hydrogen atom. The solution of the Schrödinger equation for hydrogen provides key material in any course on quantum mechanics. The interpretation of the wave function in coordinate space as a probability amplitude, transforms its absolute square to the probability density to find the electron at the corresponding location. This observable has never been measured, but forms the cornerstone of the interpretation of nonrelativistic quantum mechanics. The Schrödinger equation for hydrogen can be solved in momentum space as well, and the ground-state wave function in atomic units is given by

$$\phi_{1s}(\boldsymbol{p}) = \frac{2^{3/2}}{\pi} \frac{1}{(1+p^2)^2}. \quad (7.51)$$

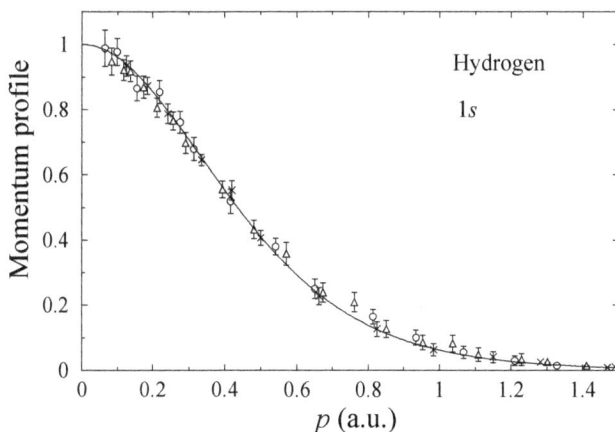

Fig. 7.1 Comparison of the normalized $(e, 2e)$ cross section (momentum profile) for hydrogen with the square of the $1s$ wave function in momentum space, adapted from [Lohmann and Weigold (1981)]. The solid line represents $(1 + p^2)^{-4}$. The measurements were performed at 1200 eV (crosses), 800 eV (circles), and 400 eV (triangles).

An $(e, 2e)$ experiment on hydrogen was performed by [Lohmann and Weigold (1981)]. As discussed in the previous section, the cross section for the process is proportional to the removal probability of an electron with momentum p under the right kinematic conditions. For hydrogen, this becomes the square of Eq. (7.51), as shown below. The final state in Eq. (7.49) for the removal of the $1s$ electron corresponds to the vacuum. The removal amplitude is therefore given by

$$\langle 0| \, a_{\boldsymbol{p}} \, |n = 1, \ell = 0\rangle = \langle \boldsymbol{p} \, |n = 1, \ell = 0\rangle = \phi_{1s}(\boldsymbol{p}), \qquad (7.52)$$

neglecting spin, which corresponds exactly to Eq. (7.51). The ability of the $(e, 2e)$ reaction to extract the square of this wave function, is demonstrated in Fig. 7.1. The cross section was obtained at several incident energies, all high enough to ensure that the PWIA result accurately describes the reaction. When the appropriate electron–electron (Mott) cross section is divided out, the outcome should become independent of energy. The comparison with the momentum profile, given by the square of $(1 + p^2)^{-2}$ [see Eq. (7.51)], convincingly demonstrates the correctness of this interpretation. The $(e, 2e)$ experiments on the hydrogen atom therefore come as close as practically possible to measuring the (square) of the electron wave function.

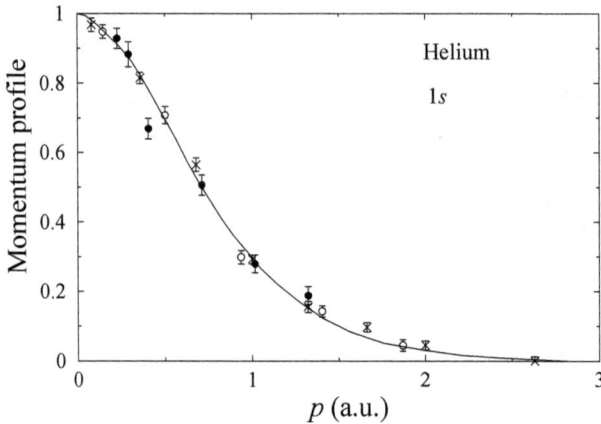

Fig. 7.2 Comparison of the $(e, 2e)$ data for helium at different energies, with theoretical DWIA calculations based on the Hartree–Fock $1s$ wave function in momentum space. The theoretical results at different energies cannot be distinguished. The experimental data were measured at 824.5 eV (crosses), 424.5 eV (open circles), and 224.5 eV (filled circles) [Hood *et. al.* (1973)].

For the helium atom, a description in terms of the independent-particle model is very successful. As discussed in Ch. 3, helium is a closed-shell system corresponding to the $1s^2$ configuration. In Ch. 10 we will develop the correct many-particle description of the mean-field or independent-particle model. The inclusion of the average effect of the two-body interaction in the description of the sp properties, is referred to as the Hartree–Fock method. The resulting sp Hamiltonian then also includes the average contribution of the electron–electron interaction, but still leads to an independent-particle description. The interpretation of the results of the $(e, 2e)$ experiment does not change when applied to this system. Data for the $(e, 2e)$ reaction on helium leading to He$^+$ in its ground state are shown in Fig. 7.2. The analysis of the reaction is more complicated than for hydrogen, since several other reactions like elastic and inelastic scattering, can take place. It is also possible that ionization occurs with other residual ion states. These processes can be represented by complex potentials that describe the elastic scattering of an electron from the atom and the residual ion [Furness and McCarthy (1973)], and are therefore constrained by other experimental data. The incoming and outgoing electrons are thus subject to the corresponding distortions, which lead to a similar description of the reaction as in the PWIA, with plane waves replaced by these distorted ones.

The resulting distorted wave impulse approximation (DWIA) is often used for describing the data. Such DWIA calculations still contain the removal amplitude $\langle \Psi_n^{N-1} | a_{\boldsymbol{p}} | \Psi_0^N \rangle$ as an important ingredient. Again, when the reaction is analyzed for different incident energies, the removal amplitude should not depend on the energy. Agreement with the data signals that the interpretation of the reaction mechanism is correct and that the proper removal amplitude was employed. More details of the description of the $(e, 2e)$ reaction can be found in [McCarthy and Weigold (1991)]. In Fig. 7.2 the distorted momentum distribution as a function of missing momentum exhibits no dependence on the incident energy, signaling that distortion effects are small for the employed kinematical conditions. Moreover, the data are perfectly described by the DWIA calculation based on the Hartree–Fock $1s$ wave function for He.

The agreement of the calculation with the data in Fig. 7.2, involves the square of the momentum space wave function that was normalized to 1. We will demonstrate in Sec. 9.5 that the removal amplitude $\langle \Psi_n^{N-1} | a_{\boldsymbol{p}} | \Psi_0^N \rangle$ obeys a Schrödinger-like equation with an energy-dependent potential, which is referred to as the self-energy. In Ch. 11 such an energy-dependent potential will be studied for atoms. The presence of this energy dependence changes the normalization of the removal amplitude to a value smaller than 1. The normalization factor is referred to as the spectroscopic factor,

$$ S = \int d\boldsymbol{p} \, \left| \langle \Psi_n^{N-1} | a_{\boldsymbol{p}} | \Psi_0^N \rangle \right|^2 . \tag{7.53} $$

The independent-particle description generates spectroscopic factors that are either 1 or 0, depending on whether the state is occupied or not.

For other closed-shell atoms, the spectroscopic factor for the removal of the last valence electron becomes a bit smaller than 1. The $(e, 2e)$ reaction on neon yields a removal probability of the valence $2p$ electron of 0.92 [Samardzic *et al.* (1993)]. Two additional fragments, each carrying 0.04, for $2p$ removal are found at higher energy. Stronger fragmentation is observed for the $2s$ removal in this atom. The sum of all the fragments adds up to 1 for both the $2p$ and $2s$ orbit, so that the occupation of these levels remains 1, according to the first line of Eq. (7.16). These features persist for heavier closed-shell atoms. This is illustrated in Fig. 7.3 for the removal of the valence $3p$ orbit in argon, described by an appropriate Hartree–Fock wave function. The shape of the solid curve in the figure describes the data only, when the theoretical result is multiplied by a spectroscopic factor of 0.95 [McCarthy *et al.* (1989)]. The interpretation is

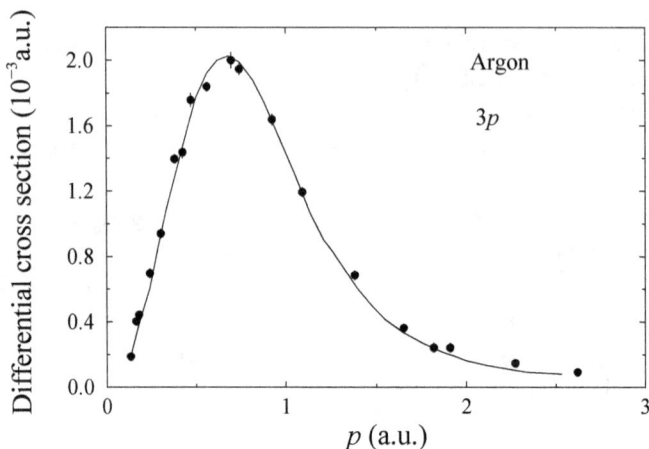

Fig. 7.3 Differential $(e, 2e)$ cross section for the $(e, 2e)$ reaction on argon for the $3p$ orbit at an incident energy of 1500 eV adapted from [McCarthy *et al.* (1989)]. The solid line corresponds to the DWIA calculation using the Hartree–Fock wave function and multiplied with a spectroscopic factor of 0.95.

that the removal probability for the (Hartree–Fock) $3p$ electron from argon is only 95%. Since there is a sum rule for the total strength [see Eqs. (7.16) – (7.18)], the remaining 5% must be located elsewhere. In a more recent experiment [Brunger *et al.* (1999)], three more fragments were indeed found with a combined strength of 0.05. The occupation number of the $3p$ orbit in argon consequently remains 1, as in the independent-particle model. The choice of the Hartree–Fock $3p$ wave function leads to a correct description of the momentum dependence of the cross section. Using a different wave function leads to an incorrect momentum dependence, and typically a smaller spectroscopic factor. The analysis of the $(e, 2e)$ [and $(e, e'p)$] reaction thus relies on employing the sp wave functions — to be interpreted as the removal amplitudes — that yield a correct description of the shape of the cross section. Such wave functions, one could say, are therefore experimentally determined.

The fragmentation of strength is easy to understand when all fragments remain below the Fermi energy, as for closed-shell atoms. This can occur when one-particle–two-hole (1p2h) states[3] mix with the one-hole configuration corresponding to the removal of a valence particle in a closed-shell system. If the mixing is weak and the energies of the 1p2h states are far

---

[3] In general, n-particle–m-hole will be denoted by $npmh$.

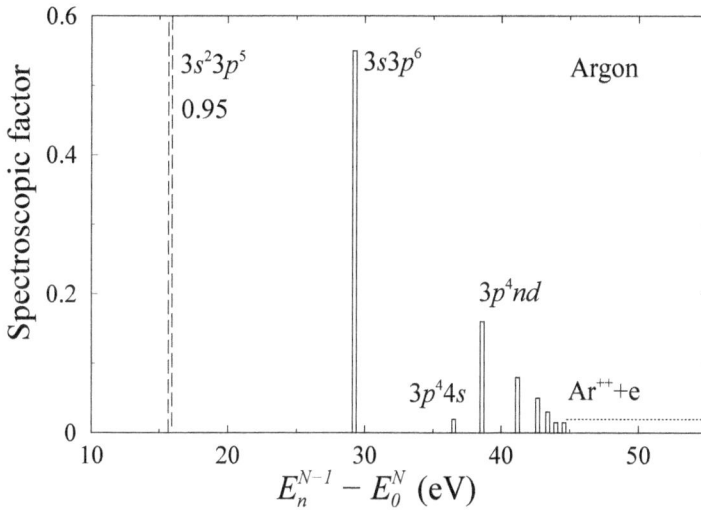

Fig. 7.4 Spectroscopic factors, indicated by bars, for the $3p$ and $3s$ removal in argon [Brunger *et al.* (1999)]. The dominant ion state configurations are indicated. The $3p^4nd$ states run from $n = 3$ to the unresolved 7 and 8 at 44 eV. The last indicated fragment corresponds to the Rydberg series. The dashed fragment denotes the $3p$ spectroscopic factor of 0.95 (not shown to scale) (see Fig. 7.3). The dotted horizontal line represents the sp strength in the Ar^{++} continuum (above 45 eV), corresponding to 0.08.

from the valence hole state, only small hole fragments will come to reside at the energies corresponding to the predominantly 1p2h states. Examples are provided by the fragmentation pattern of the $2p$ orbit in neon and the $3p$ in argon. When the hole state is more deeply bound, it is possible that it mixes more strongly with 1p2h states that are quite nearby in energy. Such a situation is illustrated in Fig. 7.4 for the removal probability of the $3s$ electron from argon as a function of $E_n^{N-1} - E_0^N$. The spectroscopic factor for the main $3p$ fragment is also shown (not to scale), to indicate the location of the Ar^+ ground state. The largest $3s$ fragment is located at 29.24 eV, with a spectroscopic factor of 0.55. The dominant 1p2h configurations at the location of the other $3s$ fragments, are indicated in the figure. The horizontal dashed line indicates the 8% strength that resides at energies, corresponding to Ar^{++} plus one electron in the continuum.

The analysis of the $(e, 2e)$ reaction on argon also shows that the results for the spectroscopic factors are independent of the incident electron energy [McCarthy *et al.* (1989)]. It is therefore permissible to infer that the removal probabilities that are experimentally extracted, correspond to the theoretical quantities associated with the numerator of the sp propagator. The sum of the spectroscopic factors for the $3s$ removal is again 1. The resulting picture of a closed-shell atom consequently retains some features of the independent-particle model, where all the sp strength is concentrated in levels that are completely full or empty. The corresponding many-particle description is studied in Ch. 10 and confirms that many atomic properties can be explained in the so-called Hartree–Fock approximation.

The observed fragmentation of the sp strength however, requires consideration of the mixing process between 1p2h and the valence hole states, as indicated above. This mixing process has little influence on the occupation numbers. For closed-shell atoms, occupation numbers jump from 1 to 0, when the Fermi energy is crossed, even including the effect of the interactions. This is no longer true for nuclei, as will be discussed in the next section. The $(e, 2e)$ reaction has also been successfully applied to molecules [McCarthy and Weigold (1991)]. More recent applications of this reaction have focused on the possibility to extract spectroscopic information for solids [Vos and McCarthy (1995)].

7.8   Discussion of $(e, e'p)$ data for nuclei

For nuclei, the $(e, e'p)$ reaction can be used to study the hole spectral function. Since the electron interacts weakly with the nucleons, it is an ideal probe to study nuclei. The dominant operators that excite the nucleus in electron scattering, have a one-body character. An example is the charge density operator which has a similar form as Eq. (7.43). Such operators can only change the quantum numbers of one particle and therefore yield excited states in which a particle is removed from the correlated ground state and placed into an unoccupied state. When the incident electron has enough energy, the removed proton can be sufficiently energetic so that the *Impulse Approximation* applies. If the electron transfers a substantial amount of energy, the resulting excited state is expected to be dominated by a simple particle-hole state. It corresponds to the outgoing particle, suitably influenced by the surrounding medium, and the valence hole that is selected by the kinematics of the reaction. It is assumed that additional interactions between this particle and hole state play no role (see also Sec. 21.5). The latter is not true at low excitation energy, where these interactions lead to collective states. The reaction description therefore corresponds to the DWIA, discussed in the previous section. The outgoing proton is subject

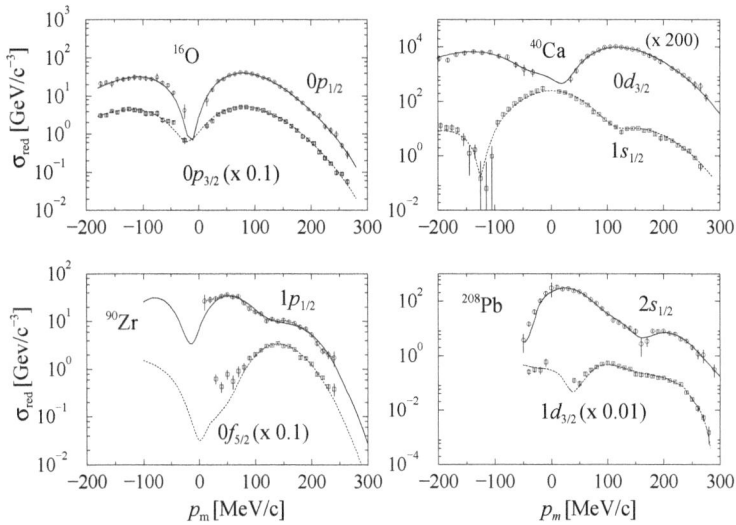

Fig. 7.5 Momentum profiles as a function of missing momentum p_m, for valence holes in several closed-shell nuclei. Appropriate scale factors have been applied to allow the representation of the data for different orbitals. The experiments were performed at the NIKHEF accelerator in Amsterdam [Lapikás (1993)].

to an optical potential that is well known from elastic scattering data. Here too, the criterion for a successful description of the reaction, requires that the extracted spectroscopic factors are independent of the incident energy of the electron. The details of the theoretical description of the $(e, e'p)$ reaction can be found in [Boffi *et al.* (1996)]. The experimental data have been reviewed in [de Witt Huberts (1990); Dieperink and de Witt Huberts (1990); Pandharipande *et al.* (1997)].

The momentum dependence of the $(e, e'p)$ cross section for a specific final state, is again dominated by the sp wave function associated with the corresponding orbital. The experimental analysis is performed by slightly adjusting the parameters of the Woods–Saxon potential, discussed in Ch. 3, to fit the shape of the cross section. For nuclei it is always necessary to reduce the theoretical cross section by a spectroscopic factor substantially less than 1, to obtain agreement with the data. As for atoms, these spectroscopic factors correspond to the removal probabilities that are contained in the numerator of the sp propagator. Examples of this type of analysis for several closed-shell nuclei, are shown in Fig. 7.5. So-called reduced cross sections are plotted, which have been divided by the elementary electron–

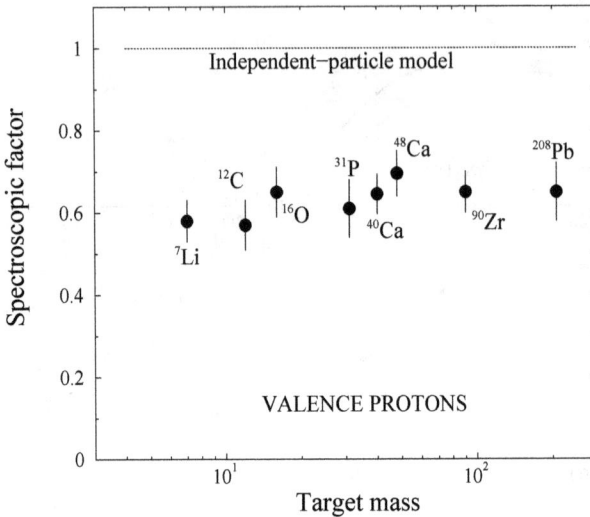

Fig. 7.6   Spectroscopic factors from the $(e, e'p)$ reaction as a function of target mass. The dotted line with a height of 1, illustrates the prediction of the independent-particle model. Data have been obtained at the NIKHEF accelerator in Amsterdam [Lapikás (1993)].

proton cross section at the appropriate kinematic conditions. The missing momentum can also have negative values when it is directed opposite to the momentum transferred to the target. A correct description of the reaction requires a good fit at all values of this quantity.

Figure 7.5 demonstrates that the shapes of the valence nucleon wave functions accurately describe the observed cross sections. Such wave functions have been employed for years in nuclear-structure calculations, which have relied on the independent-particle model. The description of the data in Fig. 7.5 however, requires a significant departure of the independent-particle model, with regard to the integral of the square of these wave functions. Indeed, the spectroscopic factors, necessary to obtain the solid curves, are substantially less than 1. Similar spectroscopic factors are extracted for nuclei all over the periodic table.[4] A compilation for the spectroscopic factor of the last valence orbit for different nuclei, adapted from [Lapikás (1993)], is shown in Fig. 7.6. The results in Fig. 7.6 indicate that there is an essentially global reduction of the sp strength of about 35% for these valence holes in most nuclei. Such a substantial deviation

---

[4]Most experiments have been performed on closed-shell nuclei.

from the prediction of the independent-particle model, requires a detailed explanation on the basis of the correlations that dominate in nuclei. We will explore this issue in Chs. 16 and 17.

An additional feature, observed in the $(e, e'p)$ reaction, is the fragmentation pattern of the more deeply bound orbitals in nuclei. It exhibits single isolated peaks only in the immediate vicinity of the Fermi energy, whereas for more deeply bound states a stronger fragmentation of the strength is observed with increasing distance from ε_F. This is beautifully illustrated by the ^{208}Pb$(e, e'p)$ data from [Quint (1988)], shown in Fig. 7.7. Whereas the $2s\frac{1}{2}$ orbit exhibits a single peak, there is a substantial fragmentation of the $0f$ strength, which corresponds to the most deeply bound strength considered. Intermediate results are extracted for orbits in between these two extremes, as illustrated in Fig. 7.7. Additional information about the occupation number of the $2s\frac{1}{2}$ valence orbit is also available. By analyzing elastic electron scattering cross sections of neighboring nuclei [Wagner (1986)], the occupation number for the $2s\frac{1}{2}$ proton orbit of 0.75 is extracted, which is about 10% larger than the spectroscopic factor [Grabmayr (1992)]. An occupation number less than 1, requires a different explanation than for the observed pattern of fragmentation. The latter pattern can be understood on the basis of the substantial mixing of the valence hole states with 1p2h states. In the case of atoms, it is permissible to continue to treat the ground state as a Slater determinant, even in the presence of electron-electron interactions. Such a treatment is not valid for nuclei, since the mutual interaction of nucleons is much stronger, particularly at short interparticle distances. Indeed, the repulsive interaction will reduce the wave function of the relative motion of two nucleons substantially. This reduction requires the admixture of high-momentum components in the relative wave function, corresponding to states at high excitation energy. The short-range repulsion of the interaction is therefore capable to admix high-lying 2p2h states into the correlated ground state $|\Psi_0^A\rangle$ of the nucleus.[5] These admixtures lead to a much more complicated ground state which includes 2p2h and additional $npnh$ components. The removal of a valence particle is not possible from these contributions to the correlated ground state, leading to a reduced occupation number. The depletion of the Fermi sea must of course be accompanied by the occupation of states that are empty in the independent-particle description, in order to conserve the total number

---

[5]We will employ the notation A for total particle number in the case of nuclei or nuclear matter, where $A = Z + N$. For other systems N will continue to denote the total number of particles.

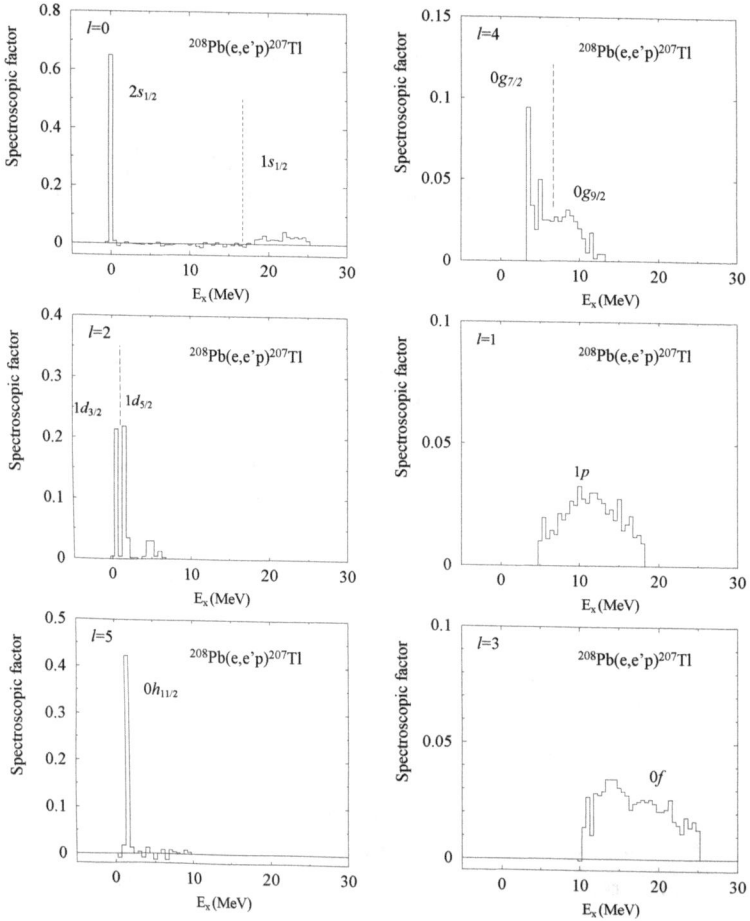

Fig. 7.7 Results for the spectroscopic strength for ^{208}Pb as a function of excitation energy, E_x, in ^{207}Tl, obtained from the $(e,e'p)$ reaction, adapted from [Quint (1988)]. The spectroscopic factor for the valence $2s\frac{1}{2}$ has been adjusted to 0.65 in accord with the analysis of [Sick and de Witt Huberts (1991)]. No error bars are shown, to emphasize the character of the observed fragmentation patterns.

of particles. The importance of short-range correlations suggests that the occupation of high-momentum states may figure prominently in accounting for all the particles in the nucleus.

One of the last $(e,e'p)$ experiments performed at the NIKHEF facility before it was decommissioned, explored the removal of all the protons in the energy and momentum domain, corresponding to the independent-particle

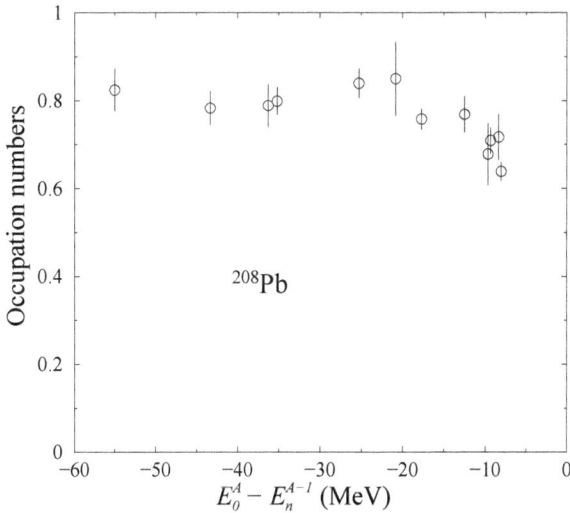

Fig. 7.8   Occupation numbers from the ^{208}Pb$(e, e'p)$ reaction. A global depletion for all the protons that are fully occupied in the independent-particle model, is clearly visible. Data have been obtained at the NIKHEF accelerator in Amsterdam [van Batenburg (2001)].

model. The experiment was performed on ^{208}Pb [van Batenburg (2001)]. The complete energy and momentum dependence of the cross section was analyzed in terms of the contribution of all the proton orbits occupied in the independent-particle model. For this purpose, energy distributions like those in Fig. 7.7 were suitably parametrized and combined with momentum profiles from a standard Woods–Saxon potential [see Eq. (3.25)]. As fit parameters to the data, the overall occupation numbers associated with these orbits, were employed. The resulting occupation numbers are shown in Fig. 7.8, plotted as a function of the sp energies of the aforementioned Woods–Saxon potential. We note the striking depletion of the Fermi sea by about 20%, which clearly distinguishes the nuclear shell occupations from the atomic ones. The depletion of the proton orbits increases when the sp energy approaches the Fermi energy. This feature is also encountered in other many-body systems, like the electron gas and nuclear matter, as discussed in Ch. 16. Many of the results, illustrated here, will be addressed again when theoretical calculations for finite nuclei are discussed in Ch. 17. The global nature of the depletion of the occupation numbers in ^{208}Pb will also appear as a feature observed in calculations of the momentum distribution of nuclear matter, presented in Ch. 16.

The presentation of a sample of experimental data from the $(e, 2e)$ and $(e, e'p)$ reactions, serves to emphasize the physical content of the sp propagator, defined in this chapter. We emphasize that their availability provides a crucial tool to assess the quality of the theoretical description of these systems. In case of disagreement between experiment and theory, it is often possible to devise improved theoretical approaches, incorporating some of the physics that is elucidated by the data.

7.9 Exercises

(1) Perform all the steps for bosons, which were discussed in this chapter for the case of fermions, pertaining to the single-particle propagator. In particular, obtain the Lehmann representation and the result for the ground-state energy (and expectation values of one-body operators).
(2) Analyze the energy sum rule, Eq. (7.26), in the presence of three-body interactions.
(3) Study the large-energy limit of Eq. (7.10). Assume a discrete and finite sp basis and interpret the sp labels as matrix indices, $G(\alpha, \beta; E) \equiv [G(E)]_{\alpha,\beta}$. Show that for $E \to \infty$,

$$[G(E)] \to \frac{1}{E - [\varepsilon]},$$

where the energy matrix $[\varepsilon]$ is given by the following commutator–anticommutator combination.

$$[\varepsilon]_{\alpha,\beta} = \left\langle \Psi_0^N \right| \{a_\alpha, [\hat{H}, a_\beta^\dagger]\} \left| \Psi_0^N \right\rangle.$$

Chapter 8

Perturbation expansion of the single-particle propagator

In this chapter, several steps are executed, which make it possible to express the exact propagator in terms of the noninteracting one and the two-body interaction. The technical details of the derivations are a bit tedious, but the end result allows an intuitive and practical analysis of the individual contributions in perturbation theory in terms of the so-called Feynman diagrams. A systematic survey of all these contributions then leads to the Dyson equation discussed in Ch. 9. In the present chapter, Sec. 8.1 summarizes the relevant material from App. A concerning time evolution in the interaction picture. The expansion of the propagator in terms of the interaction is obtained in Sec. 8.2. The evaluation of individual contributions to this expansion, as an expectation value with respect to the noninteracting ground state, is greatly facilitated by Wick's theorem, discussed in Sec. 8.4. The motivation for deriving it is provided in Sec. 8.3. All the resulting expressions contributing to the propagator can be represented pictorially by diagrams, as discussed in Sec. 8.5. Simple rules providing a dictionary between the diagrams and corresponding mathematical expression in terms of known quantities, are given in Sec. 8.6. Diagrams are presented both in time and energy formulation in this section. Additional rules are developed for systems, where it is advisable to treat direct and exchange matrix elements together.

8.1 Time evolution in the interaction picture

As discussed in App. A, it is convenient to separate the simple time-dependence of the time-evolution operator associated with $\hat{H}_0$, from the full time-evolution operator, by introducing the interaction picture. For state kets this is accomplished by defining

$$|\Psi_I(t)\rangle = \exp\left\{\frac{i}{\hbar}\hat{H}_0 t\right\}|\Psi_S(t)\rangle, \qquad (8.1)$$

where the subscripts I and S refer to the interaction and Schrödinger picture, respectively. Note that the latter picture refers to the conventional description of the time dependence in quantum mechanics. The corresponding Schrödinger equation in the interaction picture then acquires the following form, as shown in App. A,

$$i\hbar\frac{\partial}{\partial t}|\Psi_I(t)\rangle = \hat{H}_1(t)|\Psi_I(t)\rangle, \qquad (8.2)$$

with

$$\hat{H}_1(t) = \exp\left\{\frac{i}{\hbar}\hat{H}_0 t\right\}\hat{H}_1 \exp\left\{-\frac{i}{\hbar}\hat{H}_0 t\right\}. \qquad (8.3)$$

Time evolution in the interaction picture is governed by the operator that connects the interaction picture kets at different times according to

$$|\Psi_I(t)\rangle = \hat{\mathcal{U}}(t,t_0)|\Psi_I(t_0)\rangle, \qquad (8.4)$$

where the I subscript has been suppressed in the operator, but a special symbol $\mathcal{U}$ was introduced. At $t = t_0$, obviously

$$\hat{\mathcal{U}}(t_0, t_0) = 1. \qquad (8.5)$$

The equation of motion for this operator is obtained by combining Eqs. (8.2) and (8.4) with the following result

$$i\hbar\frac{\partial}{\partial t}\hat{\mathcal{U}}(t,t_0) = \hat{H}_1(t)\hat{\mathcal{U}}(t,t_0). \qquad (8.6)$$

According to App. A, one may integrate this equation formally and iterate it to all orders to yield

$$\hat{\mathcal{U}}(t,t_0) = \sum_{n=0}^{\infty}\left(\frac{-i}{\hbar}\right)^n \frac{1}{n!}\int_{t_0}^{t}dt_1 \int_{t_0}^{t}dt_2.. \int_{t_0}^{t}dt_n \; \mathcal{T}\left[\hat{H}_1(t_1)\hat{H}_1(t_2)..\hat{H}_1(t_n)\right], \qquad (8.7)$$

where the $\mathcal{T}$-operation is extended to order the $\hat{H}_1$ operator with the latest time farthest to the left, and so on.

8.2 Perturbation expansion in the interaction

The expression for the time-evolution operator in Eq. (8.7) can be used to write the correlated sp propagator as a sum of known quantities. Each of these terms is obtained by taking an appropriate expectation value with respect to the noninteracting ground state. It is convenient to consider the particle and hole part of the correlated sp propagator separately. In particular, it can be shown [Mattuck (1992)] that for the particle part of the sp propagator

$$G_+(\alpha, \beta; t - t') = \lim_{T' \to -\infty(1-i\eta)} \lim_{T \to +\infty(1-i\eta)} Q(\alpha, \beta; T, T', t - t'), \quad (8.8)$$

where

$$Q = -\frac{i}{\hbar} \frac{\langle \Phi_0^N | \, \mathcal{T} \left[\hat{\mathcal{U}}(T, T') a_{\alpha_I}(t) a_{\beta_I}^\dagger(t') \right] | \Phi_0^N \rangle}{\langle \Phi_0^N | \hat{\mathcal{U}}(T, T') | \Phi_0^N \rangle}. \quad (8.9)$$

The choice of the particle part of G implies that $t > t'$. In addition, considering the limits in Eq. (8.8), one has

$$T > t > t' > T'. \quad (8.10)$$

Note that the particle removal and addition operators, appearing explicitly and implicitly (inside $\hat{\mathcal{U}}$) in Eq. (8.8), are given in the interaction picture. The original definition of G involves the corresponding operators in the Heisenberg picture and reads

$$G_+(\alpha, \beta; t - t') = -\frac{i}{\hbar} \theta(t - t') \langle \Psi_0^N | \, a_{\alpha_H}(t) a_{\beta_H}^\dagger(t') | \Psi_0^N \rangle. \quad (8.11)$$

In order to show that Eq. (8.8) is correct, we will proceed backwards from Eq. (8.9). First observe that

$$\hat{\mathcal{U}}(T, T') = \hat{\mathcal{U}}(T, t) \hat{\mathcal{U}}(t, t') \hat{\mathcal{U}}(t', T') \quad (8.12)$$

representing the group property. Inserting this result in Eq. (8.9), one can move the addition and removal operators under the $\mathcal{T}$ sign to their location suggested by Eq. (8.12). The latter procedure involves no change in sign, since it always involves an even number of interchanges. At this point, the $\mathcal{T}$ symbol can be dropped, resulting in

$$Q = -\frac{i}{\hbar} \frac{\langle \Phi_0^N | \hat{\mathcal{U}}(T, t) a_{\alpha_I}(t) \hat{\mathcal{U}}(t, t') a_{\beta_I}^\dagger(t') \hat{\mathcal{U}}(t', T') | \Phi_0^N \rangle}{\langle \Phi_0^N | \hat{\mathcal{U}}(T, T') | \Phi_0^N \rangle}. \quad (8.13)$$

Equation (A.11) can now be used for the addition and removal operators, and Eq. (A.22) to express $\hat{\mathcal{U}}$ in terms of the regular time-evolution operator $\hat{U}_S$ given in Eq. (A.4), to rewrite Q in the following way

$$Q = -\frac{i}{\hbar} \frac{e^{\frac{i}{\hbar}E^N_{\Phi_0}(T-T')} \left\langle \Phi^N_0 \right| \hat{U}_S(T-t)a_\alpha \hat{U}_S(t-t')a^\dagger_\beta \hat{U}_S(t'-T') \left| \Phi^N_0 \right\rangle}{e^{\frac{i}{\hbar}E^N\Phi_0(T-T')} \left\langle \Phi^N_0 \right| \hat{U}_S(T-T') \left| \Phi^N_0 \right\rangle}.$$

(8.14)

Then we apply Eq. (A.4) again to write the particle addition and removal operators in the Heisenberg picture, so that

$$Q = -\frac{i}{\hbar} \frac{\left\langle \Phi^N_0 \right| e^{-\frac{i}{\hbar}\hat{H}T} a_{\alpha_H}(t)a^\dagger_{\beta_H}(t') e^{\frac{i}{\hbar}\hat{H}T'} \left| \Phi^N_0 \right\rangle}{\left\langle \Phi^N_0 \right| e^{-\frac{i}{\hbar}\hat{H}T} e^{\frac{i}{\hbar}\hat{H}T'} \left| \Phi^N_0 \right\rangle}.$$

(8.15)

Inserting the completeness relation for the exact Hamiltonian in terms of states in the N-particle system at appropriate places generates

$$Q = -\frac{i}{\hbar} \frac{\sum \langle \Phi^N_0 | \Psi^N_n \rangle \langle \Psi^N_m | \Phi^N_0 \rangle e^{-\frac{i}{\hbar}(E^N_n T - E^N_m T')} \left\langle \Psi^N_n \right| a_{\alpha_H}(t)a^\dagger_{\beta_H}(t') \left| \Psi^N_m \right\rangle}{\sum \langle \Phi^N_0 | \Psi^N_k \rangle \langle \Psi^N_k | \Phi^N_0 \rangle e^{-\frac{i}{\hbar}(E^N_k T - E^N_k T')}}.$$

(8.16)

It is now possible to perform the limits that are given in Eq (8.8). As $T \to \infty$ (or T'), the quantity ηT will be assumed to go to ∞. Accordingly, all exponentials will decay to zero. The slowest decay corresponds to the terms with the lowest energy in the exponentials, ensuring that those survive. From the ensuing cancellation between numerator and denominator, we get the desired expression Eq. (8.11) when taking this limit for Eq. (8.8). The expansion for G_+ is thus obtained by inserting Eq. (8.7) for $\hat{\mathcal{U}}$, which finally yields

$$G(\alpha, \beta; t-t') = -\frac{i}{\hbar} \sum_n^\infty \left(\frac{-i}{\hbar}\right)^n \frac{1}{n!} \int_{-\infty(1-i\eta)}^{+\infty(1-i\eta)} dt_1 ... \int_{-\infty(1-i\eta)}^{+\infty(1-i\eta)} dt_n$$

(8.17)

$$\times \left\langle \Phi^N_0 \right| \mathcal{T}\left[\hat{H}_1(t_1)...\hat{H}_1(t_n)a_{\alpha_I}(t)a^\dagger_{\beta_I}(t')\right] \left| \Phi^N_0 \right\rangle$$

$$\Big/ \sum_m^\infty \left(\frac{-i}{\hbar}\right)^m \frac{1}{m!} \int_{-\infty(1-i\eta)}^{+\infty(1-i\eta)} dt'_1 ... \int_{-\infty(1-i\eta)}^{+\infty(1-i\eta)} dt'_m \left\langle \Phi^N_0 \right| \mathcal{T}\left[\hat{H}_1(t'_1)...\hat{H}_1(t'_m)\right] \left| \Phi^N_0 \right\rangle.$$

Since the result also holds for the hole part of the propagator, the $+$ has been dropped in Eq. (8.17). The critical point in the last step of this derivation is the nonvanishing of $\langle \Phi^N_0 | \Psi^N_0 \rangle$, expressing the assumption that a nonvanishing overlap exists between the simple state $\left| \Phi^N_0 \right\rangle$ and the correlated ground state $\left| \Psi^N_0 \right\rangle$. No specific properties of the sp propagator G

were used in obtaining Eq. (8.17). The procedure can therefore be repeated for the expectation value of an arbitrary operator in the Heisenberg picture

$$\langle \Psi_0^N | \hat{O}_H(t) | \Psi_0^N \rangle = \sum_{n}^{\infty} \left(\frac{-i}{\hbar} \right)^n \frac{1}{n!} \int_{-\infty(1-i\eta)}^{+\infty(1-i\eta)} dt_1 \ldots \int_{-\infty(1-i\eta)}^{+\infty(1-i\eta)} dt_n \qquad (8.18)$$

$$\times \langle \Phi_0^H | \mathcal{T} \left[\hat{H}_1(t_1)\ldots\hat{H}_1(t_n)\hat{O}_I(t) \right] | \Phi_0^N \rangle$$

$$/ \sum_{m}^{\infty} \left(\frac{-i}{\hbar} \right)^m \frac{1}{m!} \int_{-\infty(1-i\eta)}^{+\infty(1-i\eta)} dt'_1 \ldots \int_{-\infty(1-i\eta)}^{+\infty(1-i\eta)} dt'_m \langle \Phi_0^N | \mathcal{T} \left[\hat{H}_1(t'_1)\ldots\hat{H}_1(t'_m) \right] | \Phi_0^N \rangle.$$

8.3 Lowest-order contributions and diagrams

Equation (8.17) for the sp propagator does not yet clarify how to generate each relevant contribution in the perturbation expansion, since important cancellations between numerator and denominator occur. This goal can be achieved by employing a technique, referred to as Wick's theorem, discussed in the next section. As motivation we consider a few low-order terms, contributing to the numerator and denominator of Eq. (8.17). We simplify the notation by eliminating the I subscript from the particle addition and removal operators occurring in Eq. (8.17). No confusion can arise since the Heisenberg picture operators will still be labeled by the subscript H.

Consider the $n = 0$ term in the numerator

$$n = 0 \rightarrow -\frac{i}{\hbar} \langle \Phi_0^N | \mathcal{T} \left[a_\alpha(t)a_\beta^\dagger(t') \right] | \Phi_0^N \rangle = G^{(0)}(\alpha, \beta; t - t'), \qquad (8.19)$$

which just yields the noninteracting propagator, as given explicitly in Eq. (7.32). This result is physically reasonable since without interaction one should obtain the noninteracting result for the propagator. The first-order contribution to the numerator in Eq. (8.17) requires evaluating

$$n = 1 \rightarrow \left(\frac{-i}{\hbar} \right)^2 \int_{-\infty(1-i\eta)}^{\infty(1-i\eta)} dt_1 \langle \Phi_0^N | \mathcal{T} \left[\hat{H}_1(t_1)a_\alpha(t)a_\beta^\dagger(t') \right] | \Phi_0^N \rangle \qquad (8.20)$$

$$= \left(\frac{-i}{\hbar} \right)^2 \int_{-\infty(1-i\eta)}^{\infty(1-i\eta)} dt_1 \frac{1}{2} \sum_{\gamma\delta\epsilon\theta} (\gamma\delta|V|\epsilon\theta)$$

$$\times \langle \Phi_0^N | \mathcal{T} \left[a_\gamma^\dagger(t_1)a_\delta^\dagger(t_1)a_\theta(t_1)a_\epsilon(t_1)a_\alpha(t)a_\beta^\dagger(t') \right] | \Phi_0^N \rangle$$

$$- \left(\frac{-i}{\hbar} \right)^2 \int_{-\infty(1-i\eta)}^{\infty(1-i\eta)} dt_1 \sum_{\gamma\delta} \langle \gamma| U |\delta \rangle \langle \Phi_0^N | \mathcal{T} \left[a_\gamma^\dagger(t_1)a_\delta(t_1)a_\alpha(t)a_\beta^\dagger(t') \right] | \Phi_0^N \rangle,$$

where $\hat{H}_1$ has been inserted with the inclusion of a contribution from an auxiliary one-body potential, in addition to the two-body interaction

$$\hat{H}_1(t_1) = \frac{1}{2} \sum_{\gamma\delta\epsilon\theta} (\gamma\delta|V|\epsilon\theta)\, a_\gamma^\dagger(t_1) a_\delta^\dagger(t_1) a_\theta(t_1) a_\epsilon(t_1)$$

$$- \sum_{\gamma\delta} \langle\gamma|\,U\,|\delta\rangle\, a_\gamma^\dagger(t_1) a_\delta(t_1). \tag{8.21}$$

For an explicit calculation of the $\hat{U}$ term in Eq. (8.20) for example, a choice for the time-ordering of t and t' must be made. Subsequently, the expectation value of the four operators with respect to the noninteracting ground state, must be evaluated. The time dependence of the particle addition and removal operators is given by Eqs. (A.16) and (A.15), respectively. So for $t > t'$ and the part of the integration over t_1 for which $t' > t_1$, one requires the calculation of

$$\langle \Phi_0^N |\, a_\alpha a_\beta^\dagger a_\gamma^\dagger a_\delta\, | \Phi_0^N \rangle$$

$$= \theta(\alpha - F)\theta(F - \delta) \langle \Phi_0^N |\, (\delta_{\alpha,\beta} - a_\beta^\dagger a_\alpha)(\delta_{\gamma,\delta} - a_\delta a_\gamma^\dagger)\, | \Phi_0^N \rangle$$

$$= \theta(\alpha - F)\theta(F - \delta) \Big(\underbrace{\delta_{\alpha,\beta}\delta_{\gamma,\delta}}_{a)} - \underbrace{\delta_{\beta,\delta}\delta_{\alpha,\gamma}}_{b)} \Big). \tag{8.22}$$

The same strategy as in Ch. 3 has been applied, which involves moving operators that give zero when acting on $|\Phi_0^N\rangle$ to the right, and operators that have the same effect on $\langle\Phi_0^N|$ to the left. Before taking these steps, we assert that no contribution is obtained unless both $\alpha > F$ and $F > \delta$, for the same reasons. The first term in Eq. (8.22) leads to

$$U_a \Rightarrow - \left(\frac{-i}{\hbar}\right)^2 \int_{-\infty(1-i\eta)}^{t'} dt_1 \sum_{\gamma\delta} \langle\gamma|\,U\,|\delta\rangle \tag{8.23}$$

$$\times \left\{ \frac{i}{\hbar}\theta(t_1^+ - t_1)\delta_{\gamma,\delta}\theta(F - \gamma)e^{\frac{i}{\hbar}\varepsilon_\delta(t_1^+ - t_1)} \right\} \hbar$$

$$\times \left\{ -\frac{i}{\hbar}\theta(t - t')\delta_{\alpha,\beta}\theta(\alpha - F)e^{-\frac{i}{\hbar}\varepsilon_\alpha(t-t')} \right\} \hbar$$

$$= \left\{ \int_{-\infty(1-i\eta)}^{t'} dt_1 \sum_{\gamma\delta} \langle\gamma|\,U\,|\delta\rangle\, G_-^{(0)}(\delta,\gamma; t_1 - t_1^+) \right\} G_+^{(0)}(\alpha,\beta; t-t').$$

This contribution has been written such that particle and hole parts of noninteracting propagators can be identified, even if the time arguments

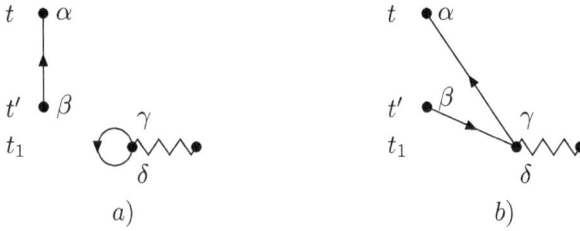

Fig. 8.1   Diagrammatic representation of Eqs. (8.23) and (8.24), respectively.

are identical. The order of the operators for this $G_-^{(0)}$ term is dictated by the $a^\dagger a$ form of $\hat{U}$, uniquely determining the sequence of the times involved. Similar steps for the second term yield

$$U_b \Rightarrow -\theta(t-t') \int_{-\infty(1-i\eta)}^{t'} dt_1 \sum_{\gamma\delta} \langle \gamma | U | \delta \rangle \, G_+^{(0)}(\alpha, \gamma; t - t_1) G_-^{(0)}(\delta, \beta; t_1 - t').$$

$$(8.24)$$

These contributions have a simple graphical representation shown in Fig. 8.1. In each picture the three times t, t' and t_1 are indicated with time increasing in the vertical direction. Such diagrams are therefore referred to as "time-ordered". Each propagator term is represented by a line with an arrow that starts at its second argument, accordingly labeled with appropriate sp quantum numbers, and ends at its first (also labeled). The beginnings and ends are indicated by a dot to which, where appropriate, a zigzag line ending in another black dot can be attached, representing the action of $\hat{U}$. The interactions are also labeled with corresponding sp quantum numbers where the final state is shown above, and the initial state below the zigzag. Diagram $a)$ in Fig. 8.1 is "disconnected" since it consists of two separate pieces that appear in product form in Eq. (8.23). This usage of language implies that the second term is "connected" since all parts are linked to each other and no factorization occurs in Eq. (8.24). The propagator in diagram $a)$ closing on itself, starts and ends at the $\hat{U}$ vertex. It is the result of addition and removal operators from the same interaction operator "contracting" with each other. Since interactions always have addition operators to the left of removal operators, such contributions involve the hole parts of the corresponding sp propagator.

Diagram $a)$ can now be read as follows: before time t_1 the system is represented by the ground state of the noninteracting system $\left| \Phi_0^N \right\rangle$ which is not explicitly shown. At time t_1 the $\hat{U}$ interaction acts by removing a particle from this state, simultaneously returning it, necessarily in the same sp level. At $t' > t_1$ a particle with quantum numbers β is added to the noninteracting ground state; at a later time t, this particle is removed again, leaving the system in the ground state of the noninteracting system. In turn, this implies that $\alpha = \beta$ since these quantum numbers correspond to the basis associated with H_0. Diagram b) represents a different physical process: before t_1 the system is in the noninteracting ground state; at t_1 the $\hat{U}$ interaction now creates a particle–hole state with the "particle" in an unoccupied state (γ) and the "hole" indicated by δ. The particle line is drawn with the arrow pointing towards increasing time, whereas the arrow of the hole line points towards decreasing time. At time t', this hole is filled again by adding a particle with quantum numbers β (equating δ and β). It can therefore also be considered as a particle going backward in time from t' to t_1, just as an antiparticle in relativistic field theory. From t' to t one particle propagates above the noninteracting ground state until, at t, it is removed, returning the system to its original state. The complete equivalence of the diagrams and the expressions given in Eqs. (8.23) and (8.24) still requires a set of simple rules that will be discussed in more detail in Sec. 8.6.

Other diagrams can be generated for different time-orderings requiring similar evaluations of the expectation value for the resulting order of particle addition and removal operators. The procedure can be applied to the two-body operator $\hat{V}$ as well. This tedious process can then be continued for higher-order terms contributing to the numerator of Eq. (8.17). Similar pictures can also be generated by evaluating the denominator of that equation, order by order. Clearly, in each order, a set of noninteracting propagators will be generated, complemented by a product of matrix elements of $\hat{U}$ and $\hat{V}$ terms. In addition, relevant integrations over time variables and summations over internal sp quantum numbers are to be performed. Wick's theorem can be used to avoid all this work, as shown in the next section.

8.4   Wick's theorem

To establish Wick's theorem, we will follow the presentation of [Fetter and Walecka (1971)]. We introduce the notion of the normal-ordered product in which the operators are organized in such a way that the ones that give zero when acting on $\left|\Phi_0^N\right\rangle$ are placed to the right of those that don't. Normal ordering therefore rearranges a product of operators with the property that the usual interchanges of fermion operators are accompanied by appropriate minus signs, as in time-ordered products. For $a_\alpha(t)$ and $a_\beta^\dagger(t')$ this normal ordering yields

$$N\left[a_\alpha(t)a_\beta^\dagger(t')\right] = \begin{cases} -a_\beta^\dagger(t')a_\alpha(t) & \alpha > F, \ \beta > F \\ a_\alpha(t)a_\beta^\dagger(t') & \alpha > F, \ \beta < F \\ a_\alpha(t)a_\beta^\dagger(t') & \alpha < F, \ \beta > F \\ a_\alpha(t)a_\beta^\dagger(t') & \alpha < F, \ \beta < F, \end{cases} \qquad (8.25)$$

where the order in the second and third term could be changed (with an attendant extra sign) since both operators have the same effect. In the first and last term in Eq. (8.25) the sequence is essential. In words, one simply moves removal operators of empty states to the right, addition operators of those same states to the left, removal operators of occupied states to the left, and, finally, addition operators of those states to the right. The extension to normal ordered products of other or more operators is defined in a similar fashion. It is now possible to define the contraction of two operators as the difference between their time-ordered and normal-ordered product. This contraction is identified by two identical symbols (*e.g.* bullets) used as superscripts attached to the operators involved.

$$a_\alpha(t)^\bullet a_\beta^\dagger(t')^\bullet = T\left[a_\alpha(t)a_\beta^\dagger(t')\right] - N\left[a_\alpha(t)a_\beta^\dagger(t')\right]. \qquad (8.26)$$

For $t > t'$ one then obtains

$$a_\alpha(t)^\bullet a_\beta^\dagger(t')^\bullet = a_\alpha(t)a_\beta^\dagger(t') - N\left[a_\alpha(t)a_\beta^\dagger(t')\right]. \qquad (8.27)$$

When $\alpha > F$ and $\beta > F$, this time-ordering and the definition of the normal-ordered product yields

$$a_\alpha(t)^\bullet a_\beta^\dagger(t')^\bullet = \theta(t-t')\left(a_\alpha(t)a_\beta^\dagger(t') + a_\beta^\dagger(t')a_\alpha(t)\right)\theta(\alpha-F)\theta(\beta-F)$$

$$= \theta(t-t')\left(a_\alpha a_\beta^\dagger + a_\beta^\dagger a_\alpha\right)\theta(\alpha-F)\theta(\beta-F)e^{-\frac{i}{\hbar}\varepsilon_\alpha t}e^{\frac{i}{\hbar}\varepsilon_\beta t'}$$

$$= i\hbar\left(\frac{-i}{\hbar}\right)\theta(t-t')\delta_{\alpha,\beta}\theta(\alpha-F)e^{\frac{i}{\hbar}\varepsilon_\alpha(t'-t)}$$

$$= i\hbar\, G_+^{(0)}(\alpha,\beta;t-t'). \qquad (8.28)$$

Applying the definitions of the time-ordered and normal-ordered product, it is straightforward to show that for all other combinations of quantum numbers ($\alpha > F$, $\beta < F$, *etc.*) the contractions vanish for $t > t'$. For $t < t'$, one finds only a nonvanishing contribution when both α and β

correspond to an occupied level in the noninteracting ground state with the result

$$a_\alpha(t)^\bullet a_\beta^\dagger(t')^\bullet = i\hbar\, G_-^{(0)}(\alpha,\beta;t-t'). \qquad (8.29)$$

This implies that for all quantum numbers α and β and all time orderings, we can write

$$a_\alpha(t)^\bullet a_\beta^\dagger(t')^\bullet = i\hbar\, G^{(0)}(\alpha,\beta;t-t'), \qquad (8.30)$$

where the particle or hole part is automatically selected by the choice of occupied or empty quantum numbers and the result is simply a c-number. Other pairs of operators, like two removal or two addition operators, have vanishing contractions, since the corresponding anticommutation relations yield zero.

The strategy to rearrange a time-ordered product of operators according to the properties of the individual addition and removal operators with respect to their action on the noninteracting ground state, can be accomplished in one stroke by invoking the following form of Wick's theorem:

$$\mathcal{T}\left[\hat{a}\hat{b}\hat{c}...\hat{x}\hat{y}\hat{z}\right] = N\left[\hat{a}\hat{b}\hat{c}...\hat{x}\hat{y}\hat{z}\right] + N\left[\hat{a}^\bullet \hat{b}^\bullet \hat{c}...\hat{x}\hat{y}\hat{z}\right] + N\left[\hat{a}^\bullet \hat{b}\hat{c}^\bullet ...\hat{x}\hat{y}\hat{z}\right]$$

$$+... + N\left[\hat{a}^\bullet \hat{b}\hat{c}...\hat{x}\hat{y}\hat{z}^\bullet\right] + N\left[\hat{a}\hat{b}^\bullet \hat{c}^\bullet ...\hat{x}\hat{y}\hat{z}\right] + ... + N\left[\hat{a}^\bullet \hat{b}^{\bullet\bullet}\hat{c}...^\bullet...^{\bullet\bullet}\hat{x}\hat{y}\hat{z}\right]$$

$$+... + N\left[\hat{a}^\bullet \hat{b}^{\bullet\bullet}\hat{c}^{\bullet\bullet\bullet}...\hat{x}^\circ \hat{y}^{\circ\circ}\hat{z}^{\circ\circ\circ}\right]$$

$$= N\left[\hat{a}\hat{b}\hat{c}...\hat{x}\hat{y}\hat{z}\right] + N\,[\text{sum over all possible pairs of contractions}]\,. \ (8.31)$$

A time-ordered product of operators is the sum of all normal-ordered products, formed by applying any number of contractions among the operators in all possible ways.

We still have to define the contractions inside the normal-ordered products, appearing in Eq. (8.31). Changing the order of operators inside a normal ordered product, requires an appropriate sign change according to

$$N\left[\hat{a}\hat{b}\hat{c}\hat{d}...\right] = -N\left[\hat{b}\hat{a}\hat{c}\hat{d}...\right], \qquad (8.32)$$

where $\hat{a},\hat{b},...$ represent either particle removal or addition operators. The contractions in Eq. (8.31) should therefore be interpreted in the following way

$$N\left[\hat{a}^\bullet \hat{b}\hat{c}^\bullet \hat{d}...\right] = -N\left[\hat{b}\hat{a}^\bullet \hat{c}^\bullet \hat{d}...\right] = -\hat{a}^\bullet \hat{c}^\bullet N\left[\hat{b}\hat{d}...\right]. \qquad (8.33)$$

For completeness we repeat the strategy of moving operators depending on their character. An operator like $a_\alpha^\dagger$ with $\alpha > F$ must be moved to the left and will be considered an operator of type I. If $\alpha < F$ this operator must be moved to the right and will be labeled type II. Similarly, a_α is a type I operator when $\alpha < F$, and type II when $\alpha > F$. Note that the distinction is related to the choice of sp basis as *its* lowest levels are filled in the noninteracting ground state $|\Phi_0^N\rangle$. This choice of basis is convenient and will be employed here. Such a choice is not mandatory, since removal and addition operators with other sp quantum numbers, can also be used. Ultimately, Wick's theorem requires the decomposition of these operators into type I and II, which is accomplished by a basis transformation according to Eqs. (2.58) and (2.59). The decomposition automatically yields two terms, one referring to occupied states, the other to empty states in the summations of Eqs. (2.58) and (2.59), which immediately identify the type I or II character. Normal ordering is a distributive operation, implying that the results discussed below, are also valid in another sp basis.

Several steps are involved in demonstrating the validity of Eq. (8.31). First one can prove the following: if $N\left[\hat{a}\hat{b}\hat{c}...\hat{x}\hat{y}\right]$ is a normal-ordered product and the operator $\hat{z}$ has a time earlier than any occurring in this product then

$$N\left[\hat{a}\hat{b}\hat{c}...\hat{x}\hat{y}\right]\hat{z} = N\left[\hat{a}\hat{b}\hat{c}...\hat{x}\hat{y}^\bullet\hat{z}^\bullet\right] + N\left[\hat{a}\hat{b}\hat{c}...\hat{x}^\bullet\hat{y}\hat{z}^\bullet\right] + ...$$
$$+ N\left[\hat{a}^\bullet\hat{b}\hat{c}...\hat{x}\hat{y}\hat{z}^\bullet\right] + N\left[\hat{a}\hat{b}\hat{c}...\hat{x}\hat{y}\hat{z}\right]. \tag{8.34}$$

Note the following points to prove Eq. (8.34). If $\hat{z}$ is of type II, then all contractions vanish with each operator $\hat{i}$ in the set $\hat{a}...\hat{y}$, since

$$\hat{i}^\bullet\hat{z}^\bullet = T\left[\hat{i}\hat{z}\right] - N\left[\hat{i}\hat{z}\right] = \hat{i}\hat{z} - \hat{i}\hat{z} = 0, \tag{8.35}$$

where in the last equality the earlier time of the operator $\hat{z}$ and its type II property were used. So in this particular case Eq. (8.34) follows, since one obviously has

$$N\left[\hat{a}\hat{b}\hat{c}...\hat{x}\hat{y}\right]\hat{z} = N\left[\hat{a}\hat{b}\hat{c}...\hat{x}\hat{y}\hat{z}\right]. \tag{8.36}$$

Assume that the operators $\hat{a}\hat{b}...\hat{y}$ are already normal ordered. If they are not, Eq. (8.34) may be reordered accordingly on both sides, using the sign conventions introduced in Eqs. (8.32) and (8.33). Consider first the situation where the operators $\hat{a}, \hat{b}, \hat{c}, ...\hat{x}, \hat{y}$ are all of type II, giving zero when acting on $\left|\Phi_0^N\right\rangle$, and $\hat{z}$ is of type I. The next step is an induction procedure. For two operators Eq. (8.34) is valid since

$$N\left[\hat{y}\right]\hat{z} = \hat{y}\hat{z} = T\left[\hat{y}\hat{z}\right] = \hat{y}^\bullet\hat{z}^\bullet + N\left[\hat{y}\hat{z}\right] = N\left[\hat{y}^\bullet\hat{z}^\bullet\right] + N\left[\hat{y}\hat{z}\right], \qquad (8.37)$$

where the second equality applies on account of the chosen time-ordering. The third equality invokes the definition of the contraction, and the last one is correct since a contraction is a c-number. We now assume that Eq. (8.34) is valid for n operators. To prove its correctness for $n+1$, we can multiply Eq. (8.34) by a type II operator $\hat{d}$ on the left (also with a time later than that of $\hat{z}$), yielding

$$\hat{d}N\left[\hat{a}\hat{b}\hat{c}...\hat{x}\hat{y}\right]\hat{z} = N\left[\hat{d}\hat{a}\hat{b}\hat{c}...\hat{x}\hat{y}^\bullet\hat{z}^\bullet\right] + N\left[\hat{d}\hat{a}\hat{b}\hat{c}...\hat{x}^\bullet\hat{y}\hat{z}^\bullet\right] + ...$$
$$+ N\left[\hat{d}\hat{a}^\bullet\hat{b}\hat{c}...\hat{x}\hat{y}\hat{z}^\bullet\right] + \hat{d}N\left[\hat{a}\hat{b}\hat{c}...\hat{x}\hat{y}\hat{z}\right]. \qquad (8.38)$$

Since all operators $\hat{a}\hat{b}\hat{c}..\hat{x}\hat{y}$ are of type II as well and the contraction of $\hat{d}$ with $\hat{z}$ is a c-number, $\hat{d}$ can be taken inside the normal ordering except for the last term in Eq (8.38), as $\hat{z}$ is still an operator there. For this last term, however, one can write

$$\hat{d}N\left[\hat{a}\hat{b}\hat{c}...\hat{x}\hat{y}\hat{z}\right] = N\left[\hat{d}^\bullet\hat{a}\hat{b}\hat{c}...\hat{x}\hat{y}\hat{z}^\bullet\right] + N\left[\hat{d}\hat{a}\hat{b}\hat{c}...\hat{x}\hat{y}\hat{z}\right], \qquad (8.39)$$

since

$$\hat{d}N\left[\hat{a}\hat{b}\hat{c}...\hat{x}\hat{y}\hat{z}\right] = (-1)^p\hat{d}\hat{z}\hat{a}\hat{b}\hat{c}...\hat{x}\hat{y} = (-1)^p T\left[\hat{d}\hat{z}\right]\hat{a}\hat{b}\hat{c}...\hat{x}\hat{y}$$
$$= (-1)^p\hat{d}^\bullet\hat{z}^\bullet\hat{a}\hat{b}\hat{c}...\hat{x}\hat{y} + (-1)^p N\left[\hat{d}\hat{z}\right]\hat{a}\hat{b}\hat{c}...\hat{x}\hat{y}$$
$$= (-1)^{2p}\hat{d}^\bullet\hat{a}\hat{b}\hat{c}...\hat{x}\hat{y}\hat{z}^\bullet + (-1)^{2p} N\left[\hat{d}\hat{a}\hat{b}\hat{c}...\hat{x}\hat{y}\hat{z}\right]$$
$$= N\left[\hat{d}^\bullet\hat{a}\hat{b}\hat{c}...\hat{x}\hat{y}\hat{z}^\bullet\right] + N\left[\hat{d}\hat{a}\hat{b}\hat{c}...\hat{x}\hat{y}\hat{z}\right]. \qquad (8.40)$$

We have used p to denote the number of times $\hat{z}$ has to be exchanged to place right after $\hat{d}$. This completes the proof of Eq. (8.34) for the chosen set of operators. Additional type I operators can be included by multiplying them on the left of Eq. (8.34); the additional contractions generated in this manner vanish and can therefore be included on the right side.

Equation (8.34) can be extended to include normal-ordered products, already containing contractions. This point can be understood by multiplying both sides of Eq. (8.34) with the contraction of the operators $\hat{d}$ and $\hat{w}$ (c-number) and making the appropriate sign changes on both sides which then cancel on account of the sign conventions

$$N\left[\hat{a}\hat{b}\hat{c}\hat{d}^\circ...\hat{w}^\circ\hat{x}\hat{y}\right]\hat{z} = N\left[\hat{a}\hat{b}\hat{c}\hat{d}^\circ...\hat{w}^\circ\hat{x}\hat{y}^\bullet\hat{z}^\bullet\right] + N\left[\hat{a}\hat{b}\hat{c}\hat{d}^\circ...\hat{w}^\circ\hat{x}^\bullet\hat{y}\hat{z}^\bullet\right] + ...$$
$$+ N\left[\hat{a}^\bullet\hat{b}\hat{c}\hat{d}^\circ...\hat{w}^\circ\hat{x}\hat{y}\hat{z}^\bullet\right] + N\left[\hat{a}\hat{b}\hat{c}\hat{d}^\circ...\hat{w}^\circ\hat{x}\hat{y}\hat{z}\right]. \quad (8.41)$$

This result also holds when additional contractions are included in Eq. (8.41).

With these preliminary items out of the way, Wick's theorem can now be proven by induction. Equation (8.31) is clearly correct for two operators

$$\mathcal{T}\left[\hat{a}\hat{b}\right] = \hat{a}^\bullet\hat{b}^\bullet + N\left[\hat{a}\hat{b}\right] = N\left[\hat{a}^\bullet\hat{b}^\bullet\right] + N\left[\hat{a}\hat{b}\right]. \quad (8.42)$$

Again, assume that Eq. (8.31) is valid for n operators and then multiply it by $\hat{A}$ from the right, where $\hat{A}$ has an earlier time than the others.

$$\mathcal{T}\left[\hat{a}\hat{b}\hat{c}...\hat{x}\hat{y}\hat{z}\right]\hat{A} = \mathcal{T}\left[\hat{a}\hat{b}\hat{c}...\hat{x}\hat{y}\hat{z}\hat{A}\right] \quad (8.43)$$
$$= N\left[\hat{a}\hat{b}\hat{c}...\hat{x}\hat{y}\hat{z}\right]\hat{A} + N\left[\hat{a}^\bullet\hat{b}^\bullet\hat{c}...\hat{x}\hat{y}\hat{z}\right]\hat{A} + N\left[\hat{a}^\bullet\hat{b}\hat{c}^\bullet...\hat{x}\hat{y}\hat{z}\right]\hat{A} + ...$$
$$+ N\left[\hat{a}^\bullet\hat{b}\hat{c}...\hat{x}\hat{y}\hat{z}^\bullet\right]\hat{A} + N\left[\hat{a}\hat{b}^\bullet\hat{c}^\bullet...\hat{x}\hat{y}\hat{z}\right]\hat{A} + ... + N\left[\hat{a}^\bullet\hat{b}^{\bullet\bullet}\hat{c}...^\bullet...^{\bullet\bullet}\hat{x}\hat{y}\hat{z}\right]\hat{A}$$
$$+... + N\left[\hat{a}^\bullet\hat{b}^{\bullet\bullet}\hat{c}^{\bullet\bullet\bullet}...\hat{x}^\circ\hat{y}^{\circ\circ}\hat{z}^{\circ\circ\circ}\right]\hat{A}$$
$$= N\left[\hat{a}\hat{b}\hat{c}...\hat{x}\hat{y}\hat{z}\hat{A}\right] + N\left[\text{sum over all possible pairs of contractions}\right].$$

The first equality in Eq. (8.43) is generated directly from the left-hand side of Eq.(8.31) since $\hat{A}$ has a time earlier than any of the other operators. The second equality reflects the multiplication of $\hat{A}$ on the right side of Eq. (8.31). To obtain the last equality, one uses Eq. (8.41) and its generalization with more contractions, to arrive at the last line. This completes the proof for the particular time-ordering where $\hat{A}$ is earlier than any other operator. The restriction can be removed by reordering each term in Eq. (8.43) without changing the outcome. The sign conventions introduced earlier, ensure that Wick's theorem is valid in these other situations as well. The proof of Eq. (8.31) is now complete.

The usefulness of this result which holds as an operator identity, becomes abundantly clear when the expectation value of a time-ordered product of operators is taken with respect to the noninteracting ground state $|\Phi_0^N\rangle$. Only those terms in Eq. (8.43) that contain fully contracted contributions give nonvanishing results. If there are $2n$ operators in the time-ordered product, $n!$ terms with n noninteracting propagators, and a factor

$(i\hbar)^n$ are generated, together with an overal sign, to be discussed below.

8.5   Diagrams

The introduction of a graphical representation greatly facilitates the analysis of the perturbation expansion of the sp propagator. We will start by considering the first few contributions to the numerator of Eq. (8.17). The term with $n = 1$ containing the auxiliary potential $\hat{U}$ in the numerator of Eq. (8.17), generates the following time-ordered product

$$\langle \Phi_0^N | \, \mathcal{T} \left[a_\gamma^\dagger(t_1) a_\delta(t_1) a_\alpha(t) a_\beta^\dagger(t') \right] | \Phi_0^N \rangle =$$
$$- a_\alpha(t)^\bullet a_\beta^\dagger(t')^\bullet a_\delta(t_1)^\circ a_\gamma^\dagger(t_1^+)^\circ + a_\delta(t_1)^\bullet a_\beta^\dagger(t')^\bullet a_\alpha(t)^\circ a_\gamma^\dagger(t_1^+)^\circ$$
$$= - (i\hbar)^2 \, G^{(0)}(\alpha, \beta; t - t') G^{(0)}(\delta, \gamma; t_1 - t_1^+)$$
$$+ (i\hbar)^2 \, G^{(0)}(\delta, \beta; t_1 - t') G^{(0)}(\alpha, \gamma; t - t_1), \qquad (8.44)$$

by direct application of Wick's theorem. The extra + superscript has been added to the propagator with two t_1 arguments, referring to the original ordering, as discussed in connection with Eq. (8.23). The two contributions have a graphical representation, shown in Fig. 8.2, with corresponding expressions

$$U_a \Rightarrow \left\{ \int_{-\infty(1-i\eta)}^{\infty(1-i\eta)} dt_1 \sum_{\gamma\delta} \langle \gamma | \, U \, | \delta \rangle \, G^{(0)}(\delta, \gamma; t_1 - t_1^+) \right\} G^{(0)}(\alpha, \beta; t - t')$$
$$(8.45)$$

and

$$U_b \Rightarrow - \int_{-\infty(1-i\eta)}^{\infty(1-i\eta)} dt_1 \sum_{\gamma\delta} \langle \gamma | \, U \, | \delta \rangle \, G^{(0)}(\alpha, \gamma; t - t_1) G^{(0)}(\delta, \beta; t_1 - t'). \quad (8.46)$$

It is important to realize that the time-orderings in these diagrams are not fixed. For this reason they are referred to as Feynman diagrams. It is important to keep in mind that both time-orderings $t > t'$ and $t' > t$ are represented by the same diagram. In addition, the internal time integrations range over all times so that t_1 can be in any position relative to t and t' in Fig. 8.2. By separating these different cases, all time-ordered diagrams can be generated. An example is given in Fig. 8.1.

   The first-order term in the numerator of Eq. (8.17) involving $\hat{V}$, yields a time-ordered product of three particle addition and three removal

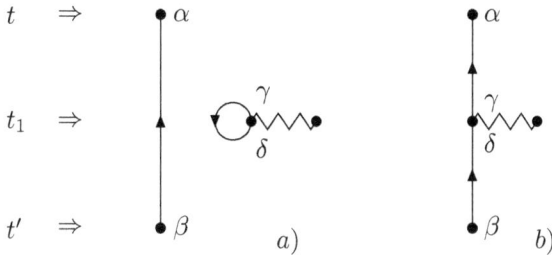

Fig. 8.2   Part *a*) shows the disconnected contribution given by Eq. (8.45).   Part *b*) represents the graphical representation of Eq. (8.46). Note that the three times indicated by horizontal arrows are not time-ordered.

operators, and therefore involves 3! terms, each containing three noninteracting propagators. As a result,

$$\hat{V}^{n=1} \Rightarrow i\hbar \int_{-\infty(1-i\eta)}^{\infty(1-i\eta)} dt_1 \, \frac{1}{2} \sum_{\gamma\delta\epsilon\theta} (\gamma\delta|V|\epsilon\theta)$$

$$\times \left\{ G^{(0)}(\alpha,\beta;t-t') \left[\underbrace{G^{(0)}(\theta,\delta;t_1-t_1^+) \, G^{(0)}(\epsilon,\gamma;t_1-t_1^+)}_{a)} \right.\right.$$

$$\left. - \underbrace{G^{(0)}(\theta,\gamma;t_1-t_1^+) \, G^{(0)}(\epsilon,\delta;t_1-t_1^+)}_{b)} \right]$$

$$+ G^{(0)}(\alpha,\gamma;t-t_1) \left[\underbrace{G^{(0)}(\theta,\beta;t_1-t') \, G^{(0)}(\epsilon,\delta;t_1-t_1^+)}_{d)} \right.$$

$$\left. - \underbrace{G^{(0)}(\theta,\delta;t_1-t_1^+) \, G^{(0)}(\epsilon,\beta;t_1-t')}_{c)} \right]$$

$$+ G^{(0)}(\alpha,\delta;t-t_1) \left[\underbrace{G^{(0)}(\theta,\gamma;t_1-t_1^+) \, G^{(0)}(\epsilon,\beta;t_1-t')}_{f)} \right.$$

$$\left.\left. - \underbrace{G^{(0)}(\theta,\beta;t_1-t') \, G^{(0)}(\epsilon,\gamma;t_1-t_1^+)}_{e)} \right] \right\}. \tag{8.47}$$

These six terms are graphically represented in Fig. 8.3. As the topology suggests, diagrams *c*) and *e*) are identical, as are *d*) and *f*). This can be verified by exchanging dummy summation variables and using the property

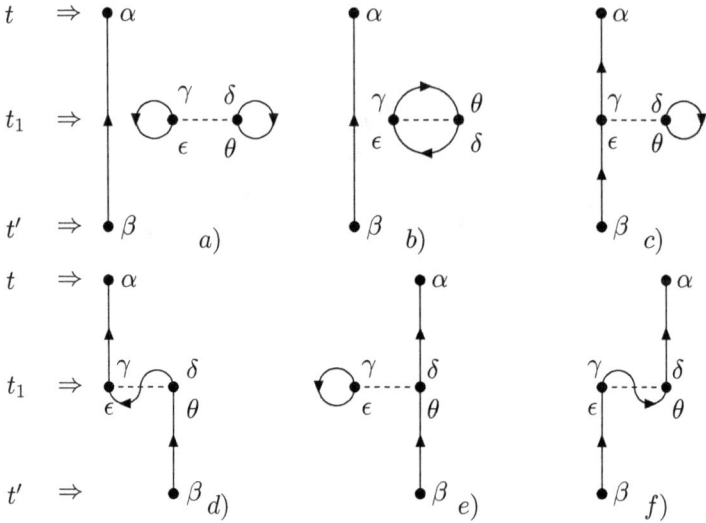

Fig. 8.3   Six diagrams corresponding to the six terms in Eq. (8.47). Note that diagrams a) and b) are disconnected and will be canceled by contributions from the denominator of Eq. (8.17).

of Eq. (2.43). It is customary to consider only diagrams c) and d) and multiply the corresponding expressions by a factor of 2. The first two diagrams are disconnected, as is the one that represents Eq. (8.45), illustrated in part a) of Fig. 8.2. We will show shortly that they cancel corresponding terms in the denominator of Eq. (8.17), leaving only connected diagrams. Anticipating this result, the true first-order contribution to the propagator can be written as

$$G^{(1)}(\alpha, \beta; t - t') = - \int dt_1 \sum_{\gamma\delta} \langle\gamma| U |\delta\rangle \, G^{(0)}(\alpha, \gamma; t - t_1) G^{(0)}(\delta, \beta; t_1 - t')$$

$$- i\hbar \int dt_1 \sum_{\gamma\delta\epsilon\theta} (\gamma\delta|V|\epsilon\theta) \qquad (8.48)$$

$$\times \Big\{ G^{(0)}(\alpha, \gamma; t - t_1) G^{(0)}(\theta, \delta; t_1 - t_1^+) G^{(0)}(\epsilon, \beta; t_1 - t')$$

$$- G^{(0)}(\alpha, \gamma; t - t_1) G^{(0)}(\epsilon, \delta; t_1 - t_1^+) G^{(0)}(\theta, \beta; t_1 - t') \Big\}.$$

The time integration limits have been suppressed in this expression. The first-order correction to the propagator therefore includes diagram b) from Fig. 8.2 and diagrams c) and d) from Fig. 8.3. It is often practical to reduce the number of diagrams even further by combining the matrix elements of the two-body interaction $\hat{V}$ as in Eq. (2.47). Using this symmetrized version, the contribution from the two-body interaction to $G^{(1)}$ can then be formulated as

$$G_{\hat{V}}^{(1)}(\alpha, \beta; t - t') = -i\hbar \int dt_1 \sum_{\gamma\delta\epsilon\theta} \langle \gamma\delta| V |\epsilon\theta \rangle \tag{8.49}$$

$$\times G^{(0)}(\alpha, \gamma; t - t_1) G^{(0)}(\theta, \delta; t_1 - t_1^+) G^{(0)}(\epsilon, \beta; t_1 - t').$$

For systems that have comparable contributions for the direct and exchange contribution in Eq. (2.47), this strategy is particularly appropriate. If it is adopted, only diagram c) in Fig. 8.3 will have to be considered in first order in $\hat{V}$. The labeling of the interactions, which is represented by a dashed line in Fig. 8.3, is usually done by reserving the two top locations for the final two-particle state and the two bottom ones for the initial state. This convention is not applied for diagram b). No ambiguity should arise however, since one can identify initial and final states by the directions of the arrows associated with the attached propagators.

The absence of the factor $\frac{1}{2}$ in front of the matrix element of $\hat{V}$ in Eq. (8.48), is due to the appearance of equivalent diagrams and remains also valid in higher order, as can be checked explicitly by the reader. In addition, for any contribution in higher order, there is an identical one from so-called topologically equivalent diagrams that differ only in the permutation of the time labels $t_1, t_2, ..., t_n$, since one can move the $\hat{H}_1(t_i)$ at will under the $\mathcal{T}$ sign. As there are $n!$ such terms, this generates a factor of $n!$ which conveniently cancels the $1/n!$ term in the numerator of Eq. (8.17).

We now turn our attention to the denominator of Eq. (8.17) to clarify the mechanism of its cancellation with terms in the numerator, as alluded to above. This denominator is sometimes written in the notation of the $\mathcal{S}$-matrix, familiar from scattering theory (see Ch. 6)

$$\langle \Phi_0^N | \hat{S} | \Phi_0^N \rangle \tag{8.50}$$

$$= \sum_{m}^{\infty} \left(\frac{-i}{\hbar} \right)^m \frac{1}{m!} \int dt_1' ... \int dt_m' \langle \Phi_0^N | \mathcal{T} \left[\hat{H}_1(t_1')...\hat{H}_1(t_m') \right] | \Phi_0^N \rangle.$$

When considering contributions to Eq. (8.50), one can proceed in a similar fashion in generating a diagrammatic representation for each order m. Clearly for $m = 0$ this contribution is 1. By direct application of Wick's theorem the first-order term yields

$$
\langle \Phi_0^N | \hat{S}^{(1)} | \Phi_0^N \rangle = \int dt_1 \sum_{\gamma\delta} \langle \gamma | U | \delta \rangle \, G^{(0)}(\delta, \gamma; t_1^+ - t_1)
$$

$$
- i\hbar \int dt_1 \, \frac{1}{2} \sum_{\gamma\delta\epsilon\theta} (\gamma\delta|V|\epsilon\theta) \left\{ G^{(0)}(\theta, \delta; t_1 - t_1^+) G^{(0)}(\epsilon, \gamma; t_1 - t_1^+) \right.
$$

$$
\left. - G^{(0)}(\theta, \gamma; t_1 - t_1^+) G^{(0)}(\epsilon, \delta; t_1 - t_1^+) \right\}. \tag{8.51}
$$

The numerator of Eq. (8.17) contains a contribution in first order of the type $G^{(0)} \times \langle \Phi_0^N | \hat{S}^{(1)} | \Phi_0^N \rangle$, as can be verified from the results of Eqs. (8.45) and (8.47). Taking the corresponding lowest order contribution to the numerators and denominator of Eq. (8.17) into account, we can therefore write schematically

$$
G(\alpha, \beta; t - t') = \left\{ G^{(0)}(\alpha, \beta; t - t') \times (1 + \langle \Phi_0^N | \hat{S}^{(1)} | \Phi_0^N \rangle + ...) \right.
$$

$$
\left. + G^{(1)}(\alpha, \beta; t - t') \times (1 + ...) + ... \right\}
$$

$$
/ \left(1 + \langle \Phi_0^N | \hat{S}^{(1)} | \Phi_0^N \rangle + ... \right). \tag{8.52}
$$

We recognize the incipient cancellation that will occur between the numerator and denominator of Eq. (8.52). When analyzing higher-order contributions, one always encounters connected terms, linked to both $a_\alpha(t)$ and $a_\beta^\dagger(t')$, possibly multiplied by disconnected ones, leading to a factorized contribution to the numerator of Eq. (8.17). This implies that in nth order, one is able to write for this numerator contribution [Abrikosov *et al.* (1975)]

$$
G_{numerator}^{(n)}(\alpha, \beta; t - t') = -\frac{i}{\hbar} \sum_{l}^{\infty} \sum_{m}^{\infty} \left(\frac{-i}{\hbar} \right)^{l+m} \delta_{n,l+m} \frac{1}{n!} \frac{n!}{l!m!}
$$

$$
\times \int dt_1 .. \int dt_m \langle \Phi_0^N | \mathcal{T} \left[\hat{H}_1(t_1)..\hat{H}_1(t_m) a_\alpha(t) a_\beta^\dagger(t') \right] | \Phi_0^N \rangle_{connected}
$$

$$
\times \int dt_{m+1} .. \int dt_n \langle \Phi_0^N | \mathcal{T} \left[\hat{H}_1(t_{m+1})..\hat{H}_1(t_n) \right] | \Phi_0^N \rangle. \tag{8.53}
$$

The result may be verified by applying Wick's theorem to both sides of this expression. The second factor contains many disconnected parts. The factor $n!/l!m!$ in Eq. (8.53) represents the number of ways to distribute the n $\hat{H}_1$ operators into the two groups. To evaluate the complete numerator, one simply has to perform the sum over n which yields

$$G_{numerator}(\alpha, \beta; t - t') = -\frac{i}{\hbar} \sum_{m}^{\infty} \left(\frac{-i}{\hbar}\right)^{m} \frac{1}{m!}$$

$$\times \int dt_1 .. \int dt_m \left\langle \Phi_0^N \middle| \mathcal{T} \left[\hat{H}_1(t_1)..\hat{H}_1(t_m)a_\alpha(t)a_\beta^\dagger(t')\right] \middle| \Phi_0^N \right\rangle_{connected}$$

$$\times \sum_{l}^{\infty} \left(\frac{-i}{\hbar}\right)^{l} \frac{1}{l!} \int dt'_1 .. \int dt'_l \left\langle \Phi_0^N \middle| \mathcal{T} \left[\hat{H}_1(t'_1)..\hat{H}_1(t'_l)\right] \middle| \Phi_0^N \right\rangle. \qquad (8.54)$$

The denominator of Eq. (8.17) can now be identified as the second factor in this expression. We therefore obtain the important result that only connected diagrams appear in the perturbation expansion of the sp propagator. This cancellation then yields the final expression for the sp propagator in terms of quantities that can, in principle, be calculated order by order from the noninteracting propagator and the Hamiltonian $\hat{H}_1$

$$G(\alpha, \beta; t - t') = -\frac{i}{\hbar} \sum_{m}^{\infty} \left(\frac{-i}{\hbar}\right)^{m} \frac{1}{m!} \int dt_1 .. \int dt_m$$

$$\times \left\langle \Phi_0^N \middle| \mathcal{T} \left[\hat{H}_1(t_1)..\hat{H}_1(t_m)a_\alpha(t)a_\beta^\dagger(t')\right] \middle| \Phi_0^N \right\rangle_{connected}. \qquad (8.55)$$

Only contributions to the sp propagator that are completely linked and connected to the operators $a_\alpha(t)$ and $a_\beta^\dagger(t')$ need be considered in this expansion. As indicated above, each term in Eq. (8.55) can be uniquely associated with a Feynman diagram.

8.6   Diagram rules

Equation (8.55) and Wick's theorem facilitate a systematic study of the sp propagator. The expansion can be depicted in a graphical manner, accompanied by a small set of rules that allow the construction of the corresponding expressions from the diagrams.

8.6.1   *Time-dependent version*

In the time-dependent formulation of the expansion, we may first consider the rules in the absence of an auxiliary potential $\hat{U}$. Only diagrams involving the two-body interaction $\hat{V}$ are encountered. The following rules apply for an mth-order contribution:

Rule 1 Draw all topologically distinct and connected diagrams
with m horizontal interaction lines for V (dashed) and $2m + 1$
directed (using arrows) Green's functions $G^{(0)}$

Rule 2 Label external points appropriately. For example, the labels
α, t and β, t' apply for Eq. (8.55)

Label each interaction with a time and sp quantum numbers

$$t \Rightarrow \quad \begin{matrix} \gamma & \delta \\ \bullet\text{-----}\bullet \\ \epsilon & \theta \end{matrix} \qquad \Rightarrow (\gamma\delta|V|\epsilon\theta)$$

For each full line one writes

$$t_i \Rightarrow \bullet\, \mu$$

$$\Rightarrow G^{(0)}(\mu, \nu; t_i - t_j)$$

$$t_j \Rightarrow \bullet\, \nu$$

Rule 3 Sum (integrate) over all internal sp quantum numbers and
integrate over all m internal times

Rule 4 Include a factor $(i\hbar)^m$ and $(-1)^F$ where F
is the number of closed fermion loops

Rule 5 Interpret equal times in a propagator as $G^{(0)}(\mu, \nu; t - t^+)$

If it is unclear whether diagrams are topologically distinct, one can
always resort to a direct application of Wick's theorem. Fermion lines
either close on themselves yielding a closed loop, or run continuously from
the external label α to β. The closing of a fermion line generates a minus
sign. The corresponding contractions can be reordered without changing
the sign into (using symbolic notation)

$$a^\dagger(t_1)^\bullet a(t_1)^{\bullet\bullet} a^\dagger(t_2)^{\bullet\bullet} a(t_2)^\circ ... a^\dagger(t_m)^{\circ\circ} a(t_m)^\bullet, \qquad (8.56)$$

requiring one additional sign change to contract the outermost operators
according to the convention. The factor $\frac{1}{2}$ appearing in the second quantized form of $\hat{V}$ can be omitted, provided only diagrams of the type c) and
d) are included for Fig. 8.3, and those of type e) and f) are discarded.
Similar considerations apply in higher order. The factor $(i\hbar)^m$ in **Rule 3**
results from the prefactor $-i/\hbar$ in Eq. (8.55), the factor $(-i/\hbar)^m$ appearing
there under the sum, and finally, a factor $(i\hbar)^{2m+1}$ from the number of contractions in mth order. The latter number corresponds to each interaction

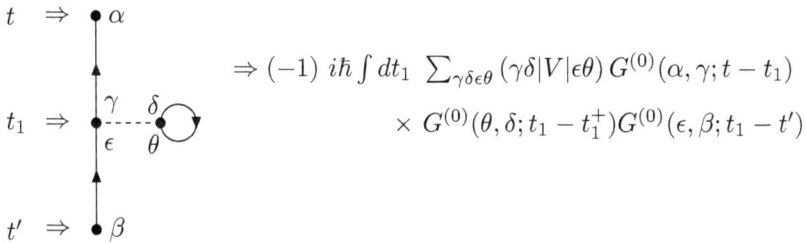

$$\Rightarrow (-1)\, i\hbar \int dt_1 \, \sum_{\gamma\delta\epsilon\theta} (\gamma\delta|V|\epsilon\theta)\, G^{(0)}(\alpha,\gamma;t-t_1)$$
$$\times\, G^{(0)}(\theta,\delta;t_1-t_1^+)G^{(0)}(\epsilon,\beta;t_1-t')$$

Fig. 8.4 Diagram V1D representing the first-order "direct" contribution from the two-body interaction V to the sp propagator in the time formulation. The minus sign in front comes from the closed fermion loop.

contributing two contractions, plus one coming from the external operators.

The first-order contributions generated by applying these rules are displayed in Figs. 8.4 and 8.5. In both diagrams, **Rule 5** applies as a result of the original ordering of the operators in the Hamiltonian. The diagram shown in Fig. 8.4 will be labeled "V1D" for the first-order "direct" contribution from V to the propagator. The exchange diagram in first-order is accordingly labeled "V1E" and shown in Fig. 8.5. In the literature one encounters different ways of drawing this exchange diagram. Here, the choice has been made to identify clearly how the propagators enter and leave the two-body interaction V. In addition, all two-body interactions are drawn horizontally to emphasize that those studied in the many-particle problem are usually static, *i.e.* occur at one time. It is possible to generate nonstatic interaction terms between particles by including higher-order contributions in the medium. Examples of such interactions will be discussed in Ch. 13.

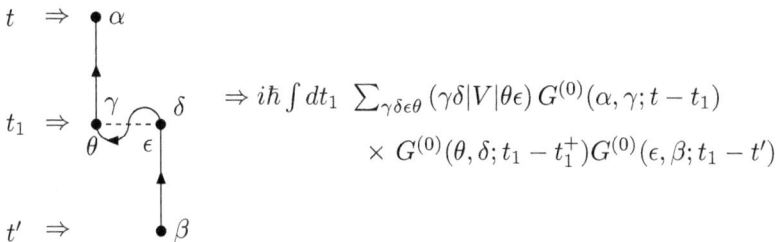

$$\Rightarrow i\hbar \int dt_1 \, \sum_{\gamma\delta\epsilon\theta} (\gamma\delta|V|\theta\epsilon)\, G^{(0)}(\alpha,\gamma;t-t_1)$$
$$\times\, G^{(0)}(\theta,\delta;t_1-t_1^+)G^{(0)}(\epsilon,\beta;t_1-t')$$

Fig. 8.5 Diagram V1E representing the first-order "exchange" contribution from the two-body interaction V.

$$t \Rightarrow \bullet \, \alpha$$

$$\Rightarrow (-1)^2 (i\hbar)^2 \int dt_1 \int dt_2 \sum_{\gamma,\delta,\epsilon,\theta} \sum_{\zeta,\xi,\lambda,\mu} G^{(0)}(\alpha,\gamma; t - t_1)$$

$$\times \, (\gamma\delta|V|\epsilon\theta) \, G^{(0)}(\epsilon, \zeta; t_1 - t_2) G^{(0)}(\theta, \delta; t_1 - t_1^+)$$

$$t_1 \Rightarrow$$

$$\times \, G^{(0)}(\mu, \xi; t_2 - t_2^+) \, (\zeta\xi|V|\lambda\mu) \, G^{(0)}(\lambda, \beta; t_2 - t')$$

$$t_2 \Rightarrow$$

$$t' \Rightarrow \bullet \, \beta$$

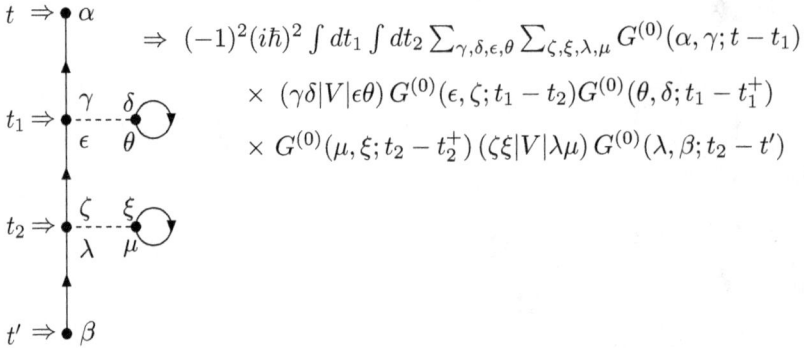

Fig. 8.6   Diagram V2a representing one of the ten second-order contributions to the sp propagator in the time formulation.

It appears more appropriate to reflect the static nature of the interactions in drawing diagrams and not use the field-theory version where the interaction lines represent propagating bosons, allowing for nonstatic interactions. The latter choice of diagrammatic representation was *e.g.* made by [Fetter and Walecka (1971)]. All diagrams up to second order in V are shown in Figs. 8.6 – 8.15, together with the corresponding expressions obtained from the rules. These diagrams clearly separate into different categories. The first four, shown in Figs. 8.6 – 8.9, are simply repeats of the first-order contributions displayed in Figs. 8.4 and 8.5. All four will be categorized as

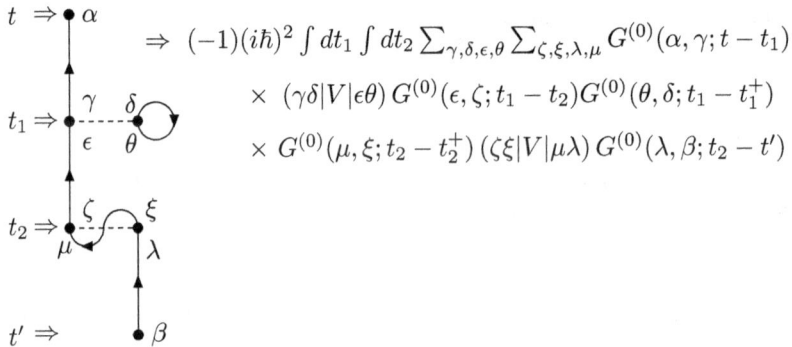

$$t \Rightarrow \bullet \, \alpha$$

$$\Rightarrow (-1)(i\hbar)^2 \int dt_1 \int dt_2 \sum_{\gamma,\delta,\epsilon,\theta} \sum_{\zeta,\xi,\lambda,\mu} G^{(0)}(\alpha,\gamma; t - t_1)$$

$$\times \, (\gamma\delta|V|\epsilon\theta) \, G^{(0)}(\epsilon, \zeta; t_1 - t_2) G^{(0)}(\theta, \delta; t_1 - t_1^+)$$

$$t_1 \Rightarrow$$

$$\times \, G^{(0)}(\mu, \xi; t_2 - t_2^+) \, (\zeta\xi|V|\mu\lambda) \, G^{(0)}(\lambda, \beta; t_2 - t')$$

$$t_2 \Rightarrow$$

$$t' \Rightarrow \qquad \bullet \, \beta$$

Fig. 8.7   Diagram V2b, representing a second-order contribution to G.

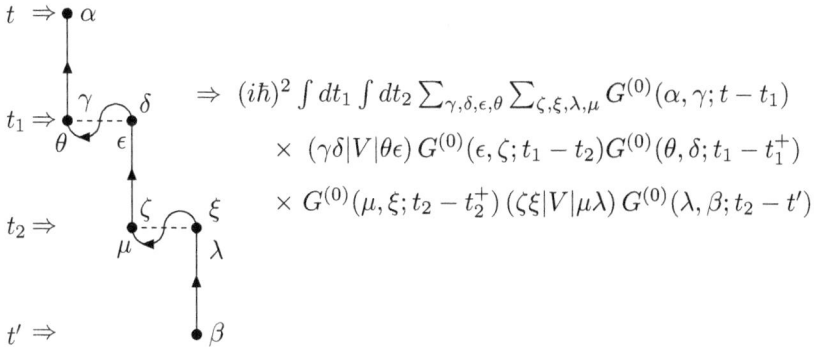

$$\Rightarrow (i\hbar)^2 \int dt_1 \int dt_2 \sum_{\gamma,\delta,\epsilon,\theta} \sum_{\zeta,\xi,\lambda,\mu} G^{(0)}(\alpha,\gamma;t-t_1)$$
$$\times \ (\gamma\delta|V|\theta\epsilon)\, G^{(0)}(\epsilon,\zeta;t_1-t_2) G^{(0)}(\theta,\delta;t_1-t_1^+)$$
$$\times \ G^{(0)}(\mu,\xi;t_2-t_2^+)\,(\zeta\xi|V|\mu\lambda)\, G^{(0)}(\lambda,\beta;t_2-t')$$

Fig. 8.8   Diagram V2c, representing a second-order contribution to G.

reducible in Ch. 9, since they can be generated by iterating lower-order contributions. For future reference these diagrams will be labeled $V2a$ through $V2d$.

The next four contributions in second order can be obtained from each other under exchange. They are labeled $V2e$ through $V2h$ and shown in Figs. 8.10 – 8.13. These terms are also related to the first-order contributions as $V2a$ through $V2d$. They are generated by replacing the propagator that leaves from and returns to V in the first-order diagrams (implying equal time arguments), by the corresponding $G^{(1)}$ represented by $V1D$ and $V1E$. Especially for the last four diagrams the topological equivalence with the

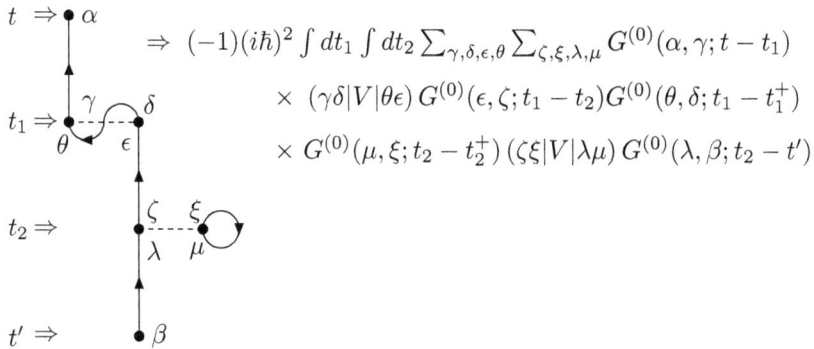

$$\Rightarrow (-1)(i\hbar)^2 \int dt_1 \int dt_2 \sum_{\gamma,\delta,\epsilon,\theta} \sum_{\zeta,\xi,\lambda,\mu} G^{(0)}(\alpha,\gamma;t-t_1)$$
$$\times \ (\gamma\delta|V|\theta\epsilon)\, G^{(0)}(\epsilon,\zeta;t_1-t_2) G^{(0)}(\theta,\delta;t_1-t_1^+)$$
$$\times \ G^{(0)}(\mu,\xi;t_2-t_2^+)\,(\zeta\xi|V|\lambda\mu)\, G^{(0)}(\lambda,\beta;t_2-t')$$

Fig. 8.9   Diagram V2d, representing a second-order contribution to G.

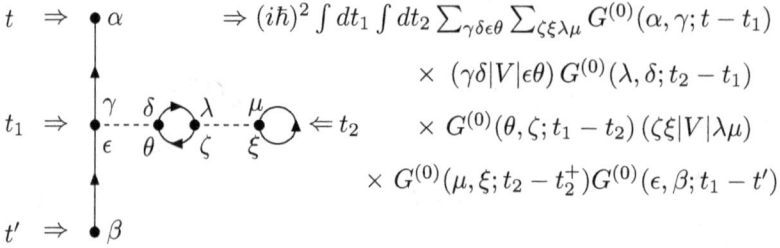

$$\Rightarrow (i\hbar)^2 \int dt_1 \int dt_2 \sum_{\gamma\delta\epsilon\theta} \sum_{\zeta\xi\lambda\mu} G^{(0)}(\alpha,\gamma;t-t_1)$$

$$\times \; (\gamma\delta|V|\epsilon\theta)\, G^{(0)}(\lambda,\delta;t_2-t_1)$$

$$\times \; G^{(0)}(\theta,\zeta;t_1-t_2)\,(\zeta\xi|V|\lambda\mu)$$

$$\times \; G^{(0)}(\mu,\xi;t_2-t_2^+)G^{(0)}(\epsilon,\beta;t_1-t')$$

Fig. 8.10   Diagram V2e, representing a second-order contribution to G.

corresponding terms, drawn according to the field-theory convention (see Fig. 9.8 in [Fetter and Walecka (1971)]), may not be immediately obvious. The present convention, clearly emphasizing the static feature of the interaction V, may be helpful in visualizing the successive dressing of internal propagators that will be explored in Ch. 9.

The final two diagrams, representing second-order contributions, are shown in Figs. 8.14 and 8.15. They are labeled by $V2i$ and $V2j$ respectively, and are also related to each other by the exchange operation. Note that there are only two such terms, as can be checked explicitly by applying the relevant algebra of Wick's theorem and relabeling of dummy indices. It will be shown in Ch. 11 that the inclusion of these types of diagrams in the propagator will cross the boundary of a mean-field description. All eight terms represented by Figs. 8.6 – 8.13 (and many more) are included in this mean-field description to be discussed in detail in Ch. 10.

Third and higher-order contributions can be obtained in a similar fashion. The reader should realize however, that in practice it is almost never

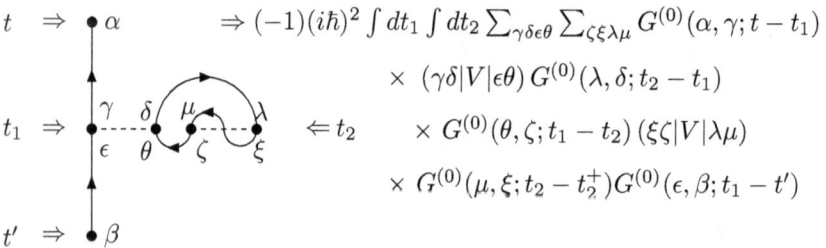

$$\Rightarrow (-1)(i\hbar)^2 \int dt_1 \int dt_2 \sum_{\gamma\delta\epsilon\theta} \sum_{\zeta\xi\lambda\mu} G^{(0)}(\alpha,\gamma;t-t_1)$$

$$\times \; (\gamma\delta|V|\epsilon\theta)\, G^{(0)}(\lambda,\delta;t_2-t_1)$$

$$\times \; G^{(0)}(\theta,\zeta;t_1-t_2)\,(\xi\zeta|V|\lambda\mu)$$

$$\times \; G^{(0)}(\mu,\xi;t_2-t_2^+)G^{(0)}(\epsilon,\beta;t_1-t')$$

Fig. 8.11   Diagram V2f, representing a second-order contribution to G.

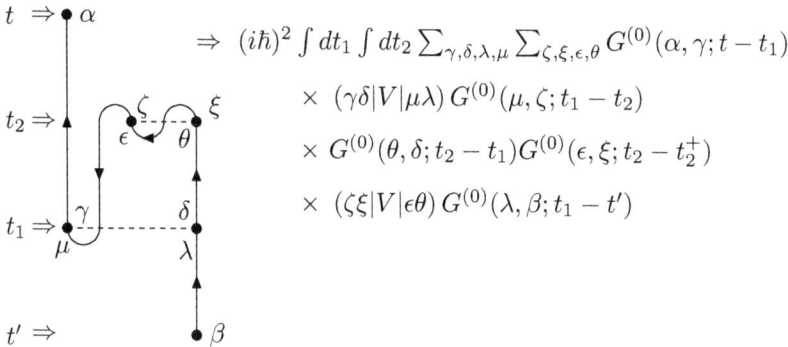

$$\Rightarrow (i\hbar)^2 \int dt_1 \int dt_2 \sum_{\gamma,\delta,\lambda,\mu} \sum_{\zeta,\xi,\epsilon,\theta} G^{(0)}(\alpha,\gamma;t-t_1)$$
$$\times (\gamma\delta|V|\mu\lambda)\, G^{(0)}(\mu,\zeta;t_1-t_2)$$
$$\times G^{(0)}(\theta,\delta;t_2-t_1)G^{(0)}(\epsilon,\xi;t_2-t_2^+)$$
$$\times (\zeta\xi|V|\epsilon\theta)\, G^{(0)}(\lambda,\beta;t_1-t')$$

Fig. 8.12   Diagram V2g, representing a second-order contribution to G.

necessary to consider such higher-order terms individually, unless they are part of infinite sums of diagrams. Nevertheless, it is useful practice to draw examples of higher-order diagrams, as will be suggested in the exercises concluding this chapter. Having learned to write down every term in the expansion of the sp propagator, it is by no means clear how to proceed making relevant approximations for a particular system under study. In fact, it will be shown in Ch. 9 that one requires infinite summations of diagrams to obtain sensible results even if the two-body interaction is quite weak.

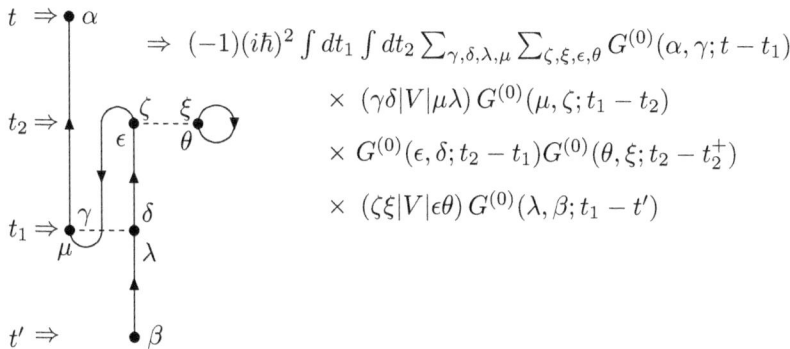

$$\Rightarrow (-1)(i\hbar)^2 \int dt_1 \int dt_2 \sum_{\gamma,\delta,\lambda,\mu} \sum_{\zeta,\xi,\epsilon,\theta} G^{(0)}(\alpha,\gamma;t-t_1)$$
$$\times (\gamma\delta|V|\mu\lambda)\, G^{(0)}(\mu,\zeta;t_1-t_2)$$
$$\times G^{(0)}(\epsilon,\delta;t_2-t_1)G^{(0)}(\theta,\xi;t_2-t_2^+)$$
$$\times (\zeta\xi|V|\epsilon\theta)\, G^{(0)}(\lambda,\beta;t_1-t')$$

Fig. 8.13   Diagram V2h, representing a second-order contribution to G.

$$t \Rightarrow \bullet \; \alpha$$

$$\Rightarrow (-1)(i\hbar)^2 \int dt_1 \int dt_2 \sum_{\gamma,\delta,\epsilon,\theta} \sum_{\zeta,\xi,\lambda,\mu} G^{(0)}(\alpha,\gamma;t-t_1)$$

$$\times (\gamma\delta|V|\epsilon\theta)\, G^{(0)}(\epsilon,\zeta;t_1-t_2)$$

$$\times G^{(0)}(\mu,\delta;t_2-t_1) G^{(0)}(\theta,\xi;t_1-t_2)$$

$$\times (\zeta\xi|V|\lambda\mu)\, G^{(0)}(\lambda,\beta;t_2-t')$$

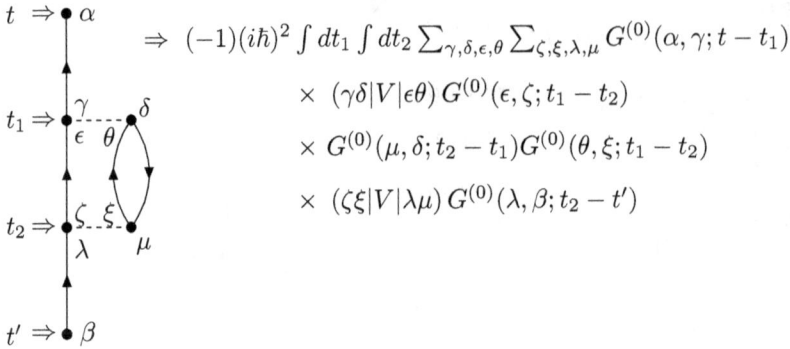

Fig. 8.14   Diagram V2i, representing a second-order contribution to G.

So far only terms involving the two-body interactions have been discussed. While these determine the general structure of the diagrams, it is sometimes necessary to introduce the auxiliary potential $\hat{U}$ as in Eq. (8.21). We then encounter additional diagrams and corresponding rules, to deal with the contributions of this one-body interaction. The inclusion of a static external potential proceeds in similar fashion. The inclusion of $\hat{U}$ does not lead to further difficulties, but requires a few additional considerations, given below. In first order, the corresponding diagram is shown in part *b*) of Fig. 8.2 and the accompanying expression is given in Eq. (8.46).

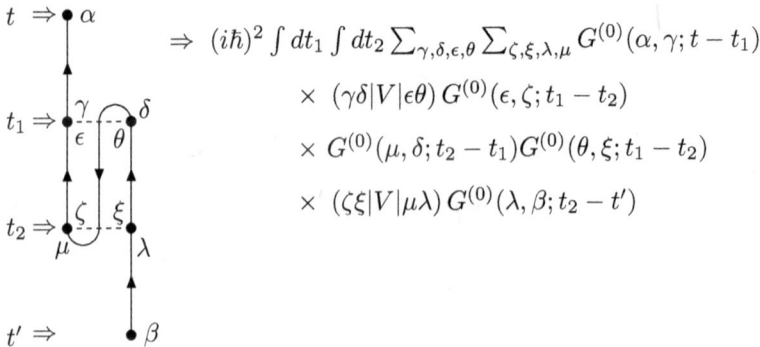

$$t \Rightarrow \bullet \; \alpha$$

$$\Rightarrow (i\hbar)^2 \int dt_1 \int dt_2 \sum_{\gamma,\delta,\epsilon,\theta} \sum_{\zeta,\xi,\lambda,\mu} G^{(0)}(\alpha,\gamma;t-t_1)$$

$$\times (\gamma\delta|V|\epsilon\theta)\, G^{(0)}(\epsilon,\zeta;t_1-t_2)$$

$$\times G^{(0)}(\mu,\delta;t_2-t_1) G^{(0)}(\theta,\xi;t_1-t_2)$$

$$\times (\zeta\xi|V|\mu\lambda)\, G^{(0)}(\lambda,\beta;t_2-t')$$

Fig. 8.15   Diagram V2j, representing a second-order contribution to G.

In second order, one encounters the contribution to Eq. (8.55) of the form

$$G^{(2)}(\alpha, \beta; t - t') = -\frac{i}{\hbar}\left(\frac{-i}{\hbar}\right)^2 \frac{1}{2!} \int dt_1 \int dt_2$$

$$\times \left\langle \Phi_0^N \middle| \mathcal{T}\left[\hat{H}_1(t_1)\hat{H}_1(t_2)a_\alpha(t)a_\beta^\dagger(t')\right] \middle| \Phi_0^N \right\rangle_{connected}. \qquad (8.57)$$

Inserting the two contributions to each $\hat{H}_1$ term requires the evaluation of the following expectation value

$$\Rightarrow \left\langle \Phi_0^N \middle| \mathcal{T}\left[\left(\hat{V}(t_1) - \hat{U}(t_1)\right)\left(\hat{V}(t_2) - \hat{U}(t_2)\right)a_\alpha(t)a_\beta^\dagger(t')\right] \middle| \Phi_0^N \right\rangle_{connected}$$

$$= \left\langle \Phi_0^N \middle| \mathcal{T}\left[\left(-\hat{U}(t_1)\right)\left(-\hat{U}(t_2)\right)a_\alpha(t)a_\beta^\dagger(t')\right] \middle| \Phi_0^N \right\rangle_{connected}$$

$$+ \left\langle \Phi_0^N \middle| \mathcal{T}\left[\hat{V}(t_1)\hat{V}(t_2)a_\alpha(t)a_\beta^\dagger(t')\right] \middle| \Phi_0^N \right\rangle_{connected}$$

$$+ 2\left\langle \Phi_0^N \middle| \mathcal{T}\left[\left(-\hat{U}(t_1)\right)\hat{V}(t_2)a_\alpha(t)a_\beta^\dagger(t')\right] \middle| \Phi_0^N \right\rangle_{connected}, \qquad (8.58)$$

where the factor of 2 in the last term originates from relabeling the dummy time integration variables in one of the $\hat{U}\hat{V}$ products. The factor of 2 arises also in the other two contributions to Eq. (8.58) resulting from the two identical terms associated with interchanging the two internal time variables t_1 and t_2. In second order, the corresponding $1/2!$ factor in Eq. (8.55) is cancelled as well, when $\hat{U}$ terms are included. This result is repeated for all higher-order terms. Two additional rules then have to be added to include these $\hat{U}$ terms. In particular, when k such $\hat{U}$ contributions appear in the diagram one has to:

Rule 6 Label each U according to

$$\Rightarrow t_i \qquad \overset{\alpha}{\underset{\beta}{\bullet\!\!\wedge\!\!\wedge\!\!\wedge\!\!\bullet}} \qquad \Rightarrow \langle\alpha| U |\beta\rangle$$

Rule 7 Include a factor $(-1)^k$ and k additional propagators $G^{(0)}$

When k $\hat{U}$ contributions appear in a diagram, it leads to k additional contractions, representing a factor $(i\hbar)^k$ since $\hat{U}$ is a one-body operator. Together with the factor $(-1)^k(-i/\hbar)^k$ this yields the factor $(-1)^k$ quoted in **Rule 7**. It is straightforward to generate the additional diagrams when these $\hat{U}$ terms are included. They will be analyzed in the next chapter, where the inclusion of $\hat{U}$ occurs quite naturally. A one-body external field

$\hat{U}_{ext}$ will generate similar terms but doesn't require the factor $(-1)^k$ in **Rule 7**. It is also wise in this case to select another symbol than the "zigzag" chosen for $\hat{U}$.

A different diagram strategy may be employed, when the relative importance of direct and exchange terms is similar. This happens in nuclear problems, for example. It is then convenient to employ the symmetrized version of the two-body interaction as in Eq. (2.46), which involves the combination of direct and exchange matrix elements of the interaction as in Eq. (2.47). This result can be obtained by considering the unsymmetrized diagrams, as studied in Figs. 8.4 and 8.5 for the first-order contribution, and all second-order terms displayed in Figs. 8.6 – 8.15. By relabeling the second term involving V in Eq. (8.48) and combining the two, the first-order term is given by Eq. (8.49). Clearly, when the antisymmetrized version of $\hat{V}$ is used, we only need to consider the direct diagram of the first-order contribution shown in Fig. 8.4. The same result is generated by applying Wick's theorem using the symmetrized version of $\hat{V}$ and applying appropriate relabeling of dummy variables. In second order, one can add the first four terms shown in Figs. 8.6 – 8.9 accordingly, by replacing in the expression for $V2a$, the two matrix elements of V by the antisymmetrized terms and only keeping Fig. 8.6. Similarly, we can collect the next four terms (Figs. 8.10 – 8.13) of the unsymmetric version and use the symmetrized matrix elements keeping only diagram $V2e$. The last two second-order diagrams shown in Figs. 8.14 and 8.15 can be combined, but require a factor of $\frac{1}{2}$ to get the correct expression. These changes lead to a modified and a new diagram rule when antisymmetrized matrix elements of the interaction are used.

Rule 1′ Draw only all topologically distinct and connected, direct
diagrams with m horizontal interaction lines for V (dashed)
and $2m + 1$ directed (using arrows) Green's functions $G^{(0)}$

Rule 8 Include a factor $\frac{1}{2}$ for each pair of equivalent lines, which
both start at the same interaction and end at another

The notion of equivalent lines introduced in **Rule 8** originates from the restriction encountered when a pair of lines start and end at the same interaction. This leads to only two diagrams of the type $V2i$ and $V2j$, whereas four diagrams occur of the type $V2a–V2d$ and $V2e–V2h$, respectively. This feature also appears in higher-order terms, resulting in the new **Rule 8** in the symmetrized version. As discussed above, certain parts of higher-order diagrams contain expressions that correspond exactly to lower-order terms. This is very helpful in resumming contributions to the perturbation expansion of the sp propagator. A systematic presentation is pursued in Ch. 9.

8.6.2 *Energy formulation*

For formal manipulations and the development of the perturbation expansion, it is useful to employ the time formulation for the propagator. However, as demonstrated in Ch. 6, for practical results it is usually preferable to apply the energy formulation. The relevant FT can be found in Eq. (7.9) leading to the important Lehmann representation of the sp propagator. A corresponding FT can then be performed on all the contributions to Eq. (8.55). It is therefore clear that a similar diagrammatic framework in the energy formulation can be constructed. For the noninteracting propagator we use Eq.(7.35) or directly Fourier transform $G^{(0)}(\alpha, \beta; t-t')$ according to

$$G^{(0)}(\alpha, \beta; E) = \int_{-\infty}^{\infty} d(t - t') \, e^{\frac{i}{\hbar}E(t-t')} G^{(0)}(\alpha, \beta; t - t')$$

$$= \delta_{\alpha,\beta} \left\{ \frac{\theta(\alpha - F)}{E - \varepsilon_\alpha + i\eta} + \frac{\theta(F - \alpha)}{E - \varepsilon_\alpha - i\eta} \right\}. \tag{8.59}$$

The results of Eq. (8.44) were generated with the integration limits $-\infty(1 - i\eta)$ and $\infty(1 - i\eta)$. Employing the integral representation of the step function, as in Eq. (7.9), already removes the unwanted contributions when the difference between the time limits approaches ∞, as discussed in [Mattuck (1992)]. For this reason we can use the integration limits given in Eq. (8.59). While it is possible to Fourier transform each contribution directly, a useful strategy is to consider the inverse transform for all of the time-dependent unperturbed propagators in every term. The inverse transform is given by

$$G^{(0)}(\alpha, \beta; \tau) = \int_{-\infty}^{\infty} \frac{dE}{2\pi\hbar} \, e^{-iE\tau/\hbar} \, G^{(0)}(\alpha, \beta; E). \tag{8.60}$$

Inserting Eq. (8.59) into Eq. (8.60) yields the proper expression for the noninteracting propagator in the time formulation [see Eq. (7.32)]. In order to obtain this outcome, one has to extend the energy integral in Eq. (8.60) to complex contour integrals in the lower half (for the particle part of the propagator) and upper half (for the hole part). Application of the residue theorem then provides the correct result. For the special case of noninteracting propagators with equal time arguments, we interpret

$$
G^{(0)}(\alpha, \beta; t - t^+) = \int_{-\infty}^{\infty} \frac{dE}{2\pi\hbar} \, e^{-iE0^-/\hbar} \, G^{(0)}(\alpha, \beta; E)
$$

$$
= \int_{C\uparrow} \frac{dE}{2\pi\hbar} \, G^{(0)}(\alpha, \beta; E), \tag{8.61}
$$

where the symbol $C \uparrow$ indicates a contour integral involving the real axis and closed in the upper half plane. In fact, the presence of the 0^- in the first line of Eq. (8.61) forces the contour to be closed in the upper half of the complex energy plane and therefore only picks up the contribution of the hole part.

With these preliminary considerations it is now possible to Fourier transform each expression for the diagrams in the time formulation, thereby obtaining corresponding ones in the energy formulation. Subsequently, replacing each unperturbed propagator by appropriate expressions, according to Eq. (8.60) or (8.61), leads to effortless time integrations. As an example, consider the FT of the first-order contribution in Eq. (8.49) and Fig. 8.4, using the symmetrized convention,

$$
G_{\hat{V}}^{(1)}(\alpha, \beta; E) = \int_{-\infty}^{\infty} d(t - t') \, e^{\frac{i}{\hbar}E(t-t')} G_{\hat{V}}^{(1)}(\alpha, \beta; t - t')
$$

$$
= -i\hbar \int_{-\infty}^{\infty} d(t - t') \, e^{\frac{i}{\hbar}E(t-t')} \int_{-\infty}^{\infty} dt_1 \sum_{\gamma\delta\epsilon\theta} \langle \gamma\delta | V | \epsilon\theta \rangle
$$

$$
\times \left\{ \int_{-\infty}^{\infty} \frac{dE_1}{2\pi\hbar} \, e^{-iE_1(t-t_1)/\hbar} \, G^{(0)}(\alpha, \gamma; E_1) \right\}
$$

$$
\times \left\{ \int_{C\uparrow} \frac{dE'}{2\pi\hbar} \, G^{(0)}(\theta, \delta; E') \right\}
$$

$$
\times \left\{ \int_{-\infty}^{\infty} \frac{dE_2}{2\pi\hbar} \, e^{-iE_2(t_1-t')/\hbar} \, G^{(0)}(\epsilon, \beta; E_2) \right\}
$$

$$
= -i \sum_{\gamma\delta\epsilon\theta} \langle \gamma\delta | V | \epsilon\theta \rangle \, G^{(0)}(\alpha, \gamma; E)
$$

$$
\times \left\{ \int_{C\uparrow} \frac{dE'}{2\pi} \, G^{(0)}(\theta, \delta; E') \right\} G^{(0)}(\epsilon, \beta; E). \tag{8.62}
$$

The last result is obtained by first performing the integration over t_1 yielding a factor $\delta(E_1 - E_2)$, then integrating over E_2, and finally, executing the

$$E \quad \bullet\, \alpha$$

$$\Rightarrow \sum_{\gamma\delta} G^{(0)}(\alpha, \gamma; E)$$

$$\times \; -i \sum_{\epsilon\theta} \langle \gamma\epsilon | V | \delta\theta \rangle \int_{C\uparrow} \frac{dE'}{2\pi} G^{(0)}(\theta, \epsilon; E')$$

$$\times \; G^{(0)}(\delta, \beta; E)$$

$$E \quad \bullet\, \beta$$

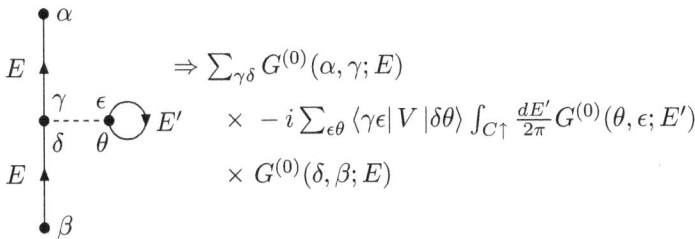

Fig. 8.16   Diagram V1DE for the symmetrized version in the energy formulation.

integration over $t-t'$. While it is clear that the structure of the diagram remains intact, there are some changes with respect to labeling. Unperturbed propagators now must be labeled by a single energy. The above example also clarifies that such labeling is consistent with energy conservation by all two-body interactions. The arrows of the propagators can then be used to represent the flow of energy in each diagram so that every interaction has the same energy coming in as flowing out. The diagram representing the first-order contribution in the energy formulation is shown in Fig. 8.16. If only diagrams with V are considered, then for an mth-order diagram we have originally m time integrations (internal times) plus the external one over $t - t'$. Each of these leads to an energy conserving δ-function. Replacing each time-dependent $G^{(0)}$ had already provided $2m + 1$ factors of $2\pi\hbar$ in the denominator. From these factors $m + 1$ are used for time integrals, leaving m independent energy integrations and a corresponding number of factors of $2\pi\hbar$ in the denominator. When U terms are included, nothing changes since for $k\,U$ terms there are k extra time integrations and propagators, and hence all factors of $2\pi\hbar$ cancel. Note that the example, illustrated above, involved a closed loop. For such loops one simply obtains an independent energy integration (to be closed in the upper half of the corresponding energy plane) which doesn't disturb the energy flow pattern.

The resulting diagram rules mimick closely those obtained for the time formulation. They are written below for the symmetrized version and illustrated up to second order by the diagrams shown in Figs. 8.16 – 8.19.

Rule 1 Draw all topologically distinct (direct) and connected diagrams with m horizontal interaction lines for V (dashed) and $2m + 1$ directed (using arrows) Green's functions $G^{(0)}$

$$\Rightarrow \sum_{\gamma\delta} G^{(0)}(\alpha,\gamma;E)$$
$$\times (-1)^2 i^2 \sum_{\epsilon,\zeta} \sum_{\lambda,\theta} \int_{C\uparrow} \frac{dE'}{2\pi} \langle \gamma\lambda | V | \epsilon\theta\rangle G^{(0)}(\theta,\lambda;E')$$
$$\times G^{(0)}(\epsilon,\zeta;E) \sum_{\xi,\mu} \int_{C\uparrow} \frac{dE''}{2\pi} \langle \zeta\xi | V | \delta\mu\rangle G^{(0)}(\mu,\xi;E'')$$
$$\times G^{(0)}(\delta,\beta;E)$$

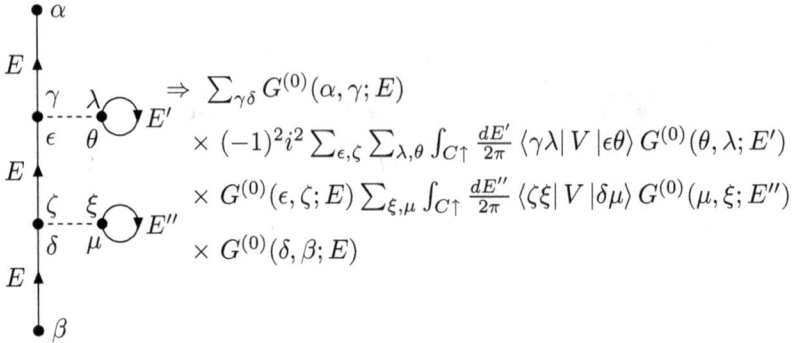

Fig. 8.17   Diagram V2a in the symmetrized version of the energy formulation.

Rule 2 Label external points only with sp quantum numbers,
   e.g. α and β
   Label each interaction with sp quantum numbers

$$\Rightarrow \langle \alpha\beta | V | \gamma\delta\rangle = (\alpha\beta|V|\gamma\delta) - (\alpha\beta|V|\delta\gamma)$$

For each arrow line one writes

$$\Rightarrow G^{(0)}(\mu,\nu;E)$$

$$\Rightarrow \sum_{\gamma\delta} G^{(0)}(\alpha,\gamma;E) \times i^2 \sum_{\epsilon\theta} \sum_{\lambda\zeta} \int_{C\uparrow} \frac{dE'}{2\pi}$$
$$\times \langle \gamma\epsilon | V | \delta\theta\rangle G^{(0)}(\lambda,\epsilon;E') G^{(0)}(\theta,\zeta;E')$$
$$\times \sum_{\mu\zeta} \int_{C\uparrow} \frac{dE''}{2\pi} \langle \zeta\xi | V | \lambda\mu\rangle G^{(0)}(\mu,\xi;E'')$$
$$\times G^{(0)}(\delta,\beta;E)$$

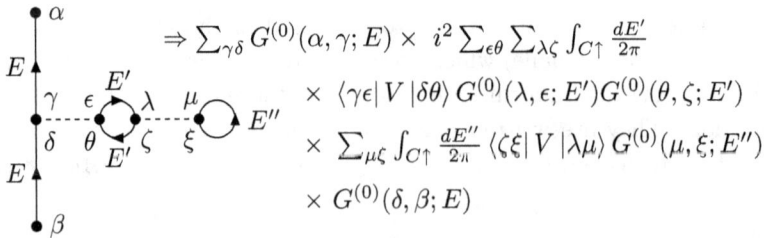

Fig. 8.18   Diagram V2e for the symmetrized case in the energy formulation.

$$\Rightarrow \sum_{\gamma\delta} G^{(0)}(\alpha,\gamma;E)$$

$$\times (-1)i^2 \tfrac{1}{2} \int \tfrac{dE_1}{2\pi} \int \tfrac{dE_2}{2\pi} \sum_{\lambda,\epsilon,\theta} \sum_{\zeta,\xi,\mu} \langle \gamma\lambda| V |\epsilon\theta\rangle$$

$$\times G^{(0)}(\epsilon,\zeta;E_1) G^{(0)}(\mu,\lambda;E_1+E_2-E)$$

$$\times G^{(0)}(\theta,\xi;E_2) \langle \zeta\xi| V |\delta\mu\rangle$$

$$\times G^{(0)}(\delta,\beta;E)$$

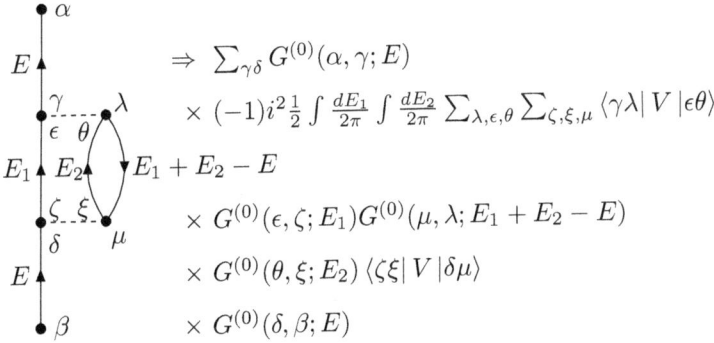

Fig. 8.19 Diagram V2i for the symmetrized version in the energy formulation.

but in such a way that energy is conserved for each V

Rule 3 Sum (integrate) over all internal sp quantum numbers and integrate over all m internal energies

For each closed loop an independent energy integration occurs over the contour $C \uparrow$

Rule 4 Include a factor $(i/2\pi)^m$ and $(-1)^F$ where F is the number of closed fermion loops

Rule 5 Include a factor of $\tfrac{1}{2}$ for each equivalent pair of lines

The diagrams shown in Figs. 8.16 – 8.19 have been purposely accompanied by expressions that emphasize their structure. In particular, it is clear that in the energy formulation, all diagrams can be written with an unperturbed propagator $G^{(0)}(\alpha,\gamma;E)$ at the top, and $G^{(0)}(\delta,\beta;E)$ at the bottom (while summing over γ and δ). The same structure of the sp propagator was encountered in Ch. 6 for the one-body problem, and will be employed in Ch. 9 to organize the perturbation expansion.

So far, only terms involving the two-body interaction are included. Additional diagrams with U terms can be easily added by noting that on account of their one-body character, the energy associated with the incoming propagator must be equal to that of the outgoing one. Similar statements (and rules) hold when an external potential is included. When in a diagram k U contributions are involved, we have to add the following rules:

Rule 6 Label each U according to

$$\Rightarrow \langle \alpha | U | \beta \rangle$$

Rule 7 Include a factor $(-1)^k$ and k additional propagators $G^{(0)}$

By now it should be clear how to generate the rules for the unsymmetrized version of the diagrams in the energy formulation.

8.7   Exercises

(1) Evaluate the expectation value of the first-order contribution to the numerator in Eq. (8.17) of the $\hat{U}$ term for the time-orderings not considered here. Construct the corresponding time-ordered diagrams for all possibilities.
(2) Evaluate the expectation value of the first-order contribution to the numerator in Eq. (8.17) for the $\hat{V}$ term for all possible time-orderings. Construct the corresponding time-ordered diagrams in all these cases.
(3) Construct all Feynman diagrams in third order, including only contributions of the two-body interaction, and write down the corresponding expressions from the diagram rules (in the unsymmetrized version).
(4) Same as the previous problem but now in the energy formulation and the symmetrized version.

Chapter 9

Dyson equation and self-consistent Green's functions

The results of Ch. 8 represent an important link between the sp propagator of a correlated system, and the two known ingredients provided by the two-body interaction $\hat{V}$ and the noninteracting ground state $\left|\Phi_0^N\right\rangle$. The latter state may require the introduction of an auxiliary one-body potential $\hat{U}$. We assume the relevant sp problem of $H_0 = T + U$ to be solvable. The lowest levels of this Hamiltonian may therefore be filled in accordance with the total number of particles and the Pauli principle. The corresponding noninteracting ground state is represented by $\left|\Phi_0^N\right\rangle$, as discussed in Chs. 3 and 5. In Ch. 8, the complete perturbation expansion of the exact propagator was established in terms of known quantities. A proper way to select contributions leading to a meaningful description was, however, not obtained.

It is the purpose of this chapter to develop an organized approach based on perturbation theory, to describe physically interesting many-particle systems. All these require a treatment that goes beyond the usual perturbation theory developed for the sp problem [Messiah (1999)]. Indeed, it is important to note that adding the first-order contribution $G_V^{(1)}$, given by Eq. (8.62), to the noninteracting propagator $G^{(0)}$, does not represent a useful approach to the problem even if the two-body interaction $\hat{V}$ would be small somehow. The reason for this inadequacy is that the resulting approximation does not have important properties that pertain to the exact propagator. For example, the sum of $G^{(0)}$ and $G_V^{(1)}$ does not have a Lehmann representation and can therefore not be interpreted as containing information describing the removal and addition probabilities of particles with respect to the ground state of the system. Also, the energies of the states with one added or removed particle cannot be extracted from such an approximation. This becomes clear when one realizes that the diagonal elements of $G^{(1)}$ have a double pole at the sp energy corresponding to the one of $G^{(0)}$, whereas the exact sp propagator has simple poles (at different energies). To obtain approximations that share such features, it is necessary to reorganize the perturbation expansion in such a way that it automatically sums infinite sets of diagrams. The relevant analysis is described in Sec. 9.1 and leads to the so-called Dyson equation with the introduction of the self-energy of a particle in the medium.

At this stage, one has a tool in hand to generate interesting descriptions of the sp propagator by making approximations to the self-energy. There is however, one more ingredient missing in this strategy. It is related to the notion that the evaluation of the self-energy, as presented in Ch. 8, involves the use of noninteracting propagators. Physically it makes more sense to let the particle, considered explicitly in the Dyson equation, interact with particles in the medium. In turn, these also experience the same correlations as one is trying to include for the particle under study. This democratic notion leads to the important concept of self-consistency between the solution of the Dyson equation and the ingredients which make up the corresponding self-energy. The concept is best developed formally by considering the equations of motion for the sp propagator as presented in Sec. 9.3. This study reveals a dynamic coupling between the sp propagator and the two-particle propagator in the medium, presented in Sec. 9.4. The perturbation expansion of the two-particle propagator can be analyzed in exactly the same way as was done for the sp propagator. This leads to the introduction of the vertex function, which can be thought of as the effective interaction between fully correlated particles in the medium, described by exact sp propagators. The results are combined at the end of Sec. 9.4 to obtain the self-energy of a particle in terms of this vertex function.

The Dyson equation can be regarded as the Schrödinger equation of a particle in the medium, subject to the self-energy as the potential. This interpretation is further developed in Sec. 9.5 and the relation with the analysis of experimental data from particle knockout experiments is emphasized (see also Ch. 7). At this point, the stage is set to study many-particle systems of interest, since all relevant ingredients like the Dyson equation and self-consistency are available. It is then possible to choose approximation schemes based on information concerning the two-body propagator in the medium. This information is provided by considering relevant experimental data, sometimes in the form of two-particle scattering results. The simplest case, involving a rather weak interaction, generates the Hartree–Fock scheme to be discussed in detail in Ch. 10. Its extension to include the next higher-order contribution is discussed in Ch. 11. Systems with stronger correlations require other schemes, involving infinite summations as relevant approximations to the effective two-body interaction in the medium. These applications will be discussed in subsequent chapters.

9.1 Analysis of perturbation expansion, self-energy, and Dyson's equation

In the last section of Ch. 8, the energy formulation of the expansion of the propagator was introduced. From the discussion in the last part of Sec. 8.6.2, one can infer that it is possible to obtain a diagrammatic representation of the propagator as shown in Fig. 9.1. It introduces the convention that the exact sp propagator is represented by two parallel, arrowed lines. The observation that any term in the perturbation expansion of G, except in zero-order, has a noninteracting propagator at the top and the bottom of the diagram, makes it possible to introduce the self-energy Σ, which represents the sum of all the intermediate contributions. The decomposition of the sp propagator in the noninteracting propagator $G^{(0)}$ and the sum of the other terms defining the self-energy, is graphically represented in Fig. 9.1. The expressions for the lowest-order contributions to the propagator in the energy formulation displayed in Figs. 8.16 – 8.19, allow for an immediate identification of the corresponding contributions to the self-energy. In Fig. 9.2 the first-order term of the self-energy is generated from Fig. 8.16 by removing the top and bottom noninteracting propagators.

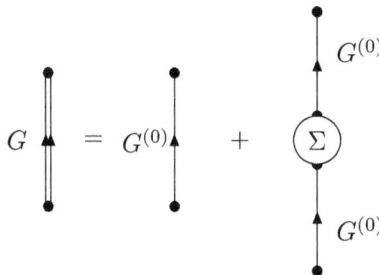

Fig. 9.1   Diagrammatic representation of the sp propagator introducing the reducible self-energy Σ.

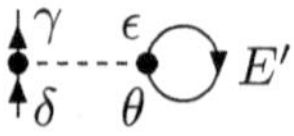

$$\gamma \quad \epsilon \atop \delta \quad \theta \;\; E' \qquad \Rightarrow \quad -i \sum_{\epsilon\theta} \langle \gamma\epsilon | \, V \, | \delta\theta \rangle \int_{C\uparrow} \tfrac{dE'}{2\pi} G^{(0)}(\theta, \epsilon; E')$$

Fig. 9.2   Diagram SE1 for the self-energy in first order.

Note that the symmetrized version of the diagram is employed and therefore both the direct and exchange contribution of V are included. In Figs. 9.3 – 9.5, similar expressions for the self-energy are obtained in second order, by applying the same procedure to the sp propagator diagrams shown in Figs. 8.17 – 8.19. In all self-energy diagrams, small arrowed lines have been included to indicate the location where the noninteracting propagators should be attached to generate the corresponding contribution to the sp propagator. In addition, the arrows act as a reminder that the energy flow is still represented by the same energy E going in and out.

This process of clipping the top and bottom noninteracting propagators can obviously be continued for all higher-order contributions, leading to an unambiguous definition of the self-energy as illustrated in Fig. 9.1. Additional terms in first and second order occur when the auxiliary sp potential $\hat{U}$ is employed. These diagrams are illustrated in Fig. 9.6. Based on the rules developed for the diagrams of the sp propagator, it is straightforward to generate the self-energy expressions given in Figs. 9.6a) – 9.6e).

It is now possible to divide the self-energy contributions, shown in Figs. 9.2 – 9.6, into two categories. The first contains terms that are called irreducible. The sum of all these contributions to the self-energy including all the higher-order terms, is denoted by Σ^*. The diagrams depicted in Figs. 9.2, 9.4, 9.5, 9.6a), and 9.6b) belong to this category. The word

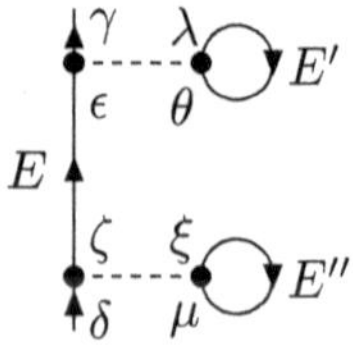

$$
\begin{aligned}
&\gamma \quad \lambda \atop \epsilon \quad \theta \;\; E' \\
E \;\; &\qquad\qquad \Rightarrow \;\; (-1)^2 i^2 \sum_{\epsilon,\zeta} \sum_{\lambda,\theta} \int_{C\uparrow} \tfrac{dE'}{2\pi} \langle \gamma\lambda | \, V \, | \epsilon\theta \rangle \, G^{(0)}(\theta, \lambda; E') \\
&\zeta \quad \xi \atop \delta \quad \mu \;\; E'' \qquad \times \; G^{(0)}(\epsilon, \zeta; E) \sum_{\xi,\mu} \int_{C\uparrow} \tfrac{dE''}{2\pi} \langle \zeta\xi | \, V \, | \delta\mu \rangle \, G^{(0)}(\mu, \xi; E'')
\end{aligned}
$$

Fig. 9.3   Diagram SE2a for the self-energy in second order.

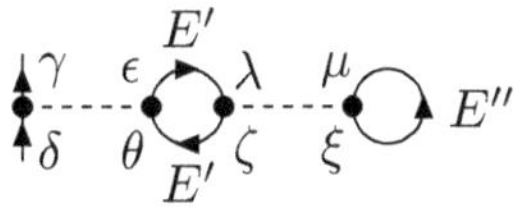

$$\Rightarrow i^2 \sum_{\epsilon\theta} \sum_{\lambda\zeta} \int_{C\uparrow} \frac{dE'}{2\pi}$$

$$\times \langle \gamma\epsilon|\, V\, |\delta\theta \rangle\, G^{(0)}(\lambda,\epsilon;E')G^{(0)}(\theta,\zeta;E')$$

$$\times \sum_{\mu\xi} \int_{C\uparrow} \frac{dE''}{2\pi} \langle \zeta\xi|\, V\, |\lambda\mu \rangle\, G^{(0)}(\mu,\xi;E'')$$

Fig. 9.4   Diagram SE2e for the self-energy in second order.

irreducible means here that such diagrams do not contain two (or more) parts that are only connected by an unperturbed sp propagator $G^{(0)}$. All other contributions to the self-energy are called reducible. Together with the irreducible ones they comprise all contributions to Σ. Analysis of the structure of the diagrams contributing to the sp propagator, makes it clear that the irreducible self-energy suffices to obtain the propagator. The corresponding diagrammatic result is shown in Fig. 9.7. The figure illustrates how successive iterations of the irreducible self-energy Σ^*, linked by the unperturbed propagator $G^{(0)}$, will generate all terms contributing to the sp propagator. The irreducible self-energy diagrams like Figs. 9.2 *etc.* contribute to the sp propagator in the term with one insertion of the irreducible self-energy on the right side of Fig. 9.7. The second-order reducible self-energy diagrams like Figs. 9.3, *etc.* appear in the next term with two irreducible self-energy insertions. Higher-order self-energy contributions distribute themselves over the terms, schematically indicated in Fig. 9.7, in a similarly unique fashion.

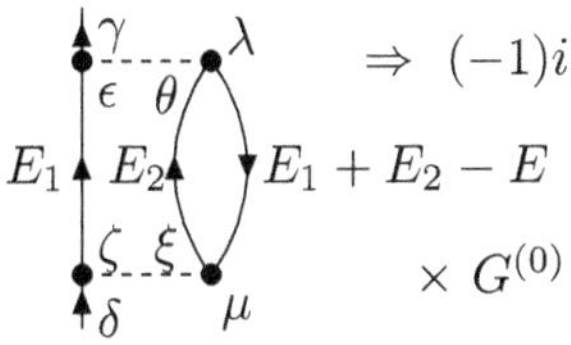

$$\Rightarrow (-1)i^2 \frac{1}{2} \int \frac{dE_1}{2\pi} \int \frac{dE_2}{2\pi} \sum_{\lambda,\epsilon,\theta} \sum_{\zeta,\xi,\mu} \langle \gamma\lambda|\, V\, |\epsilon\theta \rangle$$

$$\times G^{(0)}(\epsilon,\zeta;E_1)G^{(0)}(\mu,\lambda;E_1+E_2-E)$$

$$\times G^{(0)}(\theta,\xi;E_2)\, \langle \zeta\xi|\, V\, |\delta\mu \rangle$$

Fig. 9.5   Diagram SE2i for the self-energy in second order.

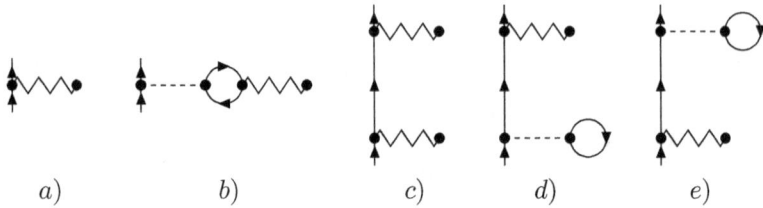

Fig. 9.6 Additional diagrams contributing to the self-energy up to second order when an auxiliary potential U is employed.

This analysis indicates that all contributions to the sp propagator can be obtained from the irreducible ones by summing the following expression:

$$
\begin{aligned}
G(\alpha, \beta; E) &= G^{(0)}(\alpha, \beta; E) \\
&+ \sum_{\gamma, \delta} G^{(0)}(\alpha, \gamma; E) \Sigma^*(\gamma, \delta; E) G^{(0)}(\delta, \beta; E) \\
&+ \sum_{\gamma, \delta, \epsilon, \theta} G^{(0)}(\alpha, \gamma; E) \Sigma^*(\gamma, \epsilon; E) G^{(0)}(\epsilon, \theta; E) \Sigma^*(\theta, \delta; E) G^{(0)}(\delta, \beta; E) \\
&+ .....
\end{aligned}
$$

(9.1)

Equation (9.1) exactly represents the diagrams shown in Fig. 9.7. In Ch. 6 such an equation was encountered for the propagator in the sp problem. Possible resummations of the corresponding Eq. (6.19) for the operator form of G were discussed there. Identical resummations will be considered here for Eq. (9.1). To visualize the resummation strategy, several lines have been drawn in Fig. 9.7. Consider first the short-dashed lines; they help identify two ways of obtaining the so-called Dyson equation for the sp propagator. Indeed, by identifying all terms below the short-dashed line with the positive slope as the sum of all contributions to the sp propagator, one may rewrite Eq. (9.1) according to

$$
G(\alpha, \beta; E) = G^{(0)}(\alpha, \beta; E) + \sum_{\gamma, \delta} G^{(0)}(\alpha, \gamma; E) \Sigma^*(\gamma, \delta; E) G(\delta, \beta; E). \quad (9.2)
$$

Alternatively, we identify all contributions above the short-dashed line with the negative slope in Fig. 9.7, with the full sp propagator, so that

$$
G(\alpha, \beta; E) = G^{(0)}(\alpha, \beta; E) + \sum_{\gamma, \delta} G(\alpha, \gamma; E) \Sigma^*(\gamma, \delta; E) G^{(0)}(\delta, \beta; E). \quad (9.3)
$$

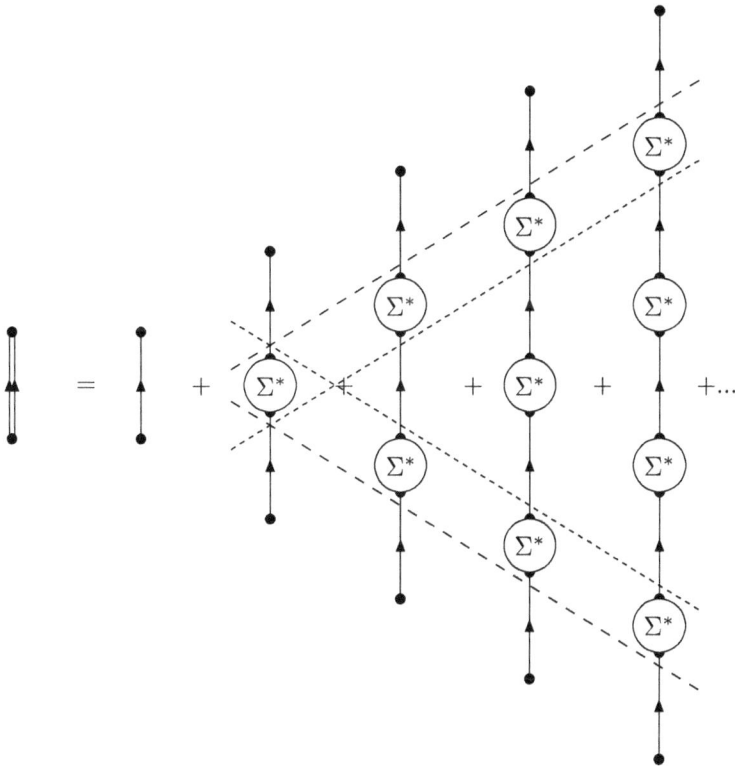

Fig. 9.7 Decomposition of the sp propagator in terms of irreducible self-energy contributions.

Equation (9.2) is illustrated diagrammatically in Fig. 9.8. A similar diagrammatic representation can be obtained for Eq. (9.3) by interchanging $G^{(0)}$ and G in the second term on the right side in Fig. 9.8. As in the sp problem, the infinite summation form of Eqs. (9.2) and (9.3) makes it possible to construct eigenvalue equations with discrete solutions for the energy. The nonperturbative aspect of the Dyson equation generates approximate solutions to the sp propagator which can be interpreted in the same way as the exact one. This includes the presence of simple poles at the approximate energies of states with one particle more or less than the ground state (with respect to the approximate energy of the ground state). The numerator of this approximate propagator then contains corresponding approximate addition and removal amplitudes, as the exact propagator (see the discussion in Ch. 7).

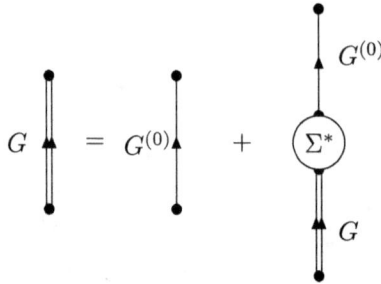

Fig. 9.8   Diagrammatic representation of the sp propagator in terms of the irreducible self-energy Σ^* and the noninteracting propagator $G^{(0)}$ representing Eq. (9.2).

Comparing the reducible self-energy, represented in Fig. 9.1, with the expansion shown in Fig. 9.7, clarifies that it is the sum of all terms inside the boundaries of the two long-dashed lines in that figure. A similar result was obtained in Ch. 6 for the $\mathcal{T}$-matrix in the sp case. This identification leads to

$$\Sigma(\gamma, \delta; E) = \Sigma^*(\gamma, \delta; E)$$
$$+ \sum_{\epsilon, \theta} \Sigma^*(\gamma, \epsilon; E) G^{(0)}(\epsilon, \theta; E) \Sigma^*(\theta, \delta; E)$$
$$+ \sum_{\epsilon, \theta, \zeta, \xi} \Sigma^*(\gamma, \epsilon; E) G^{(0)}(\epsilon, \theta; E) \Sigma^*(\theta, \zeta; E) G^{(0)}(\zeta, \xi; E) \Sigma^*(\xi, \delta; E)$$
$$+ \dots \tag{9.4}$$

Equation (9.4) can also be summed in two ways, reminiscent of the Dyson equation with its two equivalent forms given by Eqs. (9.2) and (9.3). Making use of the symmetry of Fig. 9.7, we find

$$\Sigma(\gamma, \delta; E) = \Sigma^*(\gamma, \delta; E) + \sum_{\epsilon, \theta} \Sigma^*(\gamma, \epsilon; E) G^{(0)}(\epsilon, \theta; E) \Sigma(\theta, \delta; E) \tag{9.5}$$

or

$$\Sigma(\gamma, \delta; E) = \Sigma^*(\gamma, \delta; E) + \sum_{\epsilon, \theta} \Sigma(\gamma, \epsilon; E) G^{(0)}(\epsilon, \theta; E) \Sigma^*(\theta, \delta; E). \tag{9.6}$$

They are the equivalent of the Lippmann–Schwinger equation for the $\mathcal{T}$-matrix in the case of a sp problem [see Eqs. (6.20) and (6.21)] and can therefore be used for problems involving a continuous spectrum. The irreducible self-energy Σ^* thus plays a similar role to the potential in the sp problem. Note however, that in the many-particle problem, the influence of the medium leads to an energy-dependent complex potential represented by the irreducible self-energy.

9.2 Equation of motion method for propagators

The present analysis of the diagrammatic expansion introduces the concept of the self-energy, but it does not yet provide a clear strategy for making approximations that realistically describe the correlations present in the system. The importance of the Dyson equation is related to the infinite summation it represents. The latter allows for results that are not possible to obtain using order by order summation of perturbation contributions. An example is provided by the possibility of generating bound states from a noninteracting propagator representing a continuous spectrum.

An algebraic method for deriving the Dyson equation also exists [Abrikosov *et al.* (1975)]. It gives a better insight into the possible strategies available for dealing with the most important correlations in the system and subsequently, taking those correlations into account in the self-energy. The approach starts with the equation of motion for the sp propagator. It requires a return to the time formulation. The study of the time derivative of the sp propagator is facilitated by considering the corresponding derivatives of the addition and removal operator in the Heisenberg picture [see Eq. (A.38)]. For the removal operator for example, we find

$$i\hbar\frac{\partial}{\partial t}a_{\alpha_H}(t) = \left[a_{\alpha_H}(t), \hat{H}\right] = \exp\{i\hat{H}t/\hbar\}\left[a_\alpha, \hat{H}\right]\exp\{-i\hat{H}t/\hbar\}. \quad (9.7)$$

The Hamiltonian $\hat{H}$ includes the auxiliary potential in $\hat{H}_0$ and will therefore be decomposed according to

$$\hat{H} = \hat{H}_0 - \hat{U} + \hat{V}. \quad (9.8)$$

Using the sp basis that diagonalizes H_0, one has

$$\hat{H}_0 = \sum_\gamma \varepsilon_\gamma a_\gamma^\dagger a_\gamma. \quad (9.9)$$

The three commutators required for Eq. (9.7) then yield

$$\left[a_\alpha, \hat{H}_0\right] = \varepsilon_\alpha a_\alpha, \quad (9.10)$$

$$\left[a_\alpha, \hat{U}\right] = \sum_\delta \langle\alpha|\,U\,|\delta\rangle\, a_\delta \tag{9.11}$$

using the conjugate of Eq. (2.34), and

$$\left[a_\alpha, \hat{V}\right] = \frac{1}{2}\sum_{\delta\zeta\theta} \langle\alpha\delta|\,V\,|\theta\zeta\rangle\, a_\delta^\dagger a_\zeta a_\theta \tag{9.12}$$

applying the conjugate of Eq. (2.42) for the symmetrized version of $\hat{V}$ given in Eq. (2.46). Inserting the results of Eqs. (9.10) – (9.12) in Eq. (9.7) yields

$$i\hbar\frac{\partial}{\partial t}a_{\alpha_H}(t) = \varepsilon_\alpha a_{\alpha_H}(t) - \sum_\delta \langle\alpha|\,U\,|\delta\rangle\, a_{\delta_H}(t)$$

$$+ \frac{1}{2}\sum_{\delta\zeta\theta} \langle\alpha\delta|\,V\,|\theta\zeta\rangle\, a_{\delta_H}^\dagger(t)a_{\zeta_H}(t)a_{\theta_H}(t). \tag{9.13}$$

It is now possible with the help of Eq. (9.13) to establish the time derivative of the sp propagator, but first we use the step function decomposition of the time-ordering operation to write

$$i\hbar\frac{\partial}{\partial t}G(\alpha,\beta;t-t') = \frac{\partial}{\partial t}\langle\Psi_0^N|\,T[a_{\alpha_H}(t)a_{\beta_H}^\dagger(t')]\,|\Psi_0^N\rangle \tag{9.14}$$

$$= \langle\Psi_0^N|\,\frac{\partial}{\partial t}\left\{\theta(t-t')a_{\alpha_H}(t)a_{\beta_H}^\dagger(t') - \theta(t'-t)a_{\beta_H}^\dagger(t')a_{\alpha_H}(t)\right\}|\Psi_0^N\rangle.$$

Evaluating all the time derivatives contributing to Eq. (9.14), and substituting Eq. (9.13), one finds

$$i\hbar\frac{\partial}{\partial t}G(\alpha,\beta;t-t') = \delta(t-t')\delta_{\alpha,\beta} + \langle\Psi_0^N|\,T[\frac{\partial a_{\alpha_H}(t)}{\partial t}a_{\beta_H}^\dagger(t')]\,|\Psi_0^N\rangle \tag{9.15}$$

$$= \delta(t-t')\delta_{\alpha,\beta} + \varepsilon_\alpha G(\alpha,\beta;t-t') - \sum_\delta \langle\alpha|\,U\,|\delta\rangle\,G(\delta,\beta;t-t')$$

$$+ \frac{-i}{2\hbar}\sum_{\delta\zeta\theta} \langle\alpha\delta|\,V\,|\theta\zeta\rangle\,\langle\Psi_0^N|\,T[a_{\delta_H}^\dagger(t)a_{\zeta_H}(t)a_{\theta_H}(t)a_{\beta_H}^\dagger(t')]\,|\Psi_0^N\rangle.$$

Equation (9.15) represents the first step of a hierarchy in which the $N+1$-particle propagator is related to the N-particle propagator [Martin and Schwinger (1959); Migdal (1967)]. In the present example, the coupling is established between the sp and the two-particle propagator, contained in the last line of Eq. (9.15). This two-particle propagator is in turn related to the three-particle propagator, *etc.* Before continuing the construction of the irreducible self-energy, it is necessary to analyze the diagrammatic content of the two-particle propagator, as developed in the next section.

9.3 Two-particle propagator, vertex function, and self-energy

The two-particle (tp) propagator is defined in analogy with the sp propagator [Eq. (7.1)] and given by

$$G_{II}(\alpha t_\alpha, \beta t_\beta, \gamma t_\gamma, \delta t_\delta) = -\frac{i}{\hbar} \langle \Psi_0^N | \mathcal{T}[a_{\beta_H}(t_\beta)a_{\alpha_H}(t_\alpha)a_{\gamma_H}^\dagger(t_\gamma)a_{\delta_H}^\dagger(t_\delta)] | \Psi_0^N \rangle. \tag{9.16}$$

The steps taken for the sp propagator, leading to Eq. (8.55), may now be repeated for the tp propagator. In attaining this expression for the tp propagator, the Heisenberg picture addition and removal operators have to be replaced by corresponding interaction picture operators. The resulting expectation value is taken with respect to the noninteracting ground state $|\Phi_0^N\rangle$. We then apply Wick's theorem to the equivalent of Eq. (8.17), for every term in the perturbation expansion. This again reveals a cancellation between the numerator and the denominator, leading to a corresponding set of connected contributions (diagrams). The result may be written as

$$G_{II}(\alpha t_\alpha, \beta t_\beta, \gamma t_\gamma, \delta t_\delta) = -\frac{i}{\hbar} \sum_m^\infty \left(\frac{-i}{\hbar}\right)^m \frac{1}{m!} \int dt_1 .. \int dt_m \tag{9.17}$$

$$\times \langle \Phi_0^N | \mathcal{T}\left[\hat{H}_1(t_1)..\hat{H}_1(t_m)a_\beta(t_\beta)a_\alpha(t_\alpha)a_\gamma^\dagger(t_\gamma)a_\delta^\dagger(t_\delta)\right] | \Phi_0^N \rangle_{connected}.$$

and is indeed the equivalent of Eq. (8.55). The details of the intermediate steps require only minor changes and will be left to the reader. The notion of connected diagrams in the context of Eq. (9.17) requires a little clarification that follows.

In zero order, one obtains the noninteracting tp propagator

$$G_{II}^{(0)}(\alpha t_\alpha, \beta t_\beta, \gamma t_\gamma, \delta t_\delta) = -\frac{i}{\hbar} \langle \Phi_0^N | \mathcal{T}[a_\beta(t_\beta)a_\alpha(t_\alpha)a_\gamma^\dagger(t_\gamma)a_\delta^\dagger(t_\delta)] | \Phi_0^N \rangle$$

$$= i\hbar \left[G^{(0)}(\alpha, \gamma; t_\alpha - t_\gamma)G^{(0)}(\beta, \delta; t_\beta - t_\delta)\right.$$

$$\left. - G^{(0)}(\alpha, \delta; t_\alpha - t_\delta)G^{(0)}(\beta, \gamma; t_\beta - t_\gamma)\right]. \tag{9.18}$$

This combination of unperturbed sp propagators is shown diagrammatically in Fig. 9.9. Also here, no time-ordering is assumed since we are dealing again with Feynman diagrams. Clearly, "disconnected" should not apply to the two noninteracting propagators shown in Fig. 9.9. Similarly, higher-order contributions, which have attachments to these lines, but do not link them, are still "connected" as long as there are no other disconnected parts.

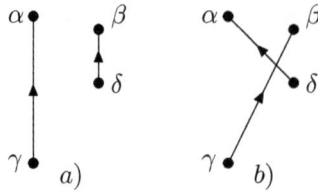

Fig. 9.9 The two contributions to the noninteracting tp propagator in the time formulation as given by Eq. (9.18).

For the analysis of the two-particle propagator in higher order, it is useful to rewrite the $\hat{V}$ contribution in $\hat{H}_1$ in a more general way. It allows for a generalization of $\hat{V}$ that includes these higher-order corrections. By including additional time integrals, one can rewrite the interaction picture $\hat{V}$ as follows

$$\hat{V}(t_1) = \frac{1}{4} \sum_{\alpha\beta\gamma\delta} \langle\alpha\beta| V |\gamma\delta\rangle \, a_\alpha^\dagger(t_1) a_\beta^\dagger(t_1) a_\delta(t_1) a_\gamma(t_1)$$

$$= \int dt_2 \int dt_3 \int dt_4 \frac{1}{4} \sum_{\alpha\beta\gamma\delta} \langle\alpha\beta| V(t_1, t_2, t_3, t_4) |\gamma\delta\rangle$$

$$\times a_\alpha^\dagger(t_1) a_\beta^\dagger(t_2) a_\delta(t_4) a_\gamma(t_3), \tag{9.19}$$

where

$$\langle\alpha\beta| V(t_1, t_2, t_3, t_4) |\gamma\delta\rangle = \delta(t_1 - t_2)\delta(t_2 - t_3)\delta(t_3 - t_4) \langle\alpha\beta| V |\gamma\delta\rangle. \tag{9.20}$$

The analysis of the first and higher-order contributions to Eq. (9.17) can now proceed. Using the formulation of $\hat{V}$ given in Eq. (9.19), the corresponding first-order contribution yields

$$G_{II}^{(1)}(\alpha t_\alpha, \beta t_\beta, \gamma t_\gamma, \delta t_\delta) \Rightarrow$$

$$\left(\frac{-i}{\hbar}\right)^2 \int dt_\epsilon \int dt_\zeta \int dt_\eta \int dt_\theta \frac{1}{4} \sum_{\epsilon\zeta\eta\theta} \langle\epsilon\zeta| V(t_\epsilon, t_\zeta, t_\eta, t_\theta) |\eta\theta\rangle$$

$$\langle\Phi_0^N| \mathcal{T} \left[a_\epsilon^\dagger(t_\epsilon) a_\zeta^\dagger(t_\zeta) a_\theta(t_\theta) a_\eta(t_\eta) a_\beta(t_\beta) a_\alpha(t_\alpha) a_\gamma^\dagger(t_\gamma) a_\delta^\dagger(t_\delta) \right] |\Phi_0^N\rangle$$

$$= (i\hbar)^2 \int dt_\epsilon \int dt_\zeta \int dt_\eta \int dt_\theta \sum_{\epsilon\zeta\eta\theta} \langle\epsilon\zeta| V(t_\epsilon, t_\zeta, t_\eta, t_\theta) |\eta\theta\rangle$$

$$\times G^{(0)}(\alpha, \epsilon; t_\alpha - t_\epsilon) G^{(0)}(\beta, \zeta; t_\beta - t_\zeta)$$

$$\times G^{(0)}(\eta, \gamma; t_\eta - t_\gamma) G^{(0)}(\theta, \delta; t_\theta - t_\delta). \tag{9.21}$$

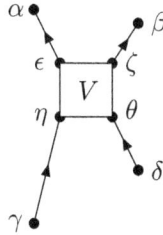

Fig. 9.10   First-order connected contribution to the tp propagator, linking two noninteracting propagators in the time formulation.

In establishing this result, Wick's theorem and the symmetry of $\hat{V}$ was used while only the connected contributions were kept. In addition, those contributions that link the interaction to one of the sp propagators, (corresponding to a self-energy insertion) have been suppressed. Diagrammatically, one may replace the dashed line for $\hat{V}$ by a box to represent the additional time arguments, thus anticipating the subsequent discussion of higher-order terms. Such a diagrammatic representation of Eq. (9.21) is given in Fig. 9.10.

In the analysis of the sp propagator two types of diagrams were encountered. The first kind contained the diagram representing $G^{(0)}$. The second contained all other connected diagrams involving higher-order self-energy insertions, as illustrated in Fig. 9.1. The tp propagator also contains two types of diagrammatic contributions. The first group includes the diagram with two noninteracting sp propagators shown in Fig. 9.9. In higher order, additional terms are generated, which contribute to the same group. These terms insert all possible self-energy corrections to these noninteracting propagators, but never link the two. The sum of all these contributions generalizes the noninteracting propagators in Fig. 9.9 to exact ones. The extension of Fig. 9.9 is shown in part a) of Fig. 9.11. Note that the dressing will include both the generalization of part a) and b) of Fig. 9.9.

The other group of diagrams in higher order, generalize the first-order contribution shown in Fig. 9.10. Each of the four noninteracting propagators will receive all possible self-energy insertions, turning them into exact sp propagators. This however, is not the only possible extension of Fig. 9.10. In addition to dressing the propagators, more complicated connections appear, which link the incoming two propagators, with the two outgoing ones. Examples of such generalizations are shown in Fig. 9.12.

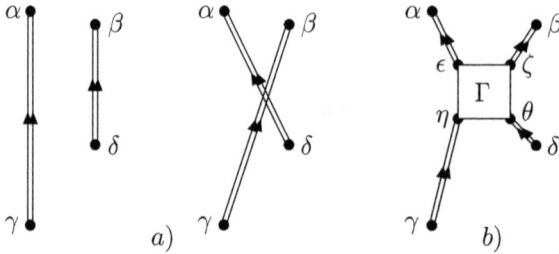

Fig. 9.11   Two contributions to the exact tp propagator in the time formulation. In part $a)$ the dressed, but noninteracting, tp propagator is shown including both direct and exchange contributions. In part $b)$ the four-point vertex function Γ is introduced to represent the sum of all higher-order contributions, generalizing Fig. 9.10.

In this figure the usual dashed line for the interaction V has been used to emphasize the actual time structure of these diagrams. Also, no additional insertions were included in the four external sp propagators. In the diagrams shown in Figs. 9.12$a)$ and 9.12$b)$, the interaction between the two incoming and two outgoing sp propagators, is characterized by two times, whereas the corresponding interaction in Fig. 9.12$c)$ has four different times. The latter term is an example illustrating the necessity to generalize V to a four-point vertex function Γ, when higher-order contributions are taken into account. The four-point vertex function includes all possible terms that connect the two incoming lines with the two outgoing ones. All intermediate sp propagators will correspondingly become fully dressed, as well as the four external ones in the diagrams shown in Fig. 9.12. The same holds for all other higher-order contributions.

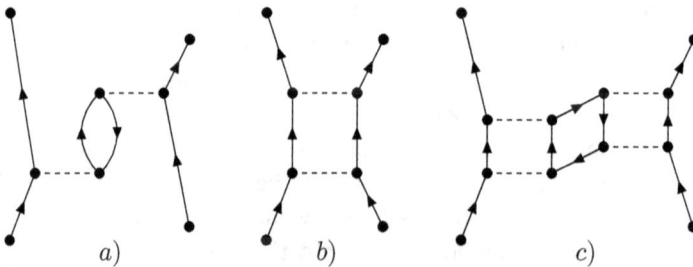

Fig. 9.12   Higher-order connected contributions to the tp propagator which generalize the first-order term in Fig. 9.10, to the four-point vertex function.

By replacing the unperturbed sp propagators with dressed ones, and replacing V by the sum of all diagrams that connect these particle lines, represented by the box labeled Γ in Fig. 9.11, one obtains the other group of contributions to G_{II} [Abrikosov *et al.* (1975)]. Γ is referred to as the four-point vertex function, since it has four external points. This quantity can be considered as the effective interaction between dressed particles in the medium. Often experimental information is available about some of the features of Γ, helping to devise approximation schemes. It is now possible to summarize the discussion by writing G_{II} in terms of dressed sp propagators and Γ as follows

$$G_{II}(\alpha t_\alpha, \beta t_\beta, \gamma t_\gamma, \delta t_\delta)$$
$$= i\hbar[G(\alpha, \gamma; t_\alpha - t_\gamma)G(\beta, \delta; t_\beta - t_\delta) - G(\alpha, \delta; t_\alpha - t_\delta)G(\beta, \gamma; t_\beta - t_\gamma)]$$
$$+ (i\hbar)^2 \int dt_\epsilon \int dt_\zeta \int dt_\eta \int dt_\theta \sum_{\epsilon\zeta\eta\theta} G(\alpha, \epsilon; t_\alpha - t_\epsilon)G(\beta, \zeta; t_\beta - t_\zeta)$$
$$\times \langle \epsilon\zeta | \Gamma(t_\epsilon, t_\zeta, t_\eta, t_\theta) | \eta\theta \rangle \, G(\eta, \gamma; t_\eta - t_\gamma)G(\theta, \delta; t_\theta - t_\delta). \tag{9.22}$$

The result is shown diagrammatically in Fig. 9.11. It is useful to transform Eq. (9.22) to the energy formulation for later applications. The first two terms give the independent, or free, propagation of pairs of particles, described by exact sp propagators, yielding

$$G_{II}^f(\alpha E_\alpha, \beta E_\beta, \gamma E_\gamma, \delta E_\delta)$$
$$= \int dt_\alpha \int dt_\beta \int dt_\gamma \int dt_\delta \; e^{\frac{i}{\hbar}E_\alpha t_\alpha} e^{\frac{i}{\hbar}E_\beta t_\beta} e^{-\frac{i}{\hbar}E_\gamma t_\gamma} e^{-\frac{i}{\hbar}E_\delta t_\delta}$$
$$\times i\hbar \; [G(\alpha, \gamma; t_\alpha - t_\gamma)G(\beta, \delta; t_\beta - t_\delta) - G(\alpha, \delta; t_\alpha - t_\delta)G(\beta, \gamma; t_\beta - t_\gamma)]$$
$$= 2\pi\hbar \; \delta(E_\alpha + E_\beta - E_\gamma - E_\delta) \; i\hbar \; [2\pi\hbar \; \delta(E_\alpha - E_\gamma) \; G(\alpha, \gamma; E_\alpha) \; G(\beta, \delta; E_\beta)$$
$$- 2\pi\hbar \; \delta(E_\beta - E_\gamma) \; G(\alpha, \delta; E_\alpha) \; G(\beta, \gamma; E_\beta)]. \tag{9.23}$$

The energy conserving δ-function is put up front, since it will also appear for the FT of the remaining term in Eq. (9.22). To perform the latter transform, it is helpful to note that the transform of Eq. (9.20) is given by

$$\langle \alpha\beta | V(E_\alpha, E_\beta, E_\gamma, E_\delta) | \gamma\delta \rangle = \int dt_1 \int dt_2 \int dt_3 \int dt_4$$
$$\times e^{\frac{i}{\hbar}E_\alpha t_1} e^{\frac{i}{\hbar}E_\beta t_2} e^{-\frac{i}{\hbar}E_\gamma t_3} e^{-\frac{i}{\hbar}E_\delta t_4} \langle \alpha\beta | V(t_1, t_2, t_3, t_4) | \gamma\delta \rangle$$
$$= 2\pi\hbar \; \delta(E_\alpha + E_\beta - E_\gamma - E_\delta) \langle \alpha\beta | V | \gamma\delta \rangle. \tag{9.24}$$

Since the vertex function conserves the energy too, the same prefactor and

δ-function can be factored out, leading to the definition

$$\langle\alpha\beta|\,\Gamma(E_\alpha, E_\beta, E_\gamma, E_\delta)\,|\gamma\delta\rangle \tag{9.25}$$
$$\equiv 2\pi\hbar\,\delta(E_\alpha + E_\beta - E_\gamma - E_\delta)\,\langle\alpha\beta|\,\Gamma(E_\alpha, E_\beta, E_\gamma, E_\alpha + E_\beta - E_\gamma)\,|\gamma\delta\rangle\,.$$

Using Eq. (9.25), it is straightforward to perform the FT of the term in Eq. (9.22) containing the vertex function, so that one finally arrives at

$$
\begin{aligned}
G_{II}(\alpha E_\alpha, \beta E_\beta, \gamma E_\gamma, \delta E_\delta) &= \int dt_\alpha \int dt_\beta \int dt_\gamma \int dt_\delta \\
&\quad \times e^{\frac{i}{\hbar}E_\alpha t_\alpha} e^{\frac{i}{\hbar}E_\beta t_\beta} e^{-\frac{i}{\hbar}E_\gamma t_\gamma} e^{-\frac{i}{\hbar}E_\delta t_\delta}\, G_{II}(\alpha t_\alpha, \beta t_\beta, \gamma t_\gamma, \delta t_\delta) \\
&= 2\pi\hbar\,\delta(E_\alpha + E_\beta - E_\gamma - E_\delta) \\
&\quad \times \big\{ i\hbar\,2\pi\hbar\,[\delta(E_\alpha - E_\gamma)\,G(\alpha, \gamma; E_\alpha)\,G(\beta, \delta; E_\beta) \\
&\quad - \delta(E_\beta - E_\gamma)\,G(\alpha, \delta; E_\alpha)\,G(\beta, \gamma; E_\beta)] \\
&\quad + (i\hbar)^2 \sum_{\epsilon\zeta\eta\theta} G(\alpha, \epsilon; E_\alpha)\,G(\beta, \zeta; E_\beta)G(\theta, \delta; E_\alpha + E_\beta - E_\gamma) \\
&\quad \times \langle\epsilon\zeta|\,\Gamma(E_\alpha, E_\beta, E_\gamma, E_\alpha + E_\beta - E_\gamma)\,|\eta\theta\rangle\,\,G(\eta, \gamma; E_\gamma)\,\big\}. \tag{9.26}
\end{aligned}
$$

This result will be used in the analysis of Sec. 14.6.

9.4   Dyson equation and the vertex function

We now come back to Eq. (9.15) to complete the analysis of the equation of motion for the propagator G, leading to an important relation between the vertex function Γ and the irreducible self-energy Σ^*. The expression for the tp propagator in Eq. (9.22) can be inserted into Eq. (9.15), yielding

$$
\begin{aligned}
i\hbar\frac{\partial}{\partial t}G(\alpha, \beta; t - t') &= \delta(t - t')\delta_{\alpha\beta} + \varepsilon_\alpha G(\alpha, \beta; t - t') \\
&\quad - \sum_{\delta} \langle\alpha|\,U\,|\delta\rangle\,G(\delta, \beta; t - t') \\
&\quad - i\hbar \sum_{\delta\zeta\theta} \langle\alpha\delta|\,V\,|\theta\zeta\rangle\,G(\zeta, \delta; t - t^+)G(\theta, \beta; t - t') \\
&\quad - \frac{1}{2}(i\hbar)^2 \sum_{\delta\zeta\theta} \sum_{\kappa\lambda\mu\nu} \int dt_\kappa \int dt_\lambda \int dt_\mu \int dt_\nu\, \langle\alpha\delta|\,V\,|\theta\zeta\rangle \\
&\quad \times G(\theta, \kappa; t - t_\kappa)G(\zeta, \lambda; t - t_\lambda)G(\nu, \delta; t_\nu - t) \\
&\quad \times \langle\kappa\lambda|\,\Gamma(t_\kappa, t_\lambda, t_\mu, t_\nu)\,|\mu\nu\rangle\,G(\mu, \beta; t_\mu - t'), \tag{9.27}
\end{aligned}
$$

using the symmetry of V and Γ under exchange. We return to the energy formulation, to show that Eq. (9.27) is equivalent to the Dyson equation. It is then useful to consider the inverse FT of all the contributions to Eq. (9.27) separately. First the time derivative of G can be written as

$$i\hbar \frac{\partial}{\partial t} G(\alpha, \beta; t - t') = \int \frac{dE}{2\pi\hbar} e^{-\frac{i}{\hbar}E(t-t')} \{E\, G(\alpha, \beta; E)\}, \qquad (9.28)$$

employing Eq. (8.60). The term with the δ-function yields

$$\delta(t - t')\delta_{\alpha,\beta} = \int \frac{dE}{2\pi\hbar} e^{-\frac{i}{\hbar}E(t-t')} \{\delta_{\alpha,\beta}\}. \qquad (9.29)$$

Continuing with the next two contributions, one finds

$$\varepsilon_\alpha G(\alpha, \beta; t - t') - \sum_\delta \langle \alpha | U | \delta \rangle\, G(\delta, \beta; t - t') \qquad (9.30)$$

$$= \int \frac{dE}{2\pi\hbar} e^{-\frac{i}{\hbar}E(t-t')} \left\{ \varepsilon_\alpha G(\alpha, \beta; E) - \sum_\delta \langle \alpha | U | \delta \rangle\, G(\delta, \beta; E) \right\}.$$

The first term with a two-body interaction reads

$$-i\hbar \sum_{\delta\zeta\theta} \langle \alpha\delta | V | \theta\zeta \rangle\, G(\zeta, \delta; t - t^+) G(\theta, \beta; t - t') \qquad (9.31)$$

$$= \int \frac{dE}{2\pi\hbar} e^{-\frac{i}{\hbar}E(t-t')} \left\{ -i \sum_{\delta\zeta\theta} \langle \alpha\delta | V | \theta\zeta \rangle \int_{C\uparrow} \frac{dE'}{2\pi} G(\zeta, \delta; E') G(\theta, \beta; E) \right\},$$

where Eq. (8.61) has been used for the sp propagator with the equal time arguments. The last term in Eq. (9.27) can be transformed according to

$$-\frac{1}{2}(i\hbar)^2 \sum_{\delta\zeta\theta} \sum_{\kappa\lambda\mu\nu} \int dt_\kappa \int dt_\lambda \int dt_\mu \int dt_\nu\, \langle \alpha\delta | V | \theta\zeta \rangle$$

$$\times\, G(\theta, \kappa; t - t_\kappa) G(\zeta, \lambda; t - t_\lambda) G(\nu, \delta; t_\nu - t)$$

$$\times\, \langle \kappa\lambda | \Gamma(t_\kappa, t_\lambda, t_\mu, t_\nu) | \mu\nu \rangle\, G(\mu, \beta; t_\mu - t')$$

$$= \int \frac{dE}{2\pi\hbar} e^{-\frac{i}{\hbar}E(t-t')} \left\{ \frac{1}{2} \sum_{\delta\zeta\theta} \sum_{\kappa\lambda\mu\nu} \langle \alpha\delta | V | \theta\zeta \rangle \right.$$

$$\times \int \frac{dE_1}{2\pi} \int \frac{dE_2}{2\pi} G(\theta, \kappa; E_1) G(\zeta, \lambda; E_2) G(\nu, \delta; E_1 + E_2 - E)$$

$$\left. \times\, \langle \kappa\lambda | \Gamma(E_1, E_2, E, E_1 + E_2 - E) | \mu\nu \rangle\, G(\mu, \beta; E) \right\}, \qquad (9.32)$$

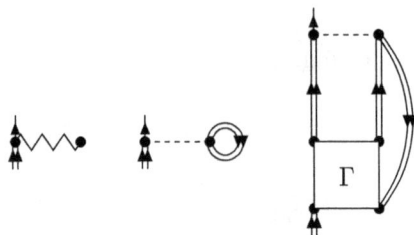

Fig. 9.13   Diagrams representing the irreducible self-energy as given by Eq. (9.34).

where Eq. (9.25) has been used. The combination of Eqs. (9.28) - (9.32) demonstrates that Eq. (9.27) can be written as the inverse FT of an expression, where several factors multiply $G(\alpha,\beta;E)$. Adding these factors and dividing this expression by the sum, one arrives at the inverse FT after some minor relabeling of dummy indices

$$G(\alpha,\beta;E) = G^{(0)}(\alpha,\beta;E) + \sum_{\gamma,\delta} G^{(0)}(\alpha,\gamma;E)\Sigma^*(\gamma,\delta;E)G(\delta,\beta;E). \quad (9.33)$$

This result obviously is identical with the Dyson equation, when one identifies the irreducible self-energy with

$$\Sigma^*(\gamma,\delta;E) = -\langle\gamma|U|\delta\rangle - i\int_{C\uparrow}\frac{dE'}{2\pi}\sum_{\mu,\nu}\langle\gamma\mu|V|\delta\nu\rangle G(\nu,\mu;E')$$

$$+\frac{1}{2}\int\frac{dE_1}{2\pi}\int\frac{dE_2}{2\pi}\sum_{\epsilon,\mu,\nu,\zeta,\rho,\sigma}\langle\gamma\mu|V|\epsilon\nu\rangle G(\epsilon,\zeta;E_1)G(\nu,\rho;E_2)$$

$$\times G(\sigma,\mu;E_1+E_2-E)\langle\zeta\rho|\Gamma(E_1,E_2;E,E_1+E_2-E)|\delta\sigma\rangle. \quad (9.34)$$

Equation (9.34) is diagrammatically shown in Fig. 9.13. In this figure the incoming line at the bottom of each self-energy diagram is represented by a short double line to signify a dressed propagator, whereas the outgoing line corresponds to a noninteracting propagator denoted by a single line. These last two contributions to the irreducible self-energy can easily be identified as the product of the dressed, but noninteracting, propagators in G_{II}, giving rise to the second term in Eq. (9.34) (middle diagram) and the contribution containing Γ yielding the last term (and last diagram). The first one refers to the auxiliary sp potential. An equivalent expression can be developed by starting the study of the equation of motion of G with the time derivative with respect to t' (see also Ch. 21). In that case,

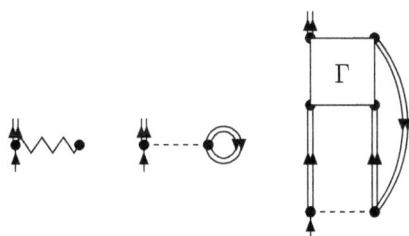

Fig. 9.14   Diagrams representing the irreducible self-energy as obtained by considering the equation of motion of G as a function of t'.

the alternative form of the Dyson equation emerges [Eq. (9.3)] and the corresponding self-energy is shown in Fig. 9.14.

The present formulation is important, since together with the Dyson equation itself, it provides a nonlinear description of the many-particle problem. At the same time, some very intuitive notions have been developed. These relate to the idea of a particle with modified properties in the medium (Dyson equation), which stem from its interaction with the others as given by Eq. (9.34). The interaction in turn takes place between particles immersed in the medium, and therefore involves dressed ones. The nonlinearity is visible in the Dyson equation [Eq. (9.2)] for the sp propagator. It includes the self-energy [Eq. (9.34)], which contains sp propagators that solve the Dyson equation. Self-consistency is therefore essential and a plausible strategy in developing calculational schemes. One may reasonably assume that for stronger correlations in the system, this self-consistency concept or, equivalently, the degree of nonlinearity, will be more important. Using the structure of the theory as outlined above, it becomes possible to develop nonlinear approximation schemes, which take the dominant physical characteristics of the system into account. By identifying suitable approximations to G_{II}, we have through Eq. (9.34) an appropriate calculational scheme that takes the corresponding physics into account. In many cases, the interaction between the particles V dictates a certain minimum approximation to have a chance of realistically describing the many-body system under study. In other instances, the size of the system and the form of the interaction, combine to dictate a "minimum" approximation scheme. In its simplest form, the diagrammatic version of the Hartree–Fock method is generated, as discussed in the next chapter.

9.5   Schrödinger-like equation from the Dyson equation

It is possible to demonstrate that the Dyson equation generates a Schrödinger-like equation, as was promised in Ch. 7 where experimental data, related to removal probabilities were presented. Indeed, we only have to follow the same steps as in Sec. 6.3. We will discuss here the case when the spectrum for the $N \pm 1$ systems near the Fermi energy involves discrete bound states, which mostly applies to finite system like atoms and nuclei. An appropriate form of the Lehmann representation looks like Eq. (6.25) (allowing for hole propagation) and is given by

$$
\begin{aligned}
G(\alpha, \beta; E) = & \sum_m \frac{\langle \Psi_0^N | a_\alpha | \Psi_m^{N+1} \rangle \langle \Psi_m^{N+1} | a_\beta^\dagger | \Psi_0^N \rangle}{E - (E_m^{N+1} - E_0^N) + i\eta} \\
& + \int_{\varepsilon_T^+}^\infty d\tilde{E}_\mu^{N+1} \frac{\langle \Psi_0^N | a_\alpha | \Psi_\mu^{N+1} \rangle \langle \Psi_\mu^{N+1} | a_\beta^\dagger | \Psi_0^N \rangle}{E - \tilde{E}_\mu^{N+1} + i\eta} \\
& + \sum_n \frac{\langle \Psi_0^N | a_\beta^\dagger | \Psi_n^{N-1} \rangle \langle \Psi_n^{N-1} | a_\alpha | \Psi_0^N \rangle}{E - (E_0^N - E_n^{N-1}) - i\eta} \\
& + \int_{-\infty}^{\varepsilon_T^-} d\tilde{E}_\nu^{N-1} \frac{\langle \Psi_0^N | a_\beta^\dagger | \Psi_\nu^{N-1} \rangle \langle \Psi_\nu^{N-1} | a_\alpha | \Psi_0^n \rangle}{E - \tilde{E}_\nu^{N-1} - i\eta},
\end{aligned}
\tag{9.35}
$$

where the continuum energy spectrum for the $N \pm 1$ systems has been included, and the corresponding energy thresholds are denoted by $\varepsilon_T^\pm$. A change of integration variable was also used to obtain this form of the Lehmann representation, introducing the integration variables for the continuum energies in the form $\tilde{E}_\mu^{N+1} = E_\mu^{N+1} - E_0^N$ and $\tilde{E}_\nu^{N-1} = E_0^N - E_\nu^{N-1}$, respectively. For the unperturbed propagator one will encounter sp energies associated with H_0, that are different from those of G, for any approximation made to the self-energy. This feature can be used to take the appropriate limits of the Dyson equation, in complete analogy with Sec. 6.3. The only difference that must be considered is associated with the energy dependence of the self-energy. By exploring the equations of motion of the two-body propagator, it is possible to show that a Lehmann representation exists for the exact self-energy that has different poles from the one for G. In the subsequent chapters we will introduce approximation schemes to the self-energy that conform to this property. As a result, one may proceed with taking the following limit of the Dyson equation involving the hole part of the propagator without generating contributions from the poles in the self-energy, or the noninteracting propagator

$$
\lim_{E \to \varepsilon_n^-} (E - \varepsilon_n^-) \left\{ G(\alpha, \beta; E) \right.
$$
$$
\left. = G^{(0)}(\alpha, \beta; E) + \sum_{\gamma\delta} G^{(0)}(\alpha, \gamma; E) \, \Sigma^*(\gamma, \delta; E) \, G(\delta, \beta; E) \right\}. \quad (9.36)
$$

The short-hand notation

$$
\varepsilon_n^- = E_0^N - E_n^{N-1} \quad (9.37)
$$

has been introduced here. As for the sp problem, this limit process generates an eigenvalue equation of the following kind (in complete analogy with the development in Ch. 6)

$$
z_\alpha^{n-} = \sum_{\gamma,\delta} G^{(0)}(\alpha, \gamma; \varepsilon_n^-) \, \Sigma^*(\gamma, \delta; \varepsilon_n^-) \, z_\delta^{n-}, \quad (9.38)
$$

where

$$
z_\alpha^{n-} = \left\langle \Psi_n^{N-1} \middle| a_\alpha \middle| \Psi_0^N \right\rangle. \quad (9.39)
$$

Since the Dyson equation can be written in a sp basis different from the one associated with H_0, one may choose the coordinate representation. Using $\boldsymbol{r}$ for the position and m for the spin projection in solving Eq. (9.38), one finds

$$
z_{\boldsymbol{r}m}^{n-} = \sum_{m_1,m_2} \int d^3 r_1 \int d^3 r_2 \; G^{(0)}(\boldsymbol{r}m, \boldsymbol{r}_1 m_1; \varepsilon_n^-) \, \Sigma^*(\boldsymbol{r}_1 m_1, \boldsymbol{r}_2 m_2; \varepsilon_n^-) \, z_{\boldsymbol{r}_2 m_2}^{n-}.
$$
$$
\quad (9.40)
$$

To obtain this result, the unperturbed propagator, $G^{(0)}$, and the self-energy, Σ^*, require a sp basis transformation on both indices, as originally the basis associated with H_0 was employed. Equation (9.40) can be rearranged by inverting the unperturbed propagator according to

$$
\sum_m \int d^3 r \, \langle \boldsymbol{r}'m' | \varepsilon_n^- - H_0 | \boldsymbol{r}m \rangle \, G^{(0)}(\boldsymbol{r}m, \boldsymbol{r}_1 m_1; \varepsilon_n^-) = \delta_{m',m_1} \delta(\boldsymbol{r}' - \boldsymbol{r}_1),
$$
$$
\quad (9.41)
$$

using Eq.(7.39). The corresponding operation on $z_{\boldsymbol{r}m}^{n-}$ generates

$$
\sum_m \int d^3 r \, \langle \boldsymbol{r}'m' | \varepsilon_n^- - H_0 | \boldsymbol{r}m \rangle \, z_{\boldsymbol{r}m}^{n-} = \left\{ \varepsilon_n^- + \frac{\hbar^2 \nabla'^2}{2m} - U(\boldsymbol{r}') \right\} z_{\boldsymbol{r}'m'}^{n-}, \quad (9.42)
$$

where U is assumed to be local and spin-independent for simplicity. Combining these results, yields the explicit cancellation of the auxiliary potential

U and the following equation

$$-\frac{\hbar^2\nabla^2}{2m}z_{rm}^{n-} + \sum_{m_1}\int d^3r_1\, \Sigma'^*(rm,r_1m_1;\varepsilon_n^-)z_{r_1m_1}^{n-} = \varepsilon_n^- z_{rm}^{n-}, \qquad (9.43)$$

where the notation Σ'^* has been used to signify that the U contribution has been removed. Equation (9.43) has the form of a Schrödinger equation with a nonlocal potential, which is represented by the self-energy Σ'^*. Note that an eigenvalue ε_n^- can only be generated when it coincides with the energy argument of the self-energy. An important difference with the ordinary Schrödinger equation is related to the normalization of the quasihole "eigenfunctions" z_{rm}^{n-}. The appropriate normalization condition is obtained by performing the same steps that lead to Eq. (6.39). The result is most conveniently expressed in terms of the sp state which corresponds to the quasihole wave function z_{rm}^{n-}. In other words, one can use the eigenstate which diagonalizes Eq. (9.43), to express the normalization condition. Assigning the notation α_{qh} to the sp state, one finds the spectroscopic factor

$$S = |\, z_{\alpha_{qh}}^{n-}\,|^2 = \left(1 - \frac{\partial\Sigma'^*(\alpha_{qh},\alpha_{qh};E)}{\partial E}\bigg|_{\varepsilon_n^-}\right)^{-1}. \qquad (9.44)$$

The subscript qh refers to the quasihole nature of this state and the fact that for those very near to the Fermi energy with quantum numbers corresponding to fully occupied mean-field levels, the normalization yields a number of order 1. As discussed in Ch. 7, the quasihole eigenfunctions and related spectroscopic factors are experimentally accessible in $(e, 2e)$ and $(e, e'p)$ reactions.

9.6   Exercises

(1) Determine the expressions for the self-energy contributions in Fig. 9.6 using the energy formulation.
(2) Generate all self-energy diagrams for the self-energy in the unsymmetrized version and determine the corresponding expressions.
(3) Perform the steps that lead to the diagrammatic version of the irreducible self-energy, shown in Fig. 9.14. Start by considering the derivative of $G(\alpha,\beta;t-t')$ with respect to t'.
(4) Determine the form of the irreducible self-energy in the time formulation, starting from Eq. (9.27).
(5) Derive Eqs. (9.31), (9.32), and (9.44).

Chapter 10

Mean-field or Hartree–Fock approximation

In the previous chapter, the formulation of many-body theory in terms of self-consistent Green's functions was developed. This chapter deals with the implementation of the theory in lowest order, which is equivalent to the so-called mean-field or Hartree–Fock (HF) approximation. The formal HF equations are derived in Sec. 10.1 together with details of the HF propagator. These results are then immediately contrasted with the conventional derivation of the HF equations, by means of the variational principle. Section 10.1 also contains a formulation in coordinate space, together with a discussion of restricted and unrestricted versions. In Sec. 10.2, the application to atoms is discussed. The section focuses on closed-shell atoms, a comparison with experimental data, and deals with relevant numerical details. Suggested steps of a numerical procedure for solving the equations are also outlined. The application to molecules is sketched in Sec. 10.3. It includes a brief discussion of the Born–Oppenheimer approximation, the use of finite, discrete basis sets, and a discussion of the hydrogen molecule. Infinite systems are reviewed in Sec. 10.4. We conclude this chapter with HF calculations of the electron gas (Sec. 10.5) and nuclear matter (Sec. 10.6).

10.1   The Hartree–Fock formalism

10.1.1   *Derivation of the Hartree–Fock equations*

The HF equations will be derived here in the general context established in Sec. 3.1, for a Hamiltonian

$$\hat{H} = \hat{T} + \hat{V} = \left(\hat{T} + \hat{U}\right) + \left(\hat{V} - \hat{U}\right), \qquad (10.1)$$

which includes an appropriately chosen (but in principle arbitrary) auxiliary potential $\hat{U}$. The sp basis that we use is the one where the noninteracting Hamiltonian

$$\hat{H}_0 = \hat{T} + \hat{U} = \sum_\alpha \varepsilon_\alpha a_\alpha^\dagger a_\alpha \qquad (10.2)$$

is diagonal. The unperturbed sp propagator $G^{(0)}$ therefore reads

$$G^{(0)}(\alpha, \beta; E) = \delta_{\alpha,\beta} \left[\frac{\theta(\alpha - F)}{E - \varepsilon_\alpha + i\eta} + \frac{\theta(F - \alpha)}{E - \varepsilon_\alpha - i\eta} \right]. \qquad (10.3)$$

We start by considering the general expression in Eq. (9.34) for the (irreducible) self-energy Σ^* in terms of the 4-point vertex function Γ. The simplest thing to do at this stage is to set $\Gamma = 0$. Note that this corresponds, according to Eq. (9.22), to replacing the tp propagator G_{II} with the antisymmetrized product of two sp propagators, an approximation which is exact for a non-interacting system. Clearly, this implies that for the system under study, the tp propagator is dominated by the noninteracting contribution in Eq. (9.22). Having made the present choice of self-energy, it will be shown shortly that it leads to a mean-field description. The quality of the approximation for the self-energy can be tested by comparing the calculations with the relevant experimental data. The results in Sec. 10.2.2 *e.g.* indicate that it is a reasonable description for electrons in atoms, as was anticipated in Ch. 3.

Setting $\Gamma = 0$ leads to the so-called HF approximation with the following self-energy [see Eq. (9.34)]

$$\Sigma^{HF}(\gamma, \delta; E) = -\langle \gamma | U | \delta \rangle - i \int_{C\uparrow} \frac{dE'}{2\pi} \sum_{\mu\nu} \langle \gamma\mu | V | \delta\nu \rangle \, G^{HF}(\nu, \mu; E').$$

$$(10.4)$$

In keeping with the self-consistent formulation of Sec. 9.4, the HF propagator G^{HF} appearing in Eq. (10.4) is *not* the noninteracting propagator $G^{(0)}$, but rather the solution of the corresponding Dyson equation

$$G^{HF}(\alpha, \beta; E) = G^{(0)}(\alpha, \beta; E) + \sum_{\gamma\delta} G^{(0)}(\alpha, \gamma; E)\Sigma^{HF}(\gamma, \delta)G^{HF}(\delta, \beta; E).$$

$$(10.5)$$

Here the energy argument of Σ^{HF} has been dropped; this is appropriate, since inspection of Eq. (10.4) clearly shows that the HF self-energy has no E-dependence. The diagrammatic equivalent of Eq. (10.5) is shown in Fig. 10.1a). It is evident that a particular, infinite class of self-energy

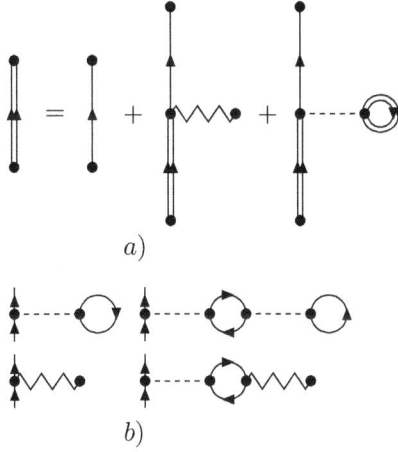

Fig. 10.1 Part *a*) shows the diagrammatic representation of the Dyson equation in the HF approximation. In part *b*) all diagrams up to second order contributing to the HF (irreducible) self-energy are displayed.

diagrams is retained in the HF self-energy, of which the lowest-order ones are shown in Fig. 10.1*b*). We emphasize that the symmetrized version of the diagram method is employed, so that both a direct and an exchange contribution are implied for each interaction V.

Further analysis of the HF self-energy Σ^{HF} in Eq. (10.4) requires the energy-dependence of the (as yet unknown) HF propagator, but we may assume that it has the same simple pole structure as the exact propagator, and write its Lehmann representation (see Sec. 7.2) as

$$G^{HF}(\alpha, \beta; E) = \sum_m \frac{z_\alpha^{m+} z_\beta^{m+*}}{E - \varepsilon_m^+ + i\eta} + \sum_n \frac{z_\alpha^{n-} z_\beta^{n-*}}{E - \varepsilon_n^- - i\eta}. \quad (10.6)$$

The (approximate) z amplitudes are defined in analogy to Eq. (9.39) by

$$z_\alpha^{n-} = \langle \Psi_n^{N-1} | a_\alpha | \Psi_0^N \rangle \quad (10.7)$$

and

$$z_\alpha^{n+} = \langle \Psi_0^N | a_\alpha | \Psi_n^{N+1} \rangle, \quad (10.8)$$

respectively. The energies $\varepsilon^\pm$ are defined in accord with Eq. (9.37) and given by

$$\varepsilon_n^- = E_0^N - E_n^{N-1} \quad (10.9)$$

and

$$\varepsilon_n^+ = E_n^{N+1} - E_0^N. \tag{10.10}$$

The assumption implied by the form of Eq. (10.6) will be correct if it is subsequently found that the Dyson equation is indeed solved by this form of the HF propagator.

Using Eq. (10.6) the HF self-energy can be readily evaluated:

$$\Sigma^{HF}(\gamma, \delta) = - \langle \gamma | U | \delta \rangle + \sum_{\mu\nu} \langle \gamma\mu | V | \delta\nu \rangle \sum_n z_\nu^{n-} z_\mu^{n-*}. \tag{10.11}$$

Recalling the relationships of Sec. 7.4 between the one-body density matrix of the system and the removal amplitudes in the sp propagator [see Eq. (7.20)], the HF one-body density matrix $n_{\mu\nu}^{HF}$ may be defined as

$$n_{\mu\nu}^{HF} = \sum_n z_\nu^{n-} z_\mu^{n-*}, \tag{10.12}$$

and the HF self-energy is written in a transparent form as

$$\Sigma^{HF}(\gamma, \delta) = - \langle \gamma | U | \delta \rangle + \sum_{\mu\nu} \langle \gamma\mu | V | \delta\nu \rangle \, n_{\mu\nu}^{HF}. \tag{10.13}$$

This expression shows that the HF self-energy really represents a "mean field" or average potential, in the sense that it contains the interparticle interaction V, averaged over the one-body density matrix. The latter quantity takes into account the occupancies of the different sp orbitals in the ground state.

The fact that Σ^{HF} is just a static (energy-independent) sp potential, also implies that the Dyson equation Eq. (10.5) is equivalent to an independent-particle problem. The bound sp eigenstates of $H_0 + \Sigma^{HF}$ can therefore be obtained by transforming the Dyson equation to an eigenvalue equation, using the standard limit procedure of Secs. 6.3 and 9.5. The limit

$$\lim_{E \to \varepsilon_n^-} (E - \varepsilon_n^-) \left\{ G^{HF} = G^{(0)} + G^{(0)} \, \Sigma^{HF} \, G^{HF} \right\}, \tag{10.14}$$

leads to the HF eigenvalue equations

$$\sum_\delta \left\{ \varepsilon_\delta \delta_{\alpha,\delta} - \langle \alpha | U | \delta \rangle + \sum_{\mu\nu} \langle \alpha\mu | V | \delta\nu \rangle \, n_{\mu\nu}^{HF} \right\} z_\delta^{n-} = \varepsilon_n^- z_\alpha^{n-}. \tag{10.15}$$

The HF equations determine the unknown removal energies ε_n^- and amplitudes z_α^{n-}. The latter should be normalized to unity

$$\sum_\alpha |z_\alpha^{n-}|^2 = 1, \tag{10.16}$$

because of the energy-independence of the HF self-energy (see Sec. 6.3). The particle-number sum rule

$$\sum_{n\alpha} |z_\alpha^{n-}|^2 = \sum_\alpha n_{\alpha\alpha}^{HF} = N \tag{10.17}$$

then dictates that there are exactly N removal states in the Lehmann representation of Eq. (10.6), which should be identified with the N lowest-energy solutions z_α^{n-} $(n = 1, \ldots, N)$ of the HF equations. This is just what we expect for the propagator of an independent-particle system as discussed in Sec. 7.5.

While the HF eigenvalue equations in Eq. (10.15) may *look* like a sp Schrödinger equation corresponding to a noninteracting many-body system, there is one complication. Due to the self-consistency condition Eq. (10.12) they are actually nonlinear in the amplitudes z_α^{n-} and therefore usually solved by iteration. Starting from the N lowest-energy sp orbitals $\alpha = 1, \ldots, N$ of H_0 one sets

$$\text{(Initial guess)} \quad \rightarrow \quad z_\alpha^{n-} = \delta_{n,\alpha}, \tag{10.18}$$

using this first approximation to construct the corresponding self-energy Σ^{HF} through Eq. (10.11). Then the N lowest-energy solutions of the Hamiltonian $H_0 + \Sigma^{HF}$ are determined (which is equivalent to solving a noninteracting system), yielding new z_α^{n-}. This cycle is repeated until convergence is achieved and the amplitudes z_α^{n-} no longer change during successive iterations. A more detailed look at practical methods for solving the HF equations in the case of atoms can be found in Sec. 10.2.

Finally we note that the derivation up to now, has used the sp basis that diagonalizes the Hamiltonian $H_0 = T + U$, *i.e.*

$$\langle \alpha | T | \delta \rangle + \langle \alpha | U | \delta \rangle = \varepsilon_\delta \, \delta_{\alpha,\delta}. \tag{10.19}$$

Substituting this result into Eq. (10.15) we see that the matrix elements of the auxiliary potential U cancel, and the HF equations become

$$\sum_\delta \left\{ \langle \alpha | T | \delta \rangle + \sum_{\mu\nu} \langle \alpha\mu | V | \delta\nu \rangle \, n_{\mu\nu}^{HF} \right\} z_\delta^{n-} = \varepsilon_n^- \, z_\alpha^{n-}. \tag{10.20}$$

They are therefore independent of the auxiliary field U — as it should be in the framework of self-consistent Green's function theory — and Eq. (10.20) represents the HF equations in an arbitrary sp basis. Of course, a suitable choice of the auxiliary field may speed up convergence of the iterative solution. In the nuclear case, *e.g.*, one can start with an auxiliary potential which provides localization of the nucleons, as discussed in Sec. 3.1. The HF method was first studied by [Hartree (1928); Slater (1930); Fock (1930)].

10.1.2   *The Hartree–Fock propagator*

The HF self-consistency problem consists of the determination of the removal amplitudes z_α^{n-} and energies ε_n^- $(n = 1, \ldots, N)$ which solve Eqs. (10.12) and (10.20). After this is achieved we can construct the remaining eigenstates $(n = N + 1, N + 2, \ldots)$ in Eq. (10.20) that have higher sp energies. These correspond to the particle addition amplitudes z_β^{m+} and energies ε_m^+ in Eq. (10.6), so we write

$$\sum_\delta \left\{ \langle \alpha | \, T \, | \delta \rangle + \sum_{\mu\nu} \langle \alpha\mu | \, V \, | \delta\nu \rangle \, n_{\mu\nu}^{HF} \right\} z_\delta^{m+} = \varepsilon_m^+ z_\alpha^{m+}. \qquad (10.21)$$

Note that Eq. (10.21) is an ordinary eigenvalue problem since the one-body density matrix given by

$$n_{\mu\nu}^{HF} = \sum_{n=1}^{N} z_\nu^{n-} z_\mu^{n-*}, \qquad (10.22)$$

is determined by the removal amplitudes and therefore fixed. No additional self-consistency steps are thus required.

The construction of the HF propagator of Eq. (10.6) is now complete, and all physical observables contained in the sp propagator (see Sec. 7.3 – 7.4) can be evaluated in the HF approximation. We observe *e.g.*, that the excited states $\left| \Psi_n^{N-1} \right\rangle$ in the $N - 1$-particle system, which can be reached by the removal of one particle from the N-particle ground state $\left| \Psi_0^N \right\rangle$, have a spectrum

$$\text{(HF:)} \qquad E_0^N - E_n^{N-1} = \varepsilon_n^-, \qquad (10.23)$$

and removal amplitudes

$$\text{(HF:)} \qquad \left\langle \Psi_n^{N-1} \right| a_\alpha \left| \Psi_0^N \right\rangle = z_\alpha^{n-}. \qquad (10.24)$$

The interpretation of the HF sp energies ε_n^- as removal energies follows naturally from the propagator formulation. It is less obvious in a variational context (see Sec. 10.1.3), where it is called Koopman's theorem. For the excited states of the $N+1$ particle system we have likewise for the spectrum

$$\text{(HF:)} \qquad E_n^{N+1} - E_0^N = \varepsilon_n^+, \qquad (10.25)$$

and the addition amplitudes

$$\text{(HF:)} \qquad \left\langle \Psi_0^N \middle| a_\alpha \middle| \Psi_n^{N+1} \right\rangle = z_\alpha^{n+}. \qquad (10.26)$$

The hole part of the spectral function defined in Eq. (7.11) reads

$$\text{(HF:)} \qquad S_h(\alpha; E) = \sum_{n=1}^{N} |z_\alpha^{n-}|^2 \delta(E - \varepsilon_n^-), \qquad (10.27)$$

and the corresponding mean removal energy is

$$\text{(HF:)} \qquad \overline{R}_{HF} = \sum_\alpha \int_{-\infty}^{\varepsilon_F^-} dE \, E \, S_h(\alpha; E) = \sum_{n=1}^{N} \varepsilon_n^-. \qquad (10.28)$$

The HF result for the ground-state energy then follows from Eq. (7.26)

$$\text{(HF:)} \qquad E_0^N = \frac{1}{2} \left\{ \sum_{\alpha\beta} \langle \alpha| T |\beta\rangle \, n_{\alpha\beta}^{HF} + \sum_{n=1}^{N} \varepsilon_n^- \right\},$$

$$= \frac{1}{2} \left\{ \overline{T}_{HF} + \overline{R}_{HF} \right\}. \qquad (10.29)$$

Obviously, $\overline{T}_{HF}$ represents the expectation value of the one-body part of the Hamiltonian (the kinetic energy and, if present, the external potential), since according to Eq. (7.19) we have

$$\text{(HF:)} \qquad \left\langle \Psi_0^N \middle| \hat{T} \middle| \Psi_0^N \right\rangle = \sum_{\alpha\beta} \langle \alpha| T |\beta\rangle \, n_{\alpha\beta}^{HF} = \overline{T}_{HF}. \qquad (10.30)$$

The contribution of the two-body interaction $\hat{V}$ to the ground-state energy can now be expressed in terms of the HF quantities $\overline{R}_{HF}$ and $\overline{T}_{HF}$

$$\text{(HF:)} \qquad \left\langle \Psi_0^N \middle| \hat{V} \middle| \Psi_0^N \right\rangle = \overline{V}_{HF} = E_0^N - \overline{T}_{HF}$$

$$= \frac{1}{2}(\overline{R}_{HF} - \overline{T}_{HF}). \qquad (10.31)$$

An alternative expression for $\overline{R}_{HF} = \sum_{n=1}^{N} \varepsilon_n^-$ is obtained by multiplying Eq. (10.20) with z_α^{n-*}, followed by a summation over α and over $n = 1, \ldots, N$. The resulting expression,

$$\sum_{n=1}^{N} \varepsilon_n^- = \sum_{\alpha\delta} \left\{ \langle \alpha | T | \delta \rangle + \sum_{\mu\nu} \langle \alpha\mu | V | \delta\nu \rangle \, n_{\mu\nu}^{HF} \right\} n_{\alpha\delta}^{HF} \quad (10.32)$$

$$= \overline{T}_{HF} + \sum_{\alpha\beta\mu\nu} \langle \alpha\mu | V | \beta\nu \rangle \, n_{\mu\nu}^{HF} n_{\alpha\beta}^{HF}, \quad (10.33)$$

when substituted in Eq. (10.31), implies that

$$\overline{V}_{HF} = \frac{1}{2} \sum_{\alpha\beta\mu\nu} \langle \alpha\mu | V | \beta\nu \rangle \, n_{\mu\nu}^{HF} n_{\alpha\beta}^{HF}. \quad (10.34)$$

As a consequence, the HF ground-state energy can also be expressed as

$$\text{(HF:)} \qquad E_0^N = \overline{R}_{HF} - \overline{V}_{HF} \quad (10.35)$$

$$= \sum_{n=1}^{N} \varepsilon_n^- - \frac{1}{2} \sum_{\alpha\beta\mu\nu} \langle \alpha\mu | V | \beta\nu \rangle \, n_{\mu\nu}^{HF} n_{\alpha\beta}^{HF}. \quad (10.36)$$

Clearly, E_0^N is *not* just the sum of the HF sp energies of the occupied orbitals, as it would be for an independent-particle problem. The correction term $-\overline{V}_{HF}$ in Eq. (10.36) is sometimes called the rearrangement energy.

Up to now the HF propagator $G^{HF}(\alpha, \beta; E)$ and all related quantities have been expressed in an arbitrary sp basis. However, once the removal amplitudes z_α^{n-} have been fixed, Eqs. (10.20) and (10.21) constitute an (Hermitian) eigenvalue problem. The solutions $z_\alpha^{n\pm}$ therefore define an orthonormal basis set of HF sp states

$$\left| n^\pm \right\rangle = \sum_\alpha z_\alpha^{n\pm} \left| \alpha \right\rangle. \quad (10.37)$$

For clarity, we will continue to label the HF sp states with Roman letters, but we drop the ($\pm$) superscripts. From the propagator context it is obvious that the HF sp states corresponding to the N lowest sp energies (the *hole* states) should be interpreted as representing removal amplitudes. The HF sp states with higher-lying energies (the *particle* states) correspond to addition amplitudes.

Expressed in the HF basis, the HF one-body density matrix is simply (using the step function for the case when its argument is true or false)

$$n_{ij}^{HF} = \delta_{i,j}\, \theta(1 \leq i \leq N), \quad (10.38)$$

and the defining equation [see Eqs. (10.20) and (10.21)] for the HF basis becomes

$$\langle n| T |m\rangle + \sum_{i=1}^{N} \langle ni| V |mi\rangle = \delta_{m,n}\varepsilon_n.$$

(10.39)

The HF propagator is also diagonal in the HF basis,

$$G^{HF}(m,n;E) = \delta_{m,n} \left[\frac{\theta(n > N)}{E - \varepsilon_n + i\eta} + \frac{\theta(1 \leq n \leq N)}{E - \varepsilon_n - i\eta} \right].$$

(10.40)

Eq. (10.40) is recognized as the propagator of a noninteracting system with sp Hamiltonian

$$\hat{H}_{HF} = \sum_n \varepsilon_n a_n^\dagger a_n$$

(10.41)

$$= \sum_{mn} \langle n| T |m\rangle a_n^\dagger a_m + \sum_{nm} \left(\sum_{i=1}^{N} \langle ni| V |mi\rangle \right) a_n^\dagger a_m,$$

(10.42)

and this would seem to imply that the HF ground state can be identified with the Slater determinant

$$(\text{HF:}) \qquad |\Psi_0^N\rangle \to |\Phi_{HF}^N\rangle = \prod_{i=1}^{N} a_i^\dagger |0\rangle.$$

(10.43)

It is readily checked that Eqs. (10.30) and (10.34) are indeed consistent with the HF ground state in Eq. (10.43), *i.e.*

$$\overline{T}_{HF} = \langle \Phi_{HF}^N | \hat{T} | \Phi_{HF}^N \rangle$$
$$\overline{V}_{HF} = \langle \Phi_{HF}^N | \hat{V} | \Phi_{HF}^N \rangle.$$

(10.44)

That this interpretation is correct is also borne out by the variational derivation in the next section.

10.1.3 *Variational content of the HF approximation*

The eigenstates for a system of N noninteracting fermions have been discussed in Section 3.1. They have the simple form of antisymmetrized product states and may generically be written as

$$|\Phi^N\rangle = \prod_{i=1}^{N} a_{h_i}^\dagger |0\rangle,$$

(10.45)

in terms of N orthonormal occupied sp orbitals $a_{h_i}^\dagger$. The orthonormality condition ensures a proper normalization, $\langle \Phi^N | \Phi^N \rangle = 1$.

The product states in Eq. (10.45), also called independent-particle states or Slater determinants, of course far from exhaust the complete N-particle Fock space. Nevertheless, there is considerable freedom in the set of all $|\Phi^N\rangle$, since the shape of the occupied sp orbitals can be chosen at will. If we now consider an interacting system with Hamiltonian $\hat{H} = \hat{T} + \hat{V}$, we may approximate the exact interacting ground state by determining the independent-particle state $|\Phi^N\rangle$ which minimizes the expectation value $\langle \Phi^N | \hat{H} | \Phi^N \rangle$ of the Hamiltonian. In the case of weak interparticle interactions $\hat{V}$ this is usually a good starting point.

The expectation value is easily evaluated as

$$ E = \langle \Phi^N | \hat{H} | \Phi^N \rangle = \sum_{i=1}^{N} \langle h_i | T | h_i \rangle + \frac{1}{2} \sum_{i,j=1}^{N} \langle h_i h_j | V | h_i h_j \rangle, \qquad (10.46) $$

in terms of the occupied sp orbitals. Expanding the unknown occupied orbitals in terms of a fixed sp basis,

$$ a_{h_i}^\dagger = \sum_\alpha z_{i\alpha} a_\alpha^\dagger, \qquad (10.47) $$

the energy becomes

$$ E = \sum_{i=1}^{N} \sum_{\alpha\beta} z_{i\alpha}^* z_{i\beta} \langle \alpha | T | \beta \rangle + \frac{1}{2} \sum_{i,j=1}^{N} \sum_{\alpha\beta\gamma\delta} z_{i\alpha}^* z_{j\beta}^* z_{i\gamma} z_{j\delta} \langle \alpha\beta | V | \gamma\delta \rangle. \quad (10.48) $$

This expression should be minimized with respect to variations in the expansion coefficients $z_{i\alpha}$, subject to the orthonormalization constraints for the occupied sp orbitals,

$$ \sum_\alpha z_{j\alpha}^* z_{i\alpha} = \delta_{i,j}. \qquad (10.49) $$

The condition for a constrained extremum reads

$$ \frac{\partial}{\partial z_{k\alpha}^*} \left[E - \sum_{i,j=1}^{N} \varepsilon_{ij} \sum_\alpha z_{j\alpha}^* z_{i\alpha} \right] = 0, \qquad (10.50) $$

where the Lagrange multipliers ε_{ij} form a Hermitian matrix. Working out the derivative, yields the set of nonlinear equations

$$\sum_{\beta} \langle \alpha | T | \beta \rangle z_{i\beta} + \sum_{\beta\gamma\delta} \langle \alpha\beta | V | \gamma\delta \rangle \left(\sum_{j=1}^{N} z_{j\beta}^* z_{j\delta} \right) z_{i\gamma} = \sum_{j=1}^{N} \varepsilon_{ij} z_{j\alpha}, \quad (10.51)$$

which should be solved together with the constraints in Eq. (10.49).

Without loss of generality we may assume the matrix ε_{ij} to be diagonal. If a solution is found where it is not, one can consider a unitary mixing of the z_i-vectors,

$$z_{i\alpha} = \sum_{j=1}^{N} U_{ij} z'_{j\alpha},$$

$$z'_{i\alpha} = \sum_{j=1}^{N} U_{ji}^* z_{j\alpha}. \quad (10.52)$$

The expression in Eq. (10.48) for the energy is invariant under such a transformation, so the set of $z_{i\alpha}$ is only determined up to a unitary transformation by the minimalization problem. The underlying solution, of course, is always the same, since the independent-particle state in Eq. (10.45) does not change (apart from a global phase) under a unitary mixing of the occupied orbitals. Equation (10.51) transforms as

$$\sum_{\beta} \langle \alpha | T | \beta \rangle z'_{i\beta} + \sum_{\beta\gamma\delta} \langle \alpha\beta | V | \gamma\delta \rangle \left(\sum_{j=1}^{N} z'^*_{j\beta} z'_{j\delta} \right) z'_{i\gamma} = \sum_{j=1}^{N} \varepsilon'_{ij} z'_{j\alpha}, \quad (10.53)$$

where

$$\varepsilon'_{ij} = \sum_{k,l=1}^{N} U_{ki}^* \varepsilon_{kl} U_{lj} = \left[U^\dagger \varepsilon U \right]_{ij}. \quad (10.54)$$

From Eq. (10.54) it follows that U can be chosen such that $\varepsilon'_{ij} = \delta_{i,j}\varepsilon_i$ is diagonal. This is the so-called canonical representation of the HF basis, and with this choice Eq. (10.53) becomes identical to the HF equations (10.20) derived in Sec. 10.1.

The variational nature of the HF ground state has two important consequences. First, the HF ground-state energy in Eq. (10.29) is always larger than the exact ground-state energy, as it is the minimal expectation value of the Hamiltonian with respect to a *restricted* class (the Slater determinants) of N-particle wave functions. A second consequence is Brillouin's theorem, which can be formulated as

$$\langle \Phi^N_{HF} | \hat{H} a^\dagger_p a_h | \Phi^N_{HF} \rangle = 0, \tag{10.55}$$

where h (hole) labels denote HF sp states that are occupied in Φ^N_{HF}, and p (particle) labels denote unoccupied sp states. Note that the Slater determinants

$$|\Phi^N_{ph}\rangle = a^\dagger_p a_h |\Phi^N_{HF}\rangle, \tag{10.56}$$

formed by replacing a hole with a particle state in the HF determinant, are called one-particle−one-hole (1p1h) excitations. Brillouin's theorem asserts that the HF ground state is stable with respect to such 1p1h excitations, which can therefore be regarded as first approximations to the excited states of the N-particle system. To prove the theorem it is sufficient to note that small variations of the occupied HF orbitals $a^\dagger_h$ are by necessity of the form

$$\delta a^\dagger_h = \sum_p \eta_{ph} a^\dagger_p, \tag{10.57}$$

since they have to be orthogonal to all hole states. The corresponding variation in Φ^N_{HF} can therefore be written as

$$|\delta \Phi^N_{HF}\rangle = \delta \left\{ \prod_h a^\dagger_h |0\rangle \right\} = \sum_h \left(\delta a^\dagger_h \right) a_h |\Phi^N_{HF}\rangle = \sum_{ph} \eta_{ph} a^\dagger_p a_h |\Phi^N_{HF}\rangle. \tag{10.58}$$

Since the energy is an extremum with respect to such variations, we have

$$0 = \langle \Phi^N_{HF} | \hat{H} | \delta \Phi^N_{HF} \rangle = \sum_{ph} \eta_{ph} \langle \Phi^N_{HF} | \hat{H} a^\dagger_p a_h | \Phi^N_{HF} \rangle, \tag{10.59}$$

for arbitrary coefficients η_{ph}, and the theorem in Eq. (10.55) results.

Alternatively, one can show (Exercise (1) of this chapter) by direct evaluation of the matrix element in second quantization that

$$\langle \Phi^N_{HF} | \hat{H} a^\dagger_p a_h | \Phi^N_{HF} \rangle = \langle h| T |p\rangle + \sum_{h'} \langle hh'| V |ph'\rangle$$

$$= \langle h| H_{HF} |p\rangle = 0, \tag{10.60}$$

where the zero result follows from the fact that $\hat{H}_{HF}$ is diagonal in the HF sp basis.[1]

---

[1] Brillouin's theorem does *not* imply that 1p1h excitations are absent when expanding the exact ground state Ψ^N_0 in a series of 1p1h, 2p2h,... excitations on the HF ground state: they can still be mixed in through the coupling between 1p1h and 2p2h states.

The interpretation of the HF energies ε_h of the occupied sp states follows from the observation (Exercise (2) of this chapter),

$$\varepsilon_h = \langle \Phi_{HF}^N | \hat{H} | \Phi_{HF}^N \rangle - \langle \Phi_{HF}^N | a_h^\dagger \hat{H} a_h | \Phi_{HF}^N \rangle. \tag{10.61}$$

This is Koopman's theorem, which states that the ε_h can be identified as removal energies [see Eq. (10.23)], if one identifies the one-hole states $|\Phi_h^{N-1}\rangle = a_h |\Phi_{HF}^N\rangle$ as approximate eigenstates in the $N-1$-system.

10.1.4 *HF in coordinate space*

We consider here a general system of spin-$\frac{1}{2}$ fermions in a local external sp potential $U(\mathbf{r})$ and interacting with a local and spin-independent tp potential $V(\mathbf{r}_1 - \mathbf{r}_2)$ (this is applicable to electrons in atoms or molecules). As the HF equations in Sec. 10.1.1 are valid in a general sp basis, the coordinate-space representation is easily derived from Eq. (10.20) by taking sp labels $\alpha \equiv \mathbf{r} m_s$. We also introduce the more familiar wave-function form for the removal amplitude

$$z_{\mathbf{r} m_s}^n = \phi_n(\mathbf{r}, m_s). \tag{10.62}$$

Recalling [see Eq. (2.54)] the matrix elements of the kinetic energy operator in coordinate space, the first term in Eq. (10.20) becomes

$$\sum_{m_s'} \int d\mathbf{r}' \langle \mathbf{r} m_s | T | \mathbf{r}' m_s' \rangle \, \phi_n(\mathbf{r}', m_s') = -\frac{\hbar^2}{2m} \nabla^2 \phi_n(\mathbf{r}, m_s). \tag{10.63}$$

The second term in Eq. (10.20) involves the tp interaction. For the present local and spin-independent tp interaction the direct matrix element is

$$(\mathbf{r}_1 m_{s_1}, \mathbf{r}_2 m_{s_2} | V | \mathbf{r}_3 m_{s_3}, \mathbf{r}_4 m_{s_4}) \tag{10.64}$$
$$= \delta_{m_{s_1}, m_{s_3}} \delta_{m_{s_2}, m_{s_4}} \delta(\mathbf{r}_1 - \mathbf{r}_3)\delta(\mathbf{r}_2 - \mathbf{r}_4) V(\mathbf{r}_1 - \mathbf{r}_2).$$

Upon substitution, the HF equations (10.20) therefore read

$$\varepsilon_n \phi_n(\mathbf{r}, m_s) = -\frac{\hbar^2}{2m} \nabla^2 \phi_n(\mathbf{r}, m_s)$$
$$+ \left[\int d\mathbf{r}' V(\mathbf{r} - \mathbf{r}') \sum_{m_s'} n^{HF}(\mathbf{r}' m_s', \mathbf{r}' m_s') \right] \phi_n(\mathbf{r}, m_s)$$
$$- \sum_{m_s'} \int d\mathbf{r}' V(\mathbf{r} - \mathbf{r}') n^{HF}(\mathbf{r}' m_s', \mathbf{r} m_s) \phi_n(\mathbf{r}', m_s'), \tag{10.65}$$

where

$$n_{HF}(\boldsymbol{r}'m'_s, \boldsymbol{r}m_s) = \sum_{n=1}^{N} \phi_n(\boldsymbol{r}, m_s)\phi_n^*(\boldsymbol{r}', m'_s) \qquad (10.66)$$

is the HF one-body density matrix.

In Eq. (10.65) the first term involving the tp interaction is called the direct or Hartree contribution to the mean field. It can be written in terms of a local potential

$$v_H(\boldsymbol{r}) = \int d\boldsymbol{r}' n^{HF}(\boldsymbol{r}')V(\boldsymbol{r} - \boldsymbol{r}') \qquad (10.67)$$

which represents the tp interaction averaged over the HF density

$$n^{HF}(\boldsymbol{r}) = \sum_{m_s} n^{HF}(\boldsymbol{r}m_s, \boldsymbol{r}m_s). \qquad (10.68)$$

The second term is the exchange or Fock contribution, and is obviously nonlocal in coordinate (and spin) space.

It may be surprising that the Hartree potential $v_H(\boldsymbol{r})$ contains the total density of the N-particle system; in a mean-field picture one would expect a particle moving in orbital ϕ_n to interact with the $N-1$ other particles in orbitals ϕ_i ($i \neq n$), and not with itself. In fact, the HF approximation is free from such spurious self-interaction. This can be seen by isolating in Eq. (10.65) the contribution from ϕ_n to the HF one-body density matrix: the Hartree and Fock terms cancel each other.

10.1.5   *Unrestricted and restricted Hartree–Fock*

At this point it is useful to analyze the spin dependence of the HF equations. The Hamiltonian we adopted has no spin dependence and obviously commutes with the total spin operator $\hat{\boldsymbol{S}}$ corresponding to $\sum_{i=1}^{N} \boldsymbol{s}_i$ for N particles. Yet we see that the general HF equations (10.65) and (10.66) derived from this Hamiltonian do not reflect this symmetry: in principle both spin-up and spin-down components may be present in the HF sp states

$$\phi_n(\boldsymbol{r}, m_s) = \delta_{m_s, +\frac{1}{2}} u_n(\boldsymbol{r}) + \delta_{m_s, -\frac{1}{2}} v_n(\boldsymbol{r}), \qquad (10.69)$$

because the Fock term mixes them. The HF Slater determinant built with such sp states, would not even be an eigenstate of the $\hat{S}_z$ operator; a typical example where the mean-field approximation may break a symmetry of the exact Hamiltonian.

In fact, the most general two-component sp states in Eq. (10.69) are almost never used in a molecular context. One imposes from the beginning, that each HF sp state is either a pure spin-up or a pure spin-down state. The two types of HF sp states corresponding to $m_s = \pm\frac{1}{2}$ will be denoted by $\phi_n^{(m_s)}$, where $n = 1, \ldots, N^{(m_s)}$ so that $\sum_{m_s} N^{(m_s)} = N$. The single spin component structure implies

$$\langle \boldsymbol{r}m_s' | \phi_n^{(m_s)} \rangle = \delta_{m_s, m_s'} \phi_n^{(m_s)}(\boldsymbol{r}). \qquad (10.70)$$

A Slater determinant with such sp states is an eigenstate of the $\hat{S}_z$ operator (but not necessarily of $\hat{S}^2$) with eigenvalue $S_z = \frac{1}{2}[N^{(+\frac{1}{2})} - N^{(-\frac{1}{2})}]$, and the resulting one-body density matrix is diagonal in spin,

$$n^{HF}(\boldsymbol{r}'m_s', \boldsymbol{r}m_s) = \delta_{m_s, m_s'} n_{HF}^{(m_s)}(\boldsymbol{r}', \boldsymbol{r}) = \delta_{m_s, m_s'} \sum_{n=1}^{N^{(m_s)}} \phi_n^{(m_s)}(\boldsymbol{r})\phi_n^{(m_s)*}(\boldsymbol{r}').$$

$$(10.71)$$

Using the ansatz (10.70), the HF equations for the spin-up and spin-down type of orbitals become

$$\varepsilon_n^{(m_s)} \phi_n^{(m_s)}(\boldsymbol{r}) = -\frac{\hbar^2}{2m} \boldsymbol{\nabla}^2 \phi_n^{(m_s)}(\boldsymbol{r}) + v_H(\boldsymbol{r})\phi_n^{(m_s)}(\boldsymbol{r})$$

$$- \int d\boldsymbol{r}' \, V(\boldsymbol{r} - \boldsymbol{r}')n_{HF}^{(m_s)}(\boldsymbol{r}', \boldsymbol{r})\phi_n^{(m_s)}(\boldsymbol{r}'). \qquad (10.72)$$

This incarnation of HF goes under the name of unrestricted HF or UHF (though it is not the most general mean-field treatment), in order to distinguish it from restricted HF discussed below. Note that in Eq. (10.72) the Fock term acts only between same-spin particles, and that the only coupling between the two spin types occurs through the total density in the Hartree term.

In restricted HF or RHF, one assumes that the HF sp states come in pairs of opposite spin, both members of the pair having the same spatial wave function. RHF is eminently suitable for most neutral molecules, which have an even number of electrons and a spin-singlet ($S = 0$) ground state. In this case the ground state is well-approximated by a closed-shell configuration: a single Slater determinant consisting of $N/2$ spatial orbitals which are doubly occupied by a spin-up and a spin-down electron. Such a RHF closed-shell configuration is then automatically an $S = 0$ eigenstate of $\hat{S}^2$. The RHF ansatz

$$\phi_n^{(m_s)}(\boldsymbol{r}) = \phi_n(\boldsymbol{r}), \quad \text{for } n = 1, \ldots, N/2 \text{ and } m_s = \pm\frac{1}{2}, \qquad (10.73)$$

when substituted in Eq. (10.72), leads to the RHF equations for the spatial
orbitals

$$\varepsilon_n \phi_n(\mathbf{r}) = -\frac{\hbar^2}{2m} \boldsymbol{\nabla}^2 \phi_n(\mathbf{r}) + v_H(\mathbf{r}) \phi_n(\mathbf{r})$$

$$- \frac{1}{2} \int d\mathbf{r}' \, V(\mathbf{r} - \mathbf{r}') n_{HF}(\mathbf{r}', \mathbf{r}) \phi_n(\mathbf{r}'), \qquad (10.74)$$

where the spin-integrated one-body density matrix is related to Eq. (10.66)
as

$$n_{HF}(\mathbf{r}', \mathbf{r}) = \sum_{m_s} n^{HF}(r'm_s, rm_s) = 2 \sum_{n=1}^{N/2} \phi_n(\mathbf{r}) \phi_n^*(\mathbf{r}'). \qquad (10.75)$$

There is of course a trade-off to be made going from the most general
Eq. (10.69) to UHF and RHF: less restrictions on the allowed HF sp states
means a lower HF energy, but also worse symmetry properties of the HF
ground state. The UHF Eqs. (10.72) are commonly used for situations with
unpaired electrons (*e.g.* ionic or radical molecular species with a non-singlet
ground state), and can be combined with projection techniques to cure possible spin contamination of the HF ground state. However, when a serious
break-down of RHF stability occurs (as happens *e.g.* in the molecular dissociation limit discussed in Sec.10.3.3), this usually signals the inadequacy
of a description in terms of a single Slater determinant, and it is better to
give up such a starting point altogether.

For electrons in atoms there is a higher symmetry, since the external
potential is spherically symmetric and the Hamiltonian also commutes with
the total orbital angular momentum operator $\hat{\boldsymbol{L}}$ corresponding to $\sum_i \hat{\boldsymbol{\ell}}_i$.
Similar considerations can be made as in the molecular case. In unrestricted
HF the radial parts of the HF sp states are allowed to depend on the
projection quantum numbers m_ℓ and m_s, and the HF ground state is an
eigenstate of $\hat{L}_z$ and $\hat{S}_z$, but not of $\hat{\boldsymbol{L}}^2$ and $\hat{\boldsymbol{S}}^2$. Restricted HF, where the
radial part of the HF sp states does not depend on m_ℓ and m_s, provides a
good approximation to describe the $L = S = 0$ ground state of closed-shell
atoms. All $n\ell$-shells are assumed to be filled with $2(2\ell + 1)$ electrons, and
the HF Slater determinant is an $L = S = 0$ eigenstate.

10.2 Atoms

10.2.1 Closed-shell configurations

The HF equations for an atomic closed-shell configuration can be derived from the RHF equations in Eq. (10.74) with the spherical ansatz

$$\phi_i(\mathbf{r}) = \varphi_{n\ell}(r)Y_{\ell m_\ell}(\hat{\mathbf{r}}), \tag{10.76}$$

for the HF sp states. Multiplying Eq. (10.74) with $Y^*_{\ell m_\ell}(\hat{\mathbf{r}})$ and integrating over $\hat{\mathbf{r}}$ leads to equations for the radial wave functions $\varphi_{n\ell}(r)$

$$\varepsilon_{n\ell}\varphi_{n\ell}(r) = \int d\hat{\mathbf{r}}\, Y^*_{\ell m_\ell}(\hat{\mathbf{r}})\left\{\left[-\frac{1}{2}\boldsymbol{\nabla}^2 - \frac{Z}{r} + v_H(r)\right]\varphi_{n\ell}(r)Y_{\ell m_\ell}(\hat{\mathbf{r}})\right.$$
$$\left. -\frac{1}{2}\int d\mathbf{r}'\,\frac{n_{HF}(\mathbf{r}',\mathbf{r})}{|\mathbf{r}-\mathbf{r}'|}\varphi_{n\ell}(r')Y_{\ell m_\ell}(\hat{\mathbf{r}}')\right\}, \tag{10.77}$$

provided the right-hand side of Eq. (10.77) is independent of m_ℓ. This independence is obvious for the kinetic and central potential term, *e.g.*

$$\boldsymbol{\nabla}^2\left[\varphi_{n\ell}(r)Y_{\ell m_\ell}(\hat{\mathbf{r}})\right] = \left(\frac{1}{r}\frac{\partial^2}{\partial r^2}r - \frac{\ell(\ell+1)}{r^2}\right)\varphi_{n\ell}(r)Y_{\ell m_\ell}(\hat{\mathbf{r}}), \tag{10.78}$$

but the Hartree and Fock terms require a bit more explanation.

Since each $(n\ell)$-shell is fully occupied, the spin-integrated one-body density matrix in Eq. (10.75) can be expressed as

$$n_{HF}(\mathbf{r}',\mathbf{r}) = 2\sum_{n\ell}\varphi_{n\ell}(r)\varphi_{n\ell}(r')\sum_{m_\ell=-\ell}^{\ell}Y_{\ell m_\ell}(\hat{\mathbf{r}})Y^*_{\ell m_\ell}(\hat{\mathbf{r}}')$$
$$= 2\sum_{n\ell}\varphi_{n\ell}(r)\varphi_{n\ell}(r')\frac{2\ell+1}{4\pi}P_\ell(\cos\omega), \tag{10.79}$$

where $P_\ell(x)$ is the Legendre polynomial of order ℓ, and ω is the angle between $\hat{\mathbf{r}}$ and $\hat{\mathbf{r}}'$. The electron density is the diagonal part of $n_{HF}(\mathbf{r},\mathbf{r}')$ and becomes [note that $P_\ell(1) = 1$]

$$n^{HF}(\mathbf{r}) = n_{HF}(\mathbf{r},\mathbf{r}) = \frac{1}{4\pi}\sum_{n\ell}2(2\ell+1)\varphi_{n\ell}^2(r). \tag{10.80}$$

The closed-shell density in Eq. (10.80) is spherically symmetric, only depending on $r = |\mathbf{r}|$. We now evaluate the Hartree potential [Eq. (10.67)]

$$v_H(\mathbf{r}) = \int d\mathbf{r}'\,\frac{n_{HF}(r')}{|\mathbf{r}-\mathbf{r}'|}. \tag{10.81}$$

To work out the angular integration, one needs the following expansion for the reciprocal distance between $\boldsymbol{r}$ and $\boldsymbol{r}'$

$$\frac{1}{|\boldsymbol{r}-\boldsymbol{r}'|} = \sum_{L=0}^{\infty} \frac{r_<^L}{r_>^{L+1}} P_L(\cos\omega), \tag{10.82}$$

where $r_<$ is the smaller and $r_>$ is the larger of the pair (r, r'). Only the $L = 0$ contribution in Eq. (10.82) survives the angular integration in Eq. (10.81), because of the orthogonality properties of the Legendre polynomials

$$\int d\hat{\boldsymbol{r}}' \, P_L(\cos\omega) = 2\pi \int_{-1}^{+1} dx \, P_L(x) = 4\pi\delta_{L,0}. \tag{10.83}$$

As a consequence, the Hartree potential is spherically symmetric

$$v_H(r) = 4\pi \int dr' r'^2 \frac{n_{HF}(r')}{r_>}, \tag{10.84}$$

and the Hartree term in Eq. (10.77) is seen to be independent of m_ℓ.

Finally, the Fock term in Eq. (10.77) becomes, with the aid of expressions (10.79) and (10.82),

$$(\hat{v}_F\varphi_{n\ell})(r) = \frac{1}{2}\int d\hat{\boldsymbol{r}}\, Y^*_{\ell m_\ell}(\hat{\boldsymbol{r}}) \int d\boldsymbol{r}' \frac{n_{HF}(\boldsymbol{r}',\boldsymbol{r})}{|\boldsymbol{r}-\boldsymbol{r}'|} \varphi_{n\ell}(r') Y_{\ell m_\ell}(\hat{\boldsymbol{r}}')$$

$$= \sum_{n'\ell'} \varphi_{n'\ell'}(r) \sum_{L=0}^{\infty} \int dr' r'^2 \varphi_{n'\ell'}(r') \varphi_{n\ell}(r') \frac{r_<^L}{r_>^{L+1}} C_{\ell\ell'L}, \tag{10.85}$$

where the angular integrations have been combined into coefficients

$$C_{\ell\ell'L} = \sum_{m'_\ell M_L} \frac{4\pi}{2L+1} \int d\hat{\boldsymbol{r}}\, Y^*_{\ell m_\ell}(\hat{\boldsymbol{r}}) Y_{LM_L}(\hat{\boldsymbol{r}}) Y_{\ell'm'_\ell}(\hat{\boldsymbol{r}})$$

$$\int d\hat{\boldsymbol{r}}'\, Y^*_{\ell'm'_\ell}(\hat{\boldsymbol{r}}') Y^*_{LM_L}(\hat{\boldsymbol{r}}') Y_{\ell m_\ell}(\hat{\boldsymbol{r}}'). \tag{10.86}$$

The integrated product of three spherical harmonics is a real number, which can be expressed as

$$\int d\hat{\boldsymbol{r}}\, Y^*_{\ell m_\ell}(\hat{\boldsymbol{r}}) Y_{LM_L}(\hat{\boldsymbol{r}}) Y_{\ell'm'_\ell}(\hat{\boldsymbol{r}}) = \frac{\sqrt{(2\ell+1)(2\ell'+1)(2L+1)}}{\sqrt{4\pi}} \tag{10.87}$$

$$\times (-1)^{m_\ell} \begin{pmatrix} \ell & L & \ell' \\ -m_\ell & M_L & m'_\ell \end{pmatrix} \begin{pmatrix} \ell & L & \ell' \\ 0 & 0 & 0 \end{pmatrix},$$

in terms of the $3j$-symbols of standard angular momentum algebra (see App. B). This implies that the summation over L in Eq. (10.85) is restricted to the finite interval $|\ell - \ell'| \leq L \leq \ell + \ell'$. Using the normalization property

$$\sum_{m'_\ell M_L} \begin{pmatrix} \ell & L & \ell' \\ -m_\ell & M_L & m'_\ell \end{pmatrix}^2 = \frac{1}{2\ell+1} \tag{10.88}$$

one arrives at

$$C_{\ell \ell' L} = (2\ell'+1) \begin{pmatrix} \ell & L & \ell' \\ 0 & 0 & 0 \end{pmatrix}^2, \tag{10.89}$$

which is indeed independent of m_ℓ.

Combining all of the above results, the HF equations for an atomic closed-shell configuration become

$$\varepsilon_{n\ell}\varphi_{n\ell}(r) = \left\{ -\frac{1}{2}\left[\frac{1}{r}\frac{\partial^2}{\partial r^2}r - \frac{\ell(\ell+1)}{r^2}\right] - \frac{Z}{r} + v_H(r) \right\} \varphi_{n\ell}(r)$$
$$- (\hat{v}_F\varphi_{n\ell})(r). \tag{10.90}$$

This represents a set of non-linear integro-differential equations, which can be solved by a variety of methods.

Near $r = 0$ the HF orbitals have the usual behavior for a central potential problem, i.e. $\varphi_{n\ell}(r) \sim r^\ell$. The asymptotic behavior $(r \to \infty)$ is a bit more tricky [Handy et al. (1969)], due to the presence of the non-local Fock potential and the long range of the Coulomb force. One can show that for occupied HF orbitals the asymptotic potential behaves as $(N - Z - 1)/r + w(r)$, where Z is the central charge, N is the number of electrons, and $w(r)$ is a residual contribution, decreasing faster than $1/r$. Moreover, all occupied orbitals have the same decay, $\varphi_{n\ell}(r) \sim e^{-\kappa r}$, where $\kappa = \sqrt{2\varepsilon}$ is determined by the largest occupied HF sp energy ε. For unoccupied orbitals there is no cancellation between the Hartree and Fock contributions. The asymptotic potential is less attractive and behaves as $(N - Z)/r + w(r)$. As a result, HF typically cannot predict bound unoccupied states for neutral atoms $(N = Z)$.

In Fig. 10.2, the occupied HF sp orbitals in the Ne atom are compared with the corresponding hydrogenic orbitals. The most deeply bound orbitals ($1s$) are very similar. In contrast, the HF valence orbitals $2s$ and $2p$ are pushed outward compared to the hydrogenic $2s$ and $2p$, because the central charge is screened by the $1s$ electrons.

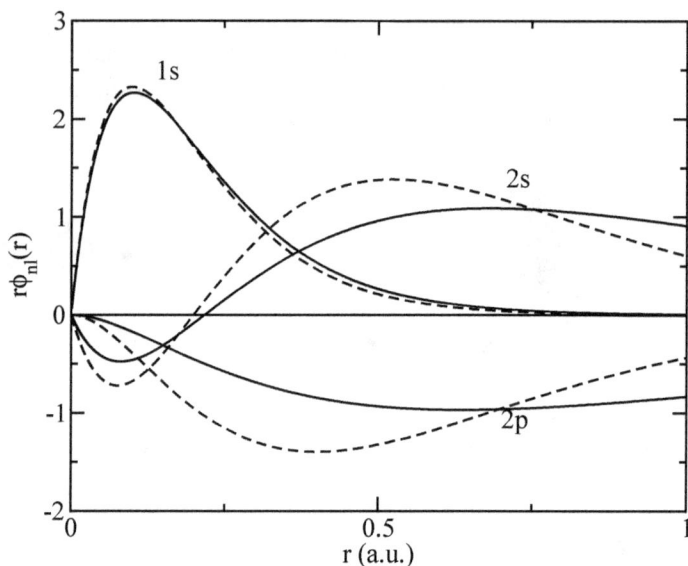

Fig. 10.2   The HF $1s$, $2s$, and $2p$ orbitals in the Ne atom ($Z = 10$), represented by full lines, are compared with the corresponding hydrogenic orbitals (dashed lines).

10.2.2   *Comparison with experimental data*

In Table 10.1, experimental results for the binding energy and the removal energies in a number of $L = S = 0$ closed-shell atoms are compared with the HF predictions. It is clear that HF in atomic systems is a good starting point, which is able to explain the bulk of the binding energy. In accordance with Koopman's theorem, the removal energies are in reasonable agreement with the data as well. Results are somewhat worse for Be and Mg, for which a representation as a pure closed-shell system is less adequate. Yet there is little reason to be smug about the performance of HF. The total energy, *e.g.*, is dominated by the rather inert core electrons. Chemical binding between atoms however, is determined by the valence electrons and is sensitive to small energy differences. In electronic systems the deviations from HF can therefore be crucial, and one often defines the *correlation energy* as the difference between the exact and the HF energy.

The spectroscopic factors associated with the removal states also point to the presence of small but nonnegligible deviations from the HF picture. As discussed in Ch. 7, experimentally one finds in $(e, 2e)$ reactions spectro-

Table 10.1   Hartree–Fock results and experimental data.

		Removal energies		Total energy	
		HF	Exp.	HF	Exp.
He	$1s$	-0.918	-0.9040	-2.862	-2.904
Be	$1s$	-4.733	-4.100	-14.573	-14.667
	$2s$	-0.309	-0.343		
Ne	$1s$	-32.77	-31.70	-128.547	-128.928
	$2s$	-1.930	-1.782		
	$2p$	-0.850	-0.793		
Mg	$1s$	-49.03	-47.91	-199.615	-200.043
	$2s$	-3.768	-3.26		
	$2p$	-2.283	-1.81		
	$3s$	-0.253	-0.2811		
Ar	$1s$	-118.6	-117.87	-526.818	-527.549
	$2s$	-12.32	-12.00		
	$2p$	-9.571	-9.160		
	$3s$	-1.277	-1.075		
	$3p$	-0.591	-0.579		

Hartree–Fock results for a number of $L = S = 0$ atoms, compared with experimental data. All energies are in atomic units (Hartree). Experimental removal energies taken from [Martin *et al.* (2002)], total energies from [Veillard and Clementi (1968)].

scopic factors $S \approx 0.90 - 0.95$ for the valence states, whereas HF predicts $S = 1$. The $(e, 2e)$ reaction also exhibits a considerable fragmentation of more deeply bound sp states, like the $3s$ in argon, which is not described by the HF results. To explain these discrepancies higher-order contributions to the self-energy must be included, as discussed in Ch. 11.

10.2.3   *Numerical details*

In a numerical solution of the HF equations (10.90), one usually chooses a grid of radial points r_i, in order to convert the continuous equations to a discrete (matrix) problem. Since the sp wave functions vary much more rapidly near the central charge (the region of small r) than in the tail region, an equidistant r_i grid is not practical. Defining a new independent variable through a logarithmic transformation $u = \ln(r)$ is therefore quite useful in atomic problems. The u-variable, which has a range $-\infty < u < +\infty$, can then be sampled on an equidistant grid. In practice, one takes a finite interval $[u_{min}, u_{max}]$, supplemented with suitable boundary conditions.

In combination with new dependent variables, $h_{n\ell}(u)$ is defined as

$$h_{n\ell}(u) = r^{\frac{3}{2}} \varphi_{n\ell}(r), \qquad (10.91)$$

the HF equations (10.90) transform as

$$\varepsilon_{n\ell} h_{n\ell}(u) = \left\{ -\frac{1}{2} \left[\frac{1}{r} \frac{\partial^2}{\partial u^2} \frac{1}{r} - \frac{(\ell + \frac{1}{2})^2}{r^2} \right] - \frac{Z}{r} + v_H(u) \right\} h_{n\ell}(u)$$
$$- (\hat{v}_F h_{n\ell})(u). \qquad (10.92)$$

The sp wave functions $h_{nl}(u)$ are now normalized as

$$\int_{-\infty}^{\infty} du\, h_{n\ell}(u) h_{n'\ell'}(u) = \delta_{n,n'} \delta_{\ell,\ell'}. \qquad (10.93)$$

The Hartree term in Eq. (10.92) reads

$$v_H(u) = \int du' \frac{1}{r_>} \sum_{n'\ell'} 2(2\ell' + 1) h_{n'\ell'}^2(u'), \qquad (10.94)$$

and the Fock term can be written as

$$(\hat{v}_F h_{n\ell})(u) = \sum_{n'\ell'} h_{n'\ell'}(u) \sum_L \int du' \frac{r_<^L}{r_>^{L+1}} C_{\ell\ell'L} h_{n'\ell'}(u') h_{n\ell}(u').$$
$$= \int du' v_F^{(\ell)}(u, u') h_{n\ell}(u'), \qquad (10.95)$$

in terms of a nonlocal potential

$$v_F^{(\ell)}(u, u') = \sum_{n'\ell'} h_{n'\ell'}(u) \sum_L \frac{r_<^L}{r_>^{L+1}} C_{\ell\ell'L} h_{n'\ell'}(u'). \qquad (10.96)$$

Tackling the nonlinear Eqs. (10.92) with iterative methods implies that one has to repeatedly solve equations of the form

$$\left[\hat{T} - \frac{Z}{r} + v_H(u) - \hat{v}_F \right] h(u) = \varepsilon h(u), \qquad (10.97)$$

where $\hat{T}$ is the kinetic energy, and the local Hartree potential $v_H(u)$ and the nonlocal Fock potential $v_F^{(\ell)}(u, u')$ are fixed by the sp wave functions $\{h_{nl}^{old}\}$ of the previous iteration step. The eigenvalue problem in Eq. (10.97) then determines both the new energy ε and the new wave function $h(u)$. After transposing Eq. (10.97) on the discrete u-grid, this is simply a matter of matrix diagonalization. However, the presence of the nonlocal Fock potential $v_F^{(\ell)}(u, u')$ would require diagonalization of a matrix with a dimension equal to the number of grid points. It is technically much easier to treat the Fock term as an inhomogeneous term. This can be done by replacing Eq. (10.97) with the following consistency loop:

$$h^{(1)}(u) = h^{old}(u) \quad \text{(Initialization)} \tag{10.98}$$

$$y^{(i)}(u) = (\hat{v}_F h^{(i)})(u) \tag{10.99}$$

$$\varepsilon^{(i)} = \int du\, h^{(i)}(u) \left[\left(\hat{T} - \frac{Z}{r} + v_H(u) \right) h^{(i)}(u) - y^{(i)}(u) \right] \tag{10.100}$$

$$\left(\hat{T} - \frac{Z}{r} + v_H(u) - \varepsilon^{(i)} \right) x^{(i)}(u) = y^{(i)}(u) \tag{10.101}$$

$$h^{(i+1)}(u) = x^{(i)}(u) / \sqrt{\int du' |x^{(i)}(u')|^2}. \tag{10.102}$$

After convergence of the sequence $(h^{(1)}, h^{(2)}, \ldots)$, this is obviously equivalent to Eq. (10.97). The advantage is that Eq.(10.101) only involves the kinetic term and purely local contributions. On a grid, this gives rise to a tridiagonal matrix structure, and the resulting linear system can be easily solved by recursion techniques (see also the computer exercise in Sec.10.2.4).

10.2.4 *Computer exercise*

Reproduce the HF results in Table 10.1 using the ideas in Sec. 10.2.3, combined with the (quick and dirty) discretization techniques explained below.

• Introduce boundaries $r_{min} = 10^{-7}/Z$ and $r_{max} = 25$ for the radial distance r, and a corresponding grid for $u = \ln(r)$,

$$u_i = c + (i-1)\Delta, \ i = 1, \ldots, M, \tag{10.103}$$

where $M \approx 1000$, $c = \ln(r_{min})$ and

$$\Delta = \frac{1}{M-1} \ln \left(\frac{r_{max}}{r_{min}} \right). \tag{10.104}$$

• For integrations use the crude approximation

$$\int du\, G(u) \to \Delta \sum_{i=1}^{M} G(u_i). \tag{10.105}$$

As a result, integral transformations of the type

$$F(u) = \int du' \frac{r_<^L}{r_>^{L+1}} G(u'), \qquad (10.106)$$

which appear in the Hartree and Fock term, can be done in only $\mathcal{O}(M)$ operations by means of the following recursion relations:

$$V_0 = 0; \quad V_i = V_{i-1} + G(u_i)r_i^L, \quad i = 1, \dots, M \qquad (10.107)$$

$$W_0 = 0; \quad W_i = W_{i-1} + G(u_i)/r_i^{L+1}, \quad i = 1, \dots, M \qquad (10.108)$$

$$F(u_i) = \Delta \left[\frac{V_i}{r_i^{L+1}} + r_i^L (W_M - W_i) \right], \quad i = 1, \dots, M. \qquad (10.109)$$

• Approximate the second-order differential operator in Eq.(10.92) on the grid as

$$\left(\frac{1}{r} \frac{\partial^2}{\partial u^2} \frac{1}{r} \right) h(u) \rightarrow \left(-\frac{2h(u_i)}{r_i^2} + \frac{h(u_{i+1})}{r_i r_{i+1}} + \frac{h(u_{i-1})}{r_i r_{i-1}} \right) \frac{1}{\Delta^2}, \qquad (10.110)$$

which gives rise to a symmetric tridiagonal matrix. Corrections at the boundaries can be neglected.

• If A_{ij} is a symmetric tridiagonal matrix with nonzero elements

$$A_{ii} = a_i, \ i = 1, \dots, M; \quad A_{i,i+1} = A_{i+1,i} = b_i, \ i = 1, \dots, M-1, \qquad (10.111)$$

then linear systems of the type

$$\sum_{j=1}^{M} A_{ij} X_j = Y_i, \ i = 1, \dots, M, \qquad (10.112)$$

can be solved for X_i by means of the following recursion relations:

$$U_M = a_M; \quad U_i = a_i - \frac{b_i^2}{U_{i+1}}, \ i = M-1, \dots, 1 \qquad (10.113)$$

$$V_M = \frac{Y_M}{U_M}; \quad V_i = (Y_i - b_i V_{i+1})/U_i, \ i = M-1, \dots, 1 \qquad (10.114)$$

$$X_1 = V_1; \quad X_i = V_i - \frac{b_{i-1}}{U_i} X_{i-1}, \ i = 2, \dots, M. \qquad (10.115)$$

• Between successive iterations an orthogonalization step should be performed for the sp wave functions having the same ℓ. For a given set of K nonorthogonal vectors $(X^{(1)}, X^{(2)}, \dots, X^{(K)})$ construct the $K \times K$ overlap matrix

$$S_{\mu\nu} = \sum_{i=1}^{M} X_i^{(\mu)} X_i^{(\nu)}, \tag{10.116}$$

and define the new orthonormal set $(X'^{(1)}, X'^{(2)}, \ldots, X'^{(K)})$ as

$$X'^{(\mu)} = \sum_{\nu=1}^{K} [S^{-\frac{1}{2}}]_{\mu\nu} X^{(\nu)}. \tag{10.117}$$

- The complete program flow should look like this:

1. Initialize the sp wave functions $\{h_{nl}^{old}(u)\}$ with hydrogenic orbitals.
2. With a fixed set of $\{h_{nl}^{old}(u)\}$:
(a) For each nl: construct a new orbital $h_{nl}(u)$ as a solution of Eq. (10.92), with v_H and $\hat{v}_F$ evaluated, using the set $\{h_{nl}^{old}(u)\}$. To do this, iterate the consistency loop in Eqs. (10.99) – (10.102) until $h_{nl}(u)$ has converged. Supplement Eq. (10.101) with an extra statement, setting the sp energy ε to zero if the value, obtained through Eq. (10.101), is positive. This is needed to guide the solution through the rough terrain of the first few iterations.
3. Orthogonalize the set of new $\{h_{nl}(u)\}$.
4. Monitor the convergence of all wave functions. If needed, set $\{h_{nl}^{old}(u)\} = \{h_{nl}(u)\}$ and repeat step 2.

While the above implementation has the advantage of simplicity, it should be noted that speed and accuracy can be greatly enhanced using more sophisticated numerical techniques. A good book on dedicated coordinate-space HF programs is [Froese Fischer et al. (1997)]; the corresponding atomic HF code is also publicly available on the internet.

10.3   Molecules

10.3.1   *Molecular problems*

The total (electrostatic) Hamiltonian for a molecular system consisting of M atomic nuclei and N electrons is

$$H = -\sum_{A=1}^{M} \frac{1}{2M_A} \nabla_A^2 + \sum_{A<B=1}^{M} \frac{Z_A Z_B}{|\mathbf{r}_A - \mathbf{r}_B|} \tag{10.118}$$
$$-\frac{1}{2} \sum_{i=1}^{N} \nabla_i^2 + \sum_{i<j=1}^{N} \frac{1}{|\mathbf{r}_i - \mathbf{r}_j|} - \sum_{i=1}^{N} \sum_{A=1}^{M} \frac{Z_A}{|\mathbf{r}_i - \mathbf{r}_A|},$$

where $M_A, Z_A, \boldsymbol{r}_A$ $(A = 1, \ldots, M)$ are the mass, charge and position vectors of the nuclei (considered as point particles), and $\boldsymbol{r}_i$ $(i = 1, \ldots, N)$ are the electron coordinates.

The nuclear masses are much larger than the electron mass, and the velocities of the nuclei are consequently much smaller than the electronic velocities. It is therefore usually a very good approximation [Born and Oppenheimer (1927)] to assume that the electrons move in a static field, generated by the nuclei at fixed positions $\boldsymbol{r}_A, \boldsymbol{r}_B, \ldots$. This is the so-called Born–Oppenheimer approximation, which decouples the nuclear and electronic motion into separate problems.

First one has to find the electronic ground state for fixed nuclear positions $\{\boldsymbol{r}_A\}$ which are considered as external parameters for the electronic problem. Let's denote with $\Psi_{el}(\{\boldsymbol{r}_A\})$ the ground state of the many-electron Hamiltonian

$$H_{el}(\{\boldsymbol{r}_A\}) = -\frac{1}{2}\sum_{i=1}^{N} \nabla_i^2 + \sum_{i<j=1}^{N} \frac{1}{|\boldsymbol{r}_i - \boldsymbol{r}_j|} - \sum_{i=1}^{N}\sum_{A=1}^{M} \frac{Z_A}{|\boldsymbol{r}_i - \boldsymbol{r}_A|}, \quad (10.119)$$

corresponding to the ground-state energy

$$E_{el}(\{\boldsymbol{r}_A\}) = \langle \Psi_{el}(\{\boldsymbol{r}_A\})| \hat{H}_{el}(\{\boldsymbol{r}_A\}) |\Psi_{el}(\{\boldsymbol{r}_A\})\rangle. \quad (10.120)$$

The motion of the nuclei is then governed by the nuclear Hamiltonian

$$H_{nuc} = -\sum_{A=1}^{M} \frac{1}{2M_A} \nabla_A^2 + \sum_{A<B=1}^{M} \frac{Z_A Z_B}{|\boldsymbol{r}_A - \boldsymbol{r}_B|} + E_{el}(\{\boldsymbol{r}_A\}), \quad (10.121)$$

where the electronic ground-state energy $E_{el}(\{\boldsymbol{r}_A\})$ plays the role of a (3M-dimensional) potential surface for the nuclei. The Hamiltonian (10.121) can be treated classically or quantummechanically, and completely determines the rotational and vibrational properties of the molecule, and the dynamical behavior in chemical reactions.

In a general situation, a complete scan of the multidimensional potential surface is out of the question, but it is possible to perform geometry optimization, *i.e.* to find the set of nuclear positions $\{\boldsymbol{r}_A\}$ which minimize the potential energy

$$V_{nuc}(\{\boldsymbol{r}_A\}) = \left(E_{el}(\{\boldsymbol{r}_A\}) + \sum_{A<B=1}^{M} \frac{Z_A Z_B}{|\boldsymbol{r}_A - \boldsymbol{r}_B|} \right). \quad (10.122)$$

This is the classical equilibrium geometry, which determines the spatial structure of the molecule (bond lengths and bond angles). Moreover, it provides a good starting point to describe intramolecular motion in terms of small departures from the equilibrium geometry, using the standard normal mode approximation.

10.3.2 *Hartree–Fock with a finite discrete basis set*

Finding the electronic structure in the complicated external potential of the nuclear charges can be exceedingly difficult. Except for some simple cases like diatomic molecules, even the single-electron problem is impossible to solve in coordinate space. One therefore introduces a finite number of (well-chosen) basis functions of known analytic form, usually resembling the orbitals of the isolated atoms present in the molecule. The adopted basis set, which determines the sp space in the problem, is by nature nonorthogonal, as it contains atomic orbitals centered on different nuclei.

Expanding the RHF spatial orbitals $\phi_n(\boldsymbol{r})$ of Eq.(10.74) in such a nonorthogonal basis set $\{\zeta_\alpha(\boldsymbol{r})\}$, one gets

$$\phi_n(\boldsymbol{r}) = \sum_\alpha x_{n\alpha}\zeta_\alpha(\boldsymbol{r}), \qquad (10.123)$$

where we may assume (without loss of generality) that all spatial electron orbitals are real functions. The RHF equations become, after projection on $\zeta_\beta(\boldsymbol{r})$,

$$\varepsilon_n \sum_\alpha x_{n\alpha} S_{\beta\alpha} = \sum_\alpha x_{n\alpha}[H_{HF}]_{\beta\alpha}. \qquad (10.124)$$

The nonorthogonality of the basis set is reflected in the presence of the overlap matrix

$$S_{\beta\alpha} = \int d\boldsymbol{r}\, \zeta_\alpha(\boldsymbol{r})\zeta_\beta(\boldsymbol{r}), \qquad (10.125)$$

and the matrix elements of the HF Hamiltonian

$$
\begin{aligned}
[H_{HF}]_{\beta\alpha} = &\int d\boldsymbol{r}\, \zeta_\beta(\boldsymbol{r}) \left(-\frac{\nabla^2}{2} - \sum_A \frac{Z_A}{|\boldsymbol{r} - \boldsymbol{r}_A|} \right) \zeta_\alpha(\boldsymbol{r}) \\
&+ \sum_{\gamma\delta} x_{n\gamma}x_{n\delta} \int \frac{d\boldsymbol{r}\,d\boldsymbol{r}'}{|\boldsymbol{r} - \boldsymbol{r}'|} \left[2\zeta_\alpha(\boldsymbol{r})\zeta_\beta(\boldsymbol{r})\zeta_\gamma(\boldsymbol{r}')\zeta_\delta(\boldsymbol{r}') \right. \\
&\left. - \zeta_\alpha(\boldsymbol{r}')\zeta_\beta(\boldsymbol{r})\zeta_\gamma(\boldsymbol{r})\zeta_\delta(\boldsymbol{r}') \right],
\end{aligned}
\qquad (10.126)
$$

can be expressed in terms of spatial integrals involving only the basis functions. The equations (10.124) are called the Roothaan equations [Roothaan (1951)], and represent a set of nonlinear algebraic equations in the unknown expansion parameters $x_{n\alpha}$. Obviously, the Roothaan equations with any finite basis set will yield a HF energy higher than the exact HF energy, usually called the *HF limit* in this context. However, quite reliable results can be obtained by a careful choice of a limited number of basis functions.

A basis function centered on atom A has the generic from

$$\zeta(\boldsymbol{r}) = \varphi(\boldsymbol{r} - \boldsymbol{r}_A). \tag{10.127}$$

Two choices are commonly adopted for the shape of the atomic orbital φ:

$$\varphi(\boldsymbol{r}) = P(x, y, z)e^{-\kappa r} \text{ (Slater type)} \tag{10.128}$$

$$\varphi(\boldsymbol{r}) = P(x, y, z)e^{-\alpha r^2} \text{ (Gaussian type).} \tag{10.129}$$

Here $P(x, y, z)$ is a polynomial, representing the symmetry character (ℓ value) of the orbital. The global shape is either exponential (Slater type) or Gaussian. The Slater type is more physical, as atomic orbitals do decrease exponentially, and have a cusp at the position of the central charge. The Gaussian type therefore has an unphysical behavior, both at $r = 0$ and at large r.

In practice, the numerical advantage of Gaussian-type wave functions is so great, that they are used in most of the present-day applications in quantum chemistry. This is because their use allows a fast computation of the matrix elements of the Coulomb interaction, even when orbitals centered on four different nuclei are involved. Moreover, the unphysical behavior of the single Gaussian orbital in Eq. (10.129) can be largely overcome by representing the orbital by a sum of (minimally three) Gaussians with different exponents, chosen so as to mimic a Slater-type orbital (see *e.g.* [Szabo and Ostlund (1989)]).

Tremendous effort has been put into the determination of optimum basis sets for molecular calculations. Together with the availability of increased computing power, such techniques have led to the present-day ability of performing ab-initio modeling of quite complicated molecules, within HF or within the framework of density functional theory. An example is shown in Fig. 10.3, representing the electron density of the buckminsterfullerene C_{60} molecule, obtained in the HF approximation using a standard Gaussian basis.

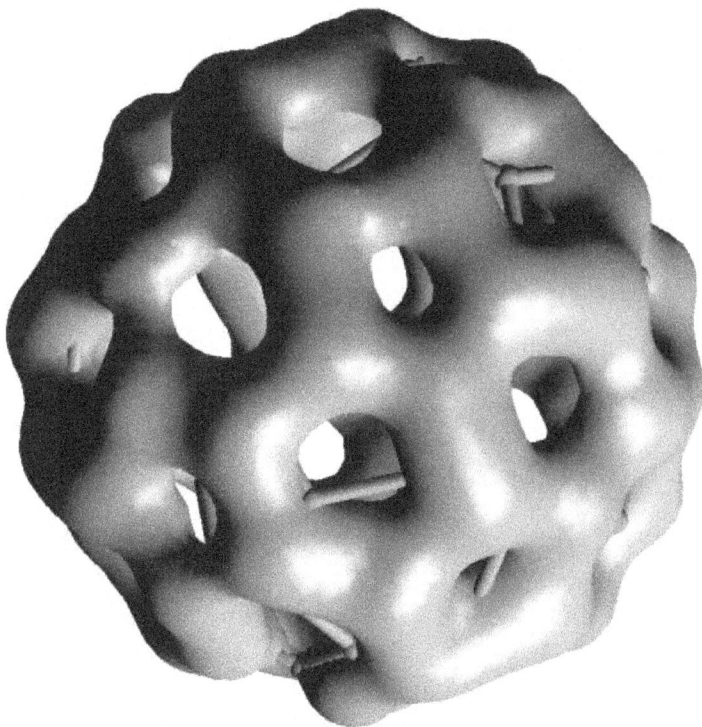

Fig. 10.3   A surface of equal electron density in the C_{60} buckminsterfullerene molecule.

10.3.3   *The hydrogen molecule*

As an example of molecular problems we discuss a highly simplified — but very instructive — model of the hydrogen molecule. The two protons P_1 and P_2 are at fixed positions $\boldsymbol{R}_1$ and $\boldsymbol{R}_2$ (Born–Oppenheimer approximation). The two-electron ground state will be a spin singlet, and the $S = 0$ spin part from the wave function can be split off

$$\Psi(\boldsymbol{r}_1 m_{s_1}, \boldsymbol{r}_2 m_{s_2}) = \Phi(\boldsymbol{r}_1, \boldsymbol{r}_2) \frac{1}{\sqrt{2}} \left(\delta_{m_{s_1},+\frac{1}{2}} \delta_{m_{s_2},-\frac{1}{2}} - \delta_{m_{s_1},-\frac{1}{2}} \delta_{m_{s_2},+\frac{1}{2}} \right).$$

(10.130)

Since the spin part is antisymmetric, the spatial part $\Phi(\boldsymbol{r}_1, \boldsymbol{r}_2)$ must be symmetric under $\boldsymbol{r}_1 \leftrightarrow \boldsymbol{r}_2$ interchange.

The finite sp basis set we employ, consists of atomic 1s orbitals centered on either one of the two protons P_1 and P_2, *i.e.*

$$\phi_i(\boldsymbol{r}) = \varphi_{1s}(\boldsymbol{r} - \boldsymbol{R}_i). \tag{10.131}$$

The following three symmetric combinations can be formed

$$\phi_1(\boldsymbol{r}_1)\phi_1(\boldsymbol{r}_2); \quad \phi_2(\boldsymbol{r}_1)\phi_2(\boldsymbol{r}_2); \quad [\phi_1(\boldsymbol{r}_1)\phi_2(\boldsymbol{r}_2) + \phi_2(\boldsymbol{r}_1)\phi_1(\boldsymbol{r}_2)]. \tag{10.132}$$

Because of an additional spatial symmetry (the mirror plane between the two protons), the two-electron wave function should be symmetric under $P_1 \leftrightarrow P_2$ interchange as well. Consequently there are only two two-electron states in the problem

$$\Phi_I(\boldsymbol{r}_1, \boldsymbol{r}_2) = \phi_1(\boldsymbol{r}_1)\phi_1(\boldsymbol{r}_2) + \phi_2(\boldsymbol{r}_1)\phi_2(\boldsymbol{r}_2), \tag{10.133}$$

$$\Phi_{II}(\boldsymbol{r}_1, \boldsymbol{r}_2) = \phi_1(\boldsymbol{r}_1)\phi_2(\boldsymbol{r}_2) + \phi_2(\boldsymbol{r}_1)\phi_1(\boldsymbol{r}_2). \tag{10.134}$$

The Φ_I configuration is called *ionic*, since both electrons are on the same atom, and has a higher energy than the Φ_{II} configuration. The ground state will nevertheless be a mixture of Φ_I and Φ_{II}, due to the interaction between both configurations. In the dissociation limit (for large distances $|\boldsymbol{R}_1 - \boldsymbol{R}_2|$ between the two nuclei) we do expect to reach a pure Φ_{II} configuration, because it is energetically more favorable to have two isolated neutral hydrogen atoms, than a H^- negative ion.

Of course we have the freedom to use, instead of the Φ_I and Φ_{II} two-electron states, any pair of linear independent combinations. Since Φ_I and Φ_{II} have equal norm, their sum and difference will be orthogonal

$$\Phi_B(\boldsymbol{r}_1, \boldsymbol{r}_2) = \Phi_I(\boldsymbol{r}_1, \boldsymbol{r}_2) + \Phi_{II}(\boldsymbol{r}_1, \boldsymbol{r}_2) \tag{10.135}$$

$$\Phi_A(\boldsymbol{r}_1, \boldsymbol{r}_2) = \Phi_I(\boldsymbol{r}_1, \boldsymbol{r}_2) - \Phi_{II}(\boldsymbol{r}_1, \boldsymbol{r}_2). \tag{10.136}$$

Looking in detail at the spatial two-electron wave functions Φ_B and Φ_A, one can easily check that they have the structure of a doubly occupied molecular sp orbital

$$\Phi_B(\boldsymbol{r}_1, \boldsymbol{r}_2) = \phi_b(\boldsymbol{r}_1)\phi_b(\boldsymbol{r}_2), \quad \Phi_A(\boldsymbol{r}_1, \boldsymbol{r}_2) = \phi_a(\boldsymbol{r}_1)\phi_a(\boldsymbol{r}_2). \tag{10.137}$$

The sp orbitals

$$\phi_b(\boldsymbol{r}) = \phi_1(\boldsymbol{r}) + \phi_2(\boldsymbol{r}), \quad \phi_a(\boldsymbol{r}) = \phi_1(\boldsymbol{r}) - \phi_2(\boldsymbol{r}), \tag{10.138}$$

are the bonding and antibonding orbital, familiar from the description of the single-electron hydrogen molecular ion H_2^+. The "exact" solution of the two-electron problem in this two-dimensional model space, then follows from diagonalization of the Hamiltonian matrix in the (Φ_A, Φ_B) tp basis. The ground-state energy is

$$E_0 = \frac{E_A + E_B}{2} - \sqrt{\left(\frac{E_A - E_B}{2}\right)^2 + \Delta^2}, \qquad (10.139)$$

where

$$E_A = \frac{\langle \Phi_A | \hat{H} | \Phi_A \rangle}{\langle \Phi_A | \Phi_A \rangle}, \quad E_B = \frac{\langle \Phi_B | \hat{H} | \Phi_B \rangle}{\langle \Phi_B | \Phi_B \rangle}, \qquad (10.140)$$

$$\Delta = \frac{\langle \Phi_A | \hat{H} | \Phi_B \rangle}{\sqrt{\langle \Phi_A | \Phi_A \rangle \langle \Phi_B | \Phi_B \rangle}}. \qquad (10.141)$$

In order to obtain the RHF results, we would now have to solve the Roothaan Eqs. (10.126) in the adopted sp space by means of an iterative procedure. Fortunately, the present model is so simple that we can skip this. The sp space contains merely ϕ_1 and ϕ_2, and the only spatial sp wave functions, which are compatible with the symmetry requirements, are precisely the bonding and antibonding combinations of Eq. (10.138). These must therefore coincide with the RHF sp basis, the bonding orbital being the doubly occupied orbital, and the antibonding orbital being unoccupied. As a consequence, the RHF ground state is the Φ_B configuration of Eq. (10.135), which together with the $S = 0$ spin wave function can be written as a single Slater determinant, and the RHF energy is given by E_B in Eq. (10.140). This implies immediately that the molecular dissociation limit, where the spatial wave function is

$$\Phi_{II}(\boldsymbol{r}_1, \boldsymbol{r}_2) = [\Phi_B(\boldsymbol{r}_1, \boldsymbol{r}_2) - \Phi_A(\boldsymbol{r}_1, \boldsymbol{r}_2)], \qquad (10.142)$$

cannot be described in RHF, since Φ_{II} is an equal mixture of two closed-shell Slater determinants Φ_A and Φ_B.

For a numerical illustration, we approximate the atomic $1s$ orbital by means of a single Gaussian[2]

$$\varphi_{1s}(\boldsymbol{r}) = e^{-\alpha r^2}, \qquad (10.143)$$

with $\alpha = 0.42$. The matrix elements E_B, E_A and Δ in Eq. (10.140), which are functions of the internuclear distance $R = |\boldsymbol{R}_1 - \boldsymbol{R}_2|$, can all be

---

[2]A single Gaussian is sufficient for illustrative purposes, but should of course never be used in genuine calculations.

expressed in terms of exponentials and the function

$$\mathrm{f}(x) = \int_0^1 du\, e^{-xu^2} = \frac{\sqrt{\pi}}{2\sqrt{x}}\mathrm{Erf}(\sqrt{x}),$$ (10.144)

related to the error function

$$\mathrm{Erf}(x) = \frac{2}{\sqrt{\pi}} \int_0^x dy\, e^{-y^2}.$$ (10.145)

Taking symmetries of the spatial integrals into account, the one-electron integrals needed for the evaluation of E_B, E_A and Δ are

$$\langle \phi_1 | \phi_2 \rangle = \int d\mathbf{r}\, e^{-\alpha(\mathbf{r}-\mathbf{R}_1)^2} e^{-\alpha(\mathbf{r}-\mathbf{R}_2)^2}$$

$$= \left(\frac{\pi}{2\alpha}\right)^{\frac{3}{2}} e^{-\frac{\alpha}{2}R^2}$$ (10.146)

$$\langle \phi_1 | -\frac{1}{2}\nabla^2 |\phi_2 \rangle = \left(\frac{\pi}{2\alpha}\right)^{\frac{3}{2}} \frac{\alpha}{2}(3-\alpha R^2)e^{-\frac{\alpha}{2}R^2}$$ (10.147)

$$\langle \phi_1 | \frac{1}{|\mathbf{r}-\mathbf{R}_1|} |\phi_2 \rangle = \frac{\pi}{\alpha}e^{-\frac{\alpha}{2}R^2}\mathrm{f}(2\alpha R^2),$$ (10.148)

and the two-electron integrals are

$$(\phi_1\phi_1| V |\phi_2\phi_2) = \int d\mathbf{r}_1 d\mathbf{r}_2 e^{-\alpha(\mathbf{r}_1-\mathbf{R}_1)^2} e^{-\alpha(\mathbf{r}_2-\mathbf{R}_1)^2}$$

$$\frac{1}{|\mathbf{r}_1-\mathbf{r}_2|}e^{-\alpha(\mathbf{r}_1-\mathbf{R}_2)^2}e^{-\alpha(\mathbf{r}_2-\mathbf{R}_2)^2}$$

$$= \frac{1}{4}\left(\frac{\pi}{\alpha}\right)^{\frac{5}{2}} e^{-\alpha R^2}$$ (10.149)

$$(\phi_1\phi_2| V |\phi_1\phi_2) = \frac{1}{4}\left(\frac{\pi}{\alpha}\right)^{\frac{5}{2}} \mathrm{f}(\alpha R^2)$$ (10.150)

$$(\phi_1\phi_1| V |\phi_1\phi_2) = \frac{1}{4}\left(\frac{\pi}{\alpha}\right)^{\frac{5}{2}} e^{-\frac{\alpha}{2}R^2}\mathrm{f}(\frac{\alpha}{4}R^2).$$ (10.151)

The potential energy V of the H_2 molecule is the sum of the electronic ground-state energy E_0 and the Coulomb repulsion between the protons

$$V(R) = E_0(R) + \frac{1}{R}.$$ (10.152)

The dissociation curve $D(R)$, defined as

$$D(R) = V(R) - 2E(\text{H-atom}),$$ (10.153)

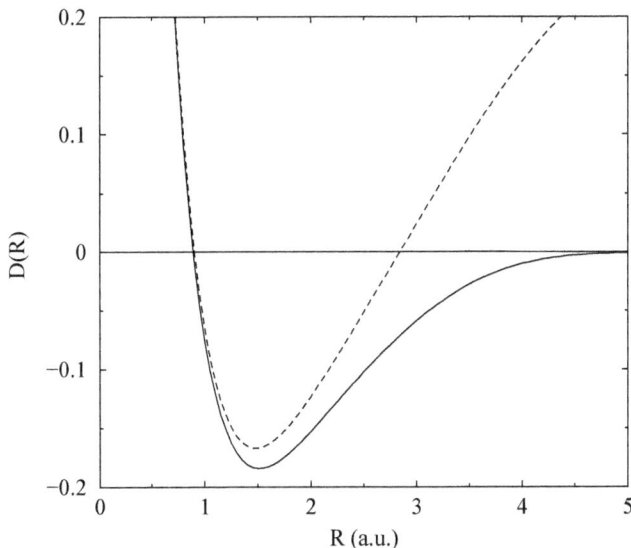

Fig. 10.4   Dissociation curves for the H_2 molecule with the minimal basis set of Gaussian form of Eqs. (10.131) and (10.143). Full and dashed line correspond to the exact solution and the RHF approximation, respectively.

is the potential energy minus the energy of the dissociation products (in this case twice the energy of the H-atom). The position R_e of the minimum in $V(R)$ or $D(R)$ determines the equilibrium geometry, *i.e.* the bond length of the H_2 molecule. The depth of the minimum, $D_e = D(R_e)$, is the dissociation energy.

The dissociation curve obtained in the present model, using Eqs. (10.129) and (10.140), and the $1s$ Gaussian orbital in Eq. (10.143), is shown in Fig. 10.4. It contains all the essential features of a more exact treatment: strong repulsion as $R \to 0$, a minimum corresponding to the equilibrium geometry of bound H_2, and a correct dissociation limit $D(R) \to 0$ as $R \to \infty$. Note that the use of a single Gaussian (having no cusp and bad asymptotics) for the $1s$ orbital, leads to an energy of the hydrogen atom

$$E(\text{H-atom}) = \frac{\langle \phi | -\frac{\nabla^2}{2} - \frac{1}{r} | \phi \rangle}{\langle \phi | \phi \rangle} = \frac{3\alpha}{2} - 2\sqrt{\frac{2\alpha}{\pi}} = -0.404, \qquad (10.154)$$

which deviates 20% from the exact value, $E_H = -0.500$. Since similar errors are made for the H_2 electronic energy, the final result for the dissociation

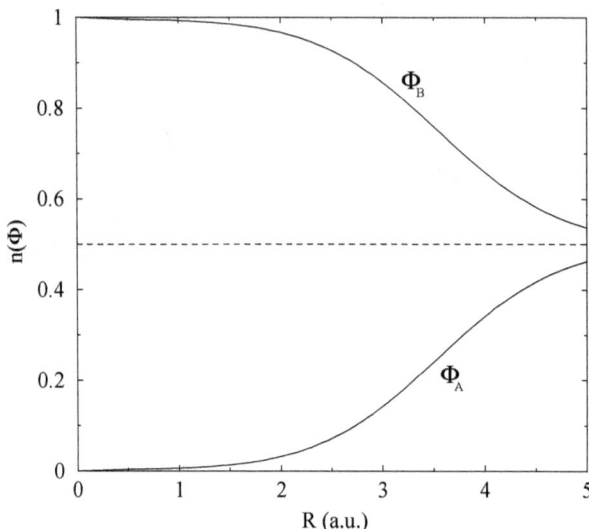

Fig. 10.5 Composition of the H_2 ground state corresponding to Eq. (10.139). In order to guide the eye, the 50% curve is indicated by the dashed line.

curve is very reasonable: a bond length $R_e = 1.51$ and dissociation energy $D_e = -0.184$, to be compared with the exact values[3] $R_e = 1.40$ and $D_e = -0.175$.

Also shown in Fig. 10.4 is the RHF dissociation curve

$$D^{RHF}(R) = E_B(R) + \frac{1}{R} - 2E(\text{H-atom}), \qquad (10.155)$$

obtained in the present model. Since $D^{RHF}(R)$ does not vanish for large R, RHF has an incorrect dissociation limit. This can be traced back to the fact that, in the $R \to \infty$ limit, $E_B(R)$ reduces to the energy of the H^- ion. Near the equilibrium distance R_e, however, the RHF curve provides a good approximation to the full $D(R)$. This is also seen in Fig. 10.5, where the occupation of the Φ_B and Φ_A configurations in the ground state are plotted. Up to $R = 2.5$ the true ground state is for more than 90% the RHF configuration Φ_B. For large R, the ground state goes to Φ_{II}, which is an equal mixture of Φ_A and Φ_B and cannot be described in RHF.

---

[3]The experimental value for the dissociation energy of H_2 is $D_0 = -0.164$. The difference with the equilibrium value $D_e = -0.175$ comes mainly from the zero-point energy of the vibrational mode.

10.4 Hartree–Fock in infinite systems

In the case of homogeneous infinite systems, a considerable simplification occurs: from the discussion on the Fermi gas in Sec. 5.1 it is clear that only plane-wave states of definite momentum are appropriate sp states in a translationally invariant system. Determination of the HF orbitals is therefore not needed, because the HF basis coincides with the plane-wave basis. In fact, it is easy to see that also the exact sp propagator[4]

$$G(\boldsymbol{p}, \boldsymbol{p}'; E) = \langle \Psi_0^N | a_{\boldsymbol{p}} \frac{1}{E - (\hat{H} - E_0^N) + i\eta} a_{\boldsymbol{p}'}^\dagger | \Psi_0^N \rangle$$
$$+ \langle \Psi_0^N | a_{\boldsymbol{p}'}^\dagger \frac{1}{E + (\hat{H} - E_0^N) - i\eta} a_{\boldsymbol{p}} | \Psi_0^N \rangle, \quad (10.156)$$

is automatically diagonal in momentum space, because total momentum $\boldsymbol{P} = \sum_{i=1}^N \boldsymbol{p}_i$ is conserved by the Hamiltonian

$$[\hat{H}, \hat{\boldsymbol{P}}] = 0, \quad (10.157)$$

and the ground state of the system has definite (zero) total momentum. Therefore the propagator in Eq. (10.156) is only nonzero when the same momentum is added and removed from the ground state

$$G(\boldsymbol{p}, \boldsymbol{p}'; E) \equiv \delta_{\boldsymbol{p}, \boldsymbol{p}'} G(p; E). \quad (10.158)$$

We introduce the notation $G(p; E)$ for the diagonal part of the sp propagator. Note too that because of the isotropy of the system (there is no preferred direction in space), the sp propagator can only depend on the magnitude $p = |\boldsymbol{p}|$ of the sp momentum $\boldsymbol{p}$. As an example, the propagator for a noninteracting system with Hamiltonian

$$\hat{H}_0 = \hat{T} + \hat{U} = \sum_{\boldsymbol{p}} \frac{p^2}{2m} a_{\boldsymbol{p}}^\dagger a_{\boldsymbol{p}} + \sum_{\boldsymbol{p}} U(p) a_{\boldsymbol{p}}^\dagger a_{\boldsymbol{p}}, \quad (10.159)$$

becomes

$$G^{(0)}(p; E) = \frac{\theta(p - p_F)}{E - \varepsilon(p) + i\eta} + \frac{\theta(p_F - p)}{E - \varepsilon(p) - i\eta}, \quad (10.160)$$

with sp energy $\varepsilon(p) = p^2/2m + U(p)$. From Eq. (9.2) it follows as well that the (irreducible) self-energy is diagonal in momentum space. As a

---

[4]For brevity of notation the discussion is restricted to fermions of a single species, but the considerations in Sec. 10.4 hold in general.

consequence, the Dyson equation becomes an algebraic relationship

$$G(p; E) = G^{(0)}(p; E) + G(p; E)\Sigma^*(p; E)G(p; E). \tag{10.161}$$

The derivation of the HF approximation for homogeneous infinite systems then becomes extremely simple. Since the HF basis coincides with the plane-wave basis, the HF one-body density matrix is according to Eq. (10.38)

$$n_{p',p}^{HF} = \delta_{p,p'}\theta(p_F - p), \tag{10.162}$$

equal to the Fermi gas step function.

As a result, the HF self-energy in Eq. (10.13) is given by

$$\Sigma^{HF}(p) = -U(p) + \sum_{p'} \theta(p_F - p') \langle pp' | V | pp' \rangle, \tag{10.163}$$

and the HF propagator is

$$G^{HF}(p; E) = \frac{\theta(p - p_F)}{E - \varepsilon^{HF}(p) + i\eta} + \frac{\theta(p_F - p)}{E - \varepsilon^{HF}(p) - i\eta}, \tag{10.164}$$

where

$$\varepsilon^{HF}(p) = \left(\frac{p^2}{2m} + U(p)\right) + \Sigma^{HF}(p)$$

$$= \frac{p^2}{2m} + \sum_{p'} \theta(p_F - p') \langle pp' | V | pp' \rangle, \tag{10.165}$$

is the HF sp energy, independent of the auxiliary potential $U(p)$.

The total energy is, according to Eq. (10.29),

$$E^{HF} = \frac{1}{2} \sum_{p} \theta(p_F - p) \left[\frac{p^2}{2m} + \varepsilon^{HF}(p)\right]$$

$$= T_{FG} + \frac{1}{2} \sum_{pp'} \theta(p_F - p)\theta(p_F - p') \langle pp' | V | pp' \rangle, \tag{10.166}$$

where $T_{FG} = \frac{3}{5} \frac{p_F^2}{2m}$ is the free Fermi-gas kinetic energy. Upon inspection, the same outcome is also obtained in first-order perturbation theory, with the kinetic energy $\hat{T}$ as unperturbed Hamiltonian, and the interaction $\hat{V}$ acting as perturbation. This is not surprising, since the translational invariance ensures that the HF ground state and the Fermi-gas ground state coincide

$$|\Phi_{FG}\rangle = |\Phi_{HF}\rangle = \prod_{p<p_F} a_p^\dagger |0\rangle \,. \tag{10.167}$$

The resulting first-order correction to the energy of the free Fermi gas is then

$$\langle \Phi_{FG}| \hat{V} |\Phi_{FG}\rangle = \frac{1}{2} \sum_{pp'} \theta(p_F - p)\theta(p_F - p') \langle pp'| V |pp'\rangle \,. \tag{10.168}$$

The HF energy in an infinite system therefore does not go beyond first-order perturbation theory, in contrast to finite systems where the determination of the shape of the HF orbitals already includes terms to all orders.

10.5   Electron gas

We assume, for simplicity, that we are dealing with the spin-unpolarized electron gas, where equal numbers of spin-up and spin-down electrons are present, or $\rho^{(m_s=+\frac{1}{2})} = \rho^{(m_s=-\frac{1}{2})} = \frac{1}{2}\rho$. The generalization to the spin-polarized case is straightforward (see Exercise (4) of this chapter). Since the electron-gas Hamiltonian conserves spin, the sp propagator and self-energy are diagonal in the spin projection quantum number m_s.

According to Eq. (10.163) the HF self-energy for the electron gas is

$$\Sigma^{HF}(p_1 m_{s_1}) = \sum_{p_2 m_{s_2}} \theta(p_F - p_2) \langle p_1 m_{s_1}, p_2 m_{s_2}| V' |p_1 m_{s_1}, p_2 m_{s_2}\rangle \,,$$
$$\tag{10.169}$$

with Fermi momentum $p_F = [3\pi^2\rho]^{\frac{1}{3}}$.

The antisymmetrized matrix element in Eq. (10.169) is given by

$$\langle p_1 m_{s_1}, p_2 m_{s_2}| V' |p_1 m_{s_1}, p_2 m_{s_2}\rangle = (p_1 m_{s_1}, p_2 m_{s_2}| V' |p_1 m_{s_1}, p_2 m_{s_2})$$
$$- (p_1 m_{s_1}, p_2 m_{s_2}| V' |p_2 m_{s_2}, p_1 m_{s_1})$$
$$= -\delta_{m_{s_1}, m_{s_2}} \frac{4\pi}{V} \frac{1}{|p_1 - p_2|^2} \,. \tag{10.170}$$

The first (direct or Hartree) term in Eq. (10.170) is zero, according to the definition in Eq. (5.28) for V' (no relative momentum is transferred). Only the second (exchange or Fock) term remains and the HF self-energy becomes

$$\Sigma^{HF}(p_1) = -\sum_{p_2} \theta(p_F - p_2)\frac{4\pi}{V}\frac{1}{|\boldsymbol{p}_1 - \boldsymbol{p}_2|^2}, \tag{10.171}$$

which (as was also clear from symmetry arguments) is independent of m_s.

Taking the thermodynamic limit and replacing the discrete sums over momentum with integrations leads to

$$\begin{aligned}\Sigma^{HF}(p_1) &= -\frac{V}{(2\pi)^3}\frac{4\pi}{V}\int d\boldsymbol{p}_2\,\theta(p_F - p_2)\frac{1}{|\boldsymbol{p}_1 - \boldsymbol{p}_2|^2}\\ &= -\frac{1}{\pi}\int_0^{p_F} dp_2\, p_2^2 \int_{-1}^{+1} dx\frac{1}{p_1^2 + p_2^2 - 2xp_1p_2}\\ &= -\frac{p_F}{\pi u_1}\int_0^1 du_2\, u_2\ln\left|\frac{u_2 + u_1}{u_2 - u_1}\right|,\end{aligned} \tag{10.172}$$

where the angular integration has been performed and dimensionless variables $u_2 = p_2/p_F$ and $u_1 = p_1/p_F$ were introduced. The remaining integral in Eq. (10.172) can be calculated using the standard primitives

$$x^n\ln|x| = \frac{d}{dx}\left[\frac{x^{n+1}}{n+1}\left(\ln|x| - \frac{1}{n+1}\right)\right], \tag{10.173}$$

and after some algebra, the HF exchange potential becomes

$$\Sigma^{HF}(p) = -\frac{p_F}{\pi}\left(1 + \frac{1 - u^2}{2u}\ln\left|\frac{1 + u}{1 - u}\right|\right). \tag{10.174}$$

Special values are $\Sigma^{HF}(0) = -2\frac{p_F}{\pi}$ and $\Sigma^{HF}(p_F) = -\frac{p_F}{\pi}$. The HF sp energies $\varepsilon^{HF}(p)$ are obtained by adding the kinetic energy contribution

$$\varepsilon^{HF}(p) = \frac{p^2}{2} + \Sigma^{HF}(p). \tag{10.175}$$

The HF potential $\Sigma^{HF}(p)$ and sp energy $\varepsilon^{HF}(p)$ are plotted in Fig. 10.6 as a function of $u = p/p_F$. The HF potential is negative, monotonically increasing, and goes to zero like

$$\Sigma^{HF}(p) \to -\frac{2}{3}\frac{p_F}{\pi}\frac{1}{u^2}, \tag{10.176}$$

in the $p \to \infty$ limit. Its main effect is a lowering of the sp energies $\varepsilon^{HF}(p)$ compared to the free sp spectrum.

The HF ground-state energy is, according to Eq. (10.29),

$$E^{HF} = \frac{1}{2} \sum_{\boldsymbol{p}m_s} \theta(p_F - p) \left[\frac{p^2}{2} + \varepsilon^{HF}(p) \right] = T_{FG} + E_x, \qquad (10.177)$$

and consists of the Fermi-gas kinetic energy T_{FG}

$$T_{FG} = N \frac{3}{10} p_F^2, \qquad (10.178)$$

to which the (negative) exchange energy E_x has been added

$$E_x = \frac{1}{2} \sum_{\boldsymbol{p}m_s} \theta(p_F - p) \Sigma^{HF}(p) = N \left(-\frac{3}{4\pi} p_F \right). \qquad (10.179)$$

Obtaining the last equality again requires the use of the primitives (10.173).

In Fig. 10.7, the density dependence of the HF energy for the spin-unpolarized electron gas

$$E^{HF}/N = \frac{3}{10} p_F^2 - \frac{3}{4\pi} p_F \qquad (10.180)$$

is displayed. Following convention, we do this as a function of

$$r_s = \left[\frac{3}{4\pi\rho} \right]^{\frac{1}{3}} = \left[\frac{9\pi}{4} \right]^{\frac{1}{3}} \frac{1}{p_F} \qquad (10.181)$$

being the radius (in atomic units) of a sphere containing one electron. Note that because of the variational nature of HF, the exact energy lies below this curve, so the electron gas is definitely bound for densities lower than $r_s \approx 2.5$. This may seem surprising in view of the repulsive Coulomb force between the electrons; the repulsion is globally compensated, however, by the attractive interaction with the positive background. The residual effect is dominated by the attractive exchange potential, which provides an explanation for the cohesion energy of metals, where the valence electrons move in the (lattice) background of the positive ions. As a matter of fact, the equilibrium point in the HF description of the electron gas corresponds to the minimum at $r_s = 4.82$ and $E/N = -0.0475$ in the energy versus density curve in Fig. 10.7. It roughly agrees with the experimental situation in solid Na with $r_s = 3.96$ and cohesion energy $E/N = -0.0415$.

The HF energy becomes exact in the limit of large densities, as can be seen on the basis of simple scaling arguments: taking the Fermi momentum as the natural scale, the electron gas Hamiltonian in Eq. (5.27) can be

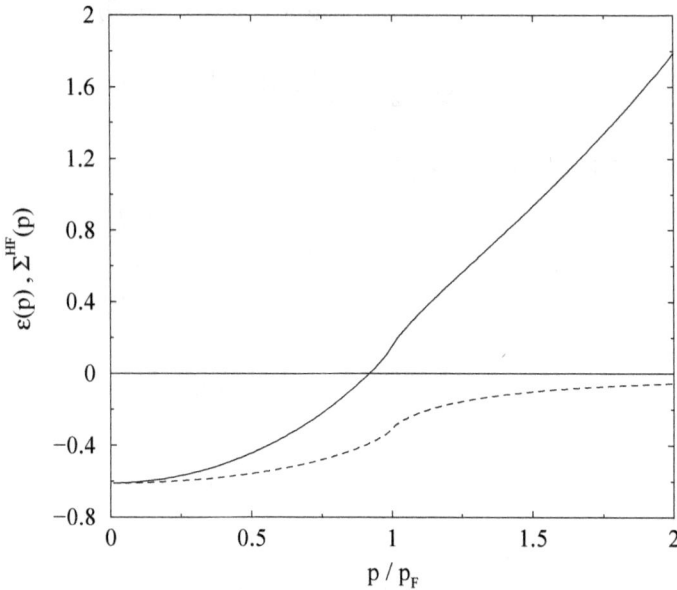

Fig. 10.6   HF potential (dashed line) and corresponding sp energy in the electron gas
with $p_F = 1$.

rewritten as

$$\hat{T} = p_F^2 \left[\sum_{\boldsymbol{p}m_s} \frac{(p/p_F)^2}{2} a_{\boldsymbol{p}m_s}^\dagger a_{\boldsymbol{p}m_s} \right] \tag{10.182}$$

$$\hat{V}' = p_F \left[\frac{1}{2} \sum_{\substack{\boldsymbol{P},\boldsymbol{p}\neq\boldsymbol{p}' \\ m_{s_1} m_{s_2}}} \frac{4\pi}{(Vp_F^3)} \frac{1}{[(\boldsymbol{p}/p_F) - (\boldsymbol{p}'/p_F)]^2} \right.$$

$$\left. \times a_{\frac{1}{2}\boldsymbol{P}+\boldsymbol{p},m_{s_1}}^\dagger a_{\frac{1}{2}\boldsymbol{P}-\boldsymbol{p},m_{s_2}}^\dagger a_{\frac{1}{2}\boldsymbol{P}-\boldsymbol{p}',m_{s_2}} a_{\frac{1}{2}\boldsymbol{P}+\boldsymbol{p}',m_{s_1}} \right],$$

where the operators inside the brackets contain only dimensionless quantities. As a consequence, in the limit of large density ($p_F \to \infty$), the kinetic
term dominates and the system approaches a free Fermi gas. The Coulomb
interaction acts as a small perturbation, and first-order perturbation theory leads precisely to the HF result (see Sec. 10.4). One can show that
all higher-order terms in a perturbation expansion in powers of $\hat{V}$ are divergent, and that an infinite set of higher-order terms must be summed in

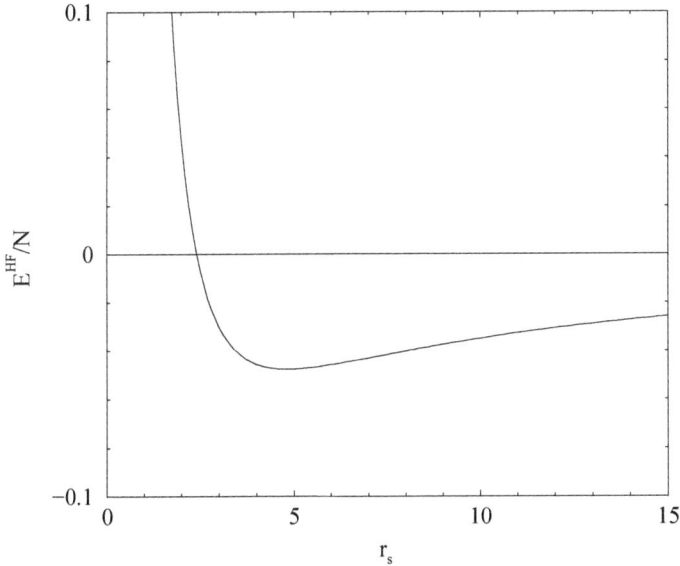

Fig. 10.7 Binding energy per particle for the electron gas, in the HF approximation.

order to get a finite result. In a large-density expansion of the electron-gas energy the dominant term beyond HF is in fact logarithmic, further discussed in Sec. 14.4.

10.6 Nuclear matter

The HF approximation fails miserably for nuclear matter, when realistic nucleon–nucleon (NN) potentials with strong short-range repulsion are used. It is nevertheless instructive to derive the HF expressions, as it is an ingredient in more elaborate theories (and provides a good exercise in handling isospin and spin degrees of freedom).

We consider symmetric (equal numbers of protons and neutrons) and spin-unpolarized nuclear matter, which is a Fermi gas with degeneracy equal to four. The expression for the HF self-energy is

$$
\Sigma^{HF}(p_1) \tag{10.183}
$$
$$
= \sum_{\boldsymbol{p}_2 m_{s_2} m_{t_2}} \theta(p_F - p_2) \langle \boldsymbol{p}_1 m_{s_1} m_{t_1}, \boldsymbol{p}_2 m_{s_2} m_{t_2}| V |\boldsymbol{p}_1 m_{s_1} m_{t_1}, \boldsymbol{p}_2 m_{s_2} m_{t_2} \rangle .
$$

The NN interaction is usually known in the form of a partial-wave decomposition. Introducing center-of-mass momentum $\boldsymbol{P}$ and relative momentum $\boldsymbol{p}$, we have

$$\langle \boldsymbol{P}p(LS)JT| V |\boldsymbol{P}'p'(L'S')J'T'\rangle = \delta_{\boldsymbol{P},\boldsymbol{P}'}\delta_{J,J'}\delta_{T,T'}\delta_{S,S'}\frac{4\pi}{V}V_{LL'}^{STJ}(p,p').$$
(10.184)

These matrix elements correspond to the two-nucleon basis defined in Eq. (4.20), including the coupling to total spin S, isospin T, and total internal angular momentum J. With the aid of the basis transformation in Eq. (4.20) one can express the HF sp potential as

$$\Sigma^{HF}(p_1) = \frac{1}{16\pi^3}\int d\boldsymbol{p}_2\theta(p_F - p_2)\sum_{LST}{}'(2T+1)(2J+1)$$
$$\times V_{LL}^{STJ}(|\boldsymbol{p}_1 - \boldsymbol{p}_2|,|\boldsymbol{p}_1 - \boldsymbol{p}_2|),$$
(10.185)

where the primed summation is restricted to odd values of $L+S+T$. The HF ground-state energy for nuclear matter can be evaluated using Eq. (10.185), and typically does not even predict binding for nuclear matter. For the CDBonn potential [Machleidt (2001)] *e.g.*, one finds an energy per nucleon of 4.64 MeV, and for the old Reid Soft Core potential [Reid (1968)] (which has a much stronger repulsive core) even 176.25 MeV.

The underlying reason is the Slater-determinant nature of the HF ground state, which allows two nucleons to come close together and experience the strong short-range repulsion of the NN potential. A more reasonable description requires the inclusion of additional short-range correlations between the nucleons, which is the subject of Sec. 16.3.1.

10.7 Exercises

(1) Prove Brillouin's theorem in Eq. (10.60) by direct calculation of the matrix element $\langle \Phi_{HF}^N| \hat{H}a_p^\dagger a_h |\Phi_{HF}^N\rangle$, with $|\Phi_{HF}^N\rangle = \prod_h a_h^\dagger |0\rangle$ and

$$\hat{H} = \sum_{\alpha\gamma}\langle\alpha| T |\gamma\rangle a_\alpha^\dagger a_\gamma + \frac{1}{4}\sum_{\alpha\beta\gamma\delta}\langle\alpha\beta| V |\gamma\delta\rangle a_\alpha^\dagger a_\beta^\dagger a_\delta a_\gamma.$$

(2) Prove Koopman's theorem in Eq. (10.61) by direct calculation of the matrix element $\langle \Phi_{HF}^N| a_h^\dagger \hat{H}a_h |\Phi_{HF}^N\rangle$.
(3) Generate the dissociation curve in Fig. 10.4 for the H_2 molecule using the two-dimensional model in Sec. 10.3.3.

(4) For the spin-polarized electron gas one can define the asymmetry parameter ξ as

$$\xi = \frac{\rho^{(+\frac{1}{2})} - \rho^{(-\frac{1}{2})}}{\rho},$$

where $\rho^{(m_s)}$ is the density of the spin-up or spin-down electrons and $\rho = \sum_{m_s} \rho^{(m_s)}$ is the total electron density. Show that:

(a) The Fermi momenta of the spin-up and spin-down electrons are

$$p_F^{(m_s)} = [6\pi^2 \rho^{(m_s)}]^{\frac{1}{3}} = (1 \pm \xi)^{\frac{1}{3}} p_F,$$

where $p_F = [3\pi^2 \rho]^{\frac{1}{3}}$ is the Fermi momentum of the unpolarized electron gas at the same density.

(b) The Fermi-gas kinetic energy is

$$T_{FG}(\xi) = T_{FG}(0)\frac{1}{2}[(1 + \xi)^{\frac{5}{3}} + (1 - \xi)^{\frac{5}{3}}],$$

where $T_{FG}(0)$ is the kinetic energy in Eq. (10.178) for the unpolarized electron gas at the same density.

(c) The HF self-energy is

$$\Sigma^{HF(m_s)}(\xi; p) = -\frac{p_F}{\pi}\left((1 \pm \xi)^{\frac{1}{3}} + \frac{(1 \pm \xi)^{\frac{2}{3}} - u^2}{2u} \ln\left| \frac{(1 \pm \xi)^{\frac{1}{3}} + u}{(1 \pm \xi)^{\frac{1}{3}} - u} \right| \right),$$

where $u = p/p_F$.

(d) The HF exchange energy is

$$E_x(\xi) = E_x(0)\frac{1}{2}[(1 + \xi)^{\frac{4}{3}} + (1 - \xi)^{\frac{4}{3}}],$$

where $E_x(0)$ is the exchange energy in Eq. (10.179) for the unpolarized electron gas at the same density.

(5) Derive the expression in Eq. (10.185) for the HF potential in nuclear matter.

Chapter 11

Beyond the mean-field approximation

In this chapter, we examine correlation effects beyond the mean-field or HF picture, by considering the next higher-order contribution to the self-energy of a particle in the medium. We recall that in the self-consistent formulation of Ch. 9 that the self-energy is related to the vertex function Γ through the master equation Eq. (9.34). The HF approximation in Ch. 10, was obtained directly by setting the vertex function $\Gamma = 0$. In terms of diagrams, we had to consider the lowest-order contributions to the self-energy in Fig. 9.2 [and Fig. 9.6a) if an auxiliary potential U is employed]. The self-consistent treatment of these diagrams, indicated by the Dyson equation in Fig. 10.1, was shown to be equivalent to the HF formalism.

The HF self-energy does not depend on energy, and on several occasions we pointed out the shortcomings of such an approach: some specific features of interacting many-body systems cannot be explained by a static self-energy, but require explicit energy-dependence. The latter appears in the self-energy for the first time when contributions of second order in the interaction are considered. These are listed in Figs. 9.3 – 9.5 and in Fig. 9.6b) – e). Upon inspection, the diagrams in Fig. 9.3 and Fig. 9.6c) – e) are all reducible, whereas the diagrams in Fig. 9.4 and Fig. 9.6b) are already generated by the HF approximation. So the diagram in Fig. 9.5 — sometimes called *the* second-order diagram, or the Born approximation to the self-energy — is the only genuinely new contribution.

Below, we deal exclusively with this second-order diagram, mainly because its energy-dependence is archetypical for all higher-order contributions and therefore warrants a careful analysis.

11.1   The second-order self-energy

The second-order diagram was evaluated in the caption of Fig. 9.5, on the
basis of the Feynman rules for the sp propagator developed in Ch. 8. The
intermediate fermion lines in the diagram of Fig. 9.5 represent sp propagators $G^{(0)}$ corresponding to a noninteracting system with Hamiltonian
$\hat{H}_0 = \hat{T} + \hat{U}$, where $\hat{U}$ is an arbitrary auxiliary potential.

In keeping with our self-consistent formulation, we should now replace
the propagators $G^{(0)}$ with interacting propagators in the self-energy

$$\Sigma^{(2)}(\gamma, \delta; E) = -\frac{1}{2} \int \frac{dE_1}{2\pi i} \int \frac{dE_2}{2\pi i} \sum_{\lambda, \epsilon, \nu} \sum_{\zeta, \xi, \mu} \langle \gamma\lambda | V | \epsilon\nu \rangle \langle \zeta\xi | V | \delta\mu \rangle$$
$$\times\, G(\epsilon, \zeta; E_1) G(\nu, \xi; E_2) G(\mu, \lambda; E_1 + E_2 - E). \quad (11.1)$$

Note that G is the sp propagator that solves the Dyson equation illustrated
in Fig. 11.1

$$G(\alpha, \beta; E) = G^{(0)}(\alpha, \beta; E) + \sum_{\gamma\delta} G(\alpha, \gamma; E)\Sigma(\gamma, \delta; E)G^{(0)}(\delta, \beta; E). \quad (11.2)$$

The irreducible self-energy

$$\Sigma(\gamma, \delta; E) = -\langle \gamma | U | \delta \rangle + \Sigma^{(1)}(\gamma, \delta) + \Sigma^{(2)}(\gamma, \delta; E), \quad (11.3)$$

now contains the second-order self-energy $\Sigma^{(2)}$ from Eq. (11.1), in addition
to the static first-order contribution

$$\Sigma^{(1)}(\gamma, \delta) = -i \int_{C\uparrow} \frac{dE'}{2\pi} \sum_{\mu\nu} \langle \gamma\mu | V | \delta\nu \rangle\, G(\nu, \mu; E'), \quad (11.4)$$

already analyzed in Ch. 10 [see Eq. (10.11)].

The self-energy[1] in Eq. (11.3) corresponds precisely to the expression
one obtains by setting in Eq. (9.34) the vertex function Γ equal to the free
interparticle interaction V,

$$\langle \zeta\rho | \Gamma(E_1, E_2; E_3, E_4) | \delta\sigma \rangle \equiv \langle \zeta\rho | V | \delta\sigma \rangle. \quad (11.5)$$

The approximation in Fig. 11.1 thus corresponds to replacing the dressed
interaction Γ (which includes all in-medium scattering processes) with the
free interaction V, and hence is sometimes called the Born approximation
in this context.

---

[1]Note that we have dropped the $*$ superscript, since all self-energies in this chapter
correspond to irreducible terms.

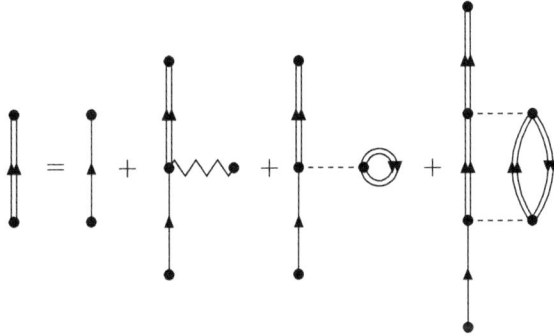

Fig. 11.1   Diagrammatic representation of the second-order Dyson equation.

By now it should be clear that the self-consistent formulation in Eqs. (11.1) – (11.4) is independent of the auxiliary potential U, since the first term in Eq. (11.3) cancels the U-dependence of the unperturbed propagator $G^{(0)}$. Also note that the first-order contribution in Eq. (11.4) will, in general, be different from the HF mean field, since the propagator G solves the second-order Dyson equation in Eq. (11.2) rather than the HF equation (10.5).

For a further analysis of the self-energy in Eq. (11.1) the same procedure as for the HF self-energy in Sec. 10.1.1 can be used: we introduce the Lehmann representation of the sp propagator G as

$$G(\alpha, \beta; E) = \sum_m \frac{z_\alpha^{m+} z_\beta^{m+*}}{E - \varepsilon_m^+ + i\eta} + \sum_n \frac{z_\alpha^{n-} z_\beta^{n-*}}{E - \varepsilon_n^- - i\eta}, \qquad (11.6)$$

and evaluate the double energy integration in Eq. (11.1) by complex contour integration.

The integrals[2] one encounters are of the form

$$I(E) = \int_{-\infty}^{+\infty} \frac{dE'}{2\pi i} \left(\frac{F_1}{E' - f_1 + i\eta} + \frac{B_1}{E' - b_1 - i\eta} \right)$$

$$\times \left(\frac{F_2}{E' - E - f_2 + i\eta} + \frac{B_2}{E' - E - b_2 - i\eta} \right). \qquad (11.7)$$

The integration contour along the real axis can be closed by including a large semicircle in the upper or lower complex E' half plane. Since the integrand behaves as $|E'|^{-2}$ for $|E'| \to +\infty$, such a semicircle (in the limit of infinite radius) yields a vanishing contribution to the integral, and its

---

[2]The contour integrals required in this book, are seldom more complicated.

inclusion does not change the result. The product in Eq. (11.4) contains four terms. However, the $F_1 F_2$ and $B_1 B_2$ terms have two poles in the same (upper or lower) half plane and do not contribute to the integral, as can be seen by closing the contour with a semicircle in the opposite half plane. The $F_1 B_2$ and $F_2 B_1$ terms have a pole in both half planes and contribute, according to the residue theorem, as

$$I(E) = \frac{F_1 B_2}{E - (f_1 - b_2) + i\eta} - \frac{B_1 F_2}{E + (f_2 - b_1) - i\eta}. \tag{11.8}$$

The self-energy in Eq. (11.1) is now easily evaluated by repeated use of Eq. (11.8) and reads

$$\Sigma^{(2)}(\gamma, \delta; E) = \frac{1}{2} \sum_{\lambda, \epsilon, \nu} \sum_{\zeta, \xi, \mu} \langle \gamma\lambda| V |\epsilon\nu\rangle \langle \zeta\xi| V |\delta\mu\rangle \tag{11.9}$$

$$\times \left(\sum_{m_1 m_2 n_3} \frac{z_\epsilon^{m_1+} z_\zeta^{m_1+^*} z_\nu^{m_2+} z_\xi^{m_2+^*} z_\mu^{n_3-} z_\lambda^{n_3^*}}{E - (\varepsilon_{m_1}^+ + \varepsilon_{m_2}^+ - \varepsilon_{n_3}^-) + i\eta} \right.$$

$$\left. + \sum_{n_1 n_2 m_3} \frac{z_\epsilon^{n_1-} z_\zeta^{n_1-^*} z_\nu^{n_2-} z_\xi^{n_2-^*} z_\mu^{m_3+} z_\lambda^{m_3+^*}}{E + (\varepsilon_{m_3}^+ - \varepsilon_{n_1}^- - \varepsilon_{n_2}^-) - i\eta} \right).$$

We recall (see Sec. 7.3) that the poles of the sp propagator in Eq. (11.6) belong either to the particle addition (ε_m^+) or removal (ε_n^-) domain

$$\forall m, n : \varepsilon_n^- \leq \varepsilon_F^- < \varepsilon_F < \varepsilon_F^+ \leq \varepsilon_m^+, \tag{11.10}$$

which are separated by the Fermi energy

$$\varepsilon_F = \frac{1}{2}[\varepsilon_F^- + \varepsilon_F^+]. \tag{11.11}$$

As a consequence, the poles appearing in the second-order self-energy of Eq. (11.9) obey the inequalities

$$\forall m_i, n_i : \varepsilon_{n_1}^- + \varepsilon_{n_2}^- - \varepsilon_{m_3}^+ < \varepsilon_F < \varepsilon_{m_1}^+ + \varepsilon_{m_2}^+ - \varepsilon_{n_3}^-. \tag{11.12}$$

This feature also holds in higher orders. In general, the energy-dependent part of the self-energy has the same analytic structure as the sp propagator G: a sum of simple poles, shifted slightly off the real axis into the lower (upper) half plane for poles corresponding to the addition (removal) domain. Similarly to the propagator in Eq. (9.35), the self-energy contains in many cases (whenever the energy spectra of the $N-1$ or $N+1$ systems have a continuous part) branch-cuts starting at some threshold energy, in addition to a set of isolated simple poles.

11.2 Solution of the Dyson equation

Before proceeding in Sec.11.5 with the full self-consistent treatment indicated in Fig. 11.1, it is instructive to examine how the Dyson equation is modified by the presence of energy-dependent terms in the self-energy. To simplify matters we evaluate the self-energy in Eq. (11.3) with HF propagators and set the auxiliary potential U equal to the HF potential [*i.e.* $G^{(0)} \equiv G^{HF}$ in Eq. (11.2)]. In fact, this can be considered as a first iteration step when solving the full self-consistency problem. We also choose the HF sp basis, so the HF propagator

$$G^{HF}(\alpha, \beta; E) = \delta_{\alpha,\beta} \left[\frac{\theta(\alpha - F)}{E - \varepsilon_\alpha + i\eta} + \frac{\theta(F - \alpha)}{E - \varepsilon_\alpha - i\eta} \right], \qquad (11.13)$$

is diagonal in the sp labels.

To find the corresponding second-order self-energy we simply have to set

$$z_\alpha^{m+} = \delta_{m,\alpha}\theta(\alpha - F); \quad z_\alpha^{n-} = \delta_{n,\alpha}\theta(F - \alpha), \qquad (11.14)$$

in the general expression (11.9). The resulting self-energy reads

$$\Sigma^{(2)}(\gamma, \delta; E) = \frac{1}{2} \sum_{\lambda, \epsilon, \nu} \langle \gamma\lambda | V | \epsilon\nu \rangle \langle \epsilon\nu | V | \delta\lambda \rangle$$

$$\times \left(\frac{\theta(\epsilon - F)\theta(\nu - F)\theta(F - \lambda)}{E - (\varepsilon_\epsilon + \varepsilon_\nu - \varepsilon_\lambda) + i\eta} + \frac{\theta(F - \epsilon)\theta(F - \nu)\theta(\lambda - F)}{E + (\varepsilon_\lambda - \varepsilon_\epsilon - \varepsilon_\nu) - i\eta} \right), (11.15)$$

or, in a more compact notation

$$\Sigma^{(2)}(\gamma, \delta; E) = \frac{1}{2} \left(\sum_{p_1 p_2 h_3} \frac{\langle \gamma h_3 | V | p_1 p_2 \rangle \langle p_1 p_2 | V | \delta h_3 \rangle}{E - (\varepsilon_{p_1} + \varepsilon_{p_2} - \varepsilon_{h_3}) + i\eta} \right.$$

$$\left. + \sum_{h_1 h_2 p_3} \frac{\langle \gamma p_3 | V | h_1 h_2 \rangle \langle h_1 h_2 | V | \delta p_3 \rangle}{E + (\varepsilon_{p_3} - \varepsilon_{h_1} - \varepsilon_{h_2}) - i\eta} \right), (11.16)$$

where labels identifying particle (p) and hole (h) states in the HF approximation have been introduced. We will proceed to examine the solution G of the equation

$$G(\alpha, \beta; E) = G^{HF}(\alpha, \beta; E) + \sum_{\gamma\delta} G(\alpha, \gamma; E)\Sigma^{(2)}(\gamma, \delta; E)G^{HF}(\delta, \beta; E).$$

$$(11.17)$$

11.2.1 *Diagonal approximation*

It is clear from Eq. (11.16) that the second-order self-energy in principle has non-diagonal contributions, even when evaluated with the diagonal HF sp propagator. However, in some cases it is a good approximation to neglect the off-diagonal terms. This happens *e.g.* in closed-shell nuclei, where off-diagonal elements would require mixing between major shells having a large energy separation.

Within this diagonal approximation, the self-energy (11.16) reads

$$\Sigma^{(2)}(\alpha; E) = \frac{1}{2} \left(\sum_{p_1 p_2 h_3} \frac{|\langle \alpha h_3| V |p_1 p_2\rangle|^2}{E - (\varepsilon_{p_1} + \varepsilon_{p_2} - \varepsilon_{h_3}) + i\eta} \right.$$
$$\left. + \sum_{h_1 h_2 p_3} \frac{|\langle \alpha p_3| V |h_1 h_2\rangle|^2}{E + (\varepsilon_{p_3} - \varepsilon_{h_1} - \varepsilon_{h_2}) - i\eta} \right), \quad (11.18)$$

and the Dyson equation (11.17) becomes

$$G(\alpha; E) = G^{HF}(\alpha; E) + G(\alpha; E)\Sigma^{(2)}(\alpha; E)G^{HF}(\alpha; E). \quad (11.19)$$

The latter has a simple algebraic solution,

$$G(\alpha; E) = \frac{1}{\frac{1}{G^{HF}(\alpha; E)} - \Sigma^{(2)}(\alpha; E)} = \frac{1}{E - \varepsilon_\alpha - \Sigma^{(2)}(\alpha; E)}. \quad (11.20)$$

For the last identity, we used the inverse of the HF propagator in Eq. (11.13) [the infinitesimal $\pm i\eta$ are irrelevant when they do not appear in the denominator of a pole term]

$$\frac{1}{G^{HF}(\alpha; E)} = E - \varepsilon_\alpha. \quad (11.21)$$

Extracting physical information from the sp propagator in general requires the knowledge of its poles and residues (see Sec. 7.3). We will assume throughout Sec. 11.2 that the self-energy $\Sigma^{(2)}$ has poles at a set of discrete energies (*i.e.* a set of isolated simple poles), and treat the case when branch-cuts are present in Sec. 11.3, where infinite Fermi systems are discussed. Of course, most realistic finite systems have branch-cuts as well, but since practical calculations are usually performed by introducing a finite and discrete sp basis, the self-energy is then automatically restricted to a discrete pole structure.

For the propagator $G(\alpha; E)$ given by the formal solution in Eq. (11.20), the discrete poles $E_{n\alpha}$ obviously correspond to the roots of the nonlinear equation

$$E_{n\alpha} = \varepsilon_\alpha + \Sigma^{(2)}(\alpha; E_{n\alpha}), \qquad (11.22)$$

with $\Sigma^{(2)}(\alpha; E)$ defined in Eq. (11.18). The residue $R_{n\alpha}$ at the pole $E_{n\alpha}$ of the propagator follows from

$$R_{n\alpha} = \lim_{E \to E_{n\alpha}} (E - E_{n\alpha}) G(\alpha; E) = \lim_{E \to E_{n\alpha}} \frac{E - E_{n\alpha}}{E - \varepsilon_\alpha - \Sigma^{(2)}(\alpha; E)}$$

$$= \left(1 - \frac{d\Sigma^{(2)}(\alpha; E)}{dE} \bigg|_{E=E_{n\alpha}} \right)^{-1}. \qquad (11.23)$$

Note that, when determining the roots of Eq. (11.22), the infinitesimal $\pm i\eta$ appearing in the denominator of Eq. (11.18) can simply be omitted. In principle the $\pm i\eta$ generate, according to Eq. (7.15), an imaginary part of the self-energy, consisting of a sequence of δ-functions located at the discrete poles of the self-energy. Since discrete solutions to Eq. (11.22) cannot coincide with a pole of the self-energy, the δ-functions do not influence the position of the roots of Eq. (11.22), which are all real. This does not hold when the self-energy has a continuous distribution of poles (which is equivalent to a branch cut), as will be clarified in the discussion on infinite systems in Sec. 11.3.

To gain insight into the location of the roots of Eq. (11.22), a graphical solution of the Dyson equation is often helpful. In Fig. 11.2 the energy-dependence of the self-energy $\Sigma^{(2)}(\alpha; E)$ of Eq. (11.18) is shown. The case on display is for a typical confined finite system, having a discrete HF sp spectrum. The hole and particle HF energies are separated by the particle-hole gap, which has a width $\Delta = \varepsilon_p^{min} - \varepsilon_h^{max}$ and is centered on the HF Fermi energy

$$\varepsilon_F = \frac{1}{2}(\varepsilon_p^{min} + \varepsilon_h^{max}). \qquad (11.24)$$

Since the poles in Eq. (11.18) all have positive residues, $\Sigma^{(2)}(\alpha; E)$ is monotonically decreasing where defined. There is a sequence of simple poles in the addition domain, located at the unperturbed HF 2p1h energies, and another sequence in the removal domain, located at (minus) the unperturbed HF 1p2h energies. The poles of the addition and removal sequence are separated by a gap of (at least) three times the HF particle–hole gap.

The roots of Eq. (11.22) are simply the intersection points of the self-energy $\Sigma^{(2)}(\alpha; E)$ with the straight line $E - \epsilon_\alpha$. It is obvious from the graph

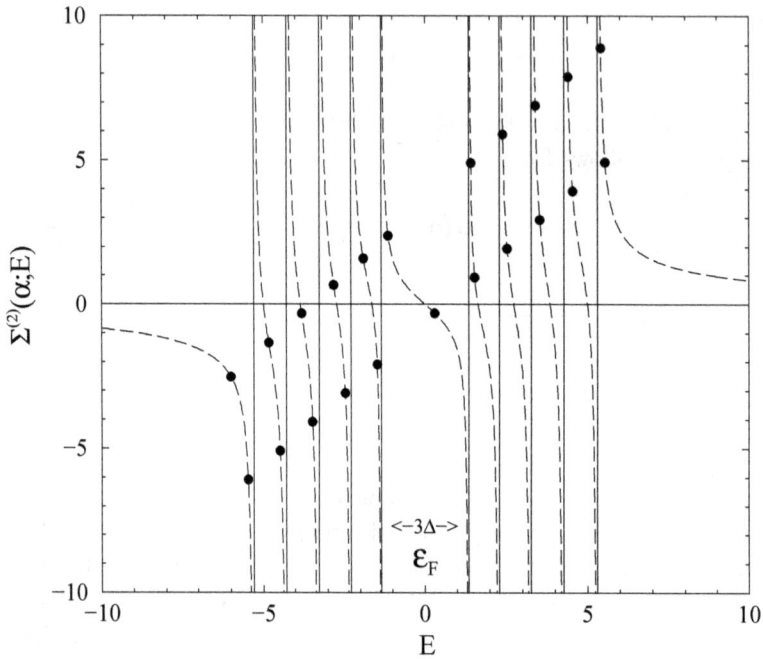

Fig. 11.2 Graphical solution of Eq. (11.23). The second-order self-energy $\Sigma^{(2)}(\alpha; E)$ of Eq. (11.18) is indicated by the dashed line. The roots of Eq. (11.22) are given by the intersections points with the straight line $E - \varepsilon_\alpha$, drawn here as dots, for two values of ε_α.

in Fig. 11.2 that between any two successive poles of the self-energy a root is located. In addition, there is a root to the left and right of the sequence of self-energy poles. When a finite sp basis set is used, this implies that a self-energy having D poles leads to a sp propagator with $D + 1$ poles.

The interpretation of these roots should by now be straightforward. The poles $E_{n\alpha}$ in the removal domain (below the Fermi energy) must be interpreted as approximate energies of the eigenstates in the $N - 1$ system

$$E_{n\alpha} \approx E_0^N - E_n^{N-1}, \tag{11.25}$$

that can be obtained by removing a particle in the sp state α from the N-particle ground state. The residue then corresponds to the (squared) removal amplitude,

$$R_{n\alpha} \approx |\langle \Psi_n^{N-1} | a_\alpha | \Psi_0^N \rangle|^2. \tag{11.26}$$

Similarly, the poles $E_{n\alpha}$ in the addition domain (above the Fermi energy) correspond to eigenstates in the $N+1$ system

$$E_{n\alpha} \approx E_n^{N+1} - E_0^N, \qquad (11.27)$$

having addition amplitudes

$$R_{n\alpha} \approx |\langle \Psi_n^{N+1} | a_\alpha^\dagger | \Psi_0^N \rangle |^2. \qquad (11.28)$$

Note that since $d\Sigma^{(2)}(\alpha; E)/dE < 0$, the residues $R_{n\alpha}$ that follow from Eq. (11.23) obey

$$0 \le R_{n\alpha} \le 1, \qquad (11.29)$$

in accordance with their relation to the physical addition or removal amplitudes.

All of this means that adding the energy-dependent second-order self-energy to the static HF self-energy produces quite dramatic effects. The removal from the ground state of a particle in an occupied HF sp state α no longer leads to a unique $N-1$ state, as in HF, but rather to a large number of $N-1$ states, each having a finite removal amplitude. Moreover, the removal from the ground state of a particle in an unoccupied HF sp state α, clearly impossible in HF, is now allowed. Similar statements hold in the addition domain. Of course, any more sophisticated treatment of the self-energy will also include these fragmentation effects on the sp strength. Experimental information on physical spectral functions indicate that such features are indispensable for a meaningful comparison with data, as discussed in Secs. 7.7 and 7.8.

As a final remark on Fig. 11.2, we note that if the unperturbed sp energy ε_α is not too far removed from the Fermi energy, the root of Eq. (11.22) lying in the interval which separates the removal and addition domain, has a special character. Since the self-energy $\Sigma^{(2)}(\alpha; E)$ has no poles in this interval, the energy derivative is relatively small here, and as a consequence the residue corresponding to the solution will be quite close to (but still smaller than) unity. Such a solution represents a quasiparticle or quasihole excitation in a finite system, and corresponds to a $N \pm 1$ eigenstate which has a rather pure sp character. On the other hand, if the sp energy ε_α is far from the Fermi energy, it is in a region where the density of 2p1h or 1p2h states is high, and the strength of this sp state will be strongly fragmented over many $N-1$ states. The different fragmentation pattern observed for valence holes and deeply-bound hole states in finite nuclei is readily understood by these elementary considerations, and explains the qualitative behavior of the data discussed in Sec. 7.8.

11.2.2   *Link with perturbation theory*

As shown in Ch. 10, the HF formalism assumes that the N-particle ground state is the HF Slater determinant

$$(\text{HF:})\ |\Psi_0^N\rangle \approx |\Phi_{HF}^N\rangle = \prod_{h=1}^{N} a_h^\dagger |0\rangle, \tag{11.30}$$

and that the eigenstates of the $N+1$ system are simple 1p excitations

$$(\text{HF:})\ |\Psi_p^{N+1}\rangle \approx a_p^\dagger |\Phi_{HF}^N\rangle. \tag{11.31}$$

Corrections to this picture can be obtained by allowing admixtures with 2p1h excitations

$$|\Psi_p^{N+1}\rangle \approx x\, a_p^\dagger |\Phi_{HF}^N\rangle + \sum_{p_1<p_2}\sum_{h_3} X_{p_1 p_2 h_3} a_{p_1}^\dagger a_{p_2}^\dagger a_{h_3} |\Phi_{HF}^N\rangle. \tag{11.32}$$

The eigenstates in this basis are found by diagonalization of the Hamiltonian matrix

$$\begin{pmatrix} \varepsilon_p & \mathcal{V} \\ \mathcal{V}^\dagger & \mathcal{E} \end{pmatrix} \begin{pmatrix} x \\ X \end{pmatrix} = E \begin{pmatrix} x \\ X \end{pmatrix}, \tag{11.33}$$

where $\mathcal{V}$ contains the coupling between 1p and 2p1h configurations

$$\mathcal{V}_{p_1 p_2 h_3} = \langle \Phi_{HF}^N | a_p \hat{H} a_{p_1}^\dagger a_{p_2}^\dagger a_{h_3} |\Phi_{HF}^N\rangle = \langle ph_3| V |p_1 p_2\rangle. \tag{11.34}$$

Under the assumption (typical in perturbation theory) that the 2p1h configurations do not interact among themselves, we have

$$\mathcal{E}_{p_1 p_2 h_3,\, p_1' p_2' h_3'} = \langle \Phi_{HF}^N | a_{h_3'}^\dagger a_{p_2} a_{p_1} \hat{H} a_{p_1}^\dagger a_{p_2}^\dagger a_{h_3} |\Phi_{HF}^N\rangle$$

$$\approx \delta_{p_1,p_1'}\delta_{p_2,p_2'}\delta_{h_3,h_3'}(\varepsilon_{p_1} + \varepsilon_{p_2} - \varepsilon_{h_3}). \tag{11.35}$$

Elimination of X from Eq. (11.33) then leads to

$$X = \frac{1}{E - \mathcal{E}} \mathcal{V}^\dagger x, \tag{11.36}$$

$$Ex = \left(\varepsilon_p + \mathcal{V} \frac{1}{E - \mathcal{E}} \mathcal{V}^\dagger \right) x. \tag{11.37}$$

According to Eq. (11.37), the eigenenergies for states with $x \neq 0$ [*i.e.* states that have a nonvanishing overlap with the $1p$ state $a_p^\dagger |\Phi_{HF}^N\rangle$] are therefore the roots of

$$E = \varepsilon_p + \sum_{p_1 < p_2, h_3} \frac{|\langle p h_3 | V | p_1 p_2 \rangle|^2}{E - (\varepsilon_{p_1} + \varepsilon_{p_2} - \varepsilon_{h_3})}. \tag{11.38}$$

This is seen to be analogous to Eq. (11.22), but only the forward (2p1h) term of the self-energy in Eq. (11.18) is generated, as a consequence of the unperturbed HF ground state (11.30) appearing in the ansatz of Eq. (11.32). In contrast, the Green function formalism which led to Eq. (11.22) automatically builds in ground-state correlations, in a single coherent framework for both the $N + 1$ and $N - 1$ excited states [Blaizot and Ripka (1986)].

11.2.3   *Sum rules*

The analytic structure of Eqs. (11.18) – (11.22), generated by the second-order diagram, is in fact quite general and also appears in more complicated cases, when higher-order diagrams or an infinite resummation of a subclass of diagrams are included in the self-energy. Hence, it is worthwhile to examine the properties of the generical expression

$$G(E) = \frac{1}{E - \varepsilon - \Sigma(E)}, \tag{11.39}$$

where both the propagator $G(E)$ and the self-energy $\Sigma(E)$ are a sum of simple poles

$$\Sigma(E) = \sum_n \frac{s_n}{E - \omega_n}, \qquad G(E) = \sum_N \frac{S_N}{E - \Omega_N}. \tag{11.40}$$

Analogous to Eq. (11.22), the Ω_N are the roots of

$$\Omega_N - \varepsilon = \sum_n \frac{s_n}{\Omega_N - \omega_n}, \tag{11.41}$$

whereas the sp strength at Ω_N follows [see Eq. (11.23)] from

$$S_N = \frac{1}{1 + \sum_n \frac{s_n}{(\Omega_N - \omega_n)^2}}. \tag{11.42}$$

However, without explicitly solving Eq. (11.41) it is possible to derive sum rules which relate the distribution of sp strength (Ω_N, S_N) with the self-energy strength (ω_n, s_n) and which are often useful as a numerical check on calculated spectral functions.

Considering the leading order in a $(1/E)$ expansion of Eq. (11.39) immediately leads to

$$E \to \infty \Rightarrow \quad \sum_N S_N = 1. \tag{11.43}$$

This means, in accordance with the fundamental sum rule in Eq. (7.18), that the summed sp strength is always equal to unity, irrespective of the values of the poles and residues appearing in the self-energy.

If we let the energy E in Eq. (11.39) coincide with one of the poles ω_n of the self-energy we get

$$E \to \omega_n \Rightarrow \quad 0 = \sum_N \frac{S_N}{\Omega_N - \omega_n}, \tag{11.44}$$

whereas the limit $E \to \Omega_N$ leads back to Eq. (11.41). Combining Eq. (11.44) with Eq. (11.41) leads to the result

$$\sum_N S_N(\Omega_N - \varepsilon) = \sum_N \sum_n \frac{S_N s_n}{\Omega_N - \omega_n} = 0, \tag{11.45}$$

which implies that the centroid of the sp strength distribution is the sp energy ε.

In a similar way, general recursion formulas for the higher-order central moments of the sp strength distribution are obtained. Defining

$$M_k = \sum_N (\Omega_N - \varepsilon)^k S_N; \quad m_k = \sum_n (\omega_n - \varepsilon)^k s_n, \tag{11.46}$$

we have according to Eqs. (11.43) and (11.45) that $M_0 = 1$ and $M_1 = 0$, and for $k \geq 2$ we can derive

$$
\begin{aligned}
M_k &= \sum_N S_N(\Omega_N - \varepsilon)^{k-1} \sum_n \frac{s_n}{\Omega_N - \omega_n} \\
&= \sum_n s_n \sum_N S_N \left[\frac{(\Omega_N - \varepsilon)^{k-1} - (\omega_n - \varepsilon)^{k-1}}{(\Omega_N - \varepsilon) - (\omega_n - \varepsilon)} + \frac{(\omega_n - \varepsilon)^{k-1}}{\Omega_N - \omega_n} \right] \\
&= \sum_{l=0}^{k-2} M_l m_{k-l-2}.
\end{aligned}
\tag{11.47}
$$

In particular we find $M_2 = m_0$, *i.e.* the second central moment, describing the width of the spectral function, is equal to the summed self-energy strength.

The moments of negative order M_{-k} (or inverse energy-weighted sum rules) can be obtained by recursion as well. Considering, for $k \geq 0$,

$$M_{-k} = \sum_N \frac{S_N}{(\Omega_N - \varepsilon)^k} = \sum_N \frac{S_N}{(\Omega_N - \varepsilon)^{k+1}} \sum_n \frac{s_n}{\Omega_N - \omega_n}, \qquad (11.48)$$

and using the identity

$$\frac{1}{x^{k+1}(x-y)} = \frac{1}{y^{k+1}(x-y)} - \sum_{l=1}^{k+1} \frac{1}{y^l x^{k-l+2}}, \qquad (11.49)$$

with $x = \Omega_N - \varepsilon$ and $y = \omega_n - \varepsilon$, one arrives at

$$M_{-(k+1)} = -\frac{1}{m_{-1}} \left(M_{-k} + \sum_{l=1}^{k} M_{-l} m_{-(k-l+1)} \right). \qquad (11.50)$$

11.2.4 *General (nondiagonal) self-energy*

We now drop the simplification introduced in Sec. 11.2.1 and include also the nondiagonal contributions to the self-energy in Eq. (11.16). We will again assume that the self-energy has a set of isolated simple poles and omit the $\pm i\eta$ in the denominator. The Dyson equation can then generically be rewritten as (we use square brackets to emphasize the matrix structure)

$$[G(E)] = (E - [\varepsilon] - [\Sigma(E)])^{-1}, \qquad (11.51)$$

where both the propagator and the self-energy are sums over discrete simple poles

$$[G(E)] = \sum_N \frac{[S_N]}{E - \Omega_N}, \quad [\Sigma(E)] = \sum_n \frac{[s_n]}{E - \omega_n}. \qquad (11.52)$$

Note that all matrices have sp labels as indices, and that the matrices $[S_N]$ and $[s_n]$ are hermitian and positive (having real eigenvalues ≥ 0).

The analysis proceeds in much the same way as in Sec. 11.2.1, but some care has to be taken because of the possibly noncommuting matrix quantities. From Eq. (11.51) it is clear that a pole Ω_N of the propagator is a zero eigenvalue appearing in the nonlinear eigenvalue equation

$$\Omega_N X_N = ([\varepsilon] + [\Sigma(\Omega_N)]) X_N. \qquad (11.53)$$

We allow for the possibility that Ω_N is a degenerate eigenvalue, and write the spectral decomposition as

$$[\varepsilon] + [\Sigma(\Omega_N)] = \Omega_N[P_N] + \sum_M \lambda_M [P_M], \qquad (11.54)$$

where the λ_M ($\neq \Omega_N$) are the other eigenvalues of ($[\varepsilon] + [\Sigma(\Omega_N)]$) and the matrices $[P_N]$ and $[P_M]$ are projection operators on the corresponding eigenspaces.

The residue of the propagator $[G(E)]$ at the pole Ω_N is given by

$$[S_N] = \lim_{E \to \Omega_N} (E - \Omega_N)[G(E)]$$

$$= \lim_{\eta \to 0} \eta \left(\Omega_N + \eta - [\varepsilon] - [\Sigma(\Omega_N)] - \eta[\Sigma'(\Omega_N)] \right)^{-1}, \quad (11.55)$$

where the energy-derivative of the self-energy is the negative hermitian matrix

$$[\Sigma'(E)] = -\sum_n \frac{[s_n]}{(E - \omega_n)^2}. \qquad (11.56)$$

Using the spectral decomposition Eq. (11.54) one can verify that the matrix

$$[A] = \Omega_N + \eta - [\varepsilon] - [\Sigma(\Omega_N)] = \eta[P_N] + \sum_M (\Omega_N - \lambda_M + \eta)[P_M] \quad (11.57)$$

is invertible for $\eta \neq 0$, the inverse being

$$[A]^{-1} = \frac{1}{\eta}[P_N] + \sum_M \frac{1}{\Omega_N - \lambda_M + \eta}[P_M]. \qquad (11.58)$$

The residue $[S_N]$ in Eq. (11.55), rewritten in terms of $[A]^{-1}$, reads

$$[S_N] = \lim_{\eta \to 0} \eta \left(1 - \eta[A]^{-1}[\Sigma'(\Omega_N)] \right)^{-1} [A]^{-1}. \qquad (11.59)$$

The limit $\eta \to 0$ can now be safely taken, and since

$$\lim_{\eta \to 0} \eta[A]^{-1} = [P_N], \qquad (11.60)$$

the residue matrix at the pole Ω_N is

$$[S_N] = (1 - [P_N][\Sigma'(\Omega_N)])^{-1}[P_N]. \qquad (11.61)$$

It is now always possible to choose eigenvectors $X_{N\nu}$, corresponding to a degenerate eigenvalue Ω_N in Eq. (11.53), in such a way that $[P_N] =$

$\sum_\nu X_{N\nu} X^\dagger_{N\nu}$ and $[\Sigma'(\Omega_N)]$ is diagonal in the $[P_N]$-subspace

$$X^\dagger_{N\nu}[\Sigma'(\Omega_N)]X_{N\nu'} = -\delta_{\nu,\nu'}u^2_{N\nu}. \tag{11.62}$$

Using this basis, Eq. (11.61) can be reexpressed as

$$[S_N] = \sum_\nu \frac{1}{1 + u^2_{N\nu}} X_{N\nu}X^\dagger_{N\nu}, \tag{11.63}$$

which clearly shows that the residue matrix $[S_N]$ is indeed a positive hermitian matrix.

At this point it may be helpful to mention that the above analysis is given primarily for the purpose of mathematical completeness. In the majority of practical applications, the eigenvalues Ω_N in Eq. (11.53) are nondegenerate, *i.e.* $[P_N] = X_N X^\dagger_N$, and Eq. (11.61) simplifies to

$$[S_N] = \frac{1}{1 - X^\dagger_N[\Sigma'(\Omega_N)]X_N} X_N X^\dagger_N. \tag{11.64}$$

Graphically, the situation is also a bit more complicated than in the diagonal case. The poles of the propagator in Eq. (11.51) can be found by plotting the eigenvalue curves $\lambda_\nu(E)$ of the matrix $[\varepsilon] + [\Sigma(E)]$ as a function of E, and determining the intersection points $\lambda_\nu(E) = E$. As an example, the eigenvalue curves of the matrix

$$\varepsilon_\alpha \delta_{\alpha,\beta} + \sum_{n=1}^{5} \frac{[s_n]_{\alpha\beta}}{E - \omega_n} \tag{11.65}$$

are shown in Fig. 11.3 as a function of energy. The dimension of the matrix (corresponding to the dimension of the sp space) is 4. The residue matrices $[s_n]$ at the poles ω_n have rank one, except for the pole at $\omega_3 = 0$ which has rank 2.

The eigenvalue curves in Fig. 11.3 are all monotonously decreasing where defined. This is easily understood by realizing that the energy derivatives follow from first-order perturbation theory

$$\lambda'_\nu(E) = X^\dagger_\nu(E)[\Sigma'(E)]X_\nu(E) = -\sum_n \frac{X^\dagger_\nu(E)[s_n]X_\nu(E)}{(E - \omega_n)^2} < 0. \tag{11.66}$$

Also note the curious behavior of the $\lambda_\nu(E)$ near a pole ω_n of the self-energy. Some of the eigenvalues curves have an asymptote, whereas others are regular at ω_n. It turns out that, if d_n is the rank of the residue matrix

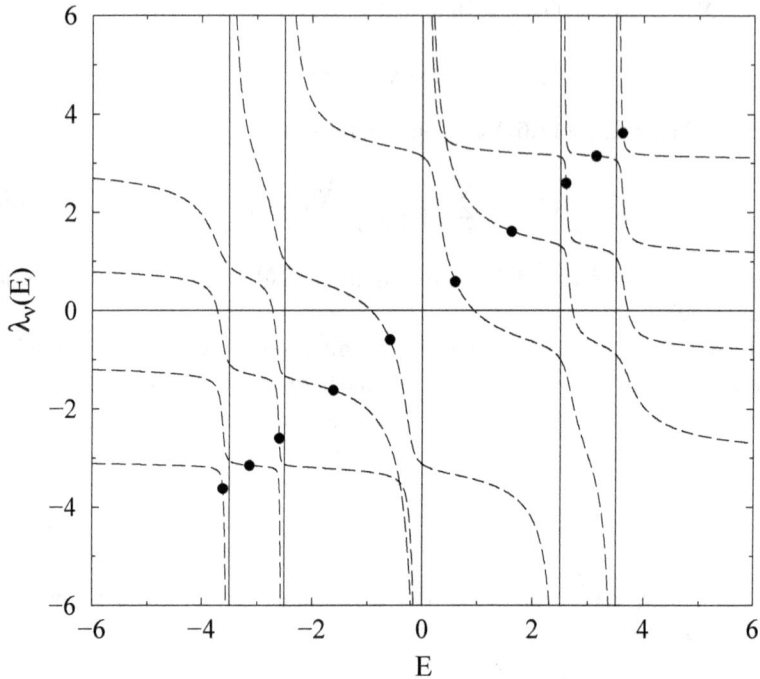

Fig. 11.3   Energy-dependence of the four eigenvalues $\lambda_\nu(E)$ of the nondiagonal second-order self-energy as in Eq. (11.65). The dots correspond to the roots of the equation $E = \lambda_\nu(E)$.

$[s_n]$ in the self-energy, then d_n of the eigenvalue curves have an asymptote at $E = \omega_n$. This can be seen by introducing the spectral decomposition

$$[\varepsilon] + [\Sigma(E)] = \sum_\nu \lambda_\nu(E)[P_\nu(E)] \tag{11.67}$$

and expressing $[s_n]$ as

$$
\begin{aligned}
[s_n] &= \lim_{E \to \omega_n} (E - \omega_n)[\Sigma(E)]) \\
&= \lim_{E \to \omega_n} \sum_\nu \{(E - \omega_n)\lambda_\nu(E)\} [P_\nu(E)].
\end{aligned} \tag{11.68}
$$

It follows that one of the eigenvalues $\lambda_\nu(E)$, say $\lambda_0(E)$, must behave as $\lambda_0(E) \to \frac{C}{E - \omega_n}$ in the limit $E \to \omega_n$, whereas the other eigenvalues are regular at ω_n. The eigenspace corresponding to the singular eigenvalue λ_0

then has a projection operator $[P_0] = \frac{1}{\tilde{C}}[s_n]$, its dimension being equal to the rank of $[s_n]$.

From the graphical analysis it is clear that the total number of poles Ω_N of the propagator in Eq. (11.51) is $M + \sum_n d_n$, where M is the dimension of the sp space. Applied to the second-order self-energy, $\sum_n d_n$ is the dimension of the combined 2p1h and 1p2h space, where the d_n take possible degeneracies in the spectrum of 2p1h or 1p2h energies into account. The number of poles in the propagator therefore agrees with what we expect from an eigenvalue problem describing the mixing of M sp states with the 2p1h and 1p2h states.

11.3 Second order in infinite systems

The second-order problem treated in Sec. 11.2 can be repeated for (homogeneous) infinite systems. Taking the thermodynamic limit then leads to a vanishing particle–hole gap and the appearance of branch-cuts in the second-order self-energy. The analysis of the resulting energy-dependence is used here to introduce, without mathematical rigor, some properties of the exact propagator and self-energy.

11.3.1 *Dispersion relations*

In the case of a homogeneous infinite system the sp propagator and self-energy are automatically diagonal in the plane-wave basis, as was discussed in Sec. 10.4. The second-order self-energy in Eq. (11.18) becomes[3]

$$\Sigma^{(2)}(p; E) = \frac{1}{2} \sum_{\boldsymbol{p}_1 \boldsymbol{p}_2 \boldsymbol{p}_3} |\langle \boldsymbol{p}\boldsymbol{p}_3| V |\boldsymbol{p}_1 \boldsymbol{p}_2 \rangle|^2 \times \qquad (11.69)$$

$$\left(\frac{\theta(p_1 - p_F)\theta(p_2 - p_F)\theta(p_F - p_3)}{E - [\varepsilon(p_1) + \varepsilon(p_2) - \varepsilon(p_3)] + i\eta} + \frac{\theta(p_F - p_1)\theta(p_F - p_2)\theta(p_3 - p_F)}{E + [\varepsilon(p_3) - \varepsilon(p_1) - \varepsilon(p_2)] - i\eta} \right).$$

The $\varepsilon(p)$ represent HF sp energies, and $\varepsilon_F = \varepsilon(p_F)$ is the HF approximation to the Fermi energy.

The interaction matrix elements in Eq. (11.69) have the form

$$\langle \boldsymbol{p}_1 \boldsymbol{p}_2| V |\boldsymbol{p}_3 \boldsymbol{p}_4 \rangle = \frac{(2\pi\hbar)^3}{V} \delta_{\boldsymbol{p}_1 + \boldsymbol{p}_2, \boldsymbol{p}_3 + \boldsymbol{p}_4} w(\boldsymbol{p}_1, \boldsymbol{p}_2, \boldsymbol{p}_3, \boldsymbol{p}_4), \qquad (11.70)$$

---

[3] For notational simplicity we consider fermions of a single species. The inclusion of spin or isospin degrees of freedom does not change the present considerations.

where w is a continuous function of the sp momenta, V is the normalization volume, and momentum conservation is expressed through the Kronecker-δ.

At this point the thermodynamic limit can be taken, resulting in

$$\Sigma^{(2)}(p; E) = \Sigma_+^{(2)}(p; E) + \Sigma_-^{(2)}(p; E), \qquad (11.71)$$

with

$$\Sigma_+^{(2)}(p; E) = \frac{1}{2}\int d\boldsymbol{p}_1 \int d\boldsymbol{p}_2 \int d\boldsymbol{p}_3\ \delta(\boldsymbol{p}_1 + \boldsymbol{p}_2 - \boldsymbol{p}_3 - \boldsymbol{p})$$
$$\times\ |w(\boldsymbol{p}, \boldsymbol{p}_i)|^2 \frac{\theta(p_1 - p_F)\theta(p_2 - p_F)\theta(p_F - p_3)}{E - [\varepsilon(p_1) + \varepsilon(p_2) - \varepsilon(p_3)] + i\eta}, \qquad (11.72)$$

$$\Sigma_-^{(2)}(p; E) = \frac{1}{2}\int d\boldsymbol{p}_1 \int d\boldsymbol{p}_2 \int d\boldsymbol{p}_3\ \delta(\boldsymbol{p}_1 + \boldsymbol{p}_2 - \boldsymbol{p}_3 - \boldsymbol{p})$$
$$\times\ |w(\boldsymbol{p}, \boldsymbol{p}_i)|^2 \frac{\theta(p_F - p_1)\theta(p_F - p_2)\theta(p_3 - p_F)}{E + [\varepsilon(p_3) - \varepsilon(p_1) - \varepsilon(p_2)] - i\eta}. \qquad (11.73)$$

As a consequence of replacing discrete summations with integrations over continuous sp momenta, the discrete poles in Eq. (11.69) have merged into the branch cuts which appear in Eqs. (11.72) and (11.73). The energy-dependent self-energy $\Sigma^{(2)}(p; E)$ is therefore an inherently complex quantity, and the real and imaginary parts are found by direct application of Eq. (7.15).

As an example, consider the real part of $\Sigma_+^{(2)}(p; E)$

$$\text{Re}\Sigma_+^{(2)}(p; E) = \frac{1}{2}\mathcal{P}\int d\boldsymbol{p}_1 \int d\boldsymbol{p}_2 \int d\boldsymbol{p}_3\ \delta(\boldsymbol{p}_1 + \boldsymbol{p}_2 - \boldsymbol{p}_3 - \boldsymbol{p})$$
$$\times\ |w(\boldsymbol{p}, \boldsymbol{p}_i)|^2 \frac{\theta(p_1 - p_F)\theta(p_2 - p_F)\theta(p_F - p_3)}{E - [\varepsilon(p_1) + \varepsilon(p_2) - \varepsilon(p_3)]}. \qquad (11.74)$$

This is a continuous function of p and E, which can be calculated directly from Eq. (11.74) by performing the integration over the sp momenta. The singularities in the denominator, which appear for $E > \varepsilon_F$, should then be regulated by the principal-value procedure, as indicated by the $\mathcal{P}$ symbol.

In many cases however, it is more convenient to consider only the imaginary part

$$\text{Im}\Sigma_+^{(2)}(p; E) = -\pi\frac{1}{2}\int d\boldsymbol{p}_1 \int d\boldsymbol{p}_2 \int d\boldsymbol{p}_3\ \delta(\boldsymbol{p}_1 + \boldsymbol{p}_2 - \boldsymbol{p}_3 - \boldsymbol{p})$$
$$\times\ |w(\boldsymbol{p}, \boldsymbol{p}_i)|^2\theta(p_1 - p_F)\theta(p_2 - p_F)\theta(p_F - p_3)$$
$$\times\ \delta\left(E - [\varepsilon(p_1) + \varepsilon(p_2) - \varepsilon(p_3)]\right). \qquad (11.75)$$

This is again a continuous function of p and E, and the presence of the additional (energy-conserving) δ-function makes the imaginary part somewhat easier to calculate than the real part. Once the former is known, the latter can be obtained by the following *dispersion relation*,

$$\text{Re}\Sigma_{+}^{(2)}(p; E) = -\frac{1}{\pi} \mathcal{P} \int dE' \, \frac{\text{Im}\Sigma_{+}^{(2)}(p; E')}{E - E'}, \qquad (11.76)$$

as may be readily verified from Eq. (11.74).

From Eq. (11.75) one also concludes that $\text{Im}\Sigma_{+}^{(2)}(p; E)$ is negative, and vanishes for $E < \varepsilon_F$ due to the presence of the energy-conserving δ-function. Similar relations hold for $\text{Im}\Sigma_{+}^{(2)}(p; E)$ and we may summarize:

$$\text{Im}\Sigma_{+}^{(2)}(p; E) = 0, \quad \text{Im}\Sigma_{-}^{(2)}(p; E) \geq 0 \ \text{ for } E < \varepsilon_F,$$
$$\text{Im}\Sigma_{-}^{(2)}(p; E) = 0, \quad \text{Im}\Sigma_{+}^{(2)}(p; E) \leq 0 \ \text{ for } E > \varepsilon_F. \qquad (11.77)$$

The dispersion relation obeyed by $\Sigma_{-}^{(2)}$ reads

$$\text{Re}\Sigma_{-}^{(2)}(p; E) = \frac{1}{\pi} \mathcal{P} \int dE' \, \frac{\text{Im}\Sigma_{-}^{(2)}(p; E')}{E - E'}, \qquad (11.78)$$

and may be combined with Eq. (11.76) into the single relation[4]

$$\text{Re}\Sigma^{(2)}(p; E) = \frac{1}{\pi} \mathcal{P} \int dE' \, \frac{|\text{Im}\Sigma^{(2)}(p; E')|}{E - E'}. \qquad (11.79)$$

A characteristic behavior for the real and imaginary part of the self-energy, connected by a dispersion relation, is shown in Fig. 11.4. The energy-dependence of the second-order self-energy in Eqs. (11.76) – (11.79) is typical for *all* contributions to the (irreducible) self-energy, as discussed further in Sec. 11.4.4.

11.3.2 *Behavior near the Fermi energy*

The analysis of the second-order self-energy in Sec. 11.2.1 made it clear that for finite systems having a particle–hole gap, $\Sigma^{(2)}$ has no poles (or equivalently: $\text{Im}\Sigma^{(2)} = 0$) in a region centered on the HF Fermi energy (11.24). In normal infinite Fermi systems there is no gap, but the imaginary part of the self-energy $\Sigma^{(2)}$ still vanishes at ε_F, as can be seen (by continuity

---

[4]The self-energy in a finite discrete sp basis (as in Sec. 11.2) also obeys these dispersion relations, but only in a trivial way: the imaginary part is a discrete sum over energy δ-functions with no connection to an integration over continuous sp labels.

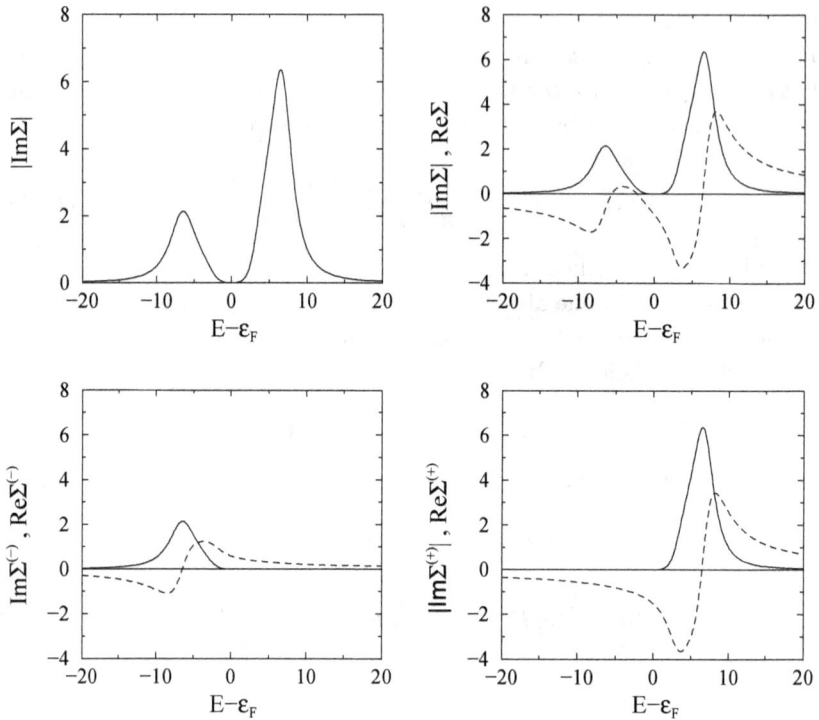

Fig. 11.4 Top left panel: illustrative energy dependence of the imaginary part of the self-energy $|\text{Im}\Sigma^{(2)}(p;E)|$. In the other panels, the corresponding real parts from the dispersion relations in Eqs. (11.76), (11.78), and (11.79), are displayed as well (dashed lines).

requirements) from the energy dependence in Eq. (11.77). In fact, the energy-dependence of $\text{Im}\Sigma^{(2)}$ near ε_F is entirely governed by phase-space restrictions, and it is easy to show that for $E \to \varepsilon_F$

$$|\text{Im}\Sigma^{(2)}(p;E)| \to C_p(E-\varepsilon_F)^2, \qquad (11.80)$$

where C_p is constant [Luttinger (1961b)].

As an example, we investigate in Eq. (11.75) the limit $E \overset{>}{\to} \varepsilon_F$. Taking the step functions into account the sp energies read

$$\varepsilon(p_1) = \varepsilon_F + u_1, \quad \varepsilon(p_2) = \varepsilon_F + u_2, \quad \varepsilon(p_3) = \varepsilon_F - u_3, \qquad (11.81)$$

where $u_i > 0$ $(i = 1, 2, 3)$. The energy-conserving δ-function in Eq. (11.75), with $E = \varepsilon_F + u$, requires that $u = u_1 + u_2 + u_3$. The limit $u \overset{>}{\to} 0$ then clearly implies that all $u_i \overset{>}{\to} 0$, and the magnitudes $p_i = |\boldsymbol{p}_i|$ of the internal sp momenta in Eq. (11.75) are by necessity close to p_F. To lowest order in u, we can therefore set $p_i = p_F$ in the interaction w, leaving w as a function of the angles only. Moreover, for $p < 3p_F$ it is always possible to find angles fulfilling momentum conservation $\boldsymbol{p}_1 + \boldsymbol{p}_2 = \boldsymbol{p} + \boldsymbol{p}_3$ with magnitudes $p_i = p_F$ for the internal momenta.[5] As a result, for $u \overset{>}{\to} 0$ the angular part of the integration in Eq. (11.75) is just a proportionality factor

$$-\frac{\pi}{2} \int d\Omega_{p_1} \int d\Omega_{p_2} \int d\Omega_{p_3} \, |w(\boldsymbol{p}, \boldsymbol{p}_i)|^2 \delta(\boldsymbol{p}_1 + \boldsymbol{p}_2 - \boldsymbol{p}_3 - \boldsymbol{p}), \qquad (11.82)$$

evaluated with $p_i = p_F$, and the energy-dependence is contained in the phase-space integral

$$F(u) = p_F^6 \int_{p_F}^{+\infty} dp_1 \int_{p_F}^{+\infty} dp_2 \int_0^{p_F} dp_3 \, \delta(u - u_1 - u_2 - u_3). \qquad (11.83)$$

We may assume that the sp energy $\varepsilon(p)$ is monotonous near ε_F, and change the integration variables from p_i to u_i. The result

$$\begin{aligned}
F(u) &= \frac{p_F^6}{[\varepsilon'(p_F)]^3} \int_0^u du_1 \int_0^u du_2 \int_0^u du_3 \, \delta(u - u_1 - u_2 - u_3) \\
&= \frac{p_F^6}{[\varepsilon'(p_F)]^3} \frac{1}{2} u^2,
\end{aligned} \qquad (11.84)$$

proves the quadratic dependence of $\mathrm{Im}\Sigma_+^{(2)}$ on $u = E - \varepsilon_F$ near the Fermi energy. The same reasoning, applied to $\mathrm{Im}\Sigma_-^{(2)}$, also leads to the quadratic behavior in Eq. (11.80), with the same proportionality constant C_p. As shown by [Luttinger (1961b)], similar arguments hold to all orders in perturbation theory, and the property (11.80) can be extended to the exact self-energy of a normal Fermi system.

11.3.3 *Spectral function*

The spectral function is in general determined (see Sec. 7.3) by the imaginary part of the propagator

$$S(p; E) = \frac{1}{\pi} |\mathrm{Im}G(p; E)|. \qquad (11.85)$$

---

[5]This is not true for one-dimensional systems, and in this case $\mathrm{Im}\Sigma$ would be proportional to $(E - \varepsilon_F)$ [Luttinger (1961b)].

In the static HF approximation the propagator in Eq. (10.164) therefore implies that the spectral function is a single delta-peak

$$S^{HF}(p; E) = \delta\left(E - \varepsilon(p)\right), \tag{11.86}$$

located at the HF sp energy $\varepsilon(p)$.

In analogy with the previous discussion on finite systems, we expect that dynamic (energy-dependent) contributions to the self-energy will provide a broadening of this peak. Indeed, one finds in normal Fermi systems that the spectral function (at least for sp momenta close to p_F) is dominated by a quasiparticle peak with a certain finite width. The latter represents a region of "most likely" energies of the residual system when a particle with momentum p is added or removed, and which can be considered as a remnant of free-particle propagation when the interparticle interaction is turned on.

To see how this works in practice, we turn again to the second-order problem. From the Dyson equation

$$G(p; E) = G^{HF}(p; E) + G^{HF}(p; E)\Sigma^{(2)}(p; E)G(p; E) \tag{11.87}$$

and its algebraic solution [see Eqs. (11.19) and (11.20)]

$$G(p; E) = \frac{1}{E - \varepsilon(p) - \Sigma^{(2)}(p; E)}, \tag{11.88}$$

the spectral function is immediately obtained as

$$S(p; E) = \frac{1}{\pi}|\mathrm{Im}G(p; E)| = \frac{\frac{1}{\pi}|\mathrm{Im}\Sigma^{(2)}(p; E)|}{[E - \varepsilon(p) - \mathrm{Re}\Sigma^{(2)}(p; E)]^2 + [\mathrm{Im}\Sigma^{(2)}(p; E)]^2}. \tag{11.89}$$

Obviously $S(p; E)$ is a continuous function of p and E and, due to the magic of complex analytic functions, it automatically obeys the sum rule [see Eq. (11.43)]

$$\int dE \ S(p; E) = 1, \tag{11.90}$$

when $\mathrm{Im}\Sigma^{(2)}$ and $\mathrm{Re}\Sigma^{(2)}$ are connected by the dispersion relation in Eq. (11.79).

Near the Fermi energy, $\mathrm{Im}\Sigma^{(2)}(p; E)$ varies proportional to $(E - \varepsilon_F)^2$ (see Sec. 11.3.2) and hence is quite small. Upon inspection of Eq. (11.89), the same must hold for the spectral function *unless* the term $[E - \varepsilon(p) - \mathrm{Re}\Sigma^{(2)}(p; E)]^2$ in the denominator becomes simultaneously small. We therefore expect $S(p; E)$ to be peaked near the energy $E_Q(p)$ which is a root of

$$E_Q(p) = \varepsilon(p) + \mathrm{Re}\Sigma^{(2)}(p; E_Q(p)). \qquad (11.91)$$

Eq. (11.91) is an implicit relation between the sp momentum p and energy E, and defines the quasiparticle spectrum[6] $E_Q(p)$ of the present approximation.

At this point the lack of self-consistency in Eq. (11.87) gives rise to a contradiction. As the quasiparticle peak can be thought to evolve adiabatically from the noninteracting spectral function, it is clear that for $p < p_F$ the peak must contribute to the removal strength and thus $E_Q(p) < \varepsilon_F$. Likewise, for $p > p_F$ one must have $E_Q(p) > \varepsilon_F$. So the Fermi energy, which separates the removal and addition domain, plays a double role in the exact theory: first, the imaginary part of the self-energy switches sign at ε_F, and second, the Fermi energy is equal to the quasiparticle energy at the Fermi momentum, $\varepsilon_F = E_Q(p_F)$.

In the present non-self-consistent approximation these definitions are clearly incompatible: the self-energy is evaluated with HF propagators, and switches sign at the HF Fermi energy $\varepsilon(p_F)$, which will in general be different from the Fermi energy $E_Q(p_F)$ defined in Eq. (11.91). Obviously, only the self-energy determined in a self-consistent approach will share this property with the exact self-energy.

11.4 Exact self-energy in infinite systems

After the preceding study of the second-order self-energy, we are now in a position to discuss the energy-dependence of the exact (irreducible) self-energy, as expressed in Eq. (9.34) or diagrammatically in Fig. 9.13. For simplicity the auxiliary potential U is omitted here, as its inclusion presents no real difficulty.

---

[6]This is the most appropriate definition of the quasiparticle spectrum, but other definitions exist, see *e.g.* the review papers [Mahaux *et al.* (1985); Mahaux and Sartor (1991)].

11.4.1   *General considerations*

We begin with some general considerations that also hold for finite systems. The self-energy can usually be split up[7] according to

$$\Sigma(\gamma, \delta; E) = \Sigma_s(\gamma, \delta) + \Sigma_d(\gamma, \delta; E), \tag{11.92}$$

where the static part Σ_s (the dynamic part Σ_d) corresponds to the second (third) term in Eq. (9.34), or to the second (third) diagram in Fig. 9.13.

The static self-energy Σ_s in Eq. (11.92) is real and independent of energy. In terms of diagrams, Σ_s contains all possible contributions to the self-energy where the external propagators are attached to the same interaction line. According to Eqs. (10.12) – (10.13), Σ_s equals

$$\Sigma_s(\gamma, \delta) = \sum_{\mu\nu} \langle \gamma\mu | V | \delta\nu \rangle \, n_{\mu\nu}, \tag{11.93}$$

which is the expression for the HF potential, but evaluated with the *exact* one-body density matrix $n_{\mu\nu}$ of the system.

The energy-dependence of the dynamic self-energy Σ_d has been worked out in Sec. 11.1, upon replacement of the vertex function Γ with the tp interaction. The exact Γ is itself energy-dependent, but the structure of Σ_d remains the same: a sum of simple poles, possibly merged into branch cuts, shifted slightly off the real axis into the lower (upper) half plane for poles corresponding to the addition (removal) domain. As a consequence, the dynamic part vanishes at large energy, and only the static part remains

$$\Sigma_s(\gamma, \delta) = \lim_{E \to \infty} \Sigma(\gamma, \delta; E). \tag{11.94}$$

11.4.2   *Self-energy and spectral function*

We now continue to discuss the case of (homogeneous) infinite systems. The static part of the self-energy

$$\Sigma_s(p) = \sum_{p'} n(p') \langle \boldsymbol{pp'} | V | \boldsymbol{pp'} \rangle, \tag{11.95}$$

is expressed in terms of the exact momentum distribution $n(p)$, which

---

[7]In some cases, *e.g.* a hard-core potential becoming infinitely repulsive when the interparticle distance is smaller than some finite hard-core radius, the separate terms in Eq. (11.92) are both divergent; in that case the dispersion relation in Eq. (11.96) must be replaced by so-called subtracted dispersion relations.

replaces the Fermi-gas step function $\theta(p_F - p)$, appearing in the HF result of Eq. (10.163).

As in Sec. 11.3.1, the dynamic self-energy $\Sigma_d(p; E)$ is a complex function, whose analytic properties can simply be taken over from Eqs. (11.76) – (11.79), by replacing $\Sigma^{(2)}$ with Σ_d. The dispersion relation (11.79), *e.g.*, becomes

$$\text{Re}\Sigma(p; E) = \Sigma_s(p) + \frac{1}{\pi}P\int dE' \frac{|\text{Im}\Sigma(p; E')|}{E - E'}, \tag{11.96}$$

where one should take into account that

$$\text{Re}\Sigma(p; E) = \Sigma_s(p) + \text{Re}\Sigma_d(p; E). \tag{11.97}$$

From the results in Sec. 11.3.2 it can be inferred that, as $E \to \varepsilon_F$,

$$|\text{Im}\Sigma(p; E)| \to C_p(E - \varepsilon_F)^2, \tag{11.98}$$

with ε_F the exact Fermi energy. Since the spectral function can be expressed as [see Eq. (11.89)]

$$S(p; E) = \frac{\frac{1}{\pi}|\text{Im}\Sigma(p; E)|}{[E - \frac{p^2}{2m} - \text{Re}\Sigma(p; E)]^2 + [\text{Im}\Sigma(p; E)]^2}, \tag{11.99}$$

also $S(p; E)$ [for $p \neq p_F$] vanishes quadratically at ε_F,

$$S(p; E) \sim (E - \varepsilon_F)^2 \text{ as } E \to \varepsilon_F. \tag{11.100}$$

11.4.3 *Quasiparticles*

We again expect, for the same reasons as explained in Sec. 11.3.3, that the spectral function $S(p; E)$ is peaked near the root of the quasiparticle equation

$$E_Q(p) = \varepsilon_s(p) + \text{Re}\Sigma_d(p; E_Q(p)), \tag{11.101}$$

where $\varepsilon_s(p) = \frac{p^2}{2m} + \Sigma_s(p)$. In a normal Fermi system, the self-energy $\Sigma(p; E)$ is smooth and can be linearized around $p = p_F$ and $E = \varepsilon_F$. One can then show that, at least for momenta p near p_F, Eq. (11.101) has a unique quasiparticle root $E_Q(p)$, which coincides with the Fermi energy at $p = p_F$, *i.e.*

$$E_Q(p_F) = \varepsilon_F = \varepsilon_s(p_F) + \text{Re}\Sigma_d(p_F; \varepsilon_F). \tag{11.102}$$

The behavior for $E \to E_Q(p)$ of the spectral function $S(p; E)$ in Eq. (11.99) is easily analyzed. A first-order expansion of $\mathrm{Re}\Sigma_d(p; E)$ in $[E - E_Q(p)]$ leads to

$$E - \varepsilon_s(p) - \mathrm{Re}\Sigma_d(p; E) \to [E - E_Q(p)]\left(1 - \frac{\partial \mathrm{Re}\Sigma_d(p; E)}{\partial E}\right)_{E=E_Q(p)}. \tag{11.103}$$

Defining the quasiparticle strength $Z_Q(p)$ as

$$Z_Q(p) = \left(1 - \frac{\partial \mathrm{Re}\Sigma_d(p; E)}{\partial E}\bigg|_{E=E_Q(p)}\right)^{-1}, \tag{11.104}$$

the spectral function in Eq. (11.99) can be decomposed as

$$S(p; E) = S_Q(p; E) + S_B(p; E), \tag{11.105}$$

where the strongly varying part (the quasiparticle peak) is isolated in $S_Q(p; E)$

$$S_Q(p; E) = \frac{\frac{1}{\pi} Z_Q^2(p) |\mathrm{Im}\Sigma(p; E_Q(p))|}{[E - E_Q(p)]^2 + [Z_Q(p)\mathrm{Im}\Sigma(p; E_Q(p))]^2}, \tag{11.106}$$

and the remainder $S_B(p; E)$, called the background contribution, is slowly varying near $E_Q(p)$.

The energy-dependence of $S_Q(p; E)$ in Eq. (11.106) has the well-known Breit–Wigner shape

$$\mathcal{L}_\Delta(E) = \frac{\Delta/\pi}{E^2 + \Delta^2}, \tag{11.107}$$

which is a distribution normalized to unity and having a full width at half maximum equal to 2Δ. We may rewrite

$$S_Q(p; E) = Z_Q(p)\mathcal{L}_{\Delta_Q(p)}(E - E_Q(p)) \tag{11.108}$$

and interpret Eq. (11.105) as a Breit–Wigner shaped quasiparticle peak with strength $Z_Q(p)$ and width parameter

$$\Delta_Q(p) = Z_Q(p)|\mathrm{Im}\Sigma(p, E_Q(p))|, \tag{11.109}$$

superimposed on a smooth background.

In Fig. 11.5, the typical energy-dependence of the spectral function for various sp momenta is shown. Note that, as $p \to p_F$, the quasiparticle

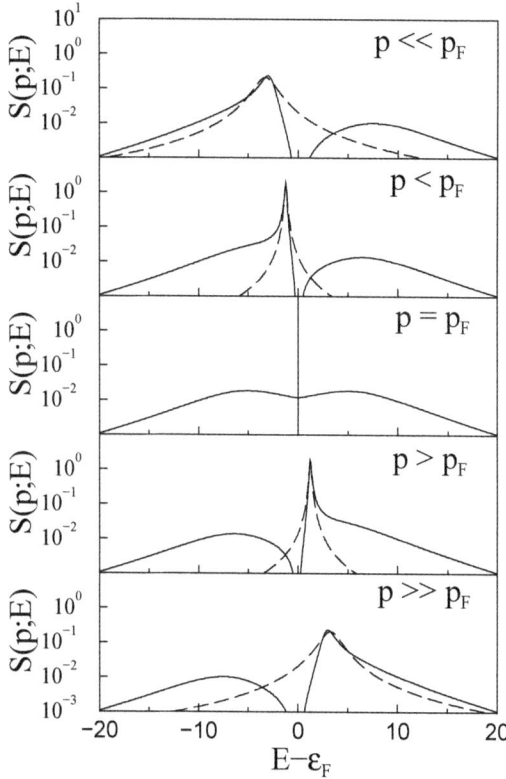

Fig. 11.5 Evolution of the energy-dependence of the spectral function $S(p; E)$ for increasing values of the sp momentum p. The dashed line represents the Breit–Wigner approximation in Eq. (11.108). The middle panel is for $p = p_F$, where the quasiparticle peak is a δ-spike, indicated by the vertical line at ε_F.

energy approaches the Fermi energy, $E_Q(p) \to \varepsilon_F$, and the imaginary part of the self-energy vanishes

$$\mathrm{Im}\Sigma(p; E_Q(p)) \to \mathrm{Im}\Sigma(p; \varepsilon_F) \to 0. \qquad (11.110)$$

From Eq. (11.109) it follows that the quasiparticle peaks become narrower as we approach the Fermi momentum, and at $p = p_F$ the peak has zero width, or $\Delta_Q(p_F) = 0$. Since $\lim_{\Delta \to 0} \mathcal{L}_\Delta(E) = \delta(E)$, the quasiparticle peak for $p = p_F$ is a δ-spike located at the Fermi energy, with strength $Z_F = Z_Q(p_F)$. The decomposition of the spectral function in a quasiparticle peak and background, is only exact and unambiguous at $p = p_F$. The background contribution is the smooth curve in the middle panel of Fig. 11.6 and can be expressed as

$$S_B(p_F; E) = \frac{\frac{1}{\pi}|\text{Im}\Sigma(p_F; E)|}{[E - \frac{p_F^2}{2m} - \text{Re}\Sigma(p_F; E)]^2 + [\text{Im}\Sigma(p_F; E)]^2}. \tag{11.111}$$

Note that $S_B(p_F; E)$ is in general nonzero at $E = \varepsilon_F$, as can be seen by expanding the numerator and denominator of Eq. (11.111) in $u = (E - \varepsilon_F)$. To leading order in u one gets

$$S_B(p_F; E) = \frac{\frac{1}{\pi}C_F u^2 + \mathcal{O}(u^3)}{(u/Z_F)^2 + \mathcal{O}(u^3)}, \tag{11.112}$$

and $S_B(p_F; \varepsilon_F) = \frac{1}{\pi}C_F Z_F^2$.

11.4.4 *Migdal–Luttinger theorem*

The so-called Migdal–Luttinger theorem [Migdal (1957); Luttinger (1961a)] states that the momentum distribution $n(p)$ in a normal Fermi liquid has a discontinuity at $p = p_F$ with magnitude equal to Z_F,

$$\lim_{\eta \overset{>}{\to} 0} [n(p_F - \eta) - n(p_F + \eta)] = Z_F. \tag{11.113}$$

This can now be understood from the relation

$$n(p) = \int_{-\infty}^{\varepsilon_F} dE \; S(p; E), \tag{11.114}$$

and the behavior of the spectral functions in Fig. 11.5. A decomposition of $S(p; E)$ in Eq. (11.114), according to Eq. (11.105), leads to $n(p) = n_Q(p) + n_B(p)$, where

$$n_B(p) = \int_{-\infty}^{\varepsilon_F} dE \; S_B(p; E) \tag{11.115}$$

is a smooth function of p, which cannot contribute to the discontinuity. The contribution of the quasiparticle peak is

$$n_Q(p) = \int_{-\infty}^{\varepsilon_F} dE \; S_Q(p; E) = Z_Q(p) \int_{-\infty}^{\varepsilon_F} dE \; \mathcal{L}_{\Delta_Q(p)}(E - E_Q(p)). \tag{11.116}$$

In the limit $p \to p_F$ the peak position $E_Q(p)$ approaches ε_F. However, the width $\Delta_Q(p)$ of the peak decreases *quadratically* like $[E_Q(p) - \varepsilon_F]^2$. As a consequence, the integration in Eq. (11.116) will, for $p \to p_F$, sample the peak either completely [for $p < p_F$ and $E_Q(p) < \varepsilon_F$], or not at all [for $p > p_F$ and $E_Q(p) > \varepsilon_F$], *i.e.*

$$n_Q(p) \rightarrow Z_Q(p)\theta(p_F - p), \tag{11.117}$$

and the theorem follows.

11.4.5 *Quasiparticle propagation and lifetime*

A further interpretation of the width $\Delta_Q(p)$ of the quasiparticle peak follows from the fact that the spectral function $S_Q(p; E)$ in Eq. (11.108) can be derived, through the relation $S_Q(p; E) = \frac{1}{\pi}|\mathrm{Im}G_Q(p; E)|$, from the following quasiparticle propagator

$$G_Q(p; E) = \frac{Z_Q(p)}{E - E_Q(p) + i\Delta_Q(p)\mathrm{sgn}(p - p_F)}. \tag{11.118}$$

In Eq. (11.118) the structure of a noninteracting propagator (10.160) can be recognized, but with a reduced strength $Z_Q(p) < 1$ and a *complex* sp energy $E_Q(p) \pm i\Delta_Q(p)$. Going back to the time domain by a FT

$$G_Q(p; t) = \frac{1}{2\pi\hbar} \int dE e^{-\frac{i}{\hbar}Et} G_Q(p; E), \tag{11.119}$$

$$= \frac{Z_Q(p)}{i\hbar} e^{-\frac{i}{\hbar}E_Q(p)t} e^{-\Delta_Q(p)\frac{|t|}{\hbar}} [\theta(p - p_F)\theta(t) - \theta(p_F - p)\theta(-t)],$$

we can see that the imaginary part of the sp energy introduces an exponential decay in time of the propagator. That is, the excitation created by the addition or removal of a particle has a finite lifetime, which is inversely proportional to the width $\Delta_Q(p)$. This is in contrast to a noninteracting system, or the mean-field treatment of an interacting system, where these sp excitations are eigenstates in the $N \pm 1$ system, and hence have an infinite lifetime. The quasiparticles in an interacting system are the remnants of this free-particle propagation. While the quasiparticle excitations are only unambiguously defined (and long-lived) near the Fermi momentum and energy, it is usually possible to extend the concept to all sp momenta. Some care must be taken for momenta far from p_F, *e.g.* multiple roots may appear in Eq. (11.101), or the definition of the quasiparticle strength in Eq. (11.104) may break down.

The fact that at the Fermi surface the propagator is completely dominated by its quasiparticle contribution, also helps to understand why the equality in Eq. (11.102) must hold: the reasoning in Sec. 11.3.2 can be repeated for the second-order diagram, evaluated with the *exact* propagator. This represents the leading energy-dependent term in a so-called skeleton diagram expansion of the exact self-energy in terms of the propagator. The role of the HF sp energies in Sec. 11.3.2 is now played by the exact quasiparticle energies $E_Q(p)$, and $\mathrm{Im}\Sigma(p; E)$ therefore vanishes at $E = \varepsilon_F = E_Q(p_F)$.

11.5   Self-consistent treatment of $\Sigma^{(2)}$

We now return to the fully self-consistent treatment of the second-order self-energy, as contained in Eqs. (11.1) – (11.4) and indicated diagrammatically in Fig. 11.1. For clarity the discussion is restricted, as in Sec. 11.2.1, to the case of a finite system with discrete poles and a diagonal approximation for the sp propagator

$$G(\alpha; E) = \sum_m \frac{|z_\alpha^{m+}|^2}{E - \varepsilon_{ma}^+ + i\eta} + \sum_n \frac{|z_\alpha^{n-}|^2}{E - \varepsilon_{na}^- - i\eta}. \qquad (11.120)$$

This covers most of the practical applications that have been performed for nuclei and atoms.

The second-order self-energy in Eq. (11.3) now reads

$$\Sigma^{(2)}(\alpha; E) = \frac{1}{2} \sum_{\lambda,\epsilon,\nu} |\langle \alpha\lambda| V |\epsilon\nu\rangle|^2 \qquad (11.121)$$

$$\times \left(\sum_{m_1 m_2 n_3} \frac{|z_\epsilon^{m_1+}|^2 |z_\nu^{m_2+}|^2 |z_\lambda^{n_3-}|^2}{E - (\varepsilon_{m_1\epsilon}^+ + \varepsilon_{m_2\nu}^+ - \varepsilon_{n_3\lambda}^-) + i\eta} \right.$$

$$\left. + \sum_{n_1 n_2 m_3} \frac{|z_\epsilon^{n_1-}|^2 |z_\nu^{n_2-}|^2 |z_\lambda^{m_3+}|^2}{E + (\varepsilon_{m_3\lambda}^+ - \varepsilon_{n_1\epsilon}^- - \varepsilon_{n_2\nu}^-) - i\eta} \right).$$

The (static) first-order self-energy in Eq. (11.3) is given by

$$\Sigma^{(1)}(\alpha) = \sum_\beta \langle \alpha\beta| V |\alpha\beta\rangle \left(\sum_n |z_\beta^{n-}|^2 \right), \qquad (11.122)$$

and can be absorbed into new sp energies $\varepsilon_\alpha^{(s)}$, by rewriting the Dyson equation (11.2) as

$$\frac{1}{G(\alpha; E)} = \frac{1}{G^{(0)}(\alpha; E)} - \Sigma(\alpha; E) = (E - \varepsilon_\alpha^{(0)}) - \Sigma(\alpha; E)$$

$$= (E - \varepsilon_\alpha^{(s)}) - \Sigma^{(2)}(\alpha; E), \qquad (11.123)$$

where

$$\varepsilon_\alpha^{(0)} = \langle \alpha | \, T \, | \alpha \rangle + \langle \alpha | \, U \, | \alpha \rangle$$
$$\Sigma(\alpha; E) = - \langle \alpha | \, U \, | \alpha \rangle + \Sigma^{(1)}(\alpha) + \Sigma^{(2)}(\alpha; E)$$
$$\varepsilon_\alpha^{(s)} = \langle \alpha | \, T \, | \alpha \rangle + \Sigma^{(1)}(\alpha). \tag{11.124}$$

Based on the results of Sec. 11.2.1, it should also present no difficulty to write down the corresponding set of equations for the unknown poles $\varepsilon_{m\alpha}^+$, $\varepsilon_{n\alpha}^-$ and residues z_α^{m+}, z_α^{n-}, which appear in Eq. (11.120). The poles correspond to the roots $E_{n\alpha}$ of the equation

$$E_{n\alpha} = \varepsilon_\alpha^{(s)} + \Sigma^{(2)}(\alpha; E_{n\alpha}), \tag{11.125}$$

where the $E_{n\alpha}$ smaller (greater) than the Fermi energy are removal energies $\varepsilon_{n\alpha}^-$ (addition energies $\varepsilon_{m\alpha}^+$), and the corresponding residues $R_{n\alpha}$ follow from Eq. (11.23).

However, unlike the treatment in Sec. 11.2.1, the self-energy should now be consistent with the sp propagator, and $\Sigma^{(1)}$ and $\Sigma^{(2)}$ themselves depend on the poles and residues of the propagator. Just as in the HF case the resulting nonlinear equations can be solved iteratively, by starting with a guess for the propagator, using it to evaluate the self-energy in Eqs. (11.121) – (11.122), and applying the Dyson equation in Eq. (11.125) to construct an updated propagator.

Note that strictly speaking, when applying a finite and discrete sp basis set, an exact solution does not exist because of the dimensionality arguments presented in Sec. 11.2.1: if the propagator of a certain iteration step has M poles, the self-energy constructed with it has [of the order of] $\mathcal{O}(M^3)$ poles, since $\Sigma^{(2)}$ contains three propagators. Consequently the Dyson equation will lead to a new propagator also having $\mathcal{O}(M^3)$ poles, and convergence cannot be achieved. The difficulty arises because the number of poles in the propagator and self-energy should in principle be restricted by the dimension of the Fock space corresponding to N fermions. In an expansion of the exact self-energy there will be cancellations, due to the Pauli principle, between diagrams containing an intermediate state with more than N simultaneous hole-lines. A truncation to the second-order self-energy spoils such delicate cancellations. In practice all this is not too important: when iterating the second-order equations the increase in the number of poles is limited to regions far from the Fermi energy, where the density of states is high. In that case individual states in the $N \pm 1$

system cannot be resolved anyway, and one is rather interested in averaged distributions of sp strength.

The simplest way to circumvent the dimensionality problem, mentioned above, is the single-pole (or quasiparticle) approximation, where only the dominant solution of Eq. (11.125) [having the largest residue or sp strength] is retained as a single pole in the updated propagator $G(\alpha; E)$. In order to include the width of a realistic spectral function, one has to go beyond the single-pole approximation. Basically two methods are in use to cut down the number of poles after having obtained a set of energies $E_{n\alpha}$ and residues $R_{n\alpha}$ of Eq. (11.125). One is to divide the energy axis into a number or narrow bins, sum the strength $R_{n\alpha}$ in each bin, and update the propagator $G(\alpha; E)$ by taking the center and the summed strength of each bin as new poles and residues. Alternatively, one replaces the spectral distribution $(E_{n\alpha}, R_{n\alpha})$ by a smaller number of poles, chosen so as to reproduce the lowest-order energy-weighted moments (see Sec. 11.2.3) of the spectral function. Both methods have been used to describe properties of atoms and nuclei.

11.5.1 *Schematic model*

The following model problem is quite transparent and useful to illustrate the additional effects caused by a self-consistent treatment of the second-order diagram. We consider in Eq. (11.121) M particle states p_i and M hole states h_i, with sp energies $\varepsilon_{h_i} = -\varepsilon_{p_i}$. The sp energies are kept fixed (*i.e.* the first order self-energy $\Sigma^{(1)}$ is neglected), and we assume a constant interaction strength

$$| \langle \alpha\beta| V |\gamma\delta \rangle |^2 = |v|^2. \qquad (11.126)$$

These model assumptions lead to a self-energy which is state-independent and antisymmetric, $\Sigma(-E) = -\Sigma(E)$, and to exact particle-hole symmetry, $G(p_i; E) = -G(h_i; -E)$.

In the example below we took $M = 6$, $|v| = 0.75$ MeV, and

$$\varepsilon_{p_i} = 2, 3, 4, 8, 9, 10 \text{ MeV, for } i = 1,\dots,6, \qquad (11.127)$$

mimicking a nuclear shell structure with two main shells above and below the Fermi level. Eqs. (11.121) and (11.125) were solved iteratively, with a division of the energy axis into 0.1 MeV wide bins.

Figure 11.6 contains the changes that occur in the self-energy and spectral functions when the system is iterated to convergence. The histograms

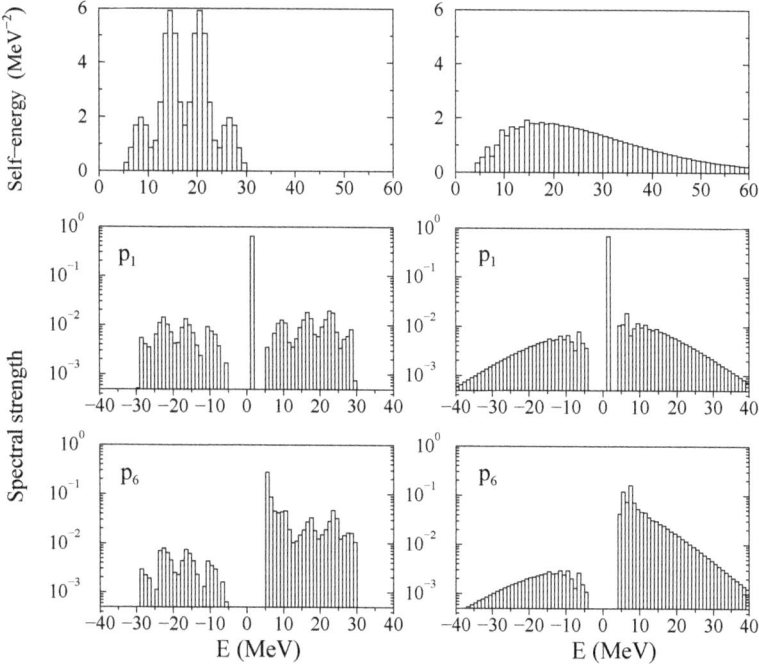

Fig. 11.6 First iteration result (left panels) and the converged self-consistent result (right panels) for the self-energy strength and the spectral functions in the model problem of Sec. 11.5.1.

shown, are the self-energy strength $\frac{1}{\pi}|\mathrm{Im}\Sigma(E)|$ and the spectral functions $S(p_1; E)$ and $S(p_6; E)$, integrated over 1 MeV wide histogram bins. The first-iteration self-energy, which is evaluated with unperturbed propagators, simply reflects the regular shell structure of the 2p1h density of states. The propagator that solves the corresponding Dyson equation shows the fragmentation of sp strength as discussed in Sec. 11.2.1: the spectral function for the p_1 valence state has a dominant, isolated quasiparticle peak carrying most (64 %) of the sp strength. In addition a background appears with sp strength located near the unperturbed 2p1h and 1p2h energies, with the strength near the 1p2h energies indicating a nonzero occupation of p_1 in the correlated ground state. For the p_6 state, which is farther from the Fermi energy, stronger fragmentation occurs, and the main peak is less dominant.

During subsequent iterations, this fragmentation of sp strength is included in the propagators that build up the self-energy. When the solution has converged, the resulting self-energy has drastically different features. Its strength is no longer concentrated exclusively at the 2p1h and 1p2h energies, but is spread out to larger excitation energies, and the minima corresponding to the shell-structure are filled up. This is due to so-called many-body damping: the self-consistent "second-order" self-energy actually contains higher-order diagrams, allowing the sp strength to be distributed over more complicated configurations. At high energies, where the density of states is large, the shell structure of the first iteration self-energy has therefore completely vanished. Near the Fermi energy however, the shell-structure peaks remain, as the isolated quasiparticle peaks of the valence states dominate in this region.

The self-consistent spectral functions show similar features: for the p_1 valence state the isolated quasiparticle peak remains, whereas the higher-lying p_6 state has a broad distribution representing the p_6 quasiparticle state. Such features agree with the data discussed in Sec. 7.8. In addition a smooth background is present in all spectral functions, stretching out to high excitation energies in both the $N+1$ and $N-1$ system.

11.5.2 *Nuclei*

A direct calculation of the second-order diagram using realistic nucleon–nucleon interactions is not meaningful. As explained in Sec. 10.6, the strong short-range repulsion of the bare force must first be neutralized by preventing nucleons in the correlated system from coming close together. In principle this must be achieved through the construction of an effective NN interaction (see Sec. 15.2) consistent with the sp propagator.

Nevertheless, self-consistent second-order calculations have been performed for some nuclei, using approximate effective interactions, *e.g.* parametrizations of the Skyrme-type [Waroquier *et al.* (1987)], which are suitable to describe NN correlations in a limited model space. As an example we compare in Fig. 11.7 the spectral function for proton removal from ^{48}Ca with experimental data derived from $(e, e'p)$ reactions. In the independent-particle model the proton hole orbitals $0s\frac{1}{2}$, $0p\frac{3}{2}$, $0p\frac{1}{2}$, $0d\frac{5}{2}$, $0d\frac{3}{2}$ and $1s\frac{1}{2}$ (listed in order of increasing sp energy) are assumed to be completely filled in the ground state of the (doubly-closed shell) nucleus ^{48}Ca. The second-order calculation of [Van Neck *et al.* (1991)], is able to describe the global evolution of the fragmentation pattern, going from the valence hole states to the deeply-bound orbitals. Especially for the $\ell = 1$ deep-hole states, the self-consistent formulation (as opposed to the first

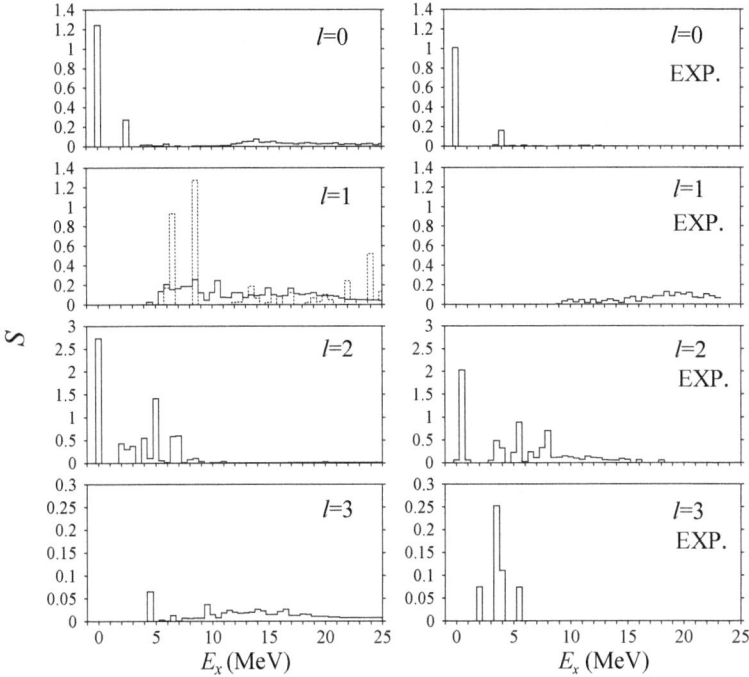

Fig. 11.7 Spectral function for proton removal from ^{48}Ca. Left panel: result of a self-consistent second-order calculation (see text). For the deeply bound $0p$ states, the first-iteration result is also shown (dotted line). Right panel: experimental spectral function adapted from [Kramer (1990)], obtained in a study of the ^{48}Ca$(e, e'p)$ reaction.

iteration result also shown in Fig. 11.7) clearly improves the description.

A more detailed study reveals several discrepancies, *e.g.* all spectroscopic factors appear to be overestimated. This is partially due to the neglect of short-range correlations: in principle the energy-dependence of the true effective interaction would provide an additional depletion of about 10%. Another reason for the discrepancies is that the second-order self-energy is just too simple: it can indeed be shown that more sophisticated approximations for the vertex function [Rijsdijk *et al.* (1992)], which include the coupling of sp degrees of freedom to low-lying collective states, provide significant improvement for both the location and the fragmentation of sp strength. This is particularly true for the $\ell = 3$ removal strength, originating from the proton $0f\frac{7}{2}$ and $0f\frac{5}{2}$ orbitals, which lie just above the Fermi energy and are unoccupied in a mean-field treatment. Short-range correlations, not included here, also provide a global depletion of the sp

Table 11.1   Second-order results and experimental data.

		Removal energies			Spectroscopic factors		
		HF	$\Sigma^{(2)}$	Exp.	HF	$\Sigma^{(2)}$	Exp.
He	$1s$	−0.918	−0.906	−0.9040	1.0	0.972	
Be	$1s$	−4.733	−4.620	−4.100	1.0	0.873	
	$2s$	−0.309	−0.320	−0.343	1.0	0.950	
Ne	$2s$	−1.930	−1.750	−1.782	1.0	0.876	0.85(2)
	$2p$	−0.850	−0.763	−0.793	1.0	0.904	0.92(2)
Mg	$2p$	−2.283	−2.146	−1.81	1.0	0.882	
	$3s$	−0.253	−0.274	−0.2811	1.0	0.962	
Ar	$3s$	−1.277	−1.159	−1.075	1.0	0.876	0.55(1)
	$3p$	−0.591	−0.585	−0.579	1.0	0.938	0.92(2)

Second-order results (taken from [Van Neck *et al.* (2001a); Peirs *et al.* (2002)]) for a number of $L = S = 0$ atoms, compared with experimental data [Samardzic *et al.* (1993); McCarthy *et al.* (1989); Brunger *et al.* (1999)]. All energies are in atomic units (Hartree).

strength below the Fermi energy. A detailed discussion of the influence of long and short-range correlations on the spectral strength distribution in nuclei is given in Ch. 17.

11.5.3   *Atoms*

It was already clear from the quality of the HF results in Sec. 10.2 that electrons in atoms constitute a rather weakly correlated many-body system. Including the second-order self-energy and performing the self-consistent calculation, one typically recovers more than 90% of the correlation energy [Dahlen and von Barth (2004); Van Neck *et al.* (2001a)].[8] Additional results for the closed-shell atoms are collected in Table 11.1, where the removal energy and spectroscopic strength for the hole orbitals nearest to the Fermi energy are listed. The second-order diagram obviously generates a shift of the HF removal energies to the experimental ones. The spectroscopic factors, typically 95 % for the highest occupied state and somewhat smaller for the next, are also in better agreement with $(e, 2e)$ results.

The numbers in Table 11.2 for the open *p*-shell (second row) atoms B, C, N, O, and F, were obtained using a spin-unrestricted formalism and a suitable angular averaging of the self-energy [Peirs *et al.* (2002)]. As is evident from Table 11.2, the electron correlations contained in the second-order

---

[8]The sp basis set for the unoccupied HF states in [Van Neck *et al.* (2001a)] was not sufficient to ensure full convergence for the total energy.

Table 11.2   Second-order results for open p-shell atoms.

	Ionization energies			Electron affinities		
	HF	$\Sigma^{(2)}$	Exp.	HF	$\Sigma^{(2)}$	Exp.
B	0.311	0.305	0.305	–	0.008	0.010
C	0.435	0.415	0.414	–	0.046	0.046
N	0.571	0.537	0.534	–	–	–
O	0.510	0.484	0.500	–	0.032	0.054
F	0.674	0.619	0.640	–	0.126	0.125

Second-order results (taken from [Peirs *et al.* (2002)]) for the open p-shell atoms, compared with experimental data. All energies are in atomic units (Hartree).

diagram also lead to a reasonable, simultaneous, description of ionization energies $I = E_0^{N-1} - E_0^N$ and electron affinities $A = E_0^N - E_0^{N+1}$. Note that HF for a neutral atom does not lead to bound unoccupied HF sp levels. As a consequence the stable negative ions B$^-$, C$^-$, O$^-$ and F$^-$ cannot be described with HF (on the neutral atom), but explicitly require correlations beyond HF. The second-order diagram correctly predicts the existence of these negative ions, as well as the absence of a stable N$^-$ ion, which is indeed not observed experimentally.

The results in Tables 11.1 – 11.2 are obviously an improvement over the HF results, but still show some serious deviations from the experimental situation. A case in point is the 3s level in Ar, which is experimentally known (see Fig. 7.4) to be strongly fragmented, with the main fragment carrying only 55% of the strength, but which remains rather pure in the self-consistent second-order calculation. This discrepancy can be traced back to the absence of screening diagrams in the second-order self-energy [Amusia and Kheifets (1985)]. In the above calculations the HF spectrum for the unoccupied sp states has only a continuum part, and had to be discretized by adding a confining potential at large distances. The resulting HF particle-hole spectrum is obviously but a poor representation of the true low-lying excitation spectrum of the neutral atom, which is dominated by Rydberg series of the type shown in Fig. 3.4. An improvement in the $N-1$ (removal) domain of the spectral function is therefore intimately connected with a more realistic description of the excited states in the N-electron system, as will be further discussed in Sec. 13.5.

11.6 Exercises

(1) Use the dispersion relation Eq. (11.96) to prove the "once subtracted" dispersion relation

$$\text{Re}\Sigma(p; E) = \text{Re}\Sigma(p; E_0) + (E - E_0)\frac{1}{\pi}P\int dE' \; \frac{|\text{Im}\Sigma(p; E')|}{(E - E')(E' - E_0)},$$

for arbitrary energies E, E_0.

(2) Given that $\varepsilon_F = 0$ and

$$\text{Im}\Sigma_+^{(2)}(p; E) = -c\theta(E)\frac{E^2}{E^4 + \Delta^4},$$

use the dispersion relation Eq. (11.76) to show that

$$\text{Re}\Sigma_+^{(2)}(p; E) = \frac{c}{E^4 + \Delta^4}\left(\frac{1}{\pi}E^2\ln(|\frac{E}{\Delta}|) - \frac{1}{4\Delta}[\Delta^3 + \sqrt{2}E(\Delta^2 - E^2)]\right).$$

Assuming symmetry around the Fermi energy

$$\text{Im}\Sigma_-^{(2)}(p; E) = -\text{Im}\Sigma_+^{(2)}(p; -E),$$

calculate $\text{Re}\Sigma_-^{(2)}(p; E)$, the total real part $\text{Re}\Sigma^{(2)}(p; E)$, and generate figures similar to Fig. 11.4.

Calculate the corresponding spectral function $S(p; E)$ through Eq. (11.89) and check by numerical integration that the normalization condition in Eq. (11.90) holds for arbitrary sp energies $\varepsilon(p)$.

(3) Reproduce the first iteration results in Fig. 11.6 for the schematic model of Sec. 11.5.1. Use a root-finding algorithm (bisection is the most convenient) to find numerically the solutions of Eq. (11.125) in each interval between successive poles of the self-energy.

Chapter 12

Interacting boson systems

In a fermion system, any sp state can be maximally occupied by one particle because of the Pauli principle. The ground state of a noninteracting fermion system is therefore characterized by filling the Fermi sea consisting of the N sp states with lowest energy. As is clear from Chs. 10 – 11, this feature persists for normal systems, even in the presence of interparticle forces. That is, N sp states have occupations smaller than, but of order, unity, whereas the other sp states have a small but nonzero occupation probability. Boson systems behave rather differently from analogous fermion systems with similar external fields and/or interparticle forces. The many-boson wave function is symmetric under permutation of particle coordinates, and multiple occupation of the same sp level is allowed. Hence a noninteracting system of N bosons in its ground state will have all particles in the sp state corresponding to the lowest energy, and the occupation number corresponding to this state is N. Such a *macroscopic* occupation (proportional to the particle number) of a particular sp state is called Bose–Einstein condensation (BEC). The condensed phase is the ground state of the many-boson system at zero temperature. However, at finite temperature, the particular role played by the lowest-energy sp state can disappear, as other sp states acquire nonzero occupations due to thermal fluctuations. The critical temperature for BEC in a noninteracting Bose gas was examined in Sec. 5.6.

This chapter deals with the inclusion of interparticle interactions in order to describe realistic Bose systems at $T=0$. The formalism of Green's functions, developed so far, is well suited for this, but requires some modifications to deal with the special role of the condensate. Sec. 12.1 begins with some considerations about the general structure of the boson sp propagator in noninteracting and interacting systems, and the definition of the condensate orbital in nonuniform systems. The Bose condensate acts as a reservoir, and particles can be easily removed or added. In Sec. 12.2 we explain why this leads to fundamental difficulties with the perturbation theory developed in Ch. 8, as Wick's theorem breaks down in the presence of a condensate. However, the counterpart of the HF approximation for fermions can easily be constructed by replacing the boson problem with an equivalent fermion one. The resulting mean-field or Hartree–Bose (HB) theory is developed in Sec. 12.3. Since most realistic interactions are too strong for HB to be applicable, one needs to consider an effective in-medium interaction. As shown in Sec. 12.4, this turns out to be very simple for dilute (low-density) systems, since the details of the interaction do not matter, and the interaction can be replaced with a zero-range pseudo-potential, only dependent on the S-wave scattering length. This replacement leads to the Gross–Pitaevskii equation, which is discussed extensively in Sec. 12.4 in the context of BEC in ultracold vapors of bosonic atoms.

12.1   General considerations

The boson Hamiltonian has the usual form $\hat{H} = \hat{T} + \hat{V}$, where $\hat{T}$ is the one-body part of the Hamiltonian

$$\hat{T} = \sum_{\alpha\beta} \langle \alpha | T | \beta \rangle \, a_\alpha^\dagger a_\beta, \tag{12.1}$$

containing the kinetic energy and (if any) the external potential, and $\hat{V}$ is the tp interaction

$$\hat{V} = \frac{1}{4} \sum_{\alpha\beta\gamma\delta} \langle \alpha\beta | V | \gamma\delta \rangle \, a_\alpha^\dagger a_\beta^\dagger a_\delta a_\gamma. \tag{12.2}$$

The angular brackets in Eq. (12.2) denote symmetrization, *i.e.*

$$\langle \alpha\beta | V | \gamma\delta \rangle = (\alpha\beta | V | \gamma\delta) + (\alpha\beta | V | \delta\gamma), \tag{12.3}$$

in terms of direct matrix elements.

12.1.1   *Boson single-particle propagator*

In Eq. (7.1) the sp propagator was generally defined (for bosons as well as fermions), as an expectation value in the exact N-particle ground state $|\Psi_0^N\rangle$

$$i\hbar G(\alpha, \beta; t, t') = \langle \Psi_0^N | T[a_{\alpha_H}(t) a_{\beta_H}^\dagger(t')] | \Psi_0^N \rangle, \tag{12.4}$$

of a time-ordered product $\mathcal{T}[\ldots]$ of a particle removal operator $a_{\alpha_H}(t)$ and an addition operator $a^\dagger_{\beta_H}(t')$, both in the Heisenberg picture [see Eqs. (7.3) and (7.4)]. Unless mentioned otherwise we will assume the exact ground state to be normalized to unity.

In the boson case, the time-ordering is simply an interchange of the operators to get the later time to the left (without adding extra minus signs). Without explicit time-dependence in the Hamiltonian, the propagator in Eq. (12.4) depends only on the time difference $\tau = t - t'$, and can be written as

$$
\begin{aligned}
i\hbar G(\alpha,\beta;\tau) &= \left\langle \Psi_0^N \middle| \theta(\tau) a_\alpha e^{-\frac{i}{\hbar}(\hat{H}-E_0^N)\tau} a^\dagger_\beta + \theta(-\tau) a^\dagger_\beta e^{\frac{i}{\hbar}(\hat{H}-E_0^N)\tau} a_\alpha \middle| \Psi_0^N \right\rangle \\
&= \theta(\tau) \sum_m e^{-\frac{i}{\hbar}(E_m^{N+1}-E_0^N)\tau} \left\langle \Psi_0^N \middle| a_\alpha \middle| \Psi_m^{N+1} \right\rangle \left\langle \Psi_m^{N+1} \middle| a^\dagger_\beta \middle| \Psi_0^N \right\rangle \\
&\quad + \theta(-\tau) \sum_n e^{\frac{i}{\hbar}(E_n^{N-1}-E_0^N)\tau} \left\langle \Psi_0^N \middle| a^\dagger_\beta \middle| \Psi_n^{N-1} \right\rangle \left\langle \Psi_n^{N-1} \middle| a_\alpha \middle| \Psi_0^N \right\rangle, \quad (12.5)
\end{aligned}
$$

where the last identity is obtained by inserting a complete set of $N \pm 1$ eigenstates of $\hat{H}$. The propagator in the energy representation then follows by means of a FT

$$
\begin{aligned}
G(\alpha,\beta;E) &= \int_{-\infty}^{+\infty} d\tau\, e^{\frac{i}{\hbar}E\tau} G(\alpha,\beta;\tau) \\
&= \left\langle \Psi_0^N \middle| a_\alpha \frac{1}{E - \hat{H} + E_0^N + i\eta} a^\dagger_\beta - a^\dagger_\beta \frac{1}{E + \hat{H} - E_0^N - i\eta} a_\alpha \middle| \Psi_0^N \right\rangle \\
&= \sum_m \frac{\left\langle \Psi_0^N \middle| a_\alpha \middle| \Psi_m^{N+1} \right\rangle \left\langle \Psi_m^{N+1} \middle| a^\dagger_\beta \middle| \Psi_0^N \right\rangle}{E - (E_m^{N+1} - E_0^N) + i\eta} \\
&\quad - \sum_n \frac{\left\langle \Psi_0^N \middle| a^\dagger_\beta \middle| \Psi_n^{N-1} \right\rangle \left\langle \Psi_n^{N-1} \middle| a_\alpha \middle| \Psi_0^N \right\rangle}{E + (E_n^{N-1} - E_0^N) - i\eta}. \quad (12.6)
\end{aligned}
$$

12.1.2 *Noninteracting boson propagator*

For a noninteracting boson system ($\hat{H} = \hat{T}$) the ground-state wave function is the simple product state

$$
\left| \Phi_0^N \right\rangle = \frac{1}{\sqrt{N!}} \left(a_0^\dagger \right)^N |0\rangle \quad (12.7)
$$

where $a_0^\dagger$ is the addition operator for the condensate, *i.e.* the sp eigenstate of $\hat{T} = \sum_\alpha \varepsilon_\alpha a_\alpha^\dagger a_\alpha$ with lowest energy ε_α.

The corresponding noninteracting sp propagator is easily derived from Eq. (12.5), using the basic boson commutation rules of Sec. 2.2. The propagator is diagonal in the sp basis of the eigenstates of $\hat{T}$ and reads

$$i\hbar G^{(0)}(\alpha, \beta; \tau) = \delta_{\alpha\beta}e^{-\frac{i}{\hbar}\varepsilon_\alpha\tau}\{\theta(\tau)(N\delta_{\alpha,0}+1) + \theta(-\tau)N\delta_{\alpha,0}\}$$
$$= \delta_{\alpha\beta}e^{-\frac{i}{\hbar}\varepsilon_\alpha\tau}\{\theta(\tau) + N\delta_{\alpha,0}\}. \tag{12.8}$$

Note that propagation in a noncondensate state only has a forward going component, whereas propagation in the condensate can go both forward and backward.

In the energy representation of Eq. (12.6) the noninteracting propagator becomes

$$G^{(0)}(\alpha, \beta; E) = \delta_{\alpha\beta}\left\{\frac{N\delta_{\alpha,0}+1}{E-\varepsilon_\alpha+i\eta} - \frac{N\delta_{\alpha,0}}{E-\varepsilon_\alpha-i\eta}\right\}$$
$$= \delta_{\alpha\beta}\left\{\frac{1}{E-\varepsilon_\alpha+i\eta} - 2i\pi N\delta_{\alpha,0}\delta(E-\varepsilon_0)\right\}. \tag{12.9}$$

The noninteracting ground-state energy is

$$E_0^N = \langle\Phi_0^N|\hat{T}|\Phi_0^N\rangle = N\varepsilon_0, \tag{12.10}$$

and the chemical potential μ of the system

$$\mu = E_0^N - E_0^{N-1} = E_0^{N+1} - E_0^N = \varepsilon_0, \tag{12.11}$$

coincides with ε_0.

12.1.3 *The condensate in an interacting Bose system*

For a finite, but large, number of N bosons the normal situation (*i.e.* when perturbation theory from the noninteracting system can be applied) implies that the boson propagator in Eq. (12.6) is dominated by the contribution from the $N\pm1$ ground states, as these have an overlap with normalization $S_{c\pm}$ of order $\mathcal{O}(N)$. More explicitly, the corresponding addition and removal amplitudes can be written as

$$\langle\Psi_0^N|a_\alpha|\Psi_0^{N+1}\rangle = \sqrt{S_{c+}}\,\phi_\alpha^{c+}$$
$$\langle\Psi_0^{N-1}|a_\alpha|\Psi_0^N\rangle = \sqrt{S_{c-}}\,\phi_\alpha^{c-}, \tag{12.12}$$

where the sp wave functions $\phi^{c\pm}$ referring to the condensate, are normalized to unity: $\sum_\alpha |\phi_\alpha^{c+}|^2 = \sum_\alpha |\phi_\alpha^{c-}|^2 = 1$. The other contributions to the propagator are of order $\mathcal{O}(1)$ or smaller.

For finite N one has in general $S_{c+} \neq S_{c-}$ and $\phi^{c+} \neq \phi^{c-}$. In the limit of large N, which is the situation of most interest, the properties of the system can hardly change when a single particle is added. The differences between the $\phi^{c\pm}$ and the $S_{c\pm}$ are negligible (being of higher order in a $1/N$ expansion) and one has

$$\left\langle \Psi_0^N \middle| a_\alpha \middle| \Psi_0^{N+1} \right\rangle \approx \left\langle \Psi_0^{N-1} \middle| a_\alpha \middle| \Psi_0^N \right\rangle \approx \sqrt{S_c}\phi_\alpha^c, \qquad (12.13)$$

$$E_0^{N+1} - E_0^N \approx E_0^N - E_0^{N-1} \approx \mu \qquad (12.14)$$

where μ is the chemical potential. The dominant part of the propagator in Eq. (12.6) then becomes

$$G(\alpha, \beta; E) \approx -2\pi i S_c \phi_\alpha^c \phi_\beta^{c*} \delta(E - \mu), \qquad (12.15)$$

analogous to the structure of the noninteracting propagator in Eq. (12.9).

However, rather than using the ϕ^{c+} or ϕ^{c-} overlap functions, the condensate orbital is often defined as the natural orbital with the largest eigenvalue [Yang (1962)]. The natural orbitals $\phi^{(i)}$ are the orthogonal sp wave functions that diagonalize the density matrix, *i.e.*

$$n_{\beta\alpha} = \left\langle \Psi_0^N \middle| a_\beta^\dagger a_\alpha \middle| \Psi_0^N \right\rangle = \sum_i d^{(i)} \phi_\alpha^{(i)} \phi_\beta^{(i)*}. \qquad (12.16)$$

Normally one[1] of the occupation numbers $d^{(i)}$, say for $i=0$, is of order $\mathcal{O}(N)$. The associated orbital $\phi^{(0)}$ is the condensate orbital, and $N_c = d^{(0)}$ is the condensate occupation. In the large-N limit one can again expect that to leading order, $\phi^{(0)} \approx \phi^c$ and $N_c \approx S_c$. Differences between ϕ^{c-} and the natural orbital $\phi^{(0)}$ were studied numerically for droplets of a few hundred ^{4}He atoms by [Lewart *et al.* (1988)] and found to be small; they vanish in the thermodynamic limit and at the mean-field level, but the precise relation appears not to have been studied in detail [Dalfovo *et al.* (1999)].

---

[1] In some special cases more than one natural orbital with an occupation of order $\mathcal{O}(N)$ occurs; the condensate is then said to be fragmented. See *e.g.* [Nozières and Saint James (1982)] and [Baym (2001)].

12.1.4 Equations of motion

The equation of motion obeyed by the boson propagator can be obtained, as in Sec. 9.2, by considering the time derivative of Eq. (12.4)

$$i\hbar\frac{\partial}{\partial t}G(\alpha,\beta;t-t') = \delta(t-t')\delta_{\alpha,\beta} \quad (12.17)$$

$$+ \langle\Psi_0^N|\,\theta(t-t')\frac{\partial a_{\alpha_H}(t)}{\partial t}a_{\beta_H}^\dagger(t') + \theta(t'-t)a_{\beta_H}^\dagger(t')\frac{\partial a_{\alpha_H}(t)}{\partial t}\,|\Psi_0^N\rangle.$$

The first term on the right in Eq. (12.17) arises from the derivative of the step function

$$\frac{d}{dt}\theta(t-t') = \delta(t-t') = -\frac{d}{dt}\theta(t'-t), \quad (12.18)$$

and application of the equal-time boson commutation relation

$$\langle\Psi_0^N|\,a_{\alpha_H}(t)a_{\beta_H}^\dagger(t) - a_{\beta_H}^\dagger(t)a_{\alpha_H}(t))\,|\Psi_0^N\rangle = \delta_{\alpha,\beta}. \quad (12.19)$$

The equation of motion for the boson removal operator in the Heisenberg picture reads

$$i\hbar\frac{\partial a_{\alpha_H}(t)}{\partial t} = [a_{\alpha_H}(t),\hat{H}] \quad (12.20)$$

$$= \sum_\gamma \langle\alpha|\,T\,|\gamma\rangle\,a_{\gamma_H}(t) + \frac{1}{2}\sum_{\gamma\delta\epsilon}\langle\alpha\gamma|\,V\,|\delta\epsilon\rangle\,a_{\gamma_H}^\dagger(t)a_{\epsilon_H}(t)a_{\delta_H}(t).$$

Substitution of Eq. (12.20) into Eq. (12.17), immediately leads to

$$i\hbar\frac{\partial}{\partial t}G(\alpha,\beta;t-t') = \delta(t-t')\delta_{\alpha,\beta} + \sum_\gamma\langle\alpha|\,T\,|\gamma\rangle\,G(\gamma,\beta;t,t') + P_2, \quad (12.21)$$

where P_2 represents the terms containing the interaction $\hat{V}$

$$P_2 = \frac{1}{2i\hbar}\sum_{\gamma\delta\epsilon}\langle\alpha\gamma|\,V\,|\delta\epsilon\rangle\left\{\theta(t-t')a_{\gamma_H}^\dagger(t)a_{\epsilon_H}(t)a_{\delta_H}(t)a_{\beta_H}^\dagger(t')\right.$$

$$\left. + \theta(t'-t)a_{\beta_H}^\dagger(t')a_{\gamma_H}^\dagger(t)a_{\epsilon_H}(t)a_{\delta_H}(t)\right\}. \quad (12.22)$$

Upon inspection P_2 is seen to involve the two-particle propagator G_{II}

$$i\hbar G_{II}(\alpha t_\alpha,\beta t_\beta,\gamma t_\gamma,\delta t_\delta)$$

$$= \langle\Psi_0^N|\,T[a_{\beta_H}(t_\beta)a_{\alpha_H}(t_\alpha)a_{\gamma_H}^\dagger(t_\gamma)a_{\delta_H}^\dagger(t_\delta)]\,|\Psi_0^N\rangle, \quad (12.23)$$

allowing to rewrite Eq. (12.21) as

$$i\hbar\frac{\partial}{\partial t}G(\alpha,\beta;t-t') = \delta(t-t')\delta_{\alpha,\beta} + \sum_{\gamma}\langle\alpha|T|\gamma\rangle\,G(\gamma,\beta;t,t')$$

$$+\frac{1}{2}\sum_{\gamma\delta\epsilon}\langle\alpha\gamma|V|\delta\epsilon\rangle\,G_{II}(\delta t,\epsilon t,\beta t',\gamma t^+). \qquad (12.24)$$

Clearly, the equation of motion (12.24) for the sp propagator connects it with the two-particle propagator G_{II}. In a similar way equations can be derived connecting G_{II} with the three-particle propagator, *etc.*

12.2 Perturbation expansions and the condensate

12.2.1 *Breakdown of Wick's theorem*

Up to now the treatment of Bose systems has proceeded in complete analogy to the Fermi case. Obviously, the next thing to do would be to repeat the analysis of Sec. 8.2, set up a perturbation expansion in powers of the interaction $\hat{V}$, and arrive at an expression for the exact boson propagator as given by Eq. (8.17), in which only the noninteracting ground state appears.

In the fermionic case it was possible to use Wick's theorem (the true engine of propagator perturbation theory) in order to evaluate the different contributions in the perturbation series in terms of noninteracting sp propagators. In the case of bosons however, a direct application of Wick's theorem is no longer possible.

As one recalls from Sec. 8.4, Wick's theorem requires that any product of sp removal and addition operators can be put in normal order, leading to a vanishing expectation value in the noninteracting ground state. For the noninteracting boson ground state in Eq. (12.7) this clearly cannot be done, since neither of

$$a_0\left|\Phi_0^N\right\rangle = \sqrt{N}\left|\Phi_0^{N-1}\right\rangle, \qquad a_0^{\dagger}\left|\Phi_0^N\right\rangle = \sqrt{N+1}\left|\Phi_0^{N+1}\right\rangle, \qquad (12.25)$$

are zero. In fact, both are rather big if N is large.

As a consequence, the condensate orbital must receive special treatment (compared to the other sp states), before the usual perturbative machinery can be used. How this is done in a genuine boson perturbation theory, the Bogoliubov formalism, will be presented at length in Ch. 18. For the purpose of quickly generating the boson mean-field equation, we will first explain the conversion of the boson problem into an equivalent fermion problem [Brandow (1971); Jackson *et al.* (1982); Wettig and Jackson (1996)], an alternative treatment that is sometimes useful.

12.2.2 *Equivalent fermion problem*

In the discussion on the hydrogen molecule in Sec. 10.3.3, it was noted that the two-electron ground state, having total spin $S = 0$, could be factorized as

$$\Psi(\boldsymbol{r}_1 m_{s_1}, \boldsymbol{r}_2 m_{s_2}) = \Phi(\boldsymbol{r}_1, \boldsymbol{r}_2)\Xi(m_{s_1}, m_{s_2}). \tag{12.26}$$

The spin part represents a Slater determinant, given by

$$\Xi(m_{s_1}, m_{s_2}) = \frac{1}{\sqrt{2}}\left(\delta_{m_{s_1}, +\frac{1}{2}}\delta_{m_{s_2}, -\frac{1}{2}} - \delta_{m_{s_1}, -\frac{1}{2}}\delta_{m_{s_2}, +\frac{1}{2}}\right), \tag{12.27}$$

and already carries the full fermion antisymmetry. Consequently, the spatial part $\Phi(\boldsymbol{r}_1, \boldsymbol{r}_2)$ must be symmetric under $\boldsymbol{r}_1 \leftrightarrow \boldsymbol{r}_2$ interchange, and represents a wave function for two spinless *bosons*.

This observation can be easily generalized and extended to an arbitrary number of particles. Considering a system of N interacting bosons, the conversion to a fermion problem is achieved by introducing an additional (fictitious) quantum number λ that can take on N values. Each original boson sp state α is now N-fold degenerate and generates N fermion sp states $(\alpha\lambda_\alpha)$ with $\lambda_\alpha = 1, \ldots, N$. The fermion Hamiltonian is taken to be diagonal in λ, *i.e.* for the sp part we have

$$\langle\alpha\lambda_\alpha| T |\beta\lambda_\beta\rangle = \langle\alpha| T |\beta\rangle \delta_{\lambda_\alpha, \lambda_\beta}, \tag{12.28}$$

and likewise for the tp interaction

$$(\alpha\lambda_\alpha, \beta\lambda_\beta| V |\gamma\lambda_\gamma, \delta\lambda_\delta) = (\alpha\beta| V |\gamma\delta) \delta_{\lambda_\alpha\lambda_\gamma}\delta_{\lambda_\beta\lambda_\delta}. \tag{12.29}$$

The noninteracting ground state of this N-fermion problem is given by

$$|\Phi_0^N\rangle = \prod_{\lambda=1}^{N} a_{0\lambda}^\dagger |0\rangle, \tag{12.30}$$

where the fermion operators $a_{0\lambda}^\dagger$ all refer to the underlying boson sp state $a_0^\dagger$ of Sec. 12.1.2. The $|\Phi_0^N\rangle$ defined in Eq. (12.30) is a closed-shell or singlet state in the λ quantum number, and factorizes into the product

$$\langle\alpha_1\lambda_1,\ldots,\alpha_N\lambda_N|\Phi_0^N\rangle = \left[\prod_{i=1}^{N}\delta_{\alpha_i,0}\right]\frac{1}{\sqrt{N!}}\mathrm{Det}[\delta_{i,\lambda_j}]_{i,j=1,\ldots,N}. \quad (12.31)$$

It has a symmetric part, which represents the noninteracting boson ground state, and an antisymmetric Slater determinant, containing the λ-dependence of the closed-shell configuration. When the interaction is turned on, the ground state of the *interacting* fermion problem remains a λ-singlet state, and still factorizes into the product of the closed-shell Slater determinant and of the true, symmetric N-boson ground state.

As far as perturbation theory and diagrams are concerned, the rules of Sec. 8.6 remain unchanged, but we now have to include λ in the summation over internal quantum numbers. Since λ is conserved at each vertex, this simply amounts to a factor N for each closed fermion loop in the diagram. The only other modification is that a sum over occupied fermion states must be restricted to the condensate orbital.

12.3 Hartree–Bose approximation

12.3.1 *Derivation of the Hartree–Bose equation*

As an example of the equivalent fermion technique in Sec. 12.2.2, we consider the HF self-energy for fermions

$$\Sigma^{HF}(\gamma,\delta) = \sum_{h=1}^{N}\left[\langle\gamma h|\,V\,|\delta h\rangle - \langle\gamma h|\,V\,|h\delta\rangle\right], \quad (12.32)$$

as given by Eq. (10.11), with h referring to the occupied sp (hole) states in the HF determinant, *i.e.* $|h\rangle = \sum_\mu z_\mu^h|\mu\rangle$. The boson counterpart, the HB mean-field, is derived according to the prescriptions in Sec. 12.2.2 by the introduction of λ-quantum numbers and the restriction of the hole summation to the contribution from the condensate orbital c

$$\Sigma^{HB}(\gamma\lambda_\gamma,\delta\lambda_\delta) = \sum_{\lambda_c}\left[\langle\gamma\lambda_\gamma,c\lambda_c|\,V\,|\delta\lambda_\delta,c\lambda_c\rangle\right.$$
$$\left. - \langle\gamma\lambda_\gamma,c\lambda_c|\,V\,|c\lambda_c,\delta\lambda_\delta\rangle\right]. \quad (12.33)$$

Taking the λ-independence of V into account [see Eq. (12.29)] one has

$$\Sigma^{HB}(\gamma\lambda_\gamma,\delta\lambda_\delta) = \Sigma^{HB}(\gamma,\delta)\delta_{\lambda_\gamma,\lambda_\delta}, \quad (12.34)$$

and performing the summation over λ_c in Eq. (12.33) results in

$$\Sigma^{HB}(\gamma, \delta) = N\,(\gamma c|\,V\,|\delta c) - (\gamma c|\,V\,|c\delta)\,. \tag{12.35}$$

Note that the first term in Eq. (12.35) has a factor N and the second does not, in agreement with the diagram rules at the end of Sec. 12.2.2 and the fact that the Hartree diagram in Fig. 8.4 has one closed loop, whereas the exchange (Fock) diagram in Fig. 8.5 has none.

The (as yet unknown) condensate orbital,

$$|c\rangle = \sum_\mu z_\mu^c\,|\mu\rangle\,, \tag{12.36}$$

then follows as the lowest-energy solution of the HB equation

$$\sum_\delta \left\{ \langle\gamma|\,T\,|\delta\rangle + \Sigma^{HB}(\gamma, \delta) \right\} z_\delta^c = \varepsilon_c z_\gamma^c, \tag{12.37}$$

which is the boson counterpart of Eq. (10.20) determining the occupied HF sp states. Like HF, this is a *self-consistent* approximation: the HB mean field following from Eqs. (12.35) – (12.36)

$$\Sigma^{HB}(\gamma, \delta) = \sum_{\mu\nu} \left\{ N\,(\gamma\mu|\,V\,|\delta\nu) - (\gamma\mu|\,V\,|\nu\delta) \right\} z_\mu^{c*} z_\nu^c, \tag{12.38}$$

represents the interaction averaged over the condensate density, and itself determines the condensate orbital through Eq. (12.37). Note that the orbital $|c\rangle$ in Eq. (12.36) is normalized to unity, $\sum_\mu |z_\mu^c|^2 = 1$.

The HB mean-field term in Eq. (12.37) can also be written as

$$\sum_\delta \Sigma^{HB}(\gamma, \delta) z_\delta^c = \sum_\delta W_{HB}(\gamma, \delta) z_\delta^c \tag{12.39}$$

with

$$W_{HB}(\gamma, \delta) = (N-1) \sum_{\mu\nu} (\gamma\mu|\,V\,|\delta\nu)\, z_\mu^{c*} z_\nu^c. \tag{12.40}$$

An equivalent form of the HB equation determining the condensate wave function, reads

$$\sum_\delta \left\{ \langle\gamma|\,T\,|\delta\rangle + W_{HB}(\gamma, \delta) \right\} z_\delta^c = \varepsilon_c z_\gamma^c. \tag{12.41}$$

12.3.2 Hartree–Bose ground-state energy

The HB ground-state energy may be similarly derived from the HF energy for fermions, given by the two equivalent expressions in Eqs. (10.29) and (10.36)

$$
\text{(HF:)} \quad E_0^N = \sum_h \langle h| T |h\rangle + \frac{1}{2} \sum_{h_1,h_2} [(h_1 h_2| V |h_1 h_2)
$$

$$
- (h_1 h_2| V |h_2 h_1)] \quad (12.42)
$$

$$
\text{(HF:)} \quad E_0^N = \frac{1}{2} \sum_h [\langle h| T |h\rangle + \varepsilon_h]. \quad (12.43)
$$

Restricting the summations over h to the condensate contribution and introducing λ quantum numbers leads to the HB energy

$$
\text{(HB:)} \quad E_0^N = \sum_{\lambda_c} \langle c\lambda_c| T |c\lambda_c\rangle + \frac{1}{2} \sum_{\lambda_{c_1},\lambda_{c_2}} [(c\lambda_{c_1}, c\lambda_{c_2}| V |c\lambda_{c_1}, c\lambda_{c_2})
$$

$$
- (c\lambda_{c_1}, c\lambda_{c_2}| V |c\lambda_{c_2}, c\lambda_{c_1})] \quad (12.44)
$$

$$
\text{(HB:)} \quad E_0^N = \frac{1}{2} \sum_{\lambda_c} [\langle c\lambda_c| T |c\lambda_c\rangle + \varepsilon_{c\lambda_c}]. \quad (12.45)
$$

After the λ summation is performed (note that ε_c is independent of λ_c), one arrives at the corresponding expressions for the ground-state energy in the HB approximation

$$
\text{(HB:)} \quad E_0^N = N \langle c| T |c\rangle + \frac{N(N-1)}{2} (cc| V |cc) \quad (12.46)
$$

$$
\text{(HB:)} \quad E_0^N = \frac{N}{2} (\langle c| T |c\rangle + \varepsilon_c). \quad (12.47)
$$

12.3.3 Physical interpretation

The HB approximation clearly assumes that all N particles are in the condensate sp state. Once Eq. (12.37) has been solved and the z_μ^c are determined, the mean field $\Sigma^{HB}(\gamma,\delta)$ generated by the condensate, is fixed and can be considered as an additional sp potential determining a noninteracting boson system. Equation (12.37) then has (apart from the condensate) other solutions

$$
\sum_\delta \{\langle \gamma| T |\delta\rangle + \Sigma^{HB}(\gamma,\delta)\} z_\delta^n = \varepsilon_n z_\gamma^n, \quad (12.48)
$$

with $\varepsilon_n > \varepsilon_c$, which describe the sp excitations of the system and, together with the condensate, constitute the HB sp basis. Since all particles are in the condensate, these excitations $(n \neq c)$ can only be made by adding a particle to the ground state, whereas a particle in the condensate can be both added to, and removed from, the ground state.

In accordance with Eqs. (12.6) and (12.9) the HB solutions with $n \neq c$ should therefore be interpreted as addition amplitudes to the excited $N+1$ states

$$\text{(HB:)} \quad z_\alpha^n = \left\langle \Psi_0^N \middle| a_\alpha \middle| \Psi_n^{N+1} \right\rangle, \quad \varepsilon_n = E_n^{N+1} - E_0^N, \quad (12.49)$$

whereas the condensate z^c is both a removal and an addition amplitude

$$\text{(HB:)} \quad z_\alpha^c = \frac{1}{N} \left\langle \Psi_0^{N-1} \middle| a_\alpha \middle| \Psi_0^N \right\rangle = \frac{1}{N+1} \left\langle \Psi_0^N \middle| a_\alpha \middle| \Psi_0^{N+1} \right\rangle, \quad (12.50)$$

and the sp energy ε_c coincides with the HB chemical potential

$$\text{(HB:)} \quad \varepsilon_c = E_0^N - E_0^{N-1} = E_0^{N+1} - E_0^N = \mu. \quad (12.51)$$

12.3.4 *Variational content*

The mean-field description for fermion systems, as shown in Sec. 10.1.3, can be obtained by a variational search for the noninteracting wave function (the HF Slater determinant) which minimizes the energy expectation value. Following the same procedure for the boson case, we should now find the condensate sp orbital for which the product state

$$\left| \Phi_0^N \right\rangle = \frac{(a_c^\dagger)^N}{\sqrt{N!}} \left| 0 \right\rangle \quad (12.52)$$

has minimal energy $E = \left\langle \Phi_0^N \middle| \hat{T} + \hat{V} \middle| \Phi_0^N \right\rangle$. With the boson second-quantization results of Sec. 2.2 the energy can be worked out as

$$E = N \left\langle c \middle| T \middle| c \right\rangle + \frac{N(N-1)}{2} \left(cc \middle| V \middle| cc \right). \quad (12.53)$$

Upon expansion of the unknown condensate orbital in a fixed sp basis

$$a_c^\dagger = \sum_\mu z_\mu^c a_\mu^\dagger, \quad (12.54)$$

the energy per particle E/N is written as

$$E/N = \sum_{\mu\nu} \left\langle \mu \middle| T \middle| \nu \right\rangle z_\mu^{c*} z_\nu^c + \frac{N-1}{2} \sum_{\kappa\lambda\mu\nu} z_\mu^{c*} z_\nu^c z_\kappa^{c*} z_\lambda^c \left(\mu\kappa \middle| V \middle| \nu\lambda \right). \quad (12.55)$$

Minimizing this expression with respect to variations in the expansion coefficients z_μ^c, subject to the normalization constraint $\sum_\mu |z_\mu^c|^2 = 1$, leads to

$$\varepsilon_c z_\mu^c = \sum_\nu \left\{ \langle \mu | T | \nu \rangle + (N-1) \left(\sum_{\kappa\lambda} (\mu\kappa | V | \nu\lambda) z_\kappa^{c*} z_\lambda^c \right) \right\} z_\nu^c, \qquad (12.56)$$

which is identical to the previously derived HB result in Eqs. (12.40) – (12.41).

The HB energy in Eq. (12.53) coincides with Eq. (12.46). Multiplying Eq. (12.56) with z_μ^{c*} and summing over μ one has

$$\varepsilon_c = \langle c | T | c \rangle + (N-1) (cc | V | cc). \qquad (12.57)$$

Elimination of the potential energy from Eqs. (12.53) and (12.57) then leads to Eq. (12.47). Note that the variational nature of the HB ground state ensures that the HB energy is a strict upper bound to the exact energy.

The interpretation in Eq. (12.51) of ε_c as the chemical potential in the HB approximation can be confirmed by considering the derivative of the HB energy (12.46) with respect to particle number, which for large N can be treated as a continuous variable. One finds (for $N \to \infty$)

$$\mu = \frac{\partial E_0^N}{\partial N} = [\langle c | T | c \rangle + N (cc | V | cc)] + R_2 + R_2^*, \qquad (12.58)$$

where the first (bracketed) term is the HB sp energy ε_c of Eq. (12.57) and the remaining terms can be written as

$$R_2 = N \langle \frac{\partial c}{\partial N} | T | c \rangle + N^2 (\frac{\partial c}{\partial N} c | V | cc)$$
$$= N \langle \frac{\partial c}{\partial N} | \hat{T} + \hat{W}_{HB} | c \rangle = N\varepsilon_c \langle \frac{\partial c}{\partial N} | c \rangle. \qquad (12.59)$$

As a result, $R_2 + R_2^* = N\varepsilon_c \frac{\partial}{\partial N} \langle c | c \rangle = 0$ and Eq. (12.58) implies $\mu = \varepsilon_c$.

12.3.5 *Hartree–Bose expressions in coordinate space*

The coordinate space representation of the HB equation can be derived straightforwardly using the results of Sec. 10.1.4. For a system of spinless bosons in a local external potential $U(\boldsymbol{r})$ and interacting with a local tp potential

$$(\boldsymbol{r}_1\boldsymbol{r}_2 | V | \boldsymbol{r}_3\boldsymbol{r}_4) = \delta(\boldsymbol{r}_1 - \boldsymbol{r}_3)\delta(\boldsymbol{r}_2 - \boldsymbol{r}_4)V(\boldsymbol{r}_1 - \boldsymbol{r}_2), \qquad (12.60)$$

one simply takes sp labels $\alpha \equiv r$ in Eqs. (12.40) – (12.41). Introducing the wave-function notation $\phi_c(\boldsymbol{r}) \equiv z_{\boldsymbol{r}}^c$ for the condensate orbital, the HB potential in Eq. (12.40) reads

$$W_{HB}(\boldsymbol{r}) = (N-1) \int d\boldsymbol{r}' \, V(\boldsymbol{r}-\boldsymbol{r}')|\phi_c(\boldsymbol{r}')|^2, \qquad (12.61)$$

and the HB equation (12.41) has the form

$$\left[-\frac{\hbar^2}{2m}\nabla^2 + U(\boldsymbol{r}) + W_{HB}(\boldsymbol{r}) \right] \phi_c(\boldsymbol{r}) = \varepsilon_c \phi_c(\boldsymbol{r}). \qquad (12.62)$$

The coordinate space representation of the HB ground state in Eq. (12.52) is

$$\langle \boldsymbol{r}_1, \boldsymbol{r}_2, .., \boldsymbol{r}_N | \Phi_0^N \rangle = \phi_c(\boldsymbol{r}_1)\phi_c(\boldsymbol{r}_2)..\phi_c(\boldsymbol{r}_N), \qquad (12.63)$$

and the condensate orbital $\phi_c(\boldsymbol{r})$ minimizes the energy functional in Eq. (12.55)

$$\begin{aligned} E_0^{HB} = \int d\boldsymbol{r}\phi_c^*(\boldsymbol{r}) &\left[-\frac{\hbar^2\nabla^2}{2m} + U(\boldsymbol{r}) \right. \\ &\left. + \frac{N(N-1)}{2} \int d\boldsymbol{r}' \, V(\boldsymbol{r}-\boldsymbol{r}')|\phi_c(\boldsymbol{r}')|^2 \right] \phi_c(\boldsymbol{r}), \qquad (12.64) \end{aligned}$$

under the constraint $\int d\boldsymbol{r}|\phi_c(\boldsymbol{r})|^2 = 1$.

12.4   Gross–Pitaevskii equation for dilute systems

12.4.1   *Pseudopotential*

The HB Eq. (12.62) is only applicable for weak and nonsingular interactions. This is not fulfilled in most practical applications, *e.g.* in BEC of ultracold vapors of alkali atoms,[2] which have already been introduced in Sec. 5.6.2, excluding the effects of the atom–atom interactions. The interatomic interaction, usually modeled by a Lennard–Jones type of potential, has a strong repulsive core when the atoms come close together and the electron clouds start to overlap. The integrand in Eq. (12.61) then becomes very large or diverges as $|\boldsymbol{r} - \boldsymbol{r}'| \to 0$.

---

[2]The trapped atomic vapors made in the laboratory are actually in a metastable state: at these ultracold temperatures the true (thermodynamically stable) ground state would be a solid. Because of the diluteness of the vapor however, the collision rate is so low that it takes a time of the order of seconds or minutes, for the atoms to cluster and reach thermodynamic equilibrium, which is long enough to do experiments.

Obviously, this is again a shortcoming of the mean-field wave function in Eq. (12.63), since the true N-boson wave function will tend to vanish whenever two atoms are closer together than the radius of the repulsive core. To include these short-range correlations between two particles, it is necessary to go beyond the mean-field approximation and construct a more realistic tp propagator. A similar problem occurs in the fermion case and will be extensively discussed in Ch. 15. The relevant class of diagrams consists of the repeated scattering of two particles in the presence of the medium. The bare interaction V, which is singular at short distances, can then be replaced by the resulting in-medium effective interaction Γ, which is well behaved. Apart from medium corrections, the relation between V and Γ is therefore the same as the well-known relation, studied in Sec. 6.4, between V and the T-matrix for scattering of two particles in free space.

In the limit of a very dilute system however, the difference between scattering in the medium and in free space is negligible: the collisions happen almost in vacuum since very few other particles are around. In Sec. 18.6 we will look more closely at the effect of medium corrections; for the moment it makes good sense to replace the bare potential by the (free-space) T-matrix.

Moreover, at ultracold temperatures the bosons move overwhelmingly with momenta close to zero. As a result, only S-wave scattering survives, and the T-matrix becomes independent of the scattering momenta. In fact, it can be characterized by just one number, the S-wave scattering length a, which is the only parameter needed to describe ultracold dilute Bose (and Fermi) gases.

The simplest way to incorporate these physical arguments is to replace the bare interaction by a so-called pseudopotential of zero range

$$V(\boldsymbol{r} - \boldsymbol{r}') \to g\delta(\boldsymbol{r} - \boldsymbol{r}'), \tag{12.65}$$

where the strength of the δ-function is related to the scattering length a of the true potential V as

$$g = \frac{4\pi\hbar^2 a}{m}. \tag{12.66}$$

Before applying Eqs. (12.65) – (12.66) to the HB formalism, we will justify the form of this pseudopotential in the next Secs. 12.4.2 – 12.4.3, which contain a brief overview of scattering length theory and the low-energy limit of the T-matrix.

12.4.2   *Quick reminder of low-energy scattering*

Phase shifts and scattering length

For identical particles of mass m and a central tp potential $V(|\mathbf{r}_1 - \mathbf{r}_2|)$, the relative scattering wave function $\psi_{\ell k}(r)$ with angular momentum ℓ and energy $E = \frac{\hbar^2 k^2}{m}$ is the solution of

$$\left(-\frac{\hbar^2}{m} \left[\frac{1}{r} \frac{d^2}{dr^2} r - \frac{\ell(\ell+1)}{r^2} \right] + V(r) - E \right) \psi_{\ell k}(r) = 0, \qquad (12.67)$$

which is regular at the origin (see also Ch. 6).

The phase shift $\delta_{\ell k}$ is defined through the asymptotic behavior of $\psi_{\ell k}(r)$: assuming that for $r \to \infty$ the potential $V(r)$ drops sufficiently fast, the wave function $\psi_{\ell k}(r)$ becomes (outside the range of V) a linear combination of the free solutions

$$\psi_{\ell k}(r) \to C_{\ell k}[\cos \delta_{\ell k}\, \mathrm{j}_\ell(kr) - \sin \delta_{\ell k}\, \mathrm{n}_\ell(kr)], \qquad (12.68)$$

where $C_{\ell k}$ is a constant. The free solutions $\mathrm{j}_\ell(x)$ and $\mathrm{n}_\ell(x)$ are spherical Bessel and Neumann functions respectively, with asymptotic behavior

$$\mathrm{j}_\ell(x) \to \frac{\sin(x - \ell\frac{\pi}{2})}{x}, \qquad \mathrm{n}_\ell(x) \to -\frac{\cos(x - \ell\frac{\pi}{2})}{x}. \qquad (12.69)$$

In the low-energy limit $k \to 0$ it can be shown, under quite general conditions for $V(r)$ (see *e.g.* [Landau and Lifshitz (1977)]), that the phase shift of the ℓth partial wave has a leading term proportional to

$$\delta_{\ell k} \sim k^{2\ell+1}[1 + \mathcal{O}(k^2)], \qquad (12.70)$$

and only S-wave scattering ($\ell = 0$) survives. The *scattering length* a, defined as

$$-\frac{1}{a} = \lim_{k \to 0} k \cot \delta_{0k}, \qquad (12.71)$$

therefore completely characterizes the potential $V(r)$ in the limit of zero-momentum scattering.[3]

---

[3] More generally one can expand $k \cot \delta_{0k} = -\frac{1}{a} + \frac{r_0}{2} k^2 + \dots$. The next-order term, containing the effective range r_0, is not needed here.

Square-well potential

As an example we consider a square-well potential of radius R, *i.e.* we take $V(r) = V_0\theta(R - r)$ in Eq. (12.67) and examine the solution for $0 < E <<|V_0|$. In the inner region $r < R$ the solution is

$$\psi_{\ell k}(r) = j_\ell(k_v r), \qquad \text{with } k_v = \sqrt{m(E - V_0)}/\hbar \qquad (12.72)$$

for an attractive potential ($V_0 < 0$), or

$$\psi_{\ell k}(r) = i_\ell(k_v r), \qquad \text{with } k_v = \sqrt{m(V_0 - E)}/\hbar \qquad (12.73)$$

for a repulsive potential ($V_0 > 0$). In the latter case $i_\ell(x) = (-i)^\ell j_\ell(ix)$ is the modified spherical Bessel function, regular at the origin.

Smoothly joining at $r = R$ with the asymptotic solution in Eq. (12.68) yields

$$\cot \delta_{\ell k} = \frac{n_\ell(kR)k_v\gamma_{\ell k} - kn'_\ell(kR)}{j_\ell(kR)k_v\gamma_{\ell k} - kj'_\ell(kR)}, \qquad (12.74)$$

where $j'_\ell(x) = dj_\ell(x)/dx$, $n'_\ell(x) = dn_\ell(x)/dx$, and $\gamma_{\ell k}$ is the logarithmic derivative of the inner solution at $r = R$

$$\gamma_{\ell k} = \frac{j'_\ell(k_v R)}{j_\ell(k_v R)} \text{ for } V_0 < 0 \; ; \quad \gamma_{\ell k} = \frac{i'_\ell(k_v R)}{i_\ell(k_v R)} \text{ for } V_0 > 0. \quad (12.75)$$

The leading term in the low-energy limit $k \to 0$ of Eq. (12.74) can easily be obtained from the behavior of $j_\ell(x)$ and $n_\ell(x)$ at small values of the argument x

$$j_\ell(x) \to \frac{x^\ell}{(2\ell + 1)!!}, \qquad n_\ell(x) \to -\frac{(2\ell - 1)!!}{x^{\ell+1}}, \qquad (12.76)$$

where $(2\ell + 1)!! = 1 \cdot 3 \ldots \cdot (2\ell + 1)$, and from the observation that the quantity $k_v\gamma_{\ell k}$ in Eq. (12.74) can be replaced with its value for $k = 0$ since

$$k_v \to k_0 + \mathcal{O}(k^2), \qquad \text{with } k_0 = \sqrt{m|V_0|}/\hbar. \qquad (12.77)$$

Hence the leading term in k is given by

$$\cot \delta_{\ell k} \to -\frac{1}{(kR)^{2\ell+1}}(2\ell + 1)!!(2\ell - 1)!!\frac{\gamma_{\ell 0} + \frac{\ell+1}{k_0 R}}{\gamma_{\ell 0} - \frac{\ell}{k_0 R}}, \qquad (12.78)$$

in accordance with Eq. (12.70). Note that the phase shifts (and hence the scattering length) are always defined, no matter how strong the interaction.

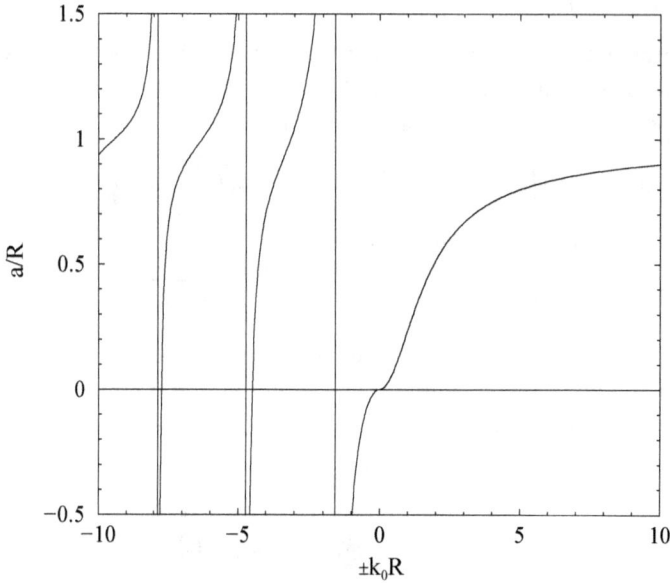

Fig. 12.1   Scattering length a of a square-well potential $V(r) = V_0\theta(R-r)$. On the plot a/R is shown as a function of $\mathrm{sign}(V_0)(k_0 R)$ [see Eq. (12.79)].

The special situation of a hard-sphere potential can be generated by taking the limit $V_0 \to +\infty$, with $\gamma_{\ell 0} \to \infty$ in Eq. (12.78).

For the dominant S-wave scattering one has $j_0(x) = [\sin x/x]$ and $i_0(x) = [\sinh x/x]$ in Eq. (12.75), and as a consequence $\gamma_{00} = \cot(\mathrm{h})(k_0 R) - \frac{1}{k_0 R}$, where the hyperbolic function must be chosen for $V_0 > 0$. The scattering length then follows from Eqs. (12.71) and (12.78)

$$a = R\left(1 - \frac{\tan(k_0 R)}{k_0 R}\right) \qquad \text{for } V_0 < 0$$

$$a = R\left(1 - \frac{\tanh(k_0 R)}{k_0 R}\right) \qquad \text{for } V_0 > 0. \qquad (12.79)$$

In Fig. 12.1, the scattering length a in Eq. (12.79) is shown as a function of V_0. For small $V_0 \approx 0$ one has positive (negative) a for a repulsive (attractive) interaction. As V_0 becomes more attractive, the potential starts to support a bound state at $k_0 R = \pi/2$. The scattering length has an asymptote, reflecting a zero-energy resonance in the cross section. This behavior is repeated at the values of $-|V_0|$ where further bound states appear. Near a resonance, all values can be obtained for the scattering length.

Values for scattering lengths in atomic collisions

Recent experimental values for a are 2.75 nm for ^{23}Na, 5.77 nm for ^{87}Rb, -1.45 nm for ^{7}Li, $\sim$ -23 nm for ^{85}Rb, and 3.45 nm for ^{133}Cs, where positive (negative) a correspond to repulsive (attractive) interactions in low-energy collisions. As will be shown in Sec. 12.4.5, a positive scattering length a leads to stable condensates, whereas atomic condensates with negative a are inherently unstable against collapse. In the presence of a confining potential, a metastable condensate with negative a can nevertheless exist, but with a limited number of condensed particles.

It is sometimes possible to tune the scattering length by applying a strong magnetic field, and exploiting the atomic hyperfine structure. The so-called Feshbach resonance is a coupled-channel effect, occurring when two atoms in the lowest-energy hyperfine state collide, and the interatomic potential in other hyperfine states supports a bound state near zero energy. Since these have different spin configurations and magnetic moments, varying the strength of the magnetic field can tune this bound-state energy into resonance with the energy of the colliding atoms. When the field strength is swept over a Feshbach resonance, a behaves as in Fig. 12.1.

This manipulation of a by means of a Feshbach resonance has been experimentally confirmed, when a ^{85}Rb condensate was suddenly brought from the regime of positive a to the unstable regime of negative a, by changing the magnetic field strength. The condensate collapsed on itself and then blew off a large fraction off its mass [Cornish *et al.* (2000)].[4]

12.4.3 The $\mathcal{T}$-matrix

Low-energy approximation

The $\mathcal{T}$-matrix has already been discussed in Sec. 6.4, in the context of a particle scattering off an external potential. For the present scattering of two identical particles in free space we must simply replace the mass m with the reduced mass $m/2$ in the expressions of Sec. 6.4.

In order to examine the low-energy limit we recall the integral equation [see Eq. (6.60)] for the general off-shell (energy-dependent) $\mathcal{T}$-matrix $\langle k'| \mathcal{T}^\ell(E) |k \rangle$ in the ℓth partial wave. The half-on-shell $\mathcal{T}$-matrix can be defined by setting the energy equal to the on-shell energy of the incoming

---

[4]The inventors of this phenomenon — presumably also fond of Latino dancing — dubbed it a Bose–nova explosion, in analogy to an astrophysical supernova.

state

$$\langle k'| \tilde{T}^\ell |k\rangle = \langle k'| T^\ell (E = \frac{\hbar^2 k^2}{m}) |k\rangle , \tag{12.80}$$

and the integral equation for $\tilde{T}$ is given by

$$\langle k'| \tilde{T}^\ell |k\rangle = \langle k'| V^\ell |k\rangle + \frac{m}{\hbar^2} \int_0^{+\infty} dq\, q^2 \frac{\langle k'| V^\ell |q\rangle \langle q| \tilde{T}^\ell |k\rangle}{k^2 - q^2 + i\eta}. \tag{12.81}$$

On the diagonal $(k = k')$ are the elements of the on-shell (physical) T-matrix, which are related to the phase shifts by [see also Eq. (6.74)]

$$\langle k| \tilde{T}^\ell |k\rangle = -\frac{2\hbar^2}{m\pi k} e^{i\delta_{\ell k}} \sin\delta_{\ell k}. \tag{12.82}$$

From the discussion of the phase shifts in Sec. 12.4.2, it follows that the low-energy limit of the on-shell T-matrix is simply

$$\langle k| \tilde{T}^\ell |k\rangle \to \frac{2\hbar^2}{m\pi} a\, \delta_{\ell,0}, \qquad \text{for } k \to 0, \tag{12.83}$$

i.e. vanishing for $\ell \neq 0$ and given by a k-independent constant for $\ell = 0$.

The low-energy limit of the half-on-shell T-matrix must, in principle, be determined by examining the limit $k, k' \to 0$, with $k \neq k'$, in Eq. (12.81). However, it is clear that it is fully controlled by the on-shell behavior (12.83), since an expansion of $\langle k'| \tilde{T}^\ell |k\rangle$ around $k = 0$, $k' = 0$, has a leading term which is necessarily the zero-energy on-shell value. In general, one therefore finds

$$\langle k'| \tilde{T}^\ell |k\rangle \to \frac{2\hbar^2}{m\pi} a\delta_{\ell,0} \tag{12.84}$$

as $k \to 0$, $k' \to 0$, with $k \neq k'$.

The S-wave dominance also implies isotropic scattering. The half-on-shell T-matrix in a plane-wave basis is

$$\langle \boldsymbol{k}'| \tilde{T} |\boldsymbol{k}\rangle = \sum_{\ell m \ell' m'} \langle k'\ell'm'| \tilde{T} |k\ell m\rangle\, Y_{\ell'm'}(\hat{\boldsymbol{k}}')Y^*_{\ell m}(\hat{\boldsymbol{k}})$$

$$= \sum_\ell \langle k'| \tilde{T}^\ell |k\rangle \frac{2\ell + 1}{4\pi} P_\ell(\omega) \tag{12.85}$$

where rotational invariance

$$\langle k'\ell'm'| \tilde{T} |k\ell m\rangle = \delta_{\ell,\ell'}\delta_{m,m'} \langle k'| \tilde{T}^\ell |k\rangle \tag{12.86}$$

has been used. Substituting Eq. (12.84) into Eq. (12.4) one finds the low-energy limit

$$\langle k' | \tilde{T} | k \rangle \to \frac{\hbar^2}{2\pi^2 m} a, \tag{12.87}$$

which is independent of the angle between k and k'. The approximation in Eq. (12.87) will be employed in subsequent work as the effective interaction in momentum space for dilute Bose systems.

Contact force

The corresponding effective interaction in coordinate space is obtained from Eq. (12.87) in the usual manner

$$(r_1' r_2' | \tilde{T} | r_1 r_2) = \int \frac{dk_1}{(2\pi)^{3/2}} \frac{dk_2}{(2\pi)^{3/2}} \frac{dk_1'}{(2\pi)^{3/2}} \frac{dk_2'}{(2\pi)^{3/2}} \tag{12.88}$$
$$\times e^{i(k_1 \cdot r_1 + k_2 \cdot r_2 - k_1' \cdot r_1' - k_2' \cdot r_2')} (k_1' k_2' | \tilde{T} | k_1 k_2).$$

Introducing CM and relative momenta

$$K = k_1 + k_2, \ K' = k_1' + k_2'; \ k = \frac{k_1 - k_2}{2}, \ k' = \frac{k_1' - k_2'}{2}, \tag{12.89}$$

and applying momentum conservation, one has

$$(k_1' k_2' | \tilde{T} | k_1 k_2) = \delta(K - K') \langle k' | \tilde{T} | k \rangle. \tag{12.90}$$

Substitution of the low-energy limit (12.87) then results in

$$(r_1' r_2' | \tilde{T} | r_1 r_2) = \frac{4\pi \hbar^2 a}{m} \delta(r_1 - r_1') \delta(r_2 - r_2') \delta(r_1 - r_2). \tag{12.91}$$

This has the form of a local zero-range or contact interaction

$$\tilde{T}(r_1 - r_2) = g\delta(r_1 - r_2), \tag{12.92}$$

with $g = \frac{4\pi \hbar^2 a}{m}$, thereby justifying the replacement in Eqs. (12.65) – (12.66).

Ultraviolet divergencies

The pseudopotential in Eq. (12.92) was generated by a FT, involving *all* relative momenta, of the constant $\mathcal{T}$-matrix in Eq. (12.87), which is an approximation only valid for small relative momenta. Use of Eq. (12.92) as an effective interaction therefore causes no problems at the HB mean-field level of Eq. (12.61), since only small momenta are sampled in the HB ground state. However, when a contact force is used to calculate a ground state quantity in higher orders of perturbation theory, the integrals over momentum k of intermediate states will typically be ultraviolet divergent (as $k \to \infty$). This is to be expected, since a constant $\mathcal{T}$-matrix (which is an approximation for small k) couples with equal ease small and very large momenta, whereas the exact $\mathcal{T}$-matrix would provide a natural cut-off at large momenta.

It is nevertheless still possible to perform higher-order calculations with a simple contact force (instead of the complicated exact $\mathcal{T}$-matrix) by means of renormalization techniques. The same divergencies, occurring in the context of many-body perturbation theory, are already present at the level of free-space scattering from a general contact force

$$V_\delta(\mathbf{r} - \mathbf{r}') = g\delta(\mathbf{r} - \mathbf{r}'), \qquad (12.93)$$

where g is now considered to be a free parameter. The corresponding (half-on-shell) $\tilde{\mathcal{T}}_\delta$ should follow from

$$\langle \mathbf{k}'| \, \tilde{\mathcal{T}}_\delta \, |\mathbf{k}\rangle = \langle \mathbf{k}'| \, V_\delta \, |\mathbf{k}\rangle + \frac{m}{\hbar^2} \int dq \, \frac{\langle \mathbf{k}'| \, V_\delta \, |\mathbf{q}\rangle \langle \mathbf{q}| \, \tilde{\mathcal{T}}_\delta \, |\mathbf{k}\rangle}{k^2 - q^2 + i\eta} \qquad (12.94)$$

$$= \langle \mathbf{k}'| \, V_\delta \, |\mathbf{k}\rangle + \frac{m}{\hbar^2} \int dq \, \frac{\langle \mathbf{k}'| \, V_\delta \, |\mathbf{q}\rangle \langle \mathbf{q}| \, V_\delta \, |\mathbf{k}\rangle}{k^2 - q^2 + i\eta} + \dots, \qquad (12.95)$$

where the last line gives the formal series expansion in powers of V_δ. Since $\langle \mathbf{k}'| \, V_\delta \, |\mathbf{k}\rangle = g/(2\pi)^3$ is a constant, the integral equation (12.94) has no solution, unless an upper bound q_{max} for the momentum integration is introduced. Taking the limit $\mathbf{k}, \mathbf{k}' \to 0$, one can now demand that V_δ has the same scattering length a as the true interparticle interaction V, and Eq. (12.95) becomes

$$\frac{4\pi\hbar^2 a}{m} = g - \frac{mg^2}{2\pi\hbar^2} \int_0^{q_{max}} dq + \dots \qquad (12.96)$$

Restricting the series to the first term on the right-hand side corresponds to the Born approximation and reproduces Eq. (12.66). The higher-order terms in the series are divergent when $q_{max} \to \infty$, and can be used to cancel similar divergencies occurring in many-body perturbation theory to the same order. It is then possible to eliminate g, take the limit $q_{max} \to \infty$, and express the result in terms of the physical scattering length a. An example of this will be given in Sec. 18.7.

General $\mathcal{T}$-matrix

The scattering solutions which behave asymptotically as an incoming plane wave

$$\psi_{\boldsymbol{k}}(\boldsymbol{r}) = \sum_{\ell m} 4\pi i^{\ell} e^{i\delta_{\ell k}} \psi_{\ell k}(r) Y_{\ell m}(\hat{k}) Y_{\ell m}^{*}(\hat{r}) \qquad (12.97)$$

$$\to e^{i\boldsymbol{k}\cdot\boldsymbol{r}} - \frac{2\pi^{2}m}{\hbar^{2}} \langle k\hat{r}| \tilde{\mathcal{T}} |\boldsymbol{k}\rangle \frac{e^{ikr}}{r}, \qquad (12.98)$$

form a complete set if the interaction has no bound states. In that case it is possible to express the general (energy-dependent) $\mathcal{T}$-matrix solely in terms of the half-on-shell $\mathcal{T}$-matrix elements. Starting from Eq. (6.20) in a plane-wave basis

$$\langle \boldsymbol{k}'| \mathcal{T}(E) |\boldsymbol{k}\rangle = \langle \boldsymbol{k}'| V |\boldsymbol{k}\rangle + \langle \boldsymbol{k}'| V \frac{1}{E - H + i\eta} V |\boldsymbol{k}\rangle, \qquad (12.99)$$

and inserting the complete set of scattering solutions, leads to

$$\langle \boldsymbol{k}'| \mathcal{T}(E) |\boldsymbol{k}\rangle = \langle \boldsymbol{k}'| V |\boldsymbol{k}\rangle + \int d\boldsymbol{q} \, \frac{\langle \boldsymbol{k}'| V |\psi_{\boldsymbol{q}}\rangle \langle \psi_{\boldsymbol{q}}| V |\boldsymbol{k}\rangle}{E - \frac{\hbar^{2}q^{2}}{m} + i\eta}. \qquad (12.100)$$

Since $\langle \boldsymbol{k}| V |\psi_{\boldsymbol{q}}\rangle = \langle \boldsymbol{k}| \tilde{\mathcal{T}} |\boldsymbol{q}\rangle$ [this follows *e.g.* from the Lippmann–Schwinger Eq. (6.52)] one has

$$\langle \boldsymbol{k}'| \mathcal{T}(E) |\boldsymbol{k}\rangle = \langle \boldsymbol{k}'| V |\boldsymbol{k}\rangle + \int d\boldsymbol{q} \, \frac{\langle \boldsymbol{k}'| \tilde{\mathcal{T}} |\boldsymbol{q}\rangle \langle \boldsymbol{k}| \tilde{\mathcal{T}} |\boldsymbol{q}\rangle^{*}}{E - \frac{\hbar^{2}q^{2}}{m} + i\eta}, \qquad (12.101)$$

and subtracting the same equation for $E = \frac{\hbar^{2}k^{2}}{m}$ leads to the desired expression

$$\langle \boldsymbol{k}'| \mathcal{T}(E) |\boldsymbol{k}\rangle = \langle \boldsymbol{k}'| \tilde{\mathcal{T}} |\boldsymbol{k}\rangle + \frac{m}{\hbar^{2}} \int d\boldsymbol{q} \, \langle \boldsymbol{k}'| \tilde{\mathcal{T}} |\boldsymbol{q}\rangle \langle \boldsymbol{k}| \tilde{\mathcal{T}} |\boldsymbol{q}\rangle^{*}$$

$$\times \left(\frac{1}{\frac{mE}{\hbar^{2}} - q^{2} + i\eta} - \frac{1}{k^{2} - q^{2} + i\eta} \right), \qquad (12.102)$$

which will be of use in later applications.

12.4.4 *Gross–Pitaevskii equation*

A system can be considered dilute when the average interparticle spacing $\rho^{-1/3}$ is large, compared to the magnitude $|a|$ of the scattering length or, equivalently, when $\rho|a|^3 << 1$. The preceding Sec. 12.4.3 made it plausible that, even when the bare interaction V is strong, the HB mean-field formalism can be applied to dilute systems by simply replacing the bare interaction with the pseudo-potential, according to Eqs. (12.65) - (12.66).

With this replacement the HB potential in Eq. (12.61) becomes

$$W_{HB}(\boldsymbol{r}) = g(N-1)|\phi_c(\boldsymbol{r})|^2 \approx gN|\phi_c(\boldsymbol{r})|^2. \qquad (12.103)$$

Note that from now on we will approximate $N-1$ by N in the HB expressions, which is convenient and harmless, since we will be dealing with large numbers of particles.

The HB equation (12.62) simplifies to

$$\left[-\frac{\hbar^2}{2m}\nabla^2 + U(\boldsymbol{r}) \right] \phi_c(\boldsymbol{r}) + gN|\phi_c(\boldsymbol{r})|^2\phi_c(\boldsymbol{r}) = \mu\phi_c(\boldsymbol{r}), \qquad (12.104)$$

where we explicitly made the identification $\mu = \varepsilon_c$ of Eq. (12.51). Eq. (12.104), which has the form of a "nonlinear" sp Schrödinger equation, is called the (time-independent) Gross–Pitaevskii (GP) equation.

The GP condensate orbital $\phi_c(\boldsymbol{r})$ also minimizes, according to Eq. (12.64), the following energy functional

$$E_0^{GP}/N = \int d\boldsymbol{r} \left(\frac{\hbar^2}{2m}|\nabla\phi_c(\boldsymbol{r})|^2 + U(\boldsymbol{r})|\phi_c(\boldsymbol{r})|^2 + \frac{gN}{2}|\phi_c(\boldsymbol{r})|^4 \right), \quad (12.105)$$

under the constraint $\int d\boldsymbol{r}|\phi_c(\boldsymbol{r})|^2 = 1$.

It is easy to generalize Eq. (12.104) to the time-dependent version

$$\left[-\frac{1}{2m}\nabla^2 + U(\boldsymbol{r};t) \right] \phi_c(\boldsymbol{r};t) + gN|\phi_c(\boldsymbol{r},t)|^2\phi_c(\boldsymbol{r};t) = i\hbar\frac{\partial}{\partial t}\phi_c(\boldsymbol{r};t),$$
$$(12.106)$$

when the external potential explicitly depends on time. Equations (12.104) and (12.106) are able to explain most of the present data on atomic BEC.

12.4.5 *Confined bosons in harmonic traps*

The first observation of atomic BEC in 1995 has triggered many new experimental and theoretical developments. We will only cover the basic physics of the BEC ground state in a confining trap, and refer to a review paper and recent books [Dalfovo *et al.* (1999); Pethick and Smith (2002); Pitaevskii and Stringari (2003)] for more material and additional references.

Estimate of interaction effects

For the magnetic traps used in atomic BEC experiments, the confining potential $U(r)$ in Eq. (12.104) is well approximated by a harmonic oscillator (HO) potential

$$U(r) = \frac{m}{2}(\omega_x^2 x^2 + \omega_y^2 y^2 + \omega_z^2 z^2).$$ (12.107)

In the most general case, it is completely anisotropic ($\omega_x \neq \omega_y \neq \omega_z$), but usually has cylindrical symmetry ($\omega_x = \omega_y$), with $\omega_x = \omega_z$ corresponding to a spherically symmetric, $\omega_x > \omega_z$ to an elongated (cigar-shaped) and $\omega_x < \omega_z$ to a flattened (pancake-shaped) trap. The behavior of noninteracting atoms in such magnetic traps has already been discussed in Sec. 5.6.2. We now examine what changes occur when the interatomic interaction is included at the level of the GP equation (12.104).

It should be stressed that, even when the system is dilute in the sense that $\rho|a|^3$ is small, the effects of the interaction can still be large. To make an estimate, let us assume that the condensate wave function $\phi_c(r)$ can be approximated by the noninteracting HO ground state $\phi_{000}(r)$ of Eq. (5.55). The central density is given by

$$\rho(0) = N|\phi_c(0)|^2 \approx N \left(\frac{m\omega_{HO}}{\pi\hbar} \right)^{3/2} = \frac{N}{\pi^{3/2} a_{HO}^3},$$ (12.108)

where the the average oscillator frequency ω_{HO} and length a_{HO} have been defined in Eq. (5.56) and Eq. (5.59), respectively. Typical values for experimental conditions are $10^3 < N < 10^6$, $|a| \sim 10^{-9}$m and $a_{HO} \sim 10^{-6}$m, which leads to values $10^{-6} < \rho|a|^3 < 10^{-3}$, indeed corresponding to very dilute systems. The interatomic interaction generates the mean-field potential in the GP equation, which in the center of the trap has a value

$$W_{HB}(0) = gN|\phi_c(0)|^2 \approx N \frac{a}{a_{HO}^3} \frac{4\hbar^2}{m\sqrt{\pi}},$$ (12.109)

and which should be compared to the HO energy scale $\hbar\omega_{HO} = \frac{\hbar^2}{ma_{HO}^2}$. The ratio is seen to be proportional to the dimensionless quantity

$$u = N \frac{a}{a_{HO}},$$ (12.110)

and can be taken as a measure of the strength of the interaction effects. Filling in the above values for N, $|a|$, and a_{HO} one finds $1 < |u| < 10^3$, *i.e.* typically exceeding unity and one can expect strong deviations of the GP

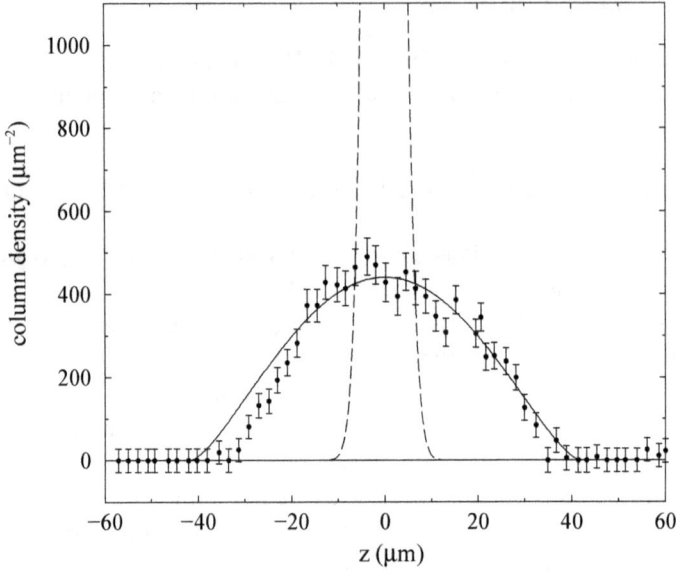

Fig. 12.2 Column density $\bar\rho(z) = \int dy \rho(0,y,z)$ of a condensate of 8×10^4 ^{23}Na atoms ($m = 21.415$ GeV/$c^2 \approx 23$ a.m.u.) in a harmonic trap with $\omega_x = \omega_y = 2050$ rad/s and $\omega_z = 170$ rad/s. The experimental atomic density shown in the figure is proportional to the optical density data by [Hau *et al.* (1998)], who directly observed the condensate as an absorption image by shining through the condensate with near-resonant laser light. With $a = 2.75$ nm and a_{HO}=1.76 μm one has $u \approx 125$. The experimental data are compared with the HO prediction (dashed line) and the GP prediction (solid line).

condensate density from the noninteracting Gaussian profile. An example is shown in Fig. 12.2 where an experimental condensate density with $u \approx 125$ (see caption) is compared to the HO and the GP ones. The interaction effects, included in the GP approach, reduce the central density by a factor of 12 compared to that of the noninteracting HO. The GP result is in very good agreement with the experimental density profile.

Scaling relations

The GP energy in Eq. (12.105) with the HO external potential of Eq. (12.107), can be split up as $E_0^{GP} = T + U + E_{int}$ in terms of its kinetic, HO and interaction parts. Since E_0^{GP} is stationary with respect to small variations in $\phi_c(\boldsymbol{r})$, one can *e.g.* consider a scaled replica ($\xi > 0$)

$$\phi_\xi(\boldsymbol{r}) = \xi^{3/2}\phi_c(\xi\boldsymbol{r}), \qquad (12.111)$$

which is a normalized wave function. Evaluating the energy functional in Eq. (12.105) with $\phi_\xi(\boldsymbol{r})$ instead of $\phi_c(\boldsymbol{r}) \equiv \phi_{\xi=1}(\boldsymbol{r})$, it is easy to see that all components have simple scaling properties, and the energy becomes

$$E_0^{GP}(\xi) = \xi^2 T + \frac{1}{\xi^2} U + \xi^3 E_{int}. \qquad (12.112)$$

The energy in Eq. (12.112) must be minimal for the GP solution with $\xi = 1$, and one has the relation

$$0 = \left.\frac{\partial E_0^{GP}(\xi)}{\partial \xi}\right|_{\xi=1} = 2T - 2U + 3E_{int} \qquad (12.113)$$

between the GP kinetic, HO, and interaction energy. Similar relations can be derived by considering scaling in the x, y or z direction [see Exercise (1)].

Attractive interactions

It is also obvious from Eq. (12.112) that for attractive interactions, a solution of the GP equation cannot represent the *global* minimum of the GP energy functional: the negative interaction energy $E_{int} < 0$ for $g < 0$ implies that $E_0^{GP}(\xi) \to -\infty$ for $\xi \to +\infty$. The system is globally unstable against collapse, as it can always find a lower energy by shrinking in size and increasing the density.[5] It is possible however, to trap the system in a local minimum, and condensates of atoms with negative scattering lengths are routinely observed. A local minimum for $a < 0$ clearly requires the presence of the confining trap potential [second term in Eq. (12.112)]. At the same time, a collapsing condensate must have its increase in negative interaction energy, at some point balanced by a rise in the kinetic energy due to its decreasing size. This turns out to be impossible when the coefficient in front of the interaction energy becomes too large in magnitude or, equivalently, when the number of particles exceeds some critical value. All this can be neatly visualized by studying a spherical trap and a one-parameter family of wave functions, *e.g.* Gaussians with variable width [Fetter (1998)],

$$\psi_\xi(\boldsymbol{r}) = \frac{\xi^{3/2}}{\pi^{3/4}} e^{-\xi^2 r^2 / 2}. \qquad (12.114)$$

---

[5]Physically, of course, the condensate will be destroyed by recombination processes, as the density increases.

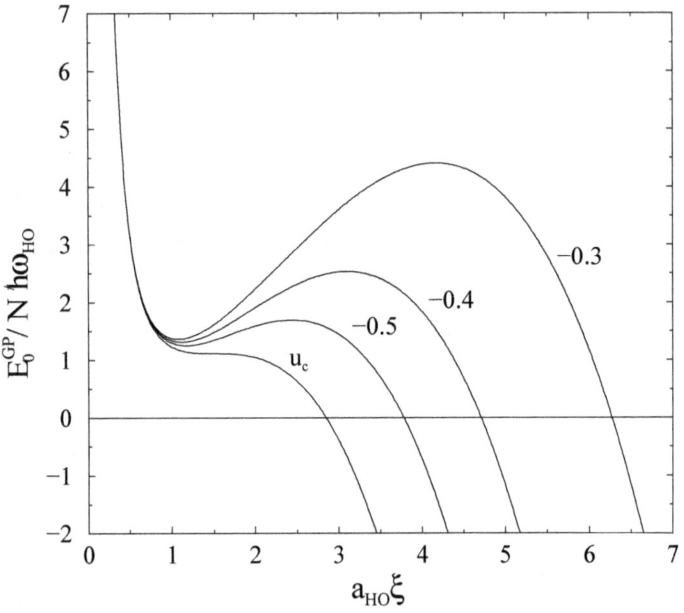

Fig. 12.3 Total energy $E_0^{GP}(\xi)/(N\hbar\omega_{HO})$ in a spherical trap with the Gaussian ansatz of Eq. (12.114), as a function of $a_{HO}\xi$, for the values $u = $ -0.3, -0.4, -0.5 , and for $u = u_c$.

Evaluating the different components of the energy functional (12.105) with the Gaussian ansatz $\psi_\xi(r)$ leads to

$$\frac{T(\xi)}{N\hbar\omega_{HO}} = \frac{3}{4}a_{HO}^2\xi^2, \quad \frac{U(\xi)}{N\hbar\omega_{HO}} = \frac{3}{4}\frac{1}{a_{HO}^2\xi^2},$$

$$\frac{E_{int}(\xi)}{N\hbar\omega_{HO}} = \frac{u}{\sqrt{2\pi}}a_{HO}^3\xi^3. \tag{12.115}$$

In Fig. 12.3, the total energy $E_0^{GP}(\xi)$ is shown as a function of ξ for various values of $u < 0$. For small $|u|$ one observes the presence of a local minimum, corresponding to a condensate of smaller size and higher density than the HO one. The minimum becomes shallower as $|u|$ increases, and vanishes completely at a critical strength $|u_c| = 2\sqrt{2\pi}5^{-5/4} \approx 0.671$. The exact GP solution (not restricted to a Gaussian ansatz) has the same behavior, but with a critical strength $|u| \approx 0.575$ (see Sec. 12.4.6).

Thomas–Fermi limit

Apart from the noninteracting limit ($u \to 0$), the GP equation (12.104) has a new interesting regime when $u \to +\infty$ in case of repulsive interactions. This so-called Thomas–Fermi[6] (TF) or large-N limit is exactly solvable. Considering the three terms on the left-hand side of Eq. (12.104), it is clear that both the kinetic term and the repulsive interaction term try to expand the atomic cloud, whereas the confining potential acts as a counterbalance by providing the necessary repelling force to the expansion. In the TF limit the interaction term completely dominates the expansion, and one can neglect the kinetic energy term in Eq. (12.104). As a result, one has a purely algebraic equation

$$\{U(\boldsymbol{r}) + gN|\phi_c(\boldsymbol{r})|^2 - \mu\} \, \phi_c(\boldsymbol{r}) = 0, \qquad (12.116)$$

having the simple solution

$$\sqrt{gN} \, \phi_c^{TF}(\boldsymbol{r}) = \theta\left(\mu - U(\boldsymbol{r})\right) \sqrt{\mu - U(\boldsymbol{r})}. \qquad (12.117)$$

Substitution of Eq. (12.117) in the normalization condition

$$1 = \int d\boldsymbol{r} \; |\phi_c^{TF}(\boldsymbol{r})|^2 = \frac{1}{gN\omega_{HO}^3} \left(\frac{2}{m}\right)^{3/2} 4\pi \int_0^{\sqrt{\mu}} dR \; R^2(\mu - R^2), \qquad (12.118)$$

allows to calculate the corresponding chemical potential

$$\mu_{TF} = (15u)^{2/5} \frac{\hbar\omega_{HO}}{2} \qquad (12.119)$$

which is much larger than the HO quantum in the present limit $u \to +\infty$. The interaction energy E_{int} can now be easily obtained

$$
\begin{aligned}
E_{int}^{TF} &= \frac{1}{2g} \int d\boldsymbol{r} \; \theta\left(\mu - U(\boldsymbol{r})\right) \left[\mu - U(\boldsymbol{r})\right]^2 \\
&= \frac{2}{7} N\mu_{TF},
\end{aligned}
\qquad (12.120)
$$

and likewise for the trap energy

$$U_{TF} = \frac{3}{7} N\mu_{TF}. \qquad (12.121)$$

---

[6]In analogy to the Thomas–Fermi approximation for the electron density in atoms, where the kinetic term is replaced by the kinetic energy of the noninteracting electron gas at the local density. Since the kinetic energy of a noninteracting boson gas is zero, the kinetic term is simply omitted in the present application.

The TF condensate density, according to Eqs. (12.117) and (12.119), obviously has a sharp boundary, given by the ellipsoid

$$2\mu_{TF}/m = \omega_x^2 x^2 + \omega_y^2 y^2 + \omega_z^2 z^2, \tag{12.122}$$

beyond which the density is strictly zero. The volume of the ellipsoid is equal to $\frac{4}{3}\pi a_{HO}^3 (15u)^{3/5}$ and grows with increasing N as $N^{3/5}$. This is compensated by a decrease in the normalized density $\rho(\boldsymbol{r})/N$, e.g. the central density $\rho_{TF}(0)/N = \mu_{TF}/(gN)$ decreases for large N as $N^{-3/5}$.

The TF density provides an easy and accurate approximation to the exact GP density for systems with a large number of atoms, except near the boundary, where the unphysical sharp edge of the TF density is rounded off in the true GP density. In fact, the kinetic energy calculated with the TF expression (12.117) diverges, since the normal derivative at the boundary becomes infinite. This has been analyzed for spherical traps in [Dalfovo *et al.* (1996)], where it was shown that this divergency can be eliminated by a universal extrapolation of the TF density near the boundary. It turns out that the kinetic energy in the $u \to +\infty$ limit has a leading term proportional to $T \sim N(\ln u)u^{-2/5}$, vanishingly small compared to the other contributions U and E_{int} which have a leading term [see Eqs.(12.120) – (12.121)] proportional to $\sim Nu^{2/5}$ in the TF limit.

Of interest is also the change in shape of the condensate (in case of an anisotropic trap), when going from the noninteracting to the TF limit. The shape can be defined through the root-mean-square radii R_x, R_y, R_z with $R_x = [\int d\boldsymbol{r} \; x^2 |\phi_c(\boldsymbol{r})|^2]^{1/2}$ and similarly for R_y and R_z. In the noninteracting limit with the HO wave function (5.55), one has $R_x^2 = \hbar/(2m\omega_x)$ and the shape is determined by

$$R_x \sqrt{\omega_x} = R_y \sqrt{\omega_y} = R_z \sqrt{\omega_z}. \tag{12.123}$$

The TF wavefunction in Eq. (12.117) on the other hand, leads to

$$R_x^{TF} = \frac{\omega_{HO}}{\omega_x} \frac{1}{\sqrt{7}} (15u)^{1/5} a_{HO} \tag{12.124}$$

and therefore

$$R_x^{TF} \omega_x = R_y^{TF} \omega_y = R_z^{TF} \omega_z. \tag{12.125}$$

Table 12.1 GP results for spherical traps.

u	μ	μ_{TF}	E_0/N	E_0^{TF}/N	T/N	U/N	E_{int}/N
0	1.500	–	1.500	–	0.7500	0.7500	0
1	2.066	1.477	1.811	1.055	0.5876	0.9692	0.2544
10	4.016	3.710	3.072	2.650	0.3561	1.772	0.9440
10^2	9.473	9.320	6.875	6.657	0.1898	4.087	2.598
10^3	23.48	23.41	16.83	16.72	9.615×10^{-2}	10.08	6.655
10^4	58.84	58.80	42.05	42.00	4.704×10^{-2}	25.22	16.78
10^5	147.7	147.7	105.5	105.5	2.238×10^{-2}	63.31	42.19
10^6	371.0	371.0	265.0	265.0	1.040×10^{-2}	159.0	106.0
-0.5	0.8497	–	1.238	–	1.105	0.5220	-0.3886
$-|u_c|$	0.3640	–	1.168	–	1.589	0.3830	-0.8042

Chemical potential μ and the GP ground-state energy per particle E_0/N, obtained by solving the spherical GP equation (12.128) for various values of the interaction strength u, compared with the corresponding Thomas–Fermi approximations μ_{TF} and E_0^{TF} (where the TF kinetic energy has been set equal to zero). Also listed are the components of the GP ground-state energy: the kinetic energy T/N, trap energy U/N and interaction energy E_{int}/N.

12.4.6   *Numerical solution of the GP equation*

The coordinate space GP Eq. (12.104) can be solved using grid-based techniques. We only examine the simple case of a spherical trap

$$\left[-\frac{\hbar^2}{2m}\nabla^2 + \frac{1}{2}m\omega_{HO}^2 r^2 \right] \phi_c(\mathbf{r}) + gN|\phi_c(\mathbf{r})|^2\phi_c(\mathbf{r}) = \varepsilon_c\phi_c(\mathbf{r}), \quad (12.126)$$

which already allows most of the relevant observations. The lowest-energy solution of Eq. (12.126) will be spherically symmetric, and can be rewritten as

$$\phi_c(\mathbf{r}) = \frac{1}{\sqrt{4\pi}}a_{HO}^{-3/2}\varphi(w)/w \quad (12.127)$$

in terms of the dimensionless variable $w = r/a_{HO}$ and a new function $\varphi(w)$ normalized as $1 = \int dw|\varphi(w)|^2$.

With this replacement, Eq.(12.126) transforms into

$$-\frac{1}{2}\varphi''(w) + \frac{1}{2}w^2\varphi(w) + u[|\varphi(w)|^2/w^2]\varphi(w) = \tilde{\mu}\varphi(w) \quad (12.128)$$

with $\tilde{\mu} = \mu/(\hbar\omega_{HO})$, and everything has been expressed in terms of the HO length and energy scales. In the representation of Eq. (12.128) it is obvious that the only relevant parameter is the interaction strength u defined in Eq. (12.110).

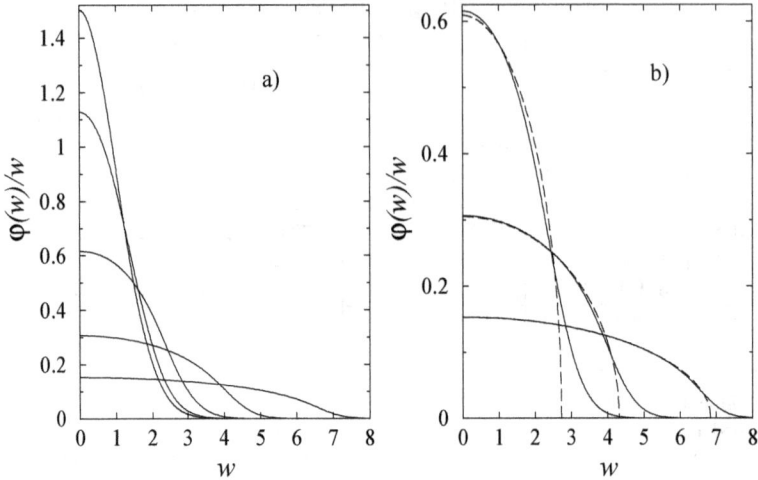

Fig. 12.4   GP wave functions for a condensate with repulsive interactions in a spherical trap. The wave functions are obtained by solving Eq. (12.128) for various values of u. Left panel: $\varphi(w)/w$ for $u = 0, 1, 10, 100, 1000$, where the central value decreases with increasing u. Right panel: comparison of the GP wave functions (full line) with the TF approximation (dashed line) for $u = 10, 100, 1000$.

Figure 12.4(a) shows how the solution of Eq. (12.128) evolves as a function of $u > 0$ (repulsive case), while the energy and its various components are listed in Table 12.1. When u increases, the atomic cloud reacts by expanding and becoming less dense so as to minimize the increase in the repulsive interaction energy E_{int}. As a consequence of the growing size, the trap energy U increases and the kinetic energy decreases. The exact GP wave function is compared to the TF approximation in Fig. 12.4(b). One observes that already at $u \approx 100$ the TF approximation is quite good, except near the TF edge. For attractive interactions ($u < 0$) the opposite happens: compared to the noninteracting HO wave function the system contracts and increases its central density so as to optimize the negative interaction energy, as illustrated in Fig. 12.5. As a consequence, the kinetic energy increases and the trap energy decreases. At a critical value $|u_c| \approx 0.5750$, corresponding to a critical particle number $N_c = |u_c|\frac{a_{HO}}{a}$, the interaction energy becomes too dominant and the spherical GP equation no longer has a solution. An overview of the energy and the chemical potential as a function of u is provided by Fig. 12.6.

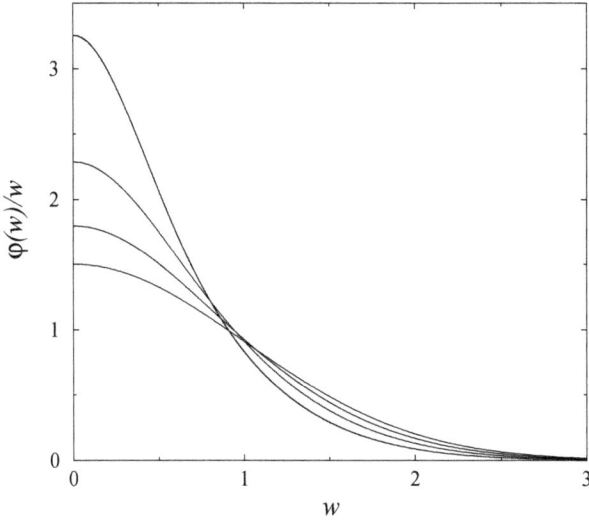

Fig. 12.5 GP wave functions for a condensate with attractive interactions in a spherical trap. The wave functions $\varphi(w)/w$ are obtained by solving Eq. (12.128) for $u = 0$, $-0.3, -0.5$ and the critical value $u = -|u_c| \approx -0.5750$. The central value rises with increasing $|u|$.

12.4.7 *Computer exercise*

Reproduce the results for spherical traps in Sec. 12.4.6 by numerical so-lution of Eq. (12.128). The simple discretization techniques developed in Sec. 10.2.4 can again be put to good use here. In this case an equidistant grid

$$w_i = (i - \frac{1}{2})\Delta, \qquad i = 1, \dots, M \tag{12.129}$$

works fine. Use $M \approx 1000$ and a cut-off at $w_{max} = 5$ or, for $u > 0$, at $w_{max} = 1.5 R_{TF}$ if the latter value is larger. Here $R_{TF} = (15u)^{1/5}$ is the TF edge.

The second-order derivative in Eq. (12.128) can be approximated on the grid as

$$\varphi''(w_i) \rightarrow (-2\varphi(w_i) + \varphi(w_{i-1}) + \varphi(w_{i+1})) \frac{1}{\Delta^2}, \tag{12.130}$$

for $i > 1$, with the boundary condition at the first point

$$\varphi''(w_1) \rightarrow (-3\varphi(w_1) + \varphi(w_2)) \frac{1}{\Delta^2}, \tag{12.131}$$

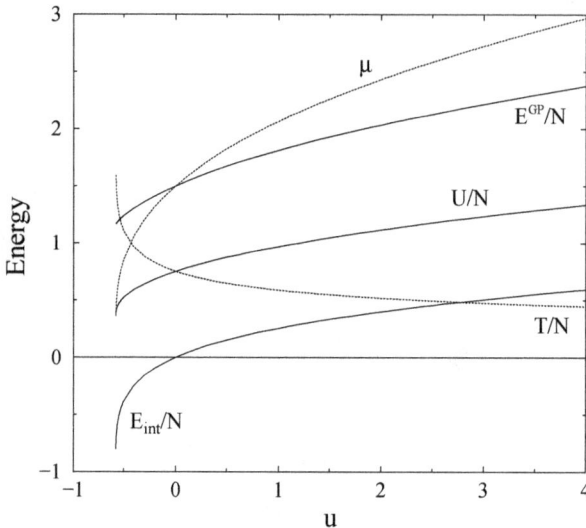

Fig. 12.6 Overview of the GP total energy E^{GP}/N, the chemical potential μ, the kinetic energy T/N, trap energy U/N and interaction energy E_{int}/N, as a function of the interaction strength u in a spherical trap.

in order to select solutions odd in w.

The nonlinearity is treated by iteration, choosing as an initial guess for $u < 0$ and small values of $u > 0$ the noninteracting HO wave function

$$\varphi^{(0)}(w) \sim w e^{-w^2/2}, \tag{12.132}$$

or for $u > 0$ the TF approximation

$$\varphi^{(0)}(w) \sim w\theta(R_{TF} - w)\sqrt{R_{TF}^2 - w^2}, \tag{12.133}$$

where $\varphi^{(0)}(w)$ should be normalized on the grid as

$$1 = \sum_{i=1}^{M} \Delta |\varphi^{(0)}(w_i)|^2. \tag{12.134}$$

Choosing the initial guess of Eq. (12.124) will lead to faster convergence for all but the smallest values of $u > 0$.

The iteration cycle looks like:

1. Use $\varphi^{(0)}$ to evaluate the HB potential $-u[|\varphi^{(0)}(w)|^2/w^2]$ in Eq. (12.128). This leads to an eigenvalue equation for a symmetric tridiagonal matrix A_{ij}.

2. An improved wave function $\varphi^{(1)}$ can be found by solving the linear system

$$\sum_{j=1}^{M} A_{ij}\varphi^{(1)}(w_j) = \varphi^{(0)}(w_i), \quad i = 1, \ldots, M, \tag{12.135}$$

by means of the recursion formula in Sec. 10.2.4, and then normalizing the solution $\varphi^{(1)}$ to unity according to Eq. (12.134).

3. At this point one can monitor the difference between $\varphi^{(0)}$ and $\varphi^{(1)}$ and, if needed, set $\varphi^{(0)} = \varphi^{(1)}$ and go back to step 1.

Note that at each iteration step the weight of the lowest-energy solution of the current eigenvalue equation is increased in $\varphi^{(1)}$, as compared to $\varphi^{(0)}$. While not optimal, the above method converges with minimal overhead and reasonably fast to the desired solution of the nonlinear GP equation in Eq. (12.128). One may also convince oneself that for values of $u < -|u_c|$ the method fails to converge.

12.5 Exercises

(1) Write the GP energy of Eq. (12.105) in an anisotropic harmonic trap as

$$E_0^{GP} = (T_x + T_y + T_z) + (U_x + U_y + U_z) + E_{int}$$

with the evident notation

$$T_x = N \int dr \, \frac{\hbar^2}{2m} \left| \frac{\partial \phi_c}{\partial x}(r) \right|^2$$
$$U_x = N \int dr \, \frac{m\omega_x^2 x^2}{2} |\phi_c(r)|^2,$$

etc. Introduce [analogous to Eq. (12.111)] a scaling of the x-coordinate only, *i.e.* $\phi_\xi(x, y, z) = \xi^{1/2}\phi(\xi x, y, z)$, and show that the GP solution obeys $2T_x - 2U_x + E_{int} = 0$.

(2) Within the Gaussian model of Eq. (12.114), find the critical attraction strength u_c beyond which an atomic condensate in a spherical trap becomes unstable against collapse. Note that for the critical strength u_c the local minimum at ξ_{min} has zero curvature, which can be translated by requiring both the first and second derivative of $E_0^{GP}(\xi)$ to vanish at ξ_{min}. The resulting solutions are $\xi_{min}a_{HO} = 5^{1/4}$ and $u_c = -2\sqrt{2\pi}5^{-5/4} \approx 0.671$.

Chapter 13

Excited states in finite systems

After studying various approaches to describe the sp propagator in a many-fermion system in Chs. 10 and 11, it is now time to discuss the description of excited states of the system with N particles. In Sec. 13.1, the relevant limit of the tp propagator G_{II}, appropriate for the calculation of these excited states, is introduced. It is possible to develop all the details of the corresponding diagrammatic expansion of this so-called particle–hole (ph) or polarization propagator. Assuming that the reader is now sufficiently familiar with the procedure, a somewhat different approach will be applied. We will identify in Sec. 13.2 the minimal set of diagrams generating a pole structure, similar to the one of the exact polarization propagator. The resulting approximation is widely used, and while the name is not too illuminating at first, it is historically called the random phase approximation (RPA).[1] In Sec. 21.4, the RPA will be linked to the Hartree–Fock description of the sp propagator. In Sec. 13.3, a simple model is employed to illustrate the scope of the RPA in a finite system. Many of the observed properties of excited states in many-fermion systems can be interpreted in terms of this schematic model. Especially, the appearance of collective states in relation to the character of the ph interaction, whether on average attractive or repulsive, can be illustrated fruitfully. Collective features also appear in the transition strength to the ground state. The energy-weighted sum rule for the RPA transition strength is the subject of Sec. 13.4. Excited states of atoms are discussed in Sec. 13.5. The summation of diagrams, involved in calculating excited states, can be applied to the determination of the so-called correlation energy as well. This topic is briefly discussed

---

[1]This term was first introduced in the study of the electron gas, treating the collective properties associated with the divergent Coulomb interaction [Bohm and Pines (1953); Pines (1953)].

in Sec. 13.6. The chapter concludes with the derivation in Sec. 13.7 of the RPA equations in the angular momentum coupled representation.

13.1 Polarization propagator

The relevant limit of the tp propagator to study excited states in many-fermion systems, is given by

$$G_{ph}(\alpha, \beta^{-1}; \gamma, \delta^{-1}; t - t') \equiv \lim_{t_\beta \to t^+} \lim_{t_\gamma \to t'^+} G_{II}(\alpha t, \overline{\delta} t', \overline{\beta} t_\beta, \gamma t_\gamma)$$

$$= -\frac{i}{\hbar} \left\langle \Psi_0^N \right| T[a_{\overline{\beta}_H}^\dagger(t) a_{\alpha_H}(t) a_{\gamma_H}^\dagger(t') a_{\overline{\delta}_H}(t')] \left| \Psi_0^N \right\rangle . \qquad (13.1)$$

The "bar" over sp quantum numbers indicates the time-reversed state, *i.e.*

$$\mathcal{T} \left| \alpha \right\rangle = \left| \overline{\alpha} \right\rangle , \qquad (13.2)$$

where $\mathcal{T}$ is the time-reversal operator.[2] It is useful to emphasize that its form depends on the sp basis that is chosen [Sakurai (1994); Messiah (1999)] and requires the selection of a particular phase convention. For a particle with momentum $\boldsymbol{p}$, spin $\frac{1}{2}$ (not explicitly shown), and spin projection m_s, we employ the convention

$$\mathcal{T} \left| \boldsymbol{p}, m_s \right\rangle \equiv \left| \overline{\boldsymbol{p}, m_s} \right\rangle = (-1)^{\frac{1}{2} + m_s} \left| -\boldsymbol{p}, -m_s \right\rangle . \qquad (13.3)$$

It implies that in this case, time reversal is equivalent to the product of the parity operator and the spin-space operator $(i\sigma_y)$, corresponding to a rotation by π around the y-axis [Sakurai (1994)]. Note that we follow the phase convention in Eq. (13.3) of [Bohr and Mottelson (1998)].

Whenever appropriate, the relevant phase convention and character of the time-reversal operator will be explicitly given. For the more general expressions in this chapter, time-reversed states are simply indicated with a bar over sp quantum numbers, and have the property (for fermions) that

$$\mathcal{T} \left| \overline{\alpha} \right\rangle = \left| \overline{\overline{\alpha}} \right\rangle = - \left| \alpha \right\rangle . \qquad (13.4)$$

For later applications we also need to be able to properly couple ph excitations to good total angular momentum states. It is then convenient to introduce operators b and $b^\dagger$, where the operator

$$b_\alpha^\dagger = a_{\overline{\alpha}} \qquad (13.5)$$

---

[2]Not to be confused with the scattering quantity or the time-ordering operation.

produces a hole with sp quantum numbers α, corresponding to the removal of a particle in the time-reversed state $\bar{\alpha}$. In the basis specified in Eq. (13.3) *e.g.*, one has

$$b^{\dagger}_{\boldsymbol{p},m_s} \equiv a_{\overline{\boldsymbol{p},m_s}} = (-1)^{\frac{1}{2}+m_s} a_{-\boldsymbol{p},-m_s}. \tag{13.6}$$

The presence in Eq. (13.1) of only two times in this limit of the tp propagator, implies that only one energy variable is required upon FT. In order to prepare for FT, one may substitute the explicit form of the Heisenberg operators [see Eqs. (7.3) and (7.4)] and the definition of the time-ordering operation in terms of step functions into Eq. (13.1). Inserting complete sets of N-particle states at appropriate places then yields the following expression for G_{ph} in the usual way

$$G_{ph}(\alpha, \beta^{-1}; \gamma, \delta^{-1}; t-t') = -\frac{i}{\hbar} \langle \Psi^N_0 | a^{\dagger}_{\bar{\beta}} a_{\alpha} | \Psi^N_0 \rangle \langle \Psi^N_0 | a^{\dagger}_{\gamma} a_{\bar{\delta}} | \Psi^N_0 \rangle \tag{13.7}$$

$$- \frac{i}{\hbar} \left\{ \sum_{n\neq 0} \theta(t-t') e^{\frac{i}{\hbar}(E^N_0-E^N_n)(t-t')} \langle \Psi^N_0 | a^{\dagger}_{\bar{\beta}} a_{\alpha} | \Psi^N_n \rangle \langle \Psi^N_n | a^{\dagger}_{\gamma} a_{\bar{\delta}} | \Psi^N_0 \rangle \right.$$

$$\left. + \sum_{n\neq 0} \theta(t'-t) e^{\frac{i}{\hbar}(E^N_0-E^N_n)(t'-t)} \langle \Psi^N_0 | a^{\dagger}_{\gamma} a_{\bar{\delta}} | \Psi^N_n \rangle \langle \Psi^N_n | a^{\dagger}_{\bar{\beta}} a_{\alpha} | \Psi^N_0 \rangle \right\}.$$

The contribution of the ground state has been explicitly isolated in the first term in Eq. (13.7), and involves matrix elements of the one-body density operator [Eq. (7.21)] that are already contained in the sp propagator. It is conventional to introduce the so-called polarization propagator, which only includes the contribution of excited states ($n \neq 0$) in Eq. (13.7)

$$\Pi(\alpha, \beta^{-1}; \gamma, \delta^{-1}; t-t') = G_{ph}(\alpha, \beta^{-1}; \gamma, \delta^{-1}; t-t')$$

$$+ \frac{i}{\hbar} \langle \Psi^N_0 | a^{\dagger}_{\bar{\beta}} a_{\alpha} | \Psi^N_0 \rangle \langle \Psi^N_0 | a^{\dagger}_{\gamma} a_{\bar{\delta}} | \Psi^N_0 \rangle. \tag{13.8}$$

With this preparation one can now perform the FT of the polarization propagator to obtain its Lehmann representation

$$\Pi(\alpha, \beta^{-1}; \gamma, \delta^{-1}; E) = \sum_{n\neq 0} \frac{\langle \Psi^N_0 | a^{\dagger}_{\bar{\beta}} a_{\alpha} | \Psi^N_n \rangle \langle \Psi^N_n | a^{\dagger}_{\gamma} a_{\bar{\delta}} | \Psi^N_0 \rangle}{E - (E^N_n - E^N_0) + i\eta}$$

$$- \sum_{n\neq 0} \frac{\langle \Psi^N_0 | a^{\dagger}_{\gamma} a_{\bar{\delta}} | \Psi^N_n \rangle \langle \Psi^N_n | a^{\dagger}_{\bar{\beta}} a_{\alpha} | \Psi^N_0 \rangle}{E + (E^N_n - E^N_0) - i\eta}, \tag{13.9}$$

by employing the integral formulation of the step functions given by Eq. (6.7). The polarization propagator, like the sp propagator, incorporates important information in its denominator, related to the position of excited states, here of the N-particle system. The numerator contains transition amplitudes, connecting the ground state with those excited states. In the case of a one-body excitation operator (summing over $\bar{\beta}$ is equivalent to summing over β)

$$\hat{O} = \sum_{\alpha\beta} \langle \alpha | O | \bar{\beta} \rangle \, a_\alpha^\dagger a_{\bar{\beta}}, \tag{13.10}$$

the transition probability from the ground state to an excited state reads

$$\left| \langle \Psi_n^N | \hat{O} | \Psi_0^N \rangle \right|^2 = \sum_{\alpha\beta} \sum_{\gamma\delta} \langle \gamma | O | \bar{\delta} \rangle \langle \alpha | O | \bar{\beta} \rangle^*$$

$$\times \langle \Psi_n^N | a_\gamma^\dagger a_{\bar{\delta}} | \Psi_0^N \rangle \langle \Psi_n^N | a_\alpha^\dagger a_{\bar{\beta}} | \Psi_0^N \rangle^*. \tag{13.11}$$

The result demonstrates that the numerator of the first term in Eq. (13.9) contains the relevant transition amplitudes for a given state n to evaluate the transition probability. Examples of cases where one-body excitation mechanisms play an important role, include inelastic electron scattering off nuclei, and neutron scattering off condensed-matter systems. Note that the information content in the second term of Eq. (13.9) is equivalent to that in the first term.

It is instructive to evaluate the noninteracting limit of the polarization propagator. The limit can be directly obtained from Eq. (13.8) by replacing $\hat{H}$ by $\hat{H}_0$ and $|\Psi_0^N\rangle$ by the noninteracting ground state $|\Phi_0^N\rangle$,

$$\Pi^{(0)}(\alpha, \beta^{-1}; \gamma, \delta^{-1}; t - t') = G_{ph}^{(0)}(\alpha, \beta^{-1}; \gamma, \delta^{-1}; t - t')$$

$$+ \frac{i}{\hbar} \langle \Phi_0^N | a_{\bar{\beta}}^\dagger a_\alpha | \Phi_0^N \rangle \langle \Phi_0^N | a_\gamma^\dagger a_{\bar{\delta}} | \Phi_0^N \rangle. \tag{13.12}$$

By employing the sp basis in which H_0 is diagonal, the states that can be excited also correspond to eigenstates of $\hat{H}_0$, as discussed in Sec. 3.1, with the following result

$$\hat{H}_0 \, a_\gamma^\dagger a_{\bar{\delta}} | \Phi_0^N \rangle = \theta(\gamma - F)\theta(F - \delta) \left\{ \varepsilon_\gamma - \varepsilon_\delta + E_{\Phi_0^N} \right\} a_\gamma^\dagger a_{\bar{\delta}} | \Phi_0^N \rangle, \tag{13.13}$$

where Kramer's degeneracy has been used in putting $\varepsilon_{\bar{\delta}} = \varepsilon_\delta$. One may leave out the bar over quantum numbers in the step functions for the same reason. Rewriting the interaction picture operators, according to

Eqs. (A.15) and (A.16), and evaluating the relevant expectation values with respect to the noninteracting ground state, yields

$$\Pi^{(0)}(\alpha, \beta^{-1}; \gamma, \delta^{-1}; t - t')$$

$$= -\frac{i}{\hbar} \left\{ \theta(t - t')\theta(\alpha - F)\theta(F - \beta)\delta_{\alpha,\gamma}\delta_{\beta,\delta}e^{-i(\varepsilon_\alpha - \varepsilon_\beta)(t-t')/\hbar} \right.$$

$$\left. + \theta(t' - t)\theta(F - \alpha)\theta(\beta - F)\delta_{\alpha,\gamma}\delta_{\beta,\delta}e^{-i(\varepsilon_\beta - \varepsilon_\alpha)(t'-t)/\hbar} \right\}, \quad (13.14)$$

where time-reversed quantum numbers can again be suppressed. The first term in Eq. (13.14) corresponds to the independent propagation of a particle with quantum numbers α (γ) from t' to t, and a hole with β (δ) from t to t'. The second term exchanges the role of t and t', as well as that of the sp quantum numbers, and can be referred to as independent hole–particle (hp) propagation. This is illustrated in Fig. 13.1, using time-ordered diagrams. It is customary to refer to part a) of Fig. 13.1 as forward propagation, and part b) as backward propagation. The graphical representation in terms of noninteracting sp propagators is totally appropriate, since a look at Eq. (7.32) demonstrates that up to a constant the two terms in Eq. (13.14) correspond exactly to such a product, *i.e.*

$$\Pi^{(0)}(\alpha, \beta^{-1}; \gamma, \delta^{-1}; t - t') = -i\hbar G^{(0)}(\alpha, \gamma; t - t')G^{(0)}(\overline{\delta}, \overline{\beta}; t' - t). \quad (13.15)$$

Note that when evaluating Eq. (13.15) one may put to zero the terms containing a product of step functions with opposite time arguments.

Using the integral representation [see Eq. (6.7)] of the step functions in Eq. (13.14), it is straightforward to evaluate the FT of $\Pi_{ph}^{(0)}$

$$\Pi^{(0)}(\alpha, \beta^{-1}; \gamma, \delta^{-1}; E)$$

$$= \delta_{\alpha,\gamma}\delta_{\beta,\delta} \left\{ \frac{\theta(\alpha - F)\theta(F - \beta)}{E - (\varepsilon_\alpha - \varepsilon_\beta) + i\eta} - \frac{\theta(F - \alpha)\theta(\beta - F)}{E + (\varepsilon_\beta - \varepsilon_\alpha) - i\eta} \right\}, \quad (13.16)$$

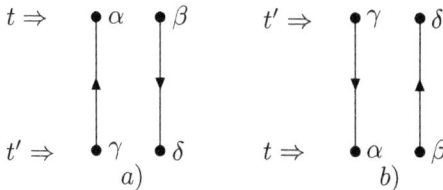

Fig. 13.1 Diagrammatic representation of the ph and hp propagation terms in Eq. (13.14), representing the unperturbed polarization propagator in the time-ordered formulation.

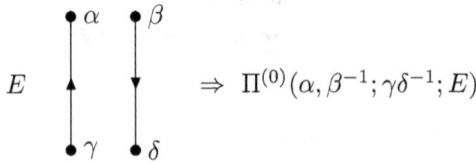

Fig. 13.2 Diagrammatic representation of the unperturbed polarization propagator in the energy formulation. This Feynman diagram includes both ph and hp propagation terms as in Eq. (13.16).

shown graphically in Fig. 13.2. The simple structure of the noninteracting polarization propagator confirms that the location of its poles corresponds to ph states obtained by removing a particle from an occupied level and placing it in an empty level of H_0. Also note that the location of the poles in Eq. (13.16) is symmetric around $E = 0$, which is a reminder that a pair of fermion operators have certain properties in common with bosons (see Ch. 18). The numerator of Eq. (13.16) illustrates how the ph pair, added to the noninteracting ground state, must propagate without change for the chosen pair of quantum numbers. In general the exact polarization propagator in Eq. (13.9) does not share this property, and may have nondiagonal contributions, as well as several excited states linked to the ground state by adding a ph pair with quantum numbers $\gamma\delta^{-1}$.

The product of sp propagators appearing in the time representation of Eq. (13.15) becomes a convolution product, when performing a FT to the energy representation. It is a useful exercise to confirm that the noninteracting polarization propagator in Eq. (13.16) can also be obtained by considering the following convolution of noninteracting sp propagators

$$\Pi^{(0)}(\alpha, \beta^{-1}; \gamma, \delta^{-1}; E) = \int \frac{dE'}{2\pi i} \, G^{(0)}(\alpha, \gamma; E + E') G^{(0)}(\bar{\delta}, \bar{\beta}; E')$$

$$\equiv \delta_{\alpha,\gamma}\delta_{\beta,\delta}\Pi^{(0)}(\alpha, \beta^{-1}; E). \tag{13.17}$$

The integral in Eq. (13.17) for each of the four terms, representing the product of the two noninteracting sp propagators, can be evaluated by considering the distribution of the poles as a function of E'. When both poles occur on the same side of the real E'-axis, the contour integral can be closed in the opposite half, thereby rendering the corresponding contribution zero. Consequently only two terms survive the convolution in Eq. (13.17), representing ph or hp contributions as shown explicitly in Eq. (13.16). It is customary to refer to the first part of Eq. (13.16) as the forward-going term, and the second one as backward-going. Nevertheless, both terms are represented by only one picture in the usual Feynman-diagram convention, indicated in Fig. 13.2.

13.2 Random Phase Approximation

Higher-order contributions to the polarization propagator can be evaluated by straightforward application of Wick's theorem. The relevant connected contributions have already been discussed in Sec. 9.3 for the more general case of the four-time tp propagator. Likewise here, there are terms that represent an interaction between the initial and final ph state, and terms that dress the noninteracting sp propagators. Both types of corrections appear in the first-order contribution given by

$$\Pi^{(1)}(\alpha, \beta^{-1}; \gamma, \delta^{-1}; t - t') = \left(\frac{-i}{\hbar}\right)^2 \int_{-\infty}^{\infty} dt_1 \frac{1}{4} \sum_{\kappa\lambda\mu\nu} \langle \kappa\lambda| V |\mu\nu\rangle \quad (13.18)$$

$$\times \left\langle \Phi_0^N \left| \mathcal{T} \left[a_\kappa^\dagger(t_1) a_\lambda^\dagger(t_1) a_\nu(t_1) a_\mu(t_1) a_{\bar{\beta}}^\dagger(t) a_\alpha(t) a_\gamma^\dagger(t') a_{\bar{\delta}}(t') \right] \right| \Phi_0^N \right\rangle,$$

in the time formulation. If the HF basis is employed, these self-energy terms will vanish in first order. A scheme to calculate excited states, using noninteracting sp propagators, is therefore obtained by keeping only those contributions to Eq. (13.18) which link the propagators, and by discarding those with first-order self-energy insertions (as well as the truly disconnected terms). Using the symmetrized version of V, then yields

$$\Pi^{(1)}(\alpha, \beta^{-1}; \gamma, \delta^{-1}; t - t') \to (i\hbar)^2 \int_{-\infty}^{\infty} dt_1 \sum_{\kappa\lambda\mu\nu} \langle \kappa\lambda| V |\mu\nu\rangle \quad (13.19)$$

$$\times G^{(0)}(\alpha, \kappa; t - t_1) G^{(0)}(\mu, \bar{\beta}; t_1 - t) G^{(0)}(\nu, \gamma; t_1 - t') G^{(0)}(\bar{\delta}, \lambda; t' - t_1),$$

illustrated in Fig. 13.3. For some systems it is useful to separate the direct and exchange contribution of the interaction V, and treat them separately. Both terms are displayed in Fig. 13.3. The physical interpretation of Eq. (13.19) is now clear. Taking into account that the unperturbed sp propagators in Eq. (13.19) are diagonal in the sp basis, one finds for the particular time-ordering $t > t_1 > t'$ that a ph pair is added at time t' with quantum numbers $\gamma\delta^{-1}$, and then propagates to t_1. At this time an interaction changes the propagation to a ph pair with quantum numbers $\alpha\beta^{-1}$ which ends at t when the pair is removed. Since V contains both direct and

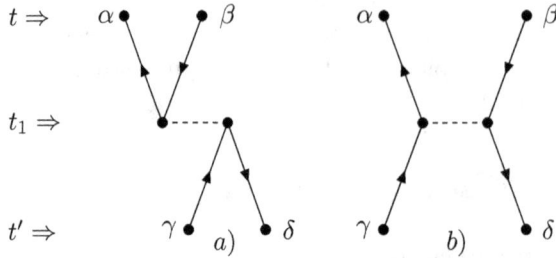

Fig. 13.3 Diagrammatic representation of the first-order correction to the polarization propagator in the time formulation. The time ordering $t > t_1 > t'$ has been chosen for this picture. In part $a)$ the direct contribution of the matrix element of the interaction is employed, whereas in part $b)$ the exchange part. The labels for the interaction have been suppressed for clarity.

exchange contributions, different diagrams can be associated with these two processes as shown in Fig. 13.3.

It is now convenient to introduce the ph version of the two-body matrix element of V as follows

$$\langle \alpha\beta^{-1} | V_{ph} | \gamma\delta^{-1} \rangle \equiv \langle \alpha\bar{\delta} | V | \bar{\beta}\gamma \rangle. \tag{13.20}$$

The definition emphasizes the physical process, illustrated in Fig. 13.3, in which the interaction connects one ph pair with another. Using Eq. (13.20) and performing a FT of Eq. (13.19), one obtains the first-order term in the energy formulation. The transformation is facilitated by using the inverse transforms of the sp propagators given by Eq. (8.60). If the definition in Eq. (13.17) is employed, this FT simplifies to

$$\Pi^{(1)}(\alpha, \beta^{-1}; \gamma, \delta^{-1}; E) = \Pi^{(0)}(\alpha, \beta^{-1}; E) \langle \alpha\beta^{-1} | V_{ph} | \gamma\delta^{-1} \rangle \Pi^{(0)}(\gamma, \delta^{-1}; E). \tag{13.21}$$

The corresponding diagrammatic representation is shown in Fig. 13.4, which is naturally similar to the one shown in Fig. 13.3. The analysis of this expression [or the equivalent one in the time formulation, see Eq. (13.19)], shows that there are four different terms. In the language of the diagrams shown in Fig. 13.1, these correspond to forward-forward, forward-backward, backward-forward, and backward-backward contributions. The Feynman diagrams in Fig. 13.4 represent all four, whereas Fig. 13.3 represents the forward-forward term in the time formulation. The pole structure of these first-order terms (in the interaction V_{ph}), clearly does not reflect the Lehmann representations of either the exact [Eq. (13.9)] or unperturbed propagator [Eq. (13.16)], since the expression in Eq. (13.21) contains a double pole on the diagonal. This problem has been encountered earlier when analyzing higher-order contributions to the sp propagator. In that case, the analysis produced the Dyson equation which generates for appropriate choices of the self-energy, like those discussed in Chs. 10 and 11. Such a formulation incorporates the correct analytic properties associated with the Lehmann representation. The resulting Dyson equation is therefore a more appropriate vehicle to calculate sp properties. In addition, it may be interpreted as a Schrödinger equation for a particle in the medium. Such an analysis also holds for the polarization propagator, which requires a similar infinite summation, likewise leading to a Schrödinger-like equation for ph propagation. One must therefore regard the sum of all contributions, in which V_{ph} is iterated with $\Pi^{(0)}$, as a minimum to obtain a polarization propagator, which can be interpreted with a valid (though approximate) Lehmann representation. The analogy then requires the identification of V_{ph} for example, with the lowest-order self-energy, and $\Pi^{(0)}$ with $G^{(0)}$. The summation of all these terms can be accomplished by rewriting the right-hand side of Eq. (13.21) as follows

$$\Pi^{(0)}(\alpha,\beta^{-1};E)\left\langle\alpha\beta^{-1}\right|V_{ph}\left|\gamma\delta^{-1}\right\rangle\Pi^{(0)}(\gamma,\delta^{-1};E) = \tag{13.22}$$
$$\Pi^{(0)}(\alpha,\beta^{-1};E)\sum_{\epsilon\theta}\left\langle\alpha\beta^{-1}\right|V_{ph}\left|\epsilon\theta^{-1}\right\rangle\Pi^{(0)}(\epsilon,\theta^{-1};\gamma,\delta^{-1};E)$$
$$\rightarrow\Pi^{(0)}(\alpha,\beta^{-1};E)\sum_{\epsilon\theta}\left\langle\alpha\beta^{-1}\right|V_{ph}\left|\epsilon\theta^{-1}\right\rangle\Pi^{RPA}(\epsilon,\theta^{-1};\gamma,\delta^{-1};E).$$

When the last expression is added to the noninteracting polarization propagator, one obtains the corresponding approximation to the exact polar-

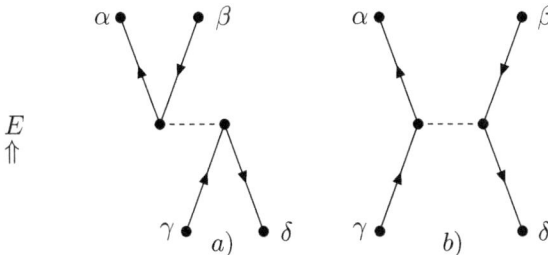

Fig. 13.4 Diagrammatic representation of the first-order correction to the polarization propagator in the energy formulation for both the direct *a*) and exchange *b*) process. The energy E labels both diagrams, as indicated on the left side by the double arrow.

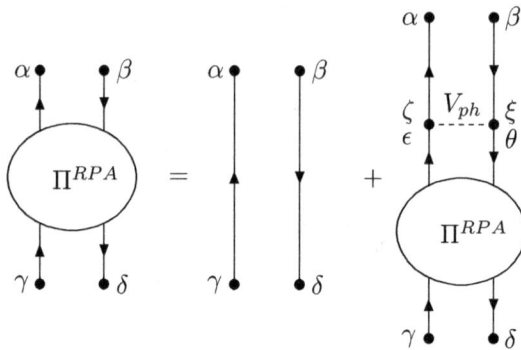

Fig. 13.5   Diagrammatic representation of the second equality in Eq. (13.24). Note that the dashed line represents V_{ph}, as defined in Eq. (13.20).

ization propagator, in which the ph interaction V_{ph} is iterated to all orders with the noninteracting polarization propagator. This is in complete analogy with the Dyson equation that iterates the lowest-order self-energy contribution. The present approximation yields the following equivalent formulations [see Eq. (13.17)]

$$\Pi^{RPA}(\alpha, \beta^{-1}; \gamma, \delta^{-1}; E) = \Pi^{(0)}(\alpha, \beta^{-1}; \gamma, \delta^{-1}; E) \qquad (13.23)$$
$$+ \Pi^{(0)}(\alpha, \beta^{-1}; E) \sum_{\epsilon\theta} \langle \alpha\beta^{-1} | V_{ph} | \epsilon\theta^{-1} \rangle \Pi^{RPA}(\epsilon, \theta^{-1}; \gamma, \delta^{-1}; E),$$

and

$$\Pi^{RPA}(\alpha, \beta^{-1}; \gamma, \delta^{-1}; E) = \Pi^{(0)}(\alpha, \beta^{-1}; \gamma, \delta^{-1}; E) \qquad (13.24)$$
$$+ \sum_{\epsilon\theta\zeta\xi} \Pi^{(0)}(\alpha, \beta^{-1}; \zeta, \xi^{-1}; E) \langle \zeta\xi^{-1} | V_{ph} | \epsilon\theta^{-1} \rangle \Pi^{RPA}(\epsilon, \theta^{-1}; \gamma, \delta^{-1}; E).$$

Successively replacing Π^{RPA} on the right side of Eq. (13.23) [or Eq. (13.24)] by the given expression, generates all higher-order terms, in which unperturbed ph propagation is interrupted zero, one, two, three, *etc.* many times by the action of V_{ph}. This particular sum of terms is known in the many-body literature as the random phase approximation (RPA) and the corresponding label has been used in Eqs. (13.22) – (13.24). The diagrammatic representation of Eq. (13.24) is given in Fig. 13.5.

   The diagrams that emerge from this approximation to the polarization propagator have various names in the literature, ring, bubble, or sausage diagrams. This name calling becomes clear when the direct contribution to V_{ph} is used to generate the higher-order terms implied by Eq. (13.24). Examples of such diagrams are shown in Fig. 13.6. The bubbles or rings emerge when the direct part of the interaction is used to connect the unperturbed ph propagation as shown in Fig. 13.6. It should be reiterated that the diagrams displayed in Fig. 13.6 represent Feynman diagrams in the energy formulation. Each bubble, for example, represents the sum of a forward- and a backward-going term corresponding to the first or second term of Eq. (13.16), respectively. The existence of both terms implies an interplay between these components when Eq. (13.24) is solved. It generates the possibility of intermediate states, in which many ph states are present at the same time. If no backward-going contributions are included, only one ph pair propagates at each time. An example of a term with more intermediate ph states is generated by taking one of the bubbles in Fig. 13.6 and flipping it upward. These contributions imply that some higher-order terms with more ph intermediate states are present. Others are however neglected, in particular those in which particles (or holes) in different parts are exchanged with each other. It has been argued that these Pauli exchange terms add up with random phases and might therefore be rather small, hence the appearance of the name random phase approximation. General features of the solutions of the RPA equation for the polarization propagator in finite systems, will be presented in the next section.

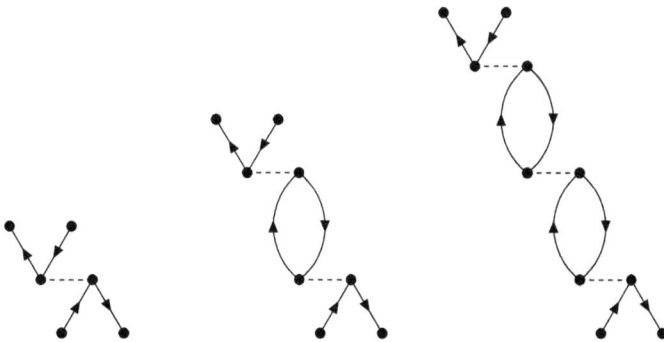

Fig. 13.6 Diagrammatic representation of a few higher-order diagrams generated by Eq. (13.24) when only the direct contribution to V_{ph} is employed, shown in Fig. 13.4. Note that the dashed line in this figure represents the direct contribution to V_{ph}, as defined in Eq. (13.20).

13.3   RPA in finite systems and the schematic model

The solution to Eq. (13.23) starts by assuming that Π^{RPA} also has a
Lehmann representation, just as the unperturbed and the exact polarization propagator. This assumption implies that the transition amplitudes
and excitation energies, in principle, require a distinct notation, since they
refer to a specific approximation (RPA) to the polarization propagator. It
is convenient to define

$$\mathcal{X}_{\alpha\beta}^n \equiv \left\langle \Psi_n^N \middle| a_\alpha^\dagger a_{\overline{\beta}} \middle| \Psi_0^N \right\rangle^* , \tag{13.25}$$

$$\mathcal{Y}_{\alpha\beta}^n \equiv \left\langle \Psi_n^N \middle| a_{\overline{\beta}}^\dagger a_\alpha \middle| \Psi_0^N \right\rangle^* = -\mathcal{X}_{\overline{\beta}\overline{\alpha}}^n , \tag{13.26}$$

which are related under time reversal, and

$$\varepsilon_n^\pi \equiv E_n^N - E_0^N , \tag{13.27}$$

keeping in mind that these quantities refer to the RPA description of the
polarization propagator. With this notation the Lehmann representation
becomes

$$\Pi^{RPA}(\alpha, \beta^{-1}; \gamma, \delta^{-1}; E) = \sum_{n\neq 0} \frac{\mathcal{X}_{\alpha\beta}^n (\mathcal{X}_{\gamma\delta}^n)^*}{E - \varepsilon_n^\pi + i\eta} - \sum_{n\neq 0} \frac{(\mathcal{Y}_{\alpha\beta}^n)^* \mathcal{Y}_{\gamma\delta}^n}{E + \varepsilon_n^\pi - i\eta}. \tag{13.28}$$

In the case of a finite system, it is natural to consider bound excited states,
and consequently, the summation in Eq. (13.28) indeed involves some discrete states[3] As a result, this Lehmann representation includes simple poles.
For simplicity we assume that all excited states correspond to discrete excitation energies. With this assumption it is possible to exploit a technique
that was introduced in Ch. 6 (and used several times since) to solve propagator equations with discrete poles. The procedure involves the calculation
of

$$\lim_{E \to \varepsilon_n^\pi} (E - \varepsilon_n^\pi)\left\{ \Pi^{RPA} = \Pi^{(0)} + \Pi^{(0)} \, V_{ph} \, \Pi^{RPA} \right\}, \tag{13.29}$$

where Eq. (13.24) has been rendered in a schematic fashion. As in Ch. 6, one
proceeds by considering the limits for the three terms in Eq. (13.29). Using
the Lehmann representation for Π^{RPA}, immediately yields the product of

---

[3]Unbound excited states, which behave asymptotically as a free particle on top of
a bound $N - 1$ state, can be described by "continuum RPA" techniques. For atoms,
see [Amusia (1990)] and the brief discussion at the end of Sec. 13.5. For applications in
nuclei, see *e.g.* [Ryckebusch *et al.* (1988)].

two transition amplitudes, associated with excited state n for the left hand side of Eq. (13.29). The limit produces no contribution when it is taken for the first term on the right side which involves $\Pi^{(0)}$. This conclusion can be drawn whenever the interaction V_{ph} is nonvanishing, since its action will imply that the RPA excitation energies will differ from the unperturbed ph ones. Using similar arguments for the final term, we arrive at the following eigenvalue equation, after cancelling a common factor $(\mathcal{X}_{\gamma\delta}^n)^*$ on both sides

$$\mathcal{X}_{\alpha\beta}^n = \Pi^{(0)}(\alpha, \beta^{-1}; \varepsilon_n^\pi) \sum_{\epsilon\theta} \langle \alpha\beta^{-1} | V_{ph} | \epsilon\theta^{-1} \rangle \mathcal{X}_{\epsilon\theta}^n. \qquad (13.30)$$

The summation over the quantum numbers ϵ and θ is restricted to either ph or hp combinations, as can be inferred from the contributions that are generated by iteration employing $\Pi^{(0)}$. The external quantum numbers α and β similarly can correspond to a ph or a hp combination. In the case that $\alpha > F > \beta$, one may write Eq. (13.30) as follows

$$\{\varepsilon_n^\pi - (\varepsilon_\alpha - \varepsilon_\beta)\} \mathcal{X}_{\alpha\beta}^n = \sum_{\epsilon\theta} \langle \alpha\beta^{-1} | V_{ph} | \epsilon\theta^{-1} \rangle \mathcal{X}_{\epsilon\theta}^n, \qquad (13.31)$$

whereas in the case $\alpha < F < \beta$ one has

$$\{\varepsilon_n^\pi + (\varepsilon_\beta - \varepsilon_\alpha)\} \mathcal{X}_{\alpha\beta}^n = -\sum_{\epsilon\theta} \langle \alpha\beta^{-1} | V_{ph} | \epsilon\theta^{-1} \rangle \mathcal{X}_{\epsilon\theta}^n. \qquad (13.32)$$

The normalization condition for each solution n (with $\varepsilon_n^\pi > 0$),

$$\sum_{\alpha>F>\beta} |\mathcal{X}_{\alpha\beta}^n|^2 - \sum_{\alpha<F<\beta} |\mathcal{X}_{\alpha\beta}^n|^2 = 1, \qquad (13.33)$$

can then be obtained in the usual fashion explained in Sec. 6.3.

Equations (13.31) and (13.32) together form a nonhermitian eigenvalue problem. In many cases it is appropriate to consider a truncated ph space with finite dimension D so that this eigenvalue problem has the double dimension 2D. The nonhermiticity has important consequences, since there is no guarantee that all eigenvalues ε_n^π will be real. In order to illustrate the circumstances of the appearance of complex eigenvalues and their physical relevance, it is quite instructive to consider the simplified case of a separable ph interaction.

Assume that the ph interaction is separable in the following way

$$\langle \alpha\beta^{-1} | V_{ph} | \epsilon\theta^{-1} \rangle = \lambda Q_{\alpha\beta} Q_{\epsilon\theta}^*, \qquad (13.34)$$

where λ is a coupling constant and $|Q_{\alpha\beta}| = |Q_{\beta\alpha}|$. Substitution of Eq. (13.34) into Eqs. (13.31) and (13.32) then yields

$$\{\varepsilon_n^\pi - (\varepsilon_\alpha - \varepsilon_\beta)\}\, \mathcal{X}_{\alpha\beta}^n = \lambda Q_{\alpha\beta} \sum_{\epsilon\theta} Q_{\epsilon\theta}^* \mathcal{X}_{\epsilon\theta}^n, \qquad (13.35)$$

for $\alpha > F > \beta$, and

$$\{\varepsilon_n^\pi + (\varepsilon_\beta - \varepsilon_\alpha)\}\, \mathcal{X}_{\alpha\beta}^n = -\lambda Q_{\alpha\beta} \sum_{\epsilon\theta} Q_{\epsilon\theta}^* \mathcal{X}_{\epsilon\theta}^n \qquad (13.36)$$

for the case $\alpha < F < \beta$. This implies that

$$\mathcal{X}_{\alpha\beta}^n = \mathcal{N} \frac{Q_{\alpha\beta}}{\varepsilon_n^\pi - (\varepsilon_\alpha - \varepsilon_\beta)} \qquad (13.37)$$

for $\alpha > F > \beta$, but

$$\mathcal{X}_{\alpha\beta}^n = -\mathcal{N} \frac{Q_{\alpha\beta}}{\varepsilon_n^\pi - (\varepsilon_\alpha - \varepsilon_\beta)}, \qquad (13.38)$$

when $\alpha < F < \beta$. The constant $\mathcal{N}$ reads

$$\mathcal{N} = \lambda \sum_{\epsilon\theta} Q_{\epsilon\theta}^* \mathcal{X}_{\epsilon\theta}^n, \qquad (13.39)$$

where the sum extends over both ph and hp combinations. One may now insert the solutions for the transition amplitudes given by Eqs. (13.37) and (13.38), into Eq. (13.39) to obtain

$$\frac{1}{\lambda} = \sum_{\epsilon > F > \theta} \frac{|Q_{\epsilon\theta}|^2}{\varepsilon_n^\pi - (\varepsilon_\epsilon - \varepsilon_\theta)} - \sum_{\epsilon < F < \theta} \frac{|Q_{\epsilon\theta}|^2}{\varepsilon_n^\pi - (\varepsilon_\epsilon - \varepsilon_\theta)}, \qquad (13.40)$$

after cancelling the common factors on both sides and dividing by λ. The only unknown quantities in this equation are the eigenvalues ε^π that yield its solution. The number of these solutions corresponds to 2D. The properties of the eigenvalues can be understood by plotting the right hand side of Eq. (13.40) as a function of the energy variable ε^π. Assuming, for illustration purposes, only 3 ph states, a corresponding plot is given in Fig. 13.7. The structure of this plot is very informative and follows a specific pattern. The location of the poles in Eq. (13.40) is denoted by the vertical asymptotes, which are located symmetrically about 0 at $\pm$ the ph energies. At negative energies, smaller than the location of the leftmost pole, the second sum in Eq. (13.40) dominates (since its poles are closer) and yields a positive contribution, which increases from zero to ∞ when the leftmost pole is reached. Beyond the first pole, but before the next one, the function

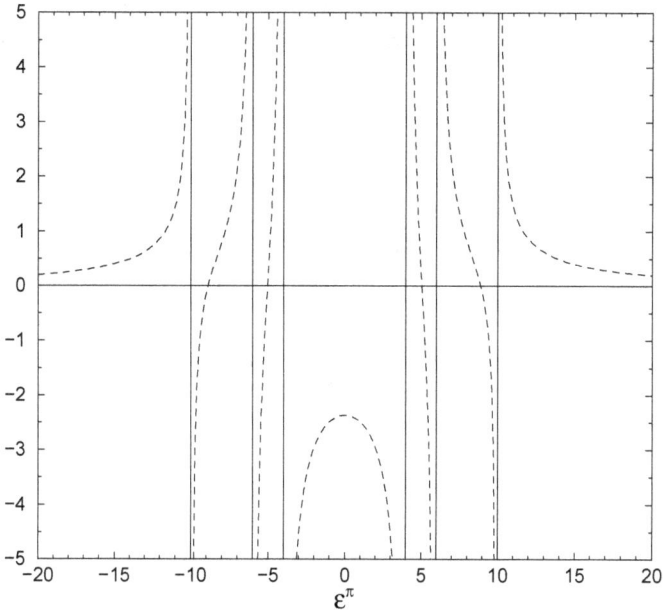

Fig. 13.7 Graphical representation of the right hand side of Eq. (13.40).

traverses all values, starting from $-\infty$ to ∞. Such behavior continues for any number of discrete poles, until one passes the last pole at negative energies. Towards the first pole at positive energies, the function approaches $-\infty$. After it and before the second one, the function goes from ∞ to $-\infty$, and so on. After the last pole the function is again positive definite, and approaches zero for large energies. One can also verify that the maximum of the function at zero energy is negative. The location of the eigenvalues is very easy to obtain graphically by drawing the horizontal straight line corresponding to $1/\lambda$, which represents the left side of Eq. (13.40). The eigenvalues are found at the energies where this straight line intersects with the dashed curve in Fig. 13.7. It is useful to distinguish between a repulsive ($\lambda > 0$) and attractive interaction ($\lambda < 0$). In the former case, the solutions at positive energy will initially be found between the unperturbed ph energies. These first D-1 solutions are therefore trapped between them. The last solution is found above the final ph energy. In the limit of a weak interaction, $1/\lambda$ becomes large and all solutions will tend to the original ph energies. If the interaction is very strong, the straight line representing

$1/\lambda$, will approach zero and the energy of the last solution may become very large. The latter state will have very collective features, to be discussed below in more detail.

For an attractive interaction, one also finds D-1 solutions in between the unperturbed ph energies. For a small value of $|\lambda|$ another solution will be found below the lowest unperturbed energy. However, if the strength of $|\lambda|$ is increased, this solution will tend to zero excitation energy. Increasing $|\lambda|$ further, the solution will disappear when the straight line representing $1/\lambda$, no longer crosses the dashed line between zero and the first ph energy. At this critical point two complex eigenvalues (a conjugate pair) will emerge. The appearance of these complex eigenvalues, identifies a characteristic instability that is inherent in the RPA eigenvalue problem. The resulting time evolution of these solutions has a component that will exponentially increase and cannot represent a true excited state. Indeed, it indicates that the ground state is unstable with respect to this type of collective excitation. The situation can often be repaired by considering the contribution of such collective excitations to the self-energy as discussed in Chs. 16 and 17.

The instability does not appear if the backward-going part of the unperturbed ph propagator is neglected in the eigenvalue problem. This simply corresponds to eliminating the summation over hp states in Eqs. (13.35), (13.36), and (13.40). In that case, the poles at negative energy in Fig. 13.7 disappear and the dashed line approaches zero from below, at negative energies. The lowest solution to the eigenvalue problem for an attractive interaction can then be found at a negative excitation energy, a physically unrealistic situation. This approximation to the polarization propagator is known as the Tamm–Dancoff approximation (TDA). It is completely equivalent to a diagonalization of the Hamiltonian in the chosen basis of ph states, as can be inferred from Eq. (13.35) or the more general Eq. (13.31). We note that the contribution of the backward-going terms becomes as important as the forward-going terms, when ε^π approaches zero, which is reflected in the values of the corresponding $\mathcal{X}$ coefficients.

The character of the RPA eigenvectors, associated with the solutions discussed above, can be inferred by considering Eqs. (13.37) and (13.38). While these solutions still require normalization, it is clear that for a weak interaction, when the eigenvalues remain close to the unperturbed ones, the eigenvector will be dominated by the corresponding unperturbed ph state belonging to that energy. In addition, the admixture of hp amplitudes is also small, since the corresponding amplitudes [see Eq. (13.38)] are even further removed in energy. These amplitudes are strictly zero for the unperturbed case.

Explicit results for the collective state can be generated when one takes the limit in which all unperturbed ph energies are degenerate. Defining

$$\mathcal{C} = \sum_{\epsilon > F > \theta} |Q_{\epsilon\theta}|^2 = \sum_{\epsilon < F < \theta} |Q_{\epsilon\theta}|^2 \qquad (13.41)$$

and denoting the degenerate ph energy by ε_{ph}, the eigenvalue problem simplifies to

$$\frac{1}{\lambda} = \mathcal{C} \left\{ \frac{1}{\varepsilon^\pi - \varepsilon_{ph}} - \frac{1}{\varepsilon^\pi + \varepsilon_{ph}} \right\}. \qquad (13.42)$$

Again, D-1 of the positive-energy solutions will remain trapped at $\varepsilon_n^\pi = \varepsilon_{ph}$. They have $\mathcal{X}_{\alpha\beta}^n = 0$ for $\alpha < F < \beta$, whereas the $\alpha > F > \beta$ amplitudes are orthogonal to the Q-amplitudes

$$\sum_{\alpha > F > \beta} Q_{\alpha\beta}^* \mathcal{X}_{\alpha\beta}^n = 0. \qquad (13.43)$$

The remaining solution has an energy ε_c^π corresponding to the positive root of Eq. (13.42),

$$\varepsilon_c^\pi = \left[2\lambda \mathcal{C} \varepsilon_{ph} + \varepsilon_{ph}^2 \right]^{1/2}. \qquad (13.44)$$

This collective state, which moves up or down from the unperturbed ph energy ε_{ph} according to the sign of λ, can have a very different energy from ε_{ph}, depending on whether $|\lambda|$ is large or not. Only in the case $\lambda < 0$ can an imaginary root appear, since $\mathcal{C}$ is a positive definite quantity. The amplitudes of the collective state are given by Eqs. (13.37) and (13.38)

$$\mathcal{X}_{ph}^c = \mathcal{N} \frac{Q_{ph}}{\varepsilon_c^\pi - \varepsilon_{ph}}; \quad \mathcal{X}_{hp}^c = -\mathcal{N} \frac{Q_{hp}}{\varepsilon_c^\pi + \varepsilon_{ph}}, \qquad (13.45)$$

where the normalization constant follows from Eq. (13.33) and can be worked out explicitly as

$$|\mathcal{N}| = \lambda \sqrt{\frac{\mathcal{C} \varepsilon_{ph}}{\varepsilon_c^\pi}}. \qquad (13.46)$$

The notion of a collective state can be further elaborated on by considering the transition probability for the operator $\hat{Q}$ that sets up the schematic interaction

$$\hat{Q} = \sum_{\alpha\beta} Q_{\alpha\beta} a_\alpha^\dagger a_{\bar\beta}. \qquad (13.47)$$

The corresponding transition amplitude to an excited state n reads

$$\langle \Psi_n^N | \hat{Q} | \Psi_0^N \rangle = \sum_{\alpha\beta} Q_{\alpha\beta} (\mathcal{X}_{\alpha\beta}^n)^*, \qquad (13.48)$$

and vanishes for a noncollective state, by virtue of Eq. (13.43). The transition probability to the collective state with energy ε_c^π in Eq. (13.44), is nonzero and given by

$$\left| \langle \Psi_c^N | \hat{Q} | \Psi_0^N \rangle \right|^2 = \frac{\varepsilon_{ph}}{\varepsilon_c^\pi} \mathcal{C}, \qquad (13.49)$$

resulting directly from Eqs. (13.46) and (13.39). In this extreme limit *all* the transition strength combines into one collective state. In particular in the case of strong attraction when ε_c^π approaches zero, one may have an extremely large transition probability. Nuclei with even numbers of protons and neutrons in the middle of major shells, exhibit enhancements of the quadrupole transition probability that exceed the sp estimate by more than the number of nucleons. Indeed, it is clear from Eq. (13.49) that the size of this constructive interference is not limited, in principle.

13.4 Energy-weighted sum rule

The RPA transition strengths can also be discussed in the context of the so-called energy-weighted sum rule (EWSR). For some operator $\hat{Q}$ this may be expressed by the following quantity

$$S(\hat{Q}) = \sum_n (E_n^N - E_0^N) \left| \langle \Psi_n^N | \hat{Q} | \Psi_0^N \rangle \right|^2. \qquad (13.50)$$

If $\hat{Q}$ is Hermitian, one can express the result as follows

$$S(\hat{Q}) = \frac{1}{2} \langle \Psi_0^N | \left[\hat{Q}, \left[\hat{H}, \hat{Q} \right] \right] | \Psi_0^N \rangle, \qquad (13.51)$$

upon replacing the energies occurring in Eq. (13.50) by the Hamiltonian in the appropriate places. Equation (13.51) expresses the EWSR in terms of the ground-state expectation value of a double commutator.

It is sometimes stated that RPA fulfills the EWSR; this holds in the sense that for a genuine HF+RPA calculation (with the same underlying tp interaction) the RPA result for Eq. (13.50) equals the expectation value of Eq. (13.51), evaluated with the HF ground state [Thouless (1972)]. For a hermitian operator $\hat{Q} = \sum_{\alpha\beta} Q_{\alpha\beta} a_\alpha^\dagger a_{\bar{\beta}}$, where $Q_{\alpha\beta} = \langle \alpha | Q | \bar{\beta} \rangle = -Q_{\bar{\beta}\bar{\alpha}}^*$, this implies that

$$\sum_{n>0} \varepsilon_n^\pi \left| \sum_{\alpha\beta} Q_{\alpha\beta} (\mathcal{X}_{\alpha\beta}^n)^* \right|^2 = \frac{1}{2} \langle \Phi_{HF}^N | \left[\hat{Q}, \left[\hat{H}, \hat{Q} \right] \right] | \Phi_{HF}^N \rangle , \qquad (13.52)$$

where the sum on the left side is over all positive energy RPA solutions.

In order to analyze this property, it is useful to rewrite the RPA eigen-value problem of Eqs. (13.31) and (13.32) in a standard matrix notation. One hereby prefers to have $\mathcal{X}_{\alpha\beta}^n$ amplitudes with $\alpha < F < \beta$ replaced by $\mathcal{Y}_{\alpha\beta}^n$ variables with $\alpha > F > \beta$, as introduced in Eq. (13.26). This allows to have only ph labels as matrix indices, and the eigenvalue problem in terms of $\mathcal{X}_{ph}^n$ and $\mathcal{Y}_{ph}^n$ amplitudes reads

$$\begin{pmatrix} A & B \\ B^* & A^* \end{pmatrix} \begin{pmatrix} \mathcal{X} \\ \mathcal{Y} \end{pmatrix} = \varepsilon^\pi \begin{pmatrix} 1 & 0 \\ 0 & -1 \end{pmatrix} \begin{pmatrix} \mathcal{X} \\ \mathcal{Y} \end{pmatrix}. \qquad (13.53)$$

Note that Eq. (13.4) has been used to write the equivalent form (13.53) of the RPA equations. The hermitian matrix A has elements

$$A_{ph,p'h'} = \delta_{p,p'} \delta_{h,h'} (\varepsilon_p - \varepsilon_h) + \langle p\bar{h'} | V | \bar{h}p' \rangle , \qquad (13.54)$$

whereas the symmetric matrix B is given by

$$B_{ph,p'h'} = \langle pp' | V | \bar{h}\,\bar{h'} \rangle . \qquad (13.55)$$

One can easily see that if $\{\varepsilon_n^\pi, \mathcal{X}^n, \mathcal{Y}^n\}$ is a solution of Eq. (13.53) with ε_n^π real, then so is $\{-\varepsilon_n^\pi, (\mathcal{Y}^n)^*, (\mathcal{X}^n)^*\}$; this reflects the structure of the polarization propagator in Eq. (13.28). Other properties of this type of eigenvalue problem are discussed more extensively in Sec. 19.3.1. For the present purpose we assume that the Hamiltonian matrix on the left side of Eq. (13.53) is positive-definite. This ensures that complex instabilities are absent and that all eigenvalues are real and come in pairs $\pm\varepsilon_n^\pi$. The corresponding eigenvectors can be normalized [following Eq. (13.33)] as

$$(\mathcal{X}^n)^\dagger \mathcal{X}^{n'} - (\mathcal{Y}^n)^\dagger \mathcal{Y}^{n'} = \delta_{n,n'} \mathrm{sgn}(\varepsilon_n^\pi). \qquad (13.56)$$

With this normalization convention, completeness of the solutions is ex-pressed as

$$\begin{pmatrix} 1 & 0 \\ 0 & -1 \end{pmatrix} = \sum_n \mathrm{sgn}(\varepsilon_n^\pi) \begin{pmatrix} \mathcal{X}^n \\ -\mathcal{Y}^n \end{pmatrix} \left((\mathcal{X}^n)^\dagger - (\mathcal{Y}^n)^\dagger \right), \qquad (13.57)$$

whereas the decomposition of the Hamiltonian matrix becomes

$$
\begin{pmatrix} A & B \\ B^* & A^* \end{pmatrix} = \sum_n |\varepsilon_n^\pi| \begin{pmatrix} \mathcal{X}^n \\ -\mathcal{Y}^n \end{pmatrix} \left((\mathcal{X}^n)^\dagger - (\mathcal{Y}^n)^\dagger \right).
\tag{13.58}
$$

Using the matrix notation introduced above, the RPA result for the left side of Eq. (13.52) can now be written as

$$
S(\hat{Q}) = \sum_{n>0} \varepsilon_n^\pi \left| \sum_{ph} Q_{ph}(\mathcal{X}_{ph}^n)^* + \sum_{ph} Q_{ph}^*(\mathcal{Y}_{ph}^n)^* \right|^2
$$

$$
= \sum_{n>0} \varepsilon_n^\pi \left| \left((\mathcal{X}^n)^\dagger - (\mathcal{Y}^n)^\dagger \right) \begin{pmatrix} Q \\ -Q^* \end{pmatrix} \right|^2,
\tag{13.59}
$$

and a straightforward (but tedious) calculation yields for the right side of Eq. (13.52)

$$
\langle \Phi_{HF}^N | \left[\hat{Q}, \left[\hat{H}, \hat{Q} \right] \right] | \Phi_{HF}^N \rangle
$$

$$
= \sum_{ph,p'h'} \left\{ 2Q_{ph}^* Q_{p'h'} A_{ph,p'h'} - Q_{ph} Q_{p'h'} B_{ph,p'h'}^* - Q_{ph}^* Q_{p'h'}^* B_{ph,p'h'} \right\}
$$

$$
= \left(Q^\dagger - Q^T \right) \begin{pmatrix} A & B \\ B^* & A^* \end{pmatrix} \begin{pmatrix} Q \\ -Q^* \end{pmatrix}.
\tag{13.60}
$$

Note that Brillouin's theorem (see Sec. 10.1.3) can be used to omit terms of the form $\langle \Phi_0^{HF} | \hat{H} a_p^\dagger a_{\bar{h}} | \Phi_0^{HF} \rangle$. Inserting the decomposition (13.58) into Eq. (13.60) one arrives immediately at the conclusion that the sum rule equality of Eq. (13.50) and Eq. (13.51) is fulfilled in RPA, at the level of Eq. (13.52). Note that this is not the case for the TDA result, which misses the terms involving the B matrix in the HF result of Eq. (13.60).

In some cases the double commutator appearing in Eq. (13.51) is a simple operator or even a number, and the EWSR becomes a model-independent quantity. An example is provided by the electric dipole operator in the absence of velocity-dependent interactions. For atoms, it leads to the Thomas–Reiche–Kuhn (TRK) or dipole sum rule, which arises as follows. The wavelength of the electromagnetic radiation emitted or absorbed in atomic transitions is typically much larger than the size of individual atoms. As a consequence, one can ignore the spatial dependence of the vector potential $\boldsymbol{A}(\boldsymbol{r}) \sim e^{i\boldsymbol{k}\cdot\boldsymbol{r}} \boldsymbol{\epsilon}$, where $\boldsymbol{\epsilon}$ is the polarization vector, and approximate $\boldsymbol{A}(\boldsymbol{r}) \sim \boldsymbol{\epsilon}$. This is known as the *dipole approximation*. The transition operator coupling the vector potential to the convection current of the electrons,[4] is in general given by

$$H_{int} \sim (-e) \sum_{i=1}^{Z} \frac{1}{2} \{ \mathbf{A}(\mathbf{r}_i) \cdot \mathbf{v}_i + \mathbf{v}_i \cdot \mathbf{A}(\mathbf{r}_i) \}, \tag{13.61}$$

where $\mathbf{v}_i = \frac{i}{\hbar}[H, \mathbf{r}_i]$ is the velocity operator. In the dipole approximation, the electromagnetic transition operator can therefore simply be written as

$$H_{int} \sim \frac{i}{\hbar} \boldsymbol{\epsilon} \cdot [H, \mathbf{D}], \tag{13.62}$$

in terms of the dipole operator $\mathbf{D} = (-e) \sum_{i=1}^{Z} \mathbf{r}_i$, and an atomic transition probability from state i to state f is characterized by the (dimensionless) *oscillator strength* parameter

$$F_{fi} = \frac{2m}{3\hbar^2 e^2} (E_f - E_i) |\langle f | \mathbf{D} | i \rangle|^2. \tag{13.63}$$

The summed oscillator strength (determining *e.g.* the total photoabsorption cross section) is recognized as an EWSR and can be expressed as

$$\sum_f F_{fi} = \frac{m}{3\hbar^2 e^2} \langle i | [\mathbf{D}, \cdot [H, \mathbf{D}]] | i \rangle. \tag{13.64}$$

In the case of atoms, velocity-dependent interactions are not present. The double commutator in Eq. (13.64) only receives contributions from the kinetic term in the Hamiltonian, and in fact becomes a constant since $[\mathbf{r}, \cdot [\nabla^2, \mathbf{r}]] = -6$. As a consequence, the sum rule for the total dipole strength reads

$$TRK = \sum_f F_{fi} = Z, \tag{13.65}$$

where Z is the number of electrons, and is known as the Thomas–Reiche–Kuhn sum rule. The TRK sum rule (experimentally confirmed for atoms, see *e.g.* [Piech (1964)]) is fulfilled in RPA calculations. It holds in general for any system of pointlike charges interacting with velocity-independent forces; a similar sum rule in the nuclear case receives corrections due to mesonic exchange currents.

---

[4]The current related to the intrinsic magnetic moment of the electrons can be neglected in the dipole approximation.

13.5   Excited states in atoms

The bound excited states in atoms do not have the strong collective structure discussed in Sec. 13.3, but can usually be described as rather pure ph excitations. However, a "collective" feature does appear in the form of an important change in the mean field for the sp states above the Fermi level [Amusia (1974); Amusia (1990)]. The sp energies appearing in the RPA equations (13.53) for a neutral atom are the HF sp energies corresponding to the neutral atom [see Eq. (10.39)]

$$\delta_{p,p'}\varepsilon_p = \langle p|\, T\, |p'\rangle + \sum_{h''} \langle p\overline{h''}|\, V\, |p'\overline{h''}\rangle. \tag{13.66}$$

As mentioned in Sec. 10.2.1, the HF potential for the unoccupied sp states in a neutral atom is too short-ranged and does not support bound particle orbitals. The HF ph spectrum is consequently also purely continuous and does not support Rydberg series of bound excited states. This unsatisfactory situation is cured in both the RPA and TDA, where a series of discrete excited states does appear. To see this in detail, one must consider the time-forward diagrams represented by the A-block in the RPA matrix equation. The diagonal sub-block, characterized by ph excitations with the same hole orbital, reads

$$\begin{aligned} A_{ph,p'h} &= \delta_{p,p'}(\varepsilon_p - \varepsilon_h) - \langle p\overline{h}|\, V\, |p'\overline{h}\rangle \\ &= -\delta_{p,p'}\varepsilon_h + \langle p|\, T\, |p'\rangle + \sum_{h''\neq h} \langle p\overline{h''}|\, V\, |p'\overline{h''}\rangle. \end{aligned} \tag{13.67}$$

This matrix can be diagonalized with a unitary transformation among the unoccupied orbitals

$$|p^{(h)}\rangle = \sum_{p'} U^{(h)}_{p,p'}\, |p'\rangle, \tag{13.68}$$

such that

$$\langle p^{(h)}|T|p'^{(h)}\rangle + \sum_{h''\neq h} \langle p^{(h)}\overline{h''}|V|p'^{(h)}\overline{h''}\rangle = \delta_{p,p'}\varepsilon_p^{(h)}, \tag{13.69}$$

hereby defining new sp energies $\varepsilon_p^{(h)}$ for the unoccupied orbitals. Obviously, this is equivalent to the sp eigenstates of a modified mean field, sometimes called the V^{N-1} potential, obtained by omitting in the HF mean field the contribution of the hole state under consideration. While in some cases (like in heavy nuclei) the removal of one particle hardly matters for the HF mean field, the change is crucial for electronic systems like atoms: the V^{N-1} potential is ion-like, with a potential that decays like $1/r$ and therefore supports a Rydberg series of unoccupied states, in addition to a continuum. An important part of the electronic correlations (the sum of all time-forward diagrams with the same hole orbital) can then be included simply by changing the basis of the unoccupied states. Expressed in terms of the $p^{(h)}h^{-1}$ basis the RPA matrix in Eq. (13.53) becomes

$$A_{ph,p'h'} = \delta_{p,p'}\delta_{h,h'}(\varepsilon_p^{(h)} - \varepsilon_h) + (1 - \delta_{h,h'})\langle p^{(h)}h^{-1}|V_{ph}|p'^{(h')}h'^{-1}\rangle$$
$$B_{ph,p'h'} = \langle p^{(h)}p'^{(h')}|V|\overline{h}\,\overline{h'}\rangle, \tag{13.70}$$

where in the A-matrix the interaction term must be omitted for $h = h'$ to avoid double-counting. Note that the eigenstates in Eq. (13.69) should be found in the subspace orthogonal to all occupied HF states, so that the modified mean-field operator in coordinate space should be supplemented with a projection operator. This is not needed when only one hole state with the relevant quantum numbers is present, since the $p^{(h)}$ states are automatically orthogonal to the h orbital.

As an example of this procedure we consider the lowest excited states in neon (see also Fig. 3.4), which involve a promotion of a $2p$ electron to an unoccupied orbit. In Fig. 13.8 the experimental levels are compared with the modified mean-field ph spectrum $\varepsilon_p^{(h)} - \varepsilon_h$, where (in atomic units) $\varepsilon_h = -0.850$ is the HF $2p$ energy. The $\varepsilon_p^{(h)}$ are obtained as eigenvalues of the HF mean field, where the contribution of one of the $2p$ electrons has been omitted. In order to keep spherical symmetry an average has been taken of the m_ℓ and m_s quantum numbers of the omitted $2p$ electron; this amounts to changing, from a value of 3 to 2.5, the degeneracy factors $(2\ell + 1)$ in Eq. (10.80) and $(2\ell' + 1)$ in Eq. (10.89) in the contribution of the $2p$ orbital to the Hartree and the Fock field. The bound sp energies for the unoccupied $3s, 3p, 4s, 3d$ and $4p$ levels that are generated in this way, already lead to a reasonable ph spectrum, with the right spacing as compared to the experimental spectrum. The agreement becomes even better when shifting the spectrum over $\Delta = .057$, which is equivalent to replacing the $2p$ HF energy with the experimental ionization energy[5] $\varepsilon_{2p}^{exp} = -0.793$. Finally, the degeneracy can be lifted by calculating the interaction energy

---

[5]The shift from the HF value to the experimental ionization energy should arise dynamically in a correct description of the self-energy. The second-order self-energy diagram in Sec. 11.5.3 generates a shift of about the right magnitude, but somewhat overshoots the experimental value (see Table 11.1). Still lacking is the incorporation of the present RPA correlations into the self-energy.

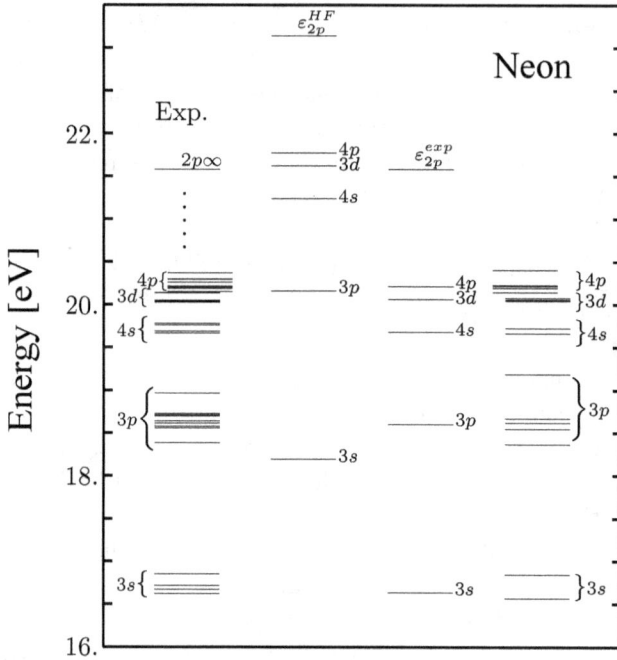

Fig. 13.8 The lowest energy levels of the Ne atom can be interpreted as the promotion of a $2p$ electron to the empty orbitals $3s, 3p, 4s, 3d, 4p...$ On the left is the experimental spectrum, taken from [Martin *et al.* (2002)]. The second column shows the modified HF ph spectrum, which is shifted in the third column so as to reproduce the experimental ionization energy. On the right the Coulomb interaction energy is added for the different LS configurations.

in each LS configuration. Note that in Eq. (13.70) the terms correcting for the omitted m_ℓ, m_s value, break the rotational symmetry. To be consistent one must take a similar spherical average, leading to the following expression

$$E_{LS} = \varepsilon_p^{(h)} - \varepsilon_h + \langle p^{(h)}h^{-1}LS|V_{ph}|p^{(h)}h^{-1}LS \rangle \qquad (13.71)$$

$$- \sum_{L'S'} \frac{(2L'+1)(2S'+1)}{4(2\ell_h+1)(2\ell_p+1)} \langle p^{(h)}h^{-1}L'S'|V_{ph}|p^{(h)}h^{-1}L'S' \rangle.$$

The ph matrix elements in Eq. (13.71) are coupled to total orbital angular momentum L and spin S; their explicit expression will be given in Sec. 13.7.

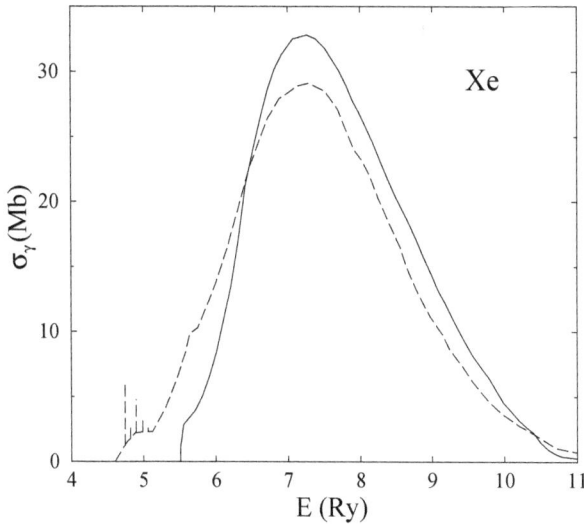

Fig. 13.9 Photoabsorption of xenon, adapted from [Amusia (1996)]. Dashed curve is the experimental cross section, the solid curve is the RPA prediction.

The additional splitting, observed in the experimental spectrum of Fig. 13.8, is due to spin-orbit coupling terms and involves configurations with good total angular momentum J.

As a final remark we note that RPA collectivity does lead to important effects when transitions to the continuum are considered. An extensive overview on ionization cross sections and atomic giant resonances can be found in [Amusia (1990)]. An example is provided by the photoabsorption spectrum of xenon in the threshold region for ionization of the $4d^{10}$ electrons, shown in Fig. 13.9. The HF prediction for the width of the giant resonance peak would be too broad; solving the continuum RPA equations leads to a concentration of the strength in reasonable agreement with the experimental cross section.

13.6   Correlation energy and ring diagrams

In Ch. 7, we have related the energy of the correlated ground state to the sp propagator by Eq. (7.26). Alternative ways to express the energy of the ground state exist. We will study one of these in the present section, since this method has been used to determine the contribution of ring diagrams to the energy of the ground state. We start with deriving an expression that is known as the Hellman–Feynman theorem, but was apparently first derived by Pauli [Pines (1962)]. Write the Hamiltonian as a function of a variable coupling constant

$$\hat{H}(\lambda) = \hat{H}_0 + \lambda \hat{H}_1 \qquad (13.72)$$

so that $\hat{H}(1) = \hat{H}$ and $\hat{H}(0) = \hat{H}_0$. We assume that for each λ, it is possible to solve

$$\hat{H}(\lambda) \left| \Psi_0^N(\lambda) \right\rangle = E_0^N(\lambda) \left| \Psi_0^N(\lambda) \right\rangle, \qquad (13.73)$$

where the eigenstate is normalized according to

$$\left\langle \Psi_0^N(\lambda) \middle| \Psi_0^N(\lambda) \right\rangle = 1. \qquad (13.74)$$

From Eq. (13.73) it is clear that

$$E_0^N(\lambda) = \left\langle \Psi_0^N(\lambda) \middle| \hat{H}(\lambda) \middle| \Psi_0^N(\lambda) \right\rangle. \qquad (13.75)$$

Taking the derivative with respect to λ yields

$$\frac{dE_0^N(\lambda)}{d\lambda} = \left\{ \frac{d}{d\lambda} \left\langle \Psi_0^N(\lambda) \middle| \right\} \hat{H}(\lambda) \middle| \Psi_0^N(\lambda) \right\rangle$$
$$+ \left\langle \Psi_0^N(\lambda) \middle| \hat{H}_1 \middle| \Psi_0^N(\lambda) \right\rangle + \left\langle \Psi_0^N(\lambda) \middle| \hat{H}(\lambda) \left\{ \frac{d}{d\lambda} \middle| \Psi_0^N(\lambda) \right\rangle \right\}$$
$$= \left\langle \Psi_0^N(\lambda) \middle| \hat{H}_1 \middle| \Psi_0^N(\lambda) \right\rangle, \qquad (13.76)$$

since the other two terms combine to

$$E_0^N(\lambda) \frac{d}{d\lambda} \left\langle \Psi_0^N(\lambda) \middle| \Psi_0^N(\lambda) \right\rangle = 0, \qquad (13.77)$$

on account of Eq. (13.74). By integrating the final expression in Eq. (13.76) from 0 to 1, we find

$$E_0^N - E_{\Phi_0^N} = \int_0^1 \frac{d\lambda}{\lambda} \left\langle \Psi_0^N(\lambda) \middle| \lambda \hat{H}_1 \middle| \Psi_0^N(\lambda) \right\rangle, \qquad (13.78)$$

which is the desired result. Equation (13.78) expresses the shift in the ground-state energy in terms of the interaction Hamiltonian $\lambda \hat{H}_1$. This implies that Eq. (13.78) is only used in special situations, *e.g.* when analytic results can be obtained, since the many-body problem must be solved for all values[6] of λ. The expectation value of the potential can also be related to the polarization propagator of Sec. 13.1 by writing

$$\langle \Psi_0^N | \hat{V} | \Psi_0^N \rangle = \frac{1}{4} \sum_{\alpha \bar{\beta} \bar{\gamma} \delta} \langle \alpha \bar{\beta} | V | \bar{\gamma} \delta \rangle \langle \Psi_0^N | a_\alpha^\dagger a_{\bar{\beta}}^\dagger a_\delta a_{\bar{\gamma}} | \Psi_0^N \rangle \tag{13.79}$$

$$= \frac{1}{4} \sum_{\alpha \bar{\beta} \bar{\gamma} \delta} \langle \alpha \bar{\beta} | V | \bar{\gamma} \delta \rangle \left\{ \langle \Psi_0^N | a_{\bar{\beta}}^\dagger a_\delta a_\alpha^\dagger a_{\bar{\gamma}} | \Psi_0^N \rangle - \delta_{\alpha\delta} \langle \Psi_0^N | a_{\bar{\beta}}^\dagger a_{\bar{\gamma}} | \Psi_0^N \rangle \right\},$$

where time-reversed states have been employed (not required) in the summations to prepare for the development in Sec. 14.4.1. For the second equality in Eq. (13.79), we moved the $a_\alpha^\dagger$ operator to the right of a_δ in the expectation values. By using the completeness of the states with N particles, the first expectation value in the last line of Eq. (13.79) can be written as

$$\langle \Psi_0^N | a_{\bar{\beta}}^\dagger a_\delta a_\alpha^\dagger a_{\bar{\gamma}} | \Psi_0^N \rangle = \sum_n \langle \Psi_0^N | a_{\bar{\beta}}^\dagger a_\delta | \Psi_n^N \rangle \langle \Psi_n^N | a_\alpha^\dagger a_{\bar{\gamma}} | \Psi_0^N \rangle \tag{13.80}$$

$$= -\frac{1}{\pi} \int_0^\infty dE \ \text{Im} \ \sum_{n \neq 0} \frac{\langle \Psi_0^N | a_{\bar{\beta}}^\dagger a_\delta | \Psi_n^N \rangle \langle \Psi_n^N | a_\alpha^\dagger a_{\bar{\gamma}} | \Psi_0^N \rangle}{E - (E_n^N - E_0^N) + i\eta} + n_{\bar{\beta}\delta} n_{\alpha\bar{\gamma}},$$

where the one-body density matrix elements were defined in Eq. (7.19). Inserting Eq. (13.80) in Eq. (13.79) and using the definition in Eq. (13.20), we obtain the relation between the potential energy and the polarization propagator, given in Eq. (13.9), in the following form

$$\langle \Psi_0^N | \hat{V} | \Psi_0^N \rangle = \frac{1}{4} \sum_{\alpha \bar{\beta} \bar{\gamma} \delta} \langle \alpha \gamma^{-1} | V_{ph} | \delta \beta^{-1} \rangle \tag{13.81}$$

$$\times \left\{ -\frac{1}{\pi} \int_0^\infty dE \ \text{Im} \ \Pi(\delta, \beta^{-1}; \alpha, \gamma^{-1}; E) + n_{\bar{\beta}\delta} n_{\alpha\bar{\gamma}} - \delta_{\alpha\delta} n_{\bar{\beta}\bar{\gamma}} \right\}.$$

Equation (13.78) then allows us to rewrite this result in terms of a coupling-constant integration. Application of Eq. (13.81) normally entails the RPA in describing Π. Since the RPA sums all ring diagrams, Eq. (13.81) shows that the corresponding energy shift contains all closed ring diagrams, on account of the presence of the V_{ph} matrix element. The correlation energy of the electron gas in the RPA is discussed in Sec. 14.4.1.

---

[6] In practical situations the integral over λ is discretized, requiring only a finite number of λ-values.

13.7 RPA in angular momentum coupled representation

It is useful to work out the RPA Eqs. (13.53) for the important case of rotationally invariant interactions and a spherical system, with ground state $|\Psi_0^N\rangle$ having total angular momentum $J = 0$. This applies *e.g.* to closed-shell nuclei and atoms. The general sp quantum number takes on the form $\alpha = am_a$, where m_a denotes the third component of the sp angular momentum j_a, and a the remaining quantum numbers. An uncoupled ph excitation operator can be written as

$$A^\dagger_{\alpha\beta} = a^\dagger_\alpha b^\dagger_\beta \qquad (13.82)$$

with $b^\dagger_\beta = (-1)^{j_b + m_b} a_{b,-m_b}$. We note that both $a^\dagger$ and $b^\dagger$ behave under rotation as spherical tensor operators (see App. B). As a result, there is no difficulty in coupling the ph excitation to good total angular momentum, yielding

$$A^\dagger_{ab}(JM) = \sum_{m_a m_b} (j_a m_a j_b m_b | JM) A^\dagger_{\alpha\beta} = [a^\dagger_a \otimes b^\dagger_b]^J_M. \qquad (13.83)$$

The coupled version of the polarization propagator in Eq. (13.9) is likewise obtained as

$$\Pi(ab^{-1}; cd^{-1}; JM) \qquad (13.84)$$
$$= \sum_{m_a m_b m_c m_d} (j_a m_a j_b m_b | JM)(j_c m_c j_d m_d | JM)\Pi(\alpha, \beta^{-1}; \gamma, \delta^{-1}; E).$$

It should be clear that only excited states with total angular momentum J contribute to the polarization propagator (13.84), which can be rewritten as

$$\Pi(ab^{-1}; cd^{-1}; JM) \qquad (13.85)$$
$$= \sum_{n\neq 0} \frac{\langle \Psi^N_{nJM} | A^\dagger_{ab}(JM) | \Psi^N_0 \rangle^* \langle \Psi^N_{nJM} | A^\dagger_{cd}(JM) | \Psi^N_0 \rangle}{E - (E^N_{nJ} - E^N_0) + i\eta}$$
$$- \sum_{n\neq 0} \frac{\langle \Psi^N_0 | A^\dagger_{ab}(JM) | \Psi^N_{nJ-M} \rangle^* \langle \Psi^N_0 | A^\dagger_{cd}(JM) | \Psi^N_{nJ-M} \rangle}{E + (E^N_{nJ} - E^N_0) + i\eta}.$$

The summation index n in Eq. (13.85) does not include the M values of the $(2J+1)$-fold degenerate multiplets. Defining the coupled version of the

RPA amplitudes in Eq. (13.25) and (13.26) as

$$
\mathcal{X}_{ab}^{nJ} = \left\langle \Psi_{nJM}^N \middle| A_{ab}^\dagger(JM) \middle| \Psi_0^N \right\rangle^* = \frac{1}{\sqrt{2J+1}} \left\langle \Psi_{nJ}^N \middle\| [a_a^\dagger \otimes b_b^\dagger]^J \middle\| \Psi_0^N \right\rangle^*,
$$

$$
\mathcal{Y}_{ab}^{nJ} = (-1)^{J-M} \left\langle \Psi_0^N \middle| A_{ab}^\dagger(JM) \middle| \Psi_{nJ-M}^N \right\rangle
$$

$$
= \frac{1}{\sqrt{2J+1}} \left\langle \Psi_0^N \middle\| [a_a^\dagger \otimes b_b^\dagger]^J \middle\| \Psi_{nJ}^N \right\rangle, \tag{13.86}
$$

the RPA polarization propagator, describing excitations with angular momentum J, reads

$$
\Pi^{RPA}(ab^{-1}; cd^{-1}; J; E) = \sum_{n \neq 0} \left(\frac{\mathcal{X}_{ab}^{nJ}(\mathcal{X}_{cd}^{nJ})^*}{E - \varepsilon_{nJ}^\pi + i\eta} - \frac{(\mathcal{Y}_{ab}^{nJ})^* \mathcal{Y}_{cd}^{nJ}}{E + \varepsilon_{nJ}^\pi - i\eta} \right). \tag{13.87}
$$

Note that the $\mathcal{X}$ and $\mathcal{Y}$ amplitudes in Eq. (13.86) are independent of M. They can be expressed in terms of reduced matrix elements, and are related by a phase

$$
\mathcal{Y}_{ab}^{nJ} = (-1)^{j_a - j_b}(-1)^J \mathcal{X}_{ba}^{nJ}. \tag{13.88}
$$

The noninteracting polarization propagator of Eq. (13.16) in the coupled representation simply becomes

$$
\Pi^{(0)}(ab^{-1}; cd^{-1}; J; E) \tag{13.89}
$$

$$
= \delta_{ac}\delta_{bd}\Pi^{(0)}(ab^{-1}; J; E) = \frac{\theta(a > F > b)}{E - \varepsilon_a + \varepsilon_b + i\eta} - \frac{\theta(a < F < b)}{E - \varepsilon_a + \varepsilon_b - i\eta},
$$

and does not depend explicitly on J [with the understanding that (j_a, j_b, J) must fulfill the triangle inequalities]. Applying the transformation (13.84) to Eq. (13.24), we also require the coupled version of the ph interaction

$$
\left\langle ab^{-1}JM \middle| V_{ph} \middle| cd^{-1}J'M' \right\rangle \tag{13.90}
$$

$$
= \sum_{m_a m_b m_c m_d} (j_a m_a j_b m_b | JM)(j_c m_c j_d m_d | J'M') \left\langle \alpha\beta^{-1} \middle| V_{ph} \middle| \gamma\delta^{-1} \right\rangle.
$$

With the definition in Eq. (13.20), the tp states in the conventional interaction matrix elements can now be coupled to good total angular momentum

$$
\left\langle \alpha\bar{\delta} \middle| V \middle| \bar{\beta}\gamma \right\rangle = \sum_{J_p M_p} (-1)^{j_b + m_b}(-1)^{j_d + m_d} \tag{13.91}
$$

$$
\times (j_a m_a j_d - m_d | J_p M_p)(j_b - m_b j_c m_c | J_p M_p) \left\langle ad\, J_p \middle| V \middle| bc\, J_p \right\rangle,
$$

where rotational invariance of the interaction has been used. Substitution of Eq. (13.91) into Eq. (13.90) results in

$$\langle ab^{-1}JM | V_{ph} | cd^{-1}J'M' \rangle = \sum_{J_p} C(abcd, JM, J'M', J_p) \langle ad J_p | V | bc J_p \rangle .$$

(13.92)

The C-factor is recognized as a geometrical recoupling coefficient

$$C(abcd, JM, J'M', J_p) = \sum_{m_a m_b m_c m_d M_p} (-1)^{j_b + m_b + j_d + m_d} (j_a m_a j_b m_b | JM)$$

$$\times (j_c m_c j_d m_d | J'M')(j_a m_a j_d - m_d | J_p M_p)(j_b - m_b j_c m_c | J_p M_p), \text{ (13.93)}$$

involving a 6j-symbol (see App. B), and is nonzero only for $J = J'$ and $M = M'$. The latter statement reflects that the ph interaction is a rotational scalar, and can only connect ph configurations with the same total angular momentum. The formula is known as the Pandya relation [Pandya (1956)]

$$\langle ab^{-1}J | V_{ph} | cd^{-1}J \rangle$$

$$= \sum_{J_p} (2J_p + 1)(-1)^{j_b + j_c + J_p} \begin{Bmatrix} j_a & j_b & J \\ j_c & j_d & J_p \end{Bmatrix} \langle ad J_p | V | bc J_p \rangle , \quad (13.94)$$

which expresses the interaction matrix elements between ph configurations in terms of the conventional tp matrix elements, in an angular momentum coupled representation.

After these preliminaries it is now straightforward to derive the RPA equation of Eq. (13.24) for the coupled polarization propagator

$$\Pi^{RPA}(ab^{-1}; cd^{-1}; J; E) = \delta_{ac}\delta_{bd}\Pi^{(0)}(ab^{-1}; J; E)$$

(13.95)

$$+ \Pi^{(0)}(ab^{-1}; J; E) \sum_{ef} \langle ab^{-1}J | V_{ph} | ef^{-1}J \rangle \Pi^{RPA}(ef^{-1}; cd^{-1}; J; E).$$

The RPA eigenvalue equation determining the amplitudes [see Eqs. (13.31) and (13.32)] becomes

$$\{\varepsilon_{nJ}^{\pi} - \varepsilon_{ph}\} \mathcal{X}_{ph}^{nJ} = \sum_{cd} \langle ph^{-1}J | V_{ph} | cd^{-1}J \rangle \mathcal{X}_{cd}^{nJ}$$

$$\{\varepsilon_{nJ}^{\pi} + \varepsilon_{ph}\} \mathcal{X}_{hp}^{nJ} = -\sum_{cd} \langle hp^{-1} | V_{ph} | cd^{-1} \rangle \mathcal{X}_{cd}^{nJ}, \quad (13.96)$$

with either $c > F > d$ or $c < F < d$. To make the symmetry of the RPA equations more transparant it is again customary to replace the $\mathcal{X}_{hp}^{nJ}$ with the $\mathcal{Y}_{ph}^{nJ}$ amplitudes, defined in Eq. (13.86), and keep only ph labels as matrix indices. The final result in matrix notation has the usual form of Eq. (13.53), with hermitian A and symmetric B submatrices given by

$$A_{ph,p'h'} = \delta_{p,p'}\delta_{h,h'}(\varepsilon_p - \varepsilon_h) + \langle ph^{-1}J| V_{ph} |p'h'^{-1}J\rangle, \qquad (13.97)$$

$$B_{ph,p'h'} = (-1)^{j_{p'}-j_{h'}}(-1)^{J} \langle ph^{-1}J| V_{ph} |h'p'^{-1}J\rangle. \qquad (13.98)$$

The above form of the RPA equations is directly applicable to the case of double-closed shell nuclei, where $j - j$ coupling is the most relevant (see Sec. 4.2). For light atoms it is more appropriate to use sp quantum numbers $\alpha = a, l_a m_{l_a}, (1/2)m_{s_a}$, and the $L - S$ coupling scheme. Since we need a separate recoupling of the orbital angular momenta and spins, the Pandya relation (13.94) must be used twice

$$\langle abf^{-1}LS| V_{ph} |cd^{-1}LS\rangle = \sum_{L_p}(2L_p + 1)(-1)^{l_b+l_c+L_p}\begin{Bmatrix} l_a & l_b & L \\ l_c & l_d & L_p \end{Bmatrix}$$

$$\times \sum_{S_p}(2S_p + 1)(-1)^{1+S_p}\begin{Bmatrix} 1/2 & 1/2 & S \\ 1/2 & 1/2 & S_p \end{Bmatrix}\langle adL_pS_p| V |bcL_pS_p\rangle. \quad (13.99)$$

Due to the spin independence of the Coulomb force, the matrix element between antisymmetric and LS-coupled tp states reads

$$\langle adL_pS_p| V |bcL_pS_p\rangle = (adL_p| V |bcL_p) \qquad (13.100)$$
$$- (-1)^{l_b+l_c+L_p}(-1)^{1+S_p} (adL_p| V |cbL_p),$$

in terms of the direct spatial matrix elements. Substituting the numerical values for the spin 6j symbols

$$\begin{Bmatrix} 1/2 & 1/2 & S \\ 1/2 & 1/2 & S_p \end{Bmatrix} \quad \begin{array}{l} (S,S_p)= \ (0,0) \ (0,1) \ (1,0) \ (1,1) \\ \Rightarrow \quad -1/2 \ \ 1/2 \ \ 1/2 \ \ 1/6 \end{array} \qquad (13.101)$$

the ph matrix element in Eq. (13.99) becomes

$$\langle ab^{-1}LS| V_{ph} |cd^{-1}LS\rangle = \sum_{L_p}(2L_p + 1)\begin{Bmatrix} l_d & l_c & L \\ l_b & l_a & L_p \end{Bmatrix} \qquad (13.102)$$

$$\times \left[2\delta_{S,0}(-1)^{l_b+l_c+L_p} (adL_p| V |bcL_p) - (adL_p| V |cbL_p) \right].$$

Explicitly, the spatial matrix element of the Coulomb force reads

$$(abL_p| V |cdL_p) = \int \frac{d\mathbf{r}_1\, d\mathbf{r}_2}{|\mathbf{r}_1 - \mathbf{r}_2|} \varphi_a(r_1)\varphi_c(r_1)\varphi_b(r_2)\varphi_d(r_2) \quad (13.103)$$

$$\times \left([Y_{l_a}(\Omega_1) \otimes Y_{l_b}(\Omega_2)]_{M_p}^{L_p} \right)^* [Y_{l_c}(\Omega_1) \otimes Y_{l_d}(\Omega_2)]_{M_p}^{L_p},$$

and with the expansion (10.82), can be worked out further as

$$(abL_p| V |cdL_p) = \sum_{L'} (-1)^{L_p+L'} \begin{Bmatrix} l_a & l_b & L_p \\ l_d & l_c & L' \end{Bmatrix} R(ac, bd, L'). \quad (13.104)$$

in terms of the coefficients

$$R(ac, bd, L) = \sqrt{2l_a+1}\sqrt{2l_b+1}\sqrt{2l_c+1}\sqrt{2l_d+1} \begin{pmatrix} l_a & l_c & L \\ 0 & 0 & 0 \end{pmatrix} \begin{pmatrix} l_b & l_d & L \\ 0 & 0 & 0 \end{pmatrix} \quad (13.105)$$

$$\times \int dr_1 r_1^2 \int dr_2 r_2^2 \frac{r_<^L}{r_>^{L+1}} \varphi_a(r_1)\varphi_c(r_1)\varphi_b(r_2)\varphi_d(r_2).$$

Exploiting the properties of the 6j symbols in App. B, the final result for Eq. (13.102) becomes quite simple

$$\langle ab^{-1}LS| V_{ph} |cd^{-1}LS \rangle = 2\delta_{S,0} \frac{(-1)^{l_b+l_d}}{2L+1} R(ab, cd, L) \quad (13.106)$$

$$- \sum_{L'} (-1)^L \begin{Bmatrix} l_c & l_d & L \\ l_b & l_a & L' \end{Bmatrix} R(ac, db, L').$$

Note that the quantities in Eq. (13.105) have the symmetries $R(ac, bd, L) = R(ca, bd, L) = R(bd, ac, L)$; the radial integrals are known as Slater integrals in atomic structure calculations.

13.8   Exercises

(1) Determine the normalization condition for the $\mathcal{X}$ amplitudes, given in Eq. (13.33), that solve the RPA eigenvalue problem.
(2) Solve the RPA eigenvalue problem numerically, employing a separable interaction, as discussed in Sec. 13.3. Vary the number of ph states and their energies, to obtain insight into the properties of this eigenvalue problem. Use the normalization condition, given in Eq. (13.33), to study the amplitudes of the collective states (adjust λ appropriately), both for an attractive and repulsive interaction.
(3) After numerically constructing (see Sec. 10.2) the HF mean field for the neon atom, generate the neon spectrum in Fig. 13.8 using the steps outlined in Sec. 13.5.
(4) Reformulate Eq. (13.81) in the form that employs a coupling constant integration.
(5) Use the angular momentum algebra of App. B to derive the expressions in Sec. 13.7.

Chapter 14

Excited states in infinite systems

The study of excited states continues in this chapter, by considering infinite systems. In Sec. 14.1, the RPA integral equation is introduced with special attention to the correct coupling of the ph states to good total spin (isospin). The calculation of the noninteracting polarization propagator is presented in Sec. 14.2. This so-called Lindhard function [Lindhard (1954)] exhibits various properties that are helpful in assessing the features of excitations in infinite systems. The important collective state of the electron gas, known as the plasmon, is discussed in Sec. 14.3. The issue of the correlation energy of the electron gas in the RPA approximation is addressed in Sec. 14.4. The response of nuclear matter with quantum numbers corresponding to the pion and rho meson, is presented in Sec. 14.5. It illustrates some of the features associated with attractive interactions in an infinite system. The presentation concludes in Sec. 14.6 with a general discussion of excited states in the limit of small momentum and excitation energy, based on an analysis by Landau, which relies on the use of exact sp propagators. It presumes further that the correlated ground state still maintains some of the properties of the Fermi gas and does not correspond to a superfluid or superconductor.

14.1 RPA in infinite systems

The polarization propagator in an infinite system with translational invariance, is diagonal in the total momentum of the ph pair. This is similar to the situation for the sp propagator, which is also diagonal in momentum space as discussed in Ch. 10. The creation of a ph pair in an infinite system corresponds, for example, to the addition of a particle with momentum $\boldsymbol{p}_\alpha$ and a hole with $\boldsymbol{p}_\beta$. When this pair is added to the noninteracting ground state, one writes

$$\left| \alpha, \beta^{-1} \right\rangle = a^{\dagger}_{\boldsymbol{p}_\alpha, m_\alpha} b^{\dagger}_{\boldsymbol{p}_\beta, m_\beta} \left| \Phi_0^N \right\rangle$$
$$= a^{\dagger}_{\boldsymbol{p}_\alpha, m_\alpha} (-1)^{\frac{1}{2}+m_\beta} a_{-\boldsymbol{p}_\beta, -m_\beta} \left| \Phi_0^N \right\rangle. \tag{14.1}$$

Since removing $-\boldsymbol{p}_\beta$ is equivalent to adding momentum $\boldsymbol{p}_\beta$ and the initial state has no momentum, the ph state has total momentum

$$\boldsymbol{Q} = \boldsymbol{p}_\alpha + \boldsymbol{p}_\beta. \tag{14.2}$$

The Hamiltonian commutes with the total momentum, so interactions will not be able to change the momentum $\boldsymbol{Q}$ (note: $\boldsymbol{Q} = \hbar\boldsymbol{q}$). As a result, the polarization propagator can be labeled with this conserved quantity, since after propagation the return to the ground state will have to correspond to the removal of another ph pair carrying the same total momentum. It is therefore convenient to rewrite the polarization propagator, given by Eq. (13.9), using the following momentum variables

$$\boldsymbol{Q}' = \boldsymbol{p}_\gamma + \boldsymbol{p}_\delta \tag{14.3}$$
$$\boldsymbol{p}' = \tfrac{1}{2}\left(\boldsymbol{p}_\gamma - \boldsymbol{p}_\delta\right), \tag{14.4}$$

corresponding to $\gamma\delta^{-1}$, with $\boldsymbol{Q}'$ being necessarily equal to the total momentum given by Eq. (14.2). Introducing also

$$\boldsymbol{p} = \tfrac{1}{2}\left(\boldsymbol{p}_\alpha - \boldsymbol{p}_\beta\right), \tag{14.5}$$

we may consider the momenta $\boldsymbol{p}$ and $\boldsymbol{p}'$ as the relative ph momenta corresponding to $\alpha\beta^{-1}$ and $\gamma\delta^{-1}$, respectively. Using this convention for the noninteracting polarization propagator, we can write

$$\Pi^{(0)}(\alpha, \beta^{-1}; \gamma, \delta^{-1}; E) \Rightarrow$$
$$\Pi^{(0)}(\boldsymbol{p}_\alpha m_\alpha, (\boldsymbol{p}_\beta m_\beta)^{-1}; \boldsymbol{p}_\gamma m_\gamma, (\boldsymbol{p}_\delta m_\delta)^{-1}; E)$$
$$\equiv \delta_{m_\alpha, m_\gamma} \delta_{m_\beta, m_\delta} \delta_{\boldsymbol{Q}, \boldsymbol{Q}'} \delta_{\boldsymbol{p}, \boldsymbol{p}'} \Pi^{(0)}(\boldsymbol{p}; \boldsymbol{Q}, E), \tag{14.6}$$

where

$$\Pi^{(0)}(\boldsymbol{p}; \boldsymbol{Q}, E) = \left\{ \frac{\theta(|\boldsymbol{p} + \boldsymbol{Q}/2| - p_F)\theta(p_F - |\boldsymbol{p} - \boldsymbol{Q}/2|)}{E - [\varepsilon(\boldsymbol{p} + \boldsymbol{Q}/2) - \varepsilon(\boldsymbol{p} - \boldsymbol{Q}/2)] + i\eta} \right.$$
$$\left. - \frac{\theta(p_F - |\boldsymbol{p} + \boldsymbol{Q}/2|)\theta(|\boldsymbol{p} - \boldsymbol{Q}/2| - p_F)}{E + [\varepsilon(\boldsymbol{p} - \boldsymbol{Q}/2) - \varepsilon(\boldsymbol{p} + \boldsymbol{Q}/2)] - i\eta} \right\}. \tag{14.7}$$

The remaining sp quantum numbers like spin (and/or isospin) have been kept explicitly in Eq. (14.6) and are denoted collectively by m_α, m_β, etc.

The Kronecker deltas identify all the relevant conserved quantities, including the relative momentum. In general, the interaction can change spins (isospins) as well as the relative momentum. So for the exact polarization propagator one may only make the replacement

$$\Pi(\alpha, \beta^{-1}; \gamma, \delta^{-1}; E) \Rightarrow \Pi(\boldsymbol{p}_\alpha m_\alpha, (\boldsymbol{p}_\beta m_\beta)^{-1}; \boldsymbol{p}_\gamma m_\gamma, (\boldsymbol{p}_\delta m_\delta)^{-1}; E)$$
$$\equiv \delta_{\boldsymbol{Q},\boldsymbol{Q}'} \Pi(\boldsymbol{p}, m_\alpha, m_\beta^{-1}; \boldsymbol{p}', m_\gamma, m_\delta^{-1}; \boldsymbol{Q}, E). \quad (14.8)$$

Such considerations also apply for the RPA approach to the polarization propagator, thus one may write Eq. (13.23) as

$$\Pi^{RPA}(\boldsymbol{p}, m_\alpha, m_\beta^{-1}; \boldsymbol{p}', m_\gamma, m_\delta^{-1}; \boldsymbol{Q}, E) = \delta_{m_\alpha, m_\gamma} \delta_{m_\beta, m_\delta} \delta_{\boldsymbol{p},\boldsymbol{p}'} \Pi^{(0)}(\boldsymbol{p}; \boldsymbol{Q}, E)$$
$$+ \Pi^{(0)}(\boldsymbol{p}; \boldsymbol{Q}, E) \sum_{\boldsymbol{p}''} \sum_{m_\epsilon m_\theta} \langle \boldsymbol{Q}, \boldsymbol{p}; m_\alpha m_\beta^{-1} | V_{ph} | \boldsymbol{Q}, \boldsymbol{p}''; m_\epsilon m_\theta^{-1} \rangle$$
$$\times \Pi^{RPA}(\boldsymbol{p}'', m_\epsilon, m_\theta^{-1}; \boldsymbol{p}', m_\gamma, m_\delta^{-1}; \boldsymbol{Q}, E), \quad (14.9)$$

where momentum conservation has been taken into account.

Uncoupled ph states correspond to Eq. (14.1) and can be written in terms of the momentum variables $\boldsymbol{Q}, \boldsymbol{p}$ according to

$$|\boldsymbol{Q}, \boldsymbol{p}; m_\alpha, m_\beta^{-1} \rangle = a^\dagger_{\boldsymbol{p}+\boldsymbol{Q}/2, m_\alpha} b^\dagger_{\boldsymbol{Q}/2-\boldsymbol{p}, m_\beta} |\Phi_0^N \rangle. \quad (14.10)$$

Specializing for now to the case of spin only, the coupled states are obtained by employing the usual Clebsch–Gordan coefficients

$$|\boldsymbol{Q}, \boldsymbol{p}; S, M_S \rangle = \sum_{m_\alpha, m_\beta} (\tfrac{1}{2} \, m_\alpha \, \tfrac{1}{2} \, m_\beta \mid S \, M_S) \, |\boldsymbol{Q}, \boldsymbol{p}; m_\alpha, m_\beta^{-1} \rangle. \quad (14.11)$$

The polarization propagator with good total spin requires the usage of the corresponding hole operators combined with the appropriate Clebsch–Gordan summations. The coupled propagator then reads

$$\Pi(\boldsymbol{p}, S, M_S; \boldsymbol{p}', S', M_S'; \boldsymbol{Q}, E) = \sum_{m_\alpha, m_\beta, m_\gamma, m_\delta} (\tfrac{1}{2} \, m_\alpha \, \tfrac{1}{2} \, m_\beta \mid S \, M_S)$$
$$\times (\tfrac{1}{2} \, m_\gamma \, \tfrac{1}{2} \, m_\delta \mid S' \, M_S') \, \Pi(\boldsymbol{p}, m_\alpha, m_\beta^{-1}; \boldsymbol{p}', m_\gamma, m_\delta^{-1}; \boldsymbol{Q}, E). \quad (14.12)$$

The possibility of changing the total spin, or its projection, has been kept open in Eq. (14.12). It depends on the character of the ph interaction. Before considering the RPA equation in more detail, it is therefore useful to construct the relevant ph matrix elements of the interaction in a coupled spin (isospin) basis. The matrix elements can be coupled to total ph spin in the following way

$$\langle Q, p; S, M_S | V_{ph} | Q, p'; S', M'_S \rangle$$
$$= \sum_{m_\alpha, m_\beta, m_\gamma, m_\delta} (\tfrac{1}{2} \, m_\alpha \, \tfrac{1}{2} \, m_\beta \mid S \, M_S) \, (\tfrac{1}{2} \, m_\gamma \, \tfrac{1}{2} \, m_\delta \mid S' \, M'_S)$$
$$\times \langle Q, p; m_\alpha, m_\beta^{-1} | V_{ph} | Q, p'; m_\gamma, m_\delta^{-1} \rangle. \tag{14.13}$$

The evaluation of the matrix element depends on the operator character of the two-body interaction V. It is instructive to evaluate Eq. (14.13) using Eq. (13.20) and express the latter matrix elements also in terms of good total spin. We will proceed by assuming that V corresponds to a local, central interaction without spin dependence but the restriction is by no means necessary, as shown in Sec. 14.5. The Coulomb interaction [see Eq. (10.182)] corresponds to such a choice, exhibiting a simple dependence on the momentum transfer. For the contribution of the direct term in Eq. (14.13), the momentum transfer is equal to Q, the conserved total ph momentum. The direct matrix element therefore does not generate any dependence on the momentum variables p and p'. Factoring out the volume for later convenience, Eq. (14.13) then yields

$$\langle Q, p; S, M_S | V_{ph} | Q, p'; S', M'_S \rangle_D$$
$$= \frac{1}{V} \sum_{m_\alpha, m_\beta, m_\gamma, m_\delta} \sum_{S_p, M_p} (\tfrac{1}{2} \, m_\alpha \, \tfrac{1}{2} \, m_\beta \mid S \, M_S) \, (\tfrac{1}{2} \, m_\gamma \, \tfrac{1}{2} \, m_\delta \mid S' \, M'_S)$$
$$\times (\tfrac{1}{2} \, m_\alpha \, \tfrac{1}{2} \, - m_\delta \mid S_p \, M_p) \, (\tfrac{1}{2} \, - m_\beta \, \tfrac{1}{2} \, m_\gamma \mid S_p \, M_p)$$
$$\times (-1)^{1/2 + m_\beta} (-1)^{1/2 + m_\delta} \, V(Q). \tag{14.14}$$

For the Coulomb interaction *e.g.* $V(Q) = 4\pi e^2 \hbar^2 / Q^2$. In arriving at Eq. (14.14), we have used that the coupled direct matrix element is diagonal in, and does not depend on, the total spin and its projection. We label the latter coupling scheme by particle-particle, to contrast it with the ph coupling in Eq. (14.13). Hence the subscript p is employed in Eq. (14.14). The m summations in Eq. (14.14) yield a so-called 6j-symbol (see App. B and Sec. 13.7) while forcing $S = S'$ and $M_S = M'_S$. The 6j-symbols occur in the recoupling of three angular momenta and some are tabulated in [Lindgren and Morrison (1982)]. The final result is given by

$$\langle Q, p; S, M_S | V_{ph} | Q, p'; S', M'_S \rangle_D$$
$$= \frac{1}{V} \delta_{S,S'} \delta_{M_S, M'_S} \sum_{S_p} (-1)^{S_p + 1} (2S_p + 1) \begin{Bmatrix} \tfrac{1}{2} & \tfrac{1}{2} & S \\ \tfrac{1}{2} & \tfrac{1}{2} & S_p \end{Bmatrix} V(Q), \tag{14.15}$$

where the summation over M_p yields the factor $2S_p + 1$. The 6j-symbols for the S, S_p combinations involve the pairs (0,0), (1,0), (0,1), and (1,1). The relevant 6j-symbols are given by, $-\frac{1}{2}, \frac{1}{2}, \frac{1}{2}$, and $\frac{1}{6}$, respectively. Inserting these and performing the sum over S_p, we finally obtain

$$\langle \boldsymbol{Q}, \boldsymbol{p}; S, M_S | V_{ph} | \boldsymbol{Q}, \boldsymbol{p}'; S', M_S' \rangle_D = \frac{1}{V} \delta_{S,S'} \delta_{M_S, M_S'} \, \delta_{S,0} \, 2 \, V(\boldsymbol{Q}), \quad (14.16)$$

showing that the direct contribution of the interaction only contributes when the ph spin $S = 0$. From a diagrammatic perspective this makes sense, since the total spin of the initial ph state must be carried over to the final one by the interaction, which does not contain spin operators, therefore selecting only the $S = 0$ channel. In the case of a spin–spin interaction, considered in Ch. 5, the direct ph matrix element only contributes when $S = 1$, as can be verified by performing similar steps.

The exchange term can be handled in similar fashion, but it yields a contribution to both $S = 0$ and 1. Combining the results and noting that only ph matrix elements survive, which are diagonal in the total spin and its projection, we find

$$\langle \boldsymbol{Q}, \boldsymbol{p} | V_{ph}^{SM_S} | \boldsymbol{Q}, \boldsymbol{p}' \rangle = \frac{1}{V} \left[\delta_{S,0} \, 2 \, V(\boldsymbol{Q}) - V(\boldsymbol{p} - \boldsymbol{p}') \right], \quad (14.17)$$

adapting the notation slightly. The exchange term of the Coulomb interaction is the sole, attractive contribution when the ph spin is 1. For other types of interactions that include spin (isospin) dependence, identical steps can be followed. In general, the total ph spin is conserved, although for a tensor interaction one must choose the quantization axis judiciously to ensure that the different projections do not mix, as discussed in Sec. 14.5.

It is now possible to present the RPA equation in the coupled spin format. By performing the relevant coupling of the ph spins, we find

$$\Pi_{SM_S}^{RPA}(\boldsymbol{p}, \boldsymbol{p}'; \boldsymbol{Q}, E) = \delta_{\boldsymbol{p}, \boldsymbol{p}'} \Pi^{(0)}(\boldsymbol{p}; \boldsymbol{Q}, E) \quad (14.18)$$
$$+ \Pi^{(0)}(\boldsymbol{p}; \boldsymbol{Q}, E) \sum_{\boldsymbol{p}''} \langle \boldsymbol{Q}, \boldsymbol{p} | V_{ph}^{SM_S} | \boldsymbol{Q}, \boldsymbol{p}'' \rangle \, \Pi_{SM_S}^{RPA}(\boldsymbol{p}'', \boldsymbol{p}'; \boldsymbol{Q}, E),$$

where the conservation of the total spin and its projection has been incorporated (also in the notation). The integral equation can be simplified considerably when the direct ph matrix element dominates the exchange term. This occurs for the Coulomb interaction, where the direct term involves the Q^{-2} divergence, whereas for the exhange term it is removed by the implied

p'' integral in Eq. (14.18). Most applications of the integral equation therefore involve simplifications, which include only the Q-dependence of the ph interaction. Equation (14.18) is then no longer an integral equation. The simplification is facilitated by considering

$$\Pi_{SM_S}^{RPA}(\boldsymbol{Q}, E) \equiv \frac{1}{V} \sum_{\boldsymbol{p}} \sum_{\boldsymbol{p}'} \Pi_{SM_S}^{RPA}(\boldsymbol{p}, \boldsymbol{p}'; \boldsymbol{Q}, E) \qquad (14.19)$$

and

$$\Pi^{(0)}(\boldsymbol{Q}, E) \equiv \frac{1}{V} \sum_{\boldsymbol{p}} \Pi^{(0)}(\boldsymbol{p}; \boldsymbol{Q}, E), \qquad (14.20)$$

where volume factors have been extracted to anticipate ensuing cancellations. Using these definitions, Eq. (14.18) can be transformed into

$$\Pi_{SM_S}^{RPA}(\boldsymbol{Q}, E) = \Pi^{(0)}(\boldsymbol{Q}, E) + \Pi^{(0)}(\boldsymbol{Q}, E) \, V_{ph}^{SM_S}(\boldsymbol{Q}) \, \Pi_{SM_S}^{RPA}(\boldsymbol{Q}, E), \quad (14.21)$$

where the interaction term no longer includes the volume dependence. Applications will be discussed in Secs. 14.3 and 14.5.

14.2   Lowest-order polarization propagator in an infinite system

Clearly, the first task, before studying Eq. (14.21), is to evaluate the quantity (sometimes called the Lindhard function) $\Pi^{(0)}(\boldsymbol{Q}, E)$. Replacing the sum over $\boldsymbol{p}$ by an integration, according to Eq. (5.6), we must calculate

$$\Pi^{(0)}(\boldsymbol{Q}, E) = \int \frac{d\boldsymbol{p}}{(2\pi\hbar)^3} \left\{ \frac{\theta(|\boldsymbol{p} + \boldsymbol{Q}/2| - p_F)\theta(p_F - |\boldsymbol{p} - \boldsymbol{Q}/2|)}{E - [\varepsilon(\boldsymbol{p} + \boldsymbol{Q}/2) - \varepsilon(\boldsymbol{p} - \boldsymbol{Q}/2)] + i\eta} \right.$$
$$\left. - \frac{\theta(p_F - |\boldsymbol{p} + \boldsymbol{Q}/2|)\theta(|\boldsymbol{p} - \boldsymbol{Q}/2| - p_F)}{E + [\varepsilon(\boldsymbol{p} - \boldsymbol{Q}/2) - \varepsilon(\boldsymbol{p} + \boldsymbol{Q}/2)] - i\eta} \right\}. \qquad (14.22)$$

Equation (14.22) can be calculated analytically, as will be illustrated here for the imaginary part explicitly. For the description of excited states, only the case $E > 0$ need be considered. Using the identity of Eq. (7.15), the first term in Eq. (14.22) solely contributes to the imaginary part

$$\text{Im } \Pi^{(0)}(\boldsymbol{Q}, E) = -\pi \int \frac{d\boldsymbol{p}}{(2\pi\hbar)^3} \qquad (14.23)$$
$$\times \theta(|\boldsymbol{p} + \boldsymbol{Q}/2| - p_F)\theta(p_F - |\boldsymbol{p} - \boldsymbol{Q}/2|)\delta(E - [\varepsilon(\boldsymbol{p} + \boldsymbol{Q}/2) - \varepsilon(\boldsymbol{p} - \boldsymbol{Q}/2)]).$$

We first consider the momentum dependence of the ph energy difference in the argument of the δ function

$$\varepsilon(\boldsymbol{p}+\boldsymbol{Q}/2) - \varepsilon(\boldsymbol{p}-\boldsymbol{Q}/2) = \frac{\boldsymbol{p}\cdot\boldsymbol{Q}}{m} = \frac{p\,Q\,\cos\theta}{m}, \tag{14.24}$$

including only kinetic energies. If $\boldsymbol{Q}$ is chosen along the z-axis, $p\cos\theta = p_z$ and energy conservation yields

$$E = \frac{p_z Q}{m}, \tag{14.25}$$

showing that it (and the accompanying imaginary part of $\Pi^{(0)}$) corresponds to a fixed value of p_z. It is useful to distinguish two cases for the total ph momentum, arising from the properties of the two step functions in Eq. (14.23). In the space corresponding to the integration variable $\boldsymbol{p}$, these conditions are represented by two spheres with radius p_F, one displaced from the origin by $\boldsymbol{Q}/2$, the other by $-\boldsymbol{Q}/2$. The step function pertaining to the hole then restricts the values of $\boldsymbol{p}$ to be inside the first sphere, while the other one only allows values outside the second sphere, as indicated in Fig. 14.1. Since these two spheres are displaced from each other by Q, they no longer overlap for $Q > 2p_F$, and the step functions allow all values of $\boldsymbol{p}$ inside the top sphere in Fig 14.1. Energy conservation in the form of Eq. (14.25), shows that a minimum and maximum energy exist, corresponding to p_z touching the bottom with $p_z = Q/2 - p_F$, and the top of the allowed sphere with $p_z = Q/2 + p_F$, respectively. These conditions translate to a nonvanishing of the imaginary part of $\Pi^{(0)}$ when

$$\frac{Q^2}{2m} - \frac{Qp_F}{m} < E < \frac{Q^2}{2m} + \frac{Qp_F}{m}. \tag{14.26}$$

It is now straightforward to perform the integrations in Eq. (14.23). The integration over the azimuth angle gives a factor of 2π, while the $\cos\theta$ integration is taken care of by the δ function, employing the following property

$$\delta\left(E - \frac{p\,Q}{m}\cos\theta\right) = \frac{m}{p\,Q}\,\delta\left(\frac{mE}{p\,Q} - \cos\theta\right). \tag{14.27}$$

The final integration over the magnitude of $\boldsymbol{p}$ is then given by

$$\text{Im}\,\Pi^{(0)}(\boldsymbol{Q}, E) = -\frac{\pi}{\hbar^3}\frac{2\pi m}{(2\pi)^3 Q}\int_{p_-}^{p_+} dp\, p = -\frac{1}{\hbar^3}\frac{m}{8\pi Q}\left[p_F^2 - \left(\frac{mE}{Q} - \frac{Q}{2}\right)^2\right], \tag{14.28}$$

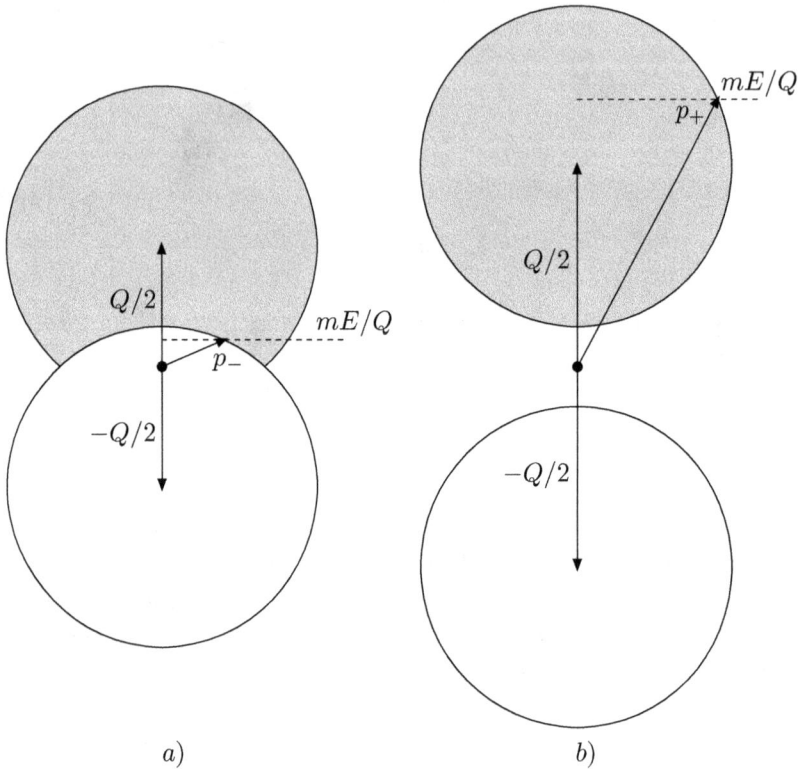

Fig. 14.1   Illustration of the constraints imposed by the step functions in Eq. (14.23). The condition $|\boldsymbol{p} + \boldsymbol{Q}/2| > p_F$ corresponds to the area outside the lower sphere in both figures, while the condition $|\boldsymbol{p} - \boldsymbol{Q}/2| < p_F$ only allows contributions from inside the top sphere. Part $a)$ illustrates the case $Q < 2p_F$, with overlapping spheres, while $b)$ is appropriate for $Q > 2p_F$, when there is no overlap. The gray area indicates the allowed region for the integration over $\boldsymbol{p}$. The dashed lines in part $a)$ and $b)$ identify possible energy values for which a nonzero imaginary part is obtained. This condition is expressed by Eq. (14.25), which shows that the only contributions to Eq. (14.23) correspond to the part of the dashed line inside the gray area of the top sphere. Limiting values of the integration variable p in Eq. (14.28) are indicated by the arrows with the corresponding labels.

where the lower limit is given by

$$p_- = \frac{mE}{Q}, \qquad (14.29)$$

and the upper limit by

$$p_+ = \left[p_F^2 - Q^2/4 + mE\right]^{1/2}. \tag{14.30}$$

This limit reflects that $|\boldsymbol{p} - \boldsymbol{Q}/2| < p_F$.

The case $Q < 2p_F$ involves two overlapping spheres, illustrated in Fig. 14.1a). For energies E, such that $\theta = 0$ is allowed (when the dashed line is above the lower sphere), the same result as in Eq. (14.28) is obtained, since one retraces identical steps. This energy domain also involves the upper limit given in Eq. (14.26). Since the excitation spectrum, associated with Eq (14.24), starts at zero energy (for $Q < 2p_F$), the remaining energy domain reads

$$0 < E < \frac{Qp_F}{m} - \frac{Q^2}{2m}, \tag{14.31}$$

where the upper limit is equal to the energy for which the dashed line in Fig. 14.1a) is no longer constrained by the lower sphere. The integrations proceed as before, except that the lower limit in Eq. (14.28) is now replaced by

$$p_- = \left[p_F^2 - Q^2/4 - mE\right]^{1/2}, \tag{14.32}$$

corresponding to $|\boldsymbol{p} + \boldsymbol{Q}/2| > p_F$. The final result for the imaginary part of the noninteracting polarization propagator then becomes (for $Q < 2p_F$)

$$\mathrm{Im}\ \Pi^{(0)}(\boldsymbol{Q}, E) = -\frac{1}{\hbar^3} \frac{m}{4\pi Q} mE. \tag{14.33}$$

A comparison of the imaginary parts for Q below and above $2p_F$ is shown in Fig. 14.2. The imaginary part of $\Pi^{(0)}$ can be interpreted as being proportional to the probability density for the absorption of momentum Q and energy E by the noninteracting Fermi sea. Its shape, as a function of energy for $Q > 2p_F$, is given by an inverted parabola, the width being proportional to the Fermi momentum. Data for inelastic electron scattering from nuclei in this high-momentum domain, can be likewise interpreted and lead to reasonable values of the Fermi momentum [Moniz *et al.* (1971)]. Still, it should be kept in mind that the response of a finite system, calculated with the appropriate sp potential well, yields quite similar shapes.

The real part of $\Pi^{(0)}$, shown in Fig. 14.3, can be calculated by straightforward integration of Eq. (14.22), as shown in [Fetter and Walecka (1971)]. An alternative procedure uses Eq. (14.23) and the identical result for the

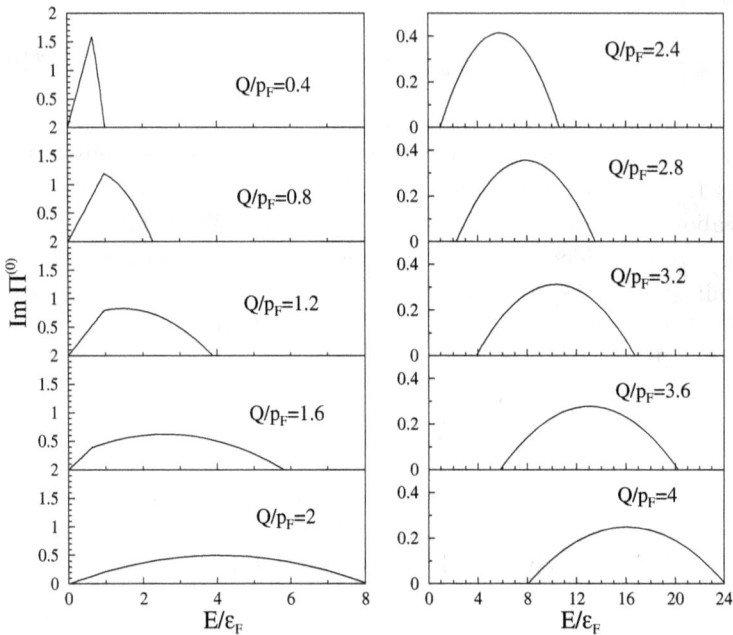

Fig. 14.2 Different shapes associated with the imaginary part of $\Pi^{(0)}$ depending on the magnitude of Q. Im $\Pi^{(0)}$ is plotted as a function of the energy in units of ε_F and is divided by the constant factor $-mp_F/(8\pi\hbar^3)$. In the left column several Q values less than $2p_F$ are considered. When the maximum is located at the end of the straight line given by Eq. (14.33), it occurs at $2 - Q/p_F$ otherwise at $(Q/p_F)^{-1}$ determined by the inverted parabola given by Eq. (14.31). The second column displays values of Q larger than $2p_F$ and has different scales for both axes as compared to the first column.

imaginary part for $E < 0$, to rewrite Eq. (14.22) in the form of a dispersion integral (see Ch. 11)

$$\Pi^{(0)}(\boldsymbol{Q}, E) = -\frac{1}{\pi} \int_{E_-}^{E_+} dE' \, \frac{\mathrm{Im}\,\Pi^{(0)}(\boldsymbol{Q}, E')}{E - E' + i\eta}$$

$$+ \frac{1}{\pi} \int_{-E_+}^{-E_-} dE' \, \frac{\mathrm{Im}\,\Pi^{(0)}(\boldsymbol{Q}, E')}{E - E' - i\eta}, \qquad (14.34)$$

where the energy limits are given by

$$E_- = \begin{cases} 0 & Q < 2p_F \\ \dfrac{Q^2}{2m} - \dfrac{Q\,p_F}{m} & Q > 2p_F \end{cases} \qquad (14.35)$$

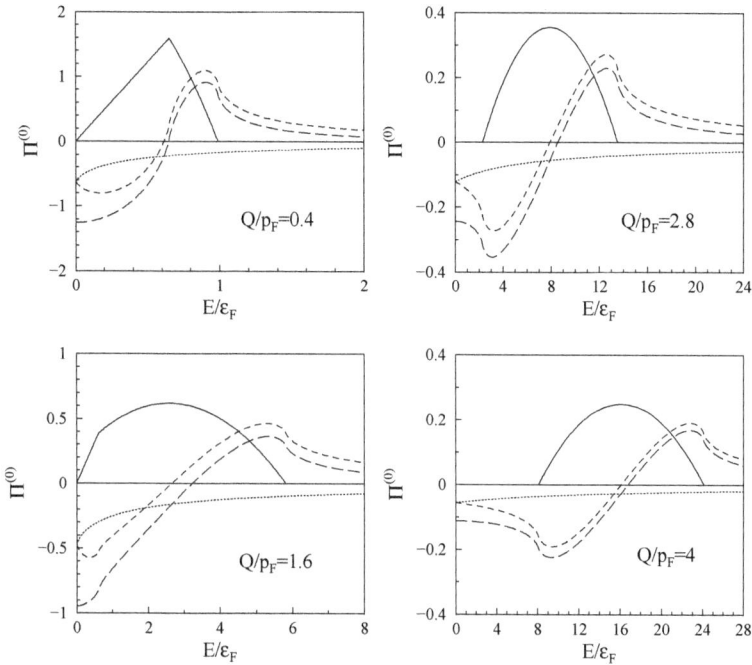

Fig. 14.3 Illustration of the contributions in Eq. (14.34) to the real part of $\Pi^{(0)}$ for four values of Q. The first term is represented by the short-dashed and the second by the dotted line. The imaginary part is also shown (full) together with the total real part (long-dashed), given by Eq. (14.37). Both the real and imaginary part of $\Pi^{(0)}$ are plotted as a function of the energy in units of ε_F. A constant factor $\pm mp_F/(8\pi\hbar^3)$ multiplies all functions, where the plus sign is for the real and the minus sign for the imaginary part of $\Pi(0)$. In the left column two Q values less than $2p_F$ are considered. The second column displays two values of Q larger than $2p_F$. Different scales are employed in the four panels.

and

$$E_+ = \frac{Q^2}{2m} + \frac{Q\, p_F}{m}, \tag{14.36}$$

for any value of Q. By simply inserting the imaginary parts of $\Pi^{(0)}$ for $E > 0$ and $E < 0$ from Eq. (14.22) into Eq. (14.34), one recovers the original equation (14.22), confirming the validity of Eq. (14.34). Using the expressions for the imaginary part obtained in Eqs. (14.28) and (14.33) one can evaluate the real part of $\Pi^{(0)}$ by employing Eq. (14.34). The resulting expression for the real part is

$$\text{Re } \Pi^{(0)}(\boldsymbol{Q}, E) = \frac{1}{\hbar^3} \frac{m p_F}{4\pi^2} \tag{14.37}$$

$$\times \left\{ -1 + \frac{p_F}{2Q} \left[1 - \left(\frac{mE}{Q p_F} - \frac{Q}{2 p_F} \right)^2 \right] \ln \left| \frac{2 Q p_F + 2mE - Q^2}{2 Q p_F - 2mE + Q^2} \right| \right.$$

$$\left. - \frac{p_F}{2Q} \left[1 - \left(\frac{mE}{Q p_F} + \frac{Q}{2 p_F} \right)^2 \right] \ln \left| \frac{2 Q p_F + 2mE + Q^2}{2 Q p_F - 2mE - Q^2} \right| \right\}.$$

A plot for the real part of $\Pi^{(0)}$ is shown in Fig. 14.3 for four values of Q/p_F, corresponding to 0.4, 1.6, 2.8 and 4. In each of the four panels the imaginary part is also indicated by the full line. The short-dashed line in every panel corresponds to the first dispersion integral in Eq. (14.34), requiring a principal value integration for energies where the imaginary part of $\Pi^{(0)}$ doesn't vanish. The dotted line corresponds to the second integration in Eq. (14.34). This contribution increases monotonically to zero with increasing energy. At $E = 0$ both terms in Eq. (14.34) yield equal contributions. The sum of these contributions is represented by the long-dashed line and corresponds to Eq. (14.37).

Special limits of the polarization propagator are required to discuss relevant applications of the RPA equation (see Sec. 14.6). For fixed Q, the limit for $E \to 0$ generates no imaginary part as illustrated in Fig. 14.2, whereas the real part is given by

$$\text{Re } \Pi^{(0)}(\boldsymbol{Q}, 0) = \frac{1}{\hbar^3} \frac{m p_F}{4\pi^2} \left\{ -1 + \frac{p_F}{Q} \left[1 - \left(\frac{Q}{2 p_F} \right)^2 \right] \ln \left| \frac{2 Q p_F - Q^2}{2 Q p_F + Q^2} \right| \right\}. \tag{14.38}$$

For fixed energy E, the limit $Q \to 0$ also produces no imaginary part, since the upper limit of the allowed energy domain, expressed by Eq. (14.36), will become smaller than any finite E in this limit. For the real part one obtains for $Q \to 0$

$$\text{Re } \Pi^{(0)}(\boldsymbol{Q}, E) \to \frac{1}{\hbar^3} \frac{m p_F}{4\pi^2} \frac{2}{3} \frac{Q^2 p_F^2}{m^2 E^2}. \tag{14.39}$$

Additional results will be discussed in Sec. 14.6.

14.3 Plasmons in the electron gas

The solution of the RPA equation for the electron gas has substantial relevance for the global properties of this system. The Coulomb interaction must be used in Eq. (14.17), yielding[1]

$$
\langle \boldsymbol{Q}, \boldsymbol{p}; S, M_S | V_{ph}^C | \boldsymbol{Q}, \boldsymbol{p}'; S', M_S' \rangle \tag{14.40}
$$
$$
= \frac{1}{V} \delta_{S,S'} \delta_{M_S,M_S'} \{ \delta_{S,0} \, 2V(\boldsymbol{Q}) - V(\boldsymbol{p} - \boldsymbol{p}') \}
$$
$$
= \delta_{S,S'} \delta_{M_S,M_S'} \frac{4\pi e^2 \hbar^2}{V} \left\{ \delta_{S,0} \frac{2}{\boldsymbol{Q}^2} - \frac{1}{|\boldsymbol{p} - \boldsymbol{p}'|^2} \right\}.
$$

Since the momentum $\boldsymbol{Q}$ is conserved, the direct term completely dominates the ph interaction at small values of Q, since the exchange contribution is integrated over in Eq. (14.18). A similar situation was encountered in the evaluation of the Fock contribution to the self-energy, as discussed in Sec. 10.5. It is therefore reasonable to expect that the implementation of the RPA equation, which neglects the exchange contribution, represents the dominant physics for small values of Q. As a result, it is permissible to use Eq. (14.21) to solve for the polarization propagator. Neglecting the exchange term, the solution to Eq. (14.21) reads

$$
\Pi_{S=0}^{RPA}(Q, E) = \frac{\Pi^{(0)}(Q, E)}{1 - 2V(Q)\Pi^{(0)}(Q, E)}, \tag{14.41}
$$

where only the magnitude of $\boldsymbol{Q}$ is kept in the notation. The factor of 2 comes from the spin matrix element for $S = 0$ states [see Eq. (14.15)]. The probability density for the absorption of momentum Q and energy E is again given by the imaginary part of Π^{RPA}. A contribution will be found in the energy domain, where the numerator has an imaginary part, according to Eqs. (14.35) and (14.36). These boundaries are plotted in Fig. 14.4 as a function of Q. The special character of the Coulomb interaction, its divergence for $Q \to 0$, requires further scrutiny. To appreciate what is happening, it is useful to remember the schematic model with attractive and repulsive ph interactions, presented in Sec. 13.3. The discussion for the repulsive case obviously applies here, and one may conjecture that a collective state could appear above the region of "trapped" ph energies, determined by the boundaries of the imaginary part of $\Pi^{(0)}$. Such a collective state,

---

[1] We will not use atomic units in this section to avoid switching notation from Secs. 14.2 and 14.5.

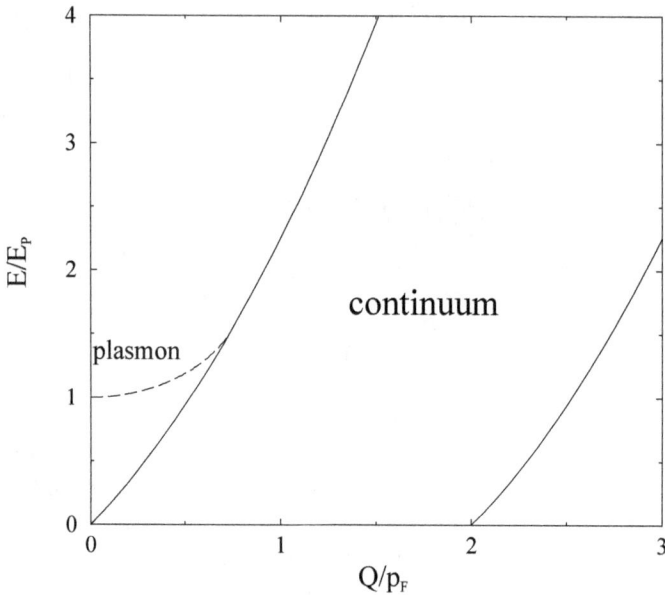

Fig. 14.4   Boundaries of the ph continuum are indicated by the full lines. The plasmon energy as a function of Q is shown by the dashed line for a value of $r_s = 2$.

usually referred to as a plasmon, will turn up in the Lehmann representation of the polarization propagator as a discrete pole. The condition for the appearance of a pole for a fixed value of Q, requires for $E > Q^2/2m + Qp_F/m$ that the denominator of Eq. (14.41) vanishes for a certain E_p

$$1 - 2V(Q) \text{ Re } \Pi^{(0)}(Q, E_p) = 0. \tag{14.42}$$

The real part indicates that in this energy domain the imaginary part of $\Pi^{(0)}$ is zero. In the language of the solution method, previously discussed for propagator equations, one proceeds by assuming a pole in the Lehmann representation for $\Pi_{S=0}^{RPA}$

$$\Pi_{S=0}^{RPA}(Q, E) = \frac{\mathcal{A}_p(Q)}{E - E_p + i\eta} - \frac{\mathcal{B}_p(Q)}{E + E_p - i\eta} + \text{continuum.} \tag{14.43}$$

Inserting this result in Eq. (14.21) and considering energies near the pole E_p, leads to the eigenvalue equation for E_p, which can be written as Eq. (14.42). Using the small Q limit for the real part of the noninteracting polarization propagator for a fixed energy E_p, given by Eq. (14.39), yields the classical plasmon energy, when the former is inserted in Eq. (14.42)

$$E_p = \left(\frac{4}{3\pi} \frac{e^2 \hbar^2 p_F^3}{m\hbar^3} \right)^{1/2} = \left(\frac{4\pi\rho e^2 \hbar^2}{m} \right)^{1/2}. \qquad (14.44)$$

Note that the Coulomb divergence has been exactly cancelled by the Q^2 in Eq. (14.39) and the density ρ has been substituted. The appearance of a collective state above the continuum in the $Q \to 0$ limit, is therefore unique for the Coulomb interaction. The root of Eq. (14.42), for nonzero values of Q, can be found by using the full expression for the real part [see Eq. (14.37)]. The location of the plasmon as a function of Q is identified in Fig. 14.4 by the dashed line for a value of $r_s = 2$. Energies are plotted in units of E_p, given by Eq. (14.44). The energy of the plasmon at $Q = 0$ can also be written as a function of r_s by dividing by ε_F

$$\frac{E_p}{\varepsilon_F} \approx 0.94\sqrt{r_s}. \qquad (14.45)$$

For $r_s = 2$ the plasmon merges into the continuum at Q_c, corresponding to a value of $Q_c/p_F \approx 0.73$ and immediately acquires a width. This feature is illustrated in Fig. 14.5 for $r_s = 1$, which yields $Q_c/p_F \approx 0.56$. Until $Q = Q_c$ the plasmon is a discrete ph state, well isolated from the continuum. The comparison for $r_s = 1$ with the noninteracting limit, indicated by the dashed lines in Fig. 14.5, shows that the plasmon carries essentially all the transition strength at small values of Q. This feature has also been encountered in Sec. 13.3 for the schematic model in a discrete ph basis, when the interaction is repulsive. The continuum part of the response in the RPA limit, is therefore not visible in the top-left panel for $Q/p_F = 0.1$. For the panel corresponding to $Q/p_F = 0.5$, the plasmon has almost reached the continuum and some transition strength already resides there, representing a sizable fraction of the sum rule strength.

The sum rule is referred to as the f-sum rule and plays an important role in analyzing the properties of the plasmon in the electron gas and in real metals. We will proceed by establishing the sum rule, and by relating it to the properties of the response of the electron gas, to an external probe that transfers momentum $\boldsymbol{Q}$ to the system. This can, for example, be achieved by the inelastic scattering of high-energy electrons, which interact sufficiently weakly with the sample, so that time-dependent perturbation theory (Fermi's golden rule) can be be applied. The relevant excitation

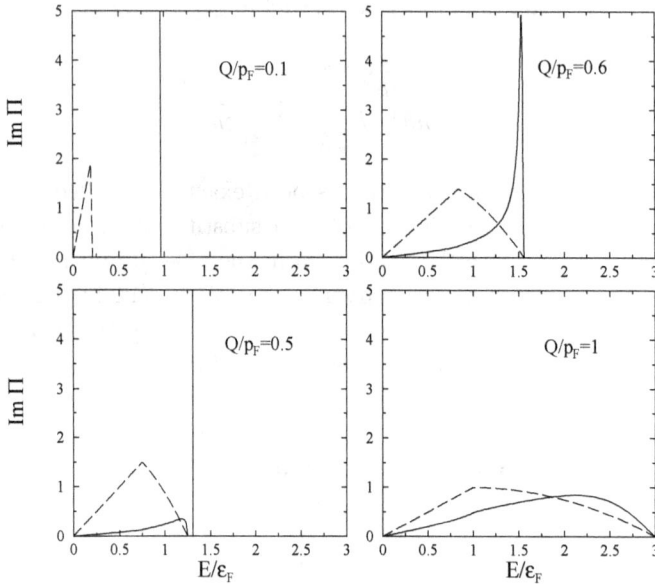

Fig. 14.5 Imaginary part of the polarization propagator for the noninteracting case (dashed lines) and RPA (full lines) for different values of Q/p_F with $r_s = 1$. The same units for the polarization propagator were used as in Fig. 14.2. The location of the plasmon energy at $Q = 0$ corresponds to $E_p/\varepsilon_F = 0.94$, given by Eq. (14.45). At $Q_c/p_F = 0.56$ the plasmon merges into the continuum, illustrated by the bottom-left panel ($Q < Q_c$) and the top-right panel ($Q > Q_c$).

operator which transfers momentum $\boldsymbol{Q}$ to the system reads[2]

$$\rho_N(\boldsymbol{Q}) = \sum_{i=1}^{N} e^{i\boldsymbol{Q}\cdot\boldsymbol{r}_i/\hbar}, \tag{14.46}$$

in first quantization. This operator was already encountered in Sec. 7.6 in the discussion of knock-out experiments. In second quantization, the corresponding operator is written as

$$\hat{\rho}(\boldsymbol{Q}) - \sum_{\boldsymbol{p}m_s} a^{\dagger}_{\boldsymbol{p}m_s} a_{\boldsymbol{p}-\boldsymbol{Q}m_s}. \tag{14.47}$$

The sum rule can be obtained by studying the energy-weighted transition

---

[2]It is tradition to use the notation $\boldsymbol{q}$ for this variable. We prefer to avoid the confusing switch here.

strength to all excited states, defined by (see Sec. 13.4)

$$S(\hat{\rho}(\boldsymbol{Q})) = \sum_n (E_n^N - E_0^N) \left| \langle \Psi_n^N | \hat{\rho}(\boldsymbol{Q}) | \Psi_0^N \rangle \right|^2. \tag{14.48}$$

The expression can be rewritten in the form of the expectation value in the ground state of a double commutator [see Eq. (13.51)]

$$S(\hat{\rho}(\boldsymbol{Q})) = \langle \Psi_0^N | \frac{1}{2} \left[\hat{\rho}^\dagger(\boldsymbol{Q}), \left[\hat{H}, \hat{\rho}(\boldsymbol{Q}) \right] \right] | \Psi_0^N \rangle. \tag{14.49}$$

It is generated by replacing the energies in Eq. (14.48) by the Hamiltonian in appropriate places. Either time reversal or parity invariance must also be employed to produce $S(\hat{\rho}(\boldsymbol{Q})) = S(\hat{\rho}(-\boldsymbol{Q}))$, noting that $\hat{\rho}^\dagger(\boldsymbol{Q}) = \hat{\rho}(-\boldsymbol{Q})$. The double commutator in Eq. (14.49) can be evaluated explicitly, when the Hamiltonian does not contain velocity-dependent interactions. The contribution of the interaction $\hat{V}$ then vanishes, since it commutes with $\hat{\rho}(\boldsymbol{Q})$. The remaining term from the kinetic energy is easiest to evaluate in first quantization, generating

$$S(\hat{\rho}(\boldsymbol{Q})) = N \frac{Q^2}{2m}. \tag{14.50}$$

The relation with the polarization propagator is obtained by rewriting the excitation operator of Eq. (14.47) as

$$\hat{\rho}(\boldsymbol{Q}) = \sqrt{2} \sum_{\boldsymbol{p}} \sum_{m_p m_h} \left(\tfrac{1}{2}\, m_p\, \tfrac{1}{2}\, m_h \mid 0\, 0 \right) a_{\boldsymbol{p}m_p}^\dagger b_{\boldsymbol{Q}-\boldsymbol{p}m_h}^\dagger, \tag{14.51}$$

which shows that it excites ph states with total spin (and projection) zero. We therefore obtain the following identity

$$S(\hat{\rho}(\boldsymbol{Q})) = 2 \sum_{\boldsymbol{p}} \sum_{\boldsymbol{p}'} \int_0^\infty dE\, E\, \left\{ -\frac{1}{\pi} \operatorname{Im} \Pi(\boldsymbol{p}, \boldsymbol{p}'; S = 0; \boldsymbol{Q}, E) \right\}, \tag{14.52}$$

using the Lehmann representation of the polarization propagator given in Eq. (13.9), adapted for the present case. A combination with Eq. (14.50) finally yields

$$S(\hat{\rho}(\boldsymbol{Q})) = \int_0^\infty dE\, E\, S_{S=0}(\boldsymbol{Q}, E) = N \frac{Q^2}{2m}. \tag{14.53}$$

The dynamic structure function for $S = 0$ is defined by

$$S_{S=0}(\boldsymbol{Q}, E) = -\frac{V}{\pi} \operatorname{Im} \Pi_{S=0}(\boldsymbol{Q}, E), \tag{14.54}$$

where $\Pi_{S=0}(\boldsymbol{Q}, E)$ can be found in Eq. (14.19). The f-sum rule in Eq. (14.53), can also be written in terms of the dielectric constant[3] $\varepsilon(\boldsymbol{Q}, E)$, which relates the Fourier transforms of the displacement field $\boldsymbol{D}(\boldsymbol{Q}, E)$ and the electric field $\boldsymbol{E}(\boldsymbol{Q}, E)$ vectors. By considering the response to an infinitesimal charge perturbation, one can show that [Pines (1963)]

$$\text{Im}\,\frac{1}{\varepsilon(\boldsymbol{Q}, E)} = -\frac{4\pi^2 e^2 \hbar^2}{Q^2} S(\boldsymbol{Q}, E). \tag{14.55}$$

Equation (14.53) can therefore be rewritten as

$$-\int_0^\infty dE\, E\, \text{Im}\,\frac{1}{\varepsilon(\boldsymbol{Q}, E)} = \frac{\pi}{2} E_p^2, \tag{14.56}$$

where Eq. (14.44) was used to introduce the classical plasmon energy.

To determine the transition strength to the plasmon excitation, it is thus necessary to calculate the numerator terms $\mathcal{A}_p(Q)$ and $\mathcal{B}_p(Q)$ in Eq. (14.43). This can be done in the standard fashion, yielding

$$\mathcal{A}_p(Q) = -\text{Re}\,\Pi^{(0)}(Q, E_p(Q)) \left\{ 2V(Q)\,\frac{\partial \text{Re}\,\Pi^{(0)}(Q, E)}{\partial E}\bigg|_{E=E_p(Q)} \right\}^{-1} \tag{14.57}$$

where $E_p(Q)$ denotes the location of the plasmon for a given Q (for which an isolated plasmon state exists) and a similar expression holds for $\mathcal{B}_p(Q)$.

When $Q \to 0$, the plasmon exhausts all the strength of the f-sum rule, according to Eq. (14.53), in the RPA method. Returning now to Fig. 14.5, we note that the plasmon in the top-left panel for $Q/p_F = 0.1$ still carries almost 100% of the strength and practically exhausts the f-sum rule. For $Q/p_F = 0.5$, shown in the bottom-left panel, already 22% of the sum-rule strength resides in the continuum, whereas for $Q/p_F = 0.6$ the plasmon has merged into the continuum and all the transition strength now resides there. The decreasing difference between the RPA and the noninteracting limit, due to the weakening Coulomb interaction with increasing Q, is illustrated in the bottom-right panel for $Q/p_F = 1$. Figure 14.5 therefore illustrates similar features to those of the schematic model with a repulsive interaction, discussed in Sec. 13.3.

The experimental results from inelastic electron scattering on metals are consistent with the global features of the properties of the plasmon, as described by the RPA method. In Fig. 14.6 inelastic electron scattering

---

[3] The sp energy $\varepsilon(\boldsymbol{p})$ is also denoted by this symbol. No confusion should arise, since the dielectric constant has an extra argument.

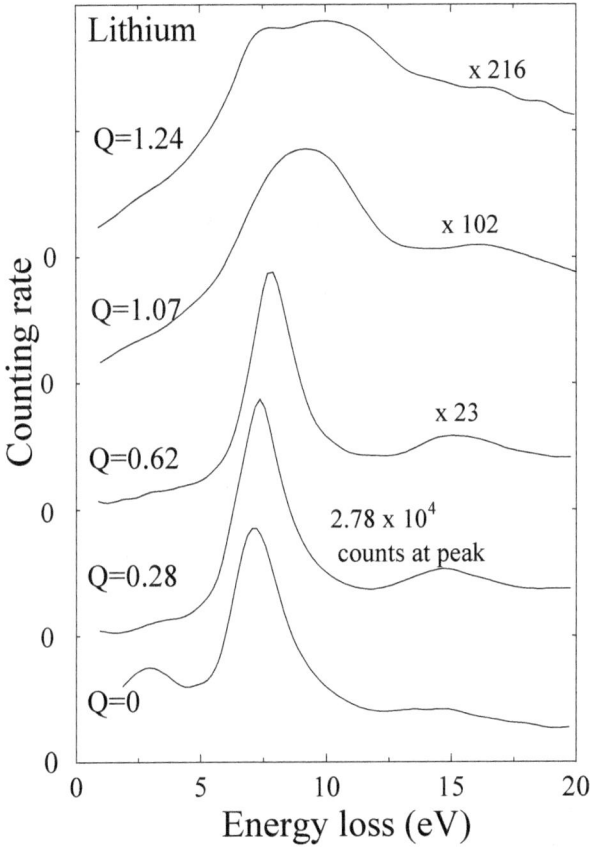

Fig. 14.6 Data for inelastic electron scattering from lithium, adapted from [Gibbons *et al.* (1976)]. The indicated momentum values are in units of Å^{-1}. The plasmon peak is sharp and moves slightly up in energy for the first three momentum values. For the last two, a substantial broadening is observed, indicating that the plasmon has merged into the continuum. All plots have been normalized to have the same maximum. The corresponding factors are also identified.

data from lithium are collected for different values of Q [Gibbons *et al.* (1976)]. For lithium the value of $r_s = 3.25$ is appropriate, yielding a Fermi energy of 4.74 eV, according to Sec. 5.2. Using these values, Eq. (14.45) produces an estimate of the plasmon energy of 8 eV, in reasonable, but not perfect, agreement with the maximum at $Q = 0$ in Fig. 14.6. The narrow peak and slight increase in the plasmon energy with larger Q are observed as well. The transition of the plasmon to the continuum is expected around

0.9 Å^{-1}. The width indeed increases dramatically for higher Q-values, as shown in Fig. 14.6. The data at zero momentum also exhibit a peak below the plasmon, corresponding to a so-called surface plasmon, associated with an excitation of the surface. Double scattering is present in the data too, reflected by the peaks at twice the plasmon energy. The general properties of the plasmon are therefore understood. Nevertheless, the width of the plasmon at small Q-values has not been explained satisfactorily and requires a more sophisticated treatment. In [DuBois (1959)] it has been shown that the plasmon acquires a width at all values of Q, when higher-order terms are taken into account. Additional experimental data related to plasmons are reviewed in [Raether (1980)].

Screening of the electron–electron interaction

An important consequence of taking the polarization of the medium into account, is reflected in the resulting modification of the electron–electron interaction. By summing all divergent contributions, the Fourier component $V(Q)$ of the Coulomb interaction becomes

$$W^{(0)}(Q, E) = V(Q) + V(Q)2\Pi^{(0)}(Q, E)V(Q) \qquad (14.58)$$
$$+ V(Q)2\Pi^{(0)}(Q, E)V(Q)2\Pi^{(0)}(Q, E)V(Q) + \dots$$
$$= V(Q) + V(Q)2\Pi^{(0)}(Q, E)W^{(0)}(Q, E) = \frac{V(Q)}{1 - 2V(Q)\Pi^{(0)}(Q, E)}.$$

The renormalization of the interaction dramatically changes the Coulomb interaction, as can be seen in the limit for small E, using Eq. (14.38). In this limit one finds

$$W^{(0)}(Q, 0) = \frac{4\pi e^2 \hbar^2}{Q^2 + p_{TF}^2 g(Q/p_F)}, \qquad (14.59)$$

where

$$p_{TF}^2 = \frac{4r_s(9\pi/4)^{1/3}}{\pi} p_F^2 \qquad (14.60)$$

and

$$g(x) = \frac{1}{2} - \frac{1}{2x}\left(1 - \frac{1}{4}x^2\right)\ln\left|\frac{1 - \frac{1}{2}x}{1 + \frac{1}{2}x}\right|. \qquad (14.61)$$

The resulting interaction is of a Yukawa type. In coordinate space it is screened with a characteristic length given by $\hbar/p_{TF}$. For energies, small compared to the plasmon energy, and momenta, small compared to p_F, the interaction is predominantly real, but in general $W^{(0)}$ is retarded with a corresponding energy dependence.

14.4 Correlation energy

14.4.1 *Correlation energy and the polarization propagator*

In Sec. 13.6, the potential energy was linked to the polarization propagator for the general case. Specializing to an infinite homogeneous system, leads to some simplifications of the corresponding expression, given in Eq. (13.81). We will first study the two contributions that contain one-body density matrix elements in Eq. (13.81). It is not necessary to use ph notation for these terms, so we will employ the usual second-quantized form of $\hat{V}$. Translational invariance suggests the use of momentum space for the sp basis. When a spin-independent, central and local interaction is employed, the second-quantized form reads

$$\hat{V} = \frac{1}{2V} \sum_{\boldsymbol{P}\boldsymbol{p}\boldsymbol{p}'} \sum_{m_1 m_2} V(\boldsymbol{p} - \boldsymbol{p}') \, a^{\dagger}_{\boldsymbol{P}/2+\boldsymbol{p}m_1} a^{\dagger}_{\boldsymbol{P}/2-\boldsymbol{p}m_2} a_{\boldsymbol{P}/2-\boldsymbol{p}'m_2} a_{\boldsymbol{P}/2+\boldsymbol{p}'m_1},$$
(14.62)

where we consider a system with spin degeneracy $\nu = 2$. The generalization to include isospin is straightforward. The corresponding result for the Coulomb interaction is given in Eq. (5.28).[4] We have used the version of $\hat{V}$ that does not employ antisymmetrized matrix elements. It slightly simplifies the following presentation, but is not necessary. No substantial effort is required to study the case for the most general two-body interaction, but this is left for the reader as an exercise.

Starting with the expression, containing two one-body density matrix elements in Eq. (13.81) (hence the subscript nn), we find

$$\langle \Psi_0^N | \hat{V} | \Psi_0^N \rangle_{nn} \Rightarrow \frac{1}{2V} \sum_{\boldsymbol{P}\boldsymbol{p}\boldsymbol{p}'} \sum_{m_1 m_2} V(\boldsymbol{p} - \boldsymbol{p}') \quad\quad\quad (14.63)$$

$$\times \langle \Psi_0^N | a^{\dagger}_{\boldsymbol{P}/2+\boldsymbol{p}m_1} a_{\boldsymbol{P}/2+\boldsymbol{p}'m_1} | \Psi_0^N \rangle \langle \Psi_0^N | a^{\dagger}_{\boldsymbol{P}/2-\boldsymbol{p}m_2} a_{\boldsymbol{P}/2-\boldsymbol{p}'m_2} | \Psi_0^N \rangle$$

$$= \frac{1}{2V} V(0) \sum_{\boldsymbol{p}_1 m_1} \langle \Psi_0^N | a^{\dagger}_{\boldsymbol{p}_1 m_1} a_{\boldsymbol{p}_1 m_1} | \Psi_0^N \rangle \sum_{\boldsymbol{p}_2 m_2} \langle \Psi_0^N | a^{\dagger}_{\boldsymbol{p}_2 m_2} a_{\boldsymbol{p}_2 m_2} | \Psi_0^N \rangle$$

$$= \frac{1}{2V} V(0) \, N^2,$$

---

[4]In the general case, the constraint $\boldsymbol{p} \neq \boldsymbol{p}'$ does not apply.

since momentum conservation implies $p = p'$. A return to individual momenta p_1 and p_2, allows the identification of the expectation values of the number operator in the last part of Eq. (14.63). Only momentum conservation and the fact that the ground state has zero total momentum have been used here. The same outcome is obtained when the correlated ground state $|\Psi_0^N\rangle$ is replaced by the noninteracting one $|\Phi_0^N\rangle$.

We proceed with the adaptation of the term with one density matrix element in Eq. (13.81). Using the n subscript for this contribution, we find

$$\langle \Psi_0^N | \hat{V} | \Psi_0^N \rangle_n \Rightarrow -\frac{1}{2V} \sum_{Ppp'} \sum_{m_1 m_2} V(p - p') \delta_{P/2+p, P/2-p'} \delta_{m_1 m_2} \qquad (14.64)$$

$$\times \langle \Psi_0^N | a^\dagger_{P/2-pm_2} a_{P/2+p'm_1} | \Psi_0^N \rangle$$

$$= -\frac{1}{2V} \sum_{Ppm_1} V(2p) \langle \Psi_0^N | a^\dagger_{P/2-pm_1} a_{P/2-pm_1} | \Psi_0^N \rangle = -\frac{1}{2V} \sum_p V(2p) \, N.$$

Again, the same ingredients were used as for Eq. (14.63), ensuring that the same result is obtained when $|\Psi_0^N\rangle$ is replaced by $|\Phi_0^N\rangle$.

Keeping the ph notation with corresponding momenta (see Sec. 14.1) for the remaining term in Eq. (13.81), we couple the ph matrix element and the polarization propagator to good total spin (and its projection). Adding and subtracting the same expression involving the noninteracting form, allows to identify the sum of the former term and the one corresponding to Eqs. (14.63) and (14.64) as the expectation value of $\hat{V}$, with respect to the noninteracting ground state $|\Phi_0^N\rangle$. Combining all these ingredients yields the expectation value of the potential energy in the correlated ground state

$$\langle \Psi_0^N | \hat{V} | \Psi_0^N \rangle = \langle \Phi_0^N | \hat{V} | \Phi_0^N \rangle - \frac{1}{4\pi} \sum_{Qpp'} \sum_{SM_S} \langle Q, p | V_{ph}^{SM_S} | Q, p' \rangle$$

$$\times \int_0^\infty dE \, \left\{ \text{Im} \, \Pi_{SM_S}(p', p; Q, E) - \delta_{p,p'} \text{Im} \, \Pi^{(0)}(p; Q, E) \right\}. \qquad (14.65)$$

Adapting the result for the energy shift, given in Eq. (13.78) with $\hat{H}_1 = \hat{V}$, we thus find the following useful expression for the correlation energy

$$E_{corr} = F_0^N - \langle \Phi_0^N | \hat{H} | \Phi_0^N \rangle \qquad (14.66)$$

$$= -\frac{1}{4\pi} \sum_{Qpp'} \sum_{SM_S} \int_0^1 \frac{d\lambda}{\lambda} \langle Q, p | \lambda V_{ph}^{SM_S} | Q, p' \rangle$$

$$\times \int_0^\infty dE \, \left\{ \text{Im} \, \Pi_{SM_S}^\lambda(p', p; Q, E) - \delta_{p,p'} \text{Im} \, \Pi^{(0)}(p; Q, E) \right\}.$$

The subtraction of the $\Pi^{(0)}$ term ensures that the correlation energy starts with contributions from second-order terms in the interaction.

14.4.2 *Correlation energy of the electron gas in RPA*

The general results of Eqs. (14.65) and (14.66) can now be applied to the electron gas by applying the RPA to the polarization propagator Π. Sec. 14.3 makes it abundantly clear that the divergent character of the Coulomb interaction can be remedied by summing all ring diagrams. A similar statement is appropriate for the diverging contributions to the potential or correlation energy, when second or higher-order ring diagram contributions are considered. The RPA approach for the electron gas incorporates this important feature and suggests further that it is permissible to neglect the exchange contribution to the ph matrix element of the Coulomb interaction in Eqs. (14.65) and (14.66). If this approximation is made, only the $S = 0$ contribution remains. Keeping the magnitude of Q in the notation, the summations over p and p' in Eq. (14.66) can be incorporated as in Eqs. (14.19) and (14.20), yielding

$$E_r = -\frac{1}{2\pi} \sum_Q \int_0^1 \frac{d\lambda}{\lambda} \lambda \frac{4\pi e^2 \hbar^2}{Q^2} \tag{14.67}$$

$$\times \int_0^\infty dE \left\{ \mathrm{Im}\, \Pi_{S=0}^{\lambda,RPA}(Q,E) - \mathrm{Im}\, \Pi^{(0)}(Q,E) \right\}$$

$$= -\frac{e^2 V}{\pi^2 \hbar} \int_0^\infty dQ \int_0^1 d\lambda \int_0^\infty dE \left\{ \mathrm{Im}\, \Pi_{S=0}^{\lambda,RPA}(Q,E) - \mathrm{Im}\, \Pi^{(0)}(Q,E) \right\}.$$

The direct ph matrix element of the Coulomb interaction [see Eq. (14.40)] was substituted in the first line of this equation. To arrive at the last expression, the remaining summation over Q has been replaced by the usual integration, with its angular part (over $\hat{Q}$) performed. The notation E_r signals that Eq. (14.67) involves the ring contribution to the correlation energy. Equation (14.67) clearly demonstrates that the ring summation eliminates the divergence of the Coulomb interaction, leaving a finite result. Indeed, keeping the lowest and first-order contribution to $\Pi^{\lambda,RPA}$, it is possible to show that Eq. (14.67) produces the logarithmic divergence, associated with the second-order ring contribution to the correlation energy (see also [Mattuck (1992)]). Equation (14.67) is purely real, as it should be. The expressions for $\mathrm{Im}\, \Pi^{(0)}$ are found in Sec. 14.2. Calculations for $\mathrm{Im}\, \Pi^{\lambda,RPA}$ can easily be performed by employing the information from

Sec. 14.3. The RPA correlation energy is compared with other approxima-
tions in Sec. 16.2.3. We note that the expectation value of the electron gas
Hamiltonian at the mean-field level was already calculated in Sec. 10.5 [see
Eq. (10.180)].

Important analytic results for the ring diagram contribution to the correlation energy can be obtained when r_s is small, as discussed in detail
in [Fetter and Walecka (1971)]. This correction represents the dominant
deviation from Eq. (10.180), when $r_s \to 0$, and was first studied by [Macke
(1950)]. The correct expression was given by [Gell-Mann and Brueckner
(1957)] (see also [Sawada *et al.* (1957)]). In atomic units it reads

$$\frac{E_{corr}}{N} = \frac{1}{2}\left[\frac{2}{\pi^2}(1-\ln 2)\ln r_s - 0.094 + ...\right]. \qquad (14.68)$$

This last expression includes the second-order exchange contribution. The
latter may be studied directly, or obtained from Eq. (14.66) by including
the $S = 1$ contribution, while including only the lowest and first-order
contribution to $\Pi_{S=1}^{\lambda,RPA}$. Terms of order $r_s \ln r_s$ were studied by [DuBois
(1959)] (see [Carr and Maradudin (1964)]). Some of the original papers on
the high-density electron gas are collected in [Pines (1962)].

14.5   Response of nuclear matter with π and ρ meson quantum numbers

In the previous section, excited states in an infinite system were studied
for a strongly repulsive interaction. For the unique case of the Coulomb
interaction, a collective state, the plasmon appears in the electron gas above
the ph continuum for a range of momenta Q, including the limit $Q \to 0$.
Application of the RPA method to other infinite systems exhibits certain
similarities for the case of repulsive interactions. Quite different behavior
is observed for an attractive interaction. In this section we will study the
excitations of nuclear matter with quantum numbers corresponding to pions
and rho mesons, covering both possibilities.

In Sec. 4.4 we encountered the basic Yukawa-type interaction, appropriate for the exchange of finite mass mesons between nucleons. A more
complete description of this exchange mechanism involves the propagation
process of *e.g.* the pion, as it travels from one nucleon to the other. Since
the pion is a boson, the energy structure of the pion propagator in free

space has a pole at $\pm$ the relativistic energy

$$E_Q = \sqrt{\hbar^2 c^2 \mu_\pi^2 + Q^2 c^2},\qquad(14.69)$$

where $\mu_\pi \hbar c = m_\pi c^2$ corresponds to the rest mass of the pion. While we will not pursue the appropriate analysis for a relativistic particle, one may point to the analogy with boson propagators to be studied in Sec. 18.3, to realize that something of the form

$$G_\pi(\boldsymbol{Q},E) \propto \frac{1}{E - \sqrt{\hbar^2 c^2 \mu_\pi^2 + Q^2 c^2} + i\eta} - \frac{1}{E + \sqrt{\hbar^2 c^2 \mu_\pi^2 + Q^2 c^2} - i\eta}$$
$$= \frac{2E_Q}{E^2 - (\hbar^2 c^2 \mu_\pi^2 + Q^2 c^2) + i\eta}\qquad(14.70)$$

must be expected. The denominator of the last result reflects the free energy-momentum relation for the pion, which is the basis for the Klein–Gordon equation. The proportionality factor in Eq. (14.70) is then simply $2E_Q$, so that

$$G_\pi(\boldsymbol{Q},E) = \frac{1}{E^2 - (\hbar^2 c^2 \mu_\pi^2 + Q^2 c^2) + i\eta}.\qquad(14.71)$$

The intrinsic quantum numbers of the pion dictate the character of the operator that couples it to a nucleon. Since the pion is a scalar particle with intrinsic negative parity, the operator must be a pseudoscalar. The only (pseudo) vectors, available for the pion and the nucleon, that yield this combination, are the momentum of the pion and the spin of the nucleon. A $\boldsymbol{\sigma} \cdot \boldsymbol{Q}$ coupling can thus be anticipated. In addition, the pion has three charge states, indicating that it has isospin $T = 1$. It implies that an operator carrying isospin 1, must be attached to each vertex as well, but in such a way that the total operator acting on the two nucleons is an isoscalar. The isospin dependence of such an operator is therefore $\boldsymbol{\tau}_1 \cdot \boldsymbol{\tau}_2$. The generalization of the Yukawa interaction for pion exchange between nucleons, can then be written as

$$V^\pi(\boldsymbol{Q},E) = \frac{f_{\pi NN}^2}{\mu_\pi^2} \frac{\boldsymbol{\sigma}_1 \cdot \boldsymbol{Q}c\, \boldsymbol{\sigma}_2 \cdot \boldsymbol{Q}c}{E^2 - (\hbar^2 c^2 \mu_\pi^2 + Q^2 c^2) + i\eta}\, \boldsymbol{\tau}_1 \cdot \boldsymbol{\tau}_2,\qquad(14.72)$$

where the coupling strength is given by $f_{\pi NN}^2/4\pi = 0.08\ \hbar c$ and the pion mass, averaged over its charge states, corresponds to 138 MeV/c^2. Equation (14.72) must be considered as a matrix element in momentum space, depending only on the momentum transfer, as discussed in Sec. 4.4. In low-energy, elastic nucleon–nucleon (NN) scattering, the Born contribution, associated with Eq. (14.72), forces the energy transfer E to be zero. In this limit (and in a similar way for $E \neq 0$), one may decompose the operator structure of the pion-exchange interaction in the following way

$$V^{\pi}(\boldsymbol{Q}, 0) = -\frac{1}{3} \frac{f_{\pi NN}^2}{\mu_{\pi}^2} \frac{3 \boldsymbol{\sigma}_1 \cdot \boldsymbol{Q} c \, \boldsymbol{\sigma}_2 \cdot \boldsymbol{Q} c - \boldsymbol{\sigma}_1 \cdot \boldsymbol{\sigma}_2 Q^2 c^2}{\hbar^2 c^2 \mu_{\pi}^2 + Q^2 c^2} \boldsymbol{\tau}_1 \cdot \boldsymbol{\tau}_2$$
$$- \frac{1}{3} f_{\pi NN}^2 \hbar^2 c^2 \frac{\boldsymbol{\sigma}_1 \cdot \boldsymbol{\sigma}_2}{\hbar^2 c^2 \mu_{\pi}^2 + Q^2 c^2} \boldsymbol{\tau}_1 \cdot \boldsymbol{\tau}_2$$
$$+ \frac{1}{3} \frac{f_{\pi NN}^2}{\mu_{\pi}^2} \boldsymbol{\sigma}_1 \cdot \boldsymbol{\sigma}_2 \, \boldsymbol{\tau}_1 \cdot \boldsymbol{\tau}_2. \tag{14.73}$$

The first term in Eq. (14.73) contains the so-called tensor force, since the operator is the scalar obtained by coupling a rank two tensor, composed out of $\boldsymbol{\sigma}_1$ and $\boldsymbol{\sigma}_2$, with a rank two tensor proportional to the spherical harmonic $Y_{2M}(\hat{\boldsymbol{Q}})$. The operator can be written as (see also App. B)

$$S_{12}(\hat{\boldsymbol{Q}}) = 3 \, \boldsymbol{\sigma}_1 \cdot \hat{\boldsymbol{Q}} \, \boldsymbol{\sigma}_2 \cdot \hat{\boldsymbol{Q}} - \boldsymbol{\sigma}_1 \cdot \boldsymbol{\sigma}_2 = \sqrt{24\pi} \, \left[[\boldsymbol{\sigma}_1 \otimes \boldsymbol{\sigma}_2]^2 \otimes Y_2 \right]_0^0$$
$$= \sqrt{24\pi} \sum_{\mu} (2 \, \mu \, 2 \, -\mu \mid 0 \, 0) \, [\boldsymbol{\sigma}_1 \otimes \boldsymbol{\sigma}_2]_{\mu}^2 \, Y_{2,-\mu}(\hat{\boldsymbol{Q}}). \tag{14.74}$$

The spin operators are coupled to a spherical tensor of rank 2 as follows

$$[\boldsymbol{\sigma}_1 \otimes \boldsymbol{\sigma}_2]_{\mu}^2 = \sum_{m_1 m_2} (1 \, m_1 \, 1 \, m_2 \mid 2 \, \mu) \, (\sigma_1)_{m_1}^1 \, (\sigma_2)_{m_2}^1, \tag{14.75}$$

explaining the use of the $\otimes$ symbol. In the last equation the spherical components of the spin operators are employed, given by $\sigma_{\pm 1}^1 = \mp 1/\sqrt{2}(\sigma_x \pm i\sigma_y)$ and $\sigma_0^1 = \sigma_z$, respectively. It is the tensor component of the nuclear interaction that is responsible for the quadrupole moment of the deuteron, since it couples the $^3S_1$ to the $^3D_1$ partial wave, leading to a D-state admixture in its ground state. The second term in Eq. (14.73) has the momentum dependence of the Yukawa interaction, discussed in Ch. 4, but also contains an explicit spin and isospin dependence. The final term in Eq. (14.73) corresponds to a δ-function upon FT to coordinate space.

The relevant ph configurations, which sample the influence of this π-exchange interaction in nuclear matter, are characterized by specific spin (and isospin) quantum numbers, as for the response to a density fluctuation, discussed in the previous section. The relevant excitation operator is written as

$$\hat{O}_{\boldsymbol{\sigma}\cdot\hat{\boldsymbol{Q}}\tau_z} = \sum_{\boldsymbol{p}} \sum_{m_s m'_s} \sum_{m_t m'_t} \langle m_s m_t | \boldsymbol{\sigma}\cdot\hat{\boldsymbol{Q}}\ \tau_z\ |m'_s m'_t\rangle\ a^\dagger_{\boldsymbol{p}m_s m_t} a_{\boldsymbol{p}-\boldsymbol{Q}m'_s m'_t},$$

$$(14.76)$$

where the z-component of the isospin operator is appropriate for symmetric nuclear matter. Choosing $\hat{\boldsymbol{Q}}$ along the z-axis, one can show in a similar way as in the previous section (see also Sec. 14.1) that the total spin and its projection of the ph states must be 1 and 0 respectively, while the total isospin equals 1. The presence of the tensor force in Eq. (14.73) (and in general when $E \neq 0$), makes the evaluation of ph matrix elements of the interaction somewhat more involved, but nevertheless straightforward [Dickhoff *et al.* (1981)] (see also App. B). These details conspire to yield the simple result that the direct contribution to the ph matrix element with $S = 1, M = 0$ and $T = 1$, is given by

$$\langle \boldsymbol{Q}, \boldsymbol{p} |\, V_{ph}^{\pi, S=1 M_S=0 T=1} \,| \boldsymbol{Q}, \boldsymbol{p}' \rangle_D = \frac{4f^2_{\pi NN}}{\mu^2_\pi}\ \frac{Q^2 c^2}{E^2 - (\hbar^2 c^2 \mu^2_\pi + Q^2 c^2) + i\eta},$$

$$(14.77)$$

with $\hat{\boldsymbol{Q}}$ along the z-axis. The tensor force of Eq. (14.73) contributes $2/3$ and the central part $1/3$ to this matrix element. The expression for the exchange contribution is more complicated and will not be derived here. It is not permissible to neglect such exchange contributions outright, as for the Coulomb interaction. Nevertheless, it is possible to represent their main effect as well as other corrections, for example related to the proper treatment of short-range correlations, by a constant interaction[5] which has the simple form

$$V_{ph}^{\prime S=1 T=1} = \frac{f^2_{\pi NN}}{\mu^2_\pi}\ g'\ \boldsymbol{\sigma}_1\cdot\boldsymbol{\sigma}_2\ \boldsymbol{\tau}_1\cdot\boldsymbol{\tau}_2.$$

$$(14.78)$$

The parameter g' is used to mimic all relevant correlations, which need to be included to correct the ph interaction in this spin-isospin channel. A plot of the Q-dependence of Eq. (14.77) combined with the matrix element of Eq. (14.78) for several values of g', is shown in Fig. 14.7 for $E = 0$. The figure clarifies that within the range of values, commonly considered for the parameter g', the effective interaction turns from repulsion to attraction rather rapidly, as a function of Q. This has interesting consequences for the Q-dependence of the response with the corresponding ph quantum numbers.

---

[5] A slight Q-dependence is obtained in detailed calculations of this quantity.

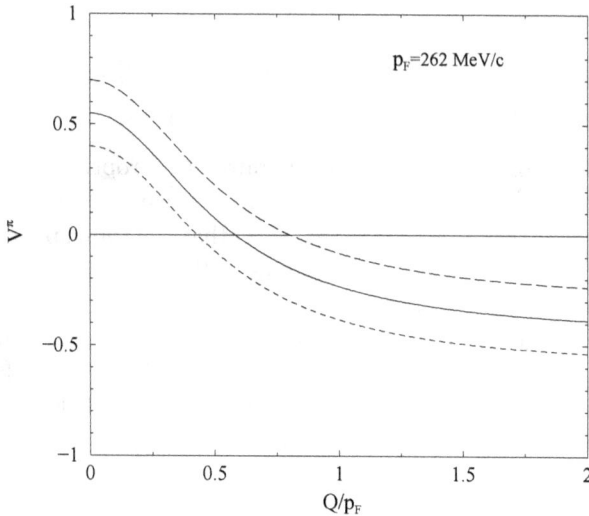

Fig. 14.7 Effective interaction (dimensionless) in the spin-isospin channel with pion quantum numbers for three values of g' corresponding to 0.4 (short-dashed), 0.55 (solid), and 0.7 (long-dashed), respectively, plotted as a function of Q/p_F. The density corresponds to normal nuclear matter, which is equivalent to $k_F = 1.33$ fm^{-1}.

For small values of Q the effective interaction is repulsive, whereas for larger values it can become quite attractive, depending on the value of g'.

Before discussing the consequences of this ph interaction in the so-called spin-longitudinal channel (spin along the quantization axis parallel to $\hat{Q}$), it is necessary to introduce one additional ingredient into the discussion. The latter can be motivated on the basis of experimental data, associated with pion–nucleon scattering. For pion kinetic energies below 300 MeV the interaction is dominated by a p-wave resonance, corresponding to an excited state of the nucleon with total spin and isospin $\frac{3}{2}$, the so-called Δ-isobar [Ericson and Weise (1988)]. In vacuum the mass of this resonance is approximately 1232 MeV/c^2 and its width, at resonance, is 115 MeV. Very successful descriptions of pion–nucleon and pion–nucleus scattering can be constructed using the Δ-isobar as a separate baryonic species, without requiring knowledge of its intrinsic structure. This is not unlike the use of nucleons in the description of nuclei and nuclear matter. It simply reflects the identification of the relevant degrees of freedom at low energy that dominate the physics (see the discussion of the NN interaction in Sec. 4.3). In the present context, the strong coupling between pions, nucleons, and

Δ-isobars can be accomodated by introducing pion-exchange interactions, which convert a nucleon into a Δ-isobar or vice versa. In complete analogy with Eq. (14.72), such an interaction reads

$$V_{NN \to \Delta N}^{\pi}(\boldsymbol{Q}, E) = \frac{f_{\pi NN} f_{\pi N\Delta}}{\mu_{\pi}^2} \frac{\boldsymbol{S}_1^{\dagger} \cdot \boldsymbol{Q}c \, \boldsymbol{\sigma}_2 \cdot \boldsymbol{Q}c}{E^2 - (\hbar^2 c^2 \mu_{\pi}^2 + \boldsymbol{Q}^2 c^2) + i\eta} \, \boldsymbol{T}_1^{\dagger} \cdot \boldsymbol{\tau}_2,$$

(14.79)

for the transition from an NN to a ΔN state with $f_{\pi N\Delta} \approx 2 f_{\pi NN}$. Note that the operators $\boldsymbol{\sigma}_1$ and $\boldsymbol{\tau}_1$ of Eq. (14.72) have been replaced by so-called transition spin and isospin operators, which carry spherical tensor rank 1 in spin and isospin space, respectively. These operators have spherical components with matrix elements between spin and isospin $\frac{1}{2}$ and $\frac{3}{2}$ states that are written as

$$\langle \tfrac{3}{2} m_\Delta | \, (\boldsymbol{S}_1^{\dagger})_\lambda^1 \, | \tfrac{1}{2} m_N \rangle = (\tfrac{1}{2} \, m_N \, 1 \, \lambda \, | \, \tfrac{3}{2} \, m_\Delta)$$

(14.80)

and

$$\langle \tfrac{3}{2} t_\Delta | \, (\boldsymbol{T}_1^{\dagger})_\lambda^1 \, | \tfrac{1}{2} t_N \rangle = (\tfrac{1}{2} \, t_N \, 1 \, \lambda \, | \, \tfrac{3}{2} \, t_\Delta).$$

(14.81)

In the opposite process, π-exchange de-excites a Δ-isobar to a nucleon. It is simply given by the adjoint of Eq. (14.79). A combination of the two processes occurs in the interaction, which takes a $N\Delta$ to a ΔN state

$$V_{N\Delta \to \Delta N}^{\pi}(\boldsymbol{Q}, E) = \frac{f_{\pi N\Delta}^2}{\mu_{\pi}^2} \frac{\boldsymbol{S}_1^{\dagger} \cdot \boldsymbol{Q}c \, \boldsymbol{S}_2 \cdot \boldsymbol{Q}c}{E^2 - (\hbar^2 c^2 \mu_{\pi}^2 + \boldsymbol{Q}^2 c^2) + i\eta} \, \boldsymbol{T}_1^{\dagger} \cdot \boldsymbol{T}_2. \quad (14.82)$$

The corresponding diagrams, representing the interactions in Eqs. (14.79) and (14.82) in ph coupling schemes, are shown in Fig. 14.8a) and b). Only direct contributions to these interactions are shown with the Δ represented by a double-arrowed line.

To include the coupling to these Δ-isobar states in the RPA description to the response, Eq. (14.21) must be generalized into a coupled problem between ph and Δh polarization propagators, expressed by

$$\Pi_N^{\pi}(\boldsymbol{Q}, E) = \Pi_N^{(0)}(\boldsymbol{Q}, E)$$

(14.83)

$$+ \, \Pi_N^{(0)}(\boldsymbol{Q}, E) \, V_{NN}^{\pi}(\boldsymbol{Q}, E) \, \Pi_N^{\pi}(\boldsymbol{Q}, E) + \Pi_N^{(0)}(\boldsymbol{Q}, E) \, V_{N\Delta}^{\pi}(\boldsymbol{Q}, E) \, \Pi_{\Delta}^{\pi}(\boldsymbol{Q}, E)$$

and

$$\Pi_{\Delta}^{\pi}(\boldsymbol{Q}, E) = \Pi_{\Delta}^{(0)}(\boldsymbol{Q}, E)$$

(14.84)

$$+ \, \Pi_{\Delta}^{(0)}(\boldsymbol{Q}, E) \, V_{\Delta N}^{\pi}(\boldsymbol{Q}, E) \, \Pi_N^{\pi}(\boldsymbol{Q}, E) + \Pi_{\Delta}^{(0)}(\boldsymbol{Q}, E) \, V_{\Delta\Delta}^{\pi}(\boldsymbol{Q}, E) \, \Pi_{\Delta}^{\pi}(\boldsymbol{Q}, E).$$

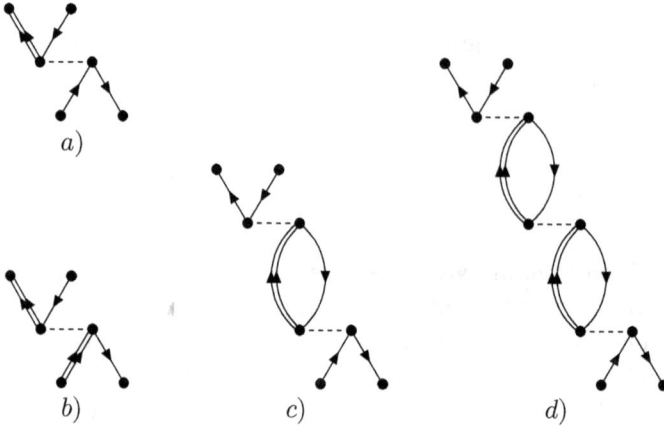

Fig. 14.8 Diagrammatic representation of ph diagrams involving Δh excitations, generated by employing the interactions of Eqs. (14.79) [a)] and (14.82) [b)]. The Δ-isobar is represented here by the customary double-arrowed line. No confusion with fully dressed sp propagators should arise, since nucleons will be treated only at the mean-field level here. Some higher-order diagrams, generated by Eq. (14.83), are shown in c) and d).

The interaction between nucleon ph states is denoted by V_{NN}^{π} and represents the sum of Eq. (14.77) and the matrix element of Eq. (14.78), which is simply given by $4f_{\pi NN}^2 g'/\mu_\pi^2$. The direct contribution to the ph-Δh matrix element has an identical energy and momentum dependence as Eq. (14.77), but the factor $4f_{\pi NN}^2$ must be replaced by $8f_{\pi NN}f_{\pi N\Delta}/3$. In analogy to Eq. (14.78), one also includes the matrix element of

$$V'^{S=1\,T=1}_{N\Delta} = \frac{f_{\pi NN}f_{\pi N\Delta}}{\mu_\pi^2}\, g'_\Delta\, \boldsymbol{S}_1^\dagger \cdot \boldsymbol{\sigma}_2\, \boldsymbol{T}_1^\dagger \cdot \boldsymbol{\tau}_2, \tag{14.85}$$

which yields $8f_{\pi NN}f_{\pi N\Delta}g'_\Delta/3V\mu_\pi^2$. A similar result is valid for the adjoint interaction, needed for $V_{\Delta N}^{\pi}$ in Eq. (14.84). Finally, the direct matrix element of Eq. (14.82), contributing to $V_{\Delta\Delta}^{\pi}$, again has the same momentum and energy dependence as Eq. (14.77), but requires the factor $16f_{\pi N\Delta}^2/9$ to replace $4f_{\pi NN}^2$ in Eq. (14.77). An interaction of the kind

$$V'^{S=1\,T=1}_{\Delta\Delta} = \frac{f_{\pi N\Delta}^2}{\mu_\pi^2}\, g'_{\Delta\Delta}\, \boldsymbol{S}_1^\dagger \cdot \boldsymbol{S}_2\, \boldsymbol{T}_1^\dagger \cdot \boldsymbol{T}_2 \tag{14.86}$$

is also included, which can be handled in the same way. The evaluation of these matrix elements requires the decomposition of the relevant interactions into central and tensor contributions as in Eq. (14.73) and the subsequent use of standard angular momentum algebra [Dickhoff *et al.* (1981)] (see also App. B).

To complete the information included in the coupled Eqs. (14.83) and (14.84), the noninteracting polarization propagator involving Δh states is required

$$\Pi_\Delta^{(0)}(\boldsymbol{Q},E) = \int \frac{d\boldsymbol{p}}{(2\pi\hbar)^3} \left\{ \frac{\theta(p_F - |\boldsymbol{p} - \boldsymbol{Q}/2|)}{E - [\varepsilon_\Delta(\boldsymbol{p} + \boldsymbol{Q}/2) - \varepsilon(\boldsymbol{Q}/2 - \boldsymbol{p})] + i\eta} \right.$$
$$\left. - \frac{\theta(p_F - |\boldsymbol{p} + \boldsymbol{Q}/2|)}{E + [\varepsilon_\Delta(\boldsymbol{Q}/2 - \boldsymbol{p}) - \varepsilon(\boldsymbol{p} + \boldsymbol{Q}/2)] - i\eta} \right\}. \qquad (14.87)$$

It is appropriate, for low-energy considerations, to neglect the width of the Δ-isobar, so that

$$\varepsilon_\Delta(\boldsymbol{p}) = \frac{\boldsymbol{p}^2}{2m_\Delta} + (m_\Delta - m)c^2 \qquad (14.88)$$

includes the kinetic energy and the mass difference with the nucleon (about 300 MeV). An important difference with Eq. (14.22) is that the Δ-isobar does not experience the Pauli principle for nucleons. The corresponding step functions are therefore absent in Eq. (14.87). The evaluation requires similar steps as for Eq. (14.22).

The polarization propagators can now be evaluated by performing the 2x2 matrix inversion that solves Eqs. (14.83) and (14.84), using the known ingredients $\Pi_N^{(0)}$, $\Pi_\Delta^{(0)}$, and the various interaction terms. Little is known about the parameter $g'_{\Delta\Delta}$ and insofar as the effects simulated by this parameter are associated with short-range correlations, one may argue that numbers similar to g' (also denoted g'_N) may be appropriate. This does not apply to the interaction that links nucleon ph with Δh states. An example of such an interaction is given by Eq.(14.79) for the pion-exchange contribution. Due to the spin and isospin $\frac{3}{2}$ character of the Δ-isobar, it is not possible to construct partial-wave matrix elements between S-wave initial and final states, when the usual particle–particle coupling is performed. Considerations related to the magnitude and relevance of the exchange matrix elements of Eq. (14.79), nevertheless suggest a value of about $g'_\Delta = \frac{1}{3}$ [Dickhoff *et al.* (1981)].

The imaginary part of Π_N^π is shown in Fig. 14.9 at normal nuclear matter density for different values of Q with, and without, the inclusion of Δh propagation. The results including only nucleons, precisely reflect the behavior of the interaction, shown in Fig. 14.7, for a value of $g' = 0.55$. For $Q/p_F = 0.2$ the interaction is sufficiently repulsive to yield a solution above the ph continuum, also indicated in the figure. This discrete state carries 71% of the energy weighted strength. The repulsive character

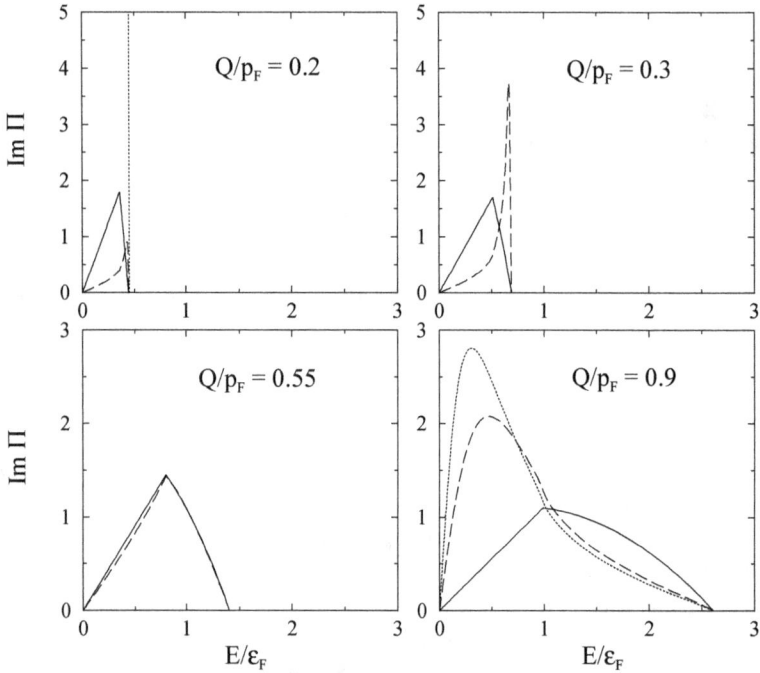

Fig. 14.9   Response of nuclear matter at a density corresponding to $p_F = 1.36$ fm^{-1} for different values of Q/p_F. The imaginary of the polarization propagator for pionic modes with nucleons only, is given by the dashed line, including also isobars (only for $Q/p_F = 0.9$) by the dotted line, and for comparison the noninteracting propagator by the full line. Where appropriate, the location of the discrete state above the ph continuum is indicated by the dotted vertical line. The imaginary part of the polarization propagator, divided by the constant factor $-mp_F/(8\pi\hbar^3)$, is plotted as a function of the energy in units of ε_F. Vertical scales are not all the same.

of the interaction is sufficiently reduced to remove this bound state at $Q/p_F = 0.3$, although the response in the continuum clearly displays the removal of strength to higher energy. At $Q/p_F = 0.55$ the interaction basically vanishes, as shown by the corresponding panel, where the response can hardly be distinguished from the noninteracting result. The interaction has turned quite attractive for $Q/p_F = 0.9$, as demonstrated by the appearance of a peak in the strength distribution at much lower energy than for the noninteracting case. The inclusion of the Δh states is particularly important for this value of Q and has been omitted in the other panels of Fig. 14.7, where it plays no role for the chosen values of the parameters. For $Q/p_F = 0.9$ however, a substantial enhancement of the response at low energy is observed, over the already enhanced one for nucleons only. The inclusion of the Δ degree of freedom in nuclear matter therefore has important consequences for excited states with pionic quantum numbers. The strength moves to lower energy at this Q-value and it is easy to visualize it moving down further with increasing density. Ultimately, a critical density occurs at which this excitation mode becomes unstable, at least when the RPA is applied. Such an instability, referred to as pion condensation, is basically identical to the appearance of complex eigenvalues when the RPA eigenvalue problem is solved, as discussed in Sec. 13.3. The possibility of pion condensation was pointed out by [Migdal (1972)]. Several scenarios have been developed, which pursue the possible consequences of collective pionic modes that develop into a new ground state of nuclear or neutron matter at higher density [Migdal (1978); Ericson and Weise (1988)].

It is also possible to study the other isovector meson, the rho meson, within the same framework. The interaction between two nucleons exchanging such a meson is written as

$$V^\rho(\boldsymbol{Q}, E) = \frac{f_{\rho NN}^2}{\mu_\rho^2} \frac{\boldsymbol{\sigma}_1 \times \boldsymbol{Q} c \cdot \boldsymbol{\sigma}_2 \times \boldsymbol{Q} c}{E^2 - (\hbar^2 c^2 \mu_\rho^2 + \boldsymbol{Q}^2 c^2) + i\eta} \boldsymbol{\tau}_1 \cdot \boldsymbol{\tau}_2. \tag{14.89}$$

In the static limit (and in a similar way for $E \neq 0$), the decomposition of the rho-exchange interaction is given by

$$V^\rho(\boldsymbol{Q}, 0) = \frac{1}{3} \frac{f_{\rho NN}^2}{\mu_\rho^2} \frac{3\boldsymbol{\sigma}_1 \cdot \boldsymbol{Q} c \, \boldsymbol{\sigma}_2 \cdot \boldsymbol{Q} c - \boldsymbol{\sigma}_1 \cdot \boldsymbol{\sigma}_2 \boldsymbol{Q}^2 c^2}{\hbar^2 c^2 \mu_\rho^2 + \boldsymbol{Q}^2 c^2} \boldsymbol{\tau}_1 \cdot \boldsymbol{\tau}_2$$
$$- \frac{2}{3} f_{\rho NN}^2 \hbar^2 c^2 \frac{\boldsymbol{\sigma}_1 \cdot \boldsymbol{\sigma}_2}{\hbar^2 c^2 \mu_\rho^2 + \boldsymbol{Q}^2 c^2} \boldsymbol{\tau}_1 \cdot \boldsymbol{\tau}_2$$
$$+ \frac{2}{3} \frac{f_{\rho NN}^2}{\mu_\rho^2} \boldsymbol{\sigma}_1 \cdot \boldsymbol{\sigma}_2 \, \boldsymbol{\tau}_1 \cdot \boldsymbol{\tau}_2. \tag{14.90}$$

The tensor component has the tendency to cancel the one from pion-exchange. By applying the same tensor algebra as for the pion, one may show that the direct matrix element of Eq. (14.89) only yields a contribution (for $\boldsymbol{Q}$ parallel to the z-axis) when $|M_S| = 1$. One therefore obtains

$$\langle \boldsymbol{Q}, \boldsymbol{p} | V_{ph}^{\rho, S=1|M_S|=1 T=1} | \boldsymbol{Q}, \boldsymbol{p}' \rangle_D = \frac{4 f_{\rho NN}^2}{\mu_\rho^2} \frac{\boldsymbol{Q}^2 c^2}{E^2 - (\hbar^2 c^2 \mu_\rho^2 + \boldsymbol{Q}^2 c^2) + i\eta}, \tag{14.91}$$

similar to Eq. (14.77).

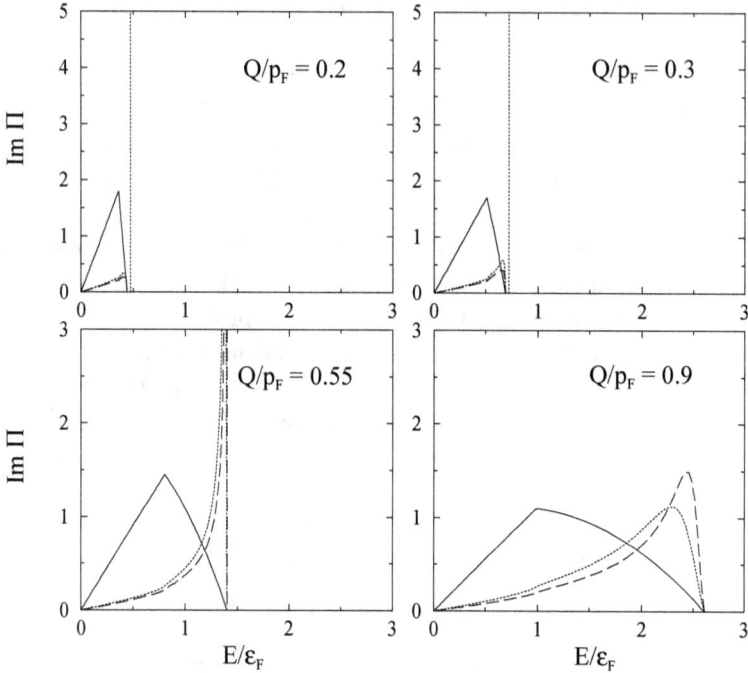

Fig. 14.10   Response of nuclear matter at a density corresponding to $p_F = 1.36$ fm^{-1} for different values of Q/p_F. The imaginary of the polarization propagator for $S = 1, |M_S| = 1, T = 1$ modes with nucleons only is given by the dashed line, including isobars by the dotted line, and for comparison the noninteracting propagator by the full line. Where appropriate the location of the discrete state above the ph continuum is indicated. The imaginary parts of the propagators are plotted as in Fig. 14.9.

The rho meson mass is about 770 MeV with the coupling constant to nucleons corresponding to $f_{\rho NN}^2/\mu_\rho^2 = 2.18 f_{\pi NN}^2/\mu_\pi^2$. This implies that at small Q, the contribution of the g' term is more important than in the pion case. The rho response in Fig. 14.10 (for the same values of Q/p_F as in Fig. 14.9) reflects this feature. The collective mode above the continuum, due to the overall repulsive interaction, now survives to a value of $Q/p_F = 0.50$, whereas it vanishes for $Q/p_F = 0.23$ in the pionic case. The amount of strength in the discrete state corresponds to 85% for $Q/p_F = 0.2$ and 78% for $Q/P_F = 0.3$. When Δ-isobars are included, these values decrease to 81% and 72%, respectively. Small changes can be observed in the continuum response in Fig. 14.10, when the Δh coupling is present. Even for $Q/p_F = 0.9$ the response testifies to the repulsive nature of the interaction in both cases, since the peak of the strength still moves to higher energies, as compared to the noninteracting case.

The marked difference between the π and ρ response function, was first studied by [Alberico *et al.* (1982)] for nuclear matter. Experimental investigations [Carey *et al.* (1984); Taddeucci *et al.* (1994); Wakasa *et al.* (1999)] of the enhancement of the spin-longitudinal (pion), as compared to the spin-transverse response (rho), have not yielded the predicted effect. When the finite size of the nucleus is taken into account, a reduced, but still significant, enhancement remains [Alberico *et al.* (1986)]. The resolution of this issue remains a subject of further investigation, since it touches on the fundamental problem of identifying signatures of the pion-exchange interaction between nucleons in the medium. The influence of Δ-isobar configurations in nuclear matter will also be discussed in the context of the saturation properties of nuclear matter (see Sec. 16.3.4).

14.6 Excitations of a normal Fermi liquid

The final section of this chapter is devoted to an analysis, based on Green's function theory, developed by Landau on the properties of Fermi liquids. We will confine ourselves mostly to a discussion of the last of a series of three seminal papers [Landau (1957a); Landau (1957b); Landau (1959)], which explored the possible excitations of the ^{3}He liquid, leading to the prediction of the so-called zero sound mode. The analysis in [Landau (1959)], employed the properties of exact sp propagators in an infinite homogeneous medium, as they have been partly explored in Ch. 11. In addition to the quasiparticle properties, discussed for fixed momentum, it is necessary to consider the properties of the exact sp propagator for both energies and momenta close the Fermi surface. To this end, we generalize the result of Eq. (11.103), by expanding the self-energy near the Fermi momentum and energy in the following way

$$\Sigma(p; E) = \Sigma(p_F; \varepsilon_F) + \left.\frac{\partial \Sigma(p; \varepsilon_F)}{\partial p}\right|_{p_F} (p - p_F) + \left.\frac{\partial \Sigma(p_F; E)}{\partial E}\right|_{\varepsilon_F} (E - \varepsilon_F).$$
$$(14.92)$$

We note that the imaginary part of the self-energy behaves as $\text{Im}\,\Sigma(p; E) \to C_p(E - \varepsilon_F)^2$ and hence does not appear in the present first-order expansion around p_F and ε_F. Including the corresponding expansion of the sp kinetic energy, one may write the exact sp propagator as follows

$$G(p; E) = \frac{1}{E - p^2/2m - \Sigma(p; E)}$$

$$\Rightarrow \frac{1}{(E - \varepsilon_F)(1 - \frac{\partial \Sigma}{\partial E}) - (\frac{p_F}{m} + \frac{\partial \Sigma}{\partial p})(p - p_F)}$$

$$= \frac{Z_F}{E - \varepsilon_F - (p - p_F)p_F/m^* \pm i\eta}. \qquad (14.93)$$

The effective mass receives distinct contributions, labeled p-mass and E-mass

$$\frac{m^*}{m} = \frac{m_p^*}{m}\frac{m_E^*}{m} \qquad (14.94)$$

which are defined by

$$\frac{m_p^*}{m} \equiv \left(1 + \frac{m}{p_F}\frac{\partial \Sigma(p; \varepsilon_F)}{\partial p}\bigg|_{p_F}\right)^{-1} \qquad (14.95)$$

and

$$\frac{m_E^*}{m} \equiv 1 - \frac{\partial \Sigma(p_F; E)}{\partial E}\bigg|_{\varepsilon_F} = Z_F^{-1}, \qquad (14.96)$$

respectively. The sign of the infinitesimal imaginary part in Eq. (14.93) is determined by whether p is above $(+)$ or below $(-)$ p_F. It is also customary to replace p_F/m^* by the Fermi velocity v_F. The form of Eq. (14.93) is appropriate for the case of a normal Fermi liquid, where there is a jump in the occupation number at p_F, given by the value of Z_F. In a superfluid or superconductor, no such jump is present in the occupation number at p_F, as discussed in Ch. 22.

Ultimately, the present analysis is intended to make statements about possible collective excitation modes in the limit of small momentum and energy transfer, the so-called Landau limit. This information is contained in the tp propagator or, equivalently, the four-point vertex function, both presented in Ch. 9. Since such collective states appear as poles in the tp propagator, they must also appear in the vertex function Γ, according to Eq. (9.26). It is therefore possible to carry out the analysis by studying the properties of Γ in the Landau limit. The additional advantage is that useful relations for the vertex function can be obtained, as explored in some of the exercises at the end of the chapter. In the Landau limit, we identify different types of contributions to Γ. Figure 14.11 illustrates the relevant

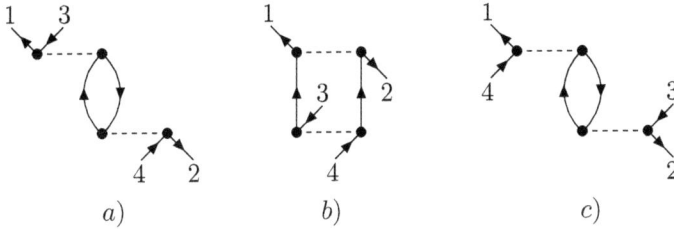

Fig. 14.11   Second-order contributions to the four-point vertex function Γ. The points labeled 1 and 2, correspond to the final state, whereas 3 and 4 refer to the initial state in the notation of Ch. 9. The direction of the corresponding short-arrowed lines is in accordance with the ph coupling representation, discussed in Secs. 13.2, 13.3, and 14.1. Diagram *a*) represents the second-order contribution to the polarization propagator, shown in Fig. 13.6 without the external propagators. Note also that for the present discussion, the propagators, inside the contributions to Γ, refer to exact ones. The interactions in these diagrams correspond to the symmetrized version and the representation has been chosen such that greatest clarity is achieved.

second-order terms in Γ, which are shown using the ph format discussed in Sec. 14.1. Since we are interested in results that identify properties of excited states of the liquid, diagram *a*) of Fig. 14.11 deserves special attention. Indeed, the corresponding term in the polarization propagator is displayed in Fig. 13.6 with external propagators attached. In the present context, the internal lines correspond to exact sp propagators, but double-arrowed lines have been suppressed for simplicity. No external lines are attached to Fig. 14.11*a*). Nevertheless, it is possible to identify this term as the one obtained by iterating a lower-order (*i.e.* first-order) contribution to Γ, with a ph propagator in complete analogy with the second-order diagram of Fig. 13.6. The labels 1 and 2 of the diagrams correspond to the final state of the corresponding contribution to Γ in the usual particle–particle (pp) formulation, employed in Ch. 9. Similarly, the labels 3 and 4 refer to the initial state.

In Sec. 13.2 we have seen that the determination of excited states requires a coupling scheme in which the lines, corresponding to labels 1 and 3, represent the final ph state, and 4 and 2 the initial one [see Eq. (13.20)]. The identification further facilitates the characterization of diagrams *b*) and *c*) in Fig. 14.11, as irreducible contributions with respect to ph propagation in this coupling scheme. For the same reason diagram *a*) is reducible with respect to the iteration of the exact, but noninteracting ph propagator, identified by the bubble with the corresponding antiparallel lines. Diagram c) also contains a bubble but the coupling scheme employed here, makes it irreducible with respect to iterating in the 1,3 or 2,4 direction. Clearly, the diagram is the exchange of diagram a) (by exchanging 3 and 4). This equivalence demonstrates that it is possible to consider diagram c) as reducible with respect to 1,4 and 3,2 ph propagation. Conventional identification in relativistic descriptions of scattering processes, refers to diagrams a), b), and c) as t-, s-, and u-channel iterations in terms of the Mandelstam variables. They identify the three four-momentum invariants, characterizing such processes [Halzen and Martin (1984)]. In the present case, 1+2 corresponds to a conserved energy (equal to 3+4) but also 1-3 or 1-4. It is convention to use the "t-channel" formulation to discuss the generation of excited states, corresponding to iterating in the 1,3 or 2,4 direction, as illustrated in Fig. 14.11a).

We proceed with the analysis, by introducing the momentum variables for the four-point function Γ. As in Eq. (9.25), we remove the energy conserving δ-function from Γ and consider three independent energy variables. Since Γ also conserves total momentum (as does the bare interaction V), one can extract a corresponding factor $\delta_{\boldsymbol{P},\boldsymbol{P'}}$ as in Eq. (4.40), and employ three independent momentum variables. We note that the interaction no longer contains the volume factor in the denominator, since it is compensated by a corresponding factor from the integration over each momentum variable. In keeping with the ph coupling for the momenta, introduced in Sec. 14.1, one can generate expressions for the integrations that need to be performed for each of the contributions, shown in Fig. 14.11. The momentum assignments for the matrix element of V in Eq. (14.9) identify the momenta associated with the labels 1-4 in the figure. The label 1 corresponds to $\boldsymbol{p}+\boldsymbol{Q}/2$, 3 to $\boldsymbol{p}-\boldsymbol{Q}/2$ (with this momentum leaving the interaction as appropriate for a hole), 4 to $\boldsymbol{p'}+\boldsymbol{Q}/2$, and, finally, 2 to $\boldsymbol{p'}-\boldsymbol{Q}/2$ (as a hole). We will choose the new energy variables associated with the three independent energies, based on those associated with the labels 1-4 by first identifying: E_1 with 1, $-E_3$ with 3, E_4 with 4, and $-E_2$ with 2. The negative signs are consistent with the hole character and indicate that these energies are flowing away from the corresponding vertices. It is now helpful to define the these three independent energy variables according to the follwoing choice

$$E_t = E_1 - E_3 = E_4 - E_2 \qquad (14.97)$$

$$E_s = E_1 + E_2 = E_3 + E_4 \qquad (14.98)$$

$$E_u = E_1 - E_4 = E_3 - E_2 \tag{14.99}$$

which can be inverted to produce

$$E_1 = \frac{1}{2}\left(E_s + E_t + E_u\right) \tag{14.100}$$

$$E_2 = \frac{1}{2}\left(E_s - E_t - E_u\right) \tag{14.101}$$

$$E_3 = \frac{1}{2}\left(E_s - E_t + E_u\right) \tag{14.102}$$

$$E_4 = \frac{1}{2}\left(E_s + E_t - E_u\right). \tag{14.103}$$

We note that in the previous sections of this chapter E_t was denoted by E. With these identifications one can construct the appropriate momentum and energy variables for the intermediate states in the three diagrams of Fig. 14.11 and thus the arguments of the two exact sp propagators. For diagram $a)$ we obtain schematically

$$a) \to \sum_{\boldsymbol{p}_t} \int \frac{dE_t'}{2\pi i} \dots G(\boldsymbol{p}_t + \boldsymbol{Q}/2; E_t' + E_t/2) \, G(\boldsymbol{p}_t - \boldsymbol{Q}/2; E_t' - E_t/2)\dots, \tag{14.104}$$

where the spin (isospin) labels (and the corresponding summations) should be included for the matrix elements of the interactions (not explicitly shown). As discussed in Ch. 11, the spin (isospin) degrees of freedom do not affect the propagators in the case of an unpolarized homogeneous system. The intermediate state in Eq. (14.104) has special singular properties when E_t and $\boldsymbol{Q}$ go to zero, to be discussed shortly in more detail. For now, it suffices to note that in this limit the arguments of both propagators coincide. Special care will be required for intermediate momenta and energies near the Fermi momentum and energy, due to the singular character exhibited by the sp propagators, as shown in Eq. (14.93).

In contrast, diagram $b)$ in Fig. 14.11 yields the schematic result

$$b) \to \sum_{\boldsymbol{p}_s} \int \frac{dE_s'}{2\pi i} \dots G(\boldsymbol{p}_s + (\boldsymbol{p} + \boldsymbol{p}')/2; E_s' + E_s/2) \, G(\boldsymbol{p}_s - (\boldsymbol{p} + \boldsymbol{p}')/2; E_s/2 - E_s')\dots, \tag{14.105}$$

which exhibits no such singular behavior in the limit when E_t and $\boldsymbol{Q}$ go to zero. This is also true for diagram c) given by

$$c) \rightarrow \sum_{\boldsymbol{p}_u} \int \frac{dE'_u}{2\pi i} \,..G(\boldsymbol{p}_u + (\boldsymbol{p} - \boldsymbol{p}')/2; E'_u + E_u/2)\, G(\boldsymbol{p}_u - (\boldsymbol{p} - \boldsymbol{p}')/2; E'_u - E_u/2)..$$

$$(14.106)$$

From the expressions in Eqs. (14.105) and (14.106), we infer that it is permissible to put E_t and $\boldsymbol{Q}$ to zero, since no singularities of the kind shown in Eq. (14.104), occur. It should be noted that although these variables do not appear explicitly, they must be included in the matrix elements of the interactions as external variables.

Similar statements hold for all higher-order contributions to Γ. These are either t-channel irreducible as diagrams b) and c) in Fig. 14.11, or reducible as diagram a). In the former case, it is permissible to put $\boldsymbol{Q}$ and E_t to zero. Having separated the contributions to Γ in this way, allows one to construct Γ by iterating the irreducible contributions to all orders in the t-channel, as in the example of diagram a). Denoting the irreducible contributions to Γ by Γ^{ir}, one obtains

$$\langle \boldsymbol{Q}, \boldsymbol{p}; \alpha\beta^{-1} | \Gamma(E_t, E_s, E_u) | \boldsymbol{Q}, \boldsymbol{p}'; \gamma\delta^{-1} \rangle \qquad\qquad (14.107)$$

$$= \langle \boldsymbol{p}; \alpha\beta^{-1} | \Gamma^{ir}(E_s, E_u) | \boldsymbol{p}'; \gamma\delta^{-1} \rangle + \sum_{\epsilon\theta} \int \frac{d\boldsymbol{p}_t}{(2\pi\hbar)^3} \int \frac{dE'_t}{2\pi i}$$

$$\times \langle \boldsymbol{p}; \alpha\beta^{-1} | \Gamma^{ir}((E_s + E_u)/2 + E'_t, (E_s + E_u)/2 - E'_t) | \boldsymbol{p}_t; \epsilon\theta^{-1} \rangle$$

$$\times G(\boldsymbol{p}_t + \boldsymbol{Q}/2; E'_t + E_t/2)\, G(\boldsymbol{p}_t - \boldsymbol{Q}/2; E'_t - E_t/2)$$

$$\times \langle \boldsymbol{Q}, \boldsymbol{p}_t; \epsilon\theta^{-1} | \Gamma(E_t, (E_s - E_u)/2 + E'_t, E'_t - (E_s - E_u)/2) | \boldsymbol{Q}, \boldsymbol{p}'; \gamma\delta^{-1} \rangle,$$

where the spin (isospin) quantum numbers are indicated by greek letters to compress the notation. Note that for Γ^{ir} only two energy arguments have been kept, corresponding to the appropriate values of $\tilde{E}_s$ and $\tilde{E}_u$.

The next several steps are all geared to transform the result of Eq. (14.107) into an equation for the collective excitations in the limit of small excitation energy and momentum transfer, where the dominant contribution of quasiparticle–quasihole propagation is singled out. The latter propagation mode corresponds to the product of exact propagators that appears in Eq. (14.107) and involves the limit of the sp propagators in Eq. (14.93). In this limit, the product of the propagators is singular, but can be evaluated explicitly. It is therefore essential to extract the singular behavior and study the energy integration contained in Eq. (14.107) involving this product of sp propagators in the Landau limit, as in Eq. (14.104). One may write a sp propagator, using the result of Eq. (14.93), for momenta in the vicinity of the Fermi momentum as [see also Eq. (11.105)]

$$G(p; E) = G_Q(p; E) + G_B(p; E)$$
$$= \frac{Z_F}{E - \varepsilon_F - (p - p_F)p_F/m^* \pm i\eta} + G_B(p; E). \quad (14.108)$$

Here the background part represents those contributions from the spectral functions, shown in Fig.11.5, that do not lead to singular contributions for the propagator. With this decomposition it is possible to perform the energy integration associated with the quasiparticle–quasihole contribution to Eq. (14.107). We write schematically

$$\int \frac{dE'_t}{2\pi i} \; f(E'_t) \; G(\boldsymbol{p}_t + \boldsymbol{Q}/2; E'_t + E_t/2) \, G(\boldsymbol{p}_t - \boldsymbol{Q}/2; E'_t - E_t/2)$$
$$= \int \frac{dE'_t}{2\pi i} \; f(E'_t) \, G_Q(\boldsymbol{p}_t + \boldsymbol{Q}/2; E'_t + E_t/2) \, G_Q(\boldsymbol{p}_t - \boldsymbol{Q}/2; E'_t - E_t/2) + ...,$$
$$(14.109)$$

where $f(E'_t)$ represents the dependence on E'_t of Γ^{ir}. Although this term also yields contributions to the energy integral, since it includes poles in both the upper and lower half of the complex E'_t-plane, they are well behaved in the Landau limit and will be considered separately. The other terms indicated by the dots in Eq. (14.109) have at least one factor involving the background part of the propagator. These terms are well behaved in the Landau limit. Each propagator in Eq. (14.109) contains a quasiparticle and a quasihole component, depending on the magnitude of $|\boldsymbol{p}_t \pm \boldsymbol{Q}/2|$, which is determined by the magnitude of $\boldsymbol{p}_t$ and its angle with respect to $\boldsymbol{Q}$. In the limit $\boldsymbol{Q} \to 0$, the magnitude of $\boldsymbol{p}_t$ will be close to p_F in order for the singular nature of G_Q to contribute. From these considerations it is clear that the product of the two quasiparticle contributions to the integral yields four terms involving pp, ph, hp, and hh products. For the pp and hh terms one realizes that the corresponding poles lie on the same side of the real E'_t-axis in the complex plane. These terms will not yield divergent contributions, since one can close the contour in the opposite half-plane, picking up contributions from the well-behaved parts of $f(E'_t)$. The only terms that require special attention are the ph and hp quasiparticle contributions. Introducing

$$E_Q(p) = \varepsilon_F + p_F \frac{p - p_F}{m^*} = \varepsilon_F + v_F(p - p_F), \quad (14.110)$$

as in Eq. (11.101), the ph product can be written as

$$\text{ph} \rightarrow \frac{Z_F \left(1 - \theta(p_F - |\boldsymbol{p}_t + \boldsymbol{Q}/2|)\right)}{E'_t + E_t/2 - E_Q(\boldsymbol{p}_t + \boldsymbol{Q}/2) + i\eta} \times \frac{Z_F \, \theta(p_F - |\boldsymbol{p}_t - \boldsymbol{Q}/2|)}{E'_t - E_t/2 - E_Q(\boldsymbol{p}_t - \boldsymbol{Q}/2) - i\eta},$$

(14.111)

where the step function identity

$$\theta(p - p_F) = 1 - \theta(p_F - p) \tag{14.112}$$

was used for the first propagator. The limit $\boldsymbol{Q} \rightarrow 0$ has already been taken for the Z-factors, since these are smooth functions of momentum. By using the identity

$$\frac{1}{E - i\eta} = 2\pi i \, \delta(E) + \frac{1}{E + i\eta} \tag{14.113}$$

which can be obtained from Eq. (7.15), one can isolate the δ-function contribution to Eq. (14.111) in the following way

$$\text{ph} \rightarrow Z_F^2 \frac{(1 - \theta(p_F - |\boldsymbol{p}_t + \boldsymbol{Q}/2|)) \, \theta(p_F - |\boldsymbol{p}_t - \boldsymbol{Q}/2|)}{E'_t + E_t/2 - E_Q(\boldsymbol{p}_t + \boldsymbol{Q}/2) + i\eta}$$
$$\times \delta(E'_t - E_t/2 - E_Q(\boldsymbol{p}_t - \boldsymbol{Q}/2)) + ... \tag{14.114}$$

The dots again indicate a nonsingular contribution, since the corresponding product of qp propagators has poles on the same side of the E'_t-axis. The integration over E'_t pertaining to the singular term can be performed with the outcome

$$\text{ph} \rightarrow 2\pi i \, Z_F^2 \frac{(1 - \theta(p_F - |\boldsymbol{p}_t + \boldsymbol{Q}/2|)) \, \theta(p_F - |\boldsymbol{p}_t - \boldsymbol{Q}/2|)}{E_t - (E_Q(\boldsymbol{p}_t + \boldsymbol{Q}/2) - E_Q(\boldsymbol{p}_t - \boldsymbol{Q}/2)) + i\eta}. \tag{14.115}$$

After establishing the singular behavior in the energy, one can also identify such behavior in the magnitude of the momentum $\boldsymbol{p}_t$, represented by Eq. (14.115). The result is obtained by noting that the step function restrictions in that equation imply $0 \le \theta_{p_t} \le \pi/2$ for the angle between $\boldsymbol{p}_t$ and $\boldsymbol{Q}$, ensuring that the magnitudes of $|\boldsymbol{p}_t \pm \boldsymbol{Q}/2|$ remain in the respective particle and hole domain. This implies that the product of the two step functions in Eq. (14.115) reduces to

$$\theta(p_F - |\boldsymbol{p}_t + \boldsymbol{Q}/2|)\theta(p_F - |\boldsymbol{p}_t - \boldsymbol{Q}/2|) = \theta(p_F - |\boldsymbol{p}_t + \boldsymbol{Q}/2|), \tag{14.116}$$

since the first step function is more restrictive when integrating over the magnitude of $\boldsymbol{p}_t$. We continue by expanding the step functions for $\boldsymbol{Q} \rightarrow 0$

in the following way

$$\theta(p_F - |\boldsymbol{p}_t \pm \boldsymbol{Q}/2|) = \theta(p_F - p_t) \pm \frac{\boldsymbol{Q}}{2} \cdot \nabla_{\boldsymbol{p}_t} \theta(p_F - p_t)$$

$$= \theta(p_F - p_t) \pm \frac{\boldsymbol{Q}}{2} \cdot \hat{\boldsymbol{p}}_t \frac{\partial}{\partial p_t} \theta(p_F - p_t)$$

$$= \theta(p_F - p_t) \mp \frac{\boldsymbol{Q}}{2} \cdot \hat{\boldsymbol{p}}_t \, \delta(p_F - p_t) \qquad (14.117)$$

so that the numerator of Eq. (14.115) reads

$$\theta(p_F - |\boldsymbol{p}_t - \boldsymbol{Q}/2|) - \theta(p_F - |\boldsymbol{p}_t + \boldsymbol{Q}/2|) = \boldsymbol{Q} \cdot \hat{\boldsymbol{p}}_t \, \delta(p_F - p_t) \qquad (14.118)$$

in the limit $\boldsymbol{Q} \to 0$. A similar expansion of the difference of the quasiparticle energies in the denominator of Eq. (14.115) yields

$$E_Q(\boldsymbol{p}_t + \boldsymbol{Q}/2) - E_Q(\boldsymbol{p}_t - \boldsymbol{Q}/2) = \frac{\boldsymbol{Q} \cdot \boldsymbol{p}_t}{m^*} = \boldsymbol{Q} \cdot \hat{\boldsymbol{p}}_t v_F. \qquad (14.119)$$

Inserting Eqs. (14.118) and (14.119) in Eq. (14.115), one finally obtains, upon performing the integration over the magnitude of $\boldsymbol{p}_t$,

$$\text{ph} \to 2\pi i \, p_F^2 Z_F^2 \frac{\boldsymbol{Q} \cdot \hat{\boldsymbol{p}}_t}{E_t - \boldsymbol{Q} \cdot \hat{\boldsymbol{p}}_t v_F + i\eta} \qquad (14.120)$$

with the restriction of the angle between 0 and $\pi/2$. An identical analysis of the hp contribution to the product of the two qp propagators yields the following complementary result for the angular range $\pi/2 \leq \theta_{p_t} \leq \pi$

$$\text{hp} \to 2\pi i \, p_F^2 Z_F^2 \frac{\boldsymbol{Q} \cdot \hat{\boldsymbol{p}}_t}{E_t - \boldsymbol{Q} \cdot \hat{\boldsymbol{p}}_t v_F - i\eta}. \qquad (14.121)$$

The last two equations indicate that there is considerable subtlety involved when the Landau limit is approached. In fact, these results depend on whether the energy E_t or the momentum $\boldsymbol{Q}$ is taken to zero first. As a consequence, the exact four-point vertex function Γ, being a function of E_t and $\boldsymbol{Q}$, is ill defined at the point $E_t = 0$, $\boldsymbol{Q} = 0$, since its value depends on how this point is approached in the (E_t, Q)-plane. This should not be a surprise: the same feature is also observed for the real part of the unperturbed polarization propagator, studied in Sec. 14.2. In the latter case, one encounters the same singularity structure and the results can be worked out exactly. We now return to Eq. (14.107) and substitute the decomposition of the product of sp propagators into its singular structure and the remaining contribution. The outcome is given by the following bulky expression

$$\langle \boldsymbol{Q}, \boldsymbol{p}; \alpha\beta^{-1} | \Gamma(E_t, E_s, E_u) | \boldsymbol{Q}, \boldsymbol{p}'; \gamma\delta^{-1} \rangle \qquad (14.122)$$

$$= \langle \boldsymbol{p}; \alpha\beta^{-1} | \Gamma^{ir}(E_s, E_u) | \boldsymbol{p}'; \gamma\delta^{-1} \rangle + p_F^2 Z_F^2 \sum_{\epsilon\theta} \int \frac{d\hat{\boldsymbol{p}}_t}{(2\pi\hbar)^3}$$

$$\times \langle \boldsymbol{p}; \alpha\beta^{-1} | \Gamma^{ir}((E_s + E_u)/2 + \varepsilon_F, (E_s + E_u)/2 - \varepsilon_F) | p_F \hat{\boldsymbol{p}}_t; \epsilon\theta^{-1} \rangle$$

$$\times \frac{\boldsymbol{Q} \cdot \hat{\boldsymbol{p}}_t}{E_t - \boldsymbol{Q} \cdot \hat{\boldsymbol{p}}_t v_F \pm i\eta}$$

$$\langle \boldsymbol{Q}, p_F \hat{\boldsymbol{p}}_t; \epsilon\theta^{-1} | \Gamma(E_t, (E_s - E_u)/2 + \varepsilon_F, (E_u - E_s)/2 + \varepsilon_F) | \boldsymbol{Q}, \boldsymbol{p}'; \gamma\delta^{-1} \rangle$$

$$+ \sum_{\epsilon\theta} \int \frac{d\boldsymbol{p}_t}{(2\pi\hbar)^3} \int \frac{dE_t'}{2\pi i}$$

$$\times \langle \boldsymbol{p}; \alpha\beta^{-1} | \Gamma^{ir}((E_s + E_u)/2 + E_t', (E_s + E_u)/2 - E_t') | \boldsymbol{p}_t; \epsilon\theta^{-1} \rangle$$

$$\times \{ G(\boldsymbol{p}_t + \boldsymbol{Q}/2; E_t' + E_t/2) G(\boldsymbol{p}_t - \boldsymbol{Q}/2; E_t' - E_t/2) \}_{reg}$$

$$\times \langle \boldsymbol{Q}, \boldsymbol{p}_t; \epsilon\theta^{-1} | \Gamma(E_t, (E_s - E_u)/2 + E_t', E_t' - (E_s - E_u)/2) | \boldsymbol{Q}, \boldsymbol{p}'; \gamma\delta^{-1} \rangle .$$

In the second term on the right side, only the integration over angles remains for a relative momentum with magnitude p_F according to Eq. (14.118). The last term in Eq. (14.122) represents all contributions from the product of propagators that do not yield singular terms.

The next step in the analysis is to sum all the contributions from the last term in Eq. (14.122). Since this involves a general principle, it can be discussed in the following generic fashion. If one encounters an integral (propagator equation) of the kind

$$X = U + U(\mathcal{G}_1 + \mathcal{G}_2)X, \qquad (14.123)$$

where X is obtained by iterating U in schematic notation with $\mathcal{G}_1$ and $\mathcal{G}_2$, representing separate contributions to the intermediate propagator (like the real and imaginary part or the separation encountered in Eq. (14.122)) the problem may be solved in two stages. First

$$Y = U + U\mathcal{G}_2 Y \qquad (14.124)$$

is solved and then X is obtained from

$$X = Y + Y\mathcal{G}_1 X. \qquad (14.125)$$

The result can be checked by comparing higher-order terms from Eq. (14.123) and the combination of Eqs. (14.124) and (14.125), also in higher order.

The application to Eq. (14.122), identifies $\mathcal{G}_1$ as the singular term which appears in the second part of the right-hand side of that equation. Only iterating the last term is however also equivalent to taking the following limit of the singular part of the propagator: first take $\boldsymbol{Q} \to 0$ and then $E_t \to 0$, in which case Eqs. (14.120) and (14.121) yield no contribution. Denoting this limit of the vertex function in the Landau limit by Γ^0, we see that the procedure that leads to Eq. (14.125) applied to Eq. (14.122) generates

$$\langle \boldsymbol{Q}, \boldsymbol{p}; \alpha\beta^{-1} | \Gamma(E_t, E_s, E_u) | \boldsymbol{Q}, \boldsymbol{p}'; \gamma\delta^{-1} \rangle = \quad\quad (14.126)$$

$$\langle \boldsymbol{p}; \alpha\beta^{-1} | \Gamma^0(E_s, E_u) | \boldsymbol{p}'; \gamma\delta^{-1} \rangle + p_F^2 Z_F^2 \sum_{\epsilon\theta} \int \frac{d\hat{\boldsymbol{p}}_t}{(2\pi\hbar)^3}$$

$$\times \langle \boldsymbol{p}; \alpha\beta^{-1} | \Gamma^0((E_s + E_u)/2 + \varepsilon_F, (E_s + E_u)/2 - \varepsilon_F) | p_F \hat{\boldsymbol{p}}_t; \epsilon\theta^{-1} \rangle$$

$$\times \frac{\boldsymbol{Q} \cdot \hat{\boldsymbol{p}}_t}{E_t - \boldsymbol{Q} \cdot \hat{\boldsymbol{p}}_t v_F + i\eta}$$

$$\langle \boldsymbol{Q}, p_F \hat{\boldsymbol{p}}_t; \epsilon\theta^{-1} | \Gamma(E_t, (E_s - E_u)/2 + \varepsilon_F, (E_u - E_s)/2 + \varepsilon_F) | \boldsymbol{Q}, \boldsymbol{p}'; \gamma\delta^{-1} \rangle .$$

The result is valid for all momentum variables $\boldsymbol{p}$, $\boldsymbol{p}'$ and energy variables E_s, E_u. Since the integral equation is restricted to intermediate states with both the particle and the hole at the Fermi surface, we may formulate a closed equation for the corresponding vertex function by considering the external momentum variables $\boldsymbol{p}$ and $\boldsymbol{p}'$ at p_F, together with placing the energies, associated with labels 1-4, on the Fermi energy. The values for E_s and E_u then become $2\varepsilon_F$ and 0, respectively. Leaving out these redundant energy variables, we obtain

$$\langle \boldsymbol{Q}, \hat{\boldsymbol{p}}; \alpha\beta^{-1} | \Gamma(E_t) | \boldsymbol{Q}, \hat{\boldsymbol{p}}'; \gamma\delta^{-1} \rangle = \langle \hat{\boldsymbol{p}}; \alpha\beta^{-1} | \Gamma^0 | \hat{\boldsymbol{p}}'; \gamma\delta^{-1} \rangle \quad (14.127)$$

$$+ p_F^2 Z_F^2 \sum_{\epsilon\theta} \int \frac{d\hat{\boldsymbol{p}}_t}{(2\pi\hbar)^3} \langle \hat{\boldsymbol{p}}; \alpha\beta^{-1} | \Gamma^0 | \hat{\boldsymbol{p}}_t; \epsilon\theta^{-1} \rangle$$

$$\times \frac{\boldsymbol{Q} \cdot \hat{\boldsymbol{p}}_t}{E_t - \boldsymbol{Q} \cdot \hat{\boldsymbol{p}}_t v_F + i\eta} \langle \boldsymbol{Q}, \hat{\boldsymbol{p}}_t; \epsilon\theta^{-1} | \Gamma(E_t) | \boldsymbol{Q}, \hat{\boldsymbol{p}}'; \gamma\delta^{-1} \rangle ,$$

where the external and internal momentum variables are only indicated by their direction, since their magnitude is equal to p_F. Various limits can be studied and we will focus on the determination of the possible pole associated with Γ in the Landau limit. The pole identifies the location of the sought after collective state and will be denoted by E_0. The first term in Eq. (14.127) is not singular near the pole of Γ and the usual eigenvalue equation emerges, although now in a form involving the vertex function

$$\left\langle \boldsymbol{Q}, \hat{\boldsymbol{p}}; \alpha\beta^{-1}\middle|\Gamma(E_t \approx E_0)\middle|\boldsymbol{Q}, \hat{\boldsymbol{p}}'; \gamma\delta^{-1}\right\rangle \tag{14.128}$$

$$= p_F^2 Z_F^2 \sum_{\epsilon\theta} \int \frac{d\hat{\boldsymbol{p}}_t}{(2\pi\hbar)^3} \left\langle \hat{\boldsymbol{p}}; \alpha\beta^{-1}\middle|\Gamma^0\middle|\hat{\boldsymbol{p}}_t; \epsilon\theta^{-1}\right\rangle$$

$$\times \frac{\boldsymbol{Q}\cdot\hat{\boldsymbol{p}}_t}{E_0 - \boldsymbol{Q}\cdot\hat{\boldsymbol{p}}_t v_F + i\eta} \left\langle \boldsymbol{Q}, \hat{\boldsymbol{p}}_t; \epsilon\theta^{-1}\middle|\Gamma(E_t \approx E_0)\middle|\boldsymbol{Q}, \hat{\boldsymbol{p}}'; \gamma\delta^{-1}\right\rangle.$$

It is convenient to take advantage of the rotational symmetry of Γ^0 in the limit $\boldsymbol{Q} \to 0$, which implies that the only dependence on the directions is of the form $\hat{\boldsymbol{p}} \cdot \hat{\boldsymbol{p}}'$. In addition, one must consider the possible spin (and isospin) degrees of freedom. We will discuss here the case of the ^{3}He liquid. One may then introduce a parametrization of the interaction on the Fermi surface of the following form

$$\begin{aligned}\Gamma^0 &= f(\hat{\boldsymbol{p}}\cdot\hat{\boldsymbol{p}}') + g(\hat{\boldsymbol{p}}\cdot\hat{\boldsymbol{p}}')\,\boldsymbol{\sigma}_1\cdot\boldsymbol{\sigma}_2 \\ &= f(\cos\theta_{pp'}) + g(\cos\theta_{pp'})\,\boldsymbol{\sigma}_1\cdot\boldsymbol{\sigma}_2 \\ &= \sum_{\ell=0}^{\infty} f_\ell\, P_\ell(\cos\theta_{pp'}) + \sum_{\ell=0}^{\infty} g_\ell\, P_\ell(\cos\theta_{pp'})\,\boldsymbol{\sigma}_1\cdot\boldsymbol{\sigma}_2. \end{aligned} \tag{14.129}$$

This representation of the interaction still requires to take the *direct* ph matrix elements of the spin-space operators. Equation (14.129) assumes that predominantly central forces operate in the ^{3}He liquid. In that case only the two operator invariants 1 and $\boldsymbol{\sigma}_1 \cdot \boldsymbol{\sigma}_2$ need be considered. In Sec. 14.1 we have seen that for a two-component fermion system with central interactions only the total ph spin S is required to characterize the different possible excitation modes. This feature is equivalent to the corresponding separation based on Eq. (14.129), since the first term will only contribute for $S = 0$ and the second only for $S = 1$, when the direct ph matrix elements are taken. In each instance, a factor of 2 must be included upon taking the appropriate matrix element, as shown in Sec. 14.1.

We now proceed by assuming that the full vertex function near E_0 can be written as

$$\left\langle \boldsymbol{Q}, \hat{\boldsymbol{p}}_t; S\middle|\Gamma(E_t \approx E_0)\middle|\boldsymbol{Q}, \hat{\boldsymbol{p}}'; S\right\rangle = \frac{\alpha(\boldsymbol{Q}, \hat{\boldsymbol{p}}; E_0, S)\alpha^*(\boldsymbol{Q}, \hat{\boldsymbol{p}}'; E_0, S)}{E_t - E_0} + \dots \tag{14.130}$$

The assumption is based on the Lehmann representation of the corresponding ph propagator, which takes on a form like the one given in Eq. (13.9). The corresponding reducible ph interaction Γ must have identical poles, noting that the vertex function in the Landau limit is also the FT of a corresponding two-time quantity. The dots in Eq. (14.130) simply represent all the other nonsingular contributions that are negligible near E_0. We will assume that we are looking for a collective state with $S = 0$, and use the corresponding notation $\alpha_0(\boldsymbol{Q}, \hat{\boldsymbol{p}})$. An identical analysis may be pursued for $S = 1$. Using Eq. (14.130), one may write Eq. (14.128) in the following way

$$\alpha_0(\boldsymbol{Q}, \hat{\boldsymbol{p}}) = 2p_F^2 Z_F^2 \int \frac{d\hat{\boldsymbol{p}}_t}{(2\pi\hbar)^3} f(\hat{\boldsymbol{p}} \cdot \hat{\boldsymbol{p}}_t) \frac{\boldsymbol{Q} \cdot \hat{\boldsymbol{p}}_t}{E_0 - \boldsymbol{Q} \cdot \hat{\boldsymbol{p}}_t v_F} \alpha_0(\boldsymbol{Q}, \hat{\boldsymbol{p}}_t), \quad (14.131)$$

where the common factor depending on $\hat{\boldsymbol{p}}'$ has been eliminated and the only remnant of the spin variables is the extra factor of 2. Even though this eigenvalue equation couples different ph orbital angular momenta on the Fermi surface, it is nevertheless useful to consider an expansion of the angular dependence in the corresponding eigenstates. We will only consider the azimuthally symmetric case, implying that α_0 depends exclusively on the angle between $\boldsymbol{Q}$ and $\hat{\boldsymbol{p}}$. The more general case is presented in [Abrikosov and Khalatnikov (1959); Baym and Pethick (1978)]. For convenience one may also take $\boldsymbol{Q}$ parallel to the z-axis. With these assumptions, α_0 can be expanded as follows

$$\alpha_0(\boldsymbol{Q}, \hat{\boldsymbol{p}}) = \sum_\ell \alpha_\ell P_\ell(\cos\theta_{Qp}), \quad (14.132)$$

where θ_{Qp} is the polar angle of $\hat{\boldsymbol{p}}$ with respect to the $\boldsymbol{Q}$-(z-)axis. Inserting this expansion and the one, given by Eq. (14.129) for f, in Eq. (14.131), one may perform the integration over the azimuth angle ϕ_t by expanding $P_\ell(\hat{\boldsymbol{p}} \cdot \hat{\boldsymbol{p}}_t)$, using the addition theorem for spherical harmonics. Since the ϕ_t-dependence is associated with the corresponding $Y_{\ell m}(\hat{\boldsymbol{p}}_t)$, the integral over ϕ_t ensures that only terms with $m = 0$ contribute. Using the relation

$$Y_{\ell m=0}(\hat{\boldsymbol{p}}) = \sqrt{\frac{2\ell + 1}{4\pi}} P_\ell(\cos\theta), \quad (14.133)$$

one finally obtains the following eigenvalue problem

$$\alpha_\ell = F_\ell \sum_{\ell'} \Omega_{\ell\ell'}(s)\alpha_{\ell'}, \quad (14.134)$$

where the orthogonality relation of the Legendre polynomials has been applied. In addition, the dimensionless constants

$$F_\ell = N_0 f_\ell, \quad (14.135)$$

with

$$N_0 = \frac{m^* p_F Z_F^2}{\pi^2 \hbar^3},\tag{14.136}$$

are introduced, referred to as *Landau parameters*. The quantities

$$s = \frac{E_0}{Q v_F} = \frac{m^* E_0}{p_F Q}\tag{14.137}$$

and

$$\Omega_{\ell\ell'}(s) = \Omega_{\ell'\ell}(s) = \frac{1}{2}\int_{-1}^{1} d(\cos\theta) P_\ell(\cos\theta) \frac{\cos\theta}{s - \cos\theta} P_{\ell'}(\cos\theta)\tag{14.138}$$

have also been introduced. The first few Ω are given by

$$\Omega_{00} = -1 - \frac{s}{2}\ln\left(\frac{s-1}{s+1}\right)\tag{14.139}$$

and

$$\Omega_{\ell 1} = -\frac{1}{3}\delta_{\ell 1} + s\Omega_{\ell 0}\tag{14.140}$$

for values of $s > 1$. Real values for these quantities are obtained only for $s > 1$. For values of $|s| < 1$, Eq. (14.138) also yields a complex contribution. We note that the eigenvalue problem in Eq. (14.134) yields solutions for $s \to -s$, just as the RPA eigenvalue problem, discussed in Sec. 13.3. Note that Ω_{00} is positive for $s > 1$ and monotonically decreasing as a function of s. If only F_0 is taken into account, this shows that F_0 must be positive in order to solve Eq. (14.134). No solution is found for negative values of F_0. The result is similar to the collective plasmon or the discrete states discussed in Sec. 14.6, which occur in energy above the ph continuum for a given value of q.

Solutions to Eq. (14.134) can be obtained if knowledge of the parameters F_ℓ is available. Landau showed that relations exist [Landau (1957a)] between the first two parameters F_0 and F_1 and the speed of ordinary (first) sound and the effective mass (specific heat), respectively. In Ch. 21 general relations between the self-energy and the irreducible ph interaction will be studied. As shown in Sec. 21.3, Landau found the following relation between the effective mass on the Fermi and the parameter F_1

$$\frac{m^*}{m} = 1 + \frac{F_1}{3}.\tag{14.141}$$

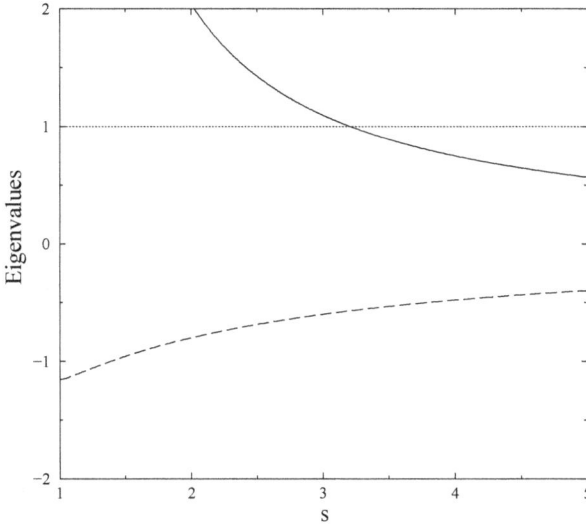

Fig. 14.12   Graphical representation of the eigenvalues (full and dashed lines) of Eq. (14.134) as a function of s using the empirical values of F_0 and F_1. The location of the collective state is identified by the crossing of the upper eigenvalue with the dotted line.

For the speed of ordinary sound c_1 he found

$$c_1^2 = \frac{p_F^2}{3m^2} \frac{1 + F_0}{1 + \frac{1}{3}F_1}. \tag{14.142}$$

Assuming the corresponding measured quantities in ^{3}He for the specific heat [Greywall (1986)] and the speed of sound [Wheatley (1975)], one may deduce values for $F_0 = 9.30$ and $F_1 = 5.40$. Assuming that higher Landau parameters are negligible, as is commonly done, one can solve Eq. (14.134) as a 2-dimensional eigenvalue problem depending on s. We find a solution of Eq. (14.134) for the value of s, denoted by s_0, for which one of these eigenvalues equals 1. The two eigenvalues are graphically shown as a function of s in Fig. 14.12, with the above values of F_0 and F_1 as input. The solution for s_0 from the intersection with 1 yields $s_0 = 3.22$, which gives the velocity of the collective excitation as

$$c_0 = s_0 v_F \tag{14.143}$$

in agreement with the measured velocity of this quantity [Abel *et al.* (1966)]. We note that the collective excitation, commonly referred to as zero sound, has a bosonic dispersion relation (see Ch.18), given by

$$E_0 = c_0 Q. \tag{14.144}$$

We emphasize that this statement pertains to the Landau limit, *i.e.* small energy and momentum transfer. Nevertheless, the empirical values of the extracted Landau parameters indicate that one may expect a continuation of the mode at higher energy and momentum transfer above the correlated ph continuum. This is indeed observed experimentally, although the presence of 2p2h states that mix with ph states, is responsible for a substantial width of the zero sound excitation at finite values of Q [Glyde *et al.* (2000a)]. While the interaction in the $S = 0$ channel is strongly repulsive, the opposite is true for the $S = 1$ channel. The corresponding lowest order Landau parameter can be related to the spin susceptibility [Landau (1957a)]. Experiment yields $G_0 = -0.710$ [Greywall (1983)]. For an undamped collective mode it is necessary that the interaction is repulsive, so for $S = 1$ one finds a behavior that corresponds more to the results, discussed in the previous section, for an attractive ph interaction. Keeping only G_0 for the interaction in the $S = 1$ channel, one may confirm the impossibility of solving Eq. (14.134) in the Landau limit, by noting that the corresponding quantity Ω_{00} becomes complex for values of $s \leq 1$.

14.7   Exercises

(1) Evaluate the double commutator in Eq. (14.49).
(2) Calculate the plasmon dispersion relation for several values of r_s.
(3) Determine the Δh-polarization propagator and perform similar calculations to those that lead to the results of Figs. 14.9 and 14.10.
(4) Use the integral equation, given in Eq. (14.127), to obtain the relation between the following limits of the vertex function ($E \to 0$ then $Q \to 0$) and the other way around, to generate the sum rule conditions on the Landau parameters.
(5) Develop the expansion of the effective ph interaction in terms of Landau parameters for nuclear matter.

Chapter 15

Excited states in $N \pm 2$ systems and in-medium scattering

In the previous chapter we have studied the excited states of the system with the same number of particles as the ground state. That discussion complements the study of the sp propagator, which probes states with one particle added or removed. In the present chapter, we will study the states that can be reached by adding or removing pairs of particles. When the sp propagator is regarded with respect to the vacuum state, one recovers the quantum mechanics of one particle (see Ch. 6). Similarly, the tp propagator studied here, can be considered for the vacuum state and then describes the quantum mechanics of two particles (fermions in this chapter). The case of two particles in free space is treated in Sec. 15.1. The subject is introduced by defining the general two-time tp propagator in the medium, and then replacing the correlated N-particle ground state by the vacuum. We will point out that the diagrammatic content of the full solution of the tp problem, corresponds to summing all the so-called ladder diagrams. Both scattering and bound-state results for two particles will be developed. When interactions have strong short-range components, a perturbative expansion is not an option and the complete sum of ladder diagrams must be included. Such diagrams also play an essential role in the medium, when dealing with two-body interactions, which exhibit such strong short-range repulsion.

Medium modifications, with respect to the interaction in free space, are considered in Sec. 15.2. These effects include the new possibility to remove pairs of particles, resulting in hole–hole (hh) propagation, and the restrictions that occur due to the Pauli principle. Summing ladder diagrams in the medium therefore leads to the combined propagation of both particle–particle (pp) and hh intermediate states. Since at low density the phase space for hh propagation is small, it has been customary to include for practical applications, only pp propagation, leading to the so-called G-matrix effective interaction, frequently used for nuclear and neutron matter calculations, further detailed in Ch. 16. We will study the consequences of medium effects for the description of the scattering process, associated with pp and hh propagation in the medium. The inclusion of hh terms may lead to difficulties, when interactions are encountered that are effectively attractive on the Fermi surface. These so-called pairing instabilities can already be identified from the behavior of the corresponding phase shifts.

Such behavior is further explored from a different perspective in Sec. 15.3. The study of the pairing instability is preceded by a presentation of the Cooper problem. This involves the addition to the Fermi sea of a pair of particles that experience a mutual attraction. We will see that it corresponds to the bound-state problem, associated with only pp propagation. The inclusion of pair removal (hh propagation), then leads to the necessity of opening up a gap in the sp spectrum to accomodate the resulting bound states that may occur for attractive interactions. This analysis therefore serves as preparation for a more in-depth treatment of pairing correlations presented in Ch. 22.

15.1   Two-time two-particle propagator

Since the full 4-time tp propagator is not required in this chapter, we take from the start the relevant two-time limit of Eq. (9.16)

$$G_{pphh}(\alpha, \alpha'; \beta, \beta'; t_1 - t_2) \equiv \lim_{t_1' \to t_1} \lim_{t_2' \to t_2} G_{II}(\alpha t_1, \alpha' t_1', \beta t_2, \beta' t_2')$$

$$= -\frac{i}{\hbar} \langle \Psi_0^N | \mathcal{T}[a_{\alpha'_H}(t_1) a_{\alpha_H}(t_1) a_{\beta_H}^\dagger(t_2) a_{\beta'_H}^\dagger(t_2)] | \Psi_0^N \rangle. \tag{15.1}$$

It is particularly suited to study excited states that can be reached by adding or removing pairs of particles in a many-fermion system. The label *pphh* is used here to include the possibility of hh propagation as well. We will refer to Eq. (15.1) as representing the pphh propagator. Here the ordering of the operators should be noted, since it inverts the order of the first pair as given by the left-hand side of the equation. The noninteracting limit is obtained in the familiar manner, *i.e.* by replacing $|\Psi_0^N\rangle$ by $|\Phi_0^N\rangle$ and by using interaction picture operators instead of the Heisenberg picture ones, employed in Eq. (15.1). The resulting noninteracting pphh propagator

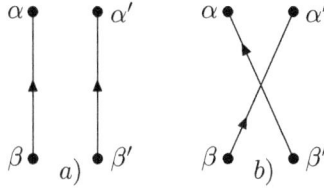

Fig. 15.1 The two contributions to the noninteracting pphh propagator in the time formulation, as given by Eq. (15.2).

reads

$$
\begin{aligned}
G^{(0)}_{pphh}(\alpha, \alpha'; \beta, \beta'; t_1 - t_2) &= -\frac{i}{\hbar} \left\langle \Phi_0^N \middle| T[a_{\alpha'}(t_1) a_\alpha(t_1) a_\beta^\dagger(t_2) a_{\beta'}^\dagger(t_2)] \middle| \Phi_0^N \right\rangle \\
&= i\hbar \left[G^{(0)}(\alpha, \beta; t_1 - t_2) G^{(0)}(\alpha', \beta'; t_1 - t_2) \right. \\
&\quad \left. - G^{(0)}(\alpha, \beta'; t_1 - t_2) G^{(0)}(\alpha', \beta; t_1 - t_2) \right],
\end{aligned}
\tag{15.2}
$$

by direct application of Wick's theorem and in agreement with the 4-time result given in Eq. (9.18). The corresponding diagrammatic representation is given in Fig. 15.1.

In this chapter we will only employ sp propagators, which have a mean-field character, as for free particles or in the HF approach. It is therefore possible to consider diagonal sp propagators, which can be written as

$$
G^{(0)}(\alpha, \alpha'; t_1 - t_2) \equiv \delta_{\alpha,\alpha'} G^{(0)}(\alpha; t_1 - t_2).
\tag{15.3}
$$

Equation (15.2) then changes into

$$
\begin{aligned}
G^{(0)}_{pphh}&(\alpha, \alpha'; \beta, \beta'; t_1 - t_2) \\
&= i\hbar \left[\delta_{\alpha,\beta} \delta_{\alpha',\beta'} - \delta_{\alpha,\beta'} \delta_{\alpha',\beta} \right] G^{(0)}(\alpha; t_1 - t_2) G^{(0)}(\alpha'; t_1 - t_2).
\end{aligned}
\tag{15.4}
$$

For practical purposes it is again useful to employ the energy formulation. Performing the usual FT, yields

$$
\begin{aligned}
G^{(0)}_{pphh}&(\alpha, \alpha'; \beta, \beta'; E) = \int_{-\infty}^{\infty} d(t_1 - t_2) \, e^{\frac{i}{\hbar} E(t_1 - t_2)} G^{(0)}_{pphh}(\alpha, \alpha'; \beta, \beta'; t_1 - t_2) \\
&= i\hbar \left[\delta_{\alpha,\beta} \delta_{\alpha',\beta'} - \delta_{\alpha,\beta'} \delta_{\alpha',\beta} \right] \int_{-\infty}^{\infty} d(t_1 - t_2) \, e^{\frac{i}{\hbar} E(t_1 - t_2)} \\
&\quad \times \int_{-\infty}^{\infty} \frac{dE_1}{2\pi\hbar} e^{-iE_1(t_1 - t_2)/\hbar} G^{(0)}(\alpha; E_1) \int_{-\infty}^{\infty} \frac{dE_2}{2\pi\hbar} e^{-iE_2(t_1 - t_2)/\hbar} G^{(0)}(\alpha'; E_2) \\
&= i\hbar \left[\delta_{\alpha,\beta} \delta_{\alpha',\beta'} - \delta_{\alpha,\beta'} \delta_{\alpha',\beta} \right] \int_{-\infty}^{\infty} \frac{dE_1}{2\pi\hbar} G^{(0)}(\alpha; E_1) G^{(0)}(\alpha'; E - E_1).
\end{aligned}
\tag{15.5}
$$

The inverse FT of the $G^{(0)}$s, given by Eq. (8.60), was used to obtain this formulation. The convolution integral in Eq. (15.5), can be evaluated by contour integration in the usual way, according to Eq. (11.8), and yields

$$G^{(0)}_{pphh}(\alpha, \alpha'; \beta, \beta'; E) \qquad\qquad\qquad\qquad\qquad (15.6)$$

$$= [\delta_{\alpha,\beta}\delta_{\alpha',\beta'} - \delta_{\alpha,\beta'}\delta_{\alpha',\beta}] \left\{ \frac{\theta(\alpha - F)\theta(\alpha' - F)}{E - \varepsilon_\alpha - \varepsilon_{\alpha'} + i\eta} - \frac{\theta(F - \alpha)\theta(F - \alpha')}{E - \varepsilon_\alpha - \varepsilon_{\alpha'} - i\eta} \right\}.$$

We define a noninteracting pphh propagator without the Kronecker deltas by

$$G^{(0)}_{pphh}(\alpha, \alpha'; \beta, \beta'; E) \equiv [\delta_{\alpha,\beta}\delta_{\alpha',\beta'} - \delta_{\alpha,\beta'}\delta_{\alpha',\beta}] \, G^{(0)}_{pphh}(\alpha, \alpha'; E). \quad (15.7)$$

Using Eq. (9.21) for the two-time case, or applying directly Wick's theorem to the first-order contribution to G_{pphh}, we will consider this contribution without the self-energy terms. As in Ch. 13 these terms are included separately. The corresponding result can be written as

$$G^{(1)}_{pphh}(\alpha, \alpha'; \beta, \beta'; t_1 - t_2) = \left(\frac{-i}{\hbar}\right)^2 \int dt \, \frac{1}{4} \sum_{\gamma\gamma'\delta\delta'} \langle \gamma\gamma' | V | \delta\delta' \rangle$$

$$\langle \Phi^N_0 | \, T \left[a^\dagger_\gamma(t) a^\dagger_{\gamma'}(t) a_{\delta'}(t) a_\delta(t) a_{\alpha'}(t_1) a_\alpha(t_1) a^\dagger_\beta(t_2) a^\dagger_{\beta'}(t_2) \right] | \Phi^N_0 \rangle$$

$$\Rightarrow (i\hbar)^2 \int dt \sum_{\gamma\gamma'\delta\delta'} \langle \gamma\gamma' | V | \delta\delta' \rangle \qquad\qquad\qquad (15.8)$$

$$\times G^{(0)}(\alpha, \gamma; t_1 - t) G^{(0)}(\alpha', \gamma'; t_1 - t) G^{(0)}(\delta, \beta; t - t_2) G^{(0)}(\delta', \beta'; t - t_2).$$

Since the present discussion is focused on two-time quantities, we will employ the static nature of the interaction, instead of the form given by Eq. (9.20). The FT of Eq. (15.8) yields the first-order contribution to the pphh propagator in the energy formulation, written in various equivalent forms

$$G^{(1)}_{pphh}(\alpha, \alpha'; \beta, \beta'; E) = G^{(0)}_{pphh}(\alpha, \alpha'; E) \, \langle \alpha\alpha' | V | \beta\beta' \rangle \, G^{(0)}_{pphh}(\beta, \beta'; E)$$

$$= G^{(0)}_{pphh}(\alpha, \alpha'; E) \, \frac{1}{2} \sum_{\gamma\gamma'} \langle \alpha\alpha' | V | \gamma\gamma' \rangle \, G^{(0)}_{pphh}(\gamma, \gamma'; \beta, \beta'; E), \quad (15.9)$$

where Eq. (15.7) can be used to verify the last equality. Equation (15.9) is graphically illustrated in Fig. 15.2. It lends itself to an immediate extension, yielding a summation of an infinite set of diagrams, in analogy to the procedure discussed in Sec. 13.2 for the polarization propagator.

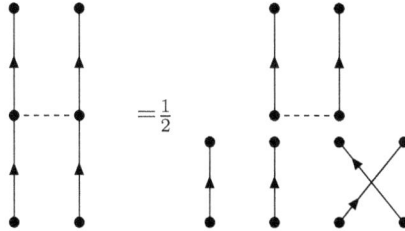

Fig. 15.2   Two equivalent representations of the first-order contribution to the pp propagator, as given by Eq. (15.9). Pairs of parallel lines, without the accompanying crossed
term, refer to $G_{pp}^{(0)}$ without the Kronecker deltas, defined in Eq. (15.7).

The extension of Eq. (15.9) exhausts all contributions to G_{pphh} for two
particles in free space, which we denote by G_{pp}. While all contributions
are not included in the case of the medium, the equation remains critically
important on account of its nonperturbative character. In the present section, we will consider the iteration of the interaction to all orders for free
particles represented by $G_{pp}^{(0)}$. Nevertheless, all results discussed here have
immediate application in the medium and can be generated by using the
corresponding $G_{pphh}^{(0)}$ in the medium. For the case of two free particles both
$|\Psi_0^N\rangle$ and $|\Phi_0^N\rangle$ become the vacuum state $|0\rangle$, since no hole propagation is
possible, and the step functions for the particles are replaced by 1, yielding

$$G_{pp}^{(0)}(\alpha,\alpha';\beta,\beta';E) = [\delta_{\alpha,\beta}\delta_{\alpha',\beta'} - \delta_{\alpha,\beta'}\delta_{\alpha',\beta}]\left\{\frac{1}{E - \varepsilon_\alpha - \varepsilon_{\alpha'} + i\eta}\right\}.$$
(15.10)

Only forward propagation occurs and the lack of holes implies that no self-
energy terms can appear in any order.

In higher order, one continues to encounter only forward-going terms
which occur sequentially after each two-body interaction. This implies that
so-called ladder diagrams are generated, where the rungs of the ladder are
represented by the interaction. The complete sum of ladder diagrams can
be obtained by combining the lowest-order contribution of Eq. (15.10) (or
Eq. (15.6) for the case of the medium) and the first-order term given by
Eq. (15.9) where the last $G_{pp}^{(0)}$ is replaced by the full sum itself as follows

$$G_{pp}(\alpha,\alpha';\beta,\beta';E) = G_{pp}^{(0)}(\alpha,\alpha';\beta,\beta';E)$$
$$+ G_{pp}^{(0)}(\alpha,\alpha';E)\frac{1}{2}\sum_{\gamma\gamma'}\langle\alpha\alpha'|V|\gamma\gamma'\rangle \; G_{pp}(\gamma,\gamma';\beta,\beta';E). \quad (15.11)$$

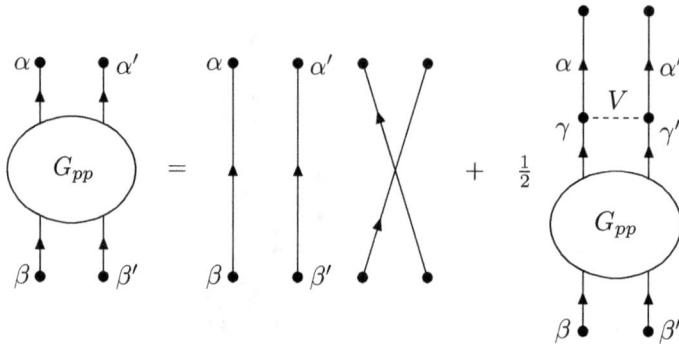

Fig. 15.3 Diagrammatic representation of the ladder equation for G_{pp} given by Eq. (15.11).

Iterating the equation generates all possible higher-order diagrams that can be obtained in perturbation theory for G_{pp} in the case of two free particles. It is therefore equivalent to the Schrödinger equation. In the medium, one still sums the same set of ladder diagrams, which form only a subset of all possible contributions. Additional terms arise in the medium in the form of self-energy insertions and more complicated ways in which the two particles can interact (see Sec. 14.6 and Ch. 9). The sum of ladder diagrams in the medium also deviates from the corresponding one in free space, due to the possibility of hh propagation between the interactions and the different sp spectrum that must be included. The solution becomes more difficult when the full dressing of the sp propagators is treated, as in fully self-consistent formulations. This will be taken up in Ch. 20.

The diagrammatic version of Eq. (15.11) is shown in Fig. 15.3. The labeling with the sp quantum numbers α and α' in the last term, identifies a contribution of $G_{pp}^{(0)}$ without Kronecker deltas, defined in Eq. (15.7). The factor of $\frac{1}{2}$ that appears in Eq. (15.11) has the following origin for higher-order contributions. For each V a factor of $\frac{1}{4}$ comes from using the symmetrized version of the interaction. For each unperturbed propagator one has two options: either one uses the version with two, or the one with four quantum numbers. The terms with two quantum numbers are generated by using the symmetry of the interaction and yield a factor of 2 for each of them. The first line of Eq. (15.9) therefore yields 1 in this case. In nth order however, we find $\frac{1}{4}^n \times 2^{n+1}$. The $\frac{1}{2}$ in Eq. (15.11) automatically generates the correct number of factors $\frac{1}{2}$ in higher order.

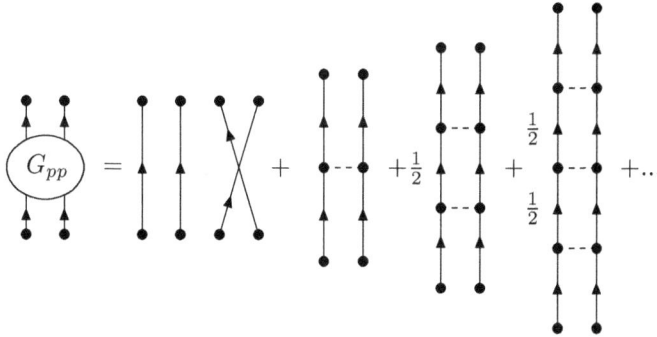

Fig. 15.4   Diagrammatic representation of the relation between G_{pp} and the corresponding vertex function, yielding Eq. (15.12).

An equivalent summation is obtained by arranging the contributions to the ladder equation for G_{pp}, according to Fig. 15.4,

$$G_{pp}(\alpha, \alpha'; \beta, \beta'; E) = G_{pp}^{(0)}(\alpha, \alpha'; \beta, \beta'; E)$$
$$+ G_{pp}^{(0)}(\alpha, \alpha'; E) \ \langle \alpha\alpha' | \Gamma_{pp}(E) | \beta\beta' \rangle \ G_{pp}^{(0)}(\beta, \beta'; E), \quad (15.12)$$

where the corresponding vertex function reads

$$\langle \alpha\alpha' | \Gamma_{pp}(E) | \beta\beta' \rangle \qquad\qquad\qquad\qquad\qquad (15.13)$$
$$= \langle \alpha\alpha' | V | \beta\beta' \rangle + \frac{1}{2} \sum_{\gamma\gamma'} \langle \alpha\alpha' | V | \gamma\gamma' \rangle \ G_{pp}^{(0)}(\gamma, \gamma'; E) \ \langle \gamma\gamma' | \Gamma_{pp}(E) | \beta\beta' \rangle.$$

We can relate Γ_{pp} back to G_{pp} by noting that all the intermediate terms in Fig. 15.4 can be resummed to G_{pp}, if an extra $\frac{1}{2}$ is included

$$\langle \alpha\alpha' | \Gamma_{pp}(E) | \beta\beta' \rangle \qquad\qquad\qquad\qquad\qquad (15.14)$$
$$= \langle \alpha\alpha' | V | \beta\beta' \rangle + \frac{1}{4} \sum_{\gamma\gamma'} \sum_{\delta\delta'} \langle \alpha\alpha' | V | \gamma\gamma' \rangle \ G_{pp}(\gamma, \gamma'; \delta, \delta'; E) \ \langle \delta\delta' | V | \beta\beta' \rangle.$$

The pole structure of G_{pp}, to be discussed shortly, is therefore identical to the one for Γ_{pp}. Such a feature was also employed in Sec. 14.6. Both resummations, given in Eqs. (15.13) and (15.14), follow similar patterns, already clarified in Ch. 6 for the sp case. The only difference occurs in the numerical factors of $\frac{1}{2}$ or $\frac{1}{4}$ that occur in the present case. The sum of ladder diagrams for the vertex function is shown in Fig. 15.5 for both Eqs. (15.13) and (15.14).

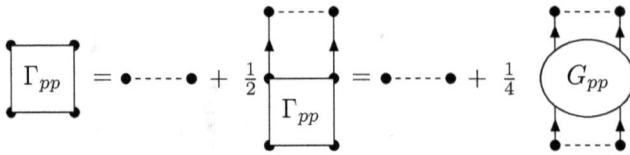

Fig. 15.5 Ladder equation for the vertex function Γ_{pp} illustrating the equivalent results contained in Eqs. (15.13) and (15.14).

15.1.1 *Scattering of two particles in free space*

As discussed in Ch. 6, the equation for the propagator, given by Eq. (15.11), is suitable for studying bound states. Many practical applications involving continuum solutions, related to the two-body scattering process, can be approached with the ladder equation for the vertex function, given by Eq. (15.13). We will first consider the latter case for scattering in free space. The natural choice is momentum or, equivalently, wave-vector space for the sp states. As usual the total wave vector (momentum) is conserved by the interaction and therefore also by the vertex function, which is normally referred to as the $\mathcal{T}$-matrix in free space.[1] The conserved total wave vector reads

$$\boldsymbol{K} = \boldsymbol{k}_\alpha + \boldsymbol{k}_{\alpha'} = \boldsymbol{k}_\beta + \boldsymbol{k}_{\beta'}, \qquad (15.15)$$

and we extract a factor $\delta_{\boldsymbol{K},\boldsymbol{K'}}/V$ ($V \to$ volume), where $\boldsymbol{K'} = \boldsymbol{k}_\beta + \boldsymbol{k}_{\beta'}$, from the matrix elements of the bare interaction and the vertex function. We define relative wave vectors, according to

$$\boldsymbol{k} = \frac{1}{2}\left(\boldsymbol{k}_\alpha - \boldsymbol{k}_{\alpha'}\right)$$

$$\boldsymbol{k'} = \frac{1}{2}\left(\boldsymbol{k}_\beta - \boldsymbol{k}_{\beta'}\right) \qquad (15.16)$$

$$\boldsymbol{q} = \frac{1}{2}\left(\boldsymbol{k}_\gamma - \boldsymbol{k}_{\gamma'}\right).$$

Transforming the summation over the relative wave vector in the resulting equation by the corresponding integration, according to Eq. (5.6), we obtain

---

[1]We will use the Γ_{pp} notation for this quantity in the present chapter.

the integral equation

$$
\langle k m_\alpha m_{\alpha'} | \Gamma_{pp}(\mathbf{K}, E) | \mathbf{k}' m_\beta m_{\beta'} \rangle = \langle k m_\alpha m_{\alpha'} | V | \mathbf{k}' m_\beta m_{\beta'} \rangle
$$

$$
+ \frac{1}{2} \sum_{m_\gamma m_{\gamma'}} \int \frac{d^3 q}{(2\pi)^3} \langle k m_\alpha m_{\alpha'} | V | \mathbf{q} m_\gamma m_{\gamma'} \rangle
$$

$$
\times G_{pp}^{(0)}(\mathbf{K}, \mathbf{q}; E) \langle \mathbf{q} m_\gamma m_{\gamma'} | \Gamma_{pp}(\mathbf{K}, E) | \mathbf{k}' m_\beta m_{\beta'} \rangle, \qquad (15.17)
$$

where the volume factor from the integration cancels with the extra volume factor in the denominator in the second term. With the present choice of variables, the noninteracting propagator in free space reads

$$
G_{pp}^{(0)}(\mathbf{K}, \mathbf{q}; E) = \frac{1}{E - \varepsilon(\frac{1}{2}\mathbf{K} + \mathbf{q}) - \varepsilon(\frac{1}{2}\mathbf{K} - \mathbf{q}) + i\eta}, \qquad (15.18)
$$

where $\varepsilon(\mathbf{k}) = \hbar^2 \mathbf{k}^2 / 2m$ represents the kinetic energy of a particle. No spin (isospin) dependence needs to be included in $G_{pp}^{(0)}$, as in Ch. 13 for the noninteracting polarization propagator. Spins (isospins) naturally require explicit consideration in the matrix elements of the interaction and the vertex function. Since E represents the total energy of the two particles, one can isolate the available energy in the center of mass, denoted by E_0, by writing

$$
E = \frac{\hbar^2 \mathbf{K}^2}{4m} + E_0 \equiv \frac{\hbar^2 \mathbf{K}^2}{4m} + \frac{\hbar^2 k_0^2}{m}, \qquad (15.19)
$$

where k_0 is introduced, which identifies the magnitude of the wave vector that corresponds to E_0. Since the sum of the sp kinetic energies in Eq. (15.18) is given by

$$
\varepsilon(\tfrac{1}{2}\mathbf{K} + \mathbf{q}) + \varepsilon(\tfrac{1}{2}\mathbf{K} - \mathbf{q}) = \frac{\hbar^2 \mathbf{K}^2}{4m} + \frac{\hbar^2 \mathbf{q}^2}{m}, \qquad (15.20)
$$

the denominator of Eq. (15.18) displays no dependence on the total wave vector and consequently, the integral equation does not depend on it either, as expected. The variable can therefore be dropped for two free particles. This is not true for particles in the medium, as will be discussed in the next section. It is helpful to make a partial wave decomposition of the scattering equation, if a numerical solution in wave-vector space is contemplated. It is also recommended if the interaction contains short-range repulsion and has a finite range, as for nucleons or atom–atom potentials. One typically proceeds to couple to good total spin (and isospin) and, for a tensor interaction, to add the orbital angular momentum of the relative motion to the total spin, to generate states with good total angular momentum. The relevant steps were given in Sec. 4.1 and can be applied to the present situation without further complications. The integral equation can then be written as

$$\langle k\ell| \, \Gamma_{pp}^{JST}(k_0) \, |k'\ell'\rangle = \langle k\ell| \, V^{JST} \, |k'\ell'\rangle \qquad (15.21)$$

$$+ \frac{m}{2\hbar^2} \sum_{\ell''} \int \frac{dq \, q^2}{(2\pi)^3} \, \langle k\ell| \, V^{JST} \, |q\ell''\rangle \, \frac{1}{k_0^2 - q^2 + i\eta} \, \langle q\ell''| \, \Gamma_{pp}^{JST}(k_0) \, |k'\ell'\rangle \, .$$

For an uncoupled channel, the diagonal (on-shell) matrix element of Γ_{pp} at wave vector k_0, is related to the phase shift. The relation is equivalent to Eq. (6.70) for the sp case. This is hardly surprising, since Eq. (15.21) is almost identical to Eq. (6.60), apart from angular momentum coupling. Indeed, the analysis given in Sec. 6.4.1 can be repeated step by step for an uncoupled channel. The only significant difference occurs in the use of twice the kinetic energy in the denominator of the unperturbed propagator in wave-vector space. The relation between the on-shell matrix element and the phase shift therefore becomes

$$\langle k_0\ell|S^{JST}(k_0)|k_0\ell\rangle = \left[1 - 2\pi i \left(\frac{mk_0}{2\hbar^2}\right) \langle k_0\ell|\Gamma_{pp}^{JST}(k_0)|k_0\ell\rangle\right] \equiv e^{2i\delta_\ell^{JST}},$$
$$(15.22)$$

where an extra factor of 2 has appeared in the denominator of the density of states as compared to the sp problem, given by Eq. (6.71). Equation (6.72) continues to hold as well

$$\tan \delta_\ell^{JST} = \frac{\text{Im} \, \langle k_0\ell|\Gamma_{pp}^{JST}(k_0)|k_0\ell\rangle}{\text{Re} \, \langle k_0\ell|\Gamma_{pp}^{JST}(k_0)|k_0\ell\rangle}. \qquad (15.23)$$

It shows, also noted in Ch. 6, that a nonvanishing imaginary part of the on-shell matrix element of Γ_{pp} is necessary for a nonvanishing phase shift.

The numerical solution of Eq. (15.21) requires knowledge of the relevant two-body matrix elements of the bare interaction, which can *e.g.* be obtained by applying the results of Sec. 4.4, together with standard angular momentum techniques (see App. B). The discretization of the integral in Eq. (15.21) is straightforward and a matrix inversion suffices to calculate the vertex function [Haftel and Tabakin (1970)]. We will mostly discuss results for nucleons here as an example of a rather complicated two-body interaction. Similarities with atom–atom interactions will be pointed out along the way. The results for the phase shifts in the $^1S_0$ channel for the potential, shown as the full line in Fig. 4.3, are given in Fig 15.6. They illustrate the attractive nature of the interaction at low energy, indicated by the

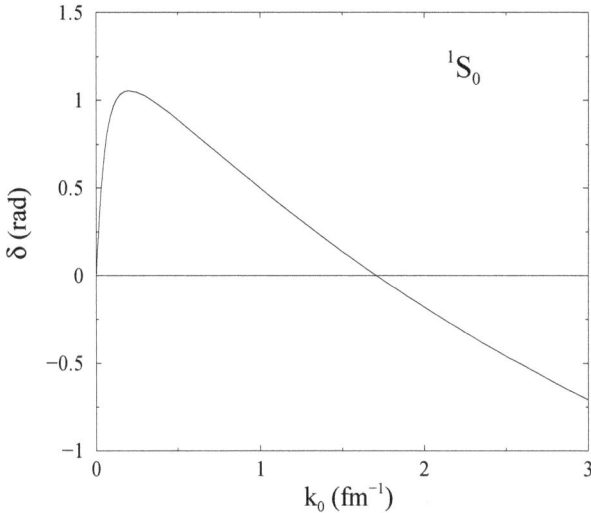

Fig. 15.6   Phase shift in radians of the $^1S_0$ nucleon–nucleon interaction as a function of
the on-shell wave vector k_0.

positive phase shifts. At higher energy the attraction turns into repulsion
(negative phase shifts), demonstrating the increasing dominance of the repulsive core. While the phase shifts were obtained numerically by working
in wave vector space, it is always illuminating to turn to coordinate space
and solve for the scattering wave function by integrating the Schrödinger
wave equation directly [see *e.g.* Eq. (12.67)]. The treatment of the repulsive
core by an all-order summation of ladder diagrams has the expected effect
on the relative wave function of two particles, *i.e.* making it vanish inside
the core. In Fig. 15.7 the free wave function is compared with the solution
of the wave equation at an energy corresponding to $k_0 = 0.25$ fm^{-1}. The
interaction is also plotted in the figure (scaled by a factor $1/100$) as the thin
dashed line, in order to visualize this feature. The enhancement of the wave
function in the domain, where the interaction is attractive, is transparent
in the figure.

The analysis of the asymptotic behavior of the scattering wave function
doesn't change appreciably in the case of coupled channels (see *e.g.* [Dick-
hoff *et al.* (1999)]). Instead of diagonal contributions in ℓ in Eq. (6.59),
nondiagonal terms are possible when a tensor interaction contributes, although only when the total spin is 1. It leads to additional, nondiagonal,

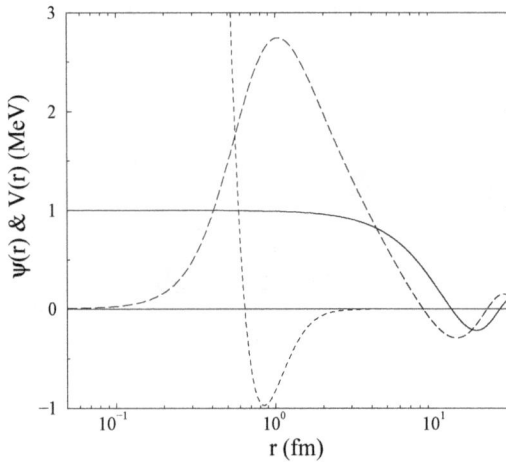

Fig. 15.7 Wave function in coordinate space for the $^1S_0$ interaction at an energy corresponding to $k_0 = 0.25$ fm^{-1}. The solid line is the free wave function $\sin(k_0 r)/k_0 r$. The correlated wave function is represented by the long-dashed line and clearly exhibits the suppression at short distances, expected for the repulsive core. The latter is indicated by the dashed line and divided by a factor 100.

$\mathcal{S}$-matrix elements, which therefore read

$$\langle k_0\ell|\mathcal{S}^{JST}(k_0)|k_0\ell'\rangle = \left[\delta_{\ell,\ell'} - 2\pi i\left(\frac{mk_0}{2\hbar^2}\right)\langle k_0\ell|\Gamma_{pp}^{JST}(k_0)|k_0\ell'\rangle\right]. \quad (15.24)$$

The unitarity of the $\mathcal{S}$-matrix and the symmetry property of the Γ_{pp}-matrix, allow the diagonalization of $\mathcal{S}$ by an orthogonal real matrix A

$$\langle k_0\ell|\mathcal{S}^{JST}(k_0)|k_0\ell'\rangle = \sum_{\alpha=1,2}\langle \ell|A^J(k_0)|\alpha\rangle e^{2i\delta_\alpha^{JST}}\langle \alpha|A^J(k_0)|\ell'\rangle, \quad (15.25)$$

where δ_α^{JST} are called the (real) eigenphase shifts. One may choose [Blatt and Biedenharn (1952)]

$$\langle \ell|A^J(k_0)|\alpha\rangle = \begin{pmatrix} \cos\epsilon^J & \sin\epsilon^J \\ -\sin\epsilon^J & \cos\epsilon^J \end{pmatrix}, \quad (15.26)$$

where ϵ^J is referred to as the mixing angle and the related mixing parameter is given by

$$\rho^J = \sin 2\epsilon^J. \quad (15.27)$$

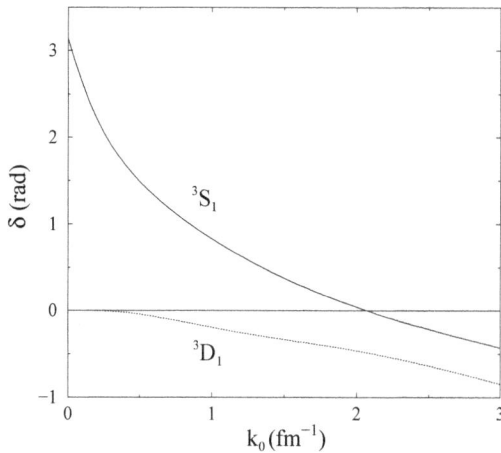

Fig. 15.8   Phase shifts in radians of the $^3S_1$-$^3D_1$ nucleon-nucleon interaction, as a function of the on-shell wave vector k_0. While substantial mixing occurs, it is customary to continue to use the original designation of $^3S_1$ and $^3D_1$ to identify the eigenphases.

Note that the three real parameters $\delta_1^{JST}, \delta_2^{JST}$, and ϵ^J can be used to represent the $\mathcal{S}$-matrix.[2] Phase shifts and mixing parameters can also be constructed from the on-shell $\mathcal{R}$-matrix elements [Haftel and Tabakin (1970)]. The $\mathcal{R}$-matrix corresponds to the solution of Eq. (15.21), when only the real part of $G_{pp}^{(0)}$ is employed. While the $^1S_0$ phase shift indicates that the interaction is almost sufficiently attractive to support a bound state, it is in the coupled $^3S_1$-$^3D_1$ channel that the sole two-nucleon bound state, the deuteron, emerges. Its presence can be inferred from the eigenphase shifts for this coupled channel, as shown in Fig. 15.8. Since at low energy the scattering wave function is orthogonal to the bound state, its presence requires the phase shift to start at π radians. In general, Levinson's theorem states that the phase shift starts at $n\pi$ for zero energy, if n bound states are present [Gottfried and Yan (2004)].

It is instructive to compare the NN phase shifts, shown in Figs. 15.6 and 15.8, with those for ^{3}He atoms, using the potential from [Aziz et al. (1979)]. The latter phase shifts are given in Fig. 15.9. From Fig. 4.4 it is clear that the interaction has a weak attraction and a huge repulsive core. This is reflected in slightly attractive phase shifts at low energy, turning

---

[2] The relation of the eigenphase shifts and corresponding mixing parameter with the so-called bar phase shifts can *e.g.* be found in [Brown and Jackson (1976)].

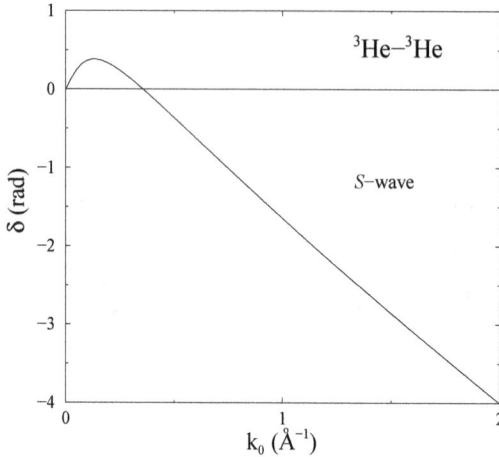

Fig. 15.9   Phase shifts in radians of the S-wave ^{3}He$-^3$He interaction, as a function of the on-shell wave vector k_0.

over into a steep dive at larger energy, reflecting the strong repulsion. We will return to the effects of this repulsion on the phase shifts in the medium in Sec. 15.2. A comparison of scattering results for nucleons in free space and in the medium will be presented there as well (see also Sec. 20.3).

15.1.2   *Bound states of two particles*

We now consider the calculation of bound states. A Lehmann representation corresponding to Eq. (15.1) can be obtained in the usual way. It includes both bound states for the $N \pm 2$ systems, as well as continuum states [see Eq. (9.35) for the sp case] and is given by

$$
\begin{aligned}
G_{pphh}(\alpha, \alpha'; \beta, \beta'; E) = &\sum_m \frac{\langle \Psi_0^N | a_{\alpha'} a_\alpha | \Psi_m^{N+2} \rangle \langle \Psi_m^{N+2} | a_\beta^\dagger a_{\beta'}^\dagger | \Psi_0^n \rangle}{E - (E_m^{N+2} - E_0^N) + i\eta} \\
&+ \int_{\varepsilon_T^+}^{\infty} d\tilde{E}_\mu^{N+2} \frac{\langle \Psi_0^N | a_{\alpha'} a_\alpha | \Psi_\mu^{N+2} \rangle \langle \Psi_\mu^{N+2} | a_\beta^\dagger a_{\beta'}^\dagger | \Psi_0^N \rangle}{E - \tilde{E}_\mu^{N+2} + i\eta} \\
&- \sum_n \frac{\langle \Psi_0^n | a_\beta^\dagger a_{\beta'}^\dagger | \Psi_n^{N-2} \rangle \langle \Psi_n^{N-2} | a_{\alpha'} a_\alpha | \Psi_0^N \rangle}{E - (E_0^N - E_n^{N-2}) - i\eta} \\
&- \int_{-\infty}^{\varepsilon_T^-} d\tilde{E}_\nu^{N-2} \frac{\langle \Psi_0^N | a_\beta^\dagger a_{\beta'}^\dagger | \Psi_\nu^{N-2} \rangle \langle \Psi_\nu^{N-2} | a_{\alpha'} a_\alpha | \Psi_0^N \rangle}{E - \tilde{E}_\nu^{N-2} - i\eta}. \quad (15.28)
\end{aligned}
$$

The continuum thresholds $\varepsilon_T^{\pm}$ and relative excitation energies are denoted by $\tilde{E}_\mu^{N+2} = E_\mu^{N+2} - E_0^N$ and $\tilde{E}_\nu^{N-2} = E_0^N - E_\nu^{N-2}$, respectively. Equation (15.28) pertains to all possible situations. Suitable modifications may occur for a particular system. In case of two free particles, for example, there are naturally no $N-2$ states and the reference state corresponds to the vacuum. Two-particle bound states can be written as

$$\left|\Psi_n^{N=2}\right\rangle = \left|\boldsymbol{K}n\right\rangle, \tag{15.29}$$

where $\boldsymbol{K}$ [see Eq. (15.15)] describes the motion of the center of mass and n is used to label the intrinsic quantum numbers. The relation with the amplitudes in the numerator of Eq. (15.28) for $\boldsymbol{K} = 0$ then reads

$$\langle 0| a_{-\boldsymbol{k}m_{\alpha'}} a_{\boldsymbol{k}m_\alpha} \left|\boldsymbol{K} = 0 \ n\right\rangle = \langle \boldsymbol{K} = 0 \ \boldsymbol{k}; m_\alpha m_{\alpha'} \left|\boldsymbol{K} = 0 \ n\right\rangle$$
$$= \psi_n(\boldsymbol{k}; m_\alpha m_{\alpha'}), \tag{15.30}$$

where $\boldsymbol{k}$ is the wave vector describing the relative motion, as in Eq. (15.16). The possibility of one (or more) bound state(s) can be studied in the standard way, by turning Eq. (15.11) into an eigenvalue problem using the Lehmann representation (15.28). As usual, the (possible) bound state appears as a discrete pole at negative energy in Eq. (15.28), whereas the noninteracting propagator in Eq. (15.18) only has poles (branch-cut) at positive energy. Using the notation of Eq. (15.30), we find

$$\frac{\hbar^2 k^2}{m}\psi_n(\boldsymbol{k}; m_\alpha m_{\alpha'}) + \frac{1}{2}\sum_{m_\gamma m_{\gamma'}}\int \frac{d^3q}{(2\pi)^3} \langle \boldsymbol{k}m_\alpha m_{\alpha'}| V \left|\boldsymbol{q}m_\gamma m_{\gamma'}\right\rangle \psi_n(\boldsymbol{q}; m_\gamma m_{\gamma'})$$
$$= E_n \ \psi_n(\boldsymbol{k}; m_\alpha m_{\alpha'}), \tag{15.31}$$

where E_n denotes the energy of the bound state and $\boldsymbol{K} = 0$ is chosen. Simple manipulations have been applied to generate the usual form of the Schrödinger equation in wave-vector space. The rotational invariance of the interaction makes it possible to write the Schrödinger equation in a basis of good total angular momentum, by coupling orbital angular momentum to total spin for noncentral forces. Since this applies to nucleons, one can also couple the states to good total isospin. The resulting eigenvalue equation in the presence of a tensor force therefore allows the coupling of different orbital angular momenta, like in the scattering problem, and is given by

$$\frac{\hbar^2 k^2}{m}\psi_n(k(\ell S)JT) + \frac{1}{2}\sum_{\ell'}\int \frac{dq\ q^2}{(2\pi)^3} \langle k\ell| V^{JST} \left|q\ell'\right\rangle \psi_n(q(\ell'S)JT)$$
$$= E_n \ \psi_n(k(\ell S)JT). \tag{15.32}$$

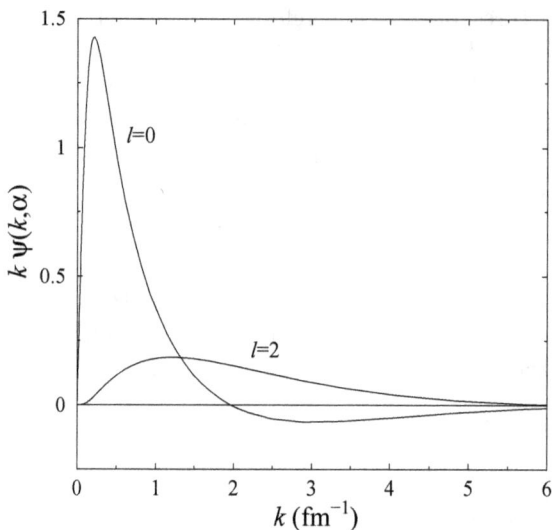

Fig. 15.10 Wave functions components in k-space multiplied by k for the $^3S_1$-$^3D_1$ deuteron bound state. The presence of the D-state component is due to the nuclear tensor interaction. The high-momentum components of the wave function are associated with the short-range repulsion of the potential.

The coupled $^3S_1$-$^3D_1$ channel exhibits a bound state for two nucleons, the deuteron, with a sizable D-state component. The tensor force is therefore crucial in generating the binding, as can be inferred by comparing the central parts of the $^1S_0$ and $^3S_1$ interaction, shown in Fig. 4.3. Since the $^1S_0$ is clearly more attractive, but does not support a bound state, it is the coupling to the ($\ell = 2$) D-state by the tensor force that is responsible for the binding of the deuteron. The solution of Eq. (15.32) can be obtained by discretizing the integral with a finite set of points and, subsequently, performing the diagonalization of the resulting matrix. Working in wave-vector space has the added benefit that it allows for the treatment of interactions that are nonlocal in coordinate space. The binding energy for the deuteron is fitted to experiment for realistic interactions. The wave functions for the S and D state components are shown in Fig. 15.10 for the Reid soft-core interaction. The D-state component of the normalized wave function corresponds to about 6.5% for this interaction [Reid (1968)]. To illustrate the effect of the short-range repulsion on the wave function, we include an illustration of these wave functions in coordinate space in

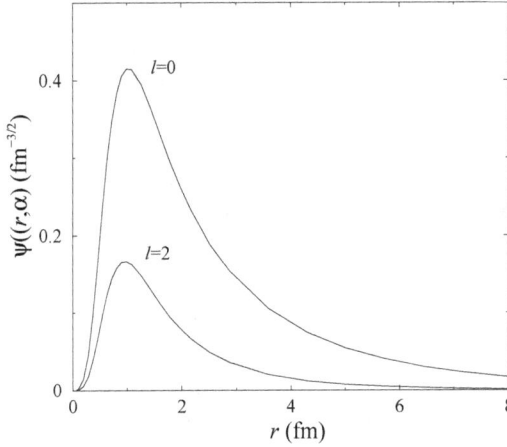

Fig. 15.11   Wave-function components in r-space for the $^3S_1$-$^3D_1$ deuteron bound state.

Fig. 15.11, by employing the following Fourier–Bessel transform

$$\psi_n(r(\ell S)JT) = \sqrt{\frac{2}{\pi}} \int_0^\infty dk k^2 \, \mathrm{j}_\ell(kr) \, \psi_n(k(\ell S)JT). \qquad (15.33)$$

While a comparison with a noninteracting wave function is not as useful as for the scattering case, it is nevertheless clear that the repulsive core suppresses the S-wave function at short distances. Proper treatment of these short-range correlations requires an all-order procedure, like solving the Schrödinger equation for two particles. Since the solution is equivalent to summing the complete set of ladder diagrams, we conclude that the same set will provide a minimum choice to deal with short-range correlations in the medium. All nuclear systems and all quantum fluids have basic interactions of this kind. Green's function or other methods based on perturbation theory, must therefore include ladder diagrams.

15.2   Ladder diagrams and short-range correlations in the medium

There are several perspectives to the summation of ladder diagrams in the medium. As already mentioned in the previous section, their summation to all orders for two free particles ensures a proper treatment of short-range correlations. This feature will be preserved in the medium, which makes ladder summations essential for systems with short-range repulsive interactions. Quantum liquids, neutron and nuclear matter, finite nuclei, all require detailed consideration of these correlations. Even for eclectrons in the electron gas it may be necessary to treat the Coulomb repulsion at short distances accordingly. Indeed, the self-consistent formulation of Sec. 9.4 suggests that a proper starting point in the medium is to approximate the vertex function by the ladder sum. Self-consistency then requires the use of fully dressed propagators. In the case of a low-density system, full dressing most likely will not involve a substantial modification of the properties of the particles from their free behavior, since the constituents of the system do not "meet" each other frequently. They are therefore not expected to experience important medium effects, associated with the Pauli principle, or changes in the location of their spectral strength. The propagation of dressed particles in solving the ladder equation, at a density where this dressing is important, will be presented in Ch. 20. The consequences of treating short-range correlations for the sp propagator, including self-consistency, are illustrated in Sec. 16.3.1.

In the present section we will focus on the solution of the ladder equation in the medium using mean-field sp propagators. It is possible to extend the analysis for the low-density limit for bosons, given in Ch. 12, to the case of fermions that interact by means of purely repulsive forces. This was first done by [Galitskii (1958)] and a full discussion of the analysis can be found in [Fetter and Walecka (1971)]. Since the fermion systems emphasized in this text, typically also have attractive interactions, leading to self-bound systems, we will forego the topic here. In addition, we will address the consequences that simple medium effects, like the Pauli principle and changes in sp energies, have on the scattering process. This is the other perspective, associated with summing ladder diagrams in an infinite system: for energies above $2\varepsilon_F$, it corresponds to the $N+2$ continuum that can be cast into an equivalent description in terms of pp scattering in the medium. For energies below $2\varepsilon_F$, hh propagation represents the scattering process of two holes in the continuum of the $N-2$ states.

The ladder equation in the medium for mean-field particles has the same form as Eq. (15.17). The essential difference is contained in the noninteracting propagator. Since we will concentrate on homogeneous infinite systems, this propagator takes the form of Eqs. (15.6) and (15.7). Employing wave vector quantum numbers, the in-medium form generalizes Eq. (15.18) for free particles to

$$G_{pphh}^{(0)}(\mathbf{K}, \mathbf{q}; E) = \frac{\theta(|\mathbf{K}/2+\mathbf{q}| - k_F)\,\theta(|\mathbf{K}/2-\mathbf{q}| - k_F)}{E - \varepsilon(\mathbf{K}/2+\mathbf{q}) - \varepsilon(\mathbf{K}/2-\mathbf{q}) + i\eta}$$
$$- \frac{\theta(k_F - |\mathbf{K}/2+\mathbf{q}|)\,\theta(k_F - |\mathbf{K}/2-\mathbf{q}|)}{E - \varepsilon(\mathbf{K}/2+\mathbf{q}) - \varepsilon(\mathbf{K}/2-\mathbf{q}) - i\eta}. \quad (15.34)$$

The presence of the medium in this propagator is expressed in the appearance of the step functions that either allow two particles to propagate when both are above the Fermi wave vector, or two holes to propagate when their wave vectors are smaller than k_F. For future reference, it is helpful to write Eq. (15.34) as a convolution of sp propagators

$$G_{pphh}^{(0)}(\mathbf{K}, \mathbf{q}; E) = i \int \frac{dE'}{2\pi}\, G^{(0)}(\mathbf{K}/2+\mathbf{q}; E/2+E')G^{(0)}(\mathbf{K}/2-\mathbf{q}; E/2-E').$$
$$(15.35)$$

This formulation yields Eq. (15.34), after employing Eq. (11.8) in the usual contour integration. The propagator of Eq. (15.34) is sometimes referred to as the Galitskii–Feynman propagator. In [Galitskii (1958)] the consequences of the medium were studied by employing the propagator given in Eq. (15.34). Equation (15.17) continues to hold, with the replacement of $G_{pp}^{(0)}$ by $G_{pphh}^{(0)}$. Also Figs. 15.3 – 15.5 are identical. The scattering equation in the medium therefore continues to generate the ladder diagrams. The ladders now also contain terms that go backward in time (Feynman diagrams), since hh propagation is included in Eq. (15.34).

The medium effect, due to the Pauli principle, introduces an important dependence on the center-of-mass wave vector $\mathbf{K}$. It can be illustrated by considering Fig. 15.12. Part a) of the figure displays the situation when $K < 2k_F$, which allows the two spheres that represent the step functions of Eq. (15.34) to overlap. The region outside both spheres, corresponds to pp propagation, whereas hh propagation can only occur when $\mathbf{q}$ is inside both spheres. This region is bounded from below by the dashed line and from above by the corresponding segment of the lower sphere. For center-of-mass momenta above $2k_F$, the situation is illustrated in part b). There is now no phase space for hh propagation. It is indeed impossible to remove two particles from the Fermi sea, each having a wave vector less than k_F, and construct a total wave vector larger than $2k_F$. The phase space dependence on the magnitude of $\mathbf{K}$ leads to an optimal situation when $K \to 0$. In that case, the two spheres in part a) overlap completely and pp propagation is represented by the phase space outside the common sphere and hh by the phase space inside. This result takes on even more significance, when

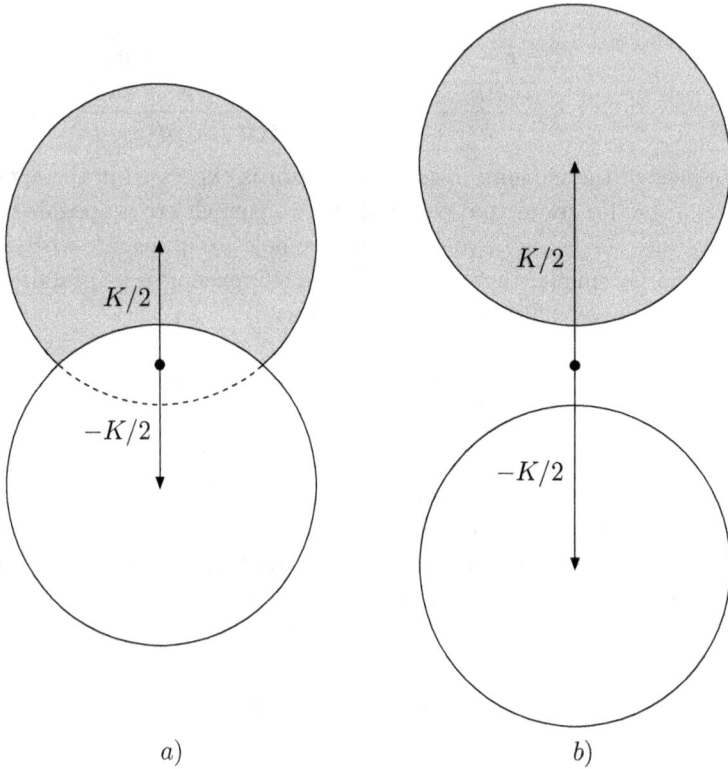

Fig. 15.12 Illustration of the contraints imposed by the pairs of step functions in Eq. (15.34). The condition $|\boldsymbol{q} \pm \boldsymbol{K}/2| > k_F$ corresponds to the area outside both the upper and lower sphere. The condition $|\boldsymbol{q} \pm \boldsymbol{K}/2| < k_F$ only allows contributions that must be inside both spheres. This area is bordered below by the dashed line in part a) which illustrates the case when $K < 2k_F$ when these two spheres overlap. Part b) is appropriate for $K > 2k_F$ when there is no overlap and therefore no hh propagation possible.

possible bound states are considered in the next section. Another influence of the medium, is the possibility that the sp spectrum $\varepsilon(\boldsymbol{q})$ deviates from the kinetic energy appropriate for particles in the Fermi gas.

15.2.1 *Scattering of mean-field particles in the medium*

We now illustrate the consequences of the presence of the medium on the properties of phase shifts. For this purpose, we assume that the sp spectrum is still described by the kinetic energy (only correct for low density), or contains an additional potential energy contribution. The latter will be a monotonically increasing function of wave vector, appropriate for nuclear systems. We consider the case when the center-of-mass wave vector is zero for simplicity. The noninteracting propagator then reduces to

$$G^{(0)}_{pphh}(\boldsymbol{K} = 0, q; E) = \frac{\theta(q - k_F)}{E - 2\varepsilon(q) + i\eta} - \frac{\theta(k_F - q)}{E - 2\varepsilon(q) - i\eta}, \qquad (15.36)$$

requiring only the magnitude of $\boldsymbol{q}$, denoted by q. To obtain the phase shifts for particles propagating in the medium with mean-field sp energies, one can follow the analysis of Sec. 6.4, although some steps require a numerical treatment. A useful reference is [Bishop *et al.* (1974)], where the introduction of the phase shift for hh propagation was first discussed in considerable detail. The sp energy $\varepsilon(q)$ can deviate from the simple kinetic energy spectrum. It may therefore yield a different relation between the energy E and the on-shell wave vector k_0 than presented in Eq. (15.19), resulting in

$$E \equiv 2\varepsilon(k_0). \qquad (15.37)$$

Nevertheless, the uniqueness of k_0, for a particular energy, is still preserved. Although one can no longer evaluate the noninteracting propagator in coordinate space completely analytically from Eq. (6.63) (even for kinetic energies), the separability of the propagator is maintained for the contribution of the pole term, as in Eq. (6.66) but possibly with a different constant prefactor. The remaining term vanishes asymptotically for r sufficiently different from r', as discussed in one of the exercises of this chapter. Consequently, one preserves the integral equation for the wave function in a partial wave basis, as in Eq. (6.68), for mean-field propagators. The only difference with free scattering, involves the use of the mean-field equivalent of the noninteracting propagator in coordinate space in Eq. (6.69). As stated above, this is due to the uniqueness of the on-shell wave vector, which guarantees that the noninteracting wave function is a plane wave or spherical Bessel function (in a partial wave basis). One can thus proceed with a similar asymptotic analysis as for free particles, yielding a corresponding definition of the phase shifts, as in Eq. (15.22) in terms of the on-shell scattering matrix. Equation (15.23) also remains valid. For coupled channels Eq. (15.24) holds, and eigenphase shifts are obtained by diagonalizing the S-matrix. The presence of a nonvanishing phase shift continues to be linked to the nonvanishing of the imaginary part of the noninteracting propagator.

For mean-field particles, the domain therefore includes all energies above $2\varepsilon(q = 0)$, which identifies the lowest energy of two occupied states.

As for particles in free space, the presence of bound states has specific consequences for the behavior of the phase shift at the corresponding thresholds in the energy variable [Bishop *et al.* (1974)]. In free space, it corresponds to zero energy (in the center of mass) and the presence of one bound state is reflected in the phase shift going to π when the scattering energy goes to zero. The relevant threshold in the medium is $2\varepsilon_F$. If the interaction is sufficiently attractive, the phase shift approaches π on both sides of $2\varepsilon_F$, to be illustrated shortly. This feature is intimately related to the presence of a pairing solution or bound-pair states in Fermi systems with attractive, effective interactions at the Fermi surface, discussed in Sec. 15.3. The phase shift can also approach $-\pi$ when a bound state below the hh continuum (*i.e.* below $2\varepsilon(q = 0)$) appears, due to a repulsive interaction. This possibility is realized in liquid ${}^3\text{He}$ at sufficiently high density when propagating mean-field particles [Bishop *et al.* (1974); Glyde and Hernadi (1983)]. Both situations (phase shifts going to π or $-\pi$) will be illustrated by employing modifications of the ${}^1S_0$ interaction of the Reid potential, as well as the actual Reid ${}^1S_0$ and ${}^3S_1$-${}^3D_1$ interactions [Reid (1968)] in the medium. If the interaction is not sufficiently attractive to yield pairing, the phase shift will be zero at $2\varepsilon_F$ on account of Eq. (15.23) and the vanishing phase space at this energy, which makes the imaginary part of the interaction Γ vanish.

The results, discussed here, involve the propagation of mean-field particles in nuclear matter at zero temperature, but are otherwise completely general. The aim is to exhibit some characteristic changes that occur in the medium for the phase shifts of $\ell = 0$ channels in the NN interaction with respect to their behavior in free space. In Fig. 15.13 the phase shift for the ${}^1S_0$ channel is shown as a function of the on-shell wave vector for various densities, and contrasted with the outcome in free space (solid line). The long-dashed, dashed, and dotted lines represent the Fermi wave vectors of 0.8, 1.36, and 1.8 fm^{-1}, respectively. For simplicity and ease of comparison a sp spectrum of kinetic energy was assumed. The on-shell wave vector was used as the plotting variable in Fig. 15.13, instead of the energy, since it allows a direct comparison between free and mean-field particles at different densities. While the nuclear interaction in the ${}^1S_0$ channel is not sufficiently attractive to generate a bound state in free space, it yields a pairing solution in a wide range of densities (see *e.g.* [Chen *et al.* (1986)]).

The presence of a pairing solution can be inferred from the behavior of

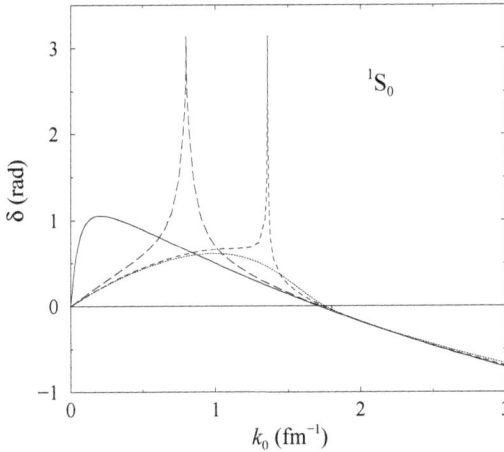

Fig. 15.13 Phase shift for the $^1S_0$ channel of the Reid potential at various densities, as a function of the on-shell wave vector. Both for free particles (solid line) and mean-field particles in the medium, corresponding to $k_F = 0.8$ (long-dashed), 1.36 (dashed), and 1.8 fm^{-1} (dotted line), a kinetic energy spectrum was used.

the phase shifts. The properties of the associated bound-pair states will be explained in the next section. Suffice it to note here that the occurrence of bound-pair states in the effective interaction, implies that the phase shift at the associated energy threshold (in this case $2\varepsilon_F$) will tend to π [Bishop et al. (1974)]. This outcome is indeed exhibited for the phase shifts for $k_F = 0.8$ and 1.36 fm^{-1}. In both cases, the phase shift on either side of $2\varepsilon_F$ (or, as in Fig. 15.13, on either side of k_F) approaches π. It is also clear from Fig. 15.13 that the phase shift tends to π more abruptly for $k_F = 1.36$ fm^{-1} than for 0.8 fm^{-1}, while it no longer does so for $k_F = 1.8$ fm^{-1}. These features illustrate the appearance and the strength of bound pair states. The latter acquire the largest binding at $k_F = 0.8$ fm^{-1}, almost no binding at 1.36 fm^{-1}, and no bound states exist at 1.8 fm^{-1}, as discussed in Sec. 15.3. Indeed, the density range for the appearance of bound-pair states, exactly tracks the appearance of a phase shift of π at k_F. This observation is commensurate with the notion that positive phase shifts in the medium near k_F ($2\varepsilon_F$), implying an attractive effective interaction, indicate the presence of bound-pair states. The general behavior around k_F is then an indication of the amount of correlation: strong attraction is indicated by a phase shift that is already large and positive quite far away from k_F.

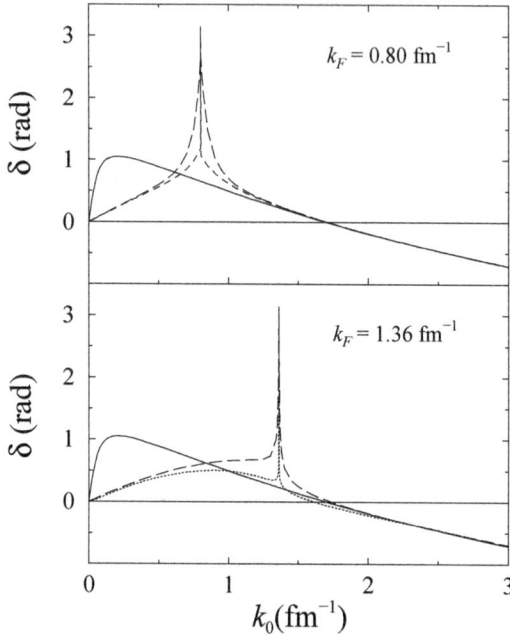

Fig. 15.14   Sensitivity of the $^1S_0$ phase shift in the medium to a gap in the sp spectrum at the Fermi momentum for $k_F = 0.8$ fm^{-1} (top panel) and the inclusion of a realistic sp spectrum for $k_F = 1.36$ fm^{-1} (bottom panel). The solid line represents the phase shift for free particles. The long-dashed line in the top panel employs the propagator $G^{(0)}_{pphh}$ with kinetic energies, whereas the dashed line includes a gap in the spectrum at k_F. In the bottom panel, the dashed line again employs Eq. (15.34) with kinetic energies at $k_F = 1.36$ fm^{-1}. The dotted line includes a realistic sp spectrum for this density.

The sensitivity of the $^1S_0$ phase shift to the sp spectrum, or a gap in the sp spectrum at k_F, is explored in Fig. 15.14. In the top panel, the kinetic energy spectrum at $k_F = 0.8$ fm^{-1} was modified by including a 7 MeV gap between sp states above and below the Fermi momentum. The gap ensures that the eigenvalues of the bound-pair states fall inside the corresponding 14 MeV gap in the two-particle spectrum (see next section) and are therefore real. When a pure kinetic energy spectrum (without gap) is used, these eigenvalues acquire complex values, indicating a pairing instability quite similar in nature to the RPA instability, discussed in Chs. 13 and 14. The dashed line in the top panel refers to the kinetic energy spectrum and the short-dashed line includes the gap in the spectrum. In order to understand these features, it is important to remember that the ladder equation, including both pp and hh propagation, is similar to an RPA summation. While the phase shift in the top panel of Fig. 15.14 still tends to π at k_F, when the gap in the sp spectrum is introduced, it is clear that a reduction of the RPA-like collectivity occurs on account of the a gap between particle and hole states, leading to less attractive phase shifts around k_F. A similar reduction of the attraction, exhibited by the phase shift, is observed in the lower panel of Fig. 15.14 for $k_F = 1.36$ fm^{-1}, when a realistic sp potential energy taken from [Vonderfecht *et al.* (1993)], is added to the kinetic energy. The dashed line in the bottom panel of Fig. 15.14 corresponds to the kinetic energy spectrum, while the dotted line includes the sp potential energy. Also in this case the average distance between particle and hole energies is enlarged by the sp potential energy, which reduces the RPA-like collectivity in the ladder equation.

Another demonstration of the connection between the behavior of the phase shift near k_F and bound-pair states, is provided in the top panel of Fig. 15.15. The panel demonstrates the behavior of the phase shifts for the same set of densities as in Fig. 15.13, but for a modified version of the Reid $^1S_0$ interaction. By multiplying the intermediate-range attraction of this interaction with 3 Yukawa terms by a factor 1.1, one generates a bound state in free space, reflected by the phase shift going to π at zero momentum in the top panel of Fig. 15.15 (solid line). The other three curves represent the same set of Fermi wave vectors as in Fig. 15.13 (again using a kinetic energy sp spectrum in the medium). Comparing the phase shifts for the two interactions at the same density, we observe, as expected, a substantially more positive phase shift for the more attractive interaction. For the highest Fermi momentum ($k_F = 1.8$ fm^{-1}), the more attractive interaction now has a phase shift of π at k_F, unlike the actual Reid $^1S_0$ interaction (shown in Fig. 15.13), demonstrating that the range of densities where bound-pair states occur, is enlarged.

It is also instructive to illustrate the condition under which the phase shift from the Galitskii–Feynman integral equation tends to $-\pi$. In [Bishop *et al.* (1974); Glyde and Hernadi (1983)] it was shown that this occurs when a hh bound state exists below the hh continuum, which reflects higher excitation energies than can be obtained by just removing two mean-field particles. Such a spectrum is generated by propagating mean-field ^{3}He atoms in the medium, interacting by means of a realistic atom–atom interaction (see Fig. 15.9). This type of interaction can be simulated by increasing the strength of the short-range repulsion of the Reid $^1S_0$ channel by a factor of ten. The phase shifts for the same set of densities, are shown in the bottom

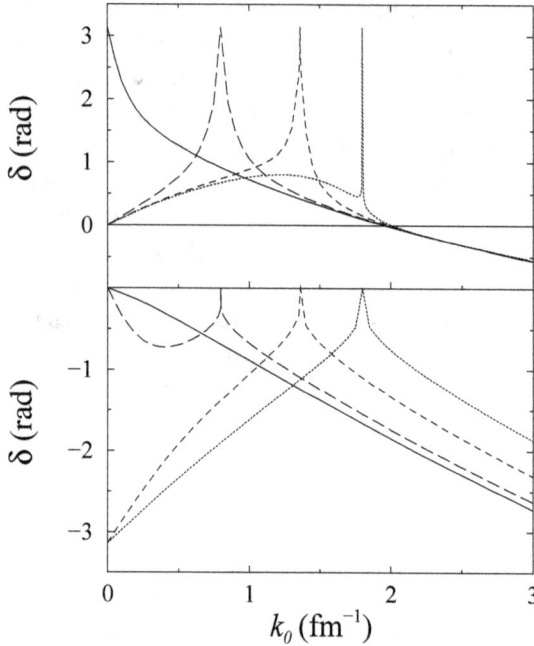

Fig. 15.15 Phase shifts obtained for modified versions of the Reid $^1S_0$ interaction. In the top panel the intermediate range attraction is increased by 10%. Phase shifts for the same densities as in Fig. 15.13 illustrate that the modified interaction yields a bound state for free particles, which is demonstrated by the corresponding phase shift going to π at zero momentum (solid line). For the highest density ($k_F = 1.8$ fm^{-1}), the phase shift at k_F tends to π, in contrast to the result shown in Fig. 15.13. In the bottom panel, the shortest-range Yukawa of the Reid $^1S_0$ interaction is multiplied by a factor of 10 to simulate an atom–atom like interaction. Results are shown for the same set of densities as in the top panel.

panel of Fig. 15.15. The phase shift for free particles now always indicates a repulsive effective interaction (negative phase shift) and is quite similar to Fig. 15.9. For $k_F = 0.8$ fm^{-1}, the phase space of the hh continuum is not yet large enough to yield a bound state, whereas for $k_F = 1.36$ and 1.8 fm^{-1} it is, yielding a phase shift of $-\pi$ at the corresponding energy threshold at zero momentum. The latter is associated with the highest two-hole excitation energy that the mean-field picture allows.

While it is numerically straightforward to generate the results of this section, the interpretation requires a better insight into the appearance of bound-pair states given in the next section.

bound state

continuum

0

Energy (arbitrary units)

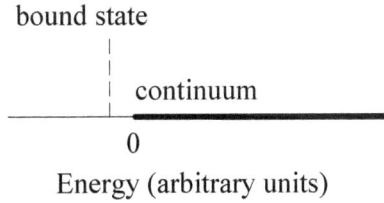

Fig. 15.16   Illustration of the spectrum of the unperturbed problem for free particles, indicated by the thick line starting at zero energy, and the possible location of a bound state at negative energy.

15.3   Cooper problem and pairing instability

Further discussion requires a reminder of the appearance of bound states for the case of two free particles. To this end, we return to Eq. (15.31), which can also be written as

$$\psi_n(\boldsymbol{k}; m_\alpha m_{\alpha'}) \tag{15.38}$$
$$= \frac{1}{E_n - \hbar^2 \boldsymbol{k}^2/m} \frac{1}{2} \sum_{m_\gamma m_{\gamma'}} \int \frac{d^3 q}{(2\pi)^3} \langle \boldsymbol{k} m_\alpha m_{\alpha'} | V | \boldsymbol{q} m_\gamma m_{\gamma'} \rangle \psi_n(\boldsymbol{q}; m_\gamma m_{\gamma'}).$$

For two electrons or two ^{3}He atoms with spin $\frac{1}{2}$, the requirement of anti-symmetry, as in Sec. 5.1, incorporates antisymmetry for spins coupled to total spin S, when the orbital angular momentum ℓ, associated with the relative motion, is such that $\ell + S$ is even. If the interaction is sufficiently attractive for $\ell = 0$, this implies that the spins must point in opposite directions, so that the total spin $S = 0$. For $\ell = 1$ the total spin must be 1 to obey the Pauli principle. Writing the eigenvalue equation in the basis with good total spin and orbital angular momentum, we find

$$\psi_n(k; \ell S) = \frac{1}{E_n - \hbar^2 \boldsymbol{k}^2/m} \frac{1}{2} \int \frac{dq\, q^2}{(2\pi)^3} \langle k | V^{\ell S} | q \rangle \psi_n(q; \ell S). \tag{15.39}$$

It is helpful to visualize the appearance of a bound state by contemplating Fig. 15.16. The continuum spectrum in the figure corresponds to the energies, associated with the propagation of noninteracting particles, and this kinetic energy spectrum therefore starts at $E = 0$. When a sufficiently attractive interaction is involved, a bound state emerges below the continuum at negative energy.

For two particles, interacting in the medium, we first treat the case when the two particles are added on top of the Fermi sea. If the total wave vector of these particles is zero, the most favorable situation for bound states as the phase space argument from the previous section shows, the noninteracting propagator reads

$$G_{pp}^{(0)}(\mathbf{K} = 0, q; E) = \frac{\theta(q - k_F)}{E - 2\varepsilon(q) + i\eta}, \tag{15.40}$$

which excludes the propagation of two holes [compare Eq. (15.36)]. The equation for propagating two particles on top of the Fermi sea is given by Eq. (15.11) in the appropriate basis, with $G_{pphh}^{(0)}$ replaced by Eq. (15.40). By considering the spectrum of $G_{pp}^{(0)}$, one realizes that it is similar in nature to the free case, illustrated in Fig. 15.16, with the threshold of the continuum now at $2\varepsilon_F$. The situation is illustrated in Fig. 15.17. The corresponding eigenvalue problem is obtained in the usual way and yields

$$\psi_C(k; \ell S) = \frac{\theta(k - k_F)}{E_C - 2\varepsilon(k)} \frac{1}{2} \int \frac{dq\ q^2}{(2\pi)^3} \langle k| V^{\ell S} |q\rangle \psi_C(q; \ell S), \tag{15.41}$$

where the subscript C has been used to identify that this problem was first considered by [Cooper (1956)] in the context of understanding superconductivity. We use the notation ψ_C for the pair addition amplitude, since it has obvious similarity to the wave function of a bound state of two particles [see Eq. (15.32)]. For a separable interaction, it is easy to show that the eigenvalue problem always has a solution for a so-called bound-pair state, when this interaction is attractive. We select a separable interaction, similar to the schematic interaction used in Ch. 13, but now of the form

$$\langle k| V^{\ell S} |q\rangle = \lambda_\ell w_\ell(k) w_\ell^*(q), \tag{15.42}$$

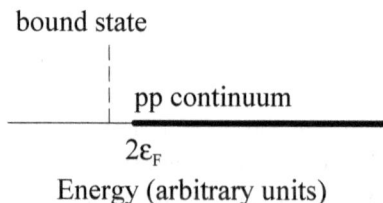

bound state

| pp continuum

$2\varepsilon_F$

Energy (arbitrary units)

Fig. 15.17 Illustration of the spectrum of the unperturbed problem for two particles on top of the Fermi sea, indicated by the thick line starting at $2\varepsilon_F$, and the possible location of a bound state below the continuum.

where S is implied by ℓ, as discussed above.   Upon substitution in Eq. (15.41) we obtain

$$\psi_C(k;\ell S) = \mathcal{N}\frac{\theta(k-k_F)w_\ell(k)}{E_C - 2\varepsilon(k)}, \tag{15.43}$$

where

$$\mathcal{N} = \frac{1}{2}\lambda_\ell \int\frac{dq\,q^2}{(2\pi)^3}w_\ell^*(q)\psi_C(q;\ell S). \tag{15.44}$$

Substituting Eq. (15.43) back in Eq. (15.41), then yields

$$\frac{1}{\lambda_\ell} = \frac{1}{2}\int\frac{dq\,q^2}{(2\pi)^3}\frac{\theta(q-k_F)|w_\ell(q)|^2}{E_C - 2\varepsilon(q)}. \tag{15.45}$$

The right-hand side of the eigenvalue equation is a negative definite quantity for $E_C < 2\varepsilon_F$. As a function of E_C, it will approach zero for $E_C \Rightarrow -\infty$. Conversely, when E_C approaches the boundary of the pp continuum the right-hand side diverges to $-\infty$. It means that for this type of separable interaction, there is always a solution to Eq (15.45) for any $\lambda_\ell < 0$. When the interaction is repulsive ($\lambda_\ell > 0$), there is no solution to Eq. (15.45) for $E_C < 2\varepsilon_F$. These arguments were also presented for excited states in finite systems in Ch 13. While in a finite system the location of the lowest excited state can still be above the ground state, in the corresponding Tamm–Dancoff approximation, in an infinite system there is an immediate peculiarity associated with the location of the energy E_C of the bound-pair state. Indeed, it should be realized that it resides in the continuum of hh states that is reached by removing two particles from the Fermi sea!

This observation immediately points out a difficulty with the inclusion of hh propagation. We first note that its inclusion in the eigenvalue equation is, strictly speaking, not allowed without first checking that the propagator equation actually leads to such a result. Postponing this examination, one may be tempted to replace Eq. (15.41) by

$$\psi_C(k;\ell S) = \frac{\theta(k-k_F)}{E_C - 2\varepsilon(k)}\frac{1}{2}\int\frac{dq\,q^2}{(2\pi)^3}\langle k|V^{\ell S}|q\rangle\,\psi_C(q;\ell S)$$

$$-\frac{\theta(k_F-k)}{E_C - 2\varepsilon(k)}\frac{1}{2}\int\frac{dq\,q^2}{(2\pi)^3}\langle k|V^{\ell S}|q\rangle\,\psi_C(q;\ell S), \tag{15.46}$$

when hh propagation is included in the unperturbed propagator $G^{(0)}_{pphh}$. Considering the spectrum of the unperturbed propagator to identify the location of possible bound-pair states, one obtains the result illustrated in

$$2\varepsilon_F$$

hh continuum ┊ pp continuum
▬▬▬▬▬▬▬┊▬▬▬▬▬▬▬▬

Energy (arbitrary units)

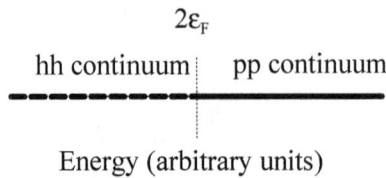

Fig. 15.18 Illustration of the spectrum of the unperturbed propagator, involving both pp and hh propagation. The pp continuum is indicated by the thick line starting at $2\varepsilon_F$ and the hh continuum is identified by the thick dashed line, ending there. The conclusion is therefore inevitable that there is no room for bound-pair states, associated with an attractive interaction. The boundary is indicated by the dotted line.

Fig. 15.18. Clearly, a bound-pair state, originating from the pp continuum, will not find a location below $2\varepsilon_F$, because the hh continuum occupies it, and vice versa. It is thus not possible to have discrete (real) eigenvalues, when the interaction is attractive. In the case of a repulsive interaction, the hh continuum is bounded on the left by the value $2\varepsilon(0)$, and for a large enough hh phase space, it is possible to find discrete hh bound states below the continuum. This was mentioned in the previous section, where the behavior of hh scattering phase shifts was discussed for an interaction with very strong short-range repulsion (see Fig. 15.15). We will see below that for an attractive interaction the eigenvalue problem of Eq. (15.46), yields a pair of conjugate complex eigenvalues, unless special precautions are taken. It signals a similar instability as was discussed in the context of excited states in the RPA approach and it is common to refer to the present one as a "pairing instability". Clearly, pairs of particles (and holes) are trying to take advantage of their mutually attractive interaction, but run into an instability, when the problem is posed in the form of Eq. (15.46). The instability is quite serious, since any time dependence involving a pair of complex conjugated eigenvalues, inevitably leads to one exploding solution, as a function of time.

Bound-pair states

Returning to the original propagator equation, we note that it is actually not possible to generate Eq. (15.46) in the usual way. The obstacle has its origin in the spectrum of the unperturbed propagator, which represents a continuum from $2\varepsilon(0)$ to $2\varepsilon_F$ for hh states, and from $2\varepsilon_F$ to ∞ for pp states. This implies that there is no room to place a bound state below the pp continuum, or above the hh continuum, since these energies already correspond to unperturbed states of the other kind. The situation is different if a gap in the sp spectrum occurs at $k = k_F$. This gap will emerge from a more complete description of pairing developed in Ch. 22. We will now employ the freedom of choosing an auxiliary potential $\hat{U}$, which includes such a gap, to illustrate some of its consequences. A simple constant energy shift Δ for all k below k_F, suffices to introduce a gap in the sp spectrum at k_F. In principle, this auxiliary potential can then be subtracted when the Dyson equation is solved, which takes the effect of bound-pair states into account. Here, we will employ the gap only to illustrate the appearance of bound-pair states, when the ladder equation is studied. The more in-depth study of pairing, leading to the famous gap also generates an important deviation from the mean-field propagator employed in the present discussion. With a gap present in the sp spectrum, a corresponding gap of 2Δ occurs between the hh and the pp continuum. It allows a legitimate derivation of the eigenvalue equation from the ladder equation for eigenvalues inside this interval. With this observation, Eq. (15.46) now becomes valid. Repeating the analysis for a separable interaction, given by Eq. (15.42), we find for the pp and hh transition amplitudes

$$\psi_{BP}(k; \ell S) = \mathcal{N}\frac{\theta(k - k_F)w_\ell(k)}{E_{BP} - 2\varepsilon(k)}, \qquad (15.47)$$

and

$$\psi_{BP}(k; \ell S) = -\mathcal{N}\frac{\theta(k_F - k)w_\ell(k)}{E_{BP} - 2\varepsilon(k)}, \qquad (15.48)$$

respectively. The label BP has been introduced to represent "bound pair." Upon substituting these results in Eq. (15.46), one obtains after a slight rearrangement

$$\frac{1}{\lambda_\ell} = \frac{1}{2}\int\frac{dq\ q^2}{(2\pi)^3}\frac{\theta(q - k_F)|w_\ell(q)|^2}{E_{BP} - 2\varepsilon(q)} - \frac{1}{2}\int\frac{dq\ q^2}{(2\pi)^3}\frac{\theta(k_F - q)|w_\ell(q)|^2}{E_{BP} - 2\varepsilon(q)}. \qquad (15.49)$$

The right side of this equation is plotted as a function of the parameter E_{BP} as the solid line in Fig. 15.19. In the figure, the dotted lines identify the boundaries of the hh and pp continuum, given by $2\varepsilon_F^-$ and $2\varepsilon_F^+$. The location of the bound-pair states is therefore confined to the domain between these two regions. Both terms on the right side of Eq. (15.49) yield negative contributions for energies E_{BP} between the hh and pp continuum. This implies that actual solutions will only be possible when the left side, *i.e.*

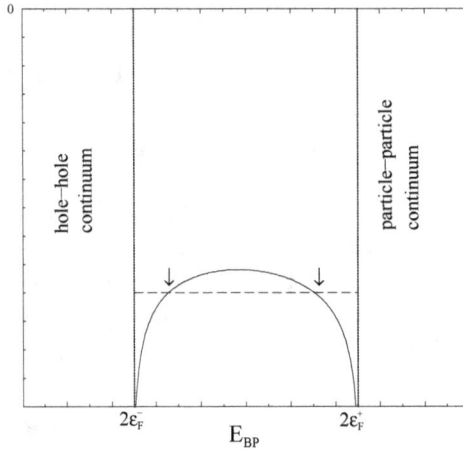

Fig. 15.19   Illustration of Eq. (15.49). The vertical dotted lines illustrate the boundaries of the hh and pp continuum. The solid line shows the dependence of the right side of Eq. (15.49) on E_{BP} between these boundaries. Finally, the dashed line represents $1/\lambda_\ell$ for a properly chosen negative value of this quantity. Bound-pair solutions are obtained where the dashed line intersects the solid line, as indicated by the arrows.

$1/\lambda_\ell$, is also negative. For a particular choice of $1/\lambda_\ell$, identified by the horizontal dashed line, these solutions correspond to the intersection points with the solid line indicated by the arrows. From the figure it is also clear that such bound states will occur even for very small absolute values of λ_ℓ, yielding solutions very close the hh and pp continuum. Increasing the magnitude of λ_ℓ from the value in the figure will ultimately generate equal eigenvalues for the hh and pp bound-pair states. Any further increase, leads again to the previously identified pairing instability, which generates two complex conjugate eigenvalues. An additional increase of the gap in the spectrum, can be used to avoid the instability and so on. This produces the important insight that obtaining real binding energies for the bound-pair states, is strongly correlated with the size of gap in the sp spectrum at the Fermi surface. It should be clear that one can always make the gap large enough, so that these bound states have real energies.

Bound-pair states in nuclear matter

An illustration is provided by the attraction in the nucleon–nucleon interaction in nuclear matter. As discussed in Sec. 15.1, the interaction in free space is sufficiently attractive in the coupled $^3S_1$-$^3D_1$ channel to sus-

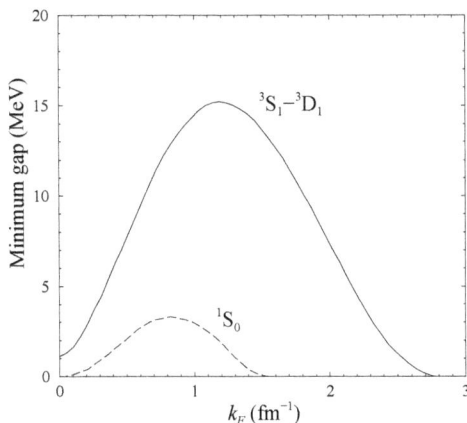

Fig. 15.20   Minimum gap between sp states at k_F necessary to avoid a pairing instability in the ladder equation as a function of k_F. For the $^3S_1$-$^3D_1$ case the limit of zero density yields half the binding energy of the deuteron.

tain one bound state. Solutions to the corresponding scattering problem in the medium, described in Sec. 15.2, indicate that the attraction is also substantial in the medium. The relevant eigenvalue problem is given by

$$\psi_{BP}(k; (\ell S)JT) \qquad (15.50)$$

$$= \frac{\theta(k - k_F)}{E_{BS} - 2\varepsilon(k)} \frac{1}{2} \sum_{\ell'} \int \frac{dq\, q^2}{(2\pi)^3} \langle k\ell | V^{JST} | q\ell' \rangle\, \psi_{BP}(q; (\ell'S)JT)$$

$$- \frac{\theta(k_F - k)}{E_{BS} - 2\varepsilon(k)} \frac{1}{2} \sum_{\ell'} \int \frac{dq\, q^2}{(2\pi)^3} \langle k\ell | V^{JST} | q\ell' \rangle\, \psi_C(q; (\ell'S)JT),$$

where a gap in the sp spectrum must be assumed, in order to derive it from the propagator equation. Equation (15.50) contains the necessary coupling of orbital angular momentum and spin, to total angular momentum J and the total isospin T. When the eigenvalue problem is studied as a function of density [Vonderfecht *et al.* (1991b)], one realizes that the gap in the sp spectrum required to avoid the pairing instability, is a sensitive function of the density. The minimum gap for the $^3S_1$-$^3D_1$ channel, but also for the $^1S_0$ channel, is plotted as a function of k_F in Fig. 15.20. The limit of zero density in the deuteron channel yields half the binding energy of the deuteron. This can be anticipated, since at very small, but nonzero, density one expects to recover the binding energy of the deuteron for the pp bound-pair state. A gap in the sp spectrum, corresponding to one half

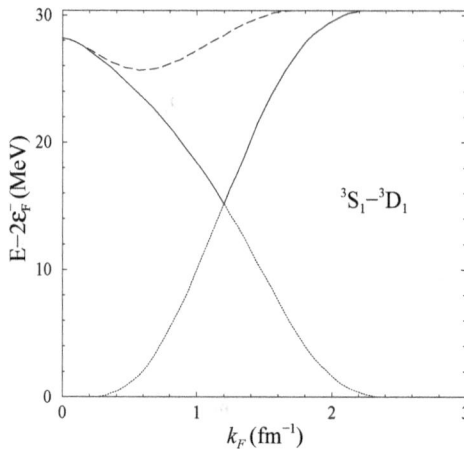

Fig. 15.21   Assuming the maximum gap between sp states found in Fig. 15.20 for $^3S_1$-$^3D_1$ coupled channel for all densities, the eigenvalues of Eq. (15.50) are plotted as a function of k_F using solid (pp) and dotted (hh) lines. Solving the eigenvalue equation without hh propagation (Cooper problem) yields the dashed curve. The limit of zero density yields the binding energy of the deuteron in both cases.

this binding energy, is then necessary to make room for the bound-pair state. Fig. 15.20 also shows that a sp gap is mandatory all the way to about $k_F = 2.8$ fm^{-1} in this channel. For the $^1S_0$ channel, which does not support a bound state in free space, it appears that the interaction in the medium, is sufficiently attractive to generate a pairing instability from $k_F = 0.1$ to about 1.5 fm^{-1}. The instability is clearly overshadowed by the much stronger one in the deuteron channel. In the latter, the largest gap required to avoid the pairing instability, occurs around $k_F = 1.2$ fm^{-1}. The required gap in the sp spectrum at this density is a little larger than 15 MeV. The presence of bound-pair states clearly signals an attractive interaction near the Fermi surface. It is therefore no coincidence that the density range, where these bound states occur for the $^1S_0$ channel, coincides with estimates of the density regime where pairing is expected in neutron matter [Chen *et al.* (1986); Baldo *et al.* (1990a)].

Keeping the gap fixed for all densities at the largest value required for the $^3S_1$-$^3D_1$ channel (see Fig. 15.20), it is possible to illustrate the behavior of the eigenvalues obtained by Eq. (15.50) as a function of k_F. The result is displayed in Fig. 15.21. The eigenvalue associated with the pp spectrum (solid line) starts at the binding of the deuteron in the limit of zero density.

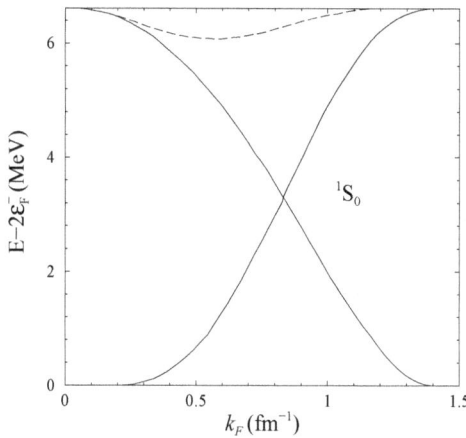

Fig. 15.22   Similar results as in Fig. 15.21 are shown for the $^1S_0$ case. Solving the eigenvalue equation without hh propagation (Cooper problem) yields the dashed curve.

The hh eigenvalue (dotted line) does not appear until a certain minimum density has been reached. As observed earlier, the pairing instability is just avoided for the density corresponding to $k_F = 1.2$ fm^{-1} where the largest binding occurs for both pp and hh bound states. For comparison, the solution to the eigenvalue problem when hh propagation is excluded, is also contained in the figure as the dashed line. The latter has the same limit at zero density, where hh propagation disappears. At higher density the difference between the pphh and pp solution, corresponding to the Cooper eigenvalue problem, becomes very large. This feature signals the important coherence, or collectivity, that is associated with an RPA-like procedure, as discussed in Chs. 13 and 14.

Similar, but less extreme, results are obtained for the $^1S_0$ channel shown in Fig. 15.22. In both cases, the amount of binding found in the simple bound-pair calculations is similar in magnitude to gaps in the sp spectrum that are found in a more advanced treatment of pairing correlations [Chen *et al.* (1986); Baldo *et al.* (1990a); Baldo *et al.* (1992)] following the Bardeen–Cooper–Schrieffer insights discussed in Ch. 22. To understand qualitatively why these interactions appear more attractive in the medium than in free space, it should be pointed out that the lowest unperturbed pp and hh states in the medium correspond to having sp states with $k = k_F$, whereas $k = 0$ for free particles. It leads to an alternate sampling of the interaction in the medium, accounting for the observed differences. This feature is particularly strong for the deuteron channel, which receives important contributions from nondiagonal terms, related to the tensor force, which couples $\ell = 0$ and 2. The coupling is increased in the medium, since finite values of k are now emphasized at low energy, whereas the $k = 0$ state in free space, has vanishing matrix elements to other states through the tensor force.

The present discussion shows that the propagation of pp and hh states with mean-field properties, leads to the possibility of bound-pair states for an attractive interaction, which can be accomodated by opening a gap in the sp spectrum. The properties of the phase shifts near $2\varepsilon_F$ for an attractive interaction, also reveal the existence of a pairing instability or bound-pair states by generating a limiting value of π. For a repulsive interaction, there is the possibility to generate a bound-pair state below the lowest energy of the hh spectrum, which corresponds to $2\varepsilon(0)$. As discussed in Sec. 15.2, the presence of such a bound state leads to the behavior of the phase shift going to $-\pi$, when the continuum boundary is reached.

15.4   Exercises

(1) Check the appearance of the factors of $\frac{1}{2}$ in Eqs. (15.11) and (15.13), by analyzing Wick's theorem in second and third order.
(2) Calculate the phase shifts of the $^1S_0$ NN interaction, given in Eq. (4.53), by applying the Numerov method in coordinate space for the range of energies, illustrated in Fig. 15.6.
(3) Obtain the same phase shifts, as in the previous problem, by employing a matrix inversion technique [Haftel and Tabakin (1970)] in wave-vector space for the $\mathcal{R}$- or $\mathcal{T}$-matrix (Γ_{pp}).
(4) Adjust the intermediate range attraction of the interaction in Eq. (4.53), to obtain a bound state at -2 MeV.
(5) Study the center-of-mass wave vector dependence of the phase shifts of the interaction in Eq. (4.53) at $k_F = 1.36$ fm^{-1}, by employing the propagator in Eq. (15.34) with kinetic sp energies. Perform an angular average of the propagator to justify the use of a partial wave basis. Compare to the case with $\boldsymbol{K} = 0$.
(6) Calculate the phase shifts for the potential of [Aziz *et al.* (1979)] in the medium (and free space, illustrated in Fig. 15.9).
(7) Analyze the eigenvalue problem of Eq. (15.45) numerically, by making a suitable choice of the function w_ℓ.

Chapter 16

Dynamical treatment of the self-energy in infinite systems

The second-order self-energy contribution was studied in Ch. 11. The consequences of incorporating this dynamic term in the solution of the Dyson equation were profound, leading to a fragmented strength distribution in the sp propagator. This is true both in finite and infinite systems. The self-consistent treatment of the second-order term already yields quite a good description of the properties of atoms. For other systems higher-order terms must be added and a second-order approach can only serve to illustrate the possibilities of a dynamical (energy-dependent) self-energy. In this chapter we will review the consequences of higher-order effects on the sp properties. Systems with prominent short-range correlations, like nuclei, neutron and nuclear matter, and quantum liquids, require the inclusion of all ladder diagrams in the vertex function. The resulting effective interaction is well-behaved, allowing a meaningful calculation of the self-energy. When special coherence is associated with certain excited states of the systems, it is necessary to treat their influence on the properties the self-energy. The simplest case involves the RPA of excited states.

In this chapter, we emphasize the sp propagator in an infinite system. We begin in Sec. 16.1 with a brief preparation concerning the use of diagrams. In Sec. 16.2.1 we study the spectral functions of the electron gas with RPA correlations. We have seen in Ch. 14 that they lead to a prominent collective state, the plasmon, at small values of q.[1] The plasmon contribution to the self-energy is indeed dominant and largely determines the properties of the particles in the electron gas, as discussed in Sec. 16.2.1. A fully self-consistent treatment of these ring (or RPA) diagrams in the self-energy has recently become available. Some of the consequences of a self-consistent approach will be presented in Sec. 16.2.2. We will also pay attention to

---

[1] We will use wave vectors in this chapter, so that $Q = \hbar q$, *etc.*

the energy per particle of the system, comparing different approximations in Sec. 16.2.3. Since a self-consistent treatment of ring diagrams has both beneficial as well as detrimental consequences when compared with data, some issues that remain unresolved will be pointed out.

An example of the inclusion of ladder diagrams in the self-energy is presented in Sec. 16.3. Nuclear matter will be studied with the apparatus, developed in Ch. 15, for dealing with short-range correlations. We will start with the calculation of the self-energy when ladder diagrams are summed in Sec. 16.3.1. Spectral functions based on the use of a pphh propagator with mean-field particles and holes, are presented in Sec. 16.3.2. The self-energy and corresponding spectral functions exhibit many illustrative examples of the effects of short-range correlations on the properties of nucleons in the medium. The pairing instabilities that are encountered, when mean-field nucleons are employed, make a self-consistent treatment all the more relevant, as discussed in Sec. 16.3.3. For a study of pairing correlations employing fully dressed nucleons we refer to Ch. 24. The related topic of scattering involving dressed particles in the medium is postponed to Ch. 20. We close the chapter with a survey of the saturation problem of nuclear matter in Sec. 16.3.4.

16.1 Diagram rules in uniform systems

The diagram rules, introduced in Chs. 8 and 9, can be used to write down expressions for the propagator and self-energy in the case of a homogeneous infinite system. Examples were worked out in Chs. 10 and 11. Simplifications arise in uniform systems on account of momentum conservation, which translates into a sp propagator that is diagonal in this variable. The propagator lines can then be labeled by the momentum (or wave vector), making sure that momentum flowing in and out of an interaction is equal. A similar feature emerged for the energy variables in the diagrams. This allows the removal of a factor, which contains the corresponding Kronecker δ divided by the volume V, contained in each two-body matrix element. In nth order in the interaction $\hat{V}$, we find n independent integrations over wave vectors. The associated n volume factors exactly cancel those from the interaction terms. In addition to being diagonal in the spin (isospin) projection, the sp propagator will not depend on this quantum number, when the ground state is spin (isospin) saturated. We can therefore label the sp propagators with wave vectors, while the beginning and end point should be marked by

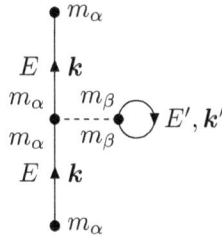

Fig. 16.1   Labels for the first-order Hartree–Fock diagram in an infinite homogeneous system in the energy formulation.

the same spin (isospin) quantum number. In lowest order, it leads to the labeling in Fig. 16.1. The resulting expression for this diagram reads

$$G_{\hat{V}}^{(1)}(k; E) = G^{(0)}(k; E)$$

$$\times \left\{ -i\frac{1}{\nu} \sum_{m_\alpha m_\beta} \int \frac{d^3 k'}{(2\pi)^3} \langle km_\alpha k' m_\beta | V | km_\alpha k' m_\beta \rangle \right.$$

$$\left. \times \int_{C\uparrow} \frac{dE'}{2\pi} G^{(0)}(k'; E') \right\} G^{(0)}(k; E). \tag{16.1}$$

An additional summation over the external spin (isospin) quantum numbers has been included, which explicitly uses the fact that the propagator does not depend on this quantity. To compensate for the summation over equal terms, the expression is divided by the degeneracy factor ν. The wave vectors in the bra and ket of the two-body interaction refer to the two in or outgoing lines, while ensuring that the total wave vector is conserved.

Higher-order self-energy diagrams in infinite systems can also be transformed, accordingly. In second order, the labeling in Fig. 16.2 applies. The corresponding expression is given by

$$\Sigma^{(2)}(k; E) = (-1)i^2 \frac{1}{2} \int \frac{dE_1}{2\pi} \int \frac{dE_2}{2\pi} \int \frac{d^3 q}{(2\pi)^3} \int \frac{d^3 k'}{(2\pi)^3} \frac{1}{\nu} \sum_{m_\alpha m_\beta m_\gamma m_\delta}$$

$$\times \langle km_\alpha k' - q/2m_\beta | V | k - qm_\gamma k' + q/2m_\delta \rangle$$
$$\times G^{(0)}(k' + q/2; E_1 + E_2) G^{(0)}(k' - q/2; E_2) G^{(0)}(k - q; E - E_1)$$
$$\times \langle k - qm_\gamma k' + q/2m_\delta | V | km_\alpha k' - q/2m_\beta \rangle. \tag{16.2}$$

We emphasize that the labeling of the internal wave vectors can be chosen to simplify the calculation. The complete self-energy, including the exact

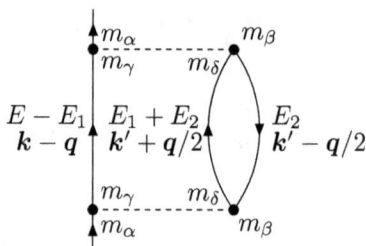

Fig. 16.2   Diagram SE2i for the self-energy in second order for an infinite homogeneous system.

vertex function of Eq. (9.34), follows a similar pattern, yielding

$$\Sigma(\boldsymbol{k}; E) = -U(\boldsymbol{k}) \tag{16.3}$$

$$-i\frac{1}{\nu}\sum_{m_\alpha m_\beta}\int\frac{d^3k'}{(2\pi)^3}\langle \boldsymbol{k}m_\alpha\boldsymbol{k}'m_\beta|\,V\,|\boldsymbol{k}m_\alpha\boldsymbol{k}'m_\beta\rangle\int_{C\uparrow}\frac{dE'}{2\pi}G(\boldsymbol{k}';E')$$

$$-i^2\frac{1}{2}\int\frac{dE_1}{2\pi}\int\frac{dE_2}{2\pi}\int\frac{d^3q}{(2\pi)^3}\int\frac{d^3k'}{(2\pi)^3}\frac{1}{\nu}\sum_{m_\alpha m_\beta m_\gamma m_\delta}$$

$$\times\,\langle \boldsymbol{k}m_\alpha\boldsymbol{k}'-\boldsymbol{q}/2m_\beta|\,V\,|\boldsymbol{k}-\boldsymbol{q}m_\gamma\boldsymbol{k}'+\boldsymbol{q}/2m_\delta\rangle$$

$$\times\,G(\boldsymbol{k}-\boldsymbol{q};E-E_1)G(\boldsymbol{k}'+\boldsymbol{q}/2;E_1+E_2)G(\boldsymbol{k}'-\boldsymbol{q}/2;E_2)$$

$$\times\,\langle \boldsymbol{k}-\boldsymbol{q}m_\gamma\boldsymbol{k}'+\boldsymbol{q}/2m_\delta|\,\Gamma(E,E_1,E_2)\,|\boldsymbol{k}m_\alpha\boldsymbol{k}'-\boldsymbol{q}/2m_\beta\rangle.$$

Several approximations to the self-energy will be discussed in the following sections. For each of these, the Dyson equation can be solved. For a uniform system, it has the simple form

$$G(k;E) = G^{(0)}(k;E) + G^{(0)}(k;E)\Sigma(k;E)G(k;E)$$

$$= \frac{E-\varepsilon(k)-\operatorname{Re}\Sigma(k;E)+i\operatorname{Im}\Sigma(k;E)}{(E-\varepsilon(k)-\operatorname{Re}\Sigma(k;E))^2+(\operatorname{Im}\Sigma(k;E))^2}, \tag{16.4}$$

where only the magnitude of k was kept, since the propagator does not depend on the direction of the vector. The explicit form of the noninteracting propagator,

$$G^{(0)}(k;E) = \frac{\theta(k-k_F)}{E-\varepsilon(k)+i\eta} + \frac{\theta(k_F-k)}{E-\varepsilon(k)-i\eta}, \tag{16.5}$$

is used to generate Eq. (16.4), together with

$$\varepsilon(k) = \frac{\hbar^2 k^2}{2m} + U(k). \tag{16.6}$$

The inclusion of U is of course optional and the results do not depend on the choice of U, when a fully self-consistent calculation is undertaken.

The spectral functions are proportional to the imaginary part of the propagator according to Eq. (11.89), [see also Eqs. (7.11) and (7.12)] and are given by

$$S_p(k; E) = \frac{-1}{\pi} \frac{\text{Im } \Sigma(k; E)}{(E - \varepsilon(k) - \text{Re } \Sigma(k; E))^2 + (\text{Im } \Sigma(k; E))^2} \tag{16.7}$$

for energies above ε_F, and by

$$S_h(k; E) = \frac{1}{\pi} \frac{\text{Im } \Sigma(k; E)}{(E - \varepsilon(k) - \text{Re } \Sigma(k; E))^2 + (\text{Im } \Sigma(k; E))^2} \tag{16.8}$$

for energies below ε_F. The connection between the sp propagator and the spectral functions can also be written in the form of a dispersion relation

$$G(k; E) = \int_{\varepsilon_F}^{\infty} dE' \frac{S_p(k; E')}{E - E' + i\eta} + \int_{-\infty}^{\varepsilon_F} dE' \frac{S_h(k; E')}{E - E' - i\eta}. \tag{16.9}$$

By taking the imaginary part, the orginal relation between the spectral functions and the imaginary part of G is recovered [see Eq. (11.89)]. The real part of G therefore has a similar connection to its imaginary part as the self-energy [see Eq. (11.96)].

The energy per particle can be calculated from the hole spectral function. The result was first obtained by [Galitskii and Migdal (1958)]. It can be generated from Eq. (7.26) by using the simplifications of the uniform system

$$\frac{E_0^N}{N} = \frac{\nu}{2\rho} \int \frac{d^3 k}{(2\pi)^3} \int_{-\infty}^{\varepsilon_F} dE \left(\frac{\hbar^2 k^2}{2m} + E \right) S_h(k; E), \tag{16.10}$$

where ρ is the density related to the Fermi wave vector by

$$\rho = \frac{\nu k_F^3}{6\pi^2}. \tag{16.11}$$

It is also possible to relate the density to the occupation number by

$$\rho = \frac{\nu}{2\pi^2} \int_0^{\infty} dk k^2 \, n(k), \tag{16.12}$$

where

$$n(k) = \int_{-\infty}^{\varepsilon_F} dE \ S_h(k; E), \tag{16.13}$$

is the occupation number. We will see in Ch. 21 that a self-consistent propagator, even when it corresponds to an approximation, guarantees particle number conservation so that Eqs. (16.11) and (16.12) should be identical. One may also separately study the kinetic energy per particle

$$\frac{T}{N} = \frac{\nu}{2\pi^2 \rho} \int_0^\infty dk k^2 \ \frac{\hbar^2 k^2}{2m} n(k). \tag{16.14}$$

The potential energy can then be derived from the difference between Eqs. (16.10) and (16.14). For normal Fermi systems at $T = 0$ the pressure P can be calculated directly from the density derivative of the energy per particle, or alternatively, on the basis of the Fermi energy $\varepsilon_F = \partial E / \partial N$:

$$P = \rho^2 \frac{\partial (E/N)}{\partial \rho} = \rho[\varepsilon_F - (E/N)]. \tag{16.15}$$

The equality follows directly from the density derivative of $E/N = E/(\rho V)$, and should hold in thermodynamically consistent many-body theories. It implies in particular that at equilibrium density (corresponding to a minimum in E/N, so at zero pressure), the Fermi energy should equal the binding energy per particle, a relation known as the Hugenholtz–van Hove theorem [Hugenholtz and van Hove (1958)].

16.2   Self-energy in the electron gas

16.2.1   *Electron self-energy in the $G^{(0)}W^{(0)}$ approximation*

The discussion of the second-order self-energy contribution in an infinite system in Sec. 11.3.1, does not apply directly to the case of the Coulomb interaction. This can be anticipated by considering the Coulomb interaction in wave-vector space, as in Eq. (14.40). In particular the direct contribution with the q^{-2} term, leads to a divergence, when included in the second-order self-energy. The direct matrix element of the Coulomb interaction in Eq. (16.2) is given by

$$\langle k m_\alpha k' - q m_\beta | V | k - q m_\gamma k' m_\delta \rangle_D \Rightarrow \delta_{m_\alpha m_\gamma} \delta_{m_\beta m_\delta} V(q), \tag{16.16}$$

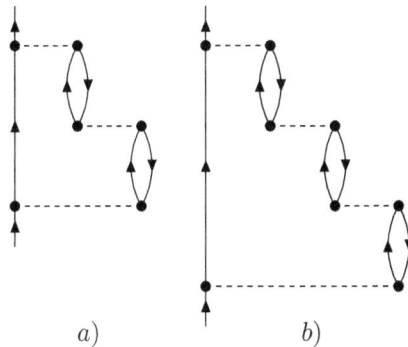

Fig. 16.3 Higher-order ring diagrams contributing to the self-energy.

where

$$V(q) = \frac{4\pi e^2}{q^2}, \tag{16.17}$$

using the result of Eq. (5.22). Performing the integral over E_2 as in Eq. (16.2), one can identify, after also integrating over $\boldsymbol{k}'$, the noninteracting polarization propagator $\Pi^{(0)}(q; E_1)$. Performing the spin summations, we find[2]

$$\Sigma_D^{(2)}(\boldsymbol{k}; E) = 2i \int \frac{dE_1}{2\pi} \int \frac{d^3q}{(2\pi)^3} V^2(q) G^{(0)}(\boldsymbol{k} - \boldsymbol{q}; E - E_1) \Pi^{(0)}(\boldsymbol{q}; E_1), \tag{16.18}$$

where $\Pi^{(0)}$ corresponds to Eq. (14.22). It is now clear that the integration over $\boldsymbol{q}$ leads to an infrared divergence, due to the presence of the q^{-4} term, while taking into account the behavior of $\Pi^{(0)}$, discussed in Sec. 14.2. The worrisome conclusion is that the second-order diagram with the direct Coulomb interaction is infinite! Such a conclusion was also reached in Sec. 14.4.2 for the correlation energy of the electron gas in second order. The divergent behavior continues in higher order. Indeed, one can replace $\Pi^{(0)}$ by $\Pi^{(0)} V \Pi^{(0)}$ to generate the third-order self-energy diagram shown in Fig. 16.3a). The interaction terms in this diagram contribute a factor q^{-6} making the divergence worse. One can repeat the procedure by making further replacements of $\Pi^{(0)}$ in the second-order self-energy, starting with $\Pi^{(0)} V \Pi^{(0)} V \Pi^{(0)}$, which leads to the fourth-order term shown in Fig. 16.3b) and so on. In each next order the divergence turns uglier (see *e.g.* [Mattuck (1992)]). This feature and its remedy have already been encountered in Sec. 14.4.2. The solution is to sum all the divergent terms, yielding the

---

[2]It is simpler to employ the unsymmetrized version of $\hat{V}$ here.

Fig. 16.4 Diagrammatic representation of the self-energy in the ring approximation. This is often referred to as the $G^{(0)}W^{(0)}$ approximation where $W^{(0)}$ corresponds to the screened Coulomb interaction of Eq. (14.58), indicated by the wiggly line.

RPA result for the polarization propagator. The procedure completely removes the unwarranted divergencies and yields the plasmon collective state at small values of q, in agreement with experimental data. It is therefore necessary to perform the same summation for the self-energy. This is instantly accomplished upon replacing $\Pi^{(0)}$ in Eq. (16.18) by the corresponding Π^{RPA} from Eq. (14.41). The sum of the divergent terms turns into the well-behaved self-energy contribution

$$\Sigma^{RPA}(\boldsymbol{k}; E) = 2i \int \frac{dE_1}{2\pi} \int \frac{d^3q}{(2\pi)^3} V^2(q) G^{(0)}(\boldsymbol{k} - \boldsymbol{q}; E - E_1) \Pi^{RPA}(\boldsymbol{q}; E_1).$$

(16.19)

The so-called ring approximation to the self-energy is thus represented by the sum of all bubble diagrams combined with the Fock term, already discussed in Sec. 10.5 (see Eq. (10.174)). This summation of self-energy diagrams is frequently represented by the diagram in Fig. 16.4 and includes the Fock contribution. The wiggly line identifies the screened Coulomb interaction $W^{(0)}$ of Eq. (14.58) that sums both the Fock contribution [Eq. (10.174)] and the terms in Eq. (16.19). The corresponding self-energy is known as the $G^{(0)}W^{(0)}$ approximation. The latter refers to a lowest-order sp propagator and the ring approximation to the screened interaction, evaluated with noninteracting sp propagators. The formulation of the Green's function method presented in Sec. 9.4, suggests a self-consistent formulation of this approach, referred to as the GW approximation. The physical argument notes that medium modifications of the particles should be included if they are important, in generating the ring approximation to the polarization propagator. This is counterbalanced by the observation that such a treatment will not conserve the f-sum rule, and thus no longer yields a good description of the plasmon. A conserving treatment, based on the ring approximation, will be discussed in Ch. 21. These issues are revisited after presenting the *GW* approximation to the self-energy in Sec. 16.2.2.

The evaluation of Eq. (16.19) is facilitated by employing the Lehmann representation of the polarization propagator, which can be written as

$$\Pi(\boldsymbol{q}, E) = -\frac{1}{\pi} \int_0^\infty dE' \frac{\text{Im } \Pi(\boldsymbol{q}, E')}{E - E' + i\eta} + \frac{1}{\pi} \int_{-\infty}^0 dE' \frac{\text{Im } \Pi(\boldsymbol{q}, E')}{E - E' - i\eta}. \quad (16.20)$$

As discussed in Ch. 14, the polarization propagator Π^{RPA} has a Lehmann representation with discrete poles, the plasmon contribution, and a continuum part, associated with the remaining ph strength. The former term is given by Eq. (14.43). It is therefore convenient to split the self-energy into a part arising from the plasmon poles and one from the remaining continuum ph pairs. Figure 14.4 illustrates the region in wave vector and energy where the two parts are nonvanishing.

If the plasmon contribution to Π^{RPA} along with $G^{(0)}$ are inserted in the expression of the self-energy [Eq. (16.19)] and the energy integration is carried out, we find the following imaginary part of the self-energy for energies above ε_F

$$\text{Im } \Sigma_p^{RPA}(\boldsymbol{k}; E) = \frac{\pi}{2} \int \frac{d^3q}{(2\pi)^3} \; \theta(|\boldsymbol{k} - \boldsymbol{q}| - k_F)\theta(q_c - q)$$

$$\times \delta\left(E - E_p(q) - \varepsilon(\boldsymbol{k} - \boldsymbol{q})\right) \left(\frac{\partial \Pi^{(0)}(\boldsymbol{q}, \tilde{E})}{\partial \tilde{E}}\Bigg|_{\tilde{E} = E_p(q)}\right)^{-1}. \quad (16.21)$$

We note that q_c is the cut-off wave vector of the plasmon branch, *i.e.* the wave vector at which it merges with the continuum. A similar result is obtained for energies below ε_F. The above expression can be further simplified by utilizing the explicit expression of the derivative of the polarization, using Eq. (14.37). The algebra is lengthy but straightforward.

Before attempting any numerical calculation of Eq. (16.21), one must consider in more detail the structure of its integrand. Whenever the limits of integration include the $q = 0$ point, it is necessary to expand the denominator of the integrand to facilitate an analytic treatment of its singular part. Consequently, whenever the limits of integration include this point, the imaginary part exhibits a logarithmic singularity. This property of the electron gas within the present approximation seems to have been first noticed in [Hedin *et al.* (1967)]. From the δ-function in the imaginary part of

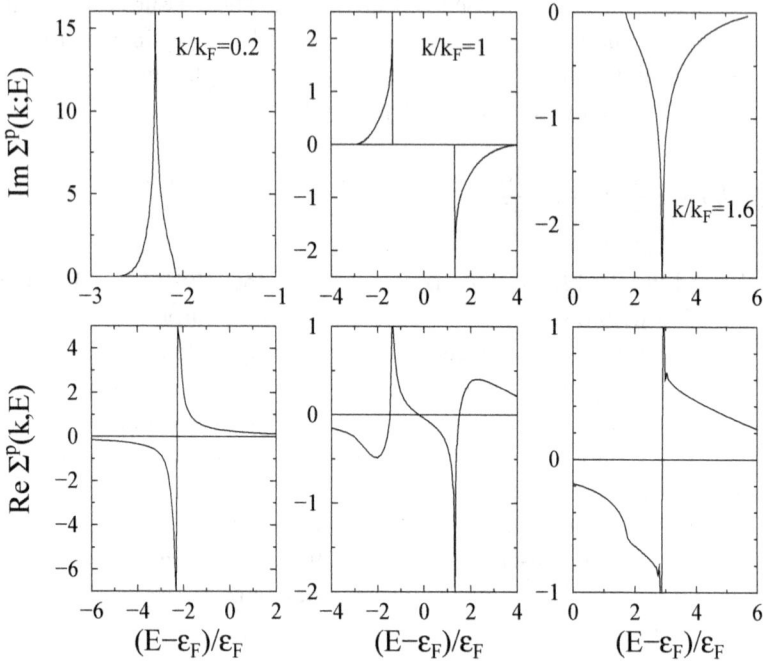

Fig. 16.5 Real and imaginary part of the self-energy for three different wave vectors, corresponding to $k/k_F = 0.2, 1$ and 1.6 at $r_s = 2$. Only the plasmon contribution to the self-energy is included in the results.

the self-energy, one can see that this occurs at energies

$$E = \pm E_p(0) + \varepsilon(\boldsymbol{k}), \qquad (16.22)$$

where the plus sign is for $k > k_F$. The behavior of the imaginary of the self-energy yields a singularity in the real part for $k = k_F$. For other values of k, the real part has a jump discontinuity at the location of the logarithmic singularity in the imaginary part.

The remaining contribution to the imaginary part of the self-energy is associated with the excitations in the ph continuum. One may employ the corresponding dispersion relation for these terms

$$\Pi_c^{RPA}(\boldsymbol{q}, E) = -\frac{1}{\pi} \int_{E_-}^{E_+} dE' \, \frac{\text{Im} \, \Pi_c^{RPA}(\boldsymbol{q}, E')}{E - E' + i\eta}$$
$$+ \frac{1}{\pi} \int_{-E_+}^{-E_-} dE' \, \frac{\text{Im} \, \Pi_c^{RPA}(\boldsymbol{q}, E')}{E - E' - i\eta}, \qquad (16.23)$$

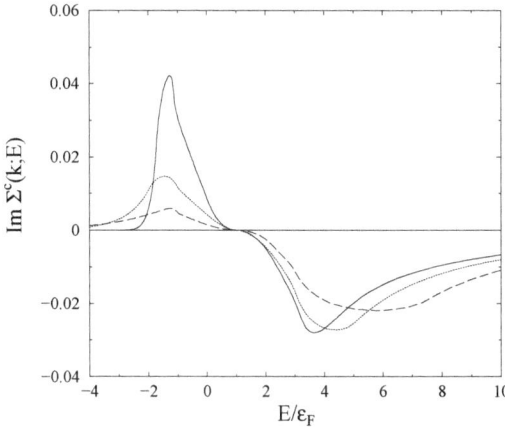

Fig. 16.6 Imaginary part of the self-energy for the same wave vectors as in Fig. 16.5. The three values $k/k_F = .2$, 1 and 1.6 are given by the solid, dotted and dashed line, respectively. Only the continuum contribution to the self-energy is included in these results.

as in Eq. (14.34). The limits $E_\pm$ of the ph continuum were discussed in Sec. 14.2. Using Eq. (16.23) in Eq. (16.19), performing the integration over E_1, and taking the imaginary part, one obtains for energies above ε_F

$$\text{Im } \Sigma_c^{RPA}(\boldsymbol{k}; E) = 2 \int \frac{d^3q}{(2\pi)^3} \, V^2(q) \tag{16.24}$$

$$\times \int_{E_-}^{E_+} dE' \, \text{Im } \Pi_c^{RPA}(q, E') \, \theta(|\boldsymbol{k} - \boldsymbol{q}| - k_F)\delta(E - E' - \varepsilon(\boldsymbol{k} - \boldsymbol{q})).$$

A similar result is generated for energies below ε_F. To complete the evaluation of the self-energy, we may employ the dispersion relation that relates the real and imaginary part of the self-energy, given in Eq. (11.79). Including the HF contribution, we find

$$\text{Re } \Sigma^{RPA}(\boldsymbol{k}; E) = \Sigma^{HF}(k) + \frac{\mathcal{P}}{\pi} \int_{-\infty}^{+\infty} dE' \, \frac{|\text{Im } \Sigma^{RPA}(\boldsymbol{k}; E')|}{E - E'}. \tag{16.25}$$

We display some results for the electron self-energy at a density corresponding to $r_s = 2$ in Fig. 16.5. The real and imaginary part of the self-energy for the three values $k/k_F = 0.2$, 1 and 1.6 are shown [Amari (1994)] (see also [Bose *et al.* (1967)]). As mentioned above, the imaginary part has a logarithmic singularity at energies $E = -E_p(0) + \varepsilon(\boldsymbol{k})$ for $k/k_F < 1$

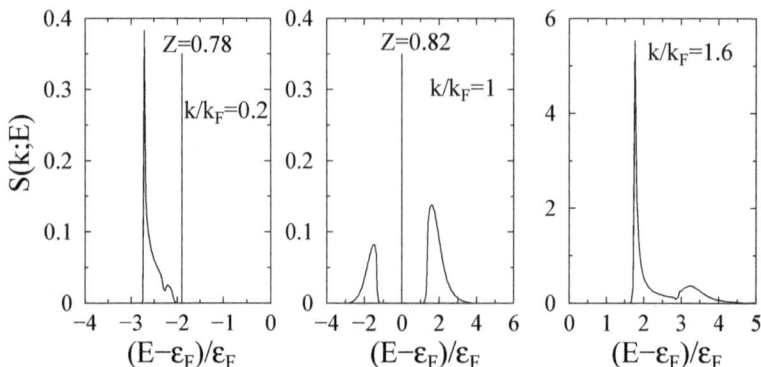

Fig. 16.7 Spectral function for the wave vectors $k/k_F = 0.2$, 1 and 1.6 at a density corresponding to $r_s = 2$. Only the plasmon contribution to the self-energy is included.

and $E = E_p(0) + \varepsilon(\boldsymbol{k})$ for $k/k_F > 1$. Also, the imaginary part vanishes for energies $-E_p(0) \leq E \leq E_p(0)$. For $k/k_F = 1$, the self-energy exhibits two sharp peaks located symmetrically about the Fermi energy. The sharp discontinuity in these two peaks implies an infinite value in the corresponding real part, when Eq. (16.25) is applied. For other wave vectors, the real part has a jump discontinuity at the position of the logarithmic singularity in the imaginary part [Hedin and Lundqvist (1969)]. The rapidly varying real parts lead to an unusual sp strength distribution, discussed below. The continuum contribution to the imaginary part of the self-energy is shown in Fig. 16.6. These contributions are much better behaved, but substantially smaller, than the plasmon terms.

The RPA self-energy has quite dramatic consequences for the sp strength distribution. Figure 16.7 shows typical results for spectral functions corresponding to the same wave vectors and density as those in Fig. 16.5. The spectral function for wave vectors deep inside the Fermi surface, has two distinct structures. Since the imaginary part of the self-energy is nonvanishing only in certain energy regions, it is possible to have multiple solutions to

$$E_Q(k) = \frac{\hbar^2 k^2}{2m} + \operatorname{Re} \Sigma^{RPA}(k; E_Q(k)) \tag{16.26}$$

outside these domains. This implies that at those energies for which Eq. (16.26) holds, the solution to the Dyson equation yields a δ-function

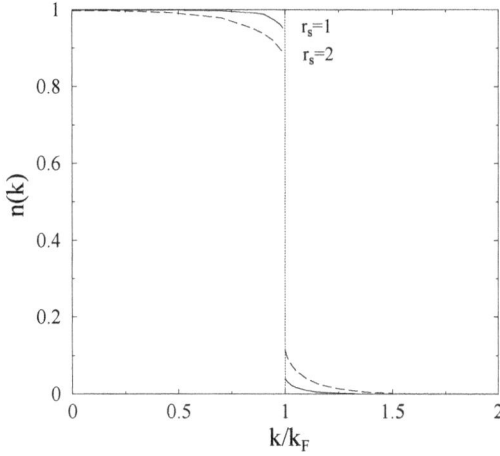

Fig. 16.8 Occupation number in the RPA. The solid curve corresponds to $r_s = 1$ and the dashed to $r_s = 2$.

term, with the strength given by

$$Z_Q(k) = \left(1 - \frac{\partial \text{Re } \Sigma^{RPA}(k; E)}{\partial E}\bigg|_{E=E_Q(k)}\right)^{-1}. \qquad (16.27)$$

This situation is almost realized for $k/k_F = 0.2$ in Fig. 16.7. The peak with the higher energy corresponds to the usual quasiparticle, whereas the second one is not present in other Fermi systems. It is referred to as the "plasmaron" peak [Hedin and Lundqvist (1969)] and can be interpreted as a resonance between a cloud of real plasmons and a hole excitation, reflecting the intermediate state of a hole and a well-defined plasmon in the self-energy. When the continuum contributions to the self-energy are included, the δ-function peaks become slightly spread out. The main peak for $k/k_F = 0.2$ contains a strength of 0.78, as indicated in Fig. 16.7. For small wave vectors, the strengths add up to about 1. Approaching the Fermi wave vector, only one peak remains, as illustrated for $k/k_F = 1$ in Fig. 16.7. The strength is 0.82 in this case. Also for values of $k/k_F > 1$, only one peak is found, as shown in Fig. 16.7 for $k/k_F = 1.6$. The occupation number is calculated by integrating the spectral function up to the Fermi energy. We illustrate results for two densities, corresponding to $r_s = 1$ and 2, in Fig. 16.8. The occupation of states above k_F increases with the parameter r_s. The jump discontinuity is equal to the quasiparticle strength or Z_F. It

Fig. 16.9 Diagrammatic form of the self-energy in the ring approximation, known as the *GW* approximation. The screened Coulomb interaction W is represented by the double wiggle and consists of summing all rings that contain the dressed sp propagator G, shown as the double line.

is given by 0.92 and 0.82 for $r_s = 1$ and $r_s = 2$, respectively.

16.2.2 *Electron self-energy in the GW approximation*

In keeping with the discussion of Ch. 9, it is natural to consider the self-consistent formulation of the inclusion of the polarization propagator in the electron self-energy. The formulation is known as the *GW* approximation, shown diagrammatically in Fig. 16.9.

We will now develop the necessary tools to study the *GW* self-energy. First we rewrite the noninteracting ph propagator in Eq. (14.7) in the form of a convolution

$$\Pi^{(0)}(\boldsymbol{k};\boldsymbol{q},E) = \int \frac{dE'}{2\pi i} G^{(0)}(\boldsymbol{k}+\boldsymbol{q}/2;E+E')G^{(0)}(\boldsymbol{k}-\boldsymbol{q}/2;E'), \quad (16.28)$$

which can be checked in the usual way. For dressed propagators we generalize to

$$\Pi^f(\boldsymbol{k};\boldsymbol{q},E) = \int \frac{dE'}{2\pi i} G(\boldsymbol{k}+\boldsymbol{q}/2;E+E')G(\boldsymbol{k}-\boldsymbol{q}/2;E'). \quad (16.29)$$

Using the spectral representation of G in Eq. (16.9), we can perform the E' integration using the appropriate contours, thereby obtaining

$$\Pi^f(\boldsymbol{k};\boldsymbol{q},E) = \int_{\varepsilon_F}^{\infty} d\tilde{E} \int_{-\infty}^{\varepsilon_F} d\tilde{E}' \frac{S_p(\boldsymbol{k}+\boldsymbol{q}/2;\tilde{E})S_h(\boldsymbol{k}-\boldsymbol{q}/2;\tilde{E}')}{E-(\tilde{E}-\tilde{E}')+i\eta}$$
$$-\int_{-\infty}^{\varepsilon_F} d\tilde{E} \int_{\varepsilon_F}^{\infty} d\tilde{E}' \frac{S_h(\boldsymbol{k}+\boldsymbol{q}/2;\tilde{E})S_p(\boldsymbol{k}-\boldsymbol{q}/2;\tilde{E}')}{E+(\tilde{E}'-\tilde{E})-i\eta}. \quad (16.30)$$

For $E > 0$ the imaginary part of Π^f can be written as

$$\mathrm{Im}\ \Pi^f(\mathbf{k}; \mathbf{q}, E)$$

$$= -\pi \int_{\varepsilon_F}^{\infty} d\tilde{E} \int_{-\infty}^{\varepsilon_F} d\tilde{E}' S_p(\mathbf{k} + \mathbf{q}/2; \tilde{E}) S_h(\mathbf{k} - \mathbf{q}/2; \tilde{E}') \delta(E - (\tilde{E} - \tilde{E}'))$$

$$= -\pi \int_{\varepsilon_F}^{E + \varepsilon_F} d\tilde{E}\ S_p(\mathbf{k} + \mathbf{q}/2; \tilde{E}) S_h(\mathbf{k} - \mathbf{q}/2; \tilde{E} - E). \tag{16.31}$$

The upper limit of the integration in the last equation, reflects the condition that the energy argument for the hole spectral function must remain smaller than ε_F. A similar result holds for $E < 0$. By integrating over $\mathbf{k}$, the dressed version of the Lindhard function emerges

$$\Pi^f(\mathbf{q}, E) = \int \frac{d^3 k}{(2\pi)^3} \Pi^f(\mathbf{k}; \mathbf{q}, E). \tag{16.32}$$

The usual dispersion relation is valid for Π^f [as in Eq.(14.34) for $\Pi^{(0)}$]

$$\Pi^f(\mathbf{q}, E) = -\frac{1}{\pi} \int_0^{\infty} dE' \frac{\mathrm{Im}\ \Pi^f(\mathbf{q}, E')}{E - E' + i\eta}$$

$$+ \frac{1}{\pi} \int_{-\infty}^{0} dE' \frac{\mathrm{Im}\ \Pi^f(\mathbf{q}, E')}{E - E' - i\eta}. \tag{16.33}$$

The imaginary part of the self-consistent Π^f, for two different values of r_s, is compared to the noninteracting term in Fig. 16.10. The difference is particularly striking, but hardly surprising. The spreading of the sp strength in the spectral functions, as discussed in the previous section and even more so for self-consistent results, leads to a reduction in the original energy domain, corresponding to noninteracting ph states. The removed strength is found in a large energy range above the usual energy threshold. For the lower density, a larger tail is calculated by [Holm and von Barth (1998)]. As in Eq. (14.41) the polarization propagator is given by

$$\Pi(q, E) = \frac{\Pi^f(q, E)}{1 - 2V(q)\Pi^f(q, E)}. \tag{16.34}$$

Employing Π^f, the screened Coulomb interaction becomes

$$W(q, E) = V(q) + V(q)2\Pi^f(q, E)V(q)$$

$$+ V(q)2\Pi^f(q, E)V(q)2\Pi^f(q, E)V(q) + ...$$

$$= V(q) + 2V(q)\left\{\Pi^f(q, E) + \Pi^f(q, E)2V(q)\Pi^f(q, E) + ...\right\}V(q)$$

$$= V(q) + 2V^2(q)\Pi(q, E), \tag{16.35}$$

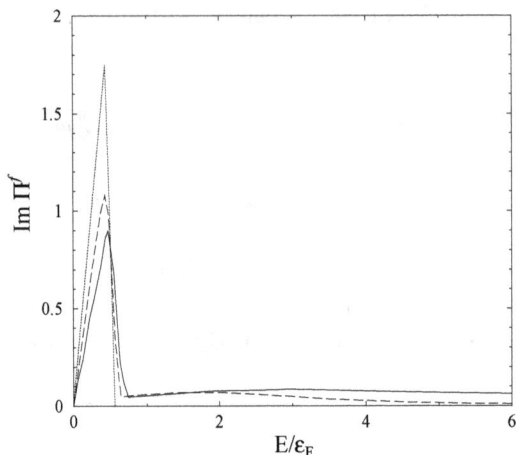

Fig. 16.10   Imaginary part of the noninteracting dressed polarization propagator Π^f for $r_s = 2$ (dashed) and $r_s = 4$ (solid) at $q/k_F = 0.25$, as a function of energy. The figure has been adapted from [Holm and von Barth (1998)]. For comparison the imaginary part of the noninteracting propagator (dotted) is included, using the same scale as in Fig. 14.2.

demonstrating that the screened interaction has the same analytic properties as Π. We can therefore write

$$W(q, E) \equiv V(q) + \Delta W(q, E) \tag{16.36}$$

$$= V(q) - \frac{1}{\pi} \int_0^{\infty} dE' \frac{\text{Im } W(q, E')}{E - E' + i\eta} + \frac{1}{\pi} \int_{-\infty}^0 dE' \frac{\text{Im } W(q, E')}{E - E' - i\eta},$$

since the imaginary part of W reads

$$\text{Im } W(q, E) = 2V^2(q)\text{Im } \Pi(q, E) \tag{16.37}$$

$$= 2V^2(q)\frac{\text{Im } \Pi^f(q, E)}{(1 - 2V(q) \text{ Re } \Pi^f(q, E))^2 + (2V(q) \text{ Im } \Pi^f(q, E))^2},$$

using Eq. (16.35). As $\Delta W = 2V^2\Pi$, Eq. (16.19) can be modified to include the dressing of the sp propagators as follows

$$\Sigma_{\Delta W}(\boldsymbol{k}; E) = i \int \frac{dE_1}{2\pi} \int \frac{d^3q}{(2\pi)^3} G(\boldsymbol{k} - \boldsymbol{q}; E - E_1)\Delta W(q; E_1). \tag{16.38}$$

When the spectral representation of G [see Eq. (16.9)] and ΔW, given in Eq. (16.36), are inserted, the integration over E_1 can be performed with the usual contour integrations, yielding

$$\Sigma_{\Delta W}(\boldsymbol{k};E) = -\frac{1}{\pi}\int\frac{d^3q}{(2\pi)^3}\int_0^\infty d\tilde{E}'\int_{\varepsilon_F}^\infty d\tilde{E}\frac{S_p(\boldsymbol{k}-\boldsymbol{q};\tilde{E})\text{Im }W(q,\tilde{E}')}{E-\tilde{E}'-\tilde{E}+i\eta}$$

$$-\frac{1}{\pi}\int\frac{d^3q}{(2\pi)^3}\int_{-\infty}^0 d\tilde{E}'\int_{-\infty}^{\varepsilon_F} d\tilde{E}\frac{S_h(\boldsymbol{k}-\boldsymbol{q};\tilde{E})\text{Im }W(q,\tilde{E}')}{E-\tilde{E}'-\tilde{E}-i\eta}.$$

$$(16.39)$$

The first term contributes an imaginary part for $E > \varepsilon_F$, of the form

$$\text{Im }\Sigma_{\Delta W}(\boldsymbol{k};E) = \int\frac{d^3q}{(2\pi)^3}\int_0^{E-\varepsilon_F} d\tilde{E}' S_p(\boldsymbol{k}-\boldsymbol{q};E-\tilde{E}')\text{Im }W(q,\tilde{E}'). \quad (16.40)$$

For energies $E < \varepsilon_F$, the second term of Eq. (16.39) yields

$$\text{Im }\Sigma_{\Delta W}(\boldsymbol{k};E) = -\int\frac{d^3q}{(2\pi)^3}\int_{E-\varepsilon_F}^0 d\tilde{E}' S_h(\boldsymbol{k}-\boldsymbol{q};E-\tilde{E}')\text{Im }W(q,\tilde{E}').$$

$$(16.41)$$

We still need the contribution to the self-energy in lowest order, corresponding to first term in Eq. (16.35). The generalization of the HF self-energy in Eq. (10.171), reads

$$\Sigma_V(k) = -\int\frac{d^3k'}{(2\pi)^3}V(\boldsymbol{k}-\boldsymbol{k}')n(k'), \quad (16.42)$$

where $n(k)$ is the occupation number, given in Eq. (16.13). Combining the last three equations, the self-energy in the *GW* approximation becomes

$$\Sigma_{GW}(\boldsymbol{k};E) = \Sigma_V(k)-\frac{1}{\pi}\int_{\varepsilon_F}^\infty dE'\frac{\text{Im }\Sigma_{\Delta W}(\boldsymbol{k};E')}{E-E'+i\eta}+\frac{1}{\pi}\int_{-\infty}^{\varepsilon_F} dE'\frac{\text{Im }\Sigma_{\Delta W}(\boldsymbol{k};E')}{E-E'-i\eta}.$$

$$(16.43)$$

With the solution of the Dyson equation [see Eq. (16.4)], the self-consistency loop is now complete.

Numerical implementation can proceed from a reasonable starting point for the spectral functions. In [Holm and von Barth (1998)], a small set of gaussians was used to represent the spectral strength

$$S(k;E) = \sum_n \frac{W_n(k)}{\sqrt{2\pi}\Gamma_n(k)}\exp\left\{-\frac{(E-E_n(k))^2}{2\Gamma_n^2(k)}\right\}, \quad (16.44)$$

where W_n, Γ_n, and E_n are parameters. A previous calculation [von Barth and Holm (1996)] in the $GW^{(0)}$ approximation, was used to initiate the parameters. The $GW^{(0)}$ approximation keeps the screened Coulomb interaction in Eq. (16.38) fixed to the original result of Eq. (14.58), but includes

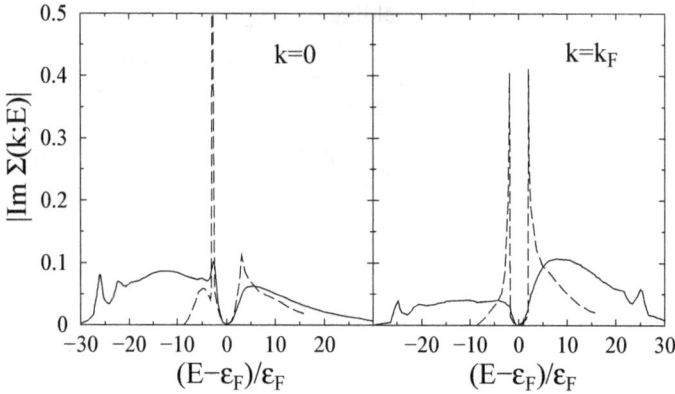

Fig. 16.11   Absolute value of the imaginary part of the self-energy for $k = 0$ (left panel) and $k = k_F$ for $r_s = 4$, as a function of $(E - \varepsilon_F)/\varepsilon_F$. The dashed lines represent the $GW^{(0)}$ approximation and the solid ones correspond to the GW result [Holm and von Barth (1998)].

a self-consistent determination of the dressed sp propagator G. Starting from Eq. (16.44), the first step is to calculate the imaginary part of Π^f in Eq. (16.32). The gaussian representation of the spectral functions allows an analytic evaluation of the convolution. Fast Fourier transform techniques can also be applied [García-González and Godby (2001)]. The real part can be obtained from the dispersion relation in Eq. (16.33). The imaginary part of ΔW is thereby determined as well [see Eq. (16.37)]. The main numerical effort is the solution for the imaginary part of $\Sigma_{\Delta W}$ in Eqs. (16.40) and (16.41). The correlated HF term in Eq. (16.42) is straightforward, computationally. The real part of the self-energy requires the dispersion integrals of Eq. (16.43). The subsequent solution of the Dyson equation yields the spectral functions, which closes the self-consistency loop. The parameters, used to fit the spectral strength in Eq. (16.44), must now be adjusted, and the whole cycle can be repeated until they no longer vary.

The most dramatic consequence of the self-consistency feature, is the disappearance of a well-defined plasmon excitation in Eq. (16.34). Indeed, the approximation to Π including dressed sp propagators, no longer fulfills the f-sum rule. Fig. 16.10 illustrates that the noninteracting Π^f contains an imaginary part in a very large energy domain. The appearance of ph strength, where the plasmon resides when noninteracting sp propagators are employed, is the main reason for its disappearance. When sp dressing is included, the response becomes a broad distribution, with a peak super-

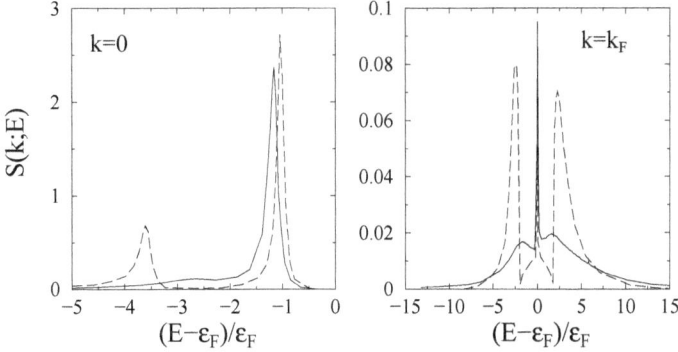

Fig. 16.12   Spectral functions for $k = 0$ (left panel) and $k = k_F$ for $r_s = 4$, as a function of $(E - \varepsilon_F)/\varepsilon_F$. The dashed lines represent the $GW^{(0)}$ approximation and the solid ones correspond to the GW result [Holm and von Barth (1998)].

imposed, but not at the location of the original plasmon [Holm and von Barth (1998)]. The absence of a well-defined plasmon excitation changes the self-energy of the electron substantially, as illustrated in Fig. 16.11 for a density corresponding to $r_s = 4$. The dashed lines reflect the intermediate $GW^{(0)}$ approximation, which yields self-energies that are not too different from the RPA (or $G^{(0)}W^{(0)}$), shown in Fig. 16.5. The solid lines in Fig. 16.11 no longer exhibit this structure, which is associated with the presence of the plasmon component of the response.

The self-consistent real and imaginary part of the self-energy lead to qualitatively different spectral functions, as demonstrated in Fig. 16.12. The satellites in the $G^{(0)}W^{(0)}$ approximation in Fig. 16.7, are retained in the partially self-consistent $GW^{(0)}$ approach, given by the dashed lines in Fig. 16.12. When full self-consistency is achieved, the satellites have completely vanished, reflecting the disappearance of the plasmon as a well-defined component of the response. Satellites are observed in Na [Steiner *et al.* (1979)], suggesting that the improved treatment of self-consistency leads to a deterioration of the description of certain features observed in experiment. The spectral functions in the GW approximation exhibit a reduction of correlations with respect to the less self-consistent implementations. This is confirmed by the properties of the quasiparticle strength, defined in Eq. (16.27), and plotted in Fig. 16.13 for the different approximations. The rising value of Z_Q with increasing self-consistency, confirms that the system becomes less correlated. At k_F, the quasiparticle strength corre-

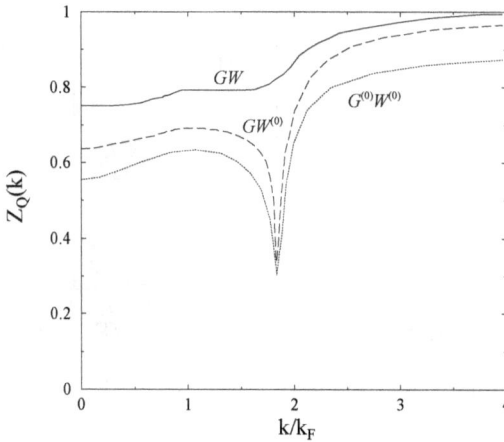

Fig. 16.13   Quasiparticle strength as a function of k/k_F in the GW (solid), $GW(0)$ (dashed), and $G^{(0)}W^{(0)}$ approximation (dotted) for $r_s = 4$ [Holm and von Barth (1998)].

sponds to the jump in the occupation number (see Sec. 11.4.4). At $r_s = 4$, it increases from 0.645 in the RPA ($G^{(0)}W^{(0)}$), via 0.702 ($GW^{(0)}$), to 0.793 in the fully self-consistent GW calculation [Holm and von Barth (1998)]. These conclusions are further strengthened by the occupation numbers, as shown in Fig. 16.14. The figure demonstrates that the depletion of the Fermi sea decreases with increasing self-consistency.

Another important observation is related to the change in the bandwidth for the various approximations. For the noninteracting electron gas it is given by the difference in energy between the $k = 0$ and $k = k_F$ sp state. In units of the Fermi energy, the bandwidth is therefore 1. When correlations are included, it can be defined by the difference between the quasiparticle energies [see Eq. (16.26)], associated with the peaks in the spectral functions, for $k = 0$ and $k = k_F$. Figure 16.12 illustrates that the effect of more self-consistency is to increase the bandwidth. A complete picture of the quasiparticle energies with different degrees of self-consistency is plotted in Fig. 16.15. Experiments on simple metals [Jensen and Plummer (1985)] suggest that the bandwidths are narrower than those obtained from band-structure calculations, employing energy-independent potentials. The $G^{(0)}W^{(0)}$ approximation exhibits narrowing (not enough), whereas the bandwidths increase substantially with increasing self-consistency. The conclusion that the properties of the self-energy do not improve for the GW approximation are therefore justified.

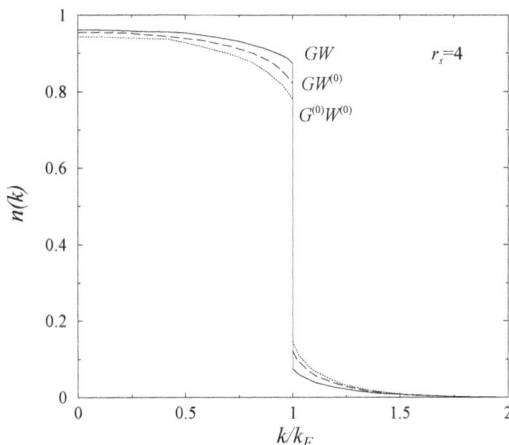

Fig. 16.14   Occupation number as a function of k/k_F in the GW (solid), $GW^{(0)}$ (dashed), and $G^{(0)}W^{(0)}$ approximation (dotted) for $r_s = 4$ [Holm and von Barth (1998)].

We are now faced with an interesting situation. On the one hand there is no doubt that self-consistency is a physically appealing concept. Furthermore, the exact formulation of the self-energy includes precisely the self-consistent sp propagators. On the other hand, essential features of the self-energy like the bandwidth and the presence of satellite structures in the sp strength, deteriorate when the self-consistent formulation is implemented. A remedy to the situation is not trivial, but we offer some remarks. First we note that improvements over the GW approximation have been proposed that are known as vertex corrections, modifying one of the vertices in Fig. 16.9 to represent higher-order exchange contributions to the self-energy. Results are not unambiguous [Mahan (1994)] and a fundamental problem with the GW approximation remains. From our studies of the properties of the electron gas, it is clear that the plasmon is a critical ingredient of the relevant physics. The screened Coulomb interaction W does not contain the correct properties, since it does not conserve the f-sum rule, let alone yield a proper plasmon. It is quite well known how to construct a (sum rule) conserving approximation for the response, for example based on the GW approach to the self-energy. We will discuss such techniques in Ch. 21. An actual implementation of this method for the electron gas is however not available. Assuming that the conserving approach yields physically relevant plasmons, it still remains to be clarified how such a description of the response can be included in the self-energy. The recent

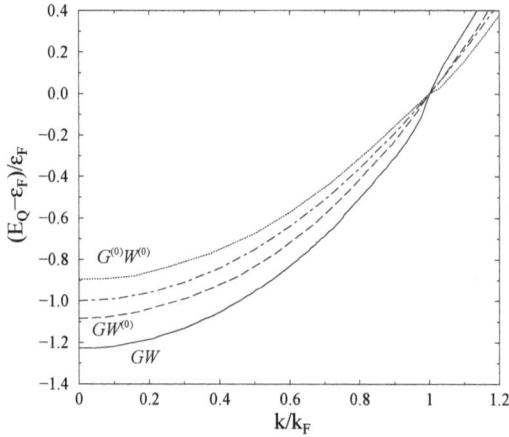

Fig. 16.15   Quasiparticle energies at $r_s = 4$, as a function of k/k_F in the GW (solid), $GW(0)$ (dashed), and $G^{(0)}W^{(0)}$ approximation (dotted) [Holm and von Barth (1998)]. The noninteracting result is given by the dashed-dot line.

development of the Faddeev technique to calculate the self-energy [Dickhoff and Barbieri (2004)], may shed some light on this problem, since it addresses the unequal treatment of the Pauli principle in the self-energy of the GW approximation. Indeed, inspection of Fig. 16.9 shows that the sp propagator G is not "Pauli aware" of the (parallel) sp propagators inside the W interaction, except for the first and last contribution in the ring summation. Improving the treatment of the self-energy in the electron is therefore an unresolved issue and requires further studies to provide a systematic improvement of the GW results.

16.2.3   *Energy per particle of the electron gas*

Finally we discuss the correlation energy of the electron gas in various approximations. First we note that an essentially exact calculation of the energy per particle of the electron gas is available employing the quantum Monte Carlo (QMC) approach of [Ceperley and Alder (1980); Ortiz *et al.* (1999); Ortiz and Ballone (1994)]. Several calculations of the energy per particle have been published for various GW implementations [Holm and von Barth (1998); García-González and Godby (2001)]. It is therefore useful to gauge the quality of these Green's function calculations by comparing with exact Monte Carlo results. The RPA correlation energy calculation, based on Sec. 14.4.2, will also be included.

Table 16.1 contains results for the negative of the so-called exchange

Table 16.1   Minus XC energies per particle (in Hartrees) for the electron gas.

r_s	1	2	4	5	10	20
QMC	0.5180	0.2742	0.1466	0.1198	0.0644	0.0344
	0.5144	0.2729	0.1474	0.1199	0.0641	0.0344
GW	0.5160(2)	0.2727(5)	0.1450(5)	0.1185(5)	0.0620(9)	0.032(1)
		0.2741	0.1465			
$GW^{(0)}$	0.5218(1)	0.2736(1)	0.1428(1)	0.1158(1)	0.0605(4)	0.030(1)
$G^{(0)}W^{(0)}$	0.5272(1)	0.2821(1)	0.1523(1)	0.1247(1)	0.0665(2)	0.0363(5)
RPA	0.5370	0.2909	0.1613	0.1340	0.0764	0.0543
$-E_x/N$	0.4582	0.2291	0.1145	0.0916	0.0458	0.0229

QMC results are from [Ceperley and Alder (1980)] (first row) and [Ortiz *et al.* (1999); Ortiz and Ballone (1994)] (second row). Second row of GW results are from [Holm and von Barth (1998)]. All other GW results are from [García-González and Godby (2001)]. Also shown are the correlation energy in RPA and the HF exchange energy per particle. Parentheses identify the numerical uncertainty in the last significant figure.

correlation (XC) energy per particle for different values of r_s. The XC energy represents the difference between the true energy of the ground state and the energy of the noninteracting system. It therefore includes the HF or correlated HF contribution [see Eq. (16.42) for the self-energy]. The former is given by Eq. (10.179) and included as the last row in Table 16.1 for reference. The QMC calculations from different groups are consistent with each other. This also holds true for the GW numbers from [Holm and von Barth (1998)] and [García-González and Godby (2001)]. The agreement between the QMC and GW results are remarkable in the limit of high density. It remains good for intermediate and low densities, although it is not a priori clear why other physical effects, not included in the GW approximation, apparently do not contribute. Partial self-consistency in the $GW^{(0)}$ approach yields slightly worse agreement. Even the $G^{(0)}W^{(0)}$ numbers are quite respectable, but less accurate. The RPA response is identical to the one used in the $G^{(0)}W^{(0)}$ method, but the energy is calculated differently, yielding a substantial discrepancy.

An important supplementary consideration is the reconstruction of the number of particles, according to Eq. (16.12), in the different GW implementations. We will see in Ch. 21 that only self-consistent approximations to the self-energy, like the full GW implementation, guarantee the conservation laws, like the one for particle number. The $G^{(0)}W^{(0)}$ approximation exhibits a negligible violation up to $r_s = 4$ [García-González and Godby (2001)]. At $r_s = 5$ the violation is 0.3 %, reaching 6.1 % for $r_s = 20$.

We are therefore confronted with a rather fundamental puzzle. While the energy per particle is quite well described in the GW approximation, it exhibits inherent failures for the self-energy, or quasiparticle properties, as discussed above. Developing an approach that reconciles these two issues, remains a formidable challenge.

16.3 Nucleon properties in nuclear matter

16.3.1 *Ladder diagrams and the self-energy*

In this section we will study the properties of the nucleon sp propagator in nuclear matter. We focus our attention on the dominant influence of short-range correlations in determining the properties of a nucleon in the medium. The correct treatment of these correlations was presented in Ch. 15. The vertex function, or effective interaction, requires the complete summation of ladder diagrams, if the two-body interaction $\hat{V}$ contains strong repulsion at small relative distance between particles. In this section we will explore the consequences for the sp propagator when the corresponding vertex function is included in the self-energy. It is convenient to formulate this treatment in the self-consistent form, originally introduced in Sec. 9.4. Since we will see that the modifications of the sp propagator are substantial with the inclusion of these correlations, one may deduce that a self-consistent treatment is important. The relevant summation of ladder diagrams for the vertex function is given by

$$
\begin{aligned}
\langle k & m_\alpha m_{\alpha'} | \, \Gamma_{pphh}(\boldsymbol{K}, E) \, | \boldsymbol{k}' m_\beta m_{\beta'} \rangle \\
&= \langle \boldsymbol{k} m_\alpha m_{\alpha'} | \, V \, | \boldsymbol{k}' m_\beta m_{\beta'} \rangle + \langle \boldsymbol{k} m_\alpha m_{\alpha'} | \, \Delta\Gamma_{pphh}(\boldsymbol{K}, E) \, | \boldsymbol{k}' m_\beta m_{\beta'} \rangle \\
&= \langle \boldsymbol{k} m_\alpha m_{\alpha'} | \, V \, | \boldsymbol{k}' m_\beta m_{\beta'} \rangle + \frac{1}{2} \sum_{m_\gamma m_{\gamma'}} \int \frac{d^3 q}{(2\pi)^3} \; \langle \boldsymbol{k} m_\alpha m_{\alpha'} | \, V \, | \boldsymbol{q} m_\gamma m_{\gamma'} \rangle \\
&\quad \times G^f_{pphh}(\boldsymbol{K}, \boldsymbol{q}; E) \, \langle \boldsymbol{q} m_\gamma m_{\gamma'} | \, \Gamma_{pphh}(\boldsymbol{K}, E) \, | \boldsymbol{k}' m_\beta m_{\beta'} \rangle , \qquad (16.45)
\end{aligned}
$$

where the first line serves to define the quantity $\Delta\Gamma_{pphh}$. The notation includes the wave vectors of the relative motion, $\boldsymbol{k}$ and $\boldsymbol{k}'$, as well as the total wave vector $\boldsymbol{K}$. Discrete quantum numbers associated with spin and isospin are indicated generically by m. The dressed version of the non-interacting propagator is employed in Eq. (16.45). The evaluation of the convolution integral corresponding to Eq. (15.5), using Eq. (16.9) for each of the sp propagators, yields

$$G^f_{pphh}(\boldsymbol{K}, \boldsymbol{q}; E) = \int_{\varepsilon_F}^{\infty} dE' \int_{\varepsilon_F}^{\infty} dE'' \frac{S_p(\boldsymbol{q} + \boldsymbol{K}/2; E') S_p(\boldsymbol{K}/2 - \boldsymbol{q}; E'')}{E - E' - E'' + i\eta}$$
$$- \int_{-\infty}^{\varepsilon_F} dE' \int_{-\infty}^{\varepsilon_F} dE'' \frac{S_h(\boldsymbol{q} + \boldsymbol{K}/2; E') S_h(\boldsymbol{K}/2 - \boldsymbol{q}; E'')}{E - E' - E'' - i\eta}.$$

$$(16.46)$$

The diagrammatic form of Eq. (16.45) is identical to Fig. 15.5. From the first equality of that figure, we recognize that the inclusion of short-range correlations in the self-energy is accomplished by replacing the exact Γ in Eq. (16.3), by the Γ_{pphh} of Eq. (16.45). The first identity of Fig. 15.5 can be used to identify the analytic structure of $\Delta\Gamma_{pphh} = \Gamma_{pphh} - V$. This information is essential to allow the proper construction of the self-energy, as shown below. Inserting the Lehmann representation of Eq. (15.28), modified for the present case, enables us to write

$$\langle k m_\alpha m_{\alpha'} | \Delta\Gamma_{pphh}(\boldsymbol{K}, E) | k' m_\beta m_{\beta'} \rangle \qquad (16.47)$$
$$= \frac{-1}{\pi} \int_{2\varepsilon_F}^{\infty} dE' \frac{\operatorname{Im} \langle k m_\alpha m_{\alpha'} | \Delta\Gamma_{pphh}(\boldsymbol{K}, E') | k' m_\beta m_{\beta'} \rangle}{E - E' + i\eta}$$
$$+ \frac{1}{\pi} \int_{-\infty}^{2\varepsilon_F} dE' \frac{\operatorname{Im} \langle k m_\alpha m_{\alpha'} | \Delta\Gamma_{pphh}(\boldsymbol{K}, E') | k' m_\beta m_{\beta'} \rangle}{E - E' - i\eta}$$
$$\equiv \langle k m_\alpha m_{\alpha'} | \Delta\Gamma_{\downarrow}(\boldsymbol{K}, E) | k' m_\beta m_{\beta'} \rangle + \langle k m_\alpha m_{\alpha'} | \Delta\Gamma_{\uparrow}(\boldsymbol{K}, E) | k' m_\beta m_{\beta'} \rangle.$$

The notation with the arrows $\downarrow$ and $\uparrow$ signals that the corresponding contribution to $\Delta\Gamma_{pphh}$ has poles in the lower or upper half of the complex energy plane. Poles in the lower half plane are associated with forward (pp) propagation or energies above $2\varepsilon_F$, whereas the opposite applies for backward (hh) terms with energies below $2\varepsilon_F$. By identifying the contribution of $\Delta\Gamma_{pphh}$ as the last term in Eq. (16.3), it is possible, after some relabeling, to write the corresponding self-energy contribution as

$$\Sigma_{\Delta\Gamma}(\boldsymbol{k}; E) = -i\frac{1}{\nu} \sum_{m_\alpha m_{\alpha'}} \int \frac{d^3 k'}{(2\pi)^3} \int \frac{dE'}{2\pi} G(\boldsymbol{k}'; E') \qquad (16.48)$$
$$\times \langle \tfrac{1}{2}(\boldsymbol{k} - \boldsymbol{k}') m_\alpha m_{\alpha'} | \Delta\Gamma_{pphh}(\boldsymbol{K}, E + E') | \tfrac{1}{2}(\boldsymbol{k} - \boldsymbol{k}') m_\alpha m_{\alpha'} \rangle.$$

The integral over E' can now be performed analytically by using Eq. (16.47) for $\Delta\Gamma_{pphh}$ and the Lehmann representation of the sp propagator, given in Eq. (16.9). As usual the contour integration can be performed, without any difficulty, by carefully considering the pole structure of $\Delta\Gamma_{pphh}$ and G.

The result of the integrations then takes the form

$$\Sigma_{\Delta\Gamma}(k;E) = \frac{1}{\nu}\sum_{m_\alpha m_{\alpha'}}\int\frac{d^3k'}{(2\pi)^3}\int_{-\infty}^{\varepsilon_F}dE' \tag{16.49}$$

$$\times \langle\tfrac{1}{2}(\boldsymbol{k}-\boldsymbol{k}')m_\alpha m_{\alpha'}|\,\Delta\Gamma_\downarrow(E+E')\,|\tfrac{1}{2}(\boldsymbol{k}-\boldsymbol{k}')m_\alpha m_{\alpha'}\rangle\,S_h(k',E')$$

$$-\frac{1}{\nu}\sum_{m_\alpha m_{\alpha'}}\int\frac{d^3k'}{(2\pi)^3}\int_{\varepsilon_F}^{\infty}dE'$$

$$\times \langle\tfrac{1}{2}(\boldsymbol{k}-\boldsymbol{k}')m_\alpha m_{\alpha'}|\,\Delta\Gamma_\uparrow(E+E')\,|\tfrac{1}{2}(\boldsymbol{k}-\boldsymbol{k}')m_\alpha m_{\alpha'}\rangle\,S_p(k',E')$$

$$\equiv \Delta\Sigma_\downarrow(k;E) + \Delta\Sigma_\uparrow(k;E).$$

The arrow notation for the self-energy is similar to the one for $\Delta\Gamma_{pphh}$. By using the dispersion relation for $\Delta\Gamma_{pphh}$ in Eq. (16.47) one may verify that the usual dispersion relation for these self-energy contributions holds. The total self-energy can then be written (discarding the auxiliary potential)

$$\Sigma(k;E) = \Sigma_V(k) - \frac{1}{\pi}\int_{\varepsilon_F}^{\infty}dE'\frac{\mathrm{Im}\ \Sigma(k;E')}{E-E'+i\eta} + \frac{1}{\pi}\int_{-\infty}^{\varepsilon_F}dE'\frac{\mathrm{Im}\ \Sigma(k;E')}{E-E'-i\eta}$$

$$= \Sigma_V(k) + \Delta\Sigma_\downarrow(k;E) + \Delta\Sigma_\uparrow(k;E), \tag{16.50}$$

where the Hartree–Fock like term is given by

$$\Sigma_V(k) = \int\frac{d^3k'}{(2\pi)^3}\frac{1}{\nu}\sum_{m_\alpha m_{\alpha'}}\langle\tfrac{1}{2}(\boldsymbol{k}-\boldsymbol{k}')m_\alpha m_{\alpha'}|\,V\,|\tfrac{1}{2}(\boldsymbol{k}-\boldsymbol{k}')m_\alpha m_{\alpha'}\rangle\,n(k').$$

$$\tag{16.51}$$

This expression can be obtained by using the Lehmann representation of the sp propagator to perform the energy integration, and subsequently replacing the integral over the hole spectral function by the occupation number.

16.3.2   *Spectral function obtained from mean-field input*

The first generation spectral functions for nuclear matter were based on the solution of the ladder equation [Eq. (16.45)], employing mean-field propagators in the construction of G_{pphh}^J. A semi-realistic interaction, based on the Reid potential [Reid (1968)], was employed in [Ramos *et al.* (1989)]. Short-range correlations were properly treated but effects of the tensor force were eliminated. The discussion below includes the full complexity of realistic interactions, including the tensor component. Employing mean-field propagators, the spectral functions take the form of δ-functions. This simplification for the self-energy, given in Eq. (16.49), yields

$$\Sigma_{\Delta\Gamma}^{(0)}(k; E) = \frac{1}{\nu} \sum_{m_\alpha m_{\alpha'}} \int \frac{d^3 k'}{(2\pi)^3} \qquad (16.52)$$

$$\times \langle \tfrac{1}{2}(\boldsymbol{k} - \boldsymbol{k}')m_\alpha m_{\alpha'} | \, \Delta\Gamma_\downarrow^{(0)}(E + \varepsilon(k')) \, | \tfrac{1}{2}(\boldsymbol{k} - \boldsymbol{k}')m_\alpha m_{\alpha'} \rangle \, \theta(k_F - k')$$

$$- \frac{1}{\nu} \sum_{m_\alpha m_{\alpha'}} \int \frac{d^3 k'}{(2\pi)^3}$$

$$\times \langle \tfrac{1}{2}(\boldsymbol{k} - \boldsymbol{k}')m_\alpha m_{\alpha'} | \, \Delta\Gamma_\uparrow^{(0)}(E + \varepsilon(k')) \, | \tfrac{1}{2}(\boldsymbol{k} - \boldsymbol{k}')m_\alpha m_{\alpha'} \rangle \, \theta(k' - k_F).$$

A superscript (0) has been attached to Σ and $\Delta\Gamma$ to identify that they are obtained from a scattering equation with mean-field propagators. As discussed in Secs. 15.2 and 15.3, a serious difficulty arises by propagating such particles, when they exhibit a continuous sp spectrum and a realistic NN interaction is employed. Indeed, so-called pairing instabilities arise in the ${}^3S_1$-${}^3D_1$ and ${}^1S_0$ partial wave channels in certain density regimes [Vonderfecht *et al.* (1991b)]. Partial self-consistency with mean-field propagators in the ladder equation can be achieved, while avoiding the above problems. This is accomplished by linking the sp potential and the full Σ in the following way [Vonderfecht *et al.* (1993)]. Below k_F, let $\varepsilon(k)$ be the first moment of $S_h(k, E)$ with respect to energy (the contribution to the total energy at that wave vector) divided by the number of particles in the system with that wave vector. This intuitive prescription can not be extended $k > k_F$, by replacing S_h with S_p, since S_p contains a significant fraction of strength up to very large energies (illustrated below). Unrealistically large values for $\varepsilon(k)$ would therefore be generated for $k > k_F$. For this reason S_p is replaced by the quasiparticle contribution S_Q to S_p. It is equivalent to setting $\varepsilon(k)$ equal to the quasiparticle energy. The sp spectrum then reads

$$\varepsilon(k) = \frac{\int dE \; E \; S_h(k, E)}{\int dE \; S_h(k, E)} \qquad k < k_F \qquad (16.53)$$

and

$$\varepsilon(k) = \frac{\int dE \; E \; S_Q(k, E)}{\int dE \; S_Q(k, E)} = E_Q(k) \qquad k > k_F. \qquad (16.54)$$

The above prescription for the sp spectrum requires calculation of the full energy dependence of Σ in each iteration step and is therefore computationally intensive. However, it naturally contains a gap large enough to prevent the pairing instabilities. The quasiparticle energy alone has no appreciable

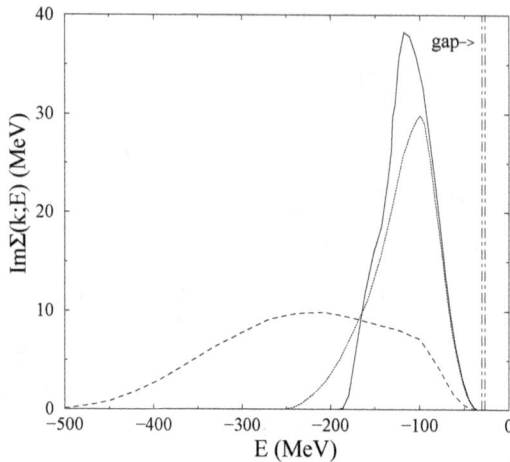

Fig. 16.16   Imaginary part of the self-energy as a function of energy below the gap for three wave vectors, $k = 0.01$ (solid), 0.51 (dotted), and 2.1 fm^{-1} (dashed). The gap reflects the different definitions of the sp spectrum for particles and holes [see Eqs. (16.53) and (16.54)].

gap, but near the Fermi wave vector there is considerable spreading of hole strength to energies below the quasiparticle energy, thereby introducing a gap at k_F when Eqs. (16.53) and (16.54) are applied.

Results from this partial self-consistency scheme, establish the main effects of short-range correlations on the sp strength distribution. In addition, they exhibit the limitations of employing mean-field sp propagators in the ladder equation and the self-energy. A characteristic property associated with the latter feature is shown in Fig. 16.16 for the imaginary part of the self-energy, calculated from Eq. (16.52). Comparing different values of k for these energies below the Fermi energy, indicates that Im$\Sigma^{\uparrow}$ becomes weak, while its energy range enlarges in a smooth manner as k increases. The energy range for this Im$\Sigma^{\uparrow}$ is determined by considering a phase space analysis, which can be based on the discussion given in Sec. 11.3.1. It shows that for each k there is a minimum energy above which 2h1p states (of mean-field type) can mix in the self-energy. So momentum conservation and the constraint on the location of 2h1p energies due to their mean-field character, are responsible for the energy range observed in Fig. 16.16. This wide range of energies will be available to the hole spectral functions discussed below.

For wave vectors near k_F one finds that the energy dependence of S_h

and S_p is dominated by the quasiparticle peak, characterizing the extent to which noninteracting features are maintained. Each wave vector has an associated quasiparticle energy which is the solution of

$$E_Q(k) = \frac{\hbar^2 k^2}{2m} + \text{Re}\Sigma(k; E_Q(k)). \tag{16.55}$$

The spectral function displays a peak at E_Q because of the vanishing term in the denominator of Eq. (11.99). The peak itself is represented by

$$S_Q(k, E) = \frac{1}{\pi} \frac{Z_Q^2(k)\,|\,W(k)\,|}{(E - E_Q(k))^2 + (Z_Q(k)W(k))^2} \tag{16.56}$$

where $W(k) = \text{Im } \Sigma(k; E_Q(k))$ and

$$Z_Q(k) = \left\{ 1 - \left(\frac{\partial \text{Re } \Sigma(k; E)}{\partial E} \right)_{E=E_Q(k)} \right\}^{-1} \tag{16.57}$$

is its strength. The first generation of these calculations have shown that near k_F the peak contains around 70% of the total sp strength, while an extra 13% is contained in the background composing the remainder of hole strength. The background is uniformly distributed across several hundred MeV below ε_F corresponding to the range of Im Σ. The range depends significantly on the value of k, as shown in Fig. 16.17. The final 17% of the strength has moved to energies greater than ε_F, including a significant high-energy tail in S_p, discussed momentarily. Farther below k_F the picture breaks down as the peak melts into the background, resulting in hole strength which is spread over a much wider range of energies. To the extent that spectral functions are described by the quasiparticle approximation, the excitations in nuclear matter are like those of a Landau Fermi liquid, illustrated in Fig. 16.17. The figure contains several hole spectral functions for wave vectors below $k_F = 1.36$ fm^{-1}. Notice that as $k \to k_F$ the peak becomes δ-function-like, due to the vanishing of ImΣ in Eq. (16.56). The infinite-lifetime character of such excitations is made possible by the loss of phase space, available to the states in Σ near the Fermi energy. This is essentially the same argument used by Landau in more general terms to develop the microscopic foundations of Fermi-liquid theory [Landau (1959)], discussed in Sec. 14.6.

In Fig. 16.18, the particle spectral function is plotted for three different wave vectors, $k = 0.79$, 1.74, and 5.04 fm^{-1}, as a function of energy. All wave vectors below k_F have the same high-energy tail as the dotted curve for $k = 0.79$ fm^{-1} in Fig. 16.18. For $k > k_F$, a quasiparticle peak,

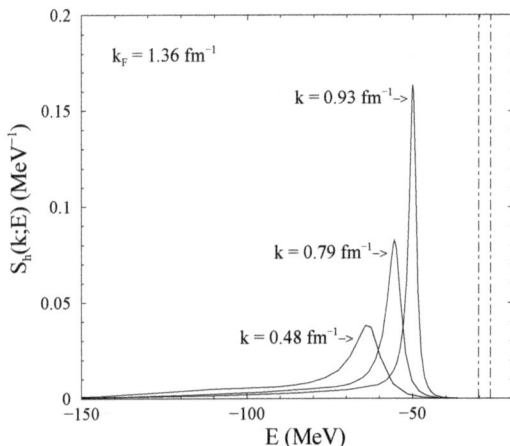

Fig. 16.17   Illustration of the decreasing width of the quasiparticle peak in the spectral function for three wave vectors below k_F given by $k = 0.48$, 0.79, and 0.93 fm^{-1}. The vertical lines indicate the position of the gap in the sp spectrum calculated from Eqs. (16.53) and (16.54).

which broadens with increasing wave vector,[3] can be observed on top of the same high-energy tail. The results therefore display a common, essentially wave vector independent, high-energy tail. The location of sp strength at high energy simply means that the interaction has sufficiently large matrix elements to compensate energy denominators encountered in the ladder equation. For this particular NN interaction [Reid (1968)], a significant amount of strength is found at high energy, as was already anticipated a long time ago [Brown (1969)].

A quantitative characterization of the missing sp strength for $k = 0.79$ fm^{-1}, shows that the integrated particle strength accounts for 17% of the sp strength. This is in agreement with the sum rule, since the integrated hole strength provides 83% of sp strength. The strength in the interval from 100 MeV above the Fermi energy to infinity amounts to 13%, with 7% residing above 500 MeV. To understand the influence of the tensor force on this distribution, a calculation of the ladder equation was performed in which the tensor coupling in the $^3S_1$-$^3D_1$ coupled channel was switched off. The integrated sp strength above the Fermi energy then amounts to 10.5% and should be regarded as due to the influence of pure short-range

---

[3]Ultimately, at very large wave vectors, the spectral functions become those of free particles of course.

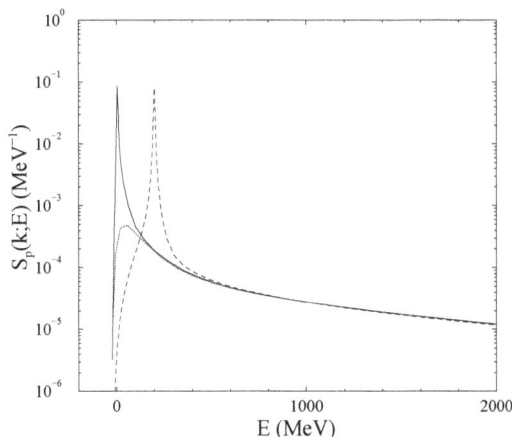

Fig. 16.18   Particle spectral functions at $k = 0.79$ (dotted), 1.74 (solid), 3.51 fm^{-1} (dashed). All three spectral functions converge to the same tail at high energy.

correlations. In [Vonderfecht *et al.* (1991a)] it is shown that the tensor force moves the additional 6.5% of strength to the first few hundred MeV above the Fermi energy. The amount of strength is consistent with other calculations of the occupation numbers, which show depletions of a similar size due to tensor correlations [Fantoni and Pandharipande (1984)].

Figure 16.19 exhibits the graph of the occupation number $n(k)$, or the number of particles in the ground state of the system with k. Near $k = 0$, $n(k)$ becomes fairly constant with a value of 0.83. From the discussion of S_p, roughly 1/3 of the 17% depletion is due to the effect of tensor correlations in the ladder equation. Another 1/3 is due the to high-energy tail in S_p at energies above 500 MeV. Other many-body methods such as Brueckner theory [Grangé *et al.* (1987)] and [Baldo *et al.* (1990b)] and correlated basis functions (CBF) theory [Benhar *et al.* (1989)], using other realistic interactions, produce very similar occupation near $k = 0$. In the work of [Grangé *et al.* (1987)] and [Baldo *et al.* (1990b)], 0.82 is reported for the Paris potential. Older CBF calculations for the Urbana v_{14} interaction, give 0.87, whereas more recent CBF numbers [Fantoni and Pandharipande (1984)] generate 0.83. All these calculations for different interactions, using different methods, give a strikingly similar value for $n(0)$. This is encouraging, since it implies that nonrelativistic many-body calculations yield stable results in the region, where one would like to compare to finite nuclei.

In contrast to $n(0)$, the occupation at k_F varies significantly between

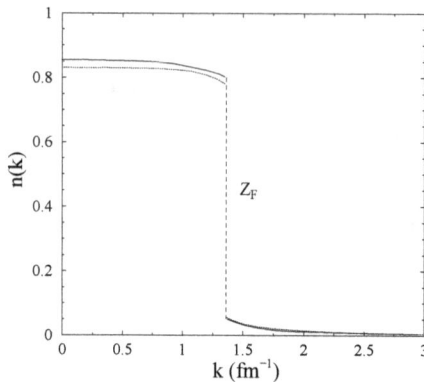

Fig. 16.19 Occupation probability for nuclear matter at equilibrium density calculated by integrating hole spectral functions obtained with mean-field propagators as input (dotted line). Self-consistent determination of the sp propagators yields the solid line.

methods. This variability is most easily expressed in terms of the discontinuity in $n(k)$ at k_F, $Z_F \equiv n(k_F^-) - n(k_F^+)$. For the results shown by the solid line in Fig. 16.19, $Z_F = 0.72$ and for the CBF calculation $Z_F = 0.70$. However, for the Paris interaction, $Z_F = 0.35$ and 0.47 has been obtained in [Grangé *et al.* (1987)] and [Baldo *et al.* (1990b)], respectively. The extra depletion in $n(k)$ as $k \to k_F$ arises from the enhanced ability of the sp state to couple to low lying 2p1h excitations, as its energy approaches these states. Also, the discontinuity depends on the level at which pairing correlations are included in the calculation. In the case of a paired system Z_F is zero, as discussed in Ch. 22.

16.3.3 *Self-consistent spectral functions*

The essential difficulty in the self-consistency procedure is the handling of the information, which describes the complete energy and wave-vector dependence of the spectral functions over their relevant domains. For interactions like the Reid potential, the energy and wave vector range needed to store the relevant information, precludes a straightforward numerical discretization of the spectral functions.

We briefly outline a method here that is able to deal with interactions like the Reid potential [Roth Stoddard (2000)], implying that it will certainly work for softer interactions. The sp spectral function is completely determined by Im $\Sigma(k; E)$. First, we note that its relation with the self-

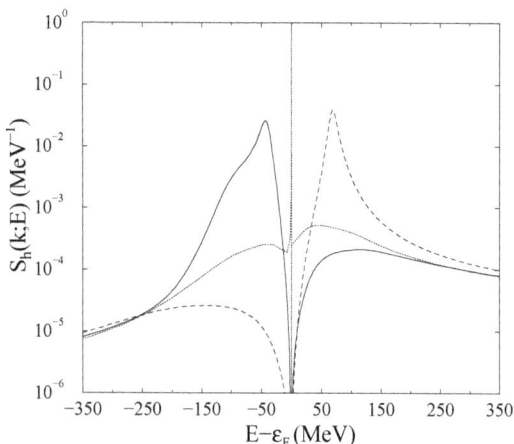

Fig. 16.20   Self-consistent spectral functions for three different wave vectors at $k_F = 1.36 \ \text{fm}^{-1}$ corresponding to 0 (full), 1.36 (dotted), and 2.1 fm^{-1} (dashed) as a function of $E - \varepsilon_F$.

energy is given by Eqs. (16.7) and (16.8). The self-energy can be written according to Eq. (16.50), which demonstrates that the spectral function can be completely expressed in terms of Im $\Sigma(k; E)$, Re $\Delta\Sigma_\downarrow(k; E)$, Re $\Delta\Sigma_\uparrow(k; E)$, and $\Sigma_V(k)$. The contributions Re $\Delta\Sigma_\downarrow(k; E)$ and Re $\Delta\Sigma_\uparrow(k; E)$ can be obtained from the Im $\Sigma(k; E)$ by performing the dispersion integrals given in Eq. (16.50). The correlated Hartree–Fock contribution, $\Sigma_V(k)$, can be calculated from the occupation numbers in a given iteration step towards self-consistency. This allows the construction of a new Fermi energy and the calculation of the next iteration for the spectral functions, closing the self-consistency loop. The analysis indicates that it is sufficient to accurately represent the imaginary part of the self-energy, for a complete determination of the spectral functions. Since this imaginary part is very well-behaved as a function of energy, Im Σ can be represented in terms of a limited set of gaussians [Roth Stoddard (2000)]. Two gaussians below the Fermi energy and two above, appear to provide sufficient flexibility for a complete representation of the imaginary part of the self-energy.

We now discuss some results of calculations that include a fully self-consistent inclusion of ladder diagrams in the self-energy. We start with the occupation numbers at a density corresponding to $k_F = 1.36 \ \text{fm}^{-1}$, shown by the dotted line in Fig. 16.19. The occupation numbers from the calculation described in Sec. 16.3.2 are given by the solid line in Fig. 16.19.

Such results can be considered as belonging to the first iteration step in the self-consistency procedure, since it is based on mean-field propagators as input. The similarity of the occupation of wave vectors below k_F between the two calculations is clear. Indeed, only a slight increase in the occupation of at most 3% is observed, when self-consistency is achieved. This implies that the corresponding depletion due to short-range correlations is still about 15% for nucleons deep in the Fermi sea. The increased occupation below k_F yields a small reduction in the occupation above k_F. This is further corroborated by the quasiparticle strength at the Fermi wave vector, since it changes only slightly from 0.72 to 0.75.

Some spectral functions are shown in Fig. 16.20 around the Fermi energy, as a function of $E - \varepsilon_F$. A comparison with the calculations, reviewed in Sec. 16.3.2, exhibits a striking difference for the strength distribution below the Fermi energy. The spectral strength in Fig. 16.20 is essentially identical at large negative energies for all wave vectors. It reflects a similar energy distribution of the imaginary part of the self-energy for these wave vectors. This could have been expected since the constraint imposed by mean-field sp energies, no longer applies and all wave vectors have the same energy domains, associated with the imaginary part of the self-energy. In the former case, the imaginary part of the self-energy has a fixed lower bound, depending on the wave vector, as shown in Fig. 16.16. The appearance of sp strength at large negative energies has important implications for the binding energy of nuclear matter, as will be discussed shortly. The spectral functions at high energy still include a common high-energy tail above ε_F for all wave vectors, which is quantitatively similar to the one shown in Fig. 16.18. Self-consistency therefore yields rather subtle changes in the properties of the spectral functions.

The special role of short-range correlations in generating saturation behavior of nuclear matter, is illustrated in Fig. 16.21. There we plot the integrand, corresponding to both terms in Eq. (16.10), as a function of wave vector, after performing the energy integral over the spectral function. Results are shown for densities, corresponding to $k_F = 1.36$ fm^{-1} and $k_F = 1.45$ fm^{-1}. At $k_F = 1.36$ fm^{-1} the high wave-vector components still provide attractive contributions, whereas for $k_F = 1.45$ fm^{-1} a changeover occurs, suggesting that at an even higher density the high wave-vector terms will provide only repulsion. From such an analysis it is clear that the expected relevance of short-range correlations in obtaining reasonable saturation properties of nuclear matter, is fully confirmed. The

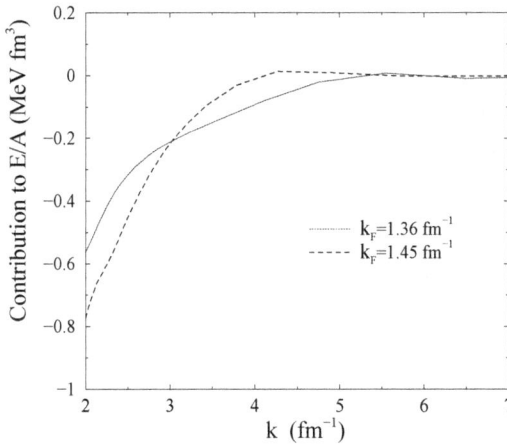

Fig. 16.21 The high wave-vector contribution to the energy per particle for $k_F = 1.36$ fm^{-1} (solid) and 1.45 fm^{-1} (dashed). It illustrates the source of the saturation process when short-range correlations are considered self-consistently.

relation of this observation to the vast body of work on the nuclear-matter saturation problem is taken up in the next section.

16.3.4 *Saturation properties of nuclear matter*

The nuclear-matter saturation problem has a long and colorful history. Many-body calculations were initiated by Brueckner. The basic idea goes back to the proper treatment of short-range correlations in the medium, which requires the solution of an in-medium scattering equation, discussed in Sec. 15.2 and Ch. 20. The solution must contain the sum of ladder diagrams, as for the scattering of nucleons in free space [Brueckner and Levinson (1955)]. Initially, only the propagation of particles above k_F was included. The method is referred to as the Brueckner–Hartree–Fock (BHF) approach. The in-medium scattering equation is solved with the noninteracting propagator of Eq. (15.34), excluding hh propagation,

$$\langle k m_\alpha m_{\alpha'} | \, G(\boldsymbol{K}, E) \, | \boldsymbol{k}' m_\beta m_{\beta'} \rangle = \langle k m_\alpha m_{\alpha'} | \, V \, | \boldsymbol{k}' m_\beta m_{\beta'} \rangle$$

$$+ \frac{1}{2} \sum_{m_\gamma m_{\gamma'}} \int \frac{d^3 q}{(2\pi)^3} \; \langle k m_\alpha m_{\alpha'} | \, V \, | \boldsymbol{q} m_\gamma m_{\gamma'} \rangle \tag{16.58}$$

$$\times \frac{\theta(|\boldsymbol{q} + \boldsymbol{K}/2| - k_F) \; \theta(|\boldsymbol{K}/2 - \boldsymbol{q}| - k_F)}{E - \varepsilon(\boldsymbol{q} + \boldsymbol{K}/2) - \varepsilon(\boldsymbol{K}/2 - \boldsymbol{q}) + i\eta} \; \langle \boldsymbol{q} m_\gamma m_{\gamma'} | \, G(\boldsymbol{K}, E) \, | \boldsymbol{k}' m_\beta m_{\beta'} \rangle .$$

The G notation for the effective interaction was introduced by [Bethe (1956)] and has since then, been referred to as the G-matrix. Equation (16.58) is also known as the Bethe–Goldstone equation. The original derivation of the corresponding linked contributions to the Brueckner theory was developed by [Goldstone (1957)]. The G-matrix interaction also obeys a dispersion relation between its real and imaginary part

$$
\langle km_\alpha m_{\alpha'} | G(\boldsymbol{K}, E) | \boldsymbol{k}' m_\beta m_{\beta'} \rangle = \langle km_\alpha m_{\alpha'} | V | \boldsymbol{k}' m_\beta m_{\beta'} \rangle \tag{16.59}
$$
$$
- \frac{1}{\pi} \int_{2\varepsilon_F}^{\infty} dE' \, \frac{\mathrm{Im} \, \langle km_\alpha m_{\alpha'} | \Delta G(\boldsymbol{K}, E') | \boldsymbol{k}' m_\beta m_{\beta'} \rangle}{E - E' + i\eta}
$$
$$
\equiv \langle km_\alpha m_{\alpha'} | V | \boldsymbol{k}' m_\beta m_{\beta'} \rangle + \langle km_\alpha m_{\alpha'} | \Delta G_\downarrow(\boldsymbol{K}, E) | \boldsymbol{k}' m_\beta m_{\beta'} \rangle \,,
$$

where we note that the imaginary part of the G-matrix only exists above $2\varepsilon_F$. The self-energy from $\Delta G_\downarrow$, can be calculated by applying Eq. (16.48). Since $\Delta G_\downarrow$ has poles in the lower half plane, only the hole part of the sp propagator yields a nonvanishing contribution. Taking a mean-field sp propagator, an equivalent result to the first term of Eq. (16.52) is obtained. The combination with the HF contribution, then gives the BHF self-energy

$$
\Sigma_{BHF}(k; E)) = \int \frac{d^3k'}{(2\pi)^3} \frac{1}{\nu} \sum_{m_\alpha m_{\alpha'}} \theta(k_F - k') \tag{16.60}
$$
$$
\times \langle \tfrac{1}{2}(\boldsymbol{k} - \boldsymbol{k}') m_\alpha m_{\alpha'} | G(\boldsymbol{k} + \boldsymbol{k}'; E + \varepsilon(\boldsymbol{k}')) | \tfrac{1}{2}(\boldsymbol{k} - \boldsymbol{k}') \, m_\alpha m_{\alpha'} \rangle \,.
$$

Using this self-energy in the Dyson equation for $k < k_F$ only produces solutions at the energies given by

$$
\varepsilon_{BHF}(k) = \frac{\hbar^2 k^2}{2m} + \Sigma_{BHF}(k; \varepsilon_{BHF}(k)), \tag{16.61}
$$

since this self-energy is real for energies less than ε_F. The sp strength at this energy requires application of Eq. (16.57). It is less than 1 and points out a difficulty inherent in an approach that includes only pp propagation, as in the G-matrix. Since the number of particles, with $k < k_F$, will not correspond to the original density, there is a violation of particle number. This can be understood from the lack of self-energy terms, which corresponds to 2h1p intermediate states that can put strength for $k > k_F$ below the Fermi energy. These contributions are missing, since hh propagation is excluded in the Bethe–Goldstone equation. One can avoid this problem by only determining a self-consistent, HF-like, sp energy, hence BHF.

A critical point in the BHF approach is encountered, when the choice of the auxiliary potential U is contemplated. Such a choice is necessary and relevant, since the final outcome will depend on this selection. A convenient (standard) choice that has often been made in the past, is given by

$$U_s(k) = \Sigma_{BHF}(k; \varepsilon_{BHF}(k)) \tag{16.62}$$

for $k < k_F$, without any self-energy contributions for $k > k_F$, *i.e.* $U_s(k) = 0$. This is a practical choice, as the Bethe–Goldstone equation needs only to be solved once for a set of energies below $2\varepsilon_F = \hbar^2 k^2/m$. One then still needs to find the self-consistent solutions to Eq. (16.61). An alternative choice of U is the so-called continuous choice where

$$U_c(k) = \Sigma_{BHF}(k; \varepsilon_{BHF}(k)) \tag{16.63}$$

for all values of k. This requires a more involved iteration scheme since the Bethe–Goldstone equation must be recalculated, as knowledge of sp energies for wave vectors above k_F is required.

For either choice, one ends up with a sp propagator of the form

$$G^{BHF}(k; E) = \frac{\theta(k - k_F)}{E - \varepsilon_{BHF}(k) + i\eta} + \frac{\theta(k_F - k)}{E - \varepsilon_{BHF}(k) - i\eta}. \tag{16.64}$$

Using this propagator in Eq. (16.10) then yields for the energy per particle

$$\frac{E_0^A}{A} = \frac{\nu}{2\rho} \int \frac{d^3k}{(2\pi)^3} \left(\frac{\hbar^2 k^2}{2m} + \varepsilon_{BHF} \right) \theta(k - k_F). \tag{16.65}$$

By applying Eq. (16.60), one can rewrite this result as

$$\frac{E_0^A}{A} = \frac{4}{\rho} \int \frac{d^3k}{(2\pi)^3} \frac{\hbar^2 k^2}{2m} + \frac{1}{2\rho} \int \frac{d^3k}{(2\pi)^3} \int \frac{d^3k'}{(2\pi)^3} \sum_{m_\alpha m_{\alpha'}} \theta(k_F - k)\theta(k_F - k')$$

$$\langle \tfrac{1}{2}(\boldsymbol{k} - \boldsymbol{k}') \, m_\alpha m_{\alpha'} | \, G(\boldsymbol{k} + \boldsymbol{k}'; \varepsilon_{BHF}(k) + \varepsilon_{BHF}(k')) \, | \tfrac{1}{2}(\boldsymbol{k} - \boldsymbol{k}') \, m_\alpha m_{\alpha'} \rangle. \tag{16.66}$$

The first term is simply the kinetic energy of the free Fermi gas with degeneracy 4, appropriate for nuclear matter. The second term can also be obtained from the potential energy contribution to Eq. (10.166) by replacing the bare interaction V by the G-matrix, further strengthening the HF character of this approach. The important replacement of V by the attractive G-matrix, gives a simple explanation of the self-binding of nuclei (nuclear matter), even though the bare interaction can be quite repulsive at short distances.

From a quantitative perspective, the BHF approximation produces binding that is typically within 10 MeV/A from the empirical volume term

Many-body theory exposed!

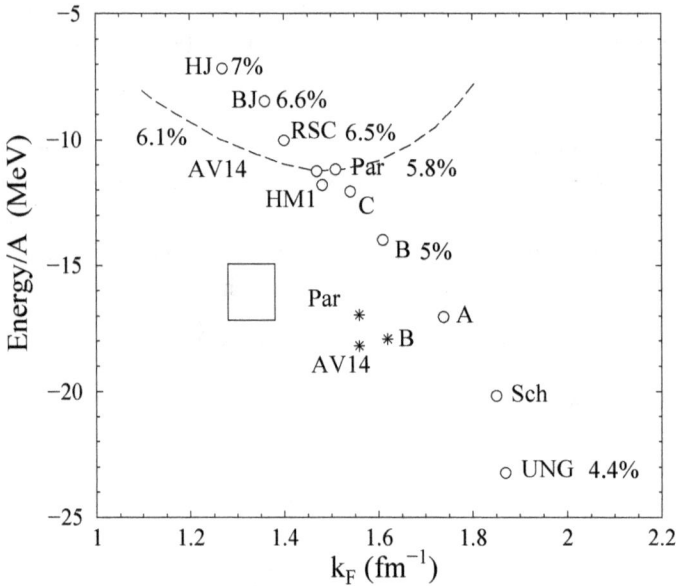

Fig. 16.22 Energy per particle as a function of the Fermi wave vector for different realistic NN interactions. The circles indicate the minima of the saturation curve for the BHF approximation when the standard choice of U is made. The symbols identify the various interactions and in some cases the D-state admixture in the deuteron wave function is given as well. The star symbols are associated with the minima of the saturation curve when three-body correlations are included. The box identifies the empirical region suggested by experimental data. For the Argonne $v14$ (AV14) interaction the saturation curve in BHF approximation is represented by the dashed line. The figure has been adapted from [Baldo (1999)].

in the mass formula in Eq. (3.38). This in itself is encouraging, since there are realistic NN interactions with a very repulsive HF contribution. In all cases, the repulsion is completely compensated by the sum of the higher-order ladder diagrams, yielding sizable binding. More problematic is that the obtained minimum as a function of density never coincides with the correct saturation point, when the amount of binding is near the empirical number. These observations are corroborated by the calculations shown in Fig. 16.22. The open circles in this figure indicate the minimum of the energy per particle for different realistic NN interactions identified by the appropriate abbreviations. For the AV14 interaction [Wiringa *et al.* (1984)] the dashed curve gives the relevant part of the complete curve. All open circles correspond to BHF calculations of the energy per particle. It was noted by [Coester *et al.* (1970)] that the minima of the BHF calculations roughly lie on a band, now called the Coester band, that does not intersect with the empirical point. Motion on this band is mainly governed by the D-state probability of the deuteron that is generated by the interaction. One finds that interactions with very strong tensor components (high D-state probability), yield low saturation densities and underbind while weak tensor forces yield high densities with overbinding. Some interactions in the figure are also labeled by this D-state probability.

From the information collected in Fig. 16.22, one may draw the conclusion that the BHF with the standard choice for U, given in Eq. (16.62) will not yield the correct empirical result. At the end of the sixties [Bäckman *et al.* (1969)] and early seventies [Pandharipande and Wiringa (1979)] several variational calculations had already indicated that it was possible to get different saturation properties for the same interaction in nuclear matter. This discrepancy generated substantial interest in studying the convergence properties of the Brueckner approach. Since most interest was focused on the energy per particle and no nucleon knockout experiments had been performed of sufficient quality to experimentally identify the limits of the mean-field approach, an expansion was developed that identifies the BHF approximation as it lowest-order contribution. This so-called hole-line expansion (sometimes referred to as Brueckner–Bethe method) emphasizes the analysis of Goldstone (time-ordered) diagrams to the energy per particle. This method first involves replacing in all diagrams every V by the well-behaved G-matrix interaction. With the choice of Eq. (16.62) one then obtains the two hole-line contribution to the energy per particle in the form of Eq. (16.66). This contribution contains the integration over two independent hole lines. Phase space arguments then suggest that all higher-order contributions involving more independent hole integrations should be ordered according to the number of independent hole lines [Brandow (1966); Day (1967)]. A similar argument was used by [Galitskii (1958)] when considering the low-density limit for strongly repulsive interactions. Indeed, an extra integration in higher order in the G-matrix over particle coordinates does not appear to give rise to a "smaller" contribution. If the density is small, however, it may be argued that the hole phase space is small and that ordering contributions in terms of the number of independent hole lines makes sense. The next contribution in the expansion includes all terms which have three independent hole integrations (the three hole-line contribution). It contains the summation of all diagrams in which three particles interact with each other any number of times by means of the G-matrix interaction [Rajamaran and Bethe (1967)]. This requires a special summation, first developed by [Faddeev (1961)] for the three-body problem. In the present case, two particles cannot repeat a G-matrix interaction since that would involve a double counting of terms. The so-called Bethe–Faddeev summation was first accomplished by [Day (1981)] for realistic interactions.

Including all three hole-line terms, moves the saturation points away from the Coester band towards the empirical region, as illustrated in Fig. 16.22 by the stars for three different realistic interactions [Baldo (1999)]. The remaining discrepancy with empirical data is still substantial, however. Nevertheless, it became clear in 1985 by the work of [Day and Wiringa (1985)] that good agreement between three-hole-line calculations and advanced variational methods for the same interactions was possible. The conclusion was further strengthened by the work of [Song *et al.* (1998)] which showed that calculations employing the continuous choice for the auxiliary potential U given by Eq. (16.63), agree extremely well with the standard choice when three-hole-line terms are included. Moreover, the continuous choice advocated by [Jeukenne *et al.* (1976)] is then already converged at the two-hole-line level. The conclusion is therefore appropriate that it is possible to obtain convergence for the energy per particle for a given realistic interaction. This observation does not say much about the quality of such calculations for other quantities like the spectral functions and occupation number, which are easier to generate using *e.g.* the Green's function method, as shown in Secs. 16.3.2 and 16.3.3.

So the nuclear saturation problem remains! Several remedies have been proposed over the years and we will now consider some of their features. The first is closely associated with the presence of excited states of the nucleon, in particular the Δ-isobar, discussed in Ch. 14. Its importance suggests that it may be necessary to include it on the same footing as the nucleon. The disadvantage of this strategy is that a lot of information is required about the interaction between nucleons and Δ-isobars for which few experimental constraints are available. An alternative strategy is to represent the influence of Δ-isobars and other nucleonic excitations by including three- and perhaps higher-body interactions. The occurrence of such interactions is inevitable if one restricts the quantum Hilbert space to nucleons. This is illustrated by considering the Δ-isobar on the same footing as the nucleon, as was done in Ch. 14. In Fig. 16.23 three nucleons initially propagate in the medium. Two of these nucleons may exchange

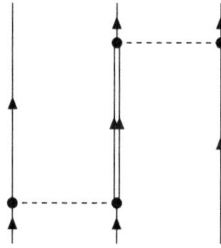

Fig. 16.23   Diagram illustrating the occurrence of three-body forces when the effects of Δ-isobar degrees of freedom are taken into account in the Hilbert space of only nucleons.

a pion which excites one of them to a Δ-isobar. After another interaction this Δ-isobar is de-excited returning to a three-nucleon situation. Only two-body mechanisms are involved in such a process, but if one restricts the Hilbert space to nucleons, it is possible to take the process of Fig. 16.23 into account by the introduction of a three-body interaction between nucleons [Fujita and Miyazawa (1957)]. More elaborate versions of this type of three-body force [Coon *et al.* (1979)] yield attractive contributions to the energy per particle [Carlson *et al.* (1983)]. Since light nuclei require additional binding beyond the contribution of two-body interactions, such three-body forces help in getting better agreement for light nuclei. In nuclear matter, however, the situation is more complicated since the Coester band properties suggest that a repulsive mechanism is needed to generate lower saturation densities that are in accord with the empirical results. For this reason an additional phenomenological repulsive three-body interaction was introduced by [Carlson *et al.* (1983)], which was then adjusted to force the correct saturation properties of nuclear matter, while also fitting the binding of light nuclei. The procedure yields an improved Hamiltonian for nuclei but gives up on a deeper insight into the saturation mechanism of nuclear matter, especially since the origin of the effective repulsion is somewhat unclear.

An alternative solution to the saturation problem has been proposed, which includes aspects of the effects of relativity [Anastasio *et al.* (1983)]. A detailed discussion is beyond the scope of the book and only a few comments will be given here for completeness. By employing a straightforward adaptation of the BHF approach, the so-called Dirac–BHF (DBHF) method gives reasonable saturation properties for nuclear matter [ter Haar and Malfliet (1986); Brockmann and Machleidt (1990); Amorim and Tjon (1992)]. The main physical effect appears to be the change of the coupling of the so-called σ-meson to the nucleons in the medium. This scalar, isoscalar meson represents to a large extent the physical exchange between nucleons of two interacting pions, coupled to zero angular momentum and isospin. Since the actual form of the scalar coupling of the σ-meson to the nucleons is essential in obtaining the saturation mechanism, it is unclear to what extent it represents the two pion-exchange processes in the medium. An additional difficulty is the necessity to deal with the properties of antiparticles and the so-called Dirac sea. Occupation numbers in this approach tend to be higher [de Jong and Lenske (1996)] than the ones discussed in the previous section, which are in better agreement with the experimental data, discussed in Sec. 7.8. A positive feature of relativistic approaches is the correct strength of the spin-orbit interaction in nuclei when the exchange of the isoscalar vector ω-meson is combined with the σ-meson terms. Further study of higher-order (three hole-line) contributions have so far not taken place to assess the convergence properties of the scheme.

Experimental data of $(e, e'p)$ reactions have stimulated yet another perspective that may be of relevance for the saturation properties of nuclear matter. A recent analysis of the $(e, e'p)$ reaction on ^{208}Pb in a wide range of missing energies (up to 100 MeV) and for missing momenta below 270 MeV/c produces information on the occupation numbers of all the deeply-bound proton orbitals, as discussed in Sec. 7.8. The data indicate that all the mean-field orbitals are depleted by the same amount of about 15% [van Batenburg (2001)]. The properties of the occupation numbers suggest that the main reason for the global depletion of these mean-field orbitals is the presence of short-range correlations, in quantitative agreement with the nuclear-matter results for occupation numbers, discussed in Secs. 16.3.2 and 16.3.3 (see also Fig. 16.19) . As we will see in the next chapter, the effect of the coupling of hole states to low-lying collective excitations in nuclei only affects occupation numbers of states in the immediate vicinity of the Fermi energy. A characteristic feature of the short-range correlations is that the depletion of the mean-field sp strength must be compensated by the admixture of a corresponding number of particles with high-momentum components. These high-momentum components have been discussed in Sec. 16.3.3 (see also Sec. 17.3). Efforts are under way to study them experimentally [Rohe et al. (2004)]. The admixture of high-momentum components must take place at energies far below the Fermi energy. The argument is based on the observation that high-momentum admixture involves self-energy terms with intermediate 2h1p states. Two holes on average will have a small total momentum leaving the intermediate particle state to compensate the external high-momentum of the self-energy. One expects these components to be centered around an energy corresponding to $2\langle\varepsilon_h\rangle - \varepsilon_p$. This expectation is completely met by the results discussed in Secs. 16.3.2 and 16.3.3. We note that the presence of high-momentum components is crucial in determining the energy per particle according to Eq. (16.10) and shown explicitly in Fig. 16.21. Results for the momentum distribution and true potential energy based on the spectral function show that enhancements as large as 200% for the kinetic and potential energy over the mean-field values can be obtained for nuclear matter [Vonderfecht *et al.* (1993)]. These large attractive contributions to the potential energy of nuclear matter are mainly from weighting the high-momentum components in the spectral function with large negative energies in Eq. (16.10). The location of high-momentum components as a function of energy is thus an important ingredient in the determination of the energy per particle as a function of density. So far, the determination of this location has relied only on quasiparticle properties in the construction of the self-energy. A self-consistent determination of the spectral function including the location of the high-momentum strength includes the dominant physics of short-range correlations (see Sec. 16.3.3). It also appears consistent with the experimental observations of the nucleon spectral function in nuclei discussed in Ch. 17 and Sec. 7.8.

We now present an argument showing that short-range correlations are the dominant factor in determining the empirical saturation density of nuclear matter. We recall that elastic electron scattering from ^{208}Pb [Frois *et al.* (1977)] accurately determines the value of the central charge density. By multiplying this number by A/Z one obtains the relevant central density of heavy nuclei, corresponding to 0.16 nucleons/fm^3 or $k_F = 1.33$ fm^{-1} (see also Sec. 3.3.1). Since the presence of nucleons at the center of a heavy nucleus is confined to s-wave nucleons, and, as argued above, their depletion is dominated by short-range correlations, one may conclude that the same is true for the actual value of the empirical saturation density of nuclear matter. While this argument is particularly appropriate for the deeply bound $0s_{\frac{1}{2}}$ and $1s_{\frac{1}{2}}$ protons, it continues to hold to a large extent for the $2s_{\frac{1}{2}}$ protons which are depleted predominantly by short-range effects (up to 15%) and by at most 10% due to long-range correlations [Sick and de Witt Huberts (1991)]. These considerations demonstrate clearly that one may expect short-range correlations to have a decisive influence on the actual value of the nuclear-matter saturation density.

The implementation of the self-consistent treatment on the spectral functions is a numerically difficult problem, especially for interactions with strong repulsive cores. A continuous scheme was employed for the calculations presented in Sec. 16.3.3. In a discrete scheme a representation of the propagator in terms of three poles [Dewulf *et al.* (2002)] was used, avoiding a full continuum solution of the ladder equation. The latter approach is equivalent to a continuous version as far as the energy per particle is concerned, since it requires a reproduction of the relevant energy-weighted moments of the hole and particle spectral function. This is substantiated by comparing the outcome of the discrete scheme with the continuous self-consistency scheme [Bożek and Czerski (2001)], employing the same NN interaction.

In Fig. 16.24 the saturation points obtained within the discrete scheme of [Dewulf *et al.* (2002)] for the updated Reid potential (Reid93), the NijmI and NijmII interaction [Stoks *et al.* (1994)] and the separable Paris interaction [Haidenbaur and Plessas (1984)] are shown [Dewulf *et al.* (2003)]. The results demonstrate an important and systematic change of the saturation properties with respect to continuous choice Brueckner–Hartree–Fock (ccBHF) calculations, leading to about 4-6 MeV less binding, and reduced values of the saturation density, closer to the empirical one. The binding energy is also shown for the continuous scheme of Sec. 16.3.3, at two densities ($k_F = 1.33$ and 1.45 fm^{-1}); the error bars are an estimate of the remaining uncertainty due to incomplete convergence and the non-self-consistent treatment of some higher order partial waves [Roth Stoddard (2000)]. A substantial shift in the saturation density for the Reid68 potential, from the ccBHF value of about 1.6 fm^{-1}, to a value below 1.45 fm^{-1}, is observed without seriously underbinding nuclear matter.

The present self-consistent treatment of short-range correlations differs in two main aspects from the ccBHF approach, the latter giving saturation properties essentially equivalent to converged three hole-line calculations [Song *et al.* (1998)]. Firstly, hole and particle lines are treated on an equal footing which is important for thermodynamic consistency. Intermediate hh propagation in the ladder diagrams is included to all orders. This feature provides, compared to ccBHF, a substantial repulsive effect in the $k < k_F$ contribution to Eq. (16.10), and comes primarily from an upward shift of the quasiparticle energy spectrum from the inclusion of $E < \varepsilon_F$ contributions to the imaginary part of the self-energy. The effect increases with density, and is the dominant factor in the observed shift of the saturation point. Secondly, the realistic spectral functions, generated

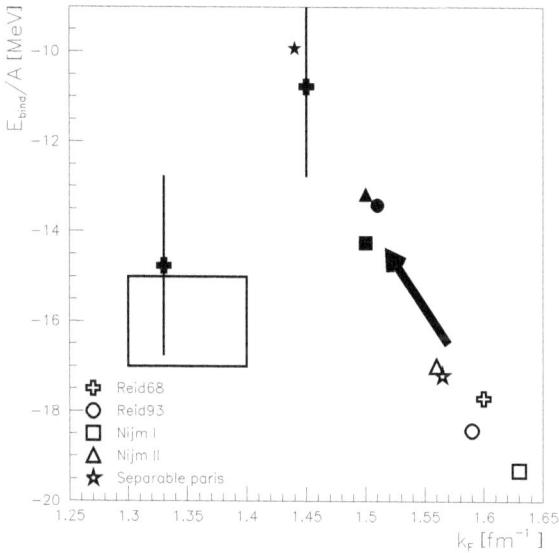

Fig. 16.24 Nuclear matter saturation points calculated with various realistic NN interactions. The open symbols refer to the continuous choice Brueckner–Hartree–Fock method. The filled symbols refer to self-consistent results and represent saturation points calculated in the discrete scheme, except for the old Reid (Reid68) interaction where the binding energy at two densities is shown in the continuous scheme.

from the self-consistent procedure outlined in Sec. 16.3.3 and used in the evaluation of the in-medium interaction Γ and self-energy Σ, are in agreement with experimental information obtained from $(e, e'p)$ reactions. For the Reid93 interaction at $k_F = 1.37$ fm^{-1} $Z_F = 0.74$ is found for the quasiparticle strength at the Fermi wave vector, whereas the hole strength for $k = 0$, integrated up to 100 MeV missing energy, equals 83%; similar values are found for the other interactions and confirmed by the calculations, discussed in the previous section for the Reid68 potential. The depletion of the quasiparticle peaks is primarily important to suppress unrealistically large pairing instabilities around normal density. The improved treatment of the high wave-vector components does affect the binding energy, through the $k > k_F$ contribution to Eq. (16.10). This feature, studied in [Dewulf et al. (2002)], provides a sizable attraction, but is smaller than the aforementioned repulsive effect. Self-consistent calculations, including pp and hh propagation, also yield much better results for the Hugenholtz–van Hove theorem [see Eq. (16.15)] than BHF numbers.

The above information indicates that a sophisticated treatment of short-range correlations lowers the ccBHF saturation densities, bringing them closer to the empirical one. It remains to be understood why apparently converged hole-line calculations [Song et al. (1998)] yield higher saturation densities. The three hole-line terms obtained in [Song et al. (1998)] seem to imply reasonable convergence properties compared to the two hole-line contribution. One may therefore assume that these calculations provide an accurate representation of the energy per particle of nuclear matter, as a function of density for nonrelativistic nucleons and two-body forces. At this point it is important to identify an underlying assumption when the nuclear-matter problem is posed. It asserts that the influence of long-range correlations in finite nuclei and nuclear matter are commensurate. It has been suggested in [Dewulf et al. (2003)] that this underlying assumption is questionable. The argument is based on the special properties of the long-range correlations associated with pion-exchange interactions as presented in Ch. 14. These attractive long-range correlations contribute at transferred wave vectors of about 1 to 2 fm^{-1}, as illustrated in Fig. 14.7. Ring diagram summations of attractive interactions yield a coherent sum (all terms are attractive), and can be calculated by expressions similar to Eq. (14.67) for the correlation energy of the electron gas. For interactions different from the Coulomb one, the integral over Q generates no contribution for small values on account of the $dQ\ Q^2$ term. The pion-exchange terms do not suffer this fate, since they occur at finite Q, and are amplified due to momentum conservation in an infinite system. It is unclear whether these terms actually generate a similar physical consequence in finite nuclei, where momentum is not a good quantum number. The nuclear matter response with pion quantum numbers, discussed in Sec. 14.5, indicates that low-lying strength should accumulate. The lack of such collective states in nuclei with pionic quantum numbers suggests that the relation between nuclear matter and finite nuclei, at least for these degrees of freedom, is far from trivial. Experimental data exhibit no enhanced response of the pion channel over that of the rho [Carey et al. (1984); Taddeucci et al. (1994); Wakasa et al. (1999)], as demanded by nuclear matter calculations (see Sec. 14.5). Given this inconsistency, it is not certain that binding-energy contributions of long-range pion-exchange terms play the same role in finite nuclei. It may therefore be necessary to excise them from nuclear-matter calculations to establish contact with finite nuclei. In other words, the original nuclear matter problem may have been ill-posed and only the effects of short-range correlations should be employed to connect the infinite with the finite system. For more details we refer to [Dewulf *et al.* (2003); Dickhoff and Barbieri (2004)].

The present overview by no means exhausts all the contributions that have been made to the subject of the nuclear-matter saturation problem. We have focused on giving an overview of some of the important historical developments and included a somewhat more in-depth discussion of those features that have come to light when using the Green's function method.

16.4 Exercises

(1) Evaluate the derivative of the real part of the polarization propagator, as required in Eq. (16.21).
(2) Determine the real part of Π^f for $r_s = 2$, based on its imaginary part, as plotted in Fig. 16.10. A simple fit to the imaginary part can be made to facilitate the numerical application of Eq. (16.33). A scaled version of the noninteracting term should adequately represent the correlated polarization propagator in the usual domain of the ph continuum. For larger energies, the remaining strength can *e.g.* be represented by a broad gaussian. Determine the polarization propagator and compare with the calculation, which is based on employing $\Pi^{(0)}(q, E)$.
(3) Verify Eq. (16.46) by performing the steps suggested in the text.
(4) The product of step functions in Eq. (16.58) is sometimes referred to as the Pauli operator Q. For practical calculations it is necessary to eliminate the dependence of Q on the angle between $\boldsymbol{K}$ and $\boldsymbol{q}$ so that a partial wave decomposition of Eq. (16.58) can be applied. This is accomplished by constructing the so-called angle-averaged Pauli operator

$$\bar{Q}(K, q) = \frac{1}{4\pi} \int d\Omega_q \, \theta(|\boldsymbol{K}/2 + \boldsymbol{q}| - k_F)\theta(|\boldsymbol{K}/2 - \boldsymbol{q}| - k_F). \quad (16.67)$$

Evaluate $\bar{Q}$ for values of $|\boldsymbol{K}|$ larger and smaller than $2k_F$. Distinguish also for different regions of $|\boldsymbol{q}|$.

Chapter 17

Dynamical treatment of the self-energy in finite systems

Some aspects of the sp propagator in a finite system have been presented in Ch. 11. The second-order approximation to the self-energy, with its self-consistent implementation, provides a good understanding of important observables of atoms. In Sec. 13.5, we identified the main missing ingredients in describing the spectral functions of atoms. In the present chapter we concentrate on the spectral functions of nuclei.

Some properties of light nuclei can nowadays be calculated in an exact manner with different techniques starting from a realistic NN interaction. An example is the application of several methods to the calculation of the ground-state energy of ^{4}He, reported in [Kamada et al. (2001)]. The low-lying states of nuclei up to $A = 10$ [Pieper et al. (2002)] can be described with the Green's function Monte Carlo method [Pieper et al. (2001)]. Calculations using the no-core shell-model approach, have yielded results for ^{12}C [Navrátil et al. (2000)]. This body of work is able to explain many aspects of the low-energy spectra of light nuclei, starting from a realistic NN interaction. Many details are further improved by including a three-body interaction between the nucleons. In all cases studied so far, the calculated energy of the ground state is always above the experimental number when only two-body interactions are included: a clear indication for the need of an overall attractive three-body force. While great advances in this area are reported, it should be pointed out that the reliance on large scale computational efforts, precludes a detailed understanding of the physics that makes nuclei "tick". Moreover, the difficulty of accounting for some of the low-lying collective states in ^{12}C [Navrátil et al. (2000)] points to the necessity for a better understanding of the physics ingredients that play a role at low energy in these nuclei. A good perspective on the relevant physical ingredients can be gained from the Green's function approach, which can be quite easily applied to heavier nuclei as well. Such insights will be developed for the sp propagator in the present chapter and for excited states in Ch. 21.

Compared to atoms, the explanation of the experimental data for nuclei, related to the sp propagator, is considerably less straightforward. Experimental data were presented in Sec. 7.8 and some qualitative features were touched on in Sec. 11.5.2. When starting from a realistic NN interaction, one can only arrive at meaningful results if short-range correlations are properly treated. The correct diagrammatic procedure was given in Ch. 16 for nuclear matter. The self-energy in a finite nucleus should therefore also include the set of ladder diagrams, with repeated interactions between parallel propagators. A completely self-consistent solution of this technically difficult problem is not yet available. We will examine various intermediate steps developed so far.

The low-energy spectral strength will be explored for "closed-shell" nuclei in Sec. 17.1. The influence of collective low-lying excitations is clarified by employing various descriptions of the intermediate states in the self-energy. The second-order method with the interaction given by the G-matrix, ensuring a proper treatment of short-range correlations, is the simplest one. Collective effects at low energy can be described by including the interaction between the intermediate 2p1h (1p2h) states in the self-energy. In Sec. 17.2, an illustrative study of open-shell systems will be presented. For nuclei with either closed proton or neutron shells, the open-shell nucleons of the other type exhibit properties that are very reminiscent of the pairing properties of superconducting or superfluid systems. The self-consistent inclusion of ladder diagrams with emphasis on their low-energy properties, can describe many of these. We will transform pphh propagation in the finite system to an RPA-like eigenvalue problem. In turn, the resulting self-energy and the corresponding solution of the Dyson equation will exhibit interesting new features that are not encountered in closed-shell systems. The interaction employed in the analysis is quite schematic, since other features associated with the sp strength are expected to exhibit many similarities with closed-shell systems.

The consequences of short-range correlations in finite systems are quite similar to those in infinite systems. Their calculation is however, technically much more difficult. We will explain some of these issues and the results of available Green's function calculations in Sec. 17.3. The chapter is concluded with a brief overview of the emerging understanding of the properties of protons in nuclei in Sec. 17.4. The summary is based both on the available experimental data associated with the $(e, e'p)$ reaction and theoretical calculations employing the Green's function method.

17.1 Influence of collective excitations at low energy

17.1.1 *Second-order effects with G-matrix interactions*

To understand the influence of the low-lying excitations, which are responsible for the observed fragmentation of the sp strength in nuclei, we employ a two-step procedure. First is the treatment of short-range correlations by the summation of ladder diagrams. The inclusion of low-energy excitations in the self-energy is expected to require only the consideration of sp orbits in the immediate vicinity of the Fermi energy. For this reason, it is convenient to exclude those states in the construction of the G-matrix interaction. The space spanned by the latter states is referred to as the model space. The effective interaction in the model space then reads

$$\langle \alpha\beta| \, G(E) \, |\gamma\delta\rangle = \langle \alpha\beta| \, V \, |\gamma\delta\rangle$$
$$+\frac{1}{2} \sum_{\sigma\tau} \langle \alpha\beta| \, V \, |\sigma\tau\rangle \, \frac{\theta(\sigma - M)\theta(\tau - M)}{E - \varepsilon_\sigma - \varepsilon_\tau} \, \langle \sigma\tau| \, G(E) \, |\gamma\delta\rangle \,. \qquad (17.1)$$

The step functions indicate that only sp states outside the configuration space M should be included in the summation. The sp quantum numbers $\alpha - \delta$ are all inside the space M. In practical calculations, it has so far been impossible to include self-energy contributions to the intermediate states. Although the energy dependence of the G-matrix is significant for the calculation of the depletion of the mostly occupied orbits, as discussed in Sec. 17.3, the early calculations of the self-energy did not treat it. Further simplifications were introduced by adopting a G-matrix, originally calculated in nuclear matter. By extracting its operator structure [Dickhoff (1983)], it is possible to employ this effective interaction inside the space M. It should be confined to reasonable excitation energies in the vicinity of the Fermi energy, [Brand *et al.* (1991)] in order to avoid the double-counting that occurs, when M becomes too large.[1] In addition, the restriction of the space M allows to avoid the treatment of the weak energy dependence of the G-matrix.

---

[1] The precise effect of the Pauli principle in the finite system is not considered in this procedure. Hence, it is wise to include only those states, which are not well described in nuclear matter, *i.e.* those near the Fermi energy, since the G-matrix is calculated with a gap in the sp spectrum according to Eq. (16.62).

The second step in the model-space approach employs the G-matrix interaction in the second-order self-energy

$$\Sigma^{(2)}(\gamma, \delta; E) = \frac{1}{2} \sum_{\epsilon \mu \nu} \langle \gamma \mu | G | \epsilon \nu \rangle \langle \epsilon \nu | G | \delta \mu \rangle \qquad (17.2)$$

$$\times \left\{ \frac{\theta(F - \epsilon)\theta(F - \nu)\theta(\mu - F)}{E - (\varepsilon_\epsilon + \varepsilon_\nu - \varepsilon_\mu) - i\eta} + \frac{\theta(\epsilon - F)\theta(\nu - F)\theta(F - \mu)}{E - (\varepsilon_\epsilon + \varepsilon_\nu - \varepsilon_\mu) + i\eta} \right\},$$

where the summation over the sp quantum numbers is restricted to the space M. A harmonic oscillator basis can be employed to approximate the mean-field sp propagators. The sp energies can be chosen such that, upon solution of the Dyson equation, they generate the experimental energies associated with the largest spectroscopic factors. The procedure should only be used for states in the immediate vicinity of the Fermi energy. For others a simple harmonic oscillator spacing with an oscillator length appropriate for the particular nucleus is adequate. Such a calculation solely fixes the position of the major fragments. Their strength and the position and strength of the other fragments are predictions.

The qualitative features of the solutions of the Dyson equation with an energy-dependent self-energy, have been discussed in Ch. 11. Applications of the self-energy in Eq. (17.2) will be presented for ^{48}Ca. In Fig. 17.1 the sp strength distribution for proton removal from ^{48}Ca in $\ell = 2$ states is compared to the available experimental data [Kramer (1990)]. The theoretical results are labeled by "Second Order" in the figure. The normalization is such that if a single state carried all the strength, the peak height would be $2j + 1$. The first peak describes $d_{\frac{3}{2}}$ removal, whereas most of the other fragments represent $d_{\frac{5}{2}}$ strength. The difference between the fragmentation of these two sp orbitals characterizes the basic features found in experiment. For an orbital which is very near the Fermi energy, like the $d_{\frac{3}{2}}$, fragmentation of strength to all 2p1h and 1p2h-like states occurs such that the distribution contains a single large fragment, and hundreds of tiny contributions spread out over the whole energy domain covered by the 2p1h and 1p2h states. The result is related to the position of the $d_{\frac{3}{2}}$ sp energy, which lies in between the 2p1h and 1p2h states with large energy denominators in Eq. (17.2), and yields a large spectroscopic factor, according to the discussion of Sec. 11.3.1 (the derivative in Eq. (11.23) is small). In the example of the second-order self-energy calculation, the theoretical strength in Fig. 17.1 for the $d_{\frac{3}{2}}$ peak is 0.73 (peak height divided by $2j + 1$), whereas experimentally it is 0.56. The total occupation number in the calculation

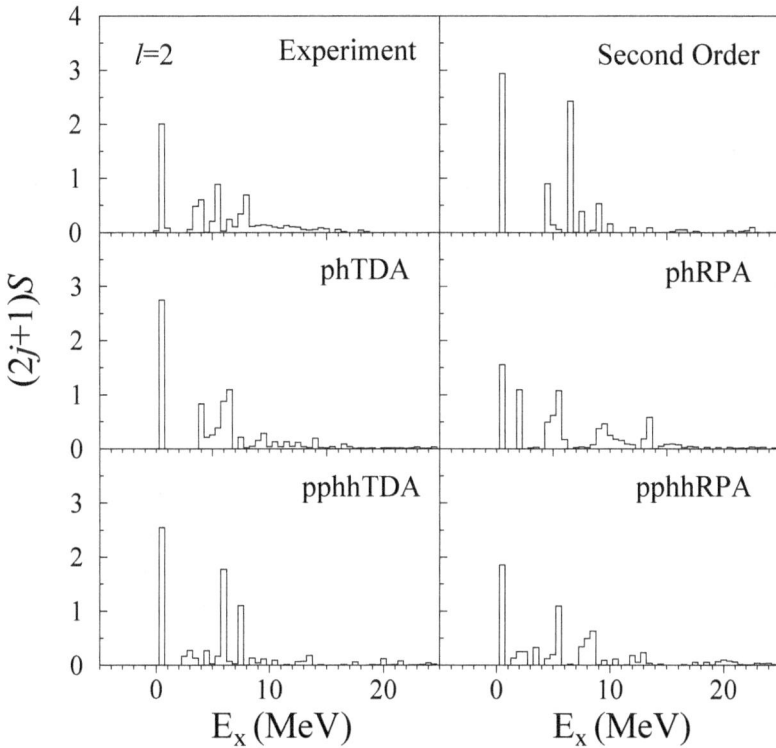

Fig. 17.1 Distribution of sp strength for $\ell = 2$ proton removal from ^{48}Ca. Experimental data are from [Kramer (1990)]. Theoretical results are displayed for various approximations to the self-energy, as discussed in the text.

is 0.89 with 0.80 residing in the experimentally accessible domain. This implies the presence of a tail of about 10% of sp strength at higher missing energy [Brand *et al.* (1991)].

For an orbital, which in mean field starts to overlap with the domain of 1p2h states, like the $d\frac{5}{2}$ state, the situation is already substantially different. The energy denominators in Eq. (17.2) are now smaller for the eigenvalues near the $d\frac{5}{2}$ sp energy. The derivative of the self-energy is typically large, resulting in reduced strengths. Consequently, the $d\frac{5}{2}$ orbital in ^{48}Ca is strongly fragmented, which is in accord with experiment. More deeply bound sp orbitals acquire even more fragmentation and are correspondingly more spread in energy [Brand *et al.* (1991)]. The calculations of the strength distribution, using the Dyson equation for a finite nucleus, are thus capable of explaining many of the qualitative features in the strength distributions that are observed in the $(e, e'p)$ reaction.

To understand the implications of the quantitative comparison of theory and experiment, it should be realized that the calculations of [Brand *et al.* (1991)] used a sp space, which included all shells below the Fermi energy and three major shells above. In this space all sp strength is assumed to reside. Consequently, the spread of sp strength resulting from solving the Dyson equation corresponds roughly to one hundred MeV above and below the Fermi energy. Note that the effect of short-range correlations at the sp level is not properly taken into account, since the coupling of the sp motion is truncated at about 100 MeV above the Fermi energy, reflecting the size of the model space. As a general conclusion of this work on ^{48}Ca, it is found that the total strength in the experimentally accessible domain in the $(e, e'p)$ reaction is overestimated by about 10–15%, while the shape of the strength distribution is already reasonably described. The background contribution to the hole strength, *i.e.* the strength which is not contained in the area of the main peak, is of the order of 10%. This is in agreement with the experimental data for the $2s\frac{1}{2}$ in ^{208}Pb, discussed in Sec. 7.8.

17.1.2 *Inclusion of collective excitations in the self-energy*

The comparison between the second-order results and experiment in Fig. 17.1 also shows that stronger fragmentation in the theoretical calculation, at low energy, is required. This can be achieved by improving the description of the intermediate 2p1h and 1p2h states in the self-energy [Rijsdijk *et al.* (1992)]. The second-order expression for the self-energy [Eq. (17.2)] can also be graphically represented by the time-ordered diagrams in Figs. 17.2a) and b). Here the lines represent the mean-field propagators discussed above. We note that additional interactions between the individual lines, which may produce collective pp, ph, or hh pairs, are missing.

An improved description of these intermediate states can thus be attempted by including the interactions between the three intermediate lines. Examples are given by the diagrams in Fig. 17.2c), d) and 17.2e), f). The simplest description of diagrams 17.2c) and d) allows the interaction between the two particles, or two holes, to act any number of times without changing the direction of the propagators. This series is summed by the TDA for the pp and hh propagator separately. For the pp propagator one

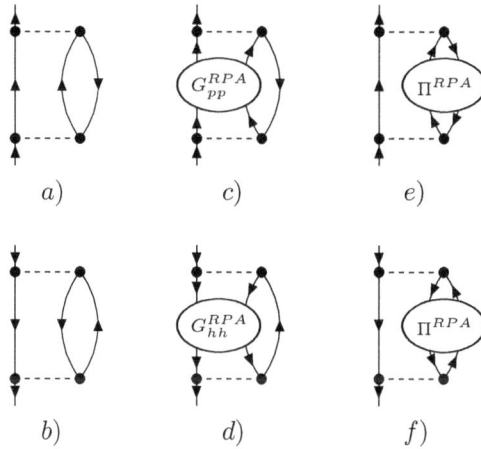

Fig. 17.2   Second-order self-energy terms represented by time-ordered diagrams $a)$ and $b)$. Extensions are made by including TDA or RPA correlations in the pp (hh) propagator (diagrams $c)$ and $d)$) or ph (hp) channel (diagrams $e)$ and $f)$). For nuclear systems, the G-matrix inside the model space replaces the dashed lines.

obtains

$$G_{pp}^{TDA}(\alpha, \beta; \gamma, \delta; E) = \sum_n \frac{T_{\alpha\beta}^{A+2,n*} T_{\gamma\delta}^{A+2,n}}{E - (E_n^{A+2} - E_0^A) + i\eta} \qquad \alpha, \beta, \gamma, \delta > F$$

(17.3)

and for the hh propagator

$$G_{hh}^{TDA}(\alpha, \beta; \gamma, \delta; E) = -\sum_m \frac{T_{\beta\alpha}^{A-2,m} T_{\delta\gamma}^{A-2,m*}}{E - (E_0^A - E_m^{A-2}) - i\eta} \qquad \alpha, \beta, \gamma, \delta < F,$$

(17.4)

according to one of the exercises at the end of the chapter. The energies $E_n^{A\pm2} - E_0^A$ are the solutions of a diagonalization of the interaction within the space of two-particle or two-hole states respectively. The coefficients $T_{\alpha\beta}^{A\pm2,n}$ are the components of the corresponding n-th eigenvector. For the interaction in this pp(hh)TDA calculation the same G-matrix interaction was adopted as in the calculation with self-energy, given by Eq. (17.2). Such an approximation should provide a reasonable estimate of the pp(hh) correlation effects in the self-energy. By replacing the noninteracting pp(hh) propagators by the pp(hh)TDA ones of Eqs. (17.3) and (17.4), one generates for the diagrams c) and d) of Fig. 17.2, the combined expression

$$\Sigma_{pphh}^{TDA}(\alpha, \beta; E) = \frac{1}{2} \left\{ \sum_{\kappa < F, n} \frac{\Gamma_{\alpha\kappa}^{A+2,n*} \Gamma_{\beta\kappa}^{A+2,n}}{E - ((E_n^{A+2} - E_0^A) - \varepsilon_\kappa) + i\eta} \right.$$

$$\left. + \sum_{\kappa > F, m} \frac{\Gamma_{\alpha\kappa}^{A-2,m} \Gamma_{\beta\kappa}^{A-2,m*}}{\omega - ((E_0^A - E_m^{A-2}) - \varepsilon_\kappa) - i\eta} \right\}, \quad (17.5)$$

with

$$\Gamma_{\alpha\kappa}^{A+2,n} = \sum_{\mu > F, \nu > F} \langle \alpha\kappa | G | \mu\nu \rangle T_{\mu\nu}^{A+2,n} \quad (17.6)$$

and

$$\Gamma_{\alpha\kappa}^{A-2,m} = \sum_{\mu < F, \nu < F} \langle \alpha\kappa | G | \mu\nu \rangle T_{\nu\mu}^{A-2,m}. \quad (17.7)$$

The Dyson equation is then solved with this approximation and the results are labeled by pphhTDA in Fig. 17.1.

Alternatively, one may consider the collectivity in the ph channel, by the summation of diagrams 17.2e) and f). Including only forward (backward) diagrams in e) f), the ph(hp)TDA propagators are given by

$$\Pi^{TDA}(\alpha, \beta; \gamma, \delta; E) = \sum_{n \neq 0} \frac{T_{\alpha\beta}^{A,n*} T_{\gamma\delta}^{A,n}}{E - (E_n^A - E_0^A) + i\eta} \qquad \alpha, \gamma > F; \beta, \delta < F,$$

$$(17.8)$$

and

$$\Pi^{TDA}(\alpha, \beta; \gamma, \delta; E) = -\sum_{m \neq 0} \frac{T_{\beta\alpha}^{A,m} T_{\delta\gamma}^{A,m*}}{E - (E_0^A - E_m^A) - i\eta} \qquad \alpha, \gamma < F; \beta, \delta > F.$$

$$(17.9)$$

It should be noted that three independent types of excitation can occur, *i.e.* neutron particle – proton hole, proton particle – neutron hole, and a mixture of neutron particle – neutron hole and proton particle – proton hole with isospin components $T_z = 1$, -1, and 0, respectively. In nuclei with even N and Z, low-lying collective states are observed with angular momentum and parity 2^+ and 3^-, which may be interpreted as surface vibrations of quadrople and octupole type. These $T_z = 0$ collective phonons are especially important as their coupling to the sp motion is an important source of fragmentation of spectral functions at low energies. Using the ph correlated propagators of Eqs. (17.8) and (17.9), the self-energy becomes

$$\Sigma_{ph}^{TDA}(\alpha,\beta;E) = \frac{1}{2}\left\{ \sum_{\mu>F,n\neq0} \frac{\Gamma_{\alpha\mu}^{A+,n*}\Gamma_{\beta\kappa}^{A+,n}}{E-(\varepsilon_\mu+(E_n^A-E_0^A))+i\eta} \right.$$

$$\left. + \sum_{\mu<F,m\neq0} \frac{\Gamma_{\alpha\mu}^{A-,m}\Gamma_{\beta\kappa}^{A-,m*}}{E-(\varepsilon_\mu+(E_0^A-E_m^A))-i\eta} \right\} \quad (17.10)$$

with

$$\Gamma_{\alpha\mu}^{A+,n} = \sum_{\nu>F,\kappa<F} \langle\alpha\kappa|G|\mu\nu\rangle T_{\nu\kappa}^{A,n} \quad (17.11)$$

and

$$\Gamma_{\alpha\mu}^{A-,m} = \sum_{\nu<F,\kappa>F} \langle\alpha\kappa|G|\mu\nu\rangle T_{\kappa\nu}^{A,m}. \quad (17.12)$$

This self-energy is subsequently used to solve the Dyson equation, and the results are labeled by phTDA in Fig. 17.1. Replacing $\Pi^{(0)}$ with Π^{TDA} or Π^{RPA} in the second-order self-energy is somewhat ambiguous when exchange diagrams are included. The factor $\frac{1}{2}$ in the second-order term reflects the pair of equivalent fermion lines in th diagram. This equivalence no longer pertains in higher order. Keeping the factor of $\frac{1}{2}$ therefore underestimates the effect of third and higher-order terms, while eliminating it, overestimates the second-order contribution. A solution to this conundrum is possible, but requires the development of additional formalism related to Faddeev's solution of the three-body problem. Such a method is reviewed in [Dickhoff and Barbieri (2004)].

It is tempting to try to add up the effects of pp and ph correlations. Unfortunately, this introduces a serious double-counting problem, which cannot be remedied by subtracting the double-counted second-order contribution. The latter term generates an incorrect energy dependence of the self-energy in certain energy regions. This invalidates the solution process of the Dyson equation. At energies close to the poles of the second-order self-energy, the slope of the total self-energy has the wrong sign. On account of the sign problem it is not possible to arrive at solutions that can be properly normalized. This can only be avoided by using self-energy contributions, which have intermediate states that can be written as propagators themselves, as in the usual Lehmann representations. All terms considered here (separately) fulfill this criterion.

A more complete description of excited states is provided by the RPA, discussed in Ch. 13 and 14 for ph states and in Ch. 15 for pp and hh states. It is well known, *e.g.* from analytic solutions of schematic models, that the collectivity of excitations is more pronounced in the RPA than in TDA, as illustrated in Sec. 13.3. For these reasons, and to obtain a better estimate of correlation effects in Σ, calculations can also be performed with the pphh and the phRPA propagators instead of the TDA ones. By proceeding in the usual manner from the material in Sec. 15.1, it is possible to arrive at the eigenvalue problem for the pphh RPA (see one of the exercises at the end of the chapter). The pphhRPA propagator can be written as

$$
\begin{aligned}
& G_{pphh}^{RPA}(\alpha, \beta; \gamma, \delta; E) \\
& = \sum_n \frac{R_{\alpha\beta}^{A+2,n*} R_{\gamma\delta}^{A+2,n}}{E - (E_n^{A+2} - E_0^A) + i\eta} - \sum_m \frac{R_{\beta\alpha}^{A-2,m} R_{\delta\gamma}^{A-2,m*}}{E - (E_0^A - E_m^{A-2}) - i\eta},
\end{aligned} \quad (17.13)
$$

where the pairs α, β and γ, δ refer to either both particle or hole states. The self-energy is then given by

$$
\begin{aligned}
\Sigma_{pphh}^{RPA}(\alpha, \beta; E) = \frac{1}{2} \Bigg\{ & \sum_{\kappa<F,n} \frac{\Delta_{\alpha\kappa}^{A+2,n*} \Delta_{\beta\kappa}^{A+2,n}}{E - ((E_n^{A+2} - E_0^A) - \varepsilon_\kappa) + i\eta} \\
& + \sum_{\kappa>F,m} \frac{\Delta_{\alpha\kappa}^{A-2,m} \Delta_{\beta\kappa}^{A-2,m*}}{E - ((E_0^A - E_m^{A-2}) - \varepsilon_\kappa) - i\eta} \Bigg\},
\end{aligned} \quad (17.14)
$$

with

$$
\Delta_{\alpha\kappa}^{A+2,n} = \sum_{\mu,\nu>F, \mu,\nu<F} \langle \alpha\kappa | G | \mu\nu \rangle R_{\mu\nu}^{A+2,n} \quad (17.15)
$$

and

$$
\Delta_{\alpha\kappa}^{A-2,m} = \sum_{\mu,\nu>F, \mu,\nu<F} \langle \alpha\kappa | G | \mu\nu \rangle R_{\nu\mu}^{A-2,m}. \quad (17.16)
$$

The energies $E_n^{A\pm2} - E_0^A$ and amplitudes $R_{\alpha\beta}^{A\pm2,n}$ are now the eigenvalues and the (RPA) normalized eigenvectors of the diagonalization of the interaction in the combined space of two-particle and two-hole states.

If, on the other hand, the ph correlations are improved, one employs the phRPA propagator

$$\Pi^{RPA}(\alpha, \beta; \gamma, \delta; E)$$
$$= \sum_{n \neq 0} \frac{R_{\alpha\beta}^{A,n*} R_{\gamma\delta}^{A,n}}{E - (E_n^A - E_0^A) + i\eta} - \sum_{m \neq 0} \frac{R_{\beta\alpha}^{A,m} R_{\delta\gamma}^{A,m*}}{E - (E_0^A - E_m^A) - i\eta}, \quad (17.17)$$

where the pairs α, γ and β, δ refer to either ph of hp combinations.[2] The corresponding self-energy reads

$$\Sigma^{RPA}(\alpha, \beta : E) = \frac{1}{2} \left\{ \sum_{\mu > F, n \neq 0} \frac{\Delta_{\alpha\mu}^{A+,n*} \Delta_{\beta\mu}^{A+,n}}{E - (\varepsilon_\mu + (E_n^A - E_0^A)) + i\eta} \right.$$
$$\left. + \sum_{\mu < F, m \neq 0} \frac{\Delta_{\alpha\mu}^{A-,m} \Delta_{\beta\mu}^{A-,m*}}{E - (\varepsilon_\mu + (E_0^A - E_m^A)) - i\eta} \right\}, \quad (17.18)$$

with

$$\Delta_{\alpha\mu}^{A+,n} = \sum_{\nu > F, \kappa < F; \nu < F, \kappa > F} \langle \alpha\kappa | G | \mu\nu \rangle R_{\nu\kappa}^{A,n} \quad (17.19)$$

and

$$\Delta_{\alpha\mu}^{A-,m} = \sum_{\nu > F, \kappa < F; \nu < F, \kappa > F} \langle \alpha\kappa | G | \mu\nu \rangle R_{\kappa\nu}^{A,m}. \quad (17.20)$$

The eigenvalues, $\pm(E_n^A - E_0^A)$, and eigenvectors $R_{\alpha\beta}^{A,n}$, are obtained by diagonalization of the interaction in the combined space of ph and hp states.

It should be noted that serious problems are encountered with these extensions of the second-order calculation. They arise, especially in phRPA, when the RPA correlations are so strong that the lowest solution for some angular momentum and parity has imaginary eigenvalues, and the amplitudes cannot be properly normalized. This happens for the 3^- phonon, which is crucial for the fragmentation of the $\frac{7}{2}^-$ strength in ^{48}Ca. As the imaginary solution can not be included in Eq. (17.20), and has to be discarded, the phRPA method is unsatisfactory here. For the same reason the lowest 1^+ states in ^{46}K en ^{50}Sc become unstable in pphh RPA and therefore must also be discarded in that method. Such problems would disappear when a self-consistent approach is adopted. This seems physically obvious:

---

[2] We have slightly modified the notation from the previous chapters to label the TDA amplitudes with T and the RPA ones with R for both the pp(hh) and ph propagators. This should not lead to any confusion.

Table 17.1   Occupation numbers for ^{48}Ca.

Shell	$\Sigma^{(2)}$	Σ^{TDA}_{ph}	Σ^{RPA}_{ph}	Σ^{TDA}_{pphh}	Σ^{RPA}_{pphh}
$0s\frac{1}{2}$	.967	.968	.963	.965	.952
$0p\frac{3}{2}$	.955	.956	.944	.950	.930
$0p\frac{1}{2}$	.951	.951	.939	.944	.920
$0d\frac{5}{2}$	.920	.925	.915	.898	.867
$0d\frac{3}{2}$	.877	.885	.891	.842	.780
$1s\frac{1}{2}$	.869	.860	.907	.818	.773
$0f\frac{7}{2}$	.060	.063	.048	.082	.120
$0f\frac{5}{2}$	.048	.044	.043	.064	.092
$1p\frac{3}{2}$	.033	.031	.036	.049	.063
$1p\frac{1}{2}$	.030	.028	.035	.042	.050
$0g\ 1d\ 2s$	.014	.014	.019	.018	.026
$0h\ 1f\ 2p$	.006	.006	.006	.007	.009
Total	20.053	20.093	20.125	20.165	20.370

Occupation numbers for relevant shells in ^{48}Ca in various approximations.

the collectivity that is generated in RPA equations, using mean-field sp propagators, will be damped when the fragmentation of the sp strength, which is induced by the coupling to collective states in the self-energy, is taken into account from the beginning. For extremely collective excitations, which are prevalent in open-shell systems, only a completely self-consistent approach seems viable, as illustrated in Sec. 17.2. One should thus keep in mind that the results with the RPA extensions, are only meant to give an indication as to how large the effect of collective correlations might be beyond the TDA description.

At low energies, the relatively pure one-hole states with valence shell quantum numbers are immediately identified by their large spectroscopic factors. For protons in ^{48}Ca this applies to the $1s\frac{1}{2}$ and $1d\frac{3}{2}$ shells. For the $\frac{1}{2}^+$ ground state of ^{47}K a spectroscopic factor of 1.07 was deduced from the $(e, e'p)$ data [Kramer (1990)] and a value of 2.26 for the $3/2^-$ state at 0.50 MeV. As discussed above, the calculation with the second-order self-energy yields about 40% larger values. Within the whole experimentally investigated energy region however, the total calculated strength for the s and d shells was only about 10% larger then deduced from the data. In Fig. 17.1 it is shown that indeed a more satisfactory fragmentation of $\ell = 2$

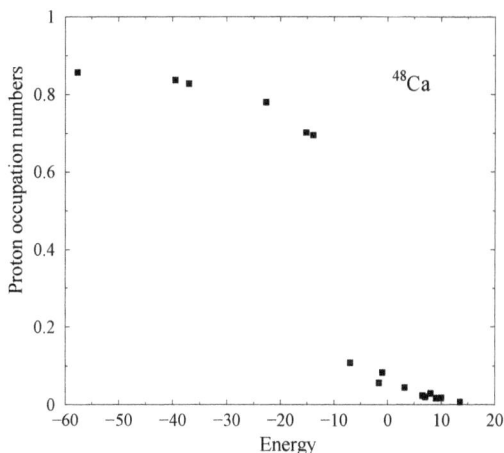

Fig. 17.3   Proton shell occupation probabilities deduced from a comparison of the present calculation and the $(e, e'p)$ data [Kramer (1990)]. The figure displays the calculated values with the pphhTDA correlations in the self-energy, multiplied by a factor 0.9 to simulate the effect of short-range correlations.

spectral strength in the low-energy region is obtained, when correlation effects beyond the second-order self-energy are included. The main peaks, which still stick out clearly in the calculation with $\Sigma^{(2)}$, are further reduced and the strength of the weaker fragments is increased by these correlations. It appears however, that the effects are overestimated by the RPA approach and underestimated in the phTDA results.

The occupation probabilities for the various orbitals can be obtained by integration of the hole spectral function [see Eq. (7.16)]. Experimentally, the hole strength has been determined within a limited energy region. For the valence $\ell = 0$ and $\ell = 2$ shells, the main portion of the total strength falls within the experimental window. In all theoretical approaches discussed here, the calculated extra hole strength at higher energies is only 5–10% of the $2j + 1$ sum rule for the 1s and only 10–15% for the 0d shell. For the more deeply bound shells a mere half, or less, of the total hole strength is within the range of the experiment, and the total calculated strength within this energy interval varies strongly with an energy shift of a few MeV in the sp energy of the orbit. A comparison of total observed and calculated strength for these orbits then becomes rather arbitrary.

As shown in Fig. 17.1, the primary effect of correlations in the self-energy is a redistribution of the spectral strength at low energies, *i.e.* at energies comparable to those of low-lying collective states in even nuclei. To what extent the total occupation numbers depend on the method of calculation is shown in Table 17.1. The occupation numbers for the deeply bound orbitals do not significantly depend on the treatment of the low-lying collective states, as was anticipated. Figure 17.1 suggests that the pphhTDA calculation yields the best description of the experimental strength distribution. A further overall reduction of about 10% of the occupation, due to short-range correlations, must also be anticipated for these nuclei, as discussed in Sec. 17.3. The overall picture that emerges when an overall reduction by short-range correlations of 10% is applied to the calculated occupation numbers with pphhTDA, is actually quite consistent and satisfactory, as illustrated in Fig. 17.3. The occupation probability of the shells just below the Fermi level is then about 75% and for the deepest lying shells about 85%. This outcome is very similar to the analysis for the case of ^{208}Pb, presented in Sec. 7.8.

For the major shells above the Fermi level the occupation probability is about 5–10% and for more remote shells only 1–2%. Also included in the table is the total number of protons. The violation of proton number is a typical occurrence when calculations are performed that are not self-consistent, as discussed in more detail in Ch. 21. A significant quantity is the jump in occupation numbers at the Fermi level. It is 0.74 for the pphhTDA. In an infinite Fermi system this jump is just the strength of the quasiparticle pole which corresponds to the spectroscopic factor, apart from the spin-factor $2j + 1$, of the states at the Fermi level. The difference between low-lying states in finite nuclei and infinite nuclear matter makes this comparison not too meaningful however, since the present results indicate that this jump is very sensitive to the treatment of collective excitations (see Table 17.1).

17.2   Self-consistent pphh RPA in finite systems

In the previous section, we have studied the influence of ladder diagrams at low energy in closed-shell nuclei. As for a homogeneous system, the summation is of course necessary to treat the core of repulsive interactions. As discussed in Sec. 15.3, the ladder diagrams are essential in describing the coherence of low-lying pair addition and removal excitations. For finite nuclei these low-lying excitations play an essential role in systems with either proton or neutron closed shells (but not both). These so-called semi-magic nuclei, have properties that are very reminiscent of pairing correlations for infinite systems [Bohr *et al.* (1958)]. We will use a simplified description of such open-shell nuclei to illustrate some of the features of a self-consistent treatment of the sp propagator, which includes the summation of ladder diagrams in a small configuration space. In turn, the sp propagators are employed to solve the ladder equation. The scheme corresponds to the self-consistent summation implied by the considerations of Sec. 9.4. We will concentrate here on the low-energy collective features of pair excitations only. The effects of short-range correlations are presented in Sec. 17.3.

For the treatment of pair addition and removal excitations we consider the Sn nuclei ($Z = 50$). The results of proton removal experiments on closed-shell nuclei exhibit important modifications of the simple shell-model picture. Nevertheless, these nuclei only have large fragments of sp strength (for removal) corresponding to the occupied states. That is no longer true for semi-magic nuclei, where it is possible to *remove* as well as *add* particles with the same sp quantum numbers with sizable probability to the ground state. This suggests that the notion of a sharp jump in occupation numbers, relevant for correlated closed-shell nuclei, becomes blurred in semi-magic ones. We will see that the Cooper-pair like excitations, obtained from summing ladder diagrams, leads to the smoothing of the Fermi surface. For a finite system the Lehmann representation of the tp propagator can be adapted from Eq. (15.28) by keeping only the discrete contributions

$$
\begin{aligned}
G_{pphh}(\alpha, \alpha'; \beta, \beta'; E) = &\sum_m \frac{\langle \Psi_0^A | a_{\alpha'} a_\alpha | \Psi_m^{A+2} \rangle \langle \Psi_m^{A+2} | a_\beta^\dagger a_{\beta'}^\dagger | \Psi_0^A \rangle}{E - (E_m^{A+2} - E_0^A) + i\eta} \\
&- \sum_n \frac{\langle \Psi_0^A | a_\beta^\dagger a_{\beta'}^\dagger | \Psi_n^{A-2} \rangle \langle \Psi_n^{A-2} | a_{\alpha'} a_\alpha | \Psi_0^A \rangle}{E - (E_0^A - E_n^{A-2}) - i\eta}.
\end{aligned}
\tag{17.21}
$$

The dressed but noninteracting approximation to this propagator can be generated by a convolution as in Eq. (15.35). For the present case it can be written in the following way

$$
\begin{aligned}
&G_{pphh}^f(\alpha, \alpha'; \beta, \beta'; E) \\
&= \sum_{m,m'} \frac{\langle \Psi_0^A | a_\alpha | \Psi_m^{A+1} \rangle \langle \Psi_m^{A+1} | a_\beta^\dagger | \Psi_0^A \rangle \langle \Psi_0^A | a_{\alpha'} | \Psi_{m'}^{A+1} \rangle \langle \Psi_{m'}^{A+1} | a_{\beta'}^\dagger | \Psi_0^A \rangle}{E - \{(E_m^{A+1} - E_0^A) + (E_{m'}^{A+1} - E_0^A)\} + i\eta} \\
&\quad - \sum_{n,n'} \frac{\langle \Psi_0^A | a_\beta^\dagger | \Psi_n^{A-1} \rangle \langle \Psi_n^{A-1} | a_\alpha | \Psi_0^A \rangle \langle \Psi_0^A | a_{\beta'}^\dagger | \Psi_{n'}^{A-1} \rangle \langle \Psi_{n'}^{A-1} | a_{\alpha'} | \Psi_0^A \rangle}{E - \{(E_0^A - E_n^{A-1}) + (E_0^A - E_{n'}^{A-1})\} - i\eta} \\
&\quad - (\gamma \longleftrightarrow \delta),
\end{aligned}
\tag{17.22}
$$

using the Lehmann representation of the sp propagator [see Eq. (7.10)] and the usual contour integrations. The ladder equation for the propagator that can be used to study the $A \pm 2$ eigenvalue problem, reads

$$G_{pphh}(\alpha, \alpha'; \beta, \beta'; E) = G_{pphh}^f(\alpha, \alpha'; \beta, \beta'; E) \qquad (17.23)$$

$$+ \frac{1}{4} \sum_{\gamma\gamma'\delta\delta'} G_{pphh}^f(\alpha, \alpha'; \gamma, \gamma'; E) \langle \gamma\gamma'| V |\delta\delta'\rangle G_{pphh}(\delta, \delta'; \beta, \beta'; E).$$

The factor $\frac{1}{4}$ appears because the full nondiagonal form of G_{pphh}^f is used instead of the clipped version, as in Eq. (15.11). Before deriving the eigen-value problem associated with Eq. (17.23), it is instructive to complete the steps that are required to construct the self-energy that includes these dia-grams. The self-energy is diagrammatically given in Fig. 9.13. We recognize that the last term in the figure merits further clarification. By consulting Fig. 15.5, it becomes clear that it is generated by closing the second contri-bution to Γ_{pphh} with a (dressed) sp propagator. In order to construct the self-energy contributions from Eq. (17.23), it is therefore useful to write the ladder equation for the vertex function [see Eq. (15.14)] as follows

$$\langle \alpha\alpha'| \Gamma_{pphh}(E) |\beta\beta'\rangle = \langle \alpha\alpha'| V |\beta\beta'\rangle + \langle \alpha\alpha'| \Delta\Gamma_{pphh}(E) |\beta\beta'\rangle \qquad (17.24)$$

$$= \langle \alpha\alpha'| V |\beta\beta'\rangle + \frac{1}{4} \sum_{\gamma\gamma'\delta\delta'} \langle \alpha\alpha'| V |\gamma\gamma'\rangle \, G_{pphh}(\gamma, \gamma'; \delta, \delta'; E) \, \langle \delta\delta'| V |\beta\beta'\rangle .$$

We will employ the second version in Fig. 15.5 of $\Delta\Gamma_{pphh}$ since the inter-mediate G_{pphh} has a Lehmann representation that can be used to perform the energy integration for the self-energy. Indeed, we find from Eqs. (17.24) and (17.21)

$$\langle \alpha\alpha'| \Delta\Gamma_{pphh}(E) |\beta\beta'\rangle = \frac{1}{4} \sum_{\epsilon\epsilon'\theta\theta'} \langle \alpha\alpha'| V |\epsilon\epsilon'\rangle \, G_{pphh}(\epsilon\epsilon'; \theta\theta'; E) \, \langle \theta\theta'| V |\beta\beta'\rangle$$

$$= \sum_n \frac{\Delta_{\alpha\alpha'}^{n+} \Delta_{\beta\beta'}^{n+\,*}}{E - (E_n^{A+2} - E_0^A) + i\eta}$$

$$- \sum_m \frac{\Delta_{\alpha\alpha'}^{m-} \Delta_{\beta\beta'}^{m-\,*}}{E - (E_0^A - E_m^{A-2}) - i\eta}, \qquad (17.25)$$

where

$$\Delta_{\alpha\alpha'}^{n+} = \frac{1}{2} \sum_{\epsilon\epsilon'} \langle \alpha\alpha'| V |\epsilon\epsilon'\rangle \langle \Psi_0^A| a_{\epsilon'} a_\epsilon |\Psi_n^{A+2}\rangle \qquad (17.26)$$

and

$$\Delta^{m-}_{\alpha\alpha'} = \frac{1}{2} \sum_{\epsilon\epsilon'} \langle \alpha\alpha' | V | \epsilon\epsilon' \rangle \langle \Psi^{A-2}_m | a_{\epsilon'} a_{\epsilon} | \Psi^A_0 \rangle . \qquad (17.27)$$

The key to making the separation of $\Delta\Gamma_{pphh}$ into forward and backward propagating parts, is thus provided by using the Lehmann representation for G_{pphh} in the last equality of Eq. (17.25).

The self-energy can now be written as (see also Fig. 9.13):

$$\Sigma^L(\gamma, \delta; E) = - \langle \gamma | U | \delta \rangle \qquad (17.28)$$
$$- i \int_{C\uparrow} \frac{dE'}{2\pi} \sum_{\mu,\nu} \langle \gamma\mu | V | \delta\nu \rangle G(\nu, \mu; E')$$
$$- i \int \frac{dE'}{2\pi} \sum_{\mu,\nu} \langle \gamma\mu | \Delta\Gamma_{pphh}(E') | \delta\nu \rangle G(\nu, \mu; E' - E),$$

where the L superscript is a reminder that only ladder diagram contributions are included. Due to the decomposition of $\Delta\Gamma_{pphh}$ into parts with poles in different halves of the complex E'-plane [see Eq. (17.25)], it is possible to perform the energy integrations for the self-energy in Eq. (17.28) explicitly, when the corresponding decomposition for the sp propagator is employed as well. Using a similar notation for the various self-energy contributions as in Sec. 16.3.1, we find

$$\Sigma^L(\gamma, \delta; E) = \Sigma^U(\gamma, \delta) + \Sigma^{HF}(\gamma, \delta) + \Delta\Sigma_\downarrow(\gamma, \delta; E) + \Delta\Sigma_\uparrow(\gamma, \delta; E), \quad (17.29)$$

where the energy-dependent contributions are given by

$$\Delta\Sigma_\downarrow(\gamma, \delta; E) = \sum_{\mu,\nu\ n,l} \frac{\Delta^{n+}_{\gamma\mu} \Delta^{n+*}_{\delta\nu} z^{l-*}_{\mu} z^{l-}_{\nu}}{E - \{(E^{A+2}_n - E^A_0) - (E^A_0 - E^{A-1}_l)\} + i\eta} \qquad (17.30)$$

and

$$\Delta\Sigma_\uparrow(\gamma, \delta; E) = \sum_{\mu,\nu\ m,k} \frac{\Delta^{m-}_{\gamma\mu} \Delta^{m-*}_{\delta\nu} z^{k+}_{\nu} z^{k+*}_{\mu}}{E - \{(E^A_0 - E^{A-2}_m) - (E^{A+1}_k - E^A_0)\} - i\eta}, \qquad (17.31)$$

respectively. The notation for the addition and removal amplitudes is familiar *e.g.* from Eqs. (9.39) and (10.8). The derivation outlined above, is quite general. Although it assumes, for simplicity, discrete sums in the different Lehmann representations, it is straightforward to include the consequences of a continuous distribution of poles, as in an infinite system. The ingredients in the self-consistent scheme based on ladder diagrams, correspond either to amplitudes related to the removal (addition) of one or two particles from the correlated ground state to the appropriate state in the $A-2$ or $A-1$ ($A+2$ or $A+1$) system, or to the associated energy differences between these states and the ground state energy E_0^A.

The use of Green's function techniques to open-shell nuclei is instructive for the following reason. We note that sp orbitals in closed-shell nuclei exhibit substantial deviations from mean-field occupation numbers (see Sec. 7.8). This raises the question whether there is any fundamental difference between closed-shell and open-shell nuclei. We have learned that the sp strength distribution is influenced by different energy scales. At low-energy, there is the mixing of sp states with collective excitations (see Sec. 17.1.2). Mixing with high-lying excitations is facilitated by the strong core in the nucleon–nucleon interaction, as discussed in Ch. 16. The latter mixing takes place on an energy scale that is much larger than the one relevant for distinguishing closed- and open-shell nuclei. Even the mixing to collective giant resonances, detailed in Ch. 21, is not expected to lead to significant differences. This expectation can be based on the smooth dependence of the position and sum-rule strength of the observed giant resonances. The true difference between open- and closed-shell nuclei, involves the extreme forms of collective behavior at low excitation energy in open-shell nuclei, like vibrations, rotations, and pairing phenomena, which are mostly absent, or at least less prevalent, in closed-shell ones.

The self-consistent treatment of ladder diagrams is also a logical extension of the discussion of bound-pair states in Sec. 15.3, since the effect of the bound states must be included in the sp propagator through solution of the Dyson equation. For illustration purposes, it is sufficient to limit the calculation to one major shell of sp orbitals. We will employ a very simple pairing force, characterized by one coupling constant G, which acts only between identical nucleons coupled to angular momentum zero. The space for the neutrons consists of the $0g\frac{7}{2}$, $1d\frac{5}{2}$, $2s\frac{1}{2}$, $1d\frac{3}{2}$, and $0h\frac{11}{2}$ shells, which are relevant for the low-lying neutron excitations in Sn nuclei.

Various simplifications of the above scheme can be applied with the restrictions to a small space and a simple interaction. For the noninteracting two-particle propagator, we can write

$$G_{pphh}^f(\alpha,\alpha';\beta,\beta';E) = (\delta_{\alpha\beta}\delta_{\alpha'\beta'} - \delta_{\alpha\beta'}\delta_{\alpha'\beta}) \qquad (17.32)$$

$$\times \left[\sum_{n+}\sum_{m+} \frac{|z_\alpha^{n+}|^2 |z_{\alpha'}^{m+}|^2}{E - \varepsilon_\alpha^{n+} - \varepsilon_{\alpha'}^{m+} + i\eta} - \sum_{n-}\sum_{m-} \frac{|z_\alpha^{n-}|^2 |z_{\alpha'}^{m-}|^2}{E - \varepsilon_\alpha^{n-} - \varepsilon_{\alpha'}^{m-} - i\eta} \right],$$

where the sp amplitudes z and energies ε correspond to solutions of the Dyson equation. The relevant Lehmann representation of the two-particle propagator reads [see also Eq. (17.21)]

$$G_{pphh}(\alpha, \alpha'; \beta\beta'; E)$$
$$= \sum_{m+} \frac{[Y_L^{m+}(\alpha\alpha')]^* Y_L^{m+}(\beta\beta')}{E - E_L^{m+} + i\eta} - \sum_{m-} \frac{Y_L^{m-}(\beta\beta')[Y_L^{m-}(\alpha\alpha')]^*}{E - E_L^{m-} - i\eta}, \quad (17.33)$$

where the two-particle amplitudes are written as

$$Y_L^{m+}(\beta\beta') = \left\langle \Psi_m^{A+2} \middle| a_\beta^\dagger a_{\beta'}^\dagger \middle| \Psi_0^A \right\rangle \quad (17.34)$$

and

$$Y_L^{m-}(\alpha\alpha') = \left\langle \Psi_0^A \middle| a_\alpha^\dagger a_{\alpha'}^\dagger \middle| \Psi_m^{A-2} \right\rangle. \quad (17.35)$$

The corresponding relative energies are denoted by

$$E_L^{m+} = E_m^{A+2} - E_0^A \quad (17.36)$$

and

$$E_L^{m-} = E_0^A - E_m^{A-2} \quad (17.37)$$

in Eq. (17.33). The subscript L serves as a reminder that the ladder approximation remains an approximate solution to the problem. Introducing the notation $\alpha = (n_a \ell_a j_a m_a) = (a m_a)$ and $\tilde{\alpha} = (n_a \ell_a j_a - m_a) = (a - m_a) = (a \tilde{m}_a)$. The Y amplitudes in Eq. (17.33) involve identical sp states that can be coupled to angular momentum zero, yielding

$$G_{pphh}^{J=0}(aa; bb; E) \quad (17.38)$$
$$= \sum_{m_\alpha} \sum_{m_\beta} \langle\, j_a \, m_a \, j_a \, \tilde{m}_a \mid 0 \; 0 \,\rangle\langle\, j_b \, m_b \, j_b \, \tilde{m}_b \mid 0 \; 0 \,\rangle G_{pphh}(\alpha, \tilde{\alpha}; \beta\tilde{\beta}; E).$$

We will then write the $J = 0$ component of Eq. (17.33) as

$$G_{pphh}(aa, bb; E) = \sum_{m+} \frac{[Y_L^{m+}(aa)]^* Y_L^{m+}(bb)}{E - E_L^{m+} + i\eta} - \sum_{m-} \frac{Y_L^{m-}(bb)[Y_L^{m-}(aa)]^*}{E - E_L^{m-} - i\eta},$$
$$(17.39)$$

suppressing the $J = 0$ label and slightly modifying the notation for the Y-amplitudes. Inserting this form of G_{pphh} and the noninteracting propagator [see Eq. (17.32)] into the ladder equation [Eq. (17.23)], and proceeding with the usual steps, one obtains an eigenvalue equation for the energies $E_L^{m\pm}$, which correspond to the excitation energies of the $(A \pm 2)$-particle system relative to the ground-state energy of the A-particle system. The resulting eigenvalue equation reads

$$
\sum_b \left[\sum_{k+} \sum_{n+} \frac{|z_{a(i)}^{k+}|^2 |z_{a(i)}^{n+}|^2}{E_{L(i+1)}^{m\pm} - \varepsilon_{a(i)}^{k+} - \varepsilon_{a(i)}^{n+}} - \sum_{k-} \sum_{n-} \frac{|z_{a(i)}^{k-}|^2 |z_{a(i)}^{n-}|^2}{E_{L(i+1)}^{m\pm} - \varepsilon_{a(i)}^{k-} - \varepsilon_{a(i)}^{n-}} \right]
$$
$$
\times \langle aa | V^{J=0} | bb \rangle \, Y_{L(i+1)}^{m\pm}(bb)
$$
$$
= Y_{L(i+1)}^{m\pm}(aa). \tag{17.40}
$$

The energies from the Dyson equation are denoted by $\varepsilon_a^{m\pm}$. A label i has been included to emphasize that one has to proceed in an iterative manner. One cycle involves diagonalizing the eigenvalue equation (17.40), then calculating the self-energy (Eq. (17.29)), and subsequently solving the Dyson equation, as discussed in Ch. 11. The process is continued until self-consistency is achieved. A practical constraint, on the number of eigenvalues of the Dyson equation that are carried over to the next iteration, must be imposed. Most of these eigenvalues carry little or no strength and it makes sense to consider the binning procedure, also discussed in Ch. 11, to keep the numerical implementation managable. The two-body matrix elements required for this calculation are written as

$$
\langle aa | V^{J=0} | bb \rangle = -\frac{1}{2} G \sqrt{(2j_a + 1)(2j_b + 1)}, \tag{17.41}
$$

where the coupling constant G can be adjusted to describe certain experimental properties and typically has a strength around 20 MeV/A.

Self-consistent eigenvalues of Eq. (17.40) are displayed in Fig. 17.4, while corresponding solutions to the Dyson equation are shown in Fig. 17.5. Note that for each value of the coupling constant G, a self-consistent result has been obtained. The solution of Eq. (17.40) can encounter a serious problem when the interaction is too attractive. If that is the case for a given G_{pphh}^f, the eigenvalues corresponding to the ground states of the $A \pm 2$ systems become complex, a well-known feature of RPA-like equations encountered in Ch. 13. Hence, the iteration procedure must take this possibility into account. If one starts the iteration by describing G_{pphh}^f in terms of noninteracting sp propagators, a critical value of G exists beyond which the above instability occurs as well. In most instances it is necessary to start the iteration procedure with correlated propagators generated for a smaller coupling constant. Figure 17.4 shows the dependence of the self-consistent eigenvalues of the ladder equation corresponding to the dominant two-particle

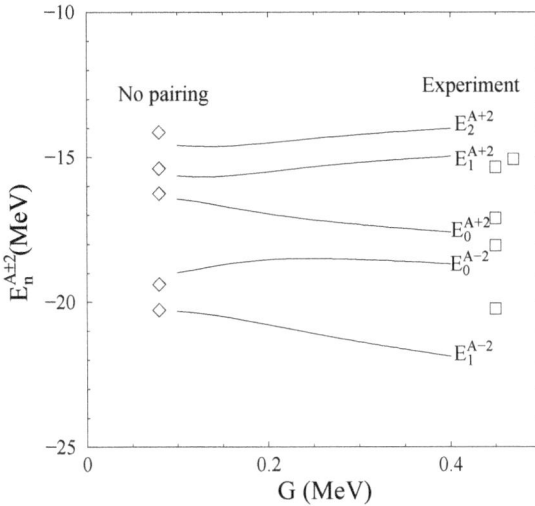

Fig. 17.4 Eigenvalue solutions of the ladder equation for the levels with the largest pair-amplitudes as a function of the coupling constant G. The position of the unperturbed pair states is indicated by the diamonds to the left of the eigenvalue curves. The squares on the right, drawn to avoid overlapping, represent the experimental data for ^{112}Sn and ^{116}Sn, respectively.

addition or removal amplitudes for ^{114}Sn. The original neutron sp energies are obtained from a standard Woods–Saxon potential, discussed in Sec. 3.3. These sp energies correspond to the diamonds in the figure (multiplied by a factor of two). The lowest experimental 0^+ energies for ^{116}Sn and ^{112}Sn are indicated by the open squares. The eigenvalues are smooth functions of the coupling constant which can be varied to investigate the convergence properties and sensitivity of the solution technique. Moreover, it is not evident that a standard value of the pairing constant is adequate for the analysis, although in practice it turns out that a value of 20 MeV/A, corresponding roughly to 0.18, will already give reasonable results for both $A\pm2$ and $A\pm1$ nuclei.

Although the self-consistent solution becomes progressively more difficult to calculate with increasing coupling constant [Yuan (1994)], the final collective bound pair states move only very slowly towards each other, as shown in Fig. 17.4. The corresponding low-lying sp fragments will therefore have to move away from each other with increasing coupling constant. This feature is illustrated in Fig. 17.5, which plots the results for the self-consistent solution of the Dyson equation for the largest fragments

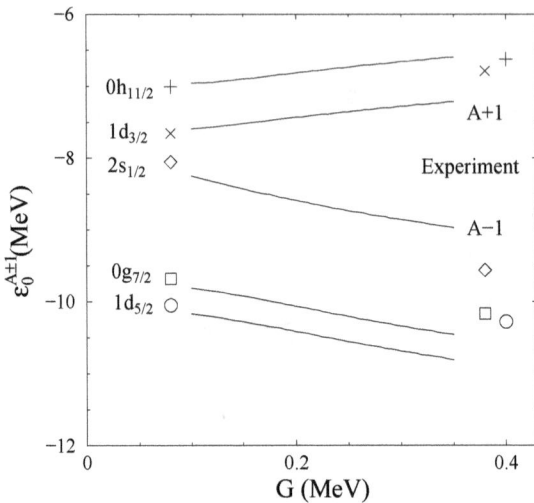

Fig. 17.5 Solutions of the Dyson equation with the largest sp strength, as a function of the coupling constant G. The position of the unperturbed energies is indicated by the symbols above the value $G = 0$ while those to the right of the curves represent the experimental data for ^{115}Sn and ^{117}Sn.

corresponding to the involved orbits. From Figs. 17.4 and 17.5, one may further conclude that it is rather easy to obtain even better agreement with the experimental data, by slightly adjusting the mean-field sp energies. Comparison with spectroscopic factors in the $A \pm 1$ nuclei yields satisfactory agreement with experimental results as well [Yuan (1994)].

Pairing in finite systems can thus be studied with the self-consistent Green's function method, when the self-consistent summation of ladder diagrams is incorporated. To illustrate the difference between the open-shell systems and closed-shell nuclei, the occupation numbers of the five neutron shells in ^{116}Sn are shown in Fig. 17.6 for a value of G corresponding to 0.35. The transition from mostly occupied to mostly empty levels, takes place in a narrow energy domain and in a considerably smoother fashion than for closed-shell nuclei, as discussed in Secs. 7.8 and 17.1.2. The jump in occupation from the $2s\frac{1}{2}$ to the $1d\frac{3}{2}$ orbit amounts to 0.38, and resembles the value of the $2s\frac{1}{2}$ quasihole fragment, which has strength 0.42. Apparently, the jump in occupation in a finite system still corresponds approximately to the strength of the last quasihole fragment. The result cannot be exact since different discrete sp quantum numbers will generate deviations, due

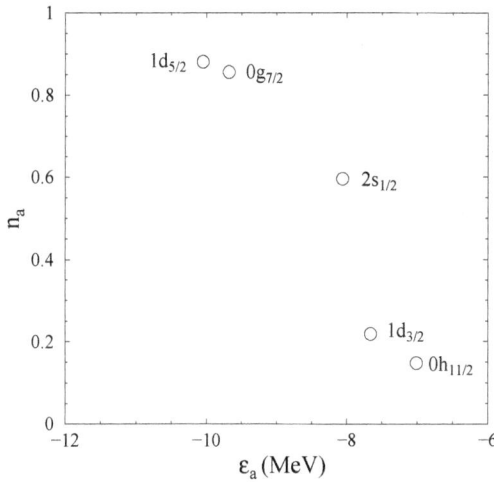

Fig. 17.6   Neutron valence shell occupation probabilities for ^{116}Sn, illustrating a smooth transition from mostly occupied to mostly empty sp levels.

to restrictions of parity and angular momentum. The smooth transition in occupation numbers in Fig. 17.6, clearly distinguishes closed- and open-shell nuclei. An even more definitive signature of the difference is provided by the size of the particle and hole fragments corresponding to the same orbit. The quasiparticle fragment of the $2s\frac{1}{2}$ in ^{116}Sn, for example, has the value 0.28 and is of similar magnitude as the quasihole fragment (0.42). Obviously, the presence of comparable particle and hole fragments close in energy, as in any system with pair correlations (see Ch. 22), represents a clear distinction between open- and closed-shell nuclei, where the corresponding fragments differ typically by an order of magnitude and lie farther apart in energy.

17.3   Short-range correlations in finite nuclei

The influence of short-range correlations can be quite well studied in nuclear matter, as discussed in Sec. 16.3.1. It is technically considerably more difficult to establish the influence of short-range correlations on the spectral function directly for finite nuclei. The main reason for this is that short-range dynamics pertains to the relative motion of two particles, whereas two-body matrix elements in the nucleus are required in the laboratory system. The transformation from relative and center-of-mass motion to individual coordinates is only straightforward with momentum or harmonic oscillator quantum numbers. Neither of these basis sets completely captures the essentials of a finite nucleus. Nevertheless, a proper treatment of short-range correlations can be obtained by constructing the G-matrix interaction, which can be used as an effective interaction in shell-model calculations in small model spaces. Only in the last decade have calculations become available, for the self-energy and corresponding spectral functions, that incorporate short-range effects.

We will sketch here the ingredients of an indirect approach that was developed by [Borromeo *et al.* (1992)] to compute the self-energy for *finite* nuclei in terms of a G-matrix, which is the solution of the Bethe–Goldstone equation for *nuclear matter*

$$\langle k\ell|\, G_{SJ_SKLT}\,|k'\ell'\rangle = \langle k\ell|\, V_{SJ_SKLT}\,|k'\ell'\rangle \tag{17.42}$$

$$+ \frac{1}{2}\sum_{\ell''}\int dk''(k'')^2 \,\langle k\ell|\, V_{SJ_SKLT}\,|k''\ell''\rangle$$

$$\frac{\bar{Q}(K,k'')}{E_{NM} - \frac{\hbar^2 K^2}{4m} - \frac{\hbar^2 k''^2}{2m}}\,\langle k''\ell''|\, G_{SJ_SKLT}\,|k'\ell'\rangle\;.$$

The variables k, k', and k'' denote the relative wave vectors between the two nucleons, ℓ, ℓ', and ℓ'' the orbital angular momenta for the relative motion, K and L the corresponding quantum numbers for the center-of-mass motion, S and T the total spin and isospin, and J_S is obtained by coupling the orbital angular momentum of the relative motion to the spin S. We note that $\bar{Q}(K,k)$ is the angle-averaged form of the product of step-functions in Eq. (16.58) that allows a partial wave decomposition of that equation (see Exercise 4 of Ch. 16). The implied sole dependence of the G-matrix on the magnitude of $\boldsymbol{K}$, the center-of-mass wave vector, also ensures that the solution of Eq. (17.42) does not depend on L. The L-label is kept however, to facilitate the recoupling to individual orbital angular momentum states, as discussed below. Equation (17.42) generates an appropriate solution of two-body short-range dynamics but the resulting matrix elements require further manipulation before becoming useful for the finite nucleus. The choices for the density of nuclear matter and the starting energy E_{NM} are not critical when the proper corrections for the finite system are taken into account. The calculation of the corresponding BHF term is not very sensitive to this choice. Furthermore, the nuclear-matter approximation is corrected by calculating the 2p1h term, displayed

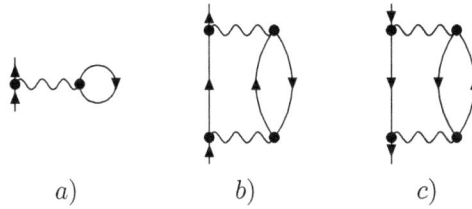

Fig. 17.7 Graphical representation of the BHF a), the 2p1h contribution b) and 1p2h term c) to the self-energy of the nucleon. The G-matrix is indicated by the wiggly line.

in Fig. 17.7b), directly for the finite system.[3] The second-order correction, which assumes harmonic oscillator states for the occupied (hole) states and plane waves for the intermediate unbound particle states, incorporates the correct energy and density dependence characteristic of a finite nucleus G-matrix. To evaluate the diagrams in Fig. 17.7, one requires matrix elements in a mixed representation of one particle in a bound harmonic oscillator while the other is in a plane wave state. Using vector bracket transformation coefficients [Balian and Brezin (1969); Wong and Clement (1972)], one can transform matrix elements from the representation in relative and center-of-mass wave vectors to individual ones in the laboratory frame, in which the two particle state is described by

$$|k_1 \ell_1 j_1 k_2 \ell_2 j_2 JT\rangle. \tag{17.43}$$

The quantum numbers k_i, ℓ_i and j_i refer to the momentum and angular momenta of particle i, whereas J and T define the total angular momentum and isospin of the two-particle state. The relevant vector bracket is given by

$$\langle k\ell KL\lambda|k_1\ell_1 k_2\ell_2\lambda\rangle, \tag{17.44}$$

where both pairs ℓ, L and ℓ_1, ℓ_2 are coupled to λ. Vector brackets can be calculated following the method of [Kung *et al.* (1979)] for example. The necessary recoupling of the angular momentum to complete the transformation from the states, employed in Eq. (17.42), to the ones in Eq. (17.43) can also be found there. Performing an integration over one of the k_i, one

---

[3]While strictly speaking the genuine BHF approach involves self-consistent sp wave functions, as in the HF approximation, the main features associated with using the G-matrix of Eq. (17.42) are approximately the same. Hence we will use the BHF abbreviation.

obtains tp states in a mixed representation

$$|n_1\ell_1j_1k_2\ell_2j_2JT\rangle = \int_0^\infty dk_1\, k_1^2 R_{n_1,\ell_1}(k_1)\, |k_1\ell_1j_1k_2\ell_2j_2JT\rangle \ . \qquad (17.45)$$

Here R_{n_1,ℓ_1} stands for the radial oscillator function, and the oscillator length is chosen to achieve an appropriate description of the bound states of ^{16}O. Using the results of Eqs. (17.42) – (17.45) the BHF approximation[4] for the self-energy (see Fig. 17.7a) can be obtained in the wave vector basis

$$\Sigma_{\ell_1j_1}^{BHF}(k_1,k_1') = \frac{1}{2(2j_1+1)}\sum_{n_2\ell_2j_2}\sum_{JT}(2J+1)(2T+1)$$

$$\times\, \langle k_1\ell_1j_1n_2\ell_2j_2|\, G_{JT}\, |k_1'\ell_1j_1n_2\ell_2j_2\rangle\,, \qquad (17.46)$$

where the summation over projection quantum numbers has been replaced by a summation over J and T. The summation over the oscillator quantum numbers is restricted to the states occupied in the independent-particle description of ^{16}O. The BHF part of the self-energy is real and does not depend on the energy.

The lowest-order terms in G, which yield an imaginary contribution to the self-energy, are represented by the diagrams displayed in Figs. 17.7b) and 17.7c), referring to intermediate 2p1h and 1p2h states. The 2p1h contribution to the imaginary part can be written as

$$\mathrm{Im}\,\Sigma_{\ell_1j_1}^{2p1h}(k_1,k_1';E) = \frac{-1}{2(2j_1+1)}\sum_{n_2\ell_2j_2}\sum_{\ell L}\sum_{JST}\int k^2dk \int K^2dK \quad (17.47)$$

$$\times\, (2J+1)(2T+1)\,\langle k_1\ell_1j_1n_2\ell_2j_2|\,G_{JT}\,|k\ell SKL\rangle$$

$$\times\, \langle k\ell SKL|\,G_{JT}\,|k_1'\ell_1j_1n_2\ell_2j_2\rangle\,\pi\,\delta\!\left(E + \varepsilon_{n_2\ell_2j_2} - \frac{\hbar^2 K^2}{4m} - \frac{\hbar^2 k^2}{m}\right),$$

where the average experimental quasihole energies $\varepsilon_{n_2\ell_2j_2}$ are used for the hole states (-47 MeV, -21.8 MeV, -15.7 MeV corresponding to $0s\frac{1}{2}$, $0p\frac{3}{2}$ and $0p\frac{1}{2}$ shells, respectively), while the energies of the particle states are given in terms of the kinetic energy only. The plane waves associated with the particle states in the intermediate states must be properly orthogonalized to the bound sp states employing the techniques discussed in [Borromeo *et al.* (1992)]. The 1p2h contribution to the imaginary part $\mathrm{Im}\,\Sigma_{\ell_1j_1}^{1p2h}(p_1,p_1';E)$ can be calculated in the same way.

---

[4]Note that the energy dependence of G is not yet included at this point.

The choice of pure kinetic energies for the particle states in calculating the imaginary parts of Im Σ^{2p1h} [Eq. (17.47)] and Im Σ^{1p2h} is not very realistic for the intermediate states at low energy. Indeed, a sizable imaginary part in Im Σ^{1p2h} is obtained only for energies E below the sp energy of the $0s\frac{1}{2}$ state. When the primary interest is to study the effects of short-range correlations, the choice appears appropriate since these involve excitations of particle states with high momenta. A different approach is required to treat the coupling to the very low-lying 2p1h and 1p2h states in an adequate way, as discussed in the Sec. 17.1. The 2p1h contribution to the real part of the self-energy can be calculated from the imaginary part, using the standard dispersion relation

$$\text{Re } \Sigma^{2p1h}_{\ell_1 j_1}(k_1, k_1'; E) = \frac{\mathcal{P}}{\pi} \int_{-\infty}^{\infty} \frac{\text{Im } \Sigma^{2p1h}_{\ell_1 j_1}(k_1, k_1'; E')}{E' - E} dE', \qquad (17.48)$$

where $\mathcal{P}$ stands for the principal value integral. A similar dispersion relation holds for Re Σ^{1p2h} and Im Σ^{1p2h}.

Since the BHF contribution Σ^{BHF} has been calculated in terms of a nuclear matter G-matrix, it already contains 2p1h terms of the kind displayed in Fig. 17.7b). In order to avoid this overcounting of the pp ladder terms, one subtracts from the real part of the self-energy a correction term (Σ_c), which just contains the 2p1h contribution calculated in nuclear matter. Summing up the various contributions one obtains the following expression for the self-energy

$$\Sigma = \Sigma^{BHF} + \Delta\Sigma \qquad (17.49)$$
$$= \Sigma^{BHF} + \left(\text{Re } \Sigma^{2p1h} - \Sigma_c + \text{Re } \Sigma^{1p2h}\right) + i\left(\text{Im } \Sigma^{2p1h} + \text{Im } \Sigma^{1p2h}\right).$$

The Dyson equation for this self-energy can now be solved in wave vector space. Energies and wave functions of the quasihole states can be determined by diagonalizing the BHF sp Hamiltonian plus $\Delta\Sigma$ in the wave vector basis. A discretization of the Schrödinger equation of Eq. (9.43) in this basis can be written as

$$\sum_{n=1}^{N} \langle k_i | \frac{\hbar^2 k_i^2}{2m} \delta_{in} + \Sigma^{BHF}_{\ell j} + \Delta\Sigma_{\ell j}(E = \varepsilon^{\ell j}_{m-}) | k_n \rangle \, w(k_n) z^{m-}_{k_n \ell j}$$
$$= \varepsilon^{\ell j}_{m-} \, z^{m-}_{k_i \ell j}, \qquad (17.50)$$

where $w(k_i)$ is the weight associated with the discrete point k_i. In this approach $\Delta\Sigma$ only contains a sizable imaginary part for energies E below the quasihole eigenvalues $\varepsilon^{\ell j}_{m-}$ ($m-$ identifying discrete solutions). The discrete solutions of Eq. (17.50) are thus separated in energy from the contribution to the hole spectral function in the continuum. The eigenvector corresponding to the discrete states yields the quasihole wave function $z_{k\ell j}^{m-}$ in wave vector space, which still needs to be normalized by the spectroscopic factor $|z_{\alpha_{qh}\ell j}^{m-}|^2$ by means of Eq. (9.44), translating into

$$|z_{\alpha_{qh}\ell j}^{m-}|^2 = \int dk \; k^2 \; |z_{k\ell j}^{m-}|^2. \tag{17.51}$$

The quasihole contribution to the spectral function contains only one contribution for each of the ℓj values that are occupied in the independent-particle description. The term diagonal in k is given by

$$S^{qh}(k\ell j; E) = |z_{k\ell j}^{m-}|^2 \, \delta(E - \varepsilon_{m-}^{\ell j}) \, . \tag{17.52}$$

In the calculations described in [Müther *et al.* (1995); Polls *et al.* (1995)], the Bethe–Goldstone equation (17.42) was solved by employing for V the one-boson-exchange potential Bonn-B, developed by [Machleidt (1989)] (Table A.2). The Pauli operator Q was approximated by the so-called angle-averaged approximation for nuclear matter with a Fermi wave vector $k_F = 1.4$ fm^{-1}. This roughly corresponds to the saturation density of nuclear matter. The starting energy E_{NM} for computing the G-matrix, Eq. (17.42), was chosen to be -10 MeV.

The square of the quasihole wave function for the $p\frac{1}{2}$ state, normalized to the spectroscopic factor, is shown in Fig. 17.8 as the solid line. For comparison the result for the BHF wave function is illustrated by the dashed line. From the comparison one can infer that at the quasihole energies no substantial change in the shape of the wave function occurs, and that the BHF wave function is a good approximation. It should be further noted that the wave function of a Woods–Saxon potential, which is constructed as the local equivalent of the BHF potential [Borromeo *et al.* (1992)], is indistinguishable from the BHF wave function. This suggests that the explicit inclusion of short-range correlations does not lead to the strong suppression of the wave function in the interior of the nucleus, as had been suggested by [Ma and Wambach (1991); Mahaux and Sartor (1991)]. These results also verify that the influence of high wave-vector components in the quasihole wave function is of minor importance. Consequently, the enhancement of the momentum distribution, as the expected signature of short-range correlations, has to come from excitations at higher missing energies.

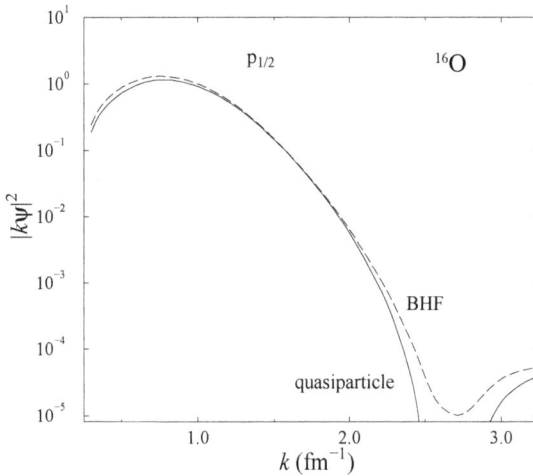

Fig. 17.8 Square of the quasihole wave function for the $p\frac{1}{2}$ state in ^{16}O (full curve), normalized to the spectroscopic factor according to Eq. (9.44), compared to the BHF result (dashed curve), adapted from [Müther and Dickhoff (1994)].

Figure 17.9 presents an example of the reduced cross section for the $(e, e'p)$ reaction on ^{16}O, leading to the bound quasihole state of ^{15}N at an excitation energy of -6.32 MeV [Polls *et al.* (1997)]. In this picture the effects of FSI have been calculated in the distorted wave impulse approximation (DWIA) [Boffi *et al.* (1996)] and the data points have been obtained at NIKHEF for the so-called parallel kinematics [Leuschner *et al.* (1994)]. Using the quasihole part of the spectral function, computed from Eq. (17.50), but adjusting the spectroscopic factor for the quasihole state contribution to fit the experimental data, one obtains the solid line of Fig. 17.9. Comparing this result with the experimental data, one finds that the calculated spectral function closely reproduces the shape of the reduced cross section as a function of the missing momentum. The absolute value for the reduced cross section can only be reproduced by employing a spectroscopic factor of 0.537, a value considerably below the one of 0.914 calculated using the properly normalized solution of Eq. (17.50) [Müther *et al.* (1995)]. An analogous result holds for the transition to the $p\frac{1}{2}$ ground state of ^{15}N for which the same analysis yields a spectroscopic factor of 0.644. For comparison, the phenomenological Woods–Saxon wave functions, adjusted to fit the shape of the reduced cross section, require spectroscopic factors ranging from 0.61 to 0.64 for the lowest $0p\frac{1}{2}$ state, and from 0.50 to 0.59 for

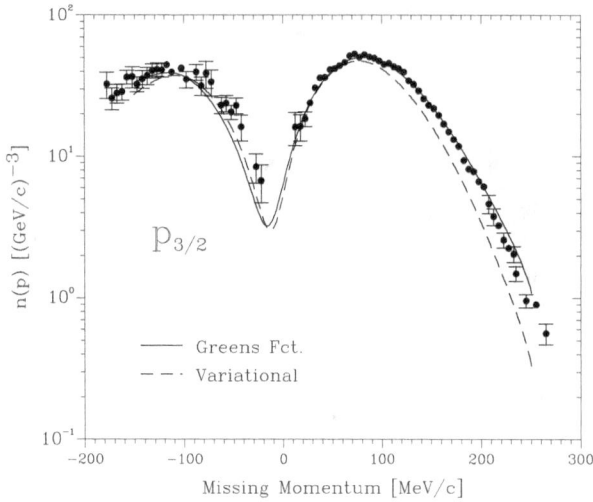

Fig. 17.9   Reduced cross section for the $^{16}O(e, e'p)$ reaction in parallel kinematics, lead-ing to the $3/2^-$ state at 6.32 MeV of the residual nucleus ^{15}N. Results of the Green's function approach (solid line) are compared to those obtained in the variational calcu-lation of [Radici *et al.* (1994)] (dashed line) and the experimental data [Leuschner *et al.* (1994)]. A spectroscopic factor of 0.537 was required for the Green's function calculation, while 0.459 has been used to for the variational one. The figure is adapted from [Polls *et al.* (1997)].

the $0p_{\frac{3}{2}}$ state, depending upon the choice of the optical potential for the outgoing proton [Leuschner *et al.* (1994)].

Figure 17.9 also contains the results for the reduced cross section derived from the variational calculation of the overlap wave function [Radici *et al.* (1994)] for the Argonne v_{14} potential [Wiringa *et al.* (1984)]. The shape of the experimental data is globally reproduced with a slightly better agreement for small negative values of p_m but with a clear underestimation at larger p_m. The overall quality of the fit is somewhat worse than for the Green's function approach and the required adjusted spectroscopic factor is 0.459, below the value of 0.537 for the latter approach. It is not clear however, whether the differences in the calculated reduced cross section are due to the use of different interactions or to the methods employed in calculating the spectral function.

Both the variational calculations of [Radici *et al.* (1994); Fabrocini and Co' (2001)] and the Green's function approach give spectroscopic factors for the p-shell orbitals of about 0.90. This value should be compared to the $\sim$0.63 deduced from the NIKHEF experiment for these orbits. The additional contribution, due to the proper consideration of the center-of-mass motion, raises the theoretical value to 0.98, worsening the agreement with data [Van Neck *et al.* (1998)]. Nevertheless, the observation that the same results are obtained with two independent many-body methods but including the same short-range physics, suggests that the effects of short-range correlations on the quasihole strength are well under control. At the same time, the discrepancy with the experiment is partly due to the emphasis on the accurate treatment of short-range correlations, and one should view the quasihole strength that has been discussed here to be due exclusively to their influence. It is clear that a considerable renormalization of the strength is to be expected, due to the coupling of the quasihole states to the low-lying collective excitations, as discussed in Sec. 17.1.2.

The continuum part of the hole spectral strength is found at higher energies and stems from the coupling to the continuum of 1p2h states. At these energies, it is useful to proceed from the BHF propagator with states $|\alpha\rangle$ that diagonalize the corresponding self-energy

$$G_{\ell j}^{(0)}(\alpha; E) = \frac{1}{E - \varepsilon_{\alpha\ell j}^{BHF} \pm i\eta}, \tag{17.53}$$

where the sign in front of the infinitesimal imaginary quantity $i\eta$ is positive (negative) depending on whether $\varepsilon_{\alpha\ell j}^{HF}$ is above or below the corresponding Fermi energy. The two-step procedure is equivalent to the solution method that was discussed in Sec. 14.6 for Eq. (14.123). The Dyson equation can then be solved by iterating the $\Delta\Sigma$ component in Eq. (17.49) of the self-energy to generate the reducible self-energy

$$\langle\alpha|\,\Sigma_{\ell j}^{red}(E)\,|\beta\rangle = \langle\alpha|\,\Delta\Sigma_{\ell j}(E)\,|\beta\rangle$$
$$+ \sum_{\gamma} \langle\alpha|\,\Delta\Sigma_{\ell j}(E)\,|\gamma\rangle\, G_{\ell j}^{(0)}(\gamma; E)\, \langle\gamma|\,\Sigma_{\ell j}^{red}(E)\,|\beta\rangle \tag{17.54}$$

and obtain the propagator from

$$G_{\ell j}(\alpha, \beta; E) = \delta_{\alpha,\beta}\, G_{\ell j}^{(0)}(\alpha; E) + G_{\ell j}^{(0)}(\alpha; E)\, \langle\alpha|\,\Sigma_{\ell j}^{red}(E)\,|\beta\rangle\, G_{\ell j}^{(0)}(\beta; E). \tag{17.55}$$

Using this representation of the Green's function, one can calculate the

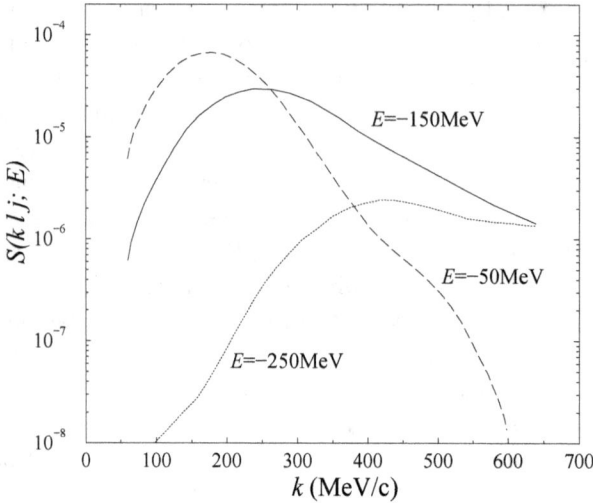

Fig. 17.10   The $p\frac{1}{2}$ spectral strength as a function of momentum at fixed energies corresponding to -50, -150, and -250 MeV, adapted from [Müther and Dickhoff (1994)]. The results demonstrate the increasing importance of high wave-vector components with higher excitation energy in the $A-1$ system.

spectral function in the wave vector basis from [see Eq. (7.11)]

$$S^c(k\ell j; E) = \frac{1}{\pi}\, \text{Im}\left(\sum_{\alpha,\beta} \langle k|\alpha\rangle_{\ell j}\; G_{\ell j}(\alpha, \beta; E)\; \langle \beta|k\rangle_{\ell j} \right). \qquad (17.56)$$

This spectral function is different from zero for energies E below the lowest sp energy of a given BHF state (with ℓj), due to the imaginary part in Σ^{red}. This contribution involves the coupling to the continuum of 1p2h states and is therefore nonvanishing only when the corresponding irreducible self-energy $\Delta\Sigma$ has a non-zero imaginary part, which is given by Eq. (17.47). The 2p1h contribution to the self-energy is responsible for the depletion of strength below the Fermi energy, moving it to high energy in the particle domain. The 1p2h term instead, is essential for the accumulation of sp strength below the Fermi energy from states (in particular those with high momenta) which are empty in the independent-particle description. The continuum contribution of Eq. (17.56) and the quasihole parts of Eq. (17.52) can be combined to generate the complete spectral function

$$S_{\ell j}(k; E) = S^c(k\ell j; E) + S^{qh}(k\ell j; E). \qquad (17.57)$$

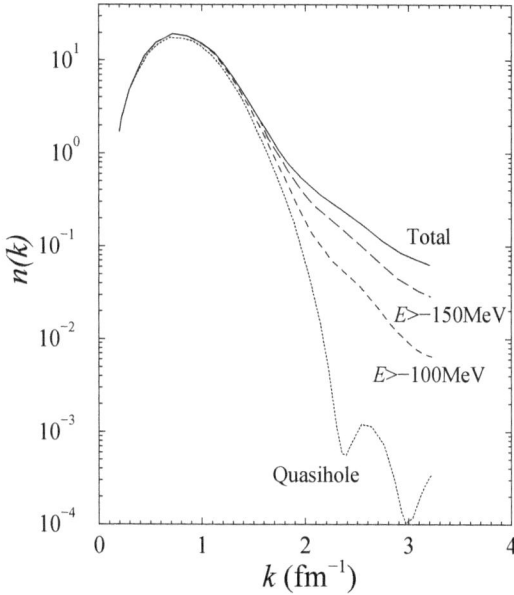

Fig. 17.11 The momentum distribution of ^{16}O, displaying the quasihole contribution and those obtained with various energy cut-offs in the integration of the spectral functions, adapted from [Müther *et al.* (1995)].

The spectral function $S^c(k\ell j; E)$ for the $p\frac{1}{2}$ quantum numbers are shown at three different energies in Fig. 17.10. The long-dashed curve corresponds to -50 MeV, the full one to -150 MeV, and finally the dotted one to -250 MeV. From these results it is clear that an important change in the momentum content of the sp strength occurs with increasing excitation energy in the $A = 15$ system. At higher excitation energy one finds more high-momentum components which are not observed in the quasihole states. This can be concluded from Fig. 17.11 where the total occupation numbers and the contribution from the quasihole states, is compared with results for other energy cuts. The required energy integration of the continuum hole strength for each k follows from Eq. (7.16).

The quasihole part corresponds to the energy domain for $(e, e'p)$ cross sections with small energy transfer, *i.e.* leading to the ground state of the final nucleus and excited states up to ≈ 20 MeV. The curve denoted by $E > -100$ MeV reflects the momentum distribution, including all states of the final nucleus up to around 80 MeV excitation energy, *etc.* As a

Table 17.2   Occupation numbers for ^{16}O.

lj	$\hat{n}^{qh}$	$\hat{n}^c(E < -100)$	$\hat{n}^c$	$\hat{n}$	$\hat{n}/(2(2j+1))$
$s\frac{1}{2}$	3.120	0.244	0.624	3.744	0.936
$p\frac{3}{2}$	7.314	0.133	0.332	7.646	0.956
$p\frac{1}{2}$	3.592	0.086	0.173	3.764	0.941
$d\frac{5}{2}$	0.0	0.106	0.234	0.234	0.020
$d\frac{3}{2}$	0.0	0.108	0.196	0.196	0.025
$f\frac{7}{2}$	0.0	0.063	0.117	0.117	0.007
$f\frac{5}{2}$	0.0	0.084	0.140	0.140	0.012
$\sum$	14.025	0.824	1.816	15.841	

Distribution of nucleons in ^{16}O, adapted from [Müther *et al.* (1995)].
Listed are the total occupation number $\hat{n}$ for various partial waves
[see Eq. (17.58)] but also the contributions from the quasihole ($\hat{n}^{qh}$)
and the continuum part ($\hat{n}^c$) of the spectral function, separately. The
continuum part with contributions originating from energies E below
-100 MeV is listed as well. The last line shows the sum of particle
numbers for all partial waves listed.

consequence, the high-momentum components of the momentum distribution due to short-range correlations can be observed mainly in knockout experiments with an energy transfer of the order of 100 MeV or more. The momentum distribution for ^{16}O calculated with different approaches and different interactions, all of which properly include the effects of short-range correlations, yield quantitatively similar results. The other approaches include: local density approximation (LDA) [Benhar *et al.* (1994)], Fermi hypernetted chain (FHNC) [Có *et al.* (94)] and Variational Monte Carlo (VMC) [Pieper *et al.* (1992)].

To understand Fig. 17.11, it is important to recall that the appearance of high-momentum components at a certain energy in the $A-1$ system is related to the self-energy contribution containing 1p2h states at that energy. From energy conservation it is then clear that at low energy it is much harder to find such states with a high-momentum particle, than at high energy. The same feature is observed in nuclear matter where the peak of the sp spectral function for momenta above k_F increases in energy as k^2. Hence, the hole strength in nuclear matter as a function of momentum, shows the same tendency shown in Fig. 17.10, *i.e.* higher momenta become more dominant at higher excitation energy.

In order to demonstrate the importance of the continuum part of the spectral functions as compared to the quasihole contribution, and to visualize the effects of correlations, we have included in Table 17.2 the particle numbers for each partial wave, including the degeneracy of the states

$$\hat{n}_{\ell j} = 2(2j+1) \int_{-\infty}^{\varepsilon_F} dE \int_0^\infty dk\, k^2 S(k\ell j; E). \qquad (17.58)$$

We have separated the contributions originating from the quasihole states and those due to the continuum [as in Eq. (17.57)]. Only 14.025 out of the 16 nucleons of ^{16}O occupy the quasihole states in this calculation (while the experimental data suggests a much smaller number). Another 1.13 nucleons are found in the 1p2h continuum with partial wave quantum numbers of the s and p shell, while an additional 0.687 nucleons are obtained from the continuum with orbital quantum numbers of the d and f shells. The distinction between quasihole and continuum contributions is somewhat artificial for the $s\frac{1}{2}$ orbital, since the coupling to low-lying 1p2h states leads to a strong fragmentation of the strength, which is observed experimentally [Mougey (1980)]. The depletion of the occupation probabilities of the hole states, indicated in Tab 17.2, is larger for the $s\frac{1}{2}$ orbit. This feature can be ascribed to the closeness of the $s\frac{1}{2}$ BHF energy to the 1p2h continuum, which yields more leakage of strength to the continuum than for the $p\frac{1}{2}$ and $p\frac{3}{2}$ quasihole states. The sum of the particle numbers listed in Table 17.2 is slightly smaller (15.841) than the number of particles in ^{16}O. This is due to the fact that only partial waves up to $\ell = 3$ were taken into account. One must keep in mind that the approach to the sp Green's function, discussed here, is not number-conserving. Indeed, the Green's functions used to evaluate the self-energy, are not determined in the required self-consistent way, as discussed in Ch. 21.

The contributions to the total energy, as derived from the energy sum rule in Eq. (7.26), are displayed in Table 17.3 for different angular momenta. The first two columns give the analogous results obtained from the solution of the BHF and BHF+2p1h terms. The latter includes the 2p1h correction to the nuclear-matter G-matrix. The BHF approach continues to describe the nucleus in terms of fully occupied sp states as in HF. However, as the sp states in BHF+2p1h are more bound, the gain in binding energy from BHF to BHF+2p1h is accompanied by a reduction of the calculated radius of the nucleon distribution. The inclusion of the 1p2h contributions to the self-energy in the complete calculation, reduces the absolute values of the quasihole energies (compare BHF+2p1h and "Total" in Table 17.3).

Table 17.3   Contributions to the energy per particle for ^{16}O.

lj	BHF			BHF+2p1h			Total		
	ϵ	t	ΔE	ϵ	t	ΔE	ϵ	t	ΔE
$s\frac{1}{2}$ qh	−36.9	11.8	−50.3	−42.6	11.9	−61.3	−34.3	11.2	−36.0
$s\frac{1}{2}$ c							−90.4	17.1	−22.9
$p\frac{3}{2}$ qh	−15.4	17.6	9.1	−20.3	19.0	−5.6	−17.9	18.1	0.4
$p\frac{3}{2}$ c							−95.2	35.2	−10.0
$p\frac{1}{2}$ qh	−11.5	16.6	10.3	−17.1	18.5	2.8	−14.1	17.2	5.5
$p\frac{1}{2}$ c							−103.6	35.9	−5.8
$\ell > 1$ c							−98.9	63.2	−12.3
E/A(MeV)		−1.9			−4.0			−5.1	
$\langle r \rangle$(fm)		2.59			2.49			2.55	

Ground-state properties of ^{16}O, adapted from [Müther *et al.* (1995)]. Listed are the energies ϵ and kinetic energies t of the quasihole states (qh) and the corresponding mean values for the continuum contribution (c), normalized to 1, for the various partial waves. Multiplying the sum: $\frac{1}{2}(t + \epsilon)$ of these mean values with the corresponding particle numbers of Table 17.2, one obtains the contribution ΔE to the energy of the ground state [as given by the energy sum rule, Eq. (7.26)]. Summing up all these contributions, and dividing by the nucleon number, yields the energy per nucleon E/A. Results are presented for the BHF, BHF+2p1h and the complete calculation (Total). The particle numbers for the qh states in BHF and BHF+2p1h are equal to the degeneracy of the states, all other occupation numbers are zero. All the energies are given in MeV.

Despite this reduction of the quasihole energies however, the total binding energy is increased as compared to BHF+2p1h. This is mainly due to the continuum part of the spectral function. Comparing various contributions to the integral in Eq. (7.26), one finds that only 37% of the total energy is due to the quasiholes Eq. (17.52). The dominating part (63%) results from the continuum part of the spectral functions, although it merely represents 1.8 nucleons (that is 11% of the total, see Table 17.2).

Summarizing, the calculation of the complete energy dependence of the hole spectral function demonstrates that the presence of high-momentum components in the nuclear ground state will show up unambiguously at high excitation energy when probed by $(e, e'p)$ reactions. These deeply bound nucleons not only generate the enhancement of the momentum distribution for momenta > 400MeV/c, depicted in Fig. 17.11, but they are essential in understanding the binding of nuclear systems. While the binding energy doesn't yield the correct experimental value of the energy per particle (-7.98 MeV/A), its origin is unambiguously related to the presence of high-momentum nucleons at large negative energies. Improvements of the calculation of the binding energy in the present scheme, suggest themselves in an obvious manner. The lack of self-consistency and the neglect of low-energy correlations are clear candidates for such extensions. In addition, three-body forces are expected to yield an important attractive contribution.

17.4 Properties of protons in nuclei

Coincidence experiments involving electron beams have played a crucial role in gathering pertinent information about the behavior of protons in the nucleus. We now proceed to summarize what has been learned in general terms about their properties. Before embarking on this overview, it is important to note that recent work on single-nucleon knockout with fast radioactive beams [Hansen and Tostevin (2003)] proposes to extend the information for valence protons to neutrons and unstable nuclei. Some promising work in this direction has been reported in [Brown *et al.* (2002); Enders *et al.* (2003)]. Indeed, there is a long tradition with hadronic probes to study spectroscopic factors in nuclei. Due to the inherent complexity of hadron-induced reactions it has been more difficult to establish absolute spectroscopic factors from corresponding experimental data. Nevertheless, it is possible to generate a consistent analysis of $(e, e'p)$ and $(d, {}^3\text{He})$ experiments, as shown in [Kramer *et al.* (2001)]. The original discrepancies between these different experiments disappear, with an improved analysis of the $(d, {}^3\text{He})$ reaction. The resulting spectroscopic factors then appear to be quite consistent with those from the $(e, e'p)$ reaction. This observation demonstrates that, with proper care, it is possible to develop valuable information from hadron-induced nucleon knockout experiments. It may therefore be possible to extend the extraction of spectroscopic factors to nuclei far off stability. This exciting new development will allow the study of the properties of nucleons in the nuclear medium in different regions of the periodic table and may provide new challenges for our theoretical understanding. Present knowledge of the properties of protons in the nucleus may be summarized as follows. The consequences of short-range correlations in nuclear systems appear to be theoretically well understood, while

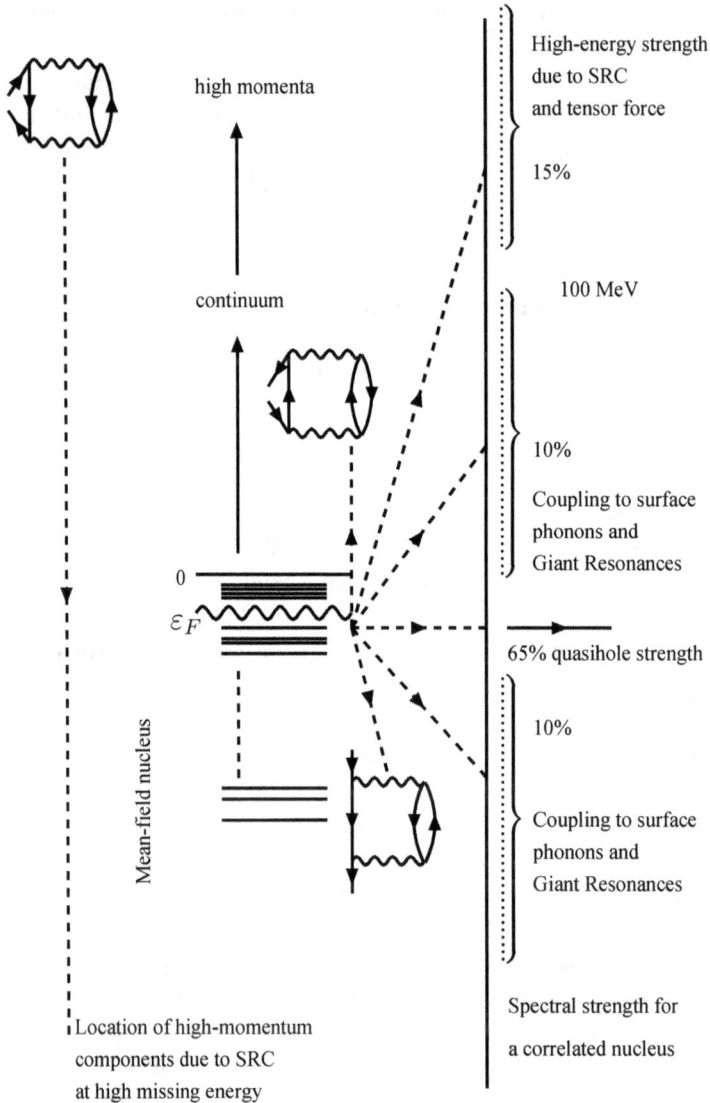

Fig. 17.12   The distribution of single-particle strength in a nucleus like ^{208}Pb. The present summary is a synthesis of experimental and theoretical work discussed in this section. A slight reduction (from 15% to 10%) of the depletion effect due to short-range correlations (SRC) must be considered for light nuclei like ^{16}O.

they are also now becoming available for experimental scrutiny [Rohe *et al.*
(2004)]. These results demonstrate that the effect of short-range correlations is two-fold. First, it involves the depletion of spectroscopic strength
from the mean-field domain as discussed in Sec. 7.8 for the experimental
data obtained for ^{208}Pb. The data in Fig. 7.8 indicate that the depletion
in heavy nuclei corresponds to about 15% for all the deeply bound proton
levels. This was predicted quite some time ago based on nuclear-matter
calculations, reviewed in Sec. 16.3.1. For lighter nuclei all theoretical work
suggests that this amount may be closer to 10%, as discussed in Sec. 17.3.
Accompanying this information is the realization that valence shells near
the Fermi energy will not contain substantial amounts of high-momentum
components. This has been experimentally confirmed and clarifies the other
role played by short-range correlations in nuclei, *i.e.* the admixtures of
high-momentum components at high missing energy that account for the
missing protons removed from the mean-field location. The location of
these high-momentum components [Rohe *et al.* (2004)] broadly conforms
with the mechanism that admixes these correlations with 1p2h states at
large missing energies, discussed in Sec. 17.3.

Being able to identify high-momentum components in addition to locat-
ing all the sp strength, associated with the mean-field orbits [van Batenburg (2001)], completes the identification of the properties of protons in
the ground state of the nucleus. The latter understanding is illustrated in
Fig. 17.12, where several generic diagrams are identified that have unique
physical consequences for the redistribution of the sp strength. The middle
column characterizes the mean-field picture that is used as a starting point
for the theoretical description. The right column identifies the location of
the sp strength of the orbits, just below the Fermi energy, when correlations are included. One may apply this picture, for example, to the $2s\frac{1}{2}$
proton orbit in ^{208}Pb. The physical mechanisms responsible for the correlated strength distribution are also identified. The strength of this orbit,
remaining at the quasihole energy, is about 65%. Long-range correlations
are responsible for the loss of 20% of the strength due to the coupling to
nearby 2p1h and 1p2h states. This loss is symmetrically distributed above
and below the Fermi energy and is physically represented by the coupling
to low-lying surface modes and higher-lying giant resonances. The resulting
occupation number of the orbit therefore corresponds to 75%. More deeply
bound nucleons have higher occupation numbers corresponding to about
85%. As discussed in Sec. 7.8, this is true for all the deep-lying orbits,

and is consistent with a global depletion due to short-range correlations of 15%. The corresponding location of this strength is identified at very high energy in the particle domain and is due to the short-range and tensor correlations induced by a realistic nucleon-nucleon interaction, as discussed in Sec. 16.3.1. The left column depicts the generic diagram that is responsible for the admixture of high-momentum components in the ground state. The energy domain of these high-momentum nucleons is at large missing energies (see Sec. 17.3).

This rather complete picture of the properties of a proton in the nucleus is unique to the field of nuclear physics. Indeed, unlike other fields with strong correlation effects, like particle or condensed matter physics, it is possible in nuclear physics to state that the properties of the constituent protons inside the nucleus are identified experimentally and understood in global theoretical terms. For atoms and molecules it is also possible to extract this information by employing the corresponding $(e, 2e)$ reaction [McCarthy and Weigold (1991); Coplan *et al.* (1994)]. This reaction generates the best possible information on the properties of individual electrons in these systems. Indeed, it was shown in 1981 that one can "measure" the square of the $1s$ wave function of the Hydrogen atom in momentum space [Lohmann and Weigold (1981)], as discussed in Sec. 7.7. Electron wave functions in the medium have been measured for a wide range of atoms and molecules [McCarthy and Weigold (1991)], as summarized in Sec. 7.7. Similar wave functions for nuclei are given in Fig. 7.5 of Sec. 7.8. The technique may also become successful in identifying the properties of electrons in solids [Vos and McCarthy (1995)].

17.5 Exercises

(1) Derive the TDA equations for the pp and hh propagators. Employ the Lehmann representation of Eqs. (17.3) and (17.4) to generate the usual eigenvalue equations from those for the propagators. Check Eq. (17.5).
(2) Do the same for the TDA equations associated with ph and hp propagation, employing Eqs. (17.8) and (17.9). Check also Eq. (17.10).
(3) Check Eqs. (17.14) and (17.18).

Chapter 18

Bogoliubov perturbation expansion for the Bose gas

In the present chapter, the description of interacting Bose systems at $T = 0$ is continued. In the mean-field treatment (Hartree–Bose or Gross–Pitaevskii) of Ch. 12 it was assumed that *all* particles are in the condensate. However, even at $T = 0$, one expects that non-condensate sp states can be occupied because of the interparticle interaction. Taking the quantum depletion of the Bose condensate into account, requires going beyond the HB or GP picture with a genuine boson perturbation theory: the Bogoliubov formalism, which is introduced in the present chapter and applied in some detail to the Bose gas. Some general remarks about this system are given in Sec. 18.1. In Sec. 18.2, the Bogoliubov prescription for dealing with the condensate, is discussed. The resulting perturbation expansion is constructed in Sec. 18.3. An important theorem concerning the chemical potential, first discussed by Hugenholtz and Pines, is the subject of Sec. 18.4. Sec. 18.5 deals with the first-order approximation of the theory, which can also be applied — as shown in Sec. 18.6 — to the Bose gas in the dilute limit, after some considerations about the low-density effective interaction. The lowest-order results are then rederived in Sec. 18.7 by the important technique of a canonical transformation to (Bogoliubov) quasiparticles.

18.1   The Bose gas

We examine in this section the homogeneous Bose gas,[1] mainly because it is conceptually simpler to introduce Bogoliubov perturbation theory when the limit of infinite particle number can be taken and particle-number non-conservation is not an issue. Moreover, the shape of the condensate orbital

---

[1]With some minor adaptations, the discussion in this chapter will mainly follow the reasoning in the pioneering paper by [Hugenholtz and Pines (1959)].

is unambiguous in a uniform system. It should be noted at the outset that no good examples are known of a weakly interacting Bose gas. The atomic BEC presented in Sec. 12.4.5 involves weakly interacting systems that are inherently nonuniform because of the trap confinement. Liquid ^{4}He, which will be examined in Sec. 19.1, is in fact a very strongly interacting system and its description requires considerable effort, going far beyond the lowest orders of perturbation theory presented here.

Just as for the Fermi gas discussed in Ch. 5, we consider N bosons (taken spinless, for simplicity) in a box of volume V with periodic boundary conditions. In the end we are interested in the thermodynamic limit, when $V \to \infty$ and $N \to \infty$ while keeping the density $\rho = N/V$ fixed. The sp states are plane waves due to translational invariance, and the Hamiltonian is given by

$$\hat{H} = \hat{T} + \hat{V} \tag{18.1}$$

$$= \sum_{\boldsymbol{p}} \frac{p^2}{2m} a_{\boldsymbol{p}}^\dagger a_{\boldsymbol{p}} + \frac{1}{2} \sum_{\boldsymbol{p}_1 \boldsymbol{p}_2 \boldsymbol{p}_3 \boldsymbol{p}_4} (\boldsymbol{p}_1 \boldsymbol{p}_2 | V | \boldsymbol{p}_3 \boldsymbol{p}_4) \, a_{\boldsymbol{p}_1}^\dagger a_{\boldsymbol{p}_2}^\dagger a_{\boldsymbol{p}_4} a_{\boldsymbol{p}_3}.$$

where the notation $\varepsilon_p = p^2/(2m)$ will be employed below. We assume a local and central interaction $V(|\boldsymbol{r}_1 - \boldsymbol{r}_2|)$ in coordinate space, so the interaction matrix elements in momentum space read

$$(\boldsymbol{p}_1 \boldsymbol{p}_2 | V | \boldsymbol{p}_3 \boldsymbol{p}_4) = \frac{1}{V} \delta_{\boldsymbol{p}_1 + \boldsymbol{p}_2, \boldsymbol{p}_3 + \boldsymbol{p}_4} W(\boldsymbol{p}_1 - \boldsymbol{p}_3), \tag{18.2}$$

where

$$W(\hbar \boldsymbol{q}) = W(\hbar q) = \int d\boldsymbol{r} \, e^{i \boldsymbol{q} \cdot \boldsymbol{r}} V(r) \tag{18.3}$$

is a real function.

According to Eq. (12.7), the ground state of the free Bose gas [only kinetic energy in Eq. (18.1)] is

$$|\Phi_0^N\rangle = \frac{1}{\sqrt{N!}} (a_0^\dagger)^N |0\rangle, \tag{18.4}$$

and the condensate of Sec. 12.1.2 is the zero-momentum plane-wave state, which is merely a constant in coordinate space

$$\phi(\boldsymbol{r}) = \langle \boldsymbol{r} | a_0^\dagger | 0 \rangle = \frac{1}{\sqrt{V}}. \tag{18.5}$$

The condensate occupation in the noninteracting ground state of Eq. (18.4) is equal to the particle number, $\langle \Phi_0^N | a_0^\dagger a_0 | \Phi_0^N \rangle = N$. When the interaction is turned on, some of the particles are lifted out of the condensate and populate non-zero momentum states. This is the so-called quantum depletion of the condensate, not to be confused with the thermal depletion discussed in Sec. 5.6.1. For the occupation of the condensate in the interacting N-boson ground state $\left| \Psi_0^N \right\rangle$ we therefore have

$$\left\langle \Psi_0^N \right| a_0^\dagger a_0 \left| \Psi_0^N \right\rangle = N_0 < N. \tag{18.6}$$

In the thermodynamic limit N_0 grows proportional to N, *i.e.* N_0 is a constant finite fraction of the total particle number, the condensate fraction $z_0 = N_0/N < 1$. The deviation of z_0 from unity is a measure of the strength of the correlation effects.

The occupation of a non-zero momentum state,

$$(\boldsymbol{p} \neq 0 :) \qquad \left\langle \Psi_0^N \right| a_{\boldsymbol{p}}^\dagger a_{\boldsymbol{p}} \left| \Psi_0^N \right\rangle = \tilde{n}(\boldsymbol{p}), \tag{18.7}$$

on the other hand, becomes constant in the thermodynamic limit, and the total particle number can be written as

$$N = N_0 + \sum_{\boldsymbol{p} \neq 0} \tilde{n}(\boldsymbol{p}) = N_0 + \frac{V}{(2\pi\hbar)^3} \int d\boldsymbol{p}\, \tilde{n}(\boldsymbol{p}). \tag{18.8}$$

In an interacting Bose gas the momentum distribution $n(\boldsymbol{p})$, normalized as

$$\rho = \int \frac{d\boldsymbol{p}}{(2\pi\hbar)^3}\, n(\boldsymbol{p}), \tag{18.9}$$

thus has the form

$$n(\boldsymbol{p}) = (2\pi\hbar)^3 z_0 \rho\, \delta(\boldsymbol{p}) + \tilde{n}(\boldsymbol{p}) \tag{18.10}$$

of a delta-spike superimposed on a continuous background. As shown by [Gavoret and Noziéres (1964)], the background $\tilde{n}(p)$ actually has a weak divergence $\sim p^{-1}$ in the limit $p \to 0$, making a negligible contribution to the integral in Eq. (18.8). Note that here and in the following, we make the assumption that the system has a meaningful thermodynamic limit, in the sense that the energy per particle remains finite as $N \to \infty$.

18.2 Bogoliubov prescription

The zero-momentum sp state is special because its occupation $N_0 = z_0 N$ is the *only* one that grows proportionally to N as $N \to \infty$. As a consequence one finds *e.g.* that to leading order in an expansion of powers of N,

$$\langle \Psi_0^N | a_0 a_0^\dagger | \Psi_0^N \rangle = 1 + N_0 \approx N_0 = \langle \Psi_0^N | a_0^\dagger a_0 | \Psi_0^N \rangle. \tag{18.11}$$

More generally

$$\langle \Psi_0^N | \hat{P} | \Psi_0^N \rangle \sim N_0^m \tag{18.12}$$

when $\hat{P}$ is a product of m addition operators $a_0^\dagger$ and m removal operators a_0 ordered in any way. Similarly, one has for the addition and removal amplitudes

$$| \langle \Psi_0^{N-1} | a_0 | \Psi_0^N \rangle | \approx | \langle \Psi_0^{N+1} | a_0^\dagger | \Psi_0^N \rangle | \approx \sqrt{N_0}, \tag{18.13}$$

and by induction it follows that

$$| \langle \Psi_0^{N-m} | (a_0)^m | \Psi_0^N \rangle | \approx N_0^{m/2}. \tag{18.14}$$

We observe in Eqs. (18.11) – (18.14) that, to leading order in N, $a_0^\dagger$ and a_0 behave just as ordinary numbers

$$a_0^\dagger \approx a_0 \rightarrow \sqrt{N_0}. \tag{18.15}$$

The replacement of the condensate operators with $\sqrt{N_0}$ in Eq. (18.15) is called the *Bogoliubov prescription*.[2]

The special treatment of the condensate in the Bogoliubov prescription lies in the fact that it is regarded as a classical field, commuting with all remaining operators. More explicitly, it decomposes the removal operator for a particle at position r according to

$$a_r = \sum_p \frac{e^{\frac{i}{\hbar} p \cdot r}}{\sqrt{V}} a_p \approx \sqrt{N_0} \phi(r) + \delta a_r. \tag{18.16}$$

The condensate wavefunction $\phi(r) = 1/\sqrt{V}$ is a number and the remainder is an operator

$$\delta a_r = \sum_{p \neq 0} \frac{e^{\frac{i}{\hbar} p \cdot r}}{\sqrt{V}} a_p, \tag{18.17}$$

which is orthogonal to $\phi(r)$,

$$\int dr \, \phi^*(r) \delta a_r = 0. \tag{18.18}$$

---

[2] The original idea of treating the zero-momentum state as a number is due to [Bogoliubov (1947)].

18.2.1 *Particle-number nonconservation*

Application of the Bogoliubov prescription to the Hamiltonian $\hat{H}$ in Eq. (18.1) leads to a new Hamiltonian $\hat{H}_B(N_0)$ given by

$$\hat{H}_B(N_0) = \hat{T} + \hat{V}_B(N_0), \tag{18.19}$$

where the parametric dependence on N_0 is made explicit. The unknown occupation of the condensate N_0 is an additional variable in the theory which must be determined by minimizing the energy with respect to it. The kinetic energy $\hat{T}$ receives no contribution from the zero-momentum state.

A useful expression for $\hat{V}_B(N_0)$ can be found by classifying the contributions to Eq. (18.1) according to the number of $\boldsymbol{p} \neq 0$ operators,

$$\hat{V}_B(N_0) = \sum_{i,j=0}^{2} \hat{V}_{i,j}(N_0), \tag{18.20}$$

where $\hat{V}_{i,j}$ contains the interaction terms in Eq. (18.1) with i non-condensate addition operators and j non-condensate removal operators. Explicitly, we find

$$V_{0,0} = \frac{1}{2} N_0^2 \left(00 \right| V \left| 00 \right), $$

$$\hat{V}_{2,0} = \frac{1}{2} N_0 \sum_{\boldsymbol{p}_1, \boldsymbol{p}_2 \neq 0} \left(\boldsymbol{p}_1 \boldsymbol{p}_2 \right| V \left| 00 \right) a_{\boldsymbol{p}_1}^\dagger a_{\boldsymbol{p}_2}^\dagger, $$

$$\hat{V}_{0,2} = \frac{1}{2} N_0 \sum_{\boldsymbol{p}_3, \boldsymbol{p}_4 \neq 0} \left(00 \right| V \left| \boldsymbol{p}_3 \boldsymbol{p}_4 \right) a_{\boldsymbol{p}_4} a_{\boldsymbol{p}_3}, $$

$$\hat{V}_{1,1} = N_0 \sum_{\boldsymbol{p}_1, \boldsymbol{p}_3 \neq 0} [\left(\boldsymbol{p}_1 0 \right| V \left| \boldsymbol{p}_3 0 \right) + \left(\boldsymbol{p}_1 0 \right| V \left| 0 \boldsymbol{p}_3 \right)] a_{\boldsymbol{p}_1}^\dagger a_{\boldsymbol{p}_3}, $$

$$\hat{V}_{2,1} = \sqrt{N_0} \sum_{\boldsymbol{p}_1, \boldsymbol{p}_2, \boldsymbol{p}_3 \neq 0} \left(\boldsymbol{p}_1 \boldsymbol{p}_2 \right| V \left| \boldsymbol{p}_3 0 \right) a_{\boldsymbol{p}_1}^\dagger a_{\boldsymbol{p}_2}^\dagger a_{\boldsymbol{p}_3}, $$

$$\hat{V}_{1,2} = \sqrt{N_0} \sum_{\boldsymbol{p}_2, \boldsymbol{p}_3, \boldsymbol{p}_4 \neq 0} \left(0 \boldsymbol{p}_2 \right| V \left| \boldsymbol{p}_3 \boldsymbol{p}_4 \right) a_{\boldsymbol{p}_2}^\dagger a_{\boldsymbol{p}_4} a_{\boldsymbol{p}_3}, $$

$$\hat{V}_{2,2} = \frac{1}{2} \sum_{\boldsymbol{p}_1, \boldsymbol{p}_2, \boldsymbol{p}_3, \boldsymbol{p}_4 \neq 0} \left(\boldsymbol{p}_1 \boldsymbol{p}_2 \right| V \left| \boldsymbol{p}_3 \boldsymbol{p}_4 \right) a_{\boldsymbol{p}_1}^\dagger a_{\boldsymbol{p}_2}^\dagger a_{\boldsymbol{p}_4} a_{\boldsymbol{p}_3}. \tag{18.21}$$

Note that in the Bose gas, the terms $\hat{V}_{1,0}$ and $\hat{V}_{0,1}$ (with only one non-condensate operator) are zero because of momentum conservation, and that $V_{0,0}$ is just a number, which we recognize as the HB energy shift in Eq. (12.46) but with the depleted occupation N_0 instead of N.

The derivative with respect to N_0 of $\hat{V}_B(N_0)$ in Eq. (18.20) also has a simple expression in terms of the $\hat{V}_{i,j}$. Since each $\hat{V}_{i,j}$ in Eq. (18.21) contains a power $N_0^{p_{i,j}}$, with $p_{i,j} = 2 - \frac{i+j}{2}$, one has the following important relation,

$$N_0 \frac{\partial}{\partial N_0} \hat{V}_B(N_0) = 2\hat{V}_B(N_0) - \sum_{i,j=0}^{2} \frac{i+j}{2} \hat{V}_{i,j}. \qquad (18.22)$$

In the new Hamiltonian $\hat{H}_B(N_0)$ of Eq. (18.19) the remaining removal operators $a_{\boldsymbol{p}}$ [with $\boldsymbol{p} \neq 0$] all annihilate the noninteracting ground state in Eq. (18.4). Consequently, the difficulty mentioned in Sec. 12.2.1 is now absent and Wick's theorem can be used to evaluate matrix elements. On the downside, Eq. (18.21) also implies that the Hamiltonian $\hat{H}_B(N_0)$ in Eq. (18.19) does not conserve the number of non-condensate particles: applying the Bogoliubov prescription to the number operator,

$$\hat{N}_B(N_0) = N_0 + \sum_{\boldsymbol{p} \neq 0} a_{\boldsymbol{p}}^\dagger a_{\boldsymbol{p}}, \qquad (18.23)$$

it is clear that $[\hat{N}_B(N_0), \hat{H}_B(N_0)] \neq 0$ and an extra condition on the expectation value of $\hat{N}_B(N_0)$ will have to be imposed to fix the total number of particles.

The Bogoliubov prescription has replaced the original problem of determining the N-boson ground state $|\Psi_0^N\rangle$ of $\hat{H}$ by that of finding the many-boson state $|\Psi_{B0}\rangle$ *and* the condensate occupation N_0 for which the expectation value

$$E_{B0}^N = \langle \Psi_{B0}| \hat{H}_B(N_0) |\Psi_{B0}\rangle \qquad (18.24)$$

is minimal, subject to the particle-number constraint

$$N = \langle \Psi_{B0}| \hat{N}_B(N_0) |\Psi_{B0}\rangle = N_0 + \sum_{\boldsymbol{p} \neq 0} \langle \Psi_{B0}| a_{\boldsymbol{p}}^\dagger a_{\boldsymbol{p}} |\Psi_{B0}\rangle. \qquad (18.25)$$

The resulting solutions depend on N and are denoted as $|\Psi_{B0}^N\rangle$ and N_{0N}.

Note that $|\Psi_{B0}^N\rangle$ is not an eigenstate of $\hat{N}$ and therefore does not coincide with the true N-boson ground state $|\Psi_0^N\rangle$. Analogous to a familiar argument from statistical mechanics however, one can expect that the fluctuation from the mean particle number

$$\Delta N = \sqrt{\langle \Psi_{B0}^N| \hat{N}_B^2(N_0) |\Psi_{B0}^N\rangle - N^2} \qquad (18.26)$$

only goes like $\sqrt{N}$, and the relative fluctuation $\Delta N/N$ vanishes as $N \to \infty$. Hence, $\left|\Psi_{B0}^N\right\rangle$ and $\left|\Psi_0^N\right\rangle$ will have the same expectation values of number-conserving operators in the thermodynamic limit. This corresponds to the well-known equivalence (in the thermodynamical limit) of the grand canonical and canonical ensembles for describing systems with a fixed number of particles. In anticipation of the discussion in Sec. 18.3 one should also keep in mind that $\left|\Psi_{B0}^N\right\rangle$ has so-called anomalous matrix elements, *e.g.*, $\left\langle\Psi_{B0}^N\right|a_{\boldsymbol{p}}^{\dagger}a_{-\boldsymbol{p}}^{\dagger}\left|\Psi_{B0}^N\right\rangle \neq 0$. An example of a many-boson wave function without fixed particle number is given in Exercise (1) at the end of this chapter.

Instead of the constrained minimum search in Eqs. $(18.24) - (18.25)$, it is far more convenient to introduce a Lagrange multiplier and perform an *unconstrained* minimization of the expectation value of

$$\hat{\Omega}_B(N_0, \mu) = \hat{H}_B(N_0) - \mu \hat{N}_B(N_0). \tag{18.27}$$

This operator is in fact identical to the thermodynamical potential of the grand-canonical ensemble at $T = 0$ [see Eq. (5.31)], with μ playing the role of the chemical potential. We now have to find the many-boson state $\left|\Psi_{B0}\right\rangle$ and N_0 for which the expectation value of Eq. (18.27) is minimal. The particle number constraint in Eq. (18.25) can be used to find μ as a function of N and eliminate μ in favor of N. Alternatively, one may decide to keep the chemical potential μ instead of N as an independent variable.

18.2.2   *The chemical potential*

We proceed by first determining, as a function of N_0 and μ, the ground state of $\hat{\Omega}_B(N_0, \mu)$ or equivalently, the state $\left|\Psi_{B0}(N_0, \mu)\right\rangle$ minimizing

$$\Omega_{B0}(N_0, \mu) = \left\langle\Psi_{B0}(N_0, \mu)\right| \hat{\Omega}_B(N_0, \mu) \left|\Psi_{B0}(N_0, \mu)\right\rangle. \tag{18.28}$$

This is a many-body problem for which the perturbative method of Ch. 8 can provide a(n approximate) solution. The variational property of the state $\left|\Psi_{B0}(N_0, \mu)\right\rangle$ then implies that with respect to small variations in N_0 or μ,

$$\frac{\partial\Omega_{B0}}{\partial N_0}(N_0, \mu) = \left\langle\Psi_{B0}(N_0, \mu)\right| \left(\frac{\partial\hat{\Omega}_B}{\partial N_0}(N_0, \mu)\right) \left|\Psi_{B0}(N_0, \mu)\right\rangle \tag{18.29}$$

and

$$\frac{\partial\Omega_{B0}}{\partial\mu}(N_0, \mu) = \left\langle\Psi_{B0}(N_0, \mu)\right| \left(\frac{\partial\hat{\Omega}_B}{\partial\mu}(N_0, \mu)\right) \left|\Psi_{B0}(N_0, \mu)\right\rangle. \tag{18.30}$$

Introducing the expectation values

$$E_{B0}(N_0, \mu) = \langle \Psi_{B0}(N_0, \mu) | \, \hat{H}_B(N_0) \, | \Psi_{B0}(N_0, \mu) \rangle, \qquad (18.31)$$

$$N_{B0}(N_0, \mu) = \langle \Psi_{B0}(N_0, \mu) | \, \hat{N}_B(N_0) \, | \Psi_{B0}(N_0, \mu) \rangle, \qquad (18.32)$$

$$U_{B0}(N_0, \mu) = \langle \Psi_{B0}(N_0, \mu) | \left(\frac{\partial \hat{V}_B}{\partial N_0}(N_0) \right) | \Psi_{B0}(N_0, \mu) \rangle, \quad (18.33)$$

and employing Eqs. (18.23) and (18.27), Eqs. (18.29) – (18.30) can be rewritten more transparently as

$$\frac{\partial \Omega_{B0}}{\partial N_0}(N_0, \mu) = U_{B0}(N_0, \mu) - \mu, \qquad (18.34)$$

$$\frac{\partial \Omega_{B0}}{\partial \mu}(N_0, \mu) = -N_{B0}(N_0, \mu). \qquad (18.35)$$

In a next step one can impose the minimum condition for variations in N_0, as well as the particle-number constraint,

$$\frac{\partial \Omega_{B0}}{\partial N_0}(N_0, \mu) = 0, \qquad (18.36)$$

$$N_{B0}(N_0, \mu) = N. \qquad (18.37)$$

The solutions μ_N and N_{0N} of Eqs. (18.36) – (18.37) then allow to express all quantities as a function of particle number N and make contact with the original problem in Eqs. (18.24) – (18.25), *i.e.*

$$\Omega_{B0}^N = \Omega_{B0}(N_{0N}, \mu_N), \quad E_{B0}^N = E_{B0}(N_{0N}, \mu_N), \dots \qquad (18.38)$$

We now derive three important relations for the chemical potential, which will be of use later on. First we calculate the total derivative

$$\frac{d\Omega_{B0}^N}{dN} = \frac{\partial \Omega_{B0}}{\partial N_0} \frac{dN_{0N}}{dN} + \frac{\partial \Omega_{B0}}{\partial \mu} \frac{d\mu_N}{dN} = -N_{B0}^N \frac{d\mu_N}{dN}, \qquad (18.39)$$

applying Eqs. (18.35) - (18.36). The partial derivatives in Eq. (18.39) are evaluated at $N_0 = N_{0N}$ and $\mu = \mu_N$. Comparing Eq. (18.39) with the result following directly from Eq. (18.27),

$$\frac{d\Omega_{B0}^N}{dN} = \frac{dE_{B0}^N}{dN} - \frac{d}{dN}(\mu_N N_{B0}^N), \qquad (18.40)$$

and using Eq. (18.37) leads to the expression for the chemical potential

$$\mu_N = \frac{dE_{B0}^N}{dN}, \qquad (18.41)$$

which is familiar from thermodynamics.

Secondly, as follows directly from Eq. (18.34), the requirement in Eq. (18.36) is equivalent to

$$\mu_N = U_{B0}^N. \tag{18.42}$$

Finally, it will also be convenient to decompose the operator $\hat{\Omega}_B$ in Eq. (18.27) as

$$\hat{\Omega}_B(N_0, \mu) = \hat{T}_\mu + \hat{V}_B(N_0) - \mu N_0, \tag{18.43}$$

where

$$\hat{T}_\mu = \sum_{\boldsymbol{p}\neq 0} [\varepsilon_p - \mu] a_{\boldsymbol{p}}^\dagger a_{\boldsymbol{p}} = \sum_{\boldsymbol{p}\neq 0} \varepsilon_{p\mu} a_{\boldsymbol{p}}^\dagger a_{\boldsymbol{p}}. \tag{18.44}$$

Since the constant term $-\mu N_0$ in Eq. (18.43) does not change the eigenstate of $\hat{\Omega}_B$, we also have

$$\mu_N = \frac{\partial}{\partial N_0} \langle \Psi_{B0}(N_0, \mu)| \hat{T}_\mu + \hat{V}_B(N_0) |\Psi_{B0}(N_0, \mu)\rangle \tag{18.45}$$

where the right-hand side is again evaluated at $\mu = \mu_N$ and $N_0 = N_{0N}$. Note that the expressions (18.41), (18.42), and (18.45) are only implicit equations for the chemical potential, since the right-hand side depends on μ_N.

18.2.3 *Propagator*

The boson propagator pertaining to the many-body problem in Eq. (18.28) can be defined exactly as in Sec. 12.1.1, but with $\hat{\Omega}_B(N_0, \mu)$ and $|\Psi_{B0}(N_0, \mu)\rangle$ playing the role of $\hat{H}$ and $|\Psi_0^N\rangle$, respectively. For notational simplicity we drop from now on the explicit dependence on N_0 and μ. The propagator is then given by

$$i\hbar G_\Omega(\boldsymbol{p}; t - t') = \langle \Psi_{B0}| \mathcal{T} \left[a_{\boldsymbol{p}\Omega}(t) a_{\boldsymbol{p}\Omega}^\dagger(t') \right] |\Psi_{B0}\rangle \tag{18.46}$$

with the Heisenberg picture defined as

$$a_{\boldsymbol{p}\Omega}^\dagger(t) = e^{\frac{i}{\hbar}\hat{\Omega}_B t} a_{\boldsymbol{p}}^\dagger e^{-\frac{i}{\hbar}\hat{\Omega}_B t}. \tag{18.47}$$

Applying the Bogoliubov prescription

$$a_{\boldsymbol{p}}^\dagger = \delta_{\boldsymbol{p},0}\sqrt{N_0} + (1 - \delta_{\boldsymbol{p},0})a_{\boldsymbol{p}}^\dagger \tag{18.48}$$

to Eq. (18.46) yields

$$i\hbar G_{\Omega B}(\boldsymbol{p}; t - t') = (1 - \delta_{\boldsymbol{p},0})i\hbar G(\boldsymbol{p}; t - t') + \delta_{\boldsymbol{p},0}N_0. \tag{18.49}$$

The energy representation of Eq. (12.6) follows by FT,

$$G_{\Omega B}(\boldsymbol{p}; E) = (1 - \delta_{\boldsymbol{p},0})G(\boldsymbol{p}; E) - 2\pi i\delta_{\boldsymbol{p},0}N_0\delta(E). \tag{18.50}$$

The last term in Eqs. (18.49) – (18.50) represents the contribution from the (depleted) condensate [see Eqs. (12.8) – (12.9)]. The noncondensate propagator $G(\boldsymbol{p}; E)$ in Eq. (18.50) can be rewritten in the usual fashion, by inserting a complete set of eigenstates of $\hat{\Omega}_B$,

$$\begin{aligned}
G(\boldsymbol{p}; E) = \sum_n &\frac{\langle \Psi_{B0}| a_{\boldsymbol{p}} |\Psi_{Bn}\rangle \langle \Psi_{Bn}| a_{\boldsymbol{p}}^\dagger |\Psi_{B0}\rangle}{E - (\Omega_{Bn} - \Omega_{B0}) + i\eta} \\
&- \sum_n \frac{\langle \Psi_{B0}| a_{\boldsymbol{p}}^\dagger |\Psi_{Bn}\rangle \langle \Psi_{Bn}| a_{\boldsymbol{p}} |\Psi_{B0}\rangle}{E + (\Omega_{Bn} - \Omega_{B0}) - i\eta}.
\end{aligned} \tag{18.51}$$

Note that $\hat{\Omega}_B$ commutes with the total momentum operator $\hat{\boldsymbol{P}}$ [see Eq. (10.157)]. Since the ground state $|\Psi_{B0}\rangle$ has zero total momentum, the eigenstates contributing to the first (second) term in Eq. (18.51) have total momentum $\boldsymbol{p}$ $(-\boldsymbol{p})$, but the dependence will not be written explicitly. As the system is isotropic we have in addition that $G(\boldsymbol{p}; E) = G(p; E)$.

Analogous to Sec. 7.4 one can show that all relevant ground-state quantities are contained in the propagator $G(p; E)$. The expectation value of any sp operator, *e.g.*, can be obtained from the momentum distribution $n(p)$ defined in Eq. (18.10),

$$n(p) = \langle \Psi_{B0}| a_{\boldsymbol{p}}^\dagger a_{\boldsymbol{p}} |\Psi_{B0}\rangle = N_0\delta_{\boldsymbol{p},0} - \int \frac{dE}{2\pi i} G(p; E)e^{i\eta E} \tag{18.52}$$

where $\eta \to 0^+$ picks up only the contributions from the second term of Eq. (18.51). The kinetic energy, *e.g.*, equals

$$\overline{T} = \sum_{\boldsymbol{p}\neq 0} \varepsilon_p n(p) = -\sum_{\boldsymbol{p}\neq 0} \int \frac{dE}{2\pi i} \varepsilon_p G(p; E)e^{i\eta E}. \tag{18.53}$$

Also the expectation value of the potential energy can be written in terms of the propagator. The following general formula is easily derived from the basic boson commutation rules in Sec. 2.2:

$$\sum_{\boldsymbol{p}} a_{\boldsymbol{p}}^\dagger [a_{\boldsymbol{p}}, \prod_{i=1}^n a_{\boldsymbol{p}_i}^\dagger] = n \prod_{i=1}^n a_{\boldsymbol{p}_i}^\dagger. \tag{18.54}$$

As a consequence we have

$$\sum_{\boldsymbol{p}\neq 0} a_{\boldsymbol{p}}^{\dagger}\,[a_{\boldsymbol{p}},\hat{V}_{i,j}] = i\hat{V}_{i,j}; \qquad \sum_{\boldsymbol{p}\neq 0}[\hat{V}_{i,j},a_{\boldsymbol{p}}^{\dagger}]a_{\boldsymbol{p}} = j\hat{V}_{i,j}. \qquad (18.55)$$

The average of both expressions in Eq. (18.55) leads to

$$\frac{1}{2}\sum_{\boldsymbol{p}\neq 0}\left(a_{\boldsymbol{p}}^{\dagger}[a_{\boldsymbol{p}},\hat{\Omega}_{B}] + [\hat{\Omega}_{B},a_{\boldsymbol{p}}^{\dagger}]a_{\boldsymbol{p}}\right) = \hat{T}_{\mu} + 2\hat{V}_{B} - N_{0}\frac{\partial}{\partial N_{0}}\hat{V}_{B}, \qquad (18.56)$$

where use has been made of Eq. (18.22) and Eq. (18.43).

The expectation value of the left-hand side of Eq. (18.56) in the state $|\Psi_{B0}\rangle$ is recognized as the mean removal energy $\overline{R}$, and can easily be expressed in terms of the propagator,

$$\begin{aligned}
\overline{R} &= \sum_{\boldsymbol{p}\neq 0} \langle\Psi_{B0}|\,a_{\boldsymbol{p}}^{\dagger}[a_{\boldsymbol{p}},\hat{\Omega}_{B}]\,|\Psi_{B0}\rangle \\
&= \sum_{\boldsymbol{p}\neq 0}\sum_{n} \langle\Psi_{B0}|\,a_{\boldsymbol{p}}^{\dagger}\,|\Psi_{Bn}\rangle\,\langle\Psi_{Bn}|\,[a_{\boldsymbol{p}},\hat{\Omega}_{B}]\,|\Psi_{B0}\rangle \\
&= \sum_{\boldsymbol{p}\neq 0}\sum_{n}(\Omega_{B0}-\Omega_{Bn})|\,\langle\Psi_{Bn}|\,a_{\boldsymbol{p}}\,|\Psi_{B0}\rangle\,|^{2} \\
&= -\sum_{\boldsymbol{p}\neq 0}\int \frac{dE}{2\pi i}\,E\,G(p;E)\mathrm{e}^{i\eta E}. \qquad (18.57)
\end{aligned}$$

Equation (18.56) therefore also implies

$$\overline{R} = \langle\Psi_{B0}|\,\hat{T}_{\mu} + 2\hat{V}_{B}\,|\Psi_{B0}\rangle - \mu N_{0} = \langle\Psi_{B0}|\,\hat{T} + 2\hat{V}_{B}\,|\Psi_{B0}\rangle - \mu N, \quad (18.58)$$

where Eqs. (18.33) and (18.42) have been employed.

The final expressions for the expectation value of $\hat{V}_{B}$, $\hat{H}_{B}$ and $\hat{\Omega}_{B}$ in terms of propagator quantities are now easily derived,

$$\langle\Psi_{B0}|\,\hat{V}_{B}\,|\Psi_{B0}\rangle = \frac{1}{2}(\mu N - \overline{T} + \overline{R}) \qquad (18.59)$$

$$E_{B0} = \langle\Psi_{B0}|\,\hat{H}_{B}\,|\Psi_{B0}\rangle = \frac{1}{2}(\mu N + \overline{T} + \overline{R}) \qquad (18.60)$$

$$\Omega_{B0} = \langle\Psi_{B0}|\,\hat{\Omega}_{B}\,|\Psi_{B0}\rangle = \frac{1}{2}(-\mu N + \overline{T} + \overline{R}). \qquad (18.61)$$

In spite of the apparent simplicity of the above equations one should keep in mind that, as the propagator depends on μ [see *e.g.* Eq. (18.51)], the same holds for $\overline{T}$ and $\overline{R}$. In principle the expression (18.60) for the energy, upon substitution of Eq. (18.41), is therefore a complicated nonlinear differential equation in $E_{B0}(N)$. In perturbative calculations (*e.g.* the dilute limit studied in Sec. 18.6) the situation brightens up considerably: the propagator quantity $\bar{T} + \bar{R}$ in Eq. (18.60) can then be evaluated with a lower-order approximation for the chemical potential. One is left with a linear first-order differential equation, determining the energy up to a quadratic solution of the homogeneous part $E_{B0} - \frac{1}{2} N \frac{dE_{B0}}{dN} = 0$. The latter solution however, corresponds to the absolute leading term in the low-density expansion and is already known from the GP treatment of Ch.12, *i.e.* Eq. (12.105) applied to the Bose gas with $U(\boldsymbol{r}) = 0$ and $\phi_c(\boldsymbol{r}) = 1/\sqrt{V}$ yields

$$E_{B0} = \frac{1}{2} g N^2 / V \quad \text{and} \quad \mu = g\rho. \qquad (18.62)$$

18.3 Bogoliubov perturbation expansion

At long last we are in a position to take advantage of the perturbative machinery developed in Ch. 8. In fact, the reasoning in that chapter can be taken over completely, with $\hat{\Omega}_B = \hat{\Omega}_0 + \hat{\Omega}_1 - \mu N_0$ in the role of the Hamiltonian, which is split up in a noninteracting piece

$$\hat{\Omega}_0 = \hat{T}_\mu \qquad (18.63)$$

defining the interaction picture, *i.e.*

$$a_{\boldsymbol{p}}^\dagger(t) = e^{\frac{i}{\hbar}\hat{\Omega}_0 t} a_{\boldsymbol{p}}^\dagger e^{-\frac{i}{\hbar}\hat{\Omega}_0 t}, \qquad (18.64)$$

and the interaction $\hat{\Omega}_1 = \hat{V}_B$. Note that the constant term $-\mu N_0$ plays no role in determining the ground state at fixed μ and N_0.

As a consequence, the following perturbation expansion for the propagator holds [see Eq. (8.55)]:

$$i\hbar G(\boldsymbol{p}; t - t') = \sum_{m=0}^\infty \frac{1}{(i\hbar)^m m!} \int dt_1 .. \int dt_m$$
$$\times \langle 0| \mathcal{T} \left[\hat{\Omega}_1(t_1)..\hat{\Omega}_1(t_m) a_{\boldsymbol{p}}(t) a_{\boldsymbol{p}}^\dagger(t') \right] |0\rangle_{connected}. \qquad (18.65)$$

The noninteracting ground state $|0\rangle$ in Eq. (18.65) is the vacuum state (without noncondensate bosons). The zero'th order term is

$$i\hbar G^{(0)}(\boldsymbol{p}; t - t') = \langle 0| \mathcal{T} \left[a_{\boldsymbol{p}}(t) a_{\boldsymbol{p}}^\dagger(t') \right] |0\rangle = \theta(t - t') e^{-\frac{i}{\hbar}\varepsilon_{p\mu}(t - t')} \qquad (18.66)$$

or, in the energy representation,

$$G^{(0)}(\boldsymbol{p}; E) = \frac{1}{E - \varepsilon_{p\mu} + i\eta}. \tag{18.67}$$

Wick's theorem can now be used in the usual manner [see Sec. 8.3] to evaluate the matrix elements appearing in Eq. (18.65), and the resulting terms are again represented by diagrams.

The diagram rules can be taken over from Sec. 8.6, with two important modifications:

(1) The boson operators in a time ordered product commute, so we don't have to worry about overall signs.

(2) The contractions in the $\mathcal{T}$-product are necessarily between noncondensate (NC) operators. A contraction will be represented by a directed NC boson line. The condensate (C) operators have been replaced with numbers in the Bogoliubov prescription, but for bookkeeping purposes they are graphically represented by short arrows entering or leaving a vertex. For the mth order contribution to the propagator $G(\boldsymbol{p}; t - t')$ in the time representation the diagram rules of Sec. 8.6 then reduce to

Rule 1: Draw all topologically distinct and connected diagrams with m horizontal interaction lines for V (represented by dashed lines). Each vertex must have one ingoing and one outgoing element, which can either be a directed NC line or a C arrow.

Rule 2: Label each interaction line with a time. Assign momenta to the internal NC propagators, making sure that momentum is conserved for every interaction V. Each C arrow carries zero momentum. Assign propagators $G^{(0)}(\boldsymbol{p}_n; t_i - t_j)$ to any directed NC line, *e.g.*

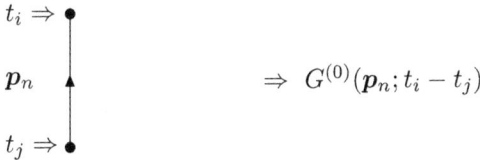

$$
\begin{array}{ll}
t_i \Rightarrow \bullet & \\
\qquad \Big\uparrow \boldsymbol{p}_n & \Rightarrow G^{(0)}(\boldsymbol{p}_n; t_i - t_j) \\
t_j \Rightarrow \bullet &
\end{array}
$$

Assign (see the fermion case) a matrix element $(\boldsymbol{p}_i \boldsymbol{p}_j | V | \boldsymbol{p}_m \boldsymbol{p}_n)$ to an interaction line, filling in zero momentum if a C arrow enters or leaves a vertex.

Rule 3: Sum over the internal momenta and integrate over the internal time variables.

Rule 4: Include an overall factor $N_0^{m_c}(i\hbar)^{m-m_c}$ where $2m_c$ is the number of C-arrows in the diagram. The number of C-arrows must obviously

Fig. 18.1   First-order diagrams for G.

be even, since the contractions remove pairwise the NC operators from the original $4m+2$ sp operators in the $\mathcal{T}$-product. The number of NC lines (including the two external ones) in the diagram is then $m_{nc} = 2m + 1 - m_c$. The power of $(i\hbar)$ follows from the expansion $(-m)$, from the definition (-1), and from the number of contractions $(+m_{nc})$.

Note that the diagrammatic expansion for bosons is somewhat simpler than in the fermion case. Looking at the structure of the contractions in Eq. (18.65), it is clear that any internal contraction within a $\hat{V}_B(t)$ operator is zero, since the NC operators are already in normal order. As a consequence, any NC line must have its start and end point at different interactions. Moreover, the NC lines have only forward propagation, so in a nonzero diagram all NC lines can be made to run in the same (upward) direction. This automatically excludes closed loops of NC lines.

The diagram rules for the propagator $G(\boldsymbol{p}; E)$ in the energy representation are obtained with a few straightforward changes:

Rule 2: Assign momenta and energies to the internal NC propagators and make sure that momentum *and* energy is conserved for each interaction V. Each C arrow of course carries zero momentum and energy. Assign a propagator $G^{(0)}(\boldsymbol{p}_n; E_i)$ to a directed NC line.

Rule 3: Sum over the internal momenta and integrate over the internal energy variables.

Rule 4: Include an overall factor $N_0^{m_c}(\frac{i}{2\pi})^{m-m_c}$. Compared to the time representation there is an additional factor $(2\pi\hbar)^{-m_{nc}}$ from the FT of the propagators, and a factor $(2\pi\hbar)^{m+1}$ from the energy-conserving δ-functions when the time integrations are performed.

The diagrams for G to first and second order in V are shown in Fig. 18.1 and Fig. 18.2. As an application of the diagram rules we evaluate a few examples. For the first-order diagrams in Fig. 18.1 one has $m = 1$, $m_c = 1$,

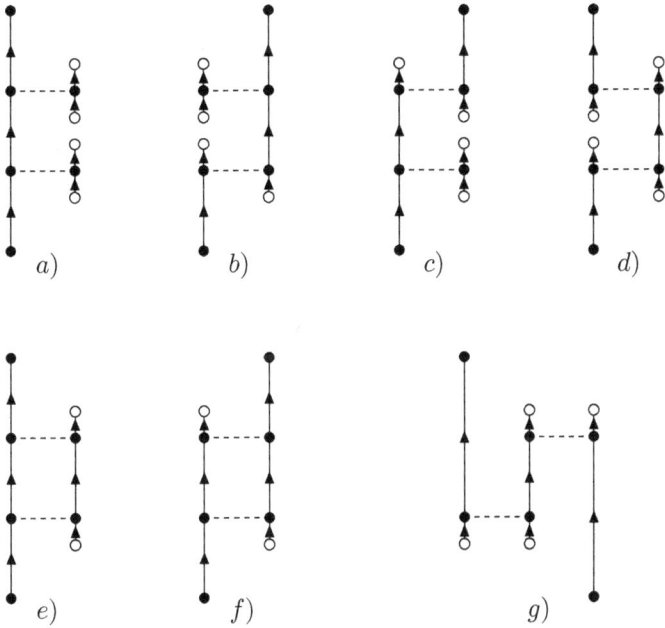

Fig. 18.2   Second-order diagrams for G.

and

$$
\begin{aligned}
G^{(1a)}(p; E) &= N_0 G^{(0)}(\boldsymbol{p}; E)\,(\boldsymbol{p}0|\,V\,|\boldsymbol{p}0)\,G^{(0)}(\boldsymbol{p}; E) \\
&= [z_0 \rho W(0)] G^{(0)}(\boldsymbol{p}; E) G^{(0)}(\boldsymbol{p}; E) \qquad (18.68) \\
G^{(1b)}(p; E) &= N_0 G^{(0)}(\boldsymbol{p}; E)\,(\boldsymbol{p}0|\,V\,|0\boldsymbol{p})\,G^{(0)}(\boldsymbol{p}; E) \\
&= [z_0 \rho W(\boldsymbol{p})] G^{(0)}(\boldsymbol{p}; E) G^{(0)}(\boldsymbol{p}; E). \qquad (18.69)
\end{aligned}
$$

For the second-order diagrams in Fig. 18.2e)f) one has $m = 2$, $m_c = 1$, and

$$
\begin{aligned}
G^{(2e)}(p; E) = N_0 \frac{i}{2\pi} \sum_{\boldsymbol{p}' \neq 0} \int dE'\, G^{(0)}(\boldsymbol{p}; E)\,(\boldsymbol{p}, 0|\,V\,|\boldsymbol{p}', \boldsymbol{p} - \boldsymbol{p}')\,G^{(0)}(\boldsymbol{p}'; E') \\
\times G^{(0)}(\boldsymbol{p} - \boldsymbol{p}'; E - E')\,(\boldsymbol{p}', \boldsymbol{p} - \boldsymbol{p}'|\,V\,|\boldsymbol{p}, 0)\,G^{(0)}(\boldsymbol{p}; E) \\
= \left[z_0 \rho \int \frac{d\boldsymbol{p}'}{(2\pi\hbar)^3} \frac{|W(\boldsymbol{p} - \boldsymbol{p}')|^2}{E - \varepsilon_{\boldsymbol{p}'} - \varepsilon_{\boldsymbol{p}-\boldsymbol{p}'} + i\eta} \right] G^{(0)}(\boldsymbol{p}; E) G^{(0)}(\boldsymbol{p}; E)
\end{aligned}
$$

$$(18.70)$$

$$G^{(2f)}(p; E) = \left[z_0\rho \int \frac{dp'}{(2\pi\hbar)^3} \frac{W(p'-p)W(-p')}{E - \varepsilon_{p'} - \varepsilon_{p-p'} + i\eta} \right] G^{(0)}(p; E)G^{(0)}(p; E).$$

$$(18.71)$$

We now investigate whether it is possible to repeat the analysis which proved so useful in the fermion case, and to identify an irreducible self-energy $\Sigma(p; E)$, in terms of which an infinite class of diagrams is easily summed. Obviously, the first-order diagrams in Fig. 18.1 are irreducible as they cannot be disconnected by cutting a propagator, and the reducible diagrams in Fig. 18.2a) $-$ d) are indeed just repetitions of Fig. 18.1a) and b). The diagrams in Fig. 18.2e), f) then define new irreducible second-order contributions. The reducible diagram in Fig. 18.2g) however, spoils the picture: it *can* be disconnected (by cutting the middle propagator) but the two resulting pieces are clearly not diagrams for G.

Instead, they are the lowest-order contributions to two novel Green's functions, the so-called anomalous propagators, which are defined as

$$i\hbar G_{12}(p; t - t') = \langle \Psi_{B0} | T[a_{p_\Omega}(t)a_{-p_\Omega}(t')] | \Psi_{B0} \rangle,$$
$$i\hbar G_{21}(p; t - t') = \langle \Psi_{B0} | T[a^\dagger_{p_\Omega}(t)a^\dagger_{-p_\Omega}(t')] | \Psi_{B0} \rangle. \qquad (18.72)$$

In the energy representation G_{12} is given by

$$G_{12}(p; E) = \langle \Psi_{B0} | a_p \frac{1}{E - (\hat{\Omega}_B - \Omega_{B0}) + i\eta} a_{-p} | \Psi_{B0} \rangle$$
$$- \langle \Psi_{B0} | a_{-p} \frac{1}{E + (\hat{\Omega}_B - \Omega_{B0}) - i\eta} a_p | \Psi_{B0} \rangle, \qquad (18.73)$$

and likewise for $G_{21}(p; E)$,

$$G_{21}(p; E) = \langle \Psi_{B0} | a^\dagger_p \frac{1}{E - (\hat{\Omega}_B - \Omega_{B0}) + i\eta} a^\dagger_{-p} | \Psi_{B0} \rangle$$
$$- \langle \Psi_{B0} | a^\dagger_{-p} \frac{1}{E + (\hat{\Omega}_B - \Omega_{B0}) - i\eta} a^\dagger_p | \Psi_{B0} \rangle. \qquad (18.74)$$

The anomalous propagators describe the process of exciting (de-exciting) two non-zero momentum particles with opposite momenta from (into) the condensate. They arise in the theory as a consequence of the nonconservation of particle number and vanish in zero'th order, when replacing $|\Psi_{B0}\rangle$ with $|0\rangle$.

Both G_{12} and G_{21} can be expanded similarly to Eq. (18.65), and are represented by diagrams with two external lines leaving or entering the diagram. We will therefore represent the exact anomalous propagators

Fig. 18.3 First order diagrams for G_{21} (left) and G_{12} (right).

graphically as

$$G_{12}(p; E) \Rightarrow \quad \boldsymbol{p}, E \quad -\boldsymbol{p}, -E \qquad\qquad G_{21}(p; E) \Rightarrow \quad \boldsymbol{p}, E \quad -\boldsymbol{p}, -E.$$

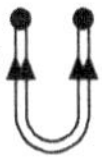

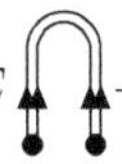

The diagram rules for G_{12} and G_{21} are exactly the same as for the normal propagator G. In Fig. 18.3 the first-order contributions are shown, which are easily evaluated as (note that $m = 1$ and $m_c = 1$):

$$G_{21}^{(1)}(p; E) = N_0 G^{(0)}(\boldsymbol{p}; E) G^{(0)}(-\boldsymbol{p}; -E) \, (00| \, V \, |\boldsymbol{p}, -\boldsymbol{p})$$
$$= [z_0 \rho W(p)] G^{(0)}(\boldsymbol{p}; E) G^{(0)}(-\boldsymbol{p}; -E) \qquad (18.75)$$

$$G_{12}^{(1)}(p; E) = N_0 \, (\boldsymbol{p}, -\boldsymbol{p}| \, V \, |00) \, G^{(0)}(\boldsymbol{p}; E) G^{(0)}(\boldsymbol{p}; E)$$
$$= z_0 \rho W(p) G^{(0)}(\boldsymbol{p}; E) G^{(0)}(-\boldsymbol{p}; -E). \qquad (18.76)$$

The second-order diagrams for G_{21} are shown in Fig. 18.4. The only irreducible diagram is the one of Fig. 18.4e); it has $m = 2$, $m_c = 1$ and its value is

$$G_{21}^{(2e)}(p; E) = N_0 \frac{i}{2\pi} \sum_{\boldsymbol{p}' \neq 0} \int dE' \, (\boldsymbol{p}', -\boldsymbol{p}'| \, V \, |\boldsymbol{p}, -\boldsymbol{p}) \, G^{(0)}(\boldsymbol{p}'; E')$$

$$\times G^{(0)}(-\boldsymbol{p}'; -E') \times (00| \, V \, |\boldsymbol{p}', -\boldsymbol{p}') \, G^{(0)}(\boldsymbol{p}; E) G^{(0)}(-\boldsymbol{p}; -E)$$

$$= \left[z_0 \rho \int \frac{d\boldsymbol{p}'}{(2\pi\hbar)^3} \frac{W(|\boldsymbol{p} - \boldsymbol{p}'|) W(p')}{-2\varepsilon_{p'} + i\eta} \right] G^{(0)}(\boldsymbol{p}; E) G^{(0)}(-\boldsymbol{p}; -E).$$

$$(18.77)$$

It is clear that the corresponding second-order diagrams for G_{12} can be obtained from the diagrams in Fig. 18.4 by rotating over 180 degrees and reversing the direction on all arrows. This corresponds to the replacement

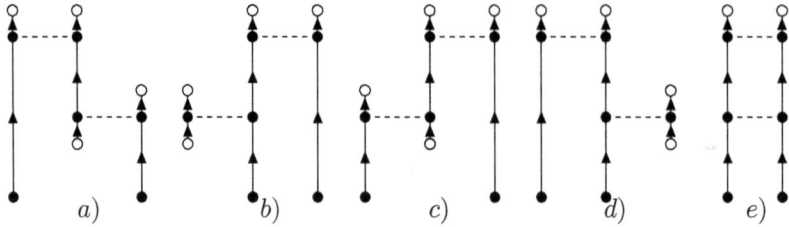

Fig. 18.4   Second-order diagrams for G_{21}.

$(\boldsymbol{p}; E) \to (-\boldsymbol{p}; -E)$, and as it obviously holds for all higher-order diagrams as well, one has

$$G_{12}(\boldsymbol{p}; E) = G_{21}(-\boldsymbol{p}; -E). \qquad (18.78)$$

In addition, Eqs. (18.73) and (18.74) imply that the anomalous propagators are even functions when both particle coordinates are interchanged,

$$G_{12}(\boldsymbol{p}; E) = G_{12}(-\boldsymbol{p}; -E); \quad G_{21}(\boldsymbol{p}; E) = G_{21}(-\boldsymbol{p}; -E) \qquad (18.79)$$

and in the Bose gas one therefore has $G_{12}(p; E) = G_{21}(p; E)$.

At this point one can define the irreducible self-energies $\Sigma_{12}(p; E)$ as the sum of all irreducible diagrams contributing to $G_{12}(\boldsymbol{p}; E)$, stripped of the external lines. An analogous definition applies to $\Sigma_{21}(p; E)$, and since $G_{12}(\boldsymbol{p}; E) = G_{21}(\boldsymbol{p}; E)$ one also has $\Sigma_{12}(\boldsymbol{p}; E) = \Sigma_{21}(\boldsymbol{p}; E)$. For reasons that will become apparent shortly, it is furthermore convenient to rename the normal propagator $G(p; E)$ as $G_{11}(p; E)$ and the irreducible normal self-energy $\Sigma(p; E)$ as $\Sigma_{11}(p; E)$. With these new ingredients, one arrives at closed expressions for G_{11} and G_{21}. The diagrammatic expansion for both these propagator terms, consists of a chain of Σ_{11}, Σ_{12}, and Σ_{21}, connected by a single NC line. In Fig. 18.5 the lowest-order terms in such an expansion are given. It is straightforward to verify by explicit order-by-order substitution that the full chain is generated by the diagrammatic equation in Fig 18.6.

The algebraic translation can be read off from Fig. 18.6 and yields

$$G_{11}(p; E) = G^{(0)}(p; E) + G^{(0)}(p; E)\Sigma_{11}(p; E)G_{11}(p; E)$$

$$+ G^{(0)}(p; E)\Sigma_{12}(p; E)G_{21}(p; E) \qquad (18.80)$$

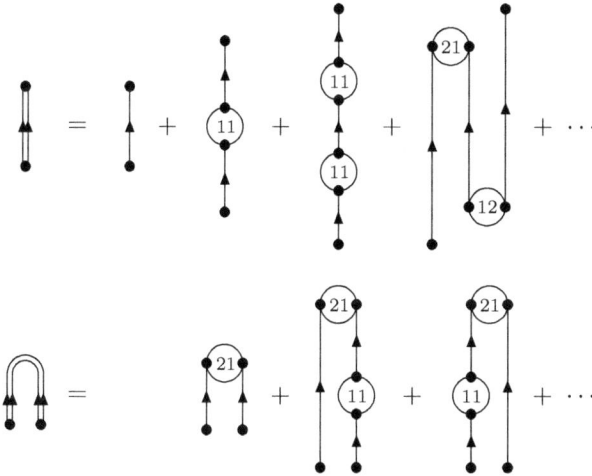

Fig. 18.5   Expansion of G_{11} and G_{21} in terms of the irreducible self-energies Σ_{11}, Σ_{12} and Σ_{21}.

and

$$G_{21}(p; E) = G^{(0)}(p; -E)\Sigma_{21}(p; E)G_{11}(p; E)$$
$$+ G^{(0)}(p; -E)\Sigma_{11}(p; -E)G_{21}(p; E). \quad (18.81)$$

These equations, first derived by [Beliaev (1958)], determine the propagator in terms of the self-energy, and represent the boson counterpart of the Dyson equation for fermions studied in Sec. 9.1.

Introducing $G_{22}(p; E) = G_{11}(p; -E)$ and $\Sigma_{22}(p; E) = \Sigma_{11}(p; -E)$ one can consider the subscripts as matrix indices and rewrite Eqs. (18.80)-(18.81) more elegantly as a 2×2 matrix equation,

$$[G(p; E)] = [G^{(0)}(p; E)] + [G^{(0)}(p; E)][\Sigma(p; E)][G(p; E)], \quad (18.82)$$

where

$$[G^{(0)}(p; E)] = \begin{bmatrix} G^{(0)}(p; E) & 0 \\ 0 & G^{(0)}(p; -E) \end{bmatrix} = \begin{bmatrix} \frac{1}{E - \varepsilon_{p\mu} + i\eta} & 0 \\ 0 & -\frac{1}{E + \varepsilon_{p\mu} - i\eta} \end{bmatrix}.$$
$$(18.83)$$

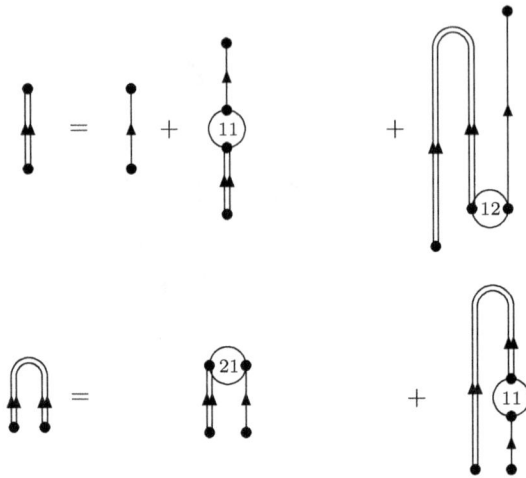

Fig. 18.6   Integral equations for G_{11} and G_{21}.

Applying a matrix inversion, yields the explicit solution of Eq. (18.82) in the following form

$$[G(p;E)] = \left([G^{(0)}(p;E)]^{-1} - [\Sigma(p;E)] \right)^{-1}$$

$$= \frac{1}{D(p;E)} \begin{bmatrix} E + \varepsilon_{p\mu} + \Sigma_{22}(p;E) & -\Sigma_{12}(p;E) \\ -\Sigma_{21}(p;E) & -E + \varepsilon_{p\mu} + \Sigma_{11}(p;E) \end{bmatrix},$$

$$\text{(18.84)}$$

where the determinantal function,

$$D(p;E) = [E - \varepsilon_{p\mu} - \Sigma_{11}(p;E)][E + \varepsilon_{p\mu} + \Sigma_{22}(p;E)] + \Sigma_{12}(p;E)\Sigma_{21}(p;E),$$

$$\text{(18.85)}$$

is easily seen to be even, *i.e.* $D(p;E) = D(p;-E)$.

18.4   Hugenholtz–Pines theorem

The Hugenholtz–Pines theorem establishes an exact relation between the chemical potential μ and the (irreducible) self-energies at zero-energy, $\Sigma_{11}(p;0)$ and $\Sigma_{12}(p;0)$, in the limit $p \to 0$ of zero momentum.

A perturbation expansion for the chemical potential μ is obtained from Eqs. (18.33) and (18.42) by considering in Eq. (8.18) the operator $\hat{A} = \partial \hat{V}_B / \partial N_0 = \partial \hat{\Omega}_1 / \partial N_0$. One immediately arrives at the expansion

$$\mu = \sum_{m=0}^{\infty} \frac{1}{m!(i\hbar)^m} \int dt_1 .. \int dt_m$$

$$\times \langle 0 | T \left[\hat{\Omega}_1(t_1)..\hat{\Omega}_1(t_m) \left(\frac{\partial \hat{\Omega}_1}{\partial N_0} \right)(0), \right] |0\rangle_{connected}, \quad (18.86)$$

where Eq. (8.18) has been used at an (arbitrary) fixed time $t = 0$. Note that the operator $\frac{\partial \hat{\Omega}_1}{\partial N_0}(0)$ in the T-product provides a fixed external point which allows to define connected and unconnected factors when contracting according to Wick's theorem. With exactly the same reasoning as in Sec. 8.5 one can show that the unconnected factors cancel the denominator, resulting in a restriction to connected diagrams as indicated in Eq. (18.86).

In a more compact notation we introduce

$$f(t_1, .., t_m; 0) = \langle 0 | T \left[\hat{\Omega}_1(t_1)..\hat{\Omega}_1(t_m) \left(\frac{\partial \hat{\Omega}_1}{\partial N_0} \right)(0) \right] |0\rangle_{connected}, \quad (18.87)$$

and rewrite the integration over $t_1..t_m$ in Eq. (18.86) as

$$I_m = \int dt_1..dt_m f(t_1, .., t_m; 0)$$

$$= \int dt_1..dt_m dt_{m+1} \delta(t_{m+1}) f(t_1, .., t_m; t_{m+1}). \quad (18.88)$$

Since $f(t_1, .., t_m; t_{m+1})$ only depends on *relative* times, it is possible to rewrite this as

$$I_m = \int d(\text{relative times of } t_1..t_{m+1}) f(t_1, .., t_m; t_{m+1})$$

$$= \int dt_1..dt_{m+1} \delta(T_{m+1}) f(t_1, .., t_m; t_{m+1}), \quad (18.89)$$

where $T_{m+1} = \frac{\sum_{i=1}^{m+1} t_i}{m+1}$. Using the symmetry of the integrand in Eq. (18.89), one therefore also has

$$I_m = \frac{1}{m+1} \frac{\partial}{\partial N_0} \int dt_1..dt_{m+1} \delta(T_{m+1}) g(t_1, .., t_{m+1}) \quad (18.90)$$

Fig. 18.7   First-order (left) and second-order (right) ground-state diagram.

where

$$g(t_1, .., t_m) = \langle 0| T \left[\hat{\Omega}_1(t_1)..\hat{\Omega}_1(t_m) \right] |0\rangle_{connected} .$$
(18.91)

The expansion for μ thus becomes

$$\mu = \frac{\partial}{\partial N_0} \left(i\hbar \sum_{m=1}^{\infty} \frac{1}{m!(i\hbar)^m} \int dt_1.. \int dt_m \delta(T_m) \right.$$

$$\left. \times \langle 0| T \left[\hat{\Omega}_1(t_1)..\hat{\Omega}_1(t_m) \right] |0\rangle_{connected} \right).$$
(18.92)

The expression following the derivative in Eq. (18.92) is recognized [see also Eq. (18.45)] as a perturbation expansion for the expectation value $K = \langle \Psi_{B0}| \hat{\Omega}_0 + \hat{\Omega}_1 |\Psi_{B0}\rangle$. The matrix elements in Eq. (18.92) can be worked out in the usual way with Wick's theorem, and the expansion for K is represented by the sum of all connected diagrams *without* external lines (ground-state diagrams).

The rules for the ground-state diagrams in the energy representation are the same as before, but the overall factor for an mth-order diagram with $2m_c$ C-arrows is given by $\frac{1}{S} N_0^{m_c} (\frac{i}{2\pi})^{m-m_c+1}$. The power of $(i/2\pi)$ is readily checked by considering the number of NC lines $m_{nc} = 2m - m_c$ and the number $(m - 1)$ of energy-conserving δ-functions arising from the m-fold time integration. Note that this leaves one time argument unrestricted, which is removed by the presence of the $\delta(T_m)$ factor in Eq. (18.92). The symmetry factor S of the diagram is the number of ways the vertices can be permuted without changing the value of the diagram (see *e.g.* [Negele and Orland (1988)]).[3]

---

[3] An appealing feature of Green functions is that one doesn't have to worry about symmetry factors in the diagrammatic expansion; they are always $S = 1$ because of the presence of the external points.

In both first and second order there is only one ground-state diagram, represented in Fig. 18.7. For the first-order diagram we have $m = 1$, $m_c = 2$ and $S = 2$ leading to a value

$$D^{(1)} = \frac{1}{2} N_0^2 \, (00| \, V \, |00) \, . \tag{18.93}$$

For the second-order diagram we have $m = 2$, $m_c = 2$ and $S = 2$, and

$$D^{(2)} = \frac{1}{2} \frac{i}{2\pi} N_0^2 \int dE \sum_{p \neq 0} \frac{|\,(p, -p|\, V \, |00)\,|^2}{(E - \varepsilon_{p\mu} + i\eta)(-E - \varepsilon_{p\mu} + i\eta)}$$

$$= -\frac{1}{2} N_0^2 \sum_{p \neq 0} \frac{|\,(p, -p|\, V \, |00)\,|^2}{2\varepsilon_{p\mu}} \, . \tag{18.94}$$

One can now generate (in a unique way) all diagrams for $\Sigma_{11}(0,0)$ by replacing in the diagrams for K two condensate arrows by an incoming and an outgoing NC line (which become the external NC lines). For the external NC lines $E = 0$ and the limit $p \to 0$ is taken, so this does not spoil the momentum and energy balance in the diagram. Note that the diagrams thus generated are automatically irreducible, since blocks connected by a single NC line cannot appear in the diagrams for K due to momentum conservation. Similarly, one obtains diagrams for $\Sigma_{12}(0,0)$ by replacing in the diagrams for K two condensate arrows by two incoming NC lines.

Let's consider a diagram of order m for K, having n_{ij} interaction lines of type $\hat{V}_{i,j}$. Since in the overall diagram as many NC particles must be added as there are removed, one has

$$\sum_{ij} (i - j) n_{ij} = 0, \tag{18.95}$$

and the number of C arrows, $2m_c$, is given by

$$m_c = \sum_{ij} [2 - \frac{1}{2}(i + j)] n_{ij} = \sum_{ij} (2 - i) n_{ij} = \sum_{ij} (2 - j) n_{ij} \, . \tag{18.96}$$

The value of the diagram for K is

$$v_K = \frac{1}{S} (\frac{i}{2\pi})^{m - m_c + 1} N_0^{m_c} D, \tag{18.97}$$

where D contains the string of interaction matrix elements and internal propagators. The corresponding contribution to the chemical potential becomes

$$v_\mu = \frac{\partial}{\partial N_0} v_K = \frac{1}{S}(\frac{i}{2\pi})^{m-m_c+1} N_0^{m_c-1} m_c D = m_c \frac{v_K}{N_0}. \tag{18.98}$$

Upon replacement of two C arrows by NC lines in the diagram for K one obtains diagrams for either $\Sigma_{11}(0,0)$ or $\Sigma_{12}(0,0)$, whose value will now be shown to be related to v_μ in Eq. (18.98). First, the mth order Σ-diagrams contain $2m_c - 2$ C-arrows and, according to Sec. 18.3, their overall factor is the same as the one in Eq. (18.98) both for the power of $(i/2\pi)$ and for the power of N_0. In addition, the evaluation of the interaction matrix elements and internal propagators in the diagram does not change under the replacement. As a consequence, all Σ-diagrams arising from a K-diagram have identical values, irrespective of which C arrows get replaced.

For the degeneracy of the diagrams we work step by step. First replace one incoming C arrow by an incoming NC line in all possible places. The number of ways this can be done is

$$\sum_{ij}(2-j)n_{ij} = m_c. \tag{18.99}$$

If the replacement is performed on an interaction line of type $\hat{V}_{i',j'}$, the resulting diagram will have one less interaction line of type $\hat{V}_{i',j'}$ and one more of type $\hat{V}_{i',j'+1}$.

If we repeat this operation we get a total degeneracy factor

$$m_c \sum_{ij}(2-j)(n_{ij} - \delta_{i,i'}\delta_{j,j'} + \delta_{i,i'}\delta_{j,j'+1}) = m_c(m_c - 1) \tag{18.100}$$

for the contribution v_{12} to $\Sigma_{12}(0,0)$ and hence

$$v_{12} = m_c(m_c - 1)\frac{v_K}{N_0}. \tag{18.101}$$

Note that when performing the replacement on all possible places of the K-diagram one generates each distinct Σ-diagram precisely S times. The expression in Eq. (18.101) still contains the $\frac{1}{S}$ factor in v_K to take care of this, and is therefore the correct contribution to $\Sigma_{12}(0,0)$.

Alternatively, we may replace as a next step an outgoing C arrow with an outgoing NC line, and get a degeneracy factor

$$m_c \sum_{ij}(2-i)(n_{ij} - \delta_{i,i'}\delta_{j,j'} + \delta_{i,i'}\delta_{j',j+1}) = m_c^2 \tag{18.102}$$

for the contribution

$$v_{11} = m_c^2 \frac{v_K}{N_0} \qquad (18.103)$$

to $\Sigma_{11}(0,0)$. Combining Eqs. (18.98), (18.101) and (18.103) leads to $v_\mu = v_{11} - v_{12}$ and since this has been derived for each specific diagram one has in general

$$\mu = \Sigma_{11}(0,0) - \Sigma_{12}(0,0), \qquad (18.104)$$

which is known as the Hugenholtz–Pines theorem. It is left as an exercise to check to second order that the above procedure indeed generates the irreducible self-energy of the corresponding order [see Exercise (2) at the end of the chapter].

The importance of the Hugenholtz–Pines theorem lies in the fact that, when considering the $p \to 0$ limit of the determinantal function $D(p; E)$ in Eq. (18.85)

$$D(0; E) = [E + \mu - \Sigma_{11}(0; E)][E - \mu + \Sigma_{11}(0; -E)] + [\Sigma_{12}(0; E)]^2, \quad (18.105)$$

the theorem ensures that $D(0; E)$ vanishes for $E = 0$. Since $D(p; E)$ appears in the denominator of Eq. (18.84), the exact boson propagator has a pole at $p = 0$, $E = 0$, and the spectrum of the elementary excitations E_p of the Bose gas is *gapless, i.e.* $E_p \to 0$ as $p \to 0$.

In fact, the spectrum vanishes linearly with p, as was shown by [Gavoret and Noziéres (1964)] to all orders in perturbation theory. Provided that the exact self-energies are regular at $p = 0$ and $E = 0$, one can see this *e.g.* by expanding the determinantal function $D(p; E) = D(p; -E)$ in Eq. (18.85) for small values of p and E as $D(p; E) \approx D(0; 0) + a^2 E^2 - b^2 p^2$. Isotropy and the even character have been used to exclude linear terms. The constant term $D(0; 0)$ then vanishes by virtue of the Hugenholtz–Pines theorem, and a linear spectrum $E_p \sim p$ results. This fact is strongly related to the appearance of superfluid properties, as will be discussed for liquid ^4He in Sec. 19.1.

18.5   First-order results

We now consider the limiting case of weak interactions, since even in lowest order the Bogoliubov perturbation theory leads to some nontrivial outcomes. It should be stressed immediately that "lowest-order" here means the expansion of the *self-energy* to first order in V. By solving the corresponding Dyson Eq. (18.82) one automatically sums a infinite class of diagrams and moves far beyond conventional first-order perturbation theory for the propagator [as considered *e.g.* in Eq. (18.68) – (18.69)]. Nevertheless, whatever the approximation for the self-energy, one knows that the corresponding chemical potential is given by the Hugenholtz–Pines theorem.

The self-energies to first order in the interaction have already been determined in Eqs. (18.68), (18.69) and (18.75):

$$\Sigma_{11}(p; E) = z_0 \rho[W(0) + W(p)] \tag{18.106}$$

$$\Sigma_{12}(p; E) = z_0 \rho W(p), \tag{18.107}$$

and are independent of energy. The chemical potential is then, according to Eq. (18.104),

$$\mu = \Sigma_{11}(0; 0) - \Sigma_{12}(0; 0) = z_0 \rho W(0). \tag{18.108}$$

The normal and anomalous Green's functions can now be derived from Eq. (18.84), with the function $D(p; E)$, appearing in the denominator, reducing to

$$D(p; E) = E^2 - \left\{[\varepsilon_p + z_0 \rho W(p)]^2 - [z_0 \rho W(p)]^2\right\} = E^2 - E_p^2. \tag{18.109}$$

The last equality defines the singularities $\pm E_p$ of the propagator, or the excitation spectrum of the elementary excitations, given by

$$E_p = \sqrt{[\varepsilon_p + z_0 \rho W(p)]^2 - [z_0 \rho W(p)]^2}. \tag{18.110}$$

In the low-momentum region, the spectrum is linear in p,

$$E_p \to p\sqrt{[z_0 \rho W(0)]/m} \tag{18.111}$$

like the dispersion relation for sound waves, with a constant velocity $\sqrt{[z_0 \rho W(0)]/m}$. At large momenta, the spectrum assumes its usual quadratic form

$$E_p \to \frac{p^2}{2m} + z_0 \rho W(p) + \mathcal{O}(p^{-2}), \tag{18.112}$$

which corresponds to almost free particles in a small potential $z_0 \rho W(p)$ arising from the interaction with the condensate. Note that Eq. (18.111) implies that for $W(0) < 0$ one has unphysical complex poles in the propagator as $p \to 0$. This reflects the instability of the present weak-interaction limit for a Bose gas with globally attractive interactions, and signals that the starting point is not valid: *e.g.* clusters will be formed, or the Bose gas will collapse and increase its density until the effective in-medium interaction is positive for small momentum transfer [Pines (1962)].

The expressions for G_{11} and G_{12} in Eq. (18.84), in terms of the E_p defined in Eq. (18.110), become

$$G_{11}(p, E) = \frac{E + \varepsilon_p + z_0\rho W(p)}{E^2 - E_p^2}$$

$$G_{12}(p, E) = -\frac{z_0\rho W(p)}{E^2 - E_p^2}. \tag{18.113}$$

This can be rewritten in a representation with simple poles as

$$G_{11}(p, E) = \frac{u_p^2}{E - E_p + i\eta} - \frac{v_p^2}{E + E_p - i\eta}$$

$$G_{12}(p, E) = \frac{-u_p v_p}{E - E_p + i\eta} - \frac{-u_p v_p}{E + E_p - i\eta} \tag{18.114}$$

where the coefficients u_p and v_p are given by

$$u_p = \left[\frac{\varepsilon_p + z_0\rho W(p) + E_p}{2E_p}\right]^{1/2}$$

$$v_p = \left[\frac{\varepsilon_p + z_0\rho W(p) - E_p}{2E_p}\right]^{1/2}, \tag{18.115}$$

and obey $u_p^2 - v_p^2 = 1$. For small momenta, one obtains to leading order

$$u_p^2 \approx v_p^2 \approx u_p v_p \approx \sqrt{z_0\rho W(0)m}/(2p). \tag{18.116}$$

For large momenta, $u_p \approx 1$ and $v_p \to \mathcal{O}(p^{-2}) \approx 0$, and the form of the free propagator is retrieved.

The fact that the low-energy excitations in Eq. (18.111) show a linear momentum dependence is one of the most outstanding differences between fermion and boson systems. In a normal Fermi system the quasiparticle spectrum near the Fermi energy is not qualitatively different from that of the free system. In a Bose system, the low-energy excitations are inherently collective because of the presence of the condensate.

18.6   Dilute Bose gas with repulsive forces

We now investigate the properties of the Bose gas in the limit of small density ρ. Since the scattering length a provides the sole other relevant dimension in the problem, the small dimensionless variable in a low-density expansion (not necessarily a power series) must be $\epsilon = \rho a^3$, and we will only be concerned with quantities to leading order in ϵ. Examining the first-order self-energy, one notes that the matrix elements $W(p)$ in Eqs. (18.106) – (18.107) become very large, or diverge for a repulsive core. This again reflects the need to replace the bare interaction with an effective one. In the dilute limit this is particularly clear: any interaction line can be extended with repeated scattering between NC lines without altering the power of N_0, so a low-density expansion in terms of the bare interaction does not make sense. Fortunately, the summation of these diagrams is quite simple. We consider the sum of the diagrams in Fig. 18.8, where Ω is the energy of the incoming pair. According to the rules of Sec. 18.3, the series reads

$$
(\boldsymbol{p}_1'\boldsymbol{p}_2'|\,\Gamma(\Omega)\,|\boldsymbol{p}_1\boldsymbol{p}_2) = (\boldsymbol{p}_1'\boldsymbol{p}_2'|\,V\,|\boldsymbol{p}_1\boldsymbol{p}_2) + \sum_{\boldsymbol{p}_3\boldsymbol{p}_4} \int dE_3 dE_4 \delta(\Omega - [E_3 + E_4])
$$

$$
\times (\boldsymbol{p}_1'\boldsymbol{p}_2'|\,V\,|\boldsymbol{p}_3\boldsymbol{p}_4) \frac{i}{2\pi} G_0(p_3; E_3) G_0(p_4; E_4)\,(\boldsymbol{p}_3\boldsymbol{p}_4|\,V\,|\boldsymbol{p}_1\boldsymbol{p}_2)
$$

$$
+ \sum_{\boldsymbol{p}_3\boldsymbol{p}_4} \sum_{\boldsymbol{p}_5\boldsymbol{p}_6} \int dE_3 dE_4 \int dE_5 dE_6 \delta(\Omega - [E_3 + E_4]) \delta(\Omega - [E_5 + E_6])
$$

$$
\times (\boldsymbol{p}_1'\boldsymbol{p}_2'|\,V\,|\boldsymbol{p}_3\boldsymbol{p}_4) \frac{i}{2\pi} G_0(p_3; E_3) G_0(p_4; E_4)\,(\boldsymbol{p}_3\boldsymbol{p}_4|\,V\,|\boldsymbol{p}_5\boldsymbol{p}_6)
$$

$$
\times \frac{i}{2\pi} G_0(p_5; E_5) G_0(p_6; E_6)\,(\boldsymbol{p}_5\boldsymbol{p}_6|\,V\,|\boldsymbol{p}_1\boldsymbol{p}_2) + \dots, \tag{18.117}
$$

and is generated by the integral equation

$$
(\boldsymbol{p}_1'\boldsymbol{p}_2'|\,\Gamma(\Omega)\,|\boldsymbol{p}_1\boldsymbol{p}_2) = (\boldsymbol{p}_1'\boldsymbol{p}_2'|\,V\,|\boldsymbol{p}_1\boldsymbol{p}_2) + \sum_{\boldsymbol{p}_3\boldsymbol{p}_4} \int dE_3 dE_4 \delta(\Omega - [E_3 + E_4])
$$

$$
\times (\boldsymbol{p}_1'\boldsymbol{p}_2'|\,V\,|\boldsymbol{p}_3\boldsymbol{p}_4) \frac{i}{2\pi} G_0(p_3; E_3) G_0(p_4; E_4)\,(\boldsymbol{p}_3\boldsymbol{p}_4|\,\Gamma(\Omega)\,|\boldsymbol{p}_1\boldsymbol{p}_2). \tag{18.118}
$$

The energy integrals in Eq. (18.118) are easily evaluated as

$$
\int dE_3 \int dE_4 \delta(\Omega - [E_3 + E_4]) G_0(p_3; E_3) G_0(p_4; E_4)
$$

$$
= \frac{-2\pi i}{\Omega - \frac{p_3^2}{2m} - \frac{p_4^2}{2m} + 2\mu + i\eta}. \tag{18.119}
$$

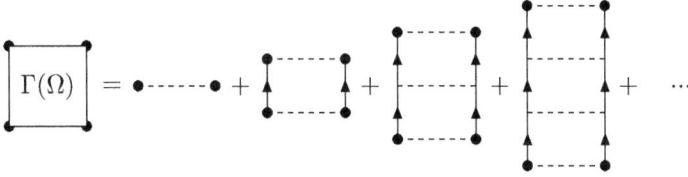

Fig. 18.8   Diagram series for the in-medium $\mathcal{T}$-matrix $\Gamma(\Omega)$.

Introducing CM and relative momenta

$$\boldsymbol{P} = \boldsymbol{p}_1 + \boldsymbol{p}_2 = \boldsymbol{p}_1' + \boldsymbol{p}_2'; \quad \boldsymbol{p} = \frac{\boldsymbol{p}_1 - \boldsymbol{p}_2}{2}; \quad \boldsymbol{p}' = \frac{\boldsymbol{p}_1' - \boldsymbol{p}_2'}{2} \qquad (18.120)$$

and splitting off the factor for overall momentum conservation,

$$(\boldsymbol{p}_1' \boldsymbol{p}_2' | \Gamma(\Omega) | \boldsymbol{p}_1 \boldsymbol{p}_2) = \frac{\delta_{\boldsymbol{p}_1 + \boldsymbol{p}_2, \boldsymbol{p}_1' + \boldsymbol{p}_2'}}{V} \Gamma(\boldsymbol{p}, \boldsymbol{p}'; \boldsymbol{P}; \Omega) \qquad (18.121)$$

the integral equation for the in-medium $\mathcal{T}$-matrix $\Gamma(\boldsymbol{p}, \boldsymbol{p}'; \boldsymbol{P}; \Omega)$ becomes

$$\Gamma(\boldsymbol{p}, \boldsymbol{p}'; \boldsymbol{P}; \Omega) = W(\boldsymbol{p} - \boldsymbol{p}') + \int \frac{d\boldsymbol{p}''}{(2\pi\hbar)^3} \frac{W(\boldsymbol{p} - \boldsymbol{p}'')\Gamma(\boldsymbol{p}'', \boldsymbol{p}'; \boldsymbol{P}; \Omega)}{\Omega - \frac{P^2}{4m} - \frac{p''^2}{m} + 2\mu + i\eta}.$$
$$(18.122)$$

We note that in the bosonic case, the in-medium $\mathcal{T}$-matrix obeys the same equation as the free-space $\mathcal{T}$-matrix in Eq. (12.100), apart from a trivial energy shift involving the chemical potential and the CM kinetic energy. It can thus immediately be expressed in terms of the free-space half off-shell $\mathcal{T}$-matrix [see Eq. (12.102)],

$$\frac{1}{(2\pi)^3} \Gamma(\boldsymbol{p}, \boldsymbol{p}'; \boldsymbol{P}; \Omega) = \langle \boldsymbol{p} | \tilde{\mathcal{T}} | \boldsymbol{p}' \rangle + \int d\boldsymbol{p}'' \langle \boldsymbol{p}' | \tilde{\mathcal{T}} | \boldsymbol{p}'' \rangle \langle \boldsymbol{p} | \tilde{\mathcal{T}} | \boldsymbol{p}'' \rangle^* \qquad (18.123)$$

$$\times \left(\frac{1}{\Omega - \frac{P^2}{4m} + 2\mu - \frac{p''^2}{m} + i\eta} - \frac{1}{\frac{p^2}{m} - \frac{p''^2}{m} + i\eta} \right).$$

In the low-energy regime of interest one has $p, p', P, \Omega \to 0$, and one can replace the matrix elements of $\tilde{\mathcal{T}}$ with the low-energy limit $g/(2\pi)^3$ of Eq. (12.87) [where $g = \frac{4\pi\hbar^2 a}{m}$]. The second term in Eq. (18.123) is then seen to scale as $g^2 \sqrt{\mu}$. Since the leading term in the low-density expansion for the chemical potential must be given by the GP result $\mu \approx \rho g$ in Eq. (18.62), one notices that $g^2 \sqrt{\mu} \sim \sqrt{\rho a^3}$. In a low-density expansion the second term therefore represents a higher-order correction to the first term in Eq. (18.123) and to leading order, the effective interaction can be treated as constant, $\Gamma \approx g$.

In the first-order self-energy diagrams of Fig. 18.1 and Fig. 18.3, one can then replace the bare interaction V with the effective interaction Γ of Fig. 18.8, leading to the ladder series for the self-energy. It is easy to infer from Fig. 18.8 that all the second-order irreducible self-energy diagrams in Fig. 18.2 and Fig. 18.4 are generated by this replacement. Moreover, contemplating the higher-order contributions to the irreducible self-energy leads to the conclusion that, in any order in V, the diagrams of the ladder series contain the minimal number (two) of condensate arrows. As a result, the third and higher-order self-energy diagrams that are not included in the ladder series are necessarily of a higher power in the density.

The upshot of all this is that the leading-order term in a low-density expansion can be simply obtained by replacing $W(p)$ with $g = 4\pi\hbar^2 a/m$ in the first-order results of the previous Sec. 18.5. In the dilute limit, Eqs. (18.110) and (18.115) then imply that

$$E_p = \sqrt{(\varepsilon_p + z_0\rho g)^2 - (z_0\rho g)^2} = z_0\rho g x \sqrt{x^2 + 2},$$

$$v_p^2 = \frac{\varepsilon_p + z_0\rho g - E_p}{2E_p} = \frac{1}{2}\left[\frac{x^2 + 1}{x\sqrt{x^2 + 2}} - 1\right],$$

$$u_p^2 = \frac{\varepsilon_p + z_0\rho g + E_p}{2E_p} = \frac{1}{2}\left[\frac{x^2 + 1}{x\sqrt{x^2 + 2}} + 1\right], \qquad (18.124)$$

where a rescaled dimensionless momentum variable $x = \frac{p}{\sqrt{2mz_0\rho g}}$ has been introduced. The depletion of the condensate is expressed in terms of the removal amplitudes [see Eqs. (18.52) and (18.114)] as

$$N - N_0 = \sum_{p \neq 0} v_p^2 = V\int \frac{d\boldsymbol{p}}{(2\pi\hbar)^3} v_p^2 = V\frac{4\pi}{(2\pi\hbar)^3}[2mz_0\rho g]^{3/2}I_1, \quad (18.125)$$

where the integral I_1 can be evaluated as

$$I_1 = \int_0^{+\infty} dx x^2 \frac{1}{2}\left[\frac{x^2 + 1}{x\sqrt{x^2 + 2}} - 1\right]$$

$$= \lim_{L \to \infty}\left(\frac{1}{4}\int_0^L d(x^2)\frac{x^2 + 1}{\sqrt{x^2 + 1}} - \frac{L^3}{6}\right) = \frac{\sqrt{2}}{6}. \qquad (18.126)$$

The condensate fraction $z_0 = N_0/N$, calculated from Eq. (18.126), reads

$$1 - z_0 = \frac{8}{3\sqrt{\pi}} z_0^{3/2} (\rho a^3)^{1/2} \tag{18.127}$$

and the resulting depletion $1 - z_0$ is seen to be small in the dilute limit $(\rho a^3) \rightarrow 0$. To leading order in ρa^3 the solution of Eq. (18.127) is given by

$$1 - z_0 = \frac{8}{3\sqrt{\pi}} (\rho a^3)^{1/2}. \tag{18.128}$$

While the kinetic energy in Eq. (18.54) and the mean removal energy in Eq. (18.58) are separately divergent (as a consequence of the contact interaction the v_p don't fall off fast enough), the ground-state energy determined by their sum is finite, and according to Eq. (18.61) one has

$$E - \frac{1}{2}\mu N = \frac{1}{2}\sum_p v_p^2 (\varepsilon_p - E_p) = V \frac{4\pi}{(2\pi\hbar)^3} \frac{[2mz_0\rho g]^{5/2}}{2m} I_2, \tag{18.129}$$

where the integral can again be evaluated as

$$I_2 = \int_0^{+\infty} dx x^2 \frac{1}{4} \left[\frac{x^2 + 1}{x\sqrt{x^2 + 2}} - 1 \right] \left(x^2 - x\sqrt{x^2 + 2} \right)$$

$$= \lim_{L \to \infty} \left(\frac{1}{8} \int_0^L d(x^2) \frac{2x^4 + 3x^2}{\sqrt{x^2 + 2}} - \frac{L^5}{10} - \frac{L^3}{12} \right) = -\frac{\sqrt{2}}{15}. \tag{18.130}$$

In accordance with the discussion at the end of Sec. 18.2.3, the resulting expression for the ground-state energy,

$$\frac{E}{N} = \frac{\mu}{2} - \frac{64\sqrt{\pi}}{15} \frac{\hbar^2}{ma^2} z_0^{5/2} (\rho a^3)^{3/2}, \tag{18.131}$$

is a simple linear differential equation in E and $\mu = dE/dN$ of the type

$$E - \frac{1}{2}N\frac{dE}{dN} = -CN^{5/2}, \tag{18.132}$$

because we have evaluated the propagator quantities with the lower-order approximation $\mu = z_0 \rho g$ from the HP theorem. Note that C is an N-independent constant and that it is permissible to keep only the leading term for $z_0 \approx 1$ in Eq. (18.131), leading to the N-dependence in Eq. (18.132). It can be solved with a power-law ansatz

$$E(N) = \frac{g}{2V} N^2 (1 + xN^{3/2}) \tag{18.133}$$

involving an unknown coefficient x. Substitution in Eq. (18.132) leads to $x = 8CV/g$, and the leading correction to the GP result $\rho g/2$ for the energy per particle is given by

$$E/N = \frac{1}{2}\rho g(1 + \frac{128}{15\sqrt{\pi}}(\rho a^3)^{1/2}), \qquad (18.134)$$

as was first proposed by [Lee and Yang (1957)]. The chemical potential then becomes

$$\mu = \rho g(1 + \frac{32}{3\sqrt{\pi}}(\rho a^3)^{1/2}). \qquad (18.135)$$

The late 1950's saw a flurry of activity on the dilute Bose gas, with other important contributions by [Lee *et al.* (1957)], [Brueckner and Sawada (1957)], and [Wu (1959)]. With similar techniques the next to leading order correction to the GP energy has been calculated as $E/N = \frac{1}{2}\rho g[1 + \frac{128}{15\sqrt{\pi}}(\rho a^3)^{1/2} + 8(4\pi/3 - \sqrt{3})\rho a^3 \ln(\rho a^3) + ..]$. The (unknown) next term goes like (ρa^3) but cannot be expressed solely in terms of the scattering length.

18.7 Canonical transformation for the Bose gas

The structure of the propagator in Eq. (18.114), containing a single removal pole for each sp momentum $\boldsymbol{p}$, is strongly reminiscent of a mean-field treatment. It is in fact possible to derive the first-order results in Sec. 18.5 directly, without applying perturbation theory. We return to the grand-canonical potential $\hat{\Omega}_B$ of Eq. (18.27) and examine in Eq. (18.21) the different terms in $\hat{V}_B$. It is clear that when the number of non-condensate particles is very small it makes sense to drop all terms involving three or four non-condensate operators, *i.e.* we neglect the terms $\hat{V}_{2,1}$, $\hat{V}_{1,2}$ and $\hat{V}_{2,2}$. The corresponding operator is

$$\hat{\Omega}_B \approx V_{0,0} - \mu N_0 + \sum_{\boldsymbol{p}\neq 0}\left\{\tilde{\varepsilon}_{\boldsymbol{p}} a_{\boldsymbol{p}}^\dagger a_{\boldsymbol{p}} + \frac{G_{\boldsymbol{p}}}{2}[a_{\boldsymbol{p}}^\dagger a_{-\boldsymbol{p}}^\dagger + a_{\boldsymbol{p}} a_{-\boldsymbol{p}}]\right\}, \qquad (18.136)$$

where we introduced

$$\tilde{\varepsilon}_{\boldsymbol{p}} = \varepsilon_{\boldsymbol{p}\mu} + z_0\rho[W(0) + W(p)]; \qquad G_{\boldsymbol{p}} = z_0\rho W(p). \qquad (18.137)$$

This operator is a quadratic form in the removal and addition operators and can therefore be diagonalized by introducing suitable linear combinations. However, it will obviously be necessary to mix removal and addition operators and to consider operators of the type

$$b_{\boldsymbol{p}}^{\dagger} = u_p a_{\boldsymbol{p}}^{\dagger} + v_p a_{-\boldsymbol{p}}; \qquad b_{\boldsymbol{p}} = u_p a_{\boldsymbol{p}} + v_p a_{-\boldsymbol{p}}^{\dagger}, \tag{18.138}$$

where the u_p and v_p are real coefficients. This is the most general linear combination, taking isotropy and momentum conservation into account. The commutators are

$$[b_{\boldsymbol{p}}, b_{\boldsymbol{p}'}] = 0; \qquad [b_{\boldsymbol{p}}, b_{\boldsymbol{p}'}^{\dagger}] = (u_p^2 - v_p^2)\delta_{\boldsymbol{p},\boldsymbol{p}'}. \tag{18.139}$$

One observes that the b-operators obey the same commutator algebra as the usual a-operators, provided that the normalization condition

$$u_p^2 - v_p^2 = 1 \tag{18.140}$$

is fulfilled. We impose this relation between the coefficients in the transformation in Eq. (18.138), which now has one degree of freedom left. The inverse transformation can easily be worked out as

$$a_{\boldsymbol{p}}^{\dagger} = u_p b_{\boldsymbol{p}}^{\dagger} - v_p b_{-\boldsymbol{p}}; \qquad a_{\boldsymbol{p}} = u_p b_{\boldsymbol{p}} - v_p b_{-\boldsymbol{p}}^{\dagger}, \tag{18.141}$$

and substitution into Eq. (18.136) yields $\hat{\Omega}_B$ in the transformed basis,

$$\hat{\Omega}_B = V_{0,0} - \mu N_0 + \sum_{\boldsymbol{p} \neq 0} \left\{ [v_p^2 \tilde{\varepsilon}_p - u_p v_p G_p] + [(u_p^2 + v_p^2)\tilde{\varepsilon}_p - 2u_p v_p G_p] b_{\boldsymbol{p}}^{\dagger} b_{\boldsymbol{p}} \right.$$

$$\left. + [-u_p v_p \tilde{\varepsilon}_p + \frac{1}{2}(u_p^2 + v_p^2)G_p](b_{\boldsymbol{p}}^{\dagger} b_{-\boldsymbol{p}}^{\dagger} + b_{\boldsymbol{p}} b_{-\boldsymbol{p}}) \right\}. \tag{18.142}$$

The last line of Eq. (18.142) contains the "off-diagonal" bb and $b^{\dagger}b^{\dagger}$ contribution, which can be made to vanish by choosing the u_p, v_p as solutions of

$$\tilde{\varepsilon}_p u_p v_p = \frac{1}{2} G_p(u_p^2 + v_p^2). \tag{18.143}$$

This equation should be solved together with the normalization constraint (18.140). It is somewhat easier to eliminate the latter by defining a new variable θ_p as

$$u_p = \cosh \theta_p; \qquad v_p = \sinh \theta_p, \tag{18.144}$$

in terms of which $u_p^2 + v_p^2 = \cosh 2\theta_p$ and $2u_p v_p = \sinh 2\theta_p$. Eq. (18.143) is now easily solved as $\theta_p = \frac{1}{2}\text{Arctanh}\,(G_p/\tilde{\varepsilon}_p)$ or

$$\cosh 2\theta_p = \frac{\tilde{\varepsilon}_p}{\sqrt{\tilde{\varepsilon}_p^2 - G_p^2}}; \quad \sinh 2\theta_p = \frac{G_p}{\sqrt{\tilde{\varepsilon}_p^2 - G_p^2}}, \tag{18.145}$$

which implies

$$u_p^2 = \frac{\tilde{\varepsilon}_p + E_p}{2E_p}, \quad v_p^2 = \frac{\tilde{\varepsilon}_p - E_p}{2E_p}, \quad \text{where } E_p = \sqrt{\tilde{\varepsilon}_p^2 - G_p^2}. \tag{18.146}$$

This solution brings Eq. (18.142) to its diagonal form,

$$\hat{\Omega}_B = V_{0,0} - \mu N_0 + \Delta\Omega + \sum_{p \neq 0} E_p\, b_p^\dagger b_p. \tag{18.147}$$

The operator part in Eq. (18.147), which describes the dynamics of the non-condensate particles, clearly has the structure of a noninteracting quasiparticle Hamiltonian with positive excitation energies E_p. Consequently, the ground state is simply the vacuum $|0\rangle$ of the b_p operators (with no quasiparticles present), whereas all excited eigenstates are of the form $(b_{p_1}^\dagger)^{n_1} (b_{p_2}^\dagger)^{n_2} .. |0\rangle$.

The occupation of the non-condensate particles in the quasiparticle vacuum follows directly from Eq. (18.141) and the fact that $b_p |0\rangle = 0$,

$$\langle 0| a_p^\dagger a_p |0\rangle = v_p^2, \tag{18.148}$$

whereas the thermodynamic potential is given by the expectation value

$$\Omega_{B0} = \langle 0| \hat{\Omega}_B |0\rangle = V_{0,0} - \mu N_0 + \Delta\Omega \tag{18.149}$$

where

$$\Delta\Omega = \sum_{p \neq 0}(v_p^2 \tilde{\varepsilon}_p - u_p v_p G_p) = \sum_{p \neq 0}\frac{1}{2}(E_p - \tilde{\varepsilon}_p) = -\sum_{p \neq 0} E_p v_p^2. \tag{18.150}$$

The quadratic approximation made in Eq. (18.136) is completely equivalent to the first-order calculations of Sec. 18.5. The dominant condensate contribution to the thermodynamic potential

$$(\Omega_{B0})_{con} = V_{0,0} - \mu N_0 = \frac{1}{2}N_0^2 W(0)/V - \mu N_0 \tag{18.151}$$

is minimized by N_0 obeying $\mu = N_0 W(0)/V = z_0 \rho W(0)$. Substituting this lowest-order solution for the chemical potential into Eq. (18.137) one sees at once that the excitation spectrum $E_p = \sqrt{\tilde{\varepsilon}_p^2 - G_p^2}$ as well as the u_p and v_p amplitudes coincide with Eqs. (18.110) and Eq. (18.115), respectively.

As a consequence the present treatment leads for dilute systems to the same condensate depletion as derived in Eq. (18.127). In order to calculate the leading correction to the ground-state energy, one can again set $z_0 = 1$ or $N_0 = N$, and Eqs. (18.149) − (18.150) lead to

$$E_{B0} = \Omega_{B0} + \mu N = V_{0,0} + \Delta\Omega = \frac{1}{2}\frac{gN^2}{V} + \sum_{\mathbf{p}\neq 0} \frac{1}{2}(E_p - \tilde{\varepsilon}_p). \qquad (18.152)$$

The second term in Eq. (18.152) is ultraviolet divergent, since $(E_p - \tilde{\varepsilon}_p) \to -\frac{m\rho^2 g^2}{p^2}$ for $p \to \infty$. In accordance with the discussion in Sec. 12.4.3 we note that this is the same divergence that appears in the second term of Eq. (12.96), and we expect a finite result when everything is expressed in terms of the physical scattering length. We therefore cancel the divergence by rewriting Eq. (18.152) as

$$E_{B0} = \left(\frac{gN^2}{2V} - \sum_{\mathbf{p}\neq 0}\frac{m\rho^2 g^2}{2p^2}\right) + \sum_{\mathbf{p}\neq 0}\left(\frac{1}{2}(E_p - \tilde{\varepsilon}_p) + \frac{m\rho^2 g^2}{2p^2}\right) \qquad (18.153)$$

$$= X_1 + X_2. \qquad (18.154)$$

The second term X_2 in Eq. (18.154) is now a convergent expression which can be evaluated [similar to Eq. (18.129)] as

$$X_2 = V\frac{4\pi}{(2\pi\hbar)^3}\frac{1}{2m}[2m\rho g]^{5/2}I, \qquad (18.155)$$

with the integral I given by

$$I = \int dx\,\frac{x^2}{2}\left(x\sqrt{x^2+2} - x^2 - 1 + \frac{1}{2x^2}\right) = \frac{4\sqrt{2}}{15}. \qquad (18.156)$$

As this outcome is finite, we must replace in Eq. (18.155) the strength of the contact force with the lowest-order term in Eq. (12.96), *i.e.* $g = 4\pi\hbar^2 a/m$.

The divergent first term X_1 in Eq. (18.154) becomes

$$X_1 = \frac{gN^2}{2V} - \frac{4\pi V}{(2\pi\hbar)^3}\frac{m\rho^2 g^2}{2}\int dp = \frac{N^2}{2V}\left(g - \frac{mg^2}{2\pi^2\hbar^2}\int\frac{dp}{\hbar}\right), \qquad (18.157)$$

and since the bracketed term is recognized as the formal expansion to *second* order of the free-space T-matrix in Eq. (12.96) it can also be replaced with $4\pi\hbar^2 a/m$.

The final result for the energy,

$$E_{B0}/N = \frac{2\pi\rho\hbar^2 a}{m}\left(1 + \frac{128}{15\sqrt{\pi}}(\rho a^3)^{1/2}\right),\qquad(18.158)$$

is in complete agreement with Eq. (18.134), which was obtained in a totally different manner as the solution of a differential equation involving the energy and the chemical potential.

The perturbative expansion of the dilute Bose gas studied in Sec. 18.6 has served as a conceptual model for the phenomenon of superfluidity in liquid ^{4}He, *e.g.* to justify the linear spectrum of low-momentum quasiparticle excitations. It does not provide a quantitative model of ^{4}He (as will be further discussed in Sec. 19.1), since this quantum liquid is neither dilute nor is the interaction purely repulsive. However, the experimental realization of BEC in dilute atomic vapors, as discussed in Sec. 12.4.5, has led to renewed interest in bosonic perturbation theory. These systems are ideally suited for a treatment along the lines of Sec. 18.6, although the presence of the confining external potential requires some modifications which will be the subject of Sec. 19.3.

18.8 Exercises

(1) Consider the many-boson state

$$|\Psi(N_0)\rangle = \sum_{N=0}^{+\infty} e^{-\frac{|\hat{N}-N_0|}{2\sqrt{N_0}}} \frac{(a_0^\dagger)^N}{\sqrt{N!}}|0\rangle.$$

Prove that as $N_0 \to \infty$, one has to leading order in N_0:

$$\langle\Psi(N_0)|\Psi(N_0)\rangle \to 2\sqrt{N_0}$$

$$\langle\hat{N}\rangle = \frac{\langle\Psi(N_0)|\hat{N}|\Psi(N_0)\rangle}{\langle\Psi(N_0)|\Psi(N_0)\rangle} \to N_0$$

$$\langle\hat{N}^2 - N_0^2\rangle = \frac{\langle\Psi(N_0)|\hat{N}^2 - N_0^2|\Psi(N_0)\rangle}{\langle\Psi(N_0)|\Psi(N_0)\rangle} \to 2N_0$$

$$\Delta N = \frac{\sqrt{\langle[\hat{N}-\langle\hat{N}\rangle]^2\rangle}}{\langle\hat{N}\rangle} \to \frac{\sqrt{2}}{\sqrt{N_0}} \to 0$$

$$\langle a_0^\dagger\rangle = \frac{\langle\Psi(N_0)|a_0^\dagger|\Psi(N_0)\rangle}{\langle\Psi(N_0)|\Psi(N_0)\rangle} \to \sqrt{N_0}.$$

Hint: Show that for $\alpha \geq 0$ and $N_0 \to +\infty$ the discrete sums can be replaced by

$$\sum_{N=0}^{+\infty} N^\alpha e^{-\frac{|N-N_0|}{\sqrt{N_0}}} \to N_0^{\frac{\alpha+1}{2}} \int_0^{+\infty} dx\, x^\alpha e^{-|x-\sqrt{N_0}|}$$

$$\to 2\sqrt{N_0} N_0^\alpha \left(1 + \frac{\alpha(\alpha-1)}{N_0} + \cdots\right).$$

(2) Check the Hugenholtz–Pines relation in Sec. 18.4 up to second order in the interaction, by taking the derivative with respect to N_0 of Eqs. (18.93) – (18.94) and comparing with the $p = 0$, $E = 0$ limit of Eqs. (18.68) and (18.71), and of Eqs. (18.75) and (18.77).

Chapter 19

Boson perturbation theory applied to physical systems

In this chapter, we explore (sometimes indirectly) applications of the bosonic perturbation theory studied in the previous chapter. Section 19.1 contains a discussion of liquid ^{4}He, whose superfluid properties can be understood using the concepts of Bose gas perturbation theory in combination with more phenomenological treatments. In Sec. 19.2, the dynamic structure function of the ^{4}He system is examined, and its asymptotic properties are derived in a general context. In Sec. 19.3, we apply the canonical transformation technique to nonuniform systems and retrieve the GP equation as a description of the depleted condensate, now supplemented with the Bogoliubov–de Gennes equations describing the dynamics of the non-condensate particles. The chapter concludes with some elements of number-conserving perturbation theory in Sec. 19.4.

19.1 Superfluidity in liquid ^{4}He

19.1.1 *The He-II phase*

The atoms of the most abundant helium isotope ^{4}He are bosonic in nature, being composed of an even number of fermions. The closed-shell $(1s^2)$ electronic configuration makes for very weak van der Waals forces. In combination with the small atomic mass this hinders the formation of clusters and the transition to a liquid. Under atmospheric pressure the gas-liquid phase transition occurs at a correspondingly low temperature $T \approx 4.2$K, and the system remains a liquid down to $T = 0$; it solidifies only under pressures of at least 25 atm.

The phase diagram, represented in Fig. 19.1, demonstrates the existence of two types of liquid ^{4}He, the so-called He-I and He-II phase. Below

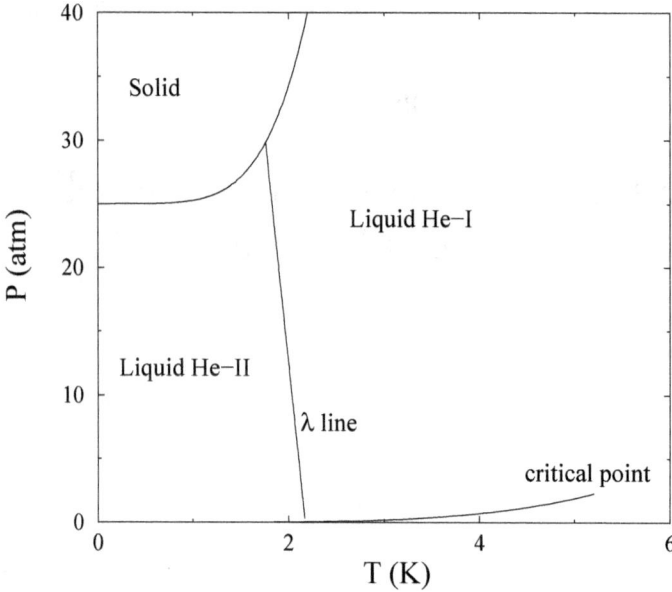

Fig. 19.1   Phase diagram of ^{4}He.

the boiling point the liquid is initially in the He-I phase and has no very remarkable properties. When cooled further it undergoes a transition to the He-II phase at a temperature $T_\lambda = 2.172$K. This is the λ-point for liquid in equilibrium with its vapor; at larger pressures the transition temperatures are somewhat lower and form the λ-line separating the He-I and He-II phase. The fact that the system undergoes a phase transition can be seen *e.g.* by the behavior of the specific heat near T_λ, which has a logarithmic singularity $\sim \ln|T-T_\lambda|$: the corresponding curve resembles the Greek letter λ from which the transition derives its name. Note that this behavior of the specific heat differs from the finite cusp predicted for the ideal Bose gas at the BEC transition (see Sec.5.6.1).

Liquid He in the He-II phase has been found to exhibit perplexing "superfluid" properties, such as the ability to flow without resistance even through very narrow tubes (zero viscosity) and an extremely high thermal conductivity. In 1938 it was suggested by [London (1938)] that superfluidity could be a manifestation of BEC of the ^{4}He atoms. The BEC temperature for an ideal Bose gas of ^{4}He atoms is 3.1 K (see Sec. 5.6), not too far from the experimental $T_\lambda \approx 2.2$ K, and the phase diagram for the fermionic isotope ^{3}He is completely different (no λ-point).

However, the strong repulsive interactions lead to difficulties in microscopic calculations, and the precise relation between BEC and superfluidity is still a subject for investigation. The fact that ^{4}He is a strongly correlated quantum liquid is witnessed by the small condensate fraction, which was estimated at about 10% [Penrose and Onsager (1956)], a value later confirmed experimentally (see *e.g.* [Sears (1983)]) by neutron and X-ray diffraction experiments. The failure of the perturbative treatment of Sec. 18.1 can also be seen from the magnitude of the dilute gas expansion parameter in liquid ^{4}He: with $\rho \approx 2 \times 10^{28} \mathrm{m}^{-3}$ and $a \approx 2 \times 10^{-10}$m one has $(\rho a^3)^{1/2} \approx 0.4$ which makes the "correction" term in Eq. (18.134) considerably larger than the leading GP term.

Ab-initio descriptions of the ^{4}He liquid with quantitative results need advanced many-body treatments using realistic He-He interactions [Aziz *et al.* (1979)]. At $T = 0$, a Green's function Monte Carlo calculation by [Kalos *et al.* (1981)] gave results in very good agreement with experimental data, both for the equation of state (energy versus density) and for structural information such as the pair distribution function $g(r)$ (see Sec. 19.1.2), the condensate fraction z_0 and the momentum distribution $n(p)$. Similar outcomes for z_0 and $n(p)$ were obtained by considering a variational ground-state wave function containing two-body and three-body correlations [Usmani *et al.* (1982); Manousakis *et al.* (1985) and (1991)]. The equation of state calculated by using diagrammatic perturbation theory at the parquet level [Jackson *et al.* (1985)] is also in reasonable agreement with the Green's function Monte Carlo numbers. For finite temperatures, a numerical study in the λ-transition range (1 K - 4 K) using path integral Monte Carlo methods [Pollock and Ceperley (1984); Ceperley and Pollock (1986)], was able to reproduce the λ-curve for the specific heat, while being consistent with the $T = 0$ Green's Function Monte Carlo calculation. Quantum Monte Carlo applications to the ^{4}He liquid (mainly the path integral method) have been reviewed in [Ceperley (1995)]. In Fig. 19.2 the calculated condensate fraction z_0 is compared with the recent experimental results of [Glyde *et al.* (2000b)]. The latter are consistent with $z_0 = 7.25 \pm 0.75$ % at $T = 0$.

19.1.2 *Phenomenological descriptions*

Most of the properties of ^{4}He can also be understood from more phenomenological theories, in combination with some ingredients of the imperfect Bose

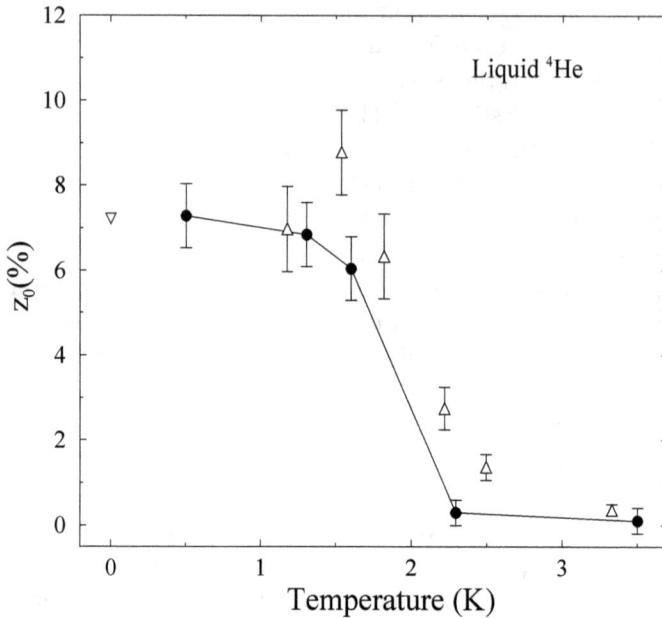

Fig. 19.2 Condensate fraction z_0 in bulk liquid ^{4}He, as a function of temperature. The filled circles are experimental data by [Glyde *et al.* (2000b)]. The up-triangles represent path integral MC calculations by [Ceperley (1995)]. The down-triangle at $T = 0$ is a diffusion MC result by [Moroni *et al.* (1997)].

gas treatment. A case in point are the two-fluid models proposed by [Tisza (1938)] and [Landau (1941) and (1947)], where the He-II phase is assumed to be a mixture of a normal and a superfluid, with densities ρ_n, ρ_s and velocities v_n, v_s, respectively. The normal fluid should then be identified with the thermal cloud (the thermally excited quasiparticle excitations) and the superfluid with the contribution of the quasiparticle vacuum. The superfluid component has no entropy, zero viscosity and is responsible for the frictionless flowing through thin capillaries; the presence of the normal component at $T > 0$ explains why a finite value is found when the viscosity is measured by dragging an object through the liquid (*e.g.* with the oscillating disk method). The two-fluid picture also leads to an extra mode for wave propagation through the liquid. In the ordinary first sound the velocities v_n and v_s of both components oscillate in phase, resulting in a density and pressure wave. The other possibility, an out of phase oscillation of v_n and v_s, leaves the density constant and leads to a temperature and entropy wave. This phenomenon is called *second sound,* and heat transfer through wave propagation (instead of diffusion) lies behind the abnormal thermal conductivity in He-II.

In a celebrated argument based on simple Galilei invariance, Landau has shown that frictionless superflow is strongly related to the shape of the excitation spectrum at low energy. A slab of superfluid flowing through a capillary with velocity v has total momentum $P = Nmv$ and kinetic energy $E = P^2/(2Nm)$. Friction occurs when part of the kinetic energy of the slab can be converted into heat by the creation of quasiparticle excitations. Let's consider a new configuration in which a quasiparticle with momentum p (in the original rest frame of the liquid) and energy E_p is present. In the new configuration the slab then has velocity $v' = v + p/(Nm)$, momentum $P' = Nmv + p = P + p$ and energy $E' = P'^2/(2Nm) + E_p = E + p \cdot v + E_p$, where the recoil energy $p^2/(2Nm)$ can be neglected. The most favorable case is for p in the opposite direction of v, and creation of a quasiparticle with momentum p is therefore allowed when $E' < E$ or $E_p < pv$. We see now that dissipation can occur only for velocities

$$v > v_c = \left(\frac{E_p}{p}\right)_{min}, \qquad (19.1)$$

where v_c is the Landau critical velocity. If the critical velocity v_c is nonzero there will be superflow for $v < v_c$. This clearly depends on the form of the excitation spectrum E_p. An ideal Bose gas has a purely quadratic spectrum, and thus would not be superfluid since Eq. (19.1) leads to $v_c = 0$. In interacting boson systems the spectrum starts linearly in p (see Sec. 18.4), and for $p \to 0$ one has $E_p = up$, where u is the macroscopic speed of sound. As a result, $v_c \leq u$ holds in any Bose gas, but the precise value of v_c depends on the detailed shape of the excitation spectrum. Indeed, the minima in E_p/p occur when $E'_p = E_p/p$ or, equivalently, whenever the tangent of the E_p versus p curve passes through the origin.

The complete dispersion curve E_p as a function of p has been accurately measured by inelastic neutron scattering. The spectrum at pressure $P \approx 0$ and $T = 1.1$ K is shown in Fig. 19.3. It is representative also for the situation at $T = 0$, since the temperature dependence of the spectrum is insignificant below $T \approx 1.5$ K. The linear part of the spectrum (phonon excitations) extends to about 0.75 Å^{-1} with a slope equal to the measured sound velocity ($v_{c,phon} \approx 240$ m/s). At larger momenta the so-called roton minimum $E_{rot} \approx 8.6$ K at $p/\hbar \approx 1.9$ Å^{-1}

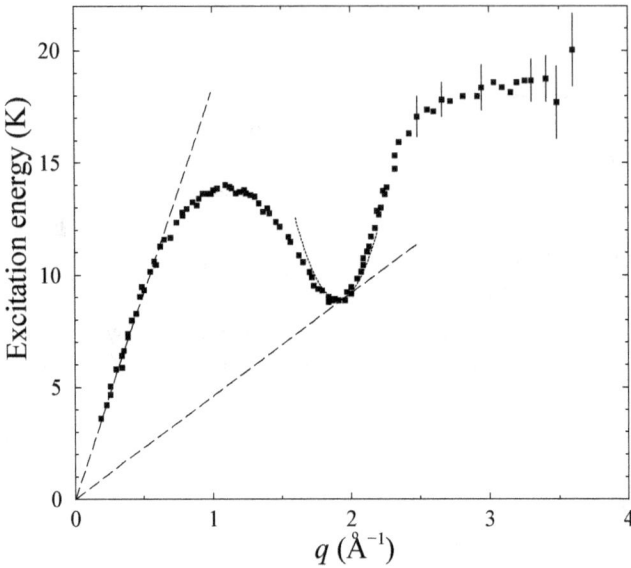

Fig. 19.3   Spectrum E_p of quasiparticle excitations in He-II at $T = 1.1$ K. Experimental data from [Woods and Cowley (1973)]. The straight lines indicate the Landau critical velocities for the phonon and roton excitations. The spectrum for $q > 2.5$ Å^{-1} has recently been measured with high accuracy (see Fig. 19.5).

appears.[1]   The precise nature of the associated excitations is still de-bated; they are usually associated with microscopic vortex rings (a local-ized rotation, hence the name[2]). In the roton region the spectrum can be approximated by a quadratic fit, and applying Eq. (19.1) leads to a Landau critical velocity $v_{c,rot} \approx 60$ m/s. This is in very good agree-ment with experimental values for the drag experienced by negative ions moving through ^{4}He with various velocities: it is vanishingly small below $v_{c,rot}$ and rises very steeply for velocities exceeding $v_{c,rot}$ [Rayfield (1966); Ellis and McClintock (1985)]. In most other situations, *e.g.* flow through thin capillaries, the observed critical velocities are orders of magnitude smaller than the roton value $v_{c,rot}$ and are determined by other mecha-nisms (the production and growth of macroscopic quantized vortices in the superfluid).

---

[1]The maximum between the phonon and roton part of the spectrum is called — rather tritely — the maxon region.

[2]More poetically, according to Feynman "*a roton is the ghost of a vanishing vortex ring*" [Feynman (1955)]

The above discussion was of necessity quite superficial. A thorough discussion on the superfluidity of ^{4}He can be found in books devoted to the subject, *e.g.* [Wilks and Betts (1987); Nozières and Pines (1990); Khalatnikov (1965)]. A good recent review article is [Griffiths *et al.* (2003)]. In the following section we take a closer look at the analysis of inclusive (n, n') scattering experiments which, apart from the dispersion curve, also allow to extract the momentum distribution and the condensate fraction.

19.2   The dynamic structure function

19.2.1   *Inclusive scattering*

The elementary neutron-He interaction is rather weak, and the (n, n') cross section for scattering off a ^{4}He target can therefore be expressed in the Born approximation,

$$\frac{d^2\sigma}{d\Omega dE_f} = \frac{\sigma_{n-He}}{4\pi} \frac{p_f}{p_i} S(Q, E). \tag{19.2}$$

The incoming (outgoing) neutron has momentum $\boldsymbol{p}_i$ ($\boldsymbol{p}_f$) and energy $E_i = p_i^2/(2m_n)$ ($E_f = p_f^2/(2m_n)$), the solid angle $d\Omega$ is the scattering direction, and σ_{n-He} is the total elementary n-He cross section. All information on the target system is contained in the dynamic structure function $S(Q, E)$, where $\boldsymbol{Q} \equiv \hbar\boldsymbol{q} = \boldsymbol{p}_i - \boldsymbol{p}_f$ is the momentum transfer and $E = E_i - E_f$ is the energy transfer to the target.

The dynamic structure function $S(Q, E)$ has already been introduced for the electron gas in Sec. 14.3 [see Eq. (14.54)], but may be defined in general as the response function to a scalar "probe" $\hat{\rho}(\boldsymbol{Q})$:

$$S(\boldsymbol{Q}, E) = \sum_n |\langle \Psi_n^N | \hat{\rho}(\boldsymbol{Q}) | \Psi_0^N \rangle|^2 \delta(E + E_0^N - E_n^N)$$

$$= \langle \Psi_0^N | \hat{\rho}^\dagger(\boldsymbol{Q}) \delta(E + E_0^N - \hat{H}) \hat{\rho}(\boldsymbol{Q}) | \Psi_0^N \rangle. \tag{19.3}$$

The operator $\rho(\boldsymbol{Q}) = \sum_i e^{i\boldsymbol{q}\cdot\boldsymbol{r}_i}$ transfers momentum $\boldsymbol{Q}$ to the system, as can be seen by its second-quantized form $\hat{\rho}(\boldsymbol{Q}) = \sum_{\boldsymbol{p}} a_{\boldsymbol{p}+\boldsymbol{Q}}^\dagger a_{\boldsymbol{p}}$; note that $\rho(-\boldsymbol{Q})$ is simply the Fourier transform of the familiar density operator $\rho(\boldsymbol{r}) = \sum_i \delta(\boldsymbol{r} - \boldsymbol{r}_i)$.

The structure factor $F(\boldsymbol{Q})$ is defined as the energy-integrated structure function $S(\boldsymbol{Q}, E)$. Using completeness, one finds immediately

$$
\begin{aligned}
F(\boldsymbol{Q}) = \int dE S(\boldsymbol{Q}, E) &= \langle \Psi_0^N | \, \hat{\rho}^\dagger(\boldsymbol{Q}) \hat{\rho}(\boldsymbol{Q}) \, | \Psi_0^N \rangle \\
&= \langle \Psi_0^N | \sum_{i,j} e^{i\boldsymbol{q}\cdot(\boldsymbol{r}_i - \boldsymbol{r}_j)} \, | \Psi_0^N \rangle .
\end{aligned} \tag{19.4}
$$

The last identity shows that $F(\boldsymbol{Q})$ is the Fourier transform of the pair correlation function $g(\boldsymbol{r})$,

$$
g(\boldsymbol{r}) = \int \frac{d\boldsymbol{q}}{(2\pi)^3} e^{i\boldsymbol{q}\cdot\boldsymbol{r}} F(\boldsymbol{Q}) = \langle \Psi_0^N | \sum_{i,j} \delta(\boldsymbol{r} - (\boldsymbol{r}_i - \boldsymbol{r}_j)) | \Psi_0^N \rangle , \tag{19.5}
$$

which represents the probability of finding two particles with relative position $\boldsymbol{r}$. For the present isotropic systems one observes of course that $S(Q, E)$, $F(Q)$ and $g(r)$ depend only on the magnitude of $\boldsymbol{Q}$ and $\boldsymbol{r}$ (as indicated in the notation), and $g(r)$ is the probability to find two particles at a relative distance r.

Provided that the interparticle potential is local, the energy-weighted sum rule of Eq. (14.50) holds with the result

$$
\int dE \, E S(Q, E) = \frac{1}{2} \langle \Psi_0^N | \, [\hat{\rho}^\dagger(\boldsymbol{Q}), [\hat{H}, \hat{\rho}(\boldsymbol{Q})]] \, | \Psi_0^N \rangle = \frac{N Q^2}{2M}, \tag{19.6}
$$

which is widely known as the f-sum rule.

Depending on the magnitude of the momentum transfer Q, the experimental structure function $S(Q, E)$ has some characteristic features in its energy dependence [Woods and Cowley (1973)]. For $0 < q < 2$ Å^{-1} there is a well-defined sharp peak (the quasiparticle[3] peak) followed by a broad background distribution extending to intermediate energies and a high-energy tail. An example is shown in Fig. 19.4. The position of the sharp peak identifies the quasiparticle energy E_Q, and the corresponding strength $N Z_Q$ is determined by energy-integration under the peak. The quasiparticle strength Z_Q diminishes rapidly beyond $q = 2$ Å^{-1}, though it is still possible (see Fig. 19.5) to track the quasiparticle peak up to $q \sim 3.6$ Å^{-1}. In addition, the broad background distribution represents multiphonon scattering. For large q ($q > 4$ Å^{-1}) the quasiparticle peak has completely disappeared,

---

[3]It may be surprising to use the quasiparticle spectrum (being a property of the sp propagator) for a description of the density response of the system. However, in condensed Bose systems the spectrum of the density fluctuations is intrinsically linked with the single-particle behavior, since for $Q \neq 0$ and to leading order in N one has $\hat{\rho}(Q) \approx a_{\boldsymbol{Q}}^\dagger a_0 + a_0^\dagger a_{-\boldsymbol{Q}} \approx \sqrt{N_0}(a_{\boldsymbol{Q}}^\dagger + a_{-\boldsymbol{Q}})$.

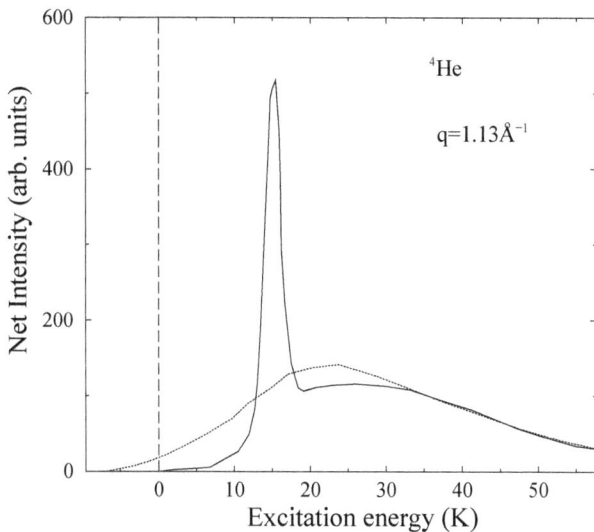

Fig. 19.4   Neutron inelastic scattering data on superfluid ^{4}He at pressure $P = 20$ bar (where $T_\lambda = 1.928$ K), adapted from [Talbot *et al.* (1988)]. The curves are proportional to the dynamic structure function $S(Q, E)$ near the maxon wave vector $q = 1.13$ Å^{-1}. The solid curve represents data at $T = 1.29$ K, well within the superfluid region, and the quasiparticle peak is clearly visible. For larger T, the quasiparticle peak decreases rapidly in intensity. It has disappeared in the dotted curve at $T = 1.90$ K, very near T_λ.

and the structure function shows only a broad peak near $E \approx Q^2/(2m)$ (see Fig. 19.6). This is characteristic of quasifree scattering and will be discussed in detail in the next section. At intermediate Q the structure of $S(Q, E)$ is more complex.

In the region of small Q the dominant quasiparticle peak exhausts the bulk of the sum rules (19.4) and (19.6). This means that $\hat{\rho}(\boldsymbol{Q}) \left| \Psi_0^N \right\rangle$ can be interpreted as an (approximate) eigenstate, with the δ-function in Eq. (19.6) generating a single peak at the corresponding eigenenergy E_Q. If one assumes that a single quasiparticle excitation exhausts the sum rules one has

$$S(Q, E) \approx NZ_Q\delta(E - E_Q), \qquad F(Q) \approx NZ_Q, \qquad \frac{Q^2}{2m} \approx E_Q Z_Q \qquad (19.7)$$

and consequently,

$$E_Q \approx \frac{Q^2 N}{2mF(Q)}, \qquad (19.8)$$

which relates the quasiparticle spectrum to the structure factor. Such a

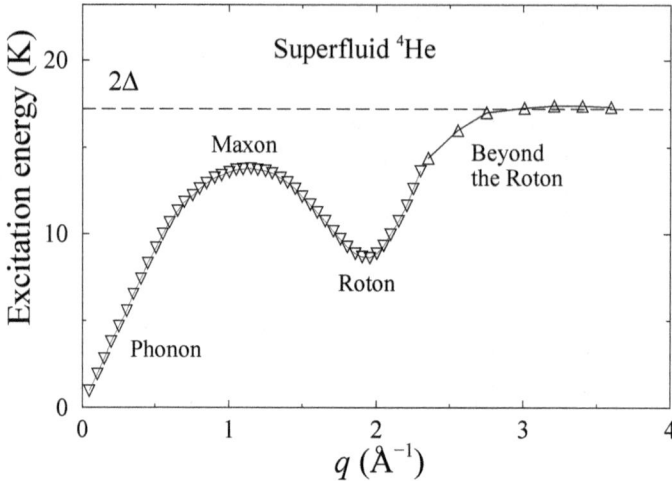

Fig. 19.5 Spectrum E_p of quasiparticle excitations in the ^{4}He liquid near zero temperature and pressure ($T = 1.35$ K and saturated vapour pressure). The down-triangles are data from [Donnelly *et al.* (1981)]. The up-triangles at larger q represent recent high-accuracy data from [Glyde *et al.* (1998)]. Note that E_p saturates at twice the roton energy E_{rot}, indicated by the dashed line. Higher excitation energies would be unstable for decay into two rotons, as predicted by [Pitaevskii (1959)].

picture provides the basis of the variational description of the excitation spectrum by [Feynman (1954)] and is able to generate roughly the shape of the dispersion curve in Fig. 19.3. A somewhat better description of the spectrum is obtained by the inclusion of backflow (three-body correlations) [Feynman and Cohen (1956)].

The analysis of the inclusive (n, n') scattering cross section at large momentum transfer allows to extract the momentum distribution of the ^{4}He atoms. The rationale behind this is quite general, and is also regularly applied to inclusive electron scattering in order to analyze *e.g.* the momentum distribution of protons in complex nuclei, or the quark momentum distribution in nucleons. In Sec. 19.2.2 we therefore discuss this topic for an arbitrary system.

19.2.2 *Asymptotic $1/Q$ expansion of the structure function*

In the large-Q limit (more precisely, $q >> 2\pi/d$ with d the average separation of the target constituents) the so-called Impulse Approximation can be used to describe the inelastic scattering process: the wavelength of the probe is so small that the constituents of the target system are seen as individual scatterers. As a result the momentum Q is transfered to only one constituent. In addition one may assume that the struck constituent, moving with large velocity, has negligible final-state interactions with the remaining $N-1$ target particles. One can then show that in this limit $S(Q, E)$ is fully determined by the momentum distribution of the constituents in the target.

To see this in detail we turn to a systematic expansion of $S(Q, E)$ in powers of $1/Q$, as derived in [Gersch *et al.* (1972)]. The time representation of the dynamic structure function is obtained by a Fourier transform of Eq. (19.3),

$$S(Q, \tau) = \int dE \, e^{-\frac{i\tau}{\hbar} E} S(Q, E)$$
$$= \langle \Psi_0^N | \hat{\rho}^\dagger(Q) e^{-\frac{i\tau}{\hbar}(\hat{H} - E_0^N)} \hat{\rho}(Q) | \Psi_0^N \rangle. \qquad (19.9)$$

Further manipulations are more convenient in the first-quantized form of the operators:

$$S(Q, \tau) = e^{\frac{i\tau}{\hbar} E_0^N} \sum_{i,j} \langle \Psi_0^N | e^{-i q \cdot r_j} e^{-\frac{i\tau}{\hbar} H} e^{i q \cdot r_i} | \Psi_0^N \rangle \qquad (19.10)$$
$$= e^{\frac{i\tau}{\hbar} E_0^N} \sum_{i,j} \langle \Psi_0^N | e^{i q \cdot (r_i - r_j)} \left[e^{-i q \cdot r_i} e^{-\frac{i\tau}{\hbar} H} e^{i q \cdot r_i} \right] | \Psi_0^N \rangle.$$

The bracketed part of the operator represents a unitary transformation on $e^{-\frac{i\tau}{\hbar} H}$, generating a momentum shift for the ith particle. This can be seen *e.g.* by applying the transformation to the momentum operator p_i, leading to $e^{-i q \cdot r_i} p_i e^{i q \cdot r_i} = p_i + Q$. As a consequence, the operator in Eq. (19.10) becomes

$$\left[e^{-i q \cdot r_i} e^{-\frac{i\tau}{\hbar} H} e^{i q \cdot r_i} \right] = e^{-\frac{i\tau}{\hbar} H(p_i \to p_i + Q)} = e^{-\frac{i\tau}{\hbar} [H + \frac{p_i \cdot Q}{m} + \frac{Q^2}{2m}]}, \qquad (19.11)$$

where we used the fact that the only momentum dependence in the Hamiltonian is contained in the kinetic energy $\sum_i p_i^2/(2m)$.

We now recall the manipulations that gave rise to the closed expression (A.32) for the evolution operator $\mathcal{U}(t, t_0)$ in the interaction picture defined in Eq. (A.22). Equations (A.22) and (A.32), with $t_0 = 0$, may be rewritten as

$$e^{-\frac{it}{\hbar}(H_0 + H_1)} = e^{-\frac{it}{\hbar} H_0} \mathcal{T} \exp\left(-\frac{i}{\hbar} \int_0^t dt' e^{\frac{it'}{\hbar} H_0} H_1 e^{-\frac{it'}{\hbar} H_0} \right), \qquad (19.12)$$

where $\mathcal{T}$ imposes the familiar time ordering in the integrals arising from the expansion of the exponential. In the derivation no special properties of the H_0 and H_1 operators were needed, and Eq. (19.12) is therefore an identity holding for arbitrary operators.

Making the substitutions $H_0 \to p_i \cdot Q/m$ and $H_1 \to H$ in Eq. (19.12), then leads to

$$e^{-\frac{i\tau}{\hbar}(H+\frac{p_i \cdot Q}{m})} = e^{-\frac{i\tau}{\hbar}\frac{p_i \cdot Q}{m}}\mathcal{T}\exp\left(\frac{-i}{\hbar}\int_0^\tau dt\, e^{\frac{it}{\hbar}\frac{p_i \cdot Q}{m}}He^{-\frac{it}{\hbar}\frac{p_i \cdot Q}{m}}\right). \quad (19.13)$$

In the integrand of Eq. (19.13) we recognize the generator for a translation $r_i \to r_i + Qt/m$, and the Hamiltonian is transformed as

$$\left[e^{\frac{it}{\hbar}\frac{p_i \cdot Q}{m}}He^{-\frac{it}{\hbar}\frac{p_i \cdot Q}{m}}\right] = H + \sum_{k(\neq i)}[V(r_i + x - r_k) - V(r_i - r_k)]$$

$$= H_i'(x), \quad (19.14)$$

in terms of a new length variable defined as $x = Qt/m$ and a vector $x = x\hat{Q}$. Note that x is the distance traveled by a particle with momentum Q and mass m in a time t.

Finally we replace in Eq. (19.13) the integration variable t with x, and introduce the corresponding quantities $\xi = Q\tau/m$, $\xi = \xi\hat{Q}$. The result is a clear separation of all Q-dependence in $S(Q,\tau)$:

$$S(Q,\tau) = e^{-iq\xi/2}\sum_{i,j}\langle\Psi_0^N|e^{iq\cdot(r_i - r_j)}e^{-\frac{i}{\hbar}p_i \cdot \xi}$$

$$\times \mathcal{T}_x\exp\left(-\frac{im}{\hbar Q}\int_0^\xi dx\,[H_i'(x) - E_0^N]\right)|\Psi_0^N\rangle \quad (19.15)$$

$$= e^{-iq\xi/2}F(Q,\xi). \quad (19.16)$$

A Fourier transform with respect to τ takes one back to the energy domain,

$$S(Q,E) = \int\frac{d\tau}{2\pi\hbar}e^{\frac{i\tau}{\hbar}E}S(Q,\tau) = \frac{m}{Q}\int\frac{d\xi}{2\pi\hbar}e^{i\xi y}F(Q,\xi), \quad (19.17)$$

where we have introduced the y-variable as

$$y = \frac{mE}{\hbar Q} - \frac{q}{2} = \frac{m}{Q\hbar}(E - \frac{Q^2}{2m}). \quad (19.18)$$

Expanding the exponential in Eq. (19.15) one can now clearly generate a series in powers of m/Q. However, Eq. (19.15) has additional Q-dependence in the factor $e^{i\boldsymbol{q}\cdot(\boldsymbol{r}_i-\boldsymbol{r}_j)}$. One therefore separates the summation over i,j in Eq. (19.15) into the *incoherent* part with $i = j$, and the coherent part with $i \neq j$. In the incoherent part one tracks a single particle which absorbs the momentum transfer of the probe, propagates, and re-emits $\boldsymbol{Q}$, according to the basic definition in Eq. (19.3). The coherent part contains the interference terms where this process involves two different particles. In the large q-limit the coherent contribution to $S(q,\omega)$ is severely suppressed as a result of the fast oscillations involved in the $e^{i\boldsymbol{q}\cdot(\boldsymbol{r}_i-\boldsymbol{r}_j)}$ factor, and one can limit oneself to the dominant incoherent contribution.

The expansion of the structure function therefore becomes

$$\frac{Q}{m}S(Q,E) = \sum_{n=0}^{\infty}\left(-\frac{im}{\hbar Q}\right)^n f_n(y), \tag{19.19}$$

where

$$f_n(y) = \int_{-\infty}^{+\infty}\frac{d\xi}{2\pi\hbar}e^{i\xi y}\sum_i\langle\Psi_0^N|e^{-\frac{i}{\hbar}\boldsymbol{p}_i\cdot\boldsymbol{\xi}}T_x\left\{\int_0^{\xi}dx_1(H_i'(\boldsymbol{x}_1) - E_0^N)\right.$$
$$\times\int_0^{x_1}dx_2(H_i'(\boldsymbol{x}_2) - E_0^N)..\int_0^{x_{n-1}}dx_n(H_i'(\boldsymbol{x}_n) - E_0^N)\right\}|\Psi_0^N\rangle. \tag{19.20}$$

It follows that in the large-Q limit, $QS(Q,E) \to mf_0(y)$ depends on the single variable y. This asymptotic property of the dynamic structure function is usually referred to as *y-scaling*. The leading term or Impulse Approximation to the structure function is given by $S_\infty(Q,E) = [m/Q]f_0(y)$ with

$$f_0(y) = \int_{-\infty}^{+\infty}\frac{d\xi}{2\pi\hbar}e^{i\xi y}\sum_i\langle\Psi_0^N|e^{-\frac{i}{\hbar}\boldsymbol{p}_i\cdot\boldsymbol{\xi}}|\Psi_0^N\rangle$$
$$= \int_{-\infty}^{+\infty}\frac{d\xi}{2\pi\hbar}\frac{V}{(2\pi\hbar)^3}\int d\boldsymbol{p}\, e^{i\xi(y-\boldsymbol{p}\cdot\hat{\boldsymbol{Q}}/\hbar)}n(\boldsymbol{p})$$
$$= \frac{N}{\rho}\int\frac{d\boldsymbol{p}}{(2\pi\hbar)^3}\delta(\hbar y - \boldsymbol{p}\cdot\hat{\boldsymbol{Q}})n(\boldsymbol{p}). \tag{19.21}$$

Here, $n(\boldsymbol{p}) = \langle\Psi_0^N|a_{\boldsymbol{p}}^\dagger a_{\boldsymbol{p}}|\Psi_0^N\rangle$ is the momentum distribution in the system and use has been made of the second-quantized form $\sum_i e^{-\frac{i}{\hbar}\boldsymbol{p}_i\cdot\boldsymbol{\xi}} \to \sum_{\boldsymbol{p}}e^{-\frac{i}{\hbar}\boldsymbol{p}\cdot\boldsymbol{\xi}}a_{\boldsymbol{p}}^\dagger a_{\boldsymbol{p}}$. The δ-function fixes the longitudinal momentum $\boldsymbol{p}\cdot\hat{\boldsymbol{Q}}$ (along the direction of $\boldsymbol{Q}$), and in the present isotropic system we get

$$f_0(y) = \frac{N}{(2\pi)^2\hbar^3\rho}\int_{\hbar|y|}^{+\infty}dp\,p\,n(p). \tag{19.22}$$

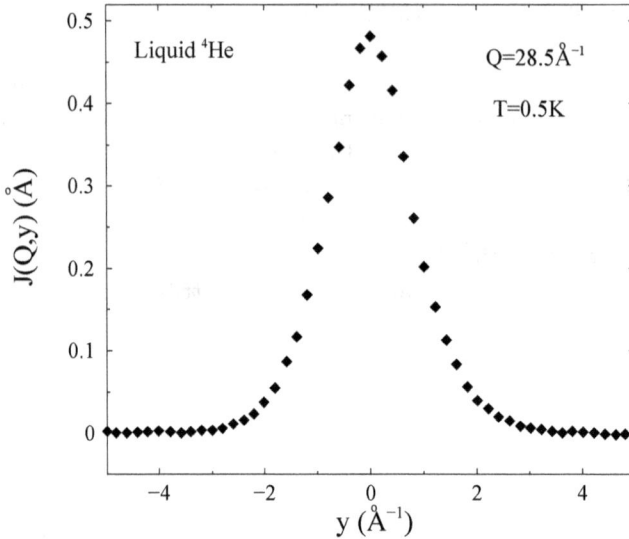

Fig. 19.6   Quasifree neutron scattering off liquid ^{4}He at $T = 0.5$ K and saturated vapour pressure. The data shown are from [Glyde *et al.* (2000b)] and represent the quantity $J(Q, y) = (\hbar/m)QS(Q, E) \approx \hbar f_0(y)$, which is (approximately) independent of Q [see Eq. (19.20)], for $Q = 28.5$ Å^{-1}. The analysis in [Glyde *et al.* (2000b)] of the y-dependence, and a comparison to temperatures just above T_λ, allows to extract the condensate fraction in Fig. 19.2, as well as the momentum distribution.

The $n = 1$ term in Eq. (19.19) represents the first correction to the Impulse Approximation and takes into account the dominant final-state interactions. The expression for $f_1(y)$ as given by Eq. (19.20) can be simplified considerably, using the fact that

$$[H'_i(\boldsymbol{x}) - E_0^N]\big|\Psi_0^N\big\rangle = \sum_{k(\neq i)} [V(\boldsymbol{r}_i - \boldsymbol{r}_k + \boldsymbol{x}) - V(\boldsymbol{r}_i - \boldsymbol{r}_k)]\big|\Psi_0^N\big\rangle \quad (19.23)$$

and that $e^{-\frac{i}{\hbar}\boldsymbol{p}_i \cdot \boldsymbol{\xi}}\phi(\boldsymbol{r}_i) = \phi(\boldsymbol{r}_i - \boldsymbol{\xi})$ acts as a generator of a translation for the i-th particle coordinate $\boldsymbol{r}_i$, which appears both in the interaction terms and in the wavefunction. Introducing the coordinate space representation of the ground-state wave function and defining the two-body density matrix as

$$\rho_2(\boldsymbol{r}_1, \boldsymbol{r}_2; \boldsymbol{r}'_1, \boldsymbol{r}'_2) = N(N-1)\int d\boldsymbol{r}_3..d\boldsymbol{r}_N \ \Psi_0^{N*}(\boldsymbol{r}'_1, \boldsymbol{r}'_2, \boldsymbol{r}_3..\boldsymbol{r}_N)$$

$$\times \Psi_0^N(\boldsymbol{r}_1, \boldsymbol{r}_2, \boldsymbol{r}_3..\boldsymbol{r}_N), \quad (19.24)$$

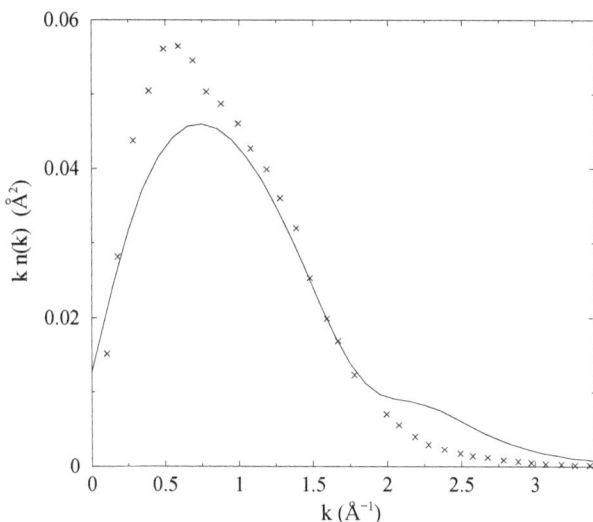

Fig. 19.7 Momentum distribution $kn(k)$ in superfluid ^{4}He (normalized as $\int dkn(k) = 1$) at the equilibrium density $\rho = 0.0219\text{Å}^{-3}$. Adapted from [Manousakis *et al.* (1985) and (1991)], with the experimental points based on [Sears (1983)].

one can show that

$$f_1(y) = \int_{-\infty}^{+\infty} \frac{d\xi}{2\pi\hbar} e^{i\xi y} \int_0^\xi dx \int dr_1 dr_2\, \rho_2(r_1 + \xi, r_2; r_1, r_2)$$
$$\times [V(r_1 - r_2 + x - \xi) - V(r_1 - r_2 - \xi)]. \tag{19.25}$$

This expression depends on the half-diagonal two-body density matrix $\rho_2(r_1, r_2; r_1', r_2)$. In general, $f_n(y)$ requires the $(n+1)$-body density matrix, diagonal except in one set of coordinates. Finally, one can change the integration variable $x \to \xi - x$ in Eq. (19.25) and exploit translational invariance of the wave function to arrive at

$$f_1(y) = \frac{N}{\rho} \int_{-\infty}^{+\infty} \frac{d\xi}{2\pi\hbar} e^{i\xi y} \int dr \rho_2(r - \xi, 0; r, 0) \int_0^\xi dx[V(r - x) - V(r - \xi)]. \tag{19.26}$$

In superfluid ^{4}He the momentum distribution has the form of Eq. (18.10). Substitution into Eq. (19.22) yields for the impulse approximation,

$$\frac{1}{N} S_\infty(Q, E) = z_0\delta\left(E - \frac{Q^2}{2m}\right) + \frac{m}{\hbar Q\rho} \int_{\hbar|y|}^{+\infty} \frac{pdp}{(2\pi\hbar)^2} \tilde{n}(p). \tag{19.27}$$

The first term arises due to the condensate and produces (in principle) a δ-peak in the spectrum of the scattered neutrons, the relative intensity of which is a measure of the condensate fraction. The second term represents Doppler-broadened quasifree scattering; the distribution is also peaked at $y = 0$ or $E = \frac{Q^2}{2m}$, and allows to derive $\tilde{n}(p)$ by taking the derivative with respect to y. In practice however, the analysis is complicated due to the final-state interactions (which provide a broadening of both the condensate peak and the background distribution) and residual coherent contributions. Fig. 19.6 contains a typical experimental result. The momentum distribution in superfluid ^{4}He (extrapolated to $T = 0$) can nevertheless be extracted from the experiments and is shown in Fig. 19.7. Also shown is a theoretical prediction by [Manousakis *et al.* (1985) and (1991)], based on a variational ground-state wave function containing two- and three-body correlations. A discussion of the liquid ^{3}He momentum distribution can be found in [Mazzanti *et al.* (2004)].

19.3 Inhomogeneous systems

The perturbation techniques for the Bose gas, developed in Sec. 18.1, can be extended to nonuniform Bose systems. We will limit the discussion on inhomogeneous systems to the first-order results for dilute systems, as is applicable to *e.g.* the atomic BEC of Sec. 12.4.5. This is the analog of the Bose gas treatment in Sec. 18.7. Making similar approximations one is again left [see Eq. (18.136)] with a thermodynamic potential, which is quadratic in the boson addition and removal operators and can be solved exactly by means of a Bogoliubov transformation to quasiparticles. For nonuniform systems this is not entirely trivial and in Sec. 19.3.1 we first discuss some mathematical properties of Bogoliubov transformations in a general setting.

19.3.1 *The bosonic Bogoliubov transformation*

The general form of a quadratic Bose Hamiltonian, expressed in an arbitrary sp basis, reads

$$\hat{H} = \sum_{\alpha\beta}\left(\varepsilon_{\alpha\beta}(a_\alpha^\dagger a_\beta + a_\beta a_\alpha^\dagger) + G_{\alpha\beta}a_\alpha^\dagger a_\beta^\dagger + G_{\alpha\beta}^* a_\beta a_\alpha\right). \qquad (19.28)$$

This may be recognized as a generalization of the Bose gas expression in Eq. (18.136), which was automatically diagonal in momentum space because of translational invariance. Note that the symmetric form in the $a^\dagger a$ and $a a^\dagger$ operators is sufficient; the difference $a_\alpha^\dagger a_\beta - a_\beta a_\alpha^\dagger = \delta_{\alpha,\beta}$ would only give rise to a constant term.

As in Sec. 18.7 we consider a general linear transformation with mixing of removal and addition operators,

$$b_i^\dagger = \sum_\gamma U_{\gamma i} a_\gamma^\dagger - V_{\gamma i} a_\gamma, \qquad b_i = \sum_\gamma U_{\gamma i}^* a_\gamma - V_{\gamma i}^* a_\gamma^\dagger. \tag{19.29}$$

The transformation coefficients are constrained by imposing boson commutation relations on the new b-operators,

$$[b_i, b_j^\dagger] = \delta_{i,j} = \sum_\gamma (U_{\gamma i}^* U_{\gamma j} - V_{\gamma i}^* V_{\gamma j}),$$

$$[b_i, b_j] = 0 = \sum_\gamma (-U_{\gamma i}^* V_{\gamma j}^* + V_{\gamma i}^* U_{\gamma j}^*). \tag{19.30}$$

We now show that for stable systems it is always possible to choose a transformation (19.29) such that the Hamiltonian takes on the simple diagonal form

$$\hat{H} = \sum_i \Lambda_i (b_i^\dagger b_i + b_i b_i^\dagger), \tag{19.31}$$

with positive energies $\Lambda_i > 0$.

It is convenient to introduce matrix notation and rewrite the Hamiltonian in Eq. (19.28) as

$$\hat{H} = \begin{pmatrix} a^\dagger & a \end{pmatrix} \begin{pmatrix} \varepsilon & G \\ G^* & \varepsilon^* \end{pmatrix} \begin{pmatrix} a \\ a^\dagger \end{pmatrix} = A^\dagger H_a A, \tag{19.32}$$

where ε is a hermitian and G a symmetric matrix in sp space, so that H_a is a hermitian matrix with doubled dimension. The transformation (19.29) is written in matrix form as

$$\begin{pmatrix} b \\ b^\dagger \end{pmatrix} = \begin{pmatrix} U^\dagger & -V^\dagger \\ -V^T & U^T \end{pmatrix} \begin{pmatrix} a \\ a^\dagger \end{pmatrix}, \quad \text{or:} \quad B = \sigma M^\dagger \sigma A, \tag{19.33}$$

where[4] $\sigma = \begin{pmatrix} 1 & 0 \\ 0 & -1 \end{pmatrix}$, $M = \begin{pmatrix} U & V^* \\ V & U^* \end{pmatrix}$, and the commutation requirements

[4]Unit and null matrices are denoted simply by 1 and 0, since no confusion is possible.

(19.30) can be expressed as

$$
\begin{pmatrix} U^\dagger & -V^\dagger \\ -V^T & U^T \end{pmatrix} \begin{pmatrix} U & V^* \\ V & U^* \end{pmatrix} = 1, \quad \text{or: } \sigma M^\dagger \sigma M = 1. \tag{19.34}
$$

It follows that $\sigma M^\dagger \sigma = M^{-1}$, and the inverse of the transformation (19.33) therefore becomes

$$
\begin{pmatrix} a \\ a^\dagger \end{pmatrix} = \begin{pmatrix} U & V^* \\ V & U^* \end{pmatrix} \begin{pmatrix} b \\ b^\dagger \end{pmatrix} \quad \text{or: } A = MB. \tag{19.35}
$$

Equation (19.35) allows to express the Hamiltonian in the b-basis as $\hat{H} = B^\dagger H_b B$, and it is easy to check that the transformed H_b keeps the symmetries of the original Hamiltonian matrix H_a, *i.e.*

$$
H_b = M^\dagger H_a M = \begin{pmatrix} U^\dagger & V^\dagger \\ V^T & U^T \end{pmatrix} \begin{pmatrix} \varepsilon & G \\ G^* & \varepsilon^* \end{pmatrix} \begin{pmatrix} U & V^* \\ V & U^* \end{pmatrix} = \begin{pmatrix} \varepsilon' & G' \\ G'^* & \varepsilon'^* \end{pmatrix}, \tag{19.36}
$$

with

$$
\begin{aligned}
\varepsilon' &= U^\dagger \varepsilon U + V^\dagger \varepsilon^* V + U^\dagger G V + V^\dagger G^* U, \\
G' &= U^\dagger G U^* + V^\dagger G^* V^* + U^\dagger \varepsilon V^* + V^\dagger \varepsilon^* U^*.
\end{aligned} \tag{19.37}
$$

Imposing on H_b the diagonal form of Eq. (19.31) implies that $G' = 0$ and $\varepsilon' = \varepsilon'^* = \Lambda$ in Eq. (19.36), with Λ a real diagonal matrix. Left-multiplying Eq. (19.36) with $\sigma M \sigma$ then leads to the equivalent diagonality condition

$$
\begin{pmatrix} \varepsilon & G \\ G^* & \varepsilon^* \end{pmatrix} \begin{pmatrix} U & V^* \\ V & U^* \end{pmatrix} = \sigma \begin{pmatrix} U & V^* \\ V & U^* \end{pmatrix} \begin{pmatrix} \Lambda & 0 \\ 0 & -\Lambda \end{pmatrix}, \quad \text{or: } H_a M = \sigma M \begin{pmatrix} \Lambda & 0 \\ 0 & -\Lambda \end{pmatrix}. \tag{19.38}
$$

Eq. (19.38) is recognized as a generalized eigenvalue problem with σ playing the role of a metric matrix: the columns of the transformation matrix M must be solutions $\begin{pmatrix} X_i \\ Y_i \end{pmatrix}$ of

$$
\begin{pmatrix} \varepsilon & G \\ G^* & \varepsilon^* \end{pmatrix} \begin{pmatrix} X_i \\ Y_i \end{pmatrix} = E_i \begin{pmatrix} 1 & 0 \\ 0 & -1 \end{pmatrix} \begin{pmatrix} X_i \\ Y_i \end{pmatrix}, \quad \text{or: } H_a \begin{pmatrix} X_i \\ Y_i \end{pmatrix} = E_i \sigma \begin{pmatrix} X_i \\ Y_i \end{pmatrix}. \tag{19.39}
$$

Note that Eq. (19.39) has the same form as the RPA equations discussed in Sec. 13.2. In particular one observes that if (X_i, Y_i, E_i) is a solution then so is $(Y_i^*, X_i^*, -E_i)$, as is also clear from the structure of the right-hand side of Eq. (19.38).

Solution for positive-definite matrices

Mathematically, it depends on the signature of the matrix H_a whether the diagonality condition in Eq. (19.38) can be fulfilled. At present we assume a physically stable system, *i.e.* the matrix H_a is positive-definite.[5] In that case it is easy to see that Eq. (19.39) is equivalent to the Hermitian eigenvalue problem $\frac{1}{E_i}\xi_i = (H_a^{-1/2}\sigma H_a^{-1/2})\xi_i$ with $\xi_i = H_a^{1/2}\binom{X_i}{Y_i}$, and consequently all eigenvalues E_i of Eq. (19.39) are real and nonzero. The corresponding eigenvectors automatically fulfill the orthogonality condition

$$X_i^\dagger X_j - Y_i^\dagger Y_j = 0, \quad \text{for } i \neq j, \tag{19.40}$$

and from the positive-definiteness of H_a in Eq. (19.39) it follows that

$$E_i(X_i^\dagger X_i - Y_i^\dagger Y_i) = \left(X_i^\dagger \; Y_i^\dagger \right) H_a \begin{pmatrix} X_i \\ Y_i \end{pmatrix} > 0. \tag{19.41}$$

Since the eigenvalues of Eq. (19.39) come in pairs $\pm\Lambda_i$ (with all $\Lambda_i > 0$), it is clear from Eq. (19.41) that the *positive-energy* solutions should be used to build up the transformation matrices U and V, in order to fulfill the commutation relation $U^\dagger U - V^\dagger V = 1$ implied by Eq. (19.34), and to determine the new b operators according to Eq. (19.33). The negative-energy solutions merely repeat this information, with the roles of b_i and $b_i^\dagger$ interchanged.

From the above one concludes that the Hamiltonian (19.28) can indeed be expressed in the diagonal form of Eq. (19.31). The diagonalization procedure in addition yields the ground state, since Eq. (19.31) may be rewritten as

$$\hat{H} = \sum_i \Lambda_i(b_i^\dagger b_i + b_i b_i^\dagger) = 2 \sum_i \Lambda_i b_i^\dagger b_i + \text{Tr}(\Lambda), \tag{19.42}$$

and represents a noninteracting system of quasiparticles with positive energies Λ_i. Obviously the vacuum state $|0_B\rangle$ of the b_i operators, defined by

$$\forall i : \; b_i |0_B\rangle = 0, \tag{19.43}$$

---

[5] In many cases H_a also has zero eigenvalues, implying a zero curvature of the energy in the direction of the corresponding eigenvector. These Goldstone modes arise whenever a continuous symmetry of the underlying physical Hamiltonian is broken in the quadratic approximation. A more general discussion for positive-semidefinite matrices is given at the end of this section.

has minimal expectation value $\langle 0_B | \hat{H} | 0_B \rangle = \text{Tr}(\Lambda)$, and is therefore the ground state of the Hamiltonian in Fock space.

The energy difference between the quasiparticle vacuum $|0_B\rangle$ and the real particle vacuum $|0\rangle$ is given by

$$\Delta E = \langle 0_B | \hat{H} | 0_B \rangle - \langle 0 | \hat{H} | 0 \rangle = \text{Tr}(\Lambda - \varepsilon) = -2 \sum_i \Lambda_i \sum_\alpha |V_{\alpha i}|^2. \quad (19.44)$$

The last identity is derived by taking the trace of the matrix identity

$$H_a = \sigma M \begin{pmatrix} \Lambda & 0 \\ 0 & \Lambda \end{pmatrix} M^\dagger \sigma = \begin{pmatrix} U & V^* \\ V & U^* \end{pmatrix} \begin{pmatrix} \Lambda & 0 \\ 0 & \Lambda \end{pmatrix} \begin{pmatrix} U^\dagger & V^\dagger \\ V^T & U^T \end{pmatrix}, \quad (19.45)$$

which follows directly from Eq. (19.38).

Canonical single-particle basis

The structure of the full Bogoliubov transformation in Eq. (19.33) and of the quasiparticle vacuum can be greatly clarified by considering the so-called canonical sp basis. Applying singular value decomposition (see *e.g.* [Golub and van Loan (1996)]) to the (complex) transformation matrices, one can write $U = W_1^\dagger u W_2$ and $V = W_3^\dagger v W_4$, where the W_i are unitary matrices and both u and v are real and positive diagonal matrices. The commutation relations in Eq. (19.34) imply that $U^\dagger U - V^\dagger V = 1$. Hence $U^\dagger U = W_2^\dagger u^2 W_2$ and $V^\dagger V = W_4^\dagger v^2 W_4$ commute, and can be brought to diagonal form by the same unitary transformation. It follows that $u^2 - v^2 = 1$, and a set of degenerate values among the u diagonal elements entails the same degeneracy block along the diagonal of v. In addition, one finds that $R_2 u^2 = u^2 R_2$ with $R_2 = W_4 W_2^\dagger$, implying that the unitary matrix R_2 is block-diagonal and therefore commutes with the positive u and v. Changing the order of the matrix factors in Eq. (19.34) yields $UU^\dagger - V^* V^T = 1$ and by the same reasoning one has that $R_1 = W_3^* W_1^\dagger$ is unitary, block-diagonal, and commutes with u and v. We can therefore write $V = W_1^T R_1^T v R_2 W_2 = W_1^T R v W_2$, with $R = R_1^T R_2$ unitary and block-diagonal, and treat each degeneracy block separately. Finally, the off-diagonal requirement in Eq. (19.34) implies that $U^\dagger V^* - V^\dagger U^* = 0$, *i.e.* $U^\dagger V^*$ must be symmetric, or equivalently, $uv(R - R^T) = 0$. This condition is automatically fulfilled for a block with $v = 0$, but in that case, of course, the matrix R is irrelevant. In a block with $v \neq 0$, R must be unitary and symmetric, and can be transformed by a real orthogonal matrix O into diagonal form $R = O^T e^{i\theta} O$, where θ is real and diagonal. Since O is real, one can construct the final canonical form $U = W_1^\dagger u W_2$ and $V = W_1^T v W_2$, by redefining $W_1 \to e^{i\frac{\theta}{2}} O W_1$ and $W_2 \to e^{i\frac{\theta}{2}} O W_2$.

With this canonical form for U and V, the full transformation in Eq. (19.33) can always be decomposed as a succession of three transformations,

$$
\begin{pmatrix} b \\ b^\dagger \end{pmatrix} = \begin{pmatrix} W_2^\dagger & 0 \\ 0 & W_2^T \end{pmatrix} \begin{pmatrix} u & -v \\ -v & u \end{pmatrix} \begin{pmatrix} W_1 & 0 \\ 0 & W_1^* \end{pmatrix} \begin{pmatrix} a \\ a^\dagger \end{pmatrix}. \tag{19.46}
$$

The first transformation (involving W_1) is an ordinary change of sp basis $a \to a'$ to the canonical basis $a'_\alpha = \sum_\gamma (W_1)_{\alpha\gamma} a_\gamma$. The second transformation $a' \to b'$ mixes particle addition and removal operators and defines the quasiparticle operators

$$
b'_\alpha = u_\alpha a'_\alpha - v_\alpha a'^\dagger_\alpha. \tag{19.47}
$$

These two transformations are sufficient to remove $b'b'$ and $b'^\dagger b'^\dagger$ terms from the Hamiltonian. The final transformation $b' \to b$ is again an ordinary unitary transformation $b_i = \sum_\alpha (W_2^*)_{\alpha i} b'_\alpha$ on the quasiparticle operators and brings the $b^\dagger b$ block in the Hamiltonian to the diagonal form of Eq. (19.31). Note that this last transformation does not involve the quasiparticle vacuum or its energy, since the defining relation (19.43) holds for any unitary transformation among the quasiparticles and $b_i |0_B\rangle = b'_\alpha |0_B\rangle = 0$.

An explicit expression for the quasiparticle vacuum is most easily found in the canonical basis. The coherent many-boson state $|\psi_\alpha\rangle = e^{\frac{w_\alpha}{2}(a'^\dagger_\alpha)^2} |0\rangle$ obeys the identity

$$
a'_\alpha |\psi_\alpha\rangle = \sum_{n=0}^\infty \frac{(w_\alpha/2)^n}{n!} [a'_\alpha, (a'^\dagger_\alpha)^{2n}] |0\rangle = w_\alpha a'^\dagger_\alpha |\psi_\alpha\rangle. \tag{19.48}
$$

Using Eq. (19.47) and Eq. (19.48) with the choice $w_\alpha = v_\alpha/u_\alpha$ one sees that $b'_\alpha |\psi_\alpha\rangle = 0$. The quasiparticle vacuum can then be expressed (apart from a normalization constant) as

$$
|0_B\rangle \sim \prod_\alpha e^{\frac{w_\alpha}{2}(a'^\dagger_\alpha)^2} |0\rangle = e^{S^\dagger} |0\rangle, \tag{19.49}
$$

which can be thought of as a coherent superposition of the bosonic pair state

$$
S^\dagger = \frac{1}{2} \sum_\alpha \frac{v_\alpha}{u_\alpha} (a'^\dagger_\alpha)^2. \tag{19.50}
$$

Finally, it is also of interest to consider the generalized density matrix

$$R = \langle 0_B | AA^\dagger | 0_B \rangle = \begin{pmatrix} 1 + \rho^* & \kappa^* \\ \kappa & \rho \end{pmatrix} \tag{19.51}$$

where the normal density matrix of the quasiparticle vacuum $\rho_{\alpha\beta} = \langle 0_B | a_\alpha^\dagger a_\beta | 0_B \rangle$ is Hermitian and $\kappa_{\alpha\beta} = \langle 0_B | a_\alpha a_\beta | 0_B \rangle$ is a symmetric matrix, the so-called anomalous density. The condition that $|0_B\rangle$ be the vacuum of the b-operators is equivalent to

$$R = M \langle 0_B | BB^\dagger | 0_B \rangle M^\dagger = M \begin{pmatrix} 1 & 0 \\ 0 & 0 \end{pmatrix} M^\dagger = \begin{pmatrix} UU^\dagger & UV^\dagger \\ VU^\dagger & VV^\dagger \end{pmatrix}, \tag{19.52}$$

and as a result $\rho = VV^\dagger$ and $\kappa = VU^\dagger$. Transforming to the canonical basis,

$$R' = \langle 0_B | A'A'^\dagger | 0_B \rangle = \begin{pmatrix} W_1 & 0 \\ 0 & W_1^* \end{pmatrix} R \begin{pmatrix} W_1^\dagger & 0 \\ 0 & W_1^T \end{pmatrix} = \begin{pmatrix} u^2 & uv \\ uv & v^2 \end{pmatrix}, \tag{19.53}$$

therefore leads to simultaneous diagonalization of both the normal density $\rho'_{\alpha\beta} = \delta_{\alpha,\beta} v_\alpha^2$ and the anomalous density $\kappa'_{\alpha\beta} = \delta_{\alpha,\beta} u_\alpha v_\alpha$.

Zero-energy solutions

In order to accommodate the presence of Goldstone modes, we still need to examine the more general case when the Hamiltonian matrix H_a in Eq. (19.39) is positive-semidefinite, *i.e.*

$$Z^\dagger H_a Z \geq 0, \quad \forall Z = \begin{pmatrix} X \\ Y \end{pmatrix}. \tag{19.54}$$

It is easy to prove that for any solution of the generalized eigenvalue problem $H_a Z = E\sigma Z$ the eigenvalue E is real; that two solutions Z_1, Z_2 with different energies $E_1 \neq E_2$ are σ-orthogonal (*i.e.* $Z_1^\dagger \sigma Z_2 = 0$); and that a set of vectors Z_i which are mutually σ-orthogonal and have nonzero σ-norm, are also linearly independent [Thouless (1961)]. The property that any solution $Z = \begin{pmatrix} X \\ Y \end{pmatrix}$ has an adjoint solution $\overline{Z} = \begin{pmatrix} Y^* \\ X^* \end{pmatrix}$ which obeys $H_a \overline{Z} = -E\sigma\overline{Z}$ still holds as well.

As a consequence, the eigenvectors Z_i corresponding to nonzero eigenvalues E_i can be treated exactly like before: they come in pairs, and it is possible to normalize them as

$$Z_i^\dagger \sigma Z_j = \text{sgn}(E_i) \delta_{i,j}. \tag{19.55}$$

The linearly independent Z_i-vectors span a linear subspace $\mathcal{L}$ of (necessarily even) dimension $2m$, with a nonhermitian projection operator

$$\mathcal{P} = \sum_{i=1}^{2m} \operatorname{sgn}(E_i) Z_i Z_i^\dagger \sigma, \tag{19.56}$$

obeying $\mathcal{P}^2 = \mathcal{P}$, $\mathcal{P}Z = Z$ for any Z in $\mathcal{L}$, and $\mathcal{P}Z = 0$ for any vector Z σ-orthogonal to $\mathcal{L}$.

In addition, we should now allow for the presence of zero-energy solutions ($H_a N = 0$). Note that any vector of the null space $\mathcal{N}$ of H_a is automatically σ-orthogonal to all nonzero eigenvectors, $N^\dagger \sigma Z_i = 0$. Diagonalizing σ in $\mathcal{N}$, it is always possible to have a basis set for $\mathcal{N}$ which is both orthogonal and σ-orthogonal, *i.e.*

$$N_m^\dagger N_n = \delta_{m,n} \text{ and } N_m^\dagger \sigma N_n = \lambda_n \delta_{m,n}. \tag{19.57}$$

We note that the adjoint vectors $\overline{N}_m$ also form an orthogonal basis in which σ is diagonal:

$$\overline{N}_m^\dagger \overline{N}_n = N_n^\dagger N_m = \delta_{m,n}; \quad \overline{N}_m^\dagger \sigma \overline{N}_n = -N_n^\dagger \sigma N_m = -\lambda_n \delta_{m,n}. \tag{19.58}$$

As a consequence, the $\lambda \neq 0$ come in pairs $(\lambda_n, -\lambda_n)$.

In principle one should therefore distinguish two types of zero-energy solutions N_n. The first type, having $\lambda_n \neq 0$, can be paired off such that $N_{-\lambda_n} = \overline{N}_{\lambda_n}$ (possibly after a unitary transformation if λ_n is degenerate). One may renormalize, $Z_n^0 = \frac{1}{|\lambda_n|^{1/2}} N_n$, to obtain

$$(Z_m^0)^\dagger \sigma Z_n^0 = \operatorname{sgn}(\lambda_n) \delta_{m,n}. \tag{19.59}$$

The normalizable zero-energy solutions Z_n^0 behave exactly like the eigenvectors Z_i with nonzero eigenvalues. Such eigenvectors Z_n^0, while mathematically possible, can usually be discarded on physical grounds: they would correspond to intrinsic excitations with zero excitation energy.

More important from a physical viewpoint are a second type of zero-energy solutions $N_{\lambda_j} \equiv P_j$, having $\lambda_j = P_j^\dagger \sigma P_j = 0$ or a zero σ-norm. These are the Goldstone modes corresponding to the generators of symmetries of the underlying Hamiltonian which are broken by the mean-field treatment. The P_j can always (possibly after a unitary transformation) be chosen self-adjoint, *i.e.* $\overline{P}_j = P_j$. Since any P_j is σ-orthogonal to all vectors previously constructed, the set of the Z_i, Z_n^0, P_j cannot be complete: otherwise one would have $\sigma P_j = 0$, which is clearly absurd. We see that in this case the solutions of the eigenvalue problem do not span the entire sp space.

In order to complete the basis, for any P_j the "missing" vector Q_j can be constructed (see [Thouless (1961); Thouless and Valatin (1962)]) as the solution of

$$H_a Q_j = -i\sigma P_j. \qquad (19.60)$$

This is solvable, since σP_j is orthogonal to the null space $\mathcal{N}$ of H_a; the vector Q_j can then be defined as the unique solution[6] which is orthogonal to $\mathcal{N}$.

Combining Eq. (19.60) with $H_a Z_i = E_i \sigma Z_i$ one finds in the standard fashion that $Q_j^\dagger \sigma Z_i = 0$ for all nonzero-energy solutions Z_i. Further manipulating Eq. (19.60), it is easy to see that the $Q_j = \overline{Q}_j$ are self-adjoint, and that they form a linearly independent set. We may therefore conclude that the vectors Z_i, P_j, Q_j span the entire space, and expand σP_i as

$$\sigma P_i = \sum_j \alpha_{ij} Q_j + \sum_m \beta_{im} Z_m, \qquad (19.61)$$

where α_{ij} is an invertible matrix. Left-multiplying Eq. (19.61) with $Q_k^\dagger \sigma$, and employing (σ-)orthogonality, leads to the result that $Q_k^\dagger \sigma Q_j = 0$, for all k, j. Finally, we can employ the fact that the matrix A_{ij} defined as

$$A_{ij} = Q_i^\dagger H_a Q_j = -i Q_i^\dagger \sigma P_j = i\overline{P}_j^\dagger \sigma \overline{Q}_i, \qquad (19.62)$$

is a real symmetric and positive-definite matrix. The matrix A_{ij} can be diagonalized with a real orthogonal transformation, which does not spoil the self-adjointness property of the P_i and Q_i. It is thus always possible to construct a set of self-adjoint P_i and Q_i which obey

$$Q_i^\dagger \sigma Q_j = P_i^\dagger \sigma P_j = 0, \qquad (19.63)$$
$$Q_i^\dagger \sigma P_j = i\Gamma_i \delta_{i,j}, \text{with } \Gamma_i > 0, \qquad (19.64)$$

and are decoupled from the excitations with nonzero energy,

$$P_i^\dagger \sigma Z_j = Q_i^\dagger \sigma Z_j = 0. \qquad (19.65)$$

Expressed in terms of operators, according to Eq. (19.29), it is clear that iP_j and $iQ_j\hbar/\Gamma_j$ represent Hermitian operators that obey the commutation relations of canonically conjugate momenta and coordinates. Each

---

[6]The solution is defined up to an arbitrary vector of the null space $\mathcal{N}$; this freedom can be used to make the Q_j σ-orthogonal to all Z_n^0 vectors. For simplicity we skip this unnecessary complication, and assume in the remainder of the discussion that no Z_n^0 solutions are present.

momentum operator corresponds to the generator of a continuous symmetry of the underlying system [Ring and Schuck (1980)], which is broken in the mean-field Hamiltonian of Eq. (19.28).

The P_i and Q_i span a linear subspace, with a projection operator

$$\mathcal{P}_0 = \sum_i [Q_i P_i^\dagger - P_i Q_i^\dagger] \frac{i}{\Gamma_i} \sigma, \qquad (19.66)$$

obeying $\mathcal{P}_0^2 = \mathcal{P}_0$, $\mathcal{P}_0 P_i = P_i$ and $\mathcal{P}_0 Q_i = Q_i$. Equation (19.56), together with $\mathcal{P}_0 \mathcal{P} = 0$, allows to write the identity operator for the complete space as

$$1 = \mathcal{P} + \mathcal{P}_0. \qquad (19.67)$$

The Hamiltonian matrix can likewise be expressed as

$$H_a = \sum_i |E_i| Z_i Z_i^\dagger \sigma + \sum_j \frac{1}{\Gamma_i} \sigma P_i P_i^\dagger \sigma. \qquad (19.68)$$

This section on the mathematical properties of bosonic Bogoliubov transformations was somewhat abstract, but can now be put to use in a description of nonuniform dilute Bose systems.

19.3.2 *Bogoliubov prescription for nonuniform systems*

We consider (as in Sec. 18.7) a dilute system of spinless bosons, with the Hamiltonian now containing a confining potential $U(\boldsymbol{r})$ in addition to the usual contact interaction between the particles,

$$\hat{H} = \int d\boldsymbol{r}\, a_{\boldsymbol{r}}^\dagger \left[-\frac{\hbar^2 \nabla^2}{2m} + U(\boldsymbol{r}) \right] a_{\boldsymbol{r}} + \frac{g}{2} \int d\boldsymbol{r}\, a_{\boldsymbol{r}}^{\dagger 2} a_{\boldsymbol{r}}^2. \qquad (19.69)$$

The Bogoliubov prescription for the boson removal and addition operators,

$$a_{\boldsymbol{r}} \to \sqrt{N_0}\, \phi(\boldsymbol{r}) + \delta a_{\boldsymbol{r}}, \qquad a_{\boldsymbol{r}}^\dagger \to \sqrt{N_0}\, \phi^*(\boldsymbol{r}) + \delta a_{\boldsymbol{r}}^\dagger, \qquad (19.70)$$

again amounts to splitting off the contribution of the (as yet unknown) condensate, which is described by a normalized wave function $\phi(\boldsymbol{r})$ with a macroscopic occupation N_0. The non-condensate part $\delta a_{\boldsymbol{r}}$ only acts in the sp space orthogonal to $\phi(\boldsymbol{r})$ [see Eq. (18.18)].

In the weakly interacting limit one has $N_0 \approx N$ and the operators $\delta a_{\boldsymbol{r}}$, $\delta a_{\boldsymbol{r}}^\dagger$ represent but small fluctuations. Applying the Bogoliubov prescription to the thermodynamic potential $\hat{\Omega} = \hat{H} - \mu\hat{N}$ and expanding to second order in the operators $\delta a_{\boldsymbol{r}}$ and $\delta a_{\boldsymbol{r}}^{\dagger}$, results in $\hat{\Omega}_B = \Omega_0 + \hat{\Omega}_1 + \hat{\Omega}_2$, where

$$\Omega_0 = N_0 \int d\boldsymbol{r}\; \phi^*(\boldsymbol{r}) \left[-\frac{\hbar^2\nabla^2}{2m} + U(\boldsymbol{r}) - \mu \right] \phi(\boldsymbol{r}) + \frac{gN_0^2}{2} \int d\boldsymbol{r}\; |\phi(\boldsymbol{r})|^4$$

$$\hat{\Omega}_1 = \sqrt{N_0} \int d\boldsymbol{r}\; \delta a_{\boldsymbol{r}}^{\dagger} \left[-\frac{\hbar^2\nabla^2}{2m} + U(\boldsymbol{r}) - \mu + N_0\, g|\phi(\boldsymbol{r})|^2 \right] \phi(\boldsymbol{r}) + \text{h.c.}$$

$$\hat{\Omega}_2 = \int d\boldsymbol{r}\; \delta a_{\boldsymbol{r}}^{\dagger} \left[-\frac{\hbar^2\nabla^2}{2m} + U(\boldsymbol{r}) - \mu + 2N_0\, g|\phi(\boldsymbol{r})|^2 \right] \delta a_{\boldsymbol{r}}$$

$$+ \frac{gN_0}{2} \int d\boldsymbol{r}\; \left[\phi^{*2}(\boldsymbol{r})(\delta a_{\boldsymbol{r}})^2 + \phi^2(\boldsymbol{r})(\delta a_{\boldsymbol{r}}^{\dagger})^2 \right]. \tag{19.71}$$

Compared to the homogeneous case in Eq. (18.19–18.20) a term $\hat{\Omega}_1$ appears which is linear in the fluctuation operators. The linear term vanishes when $\phi(\boldsymbol{r})$ obeys

$$\left[-\frac{\hbar^2\nabla^2}{2m} + U(\boldsymbol{r}) + N_0\, g|\phi(\boldsymbol{r})|^2 \right] \phi(\boldsymbol{r}) = \mu\phi(\boldsymbol{r}). \tag{19.72}$$

This is recognized as the GP equation (12.104), but takes into account the quantum depletion ($N_0 < N$) of the condensate. The choice of $\phi(\boldsymbol{r})$ as the ground-state solution of Eq. (19.72) also minimizes, for any N_0, the dominant contribution Ω_0 of the condensate to the thermodynamic potential [see Eq. (12.105)].

The remaining potential Ω_2 is quadratic in the fluctuation operators and the corresponding ground state can therefore be found using the general results of Sec. 19.3.1.

19.3.3 *Bogoliubov–de Gennes equations*

In coordinate space the Bogoliubov transformation (19.29), upon the replacement $\alpha \equiv \boldsymbol{r}$, takes on the form

$$b_i = \int d\boldsymbol{r}\; \left(U_i^*(\boldsymbol{r})\delta a_{\boldsymbol{r}} - V_i^*(\boldsymbol{r})\delta a_{\boldsymbol{r}}^{\dagger} \right),$$

$$b_i^{\dagger} = \int d\boldsymbol{r}\; \left(U_i(\boldsymbol{r})\delta a_{\boldsymbol{r}}^{\dagger} - V_i(\boldsymbol{r})\delta a_{\boldsymbol{r}} \right), \tag{19.73}$$

with the amplitudes $U_{\alpha i} \equiv U_i(\boldsymbol{r})$ and $V_{\alpha i} \equiv V_i(\boldsymbol{r})$. These diagonalize $\hat{\Omega}_B$ and are obtained as the (positive energy) solutions of the generalized eigenvalue problem of Eq. (19.39). Introducing the operator

$\hat{L} = -\hbar^2 \nabla^2/(2m) + U(\boldsymbol{r})$, the resulting equations read

$$\left[\hat{L} - \mu + 2N_0 \, g|\phi(\boldsymbol{r})|^2\right] U_i(\boldsymbol{r}) + N_0 \, g\phi^2(\boldsymbol{r})V_i(\boldsymbol{r}) = E_i U_i(\boldsymbol{r}) \qquad (19.74)$$

$$\left[\hat{L} - \mu + 2N_0 \, g|\phi(\boldsymbol{r})|^2\right] V_i(\boldsymbol{r}) + N_0 \, g\phi^{*2}(\boldsymbol{r})U_i(\boldsymbol{r}) = -E_i V_i(\boldsymbol{r}). \qquad (19.75)$$

Equations (19.74) and (19.75) are known as the bosonic Bogoliubov–de Gennes (BdG) equations, in analogy to their fermionic counterpart discussed in Ch. 22. They should be solved consistently with the GP equation (19.72),

$$\left[\hat{L} + N_0 \, g|\phi(\boldsymbol{r})|^2\right] \phi(\boldsymbol{r}) = \mu\phi(\boldsymbol{r}). \qquad (19.76)$$

The GP equation (19.76) describes the condensate. The BdG equations (19.74) and (19.75) describe the non-condensate particles: both the structure of the vacuum and the elementary excitations of the system. For fixed particle number one should of course add

$$N = N_0 + \sum_i \int d\boldsymbol{r} \, |V_i(\boldsymbol{r})|^2. \qquad (19.77)$$

Note that the condensate wave function itself $[i.e.\ U_i(\boldsymbol{r}) = \phi(\boldsymbol{r}),\ V_i(\boldsymbol{r}) = -\phi^*(\boldsymbol{r})]$ is a zero-energy solution[7] of Eqs. (19.74) and (19.75). The positive energy solutions obey the orthogonality conditions contained in Eq. (19.34)

$$\int d\boldsymbol{r} \, (U_i^*(\boldsymbol{r})U_j(\boldsymbol{r}) - V_i^*(\boldsymbol{r})V_j(\boldsymbol{r})) = \delta_{i,j},$$

$$\int d\boldsymbol{r} \, (U_i(\boldsymbol{r})V_j(\boldsymbol{r}) - V_i(\boldsymbol{r})U_j(\boldsymbol{r})) = 0, \qquad (19.78)$$

and are also orthogonal to the condensate wave function,

$$\int d\boldsymbol{r} \, (U_i^*(\boldsymbol{r})\phi(\boldsymbol{r}) + V_i^*(\boldsymbol{r})\phi^*(\boldsymbol{r})) = 0. \qquad (19.79)$$

All other results from Sec. 19.3.1 can be taken over as well, *e.g.* the "coordinate" $\Phi(\boldsymbol{r})$ conjugate to the zero-mode momentum represented by the

---

[7]The appearance of a zero-energy solution is in accordance with the previous footnote[5]: the continuous symmetry of the original Hamiltonian (19.69) is the invariance under a $U(1)$ global phase change $a_{\boldsymbol{r}}^\dagger \to e^{i\phi}a_{\boldsymbol{r}}^\dagger$, which is equivalent to particle-number conservation. This symmetry is broken by the particle-number nonconserving terms in Eq. (19.71).

condensate wave function $\phi(r)$, is found by solving [see Eq. (19.60)]

$$\left[\hat{L} - \mu + 2N_0\, g|\phi(r)|^2\right]\Phi(r) + N_0\, g\phi^2(r)\Phi^*(r) = \phi(r), \quad (19.80)$$

and is needed to complete the sp space [Lewenstein and You (1996)], according to Eq. (19.67).

There is equivalence between the Bogoliubov quasiparticle excitations and the normal modes (small amplitude oscillations) of the condensate. Consider the addition of a small time-dependent perturbation $\delta U(r,t)$ to the confining potential $U(r)$. The first-order response $\delta\phi(r,t)$ of the unperturbed time-dependent condensate wave function $\phi(r,t) = \phi(r)e^{-\frac{i}{\hbar}\mu t}$ is found by replacing $U \to U + \delta U$ and $\phi \to \phi + \delta\phi$ in the time-dependent GP equation [see Eq. (12.106)], and putting the first-order terms equal

$$\left[\hat{L} + 2N_0\, g|\phi(r)|^2\right]\delta\phi(r,t) + N_0\, g\phi^2(r)e^{-\frac{2i}{\hbar}\mu t}\delta\phi^*(r,t)$$

$$+ \delta U(r,t)\phi(r)e^{-\frac{i}{\hbar}\mu t} = i\hbar\frac{\partial}{\partial t}\left(\delta\phi(r,t)\right). \quad (19.81)$$

For a harmonic perturbation with frequency ω [i.e. $\delta U(r,t) = \delta U(r)e^{-i\omega t} + \delta U^*(r)e^{i\omega t}$], the time-dependence dictates a response of the form

$$\delta\phi(r,t) = e^{-\frac{i}{\hbar}\mu t}\left(\delta f_1(r)e^{-i\omega t} + \delta f_2^*(r)e^{i\omega t}\right). \quad (19.82)$$

Substitution into Eq. (19.81) yields the equation for the spatial parts $\delta f_1(r)$ and $\delta f_2(r)$ of the condensate response:

$$\left[\hat{L} + 2N_0\, g|\phi(r)|^2\right]\delta f_1(r) + N_0\, g\phi^2(r)\delta f_2(r) + \delta U(r)\phi(r)$$
$$= (\mu + \hbar\omega)\delta f_1(r)$$

$$\left[\hat{L} + 2N_0\, g|\phi(r)|^2\right]\delta f_2(r) + N_0\, g\phi^{*2}(r)\delta f_1(r) + \delta U(r)\phi^*(r)$$
$$= (\mu - \hbar\omega)\delta f_2(r). \quad (19.83)$$

Resonances in the linear response appear for frequencies that yield zero for the inhomogeneous part of Eq. (19.83). The final equations for the eigenfrequencies and normal modes are seen to coincide with the BdG equations.

The energies predicted by the BdG equations [see Eqs. (19.74) and (19.75)] agree quite well with the experimental frequencies of the low-lying collective modes in atomic BEC, obtained by modulating the trapping magnetic fields [Dalfovo *et al.* (1999)].

19.4 Number-conserving approach

The Bogoliubov prescription, as was mentioned in Sec. 18.2, leads to exact results only in the thermodynamic limit. While the grand-canonical ensemble treats open systems in contact with a particle reservoir, its use in describing systems with a fixed and finite particle number can be (and has been [Leggett (2001)]) frowned upon.

It is therefore of interest to investigate particle-number conserving versions of bosonic perturbation theory. Such extensions are usually performed by a non-linear transformation, which seems originally due to [Kromminga and Bolsterli (1962)]. New operators b_α are introduced as

$$b_\alpha^\dagger = a_\alpha^\dagger a_0 \, \hat{N}_0^{-1/2}$$
$$b_\alpha = \hat{N}_0^{-1/2} a_0^\dagger a_\alpha, \tag{19.84}$$

where a_0 is the removal operator for a particle in the condensate. The operator $\hat{N}_0 = a_0^\dagger a_0$ is the condensate number operator, and can be assumed to have non-zero eigenvalues in the Fock space relevant for this problem.

Obviously, the b_α operators do not change the particle number. Moreover, one can show that

$$b_\alpha^\dagger b_\beta = a_\alpha^\dagger a_\beta, \tag{19.85}$$

by using the identity

$$a_0 f(\hat{N}_0) = f(\hat{N}_0 + 1) a_0, \tag{19.86}$$

which holds for any well-behaved function $f(x)$. The identity is easily seen to hold by applying Eq. (19.86) to the eigenstates $(a_0^\dagger)^m \, |\rangle$ of $\hat{N}_0$ and replacing, as the functional argument of $f(x)$, the operator with the corresponding eigenvalue; since an arbitrary many-boson state can always be decomposed in the eigenstates of $\hat{N}_0$, Eq. (19.86) holds in general. As a result of the identity (19.86) one has *e.g.* that $a_0 \hat{N}_0^{-1} a_0^\dagger = 1$, and Eq. (19.85) follows.

The condensate operators are simply given by [see Eq. (19.84)]

$$b_0^\dagger = b_0 = \sqrt{\hat{N}_0}. \tag{19.87}$$

The non-condensate sp states will be denoted with Latin labels $a, b, \ldots$, while Greek labels $\alpha, \beta, \ldots$ are kept for the general case, *e.g.* $\hat{N} = \sum_\alpha \hat{N}_\alpha = \hat{N}_0 + \sum_a \hat{N}_a$. Using Eq. (19.86) one can show that the non-condensate b_a

operators obey the boson commutation rules

$$[b_a, b_b^\dagger] = \delta_{ab} \quad , [b_a, b_b] = 0 \quad , [b_a^\dagger, b_b^\dagger] = 0. \tag{19.88}$$

Equation (19.85) allows us to re-express the Hamiltonian in terms of the $b_\alpha^\dagger$ and b_α operators. The condensate operators, as given by Eq. (19.87), can be eliminated by replacing

$$\sqrt{\hat{N}_0} = \sqrt{N}\sqrt{1 - \frac{1}{N}\sum_a b_a^\dagger b_a}, \tag{19.89}$$

leaving only non-condensate operators. Finally the square-roots, appearing due to Eq. (19.89), are expanded in a series which can be combined with the usual perturbative expansion. The reader is referred to [Kromminga and Bolsterli (1962)] and [Grest and Rajagopal (1975)] for a treatment of the homogeneous system, and to [Gardiner (1997); Castin and Dum (1998)] for recent applications of particle-number conserving approaches in trapped dilute Bose gases.

19.5   Exercises

(1) Show that an alternative form of the Bogoliubov quasiparticle vacuum in Eq. (19.49) is given by

$$|0_B\rangle = e^{\frac{1}{2}\sum_\alpha w_\alpha[(a_\alpha'^\dagger)^2 - (a_\alpha')^2]}|0\rangle,$$

where $w_\alpha = \text{Atanh}(v_\alpha/u_\alpha)$. Hint: use the fact that $Q_\alpha = \frac{w_\alpha}{2}[(a_\alpha'^\dagger)^2 - (a_\alpha')^2]$ is antihermitian, that e^{Q_α} is unitary, and the expansion

$$e^{-Q_\alpha} a_\alpha e^{Q_\alpha} = a_\alpha - [Q_\alpha, a_\alpha] + \frac{1}{2!}[Q_\alpha, [Q_\alpha, a_\alpha]] - ..$$
$$= (\cosh w_\alpha)a_\alpha + (\sinh w_\alpha)a_\alpha^\dagger.$$

(2) Prove the commutation relations in Eq. (19.88).

Chapter 20

In-medium interaction and scattering of dressed particles

Chapter 16 illustrated the influence of short-range correlations on the sp propagator, one of their characteristic features being the depletion of the Fermi sea. The inevitable consequence is the occupation of states above the Fermi energy, which in an infinite system implies the presence of momentum components above the Fermi momentum in the ground state. The occupation of high-momentum components thereby restores the density to its proper value. The distribution of the sp strength exhibits the expected broadening of the quasiparticle peaks away from the Fermi momentum. Specific features, induced by short-range correlations in the distribution of the sp strength, include the appearance of low-momentum components in a wide energy domain above the Fermi energy. In turn, high-momentum components are found centered around a ridge determined by the characteristic 2h1p energy given by an the average two-hole energy with the high-momentum quasiparticle energy subtracted. The latter term contains a dominant kinetic energy contribution which pushes them to ever lower energy as $-\boldsymbol{p}^2/2m$ with increasing p.

It is the purpose of the present chapter to explore the consequences of these modifications of the sp propagator on the properties of the vertex function Γ. In Ch. 15, the vertex function was studied by propagating mean-field particles in the medium, when summing the ladder diagrams. We will now turn our attention to propagating particles that are dressed by the same short-range correlations that are treated at the tp level when the ladder diagrams are summed. This calculation is of course necessary when the sp propagator is calculated self-consistently, in terms of a self-energy that includes the contribution of ladder diagrams. The ladder equation must then include the full off-shell behavior of the dressed particles and

exhibits significant differences with respect to its mean-field counterpart,
discussed in Ch. 15.

Throughout this section whenever quantitative calculations are presented, the dressing of the sp propagator originates from ladder diagrams
for nuclear matter at a density corresponding to $k_F = 1.36$ fm^{-1}. We are
not aware of other discussions of the in-medium scattering process involving dressed particles. The chapter is therefore not as well digested as some
of the other topics covered in the book and can therefore be omitted at a
first reading.

The dressing of the sp propagator is qualitatively similar to that displayed in Fig. 11.5, but differs in quantitative aspects. The results of
Sec. 16.3.2 form the quantitative basis for the present discussion. Consider the occupation of k given by the integral over the spectral strength
up to ε_F [see Eq. (11.114)]. Around normal nuclear matter density the occupation number for $k < k_F$ is about 0.85 for most values of k except near
k_F, *i.e.* about 15% of the strength is depleted due to the inclusion of ladder
diagrams in the nucleon self-energy. The strength of the quasiparticle pole
at k_F [see Eq. (11.113)] is about 0.75. In Sec. 20.1 we will examine the consequences of such dressing on the propagator in wave vector space of two
noninteracting (but dressed) particles. In Sec. 20.2 we proceed to coordinate space, since this analysis sheds light on how to frame the discussion of
the scattering process in the medium when dressing is taken into account.
The scattering process will then be discussed in Sec. 20.3 and compared to
conventional descriptions.

20.1   Propagation of dressed particles in wave-vector space

For a uniform system the tp propagator defined in Eq. (15.1) is conveniently
studied by employing wave vector quantum numbers

$$G_{pphh}(\boldsymbol{k}_\alpha, \boldsymbol{k}_{\alpha'}; \boldsymbol{k}_\beta, \boldsymbol{k}_{\beta'}; t_1 - t_2) \tag{20.1}$$
$$= -\frac{i}{\hbar} \left\langle \Psi_0^A \right| \mathcal{T} \left[a_{\boldsymbol{k}_{\alpha'}}(t_1) a_{\boldsymbol{k}_\alpha}(t_1) a_{\boldsymbol{k}_\beta}^\dagger(t_2) a_{\boldsymbol{k}_{\beta'}}^\dagger(t_2) \right] \left| \Psi_0^A \right\rangle,$$

where we have suppressed additional quantum numbers like spin (and
isospin). These quantities can be reinstated where appropriate. We continue to study the ladder approximation to the tp propagator but allow for
the dressing of the particles in the medium. The diagrams contributing to
Eq. (20.1) are shown in Fig. 20.1.

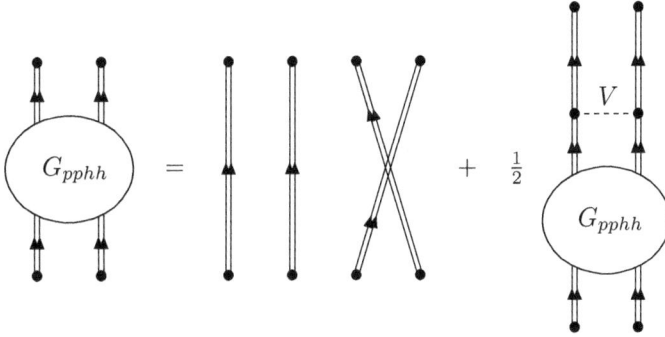

Fig. 20.1 Diagrammatic representation of the ladder equation for G_{pphh} given by the first equality of Eq. (20.5) using dressed sp propagators.

As usual, the FT of Eq.(20.1) is more relevant for practical calculations:

$$G_{pphh}(\boldsymbol{k}_\alpha, \boldsymbol{k}_{\alpha'}; \boldsymbol{k}_\beta, \boldsymbol{k}_{\beta'}; E) \qquad (20.2)$$
$$= \int_{-\infty}^{\infty} d(t_1 - t_2) e^{\frac{i}{\hbar} E(t_1 - t_2)} G_{pphh}(\boldsymbol{k}_\alpha, \boldsymbol{k}_{\alpha'}; \boldsymbol{k}_\beta, \boldsymbol{k}_{\beta'}; t_1 - t_2).$$

Since the total wave vector of the added and removed pair must be the same on account of momentum conservation, it is more appropriate to consider the propagator without the resulting δ-function as in Sec. 15.1. We continue to employ the infinite summation of diagrams, shown in Fig. 20.1, provided the proper substitution from $G_{pphh}^{(0)}$ to G_{pphh}^{f} is made. The latter term represents the free propagation of dressed particles and is given in analogy to Eq. (15.35) by

$$G_{pphh}^{f}(\boldsymbol{q}; \boldsymbol{K}, E) = i \int \frac{dE'}{2\pi} \, G(\boldsymbol{K}/2 + \boldsymbol{q}; E/2 + E') G(\boldsymbol{K}/2 - \boldsymbol{q}; E/2 - E'). \qquad (20.3)$$

The Lehmann representation of G in Eq. (16.9) can then be employed to calculate Eq. (20.3) by contour integration

$$G_{pphh}^{f}(\boldsymbol{q}; \boldsymbol{K}, E) = \int_{\varepsilon_F}^{\infty} dE' \int_{\varepsilon_F}^{\infty} dE'' \frac{S_p(\boldsymbol{K}/2 + \boldsymbol{q}; E') S_p(\boldsymbol{K}/2 - \boldsymbol{q}; E'')}{E - E' - E'' + i\eta}$$
$$- \int_{-\infty}^{\varepsilon_F} dE' \int_{-\infty}^{\varepsilon_F} dE'' \frac{S_h(\boldsymbol{K}/2 + \boldsymbol{q}; E') S_h(\boldsymbol{K}/2 - \boldsymbol{q}; E'')}{E - E' - E'' - i\eta}. \qquad (20.4)$$

The integral equation for the tp propagator can be written in a similar way

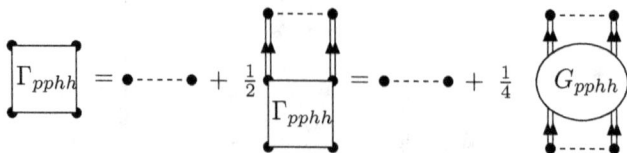

Fig. 20.2 Ladder equation for the vertex function Γ_{pphh} illustrating the equivalent results contained in Eqs. (20.7) and the dressed version of (15.14).

as in Eqs. (15.11) and (15.12)

$$G_{pphh}(\boldsymbol{k}, \boldsymbol{k}'; \boldsymbol{K}, E) = G^f_{pphh}(\boldsymbol{k}, \boldsymbol{k}'; \boldsymbol{K}, E) \tag{20.5}$$

$$+ G^f_{pphh}(\boldsymbol{k}; \boldsymbol{K}, E) \frac{1}{2} \int \frac{d^3 q}{(2\pi)^3} \langle \boldsymbol{k} | V | \boldsymbol{q} \rangle G_{pphh}(\boldsymbol{q}, \boldsymbol{k}'; \boldsymbol{K}, E)$$

$$= G^f_{pphh}(\boldsymbol{k}, \boldsymbol{k}'; \boldsymbol{K}, E) + G^f_{pphh}(\boldsymbol{k}; \boldsymbol{K}, E) \langle \boldsymbol{k} | \Gamma(\boldsymbol{K}, E) | \boldsymbol{k}' \rangle G^f_{pphh}(\boldsymbol{k}'; \boldsymbol{K}, E),$$

where

$$G^f_{pphh}(\boldsymbol{k}, \boldsymbol{k}'; \boldsymbol{K}, E) = [\delta(\boldsymbol{k} - \boldsymbol{k}') - \delta(\boldsymbol{k} + \boldsymbol{k}')] G^f_{pphh}(\boldsymbol{k}; \boldsymbol{K}, E) \tag{20.6}$$

is the noninteracting tp propagator which, both in homogeneous matter and free space, conserves the relative wave vector as expressed by the δ-functions in Eq. (20.6). The appearance of two δ-functions is associated with the exchange term contained in G^f_{pphh}. The relative wave vectors $\boldsymbol{k}$, $\boldsymbol{k}'$, and the total one $\boldsymbol{K}$ are defined in Eqs. (15.17) and (15.16), respectively. The second equality in Eq. (20.5) links the tp propagator with the vertex function, or effective interaction Γ, which contains the summation of all ladder diagrams [see Eq. (15.12)]. The integral equation for Γ can then be written [reinstating spin-like quantum numbers as in Eq. (15.17)]

$$\langle \boldsymbol{k} m_\alpha m_{\alpha'} | \Gamma_{pphh}(\boldsymbol{K}, E) | \boldsymbol{k}' m_\beta m_{\beta'} \rangle = \langle \boldsymbol{k} m_\alpha m_{\alpha'} | V | \boldsymbol{k}' m_\beta m_{\beta'} \rangle$$

$$+ \frac{1}{2} \sum_{m_\gamma m_{\gamma'}} \int \frac{d^3 q}{(2\pi)^3} \langle \boldsymbol{k} m_\alpha m_{\alpha'} | V | \boldsymbol{q} m_\gamma m_{\gamma'} \rangle$$

$$\times G^f_{pphh}(\boldsymbol{K}, \boldsymbol{q}; E) \langle \boldsymbol{q} m_\gamma m_{\gamma'} | \Gamma_{pphh}(\boldsymbol{K}, E) | \boldsymbol{k}' m_\beta m_{\beta'} \rangle. \tag{20.7}$$

This result is shown diagrammatically in Fig. 20.2 together with the dressed version of Eq. (15.14).

The new ingredient in the scattering process in the medium is the occurrence of the dressed but noninteracting propagator G^f_{pphh} appearing in Eq. (20.7). By returning to sp wave vectors for the removal or addition of individual particles, we can write $G^f_{pphh}(\boldsymbol{k}; \boldsymbol{K}, E)$ in Eq. (20.4) as

$$G^f_{pphh}(k_\alpha, k_{\alpha'}; E) = \int_{\varepsilon_F}^{\infty} dE_\alpha \int_{\varepsilon_F}^{\infty} dE_{\alpha'} \frac{S_p(k_\alpha, E_\alpha) S_p(k_{\alpha'}, E_{\alpha'})}{E - E_\alpha - E_{\alpha'} + i\eta}$$

$$- \int_{-\infty}^{\varepsilon_F} dE_\alpha \int_{-\infty}^{\varepsilon_F} dE_{\alpha'} \frac{S_h(k_\alpha, E_\alpha) S_h(k_{\alpha'}, E_{\alpha'})}{E - E_\alpha - E_{\alpha'} - i\eta}. \quad (20.8)$$

Only the magnitudes k_α and $k_{\alpha'}$ are indicated here since there is no dependence of the sp spectral functions on the direction of the sp wave vector. For the imaginary part we then find

$$\mathrm{Im}\, G^f_{pphh}(k_\alpha, k_{\alpha'}; E)$$

$$= -\pi \begin{cases} \int_{\varepsilon_F}^{\infty} dE_\alpha\, S_p(k_\alpha, E_\alpha) S_p(k_{\alpha'}, E - E_\alpha) & E > 2\varepsilon_F \\ \int_{-\infty}^{\varepsilon_F} dE_\alpha\, S_h(k_\alpha, E_\alpha) S_h(k_{\alpha'}, E - E_\alpha) & E < 2\varepsilon_F \end{cases}. \quad (20.9)$$

Employing Eq. (20.9), one can generate the following two sum rules by a change of integration variables

$$I^{II}_>(k_\alpha, k_{\alpha'}) = -\frac{1}{\pi} \int_{2\varepsilon_F}^{\infty} dE\, G^f_{pphh}(k_\alpha, k_{\alpha'}; E) = (1 - n(k_\alpha))\,(1 - n(k_{\alpha'})),$$

$$(20.10)$$

and

$$I^{II}_<(k_\alpha, k_{\alpha'}) = -\frac{1}{\pi} \int_{-\infty}^{2\varepsilon_F} dE\, G^f_{pphh}(k_\alpha, k_{\alpha'}; E) = n(k_\alpha) n(k_{\alpha'}), \quad (20.11)$$

where $n(k)$ refers to the occupation of the sp state k. The sum rule in Eq.(20.11) may yield large deviations from the free Fermi gas result when substantial correlations are present. Examples of such systems are nuclear matter, neutron matter, and liquid ^{3}He at densities corresponding to saturation. Around nuclear saturation density Eq. (20.11) deviates from 1 (valid for the Fermi gas), when $k < k_F$, by as much as 35%. For the ^{3}He liquid at saturation the deviation is at least 75%.

We start our discussion by pointing out that the scattering of dressed particles does not yield a unique on-shell wave vector as in Eq. (15.37) for mean-field and Eq. (15.19) for free particles. This is illustrated in Fig. 20.3 where the real (dashed) and imaginary part of G^f_{pphh} (solid) are shown for an energy of about 280 MeV in nuclear matter at $k_F = 1.36$ fm^{-1} with

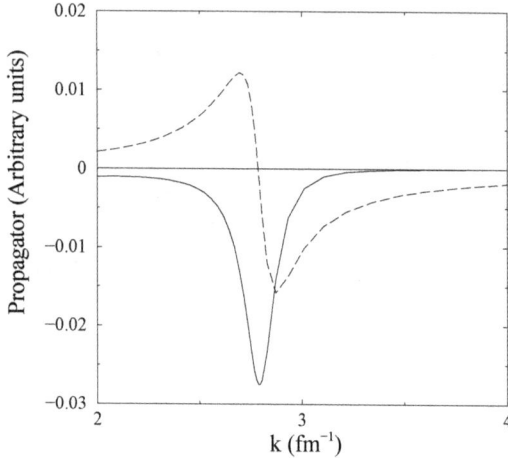

Fig. 20.3 Real (dashed) and imaginary part (solid) of the dressed noninteracting tp propagator for an on-shell wave vector of $k_0 = 2.8$ fm^{-1}. The total wave vector of the particles is zero.

$K = 0$. Equation (20.8) with $k_\alpha = k_{\alpha'}$ (equal to the relative wave vector k) can be used for this result. The imaginary part is calculated by performing the convolution integral in Eq. (20.9) and the real part is constructed by employing the usual dispersion relation between the real and imaginary parts of propagators. In the present case, it reads

$$G_{pphh}^f(\boldsymbol{k}; \boldsymbol{K}, E) = -\frac{1}{\pi} \int_{2\varepsilon_F}^{\infty} dE' \frac{\text{Im } G_{pphh}^f(\boldsymbol{k}; \boldsymbol{K}, E')}{E - E' + i\eta}$$

$$+\frac{1}{\pi} \int_{-\infty}^{2\varepsilon_F} dE' \frac{\text{Im } G_{pphh}^f(\boldsymbol{k}; \boldsymbol{K}, E')}{E - E' - i\eta}. \qquad (20.12)$$

The imaginary part of G_{pphh}^f contrasts dramatically with the δ-function of the Galitskii–Feynman propagator in Eq. (15.34). In the dressed case a broadened distribution over the relative wave vector k is found, which precludes a unique definition of an "on-shell" wave vector. One can nevertheless generate an approximate result by recalling the discussion of the sp propagator in Sec. 11.4.3. The spectral function for a sp particle (which is proportional to the imaginary part of the sp propagator) exhibits a peak at the quasiparticle energy. One may therefore expect that the peak in Im G_{pphh}^f will occur at the sum of the quasiparticle energies. This suggests that for zero total wave vector an approximate "on-shell" wave vector can

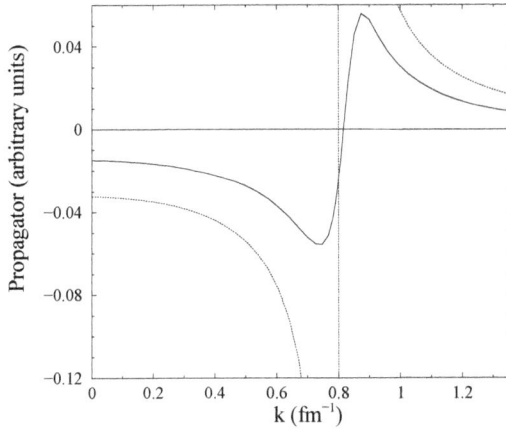

Fig. 20.4 Comparison of the real part of the mean-field (dotted) and dressed (full line) tp propagator at an on-shell wave vector of 0.8 fm^{-1} indicated by the vertical dotted line. In the Galitskii–Feynman case a realistic sp spectrum was employed.

be determined by

$$E = 2E_Q(k_0) = 2\left\{\frac{\hbar^2 k_0^2}{2m} + \mathrm{Re}\Sigma(k_0, E_Q(k_0))\right\}. \tag{20.13}$$

Except for energies deep in the Fermi sea [Dickhoff (1998)], the on-shell wave vector k_0 coincides with the location of the peak in the imaginary part of G^f_{pphh} as a function of wave vector. In Fig. 20.3 the on-shell wave vector is given by $k_0 = 2.8$ fm^{-1}, associated with the propagation of two dressed particles above the Fermi sea. Also, for the real part of the propagator there is a distinct difference between the mean-field and the dressed case. While the former propagator jumps from $+\infty$ to $-\infty$ around 2.8 fm^{-1}, the dressed one exhibits a characteristic wiggle around this energy.

For other wave vectors in Fig. 20.3, one finds a reduction factor of about 0.5 with respect to the mean-field result, while only for large values of k the real parts approach each other. The reduction is illustrated in Fig. 20.4 for the real part of G^f_{pphh}. The noninteracting propagator in Eq. (20.8) becomes the familiar Galitskii–Feynman propagator [see Eq. (15.34)], when spectral functions are inserted that are characterized by a δ-function peak of strength 1. The substantial spreading in the imaginary part of the propagator, displayed in Figs. 20.3 and 20.4, alters the conventional picture of the scattering process. Even small components with large wave vectors are present but not shown in these figures. In the partial wave decomposition the integral equation reads

$$\langle k\ell| \, \Gamma^{JST}_{pphh}(K, E) \, |k'\ell'\rangle = \langle k\ell| \, V^{JST} \, |k'\ell'\rangle \tag{20.14}$$

$$+ \frac{1}{2} \sum_{\ell''} \int \frac{dq \, q^2}{(2\pi)^3} \, \langle k\ell| \, V^{JST} \, |q\ell''\rangle \, G^f_{pphh}(q; K, E) \, \langle q\ell''| \, \Gamma^{JST}_{pphh}(K, E) \, |k'\ell'\rangle \, .$$

To arrive at this equation, one may proceed by an angle-averaging procedure of Eq. (20.12), similar to Eq. (16.67), to eliminate the dependence on the angle between the relative and total wave vectors before performing the partial wave decomposition.

For mean-field propagators one usually solves Eq. (20.14) by discretizing the integral equation, taking only the real part of the propagator into account, as for two free particles. The solution to this integral equation then yields the $\mathcal{R}$-matrix by a real matrix inversion [Haftel and Tabakin (1970)]. The contribution of the imaginary part of the propagator can then be generated algebraically using the $\mathcal{R}$-matrix elements [Trefz et al. (1985)]. Employing dressed propagators, it is more convenient to discretize the integral equation in such a way that the relevant sampling of both the real and imaginary part of the propagator occurs simultaneously, leading to a complex matrix inversion that yields Γ directly.

In order to perform the analysis of the scattering process employing dressed propagators, it will be helpful to use an analytical approximation to the noninteracting propagator in Eq. (20.8) that describes the essential new features. For this purpose, only the case of zero total wave vector of the propagating pair will be considered. The noninteracting propagator then contains equal and opposite wave vectors for the two particles (holes). Since the spectral functions do not depend on the direction of the wave vector, we can use Eq. (20.8) with $k_\alpha = k_{\alpha'} = k$. Both the absolute value of the sp and the relative wave vectors are now represented by k.

Since the imaginary part of G^f_{pphh} exhibits similar behavior as found the sp spectral functions, one may try to introduce an *ad hoc* tp self-energy for further analysis. We then attempt to write the noninteracting propagator as

$$G^f_{pphh}(k, E) = \frac{\pm 1}{E - \Sigma_{pphh}(k, E)}, \tag{20.15}$$

where the sign is determined by whether the energy E is above (+) or below (-) $2\varepsilon_F$. By assuming that this ad hoc self-energy Σ_{pphh} has a slowly varying imaginary part as a function of the relative wave vector k, one can expand the self-energy at k_0 for which

$$E \equiv \mathrm{Re}\ \Sigma_{pphh}(k_0, E). \qquad (20.16)$$

We note that this expansion is in powers of the square of the wave vector since odd powers of k cannot contribute. A complex pole approximation (CPA) to the propagator emerges by only keeping the real and imaginary part of Σ_{pphh} at k_0^2, as well as the first derivative of the real part. The resulting propagator has the form

$$G_{pphh}^{f,CPA}(k, E) = \frac{m}{\hbar^2} \frac{\pm \mathcal{C}}{k_0^2 - k^2 \pm i\gamma}, \qquad (20.17)$$

where the constant $\mathcal{C}$ is given by

$$\mathcal{C} = \frac{\hbar^2}{m} \left(\left. \frac{\partial \mathrm{Re}\ \Sigma_{pphh}}{\partial k^2} \right|_{k_0^2} \right)^{-1} \qquad (20.18)$$

and γ by

$$\gamma = |\mathrm{Im}\ \Sigma_{pphh}(k_0, E)| \left(\left. \frac{\partial \mathrm{Re}\ \Sigma_{pphh}}{\partial k^2} \right|_{k_0^2} \right)^{-1}. \qquad (20.19)$$

Typical values for $\mathcal{C} \approx 0.5$ are found at low energies, whereas it rises slowly to 1 for higher energies. This feature is closely related to the pattern of the distribution of the sp strength. The quasiparticle pole strength at k_F = 1.36 fm^{-1} is about 0.7, so for a tp propagator close to these energies a factor of $(0.7)^2$ is expected. For larger wave vectors the strength in the peak grows back to 1, yielding a propagator which is more of the mean-field, or even free-particle, kind. It is also apparent that this factor of about 0.5 can be identified from Fig. 20.4. The CPA is calculated after first numerically calculating the noninteracting propagator of the dressed particles. In Fig. 20.5 the quality of this CPA to the propagator can be judged by comparing it to the numerically generated imaginary part of the propagator, at an energy associated with an on-shell wave vector of 0.5 fm^{-1}. For the real part of the propagator a satisfactory description for wave vectors below k_F is found as well.

The CPA to the propagator cannot be used to solve the integral equation in Eq. (20.14), since it is only a good approximation close to the peak of the imaginary part. The full solution of Eq. (20.14) furthermore requires an accurate representation of the high-momentum components of the propagator in order to properly include the effect of short-range correlations in

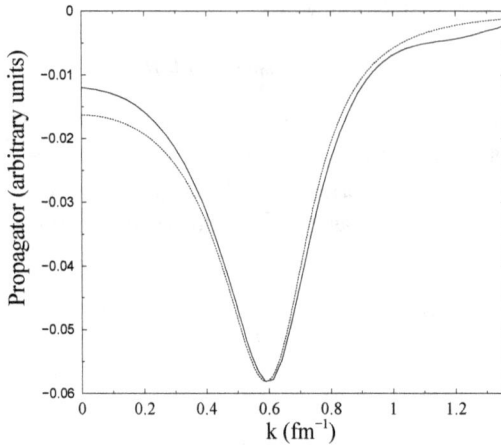

Fig. 20.5 Comparison for the imaginary part of the noninteracting tp propagator of dressed nucleons between the complete (numerical) result represented by the full line and the simple CPA [see Eq. (20.17)] given by the dotted line for values below k_F. The energy (below $2\varepsilon_F$) corresponds to an on-shell wave vector of 0.5 fm^{-1}.

the interaction or wave function. The CPA result does provide a reasonable representation of the long-range part of the propagator and therefore can be profitably used to discuss the asymptotic analysis of the scattering process.

20.2   Propagation of dressed particles in coordinate space

As discussed in Ch. 6, scattering problems require an analysis in coordinate space. The CPA to the dressed propagator can easily be Fourier transformed to coordinate space, and facilitates this study. In general, one obtains the dressed but noninteracting propagator in coordinate space by a Fourier–Bessel transform. Since we will only consider the case for total wave vector equal to zero, we have to consider

$$ G_{pphh}^{f,\ell}(r,r';E) = \frac{2}{\pi} \int_0^\infty dk \, k^2 \, \mathrm{j}_\ell(kr)\mathrm{j}_\ell(kr')G_{pphh}^{f}(k;K=0,E). \quad (20.20) $$

For free particles, the integral in Eq. (20.20) yields the product of a spherical Bessel function and one of the spherical Hankel functions with k_0, the real on-shell wave vector, as argument [see Eq. (6.66) for the sp case]. This on-shell wave vector is real since the corresponding noninteracting propagators

(Eqs. (6.41) and (15.18)) can only have a vanishing denominator for a real wave vector. Since Eq. (20.20) can be calculated by a contour integral for Eqs. (6.41) and (15.18) (at least for the long-range part), as well as for Eq. (20.17), it is clear that a nonvanishing imaginary part for the pole of Eq. (20.17), due to the presence of γ, will lead to a complex on-shell wave vector that will be denoted by κ_0. Using the CPA (Eq. (20.17)) for $E < 2\varepsilon_F$, Eq. (20.20) yields

$$G_{pphh}^{\ell,CPA}(r,r';E) = -i\mathcal{C}\frac{m}{\hbar^2}\kappa_0\, \mathrm{j}_\ell(\kappa_0 r_<)\, \mathrm{h}_\ell^*(\kappa_0 r_>). \qquad (20.21)$$

The wave-vector argument of the spherical Bessel and Hankel functions κ_0 is now complex, and its real and imaginary parts κ_0^R and κ_0^I are easily calculated from k_0 and γ by determining the zeros of the denominator of Eq. (20.17). Equation (20.21) contains the Hankel function h_ℓ^* due to the different boundary condition associated with hh propagation for energies below $2\varepsilon_F$. This leads to a pole in the upper half of the complex k-plane in contrast to the case of pp or free-particle scattering. Consequently, κ_0^I is negative for $E < 2\varepsilon_F$ and its magnitude can become as large as 0.2 to 0.3 fm^{-1}[Dickhoff *et al.* (1999)]. The propagator for $\ell = 0$ can then be written as (for $r < r'$)

$$G_{pphh}^{\ell=0,CPA}(r,r';E) \qquad (20.22)$$

$$= \frac{-i\mathcal{C}m}{2\hbar^2(\kappa_0^R + i\kappa_0^I)}\left(\frac{e^{i\kappa_0^R r}e^{-\kappa_0^I r}}{r} - \frac{e^{-i\kappa_0^R r}e^{\kappa_0^I r}}{r}\right)\frac{e^{-i\kappa_0^R r'}e^{\kappa_0^I r'}}{r'}.$$

A typical comparison between this expression and the numerical Fourier–Bessel transform of the dressed noninteracting propagator is shown in Fig. 20.6 for the imaginary part. For fixed r', corresponding to the location of the maximum in Fig. 20.6, both propagators are shown as a function of r without the factor $1/rr'$. While confirming the validity of the CPA, Fig. 20.6 also demonstrates that the propagator for dressed particles is radically different from the noninteracting or mean-field one, due to presence of damping terms related to the nonzero value of κ_0^I. For comparison, Eq. (20.22) reduces to

$$G_{pp}^{\ell=0}(r,r';E) = \frac{m}{\hbar^2 k_0}\frac{\sin(k_0 r)}{r}\frac{e^{-ik_0 r'}}{r'} \qquad (20.23)$$

for free particles. Taking *e.g.* the imaginary part, yields the product of the two $\ell = 0$ spherical Bessel functions exhibiting no damping.

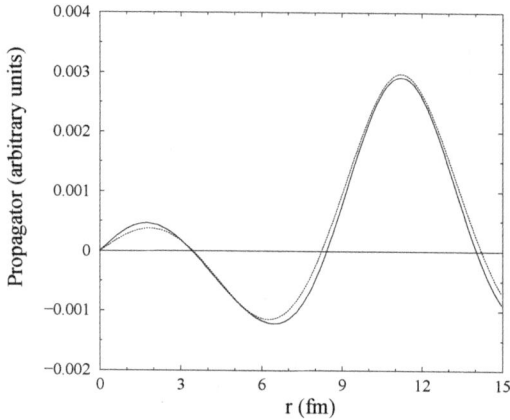

Fig. 20.6   Comparison between the CPA and the complete numerical calculation for the dressed noninteracting tp propagator in coordinate space for a value of r' for which both propagators have a maximum (around 11 fm). Indeed, for $r = r'$ the damping is least effective (see Eq. (20.22)). Shown is the imaginary part for an energy below $2\varepsilon_F$ (also used in Fig. 20.5) for the CPA propagator (dotted line) and the complete result (full line).

As noted before, there is no longer a unique on-shell wave vector. Indeed, the complex pole at κ_0 in the CPA propagator is just a simple (and approximate) representation of this feature. As a consequence, the relative wave function of the dressed particles contains a spread in wave-vector states. In turn, it must yield a localization of the wave function in coordinate space, based on the Heisenberg uncertainty principle. This has interesting physical consequences, since it means that if the separation distance between the scatterers is too large there is little probability that they will actually interact because this requires a small relative distance. Indeed, taking r' to much larger values than in Fig. 20.6, yields a negligible contribution to the noninteracting propagator near small r, where the interaction will act to modify the wave function. Clearly, the effect is governed by the size of κ_0^I, the imaginary part of the pole of the CPA, which only becomes small when the scattering energy approaches $2\varepsilon_F$. Just as in the case of sp motion, the noninteracting wave function tends to a plane wave only in this limit. Figure 20.6 would then yield a simple sine wave characterized by $\kappa_0^R \to k_F$, as in Eq. (20.23). For all other energies damping does occur sufficiently rapidly to warrant the following observation: since only the part of the wave which returns from the scattering can be affected and

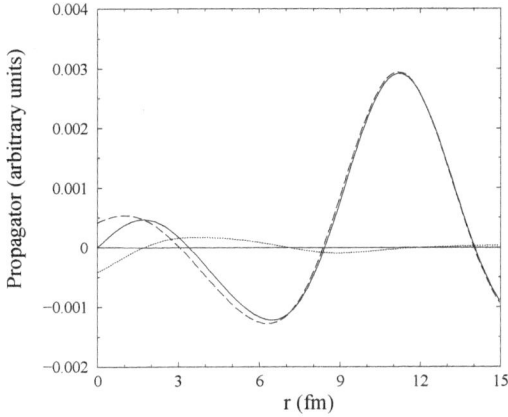

Fig. 20.7 Decomposition of the tp CPA propagator in coordinate space into in- (dashed) and outgoing wave (dotted line) for the same value of r' and energy as in Fig. 20.6. Also shown is the sum of both contributions (full line). The in- and outgoing waves correspond to the two terms in Eq. (20.21).

always decreases with increasing r, only that part of the noninteracting wave can be influenced by the scattering which is exponentially damped in r. Figure 20.7 illustrates the decomposition of the CPA propagator, shown in Fig. 20.6, in terms of the incoming and outgoing wave. For values of r' outside the range of the interaction, even a substantial modification of the outgoing wave will hardly affect the total propagator and the wave function automatically "heals", according to the value of κ_0^I, to the noninteracting one.

An asymptotic analysis of the scattering process employing the CPA propagator is also possible. By following the steps leading to Eq. (6.70) in the case when the noninteracting propagator is given by the CPA result Eq. (20.21), one obtains the asymptotic propagator in the following form (for an uncoupled channel and energy E above $2\varepsilon_F$)

$$G_{pphh}^{\ell,JST}(r,r';E) \rightarrow -i\left(\frac{m\mathcal{C}}{2\hbar^2}\right)\kappa_0 \mathrm{h}_\ell(\kappa_0 r')\Big\{\mathrm{h}_\ell^*(\kappa_0 r) + \mathrm{h}_\ell(\kappa_0 r)$$

$$\times\left[1 - 2i\frac{m\mathcal{C}}{\hbar^2}\kappa_0\int_0^\infty dr_1\ r_1^2\int_0^\infty dr_2\ r_2^2\ \langle r_1|\Gamma_l^{JST}(E)|r_2\rangle \mathrm{j}_\ell(\kappa_0 r_1)\mathrm{j}_\ell(\kappa_0 r_2)\right]\Big\}$$

$$= -i\frac{mc}{2\hbar^2}\kappa_0 \mathrm{h}_\ell(\kappa_0 r')\left\{\mathrm{h}_\ell^*(\kappa_0 r) + \mathrm{h}_\ell(\kappa_0 r)e^{2i\delta_\ell^{JST}}\right\}. \qquad (20.24)$$

A simple example for a hard-core potential will be used to illustrate some features in more detail. The term in brackets in Eq. (20.24) represents the asymptotic wave function including the effect of the potential in terms of a phase shift, as in Eq. (6.70) for free or mean-field particles. For a hard-core potential with hard-core radius r_0 one must require this correlated wave function to vanish at r_0 also in the case of [dropping the JST subscript in Eq. (20.24)]

$$0 = \mathrm{h}_\ell^*(\kappa_0 r_0) + \mathrm{h}_\ell(\kappa_0 r_0)e^{2i\delta_\ell}. \tag{20.25}$$

The boundary condition generates the following expression for the phase shifts

$$\tan \delta_\ell = \frac{\mathrm{j}_\ell(\kappa_0 r_0)}{\mathrm{n}_\ell(\kappa_0 r_0)}. \tag{20.26}$$

In the limit that κ_0^I vanishes, corresponding to free or mean-field particles, one recovers the usual ratio of spherical Bessel and Neumann functions with real arguments. For the $\ell = 0$ case one obtains

$$\tan \delta_0 = -\tan \kappa_0 r_0, \tag{20.27}$$

yielding a real

$$\delta_0^R = -\kappa_0^R r_0 \tag{20.28}$$

and an imaginary part of the phase shift

$$\delta_0^I = -\kappa_0^I r_0. \tag{20.29}$$

Somewhat surprisingly a complex phase shift appears. Its role becomes clear when one considers the asymptotic propagator explicitly. For energies E above $2\varepsilon_F$ ($\kappa_0^I > 0$) the CPA for $\ell = 0$, inserting the real and imaginary part of the phase shift according to Eqs. (20.28) and (20.29), yields

$$G_{pphh}^{\ell=0}(r, r'; E) \rightarrow i\frac{\mathcal{C}m}{2(\kappa_0^R + i\kappa_0^I)\hbar^2}\frac{1}{rr'}e^{i\kappa_0^R r'}e^{-\kappa_0^I r'}$$

$$\left\{-e^{-i\kappa_0^R r}e^{\kappa_0^I r} + e^{i\kappa_0^R(r-2r_0)}e^{\kappa_0^I(2r_0-r)}\right\}. \tag{20.30}$$

It is clear that Eq. (20.30) vanishes for $r = r_0$. Note however, that this can only be achieved by the presence of a complex phase shift. The incoming wave, given by the first term in the bracket, needs to be exactly compensated by the outgoing wave at r_0. Merely shifting the oscillatory character of the wave function by the real part of the phase shift does not suffice.

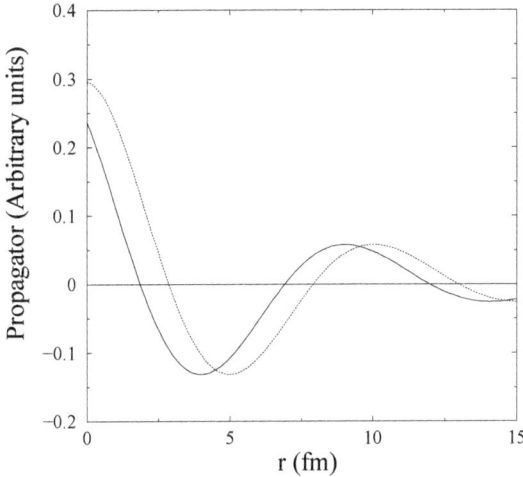

Fig. 20.8 Outgoing tp wave function, with (dotted) and without (full line) complex phase shift for a hard-core radius of 0.5 fm, corresponding to the second term in Eq. (20.30) for values of the energy and r' used in Fig. 20.6.

An additional enhancement provided by the imaginary part of the phase shift is required to achieve cancellation at r_0, since the incoming wave has a larger amplitude than the outgoing part (without the phase shift). The effect is a shift of the complete wave function as appropriate for a hard-core potential, illustrated in Fig. 20.8. The solid line represents the uncorrelated outgoing wave and the dashed line includes the real and imaginary part of the phase shift. Results for values of $k_0 = 0.6$ fm^{-1}, $\gamma = 0.2$ fm^{-1} [see Eqs. (20.16) and (20.19)] yield phase shifts of $\delta_0^R = $ -0.3 and $\delta_0^I = $ -0.08 for a hard-core radius of 0.5 fm used in Fig. 20.8.

The complete asymptotic propagator, including the complex phase shift (dashed line), and the noninteracting CPA propagator (full line) are shown in Fig. 20.9. As in Fig. 20.7 the unaffected incoming wave dominates both propagators (wave functions) while both outgoing waves (shown in Fig. 20.8) are damped exponentially. The dashed line vanishes at the hard-core radius of 0.5 fm, as required. More importantly, even though a phase shift will exist representing the effect of the scattering interaction, the asymptotic wave function nevertheless heals to the noninteracting one, as shown in Fig 20.9. This same feature is observed for the complete numerical calculation including a realistic interaction [Dickhoff *et al.* (1999)].

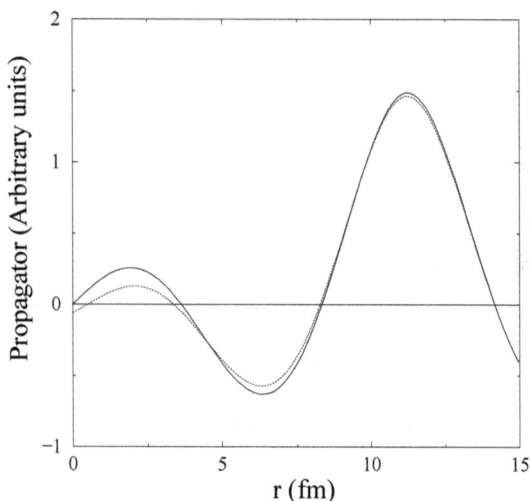

Fig. 20.9 Comparison of the imaginary part of the total noninteracting tp propagator (full line) with the asymptotic one (dotted line) for a hard-core radius of 0.5 fm for values of the energy and r' used in Fig. 20.6. The asymptotic wave function (Eq. (20.30)) does not vanish inside the hard-core radius. The exact wave function vanishes of course for $r < r_0$.

Healing properties and the nuclear shell model

The above observations allow for the resolution of an apparent paradox related to the so-called healing properties of wave functions in the medium. This property has been considered the physical justification of the mean-field-like properties observed in nuclei in the presence of strong short-range interactions. They correspond to the simple shell-model results discussed in Ch. 3. The problem is to explain them while it is known that the particles involved have very violent encounters in the medium on account of their short-range repulsion. The original discussion of the healing properties of the relative wave function of particles in the medium [Gomes *et al.* (1958)] used a Bethe–Goldstone propagator involving mean-field nucleons above k_F to arrive at the healing property of the relative wave function. The easiest way to infer this is to consider the FT of Eq. (15.34) (for $\boldsymbol{K} = 0$ for simplicity)

$$G^{(0)}_{pphh}(\boldsymbol{r}, \boldsymbol{r}'; E) = \frac{1}{(2\pi)^3} \int d^3k \; e^{i\boldsymbol{k}.(\mathrm{r}-\mathrm{r}')} G^{(0)}_{pphh}(\boldsymbol{k}; \boldsymbol{K} = 0, E). \quad (20.31)$$

For free particles the FT leads to the usual expression, which is given in Eq. (6.48) for the sp case. Excluding hh propagation but considering energies below $2\varepsilon_F$, excises the energy-conserving pole contribution. The nonsingular integrand then yields only real contributions to the FT. More importantly, the propagator vanishes like $|\boldsymbol{r} - \boldsymbol{r}'|^{-2}$. When inserted into the asymptotic analysis of the wave function, it leads to a vanishing difference between the correlated wave function and the unperturbed plane wave. This is the healing property of the wave function that is accompanied by a vanishing phase shift. The description relies heavily on the Fermi gas picture of the ground state and the assumption that only pp propagation is relevant. However, there is overwhelming experimental evidence, discussed in Sec. 7.8, that sp motion in nuclei must be described in terms of dressed nucleons with substantial fragmentation of the strength.

If the nucleons are dressed particles, a Bethe–Goldstone propagator does not suffice to generate a self-energy that will describe the sp strength distribution. Instead a Galitskii–Feynman propagator must be employed, which will generate quite a reasonable description of the sp strength, including the quasiparticle features for nucleons at the Fermi surface (see Ch. 16). The description of the scattering process is however modified by employing a Galitskii–Feynman propagator. Whereas it is possible to generate healing with a Bethe–Goldstone propagator, due to a vanishing phase shift for scattering energies below $2\varepsilon_F$, this is no longer possible with a Galitskii–Feynman propagator. A nonvanishing phase shift is found and the asymptotic relative wave function of mean-field particles does not heal. The discussion of the scattering of dressed particles in this chapter demonstrates that one automatically encounters a localization of the relative wave functions in coordinate space. The CPA analysis for hard-core scattering indicates that even with sizable phase shifts its localization leads to the desired healing property, since the part of the wave function affected by the scattering is exponentially damped. The same features are also observed for the complete numerical propagator [Dickhoff *et al.* (1999)]. Even in the presence of strong interaction processes the ensuing picture of the nuclear medium is a tranquil one, in which the dressed particles no longer remember their scattering event beyond some finite distance, and their wave functions heal to the corresponding noninteracting ones. This appears to be a satisfactory depiction of a correlated medium in which particles do not carry the information of their interaction indefinitely around, unlike a description of scattering using a Galitskii–Feynman (or Bethe–Goldstone) propagator. The sp properties of the nuclear shell model are therefore better viewed in terms of quasiparticles (holes) as in Landau's Fermi liquid analysis, provided the proper sp basis for the relevant nucleus is employed (see Sec. 17.4).

20.3   Scattering of particles in the medium

The preceding discussion has focused on an analytically solvable model. First, the CPA was developed to provide a sufficiently realistic approximation to the propagator of two dressed particles. Second, a hard-core scattering problem for such a propagator was analyzed. A generalization to the complete propagator will be outlined next. As for the CPA shown in Fig. 20.7, it is possible to separate the in and outgoing part of the (numerical) noninteracting propagator in coordinate space. This can be written schematically as

$$G_{pphh}^{f,\ell} = G_{pphh}^{f,\ell}(in) + G_{pphh}^{f,\ell}(out). \tag{20.32}$$

The separation is similar to the one presented in Fig. 20.7. The equation for the propagator in a partial wave basis is obtained in complete analogy to the steps that yield Eq. (6.64), and is given by

$$G_{pphh}^{JST}(r\ell, r'\ell'; E) = \left(1 - (-1)^{\ell+S+T}\right) \delta_{\ell,\ell'} G_{II}^{f,\ell}(r, r'; E) \tag{20.33}$$

$$+ \int_0^\infty dr_1 r_1^2 \int_0^\infty dr_2 r_2^2 \, G_{pphh}^{f,\ell}(r, r_1; E) \, \langle r_1\ell| \, \Gamma_{pphh}^{JST}(E) \, |r_2\ell'\rangle \, G_{pphh}^{f,\ell'}(r_2, r'; E).$$

This equation for the propagator can be written for an uncoupled channel as

$$G_{pphh}^{\ell,JST} = G_{pphh}^{f,\ell} + \Delta G_{pphh}^{\ell,JST}, \tag{20.34}$$

where ΔG_{pphh} contains the contribution due to the interaction Γ which can only affect the outgoing wave. By using Eq. (20.32) it is possible to identify the phase shift similar to Eq. (20.24)

$$e^{2i\delta_\ell^{JST}} = \frac{G_{pphh}^{f,\ell}(out) + \Delta G_{pphh}^{\ell,JST}}{G_{pphh}^{f,\ell}(out)}. \tag{20.35}$$

Two remarks are in order here. First, due to the localization of the propagator the phase shift must be calculated for r' not too far away from the origin, in order to generate a nonvanishing outgoing wave. Second, and relatedly, one must expect some r' dependence of this definition of the phase shift, since the dressed noninteracting propagator does not completely separate into a product of a function of r and a function of r' as in the CPA of Eq. (20.21). More importantly, these observations and the healing property of the propagator imply that a conventional derivation and definition of the cross section for the scattering process is not possible. No outgoing wave reaches asymptotically meaningful distances, with damping constants of the order of fm in the case of nuclear matter around saturation density.

Similar to the definition of the phase shift in the partial wave basis above, one may obtain the scattering amplitude by considering the CPA of the noninteracting propagator in coordinate space. Equation (6.45) can be calculated analytically, using Eq. (20.17), yielding for $E > 2\epsilon_F$

$$G_{pphh}^{f,CPA}(\boldsymbol{r}, \boldsymbol{r}'; E) = -\frac{m\mathcal{C}}{4\pi\hbar^2} \frac{e^{i\kappa_0|\boldsymbol{r}-\boldsymbol{r}'|}}{|\boldsymbol{r} - \boldsymbol{r}'|}. \tag{20.36}$$

For a derivation of the scattering amplitude, one requires the separability of this propagator as in Eq. (6.50), contingent on the condition that either r' is much larger than r or vice versa. In the present case one cannot make this assumption without running into a vanishing propagator, due to the presence of the imaginary part of κ_0. This observation does not change for the complete numerical propagator. As a result, there is no asymptotic (large distance) cross section as in the case of conventional scattering experiments. Only a local modification of the wave function is possible, with a rapid healing to the noninteracting wave. Even for energies close to $2\varepsilon_F$, where the imaginary part of κ_0 becomes small, the phase shift vanishes (or approaches π [Bishop *et al.* (1974)]) and no asymptotically significant cross section can be identified. Similar conclusions are reached for the complete numerical propagator [Dickhoff *et al.* (1999)].

The above discussion does not imply that the localized interaction between dressed particles is small. In order to provide a way to assess the strength of the interaction of dressed particles, it is convenient to generate a quantity that will yield the conventional cross section in the limit of mean-field or free-particle scattering. In addition, it is useful to obtain a similar quantity for the phase shifts to make meaningful comparisons to conventional results. Although we introduce approximations below, they do provide physically useful generalizations. The first step involves the practical observation that in most cases the imaginary part of κ_0, which characterizes the damping of the wave function, is considerably smaller than its real part, as shown for example in Fig. 20.3. Only for energies of two particles deep in the Fermi sea do the real and imaginary parts of κ_0

become comparable. Consider the identity (for $r < r'$)

$$
\begin{aligned}
-ik_0 j_\ell(k_0 r) j_\ell(k_0 r') &= \frac{-ik_0}{2} \{ j_\ell(k_0 r) h_\ell(k_0 r') + j_\ell(k_0 r) h_\ell^*(k_0 r') \} \\
&= \frac{1}{\pi} \int_0^\infty dk\, k^2\, \frac{j_\ell(kr) j_\ell(kr')}{k_0^2 - k^2 + i\eta} - \frac{1}{\pi} \int_0^\infty dk\, k^2\, \frac{j_\ell(kr) j_\ell(kr')}{k_0^2 - k^2 - i\eta} \\
&= i\frac{2}{\pi} \int_0^\infty dk\, k^2 j_\ell(kr) j_\ell(kr')\, \mathrm{Im} \left\{ \frac{1}{k_0^2 - k^2 + i\eta} \right\} \\
&\Rightarrow i\frac{2\hbar^2}{\pi m} \int_0^\infty dk\, k^2 j_\ell(kr) j_\ell(kr')\, \mathrm{Im} \left\{ G_{pp}^{(0)}(k; E) \right\},
\end{aligned}
\tag{20.37}
$$

which is valid for vanishing η in the case of particles propagating in free space. Invoking the smallness of the imaginary part of κ_0 with respect to its real part, one may heuristically write the product of the spherical Bessel functions appearing in the first line of Eq. (20.24) using the identity given in Eq. (20.37). This is appropriate for a pole in the complex wave-vector plane, not too far from the real axis ($|\kappa_0^I| \ll \kappa_0^R$ for the CPA result), but also makes sense for r not too different from r' since the damping effect is smallest there. Since the integral in Eq. (20.37) contains real spherical Bessel functions, one can use the transformation to k-space for both integrals in the first line of Eq. (20.24), yielding the asymptotic propagator for the CPA as

$$
\begin{aligned}
G_{pphh}^{\ell, JST}(r, r'; E)_{CPA} \rightarrow -i \left(\frac{mc}{2\hbar^2} \right) &\kappa_0 h_\ell(\kappa_0 r') \Big\{ h_\ell^*(\kappa_0 r) + h_\ell(\kappa_0 r) \\
&\times \left[1 + 2i \int_0^\infty dk\, kk^2\, \mathrm{Im} \left\{ G_{pphh}^f(k; E) \right\} \langle k | \Gamma_{pphh}^{\ell JST}(E) | k \rangle \right] \Big\}.
\end{aligned}
\tag{20.38}
$$

The S-matrix element (and phase shift) can thus be written for an uncoupled channel in the following way

$$
S_\ell(E) = 1 + 2i \int_0^\infty dk\, k^2\, \mathrm{Im} \left\{ G_{pphh}^f(k; E) \right\} \langle k | \Gamma_{pphh}^{\ell JST} | k \rangle \equiv e^{2i\delta_\ell^{JST}}.
\tag{20.39}
$$

It reduces to Eq. (15.22) for free or mean-field particles. Equation (20.39) can then be used to calculate phase shifts for the complete propagator. Using Eq. (20.39) implies that the phase shift δ_ℓ^{JST} remains real, a reasonable approximation at most energies considering the smallness of $|\kappa_0^I|$ compared to κ_0^R, which determines the strength of δ_ℓ^I with respect to δ_ℓ^R [see Eqs. (20.29) and (20.28)]. Consequently, the phase shifts can be fruitfully compared with calculations for mean-field or free particles.

The approximation is sensible for a pole in the complex wave-vector plane not too far from the real axis for the CPA, but makes sense for r no too different from r' in general. Using this extension of the last equality in Eq. (20.37) to the dressed propagator in the case of coupled channels, the $\mathcal{S}$-matrix element can be written like Eq. (20.39)

$$\mathcal{S}_{\ell,\ell'}^{JST}(E) = 1 + 2i \int_0^\infty dk \, k^2 \, \mathrm{Im}\left\{ G_{pphh}^f(k;E) \right\} \langle k\ell|\, \Gamma_{pphh}^{JST}(E)\, |k\ell'\rangle . \quad (20.40)$$

This reduces to the conventional formulations [see *e.g.* Eq. (15.24)] for free or mean-field particles. For coupled channels Eq. (20.40) can be used to generate phase shifts by diagonalization of the $\mathcal{S}$-matrix, as discussed in Sec. 15.1. Equation (20.40) is exact for noninteracting or mean-field particles. For dressed particles, it includes the physically reasonable expectation that the distribution over the wave vectors as contained in the imaginary part of the propagator will feature in determining the scattering process. While this approximation does not make sense at large distance scales, it provides locally, a very useful generalization of the phase shift. The corresponding "short-distance" approximation to the scattering amplitude then yields

$$f_{m_s' m_s}^S(\theta, \phi) = 4\pi \sum_{\ell\ell' J} \sum_{mm' M} i^{\ell'} (-i)^\ell Y_{\ell m_\ell}(\hat{\mathbf{r}}) Y_{\ell' m_\ell'}^*(\hat{\mathbf{z}})$$
$$(\ell m_\ell \, Sm_s|JM)(\ell'm_\ell' \, Sm_s'|JM)$$
$$\int_0^\infty dk \, k \, \mathrm{Im}\left\{ G_{pphh}^f(k;E) \right\} \langle k(\ell S)J|\, \Gamma_{pphh}(E)\, |k(\ell'S)J\rangle , \quad (20.41)$$

where a coupling to total spin S and projections m_s, m_s' for initial and final spin states has been included together with the usual decomposition in partial waves. For free or mean-field particle scattering the δ-function of the imaginary part of G_{pp}^f yields the usual formulation. For a central interaction and free particles Eq. (20.41) reduces to (suppressing spin indices)

$$f(\theta, \phi) = \sum_\ell \frac{2\ell+1}{k_0} \left\{ \frac{-mk_0\pi}{2\hbar^2} \right\} \langle k_0|\, \mathcal{T}_\ell(k_0)\, |k_0\rangle \, P_\ell(\cos\theta)$$
$$= \sum_\ell \frac{2\ell+1}{k_0} e^{i\delta_\ell} \sin \delta_\ell P_\ell(\cos\theta), \quad (20.42)$$

where the addition theorem for spherical harmonics and the δ-function of the imaginary part of the propagator have been used for the first equality and Eq. (15.22) for the second one. For the total cross section (in the neutron–proton system) one finds

$$\sigma_{tot} = \pi \sum_{S\ell\ell' J} (2J+1) \tag{20.43}$$

$$\times \left| \int_0^\infty dk \; k \; \text{Im} \left\{ G^f_{pphh}(k;E) \right\} \langle k(\ell S)J | \, \Gamma_{pphh}(E) \, | k(\ell' S)J \rangle \right|^2 ,$$

which for a central interaction and free particles, reduces to the standard result

$$\sigma_{tot} = \frac{4\pi}{k_0^2} \sum_\ell (2\ell+1) \sin^2 \delta_\ell. \tag{20.44}$$

Equation (20.43) demonstrates that a sensible cross section will be obtained for dressed particles at all energies for which a nonvanishing imaginary part of the propagator exists. For two particles deep in the Fermi sea *e.g.*, Eq. (20.43) avoids the divergence associated with the k_0^{-2} term in Eq. (20.44). The formulation of the cross section in terms of Eq. (20.43) provides a reasonable way to assess the strength of the interaction between dressed particles in the medium in terms of the square of the relevant transition matrix element (Γ), multiplied by an appropriate measure of the density of states represented by the imaginary part of the noninteracting propagator.

This density of states can be written for zero total wave vector as

$$N^{(2)}(E) = -\frac{1}{\pi} \int_0^\infty dk \; k^2 \; \text{Im} \; G^f_{pphh}(k; K=0, E), \tag{20.45}$$

where only the magnitude of the relative vector needs to be considered. For a free Fermi gas this expression reduces to

$$N_F^{(2)}(E) = \frac{m^{\frac{3}{2}} E^{\frac{1}{2}}}{2\hbar^3}. \tag{20.46}$$

Using the unique relation between the energy and the on-shell wave vector k_0 [see Eq. (15.37)] this result can be written as

$$N_F^{(2)}(k_0) = \frac{mk_0}{2\hbar^2}, \tag{20.47}$$

when only kinetic energy contributes. For mean-field particles in the medium one may have to include the effect of the sp potential energy U

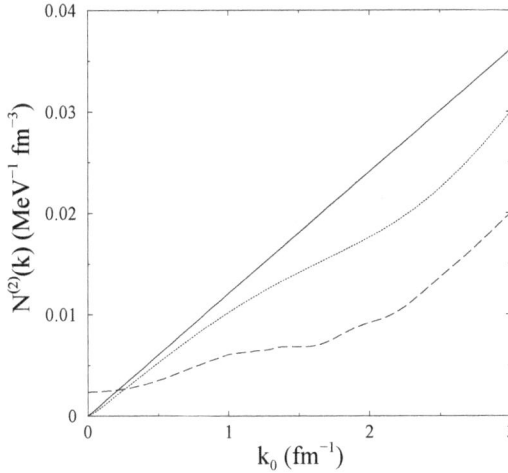

Fig. 20.10 Density of tp states as a function of the on-shell wave vector for free particles (solid line), for mean-field particles at $k_F = 1.36$ fm^{-1} including a sp spectrum U (dotted line), and for dressed particles (dashed line). All three lines correspond to zero total wave vector.

which yields the following density of states

$$N_{mf}^{(2)}(k_0) = \frac{mk_0^2}{2\hbar^2}\left(k_0 + \frac{m}{\hbar^2}\frac{\partial U}{\partial k}\bigg|_{k_0}\right)^{-1}. \qquad (20.48)$$

For purposes of comparison we consider the density of states for dressed particles also as a function of wave vector. This is achieved for zero total wave vector by determining an on-shell wave vector according to Eq. (20.13). This on-shell wave vector k_0 normally coincides with the location of the peak in the imaginary part of G^f_{pphh}, as discussed in Sec. 20.1.

In Fig. 20.10 the density of states for these different cases is presented. The dashed line represents free particles (or mean-field particles with only kinetic energy) according to Eq. (20.47). The dotted line shows the effect of a realistic sp spectrum [Vonderfecht *et al.* (1993)] for mean-field particles at $k_F = 1.36$ fm^{-1} and uses Eq. (20.48). The inclusion of the complete dressing leads to the dashed line in Fig. 20.10 based on the evaluation of Eq. (20.45). When the on-shell wave vector approaches k_F, the reduction of the density of states compared to the dotted line is given exactly by a factor Z_F^2 (about 0.5 here), representing a reduction of the strength of the quasiparticle pole at k_F for each of the particles. Figure 20.10 shows

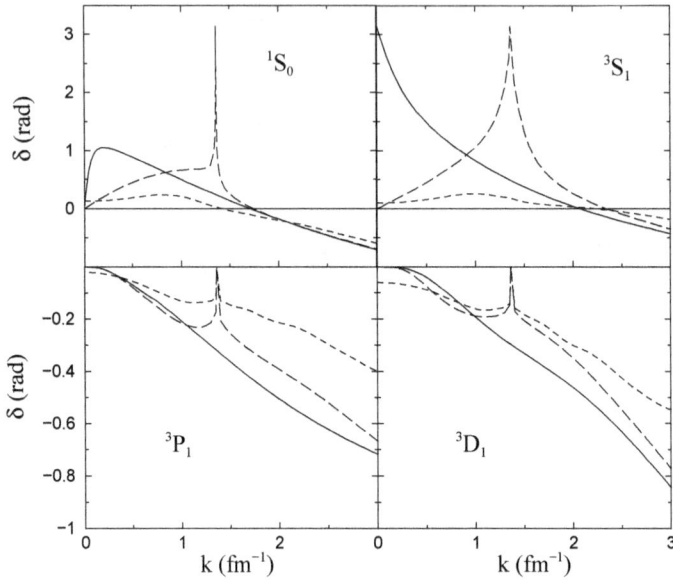

Fig. 20.11 Comparison of phase shifts for free particles (solid), mean-field particles (dashed), and dressed particles (short-dashed lines) for different partial waves. The density of the medium corresponds to $k_F = 1.36 \text{ fm}^{-1}$.

that this reduction is substantial in a large domain of wave vectors (or equivalently in a large domain of energies). For small wave vectors the dressed density of states does not go to zero, since the on-shell relation (20.13) is no longer applicable for energies below $2U(k = 0)$, while the density of states does not vanish at those energies.

The phase shifts for some of the more important partial wave channels are summarized in Fig. 20.11. A comparison is made between phase shifts for free particles (solid line), mean-field particles at $k_F = 1.36 \text{ fm}^{-1}$ (dashed line), and dressed particles (short-dashed line) at the same density for the $^1S_0$, $^3S_1$, $^3P_1$, and $^3D_1$ channels (corresponding to the different panels in Fig. 20.11) as a function of the on-shell wave vector. In general, one finds that the dressed phase shifts suggest weaker interactions, since they are either less repulsive or less attractive than in the mean-field. For the two S-wave channels the most striking feature of the dressed phase shift is the disappearance of the pairing signature for the $^1S_0$ channel and the enormous reduction of the signal in the $^3S_1$ partial wave. While the dressed $^1S_0$ phase shift is essentially zero at k_F, it is still clearly attractive at this wave vector for the $^3S_1$ channel. The calculation of the phase shift for this channel displays a slight kink close to k_F, suggesting that the phase shift may actually rise rapidly to π very close to k_F. It implies a tremendous reduction in the strength of the pairing correlations in this coupled channel, as compared to a mean-field treatment. Gaps of the order of 10 MeV have been generated for this channel in [Vonderfecht *et al.* (1991b); Baldo *et al.* (1992)] (see also Ch. 22). Clearly, the dressing of the nucleons has a strong influence on pairing. While one would expect a gap using dressed nucleons, based on the attractive effective interaction at the Fermi surface, its magnitude is drastically reduced, as suggested by the phase shift calculation illustrated in Fig. 20.11. The main ingredient in this reduction is the decrease in the density of states at $2\varepsilon_F$ when dressed nucleons are propagated. According to Fig. 20.10 the reduction is essentially the square of the strength of the quasiparticle pole at k_F, leading to a factor of about 0.5. Since pairing correlations are particularly sensitive to the density of states, it is not surprising that the strength of the pairing is substantially diminished when dressing is taken into account. It is also noteworthy that one observes a smaller negative phase shift for both S-waves at higher energy, as compared to the mean-field. A similar conclusion may be drawn by inspecting the phase shifts for the $^3P_1$ and $^3D_1$ channel in the bottom panels of Fig. 20.11. For these partial waves, which represent repulsive effective interactions, one observes a reduction of the magnitude of the phase shift as well, when dressing is included. It is important to note that for mean-field propagation Fig. 20.11 shows that the results tend to those of free particles at high energy, whereas it is not so for dressed particles. The latter indicates that the effect of the dressing extends to a large energy domain. Such an observation is not too surprising since the spreading of the sp strength, due to short-range and tensor correlations, takes place in a very large energy domain and is quite different from a local (in energy) spreading of the strength, as would be generated by a complex quasiparticle energy.

The phase shifts for dressed particles lead to the expectation that the total cross sections are substantially reduced, compared to the mean-field calculations. Neutron–proton (np) and neutron–neutron (nn) total cross sections are displayed in Fig. 20.12, and confirm this expectation. The cross sections have been obtained for free (solid), mean-field (dashed), and dressed particles (short-dashed line) by including all partial wave channels of the Reid potential with $J \leq 2$. For mean-field particles a realistic sp energy spectrum was incorporated. The effect of the pairing correlations

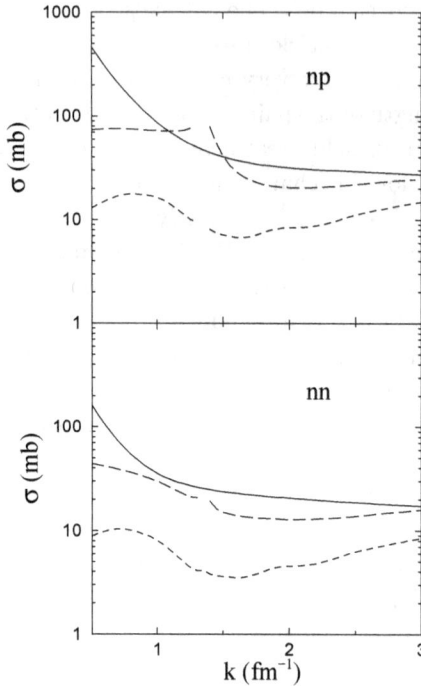

Fig. 20.12 Total neutron–proton (top) and neutron–neutron (bottom panel) cross sections for free (solid), mean-field (dashed), and dressed particles (short-dashed line) as a function of the on-shell wave vector. Gaps in the dashed and short-dashed curves reflect the difficulty in attaining accurate results near k_F when pairing occurs.

on the cross sections, yields a cusp-like behavior around k_F. As the phase shifts for mean-field particles suggest, the corresponding cross sections in the medium become essentially identical to the one in free space at high energy. Both for the np and nn total cross sections, the effect of dressing the nucleons is quite dramatic, leading to a substantial reduction of the total cross section at all energies. Indeed, on average a cross section of only about 10 mb is obtained. While this may seem a small number, it does not imply that the effective interaction in the medium has become insignificant. In addition, one should recall that the concept of asymptotic flux in the medium, representing preserved information of a scattering event deep in the medium, is not a realistic consideration when the dressing of the nucleons is significant, as it is at $k_F = 1.36$ fm^{-1}. The main ingredient accounting for the dressing is the tp density of states. Its reduction for dressed particles is to a large extent responsible for the change in the cross section. While no results are shown in Fig. 20.12 below 0.5 fm^{-1} in order to avoid the large value of the total cross sections for free particles, it should be noted that the cross sections for dressed particles smoothly go to zero when expression Eq. (20.43) is used at lower energies. This expression avoids the problem associated with Eq. (20.44) which would yield an infinite cross section for the on-shell wave vector going to zero. In addition, the latter does not generate a cross section for energies that do not yield a solution for the on-shell wave vector according to Eq. (20.13), *i.e.* for energies deep in the hh continuum.

20.4 Exercises

(1) Extend the analysis of the complex phase shift to a hard-core interaction supplemented by an attractive square well, just outside the hard core.
(2) Perform the integral in Eq. (20.31) for a Bethe–Goldstone propagator. Solve for the Bethe–Goldstone wave function for a hard-core plus attractive square well. Study the healing properties of this wave function.

Chapter 21

Conserving approximations and excited states

The necessity to go beyond the mean-field description of sp motion has attendant consequences for the approach to excited states. So far the discussion in Chs. 13 and 14 has assumed that the sp propagators that generate them, at most include the HF contribution to the self-energy. The situation changes in a qualitative manner when it becomes necessary to include higher-order terms in the self-energy. Examples have been discussed in Ch. 16 for the electron gas (ring diagrams) and nuclear matter (ladder diagrams). Even at the level of the second-order self-energy, an important change in the treatment of excited states is warranted. When an energy-dependent self-energy is required, the RPA method for excited states will no longer be adequate, at least in principle. Indeed, intermediate states in the self-energy will include 2p1h- and 2h1p- and more complicated states. When the resulting sp propagator is combined with another one to describe excited states, it establishes an automatic link to 2p2h states. Such a coupling is included, in a minimal fashion, in the RPA by the backward-going diagrams. It is however, insufficient to represent the proper mixing of 1p1h with 2p2h states. As a result, an improved description of excited states is necessary that should be consistent with relevant conservation laws at the same time.

A treatment of excited states that builds on a given approximation of the self-energy was developed by [Baym and Kadanoff (1961); Baym (1962)] for the many-body problem, and will be presented in this chapter. We will first consider the self-energy in the time formulation in the presence of a time-dependent external potential in Sec. 21.1. Thus a useful link with the response to this external field is set up. The application of the equations of motion for the sp propagator in its presence, will enable to identify an important condition on the (approximate) tp propagator.

The link between general conservation laws and the equations of motion leads to a second condition. Conserving approximations to the response, including the fulfillment of sum rules, will be studied in Sec. 21.2. Exact relations between properties of the sp propagator in an infinite system and the ph interaction on the Fermi surface are discussed in Sec. 21.3. A survey of several different approximations to the self-energy and the resulting conserving description of excited states will be presented in Sec. 21.4. A brief overview of the properties of excited states in nuclei and their theoretical description, based on the material in this chapter, will be given in Sec. 21.5.

21.1   Equations of motion and conservation laws

The Hamiltonian of the many-particle problem has the form

$$
\begin{aligned}
\hat{H} &= \hat{T} + \hat{V} \\
&= \sum_{\alpha\beta} \langle\alpha| T |\beta\rangle \, a_\alpha^\dagger a_\beta + \frac{1}{4} \sum_{\alpha\beta\gamma\delta} \langle\alpha\beta| V |\gamma\delta\rangle \, a_\alpha^\dagger a_\beta^\dagger a_\delta a_\gamma \\
&= \hat{T} + \hat{U} + \hat{V} - \hat{U} = \hat{H}_0 + \hat{H}_1 \\
&= \sum_\alpha \varepsilon_\alpha a_\alpha^\dagger a_\alpha + \hat{V} - \sum_{\alpha\beta} \langle\alpha| U |\beta\rangle \, a_\alpha^\dagger a_\beta,
\end{aligned}
\tag{21.1}
$$

with the usual separation into $\hat{H}_0 = \sum_\alpha \varepsilon_\alpha a_\alpha^\dagger a_\alpha$ and $\hat{H}_1 = \hat{V} - \hat{U}$ that involves the auxiliary potential $\hat{U}$. We will employ the sp basis that diagonalizes H_0. Experimentally, information about the ground state or excitations of the system can only be gathered by applying a time-dependent external field, which probes the density (or other degrees of freedom). The electromagnetic field is a prime example of such a probe. A field that couples directly to the "density" through a one-body operator, can be written as

$$
\hat{\phi}(t) = \sum_{\gamma\delta} \langle\gamma| \phi(\boldsymbol{x}, t) |\delta\rangle \, a_\gamma^\dagger a_\delta = \sum_{\gamma\delta} \phi_{\gamma\delta}(t) a_\gamma^\dagger a_\delta.
\tag{21.2}
$$

The field $\phi(\boldsymbol{x}, t)$ is for instance given by:

$$
\phi(\boldsymbol{x}, t) = \phi(\boldsymbol{x}) \, e^{-iEt/\hbar} + \phi^\dagger(\boldsymbol{x}) \, e^{iEt/\hbar}
\tag{21.3}
$$

but may depend on other dynamical variables like momentum, spin, and isospin *etc*. The density will oscillate with this external field and the system exhibits resonances whenever the energy E is close to an excitation energy. We will therefore proceed by examining the more general case of the system immersed in an external field, which can transfer energy and momentum (and other quantities) to it. This will allow for the description of the response in a natural way. The total Hamiltonian $\hat{H}^\phi$ now becomes explicitly time dependent:

$$\hat{H}^\phi(t) = \hat{H} + \hat{\phi}(t). \tag{21.4}$$

We start by generating an expression for the sp propagator in the presence of the external field ϕ. This requires a Heisenberg picture that is different from the usual one, since ϕ is time dependent. Consider first the time-dependent Schrödinger equation in the presence of ϕ

$$i\hbar \frac{\partial}{\partial t} |\Psi_S^\phi(t)\rangle = \hat{H}^\phi(t)|\Psi_S^\phi(t)\rangle. \tag{21.5}$$

The time-evolution operator is defined by

$$|\Psi_S^\phi(t)\rangle = \hat{U}_S^\phi(t, t_0)|\Psi_S^\phi(t_0)\rangle \tag{21.6}$$

with additional labels to distinguish it from the time-evolution operator associated with the time-independent Hamiltonian $\hat{H}$ (while dropping the one for particle number). When $\hat{H}^\phi$ doesn't commute with itself at different times, one obtains a time-evolution operator similar to Eq. (8.7)

$$\hat{U}_S^\phi(t, t_0) \tag{21.7}$$
$$= \sum_{n=0}^{\infty} \left(\frac{-i}{\hbar}\right)^n \frac{1}{n!} \int_{t_0}^t dt_1 \int_{t_0}^t dt_2 ... \int_{t_0}^t dt_n \; \mathcal{T}\left[\hat{H}^\phi(t_1)\hat{H}^\phi(t_2)...\hat{H}^\phi(t_n)\right].$$

21.1.1   *The field picture*

It is now useful to introduce the field picture, which is analogous to the interaction picture in the standard case. In the field picture we identify the time evolution generated by $\hat{H}$, the Hamiltonian without the field ϕ, as playing the role of $\hat{H}_0$ in the interaction picture. The field picture state ket can thus be defined by using Eq. (8.1) as a template

$$|\Psi_F^\phi(t)\rangle = \exp\left\{\frac{i}{\hbar}\hat{H}t\right\}|\Psi_S^\phi(t)\rangle. \tag{21.8}$$

The adjoint of $\hat{U}_S(t, 0)$, the time-evolution operator without the external field, appears here as the present equivalent of $\exp\left\{\frac{i}{\hbar}\hat{H}_0 t\right\}$, which defines the usual interaction picture. It is straightforward to derive the Schrödinger equation in the field picture with the result

$$i\hbar\frac{\partial}{\partial t}|\Psi_F^\phi(t)\rangle = \hat{\phi}_F(t)|\Psi_S^\phi(t)\rangle, \qquad (21.9)$$

where the field picture operator $\hat{\phi}_F$ is given by

$$\hat{\phi}_F(t) = \exp\left\{\frac{i}{\hbar}\hat{H}t\right\}\hat{\phi}(t)\exp\left\{-\frac{i}{\hbar}\hat{H}t\right\} \qquad (21.10)$$

in analogy to Eq. (8.3). Time evolution in the field picture is governed by the corresponding operator

$$|\Psi_F^\phi(t)\rangle = \hat{U}_F(t,t_0)|\Psi_F^\phi(t_0)\rangle. \qquad (21.11)$$

From Eq. (21.9) one therefore obtains

$$i\hbar\frac{\partial}{\partial t}\hat{U}_F(t,t_0) = \hat{\phi}_F(t)\hat{U}_F(t,t_0). \qquad (21.12)$$

Again, a formal solution can be applied here as well to yield

$$\hat{U}_F(t,t_0) \qquad (21.13)$$
$$= \sum_{n=0}^{\infty}\left(\frac{-i}{\hbar}\right)^n\frac{1}{n!}\int_{t_0}^t dt_1\int_{t_0}^t dt_2...\int_{t_0}^t dt_n\ \mathcal{T}\left[\hat{\phi}_F(t_1)\hat{\phi}_F(t_2)...\hat{\phi}_F(t_n)\right].$$

The Heisenberg picture state ket in the presence of $\phi(t)$ becomes

$$|\Psi_H^\phi(t)\rangle = (\hat{U}_S^\phi)^\dagger(t,t_0)|\Psi_S^\phi(t)\rangle = |\Psi_S^\phi(t_0)\rangle, \qquad (21.14)$$

generalizing the case of a time-independent Hamiltonian. We still find

$$i\hbar\frac{\partial}{\partial t}|\Psi_H^\phi(t)\rangle = 0. \qquad (21.15)$$

An operator in this Heisenberg picture is defined by

$$\hat{O}_H(t) = (\hat{U}_S^\phi)^\dagger(t,t_0)\hat{O}_S\hat{U}_S^\phi(t,t_0), \qquad (21.16)$$

which applies in particular to the addition and removal operators, $a^\dagger$ and a. We are now ready to consider the sp propagator in the presence of $\hat{\phi}(t)$ with respect to the Heisenberg ground state $|\Psi_0^\phi\rangle$ (H suppressed) of the combined system by

$$G^\phi(\alpha,\beta;t-t') = -\frac{i}{\hbar}\langle\Psi_0^\phi|\ \mathcal{T}\left[a_{\alpha H}(t)a_{\beta H}^\dagger(t')\right]|\Psi_0^\phi\rangle, \qquad (21.17)$$

where the H subscript for a and $a^\dagger$ refers to the time dependence given by Eq. (21.16). The expansion that was used in Eq. (8.17) for the sp propagator, can be employed to generate the analogous result for Eq. (21.17)

$$G^\phi(\alpha, \beta; t - t') = -\frac{i}{\hbar} \sum_n \left(\frac{-i}{\hbar}\right)^n \frac{1}{n!} \int_{-\infty}^{+\infty} dt_1 \ldots \int_{-\infty}^{+\infty} dt_n$$

$$\times \langle \Psi_0| \, \mathcal{T} \left[\hat{\phi}_F(t_1)\ldots\hat{\phi}_F(t_n) a_{\alpha_F}(t) a^\dagger_{\beta_F}(t')\right] |\Psi_0\rangle \qquad (21.18)$$

$$/ \sum_m \left(\frac{-i}{\hbar}\right)^m \frac{1}{m!} \int_{-\infty}^{+\infty} dt'_1 \ldots \int_{-\infty}^{+\infty} dt'_m \, \langle \Psi_0| \, \mathcal{T} \left[\hat{\phi}_F(t'_1)\ldots\hat{\phi}_F(t'_m)\right] |\Psi_0\rangle .$$

It is convenient to simplify the notation for further development

$$G^\phi(\alpha, \beta; t - t') = -\frac{i}{\hbar} \frac{\langle \Psi_0| \, \mathcal{T}[\hat{S} a_{\alpha_F}(t) a^\dagger_{\beta_F}(t')] |\Psi_0\rangle}{\langle \Psi_0| \, \mathcal{T}[\hat{S}] |\Psi_0\rangle}, \qquad (21.19)$$

where

$$\hat{S} = \exp\left\{-\frac{i}{\hbar} \int_{-\infty}^{\infty} dt' \left[\sum_{\gamma\delta} \phi_{\gamma\delta}(t') a^\dagger_{\gamma_F}(t') a_{\delta_F}(t')\right]\right\} \qquad (21.20)$$

and the time dependence corresponds to the usual Heisenberg picture linked with $\hat{H}$ and thus can be associated with the field picture time dependence of Eq. (21.10). For a very weak field one therefore obtains information about the ground state $|\Psi_0\rangle$ associated with $\hat{H}$

$$\lim_{\phi \to 0} G^\phi(\alpha, \beta; t - t') = G(\alpha, \beta; t - t'), \qquad (21.21)$$

but also about excited states, as discussed in Sec. 21.2.

21.1.2 *Equations of motion in the field picture*

Following the development in Ch. 9, we can study the equations of motion for the sp propagator that lead to the identification of the self-energy. The equivalent result to Eq. (9.15) in the presence of $\hat{\phi}$ becomes

$$i\hbar \frac{\partial}{\partial t} G^\phi(\alpha, \beta; t - t') = \delta(t - t')\delta_{\alpha\beta} + \varepsilon_\beta G^\phi(\alpha, \beta; t - t')$$

$$- \sum_\delta \langle \alpha| U |\delta\rangle \, G^\phi(\delta, \beta; t - t') + \sum_\delta \phi_{\alpha\delta}(t) \, G^\phi(\delta, \beta; t - t')$$

$$+ \frac{1}{2} \sum_{\delta\zeta\theta} \langle \alpha\delta| V |\theta\zeta\rangle \, G^\phi_{II}(\theta t, \zeta t, \delta t^+, \beta t'), \qquad (21.22)$$

where the corresponding tp propagator

$$G_{II}^{\phi}(\alpha t_{\alpha}, \beta t_{\beta}, \gamma t_{\gamma}, \delta t_{\delta}) = -\frac{i}{\hbar}\langle \Psi_0^{\phi}|\mathcal{T}[a_{\beta_H}(t_{\beta})a_{\alpha_H}(t_{\alpha})a_{\gamma_H}^{\dagger}(t_{\gamma})a_{\delta_H}^{\dagger}(t_{\delta})]|\Psi_0^{\phi}\rangle$$

$$(21.23)$$

has been employed. When the time derivative of the addition operator in the sp propagator is used, we find

$$-i\hbar\frac{\partial}{\partial t'}G^{\phi}(\alpha, \beta; t - t') = \delta(t - t')\delta_{\alpha\beta} + \varepsilon_{\alpha}G^{\phi}(\alpha, \beta; t - t')$$

$$- \sum_{\delta}\langle \delta| U |\beta\rangle \ G^{\phi}(\alpha, \delta; t - t') + \sum_{\delta}\phi_{\delta\beta}(t') \ G^{\phi}(\alpha, \delta; t - t')$$

$$+ \frac{1}{2}\sum_{\delta\zeta\theta}\langle \theta\zeta| V |\beta\delta\rangle \ G_{II}^{\phi}(\delta t', \alpha t, \theta t'^{+}, \zeta t'^{+}). \qquad (21.24)$$

As discussed in Sec. 9.4, it is possible to develop the equation of motion for the tp propagator that will couple to the three-particle propagator, *etc.* In most practical applications an approximation to G_{II} is made and the hierarchic coupling to n-particle Green's functions is terminated at $n = 2$. It is reasonable to expect that a physically meaningful approximation requires that both Eq. (21.22) and (21.24) lead to the same G^{ϕ} and that the conservation laws implied by Eq. (21.4) should be obeyed. We will investigate the first condition by transforming Eqs. (21.22) and (21.24) into the Dyson equation in the time formulation. The Dyson equation then allows for the calculation of the sp propagator, based on diagrammatic approximations that have been discussed in previous chapters.

We start by introducing the equations of motion for the sp propagator $G^{(0)}$ associated with $\hat{H}_0 = \hat{T} + \hat{U}$. These are generated in the same way as for G^{ϕ} and read

$$i\hbar\frac{\partial}{\partial t}G^{(0)}(\alpha, \beta; t - t') = \delta(t - t')\delta_{\alpha\beta} + \varepsilon_{\alpha}G^{(0)}(\alpha, \beta; t - t') \qquad (21.25)$$

and

$$-i\hbar\frac{\partial}{\partial t'}G^{(0)}(\alpha, \beta; t - t') = \delta(t - t')\delta_{\alpha\beta} + \varepsilon_{\beta}G^{(0)}(\alpha, \beta; t - t'). \qquad (21.26)$$

We proceed by defining the inverted Green's function $G^{\phi-1}$ by

$$\sum_{\gamma}\int_{-\infty}^{\infty}dt'' G^{\phi}(\alpha, \gamma; t - t'')G^{\phi-1}(\gamma, \beta; t'' - t') = \delta(t - t')\delta_{\alpha\beta} \qquad (21.27)$$

or

$$\sum_{\gamma} \int_{-\infty}^{\infty} dt'' G^{\phi^{-1}}(\alpha, \gamma; t - t'') G^{\phi}(\gamma, \beta; t'' - t') = \delta(t - t')\delta_{\alpha\beta}. \quad (21.28)$$

For $G^{(0)^{-1}}$ one has in like fashion

$$\sum_{\gamma} \int_{-\infty}^{\infty} dt'' G^{(0)}(\alpha, \gamma; t - t'') G^{(0)^{-1}}(\gamma, \beta; t'' - t') = \delta(t - t')\delta_{\alpha\beta} \quad (21.29)$$

or

$$\sum_{\gamma} \int_{-\infty}^{\infty} dt'' G^{(0)^{-1}}(\alpha, \gamma; t - t'') G^{(0)}(\gamma, \beta; t'' - t') = \delta(t - t')\delta_{\alpha\beta}. \quad (21.30)$$

Multiplying Eq. (21.25) by $G^{(0)^{-1}}$ from the right, performing the appropriate summation and integration, and using Eq. (21.29) we obtain the explicit expression

$$G^{(0)^{-1}}(\alpha, \beta; t - t') = \left(i\hbar \frac{\partial}{\partial t} - \varepsilon_\alpha \right) \delta_{\alpha\beta} \delta(t - t'). \quad (21.31)$$

Similar steps also yield the alternative relation

$$G^{(0)^{-1}}(\alpha, \beta; t - t') = \left(-i\hbar \frac{\partial}{\partial t'} - \varepsilon_\beta \right) \delta_{\alpha\beta} \delta(t - t'). \quad (21.32)$$

Employing these equations, it is possible to rewrite Eqs. (21.22) in the following way

$$G^{\phi^{-1}}(\alpha, \beta; t - t') = G^{(0)^{-1}}(\alpha, \beta; t - t') \\ + \langle \alpha | U | \beta \rangle - \phi_{\alpha\beta}(t)\delta(t - t') - \Sigma^{(A)}(\alpha, \beta; t - t') \quad (21.33)$$

with

$$\Sigma^{(A)}(\alpha, \beta; t - t') \quad (21.34)$$
$$= \frac{1}{2} \int_{-\infty}^{\infty} dt'' \sum_{\gamma\delta\zeta\theta} \langle \alpha\delta | V | \theta\zeta \rangle \, G_{II}^{\phi}(\theta t, \zeta t, \delta t^+, \gamma t'') G^{\phi^{-1}}(\gamma, \beta; t'' - t').$$

In the same way we can transform Eq. (21.24) so that

$$G^{\phi^{-1}}(\alpha, \beta; t - t') = G^{(0)^{-1}}(\alpha, \beta; t - t') + \langle \alpha | U | \beta \rangle \\ - \phi_{\alpha\beta}(t)\delta(t - t') - \Sigma^{(B)}(\alpha, \beta; t - t') \quad (21.35)$$

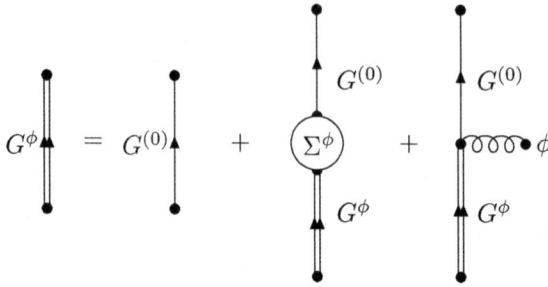

Fig. 21.1 Diagrammatic representation of the sp propagator in terms of the irreducible self-energy Σ and the noninteracting propagator $G^{(0)}$ representing Eq. (21.38).

with

$$\Sigma^{(B)}(\alpha,\beta;t-t') \qquad (21.36)$$
$$= \frac{1}{2}\int_{-\infty}^{\infty} dt'' \sum_{\gamma\delta\zeta\theta} G^{\phi^{-1}}(\alpha,\gamma;t-t'') \langle\theta\zeta|\,V\,|\beta\delta\rangle\, G^{\phi}_{II}(\delta t',\gamma t'',\theta t'^{+},\zeta t'^{+}).$$

To guarantee that both Eqs. (21.33) and (21.35) produce the identical solution for the sp propagator, it is clear that the approximation to G^{ϕ}_{II} should be such that

$$\Sigma^{(A)} = \Sigma^{(B)}. \qquad (21.37)$$

If this condition is fulfilled both equations of motion can be replaced by one equation: the Dyson equation with $\Sigma^{\phi} = \Sigma^{(A)} - U = \Sigma^{(B)} - U$

$$G^{\phi}(\alpha,\beta;t-t') = G^{(0)}(\alpha,\beta;t-t')$$
$$+ \sum_{\gamma\delta}\int_{-\infty}^{\infty} dt'' \int_{-\infty}^{\infty} dt''' G^{(0)}(\alpha,\gamma;t-t'')\Sigma^{\phi}(\gamma,\delta;t''-t''')G^{\phi}(\delta,\beta;t'''-t')$$
$$+ \sum_{\gamma\delta}\int_{-\infty}^{\infty} dt'' G^{(0)}(\alpha,\gamma;t-t'')\phi_{\gamma\delta}(t'')G^{\phi}(\delta,\beta;t''-t'). \qquad (21.38)$$

This result is shown diagrammatically in Fig. 21.1. The external field ϕ is indicated by the helical line and represents the difference between Fig. 21.1 and Fig. 9.8. Note that the presence of ϕ also changes all the internal sp propagators in the self-energy, which must therefore be labeled accordingly.

The condition of Eq. (21.37) can be checked by writing the tp propagator as in Eq. (9.22)

$$G_{II}^{\phi}(\alpha t_{\alpha}, \beta t_{\beta}, \gamma t_{\gamma}, \delta t_{\delta})$$
$$= i\hbar[G^{\phi}(\alpha, \gamma; t_{\alpha} - t_{\gamma})G^{\phi}(\beta, \delta; t_{\beta} - t_{\delta}) - G^{\phi}(\alpha, \delta; t_{\alpha} - t_{\delta})G^{\phi}(\beta, \gamma; t_{\beta} - t_{\gamma})]$$
$$+ (i\hbar)^2 \int dt_{\epsilon} \int dt_{\zeta} \int dt_{\eta} \int dt_{\theta} \sum_{\epsilon\zeta\eta\theta} G^{\phi}(\alpha, \epsilon; t_{\alpha} - t_{\epsilon})G^{\phi}(\beta, \zeta; t_{\beta} - t_{\zeta})$$
$$\times \langle \epsilon\zeta| \Gamma(t_{\epsilon}, t_{\zeta}, t_{c}\eta, t_{\theta}) |\eta\theta\rangle \, G^{\phi}(\eta, \gamma; t_{\eta} - t_{\gamma})G^{\phi}(\theta, \delta; t_{\theta} - t_{\delta}), \tag{21.39}$$

in terms of the vertex function Γ shown in Fig. 9.11. An approximation to G_{II} can obviously be formulated in terms of Γ. Equations (21.34) and (21.36) may now be represented in the diagrammatic form of Figs. 9.13 and 9.14, respectively. Checking condition Eq. (21.37) is therefore equivalent to making sure that the approximation to Γ or G_{II} yields the same diagrams for Figs. 9.13 and 9.14. This development emphasizes the importance of self-consistent propagators, as indicated by the use of the equations of motion.

21.1.3 *Conservation laws and approximations*

Conservation laws that are implied by the Hamiltonian are fulfilled by imposing another condition on the approximate sp propagator. In turn, this leads to an extra condition on the tp propagator, which is already implied by its original definition given in Eq. (21.23). The extra symmetry condition, together with Eq. (21.37), is sufficient to ensure that the sp propagator satisfies conservation laws. We proceed by studying some examples.

If the approximate sp propagator solves both equations of motion in Eqs. (21.22) and (21.24), then the number of particles is conserved. This can be shown as follows: subtracting Eq. (21.24) from (21.22), putting $\alpha = \beta$, $t' = t^{+}$ and summing over α one obtains

$$i\hbar \frac{d}{dt} \sum_{\alpha} G^{\phi}(\alpha, \alpha; t - t^{+}) = 0, \tag{21.40}$$

or

$$\frac{d}{dt} \langle \hat{N}(t) \rangle = 0. \tag{21.41}$$

Particle number conservation is thus guaranteed if the equations of motion in Eqs. (21.22) and (21.24) are satisfied. In turn, this requires for any approximation the use of self-consistent propagators.

To satisfy conservation of momentum, the approximate G_{II} has to fulfill an additional symmetry condition. We proceed by multiplying Eqs. (21.22) and (21.24) by the matrix element $\langle\beta|\,\boldsymbol{p}\,|\alpha\rangle$. Performing the subtraction, summing over α and β, putting $t' = t^+$, and using completeness of the sp states $\{|\alpha\rangle\}$ yields, after cancelling the U contribution,

$$
i\hbar\frac{d}{dt}\sum_{\alpha\beta}\langle\beta|\,\boldsymbol{p}\,|\alpha\rangle\,G^\phi(\alpha,\beta;t-t^+)
$$
$$
=\sum_{\delta\beta}\langle\beta|\,[\boldsymbol{p},\phi(\boldsymbol{x},t)]\,|\delta\rangle\,G^\phi(\delta,\beta,t-t^+)
$$
$$
+\sum_{\delta\beta}\langle\beta|\,[\boldsymbol{p},T]\,|\delta\rangle\,G^\phi(\delta,\beta;t-t^+)
$$
$$
+\frac{1}{2}\sum_{\beta\delta\theta\zeta}\langle\beta\delta|\,[\boldsymbol{p}_1,V]\,|\theta\zeta\rangle\,G^\phi_{II}(\theta t,\zeta t,\delta t^+,\beta t^+). \tag{21.42}
$$

The commutation relation of $\boldsymbol{p}$ with the kinetic energy operator vanishes. The implication is that Eq. (21.42) can be written as

$$
i\hbar\frac{d}{dt}\left[\sum_{\alpha\beta}\langle\beta|\,\boldsymbol{p}\,|\alpha\rangle\,G^\phi(\alpha,\beta;t-t^+)\right]
$$
$$
=-\sum_{\beta\delta}\langle\beta|\,(i\hbar\nabla\phi(\boldsymbol{x},t)\,|\delta\rangle\,G^\phi(\delta,\beta;t-t^+), \tag{21.43}
$$

by observing that

$$
[\boldsymbol{p},f(\boldsymbol{x})]=-i\hbar\nabla f(\boldsymbol{x}) \tag{21.44}
$$

and provided that the term

$$
-i\hbar\frac{1}{2}\sum_{\beta\delta\theta\zeta}\langle\beta\delta|\,(\nabla_1 V)\,|\theta\zeta\rangle\,G^\phi_{II}(\theta t,\zeta t,\delta t^+,\beta t^+)=0. \tag{21.45}
$$

Assuming an interaction of the form $V(|\boldsymbol{x}_1-\boldsymbol{x}_2|)$ and evaluating this expression in coordinate space, one can show that the integrand is odd upon interchange of $\boldsymbol{x}_1$ and $\boldsymbol{x}_2$ when the following symmetry property holds

$$
G^\phi_{II}(\alpha t,\beta t,\gamma t^+,\delta t^+)=G^\phi_{II}(\beta t,\alpha t,\delta t^+,\gamma t^+). \tag{21.46}
$$

The symmetry condition is certainly fulfilled by the exact tp propagator. We conclude that, if it is obeyed by the approximate tp propagator, the momentum conservation law holds, since it is possible to write

Eq. (21.43) as

$$\frac{d\langle \hat{\boldsymbol{P}}(t)\rangle}{dt} = -\hbar \langle \nabla \phi(\boldsymbol{x}, t)\rangle, \qquad (21.47)$$

which states that the change in the total momentum of the system equals the applied force. The condition stated in Eq. (21.46) is then written as

$$\langle \alpha\beta | \Gamma(t_1, t_2, t_3, t_4) | \gamma\delta\rangle = \langle \beta\alpha | \Gamma(t_2, t_1; t_4, t_3) | \delta\gamma\rangle, \qquad (21.48)$$

and both forms can be checked by inspecting the diagrammatic content of these quantities. The method may be employed to demonstrate the conservation law associated with angular momentum. A more complicated derivation is required to prove conservation of energy. Provided the conditions of Eqs. (21.37) and (21.46), or (21.48) hold, it is shown in [Kadanoff and Baym (1962)] that the energy conservation law is fulfilled.

21.2 Linear response and extensions of RPA

The discussion of the sp propagator for the electron gas, nuclei, and nuclear matter has made it clear that self-energy contributions beyond the HF approximation are essential for these systems. The need to go beyond the mean-field approximation in the self-energy has its counterpart in the description of excited states. This relation is well known in quantum field theory and is referred to under the heading of Ward identities. For the many-body system the work of [Kadanoff and Baym (1962)] has been particularly helpful in clarifying the proper way to describe excited states, once an approximation to the self-energy has been chosen. We have already seen in the previous section how an analysis of the sp propagator in the presence of an external field ϕ leads to constraints on the self-energy that guarantee the fulfillment of important conservation laws. We will now extend this discussion to the description of excited states.

A convenient starting point is again provided by the field picture introduced in Sec. 21.1. In the presence of the external field ϕ the sp propagator can be written as a perturbation expansion in powers of the external field. This expansion takes the familiar form

$$G^\phi(\alpha, \bar{\beta}; t - t') = -\frac{i}{\hbar} \frac{\langle \Psi_0 | T[\hat{S} a_{\alpha_F}(t) a^\dagger_{\bar{\beta}_F}(t')] | \Psi_0\rangle}{\langle \Psi_0 | T[\hat{S}] | \Psi_0\rangle}, \qquad (21.49)$$

where

$$\hat{S} = \exp\left\{-\frac{i}{\hbar}\int_{-\infty}^{\infty}dt'\left[\sum_{\gamma\bar{\delta}}\phi_{\gamma\bar{\delta}}(t')a^{\dagger}_{\gamma_F}(t')a_{\bar{\delta}_F}(t')\right]\right\}. \tag{21.50}$$

We have employed time-reversed states to facilitate the proper coupling to total angular momentum when necessary. In the linear response formulation the assumption is made that the external field is weak and can therefore be treated in first order. In particular, we will study the linear response of the sp propagator to the external field as expressed by Eq. (21.49). The strategy is to consider all terms linear in ϕ (in the limit $\phi \Rightarrow 0$). If ϕ were a variable, this would correspond to the first derivative of G^{ϕ} with respect to that variable. Since ϕ is a function, it is necessary to treat it as a functional derivative.

21.2.1 *Brief encounter with functional derivatives*

A functional $\mathcal{F}[\psi]$ is a mapping between the functions $\psi(x)$ of a variable x (either a single or a multiple variable), and the complex numbers. The functional derivative $\delta\mathcal{F}[\psi]/\delta\psi(x)$ is also a functional, and in addition a function of the external variable x. It is defined as the variation $\delta\mathcal{F}$ of the functional when a small change $\delta\psi$ is made,

$$\delta\mathcal{F}[\psi] = \mathcal{F}[\psi + \delta\psi] - \mathcal{F}[\psi] = \int dx \, \frac{\delta\mathcal{F}[\psi]}{\delta\psi(x)}\delta\psi(x), \tag{21.51}$$

up to first order in $\delta\psi$. This definition can be viewed as a generalization of the ordinary (partial) derivative. The usual rules of differentiation are also valid for functional derivatives [Volterra (1959); Strinati (1988)]. The following results hold

$$\frac{\delta\psi(x)}{\delta\psi(y)} = \delta(x - y), \tag{21.52}$$

$$\frac{\delta}{\delta\psi(x)}\{\mathcal{F}[\psi]\mathcal{G}[\psi]\} - \frac{\delta\mathcal{F}[\psi]}{\delta\psi(x)}\mathcal{G}[\psi] + \mathcal{F}[\psi]\frac{\delta\mathcal{G}[\psi]}{\delta\psi(x)}, \tag{21.53}$$

and

$$\frac{\delta}{\delta\psi(x)}\frac{\mathcal{F}[\psi]}{\mathcal{G}[\psi]} = \frac{1}{\mathcal{G}^2[\psi]}\left[\frac{\delta\mathcal{F}[\psi]}{\delta\psi(x)}\mathcal{G}[\psi] - \mathcal{F}[\psi]\frac{\delta\mathcal{G}[\psi]}{\delta\psi(x)}\right]. \tag{21.54}$$

Another useful relation involves a functional $\mathcal{F}[\psi; x, y]$ dependent on two external variables, as the matrix inverse of $\mathcal{G}[\psi; x, y]$ which means that

$$\int dz \; \mathcal{F}[\psi; x, z]\mathcal{G}[\psi; z, y] = \delta(x - y), \qquad (21.55)$$

for any function ψ. Taking the functional derivative $\delta/\delta\psi(u)$ in accordance with Eq. (21.53) one finds

$$\int dz \; \left\{ \frac{\delta\mathcal{F}[\psi; x, z]}{\delta\psi(u)}\mathcal{G}[\psi; z, y] + \mathcal{F}[\psi; x, z]\frac{\delta\mathcal{G}[\psi; z, y]}{\delta\psi(u)} \right\} = 0, \qquad (21.56)$$

and a (matrix) multiplication with $\mathcal{F}[\psi; y, z']$ then yields

$$\frac{\delta\mathcal{F}[\psi; x, z']}{\delta\psi(u)} = -\int dy \int dz \; \mathcal{F}[\psi; x, z]\frac{\delta\mathcal{G}[\psi; z, y]}{\delta\psi(u)}\mathcal{F}[\psi; y, z']. \qquad (21.57)$$

For a functional $\mathcal{F}[\mathcal{G}[\psi; x]]$ defined through its dependence on a functional $\mathcal{G}[\psi; x]$ with *e.g.* a single external variable, the "chain" rule is given by

$$\frac{\delta\mathcal{F}}{\delta\psi(u)} = \int dx \; \frac{\delta\mathcal{F}}{\delta\mathcal{G}[\psi; x]}\frac{\delta\mathcal{G}[\psi; x]}{\delta\psi(u)}. \qquad (21.58)$$

Equipped with this material, we can generate the results of the following section.

21.2.2 *Linear response and functional derivatives*

The functional derivative of the sp propagator, given in Eq. (21.49), with respect to ϕ can now be evaluated, employing the material of Sec. 21.2.1. The ϕ-dependence, represented by $\mathcal{S}$ in Eq. (21.49), is present in both the numerator and denominator. We can therefore apply Eq. (21.54) varying ϕ by a small amount $\delta\phi$ and keeping only the lowest-order terms to write the change in G as

$$\delta G^\phi(\alpha, \overline{\beta}; t - t') = -\frac{i}{\hbar}\frac{\langle \Psi_0| \, \mathcal{T}[\delta\hat{\mathcal{S}}a_{\alpha_F}(t)a^\dagger_{\overline{\beta}_F}(t')] \, |\Psi_0\rangle}{\langle \Psi_0| \, \mathcal{T}[\hat{\mathcal{S}}] \, |\Psi_0\rangle}$$
$$-G^\phi(\alpha, \overline{\beta}; t - t')\frac{\langle \Psi_0| \, \mathcal{T}[\delta\hat{\mathcal{S}}] \, |\Psi_0\rangle}{\langle \Psi_0| \, \mathcal{T}[\hat{\mathcal{S}}] \, |\Psi_0\rangle}. \qquad (21.59)$$

The change in $\hat{S}$ is related to $\delta\phi$ by

$$\delta\hat{S} = -\frac{i}{\hbar}\int_{-\infty}^{\infty}dt'\left[\sum_{\gamma\bar{\delta}}\delta\phi_{\gamma\bar{\delta}}(t')a_{\gamma_F}^{\dagger}(t')a_{\bar{\delta}_F}(t')\right]. \quad (21.60)$$

Upon substitution of Eq. (21.60) into (21.59) the change in G^{ϕ} can be written as

$$\delta G^{\phi}(\alpha,\bar{\beta};t-t') = \int dt''\sum_{\gamma\bar{\delta}}\delta\phi_{\gamma\bar{\delta}}(t'') \quad (21.61)$$

$$\left\{\frac{i}{\hbar}G_{II}(\alpha t,\bar{\delta}t''^{+},\bar{\beta}t',\gamma t'') + G(\alpha,\bar{\beta};t-t')G(\bar{\delta},\gamma;t''-t''^{+})\right\}.$$

The functional derivative can now be identified according to Eq. (21.51) as

$$\frac{\delta G^{\phi}(\alpha,\bar{\beta};t-t')}{\delta\phi_{\gamma\bar{\delta}}(t'')} = \frac{i}{\hbar}G_{II}(\alpha t,\bar{\delta}t''^{+},\bar{\beta}t',\gamma t'') + G(\alpha,\bar{\beta};t-t')G(\bar{\delta},\gamma;t''-t''^{+}).$$

$$(21.62)$$

The exact propagators on the right side of this equation correspond to the usual Heisenberg picture. We observe that the tp propagator is equivalent to a three-time generalization of the ph propagator introduced in Eq. (13.1). The term with the product of sp propagators provides the correction that turns the functional derivative of G^{ϕ} into the three-time generalization of the polarization propagator of Eq. (13.8)

$$\Pi(\alpha t,\beta^{-1}t';\gamma t''^{+},\delta^{-1}t'') = -i\hbar\frac{\delta G^{\phi}(\alpha,\bar{\beta};t-t')}{\delta\phi_{\gamma\bar{\delta}}(t'')}. \quad (21.63)$$

The two-time polarization propagator can be obtained by taking the following limit

$$\Pi(\alpha,\beta^{-1};\gamma,\delta^{-1};t-t') = \lim_{\tilde{t}\Rightarrow t}\Pi(\alpha t,\beta^{-1}\tilde{t};\gamma t'^{+},\delta^{-1}t'). \quad (21.64)$$

We will next derive the integral equation for the three-time polarization propagator that ensures the fulfillment of the relevant conservation laws in describing excited states. We start by rewriting Eq. (21.33) or (21.35) as

$$G^{\phi^{-1}}(\alpha,\bar{\beta};t-t') = G^{(0)^{-1}}(\alpha,\bar{\beta};t-t')$$
$$-\phi_{\alpha\bar{\beta}}(t)\delta(t-t') - \Sigma^{\phi}(\alpha,\bar{\beta};t-t'). \quad (21.65)$$

Employing Eq. (21.57), we can replace the functional derivative of G^ϕ by the one of its inverse

$$\frac{\delta G^\phi(\alpha,\overline{\beta};t-t')}{\delta\phi_{\gamma\overline{\delta}}(t'')} = -\sum_{\epsilon\zeta}\int dt_1 \int dt_2$$

$$G^\phi(\alpha,\epsilon;t-t_1)\frac{\delta G^{\phi^{-1}}(\epsilon,\overline{\zeta};t_1-t_2)}{\delta\phi_{\gamma\overline{\delta}}(t'')} G^\phi(\overline{\zeta},\overline{\beta};t_2-t'). \quad (21.66)$$

Inserting Eq. (21.65) and using the chain rule of Eq. (21.58), yields the so-called Bethe–Salpeter equation for the polarization propagator in the limit $\phi \Rightarrow 0$

$$\Pi(\alpha t,\beta^{-1}t';\gamma t''^+,\delta^{-1}t'') = \Pi^f(\alpha t,\beta^{-1}t';\gamma t''^+,\delta^{-1}t'')$$

$$+\frac{i}{\hbar}\sum_{\epsilon\zeta\eta\theta}\int dt_1\int dt_2\int dt_3\int dt_4\left[\Pi^f(\alpha t,\beta^{-1}t';\epsilon t_1,\zeta^{-1}t_2)\right.$$

$$\left.\Gamma^{ph}(\epsilon t_1,\zeta^{-1}t_2,\eta t_3,\theta^{-1}t_4)\Pi(\eta t_3,\theta^{-1}t_4;\gamma t''^+,\delta^{-1}t'')\right]. \quad (21.67)$$

The noninteracting polarization propagator is given by the product of dressed sp propagators

$$\Pi^f(\alpha t,\beta^{-1}t',\gamma t_1,\delta^{-1}t_2) = -i\hbar G(\alpha,\gamma;t-t_1)G(\overline{\delta},\overline{\beta};t_2-t') \quad (21.68)$$

and describes the independent propagation of a particle and a hole. Since the self-energy is conserving in the sense of the previous discussion, the internal propagators are self-consistent within the chosen approximation scheme. The resulting conserving approximation for the polarization propagator requires a ph interaction vertex determined by the functional derivative of the self-energy with respect to the self-consistent sp propagator. We therefore obtain

$$\Gamma^{ph}(\alpha t_1,\beta^{-1}t_2,\gamma t_3,\delta^{-1}t_4) = \frac{\delta\Sigma(\alpha,\overline{\beta};t_1-t_2)}{\delta G(\gamma,\overline{\delta};t_3-t_4)}. \quad (21.69)$$

The Bethe–Salpeter equation contains the three-time version of the polarization propagator. Hence, practical applications are not easy to implement. We will discuss some examples of the relation between the approximation for the self-energy and the corresponding description of excited states in Sec. 21.4.

Equation (21.67) can be regarded as a reformulation of the standard perturbation expansion of the polarization propagator in terms of noninteracting propagators. Instead, the first term contributing to the polarization

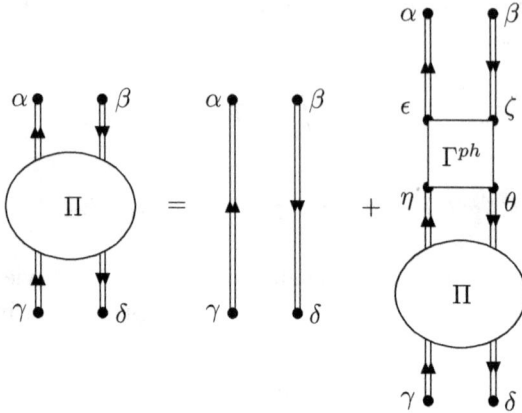

Fig. 21.2 Diagrammatic representation of Eq. (21.67) for the polarization propagator. Note that the box represents the irreducible ph interaction Γ^{ph} as defined in Eq. (21.69). The use of self-consistent or exact sp propagators is indicated by the double-line notation. The present diagrammatic form may be employed in the time formulation but can also be used with energy variables.

propagator, involves the product of exact sp propagators, as illustrated in Fig. 21.2. The advantage of the present approach over the standard perturbation expansion is that, in an approximation scheme, it automatically produces the set of diagrams, according to Eq. (21.69), that fulfill the conservation laws. The latter is based on the use of a self-energy that includes self-consistent propagators, leading to a conserving approximation to G and to its functional derivative Π [see Eq. (21.63)]. Several examples of conserving approximations and the consequences for the description of excited states are reviewed in Sec. 21.4.

21.3 Ward–Pitaevskii relations for a Fermi liquid

An important collection of identities exists that relates properties of the exact sp propagator in the vicinity of the Fermi surface with those of the corresponding irreducible ph interaction in an infinite system. Some of these results have already been employed in Sec. 14.6 when we discussed the excitations of a normal Fermi liquid in the Landau limit. They will now be presented in a more detail. For the many-body problem these identities are referred to as Ward–Pitaevskii relations. First studied for quantum electrodynamics by [Ward (1950)], they were applied to the many-body problem by [Pitaevskii (1960)]. We will follow the presentation of [Abrikosov *et al.* (1975)] for the first relation and present the others without too much detail. Slightly different presentations can be found in [Nozières (1997); Gross *et al.* (1991)].

The framework, developed so far, lends itself quite well to a derivation of the Ward–Pitaevskii relations. We therefore start with Eq. (21.61) and apply it for several different choices of the external field ϕ. By expressing the change in G in terms of the tp propagator, it is possible to rewrite the latter in terms of the four-point vertex function. Using Eq. (21.39) to replace G_{II} in Eq. (21.61), we find

$$\delta G^\phi(\alpha, \overline{\beta}; t - t') \tag{21.70}$$
$$= \int dt'' \sum_{\gamma\overline{\delta}} \delta\phi_{\gamma\overline{\delta}}(t'') G(\alpha, \gamma; t - t'') G(\overline{\delta}, \overline{\beta}; t'' - t')$$
$$- i\hbar \int dt'' \sum_{\gamma\overline{\delta}} \delta\phi_{\gamma\overline{\delta}}(t'') \int dt_\epsilon \int dt_\zeta \int dt_\eta \int dt_\theta \sum_{\epsilon\overline{\zeta}\overline{\eta}\theta} G(\alpha, \epsilon; t - t_\epsilon)$$
$$\times G(\overline{\delta}, \overline{\zeta}; t'' - t_\zeta) G(\overline{\eta}, \overline{\beta}; t_\eta - t') G(\theta, \gamma; t_\theta - t'') \left\langle \epsilon\eta^{-1} \middle| \Gamma \middle| \theta\zeta^{-1} \right\rangle,$$

where the ph notation of Eq. (13.20) has been used for the matrix element of Γ and the latter's time dependence is suppressed. We will now specialize to the momentum sp basis appropriate for infinite systems. Making relevant substitutions, we take advantage of the fact that the sp propagators are diagonal in momentum and spin variables. Employing the momentum variables introduced in Sec. 14.2, we obtain

$$\delta G^\phi(\boldsymbol{p} + \boldsymbol{Q}/2\, m_\alpha, \boldsymbol{p} - \boldsymbol{Q}/2\, \overline{m}_\beta; t - t') \tag{21.71}$$
$$= \int dt'' \left\langle \boldsymbol{p} + \boldsymbol{Q}/2\, m_\alpha \middle| \delta\phi(t'') \middle| \boldsymbol{p} - \boldsymbol{Q}/2\, \overline{m}_\beta \right\rangle$$
$$\times G(\boldsymbol{p} + \boldsymbol{Q}/2; t - t'') G(\boldsymbol{p} - \boldsymbol{Q}/2; t'' - t')$$
$$- i\hbar \int dt'' \sum_{\boldsymbol{p}' m_\gamma \overline{m}_\delta} \left\langle \boldsymbol{p}' + \boldsymbol{Q}/2\, m_\gamma \middle| \delta\phi(t'') \middle| \boldsymbol{p}' - \boldsymbol{Q}/2\, \overline{m}_\delta \right\rangle$$
$$\times \int dt_\epsilon \int dt_\zeta \int dt_\eta \int dt_\theta G(\boldsymbol{p} + \boldsymbol{Q}/2; t - t_\epsilon) G(\boldsymbol{p}' - \boldsymbol{Q}/2; t'' - t_\zeta)$$
$$\times G(\boldsymbol{p} - \boldsymbol{Q}/2; t_\eta - t') G(\boldsymbol{p}' + \boldsymbol{Q}/2; t_\theta - t'') \left\langle \boldsymbol{Q}, \boldsymbol{p}; \alpha\beta^{-1} \middle| \Gamma \middle| \boldsymbol{Q}, \boldsymbol{p}'; \gamma\delta^{-1} \right\rangle,$$

assuming a spin-$\frac{1}{2}$ fermion system. Note that we have also used conservation of the total ph momentum by Γ to arrive at this result.

The first choice for the external field is given by

$$\hat{\phi}(t) \Rightarrow \sum_{\alpha\bar{\beta}} \langle \alpha | \, \delta\phi(t) \, | \bar{\beta} \rangle \, a^{\dagger}_{\alpha} a_{\bar{\beta}} = \delta\phi(t) \hat{N}, \qquad (21.72)$$

where we consider an infinitely small field $\delta\phi(t)$ which is homogeneous in space and varies slowly in time. The matrix element of $\delta\phi$ reads

$$\langle \boldsymbol{p} + \boldsymbol{Q}/2 \, m_{\alpha} | \, \delta\phi(t'') \, | \boldsymbol{p} - \boldsymbol{Q}/2 \, \overline{m}_{\beta} \rangle = \delta_{\boldsymbol{Q},0} \delta\phi(t) \langle m_{\alpha} | \overline{m}_{\beta} \rangle \qquad (21.73)$$
$$= \delta_{\boldsymbol{Q},0} \delta\phi(t) (-1)^{\frac{1}{2} - m_{\alpha}} \delta_{m_{\alpha}, -m_{\beta}},$$

using the definition of the time-reversed spin state in Eq. (13.3). Inserting Eq. (21.73) in the second term of Eq. (21.71), generates a ph state with good total spin on the ket side of Γ, since

$$\sum_{m_{\gamma} m_{\delta}} | \tfrac{1}{2} \, m_{\gamma} \, (\tfrac{1}{2} m_{\delta})^{-1} \rangle \, (-1)^{\frac{1}{2} - m_{\gamma}} \delta_{m_{\gamma}, -m_{\delta}} = \sqrt{2} | \tfrac{1}{2} \, \tfrac{1}{2}^{-1} S = 0 M_S = 0 \rangle,$$
$$(21.74)$$

using Eq. (B.12). Since Γ conserves the total spin, the bra side of the interaction will also have $S = 0$, generating a corresponding Clebsch-Gordan coefficient. Together with the $\sqrt{2}$ factor in Eq. (21.74) the latter produces the same phase factor and spin Kronecker δ as in Eq. (21.73), which is therefore common to both terms in Eq. (21.71). We proceed to perform the transformation to energy variables, for example using

$$\delta\phi(t) = \int \frac{dE_{\phi}}{2\pi\hbar} \, e^{-\frac{i}{\hbar} E_{\phi} t} \, \delta\phi(E_{\phi}), \qquad (21.75)$$

and similar expressions for the sp propagators G and the interaction Γ (see Sec. 9.4). Employing Eq. (9.25) and eliminating the common factors, the change in G under the influence of $\hat{\phi}$ can now be written as

$$\delta G = G(\boldsymbol{p}; E) \, G(\boldsymbol{p}; E - E_{\phi}) \, \delta\phi(E_{\phi})$$
$$- iG(\boldsymbol{p}; E) \, G(\boldsymbol{p}; E - E_{\phi}) \, \delta\phi(E_{\phi}) \int \frac{dE'}{2\pi} \int \frac{d^3 p'}{(2\pi\hbar)^3}$$
$$\times \, \langle \boldsymbol{p}; S = 0 | \, \Gamma^0 \, | \boldsymbol{p}'; S = 0 \rangle \, G(\boldsymbol{p}'; E') \, G(\boldsymbol{p}'; E' - E_{\phi}). \quad (21.76)$$

As in Sec. 14.6, the limit of Γ when $\boldsymbol{Q} \Rightarrow 0$ first is denoted by Γ^0. The last form of Eq. (21.72) shows that when this operator is added to the Hamiltonian of the system, the sp propagator will be multiplied by a factor $e^{-\frac{i}{\hbar} \delta\phi(t-t')}$ in the limit $\delta\phi \Rightarrow$ constant. In the energy formulation this implies that E is replaced by $E - \delta\phi$, which shows that

$$\frac{\delta G}{\delta \phi} \Rightarrow -\frac{\partial G}{\partial E}, \qquad (21.77)$$

when $E_\phi \Rightarrow 0$. Denoting the limit of $G(\boldsymbol{p}; E)\, G(\boldsymbol{p}; E - E_\phi)$ by $\{G^2(\boldsymbol{p}; E)\}_0$ when $E_\phi \Rightarrow 0$ and using Eq. (21.77), Eq. (21.76) is written as

$$\frac{\partial G}{\partial E} = -\{G^2(\boldsymbol{p}; E)\}_0 \left[1 - i \int \frac{dE'}{2\pi} \int \frac{d^3p'}{(2\pi\hbar)^3} \, \langle \boldsymbol{p} | \Gamma^0 | \boldsymbol{p}' \rangle \, \{G^2(\boldsymbol{p}'; E')\}_0 \right],$$
$$(21.78)$$

suppressing the label $S = 0$. If we consider G near the Fermi energy and Fermi momentum, as in Eq. (14.93), we may rewrite Eq. (21.78) upon division by $\{G^2(\boldsymbol{p}; E)\}_0$ as

$$\frac{\partial G^{-1}}{\partial E} = \frac{1}{Z_F} = 1 - i \int \frac{dE'}{2\pi} \int \frac{d^3p'}{(2\pi\hbar)^3} \, \langle \boldsymbol{p} | \Gamma^0 | \boldsymbol{p}' \rangle \, \{G^2(\boldsymbol{p}'; E')\}_0, \quad (21.79)$$

implying that the magnitude of $\boldsymbol{p} \Rightarrow p_F$ and $E \Rightarrow \varepsilon_F$. By performing similar steps for different types of external perturbations, other useful relations between quasiparticle properties and the vertex function can be generated. We refer to [Abrikosov *et al.* (1975); Lifshitz and Pitaevskii (1980)] for more details and list these relations below. For a field that does not depend on time, no energy is transferred. When the particles have fictitious infinitesimal charge, the presence of a magnetic field that has a small spatial but no time dependence, leads to the substitution of the momenta of the particles from $\boldsymbol{p} \Rightarrow \boldsymbol{p} - \delta e \boldsymbol{A}/c$ and results in a small change in the Hamiltonian. Such a field transfers an infinitesimal momentum and no energy. When this momentum is taken to zero, it yields

$$\frac{\delta G}{\delta e \boldsymbol{A}/c} = -\frac{\partial G}{\partial \boldsymbol{p}}. \qquad (21.80)$$

Repeating the same steps as before, one arrives near the Fermi momentum and energy at the following identity

$$\nabla_{\boldsymbol{p}} G^{-1} = -\frac{\boldsymbol{p}}{m^* Z_F} \qquad (21.81)$$

$$= -\frac{\boldsymbol{p}}{m} + i \int \frac{dE'}{2\pi} \int \frac{d^3p'}{(2\pi\hbar)^3} \, \langle \boldsymbol{p} | \Gamma^\infty | \boldsymbol{p}' \rangle \, \frac{\boldsymbol{p}'}{m} \{G^2(\boldsymbol{p}'; E')\}_\infty,$$

where we have denoted the limit of Γ when first the energy and than the momentum is taken to zero, so $\boldsymbol{Q}/E_t \Rightarrow \infty$ in the notation of Sec. 14.6, by

Γ^∞. This notation is introduced for the product of the two sp propagators as well.

If the change in the sp propagator is considered when the system as a whole moves with a small, slowly varying velocity $\delta\boldsymbol{u}(t)$, the Hamiltonian acquires an additional term $-\delta\boldsymbol{u}\cdot\hat{\boldsymbol{P}}$, where $\hat{\boldsymbol{P}}$ is the total momentum operator of the system. The resulting identity for such a field, with the energy transfer and $\delta\boldsymbol{u}\Rightarrow 0$, is given by

$$\boldsymbol{p}\frac{\partial G^{-1}}{\partial E} = \frac{\boldsymbol{p}}{Z_F} \tag{21.82}$$

$$= \boldsymbol{p} - i\int\frac{dE'}{2\pi}\int\frac{d^3p'}{(2\pi\hbar)^3}\,\langle\boldsymbol{p}|\,\Gamma^0\,|\boldsymbol{p}'\rangle\,\boldsymbol{p}'\{G^2(\boldsymbol{p}';E')\}_0.$$

Finally, a small field that is constant in time and weakly inhomogeneous in space requires the invocation of the equilibrium condition involving the chemical potential $\mu + \delta\phi = 0$, leading to the final identity

$$\frac{\partial G^{-1}}{\partial\mu} = 1 - i\int\frac{dE'}{2\pi}\int\frac{d^3p'}{(2\pi\hbar)^3}\,\langle\boldsymbol{p}|\,\Gamma^\infty\,|\boldsymbol{p}'\rangle\,\{G^2(\boldsymbol{p}';E')\}_\infty, \tag{21.83}$$

valid for all momenta. We note that all the fields generate $S = 0$ ph states.

We are now in a position to verify Eqs. (14.141) and (14.142). In preparation we rewrite Eq. (14.126) in the limit that first $E_t \Rightarrow 0$ and then $\boldsymbol{Q}$ as follows,

$$\langle\boldsymbol{p}|\,\Gamma^\infty\,|\boldsymbol{p}'\rangle = \langle\boldsymbol{p}|\,\Gamma^0\,|\boldsymbol{p}'\rangle \tag{21.84}$$

$$-\frac{p_F Z_F^2}{m^*}\int\frac{d\hat{\boldsymbol{p}}_t}{(2\pi\hbar)^3}\,\langle\boldsymbol{p}|\,\Gamma^0\,|p_F\hat{\boldsymbol{p}}_t\rangle\,\langle p_F\hat{\boldsymbol{p}}_t|\,\Gamma^\infty\,|\boldsymbol{p}'\rangle\,,$$

adjusting the notation to the present situation and noting that the magnitude of both $\boldsymbol{p}$ and $\boldsymbol{p}'$ will be assumed equal to p_F. We also note that Eqs. (14.120) and (14.121) lead to the following relation between the opposite limits of G^2

$$\{G^2(\boldsymbol{p};E)\}_\infty = -\frac{2m^*\pi Z_F^2}{p_F}\,\delta(E - \varepsilon_F)\delta(|\boldsymbol{p}| - p_F) + \{G^2(\boldsymbol{p};E)\}_0. \tag{21.85}$$

The last two equations are useful vehicles to manipulate the four identities that were obtained for various external fields.

We proceed by inserting Eq. (21.84)[1] into Eq. (21.81), thereby

---

[1] Note that this equation holds for arbitrary values of $\boldsymbol{p}$ and $\boldsymbol{p}'$.

generating

$$-\frac{p}{m^* Z_F} + \frac{p}{m} = +i \int \frac{dE'}{2\pi} \int \frac{d^3 p'}{(2\pi\hbar)^3} \langle p| \Gamma^0 |p'\rangle \frac{p'}{m} \{G^2(p'; E')\}_\infty$$

$$+i \int \frac{dE'}{2\pi} \int \frac{d^3 p'}{(2\pi\hbar)^3} \left\{ -\frac{p_F Z_F^2}{m^*} \right\} \tag{21.86}$$

$$\times \int \frac{d\hat{p}_t}{(2\pi\hbar)^3} \langle p| \Gamma^0 |p_F\hat{p}_t\rangle \langle p_F\hat{p}_t| \Gamma^\infty |p'\rangle \frac{p'}{m} \{G^2(p'; E')\}_\infty$$

$$= +i \int \frac{dE'}{2\pi} \int \frac{d^3 p'}{(2\pi\hbar)^3} \langle p| \Gamma^0 |p'\rangle \frac{p'}{m} \{G^2(p'; E')\}_\infty$$

$$- \frac{p_F Z_F^2}{m^*} \int \frac{d\hat{p}_t}{(2\pi\hbar)^3} \langle p| \Gamma^0 |p_F\hat{p}_t\rangle \langle p_F\hat{p}_t| \left\{ \frac{-p_F\hat{p}_t}{m^* Z_F} + \frac{p_F\hat{p}_t}{m} \right\},$$

where the magnitude of p and p' are equal to p_F and Eq. (21.81) was employed in rewriting the second term after the last equality. Substitution of Eq. (21.85) into the first term yields, after some cancellations and minor rearrangement,

$$\frac{1}{m} = \frac{1}{m^*} + p_F Z_F^2 \int \frac{d\hat{p}_t}{(2\pi\hbar)^3} \langle p| \Gamma^0 |p_F\hat{p}_t\rangle \hat{p}_t. \tag{21.87}$$

Using Eq. (14.129) and the defintion of the Landau parameters, combined with the orthogonality of the Legendre polynomials, finally generates the relation between the effective mass and the Landau parameter F_1

$$\frac{m^*}{m} = 1 + \frac{F_1}{3}, \tag{21.88}$$

which was employed in Sec. 14.6.

To obtain the relation between the speed of sound and the Landau parameter F_0, we first note that the smooth dependence of Fermi velocity v_F and the pole strength Z_F on the chemical potential, or Fermi energy, allows us to rewrite the derivative of the left-hand side of Eq. (21.83) near the pole as follows

$$\frac{\partial G^{-1}}{\partial \mu} \Rightarrow \frac{p_F}{m^* Z_F} \frac{dp_F}{d\mu}. \tag{21.89}$$

Employing the right side of Eq. (21.83), while replacing Γ^∞ by Eq. (21.84) and using Eq. (21.85) as before, yields after some rearrangement

$$\frac{p_F}{m^*} \frac{dp_F}{d\mu} = \left[1 + m^* p_F Z_F^2 \int \frac{d\hat{p}_t}{(2\pi\hbar)^3} \langle p| \Gamma^0 |p_F\hat{p}_t\rangle \right]^{-1}. \tag{21.90}$$

Noting that, in general, the derivative of the density can be written as

$$\frac{dN/V}{d\mu} = \frac{8\pi p_F^2}{(2\pi\hbar)^3}\frac{dp_F}{d\mu}, \qquad (21.91)$$

we may establish that Eq. (21.90) is equivalent to the following expression for the speed of sound

$$c_1^2 = \frac{N}{m}\frac{\partial\mu}{\partial N} = \frac{p_F^2}{3m^2}\frac{1+F_0}{1+\frac{1}{3}F_1}, \qquad (21.92)$$

where the first equality represents a well-known thermodynamic relation for an infinite system. This concludes our discussion of the relation between the first two Fermi liquid parameters and important observables of an infinite Fermi system.

21.4   Examples of conserving approximations

We will now consider in some detail the relation between the conserving polarization propagator and specific choices for the self-consistent approximations to the self-energy that have been presented in previous chapters.

21.4.1   *Hartree–Fock and the RPA approximation*

The simplest self-consistent approximation to the sp propagator or self-energy produces the HF approximation. The self-consistent HF self-energy is given in the time formulation by

$$\Sigma^{HF}(\alpha,\bar{\beta};t-t') = -i\hbar\delta(t-t')\sum_{\gamma\bar{\delta}}\langle\alpha\bar{\delta}|\,V\,|\bar{\beta}\gamma\rangle\,G^{HF}(\gamma,\bar{\delta};t-t^+). \quad (21.93)$$

This result can be obtained from the expression for the first-order term in Fig. 8.4 by replacing the propagator by G^{HF}. The self-consistent HF solution of the Dyson equation with the self-energy of Eq. (21.93) generates the HF propagator. In the corresponding HF sp basis, discussed in Ch. 10, the propagator reads, after FT,

$$G^{HF}(\alpha,\bar{\beta};E) = \delta_{\alpha\bar{\beta}}\left[\frac{\theta(\alpha-F)}{E-\varepsilon_\alpha^{HF}+i\eta} + \frac{\theta(F-\alpha)}{E-\varepsilon_\alpha^{HF}-i\eta}\right]. \qquad (21.94)$$

The application of Eq. (21.69) to the HF self-energy yields the appropriate irreducible ph interaction given by

$$\Gamma^{ph}_{HF}(\alpha t_1, \beta^{-1} t_2, \gamma t_3, \delta^{-1} t_4) = \frac{\delta \Sigma^{HF}(\alpha, \overline{\beta}; t_1 - t_2)}{\delta G^{HF}(\gamma, \overline{\delta}; t_3 - t_4)}$$

$$= -i\hbar \delta(t_1 - t_2)\delta(t_1 - t_3)\delta(t_1 - t_4)\left\langle \alpha \overline{\delta} \right| V \left| \overline{\beta} \gamma \right\rangle. \qquad (21.95)$$

Insertion of this ph interaction in Eq. (21.67), generates

$$\Pi^{RPA}(\alpha, \beta^{-1}; \gamma, \delta^{-1}; t_1 - t_2) = \Pi^{(0)}(\alpha, \beta^{-1}; \gamma, \delta^{-1}; t_1 - t_2)$$

$$+ \sum_{\epsilon\zeta\eta\theta} \int dt_3\ \Pi^{(0)}(\alpha, \beta^{-1}; \epsilon\zeta^{-1}; t_1 - t_3)$$

$$\times \left\langle \epsilon \overline{\theta} \right| V \left| \overline{\zeta} \eta \right\rangle \Pi^{RPA}(\eta, \theta^{-1}; \gamma\delta^{-1}; t_3 - t_2), \qquad (21.96)$$

where

$$\Pi^{(0)}(\alpha, \beta^{-1}, \gamma, \delta^{-1}; t_1 - t_2) = -i\hbar G^{HF}(\alpha, \gamma; t_1 - t_2)G^{HF}(\overline{\delta}, \overline{\beta}; t_2 - t_1). \qquad (21.97)$$

After applying the usual FT, we recognize the RPA version of the polarization propagator, which can be written as

$$\Pi^{RPA}(\alpha, \beta^{-1}; \gamma, \delta^{-1}; E) = \Pi^{(0)}(\alpha, \beta^{-1}; \gamma, \delta^{-1}; E) \qquad (21.98)$$

$$+ \Pi^{(0)}(\alpha, \beta^{-1}; E) \sum_{\eta\theta} \left\langle \alpha\beta^{-1} \right| V_{ph} \left| \eta\theta^{-1} \right\rangle \Pi^{RPA}(\eta, \theta^{-1}; \gamma, \delta^{-1}; E),$$

where the definition of the ph interaction of Eq. (13.20) was used. The noninteracting polarization propagator in the HF basis

$$\Pi^{(0)}(\alpha, \beta^{-1}; \gamma, \delta^{-1}; E) \qquad (21.99)$$

$$= \delta_{\alpha,\gamma}\delta_{\beta,\delta} \left\{ \frac{\theta(\alpha - F)\theta(F - \beta)}{E - (\varepsilon^{HF}_\alpha - \varepsilon^{HF}_\beta) + i\eta} - \frac{\theta(F - \alpha)\theta(\beta - F)}{E + (\varepsilon^{HF}_\beta - \varepsilon^{HF}_\alpha) - i\eta} \right\},$$

was also employed to generate Eq. (21.98) in the same form as Eq. (13.23) in Ch. 13. The unique link between the RPA and the HF approximation is therefore clearly demonstrated. The results of Sec. 13.5 for excited states in atoms should thus be viewed in the present context.

21.4.2 Second-order self-energy and the particle-hole interaction

The self-consistent treatment of the second-order self-energy was documented *e.g.* in Sec. 11.5.3 for atoms. The second-order contribution to the

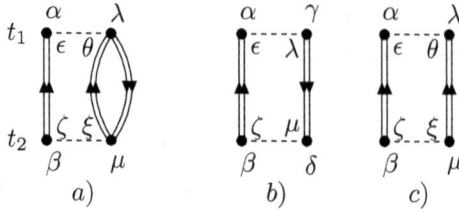

Fig. 21.3 Diagrammatic representation of the self-consistent second-order self-energy in a). The contributions to the irreducible ph interaction that are obtained by the functional derivative of a), with respect to the self-consistent sp propagator $G^{(2)}$, are given by diagrams b) and c).

self-energy in nuclei, employing the G-matrix interaction, was presented in Sec. 17.1.1. The relevant self-energy diagram is shown in Fig. 21.3. The diagram is given in the time formulation by

$$\Sigma^{(2)}(\alpha,\beta;t_1-t_2) = -(i\hbar)^2 \frac{1}{2} \sum_{\lambda\epsilon\theta}\sum_{\zeta\xi\mu} \langle\alpha\lambda|\,V\,|\epsilon\theta\rangle\,\langle\zeta\xi|\,V\,|\beta\mu\rangle$$

$$G^{(2)}(\epsilon,\zeta;t_1-t_2)G^{(2)}(\theta,\xi;t_1-t_2)G^{(2)}(\mu,\lambda;t_2-t_1), \quad (21.100)$$

where $G^{(2)}$ is the self-consistent propagator (up to second order). As illustrated explicitly in Fig. 11.1, the HF contribution is also incorporated in the self-energy. We will only consider the new contributions to the irreducible ph interaction when the second-order diagram of Fig. 21.3 is included. By applying Eq. (21.69) to Eq. (21.100), we generate the extra interaction terms

$$\Gamma^{ph}_{(2)}(\alpha t_1, \beta^{-1}t_2, \gamma t_3, \delta^{-1}t_4)$$

$$= -(i\hbar)^2 \delta(t_1-t_3)\delta(t_2-t_4) \sum_{\lambda\epsilon\zeta\mu} \langle\alpha\lambda|\,V\,|\epsilon\gamma\rangle\,\langle\zeta\bar\delta|\,V\,|\bar\beta\mu\rangle$$

$$\times G^{(2)}(\epsilon,\zeta;t_1-t_2)G^{(2)}(\mu,\lambda;t_2-t_1)$$

$$- (i\hbar)^2 \frac{1}{2}\delta(t_2-t_3)\delta(t_1-t_4) \sum_{\epsilon\theta\zeta\xi} \langle\alpha\bar\delta|\,V\,|\epsilon\theta\rangle\,\langle\zeta\xi|\,V\,|\bar\beta\gamma\rangle$$

$$\times G^{(2)}(\epsilon,\zeta;t_1-t_2)G^{(2)}(\theta,\xi;t_1-t_2). \quad (21.101)$$

Two of the three contributions from the functional derivative of Eq. (21.100) are identical, and the factor of $\frac{1}{2}$ only survives in the last term of Eq. (21.101). The interaction associated with Eq. (21.101) is displayed in Fig. 21.3. Diagram a) represents the first term and b) the second one. In this approximation the equation for the polarization propagator becomes

$$\Pi^{(2)}(\alpha\beta^{-1};\gamma,\delta^{-1};t_1-t_2) = \Pi^f(\alpha,\beta^{-1};\gamma,\delta^{-1};t_1-t_2)$$

$$+ \frac{i}{\hbar}\sum_{\epsilon\zeta\eta\theta}\int dt_3 \int dt_4 \int dt_5 \int dt_6 \Pi^f(\alpha t_1,\beta^{-1}t_1;\epsilon t_3,\zeta^{-1}t_4)$$

$$\left[\Gamma^{ph}_{HF}(\epsilon t_3,\zeta^{-1}t_4,\eta t_5,\theta^{-1}t_6) + \Gamma^{ph}_{(2)}(\epsilon t_3,\zeta^{-1}t_4,\eta t_5,\theta^{-1}t_6)\right]$$

$$\times \Pi^{(2)}(\eta t_5,\theta^{-1}t_6;\gamma t_2,\delta^{-1}t_2). \tag{21.102}$$

This equation is not yet ready for practical application, but it leads to an important extension of the RPA method.

21.4.3 *Extension of the RPA including second-order terms*

The time dependence of Eq. (21.102) is still quite formidable and we will now proceed to a manageable form, which takes the major new physical ingredient of the second-order self-energy into account [Brand *et al.* (1990)]. The latter corresponds to the fragmentation of the sp strength that becomes inevitable when there is substantial mixing between hole (or particle) states with 2h1p (or 2p1h) states. For a system where this fragmentation is important, it is unavoidable that the description of excited states requires the mixing between 1p1h and 2p2h states. It is this physical ingredient that will now be incorporated in an approximate treatment of Eq. (21.102).

By employing the HF sp basis, it is possible to write the first iteration of the Dyson equation towards the self-consistent propagator $G^{(2)}$ in the following manner (see also Ch. 11)

$$G^{(2)}(\alpha,\beta;t_1-t_2) = G^{HF}(\alpha,\beta;t_1-t_2) \tag{21.103}$$

$$+ \sum_{\gamma\delta}\int dt_3 \int dt_4\, G^{HF}(\alpha,\gamma;t_1-t_3)\Sigma^{(2)}(\gamma,\delta;t_3-t_4)G^{(2)}(\delta,\beta;t_4-t_2),$$

using HF sp propagators in the self-energy. The result of this approximation is to transform Eq. (21.102) into an expansion in which only iterations of Γ^{ph}_{HF}, $\Gamma^{ph}_{(2)}$, and $\Sigma^{(2)}$ according to Eq. (21.103) occur, all expressed in HF sp propagators. Iteration of Eq. (21.102) using only Γ^{ph}_{HF}, will generate the usual RPA diagrams. The extra Feynman diagrams up to second order that are generated in this scheme from Eq. (21.102), are shown in Fig. 21.4. Switching to the energy formulation, the contribution of the

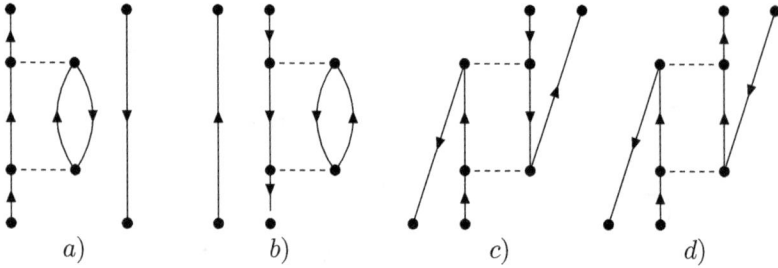

Fig. 21.4 Feynman diagrams for the polarization propagator up to second order in the interaction. The self-energy diagrams are identified in $a)$ and $b)$. The so-called ph screening diagram is displayed in $c)$. The ladder diagram is illustrated in $d)$. The sp lines represent the HF sp propagators. The sum of diagrams $a) - d)$ defines the irreducible ph interaction $\Gamma_{(2)}^{ERPA}$.

diagrams $a) - d)$ to the ph propagator Π can be written as

$$\Pi^{(2*)}(\alpha, \beta^{-1}; \gamma, \delta^{-1}; E) \qquad (21.104)$$
$$= \Pi^{(0)}(\alpha, \beta^{-1}; \alpha, \beta^{-1}; E)\Gamma_{(2*)}^{ph}(\alpha, \beta^{-1}; \gamma, \delta^{-1}; E)\Pi^{(0)}(\gamma, \delta^{-1}; \gamma, \delta^{-1}; E),$$

employing a generalized meaning (and notation) for the irreducible interaction, which now also contains self-energy terms. We thus define the irreducible ph interaction $\Gamma_{(2*)}^{ph}$. Irreducible here implies that it cannot be separated into successive ph interactions by cutting a pair of ph lines *at the same time*. Note that Eq. (21.104) contains terms that include all possible connections between forward and backward-going contributions present in $\Pi^{(0)}$. By iterating the ph interaction $\Gamma_{(2*)}^{ph}$ together with the usual ph interaction that generates the RPA, we arrive at the following approximation to Eq. (21.102)

$$\Pi^{(2)}(ab^{-1}; cd^{-1}; J; E) = \Pi^{(0)}(ab^{-1}; cd^{-1}; J; E) + \sum_{ef} \Pi^{(0)}(ab^{-1}; J; E)$$

$$\times \left\{ \langle ab^{-1}J | V_{ph} | ef^{-1}J \rangle + \Gamma_{(2*)}^{ph}(ab^{-1}; ef^{-1}; J; E) \right\}$$

$$\times \Pi^{(2)}(ef^{-1}; cd^{-1}; J; E), \qquad (21.105)$$

where we have applied the angular momentum coupling that was developed in Sec. 13.7. The equation takes the form of an inhomogeneous RPA equation with an energy-dependent interaction. The polarization propagator $\Pi^{(2)}$ can also be derived from the equation of motion method [Takayanagi *et al.* (1988)]. Besides the usual RPA contributions, it takes into account the coupling to antisymmetrized 2p2h states for the excited states (through forward-going terms) as well as for the ground state (backward-going terms), and the coupling of the particle or the hole to a 2p2h admixture in the ground state (self-energy contributions).

The diagram expansion implied by Eq. (21.105), represents only a sub-series of the terms generated by Eq. (21.102). The latter will also produce irreducible ph interactions which are of higher order in the ph vertex $\Gamma^{ph}_{(2)}$, *i.e.* they cannot be partitioned by cutting a particle and a hole line *at the same time*. They can be thought of, *e.g.*, as linking subsequent $\Gamma^{ph}_{(2)}$ interactions by means of two hole or two particle lines that do not return the polarization propagator to its simple ph component. Such higher-order irreducible ph interactions describe the coupling of the 1p1h state to more complicated configurations, in which the 2p2h state couples to npnh states with n≥3. These contributions to the ph interaction are expected to be suppressed because, due to the gap in the sp spectrum of a closed-shell system, there will be a large energy mismatch between the initial 1p1h state and the intermediate propagating npnh state (n≥3). At low energy, the coupling of 1p1h to 2p2h excitations will dominate, whereas at higher excitation energies typically no coherent excitations are observed. More-over, in this energy region it is probably most important to consider the small admixture of 1p1h states into the abundantly present 2p2h states. In practical applications [Brand *et al.* (1990)] such higher-order terms have therefore been neglected.

A remaining complication in Eqs. (21.102) – (21.105) is that one should employ the fully dressed sp propagator $G^{(2)}$ everywhere. Dressing the dia-grams of $\Gamma^{ph}_{(2*)}$ means replacing the HF sp propagators G^{HF} by $G^{(2)}$. How-ever, the second-order self-energy causes a fragmentation of the sp strength, which leads to very large numbers of 1p1h and 2p2h states in practical real-izations. Furthermore, $G^{(2)}$ no longer provides a sharp distinction between particle ($> F$) and hole ($< F$) states. Consequently, the dressed diagrams of $\Gamma^{ph}_{(2*)}$ contribute to the pp and hh amplitudes of the polarization prop-agator, requiring their explicit consideration. Such calculations have not been reported so far.

In the treatment of the sp propagator one may neglect the fragmentation of the sp strength and restrict the effect of the second-order self-energy to a shift of the HF sp energy, approximating $G^{(2)}$ by the single pole expression

$$G^{(2*)}(\alpha, \beta; E) = \delta_{\alpha\beta} \left[\frac{\theta(\alpha - F)}{E - \varepsilon^{(2*)}_\alpha + i\eta} + \frac{\theta(F - \alpha)}{E - \varepsilon^{(2*)}_\alpha - i\eta} \right]. \qquad (21.106)$$

These quasiparticle energies $\varepsilon_\alpha^{(2*)}$ are defined in a self-consistent way as the poles with the largest fragment in the sp propagator that satisfies the Dyson equation for the diagonal second-order self-energy. This means that the $\varepsilon_\alpha^{(2*)}$ correspond to the solutions with the largest strength of the equation

$$\varepsilon_\alpha^{(2*)} = \varepsilon_\alpha^{HF} + \Sigma^{(2*)}(\alpha, \alpha; E = \varepsilon_\alpha^{(2*)}). \qquad (21.107)$$

The single-pole expression of Eq. (21.106) maintains the sharp distinction between particle and hole states, so that when the HF propagators in $\Gamma_{(2*)}^{ph}(E)$ are replaced by Eq. (21.106), again only the ph and hp amplitudes have to be considered. This dressed ph interaction leads to what we will call the extended RPA (ERPA) equation, represented by Eq. (21.105) with the appropriate substitutions discussed here. We note that the independent propagation of a particle and a hole is described by the iteration of diagrams a) and b) of Fig. 21.4 in Eq. (21.105). This incorporates the desired properties that the largest ph strength is associated with the corresponding ph energy, while at the same time a fragmented ph spectrum is generated due to the coupling to 2p2h states in the self-energy terms.

We have now arrived at a conserving extension of the RPA in which the coupling to 2p2h states is explicitly accounted for, while allowing for a shift of the 1p1h and 2p2h energies with respect to the HF energies through the action of the second-order self-energy. The latter coupling can also be accomplished by calculating the matrix element between a ph state with a properly antisymmetrized 2p2h state, propagating the intermediate 2p2h state, and then returning to a ph state, which is potentially different from the original one. When comparing with the contributions from the four diagrams in Fig. 21.4, it becomes clear that **all** four terms are required to account for the Pauli principle, which reduces the number of possible 2p2h states. This illustrates the role of conserving approximations to the polarization propagator. In the present approximation one encounters fragmentation of ph strength from iterating only the self-energy terms, which cannot be described in the standard 1p1h RPA. It is this feature plus the additional interaction terms, iterated in Eq. (21.105), that lead to a treatment of sp properties and response beyond a mean-field description.

21.4.4   *Practical ingredients of ERPA calculations*

We start by discussing the procedure to solve the ERPA equations in a discrete basis, emphasizing the proper normalization of the transition amplitudes and strength functions. Equation (21.105) can be solved in the usual way by explicitly calculating the poles of $\Pi^{(2)}$. Inserting the Lehmann representation, given by Eq.(13.87), with the ph states coupled to good total angular momentum, we multiply Eq. (21.105) by $E - \varepsilon^\pi_{nJ}$ and take the limit $E \Rightarrow \varepsilon^\pi_{nJ}$. The inhomogeneous term in Eq. (21.105) vanishes and an RPA-like eigenvalue problem emerges as in Eq. (13.53), which we write as

$$
\begin{pmatrix} A(E) & B \\ B^* & A^*(E) \end{pmatrix} \begin{pmatrix} \mathcal{X}^{nJ}_{ph} \\ \mathcal{X}^{nJ}_{hp} \end{pmatrix} = \varepsilon^\pi_{nJ}(E) \begin{pmatrix} 1 & 0 \\ 0 & -1 \end{pmatrix} \begin{pmatrix} \mathcal{X}^{nJ}_{ph} \\ \mathcal{X}^{nJ}_{hp} \end{pmatrix}, \qquad (21.108)
$$

to be solved subject to the condition

$$
E = \varepsilon^\pi_{nJ}(E). \qquad (21.109)
$$

The ERPA sub-matrices contain the new ingredients and read

$$
A(E) \Rightarrow \langle p_1 h_2^{-1} J | V + \Gamma^{ph}_{(2*)}(E) | p_3 h_4^{-1} J \rangle + (\varepsilon^{HF}_{p_1} - \varepsilon^{HF}_{h_2}) \delta_{1,3} \delta_{2,4} \qquad (21.110)
$$

$$
B \Rightarrow \langle p_1 h_2^{-1} J | V + \Gamma^{ph}_{(2*)} | h_4 p_3^{-1} J \rangle \qquad (21.111)
$$

$$
A^*(E) \Rightarrow \langle h_2 p_1^{-1} J | V + \Gamma^{ph}_{(2*)}(E) | h_4 p_3^{-1} J \rangle + (\varepsilon^{HF}_{p_1} - \varepsilon^{HF}_{h_2}) \delta_{1,3} \delta_{2,4} \qquad (21.112)
$$

$$
B^* \Rightarrow \langle h_2 p_1^{-1} J | V + \Gamma^{ph}_{(2*)} | p_3 h_4^{-1} J \rangle. \qquad (21.113)
$$

Explicit formulas for all diagrams contained in $\Gamma^{ph}_{(2*)}$ are presented here for completeness [Brand *et al.* (1990)]. We have adopted the notation $p_1, p_2,$ *etc.* for sp states above the Fermi level and $h_1, h_2,$ *etc.* for sp states below. These indices represent a complete set of quantum numbers, for example given by $n_1, \ell_1, j_1, m_1,$ and m_{t_1}. Coupling to good total angular momentum is indeed appropriate since it reduces the number of states that needs to be considered and is essential in distinguishing the different types of excitation modes, as will be illustrated in Sec. 21.5. The contribution to $\Gamma^{ph}_{(2*)}$ of the ph screening diagram of Fig. 21.4c) to the ERPA A-matrix is given by

$$
\langle p_1 h_2^{-1} J | \Gamma^{ph}_{(2s)}(E) | p_3 h_4^{-1} J \rangle = \sum_{J'} \langle 12, 34; J | 13, 24; J' \rangle
$$

$$
\times \sum_{p_5 h_6} \left[\frac{\langle p_1 p_3^{-1} J' | V | p_5 h_6^{-1} J' \rangle \langle p_5 h_6^{-1} J' | V | h_2 h_4^{-1} J' \rangle}{E - (\varepsilon^{HF}_3 + \varepsilon^{HF}_5 - \varepsilon^{HF}_2 - \varepsilon^{HF}_6) + i\eta} \right.
$$

$$
\left. + \frac{\langle p_1 p_3^{-1} J' | V | h_6 p_5^{-1} J' \rangle \langle h_6 p_5^{-1} J' | V | h_2 h_4^{-1} J' \rangle}{E - (\varepsilon^{HF}_1 + \varepsilon^{HF}_5 - \varepsilon^{HF}_4 - \varepsilon^{HF}_6) + i\eta} \right]. \qquad (21.114)
$$

The angular momentum recoupling factor is discussed in App. B and reads for the present situation

$$\langle 12, 34; J | 13, 24; J' \rangle = -(-1)^{j_2 + j_3 + J + J'} (2J' + 1) \begin{Bmatrix} j_1 & j_2 & J \\ j_4 & j_3 & J' \end{Bmatrix}. \quad (21.115)$$

For closed-shell nuclei with $N = Z$, it is convenient to employ the isospin formulation. The inclusion of isospin in constructing the recoupling factor then yields a result similar to Eq. (21.115) and can be incorporated straightforwardly. The contribution of the ph screening diagram to the ERPA B-matrix is represented by

$$\langle p_1 h_2^{-1} J | \Gamma_{(2s)}^{ph} | h_3 p_4^{-1} J \rangle = \sum_{J'} \langle 12, 34; J | 13, 24; J' \rangle$$

$$\times \sum_{p_5 h_6} \left[\frac{\langle p_1 h_3^{-1} J' | V | p_5 h_6^{-1} J' \rangle \langle p_5 h_6^{-1} J' | V | h_2 p_4^{-1} J' \rangle}{-(\varepsilon_4^{HF} + \varepsilon_5^{HF} - \varepsilon_2^{HF} - \varepsilon_6^{HF})} \right.$$

$$\left. + \frac{\langle p_1 h_3^{-1} J' | V | h_6 p_5^{-1} J' \rangle \langle h_6 p_5^{-1} J' | V | h_2 p_4^{-1} J' \rangle}{-(\varepsilon_1^{HF} + \varepsilon_5^{HF} - \varepsilon_3^{HF} - \varepsilon_6^{HF})} \right]. \quad (21.116)$$

The contributions from the ph screening diagram to the ERPA A^* and B^* matrices follow from the symmetry relations [Hengeveld *et al.* (1986)]

$$\langle h_2 p_1^{-1} J | \Gamma_{(2s)}^{ph}(E) | h_4 p_3^{-1} J \rangle =$$

$$(-1)^{j_1 + j_2 + j_3 + j_4} \langle p_1 h_2^{-1} J | \Gamma_{(2s)}^{ph}(-E) | p_3 h_4^{-1} J \rangle \quad (21.117)$$

and

$$\langle h_2 p_1^{-1} J | \Gamma_{(2s)}^{ph}(E) | p_4 h_3^{-1} J \rangle =$$

$$(-1)^{j_1 + j_2 + j_3 + j_4} \langle p_1 h_2^{-1} J | \Gamma_{(2s)}^{ph}(E) | h_3 p_4^{-1} J \rangle. \quad (21.118)$$

The contribution from the ladder diagram of Fig. 21.4d) to the ERPA A-matrix is given by the expression

$$\langle p_1 h_2^{-1} J | \Gamma_{(2l)}^{ph}(E) | p_3 h_4^{-1} J \rangle = \sum_{J'} [12, 34; J | 13, 24; J']$$

$$\times \left\{ \sum_{p_5 p_6} \frac{\langle p_1 h_4 J' | V | p_5 p_6 J' \rangle \langle p_5 p_6 J' | V | p_3 h_2 J' \rangle}{E - (\varepsilon_5^{HF} + \varepsilon_6^{HF} - \varepsilon_2^{HF} - \varepsilon_4^{HF}) + i\eta} \right.$$

$$\left. + \sum_{h_5 h_6} \frac{\langle p_1 h_4 J' | V | h_5 h_6 J' \rangle \langle h_5 h_6 J' | V | p_3 h_2 J' \rangle}{E - (\varepsilon_1^{HF} + \varepsilon_3^{HF} - \varepsilon_5^{HF} - \varepsilon_6^{HF}) + i\eta} \right\}. \quad (21.119)$$

The contribution to the ERPA B-matrix from the ladder diagram of Fig. 21.4d) has the form:

$$\langle p_1 h_2^{-1} J | \Gamma_{(2l)}^{ph} \langle h_3 p_4^{-1} J | = \sum_{J'} [12, 34; J | 13, 24; J']$$

$$\times \left\{ \sum_{p_5 p_6} \frac{\langle p_1 p_4 J' | V | p_5 p_6 J' \rangle \langle p_5 p_6 J' | V | h_3 h_2 J' \rangle}{-(\varepsilon_5^{ph} + \varepsilon_6^{HF} - \varepsilon_2^{HF} - \varepsilon_3^{HF})} \right.$$

$$\left. + \sum_{h_5 h_6} \frac{\langle p_1 p_4 J' | V | h_5 h_6 J' \rangle \langle h_5 h_6 J' | V | h_3 h_2 J' \rangle}{-(\varepsilon_1^{HF} + \varepsilon_4^{HF} - \varepsilon_5^{HF} - \varepsilon_6^{HF})} \right\}. \tag{21.120}$$

The contributions to the ERPA A^* and B^*-matrix can be found by applying Eqs. (21.117) and (21.118). The angular momentum recoupling factor for the present case produces

$$[12, 34; J | 13, 24; J'] = -(2J' + 1) \begin{Bmatrix} j_1 & j_2 & J \\ j_3 & j_4 & J' \end{Bmatrix}. \tag{21.121}$$

The self-energy diagrams of Figs. 21.4a) and b) will be dealt with next. The expression for the contribution to the A-matrix when the self-energy couples to the particle line, yields

$$\langle p_1 h_2^{-1} J | \Gamma_{(2p)}^{ph}(E) | p_1 h_2^{-1} J \rangle = \frac{1}{2} \sum_{J'} \frac{2J' + 1}{2j_2 + 1}$$

$$\times \left\{ \sum_{p_4 p_5 h_6} \frac{\langle p_1 h_6 J' | V | p_4 p_5 J' \rangle^2}{E - (\varepsilon_4^{HF} + \varepsilon_5^{HF} - \varepsilon_2^{HF} - \varepsilon_6^{HF}) + i\eta} \right.$$

$$\left. + \sum_{h_4 h_5 p_6} \frac{\langle p_1 p_6 J' | V | h_4 h_5 J' \rangle^2}{\varepsilon_1^{HF} - (\varepsilon_4^{HF} + \varepsilon_5^{HF} - \varepsilon_6^{HF})} \right\}. \tag{21.122}$$

The expression for the contribution to the A-matrix when the self-energy couples to the hole line is given by:

$$\langle p_1 h_2^{-1} J | \Gamma_{(2h)}^{ph}(E) | p_1 h_2^{-1} J \rangle = \frac{1}{2} \sum_{J'} \frac{2J' + 1}{2j_2 + 1}$$

$$\times \left\{ \sum_{h_4 h_5 p_6} \frac{\langle h_2 p_6 J' | V | h_4 h_5 J' \rangle^2}{E - (\varepsilon_1^{HF} + \varepsilon_6^{HF} - \varepsilon_4^{HF} - \varepsilon_5^{HF}) + i\eta} \right.$$

$$\left. + \sum_{p_4 p_5 h_6} \frac{\langle h_2 h_6 J' | V | p_4 p_5 J' \rangle^2}{\varepsilon_2^{HF} - (\varepsilon_4^{HF} + \varepsilon_5^{HF} - \varepsilon_6^{HF})} \right\}. \tag{21.123}$$

The contribution to the ERPA A^*-matrix can be found by employing Eqs. (21.117) and (21.118).

The pole structure of the A and A^*-matrix, due to the intermediate 2p2h propagation (the B-matrices are energy independent), ensures also that solutions are obtained corresponding to predominantly 2p2h states. The dimension of the matrix that one diagonalizes is however still the same as in the standard RPA. In order to satisfy the condition of Eq. (21.109), these diagonalizations have to be performed for various values of E. The amplitudes $\mathcal{X}$ can be normalized by manipulating the inhomogeneous equation (21.105) in the usual way:

$$\sum_{p_1 h_2} |\mathcal{X}^{nJ}_{p_1 h_2}|^2 - \sum_{h_2 p_1} |\mathcal{X}^{nJ}_{h_2 p_1}|^2 = 1$$

$$+ \sum_{abcd} \mathcal{X}^{nJ*}_{ab} \mathcal{X}^{nJ}_{cd} \left\langle ab^{-1}J \right| \frac{d\Gamma^{ph}_{(2*)}(E)}{dE} \Bigg|_{E=\varepsilon^{\pi}_{nJ}} \left| cd^{-1}J \right\rangle . \qquad (21.124)$$

This shows explicitly that it is necessary for $\Gamma^{ph}_{(2*)}$ to have the correct energy dependence (containing all diagrams of Fig. 21.4) to generate the proper normalization of the excitation amplitudes. All transition amplitudes to the excited states can thus be calculated, yielding besides excitation energies, transition densities as well as cross sections. The resulting transition strength distribution can be written as

$$S^J_Q(E) = \frac{1}{2J+1} \sum_n \left| \sum_{ph} \langle p \parallel Q^J \parallel h \rangle \left\{ \mathcal{X}^{nJ}_{ph} + \mathcal{X}^{nJ}_{hp} \right\} \right|^2 \delta(E - \varepsilon^{\pi}_{nJ}),$$

$$(21.125)$$

where Q_{JM} represents an external field with multipolarity (J, M).

The density of 2p2h states, and thus the number of solutions, increases rapidly with energy so that for the calculation of high-lying excitations like the giant resonances in nuclei, the above method, although still feasible, becomes very involved. In this region a second method can be employed to calculate a continuous strength distribution directly from the polarization propagator. This is accomplished by making the energy complex: $E \Rightarrow E + i\Delta$. The value $\Delta = 0.25$ MeV can be adopted as a reasonable choice for closed-shell nuclei [Brand et al. (1990)]. In addition, a width of 1 MeV for the 2p2h states may be introduced, which can be interpreted to represent the spreading of these states into more complicated states. The polarization propagator can then be calculated directly from Eq. (21.105) by matrix inversion:

$$\Pi^{(2)}(E) = \left[\left(\Pi^{(0)}(E) \right)^{-1} - V - \Gamma^{ph}_{(2*)}(E) \right]^{-1}, \tag{21.126}$$

and the strength distribution is obtained from

$$S^J_Q(E) = -\frac{1}{\pi} \mathrm{Im} \sum_{abcd} \left[\langle a \parallel Q^J \parallel b \rangle^* \Pi^{(2)}(ab^{-1}; cd^{-1}; J; E) \langle c \parallel Q^J \parallel d \rangle \right]. \tag{21.127}$$

The width of 1 MeV is sufficiently large to yield a smooth strength distribution and sufficiently small so as not to influence it significantly. With a complete set of conserving diagrams for $\Gamma^{ph}_{(2*)}$ the response function is positive definite. These two complementary methods have been used to study both the low-lying discrete states as well as the strength distribution at higher energies in closed-shell nuclei, further discussed in Sec. 21.5.

21.4.5 *Ring diagram approximation and the polarization propagator*

The description of collective states by means of the RPA suggests that such correlations should be included in the self-energy. Examples are provided by the plasmon excitation in the electron gas, discussed in Sec. 16.2.1 and 16.2.2, and low-lying collective states in nuclei, as studied in Sec. 17.1.2. When such phenomena dominate the properties of the tp propagator, the resulting self-energy contains the sum of all ring diagrams. We will consider here the case where only the direct contributions to the RPA are included, as is appropriate for the electron gas. Keeping the notation general however, allows possible applications to finite systems. The HF contribution to the self-energy will be included together with the other ring diagram terms of the self-energy. The additional self-energy diagrams for constructing the new ingredients of the irreducible ph interaction are given by Figs. 17.2e) and f), provided self-consistent sp propagators are employed. For the electron gas this self-energy contribution can be found in Eq. (16.38). Using the general notation of the present chapter, we can write for this term

$$\Delta\Sigma^R(\alpha, \beta; t_1 - t_2) = i\hbar \sum_{\lambda\epsilon\nu} \sum_{\zeta\xi\mu} (\alpha\lambda|V|\epsilon\theta)(\zeta\xi|V|\beta\mu)$$
$$G(\epsilon, \zeta; t_1 - t_2)\Pi^{RPA}(\theta, \lambda^{-1}; \xi, \mu^{-1}; t_1 - t_2), \tag{21.128}$$

noting that the factor $\frac{1}{2}$ disappears when only direct contributions are included. The construction of the irreducible ph interaction from Eq. (21.128) employs yet again Eq. (21.69). Its application requires the formulation of Π^{RPA} according to Eq. (21.96) to exhibit its dependence on the self-consistent G, while omitting the exchange contribution from the two-body matrix element of the interaction. We note that, as before, the application of the functional derivative with respect to G is actually equivalent to cutting one of the internal G lines in the self-energy diagram. For the present case, two types of contributions are generated. The first involves the term where the G not involved in Π^{RPA} is cut. For the other, a G inside Π^{RPA} must be cut, giving rise to two separate terms. The algebra is not complicated but somewhat tedious.[2] The final result for the conserving polarization propagator is given by the following expression

$$
\begin{aligned}
\Pi(\alpha t_1, \beta^{-1}t_2; \gamma t_3^+, \delta^{-1}t_3) &= -i\hbar G(\alpha,\gamma;t_1-t_3)G(\delta,\beta;t_3-t_2) \\
&- i\hbar \sum_{\epsilon\zeta\eta\theta} \int dt_4\ G(\alpha,\epsilon;t_1-t_3)G(\zeta,\beta;t_3-t_2) \\
&\times [(\epsilon\theta|V|\zeta\eta) - (\epsilon\theta|V|\eta\zeta)]\,\Pi(\eta t_4, \theta^{-1}t_4, \gamma t_3^+, \delta^{-1}t_3) \\
&- i\hbar \sum_{\epsilon\zeta\eta\theta} \int dt_4 \int dt_5 G(\alpha,\epsilon;t_1-t_4)G(\zeta,\beta;t_5-t_2) \\
&\times \sum_{\rho\tau\xi\phi}(\epsilon\rho|V|\eta\tau)\Pi^{RPA}(\tau,\rho^{-1};\xi,\phi^{-1};t_4-t_5)(\theta\xi|V|\zeta\phi) \\
&\times \Pi(\eta t_3, \theta^{-1}t_4; \gamma t_3^+, \delta^{-1}t_3) \\
&+ \hbar \sum_{\epsilon\zeta\eta\theta} \int dt_4 \int dt_5 \int dt_6 \int dt_7 G(\alpha,\epsilon;t_1-t_4)G(\zeta,\beta;t_5-t_2) \\
&\times \Big\{ \sum_{\rho\sigma\kappa\lambda}(\epsilon\rho|W(t_4-t_6)|\sigma\eta)G(\sigma,\kappa;t_4-t_5) \\
&\times G(\lambda,\rho;t_7-t_6)(\kappa\theta|W(t_7-t_5)|\zeta\lambda) \\
&+ \sum_{\sigma\lambda\kappa\rho}(\epsilon\theta|W(t_4-t_7)|\sigma\lambda)G(\sigma,\kappa;t_4-t_5) \\
&\times G(\lambda,\rho;t_7-t_5)(\kappa\rho|W(t_6-t_5)|\zeta\eta)\Big\}\,\Pi(\eta t_6, \theta^{-1}t_7; \gamma t_3^+, \delta^{-1}t_3),
\end{aligned}
$$

$$(21.129)$$

---

[2] We note that the external field dependence of G must be included before the functional derivative is taken and ϕ is put to zero. In the case of the electron gas, this restores the presence of the bare Coulomb interaction in the irreducible ph interaction since the Hartree self-energy contribution must be included.

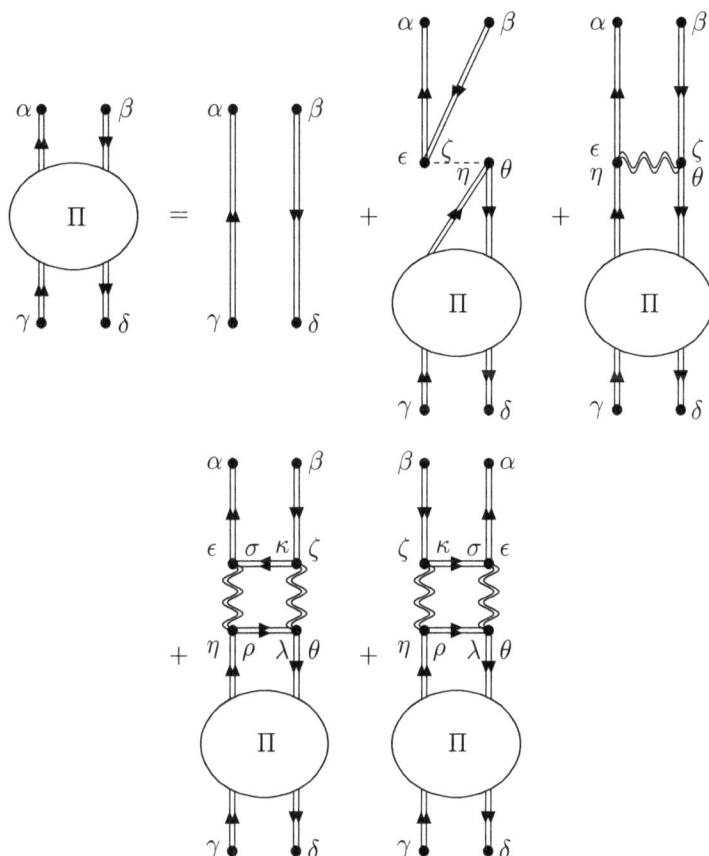

Fig. 21.5   Diagrammatic representation of Eq. (21.129). The use of self-consistent propagators is indicated by the double-line notation.

where the screened interaction

$$(\mu\lambda|W(t_1 - t_2)|\sigma\tau) = (\mu\lambda|V|\sigma\tau)\delta(t_1 - t_2)$$
$$+ \sum_{\rho\epsilon\eta\pi} (\mu\rho|V|\sigma\pi)\Pi^{RPA}(\pi, \rho^{-1}; \epsilon, \eta^{-1}; t_1 - t_2)(\epsilon\lambda|V|\eta\tau) \quad (21.130)$$

has been introduced in analogy to the screened Coulomb interaction W. As it stands, it appears that Eq. (21.129) is quite intractable. Its physical content becomes more distinct when the diagrammatic structure is viewed, as in Fig. 21.5. The last term in the top row of the figure combines the exchange of the bare interaction term with the next one in Eq. (21.129), resulting in the screened interaction W, indicated by the doubly wiggled

line. Two such screened interactions contribute to the last two terms in
Eq. (21.129). Figure 21.5 demonstrates that the self-consistent inclusion
of ring diagrams in the self-energy leads to two new ingredients of the
irreducible ph interaction. The first involves the exchange term of the
screened interaction W. For the electron gas one may therefore loosely
speak about the exchange of a plasmon-like excitation between the particle
and the hole. In this language the last two diagrams in Fig. 21.5 can be
interpreted as the coupling of the ph propagation to two such excitations
representing the second new feature. We note that the construction of the
irreducible interaction guarantees that the f-sum rule is now conserved. For
the electron gas this should lead to the restoration of the physical plasmon
excitation in the polarization propagator. We are not aware of an actual
(approximate) implementation of such a scheme.

The other self-consistent approximation to the self-energy includes the
summation of all ladder diagrams, as discussed in Sec. 16.3.3. The construction of the irreducible ph interaction proceeds by calculating the functional
derivative of this self-energy. The irreducible interaction contains the ph
form of the ladder-summed effective interaction plus the coupling to two
such interactions. The latter can be considered as the coupling to a simultaneous pp and a hh excitation. More details can be found in [Baym
and Kadanoff (1961)]. Also in this case, we are not aware of an actual
application.

21.5 Excited states in nuclei

We conclude this chapter with a brief overview of some of the properties of
excited states in nuclei. In order to appreciate the complexities of nuclear
spectra, it is helpful to consider the information that is contained in the
noninteracting polarization propagator of Eq. (21.68). Taking $t = t'$ and
performing the transformation to the energy formulation yields

$$\Pi^f(\alpha, \beta^{-1}; \gamma, \delta^{-1}; E) \tag{21.131}$$
$$= \sum_{n,k} \frac{\langle \Psi_0^A | a_{\bar{\beta}}^\dagger | \Psi_k^{A-1}\rangle \langle \Psi_0^A | a_\alpha | \Psi_n^{A+1}\rangle \, \langle \Psi_n^{A+1} | a_\gamma^\dagger | \Psi_0^A \rangle \langle \Psi_k^{A-1} | a_{\bar{\delta}} | \Psi_0^A\rangle}{E - (\varepsilon_n^+ - \varepsilon_k^-) + i\eta}$$
$$- \sum_{n,k} \frac{\langle \Psi_0^A | a_\gamma^\dagger | \Psi_k^{A-1}\rangle \langle \Psi_0^A | a_{\bar{\delta}} | \Psi_n^{A+1}\rangle \, \langle \Psi_n^{A+1} | a_{\bar{\beta}}^\dagger | \Psi_0^A \rangle \langle \Psi_k^{A-1} | a_\alpha | \Psi_0^A\rangle}{E + (\varepsilon_n^+ - \varepsilon_k^-) - i\eta},$$

Oxygen-16

Excitation Energy [MeV]

$\pi+\ \ T=0\ \ \pi-$ $T=1$ $A\pm1\ data$

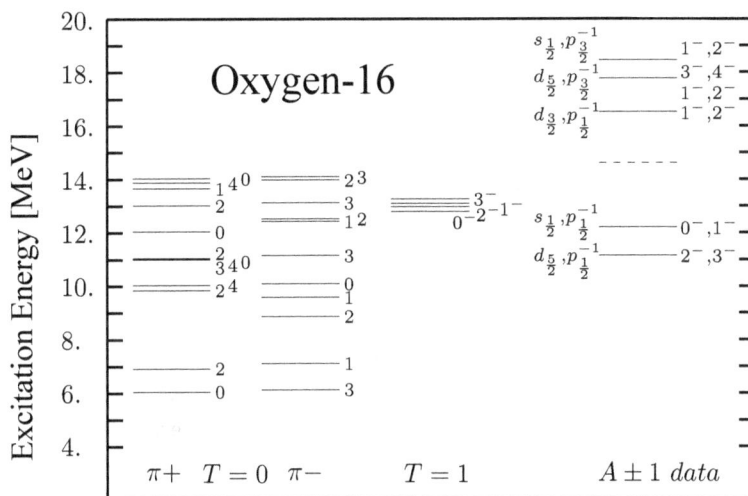

Fig. 21.6 Energy levels for the ^{16}O nucleus below 14 MeV excitation energy. The first column displays positive parity states with isospin $T=0$. The second column contains isoscalar negative parity states and the third column exhibits all isovector states. The last column employs information from neighboring $A\pm1$ nuclei to identify the poles of Π^f associated with the main sp shell states for this nucleus. The dashed line corresponds to the first pole of Π^f associated with more complicated configurations in the $A\pm1$ spectra. Data were taken from [Tilley *et al.* (1993)].

where the Lehmann representations of the individual sp propagators has been employed. The notation of Eqs. (10.23) and (10.25) has been used to identify the energy of states with one particle removed or added, with respect to the energy of the ground state of A particles. The latter state will be presumed to correspond to a closed-shell nucleus, as identified in the discussion of Sec. 3.3. Figure 21.6 illustrates the excitation spectrum of the doubly-closed shell nucleus ^{16}O up to 14 MeV, identifying the isoscalar positive and negative parity states in the first two columns, respectively. All $T=1$ states below 14 MeV are shown in the third column. The last column identifies the poles of Eq. (21.131) as they are obtained from the experimental information of the neighboring odd nuclei. The possible configurations include the $p_{\frac{1}{2}}$ and $p_{\frac{3}{2}}$ hole states combined with the $d_{\frac{5}{2}}, s_{\frac{1}{2}},$ and $d_{\frac{3}{2}}$ particle ones. These sp states correspond to large spectroscopic factors as identified by appropriate nucleon transfer reactions. The possible angular momentum and parity of these ph states are also indicated. The dashed line in the last column identifies the first pole in Π^f that is associated with smaller-strength fragments in the $A + 1$ system.

The spectrum of Fig. 21.6 is not easy to understand from the perspective of a mean-field description. Only the isovector states in the third column can easily be identified with the four states at slightly lower energy in the last column. In the RPA based on the HF approximation, the latter states can obviously be accounted for if the ph interaction is slightly repulsive. Since HF for nuclei must be replaced by at least a BHF approach, it is useful to summarize the successes and failures of the RPA for the excited states when a G-matrix ph interaction is employed [Czerski *et al.* (1986)]. Similar results have been found with phenomenological effective ph interactions. The description of the isovector states clustered around 13 MeV excitation energy is quite satisfactory, indicating that the G-matrix has the right tendency to move these states up from their unperturbed position that is based on the experimental information from the $A \pm 1$ nuclei. The RPA diagonalization of the isovector states does not lead to any significant collective enhancements, except in the 1^- case. A strong collective state is calculated around 23 MeV, documenting a substantial repulsion in the interaction, which is capable of mixing ph states in a way that is quite similar to the schematic model from Sec. 13.3. Unfortunately, while in this region there is substantial electric dipole strength observed experimentally [Dolbilkin *et al.* (1965)], it is strongly fragmented unlike the result of the RPA calculations. Other negative parity isovector states exhibit similar fragmentation of the transition strength, like the magnetic quadrupole, that cannot be obtained within the RPA framework. For isoscalar states, the G-matrix ph interaction generates one low-lying collective 3^- state but completely fails to account for all the low-lying states that are observed. This is true for the negative parity states, but even more conspicuously for the positive parity states that in the mean-field approach are only available as ph excitations that jump two major shells and are therefore not found below 20 MeV excitation energy. The only positive conclusion here is that the interaction has the correct sign for the isoscalar states with natural parity.[3]

The failure of the RPA should not come as a surprise when the fragmentation of the sp strength is taken into account. The experimental results, discussed in Sec. 7.8, and the theoretical ones of Sec. 17.1.2, leave no doubt that a dynamical description of the nucleon self-energy is necessary to produce the pattern of the sp strength. Following the presentation

---

[3]These states have quantum numbers $0^+, 1^-, 2^+, 3^-$, *etc.*

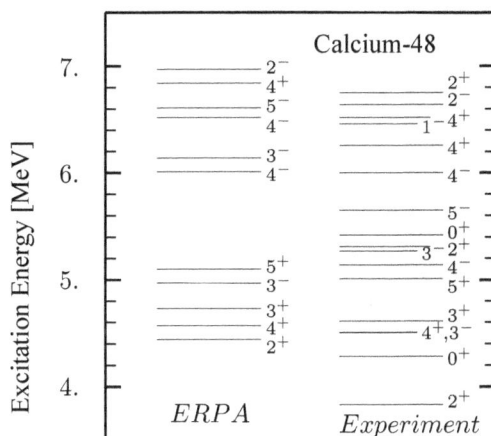

Fig. 21.7   Energy levels for the ^{48}Ca nucleus below 7 MeV excitation energy. The first column displays the results of the ERPA calculation. The second contains the experimental data taken from [Wise *et al.* (1985); Fujita *et al.* (1988); Lederer and Shirley (1978)].

of this chapter, it is thus necessary to include higher-order terms to the self-energy, which then automatically yield an effective ph interaction that takes the consequences of this fragmentation into account in the treatment of excited states. For nuclei this has been accomplished in the form of the ERPA method, as developed in Secs. 21.4.3 and 21.4.4. It is based on the second-order self-energy approximation in the G-matrix interaction and has been applied to the closed-shell nucleus ^{48}Ca and the semi-magic nucleus ^{90}Zr in [Brand *et al.* (1990)]. The latter nucleus exhibits some of the pairing properties that were discussed in Sec. 17.2.

Another deficiency of the RPA method that employs a G-matrix ph interaction, is the development of instabilities in heavier nuclei for 2^+ and 3^- states. For ^{48}Ca this occurs for the first 3^- state. A gratifying feature of the ERPA method is the disappearance of the instability, as shown by the presence of this state in Fig. 21.7 in reasonable agreement with experiment (note that the scale starts at 4 MeV excitation energy). While the level density in the ERPA at low energy is still smaller than the experimental one, the overall agreement with the angular momentum and parity of the excited states is much better than for the RPA. In general, the low-lying states are better accounted for than in the RPA calculation. The latter feature can be nicely illustrated by considering electromagnetic transition

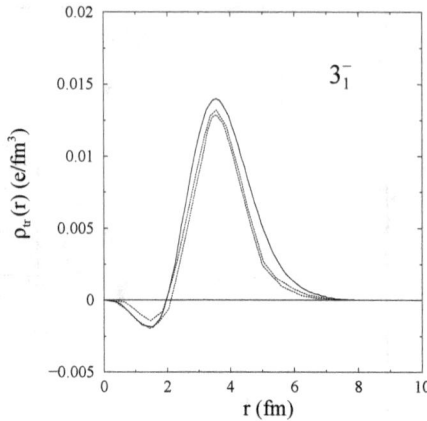

Fig. 21.8 Electromagnetic transition density from the ground state to the first 3^- in ^{48}Ca. The dotted lines indicate the experimental uncertainty. Data are adapted from [Wise *et al.* (1985)]. The solid line shows the ERPA result. Note that the RPA method generates an instability for this state.

densities.

The latter can be extracted from inelastic electron scattering to discrete final states [Heisenberg and Blok (1983)]. Electromagnetic transition densities (and currents) provide extremely useful information about the character of the excited states. To elucidate their sensitive character, it is helpful to write the second-quantized density operator in the following way

$$\hat{\rho}(\boldsymbol{r}) = \sum_{JM} \sum_{n\ell j} i^{\ell_2 - \ell_1} u_{n_1 \ell_1 j_1}(r) u_{n_2 \ell_2 j_2}(r) Y^*_{JM}(\hat{\boldsymbol{r}}) (-1)^{j_2 - \frac{1}{2} + J}$$

$$\times \frac{[\hat{j}_1][\hat{j}_2]}{[4\pi]^{\frac{1}{2}}} \begin{pmatrix} j_1 & j_2 & J \\ \frac{1}{2} & \frac{1}{2} & 0 \end{pmatrix} \left[a^{\dagger}_{n_1 \ell_1 j_1} \otimes b^{\dagger}_{n_2 \ell_2 j_2} \right]^J_M . \tag{21.132}$$

The multipole decomposition employs results and notation from Sec. 13.7 and App. B. The critical ingredients of the decomposition are the sp wave functions $u_{n\ell j}$ that determine the radial shape of the matrix element of this operator between the ground state and the excited state under study. For a simple ph state coupled to good total angular momentum, Eq. (21.132) is proportional to the product of the sp wave functions that comprise the ph state.

The transition density for the first 3^- state in ^{48}Ca is shown in Fig. 21.8. The two dotted lines indicate the uncertainty in the experimental data [Wise *et al.* (1985)]. Since the RPA is unstable, only the ERPA curve

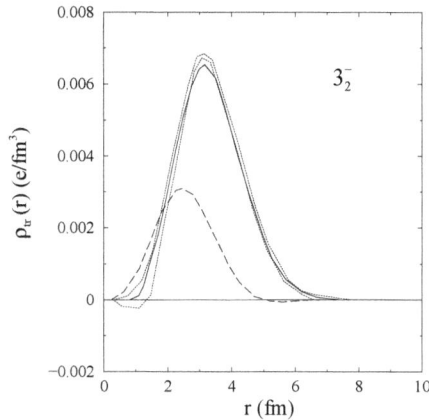

Fig. 21.9   Electromagnetic transition density from the ground state to the second 3^- in ^{48}Ca. The dotted lines indicate the experimental uncertainty. Data are adapted from [Wise *et al.* (1985)]. The solid line represents the ERPA calculation. The RPA (dashed line) underestimates the experimental data substantially.

is given, which describes the data reasonably well. The transition to the second 3^- state is illustrated in Fig. 21.9. Clearly, the RPA calculation (dashed line) for this transition density underestimates the experimental data. This outcome is not unexpected since the transition strength collects in the lowest state of the same J^π for an attractive interaction and little strength remains for other states. The ERPA method corrects for this deficiency by including the fragmentation of the sp strength, which generates the good agreement with the data, as illustrated by the solid line. Similar improvements are observed for other low-lying states when the ERPA is applied [Brand *et al.* (1990)].

The coupling to 2p2h states incorporated in the ERPA is also essential for the description of the response at higher energy. At those energies several nuclear excitation modes exist that are referred to as "giant resonances" [van der Woude (1987)]. These collective modes exhaust substantial fractions of the relevant EWSR given in Eq. (13.50) and their energy and width exhibit a smooth dependence on particle number. The giant dipole resonance (GDR) emerged when the absorption of high-energy photons was studied [Baldwin and Klaiber (1947)]. The GDR is a common feature of all nuclei and for heavier nuclei the absorption cross section has the shape of a Lorentzian. The resonance energy exhibits a simple A-dependence given by $E_{GDR} = 47.9A^{-1/4.27}$ [Berman and Fultz (1975)].

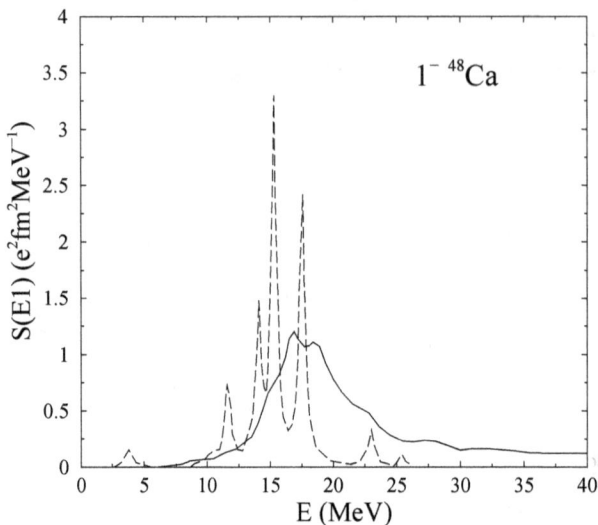

Fig. 21.10   Strength function for the electric dipole resonance in ^{48}Ca. RPA results are represented by the dashed line and are divided by a factor of 2. The ERPA calculation is illustrated by the solid line.

The width varies from 4 MeV in ^{208}Pb to 7 MeV in ^{65}Cu. An explanation of this phenomenon was provided by [Goldhaber and Teller (1948)] as a collective, dipole-like, vibration of protons against neutrons. In microscopic language, the resonance results from the coherent interplay of a substantial number of ph excitations which contribute to this $\Delta T = 1, J^\pi = 1^-$ excitation. The location of the resonance is about double the energy difference between major nuclear shells, and must therefore be associated with a repulsive ph interaction.

A typical RPA description of the GDR for ^{48}Ca is illustrated in Fig. 21.10 by the dashed line. The strength distribution is calculated by applying Eq. (21.127) when the RPA problem is solved as an inhomogeneous equation by assigning the unperturbed energies a small width. The unperturbed ph strength is concentrated around 11.5 MeV and is moved to 15 MeV by the RPA correlations. A further push to 19 MeV in the ERPA description puts the center of the strength distribution at the energy observed in neighboring nuclei and in accord with systematics. The calculated width of about 6 MeV also compares favorably with empirical information. The energy-weighted dipole strength summed up to 30 MeV yields 125%

of the classical sum rule given by $S(E1)_{class} = 14.8NZ/A$ e^{2}fm^2 [Bohr and Mottelson (1998)]. The enhancement is due to the more complicated nature of the nuclear interaction, invalidating the simple result corresponding to Eq. (13.65). In the ERPA calculation a reduction occurs to 102% of the classical value. The experimental value in ^{51}V only yields 73% suggesting that too much dipole strength is found below 30 MeV. This feature can be identified for other giant resonances like the electric quadrupole and the charge-exchange Gamow–Teller mode as well. The main distribution of the strength is in agreement with experiment but the ERPA overestimates it by 20–30%. The calculations of [Brand *et al.* (1990)] did not include the depletion effect for the sp strength due to short-range correlations that was discussed in Sec. 17.3. Since such depletions easily remove at least 10% of the strength of both the relevant particle and hole strength, their combined effect matches the discrepancy between the ERPA strength and experiment quite well. Thus, consistency is achieved with the observed and calculated sp strength of this nucleus discussed in Sec. 17.1.

The conclusion may therefore be drawn that heavier closed-shell nuclei, like ^{48}Ca, are somewhat easier to describe than the lighter $N = Z$ ones like ^{16}O. Both the sp strength and the response for the latter nucleus require a more sophisticated treatment of the self-energy and the resulting ph interaction. A first attempt at incorporating both ph and pp/hh collective phenomena in the self-energy has been developed in [Barbieri and Dickhoff (2001, 2002)]. Such an approach still yields an overestimate of the spectroscopic factors in ^{16}O but improves their overall description. The inclusion of ph phonons in the self-energy may be employed to construct irreducible ph interaction terms like those in Fig. 21.5. The presence of two-phonon coupling in the ph interaction improves the excitation spectrum of ^{16}O by introducing positive parity states of the two-phonon kind at lower excitation energy [Barbieri and Dickhoff (2003)]. Such states can be thought of as a combination of two negative parity (microscopic) RPA phonons. This development appears a promising step towards a deeper understanding of the excitation spectrum shown in Fig. 21.6.

We conclude with a brief comment on the extraction of removal probabilities and spectroscopic factors in the $(e, e'p)$ reaction. First note that the reaction predominantly acts as a one-body excitation operator to nuclear excited states. The preceding discussion of excited states of nuclei clarifies that collective correlations occur at small excitation energy, associated with 2^+ and 3^- surface modes and at somewhat higher energy as giant resonances. Theoretical and experimental studies at even higher energy suggest that no further coherence of ph states and corresponding transition strength is to be expected. It can thus be concluded that the dominant contribution to the polarization propagator at these energies is of the form given by Eq. (21.131). The latter involves dressed but noninteracting particle and hole components. It should be clear that an appropriate choice of the kinematics of the $(e, e'p)$ reaction (including high enough excitation energy), enables the analysis of the reaction in terms of a specific hole transition amplitude to a discrete final $A - 1$ state, which can be selected by the coincidence set-up, and a particle amplitude that describes the addition of a proton at high energy. The latter term corresponds to the elastic scattering of protons from the target nucleus. By employing empirical information concerning this process, constrained by such elastic scattering data, it becomes possible to extract accurate information on the removal amplitude of valence nucleons. The resulting analysis [Sick and de Witt Huberts (1991); Pandharipande *et al.* (1997)] is capable of generating absolute spectroscopic factors for nuclei with errors of about 5%.

21.6 Exercises

(1) Perform the FT of the terms displayed in Fig. 21.4 to demonstrate the validity of Eq. (21.104).
(2) Calculate the second-order term that links an initial ph state with a final (different) ph state by going through an intermediate properly antisymmetrized 2p2h state. Show that this procedure is equivalent to the A-matrix contribution of $\Gamma^{ph}_{(2*)}$. Compare also with the separate contributions from diagrams $a) - d)$ of Fig. 21.4.
(3) Check Eqs. (21.114) – (21.123).
(4) Construct the irreducible ph interaction when the self-energy includes all the direct contributions of the ring diagrams, which contain self-consistent sp propagators. Verify Eq. (21.129).
(5) Construct the irreducible ph interaction from the previous exercise by cutting one of the self-consistent propagators in the self-energy in all possible ways.
(6) Calculate the functional derivative with respect to G of the self-energy that includes all ladder diagrams containing such self-consistent propagators G. Perform this feat diagrammatically by cutting G in the self-energy in all possible ways.

Chapter 22

Pairing phenomena

In this chapter the extension of the Green's function formalism to superfluid (superconducting) systems is presented. The ground state of these systems can be characterized by a macroscopic occupation of a specific pair state. Such a possibility had already been anticipated in the treatment of Cooper pairing in Sec. 15.3. In Sec. 22.1 the general concept of condensation into a quasi-boson or fermion pair state is introduced. This has much in common with the phenomenon of Bose–Einstein condensation for identical bosons, and throughout this chapter we will emphasize the analogy. In Sec. 22.2 the Fermi gas is revisited and we introduce the concept of anomalous propagators. The perturbation expansion of the propagators, and the corresponding diagram rules, are derived in Sec. 22.3. In lowest-order this leads to the Bardeen–Cooper–Schrieffer (BCS) treatment of superconductivity, explained in Sec. 22.4. We rederive in Sec. 22.5 the BCS theory by employing the Bogoliubov–Valatin transformation to quasiparticles. This alternative formulation exposes the mean-field character and variational interpretation of BCS. Various applications, such as superconductivity in metals, the superfluid ^{3}He, and nucleon pairing in neutron stars, are briefly discussed in Sec. 22.6. In Sec. 22.7 the BCS formalism is expressed in terms of an arbitrary sp basis, allowing an extension to inhomogeneous settings. The chapter ends with a discussion of exactly solvable pairing models in Sec. 22.8, a recent topic of importance for finite systems.

22.1 General considerations

The normal perturbation expansion for identical fermions, as developed in Chs. 8 and 9, has a chance to converge only when the true ground state bears a reasonable resemblance to a noninteracting ground state (Slater

determinant). This is not guaranteed when the interparticle interaction is
sufficiently attractive. Section 15.3 already contained the notion that the
pair excitation energies in the Fermi gas [as determined from Eq. (15.46)]
can become complex, signaling an instability of the Fermi sea with respect
to the formation of the corresponding pair state (Cooper pairs). In such a
case we must drastically alter the starting point of our perturbation expansion, in order to accommodate pairing effects.

One can appreciate that such considerations hold quite generally by
contemplating the following Hamiltonian, symmetric under time reversal,

$$\hat{H} = \hat{H}_0 + \hat{V} = \sum_{\alpha>0} \varepsilon_\alpha (a_\alpha^\dagger a_\alpha + a_{\bar{\alpha}}^\dagger a_{\bar{\alpha}}) - gB^\dagger B, \qquad (22.1)$$

where $(\alpha, \bar{\alpha})$ are time-reversed states and $\sum_{\alpha>0}$ refers to a summation over
one member of each pair. In Eq. (22.1) the attractive interaction $\hat{V}$ is
assumed to be dominated by a single pair state, $B^\dagger = \sqrt{2}\sum_{\alpha>0} x_\alpha a_\alpha^\dagger a_{\bar{\alpha}}^\dagger$,
with $2\sum_{\alpha>0}|x_\alpha|^2 = 1$.

For $g = 0$ the sp Hamiltonian $\hat{H}_0$ gives rise to the noninteracting ground
state

$$|F\rangle = \prod_{h=1}^{N/2} a_h^\dagger a_{\bar{h}}^\dagger |0\rangle, \qquad (22.2)$$

where the orbitals with lowest ε are filled. For $g > 0$ the total energy for
a state $|\Psi^N\rangle$ has a negative interaction contribution, proportional to the
occupation $n_B = \langle \Psi^N| B^\dagger B |\Psi^N\rangle$ of the pair state $B^\dagger$. In some cases the
interaction term in Eq. (22.1) may become dominant, *e.g.* when the coupling
strength $g > 0$ increases, or the sp spectrum ε_α is nearly degenerate. It
is then obviously advantageous to make n_B large, by putting as many $B^\dagger$
pairs as possible in the ground state.[1] Consequently, the Fermi sea $|F\rangle$ is
a poor starting point, since it is plausible that the true ground state will
look more like a coherent superposition of the pair state $B^\dagger$, *i.e.*

$$|P\rangle \sim (B^\dagger)^{N/2} |0\rangle. \qquad (22.3)$$

As is usual for coherent phenomena, the gain in interaction energy going
from $|F\rangle$ to $|P\rangle$ can become very large and, in fact, proportional to the

---

[1]As discussed in more detail in Sec. 22.8, the pair occupation n_B cannot be larger
than $N/2$, and in general has an upper bound smaller than $N/2$ depending on the internal
structure of the pair state $B^\dagger$.

particle number. This is illustrated by the schematic example of a "structureless" pair state

$$B^\dagger = \sqrt{2/\Omega} \sum_{\alpha>0} a_\alpha^\dagger a_{\bar\alpha}^\dagger, \tag{22.4}$$

where Ω (with $\Omega \geq N$) is the total number of available sp states. It is easily shown that B, $B^\dagger$, and the number operator

$$\hat{N} = \sum_{\alpha>0} (a_\alpha^\dagger a_\alpha + a_{\bar\alpha}^\dagger a_{\bar\alpha}) \tag{22.5}$$

obey $SU(2)$-like commutation relations,

$$[B, B^\dagger] = 1 - \frac{2}{\Omega}\hat{N}, \qquad [\hat{N}, B^\dagger] = 2B^\dagger. \tag{22.6}$$

As a consequence, one can verify [see Exercise (1)] that, for integer n,

$$B(B^\dagger)^n |0\rangle = \sum_{i=1}^{n} (B^\dagger)^{i-1}[B, B^\dagger](B^\dagger)^{n-i} |0\rangle$$

$$= n\left[1 - \frac{2(n-1)}{\Omega}\right](B^\dagger)^{n-1} |0\rangle. \tag{22.7}$$

Considering the limit $N \to \infty$ with $N/\Omega \to 0$, the interaction energy in the paired state of Eq. (22.3) is given by

$$\frac{\langle P| B^\dagger B |P\rangle}{\langle P| P\rangle} \to \frac{N}{2}, \tag{22.8}$$

whereas for the Fermi sea $|F\rangle$ in Eq. (22.2) one finds $\langle F| B^\dagger B |F\rangle = N/\Omega \to 0$. In this case, the Fermi sea cannot be a meaningful starting point for a perturbative treatment of exact ground-state quantities.

The structure of $|P\rangle$ in Eq. (22.3) looks suspiciously like the Bose condensed state in Eq. (12.7), with the fermion pair operator $B^\dagger$ playing the role of the condensate boson operator, and with a macroscopic number of condensed "bosons" (fermion pairs) in the ground state. It is clear that a candidate for a modified perturbation expansion can be closely modeled on the structure of the Bogoliubov perturbation expansion in the presence of a condensate, as presented in Ch. 18.

22.2 Anomalous propagators in the Fermi gas

We study a homogeneous and unpolarized system of spin-$\frac{1}{2}$ fermions, interacting with spin-independent forces. The Hamiltonian reads

$$\hat{H} = \sum_{\boldsymbol{p}m} \varepsilon_p a^\dagger_{\boldsymbol{p}m} a_{\boldsymbol{p}m} + \frac{1}{2} \sum_{\boldsymbol{p}_i m_i} (\boldsymbol{p}_1 \boldsymbol{p}_2 | V | \boldsymbol{p}_3 \boldsymbol{p}_4) \, a^\dagger_{\boldsymbol{p}_1 m_1} a^\dagger_{\boldsymbol{p}_2 m_2} a_{\boldsymbol{p}_4 m_2} a_{\boldsymbol{p}_3 m_1},$$

(22.9)

where $m = \pm\frac{1}{2}$ denotes the spin projection. The interaction is given by

$$(\boldsymbol{p}_1 \boldsymbol{p}_2 | V | \boldsymbol{p}_3 \boldsymbol{p}_4) = \frac{1}{V} \delta_{\boldsymbol{p}_1 + \boldsymbol{p}_2, \boldsymbol{p}_3 + \boldsymbol{p}_4} W(\boldsymbol{p}_1 - \boldsymbol{p}_3), \qquad (22.10)$$

with $W(\boldsymbol{Q}) \equiv W(|\boldsymbol{Q}|)$ a real function. The formalism will be developed for the case the fermions couple to total spin $S = 0$. Pairing for $S = 1$ will not be considered explicitly here. Once it has been accepted that a pair state is condensed (*i.e.* has a macroscopic occupation in the exact ground state), it makes sense to give up particle-number conservation, and to consider the grand-canonical potential $\hat{\Omega} = \hat{H} - \mu \hat{N}$ in Fock space. Since a similar matrix structure as in Sec. 18.3 is expected, we define the normal sp propagator

$$i\hbar G_{11}(\boldsymbol{p}m; t - t') = \langle \Psi_0 | T[a_{\boldsymbol{p}m_\Omega}(t) a^\dagger_{\boldsymbol{p}m_\Omega}(t')] | \Psi_0 \rangle , \qquad (22.11)$$

as well as the complementary propagator

$$i\hbar G_{22}(\boldsymbol{p}m; t - t') = \langle \Psi_0 | T[a^\dagger_{-\boldsymbol{p}-m_\Omega}(t) a_{-\boldsymbol{p}-m_\Omega}(t')] | \Psi_0 \rangle . \qquad (22.12)$$

The operators $a_{\boldsymbol{p}m_\Omega} = e^{\frac{i}{\hbar}\hat{\Omega}t} a_{\boldsymbol{p}m} e^{-\frac{i}{\hbar}\hat{\Omega}t}$ are in the Heisenberg picture, with $\hat{\Omega}$ defining the time evolution. Moreover, we should be prepared (see later) to introduce *anomalous* propagators

$$i\hbar G_{12}(\boldsymbol{p}m; t - t') = \langle \Psi_0 | T[a_{\boldsymbol{p}m_\Omega}(t) a_{-\boldsymbol{p}-m_\Omega}(t')] | \Psi_0 \rangle ,$$

$$i\hbar G_{21}(\boldsymbol{p}m; t - t') = \langle \Psi_0 | T[a^\dagger_{-\boldsymbol{p}-m_\Omega}(t) a^\dagger_{\boldsymbol{p}m_\Omega}(t')] | \Psi_0 \rangle . \qquad (22.13)$$

With these definitions one has set up a 2×2 matrix propagator $[G(\boldsymbol{p}m; E)]$ with elements $G_{ij}(\boldsymbol{p}m; E)$, that can be formally written as

$$i\hbar [G(\boldsymbol{p}m; E)] = \langle \Psi_0 | T[A_{\boldsymbol{p}m}(t) A^\dagger_{\boldsymbol{p}m}(t')] | \Psi_0 \rangle , \qquad (22.14)$$

where

$$A_{\boldsymbol{p}m}(t) = \begin{pmatrix} a_{\boldsymbol{p}m_\Omega}(t) \\ a^\dagger_{-\boldsymbol{p}-m_\Omega}(t) \end{pmatrix} ; A^\dagger_{\boldsymbol{p}m}(t) = \begin{pmatrix} a^\dagger_{\boldsymbol{p}m_\Omega}(t) & a_{-\boldsymbol{p}-m_\Omega}(t) \end{pmatrix} . \qquad (22.15)$$

It is possible [Nambu (1960); Mattuck (1992)] to derive the perturbative expansion of the Green's function solely in terms of matrix quantities. However, we will maintain the distinction between the normal and anomalous quantities, since we prefer to closely follow the development of the boson formalism in Sec. 18.3.

The propagators in the energy representation are obtained by FT,

$$G_{11}(\boldsymbol{p}m; E) = \langle \Psi_0 | a_{\boldsymbol{p}m} \frac{1}{E - \hat{\Omega} + \Omega_0 + i\eta} a^\dagger_{\boldsymbol{p}m} + a^\dagger_{\boldsymbol{p}m} \frac{1}{E + \hat{\Omega} - \Omega_0 - i\eta} a_{\boldsymbol{p}m} | \Psi_0 \rangle$$

$$G_{12}(\boldsymbol{p}m; E) = \langle \Psi_0 | a_{\boldsymbol{p}m} \frac{1}{E - \hat{\Omega} + \Omega_0 + i\eta} a_{-\boldsymbol{p}-m} + a_{-\boldsymbol{p}-m} \frac{1}{E + \hat{\Omega} - \Omega_0 - i\eta} a_{\boldsymbol{p}m} | \Psi_0 \rangle$$

$$G_{21}(\boldsymbol{p}m; E) = \langle \Psi_0 | a^\dagger_{-\boldsymbol{p}-m} \frac{1}{E - \hat{\Omega} + \Omega_0 + i\eta} a^\dagger_{\boldsymbol{p}m} + a^\dagger_{\boldsymbol{p}m} \frac{1}{E + \hat{\Omega} - \Omega_0 - i\eta} a^\dagger_{-\boldsymbol{p}-m} | \Psi_0 \rangle$$

$$G_{22}(\boldsymbol{p}m; E) = \langle \Psi_0 | a^\dagger_{-\boldsymbol{p}-m} \frac{1}{E - \hat{\Omega} + \Omega_0 + i\eta} a_{-\boldsymbol{p}-m}$$

$$+ a_{-\boldsymbol{p}-m} \frac{1}{E + \hat{\Omega} - \Omega_0 - i\eta} a^\dagger_{-\boldsymbol{p}-m} | \Psi_0 \rangle . \tag{22.16}$$

These definitions are in complete formal analogy to the boson case, but one must keep in mind some differences as well. Firstly, minus signs now appear upon switching the arguments of the propagators in both the time ($\tau = t - t'$) and energy representation:

$$G_{22}(\boldsymbol{p}m; \tau) = -G_{11}(-\boldsymbol{p} - m; -\tau); \quad G_{22}(\boldsymbol{p}m; E) = -G_{11}(-\boldsymbol{p} - m; -E)$$

$$G_{12}(\boldsymbol{p}m; \tau) = -G_{12}(-\boldsymbol{p} - m; -\tau); \quad G_{12}(\boldsymbol{p}m; E) = -G_{12}(-\boldsymbol{p} - m; -E)$$

$$G_{21}(\boldsymbol{p}m; \tau) = -G_{21}(-\boldsymbol{p} - m; -\tau); \quad G_{21}(\boldsymbol{p}m; E) = -G_{21}(-\boldsymbol{p} - m; -E). \tag{22.17}$$

Secondly, the anomalous quantities have a somewhat different interpretation here. The so-called anomalous density

$$F(\boldsymbol{p}m) = -i\hbar G_{21}(\boldsymbol{p}m; \tau = 0) = \langle \Psi_0 | a^\dagger_{\boldsymbol{p}m} a^\dagger_{-\boldsymbol{p}-m} | \Psi_0 \rangle \tag{22.18}$$

e.g., represents the structure of the condensed fermion pair, and is therefore related to the condensate parameter $\sqrt{N_0} = \langle \Psi_0 | a_0 | \Psi_0 \rangle$ in Bose gas theory, rather than to the noncondensate bosons. Also, note the sign change when both operators in Eq. (22.18) are switched, $F(\boldsymbol{p}m) = -F(-\boldsymbol{p} - m)$. For the present isotropic system we can take

$$F(\boldsymbol{p}m) = F_p(-1)^{\frac{1}{2}+m} = F_p s_m, \tag{22.19}$$

where F_p depends only on the magnitude $p = |\boldsymbol{p}|$ and can be taken real.

The spin-dependent phase factor $s_m = (-1)^{\frac{1}{2}+m}$ is required for the sign change mentioned above.

The introduction of anomalous propagators like Eq. (22.13) for an isolated system[2] requires some explanation. We recall that in the boson case the condensate was singled out as a special sp state (Bogoliubov prescription), resulting in a potential $\hat{\Omega}_B$ that was automatically particle-number nonconserving. It was then argued that in the thermodynamic limit the same results are generated as in a particle-number conserving approach. At present we cannot do this explicitly: it is not a sp state but a pair state that is possibly condensed, due to the interparticle interactions. Nevertheless, the use of anomalous propagators makes perfect sense in the thermodynamic limit (or whenever a very large number N of particles is studied). If a Cooper-like pair state is really macroscopically occupied, the exact ground state in Fock space is automatically massively degenerate; neighboring ground states for $N \pm 2$, $N \pm 4$, ..., can be obtained without expense by adding or removing Cooper pairs, and have the same Ω (up to vanishingly small terms in $1/N$). This means that we may as well include superpositions of these states in the state $|\Psi_0\rangle$, appearing in the bra and ket of the propagator definitions. Such a state minimizes $\Omega_0 = \langle \Psi_0 | \hat{H} - \mu \hat{N} | \Psi_0 \rangle$ under the constraint $N = \langle \Psi_0 | \hat{N} | \Psi_0 \rangle$. It has no definite particle number, but is sharply peaked around N, and the relative fluctuations vanish in the thermodynamic limit $\Delta N / N \to 0$. In this limit all expectation values of particle-number conserving operators will be the same for $|\Psi_0\rangle$ and the fixed-N ground state $|\Psi_0^N\rangle$. In addition, $|\Psi_0\rangle$ has anomalous expectation values. The advantage is that $|\Psi_0\rangle$ has a graceful perturbation expansion, in particular a simple mean-field approximation (BCS), of which the ground state is a (generalized) product wave function that already captures the essentials of a coherent pair superposition like the one in Eq. (22.3). Note that this reasoning is only precise in the limit $N \to \infty$: when dealing with a rather small number of particles (*i.e.* pairing in nuclei) we must take care to correct for particle-number nonconservation.

22.3 Diagrammatic expansion in a superconducting system

Employing familiar reasoning, the general Eq. (8.18) can be applied to the time-ordered products appearing in the propagators G_{ij} of Eqs. (22.11) –

---

[2]There is of course no problem for a genuinely open system like a Josephson junction, where electrons can tunnel into and out of the system.

(22.13). The resulting perturbation series is expressed in terms of time-ordered products of the fermion operators in the interaction picture, where the free $\hat{\Omega}_0 = \hat{T} - \mu\hat{N}$ governs the time evolution. The operator products should be taken in the Fermi-gas noninteracting ground state $|\Phi_0^\mu\rangle$ at fixed chemical potential μ. Wick's theorem can be used to evaluate the matrix elements in terms of contractions, and to make a one-to-one correspondence with a suitably defined diagrammatic series.

At this point we seem to be in deep trouble, since the Fermi gas $|\Phi_0^\mu\rangle$ has fixed particle-number (with $\frac{N}{V} = \rho = \frac{(2m\mu)^{3/2}}{3\pi^2}$), and obviously does not have any anomalous expectation values, *i.e.*

$$i\hbar G_{12}^{(0)}(\boldsymbol{p}m; t - t') = \langle\Phi_0^\mu| T[a_{\boldsymbol{p}m}(t)a_{-\boldsymbol{p}-m}(t')] |\Phi_0^\mu\rangle = 0. \qquad (22.20)$$

It only has the usual normal type

$$i\hbar G_{11}^{(0)}(\boldsymbol{p}m; t - t') = \langle\Phi_0^\mu| T[a_{\boldsymbol{p}m}(t)a_{\boldsymbol{p}m}^\dagger(t')] |\Phi_0^\mu\rangle \qquad (22.21)$$
$$= e^{-\frac{i}{\hbar}\varepsilon_{p\mu}(t-t')}[\theta(t - t')\theta(\varepsilon_{p\mu}) - \theta(t' - t)\theta(-\varepsilon_{p\mu})],$$

where $\varepsilon_{p\mu} = \varepsilon_p - \mu = p^2/(2m) - \mu$. This seems to imply that anomalous contractions between aa and $a^\dagger a^\dagger$ do not contribute to Wick's theorem, and we are simply left with the perturbation series for G_{11} in a normal system.

A moment of reflection learns that this should be expected: we are trying to take the Fermi sea as a perturbative starting point for a series in powers of the interaction, whereas we strongly suspect (see Sec. 22.1) that condensation of a fermion pair state is a nonperturbative phenomenon. Fortunately we know that *self-consistent* Green's function theory is able to cope with nonperturbative phenomena, and that is indeed how the perturbation expansion of a superconducting system should be organized.

The situation is one of spontaneous symmetry breaking, and in fact very similar to that of a self-bound system like a nucleus, in which the particles are localized around the center of mass due to their mutual interactions. It is clearly impossible to generate a useful approximation for the sp propagator, by starting with the free plane-wave propagator (corresponding to the kinetic energy T) and expanding in powers of the translationally invariant interaction V. The way out is to change the noninteracting starting point to $H_0 = T + U$, by adding an auxiliary potential U localized around a fixed point in space, and subtracting the contributions of U from the interaction term. As shown in Sec. 9.5, the auxiliary potential then drops out of the calculation in a self-consistent treatment (as in HF theory), when the self-energy is evaluated with the interacting propagator.

In the present case we can imagine adding to the kinetic energy a small auxiliary potential U of the type $(a^\dagger a^\dagger + aa)$, which is explicitly number-nonconserving, and use this as the noninteracting reference system. The corresponding ground state $|\Phi_0'\rangle$ now appears in Eq. (8.18), and acts as the reference vacuum for Wick's theorem. As a consequence, also contractions between $a^\dagger a^\dagger$ and aa operators appear, and should be identified with the $G_{21}^{(0)'}$ and $G_{12}^{(0)'}$ propagators. Again, the auxiliary potential will drop out of the calculation in a self-consistent theory. In view of this, we may as well skip an explicit introduction of a particle-number nonconserving auxiliary potential: we simply keep in mind that the internal contractions in Wick's theorem can be of both the normal and anomalous type, and represent the interacting propagators of the self-consistent theory.

After these preliminaries, the diagrammatic expansion can be safely interpreted, and it is clear that its structure will bear a strong resemblance to the bosonic case of Sec. 18.3. The expansion for the propagators G_{ij} consists of all connected diagrams containing interaction lines, propagators, and two external points, with all the usual construction rules of Sec. 8.6 applying. The normal propagators G_{11} and G_{22} have one incoming and one outgoing external fermion line, and are represented graphically as follows.

$$G_{11}(\boldsymbol{p}m; E) \Rightarrow \qquad\qquad G_{22}(\boldsymbol{p}m; E) \Rightarrow$$

The anomalous propagators G_{12} and G_{21} have either two incoming or two outgoing fermion lines, and are represented graphically below.

$$G_{12}(\boldsymbol{p}m; E) \Rightarrow \qquad\qquad G_{21}(\boldsymbol{p}m; E) \Rightarrow$$

One may then define the associated (irreducible) self-energies Σ_{ij}, as the sum of the irreducible diagrams, stripped of the external lines, contributing to the propagator G_{ij}. Note again, that self-consistency is a necessity here. The internal propagators, appearing in the self-energy diagrams, should be the dressed ones: the anomalous self-energy cannot be written down in terms of the noninteracting propagator, which only contains normal pieces. This is logical, as a condensate of fermion pairs may arise solely because of the interactions, in contrast to the bosonic case where condensation occurs at the noninteracting (sp) level.

In the algebraic translation of the diagrams for $G_{ij}(\boldsymbol{p}m; E)$ some ambiguity exists, related to the fact that switching the labeling of both legs in the anomalous propagators results in a minus sign. A sign ambiguity also emerges when a normal fermion line is identified with either G_{11} or G_{22}. These ambiguities are simply resolved by looking at the spin label of the fermion lines, and restricting the propagators in the expansion to the G_{ij} with spin m. An examination of the contractions arising in the time-ordered products then leads to the conclusion that all diagram rules of Sec. 8.6 remain unchanged, apart from one additional rule: an extra minus sign should be introduced for each interaction vertex with the opposite spin $(-m)$.

Finally we should determine how to interpret equal-time contractions, or equivalently, propagators having both endpoints at the same interaction line. In this case the original order of the removal and addition operators, as they appear in the interaction, should be restored. For G_{11} this leads to the usual prescription (see Sec. 8.6.1)

$$\lim_{t\to t'} G_{11}(\boldsymbol{p}m; t - t') = G_{11}(\boldsymbol{p}m; -\eta),$$
$$\int dE\, G_{11}(\boldsymbol{p}m; E) \Rightarrow \int dE\, e^{i\eta E} G_{11}(\boldsymbol{p}m; E), \qquad (22.22)$$

in the time and energy formulation, respectively. For the G_{22} propagators the opposite holds,

$$\lim_{t\to t'} G_{22}(\boldsymbol{p}m; t - t') = G_{22}(\boldsymbol{p}m; +\eta),$$
$$\int dE\, G_{22}(\boldsymbol{p}m; E) \Rightarrow \int dE\, e^{-i\eta E} G_{22}(\boldsymbol{p}m; E). \qquad (22.23)$$

For the anomalous propagators either choice can be made, since the equal-time $a^\dagger a^\dagger$ and aa operators anticommute anyway.

As an example of the diagram rules we treat the first-order contributions to the $\Sigma_{ij}(\boldsymbol{p}m; E)$, shown in Figs. 22.1 – 22.2. Note that in keeping with the self-consistent formulation, the internal propagators are dressed ones. For $\Sigma_{11}(\boldsymbol{p}m; E)$ the direct [part a)] and exchange [part b)] diagrams of Fig. 22.1 are of course already known from the treatment of normal Fermi systems,

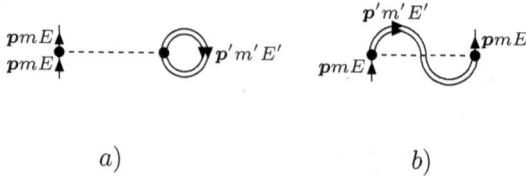

a)   b)

Fig. 22.1 First-order diagrams for the normal self-energy $\Sigma_{11}(\boldsymbol{p}m; E)$. Direct a) and exchange b) contributions.

but their evaluation provides a useful check for the present diagram rules:

$$\Sigma_{11}^D(\boldsymbol{p}m; E) = (-1)\left(\frac{i}{2\pi}\right)\sum_{\boldsymbol{p}'}\int dE'\frac{W(0)}{V}\left(e^{i\eta E'}G_{11}(\boldsymbol{p}'m; E')\right.$$
$$\left. - e^{i\eta E'}G_{22}(-\boldsymbol{p}'m; -E')\right). \tag{22.24}$$

The factors in front, (-1) (due to the closed fermion loop) and $\frac{i}{2\pi}$, come from the familiar diagram rules. In part a) of Fig. 22.1, the propagator in the loop should be interpreted as G_{11} for $m' = m$, and as G_{22} for $m' = -m$. In the latter case an extra minus sign appears (additional rule), because a vertex is present with opposite spin $(-m)$. Finally, Eqs. (22.22) and (22.23) have been used. The energy integrations can be worked out and yield

$$\Sigma_{11}^D(\boldsymbol{p}m; E) = \sum_{\boldsymbol{p}'}\frac{W(0)}{V}2n_{\boldsymbol{p}'}, \tag{22.25}$$

where

$$n_{\boldsymbol{p}} = \langle\Psi_0| a_{\boldsymbol{p}m}^\dagger a_{\boldsymbol{p}m}|\Psi_0\rangle \tag{22.26}$$

is the normal density of the system, *i.e.* the momentum distribution. For the exchange diagram (part b) of Fig. 22.1) one can only have $m' = m$ for the spin label of the internal propagator,

$$\Sigma_{11}^E(\boldsymbol{p}m; E) = \left(\frac{i}{2\pi}\right)\sum_{\boldsymbol{p}'}\int dE'\frac{W(\boldsymbol{p}-\boldsymbol{p}')}{V}e^{i\eta E'}G_{11}(\boldsymbol{p}'m; E')$$
$$= -\sum_{\boldsymbol{p}'}\frac{W(\boldsymbol{p}-\boldsymbol{p}')}{V}n_{\boldsymbol{p}'}. \tag{22.27}$$

The contributions to $\Sigma_{22}(\boldsymbol{p}m; E)$ are similar, but the external legs in Fig. 22.1 should be labeled $(-\boldsymbol{p}-m; -E)$, resulting in a global sign change.

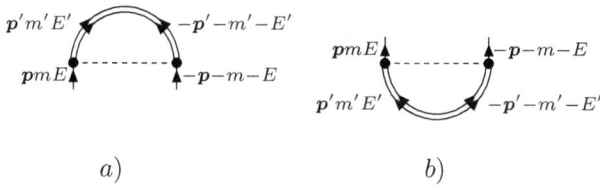

a)   b)

Fig. 22.2 First-order diagrams for the anomalous self-energy $\Sigma_{21}(\boldsymbol{p}m; E)$ in part $a)$ and $\Sigma_{12}(\boldsymbol{p}m; E)$ in part $b)$.

The first-order contribution to $\Sigma_{21}(\boldsymbol{p}m; E)$ is shown in part $a)$ of Fig. 22.2. The internal propagator has $m' = m$ and we find (the additional rule gives an extra minus sign)

$$\Sigma_{21}(\boldsymbol{p}m; E) = (-1)\left(\frac{i}{2\pi}\right)\sum_{\boldsymbol{p'}}\int dE' \frac{W(\boldsymbol{p}-\boldsymbol{p'})}{V}e^{i\eta E'}G_{21}(\boldsymbol{p'}m; E')$$

$$= \sum_{\boldsymbol{p'}}\frac{W(\boldsymbol{p}-\boldsymbol{p'})}{V}F(\boldsymbol{p'}m), \tag{22.28}$$

with the anomalous density $F(\boldsymbol{p}m)$ as defined in Eq. (22.18). Likewise, part $b)$ of Fig. 22.2 leads to

$$\Sigma_{12}(\boldsymbol{p}m; E) = (-1)\left(\frac{i}{2\pi}\right)\sum_{\boldsymbol{p'}}\int dE' \frac{W(\boldsymbol{p}-\boldsymbol{p'})}{V}e^{i\eta E'}G_{12}(\boldsymbol{p'}m; E')$$

$$= \sum_{\boldsymbol{p'}}\frac{W(\boldsymbol{p}-\boldsymbol{p'})}{V}F(\boldsymbol{p'}m). \tag{22.29}$$

The general topological structure of the diagrammatic series is the same as the Bose case of Fig. 18.5: a sequence of normal and anomalous Σ_{ij} (irreducible) self-energies, connected by the non-interacting (Fermi gas) normal propagators

$$G_{11}^{(0)}(\boldsymbol{p}m; E) = \frac{\theta(\varepsilon_{p\mu})}{E - \varepsilon_{p\mu} + i\eta} + \frac{\theta(-\varepsilon_{p\mu})}{E - \varepsilon_{p\mu} - i\eta},$$

$$G_{22}^{(0)}(\boldsymbol{p}m; E) = \frac{\theta(\varepsilon_{p\mu})}{E + \varepsilon_{p\mu} - i\eta} + \frac{\theta(-\varepsilon_{p\mu})}{E + \varepsilon_{p\mu} + i\eta}. \tag{22.30}$$

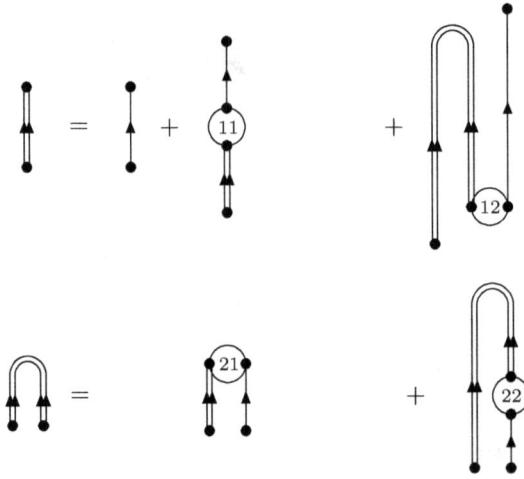

Fig. 22.3 Diagrammatic representation of the propagators $G_{ij}(\boldsymbol{pm}; E)$, summing all reducible repetitions of the irreducible self-energies $\Sigma_{ij}(\boldsymbol{pm}; E)$.

The full sequence can be summed in the way shown in Fig. 22.3. Adding the momentum, spin, and energy labels in Fig. 22.3 the corresponding algebraic Dyson-like equations becomes

$$G_{11}(\boldsymbol{pm}; E) = G_{11}^{(0)}(\boldsymbol{pm}; E) + G_{11}^{(0)}(\boldsymbol{pm}; E)\Sigma_{11}(\boldsymbol{pm}; E)G_{11}(\boldsymbol{pm}; E)$$
$$+ G_{11}^{(0)}(\boldsymbol{pm}; E)\Sigma_{12}(\boldsymbol{pm}; E)G_{21}(\boldsymbol{pm}; E)$$
$$G_{21}(\boldsymbol{pm}; E) = G_{22}^{(0)}(\boldsymbol{pm}; E)\Sigma_{21}(\boldsymbol{pm}; E)G_{11}(\boldsymbol{pm}; E)$$
$$+ G_{22}^{(0)}(\boldsymbol{pm}; E)\Sigma_{22}(\boldsymbol{pm}; E)G_{21}(\boldsymbol{pm}; E). \tag{22.31}$$

These are called the Gorkov equations [Gorkov (1958)], and determine the propagators in terms of the self-energies. One can rewrite Eq. (22.31) more elegantly as a 2×2 matrix equation,

$$[G(\boldsymbol{pm}; E)] = [G^{(0)}(\boldsymbol{pm}; E)] + [G^{(0)}(\boldsymbol{pm}; E)][\Sigma(\boldsymbol{pm}; E)][G(\boldsymbol{pm}; E)]. \tag{22.32}$$

The noninteracting matrix propagator is defined as

$$[G^{(0)}(\boldsymbol{pm}; E)] = \begin{bmatrix} G_{11}^{(0)}(\boldsymbol{pm}; E) & 0 \\ 0 & G_{22}^{(0)}(\boldsymbol{pm}; E) \end{bmatrix} \tag{22.33}$$

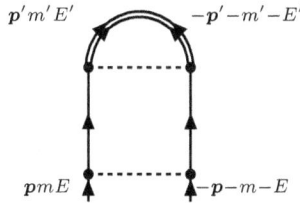

Fig. 22.4 Diagram generated by iterating the Dyson equation for the anomalous propagator in Fig. 22.2a).

and its inverse follows from Eq. (22.30)

$$[G^{(0)}(\bm{pm}; E)]^{-1} = \begin{bmatrix} E - \varepsilon_{p\mu} & 0 \\ 0 & E + \varepsilon_{p\mu} \end{bmatrix}. \tag{22.34}$$

The explicit solution of Eq. (22.32) is given by matrix inversion as

$$[G(\bm{pm}; E)] = \left([G^{(0)}(\bm{pm}; E)]^{-1} - [\Sigma(\bm{pm}; E)]\right)^{-1} \tag{22.35}$$

$$= \frac{1}{D(\bm{pm}; E)} \begin{bmatrix} E + \varepsilon_{p\mu} - \Sigma_{22}(\bm{pm}; E) & \Sigma_{12}(\bm{pm}; E) \\ \Sigma_{21}(\bm{pm}; E) & E - \varepsilon_{p\mu} - \Sigma_{11}(\bm{pm}; E) \end{bmatrix},$$

where the determinant function

$$D(\bm{pm}; E) = \{E - \varepsilon_{p\mu} - \Sigma_{11}(\bm{pm}; E)\}\{E + \varepsilon_{p\mu} - \Sigma_{22}(\bm{pm}; E)\}$$
$$- \Sigma_{12}(\bm{pm}; E)\Sigma_{21}(\bm{pm}; E), \tag{22.36}$$

identifies the singularities of the propagator.

A final issue remains to be clarified concerning the anomalous self-energy in a paired system. Its diagrammatic content is restricted to terms that generalize the lowest-order terms of Fig. 22.2a) and b) in such a way that they do not contain pp (hh) reducible contributions [Migdal (1967)]. We illustrate the point by noting that the iteration of the coupled Dyson equations yields in lowest order the diagram shown in Fig. 22.4 for the anomalous propagator in diagram a) of Fig. 22.2. We therefore recognize that the second-order ladder diagram in the limit of total momentum and energy zero is automatically produced from the coupled Dyson equations. In higher order all such ladder diagrams are generated accordingly. It is thus clear that pp reducible diagrams, like the one in Fig. 22.4, should not be included in the anomalous self-energy.

22.4   The BCS gap equation

The first-order self-consistent equations generated by this scheme are called the BCS equations, after [Bardeen *et al.* (1957)] who derived them starting from an inspired variational ansatz (see Sec. 22.5). A more involved application of the Gorkov equations will be presented in Ch. 24. Here we consider only the first-order self-energies in Fig. 22.1 and Fig. 22.2, calculated in Eqs. (22.24) – (22.29), in order to establish the basic features of the BCS gap equation. These terms are independent of energy and are given by

$$\Sigma_{11}(\boldsymbol{p}m) = -\Sigma_{22}(\boldsymbol{p}m) = \sum_{\boldsymbol{p}'} n_{\boldsymbol{p}'} \left(2 \frac{W(0)}{V} - \frac{W(\boldsymbol{p}-\boldsymbol{p}')}{V} \right) = V_p$$

$$\Sigma_{21}(\boldsymbol{p}m) = \Sigma_{12}(\boldsymbol{p}m) = \sum_{\boldsymbol{p}'} F_{\boldsymbol{p}'} s_m \frac{W(\boldsymbol{p}-\boldsymbol{p}')}{V} = \Delta_p s_m. \qquad (22.37)$$

One recognizes in Σ_{11} the familiar expression for a HF-type self-energy V_p. The anomalous Σ_{12} is called the gap function, for reasons that will soon be apparent. Note that V_p and Δ_p are expressed in terms of the (as yet unknown) normal and anomalous densities, n_p and F_p.

Substitution of Eq. (22.37) into Eq. (22.35) yields the normal and anomalous Green's functions, with the determinant function appearing in the denominator reducing to

$$D(\boldsymbol{p}m; E) = E^2 - (\varepsilon_{p\mu} + V_p)^2 - \Delta_p^2 = E^2 - E_p^2. \qquad (22.38)$$

The last equality defines the singularities $\pm E_p$ of the propagator, or the excitation spectrum of the elementary sp excitations

$$E_p = \sqrt{\chi_p^2 + \Delta_p^2}, \qquad (22.39)$$

where the modified sp energies

$$\chi_p = \varepsilon_{p\mu} + V_p, \qquad (22.40)$$

are taken relative to the chemical potential, and include the HF-like energy shift. The total expression for the propagator G_{ij} in Eq. (22.35) becomes

$$[G(\boldsymbol{p}m; E)] = \frac{1}{E^2 - E_p^2} \begin{pmatrix} E + \chi_p & \Delta_p s_m \\ \Delta_p s_m & E - \chi_p \end{pmatrix}. \qquad (22.41)$$

To retrieve a representation with simple poles, we note that

$$\frac{E}{E^2 - E_p^2} = \frac{1}{2}\left(\frac{1}{E - E_p + i\eta} + \frac{1}{E + E_p - i\eta}\right),$$

$$\frac{1}{E^2 - E_p^2} = \frac{1}{2E_p}\left(\frac{1}{E - E_p + i\eta} - \frac{1}{E + E_p - i\eta}\right), \qquad (22.42)$$

where the identification $\pm i\eta$ has been made in agreement with Eq. (22.16). As a consequence, the propagator can be written as

$$[G(\boldsymbol{pm}; E)] = \frac{\begin{pmatrix} u_p^2 & u_p v_p s_m \\ u_p v_p s_m & v_p^2 \end{pmatrix}}{E - E_p + i\eta} + \frac{\begin{pmatrix} v_p^2 & -u_p v_p s_m \\ -u_p v_p s_m & u_p^2 \end{pmatrix}}{E + E_p - i\eta}, \qquad (22.43)$$

in terms of the BCS amplitudes u_p and v_p

$$u_p^2 = \frac{1}{2}\left(1 + \frac{\chi_p}{E_p}\right), \quad v_p^2 = \frac{1}{2}\left(1 - \frac{\chi_p}{E_p}\right). \qquad (22.44)$$

Note that the amplitudes obey $u_p^2 + v_p^2 = 1$, and that the relative sign of u_p and v_p is fixed by

$$u_p v_p = \frac{\Delta_p}{2E_p}. \qquad (22.45)$$

In Fig. 22.5 we illustrate the typical behavior near the Fermi surface, of the BCS amplitudes in Eq. (22.44), assuming a constant gap Δ (independent of momentum). It is clear that the deviation from the normal Fermi-gas quantities only occurs in a region of width Δ, where, instead of the Fermi-gas discontinuity, one now has a smooth transition symmetrical around the chemical potential μ (*i.e.* $\chi_p = 0$).

The densities that follow from the propagator (22.41) are given by Eqs. (22.26) and (22.18)

$$n_p = v_p^2 = \frac{1}{2}\left(1 - \frac{\chi_p}{E_p}\right), \quad -F_p = u_p v_p = \frac{\Delta_p}{2E_p}. \qquad (22.46)$$

These equations in fact represent the self-consistency relations between the propagator (22.43) and the self-energy in Eq. (22.37). The last relation in Eq. (22.46) can also be expressed through Eq. (22.37), in the form

$$\Delta_p = -\frac{1}{2}\sum_{p'} \frac{W(\boldsymbol{p} - \boldsymbol{p'})}{V} \frac{\Delta_{p'}}{E_{p'}}, \qquad (22.47)$$

which is the celebrated BCS gap equation.

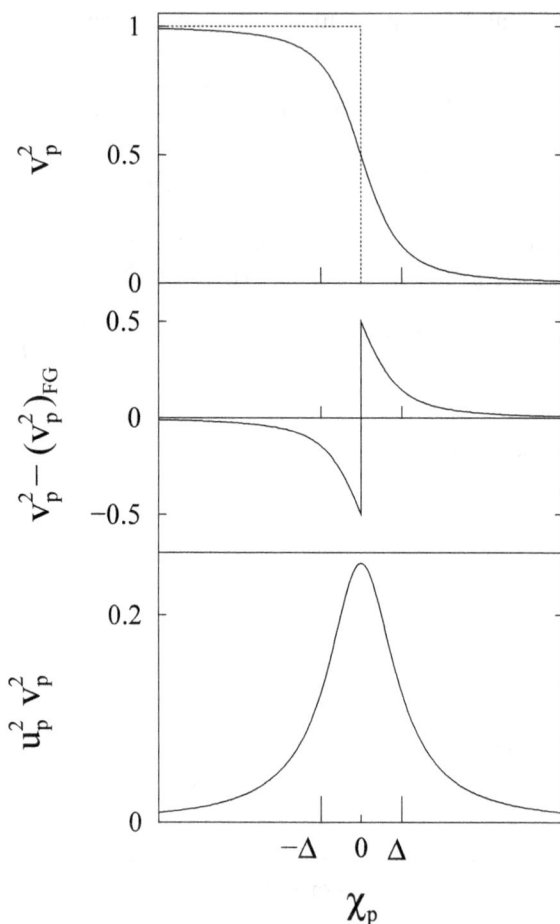

Fig. 22.5   Behavior as a function of sp energy χ_p, of the BCS occupation v_p^2 [upper panel], its deviation from the Fermi-gas step function $v_p^2 - \theta(-\chi_p)$ [middle panel], and the product $u_p^2 v_p^2$ (lower panel), as given by Eq. (22.44) and assuming a constant gap Δ.

Several important observations regarding the nature of the solutions can be made without explicitly solving the BCS equations. First, one observes that the gap equation (22.47) always has the trivial solution $\Delta_p = 0, \forall p$. This trivial (or normal) solution implies $E_p = |\chi_p|$ and, through Eq. (22.44), that the system has Fermi-gas occupations $u_p^2 = \theta(\chi_p)$, $v_p^2 = \theta(-\chi_p)$. Equation (22.37) then simply restates the HF expression for the normal self-energy Σ_{11}. On the other hand, a nontrivial (or superconducting)

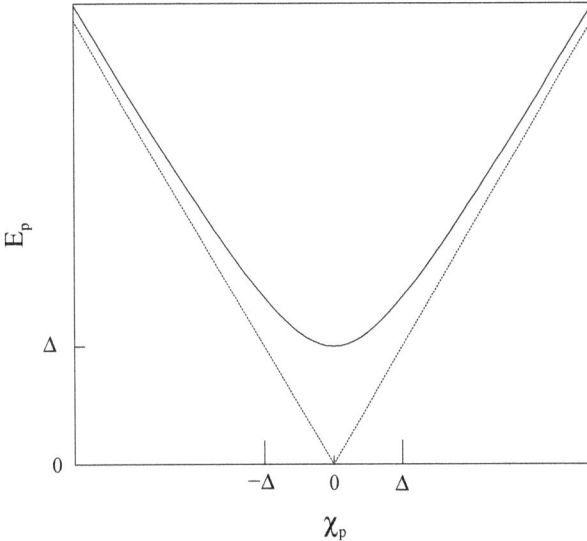

Fig. 22.6   The superconducting quasiparticle spectrum $E_p = \sqrt{\chi_p^2 + \Delta_p^2}$ as a function of sp energy χ_p, assuming a constant gap Δ. The dotted lines represent the normal quasiparticle spectrum $E_p = |\chi_p|$.

solution is characterized by a finite $|\Delta_p| > 0$. As a consequence, one has $E_p \geq |\Delta_p| > 0$, and there is a *finite gap* separating the ground state from the spectrum of sp excitations (hence the name of the gap function). Such a feature is not present in a normal infinite Fermi system, but agrees with the expectations for a system with pair condensation. The gap is then related to the energy it takes to break a condensed pair, and create a sp excitation. The shape of the quasiparticle spectrum E_p near the Fermi surface is illustrated in Fig. 22.6 for a constant gap Δ.

From the gap equation (22.47) it is also clear that a purely repulsive interaction cannot generate a superconducting solution,[3] and that the interaction must be globally attractive for the signs on the left and right to

---

[3]With "purely repulsive" we mean that the interaction, viewed as an operator in momentum space, has only positive eigenvalues. In the present example $V(\boldsymbol{r})$ is local in coordinate space, $V(\boldsymbol{r}) \sim \int d\boldsymbol{Q} \, \exp[i\boldsymbol{Q} \cdot \boldsymbol{r}/\hbar] W(\boldsymbol{Q})$, and purely repulsive means that $V(\boldsymbol{r}) > 0$. Sometimes (*e.g.* for nuclear interactions), $W(\boldsymbol{Q}) > 0$ in momentum space and still yields a solution to the BCS gap equation. By overcoming the short-range repulsion in coordinate space, the gap equation emphasizes the longer range attraction.

be balanced. Finally, we note that BCS superconductivity is a phenomenon limited to a region near the Fermi surface. The gap Δ_p in Eq. (22.47) can be appreciably different from zero only for $\chi_p \approx 0$, since one may naturally assume that the interaction $W(\boldsymbol{p} - \boldsymbol{p}')$ decreases rapidly with momentum transfer and requires $p \approx p'$, whereas the denominator $E_{p'} = \sqrt{\chi_{p'}^2 + \Delta_{p'}^2}$ increases with $|\chi_{p'}|$ and demands $\chi_{p'} \approx 0$.

In accordance with these observations, further insight can be gained by a schematic model of a weak, constant attractive interaction

$$W(\boldsymbol{p} - \boldsymbol{p}') = -\lambda\theta(c - |\chi_p|)\theta(c - |\chi_{p'}|), \qquad (22.48)$$

which is limited to a small region $R = (-c < \chi_p < c)$ around the Fermi surface. The interaction vanishes when $|\chi_p|$ exceeds some cut-off energy c, assumed to be much smaller than the chemical potential $(0 < c \ll \mu)$.

Substitution into Eq. (22.47) immediately leads to the conclusion that Δ_p vanishes outside R, and is a constant Δ inside R

$$\Delta_p = \theta(c - |\chi_p|)\Delta, \qquad (22.49)$$

with Δ a solution of

$$1 = \frac{\lambda}{2V} \sum_p \frac{\theta(c - |\chi_p|)}{\sqrt{\chi_p^2 + \Delta^2}}. \qquad (22.50)$$

For a further analytical evaluation we set the occupations appearing in the normal self-energy of Eq. (22.37) equal to the Fermi-gas step function, $v_{p'}^2 = \theta(-\chi_{p'})$, *i.e.* Eq. (22.40) is replaced by

$$\chi_p = \varepsilon_{p\mu} + \sum_{p'} \theta(-\chi_{p'}) \left(2\frac{W(0)}{V} - \frac{W(\boldsymbol{p} - \boldsymbol{p}')}{V} \right). \qquad (22.51)$$

This allows to decouple the changes to the normal self-energy from the determination of the gap, and to concentrate on the new features generated by the superconducting solution. Note that the error introduced is presumably very small: the bulk of the normal self-energy comes from the entire Fermi sea, whereas the deviation $v_{p'}^2 - \theta(-\chi_{p'})$ is nonzero only in the small region R, and is in addition antisymmetrical.

Taking the thermodynamic limit, and switching the integration variable from p to χ_p in Eq. (22.50) one obtains

$$
\begin{aligned}
1 &= \frac{\lambda}{2} \int_{-c}^{+c} d\chi \frac{D(\chi)}{\sqrt{\chi^2 + \Delta^2}} \\
&\approx \frac{\lambda D(0)}{2} \int_{-c}^{+c} \frac{d\chi}{\sqrt{\chi^2 + \Delta^2}} = \lambda D(0) \mathrm{Asinh}(c/\Delta),
\end{aligned}
\tag{22.52}
$$

where the density of states has been defined as

$$
D(\chi) = \frac{4\pi}{(2\pi\hbar)^3} p^2 \frac{dp}{d\chi}.
\tag{22.53}
$$

Since $D(\chi)$ is a smooth function, which does not change appreciably in the small interval R, it can be replaced with its value $D(0)$ at the Fermi surface. The solution of Eq. (22.52) is

$$
\Delta = \frac{c}{\sinh\left(\frac{1}{\lambda D(0)}\right)} \to 2c e^{-\frac{1}{\lambda D(0)}}.
\tag{22.54}
$$

The latter limit applies to the normal situation where $\lambda D(0) \ll 1$ and hence $\Delta \ll c$, which is consistent with the nature of this schematic model. Note that, while $\lambda \to 0$ implies $\Delta \to 0$, it is impossible to expand Δ in a power series of the interaction strength λ, and the presence of a finite gap is indeed a nonperturbative effect. Having determined the gap Δ, the BCS propagator $G_{ij}(\boldsymbol{p}m; E)$ at fixed chemical potential μ is now fully known, and various properties of the system can be calculated in the standard way. These properties should be compared to the ones that follow from the normal Fermi gas (or HF) propagator at fixed μ, *i.e.*

$$
G_{11}^{HF}(\boldsymbol{p}m; E) = \frac{\theta(\chi_p)}{E - \chi_p + i\eta} + \frac{\theta(-\chi_p)}{E - \chi_p - i\eta}; \quad G_{12}^{HF}(\boldsymbol{p}m; E) = 0.
\tag{22.55}
$$

The number of particles is obtained as

$$
\begin{aligned}
N_{BCS}(\mu) &= \sum_{\boldsymbol{p}m} \int \frac{dE}{2\pi i} e^{i\eta E} G_{11}(\boldsymbol{p}m; E) = 2 \sum_{\boldsymbol{p}} v_p^2 \\
&= N_{HF}(\mu) + 2 \sum_{\boldsymbol{p}} (v_p^2 - \theta(-\chi_p)).
\end{aligned}
\tag{22.56}
$$

The last term may be manipulated in a similar way as Eq. (22.52) yielding

$$N_{BCS}(\mu) - N_{HF}(\mu) = 2D(0)V \int_{-c}^{+c} d\chi \left(\frac{1}{2} \left(1 - \frac{\chi}{\sqrt{\chi^2 + \Delta^2}} \right) - \frac{1}{2} \left(1 - \frac{\chi}{|\chi|} \right) \right)$$

$$= D(0)V \int_{-c}^{+c} d\chi \, \chi \left(\frac{1}{|\chi|} - \frac{1}{\sqrt{\chi^2 + \Delta^2}} \right) = 0, \quad (22.57)$$

and vanishes due to the antisymmetry of the integrand. Therefore, to leading order, the BCS solution has the same number of particles as the underlying Fermi gas (or HF) state. The grand-canonical potential reads

$$\Omega_{BCS}(\mu) = \frac{1}{2} \sum_{\boldsymbol{pm}} \int \frac{dE}{2\pi i} e^{i\eta E} G_{11}(\boldsymbol{pm}; E)(\varepsilon_{p\mu} + E) = \sum_{\boldsymbol{p}} (\varepsilon_{p\mu} - E_p) v_p^2,$$
$$(22.58)$$

which should be compared to the normal solution

$$\Omega_{HF}(\mu) = \sum_{\boldsymbol{p}} (\varepsilon_{p\mu} + \chi_p) \, \theta(-\chi_p). \quad (22.59)$$

The difference can be split into two contributions

$$\Omega_{BCS}(\mu) - \Omega_{HF}(\mu) = \sum_{\boldsymbol{p}} (\varepsilon_{p\mu} + \chi_p) \left(v_p^2 - \theta(-\chi_p) \right) - \sum_{\boldsymbol{p}} (\chi_p + E_p) v_p^2$$

$$= R_1 + R_2. \quad (22.60)$$

The contribution R_1 may be rewritten using Eq. (22.40)

$$R_1 = \sum_{\boldsymbol{p}} (2\chi_p - V_p)(v_p^2 - \theta(-\chi_p))$$

$$= D(0)V \int_{-c}^{+c} d\chi \chi^2 \left(\frac{1}{|\chi|} - \frac{1}{\sqrt{\chi^2 + \Delta^2}} \right). \quad (22.61)$$

The term involving V_p has been dropped, with a similar reasoning as before: V_p is not peaked at the Fermi surface and is assumed constant in the small region R, whereas the factor $(v_p^2 - \theta(-\chi_p))$ is antisymmetric. The remainder of R_1 is clearly positive, and represents the cost in kinetic energy, for deforming the Fermi-gas step function into the BCS distribution depicted in Fig. 22.5. The remaining integrals are standard and result in

$$R_1 = D(0)V \left\{ c^2 - \left(c\sqrt{c^2 + \Delta^2} + \Delta^2 \text{Asinh}(\frac{c}{\Delta}) \right) \right\}$$

$$= -\frac{1}{2}D(0)V\Delta^2 + V\frac{\Delta^2}{\lambda} \quad (22.62)$$

where the expression (22.54) for the gap has been used, and the limit $\Delta \ll c$ was taken. The pairing energy, gained by introducing a gap in the sp spectrum, is represented by the negative contribution R_2 and reads

$$R_2 = -\sum_p (E_p + \chi_p)\frac{1}{2}\left(1 - \frac{\chi_p}{E_p}\right)$$

$$= -\frac{D(0)V}{2}\int_{-c}^{+c} d\chi \frac{\Delta^2}{\sqrt{\chi^2 + \Delta^2}} = -V\frac{\Delta^2}{\lambda}. \qquad (22.63)$$

It more than compensates for the increased kinetic energy, with a net result $R_1 + R_2 = -\frac{1}{2}D(0)V\Delta^2$ that is negative and extensive (proportional to the volume or the particle number). The shift in the potential [or the energy, by virtue of Eq. (22.57)] at fixed μ thus becomes

$$\Omega_{BCS}(\mu) - \Omega_{HF}(\mu) = E_{BCS}(\mu) - E_{HF}(\mu) = -\frac{1}{2}D(0)V\Delta^2. \qquad (22.64)$$

For the energy shift, the relevant quantity is of course the energy at fixed particle number N. However, the energy shift at fixed N is the same as the shift at fixed μ given by Eq. (22.64). This is because $N_{BCS}(\mu) = N_{HF}(\mu)$ implies that for fixed N, $\mu_{BCS}(N) = \mu_{HF}(N)$ to leading order in $1/N$. So the leading order of the chemical potential is not affected by the transition to the BCS ground state. However, the extensive quantity $N[\mu_{BCS} - \mu_{HF}]$ is again of leading order (see *e.g.* [Fetter and Walecka (1971)]), and as a consequence $\Omega_{BCS}(N) - \Omega_{HF}(N)$ differs from Eq. (22.64). This difference drops out in the more relevant energy shift, since

$$E_{BCS}(N) - E_{HF}(N) = \Omega_{BCS}(\mu_{BCS}) - \Omega_{HF}(\mu_{HF}) + N[\mu_{BCS} - \mu_{HF}] \quad (22.65)$$

$$= \Omega_{BCS}(\mu_{HF}) - \Omega_{HF}(\mu_{HF}) + \left\{\left(\frac{\partial\Omega_{BCS}}{\partial\mu}\right)_{\mu_{HF}} + N\right\}[\mu_{BCS} - \mu_{HF}].$$

As a consequence of the thermodynamic identity $d\Omega/d\mu = -N$, the last term vanishes, and we have

$$E_{BCS}(N) - E_{HF}(N) = -\frac{1}{2\rho}D(0)N\Delta^2. \qquad (22.66)$$

22.5 Canonical BCS transformation

The BCS propagator in Eq. (22.43) has a mean-field character, *i.e.* it contains a single removal pole for each sp momentum p. Just as in the boson case of Sec. 18.7, it is therefore possible to derive the BCS equation as a mean-field approximation.

We consider a general linear transformation which mixes the fermion addition and removal operators

$$b^\dagger_{pm} = u_p a^\dagger_{pm} + v_p s_m a_{-p-m}; \qquad b_{pm} = u_p a_{pm} + v_p s_m a^\dagger_{-p-m}, \qquad (22.67)$$

where the u_p and v_p are real coefficients. Note that this is consistent with isotropy, momentum conservation and time-reversal invariance. Equation (22.67) was first considered in [Bogoliubov (1958); Valatin (1958)], and is called the Bogoliubov–Valatin transformation. We demand that the transformation is canonical, *i.e.* that it should preserve the fundamental fermion anticommutation relations. From Eq. (22.67) one calculates

$$\{b_{pm}, b_{p'm'}\} = 0; \qquad \{b_{pm}, b^\dagger_{p'm'}\} = (u_p^2 + v_p^2)\delta_{p,p'}\delta_{m,m'}, \qquad (22.68)$$

and observes that the transformation is canonical, provided that the normalization condition

$$u_p^2 + v_p^2 = 1 \qquad (22.69)$$

is fulfilled. The inverse transformation is easily worked out as

$$a^\dagger_{pm} = u_p b^\dagger_{pm} - v_p s_m b_{-p-m}; \qquad a_{pm} = u_p b_{pm} - v_p s_m b^\dagger_{-p-m}, \qquad (22.70)$$

The vacuum state of the new b operators is denoted as $|BCS\rangle$, and is defined as the state for which

$$b_{pm}|BCS\rangle = 0, \qquad \forall pm. \qquad (22.71)$$

We now want to minimize the expectation value of the grand-canonical potential $\langle BCS|\hat{\Omega}|BCS\rangle$, in order to get the best possible approximation to the true ground state among all vacuum states generated by transformations of the type (22.67). The strategy consists of rewriting $\hat{\Omega}$ in terms of the $b, b^\dagger$ operators, and reshuffle to normal ordering with respect to the vacuum $|BCS\rangle$ of the b operators (*i.e.* all $b^\dagger$ to the left of all b operators). The result contains normal products of zero, two, and four $b^\dagger, b$ operators, where the fully contracted or zero-operator term is just the required $\Omega_0 = \langle BCS|\hat{\Omega}|BCS\rangle$.

In order to execute this strategy, Wick's theorem (for equal time arguments, see Sec. 8.4) is very useful. The possible contractions of the $a, a^\dagger$

operators with respect to the b-vacuum are given by the following expressions

$$a_{\boldsymbol{p}m}^{\dagger\bullet} a_{\boldsymbol{p}'m'}^{\bullet} = \langle BCS | a_{\boldsymbol{p}m}^{\dagger} a_{\boldsymbol{p}'m'} | BCS \rangle = v_p^2 \delta_{\boldsymbol{p},\boldsymbol{p}'} \delta_{m,m'},$$

$$a_{\boldsymbol{p}m}^{\bullet} a_{\boldsymbol{p}'m'}^{\dagger\bullet} = u_p^2 \delta_{\boldsymbol{p},\boldsymbol{p}'} \delta_{m,m'},$$

$$a_{\boldsymbol{p}m}^{\bullet} a_{\boldsymbol{p}'m'}^{\bullet} = u_p v_p s_m \delta_{\boldsymbol{p},-\boldsymbol{p}'} \delta_{m,-m'},$$

$$a_{\boldsymbol{p}m}^{\dagger\bullet} a_{\boldsymbol{p}'m'}^{\dagger\bullet} = -u_p v_p s_m \delta_{\boldsymbol{p},-\boldsymbol{p}'} \delta_{m,-m'}. \tag{22.72}$$

Transforming the sp term is easy:

$$\sum_{\boldsymbol{p}m} \varepsilon_{p\mu} a_{\boldsymbol{p}m}^{\dagger} a_{\boldsymbol{p}m} = \sum_{\boldsymbol{p}m} \varepsilon_{p\mu} (v_p^2 + N[a_{\boldsymbol{p}m}^{\dagger} a_{\boldsymbol{p}m}]), \tag{22.73}$$

where $N[..]$ denotes normal ordering with respect to $|BCS\rangle$. The interaction term

$$\frac{1}{2} \sum_{\boldsymbol{p}_i m_i} (\boldsymbol{p}_1 \boldsymbol{p}_2 | V | \boldsymbol{p}_3 \boldsymbol{p}_4) \, a_{\boldsymbol{p}_1 m_1}^{\dagger} a_{\boldsymbol{p}_2 m_2}^{\dagger} a_{\boldsymbol{p}_4 m_2} a_{\boldsymbol{p}_3 m_1}, \tag{22.74}$$

requires a bit more work. We note that the 4 $a, a^{\dagger}$ operators give rise to 6 singly contracted terms, but only 3 are independent. In terms of the momentum labels in Eq. (22.74), the $1^{\bullet}2^{\bullet}$ and $4^{\bullet}3^{\bullet}$ terms are hermitian conjugate, whereas the $1^{\bullet}3^{\bullet}$ and $2^{\bullet}4^{\bullet}$ terms are equal, and so are the $1^{\bullet}4^{\bullet}$ and $2^{\bullet}3^{\bullet}$ terms.

Using Eqs. (22.72) and (22.74) we therefore obtain for the singly contracted terms

$$\hat{V}_1 = -\frac{1}{2} \sum_{\boldsymbol{p}\boldsymbol{p}'m} \left\{ \frac{W(\boldsymbol{p}-\boldsymbol{p}')}{V} u_p v_p s_m \left(N[a_{\boldsymbol{p}'m}^{\dagger} a_{-\boldsymbol{p}'-m}^{\dagger}] + N[a_{-\boldsymbol{p}'-m} a_{\boldsymbol{p}'m}] \right) \right.$$

$$\left. + 2 \left(2\frac{W(0)}{V} - \frac{W(\boldsymbol{p}-\boldsymbol{p}')}{V} \right) v_p^2 \sum_m N[a_{\boldsymbol{p}'m}^{\dagger} a_{\boldsymbol{p}'m}] \right\}. \tag{22.75}$$

The doubly contracted terms arise from the previous expression by replacing the normal products by contractions, and applying a factor $\frac{1}{2}$ to correct for the fact that each double contraction has been counted twice. Adding the constant contribution from Eq. (22.73) we get

$$\Omega_0 = 2 \sum_{\boldsymbol{p}} \varepsilon_{p\mu} v_p^2 + \sum_{\boldsymbol{p}\boldsymbol{p}'} \left\{ v_p^2 v_{p'}^2 \left(2\frac{W(0)}{V} - \frac{W(\boldsymbol{p}-\boldsymbol{p}')}{V} \right) \right. \tag{22.76}$$

$$\left. + u_p u_{p'} v_p v_{p'} \frac{W(\boldsymbol{p}-\boldsymbol{p}')}{V} \right\} = \sum_{\boldsymbol{p}} \left\{ (2\varepsilon_{p\mu} + V_p) v_p^2 - u_p v_p \Delta_p \right\},$$

in terms of quantities V_p and Δ_p which depend on the u_p, v_p as

$$V_p = \sum_{p'} [2\frac{W(0)}{V} - \frac{W(p-p')}{V}]v_{p'}^2,$$

$$\Delta_p = \sum_{p'} \frac{W(p-p')}{V} u_{p'} v_{p'}. \tag{22.77}$$

The full $\hat{\Omega}$ can now be rewritten as $\hat{\Omega} = \Omega_0 + \hat{\Omega}_1 + \hat{\Omega}_2$, where

$$\hat{\Omega}_1 = \sum_{pm} \varepsilon_{p\mu} N[a_{pm}^\dagger a_{pm}] + \hat{V}_1, \tag{22.78}$$

and $\hat{\Omega}_2$ contains the uncontracted contributions of the interaction

$$\hat{\Omega}_2 = \frac{1}{2} \sum_{p_i m_i} (p_1 p_2| V |p_3 p_4) N[a_{p_1 m_1}^\dagger a_{p_2 m_2}^\dagger a_{p_4 m_2} a_{p_3 m_1}]. \tag{22.79}$$

It remains to express $\hat{\Omega}$ in terms of the $b, b^\dagger$ operators. Employing Eq. (22.70) this is straightforward, and with $\chi_p = \varepsilon_{p\mu} + V_p$ one finds[4]

$$\hat{\Omega}_1 = \sum_{pm} \{\chi_p(u_p^2 - v_p^2) + 2\Delta_p u_p v_p\} b_{pm}^\dagger b_{pm} \tag{22.80}$$

$$+ \left\{ \frac{\Delta_p}{2}(u_p^2 - v_p^2) - \chi_p u_p v_p \right\} s_m (b_{pm}^\dagger b_{-p-m}^\dagger + b_{-p-m} b_{pm}).$$

The last line of Eq. (22.80) contains the "off-diagonal" bb and $b^\dagger b^\dagger$ contribution, which can be made to vanish by choosing the u_p, v_p as solutions of

$$\chi_p u_p v_p = \frac{1}{2}\Delta_p(u_p^2 - v_p^2). \tag{22.81}$$

This equation should be solved together with the normalization constraint (22.69), and it is somewhat easier to eliminate the latter by defining a new variable θ_p as

$$u_p = \cos\theta_p; \quad v_p = \sin\theta_p, \tag{22.82}$$

---

[4]It is of course possible to rewrite $\hat{\Omega}_2$ in terms of the $b, b^\dagger$ operators as well. However, $\hat{\Omega}_2$ plays no role at the BCS mean-field level, and while it can be employed to go beyond BCS, it seems more practical to use the systematic diagrammatic expansion of Sec. 22.3. We therefore omit further consideration of $\hat{\Omega}_2$.

in terms of which $u_p^2 - v_p^2 = \cos 2\theta_p$ and $2u_p v_p = \sin 2\theta_p$. Equation (22.81) is now easily solved as $\theta_p = \frac{1}{2}\text{Arctan}\,(\Delta_p/\chi_p)$ or

$$\cos 2\theta_p = \frac{\chi_p}{\sqrt{\chi_p^2 + \Delta_p^2}}; \qquad \sin 2\theta_p = \frac{\Delta_p}{\sqrt{\chi_p^2 + \Delta_p^2}}, \qquad (22.83)$$

which implies

$$u_p^2 = \frac{E_p + \chi_p}{2E_p}, \qquad v_p^2 = \frac{E_p - \chi_p}{2E_p}, \qquad \text{where } E_p = \sqrt{\chi_p^2 + \Delta_p^2}, \qquad (22.84)$$

bringing $\hat{\Omega}_1$ to its diagonal form

$$\hat{\Omega}_1 = \sum_{pm} E_p b_{pm}^\dagger b_{pm}. \qquad (22.85)$$

At the same time we have found the transformation in Eq. (22.67) that minimizes $\Omega_0 = \langle BCS|\,\hat{\Omega}\,|BCS\rangle$, since this is equivalent to the diagonalization requirement (22.81). This can be understood by noting that a small variation in the $b, b^\dagger$ operators leads to a change in the vacuum $|\delta BCS\rangle$ which must be orthogonal to $|BCS\rangle$, and is thus of the form

$$|\delta BCS\rangle \sim b_{pm}^\dagger b_{-p-m}^\dagger |BCS\rangle. \qquad (22.86)$$

For $|BCS\rangle$ to have minimal Ω_0, it must be stable with respect to these two-quasiparticle excitations

$$0 = \langle BCS|\,\hat{\Omega} b_{pm}^\dagger b_{-p-m}^\dagger |BCS\rangle = \langle BCS|\,\hat{\Omega}_1 b_{pm}^\dagger b_{-p-m}^\dagger |BCS\rangle. \qquad (22.87)$$

Only the off-diagonal bb-term in Eq. (22.80) can contribute to Eq. (22.87), which is therefore equivalent to Eq. (22.81). It is in fact possible to derive the BCS equations by direct minimization of Ω_0 [see Exercise (2)].

A survey of the relations (22.77) and (22.84) establishes that these are completely equivalent to the BCS equations derived in Sec. 22.4. It allows to view BCS theory as a mean-field approximation: the remaining operator $\hat{\Omega}_1$ in Eq. (22.85) clearly has the structure of a noninteracting quasiparticle Hamiltonian, with positive excitation energies E_p. As a consequence, the ground state is simply the vacuum $|BCS\rangle$ (with no quasiparticles present), whereas all excited eigenstates of $\hat{\Omega}_1$ are of the form: $b_{p_1 m_1}^\dagger b_{p_2 m_2}^\dagger .. |BCS\rangle$. One may verify, using Eq. (22.67), that the state

$$b_{pm}[u_p - v_p s_m a_{pm}^\dagger a_{-p-m}^\dagger]\,|0\rangle = 0 \qquad (22.88)$$

vanishes $\forall \boldsymbol{pm}$. Since the different momenta $\boldsymbol{p}$ do not interfere, the product wave function

$$|BCS\rangle = \prod_{\boldsymbol{p}} [u_p - v_p s_m a^\dagger_{\boldsymbol{pm}} a^\dagger_{-\boldsymbol{p}-m})] \, |0\rangle \,, \qquad (22.89)$$

has the property (22.71) and can therefore be identified as the BCS vacuum. The product state is normalized to unity, since

$$\langle 0| \, [u_p - v_p s_m a_{-\boldsymbol{p}-m} a_{\boldsymbol{pm}}][u_p - v_p s_m a^\dagger_{\boldsymbol{pm}} a^\dagger_{-\boldsymbol{p}-m}] \, |0\rangle = 1 \qquad (22.90)$$

implies $\langle BCS|BCS\rangle = 1$. In the original BCS paper [Bardeen *et al.* (1957)] the explicit form of Eq. (22.89) was in fact used as a variational ansatz. The fact that $|BCS\rangle$ is a coherent pair state, is best illustrated by rewriting Eq. (22.89) in exponential form

$$|BCS\rangle = \prod_{\boldsymbol{p}} [u_p] \prod_{\boldsymbol{p}} [1 - \frac{v_p}{u_p} s_m a^\dagger_{\boldsymbol{pm}} a^\dagger_{-\boldsymbol{p}-m}] \, |0\rangle$$

$$= \prod_{\boldsymbol{p}} [u_p] \prod_{\boldsymbol{p}} [e^{-\frac{v_p}{u_p} s_m a^\dagger_{\boldsymbol{pm}} a^\dagger_{-\boldsymbol{p}-m}}] \, |0\rangle \sim e^{-B^\dagger} \, |0\rangle \,, \qquad (22.91)$$

in terms of a pair state $B^\dagger = \frac{1}{2} \sum_{\boldsymbol{pm}} \frac{v_p}{u_p} s_m a^\dagger_{\boldsymbol{pm}} a^\dagger_{-\boldsymbol{p}-m}$.

22.6   Applications

The present chapter only covers the overall structure of a Green's function treatment of superconductivity in any detail, and the following discussion is quite superficial. For in-depth treatments of specific applications we must refer to dedicated books, *e.g.* [Schrieffer (1983); de Gennes (1966)] (superconductivity), [Anderson (1997)] (high-T_c superconductors), and [Vollhardt and Wölfle (1990)] (^{3}He superfluidity).

22.6.1   *Superconductivity in metals*

Most metals, cooled below a certain critical temperature, exhibit a range of macroscopic quantum phenomena known under the heading of superconductivity. The resistance to electric current drops to zero, and a current in a superconducting ring persists without observable dissipation. Zero resistance implies that the electric field inside the material vanishes (otherwise an infinite current would result). Maxwell's equations then dictate a

constant magnetic field. In fact, the magnetic field inside a metallic superconductor is zero. The superconductor has perfect diamagnetic properties
and shows the Meissner effect: a — not too strong — magnetic field is
completely expelled from its interior [Meissner and Ochsenfeld (1933)].

After the discovery of superconductivity in mercury by Kamerlingh–
Onnes in 1911, the phenomenon remained unexplained for a long time.
It became clear that strong parallels exist with the phase transition to
superfluidity in ^{4}He, and that diamagnetism was the fundamental property. Based on the "electron superfluid" analogy, successful phenomenological theories were proposed by [London and London (1935)] and [Pippard
(1953)], and by [Ginzburg and Landau (1950)]. The Ginzburg–Landau
theory is actually a general treatment of second-order phase transitions.

For a long time it was unclear how the repulsive Coulomb force between
electrons, could be reconciled with an attraction needed for the formation of
quasibosonic electron pairs in a microscopic theory. With the observation
of the isotope effect (the fact that the critical temperature depends as
$T_c \sim M^{-\frac{1}{2}}$ on the mass M of the positive ions forming the crystal lattice
of the metal), the relevance of the interplay between the electrons and the
lattice became clear.

The quantized vibrations of the lattice around the ionic equilibrium
positions are called phonons. Electrons in metals also interact through the
exchange of virtual phonons. In pictorial terms, an electron moving through
the crystal bends the lattice somewhat towards itself (phonon creation); this
temporarily leaves a region of positive charge, which can attract another
electron (phonon absorption) before the slow-moving ions have had time
to restore the lattice. The phonon-exchange or Frölich interaction [Frölich
(1952)] is spin-independent and has the form

$$(\boldsymbol{p_1 p_2}| \, V(E) \, |\boldsymbol{p_3 p_4}) = \delta_{\boldsymbol{p_1 + p_2}, \boldsymbol{p_3 + p_4}} \frac{1}{V} \gamma^2 \frac{\Omega_{\boldsymbol{Q}}^2}{E^2 - \Omega_{\boldsymbol{Q}}^2} \theta(\Omega_D - \Omega_{\boldsymbol{Q}}), \quad (22.92)$$

where $\Omega_{\boldsymbol{Q}}$ is the phonon spectrum, γ is the electron–phonon coupling
strength, and $E = \varepsilon_{\boldsymbol{p_1}} - \varepsilon_{\boldsymbol{p_3}}$ ($\boldsymbol{Q} = \boldsymbol{p_1} - \boldsymbol{p_3}$) is the energy (momentum)
transfer. The denominator of Eq. (22.92) reflects the propagation of the
exchanged phonon; the interaction $V(E)$ is therefore retarded, as indicated
by the dependence on the energy transfer. Finally, $\Omega_D/\hbar$ is the Debye
frequency, or the maximal frequency allowed in the discrete lattice. The
interaction in Eq. (22.92) is attractive when $|E| < \Omega_{\boldsymbol{Q}} < \Omega_D$, and can dom-
inate the (screened) Coulomb repulsion when the electron–phonon coupling
strength is large enough. The attraction vanishes when $\varepsilon_{\boldsymbol{p_1}}$ and $\varepsilon_{\boldsymbol{p_3}}$ differ

by more than Ω_D (which is quite tiny, of the order of 10^{-2} eV, compared to typical kinetic and Coulomb energies of the order of eV). This clearly limits allowed particle–hole excitations to momenta within a small region around the Fermi surface.

The net result of the Coulomb repulsion and the Frölich interaction, including the in-medium charge screening, produces an effective interaction that is attractive for electrons near the Fermi surface. The first microscopic explanation of metallic superconductivity was provided by BCS [Bardeen *et al.* (1957)], using a variational ansatz [see Eq. (22.77)] and the schematic interaction of Eq. (22.48), which is essentially the static approximation ($E = 0$) of Eq. (22.92), with $\lambda = -\gamma^2$ and the cut-off energy $c = \Omega_D$. This leads to values for the gap Δ of the order of 10^{-4} eV. The condensation energy per particle [the energy gain because of pairing, see Eq. (22.66)] is very small, of the order of 10^{-8} eV: a very subtle effect, but responsible for huge changes in the macroscopic behavior.

Note that diagrammatic methods allow to go beyond the static approximation, and to study a coupled system of electrons and phonons. This is described by the Eliashberg equations [Eliashberg (1960)], which include retardation effects in the phonon exchange.

The BCS theory (and its finite temperature extension) proved to be astonishingly successful. The attractive interaction produces a condensate of Cooper pairs; the pairing occurs in states with opposite spin and momentum, and each electron can be thought to move in coherence with its partner.

One of the fundamental length scales is the *coherence length* ξ_0 that can be thought of as the spatial extent of a Cooper pair. This can be roughly estimated by realizing that in the model with a constant gap Δ, the pair wave function $F_p = \Delta/(2E_p)$ [see Eq. (22.46] has an extent (δp) in momentum space corresponding to $(\delta \chi) = \Delta$, or $(\delta p) = \Delta (dp/d\chi)_{\chi=0}$. For a free spectrum (or introducing an effective mass) one has $(dp/d\chi)_{\chi=0} = m/p_F = 1/v_F$, where v_F is the velocity at the Fermi surface. As a consequence, the spatial extent is characterized by $(\delta r) = \hbar/(\delta p) = \hbar v_F/\Delta$. A more precise calculation yields the BCS coherence length $\xi_0 = \frac{\hbar v_F}{\pi \Delta}$, which is of the order of 10^{-6} m. This should be compared to the average distance d_s between the electrons involved in pairing. The density of superconducting electrons is $\rho_s = N_s/V = 2D(0) \int_{-c}^{c} d\chi v_p^2 = 2D(0)\Omega_D$. Assuming a free spectrum yields $N_s/N = 3\Omega_D/(2\varepsilon_F)$, corresponding to d_s of the order of 10^{-9} m. This implies that the spatial extent of the Cooper pairs is of the order of a thousand times the average interparticle distance, *i.e.* a great many overlapping Cooper pairs occupy the same volume. BCS pairing is therefore called the weak-coupling limit, and is different from condensation of fermion pairs into isolated bosonic "molecules". The latter situation would arise when the attractive forces become very strong.

Up to now we have discussed normal, metallic superconductors of the so-called type-I, characterized by a complete Meissner effect. These materials must completely expel an applied magnetic field (costing energy) in order to remain superconducting, and thus cannot withstand large field strengths. The Meissner effect does not occur discontinuously at the surface; in fact the magnetic field penetrates slightly in the interior and dies out exponentially, with a length scale given by the *London penetration depth* λ_L. The latter depends on the superconducting electron density according to $\lambda_L \sim \rho_s^{-\frac{1}{2}}$.

Other superconductors (mostly alloys) belong to the type-II class, and show only a partial Meissner effect: between two critical values of the field strength, magnetic fields are not completely expelled but confined to an array of tubes with matter in the normal state (the vortex lattice). This allows the bulk of the material to remain superconducting at much larger field strengths than the type-I class. The ratio of the two length scales (penetration depth and coherence length) is crucial for the distinction between type-I or type-II behavior.

In 1986, [Bednorz and Müller (1986)] discovered "high T_c" superconductivity in a new class of ceramic materials containing stacks of CuO_2 planes, alternated by layers which act as charge reservoirs for the planes. These cuprate materials are of extreme type-II and anisotropic, and the critical temperature can be amazingly high, up to 165 K. While the role of electron pairing itself is generally accepted, the precise origin of the pairing mechanism is still hotly debated.

22.6.2 *Superfluid ^3He*

The transition to a superfluid state in the ^{3}He liquid only occurs at extremely low temperatures (below 3 mK). In fact, depending on the pressure, and the presence of a magnetic field, three superfluid phases are observed [Osheroff *et al.* (1972a); Osheroff *et al.* (1972b)]. The pair state has relative orbital angular momentum $L = 1$, and spin $S = 1$, which is energetically more favorable than the $L = 0$, $S = 0$ channel due to the very repulsive short-distance part of the interaction.

As a consequence, the ^{3}He liquid is a prime example of an anisotropic superfluid (in contrast to the isotropic singlet pairing in metals), being characterized by two vector quantities $\hat{\boldsymbol{L}}$ and $\hat{\boldsymbol{S}}$ that can exhibit long-range order. This makes for a far richer phenomenology (*e.g.* the three possible superfluid phases are distinguished by the S_z structure of the pairs), which is at present rather well understood [Leggett (1975); Leggett (2004)].

22.6.3   *Superfluidity in neutron stars*

The structure of a neutron star is schematically illustrated in Fig. 5.1. For many properties of a neutron star (*e.g.* the cooling rate through neutrino emission, after its creation in a supernova explosion), the treatment of pairing in the nucleonic quantum fluids of its interior is absolutely crucial. A detailed account can be found in a recent review by [Dean and Hjorth–Jensen (2003)], concerning nucleon pairing in both finite nuclei and fluids.

In a first step one can study the free NN interaction V, and determine the attractive components that can give rise to pairing. The momentum range in which a specific partial wave of V is attractive, corresponds roughly to a density range where such momenta are typically found near the Fermi surface. For the neutrons in the inner crust region, the density is quite low ($\rho \approx \rho_0/10$ with $\rho_0 = 0.16\,\mathrm{fm}^{-3}$), and the singlet $^1S_0$ channel will be dominant for neutron pairing. The neutron density becomes much larger in the quantum fluid interior ($\rho \approx \rho_0 - 2\rho_0$), corresponding to momenta that allow the neutrons to experience the strongly repulsive core of the $^1S_0$ component. One then expects neutron pairing predominantly in the triplet $^3P_2 - {}^3F_2$ channel, which is the most attractive $T = 1$ component in this momentum region. This global picture is indeed confirmed by microscopic calculations (*e.g.* at the BHF+BCS level).

At the same time, one should keep in mind that many calculations treat the nucleon propagator at the mean-field level, and do not include the reduced quasiparticle strength, and the corresponding reduction of the in-medium interaction strength near the Fermi surface. This effect can significantly decrease the pairing gap, and in principle requires a self-consistent treatment of the sp propagator, the BCS gap, and the in-medium interaction [Bożek (2002, 2003)]. Some recent developments are explored in Sec. 24.8.2.

Note that for realistic NN potentials, the BCS gap equation (22.47) couples different partial-wave components, and becomes more difficult to solve. Angle-averaging procedures are usually employed to simplify the problem. A recent separation method proposed in [Khodel *et al.* (1996)] allows to solve the BCS problem in a complete way [Khodel *et al.* (1998)].

22.7 Inhomogeneous systems

The BCS pairing mechanism was originally developed for electrons in metals, but is also relevant for inhomogeneous finite systems like nuclei. In this section, the BCS treatment of Sec. 22.5 will be reformulated in an arbitrary sp basis. We note the large parallels with the corresponding discussion for bosons in Sec. 19.3. We start with the Hamiltonian

$$\hat{H} = \sum_{\alpha\gamma} \varepsilon_{\alpha\gamma} a_\alpha^\dagger a_\gamma + \frac{1}{4} \sum_{\alpha\beta\gamma\delta} \langle \alpha\beta| \, V \, |\gamma\delta\rangle \, a_\alpha^\dagger a_\beta^\dagger a_\delta a_\gamma. \tag{22.93}$$

The general Bogoliubov–Valatin transformation mixes addition and removal operators

$$b_i^\dagger = \sum_\gamma U_{\gamma i} a_\gamma^\dagger + V_{\gamma i} a_\gamma, \qquad b_i = \sum_\gamma U_{\gamma i}^* a_\gamma + V_{\gamma i}^* a_\gamma^\dagger, \tag{22.94}$$

which can be reformulated in matrix notation as

$$\begin{pmatrix} b \\ b^\dagger \end{pmatrix} = \begin{pmatrix} U^\dagger & V^\dagger \\ V^T & U^T \end{pmatrix} \begin{pmatrix} a \\ a^\dagger \end{pmatrix}, \quad \text{or:} \ B = M^\dagger A, \tag{22.95}$$

where $M = \begin{pmatrix} U & V^* \\ V & U^* \end{pmatrix}$. The transformation is canonical, provided that M is unitary, *i.e.* $M^\dagger M = MM^\dagger = 1$, and the inverse transformation is simply

$$\begin{pmatrix} a \\ a^\dagger \end{pmatrix} = \begin{pmatrix} U & V^* \\ V & U^* \end{pmatrix} \begin{pmatrix} b \\ b^\dagger \end{pmatrix} \quad \text{or:} \ A = MB. \tag{22.96}$$

As in Sec. 22.5 we try to find the b-operators for which the corresponding vacuum $|0_B\rangle$ has minimal energy $E_0 = \langle 0_B| \, \hat{H} \, |0_B\rangle$. This can be done by reshuffling to normal order with respect to $|0_B\rangle$. Contractions then read

$$a_\alpha^{\dagger\bullet} a_\gamma^{\bullet} = \langle 0_B| \, a_\alpha^\dagger a_\gamma \, |0_B\rangle = \rho_{\alpha\gamma} = \rho_{\gamma\alpha}^*,$$
$$a_\alpha^{\dagger\bullet} a_\beta^{\dagger\bullet} = \langle 0_B| \, a_\alpha^\dagger a_\beta^\dagger \, |0_B\rangle = F_{\alpha\beta} = -F_{\beta\alpha}, \tag{22.97}$$

and are recognized as the normal and anomalous density, respectively. Employing Wick's theorem, the Hamiltonian is decomposed as $\hat{H} = E_0 + \hat{H}_1 + \hat{H}_2$, where the energy E_0 represents the doubly contracted term

$$E_0 = \sum_{\alpha\gamma} \varepsilon_{\alpha\gamma} \rho_{\alpha\gamma} + \frac{1}{2} \sum_{\alpha\beta\gamma\delta} \langle \alpha\beta| \, V \, |\gamma\delta\rangle \left(\rho_{\alpha\gamma}\rho_{\beta\delta} + \frac{1}{2} F_{\alpha\beta} F_{\gamma\delta}^* \right)$$
$$= \frac{1}{2} \sum_{\alpha\gamma} (\varepsilon_{\alpha\gamma} + \chi_{\alpha\gamma}) \rho_{\alpha\gamma} + \frac{1}{2} \sum_{\alpha\beta} \Delta_{\alpha\beta} F_{\alpha\beta}, \tag{22.98}$$

and the singly contracted term is

$$\hat{H}_1 = \sum_{\alpha\gamma} \chi_{\alpha\gamma} N[a_\alpha^\dagger a_\gamma] + \frac{1}{2} \sum_{\alpha\beta} \left(\Delta_{\alpha\beta} N[a_\alpha^\dagger a_\beta^\dagger] + \Delta_{\alpha\beta}^* N[a_\beta a_\alpha] \right). \quad (22.99)$$

In Eqs. (22.98) and (22.99) the HF-like sp energies χ, and the pairing field Δ, have been introduced,

$$\chi_{\alpha\gamma} = \varepsilon_{\alpha\gamma} + \sum_{\beta\delta} \langle \alpha\beta| V |\gamma\delta\rangle \rho_{\beta\delta}; \quad \Delta_{\alpha\beta} = \frac{1}{2} \sum_{\gamma\delta} \langle \alpha\beta|V|\gamma\delta\rangle F_{\gamma\delta}^*. \quad (22.100)$$

The term $\hat{H}_2$, containing normal products of four $b, b^\dagger$, is irrelevant for the further discussion, and will not be considered.

Minimizing E_0 is again equivalent to requiring $\hat{H}_1$ to be diagonal in the b-operators. Using antisymmetry ($N[a_\alpha^\dagger a_\gamma] = -N[a_\gamma a_\alpha^\dagger]$) and linearity of the normal product, $\hat{H}_1$ can be rewritten in a more symmetrical form as

$$\hat{H}_1 = \frac{1}{2} N \left[(a^\dagger \ a) \begin{pmatrix} \chi & \Delta \\ -\Delta^* & -\chi^* \end{pmatrix} \begin{pmatrix} a \\ a^\dagger \end{pmatrix} \right]$$

$$= \frac{1}{2} N[A^\dagger H_a A] = \frac{1}{2} N[B^\dagger (M^\dagger H_a M) B]. \quad (22.101)$$

Note that χ is hermitian, Δ is antisymmetric, and H_a is therefore a hermitian matrix of twice the dimension of the sp space. Diagonality of $H_b = M^\dagger H_a M$ is achieved by taking the columns of the unitary transformation matrix M as the solutions $\begin{pmatrix} U_i \\ V_i \end{pmatrix}$ of the eigenvalue problem

$$\chi U_i + \Delta V_i = E_i U_i,$$
$$\Delta^* U_i + \chi^* V_i = -E_i V_i. \quad (22.102)$$

It has the same form as the bosonic Eq. (19.39), but since Δ is now antisymmetric, Eq. (22.102) represents a simple hermitian eigenvalue problem with real eigenvalues. These come in pairs with opposite signs, since any solution (U_i, V_i, E_i) implies that $(V_i^*, U_i^*, -E_i)$ is also one. The positive-energy solutions correspond to the physical quasiparticle excitations $b_i^\dagger$, in terms of which the Hamiltonian of Eq. (22.101) becomes $\hat{H}_1 = \sum_{i>0} E_i b_i^\dagger b_i$.

The relations (22.102) are called the Hartree–Fock–Bogoliubov (HFB) equations in general, or the Bogoliubov–de Gennes equations when expressed in coordinate space. Note that the HF-like energy χ, and the pairing field Δ, still depend on the normal and anomalous densities ρ and F, which are in turn determined by the amplitudes U and V. Like in HF, the full HFB method therefore requires iterative methods for finding a self-consistent solution. For finite N, the energy minimization should be performed under the particle-number constraint $[N = \text{Trace}(\rho)]$, which is done in the usual fashion by introducing the grand-canonical potential $\hat{\Omega} = \hat{H} - \mu \hat{N}$. It amounts to replacing $\varepsilon \rightarrow \varepsilon - \mu$ in the treatment above.

The HFB method or approximations to it, have been used extensively in nuclear calculations [Bender *et al.* (2003)] to study so-called superfluid nuclei: in open-shell nuclei the attractive forces acting between like valence nucleons with nearly degenerate sp energies favor the formation of pairs with total angular momentum $J = 0$ (see *e.g.* [Heyde (1990)] and also Sec. 17.2). Since the number of nucleons is quite small, particle-number nonconservation is an important problem that needs correcting by projection techniques [Ring and Schuck (1980); Allaart *et al.* (1988)].

Canonical single-particle basis

The full Bogoliubov–Valatin transformation in Eq. (22.94) can always be decomposed into three successive transformations. This is by and large analogous to the bosonic case of Sec. 19.3, where additional details can be found. Applying singular value decomposition to the (complex) transformation matrices U and V, one can write $U = W_1^\dagger u W_2$ and $V = W_3^\dagger v W_4$, where the W_i are unitary matrices and both u and v are real and positive diagonal matrices. The unitarity requirements $U^\dagger U + V^\dagger V = 1$ and $UU^\dagger + V^* V^T = 1$ again imply that $u^2 + v^2 = 1$, and hence u and v have corresponding blocks with degenerate values. In addition, one can write $V = W_1^T R v W_2$, with R unitary and block-diagonal (*i.e.* commuting with u and v. This allows treating each degeneracy block separately.

For fermions, the additional unitarity condition now requires $UV^\dagger$ to be antisymmetric, or equivalently $uv(R - R^T) = 0$. For a block with a nonzero value of uv, the matrix R should therefore be unitary and antisymmetric. It is easy to see that R then has necessarily even dimension, with eigenvectors and eigenvalues that come in pairs $(X, e^{i\theta})$ and $(X^*, -e^{i\theta})$. Taking the real combinations $X_+ = (X + X^*)/\sqrt{2}$ and $X_- = (X - X^*)/(\sqrt{2}i)$, one can show that R can be transformed by a real orthogonal matrix O into the canonical form $R = O^T e^{i\theta} \tau O$, where τ contains 2×2 blocks of the form $\begin{pmatrix} 0 & -1 \\ 1 & 0 \end{pmatrix}$ along its diagonal, and the real-diagonal matrix θ is twofold degenerate in each such block. One is thus allowed to redefine $W_1 \rightarrow e^{i\frac{\theta}{2}} O W_1$ and $W_2 \rightarrow e^{i\frac{\theta}{2}} O W_2$ to get the final canonical form $U = W_1^\dagger u W_2$ and $V = W_1^T v \tau W_2$.

It is curious to see that fermions have this natural pairing property, where each sp state has a partner associated to it, having the same u and v. In systems with time-reversal invariance the associated partners $(\alpha, \overline{\alpha})$ are time-reversed sp states, but in general this does not have to be the case. Note that the sp states with $uv = 0$ are either fully occupied ($u = 0, v = 1$) or unoccupied ($u = 1, v = 0$). In that case, the condition that $UV^\dagger$ be antisymmetric is automatically fulfilled for the corresponding block. In the blocks with $uv = 0$ the matrix R is irrelevant, and τ can be replaced by the unit matrix.

Hence, the full transformation in Eq. (22.95) can always be decomposed as a succession of three transformations,

$$\begin{pmatrix} b \\ b^\dagger \end{pmatrix} = \begin{pmatrix} W_2^\dagger & 0 \\ 0 & W_2^T \end{pmatrix} \begin{pmatrix} u & -v\tau \\ -v\tau & u \end{pmatrix} \begin{pmatrix} W_1 & 0 \\ 0 & W_1^* \end{pmatrix} \begin{pmatrix} a \\ a^\dagger \end{pmatrix}. \qquad (22.103)$$

This is known as the Bloch–Messiah decomposition [Bloch and Messiah (1962)]. The first transformation (involving W_1) is an ordinary change of sp basis $a \to a'$ to the canonical basis $a'_\alpha = \sum_\gamma (W_1)_{\alpha\gamma} a_\gamma$. The second transformation $a' \to b'$ is a BCS transformation among the associated partners of the canonical basis,

$$b'_\alpha = u_\alpha a'_\alpha + v_\alpha a'^\dagger_{\overline{\alpha}}, \quad b'_{\overline{\alpha}} = u_\alpha a'_{\overline{\alpha}} - v_\alpha a'^\dagger_\alpha. \qquad (22.104)$$

Note that $u_\alpha = u_{\overline{\alpha}}$ and $v_\alpha = v_{\overline{\alpha}}$ are all taken to be positive here; it is sometimes easier to shift the phase difference in Eq. (22.104) into the v-amplitudes and/or the sp states.

These two transformations are sufficient to remove $b'b'$ and $b'^\dagger b'^\dagger$ terms from the Hamiltonian and to fix the quasiparticle vacuum. The final transformation $b' \to b$ is again an ordinary unitary transformation $b_i = \sum_\alpha (W_2^*)_{\alpha i} b'_\alpha$ on the quasiparticle operators and brings the $b^\dagger b$ block in the Hamiltonian to its diagonal form.

The quasiparticle vacuum and the structure of the condensed pair state, are most easily expressed in the canonical basis. The derivation then becomes identical to the Fermi gas expressions of Sec. 22.5. With the identity $b'_\alpha [u_\alpha - v_\alpha a'^\dagger_\alpha a'^\dagger_{\overline{\alpha}}] |0\rangle = 0$, one can check that the wave function

$$|0_B\rangle = \prod_{\alpha > 0} [u_\alpha - v_\alpha a'^\dagger_\alpha a'^\dagger_{\overline{\alpha}}] |0\rangle \qquad (22.105)$$

obeys $b'_\alpha |0_B\rangle = 0, \forall \alpha$; it is normalized to unity as well. The labels $\alpha > 0$ refer to the first member of each associated pair. It is also assumed that no fully occupied sp states (with $u_\alpha = 0$) are present; these are of course trivial to include as a Slater determinant. The BCS wave function can alternatively be written as $|0_B\rangle \sim e^{-B^\dagger}|0\rangle$, in terms of the condensed pair state $B^\dagger = \sum_{\alpha>0} \frac{v_\alpha}{u_\alpha} a'^\dagger_\alpha a'^\dagger_{\bar\alpha}$.

Finally, it is sometimes useful to consider the generalized density matrix

$$R = \langle 0_B | AA^\dagger | 0_B \rangle = \begin{pmatrix} 1 - \rho^* & -F^* \\ F & \rho \end{pmatrix}. \qquad (22.106)$$

The condition that $|0_B\rangle$ be the vacuum of the b-operators, is equivalent to

$$R = M \langle 0_B | BB^\dagger | 0_B \rangle M^\dagger = M \begin{pmatrix} 1 & 0 \\ 0 & 0 \end{pmatrix} M^\dagger = \begin{pmatrix} UU^\dagger & UV^\dagger \\ VU^\dagger & VV^\dagger \end{pmatrix}, \qquad (22.107)$$

and one sees that $\rho = VV^\dagger$, and $F = VU^\dagger$. Going to the canonical basis

$$R' = \langle 0_B | A'A'^\dagger | 0_B \rangle = \begin{pmatrix} W_1 & 0 \\ 0 & W_1^* \end{pmatrix} R \begin{pmatrix} W_1^\dagger & 0 \\ 0 & W_1^T \end{pmatrix} = \begin{pmatrix} u^2 & -uv\tau \\ -uv\tau & v^2 \end{pmatrix}, \qquad (22.108)$$

therefore brings both the normal density, $\rho'_{\alpha\beta} = \delta_{\alpha,\beta} v_\alpha^2$, and the anomalous density, $F'_{\alpha\beta} = u_\alpha v_\alpha \delta_{\beta,\bar\alpha}[\theta(\alpha) - \theta(\beta)]$, to its canonical form.

22.8 Exact solutions of schematic pairing problems

Two-fermion states

We first look more closely at the structure of a general two-fermion state

$$|\Psi^2\rangle = B^\dagger |0\rangle = \sum_{\alpha\beta} C_{\alpha\beta} a^\dagger_\alpha a^\dagger_\beta |0\rangle. \qquad (22.109)$$

With the same reasoning as in Sec. 22.7, one proves the existence of a natural sp basis, in terms of which the pair addition operator has the form

$$B^\dagger = \sqrt{2} \sum_{\alpha>0} x_\alpha a^\dagger_\alpha a^\dagger_{\bar\alpha}, \qquad (22.110)$$

where the x_α are real and positive (and correspond to the singular values of the antisymmetric matrix $C_{\alpha\beta}$). The pairs $(\alpha, \bar\alpha)$ are not necessarily time-reversed sp states, but are associated pairs in the general sense. The density matrix is diagonal in this sp basis, with all natural occupations $x_\alpha^2 = \langle \Psi^2 | a^\dagger_\alpha a_\alpha | \Psi^2 \rangle = \langle \Psi^2 | a^\dagger_{\bar\alpha} a_{\bar\alpha} | \Psi^2 \rangle$ being (at least) twofold degenerate.

We will assume normalized two-fermion states, *i.e.* $\langle 0| BB^\dagger |0\rangle = 2\sum_{\alpha>0} x_\alpha^2 = 1$. It is also convenient to define $x_{\bar\alpha} = -x_\alpha$ and write equivalently,

$$B^\dagger = \frac{1}{\sqrt{2}} \sum_\alpha x_\alpha a_\alpha^\dagger a_{\bar\alpha}^\dagger, \text{with } \sum_\alpha x_\alpha^2 = 1. \qquad (22.111)$$

Pairing Hamiltonians

We now revisit the problem posed in the beginning of this chapter, in the form of a pairing Hamiltonian like Eq. (22.1)

$$\hat{H} = \sum_\alpha \varepsilon_\alpha a_\alpha^\dagger a_\alpha - g B^\dagger B, \qquad (22.112)$$

expressing a competition between the tendency to fill the lowest sp levels, and the tendency to lower the energy by creating $B^\dagger$ pairs. The eigenstates can be classified according to the presence of unpaired sp states $\beta_1..\beta_m$, since the subspaces

$$|\Psi_{\beta_1..\beta_m}\rangle = a_{\beta_1}^\dagger..a_{\beta_m}^\dagger \sum_{\alpha_1..\alpha_n > 0} C_{\alpha_1..\alpha_n} a_{\alpha_1}^\dagger a_{\bar\alpha_1}^\dagger..a_{\alpha_n}^\dagger a_{\bar\alpha_n}^\dagger |0\rangle, \qquad (22.113)$$

with different $\beta_1..\beta_m$, cannot be connected by the Hamiltonian (22.112). We can therefore restrict the discussion to the "fully paired" eigenstates with $N = 2n$ particles; in the other subspaces the unpaired sp states are blocked, and simply removed from the available sp space.

For general ε_α and x_α, the Hamiltonian (22.112) cannot be solved exactly. However, in two limiting cases it is possible: a flat distribution of $x_\alpha \equiv x$ (corresponding to a schematic pairing force) with an arbitrary sp spectrum ε_α, and a flat distribution of the sp energies $\varepsilon_\alpha \equiv \varepsilon$ (corresponding to one degenerate shell) with an arbitrary pairing vector x_α. These two cases are in fact the most relevant examples of a larger class of Hamiltonians [Dukelsky *et al.* (2004)], consisting of linear combinations of a complete set of commuting operators (the integrals of motion) that are quadratic in the generators of some Lie algebra [$SU(2)$ in the present case]. The models based on $SU(2)$ were discovered quite early [Richardson (1963, 1968); Richardson and Sherman (1964); Gaudin (1976)], but were largely ignored, mainly because the resulting set of nonlinear equations is difficult to solve near a singularity. Recently they were rediscovered [Dukelsky *et al.* (2001)], and algorithms have become available to solve them without much effort [Rombouts *et al.* (2004)].

Maximal pair occupation

In the limiting case of degenerate sp energies, the Hamiltonian (22.112) is equivalent to the operator $B^\dagger B$, and the N-particle ground state has a maximal pair occupation $n_B = \langle \Psi^N | B^\dagger B | \Psi^N \rangle$. Defining the set of two-fermion operators [Pan *et al.* (1998)]

$$\phi^\dagger(y) = \frac{1}{\sqrt{2}} \sum_\alpha \frac{x_\alpha}{1 - yx_\alpha^2} a_\alpha^\dagger a_{\bar\alpha}^\dagger, \qquad (22.114)$$

where the y are complex numbers [note that $B^\dagger \equiv \phi^\dagger(0)$], one may verify the following commutator relations

$$[B, \phi^\dagger(y)] = \sum_\alpha \frac{x_\alpha^2}{1 - yx_\alpha^2}(1 + 2\eta a_\alpha^\dagger a_\alpha) = \Lambda(y) + 2\eta \sum_\alpha \frac{x_\alpha^2}{1 - yx_\alpha^2} a_\alpha^\dagger a_\alpha,$$

$$[[B, \phi^\dagger(y_1)], \phi^\dagger(y_2)] = 4\eta \frac{\phi^\dagger(y_1) - \phi^\dagger(y_2)}{y_1 - y_2}, \qquad (22.115)$$

where $\eta = -1$. It is possible to show that all fully-paired eigenstates of $B^\dagger B$ with $N = 2n$ particles have the form of a product of pair operators

$$|\Psi^N\rangle = \phi^\dagger(0)\phi^\dagger(y_1)\phi^\dagger(y_2)..\phi^\dagger(y_{n-1}) |0\rangle, \qquad (22.116)$$

with the set $\{y\} = \{y_1, .., y_{n-1}\}$ to be determined. This can be seen explicitly by acting with $B^\dagger B$ on the state (22.116), using the commutation relations in Eq. (22.115) to move B to the right where it destroys the vacuum $|0\rangle$. The resulting state can be written as

$$B^\dagger B |\Psi^N\rangle = E(\{y\}) |\Psi^N\rangle + \sum_{k=1}^{n-1} V_k(\{y\}) B^\dagger B^\dagger \prod_{i(\neq k)=1}^{n-1} \phi^\dagger(y_i) |0\rangle,$$

$$V_k(\{y\}) = \Lambda(y_k) - 4\eta \left(\frac{1}{y_k} + \sum_{i(\neq k)=1}^{n-1} \frac{1}{y_k - y_i} \right),$$

$$E(\{y\}) = \Lambda(0) + 4\eta \sum_{k=1}^{n-1} \frac{1}{y_k}. \qquad (22.117)$$

Clearly, an eigenstate is obtained for any solution $\{y\}$ of the nonlinear set of equations $V_k(\{y\}) = 0$, or:

$$\sum_\alpha \frac{x_\alpha^2}{1 - y_k x_\alpha^2} - 4\eta \left(\frac{1}{y_k} + \sum_{i(\neq k)=1}^{n-1} \frac{1}{y_k - y_i} \right) = 0, \quad k = 1, .., n - 1. \quad (22.118)$$

Equation (22.117) then yields the corresponding eigenvalue,

$$E(\{y\}) = 1 + 4\eta \sum_{k=1}^{n-1} \frac{1}{y_k}.$$ (22.119)

Of special interest [Van Neck *et al.* (2001b)] is the largest among the eigenvalues (22.119) of $B^\dagger B$, which corresponds to the maximal occupation n_B^{max} of the pair state $B^\dagger$. By virtue of Eqs. (22.118), this is seen to be a function of the singular values $\{x_\alpha\}$ of $B^\dagger$. There are two extreme situations: an uncorrelated pair state $B^\dagger = a_\alpha^\dagger a_{\bar\alpha}^\dagger$ for which $n_B^{max} = 1$; and a maximally correlated pair state (Ω is the total number of sp states)

$$B^\dagger = \sqrt{\frac{2}{\Omega}} \sum_{\alpha>0} a_\alpha^\dagger a_{\bar\alpha}^\dagger = \sqrt{\frac{1}{2\Omega}} \sum_\alpha \mathrm{sgn}(\alpha) a_\alpha^\dagger a_{\bar\alpha}^\dagger \equiv G^\dagger,$$ (22.120)

for which the spectrum can be solved by SU(2) algebra [see Eq. (22.7)] and $n_B^{max} = n[1 - \frac{2(n-1)}{\Omega}]$. For an arbitrary pair state, n_B^{max} lies between these two extremes, and $1 \le n_B^{max} \le n$ for $\Omega \gg N$.

Extension to bosons

The above exact solutions are easily extended to boson pairing. An arbitrary two-boson state has the same form as Eq. (22.109), the matrix $C_{\alpha\beta}$ now being symmetric. The canonical form for a normalized two-boson state (see Sec. 19.3) is given by

$$B^\dagger = \frac{1}{\sqrt{2}} \sum_\alpha x_\alpha (a_\alpha^\dagger)^2,$$ (22.121)

where the real and positive x_α correspond to the singular values of $C_{\alpha\beta}$, and $\sum_\alpha x_\alpha^2 = 1$. One defines similarly to Eq. (22.114) the set of two-boson operators

$$\phi^\dagger(y) = \frac{1}{\sqrt{2}} \sum_\alpha \frac{x_\alpha}{1 - yx_\alpha^2} (a_\alpha^\dagger)^2,$$ (22.122)

and applying the same manipulations, Eqs. (22.115)-(22.119) still hold with $\eta = 1$.

For bosons, an uncorrelated pair state $B^\dagger = (a_\alpha^\dagger)^2$ has a maximal occupation $n_B^{max} = n(2n - 1) = \frac{1}{2}N(N - 1)$, whereas the maximally correlated pair state

$$B^\dagger = \sqrt{\frac{1}{2\Omega}} \sum_\alpha (a_\alpha^\dagger)^2 \equiv G^\dagger, \qquad (22.123)$$

has a maximal occupation $n_B^{max} = n[1 + \frac{2(n-1)}{\Omega}]$. An arbitrary two-boson state therefore has $n \le n_B^{max} \le \frac{1}{2}N(N-1)$ for $\Omega \gg N$. Note the different behavior for fermions (bosons): the maximal occupation of a pair state is enhanced (reduced) with respect to uncorrelated pair states.

22.8.1 *Richardson–Gaudin equations*

The second special case of an exactly solvable pairing Hamiltonian (22.112) involves a nondegenerate sp spectrum ε_α, and a schematic pairing force with $B^\dagger \equiv G^\dagger$ as given by Eq. (22.120). It is convenient to introduce the following set of two-fermion operators

$$\psi^\dagger(z) = \sqrt{2} \sum_{\alpha > 0} \frac{1}{\varepsilon_\alpha - z} a_\alpha^\dagger a_{\bar\alpha}^\dagger = \frac{1}{\sqrt{2}} \sum_\alpha \mathrm{sgn}(\alpha) \frac{1}{\varepsilon_\alpha - z} a_\alpha^\dagger a_{\bar\alpha}^\dagger, \qquad (22.124)$$

where the z are complex numbers. The following commutation relations are easily verified

$$[\hat{H}_0, \psi^\dagger(z)] = [\sum_\alpha \varepsilon_\alpha a_\alpha^\dagger a_\alpha, \psi^\dagger(z)] = 2\sqrt{\Omega} G^\dagger + 2z\psi^\dagger(z),$$

$$[G, \psi^\dagger(z)] = \frac{1}{\sqrt{\Omega}} \sum_\alpha \frac{1}{\varepsilon_\alpha - z}(1 + 2\eta a_\alpha^\dagger a_\alpha) = \tilde\Lambda(y) + \frac{2\eta}{\sqrt{\Omega}} \sum_\alpha \frac{1}{\varepsilon_\alpha - z} a_\alpha^\dagger a_\alpha,$$

$$[[G, \psi^\dagger(z_1)], \psi^\dagger(z_2)] = \frac{4\eta}{\sqrt{\Omega}} \frac{\psi^\dagger(z_1) - \psi^\dagger(z_2)}{z_1 - z_2}, \qquad (22.125)$$

where $\eta = -1$. One can again show that all fully-paired eigenstates with $N = 2n$ particles have the form of a product of pair operators

$$|\Psi^N\rangle = \psi^\dagger(z_1)\psi^\dagger(z_2)..\psi^\dagger(z_n)|0\rangle, \qquad (22.126)$$

with the set $\{z\} = \{z_1, .., z_n\}$ to be determined. Acting with the Hamiltonian on the state (22.126), the commutation relations in Eq. (22.125) can be used to move $\hat{H}_0$ and G to the right, where they destroy the vacuum $|0\rangle$. One finds

$$(\hat{H}_0 - gG^\dagger G)|\Psi^N\rangle = E(\{z\})|\Psi^N\rangle + G^\dagger \sum_{k=1}^n V_k(\{z\}) \prod_{i(\neq k)=1}^n \psi^\dagger(z_i)|0\rangle,$$

$$V_k(\{z\}) = 2\sqrt{\Omega} - g\left(\tilde{\Lambda}(z_k) - 4\eta\frac{1}{\sqrt{\Omega}}\sum_{i(\neq k)=1}^{n}\frac{1}{z_k - z_i}\right),$$

$$E(\{z\}) = 2\sum_{k=1}^{n} z_i. \qquad (22.127)$$

An eigenstate is obtained for any solution $\{z\}$ of the set of nonlinear equations $V_k(\{z\}) = 0$, or

$$\frac{2\sqrt{\Omega}}{g} = \sum_{\alpha}\frac{1}{\varepsilon_\alpha - z_k} - 4\eta\sum_{i(\neq k)=1}^{n}\frac{1}{z_k - z_i}, \quad k = 1, .., n. \qquad (22.128)$$

These equations were first derived in [Richardson (1963, 1968)]. The eigenvalue $E(\{z\})$ corresponding to a solution $\{z\}$ is then represented by Eq. (22.127). For a boson Hamiltonian (22.112) with a schematic pairing force $G^\dagger$ as given by Eq. (22.123), one introduces

$$\psi^\dagger(z) = \frac{1}{\sqrt{2}}\sum_{\alpha}\frac{1}{\varepsilon_\alpha - z}(a_\alpha^\dagger)^2. \qquad (22.129)$$

Performing the same manipulations then leads to Eqs. (22.125) – (22.128), with $\eta = 1$ for bosons.

22.9 Exercises

(1) Derive the spectrum and eigenstates of the schematic pairing force $\hat{H} = B^\dagger B$, with $B^\dagger$ the pair creation operator as defined in Eq. (22.4). Use the commutator algebra of Eq. (22.6).
(2) Generate the BCS equations by minimizing Ω_0 in Eq. (22.76) with respect to the amplitudes u_p and v_p, subject to the normalization constraint $u_p^2 + v_p^2 = 1$.
(3) Calculate the energy in Eq. (22.76) starting from the BCS vacuum ansatz in Eq. (22.89).
(4) Show that a solution to the HFB equations (22.102) indeed provides an extremum for the energy in Eq. (22.98).

Chapter 23

Elastic scattering and the self-energy in finite systems

In Ch. 17 we focused on the removal part of the sp propagator in a finite system. This chapter will highlight the addition part of the sp propagator for a finite system, in particular as it applies to nuclei. The particle part of the sp propagator is related to the elastic scattering of a nucleon from the ground state of a target nucleus. It is therefore possible to generate the scattering wave function for this process with the corresponding experimentally accessible cross sections from the Dyson equation for a single nucleon. As for the discussion of the sp removal properties, we will assume that the (target) ground state of A particles corresponds to a nucleus with magic numbers for both protons and neutrons. This can be relaxed to even-even nuclei when empirical scattering potentials are considered. We thus deal with the scattering of a spin-$\frac{1}{2}$ particle (a nucleon) from a spinless particle (the A-particle ground state). We establish the formal relation between the self-energy and the elastic scattering amplitude in Sec. 23.1. In Sec. 23.2 the relation of the sp propagator with quantities like phase shifts and the scattering amplitude are further developed for the present case, based on the discussion initiated in Ch. 6 (see also Chs. 12, 15 and 20). We also collect some of the standard results of scattering theory for this problem [Joachain (1975); Gottfried and Yan (2004)] in Sec. 23.2 to facilitate the comparison of experimental data with calculations. Some empirical information, associated with the potentials that have been used to analyze elastic scattering off nuclei, is touched upon in Sec. 23.3.

The self-energy plays the crucial role of the potential the nucleon experiences when scattering elastically from the target ground state. Since inelastic scattering becomes possible at energies that permit excited states, some of the flux associated with the incoming beam must disappear from the elastic channel. The latter effect is represented by the appearance of an imaginary contribution to the self-energy at the corresponding energies. As usual, it represents the coupling to more complicated states as discussed throughout this book.

We establish in Sec. 23.4 a link with a large body of work that has been applied to the phenomenological analysis of elastic nucleon scattering in the past. The goal of this section is to employ the *framework* of self-consistent Green's functions to analyze this process. Many empirical calculations of elastic scattering have historically been performed with a complex potential, representing the effect of the interaction with the nucleons in the target ground state. Mahaux and Sartor proposed an important unifying development in a series of papers summarized in [Mahaux and Sartor (1991)]. These authors utilized the dispersion relation that links the real to the imaginary part of the self-energy with suitable approximations that connect the imaginary part below and above the Fermi energy. This approach is referred to as the dispersive optical model (DOM).

How to deal with nonlocality in the DOM potentials is taken up in Sec. 23.4.1. A standard procedure is presented that links a nonlocal potential with an equivalent local one that exhibits a smooth energy dependence. The related issue of the correct interpretation of the imaginary part of the empirical potentials and the microscopic imaginary part of the self-energy, is introduced in Sec. 23.4.2. Due to the treatment of nonlocality in the DOM approach of [Mahaux and Sartor (1991)], it is not possible to consider the DOM potentials directly as a proper self-energy to solve the Dyson equation. While not a serious issue for the description of scattering data at positive energies, it is critical when quantities that are extracted from the sp propagator below the Fermi energy are studied. It is therefore necessary to determine pertinent approximations for such quantities. We will present these DOM ingredients in Sec. 23.5. A more detailed analysis is required for the determination of occupation numbers from the DOM potentials. This issue is taken up in Sec. 23.5.1.

Results of a typical DOM analysis are presented in Sec. 23.6. In particular we will discuss a combined DOM analysis for Ca nuclei, including in particular ^{40}Ca and ^{48}Ca, to illustrate the flexiblility of this approach. The possibility to extrapolate to Ca nuclei with larger nucleon asymmetry, is utilized to make predictions for isotopes that approach the neutron dripline. This part of the chart of nuclides will be studied with future radioactive beam accelerators. Since the DOM analysis employs local potentials that do not allow a direct determination of the sp propagator, we present some calculations for a nonlocal potential in Sec. 23.6.1.

23.1   Self-energy and elastic scattering in finite systems

While it is intuitively clear that the sp propagator contains information related to the elastic scattering of particles from a target ground state, it is appropriate to provide a more solid derivation for this observation. Historically, the first attempt for such a connection was presented by [Bell and Squires (1959)]. We will follow the presentation of [Villars (1967)] and [Blaizot and Ripka (1986)]. We will not deal with the complications associated with treating the recoil of the nucleus [Redish and Villars (1970)]. Neither will we attempt to discuss this problem in terms of the more correct wave-packet formulation. The initial and final state of the target nucleus are denoted by $|\Psi_0^A\rangle$, representing an even-even nucleus with total angular momentum zero. As usual,

$$\hat{H}\,|\Psi_0^A\rangle = E_0^A\,|\Psi_0^A\rangle. \tag{23.1}$$

The states describing the projectile before and after the reaction will have the same magnitude of momentum and correspond to plane waves with momentum $\boldsymbol{p}_i$ and $\boldsymbol{p}_f$, respectively. We suppress spin (isospin) for convenience and represent the initial state of the combined projectile-target system by

$$a_{\boldsymbol{p}_i}^\dagger\,|\Psi_0^A\rangle = |\phi_i\rangle, \tag{23.2}$$

with energy

$$E_i = E_0^A + \frac{p_i^2}{2m}. \tag{23.3}$$

For the final state we write similarly

$$a_{\boldsymbol{p}_f}^\dagger\,|\Psi_0^A\rangle = |\phi_f\rangle, \tag{23.4}$$

with

$$E_f = E_0^A + \frac{p_f^2}{2m} = E_i. \tag{23.5}$$

Note that these states are not eigenstates of the Hamiltonian in the wave-packet sense, except in the distant past or future, respectively. This becomes clear by evaluating

$$[\hat{H}, a_{\boldsymbol{p}_i}^\dagger] = \frac{p_i^2}{2m}a_{\boldsymbol{p}_i}^\dagger + J_i^\dagger = \varepsilon_i a_{\boldsymbol{p}_i}^\dagger + J_i^\dagger \tag{23.6}$$

and

$$[\hat{H}, a_{\boldsymbol{p}_f}] = -\frac{\boldsymbol{p}_f^2}{2m} a_{\boldsymbol{p}_f} - J_f = -\varepsilon_f a_{\boldsymbol{p}_f} - J_f. \qquad (23.7)$$

The first term on the right in Eqs. (23.6) and (23.7) comes from the commutator with the kinetic energy operator $\hat{T}$ and the other terms can be obtained by employing Eq. (2.42), involving two addition and one removal operator for $J_i^\dagger$ and two removal and one addition operator for J_f. From Eq. (23.6) we infer

$$\begin{aligned}\hat{H}\,|\phi_i\rangle &= [\hat{H}, a_{\boldsymbol{p}_i}^\dagger]\,|\Psi_0^A\rangle + E_0^A a_{\boldsymbol{p}_i}^\dagger\,|\Psi_0^A\rangle \\ &= E_i\,|\phi_i\rangle + J_i^\dagger\,|\Psi_0^A\rangle\,,\end{aligned} \qquad (23.8)$$

and similarly from Eq. (23.7)

$$\langle\phi_f|\,\hat{H} = E_f\,\langle\phi_f| + \langle\Psi_0^A|\,J_f. \qquad (23.9)$$

The stationary scattering states are eigenstates of $\hat{H}$ and obey

$$(\hat{H} - E_i)|\Psi_i^{(+)}\rangle = 0. \qquad (23.10)$$

Inverting Eq. (23.8), while adding a solution of the homogeneous equation (23.10), yields

$$|\Psi_i^{(+)}\rangle = |\phi_i\rangle + \frac{1}{E_i - \hat{H} + i\eta} J_i^\dagger\,|\Psi_0^A\rangle\,, \qquad (23.11)$$

where the $i\eta$ signals the outgoing wave boundary condition and it can be shown that $|\Psi_i^{(+)}\rangle$ is properly normalized. Similarly, one finds

$$|\Psi_f^{(-)}\rangle = |\phi_f\rangle + \frac{1}{E_f - \hat{H} - i\eta} J_f^\dagger\,|\Psi_0^A\rangle. \qquad (23.12)$$

The S-matrix element for the transition $i \to f$ then reads [Gottfried and Yan (2004)] (see also Ch. 6)

$$\hat{S}_{fi} = \langle\Psi_f^{(-)}|\Psi_i^{(+)}\rangle. \qquad (23.13)$$

Inserting Eqs. (23.11) and (23.12) judiciously in this expression, we find

$$\begin{aligned}\langle\Psi_f^{(-)}|\Psi_i^{(+)}\rangle &= \langle\phi_f|\phi_i\rangle + \langle\phi_f|\frac{1}{E_i - \hat{H} + i\eta} J_i^\dagger\,|\Psi_0^A\rangle \\ &\quad + \langle\Psi_0^A|\,J_f\frac{1}{E_f - \hat{H} + i\eta}|\Psi_i^{(+)}\rangle.\end{aligned} \qquad (23.14)$$

Using Eq. (23.7) one may show

$$a_{\boldsymbol{p}_f} \frac{1}{E - \hat{H} \pm i\eta} = \frac{1}{E - \varepsilon_f - \hat{H} \pm i\eta} a_{\boldsymbol{p}_f} + \frac{1}{E - \varepsilon_f - \hat{H} \pm i\eta} J_f \frac{1}{E - \hat{H} \pm i\eta}. \tag{23.15}$$

With this result the second term of Eq. (23.14) can be rewritten as

$$\left\langle \Psi_0^A \middle| a_{\boldsymbol{p}_f} \frac{1}{E_i - \hat{H} + i\eta} J_i^\dagger \middle| \Psi_0^A \right\rangle = \frac{1}{E_i - E_f + i\eta} \left\langle \Psi_0^A \middle| a_{\boldsymbol{p}_f} J_i^\dagger \middle| \Psi_0^A \right\rangle$$
$$+ \frac{1}{E_i - E_f + i\eta} \left\{ \left\langle \Psi_0^A \middle| J_f \middle| \Psi_i^{(+)} \right\rangle - \left\langle \Psi_0^A \middle| J_f \middle| \phi_i \right\rangle \right\}, \tag{23.16}$$

where Eq. (23.11) has been employed as well. Inserting Eq. (23.16) in (23.14), yields after further manipulation

$$\left\langle \Psi_f^{(-)} \middle| \Psi_i^{(+)} \right\rangle = \left\langle \phi_f \middle| \phi_i \right\rangle + \frac{1}{E_i - E_f + i\eta} \left\langle \Psi_0^A \middle| \left(a_{\boldsymbol{p}_f} J_i^\dagger - J_f a_{\boldsymbol{p}_i}^\dagger \right) \middle| \Psi_0^A \right\rangle$$
$$- 2\pi i \delta(E_f - E_i) \left\langle \Psi_0^A \middle| J_f \middle| \Psi_i^{(+)} \right\rangle. \tag{23.17}$$

Using Eqs. (23.8) and (23.9), the second term on the right side can be shown to be proportional to $E_i - E_f = 0$, and therefore vanishes for all $\eta > 0$. We may therefore identify the transition matrix [see *e.g.* Eq. (6.71)] from Eq. (23.17) according to

$$\left\langle \phi_f \middle| T \middle| \phi_i \right\rangle = \left\langle \Psi_0^A \middle| J_f \middle| \Psi_i^{(+)} \right\rangle. \tag{23.18}$$

A similar derivation yields the equivalent relation

$$\left\langle \phi_f \middle| T \middle| \phi_i \right\rangle = \left\langle \Psi_f^{(-)} \middle| J_i^\dagger \middle| \Psi_0^A \right\rangle \tag{23.19}$$

noting that the condition $E_f = E_i$ must hold.

It remains to be shown how to relate these expressions to the sp propagator. For this purpose we use Eq. (23.11) to transform Eq. (23.18) into

$$\left\langle \phi_f \middle| T \middle| \phi_i \right\rangle = \left\langle \Psi_0^A \middle| \left\{ J_f, a_{\boldsymbol{p}_i}^\dagger \right\} \middle| \Psi_0^A \right\rangle$$
$$- \left\langle \Psi_0^A \middle| a_{\boldsymbol{p}_i}^\dagger J_f \middle| \Psi_0^A \right\rangle + \left\langle \Psi_0^A \middle| J_f \frac{1}{E_i - \hat{H} + i\eta} J_i^\dagger \middle| \Psi_0^A \right\rangle. \tag{23.20}$$

The second term in this relation can be rewritten by utilizing again Eq. (23.6) yielding

$$\left\langle \Psi_0^A \middle| a_{\boldsymbol{p}_i}^\dagger = \left\langle \Psi_0^A \middle| J_i^\dagger \frac{1}{E_0^A - \hat{H} - \varepsilon_i}, \tag{23.21}$$

where the denominator never vanishes. This observation follows by inserting a complete set of $A - 1$ states on the right and noting that the ground state energy E_0^A can never be equal to the sum of the energies of the two fragments. We can then write

$$\langle \phi_f | \, T \, | \phi_i \rangle = \langle \Psi_0^A | \, \{ J_f, a_{\boldsymbol{p}_i}^\dagger \} \, | \Psi_0^A \rangle - \langle \Psi_0^A | \, J_i^\dagger \frac{1}{E_0^A - \hat{H} - \varepsilon_i} J_f \, | \Psi_0^A \rangle$$

$$+ \langle \Psi_0^A | \, J_f \frac{1}{E_i - \hat{H} + i\eta} J_i^\dagger \, | \Psi_0^A \rangle . \tag{23.22}$$

It is now straightforward but somewhat tedious to show that

$$\langle \phi_f | \, T \, | \phi_i \rangle = \lim_{E \to \varepsilon_i} (\varepsilon_i - E)^2 G(\boldsymbol{p}_f, \boldsymbol{p}_i; E), \tag{23.23}$$

and furthermore, using the Dyson equation in terms of the reducible self-energy, that

$$\langle \phi_f | \, T \, | \phi_i \rangle = \langle \boldsymbol{p}_f | \, \Sigma_{red}(E = \varepsilon_i) \, | \boldsymbol{p}_i \rangle , \tag{23.24}$$

which completes the proof of the relation between the elastic scattering amplitude and the sp propagator.

23.2   Scattering formalism for spin-$\frac{1}{2}$ spin-0 scattering

The presentation of scattering in a partial wave basis in Sec. 6.4.1 can easily be adapted to represent the elastic scattering problem of a spin-$\frac{1}{2}$ nucleon from a nucleus with angular momentum and parity 0^+. Instead of starting from states which are only characterized by wave vectors, as in Sec. 6.4, it is necessary to introduce the spin degree of freedom explicitly in the basis transformation of Eq. (6.57) according to

$$|\boldsymbol{k} m_s\rangle = \sum_{\ell m_\ell} |k \ell m_\ell m_s\rangle Y_{\ell m_\ell}^*(\hat{\boldsymbol{k}})$$

$$= \sum_{\ell m_\ell} \sum_{j m_j} |k(\ell \tfrac{1}{2}) j m_j\rangle Y_{\ell m_\ell}^*(\hat{\boldsymbol{k}}) \, (\ell \, m_\ell \, \tfrac{1}{2} \, m_s | j \, m_j), \tag{23.25}$$

employing the usual Clebsch–Gordon coupling between orbital and spin angular momentum. In the corresponding basis the noninteracting propagator becomes

$$G^{(0)}(k(\ell \tfrac{1}{2}) j m_j, k'(\ell' \tfrac{1}{2}) j' m_{j'}; E) = \frac{\delta(k - k')}{k^2} \delta_{\ell, \ell'} \delta_{j, j'} \delta_{m_j, m_{j'}} G^{(0)}(k; E), \tag{23.26}$$

where $G^{(0)}(k; E)$ has the same form as in Eq. (6.58). While this statement is not correct for proton elastic scattering on account of the presence of the long-range Coulomb interaction, the current analysis is appropriate for neutron scattering. The generalization to proton elastic scattering is standard [Gottfried and Yan (2004)]. Expressing Eq. (6.19) in the angular momentum basis and assuming that the interaction is rotationally invariant but contains spin-orbit-like terms, we find

$$G_{\ell j}(k, k'; E) = \frac{\delta(k - k')}{k^2} G^{(0)}(k; E)$$
$$+ G^{(0)}(k; E) \int_0^\infty dq q^2 \langle k|\Sigma^{\ell j}(E)|q\rangle G_{\ell j}(q, k'; E)$$
$$= \frac{\delta(k - k')}{k^2} G^{(0)}(k; E) + G^{(0)}(k; E)\langle k|\Sigma^{\ell j}_{red}(E)|k'\rangle G^{(0)}(k'; E), \qquad (23.27)$$

where the required coupling of orbital and spin to total angular momentum has been performed. The equation for the reducible self-energy Σ_{red} can then be written as

$$\langle k|\Sigma^{\ell j}_{red}(E)|k'\rangle = \langle k|\Sigma^{\ell j}(E)|k'\rangle$$
$$+ \int_0^\infty dq\, q^2 \langle k|\Sigma^{\ell j}(E)|q\rangle G^{(0)}(q; E)\langle q|\Sigma^{\ell j}_{red}(E)|k'\rangle. \quad (23.28)$$

The coordinate space version of Eq. (23.27) is obtained by a double Fourier–Bessel transform

$$G_{\ell j}(r, r'; E) = \frac{2}{\pi} \int_0^\infty dk\, k^2 \int_0^\infty dk'\, k'^2 j_\ell(kr) j_\ell(k'r') G_{\ell j}(k, k'; E). \quad (23.29)$$

The transformation is similar to the one in Eq. (6.62). The result for the noninteracting part of the propagator, represented by the first term in Eq. (23.27) is still given by Eqs. (6.63) and (6.66). Equation (23.27) tranforms to

$$G_{\ell j}(r, r'; E) = G_\ell^{(0)}(r, r'; E) + \int_0^\infty dr_1 r_1^2 \int_0^\infty dr_2 r_2^2$$
$$\times G_\ell^{(0)}(r, r_1; E)\langle r_1|\Sigma^{\ell j}(E)|r_2\rangle G_{\ell j}(r_2, r'; E)$$
$$= G_\ell^{(0)}(r, r'; E) + \int_0^\infty dr_1 r_1^2 \int_0^\infty dr_2 r_2^2$$
$$\times G_\ell^{(0)}(r, r_1; E)\langle r_1|\Sigma^{\ell j}_{red}(E)|r_2\rangle G_\ell^{(0)}(r_2, r'; E) \qquad (23.30)$$

in wave vector space. The second equality can be used again to study the asymptotic behavior of the propagator outside the influence of the strong interaction associated with the spatial extent of the nuclear ground state.

For neutron scattering it is appropriate to assume that the reducible self-energy has a finite range. It is therefore possible to copy the analysis in Sec. 6.4.1 for the asymptotic form of the propagator with minor alterations. The equivalent of Eq. (6.70) for similar conditions on r and r' then reads in the asymptotic domain

$$G_{\ell j}(r, r'; E) \rightarrow = -i\frac{m}{\hbar^2}k_0 \tag{23.31}$$

$$\times h_\ell(k_0 r') \left\{ h_\ell^*(k_0 r) + h_\ell(k_0 r) \left[1 - 2\pi i \left(\frac{mk_0}{\hbar^2} \right) \langle k_0 | \Sigma_{red}^{\ell j}(E) | k_0 \rangle \right] \right\}.$$

We note that the on-shell matrix element of the $\Sigma_{red}^{\ell j}$ completely determines the asymptotic behavior, as in the sp case presented in Sec. 6.4.1 and formalized in the previous section. The term in square brackets corresponds to the relevant $\mathcal{S}$-matrix element that defines the phase shift in this more general case

$$\langle k_0 | \mathcal{S}^{\ell j}(E) | k_0 \rangle = \left[1 - 2\pi i \left(\frac{mk_0}{\hbar^2} \right) \langle k_0 | \Sigma_{red}^{\ell j}(E) | k_0 \rangle \right] \equiv e^{2i\delta_{\ell j}}, \tag{23.32}$$

where we now must allow for the possibility that the phase shift contains an imaginary part due to the presence of an imaginary part in the irreducible self-energy.

To obtain the scattering amplitude it is sufficient to trace the same steps as in Sec. 6.4, while properly accounting for the presence of spin-$\frac{1}{2}$ states. Not surprisingly, the equivalent of Eq. (6.55) becomes

$$f_{m'_s, m_s}(\theta, \phi) = -\frac{4m\pi^2}{\hbar^2} \langle k' m'_s | \Sigma_{red}(E = \frac{\hbar^2 k^2}{2m}) | k m_s \rangle. \tag{23.33}$$

It is conventional to write the matrix structure of the scattering amplitude in the following way [Joachain (1975)]

$$[f(\theta, \phi)] = \mathcal{F}(\theta)I + \boldsymbol{\sigma} \cdot \hat{\boldsymbol{n}}\mathcal{G}(\theta), \tag{23.34}$$

based on rotational invariance and parity conservation. The unit vector is given by

$$\hat{\boldsymbol{n}} = \frac{\boldsymbol{k} \times \boldsymbol{k}'}{|\boldsymbol{k} \times \boldsymbol{k}'|} = \frac{\hat{\boldsymbol{k}} \times \hat{\boldsymbol{k}}'}{\sin \theta}, \tag{23.35}$$

and $\boldsymbol{\sigma}$ is formed by the Pauli spin matrices. The relation between $\mathcal{F}$ and $\mathcal{G}$ and the phase shifts determined by Eq. (23.32), can now be worked out (see *e.g.* [Joachain (1975)]) yielding

$$\mathcal{F}(\theta) = \frac{1}{2ik} \sum_{\ell=0}^{\infty} \left[(\ell+1) \left\{ e^{2i\delta_{\ell+}} - 1 \right\} + \ell \left\{ e^{2i\delta_{\ell-}} - 1 \right\} \right] P_\ell(\cos\theta) \quad (23.36)$$

and

$$\mathcal{G}(\theta) = \frac{\sin\theta}{2k} \sum_{\ell=1}^{\infty} \left[e^{2i\delta_{\ell+}} - e^{2i\delta_{\ell-}} \right] P_\ell'(\cos\theta). \quad (23.37)$$

We employ the notation $\delta_{\ell\pm} \equiv \delta_{\ell j = \ell \pm \frac{1}{2}}$ and P_ℓ' denotes the derivative of the Legendre polynomial with respect to $\cos\theta$. The unpolarized cross section reads

$$\left(\frac{d\sigma}{d\Omega} \right)_{unpol} = |\mathcal{F}|^2 + |\mathcal{G}|^2. \quad (23.38)$$

For polarization measurements with an initially unpolarized beam one obtains a polarization along $\hat{n}$, *i.e.* perpendicular to the scattering plane characterized by

$$P(\theta) = \frac{2\mathrm{Re}\{\mathcal{F}(\theta)\mathcal{G}^*(\theta)\}}{|\mathcal{F}|^2 + |\mathcal{G}|^2}. \quad (23.39)$$

It is also common to denote this quantity by A_y. A third independent observable called the spin-rotation parameter was introduced by [Glauber and Osland (1979)]

$$Q(\theta) = \frac{2\mathrm{Im}\{\mathcal{F}(\theta)\mathcal{G}^*(\theta)\}}{|\mathcal{F}|^2 + |\mathcal{G}|^2}. \quad (23.40)$$

Important information is also contained in the total cross sections. The total elastic cross section can be derived from Eq. (23.38)

$$\sigma_{tot}^{el} = \int d\Omega \left(|\mathcal{F}|^2 + |\mathcal{G}|^2 \right). \quad (23.41)$$

Employing the partial wave expansions (23.36) and (23.37) and the orthogonality of the Legendre polynomials, we find

$$\sigma_{tot}^{el} = \frac{\pi}{k^2} \sum_{\ell=0}^{\infty} \frac{\left| (\ell+1)\left\{ e^{2i\delta_{\ell+}} - 1 \right\} + \ell \left\{ e^{2i\delta_{\ell-}} - 1 \right\} \right|^2}{2\ell+1}$$

$$+ \frac{\pi}{k^2} \sum_{\ell=0}^{\infty} \frac{\ell(\ell+1) \left| e^{2i\delta_{\ell+}} - e^{2i\delta_{\ell-}} \right|^2}{2\ell+1}. \quad (23.42)$$

We can define partial elastic cross sections such that

$$\sigma_{tot}^{el} = \sum_{\ell=0}^{\infty} \sigma_{\ell}^{el}, \tag{23.43}$$

which for a given ℓ read

$$\sigma_{\ell}^{el} = \frac{\pi}{k^2} \left[(\ell+1) \left| e^{2i\delta_{\ell+}} - 1 \right|^2 + \ell \left| e^{2i\delta_{\ell-}} - 1 \right|^2 \right] \leq \frac{4\pi(2\ell+1)}{k^2}, \tag{23.44}$$

and are limited as indicated. With complex potentials, and therefore complex phase shifts, it is possible to calculate the total reaction cross section

$$\sigma_{tot}^{r} = \sum_{\ell=0}^{\infty} \sigma_{\ell}^{r}, \tag{23.45}$$

with

$$\sigma_{\ell}^{r} = \frac{\pi}{k^2} \left[(2\ell+1) - (\ell+1) \left| e^{2i\delta_{\ell+}} \right|^2 - \ell \left| e^{2i\delta_{\ell-}} \right|^2 \right] \leq \frac{\pi(2\ell+1)}{k^2}. \tag{23.46}$$

We arrive at these results by using the optical theorem that yields the total cross section from the imaginary part of the forward scattering amplitude [Gottfried and Yan (2004)]

$$\sigma_{tot} = \sigma_{tot}^{el} + \sigma_{tot}^{r}. \tag{23.47}$$

For potentials without a spin-orbit interaction we find the following simplification

$$f(\theta) = \frac{1}{2ik} \sum_{\ell=0}^{\infty} (2\ell+1) \left[e^{2i\delta_{\ell}} - 1 \right] P_{\ell}(\cos\theta) \tag{23.48}$$

for the elastic scattering amplitude [see also Eq. (6.74)]. The total elastic cross section becomes

$$\sigma_{tot}^{el} = \frac{\pi}{k^2} \sum_{\ell=0}^{\infty} (2\ell+1) \left| e^{2i\delta_{\ell}} - 1 \right|^2. \tag{23.49}$$

For the reaction cross section we obtain

$$\sigma_{tot}^{r} = \frac{\pi}{k^2} \sum_{\ell=0}^{\infty} (2\ell+1) \left(1 - \eta_{\ell}^2 \right), \tag{23.50}$$

where we have introduced the inelasticity parameter

$$\eta_{\ell} = e^{-2 \, \text{Im} \, \delta_{\ell}}, \tag{23.51}$$

which is limited according to $0 \le \eta_\ell \le 1$. The latter limit corresponds to pure elastic scattering (no absorption). The total cross section can then be written as

$$\sigma_{tot} = \frac{2\pi}{k^2} \sum_{\ell=0}^{\infty} (2\ell + 1) \left[1 - \eta_\ell \cos \left(2\mathrm{Re}\ \delta_\ell \right) \right]. \tag{23.52}$$

The treatment of the Coulomb interaction requires a similar procedure. Instead of employing spherical Bessel functions representing the incoming and outgoing spherical waves, it is necessary to use Coulomb wave functions. Such an analysis is for example presented by [Gottfried and Yan (2004)] and can be copied directly.

23.3 Empirical information for nuclei

An enormous wealth of elastic nucleon scattering data exists. From these studies an empirical picture emerges, confirming the expectation that both the real and imaginary part of the self-energy depend on energy. The analysis of data related to targets that have an even number of protons and neutrons, often employs local potentials to represent the interaction of the nucleon with the target. When results are analyzed in this manner, it is possible to represent the so-called optical potential by a real part given by

$$\mathcal{V}(r; E) = \mathcal{U}_V(E) f(r, r_v, a_v), \tag{23.53}$$

where f has the same Woods-Saxon form as in Eq. (3.26)

$$f(r, r_v, a_v) = \left[1 + \exp \left(\frac{r - r_v A^{1/3}}{a_v} \right) \right]^{-1}. \tag{23.54}$$

The description of the experimental data requires an imaginary potential that has both a volume and a surface contribution according to

$$\mathcal{W}(r; E) = W_V(E) f(r, r_{W_v}, a_{W_v}) - 4a_{W_s} W_S(E) \frac{d}{dr} f(r, r_{W_s}, a_{W_s}). \tag{23.55}$$

It is not possible to link these potentials directly to the nucleon self-energy, since the energy dependence of the real part is not related to the imaginary part by a dispersion relation. For a fixed energy, a fit to the experimental points then involves too many parameters to be uniquely determined in practice. Nevertheless, properties of these potentials like the volume

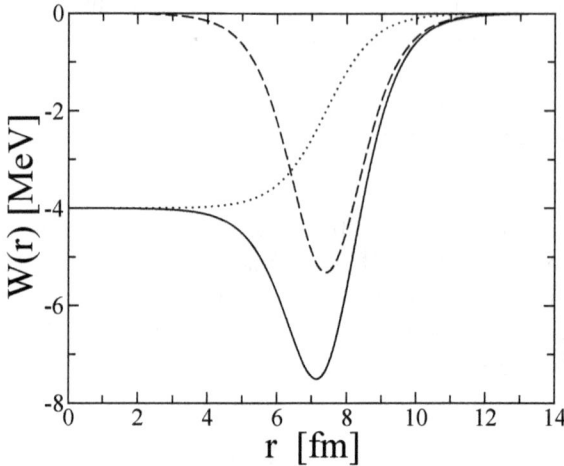

Fig. 23.1 Empirical imaginary part of the optical potential for 30 MeV proton scattering on ^{208}Pb. Dotted line corresponds to the volume absorption represented by a standard Woods-Saxon form. The surface absorption is given by the dashed line. The solid line represents the combined imaginary potential. Parameters were taken from [Bohr and Mottelson (1998)].

integral

$$J_V(E) = 4\pi \int_0^\infty r^2 \mathcal{V}(r; E) \, dr, \qquad (23.56)$$

and the root-mean-square (rms) radius

$$R_{rms}^V(E) = \left[\frac{4\pi \int_0^\infty r^4 \mathcal{V}(r; E) \, dr}{J_V(E)} \right]^{\frac{1}{2}}, \qquad (23.57)$$

and the corresponding quantities $J_W(E)$ and $R_{rms}^W(E)$, are quite well constrained by the data. A Coulomb potential must naturally be added when proton scattering is studied. Typical values of radius parameters r_i range from 1 to 1.3 fm. The diffuseness parameters have values between 0.6 and 0.7 fm. An example for the imaginary part of the potential is shown in Fig. 23.1 for 30 MeV proton scattering from ^{208}Pb. While absorption in the surface dominates at low energies, the volume contribution becomes dominant at scattering energies above 50 MeV. The depth of the real potential exhibits an almost linear dependence on energy up to 150 MeV, where it becomes very small. Such phenomenological results for the analysis of elastic nucleon scattering must still be related to the microscopic picture of the irreducible self-energy acting as the potential. In particular, it will be necessary to understand in more detail the meaning of the energy dependence of the empirical potentials. Steps in this direction are taken in the next section. Comprehensive analyses of the data in terms of optical potentials are reported in [Becchetti and Greenlees (1969); Nadasen *et al.* (1981); Schwandt *et al.* (1982); Varner *et al.* (1991); Koning and Delaroche (2003)].

23.4 Dispersive optical model analysis

The link between the empirical results of the previous section and the statement from Sec. 23.1 that the scattering amplitude for elastic nucleon scattering is proportional to the on-shell reducible self-energy at the corresponding energy, requires further strengthening. Important work establishing this connection was performed by [Mahaux and Sartor (1991)]. An essential ingredient in their development is the dispersion relation employed for the irreducible self-energy (see Eq. (11.96) for a particle in an infinite system). For a finite system, the self-energy near the Fermi energy exhibits discrete poles detailed in Chs. 11 and 17. We also treated the continuum for energies below the Fermi energy in Sec. 17.3 to calculate high-momentum components induced by short-range correlations. We have tacitly assumed however that the continuum associated with elastic scattering can be discretized for the analysis of low-energy sp strength distributions. A proper description of all relevant ingredients in the nucleon self-energy leads to the following dispersion relation

$$
\begin{aligned}
\Sigma^*(\boldsymbol{rm}, \boldsymbol{r'm'}; E) &= \Sigma_s(\boldsymbol{rm}, \boldsymbol{r'm'}) \\
&+ \sum_k \frac{\phi_k(\boldsymbol{r}, m)\phi_k^*(\boldsymbol{r'}, m')}{E - \varepsilon_k^+ + i\eta} - \frac{1}{\pi} \int_{\varepsilon_T^>}^\infty dE' \frac{\operatorname{Im} \Sigma^*(\boldsymbol{rm}, \boldsymbol{r'm'}; E')}{E - E' + i\eta} \\
&+ \sum_n \frac{\phi_n(\boldsymbol{r}, m)\phi_n^*(\boldsymbol{r'}, m')}{E - \varepsilon_n^- - i\eta} + \frac{1}{\pi} \int_{-\infty}^{\varepsilon_T^<} dE' \frac{\operatorname{Im} \Sigma^*(\boldsymbol{rm}, \boldsymbol{r'm'}; E')}{E - E' - i\eta}. \quad (23.58)
\end{aligned}
$$

The energy-independent part of the self-energy is equal to the correlated HF term shown *e.g.* diagrammatically in Fig. 9.13. The terms with discrete poles (that differ from those of G) contain residues with properties that were discussed in Sec. 11.2.4. Equation (23.58) should be compared to the Lehmann representation of the sp propagator given in Eq. (9.35).

The threshold energies $\varepsilon_T^>$ and $\varepsilon_T^<$ differ from the ones for the sp propagator. Since the elastic scattering phase shift is real below the first inelastic threshold, one can identify $\varepsilon_T^>$ with the corresponding energy.

The actual use of the dispersion relation is in its subtracted form (see the first exercise in Ch. 11). A useful choice for the subtraction energy proposed by [Mahaux and Sartor (1991)] is the average Fermi energy $\varepsilon_F = \frac{1}{2}[\varepsilon_F^- + \varepsilon_F^+]$, already considered in Eq (11.11). With this choice the once subtracted dispersion relation for the self-energy becomes

$$
\begin{aligned}
\Sigma^*(\boldsymbol{r}m, \boldsymbol{r}'m'; E) = {}& \Sigma^*(\boldsymbol{r}m, \boldsymbol{r}'m'; \varepsilon_F) \\
& + \sum_m \phi_m(\boldsymbol{r}, m)\phi_m^*(\boldsymbol{r}', m') \left\{ \frac{1}{E - \varepsilon_m^+ + i\eta} - \frac{1}{\varepsilon_F - \varepsilon_m^+ + i\eta} \right\} \\
& - \frac{1}{\pi} \int_{\varepsilon_T^>}^{\infty} dE'\, \mathrm{Im}\, \Sigma^*(\boldsymbol{r}m, \boldsymbol{r}'m'; E') \left\{ \frac{1}{E - E' + i\eta} - \frac{1}{\varepsilon_F - E' + i\eta} \right\} \\
& + \sum_n \phi_n(\boldsymbol{r}, m)\phi_n^*(\boldsymbol{r}', m') \left\{ \frac{1}{E - \varepsilon_n^- - i\eta} - \frac{1}{\varepsilon_F - \varepsilon_n^- - i\eta} \right\} \\
& + \frac{1}{\pi} \int_{-\infty}^{\varepsilon_T^<} dE'\, \mathrm{Im}\, \Sigma^*(\boldsymbol{r}m, \boldsymbol{r}'m'; E') \left\{ \frac{1}{E - E' - i\eta} - \frac{1}{\varepsilon_F - E' - i\eta} \right\}.
\end{aligned}
$$
(23.59)

The (real) self-energy at the Fermi energy was interpreted by [Mahaux and Sartor (1991)] as a generalized HF potential. This quantity should not be confused with the correlated HF contribution Σ_s, but we will employ the HF designation for the remainder of this chapter, since it is common usage. Empirical information concerning the form of this potential can then be extracted from the experience with HF calculations employing effective forces. Such calculations include parameters that are chosen in order to reproduce certain experimental properties, *e.g.* the Fermi energy. As for atoms, the HF potential is nonlocal in coordinate space. It has been customary to introduce an equivalent local potential in the sense that both bound single-particle energies and scattering phase shifts are approximately equal to the corresponding quantities of the nonlocal potential. In addition, a simple relation exists between the wave functions of these local and nonlocal solutions. It will be instructive to explore this relation in more detail, since it is widely used. In the future however, it may no longer be necessary to take advantage of the reduction in computer requirements when local potentials are utilized.

23.4.1 *Local-equivalent and nonlocal potentials*

We will illustrate the approximate relation between an equivalent local but energy-dependent potential, and a nonlocal spin-independent potential of the form $\mathcal{V}(r, r')$ that is assumed separable, in the following way

$$\mathcal{V}(r, r') = V_N \left(\tfrac{1}{2}(r + r')\right) H(|r - r'|). \tag{23.60}$$

The degree of nonlocality can be expressed by a gaussian proposed by [Perey and Buck (1962)]

$$H(|r - r'|) = \frac{1}{\pi^{\frac{3}{2}} \beta^3} \exp\left[-\left(\frac{r - r'}{\beta}\right)^2\right], \tag{23.61}$$

which is normalized to 1 when H is integrated over $r' - r$. Typical values of β that characterize the degree of nonlocality of the original potential are taken to be of the order of 1 fm. The wave equation relevant for both bound and scattering states can be written as

$$-\frac{\hbar^2 \nabla^2}{2m} \psi_N(r) + \int d^3 r' \mathcal{V}(r, r') \psi_N(r') = E \psi_N(r), \tag{23.62}$$

where E either represents continuous positive energies, or discrete negative ones, and we denote the eigenfunction of the nonlocal potential by ψ_N. We follow [Fiedeldey (1966)] for the derivation of the relation between the nonlocal and the local-equivalent potential, when the former can be written as in Eq. (23.60). Concurrently, it will become clear how to relate ψ_N to the wave function ψ_L of the local equivalent potential. We define $E \equiv \hbar^2 k^2 / 2m$ and write Eq. (23.62) in the form

$$\left(\nabla^2 + k^2\right) \psi_N(r) = I_N(r), \tag{23.63}$$

where we have identified

$$I_N(r) = \frac{2m}{\hbar^2} \int d^3 r' V_N \left(\tfrac{1}{2}(r + r')\right) H(|r' - r|) \psi_N(r'). \tag{23.64}$$

Defining

$$U_N(r) = \frac{2m}{\hbar^2} V_N(r), \tag{23.65}$$

we employ

$$s = r' - r \tag{23.66}$$

to rewrite Eq. (23.64) as

$$I_N(\boldsymbol{r}) = \int d^3 s \, U_N \left(\boldsymbol{r} + \tfrac{1}{2}\boldsymbol{s}\right) H(\boldsymbol{s}) \psi_N(\boldsymbol{r} + \boldsymbol{s})$$

$$= \left[\int d^3 s \, H(s) \exp\left(\boldsymbol{s} \cdot \left[\tfrac{1}{2}\nabla_1 + \nabla_2\right]\right)\right] U_N(\boldsymbol{r}) \psi_N(\boldsymbol{r}), \quad (23.67)$$

where ∇_1 acts solely on U_N and ∇_2 on ψ_N. We recognize the exponential form of these operators as being responsible for the translations that return us to the first line of Eq. (23.67). The FT of $H(s)$, given by

$$f(k^2) = \int d^3 s \, H(s) \exp\left(i\boldsymbol{k} \cdot \boldsymbol{s}\right) \quad (23.68)$$

allows us to write

$$I_N(\boldsymbol{r}) = f\left(-\left[\tfrac{1}{2}\nabla_1 + \nabla_2\right]^2\right) U_N(\boldsymbol{r}) \psi_N(\boldsymbol{r})$$

$$= \exp\left(\tfrac{1}{4}\beta^2 \left[\tfrac{1}{2}\nabla_1 + \nabla_2\right]^2\right) U_N(\boldsymbol{r}) \psi_N(\boldsymbol{r}), \quad (23.69)$$

where the explicit form of $H(s)$ in Eq. (23.61) has been applied. A useful strategy is to expand f about a parameter κ^2 that yields the best approximation to the wave equation for the nonlocal wave function

$$f(k^2) = f(\kappa^2) + (k^2 - \kappa^2) f'(\kappa^2) + \dots \quad (23.70)$$

Keeping only the first two terms in this expansion, we obtain

$$I_N(\boldsymbol{r}) \approx \left[f(\kappa^2) - f'(\kappa^2)\left(\left[\tfrac{1}{2}\nabla_1 + \nabla_2\right]^2 + \kappa^2\right)\right] U_N(\boldsymbol{r}) \psi_N(\boldsymbol{r}). \quad (23.71)$$

Substitution of Eq. (23.71) in Eq. (23.64) yields, after some manipulation, the approximate nonlocal wave equation in the form

$$\left(\nabla^2 + k^2\right) \psi_N(\boldsymbol{r}) \approx \frac{f U_N + (k^2 - \kappa^2) f' U_N - \tfrac{1}{4} f' \nabla^2 U_N}{1 + f' U_N} \psi_N$$

$$- \frac{f' \nabla U_N}{1 + f' U_N} \nabla \psi_N. \quad (23.72)$$

We now postulate [Austern (1965)] the existence of an equivalent local potential $U_L(\boldsymbol{r})$ and a corresponding wave function such that

$$\psi_L(\boldsymbol{r}) \stackrel{r \to \infty}{\longrightarrow} \psi_N(\boldsymbol{r}), \quad (23.73)$$

while satisfying the wave equation

$$\left(\nabla^2 + k^2\right) \psi_L(\boldsymbol{r}) = U_L(\boldsymbol{r}) \psi_L(\boldsymbol{r}). \quad (23.74)$$

Defining

$$\psi_L(\mathbf{r}) = p(\mathbf{r})\psi_N(\mathbf{r}), \tag{23.75}$$

with $p(\mathbf{r}) \to 1$ for $\mathbf{r} \to \infty$, we can rewrite Eq. (23.74) in the form

$$\left(\nabla^2 + k^2\right)\psi_N = \left(U_L - \frac{\nabla^2 p}{p}\right)\psi_N - 2\frac{\nabla p}{p} \cdot \nabla\psi_N. \tag{23.76}$$

Comparing Eq. (23.76) with Eq. (23.72) we can identify the coefficients of ψ_N and $\nabla\psi_N$. Putting these equal yields

$$\frac{\nabla p}{p} = \frac{1}{2}\frac{f'\nabla U_N}{1 + f'U_N} \tag{23.77}$$

and

$$U_L = \frac{f + (k^2 - \kappa^2)f'}{1 + f'U_N}U_N - \frac{1}{4}f'\frac{\nabla^2 U_N}{1 + f'U_N} + \frac{\nabla^2 p}{p}. \tag{23.78}$$

A suitable choice of κ^2 is made by

$$\kappa^2 = k^2 - U_L^0(\mathbf{r}), \tag{23.79}$$

with

$$U_L^0(\mathbf{r}) = f(\kappa^2)U_N(\mathbf{r}). \tag{23.80}$$

With this assumption Eq. (23.78) simplifies to

$$U_L = U_L^0 - \frac{1}{4}\frac{f'\nabla^2 U_N}{1 + f'U_N} + \frac{\nabla^2 p}{p}. \tag{23.81}$$

In addition we find starting from Eq. (23.80)

$$\nabla U_L^0 = \frac{f\nabla U_N}{1 + f'U_N} = \frac{2f}{f'}\frac{\nabla p}{p}, \tag{23.82}$$

while using Eq. (23.77). Since $f'/2f = -\beta^2/8$ for the assumed gaussian nonlocality formfactor, we arrive at

$$p(r) = \exp\left(-\tfrac{1}{8}\beta^2 U_L^0(r)\right) \tag{23.83}$$

for potentials that depend only on r. The relation between the nonlocal and local wave function then becomes

$$\psi_N(\mathbf{r}) = \exp\left(\tfrac{1}{8}\beta^2 U_L^0(r)\right)\psi_L(\mathbf{r}), \tag{23.84}$$

while

$$U_L^0(r) = \exp\left(-\tfrac{1}{4}\beta^2[k^2 - U_L^0(r)]\right)U_N(r) \qquad (23.85)$$

allows the construction of U_L^0 based on an assumed form of U_N. The local equivalent potential in Eq. (23.81) therefore has as leading contribution U_L^0 with minor corrections for Woods-Saxon representations of U_N [Fiedeldey (1966)]. The depth of the local equivalent potential corresponding to $r = 0$ will be denoted by V_{HF}. Employing Eq. (23.85) we find for the depth an energy dependence given by

$$V_{HF}(E) = \exp\left(-\tfrac{1}{4}\beta^2 2m[E - V_{HF}(E)]/\hbar^2\right)V_N(r = 0). \qquad (23.86)$$

This function is negative for all values of E, since $V_N(r = 0)$ is negative. For large negative values of E it represents a linear function and approaches zero for large positive values of E. Such behavior matches empirical results discussed in Sec. 23.3.

The energy dependence associated with local approximations to calculated HF potentials is rather weak and can also be represented by a linear function of E. Identifying the local equivalent HF potential by $\mathcal{V}_{HF}(r; E)$, we can characterize its energy dependence by an effective mass defined by

$$\frac{\tilde{m}^*(r; E)}{m} = 1 - \frac{d\mathcal{V}_{HF}(r; E)}{dE}, \qquad (23.87)$$

which represents the influence of nonlocality on the HF potential. We have encountered a corresponding quantity in infinite systems identified by m_p^* in Eq. (14.95), which characterizes the momentum dependence of the self-energy, or equivalently its nonlocal character. The relation between the solution of the wave equation for the nonlocal and local equivalent potential can now be written to good approximation in the form

$$\psi_N(\boldsymbol{r}) = \left[\frac{\tilde{m}^*(r; E)}{m}\right]^{\frac{1}{2}}\psi_L(\boldsymbol{r}), \qquad (23.88)$$

for wave functions calculated at E and employing Eq. (23.84). Outside the range of the nuclear potential these wave functions are equal and therefore yield similar results for scattering quantities. For bound states an additional constant factor must be introduced on the right side of Eq. (23.88) when both wave functions are normalized to 1.

23.4.2   *Interpretation of the empirical imaginary potentials*

The empirical treatment that replaces nonlocality by the energy dependence incorporated by Eqs. (23.86) and (23.88) has important implications for the solution of the Dyson equation. It is perfectly appropriate to analyze scattering data above the Fermi energy together with bound state information (both above and below ε_F) in terms of energy-dependent potentials like in Eq. (23.86), adjusting parameters to acquire an accurate representation of the experimental data. However, it is no longer possible to employ such empirical potentials directly as representations of the self-energy. One aspect of this problem is associated with the energy dependence of the local equivalent potential. Since it does not emerge from the dispersive contributions to the real part of the self-energy, it leads to a distortion of the normalization, as can be expected from Eq. (23.88). We will further illustrate this point in Sec. 23.6.1. Another feature that does not immediately allow the comparison of the imaginary part of the optical potential with microscopically calculated self-energies, is also associated with the treatment of nonlocality. Understanding of this issue was crucial in clarifying the apparent discrepancy between empirical results for the mean free path of a nucleon in the nucleus and theoretical calculations [Negele and Yazaki (1981)]. This is best illustrated by recalling the analysis of scattering of two nucleons in the medium that was presented in Sec. 20.2. We will therefore briefly digress to infinite nuclear matter to elucidate this argument.

In Sec. 20.2 we analyzed the progagator associated with the relative motion of two nucleons. A similar analysis can of course be carried out for the propagator of a single nucleon and follows the development that yields Eq. (14.93) now including a finite imaginary part of the self-energy. We consider the usual quasiparticle energy [see *e.g.* Eq. (11.101)], and identify at a fixed energy the momentum for which

$$E = \frac{p_0^2}{2m} + \mathrm{Re}\,\Sigma(p_0; E). \tag{23.89}$$

We expand the self-energy around this momentum. As usual it is necessary to maintain the two lowest-order terms in the expansion of the real part, while it suffices to keep only the lowest-order term for the imaginary part at energies that are relevant for the analysis of elastic scattering data. The outcome is familiar and yields a propagator with a pole in the complex momentum plane, as discussed in Sec. 20.2. Contrary to that section, we will follow [Negele and Yazaki (1981)] and expand the self-energy as a function of p (not p^2) according to

$$\Sigma(p; E) = \mathrm{Re}\, \Sigma(p_0; E) + (p - p_0) \left.\frac{\partial \mathrm{Re}\, \Sigma(p; E)}{\partial p}\right|_{p_0} + i\mathrm{Im}\, \Sigma(p_0; E). \quad (23.90)$$

Using also $p^2 = p_0^2 + 2p_0(p-p_0)$, the real part of the complex pole momentum p_R is still given by p_0. Its imaginary part however reads

$$p_I = -\mathrm{Im}\, \Sigma(p_0; E) \left(\frac{p_0}{m} + \left.\frac{\partial \mathrm{Re}\, \Sigma(p; E)}{\partial p}\right|_{p_0}\right)^{-1}. \quad (23.91)$$

The damping of the propagation is directly related to the mean free path λ by

$$\lambda = \frac{\hbar}{2p_I} = -\frac{\hbar}{2m_{p_0}^* \mathrm{Im}\, \Sigma(p_0; E)} \quad (23.92)$$

where

$$\frac{m_{p_0}^*}{m} = \left(1 + \frac{m}{p_0} \left.\frac{\partial \mathrm{Re}\, \Sigma(p; E)}{\partial p}\right|_{p_0}\right)^{-1}. \quad (23.93)$$

The important conclusion from this analysis is that a phenomenological local imaginary part of the optical potential should be compared with the equivalent of $m_{p_0}^*/m \times \mathrm{Im}\, \Sigma(p_0; E)$ and not with $\mathrm{Im}\, \Sigma(p_0; E)$. This implies for finite nuclei that

$$\mathcal{W}(r; E) = \frac{\widetilde{m}^*(r; E)}{m}\, \mathrm{Im}\, \Sigma(r; E), \quad (23.94)$$

where we anticipate the assumption that the imaginary part of the self-energy is local and the empirical imaginary part of the optical potential is denoted by $\mathcal{W}$. This realization allowed [Negele and Yazaki (1981)] to reconcile the apparent discrepancy between mean free paths, calculated from imaginary parts of the nucleon self-energy in nuclear matter, and those obtained from experimental reaction cross sections [Nadasen et al. (1981)]. If the nonlocality of the self-energy, expressed by m_p^*, is included as in Eq. (23.92), qualitative agreement is achieved. Since a local approximation is adopted in the DOM analysis, it will be necessary to take this m_p^*/m factor into account when nonlocality is explicitly included in Sec. 23.6.1.

23.5 Ingredients of the DOM analysis

Nonlocality clearly requires careful consideration when links with the nucleon self-energy are sought based on an analysis of experimental data. Another important issue is the presence of a very strong energy dependence near the Fermi energy due to the low-lying (correlated) 2p1h and 1p2h contributions in Eq. (23.58). While these discrete contributions have been discussed and calculated in Chs. 11 and 17, they have traditionally not been included in the analysis of elastic nucleon scattering, since they depend in detail on the nucleus that is being studied. In heavier nuclei the presence of many resonances and the limitation of the experimental resolution also requires that the optical-model description only reproduce the energy-averaged elastic part of the scattering matrix. This averaging can be achieved by replacing the energy E in Eq. (23.58) by $E + i\Delta$, resulting in a smooth energy (E) dependence if Δ is larger than the average separation between levels with the same angular momentum and parity. The averaging procedure does not have any calculational consequences, since the data will be fit with an ansatz that has an appropriately smooth energy dependence. Nevertheless, it should be kept in mind that resulting DOM potentials cannot represent more than an energy-averaged representation of self-energy properties.

The remaining ingredient of the DOM potential concerns the dispersive contribution involving the imaginary part of the self-energy. Since the discrete-pole contributions to the self-energy have been smoothed out, one only needs continuous contributions. These will start at a threshold closer to the Fermi energy representing also the discrete terms, but in an average sense. While the dispersive part of the self-energy, in particular its imaginary part, are in principle also nonlocal in coordinate space, it was assumed by [Mahaux and Sartor (1991)] that its equivalent momentum dependence is smooth, as born out by nuclear matter calculations. It is therefore reasonable to replace the nonlocal imaginary part of the self-energy by the local empirical potential, keeping in mind that nonlocality corrections typically lead to a linear (smooth) energy dependence that can be incorporated by adjusting the energy dependence of the resulting real part through the fit of the local equivalent HF term.

Although a lot of information about the local imaginary optical potential is available, less is known about its behavior below the Fermi energy. From nuclear matter calculations we learn that within 50 MeV of the Fermi energy there is considerable symmetry of the imaginary part of the self-energy with respect to the Fermi energy (see *e.g.* [Müther and Dickhoff (2005)]). Calculations in finite nuclei [Brand *et al.* (1991)] also exhibit this symmetry when configuration spaces are used that are not too large. An essential asymmetry will occur at large distances from the Fermi energy on account of the phase space difference between particles and holes. Since a subtracted dispersion relation is utilized, the emphasis is on the contribution of the imaginary part at lower energies. Asymmetric energy contributions therefore hardly influence results near the Fermi energy, but do have consequences for the depletion of deeply bound orbits (see Sec. 23.6).

In summary, we can represent the DOM potentials by the following contributions

$$\mathcal{V}(r; E) = \mathcal{V}_{HF}(r; E) + \Delta\mathcal{V}(r; E). \tag{23.95}$$

The corresponding local dispersive correction reads

$$\Delta\mathcal{V}(r; E) = \frac{1}{\pi}\mathcal{P}\int \mathcal{W}(r; E')\left(\frac{1}{E' - E} - \frac{1}{E' - \varepsilon_F}\right)dE'. \tag{23.96}$$

It should be noted that it is customary to employ the same sign for the imaginary part of the optical potential above and below the Fermi energy. This feature has been included in going from Eq. (23.58) to Eq. (23.96). More details of the employed parametrizations are presented in Sec. 23.6.

In the analysis with DOM potentials it has been assumed, until now, that all the nonlocality is associated with the HF component. We can therefore assume that the effective mass associated with nonlocality is given by Eq. (23.87). Since we have already explored the link between the empirical imaginary part of the optical potential and the related imaginary part of the self-energy in Eq. (23.94), we must keep in mind that if a nonlocal HF contribution is used, the dispersive correction for the real part reads

$$\Delta\mathcal{V}_N(r; E) = \frac{m}{\widetilde{m}(r; E)}\Delta\mathcal{V}(r; E). \tag{23.97}$$

It is necessary to use Eq. (23.97) when the quasiparticle strength is evaluated [Mahaux and Sartor (1991)]. The exact expression for this quantity can be found in Eq. (9.44). It is convenient to identify an effective mass contribution that reflects the true energy dependence of the self-energy. The total effective mass is defined as

$$\frac{m^*(r; E)}{m} = 1 - \frac{d\mathcal{V}(r; E)}{dE}, \tag{23.98}$$

consequently it makes sense to identify the E-mass $\overline{m}^*(r; E)$ by

$$\frac{m^*(r; E)}{m} = \frac{\widetilde{m}^*(r; E)}{m} \frac{\overline{m}^*(r; E)}{m}. \tag{23.99}$$

For nuclei we employ discrete quantum numbers labeled by n for energy, ℓ for orbital, and j for total angular momentum. We associate $E_{n\ell j}$ with the quasiparticle energy that solves the energy-dependent eigenvalue problem utilizing the real part of the DOM potential according to

$$-\frac{\hbar^2 \nabla^2}{2m} \psi_{nlj}(\boldsymbol{r}) + [\mathcal{V}_{HF}(r; E_{n\ell j}) + \Delta \mathcal{V}(r; E_{n\ell j})]\, \psi_{n\ell j}(\boldsymbol{r}) = E_{n\ell j} \psi_{n\ell j}(\boldsymbol{r}). \tag{23.100}$$

The nonlocality correction in Eq. (23.88) then takes the form

$$\overline{\psi}_{n\ell j}(\boldsymbol{r}) = \left[\frac{\widetilde{m}^*(r; E_{n\ell j})}{m}\right]^{\frac{1}{2}} \psi_{n\ell j}(\boldsymbol{r}), \tag{23.101}$$

with the left-hand side normalized to unity. Using these wave functions one can represent the characteristics of the quasiparticle properties of the sp strength distribution by a width calculated according to

$$\Gamma_{n\ell j} = -2 \frac{\int d^3 r\, \overline{\psi}^2_{n\ell j}(\boldsymbol{r})\, \mathcal{W}(r; E_{n\ell j})}{\int d^3 r\, \overline{\psi}^2_{n\ell j}(\boldsymbol{r})\, \frac{m^*(r; E_{n\ell j})}{m}}, \tag{23.102}$$

which should be compared to Eq. (11.109). Consulting the infinite matter definition of the effective mass in Eq. (14.96) and comparing it to Eq. (9.44), it is reasonable to suggest that the quasiparticle strength for discrete states, *i.e.* the spectroscopic factor, can be approximated by

$$S_{n\ell j} = \int d^3 r\, \overline{\psi}^2_{n\ell j}(\boldsymbol{r})\, \frac{m}{\overline{m}(r, E_{n\ell j})}. \tag{23.103}$$

For comparison with $(e, e'p)$ results, it is useful to consider the rms radius of valence hole states

$$R^{n\ell j}_{rms} = \sqrt{\int d^3 r\, \overline{\psi}^2_{n\ell j}(\boldsymbol{r})}. \tag{23.104}$$

23.5.1   *Occupation numbers in the DOM analysis*

The determination of occupation numbers associated with the orbits characterized by quantum numbers n, ℓ, and j is more involved [Mahaux and

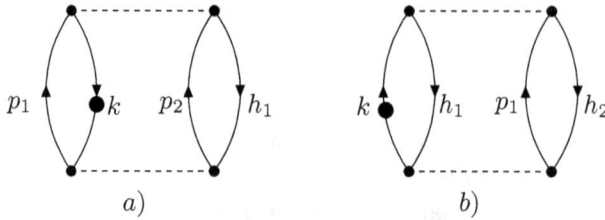

Fig. 23.2 Diagrams representing the second-order corrections to the Fermi gas momentum distribution. The action of the $a_k^\dagger a_k$ operator is represented by the dot. Part $a)$ displays the correction for momenta below k_F, while part $b)$ refers to $k > k_F$.

Sartor (1991)]. Since it is not correct to solve the Dyson equation containing the energy dependence of the HF potential that represents nonlocality, it is necessary to evaluate occupation numbers directly from perturbation theory. We refer to [Thouless (1972)] for the rules to generate the relevant diagrams and associated expressions. We discuss a few examples here that clarify the expressions proposed by [Mahaux and Sartor (1991)] for generating occupation numbers. We will develop the relevant equations for nuclear matter and subsequently produce formulations for finite nuclei. In lowest order , *i.e.* no contribution from the two-body interaction V, the momentum distribution is given by the usual step function $\theta(k - k_F)$. Corrections only appear in second order in V and are displayed in Fig. 23.2$a)$ and $b)$ for momenta below and above the Fermi momentum, respectively. The corresponding expression for this correction to the momentum distribution in second order for $k < k_F$ reads

$$\delta n_<^{(2)}(k) = -\frac{1}{2} \sum_{p_1 p_2 h_1} \frac{|\langle p_1 \, p_2 | \, V \, | k \, h_1 \rangle|^2}{\left(\varepsilon(k) + \varepsilon(h_1) - \varepsilon(p_1) - \varepsilon(p_2)\right)^2}, \qquad (23.105)$$

where, p_i refers to a particle state $(p_i > k_F)$ and h_i to a hole state $(h_i < k_F)$. We have also suppressed bold-face notation for vectors and discrete quantum numbers in the summation. A similar expression is found for $k > k_F$

$$\delta n_>^{(2)}(k) = \frac{1}{2} \sum_{p_1 h_1 h_2} \frac{|\langle k \, p_1 | \, V \, | h_1 \, h_2 \rangle|^2}{\left(\varepsilon(h_1) + \varepsilon(h_2) - \varepsilon(k) - \varepsilon(p_1)\right)^2}. \qquad (23.106)$$

The sp energies will not be specified here, but can of course represent kinetic energy only, or some form of self-consistently calculated sp spectrum.

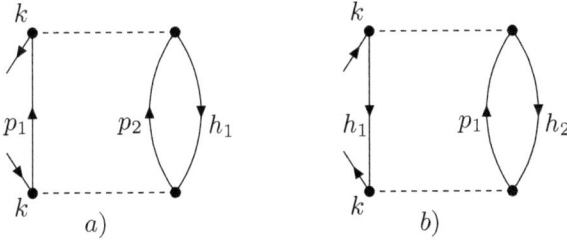

Fig. 23.3 Time-ordered diagrams for the second-order self-energy, relevant for a hole ($k < k_F$) in part a) and a particle in b).

An interesting relation exists between the corrections to the momentum distribution in second order and related second-order self-energy diagrams. Indeed the second order term with an intermediate 2p1h state can be written as

$$\Sigma_2^\downarrow(k; E) = \frac{1}{2} \sum_{p_1 p_2 h_1} \frac{|\langle p_1\, p_2 | V |k\, h_1\rangle|^2}{E + \varepsilon(h_1) - \varepsilon(p_1) - \varepsilon(p_2)}, \tag{23.107}$$

where the downward pointing arrow is a reminder that this contribution has an imaginary part above the Fermi energy. For $k < k_F$ the energy derivative of Eq. (23.107) at $\varepsilon(k)$ yields precisely Eq. (23.105). To facilitate the comparison we include in Fig. 23.3a) the time-ordered diagram that yields Eq. (23.107). For $k > k_F$ a similar result exists, related to the second-order diagram with an intermediate 1p2h state [see also Fig. 23.3b)]

$$\Sigma_2^\uparrow(k; E) = \frac{1}{2} \sum_{p_1 h_1 h_2} \frac{|\langle h_1\, h_2 | V |k\, p_1\rangle|^2}{E + \varepsilon(p_1) - \varepsilon(h_1) - \varepsilon(h_2)}. \tag{23.108}$$

The relations between the self-energy expressions and the momentum distribution can thus be summarized by

$$n(k) = 1 + \left(\frac{\partial \, \Sigma^\downarrow(k; E)}{\partial E} \right)_{E=\varepsilon(k)} \qquad k < k_F \tag{23.109}$$

and

$$n(k) = - \left(\frac{\partial \, \Sigma^\uparrow(k; E)}{\partial E} \right)_{E=\varepsilon(k)} \qquad k > k_F. \tag{23.110}$$

By taking the derivatives at the sp energies $\varepsilon(k)$ in the pertinent expression (depending on k being above or below k_F), one ensures that the corresponding momentum distributions are real. Equations (23.109) and (23.110) are not exact expressions, but yield the correct set of contributions when all ladder diagram contributions to the self-energy are included while using mean-field sp propagators [Ramos *et al.* (1991)].

Keeping in mind the nonlocality correction and using the dispersion relation for the real part of the self-energy, one may tentatively rewrite Eqs. (23.109) and (23.110) for DOM orbits in finite nuclei. For hole states, the occupation number is approximated by

$$
n_{n\ell j} = \int d^3 r \, \overline{\psi}^2_{n\ell j}(\boldsymbol{r}) \left[1 + \frac{m}{\widetilde{m}(r; E_{n\ell j})} \frac{1}{\pi} \int_{E_F}^{\infty} \frac{\mathcal{W}(r; E')}{(E' - E_{n\ell j})^2} dE' \right],
$$
(23.111)

while for particle states, the same approximation gives

$$
n_{n\ell j} = -\int d^3 r \, \overline{\psi}^2_{n\ell j}(\boldsymbol{r}) \left[\frac{m}{\widetilde{m}(r; E_{n\ell j})} \frac{1}{\pi} \int_{-\infty}^{E_F} \frac{\mathcal{W}(r; E')}{(E' - E_{n\ell j})^2} dE' \right].
$$
(23.112)

With this preparation all relevant quantities below the Fermi energy have been collected. Details of the DOM potentials and results of the fits are discussed in the next section.

23.6   Results of the DOM analysis in several nuclei

A number of DOM analyses have been reported in addition to the material reviewed in [Mahaux and Sartor (1991)]. We will discuss recent work that involves a combined DOM analysis of several nuclei simultaneously [Charity *et al.* (2006); Charity *et al.* (2007)]. Figures in this section are adapted from these publications. References to earlier DOM calculations can also be found there. The idea behind a simultaneous analysis of several isotopes of Ca nuclei is to establish the dependence of the DOM potentials on the nucleon asymmetry parameter

$$
\delta = \frac{N - Z}{N + Z}.
$$
(23.113)

Detailed knowledge of this dependence will allow the study of exotic nuclei all the way to the proton and neutron dripline, where nuclei are no longer bound. Such nuclei will be studied at future rare isotope accelerators.

We begin by presenting the functional forms of the DOM potentials that are employed in [Charity *et al.* (2006); Charity *et al.* (2007)]. The imaginary potential is composed of the sum of volume, surface, and imaginary spin-orbit components,

$$W\left(r;E\right) = -W_v\left(E\right)f(r,r_v,a_v) + 4a_sW_s\left(E\right)\frac{d}{dr}f(r,r_s,a_s) + W_{so}\left(r;E\right),$$
(23.114)

with Woods–Saxon form factors $f(r,r_i,a_i)$, as given in Eq. (23.54). Two contributions to the surface imaginary potential have been included: one dependent and the other, independent of asymmetry, *i.e.*

$$W_s(E) = W_s^0(E) + D\left(N,Z\right)W_s^1(E).$$
(23.115)

The asymmetry-dependent factor $D\left(N,Z\right)$ for protons is just

$$D^p\left(N,Z\right) = \frac{N-Z}{A},$$
(23.116)

similar to Eq. (3.27). Both surface terms have been parametrized by the same form, which is assumed symmetric about ε_F, representing the coupling to low-lying (collective) states including giant resonances. The latter form consists of the difference of two functions that cancel at large energies

$$W_s^i\left(E\right) = \omega_4(E,A_s^i,B_{s1}^i,0) - \omega_2(E,A_s^i,B_{s2}^i,C_s^i),$$
(23.117)

with

$$\omega_n(E,A_s^i,B_s^i,C_s^i) = A_s^i\,\theta\left(X\right)\frac{X^n}{X^n + (B_s^i)^n},$$
(23.118)

where $\theta\left(X\right)$ is the step function and $X = |E - \varepsilon_F| - C_s^i$. The functions ω_n are very practical to construct the imaginary potentials as there are analytical expressions for the corresponding dispersion integrals [Charity *et al.* (2007)]. Figure 23.4(a) demonstrates the relationship between the shape of $W_s^i\left(E\right)$ and its four defining parameters A_s, B_{s1}, B_{s2}, and C_s. It is important to note that the present treatment of the optical potential generates an ℓj-independent representation of the nucleon self-energy, apart from the spin-orbit contribution which is explicitly included. Details of the pole structure of the nucleon self-energy at very low energy, that depend on angular momentum and parity, are thus only treated in an average way, as discussed in the previous section.

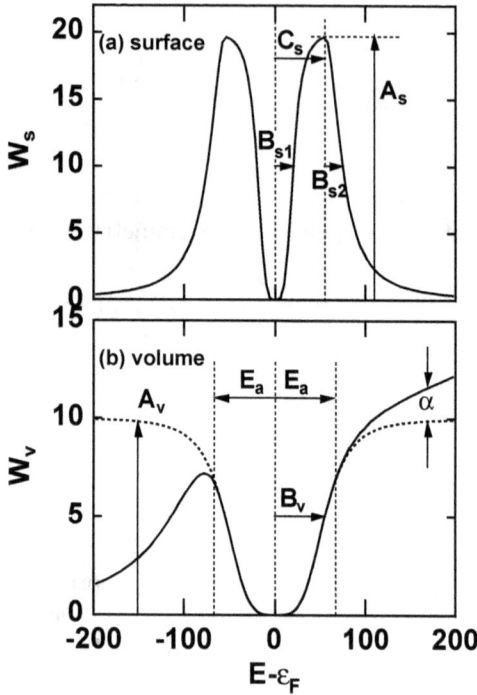

Fig. 23.4   The curves show typical energy-dependences of (a) the surface and (b) the volume imaginary potentials. In each case the meaning of the defining parameters is indicated. For the volume term in (b), the dashed curve indicates the symmetric potential, before the energy asymmetry correction ΔW_{NM} is added.

The phase space of particle levels for $E \gg \varepsilon_F$ is significantly larger than that of hole levels for $E \ll \varepsilon_F$. Thus the contributions from 2p1h states will be larger than those from 1p2h states. Hence, at energies well removed from ε_F, the form of the volume imaginary potential should no longer be symmetric about ε_F. The following form was therefore assumed

$$W_v(E) = \omega_4(E, A_v, B_v, 0) + \Delta W_{NM}(E), \qquad (23.119)$$

where $\Delta W_{NM}(E)$ is the energy-asymmetric correction modeled after nuclear-matter calculations. Apart from this correction, the parametrization is the form from [Jeukenne and Mahaux (1983)] used in many DOM analyses.

The energy-asymmetric correction was taken as

$$\Delta W_{NM}(E) = \begin{cases} \alpha \left[\sqrt{E} + \frac{(\varepsilon_F + E_a)^{3/2}}{2E} - \frac{3}{2}\sqrt{\varepsilon_F + E_a} \right] & \text{for } E - \varepsilon_F > E_a \\ -w_2(E, A_v, E_a, 0, E_a) & \text{for } E - E_F < E_a \\ 0 & \text{otherwise.} \end{cases}$$

(23.120)

This is the same form as assumed by [Mahaux and Sartor (1991)], except for $E - \varepsilon_F < E_a$, where the form is slightly different but has the advantage of making the calculation of the dispersive correction easier. The parameter α defines the magnitude of the asymmetry correction for $E > \varepsilon_F$ and Mahaux and Sartor employed a value of $\alpha=1.65$ MeV$^{1/2}$. However the value of this quantity is not constrained theoretically. For example, nuclear-matter calculations with the CDBonn [Machleidt (2001)] and ArV18 [Wiringa *et al.* (1995)] interactions predict very different behavior, as shown in Fig. 24.13. Fitting these calculated imaginary potentials with Eqs. (23.119) and (23.120), values are obtained of $\alpha \approx 0$ and $\alpha = 2.2$ MeV$^{1/2}$ for the two interactions, respectively. In the DOM analysis the larger values of α do not give good fits, otherwise the value is not well constrained. Subsequently, a fixed, intermediate value of $\alpha=0.61$ MeV$^{1/2}$ was employed. The uncertainty in α gives rise to uncertainties in the absolute values of the predicted spectroscopic factors and occupation probabilities [Mahaux and Sartor (1991); Charity *et al.* (2006)]. If W_{NM} is set to zero, a refit of the data increases the occupation numbers of bound states by about 7% [Charity *et al.* (2006)].

The dependence of $W_v(E)$ on its defining parameters (A_v, B_v, E_a, and α) is shown in Fig. 23.4(b). We note that the effect of the energy asymmetry in the volume potential is mainly confined to the deeply bound orbits. In the present description few data are available to constrain this energy asymmetry. The situation would likely improve if charge densities could be included in the fit procedure.

The HF potential is parametrized in the following way

$$\mathcal{V}_{HF}(r; E) = -V_{HF}^{vol}(E) \, f(r, r_{HF}, a_{HF}) \tag{23.121}$$

$$+ 4V_{HF}^{sur} \frac{d}{dr} f(r, r_{HF}, a_{HF}) + V_c(r) + \mathcal{V}_{so}(r; E),$$

where the Coulomb V_c and real spin-orbit $\mathcal{V}_{so}$ terms have been separated from the volume and surface components. The volume component contains the energy-dependence representing nonlocality which was assumed to be linear below the Fermi energy and given by a sum of two exponentials

Table 23.1 Parameters describing the OM potential obtained from a fit to the data.

$$B_{HF} = .415 \text{ MeV}, \ C_{HF} = 18.29 \text{ MeV}, \ D_{HF} = 0.153 \text{ MeV}$$
$$V_{HF}^{sur} = 1.81 \text{ MeV}, \ r_{HF} = 1.19 \text{ fm}, \ a_{HF} = 0.70 \text{ fm}$$

$$A_s^0 = 6.19 \text{ MeV}, \ B_{s1}^0 = 9.82 \text{ MeV}, \ B_{s2}^0 = 36.94 \text{ MeV}$$
$$C_s^0 = 58.1 \text{ MeV}$$
$$A_s^1 = 110.34 \text{ MeV}, \ B_{s1}^1 = 7.82 \text{ MeV}, \ B_{s2}^1 = 17.70 \text{ MeV}$$
$$C_s^1 = 30.0 \text{ MeV}$$
$$r_s = 1.27 \text{ fm}, \ a_s = 0.64 \text{ fm}$$

$$A_v = 6.18 \text{ MeV}, \ B_v = 47.85 \text{ MeV}$$
$$r_v = 1.38 \text{ fm}, \ a_v = 0.63 \text{ fm}$$
$$E_a = 60 \text{ MeV (fixed)}, \ \alpha = 0.61 \text{ MeV}^{1/2} \text{ (fixed)}$$

$$V_{so} = 5.51 \text{ MeV}, \ A_{so} = -4.38 \text{ MeV}, \ B_{so} = 208.6 \text{ MeV}$$
$$r_{so} = 1.058 \text{ fm}, \ a_{so} = 0.67 \text{ fm}, \ C_{so} = -4.5 \times 10^{-4}$$

$$r_C = 1.31 \text{ fm (fixed)}$$

above. The two forms are matched at the Fermi energy

$$V_{HF}^{Vol}(E) = \begin{cases} A_{HF} \exp\left[-\dfrac{B_{HF}}{A_{HF}}(E - \varepsilon_F)\right] + C_{HF} \exp\left[-\dfrac{D_{HF}}{C_{HF}}(E - \varepsilon_F)\right] \\ A_{HF} - B_{HF}(E - \varepsilon_F) + C_{HF} - D_{HF}(E - \varepsilon_F), \end{cases}$$
(23.122)

where the first line is for $E > \varepsilon_F$ and the second one for $E < \varepsilon_F$.

At high energies, OM potentials generally include an imaginary spin-orbit potential. This term is usually assumed to be zero for lower energies, implying that it is energy dependent, giving rise to a dispersive correction to the real component. The total spin-orbit potential has the form

$$\mathcal{U}_{so}(r; E) = \mathcal{V}_{so}(r, E) + i\mathcal{W}_{so}(r; E) = \Delta\mathcal{V}_{so}(r, E)$$
$$+ \left(\frac{h}{m_\pi c}\right)^2 [V_{so} - C_{so}E + iW_{so}(E)]\frac{1}{r}\frac{d}{dr}f(r, r_{so}, a_{so})\frac{\boldsymbol{\ell} \cdot \boldsymbol{s}}{2},$$
(23.123)

where $(\hbar/m_\pi c)^2 = 2.0 \text{ fm}^2$ and $\Delta\mathcal{V}_{so}$ is the dispersive correction determined from the imaginary component $\mathcal{W}_{so}$. Since the imaginary spin-orbit component is generally needed only at high energies, the following form was

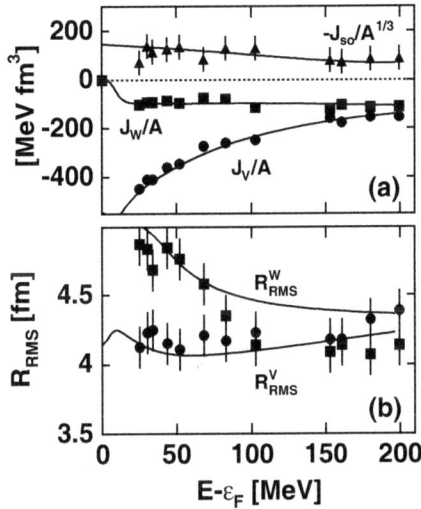

Fig. 23.5   (a) Integrated potentials and (b) rms radii obtained from combined fits to $p+^{40}$Ca elastic-scattering data and $n+^{40}$Ca total cross sections.

chosen

$$W_{so}(E) = \omega_4(E, A_{so}, B_{so}, 0). \tag{23.124}$$

The dispersive correction $\Delta V_{so}(E)$, associated with this component from Eq. (23.96), gives an approximately linear decrease in magnitude of the total real spin-orbit strength over the energy region of interest. Through the term C_{so}, the possibility of an additional linear decrease in the real spin-orbit strength due to nonlocality is also included. However, in the final fit the C_{so} term is very small (Table 23.1) suggesting that the dispersive term accounts for most of the energy dependence. Thus imposition of causality provides a natural way to account for all of the linear decrease in the magnitude of the integrated potential J_{so} observed in Fig. 23.5.

The integrated potentials and rms radii obtained from the fits are plotted as a function of $E - \varepsilon_F$ in Fig. 23.5. The inclusion of the neutron total cross sections in the fits reduced the fluctuations in all of these quantities, especially for the rms radii. A number of important trends are observed in these data. First, the spin-orbit term J_{SO} is approximately constant, while showing a small, but significant, decrease with energy. Second, for the imaginary potential, R_{rms}^W continues to decrease above $E - \varepsilon_F = 50$ MeV, where the asymmetry dependence has largely vanished. It is possible that the radial dependence of the volume term W_v is itself energy dependent or that there is an asymmetry-independent surface component that persists to larger energies. These two possibilities are indistinguishable in practice, since adding a small surface correction to a volume component is equivalent to modifying the radius of the volume component. Finally for the real potential, R_{rms}^V shows much less energy dependence and may increase at the highest energies. The latter may be an indication of a "wine-bottle" shaped potential.

The fit parameters are listed in Table 23.1. For such a large body of data (81 data sets comprising 3569 data points), the excellent agreement of the fit with just 25 free parameters provides confidence in the predictive power of the DOM calculations. Note that parameters like radii and diffuseness properties in Table 23.1 have values that are quite standard.

Comparisons of experimental and fitted elastic scattering data are shown in Figs. 23.6 and 23.7 for differential cross sections [see Eq. (23.38)] and analyzing powers [see Eq. (23.39)], respectively. The elastic proton scattering data for ^{40}Ca and ^{48}Ca are included in the fit, whereas the results for ^{42}Ca and ^{44}Ca are calculated by employing the fitted dependence of the potential on asymmetry. The description of these data is also excellent and it is clear that for protons it makes sense to consider extrapolations beyond ^{48}Ca. It is useful to realize that the concept of an empirical representation of the self-energy for nuclei that do not have shell closures makes sense as well. Clearly, the presence of more collective low-lying positive parity states in ^{42}Ca and ^{44}Ca does not spoil the description of the elastic scattering data. The rather large energy domain considered in the fits, apparently is not a problem in describing elastic scattering data either. The limit of 200 MeV is indicated by the need for a treatment of relativistic aspects of the scattering process at higher energies. By taking this wide energy perspective, an important constraint is imposed on the real part of the self-energy below the Fermi energy, since a major part of the dispersive contribution is determined by the imaginary part above the Fermi energy. We emphasize the importance of employing a subtracted dispersion relation, since it allows the real part of the self-energy at the subtraction point ε_F to be directly related to detailed empirical information. Indeed the assumed form of $\mathcal{V}_{HF}$ and the resulting fit parameters correspond nicely to expectations based on nuclear-structure information. The assumed $\ell j \pi$ indendence of the DOM potentials (apart from the spin-orbit term), does not present any serious problem in describing the elastic scattering data either.

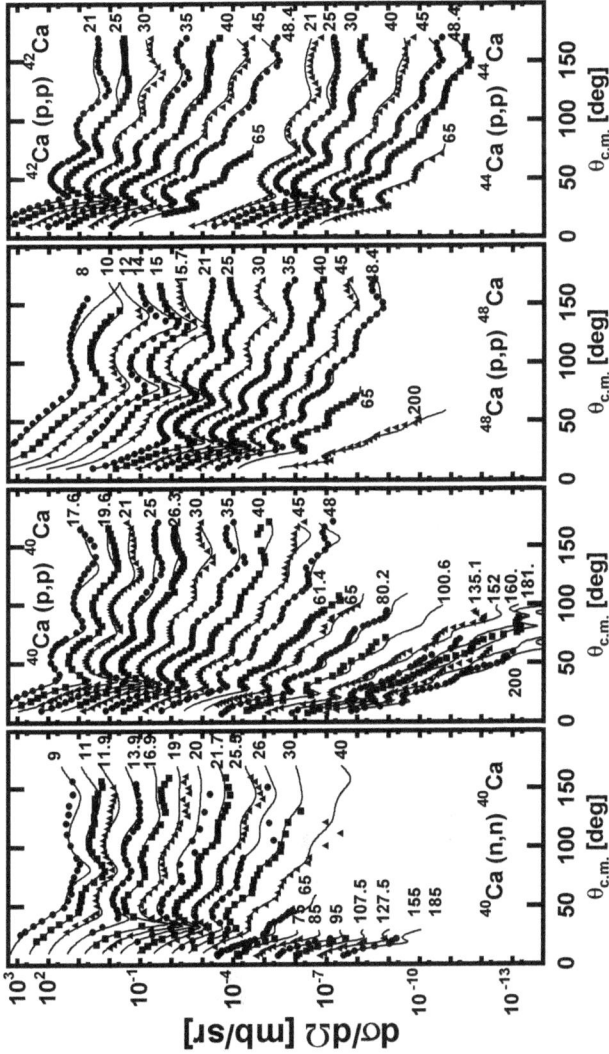

Fig. 23.6 Comparison of experimental (data) and fitted (curves) differential elastic scattering cross sections. For display purposes, both data and curves for successively larger energies are scaled down by a factor of 4. For $p+{}^{42}$Ca, the data and curves are scaled down by an additional factor of 100.

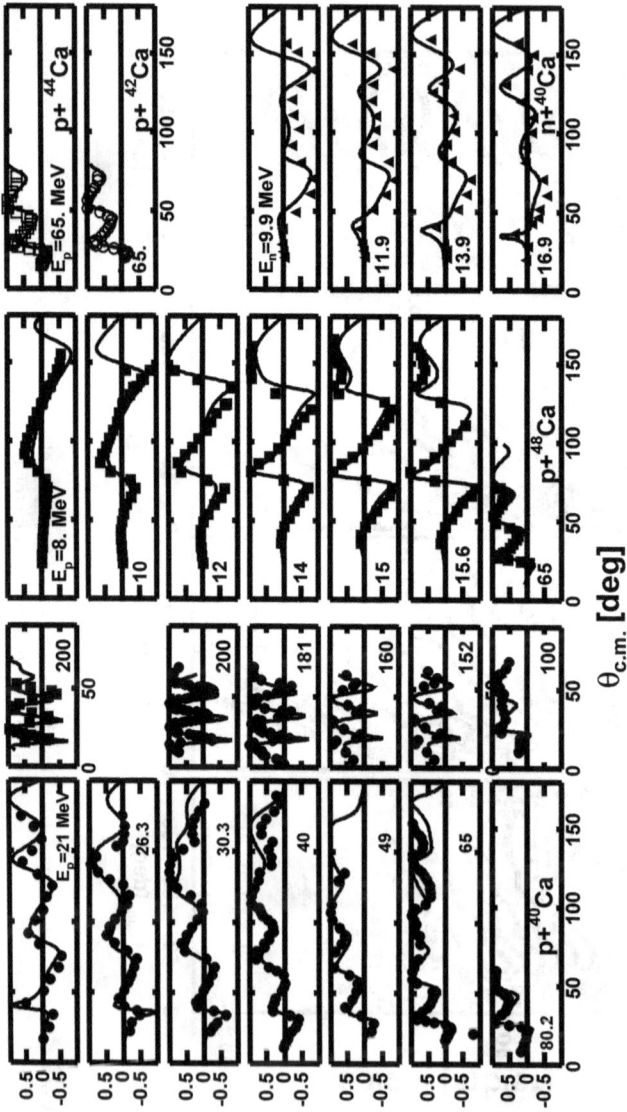

Fig. 23.7   Comparison of experimental and fitted analyzing powers. The data points depicted by the solid circles, solid squares, open circles, open squares, and triangles are for the $p+^{40}$Ca, $p+^{42}$Ca, $p+^{44}$Ca, $p+^{48}$Ca, and $n+^{40}$Ca reactions, respectively.

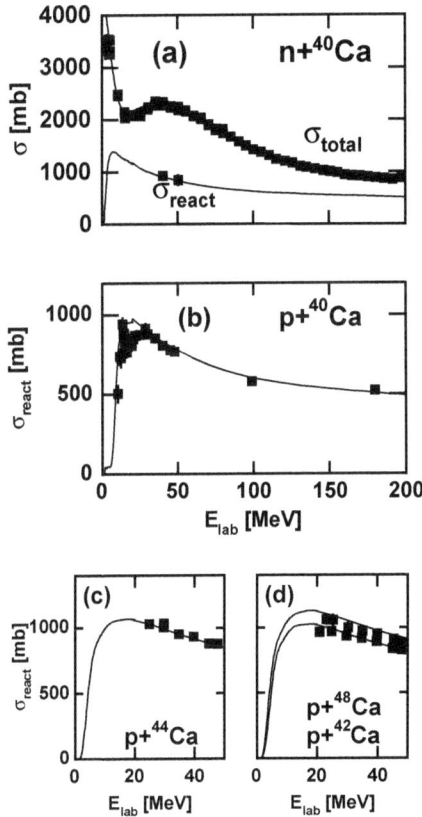

Fig. 23.8 Comparison of experimental and predicted reaction and total scattering cross sections. (a) total and reactions cross section for the $n+^{40}$Ca reaction. Reactions cross sections for (b) $p+^{40}$Ca, (c) $p+^{44}$Ca, and (d) $p+^{42,48}$Ca reactions.

Fitted reaction and total cross sections are displayed in Fig. 23.8. We emphasize the importance of the presence and the behavior of the reaction cross section which reflects the fact that the corresponding nucleon self-energy has a considerable imaginary part. Such a presence at energies above ε_F inevitably leads to a depletion of the sp strength associated originally with fully occupied orbits in the mean-field picture. We refer to the discussion in Sec. 17.4 for an overview of our present understanding of the sp strength distribution in closed-shell nuclei.

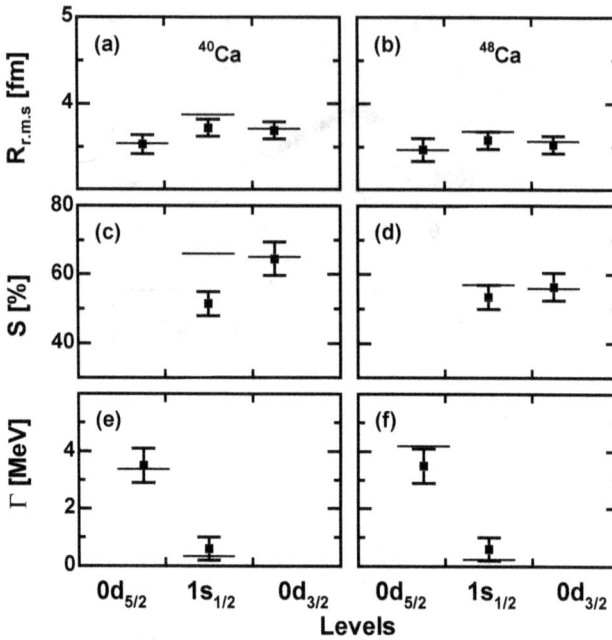

Fig. 23.9 Fitted level properties for the $0d\frac{5}{2}$, $1s\frac{1}{2}$, and $0d\frac{3}{2}$ proton hole states. The left (a,c,e) and the right (b,d,f) panels display the fits for ^{40}Ca and ^{48}Ca, respectively. The fitted quantities include (a,b) the rms radius R_{rms}, (c,d) the spectroscopic factors S expressed as a percentage of the independent-particle-model value, and (e,f) the widths Γ of these states.

Bound state information from $(e, e'p)$ data are shown in Fig. 23.9. For the $0d\frac{5}{2}$, $1s\frac{1}{2}$, and $0d\frac{3}{2}$ proton holes states in ^{40}Ca and ^{48}Ca, the widths Γ [see Eq. (23.102)], rms radii R_{rms} [Eq. (23.104)], and spectroscopic factors S [Eq. (23.103)] measured in the $(e, e'p)$ reactions at NIKHEF by [Kramer *et al.* (1989); Kramer (1990); Kramer *et al.* (2001)] are shown. The spectroscopic factor for the $1s\frac{1}{2}$ state in ^{40}Ca, although shown in the plots, was not included in the fits. More recent experiments by the same group indicate that the value of this quantity is larger than the published value [Lapikás (2005)]. The results for these bound-state quantities suggest that the assumption of an imaginary part of the potential that is symmetric with respect to the Fermi energy within about 60 MeV is a reasonable starting point, as was found by other authors for other systems [Mahaux and Sartor (1991)]. The success in describing these spectroscopic properties below the

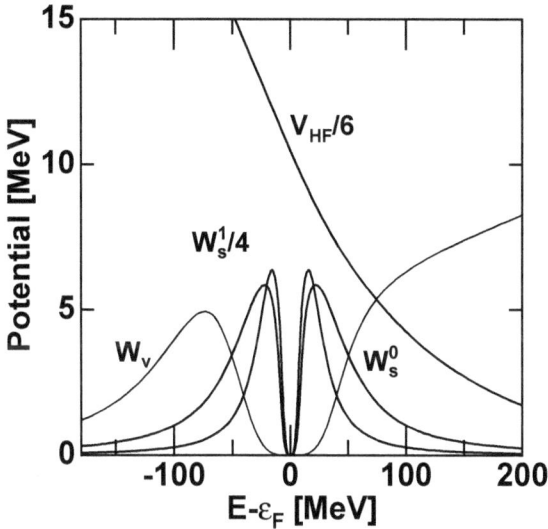

Fig. 23.10   Potentials determined from the fit. The HF potential V_{HF} is for $p+^{40}$Ca while the other potentials are system independent.

Fermi energy furthermore provides independent confirmation of the validity of the procedure to extract spectroscopic factors from the $(e, e'p)$ reaction.

The energy dependence of the potentials that provide a successful description of the experimental data, is shown in Fig. 23.10. The HF potential V_{HF} is dependent on the parameter A_{HF} which is adjusted for each system to the experimental value of ε_F. The potential plotted in Fig. 23.10 is for the $p+^{40}$Ca system. For the other systems, while the absolute potential will be different, the relative energy dependence is the same.

The two imaginary surface potentials have quite different energy dependences. The asymmetry-dependent or isovector component W_s^1 rises and falls faster with $|E - \varepsilon_F|$ than the isoscalar component W_s^0. The elastic-scattering and other positive-energy data constrain the falling part of these potentials, while the rising part is constrained by the $(e, e'p)$ data and the level energies. It is therefore essential to have both positive and negative energy data to fully constrain these potentials.

The fitted potentials are consistent with the integrated potentials and rms radii for the $p+^{40}$Ca reaction presented in Fig. 23.5, where the curves indicate the DOM predictions. The slow fall off with energy of the isoscalar surface potential W_s^0 (seen in Fig. 23.10) is responsible for the slow fall off

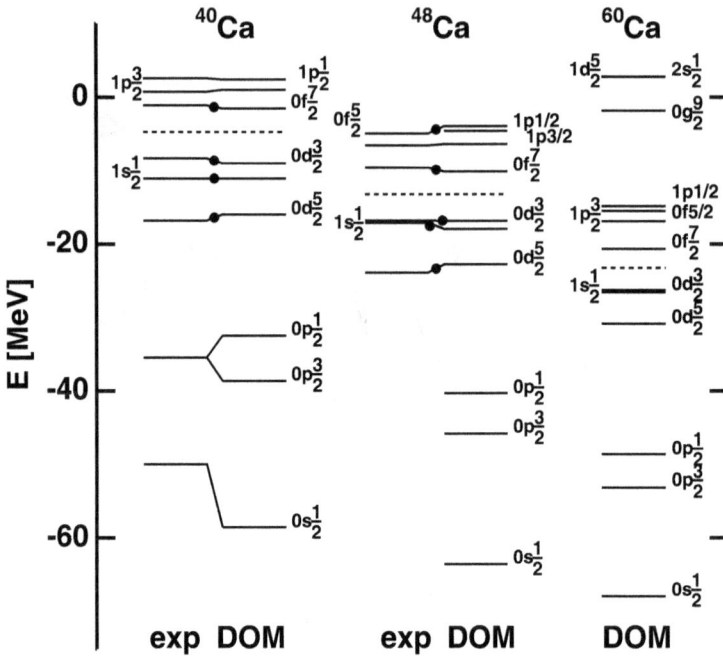

Fig. 23.11   Comparison of experimental proton single-particle levels to DOM calculations. The energies of the levels indicated with solid dots were included in the fits. The
dashed lines indicate the Fermi energy.

of the R_{rms}^W values in Fig. 23.5(b). The faster fall off for the isovector
component is required in order for the J_W values to be largely asymmetry
independent by 50 MeV, as found experimentally. The surface HF component is responsible for the small increase of R_{rms}^V at the larger energies.

   Single-particle level energies for protons in ^{40}Ca and ^{48}Ca are compared
in Fig. 23.11 with data. References to the experimental results can be found
in [Charity *et al.* (2006)]. Apart from the levels included in the fit (indicated
with the solid dots), the other known levels are well reproduced. The $0s_{1/2}$
level of ^{40}Ca is very wide and even though the DOM prediction for its energy
is low, it lies within the experimentally determined width. Extrapolating
the asymmetry dependence, it is possible to predict the proton levels in
^{60}Ca. The surface dispersive correction is large for this nucleus and its
effects are quite apparent. The levels in the immediate vicinity of ε_F are
focused closer to ε_F, increasing the density of sp levels. A reduced gap

between the particle and hole valence levels implies that the closed-shell nature of this nucleus has diminished and proton pairing may be important. The levels further from ε_F have been pushed away and as a result there are big gaps between the $0p\frac{1}{2}$ and the $0d\frac{5}{2}$ and between the $1p\frac{1}{2}$ and $0g\frac{9}{2}$.

While extrapolations for proton elastic scattering and level properties with increasing asymmetry are based on solid data input, it is clear that these extrapolations can be experimentally investigated with rare isotope beam facilities that can perform elastic proton scattering experiments in inverse kinematics. Employing such new data, will fine tune the DOM potentials, in particular their asymmetry dependence. Ultimately this procedure should allow access to proton properties near the neutron dripline. It is already clear that the asymmetry degree of freedom, as discussed here, implies that the minority species will become more correlated with increasing asymmetry. Experimental work [Gade *et al.* (2004)] involving heavy-ion knockout reactions suggests that the reduction factor, the ratio of the measured spectroscopic factor relative to the shell-model value, decreases strongly with nucleon separation energy. The latter, of course, is related to the asymmetry. Nuclear-matter calculations suggest that protons (neutrons) feel stronger (weaker) correlations with increasing neutron fraction [Frick *et al.* (2005)]. These effects are related to the increased (decreased) importance of the stronger *p-n* tensor interaction compared to the *p-p* (*n-n*) interaction for protons (neutrons) with increasing asymmetry (see Sec. 24.7, in particular Fig. 24.9).

Occupation numbers calculated according to Eqs. (23.111) and (23.112) for protons in ^{40}Ca and ^{48}Ca are displayed in Fig. 23.12. Results extrapolated for ^{60}Ca are also shown. There is some uncertainty associated with the absolute values of these probabilities. For instance the parameter α, defining the energy-asymmetry of the volume imaginary potential, is not constrained in the fits and variation of this parameter leads to scaling of the occupation probabilities [Mahaux and Sartor (1991); Charity *et al.* (2006); Charity *et al.* (2007)]. One should therefore concentrate on the relative difference in the probabilities between the two isotopes. The protons show a significant difference: Compared to ^{40}Ca, the occupation probabilities for ^{48}Ca and ^{60}Ca are further reduced from unity below the Fermi surface and further enhanced from zero above. This implies that protons indeed feel stronger correlations in the more neutron-rich nucleus. The differences are mainly confined to levels near the Fermi surface. The $0s$ and $0p$ levels show very little asymmetry dependence. Their occupation probabilities are dominated by the effects of short-range correlations. Their asymmetry

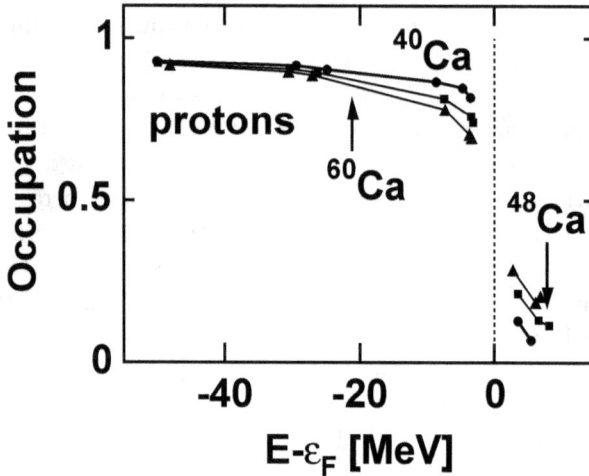

Fig. 23.12   Occupation probabilities predicted for protons in ^{40}Ca (circular points), ^{48}Ca (square points) and ^{60}Ca (triangular points).

dependence is therefore more strongly connected to any asymmetry-dependence of the volume imaginary component, which is absent in the calculations [Charity *et al.* (2007)]. We note that the DOM calculations for the proton occupation numbers in ^{48}Ca are similar to the theoretical results presented in Fig. 17.3.

The spectroscopic factors of $0d\frac{3}{2}$ protons further illustrate the increase of correlations change from 65.5%±4.8% to 56.5%±4.0% of the mean-field value in going ^{40}Ca to ^{48}Ca [Kramer *et al.* (1989); Kramer (1990); Kramer *et al.* (2001)]. A further reduction to 50% for the proton $0d\frac{3}{2}$ spectroscopic factor in ^{60}Ca is obtained from the extrapolated DOM potentials. These values are expected to further decrease with increasing asymmetry, suggesting that asymmetry dials from Z_F factors of about 2/3 for stable closed-shell nuclei to values that may approach the value of 0.2 for the strongly correlated ^{3}He liquid. Heavy-ion knockout reactions further-more suggest [Hansen and Tostevin (2003)] that the spectroscopic factors of weakly bound nucleons approach values close to the mean-field expectation. It may therefore be possible to employ the asymmetry "dial" also to study the majority species in the limit that the correlations become weak as in atoms.

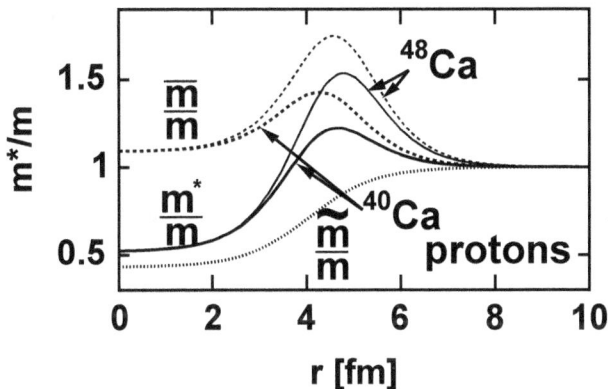

Fig. 23.13 DOM effective masses relative to the nucleon mass for protons. The momentum-dependent effective mass $\tilde{m}/m$ in ^{40}Ca is given by the dotted curve. Results for ^{48}Ca are similar. The energy-dependent effective masses are indicated by the dashed curves, while the solid curves represent the total effective masses. Thick and thin curves identify ^{40}Ca and ^{48}Ca, respectively.

Plotted in Fig. 23.13 are proton effective masses at the corresponding Fermi energy ε_F obtained from the DOM calculations. The momentum-dependent effective mass $\tilde{m}/m$ of Eq. (23.87) is shown only for ^{40}Ca. The results for ^{48}Ca are similar with a slightly larger radius. As for the occupation numbers, the proton's energy-dependent $\overline{m}/m$ defined in Eq. (23.99) and total m^*/m effective masses [see Eq. (23.98)] exhibit a significant difference between ^{40}Ca and ^{48}Ca. The larger surface imaginary component for ^{48}Ca gives rise to an increased enhancement of the effective masses in the surface.

The analysis of DOM potentials for neutrons in Ca isotopes is hampered by the lack of elastic scattering data in ^{48}Ca. The standard asymmetry dependence assumption suggests that the factor for protons in Eq. (23.116) should simply change sign for neutrons. It is shown in [Charity *et al.* (2007)] that this choice leads to imaginary potentials for neutrons near the Fermi energy that have the wrong sign and are therefore unacceptable. Various other choices are considered in [Charity *et al.* (2007)] demonstrating that elastic neutron scattering data on ^{48}Ca will be critical in determing the asymmetry dependence of DOM potentials for neutrons. Evidently this is necessary before detailed and reliable predictions of the properties of neutrons towards the neutron dripline can be made.

The *framework* of Green's functions is a critical ingredient in the DOM analysis, but there is a need for detailed theoretical calculations of the self-energy as a function of asymmetry. It is empirically determined that the proton imaginary part of the self-energy increases near the Fermi energy with increasing asymmetry. Understanding in detail which low-lying excitations are responsible for this effect is interesting and important. One possible candidate is the charge-exchange Gamow-Teller giant resonance that increases in strength with increasing asymmetry.

23.6.1 *Nonlocal HF potential in the DOM calculation*

The inclusion of nonlocality is necessary if the DOM potential is to be considered as a representation of the nucleon self-energy. The same number of parameters are required for a nonlocal potential [see Eq. (23.60)], when V_N is represented by a Woods-Saxon form, as for an local-equivalent potential with an assumed linear energy dependence. It is thus necessary to perform future studies with nonlocal potentials. This has the added advantage that all the properties that can be extracted from the propagator below the Fermi energy, can be compared to corresponding experimental data like the proton charge density and the energy per particle, leading to further experimental constraints of the DOM potentials. The solution of the Dyson equation with such a nonlocal potential to represent the HF potential for the proton $s\frac{1}{2}$ sp strength in ^{40}Ca is shown in Fig. 23.14 by the solid line. This strength is obtained by solving the propagator in coordinate space according to

$$G_{\ell j}(r, r'; E) = G_\ell^{(0)}(r, r'; E) + \int_0^\infty dr_1 r_1^2 \int_0^\infty dr_2 r_2^2 \qquad (23.125)$$
$$\times G_\ell^{(0)}(r, r_1; E)\langle r_1|\Sigma^{\ell j}(E)|r_2\rangle G_{\ell j}(r_2, r'; E),$$

where the HF part of the DOM potential is replaced by a nonlocal potential and the imaginary part is corrected by the factor given in Eq. (23.94). The corresponding dispersive real part follows a similar prescription [see Eq. (23.97)]. The hole spectral function is obtained from

$$S_{\ell j}(r; E) = \frac{1}{\pi} |\text{Im } G_{\ell j}(r, r; E)|. \qquad (23.126)$$

The spectral strength at E

$$S_{\ell j}(E) = \int_0^\infty dr \ r^2 \ S_{\ell j}(r; E) \qquad (23.127)$$

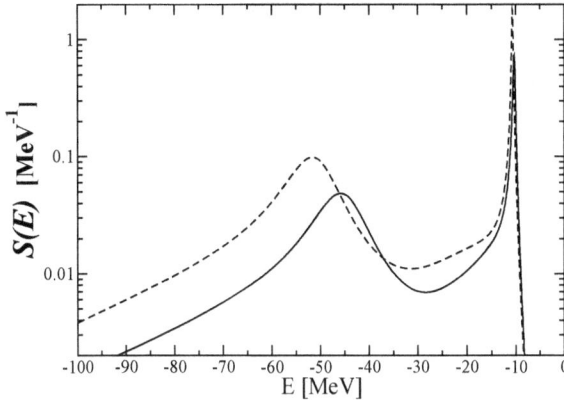

Fig. 23.14 Spectral strength for $s\frac{1}{2}$ protons in ^{40}Ca from solving the Dyson equation with the local equivalent HF potential (dashed) and nonlocal version (solid).

yields the spectroscopic strength per unit of energy in analogy to the spectroscopic factor in Eq. (7.53). The nonlocal potential was chosen as in Eq. (23.60) with a Woods-Saxon shape for V_N. The depth of the V_N was 100 MeV, the nonlocality parameter $\beta = 1.2$ fm, the radius parameter 1.25 fm, and the diffuseness 0.6 fm. These numbers were selected to put the peak of the distribution close to the position of the $1s\frac{1}{2}$ sp level, obtained from the DOM calculation. For comparison we include the solution of the Dyson equation with the original DOM potential with the energy-dependent local equivalent HF potential. The total s strength integrated up to the Fermi energy for the solid line is 1.71 corresponding to 85% of the 0s and 1s mean-field occupation. Since the DOM potential cannot be legitimately used on account of its energy dependence in the HF part, we confirm this assessment by noting that the dashed curve severely overestimates the s strength. We note that the strength contained in the peak near the Fermi energy with the nonlocal potential corresponds accurately to the $1s\frac{1}{2}$ spectroscopic factor shown in Fig. 23.9. This demonstrates the importance of properly retrofitting the strength of the imaginary part of the self-energy when nonlocal potentials are employed. We also note that it is easier with nonlocal potentials to avoid placing the first peak of the $s\frac{1}{2}$ strength distribution at a too low energy. With a linear energy dependence for the local equivalent potential one may find the corresponding peak at a very deeply bound location.

We close the discussion of the DOM results with the observation that no concerted efforts have been made to perform microscopic calculations of the nucleon self-energy at positive energy calculated for a specific nucleus. This daunting task remains a challenge for the future. We refer to a substantial body of work that was aimed at calculating the volume properties of the optical-model potential from nuclear matter calculations through some form of local-density approximation [Jeukenne *et al.* (1976)]. Such procedures have been fruitfully used in the empirical analysis of experimental data [Jeukenne *et al.* (1977)] but cannot replace microscopic calculations in finite nuclei.

23.7 Exercises

(1) Show that $|\Psi_i^{(+)}\rangle$ in Eq. (23.11) is properly normalized and confirm the validity of Eq. (23.23).
(2) Verify Eqs. (23.36), (23.37), (23.42), and (23.46).
(3) Determine the elastic differential cross section for 30 MeV proton scattering off ^{208}Pb using the parameters given in [Bohr and Mottelson (1998)] and employing an appropriate numerical procedure.
(4) Verify that Eq. (23.88) is a sensible representation of the wave function for the nonlocal potential.

Chapter 24

Finite-temperature formalism and applications

The material presented in Ch. 5 treated many-boson and fermion systems without the full consideration of the interaction between the constituent particles. Interacting systems are often studied at finite temperature, so it is important to develop the tools for the description of their quantum properties. With some minor adjustments the formalism of Chs. 7 and 8 for $T = 0$ will be equally useful to study sp propagation at finite temperature. Similar material can be found in [Kadanoff and Baym (1962); Fetter and Walecka (1971); Abrikosov *et al.* (1975); Mattuck (1992)].

In Sec. 24.1 we introduce the sp propagator at finite temperature. The formalism will be presented for fermions, leaving the treatment of bosons as exercise material at the end of the chapter. In order to develop a systematic perturbation expansion, it will be necessary to introduce the concept of imaginary time propagation, as opposed to that in real time. Ensemble averages at finite temperature involve the presence of the statistical operator which provides both an analogy with, and a deviation from, the zero-temperature case. The finite-temperature expansion in the modified interaction picture is discussed in Sec. 24.2. The same diagrams as for $T = 0$ appear in Sec. 24.3 but a Fourier summation employing so-called Matsubara sums [Matsubara 1955], is required in the energy formulation. These sums must be introduced on account of the quasiperiodic boundary conditions obeyed by the finite-temperature sp propagator.

The spectral representation of the propagator is studied in Sec. 24.4, where it is also shown that relevant ensemble averages of one-body operators together with the ensemble average of the Hamiltonian, can be calculated from the sp propagator. We briefly review the Dyson equation in Sec. 24.5 and discuss the HF approximation at finite T in 24.6. As an example of tp propagation at finite T, we study homogeneous systems with short-range correlations. The required ladder summations are reviewed in Sec. 24.7 together with a discussion of some recent results. The summation of ring diagrams at finite T follows similar developments and is not pursued here. We refer to [Fetter and Walecka (1971)] for a detailed analysis pertaining to the electron gas. Pairing at finite temperature is presented in Sec. 24.8.

24.1 Real and imaginary time single-particle propagator

Noting the results of Eqs. (5.29) – (5.33), the sp propagator at a finite temperature T has a logical definition as the ensemble average

$$G_T(\alpha, \beta; t, t') = -\frac{i}{\hbar} \left\langle T[a_{\alpha_\Omega}(t) a^{\dagger}_{\beta_\Omega}(t')] \right\rangle = -\frac{i}{\hbar} \text{Tr} \left(\hat{\rho}_G T[a_{\alpha_\Omega}(t) a^{\dagger}_{\beta_\Omega}(t')] \right),$$

(24.1)

where $\hat{\rho}_G$ is the statistical operator given in Eq. (5.29), and T affects the same time-ordering as in Eq. (7.5). Since we will use the grand-canonical description of the system, it is convenient to let time evolution be governed by

$$\hat{\Omega} = \hat{H} - \mu \hat{N},$$

(24.2)

as in Ch. 18 for bosons at $T = 0$. Particle number is only conserved on average, and as usual, μ denotes the chemical potential. In Eq. (24.1) the appropriately modified definition of the operator in the Heisenberg picture is used, where *e.g.*

$$a_{\alpha_\Omega}(t) = \exp\left\{\frac{i}{\hbar}\hat{\Omega}t\right\} a_{\alpha} \exp\left\{-\frac{i}{\hbar}\hat{\Omega}t\right\}.$$

(24.3)

Other authors continue to utilize the standard definition of the Heisenberg picture [Kadanoff and Baym (1962)].

The real-time propagator of Eq. (24.1) is not yet in a form that can be employed to generate a perturbation expansion analogous to the one for zero temperature. Since the latter result is related to the expansion of the time-evolution operator in the interaction picture, it originates from the many-body time-dependent Schrödinger equation. By recognizing the possibility to identify a similar evolution equation for the operator $\exp(-\hat{\Omega}\tau/\hbar)$, and thereby relating it to $\exp(-\beta\hat{\Omega})$, one can obtain a finite-T perturbation expansion, if one considers propagation in imaginary time (*i.e.* $it \Rightarrow \tau$).

We therefore introduce, for real τ,

$$a_{\alpha\Omega}(\tau) = \exp\left\{\hat{\Omega}\tau/\hbar\right\}a_\alpha \exp\left\{-\hat{\Omega}\tau/\hbar\right\} \tag{24.4}$$

and

$$a_{\beta\Omega}^\dagger(\tau) = \exp\left\{\hat{\Omega}\tau/\hbar\right\}a_\beta^\dagger \exp\left\{-\hat{\Omega}\tau/\hbar\right\}, \tag{24.5}$$

which are *not* adjoints of each other.

The replacement $it \Rightarrow \tau$ requires the introduction of an imaginary time-ordering operation $\mathcal{T}_\tau$ which orders according to the value of τ with the smallest argument occurring on the right, while maintaining the fermion sign change for an odd number of permutations. The temperature or imaginary-time sp propagator can now be defined (for real values of τ and τ') by

$$\begin{aligned}\mathcal{G}_T(\alpha,\beta;\tau-\tau') &= -\frac{1}{\hbar}\left\langle \mathcal{T}_\tau[a_{\alpha\Omega}(\tau)a_{\beta\Omega}^\dagger(\tau')]\right\rangle \\ &= -\frac{1}{\hbar}\operatorname{Tr}\left(\hat{\rho}_G\mathcal{T}_\tau[a_{\alpha\Omega}(\tau)a_{\beta\Omega}^\dagger(\tau')]\right),\end{aligned} \tag{24.6}$$

dropping an additional factor of i, as is customary [Fetter and Walecka (1971); Abrikosov *et al.* (1975); Mattuck (1992)]. The notation $\tau-\tau'$ in Eq. (24.6) indicates that the sp propagator only depends on the difference of its imaginary time arguments when the Hamiltonian is not an explicit function of time.

24.1.1 *Quasiperiodic boundary conditions and the Fourier series*

Setting $\tau'=0$ for simplicity, the temperature propagator in Eq. (24.6) can be written as

$$\begin{aligned}\mathcal{G}_T(\alpha,\beta;\tau) = -\frac{1}{\hbar Z_G}&\left\{\theta(\tau)\operatorname{Tr}\left(e^{-(\beta-\tau/\hbar)\hat{\Omega}}a_\alpha e^{-\tau/\hbar\hat{\Omega}}a_\beta^\dagger\right)\right. \\ &\left.-\theta(-\tau)\operatorname{Tr}\left(e^{-(\beta+\tau/\hbar)\hat{\Omega}}a_\beta^\dagger e^{\tau/\hbar\hat{\Omega}}a_\alpha\right)\right\}, \tag{24.7}\end{aligned}$$

where the grand partition function Z_G is given in Eq. (5.30). We also employed the invariance of the trace of an operator product under cyclical permutations

$$\operatorname{Tr}\left(\hat{A}\hat{B}...\hat{X}\hat{Y}\right) = \operatorname{Tr}\left(\hat{Y}\hat{A}\hat{B}...\hat{X}\right). \tag{24.8}$$

748   *Many-body theory exposed!*

In Eq. (24.7) the sum over eigenstates contained in the trace is accompanied by a weight factor $e^{-(\beta\pm\tau/\hbar)\hat{\Omega}}$. Since the eigenvalue spectrum of $\hat{\Omega}$ is in general not bounded from above, the temperature propagator is defined for arguments in the range $-\beta\hbar \leq \tau \leq \beta\hbar$, thereby ensuring convergence. Note that when the general form is employed with two independent imaginary times τ and τ' [see Eq. (24.6)], these can be restricted to the interval $0 \leq \tau, \tau' \leq \beta\hbar$ since their difference covers the relevant interval $-\beta\hbar \leq \tau - \tau' \leq \beta\hbar$.

Using Eq. (24.7), we now evaluate the temperature propagator for a negative argument $-\beta\hbar < \tau < 0$, and compare it with the value obtained at the positive argument $\tau + \beta\hbar$

$$\mathcal{G}_T(\alpha, \beta; \tau) = \frac{1}{\hbar Z_G}\text{Tr}\left(e^{-(\beta+\tau/\hbar)\hat{\Omega}}a_\beta^\dagger e^{\tau/\hbar\hat{\Omega}}a_\alpha\right) \tag{24.9}$$

$$\mathcal{G}_T(\alpha, \beta; \tau + \beta\hbar) = -\frac{1}{\hbar Z_G}\text{Tr}\left(e^{\tau/\hbar\hat{\Omega}}a_\alpha e^{-(\beta+\tau/\hbar)\hat{\Omega}}a_\beta^\dagger\right). \tag{24.10}$$

With the cyclical property of the trace one then finds the important antiperiodic property

$$\mathcal{G}_T(\alpha, \beta; \tau) = -\mathcal{G}_T(\alpha, \beta; \tau + \beta\hbar), \tag{24.11}$$

relating the propagator at positive and negative imaginary times.

The temperature propagator is a continuous function of τ defined over the interval $[-\beta\hbar, \beta\hbar]$. The boundary condition in Eq. (24.11) implies that the function can be periodically repeated (with period $2\beta\hbar$). This is sufficient to ensure an expansion as a discrete Fourier series

$$\mathcal{G}_T(\alpha, \beta; \tau) = \frac{1}{\hbar\beta}\sum_{n=-\infty}^{+\infty} e^{-iE_n\tau/\hbar}\mathcal{G}_T(\alpha, \beta; E_n) \tag{24.12}$$

containing (for fermions) all energies that are odd multiples of π/β, the so-called Matsubara energies [1]

$$E_n = \frac{(2n+1)\pi}{\beta}. \tag{24.13}$$

---

[1]Throughout this chapter the FT of the (real or imaginary) time variable is loosely referred to as "energy", and the corresponding propagator argument is denoted by "E". Since the time evolution is governed by $\hat{\Omega} = \hat{H} - \mu\hat{N}$, it is clear that these correspond to energies relative to the chemical potential μ. If confusion is possible, we will use the term absolute energy (corresponding to the Hamiltonian); so $E = E(abs.) - \mu$ for a sp quantity, $E = E(abs.) - 2\mu$ for a tp quantity.

The set of Fourier coefficients contains an equivalent amount of information as the original representation in the imaginary time domain

$$\mathcal{G}_T(\alpha,\beta;E_n) = \frac{1}{2}\int_{-\beta\hbar}^{\beta\hbar} d\tau \; e^{iE_n\tau/\hbar}\mathcal{G}_T(\alpha,\beta;\tau) = \int_0^{\beta\hbar} d\tau \; e^{iE_n\tau/\hbar}\mathcal{G}_T(\alpha,\beta;\tau),$$
(24.14)

using the boundary condition to generate the last equality. We will refer to this propagator as the temperature Green's function in the imaginary energy domain.

24.1.2 *Noninteracting temperature propagator*

As usual, a helpful illustration is provided by the noninteracting Fermi system. The Hamiltonian operator then simplifies to

$$\hat{\Omega}_0 = \hat{H}_0 - \mu\hat{N}.$$
(24.15)

Employing the notation [see also Eq. (5.40)]

$$Z_0^F = \mathrm{Tr}\left(e^{-\beta\hat{\Omega}_0}\right)$$
(24.16)

and $\hat{\rho}_G^0 = e^{-\beta\hat{\Omega}_0}/Z_0^F$, we can write the noninteracting temperature propagator as

$$\begin{aligned}
\mathcal{G}_T^{(0)}(\alpha,\beta;\tau-\tau') &= -\frac{1}{\hbar}\left\langle T_\tau[a_{\alpha\Omega_0}(\tau)a_{\beta\Omega_0}^\dagger(\tau')]\right\rangle_0 \\
&= -\frac{1}{\hbar}\mathrm{Tr}\left(\hat{\rho}_G^0 T_\tau[a_{\alpha\Omega_0}(\tau)a_{\beta\Omega_0}^\dagger(\tau')]\right).
\end{aligned}$$
(24.17)

The notation $\langle\rangle_0$ indicates the ensemble average over noninteracting systems. A similar calculation as in Eq. (A.14) demonstrates that

$$\begin{aligned}
\hbar\frac{\partial}{\partial\tau}a_{\alpha\Omega_0}(\tau) &= \left[\hat{\Omega}_0, a_{\alpha\Omega_0}(\tau)\right] \\
&= \exp\left\{\hat{\Omega}_0\tau/\hbar\right\}\left[\hat{\Omega}_0, a_\alpha\right]\exp\left\{-\hat{\Omega}_0\tau/\hbar\right\} \\
&= -\left(\varepsilon_\alpha - \mu\right)a_{\alpha\Omega_0}(\tau) = -\varepsilon_{\alpha\mu}a_{\alpha\Omega_0}(\tau),
\end{aligned}$$
(24.18)

where we assume that α and β refer to eigenstates of the sp Hamiltonian H_0 and the notation $\varepsilon_{\alpha\mu} = \varepsilon_\alpha - \mu$ is used [see Eq. (18.44)]. The solution to Eq. (24.18) is given by

$$a_{\alpha\Omega_0}(\tau) = e^{-\varepsilon_{\alpha\mu}\tau/\hbar}a_\alpha.$$
(24.19)

An identical calculation yields

$$a^\dagger_{\alpha\Omega_0}(\tau) = e^{\varepsilon_{\alpha\mu}\tau/\hbar}a^\dagger_\alpha, \tag{24.20}$$

demonstrating again that Eq. (24.19) and (24.20) are not related by Hermitian conjugation. Evaluation of Eq. (24.17) is now straightforward by employing these equations and the definition of T_τ

$$\mathcal{G}_T^{(0)}(\alpha,\beta;\tau-\tau') = -\frac{1}{\hbar}\Big\{\theta(\tau-\tau')e^{-\varepsilon_{\alpha\mu}\tau/\hbar}e^{\varepsilon_{\beta\mu}\tau'/\hbar}\langle a_\alpha a^\dagger_\beta\rangle_0$$

$$-\theta(\tau'-\tau)e^{-\varepsilon_{\alpha\mu}\tau/\hbar}e^{\varepsilon_{\beta\mu}\tau'/\hbar}\langle a^\dagger_\beta a_\alpha\rangle_0\Big\} \tag{24.21}$$

$$= -\frac{1}{\hbar}\delta_{\alpha\beta}e^{-\varepsilon_{\alpha\mu}(\tau-\tau')/\hbar}\{\theta(\tau-\tau')(1-n^0_\alpha)-\theta(\tau'-\tau)n^0_\alpha\}.$$

The last equality follows from the realization that, using eigenstates of $\hat{H}_0$ and $\hat{N}$ in the evaluation of the trace, only contributions with $\alpha=\beta$ will survive. The required ensemble averages, *e.g.* $\langle a_\alpha a^\dagger_\alpha\rangle_0 = 1 - \langle a^\dagger_\alpha a_\alpha\rangle_0 = 1 - n^0_\alpha$, have been calculated in Sec. 5.5 [see Eq. (5.42)] with the standard result

$$\langle a^\dagger_\alpha a_\alpha\rangle_0 = n^0_\alpha = \frac{1}{\exp\{\beta\varepsilon_{\alpha\mu}\}+1}. \tag{24.22}$$

Equation (24.21) for the noninteracting temperature sp propagator should be compared to Eq. (7.32) for $T=0$. In that case the distinction between the first and second term is very sharp, either α corresponds to an empty level in the ground state for the former, or to an occupied one for the latter. At finite temperature, this distinction becomes blurred so that there exists *e.g.* a partial occupation even for levels that are empty at $T=0$.

Note that the antiperiodic property of Eq. (24.11) is easily checked for the noninteracting case. For $-\hbar\beta < \tau < 0$ one has

$$\mathcal{G}_T^{(0)}(\alpha,\beta;\tau) = \frac{1}{\hbar}\delta_{\alpha\beta}e^{-\varepsilon_{\alpha\mu}\tau/\hbar}n^0_\alpha$$

$$\mathcal{G}_T^{(0)}(\alpha,\beta;\tau+\hbar\beta) = -\frac{1}{\hbar}\delta_{\alpha\beta}e^{-\varepsilon_{\alpha\mu}(\tau+\beta)/\hbar}(1-n^0_\alpha). \tag{24.23}$$

The relation $1-n^0_\alpha = e^{\beta\varepsilon_{\alpha\mu}}n^0_\alpha$ then leads automatically to Eq. (24.11).

Inserting Eq. (24.21) in (24.14) yields the noninteracting temperature Green's function in the imaginary energy domain

$$\mathcal{G}_T^{(0)}(\alpha,\beta;E_n) = \delta_{\alpha,\beta}\frac{1}{iE_n - \varepsilon_{\alpha\mu}}. \tag{24.24}$$

This simple form applies when the propagator is expressed in the sp basis of eigenstates of H_0. When the sp labels α and β refer to an arbitrary basis, the more general form of the noninteracting propagator reads

$$\mathcal{G}_T^{(0)}(\alpha, \beta; E_n) = \sum_i \frac{z_{i\alpha} z_{i\beta}^*}{iE_n - \varepsilon_{i\mu}}, \qquad (24.25)$$

where the index i labels the eigenstates $|i\rangle$ and energies ε_i of the sp Hamiltonian H_0, and $|i\rangle = \sum_\alpha z_{i\alpha} |\alpha\rangle$. We will encounter this transformation when the HF description is discussed in Sec. 24.6.

24.2   Interaction-picture expansion at finite temperature

The finite-temperature expansion requires a similar strategy as the one presented in Ch. 8 and App. A for $T = 0$. We will therefore only sketch this development while stressing the similarities with the results of Secs. 8.1 and 8.2. As usual $\hat{\Omega} = \hat{\Omega}_0 + \hat{H}_1$. The starting point is the equivalent of the Schrödinger equation given by Eq. (A.2). Indeed, one verifies directly that

$$\hbar \frac{\partial}{\partial \tau} \exp(-\hat{\Omega}\tau/\hbar) = -\hat{\Omega} \exp(-\hat{\Omega}\tau/\hbar). \qquad (24.26)$$

As for $T = 0$, the relation between the exact propagator and the noninteracting one requires the equivalent of the interaction-picture time evolution. We define the Heisenberg picture at finite T for a general operator O_S in the Schrödinger picture by

$$\hat{O}_\Omega(\tau) = \exp\left\{\hat{\Omega}\tau/\hbar\right\} \hat{O}_S \exp\left\{-\hat{\Omega}\tau/\hbar\right\} \qquad (24.27)$$

like in Eqs. (24.4) and (24.5). The corresponding interaction-picture operator reads

$$\hat{O}(\tau) = \exp\left\{\hat{\Omega}_0\tau/\hbar\right\} \hat{O}_S \exp\left\{-\hat{\Omega}_0\tau/\hbar\right\}. \qquad (24.28)$$

We can relate the two pictures by observing that

$$\hat{O}_\Omega(\tau) = \exp\left\{\hat{\Omega}\tau/\hbar\right\} \exp\left\{-\hat{\Omega}_0\tau/\hbar\right\} \hat{O}(\tau) \exp\left\{\hat{\Omega}_0\tau/\hbar\right\} \exp\left\{-\hat{\Omega}\tau/\hbar\right\}$$
$$= \hat{\mathcal{U}}(0, \tau) \hat{O}(\tau) \hat{\mathcal{U}}(\tau, 0). \qquad (24.29)$$

Unlike the $T = 0$ version [see Eq. (A.22)]

$$\hat{\mathcal{U}}(\tau, \tau') = \exp\left\{\hat{\Omega}_0\tau/\hbar\right\} \exp\left\{-\hat{\Omega}(\tau - \tau')/\hbar\right\} \exp\left\{-\hat{\Omega}_0\tau'/\hbar\right\} \qquad (24.30)$$

is not unitary, whereas the group property similar to Eq. (A.25) continues to hold. Also the boundary condition

$$\hat{\mathcal{U}}(\tau, \tau) = 1 \tag{24.31}$$

remains valid. Using Eqs. (24.26)–(24.30), we generate the evolution equation in the interaction picture at finite temperature according to

$$
\begin{aligned}
\hbar \frac{\partial}{\partial \tau} \hat{\mathcal{U}}(\tau, \tau') &= \exp\left\{ \hat{\Omega}_0 \tau / \hbar \right\} \left(\hat{\Omega}_0 - \hat{\Omega} \right) \exp\left\{ -\hat{\Omega}(\tau - \tau')/\hbar \right\} \exp\left\{ -\hat{\Omega}_0 \tau' / \hbar \right\} \\
&= \exp\left\{ \hat{\Omega}_0 \tau / \hbar \right\} \left(\hat{\Omega}_0 - \hat{\Omega} \right) \exp\left\{ -\hat{\Omega}_0 \tau / \hbar \right\} \hat{\mathcal{U}}(\tau, \tau') \\
&= -\hat{H}_1(\tau) \hat{\mathcal{U}}(\tau, \tau'),
\end{aligned}
\tag{24.32}
$$

where

$$\hat{H}_1(\tau) = \exp\left\{ \hat{\Omega}_0 \tau / \hbar \right\} \hat{H}_1 \exp\left\{ -\hat{\Omega}_0 \tau / \hbar \right\}. \tag{24.33}$$

The operator differential equation (24.32) can formally be solved in exactly the same way as Eq. (8.6) leads to Eq. (8.7)

$$\hat{\mathcal{U}}(\tau, \tau') = \sum_{n=0}^{\infty} \left(\frac{-1}{\hbar} \right)^n \frac{1}{n!} \int_{\tau'}^{\tau} d\tau_1 \ldots \int_{\tau'}^{\tau} d\tau_n \, T_{\tau} \left[\hat{H}_1(\tau_1) \ldots \hat{H}_1(\tau_n) \right]. \tag{24.34}$$

To proceed with the relation between the interacting and noninteracting temperature propagator, we rewrite Eq. (24.30) with $\tau' = 0$ as

$$\exp\left\{ -\hat{\Omega}\tau / \hbar \right\} = \exp\left\{ -\hat{\Omega}_0 \tau / \hbar \right\} \hat{\mathcal{U}}(\tau, 0). \tag{24.35}$$

With $\tau = \beta \hbar$, Eq. (24.35) yields a useful expansion of the grand partition function [see Eq. (5.30)]

$$Z_G = \mathrm{Tr}\left(e^{-\beta \hat{\Omega}} \right) = \mathrm{Tr}\left(e^{-\beta \hat{\Omega}_0} \hat{\mathcal{U}}(\hbar\beta, 0) \right) \tag{24.36}$$

$$= \sum_{n=0}^{\infty} \left(\frac{-1}{\hbar} \right)^n \frac{1}{n!} \int_0^{\hbar\beta} d\tau_1 \ldots \int_0^{\hbar\beta} d\tau_n \, \mathrm{Tr}\left\{ e^{-\beta \hat{\Omega}_0} T_{\tau} \left[\hat{H}_1(\tau_1) \ldots \hat{H}_1(\tau_n) \right] \right\}.$$

The development of Sec. 8.2 can now be largely copied. We first consider $\tau - \tau' > 0$ (the case $\tau - \tau' < 0$ is completely analogous). For $0 < \tau' < \tau < \beta$, the propagator becomes

$$\mathcal{G}_{T+}(\alpha, \beta; \tau - \tau') = -\frac{1}{\hbar} \frac{\mathrm{Tr}\left(e^{-\beta \hat{\Omega}} a_{\alpha\Omega}(\tau) a_{\beta\Omega}^{\dagger}(\tau') \right)}{\mathrm{Tr}\left(e^{-\beta \hat{\Omega}} \right)} \tag{24.37}$$

$$= -\frac{1}{\hbar} \frac{\mathrm{Tr}\left(e^{-\beta\hat{\Omega}_0}\hat{\mathcal{U}}(\hbar\beta,\tau)a_\alpha(\tau)\hat{\mathcal{U}}(\tau,\tau')a_\beta^\dagger(\tau')\hat{\mathcal{U}}(\tau',0)\right)}{\mathrm{Tr}\left(e^{-\beta\hat{\Omega}_0}\hat{\mathcal{U}}(\hbar\beta,0)\right)},$$

using Eq. (24.35) with $\tau = \hbar\beta$, and Eq. (24.29). We also employed the identity

$$e^{-\beta\hat{\Omega}}a_{\alpha_\Omega}(\tau)a_{\beta_\Omega}^\dagger(\tau') \tag{24.38}$$
$$= e^{-\beta\hat{\Omega}_0}\hat{\mathcal{U}}(\hbar\beta,0)\left[\hat{\mathcal{U}}(0,\tau)a_{\alpha_{\Omega_0}}(\tau)\hat{\mathcal{U}}(\tau,0)\right]\left[\hat{\mathcal{U}}(0,\tau')a_{\beta_{\Omega_0}}^\dagger(\tau')\hat{\mathcal{U}}(\tau',0)\right].$$

The expansion in Eq. (24.34) can then be applied to the $\hat{\mathcal{U}}$ operators, and the operator product in the numerator of Eq. (24.37) becomes (dropping the Ω_0 interaction picture subscript as in Ch. 8)

$$e^{-\beta\hat{\Omega}_0}\sum_{n=0}^{\infty}\left(\frac{-1}{\hbar}\right)^n\frac{1}{n!}\sum_{k,l,m=0}^{\infty}\delta_{n,k+l+m}\frac{n!}{k!l!m!}\int_\tau^{\hbar\beta}dx_1..\int_\tau^{\hbar\beta}dx_k$$

$$\times\,\mathcal{T}_\tau\left[\hat{H}_1(x_1)...\hat{H}_1(x_k)\right]a_\alpha(\tau)\int_{\tau'}^{\tau}dy_1..\int_{\tau'}^{\tau}dy_l\,\mathcal{T}_\tau\left[\hat{H}_1(y_1)..\hat{H}_1(y_l)\right]a_\beta^\dagger(\tau')$$

$$\times\int_0^{\tau'}dz_1..\int_0^{\tau'}dz_m\,\mathcal{T}_\tau\left[\hat{H}_1(z_1)..\hat{H}_1(z_m)\right]$$

$$= e^{-\beta\hat{\Omega}_0}\sum_{n=0}^{\infty}\left(\frac{-1}{\hbar}\right)^n\frac{1}{n!}\sum_{k,l,m=0}^{\infty}\delta_{n,k+l+m}\frac{n!}{k!l!m!}\int_\tau^{\hbar\beta}dx_1..\int_\tau^{\hbar\beta}dx_k$$

$$\times\int_{\tau'}^{\tau}dy_1..\int_{\tau'}^{\tau}dy_l\int_0^{\tau'}dz_1..\int_0^{\tau'}dz_m \tag{24.39}$$

$$\times\,\mathcal{T}_\tau\left[\hat{H}_1(x_1)...\hat{H}_1(x_k)\hat{H}_1(y_1)..\hat{H}_1(y_l)\hat{H}_1(z_1)..\hat{H}_1(z_m)a_\alpha(\tau)a_\beta^\dagger(\tau')\right].$$

Integration variables x, y, z have been introduced to represent the intermediate imaginary times, and the external operators can be shifted to the right of the global τ-ordered product. The combinatorial factor $n!/(k!l!m!)$ can now be absorbed by extending the integration boundaries to the entire interval $[0, \hbar\beta]$, generating the final result,

$$\mathcal{G}_T(\alpha,\beta;\tau-\tau') = -\frac{1}{\hbar} \tag{24.40}$$

$$\times\,\frac{\sum_{n=0}^{\infty}\left(\frac{-1}{\hbar}\right)^n\frac{1}{n!}\int_0^{\hbar\beta}d\tau_1...\int_0^{\hbar\beta}d\tau_n\left\langle\mathcal{T}_\tau\left[\hat{H}_1(\tau_1)...\hat{H}_1(\tau_n)a_\alpha(\tau)a_\beta^\dagger(\tau')\right]\right\rangle_0}{\sum_{m=0}^{\infty}\left(\frac{-1}{\hbar}\right)^m\frac{1}{m!}\int_0^{\hbar\beta}d\tau_1...\int_0^{\hbar\beta}d\tau_m\left\langle\mathcal{T}_\tau\left[\hat{H}_1(\tau_1)...\hat{H}_1(\tau_m)\right]\right\rangle_0}.$$

24.2.1 Wick's theorem at finite temperature

The ingredients in the perturbative expansion (24.40) can in principle be calculated to any order: one has to work out noninteracting ensemble averages of τ-ordered products of addition and removal operators in the interaction picture. For further manipulation of the perturbation series, it is indispensable to have a systematic way of generating the contributions and expressing them in terms of the noninteracting propagators. Consequently, one needs the counterpart of Wick's theorem (see Sec. 8.4), which proved to be so useful for the real-time propagator at $T = 0$.

A finite-temperature extension of Eq. (8.31) is indeed available. For a string $[\hat{x}_1\hat{x}_2..\hat{x}_n]$, where each $\hat{x}_i$ is an operator in the interaction picture at imaginary time τ_i, Wick's theorem states that

$$\langle \mathcal{T}_\tau [\hat{x}_1\hat{x}_2..\hat{x}_n]\rangle_0 = \text{sum of all fully contracted terms} \qquad (24.41)$$
$$= [\hat{x}_1^\bullet \hat{x}_2^{\bullet\bullet} \hat{x}_3^{\bullet\bullet} \hat{x}_4^{\bullet\bullet}..\hat{x}_n^{\bullet\bullet\bullet}] + [\hat{x}_1^\bullet \hat{x}_2^{\bullet\bullet} \hat{x}_3^{\bullet\bullet} \hat{x}_4^\bullet..\hat{x}_n^{\bullet\bullet\bullet}] + ...,$$

defining a contraction as

$$\hat{x}_i^\bullet \hat{x}_j^\bullet = \langle \mathcal{T}_\tau \hat{x}_i \hat{x}_j \rangle_0. \qquad (24.42)$$

Any $\hat{x}_i$ can be either an addition operator $a_{\alpha_i}^\dagger(\tau_i)$ or a removal operator $a_{\alpha_i}(\tau_i)$. Since the Hamiltonian conserves the number of particles, the number of addition and removal operators must be equal, requiring n to be even. Each fully contracted term is then generated by a choice of $n/2$ pairs in the string. As usual, when the partners of a contraction in a term of Eq. (24.41) are not adjacent, they can be moved through the string generating additional minus signs (for fermions). Contractions between two addition or two removal operators are zero, whereas the nontrivial contractions

$$a_{\alpha_i}^\bullet(\tau_i)a_{\alpha_j}^{\dagger\bullet}(\tau_j) = -\hbar\mathcal{G}_T^{(0)}(\alpha_i,\alpha_j;\tau_i - \tau_j) = -a_{\alpha_j}^{\dagger\bullet}(\tau_j)a_{\alpha_i}^\bullet(\tau_i) \qquad (24.43)$$

are related to the noninteracting propagator.

Note that Eq. (24.41) is a statement about c-numbers (ensemble averages), in contrast to the $T = 0$ Wick's theorem in Eq. (8.31), which holds at the operator level. The proof is therefore much simpler. It is sufficient to consider the standard ordering $\tau_1 > \tau_2 > ..\tau_n$: any other ordering would generate the same sign change to the left and right of Eq. (24.41). Now, one can always move the first operator $\hat{x}_1$ to the right of the string, and correct with anticommutator terms

$$\langle \hat{x}_1 \hat{x}_2 .. \hat{x}_n \rangle_0 = - \langle \hat{x}_2 .. \hat{x}_n \hat{x}_1 \rangle_0 + \sum_{m=2}^{n} (-1)^m \langle \hat{x}_2 .. \hat{x}_{m-1} \{ \hat{x}_1, \hat{x}_m \} \hat{x}_{m+1} \hat{x}_n \rangle_0 .$$

(24.44)

The first term on the right can be rewritten by observing that

$$\text{Tr} \left(e^{-\beta \hat{\Omega}_0} \hat{x}_2 .. \hat{x}_n \hat{x}_1 \right) = \text{Tr} \left(\hat{x}_1 e^{-\beta \hat{\Omega}_0} \hat{x}_2 .. \hat{x}_n \right),$$

(24.45)

and

$$\hat{x}_1 e^{-\beta \hat{\Omega}_0} = C_1 e^{-\beta \hat{\Omega}_0} \hat{x}_1$$

(24.46)

where $C_1 = e^{\pm \varepsilon_{\alpha_1 \mu} \beta}$, the sign $+/-$ in the exponent depending on whether $\hat{x}_1$ adds/removes a particle. We can then rewrite Eq. (24.44) as

$$\langle \hat{x}_1 \hat{x}_2 .. \hat{x}_n \rangle_0 = \sum_{m=2}^{n} (-1)^m \frac{\{ \hat{x}_1, \hat{x}_m \}}{1 + C_1} \langle \hat{x}_2 .. \hat{x}_{m-1} \hat{x}_{m+1} \hat{x}_n \rangle_0 .$$

(24.47)

Note that the anticommutator is a number and can be taken out of the average. One can now also apply Eq. (24.47), with $n = 2$, to the string $\hat{x}_1 \hat{x}_m$. Noting that $\tau_1 > \tau_m$ one finds

$$\langle \hat{x}_1 \hat{x}_m \rangle_0 = \hat{x}_1^{\bullet} \hat{x}_m^{\bullet} = \frac{\{ \hat{x}_1, \hat{x}_m \}}{1 + C_1},$$

(24.48)

allowing to identify the prefactors appearing in the sum (24.47) as contractions, *i.e.*

$$\langle \hat{x}_1 \hat{x}_2 .. \hat{x}_n \rangle_0 = \sum_{m=2}^{n} (-1)^m \hat{x}_1^{\bullet} \hat{x}_m^{\bullet} \langle \hat{x}_2 .. \hat{x}_{m-1} \hat{x}_{m+1} \hat{x}_n \rangle_0$$

$$= \sum_{m=2}^{n} \langle \hat{x}_1^{\bullet} \hat{x}_2 .. \hat{x}_{m-1} \hat{x}_m^{\bullet} \hat{x}_{m+1} .. \hat{x}_n \rangle_0 .$$

(24.49)

This takes care of all possible contractions involving $\hat{x}_1$. Obviously the process can be repeated on the strings of $n - 2$ remaining operators appearing on the right of Eq. (24.49), thereby proving the general statement (24.41) concerning Wick's theorem at finite temperature.

24.3 Diagrams at finite temperature

In view of the discussion in Sec. 24.2 it should not come as a surprise that the ensuing perturbation series for the temperature propagator is a carbon copy of the $T = 0$ version. Wick's theorem can be used to expand the ensemble averages appearing in Eq. (24.40) in terms of contractions. This in turn allows a graphical representation in the form of diagrams, identical to the one discussed in Sec. 8.5. The presence of the denominator in Eq.(24.40) ensures that only connected diagrams contribute to the sp propagator at finite temperature

$$\mathcal{G}(\alpha, \beta; \tau - \tau') = -\frac{1}{\hbar} \sum_{n=0}^{\infty} \left(\frac{-1}{\hbar}\right)^n \frac{1}{n!} \int_0^{\hbar\beta} d\tau_1 ... \int_0^{\hbar\beta} d\tau_n \qquad (24.50)$$
$$\times \left\{ \left\langle \mathcal{T}_\tau \left[\hat{H}_1(\tau_1)...\hat{H}_1(\tau_n) a_\alpha(\tau) a_\beta^\dagger(\tau') \right] \right\rangle_0 \right\}_{connected},$$

incorporating the familiar cancellation beween the numerator and denominator of Eq. (24.40). No new diagrams are generated either, greatly simplifying the effort to establish useful results for finite T. Only the translation from diagrams to analytical expressions requires some changes that are related to Eq. (24.43).

24.3.1   *Rules for time formulation*

The following rules apply for the mth order contribution to $\mathcal{G}_T(\alpha, \beta, \tau - \tau')$:

Rule 1 Draw all topologically distinct and connected diagrams
with m horizontal interaction lines for V (dashed) and $2m + 1$
directed (using arrows) Green's functions $\mathcal{G}_T^{(0)}$

Rule 2 Label the external points ($\alpha\tau$ and $\beta\tau'$) using imaginary times.
Label each interaction with an imaginary time τ_i and
sp quantum numbers

$$\tau \Rightarrow \underset{\epsilon \quad \theta}{\overset{\gamma \quad \delta}{\bullet\text{----}\bullet}} \qquad \Rightarrow (\gamma\delta|V|\epsilon\theta)$$

For each full line one writes

$$\tau_i \Rightarrow \bullet\, \alpha$$
$$\Big\uparrow \qquad \Rightarrow \mathcal{G}_T^{(0)}(\alpha, \beta; \tau_i - \tau_j)$$
$$\tau_j \Rightarrow \bullet\, \beta$$

Rule 3 Sum (integrate) over all internal sp quantum numbers and
integrate all m internal τ_i over the interval $[0, \hbar\beta]$

$$\Rightarrow (-1)(-\hbar) \int_0^{\hbar\beta} d\tau_1 \sum_{\gamma\delta\epsilon\theta} \langle\gamma\delta|V|\epsilon\theta\rangle \, \mathcal{G}_T^{(0)}(\alpha,\gamma;\tau-\tau_1)$$
$$\times \, \mathcal{G}_T^{(0)}(\theta,\delta;\tau_1-\tau_1^+)\mathcal{G}_T^{(0)}(\epsilon,\beta;\tau_1-\tau')$$

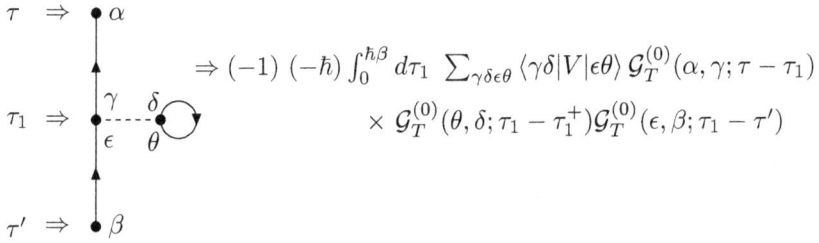

Fig. 24.1 Diagram representing the first-order contribution from the two-body interaction V to the temperature sp propagator in the imaginary-time formulation. The minus sign in front accounts for the closed fermion loop.

Rule 4 Include a factor $(-\hbar)^m$ and $(-1)^F$ where F
        is the number of closed fermion loops
Rule 5 Interpret equal imaginary times in a propagator as
        $\mathcal{G}_T^{(0)}(\alpha,\beta;\tau-\tau^+)$

    The factor $(-\hbar)^m$ in **Rule 3** results from the $-1/\hbar$ in Eq. (8.55), the $(-1/\hbar)^m$ appearing there under the sum, and finally, a factor $(-\hbar)^{2m+1}$ from the number of contractions in mth order. The latter number corresponds to each interaction contributing two contractions, plus one coming from the external operators. As before, **Rule 5** is related to the original ordering of the operators in the Hamiltonian. When an auxiliary potential $\hat{U}$ is introduced and the perturbation is $\hat{H}_1 = \hat{V} - \hat{U}$ (see the discussion in Sec 8.6.1), the following extra rules apply when k such $\hat{U}$ contributions appear in the diagram. A one-body external field $\hat{U}_{ext}$ will generate similar terms but doesn't require the factor $(-1)^k$ in **Rule 7**.

Rule 6 Label each U according to

           $\Rightarrow \tau_i$ $\Rightarrow \langle\alpha|U|\beta\rangle$

Rule 7 Include a factor $(-1)^k$ and k additional propagators $\mathcal{G}_T^{(0)}$

    The evaluation of the first-order diagram is given in Fig. 24.1 as an example. Note that antisymmetrized interactions are used.

24.3.2 *Rules for energy formulation*

It is usually preferable for practical calculations to work in the imaginary energy domain, rather than in imaginary time. This is again closely analogous to the $T = 0$ case presented in Sec. 8.6.2. The main difference is the use of the discrete Fourier series explained in Sec. 24.1.1, instead of the continuous FT employed for $T = 0$. It implies that the familiar integrations over intermediate energies should now be replaced by sums over Matsubara frequencies.

The following rules apply for the mth order contribution to $\mathcal{G}_T(\alpha, \beta, E_n)$:

Rule 1 Draw all topologically distinct (direct) and connected
diagrams with m horizontal interaction lines for V (dashed)
and $2m + 1$ directed (using arrows) Green's functions $\mathcal{G}_T^{(0)}$

Rule 2 Label external points only with sp quantum numbers,
e.g. α and β
Label each interaction with sp quantum numbers

$$
\begin{array}{cc}
\alpha & \beta \\
\bullet\!\!-\!-\!-\!-\!\bullet \\
\gamma & \delta
\end{array}
\qquad \Rightarrow \qquad \langle\alpha\beta|\,V\,|\gamma\delta\rangle = (\alpha\beta|V|\gamma\delta) - (\alpha\beta|V|\delta\gamma)
$$

For an arrow line one writes

$$
\begin{array}{l}
\bullet\ \alpha \\
\Big\uparrow E_k \quad \Rightarrow \quad \mathcal{G}_T^{(0)}(\alpha, \beta; E_k) = \delta_{\alpha,\beta}\,\frac{1}{iE_k - \varepsilon_{\alpha\mu}} \\
\bullet\ \beta
\end{array}
$$

but in such a way that energy is conserved for every V

Rule 3 Sum (integrate) over all internal sp quantum numbers and
sum over all m internal energies (which should be interpreted as
Matsubara energies); a propagator starting and ending on the
same interaction line should have a convergence factor $e^{i\eta E_k}$.

Rule 4 Include a factor $(-1/\beta)^m$ and $(-1)^F$ where F
is the number of closed fermion loops

Rule 5 Include a factor of $\frac{1}{2}$ for equivalent pairs of lines

The overall factor $(-1/\beta)^m$ arises in the following way. Taking the discrete FT of an imaginary time mth order contribution has in total $m+1$ τ integrations. Switching the $2m + 1$ propagators from imaginary time

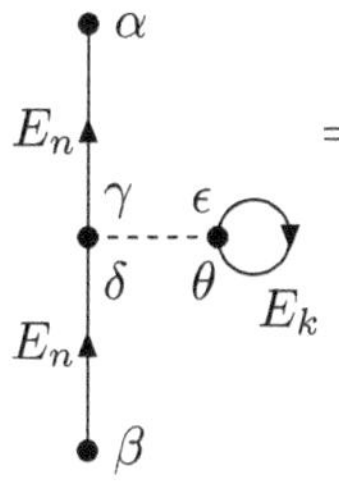

Fig. 24.2 First-order diagram in the energy formulation.

to frequency yields a factor $1/(\hbar\beta)^{2m+1}$. The τ integrations now lead to $m+1$ energy conserving Kronecker-deltas (m at the interaction lines, and one for the external points). Each supplies a factor $\hbar\beta$, so the extra factor generated by switching to the energy domain is $1/(\hbar\beta)^m$. This combines with the factor $-\hbar^m$ to $(-1/\beta)^m$, as stated.

A propagator starting and ending at the same interaction line represents a contraction in the same interaction $\hat{H}_1$, so the original "$a^\dagger a$" order should be respected. In the imaginary-time domain this part of the propagator is selected by taking $\mathcal{G}_T(\alpha\beta; \tau = 0^- = -\eta)$. Application of Eq. (24.12),

$$\mathcal{G}_T(\alpha, \beta; \tau = -\eta) = \frac{1}{\beta\hbar} \sum_{n=-\infty}^{+\infty} \mathcal{G}_T(\alpha, \beta; E_n) e^{i\eta E_n}, \qquad (24.51)$$

makes clear that the convergence factor mentioned above, should be added in the imaginary-energy domain.

When $k\,U$ contributions are involved in a diagram, we have to add the following rules:

Rule 6 Label each U according to

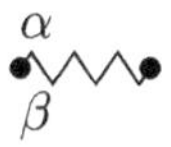

$$\Rightarrow \quad \langle \alpha |\, U\, | \beta \rangle$$

Rule 7 Include a factor $(-1)^k$ and k additional propagators $G^{(0)}$

The diagram representing the first-order contribution in the energy formulation is shown in Fig. 24.2. The (irreducible, non-HF) second-order diagram is calculated in Fig. 24.3.

24.3.3   *Energy sums*

Actual calculations require the evaluation of the infinite sums over Matsub-ara energies $E_n = (2n+1)\pi/\beta$ appearing *e.g.* in Figs. 24.2–24.3. These can be calculated by noting that the Fermi function $f(z) = 1/(1+e^{\beta z})$, viewed as an analytic function in the complex plane, has its only singularities precisely at the imaginary Matsubara energies iE_n, which are all simple poles with residue $-1/\beta$. This allows replacing the sums with contour integrals in the complex energy plane.

As an example we evaluate the first-order contribution to the self-energy by omitting the external propagators in Fig. 24.2. With the explicit expression in **Rule 2** for the noninteracting sp propagator we find

$$\Sigma^{(1)}(\gamma, \delta; E_n) = \sum_{\theta} \langle \gamma\theta | V | \delta\theta \rangle S_\theta. \qquad (24.52)$$

The energy sum reads as

$$S_\theta = \frac{1}{\beta} \sum_{m=-\infty}^{+\infty} \frac{e^{i\eta E_m}}{iE_m - \varepsilon_{\theta\mu}} = \sum_m F(iE_m), \qquad (24.53)$$

where the analytic function $F(z) = \frac{1}{\beta}\frac{e^{\eta z}}{z - \varepsilon_{\theta\mu}}$ has a pole on the real axis.

One may then consider the complex contour integral $\int_C dz\, f(z)F(z)$, counterclockwise along a large circle (the radius of which becomes infinitely large) centered at $z = 0$. When $|z| \to \infty$, the integrand behaves as $e^{z(\eta-\beta)}/z$ for Re $z > 0$ and as $e^{\eta z}/z$ for Re $z < 0$. Since $0 < \eta < \beta$, this ensures that the contour integral vanishes (exponentially) in the limit of infinite radius. Applying the residue theorem, we have

$$0 = [\text{sum of residues at poles of } f(z)F(z)]$$
$$= \sum_m \left(-\frac{1}{\beta}\right) F(iE_m) + \frac{1}{\beta}e^{\eta\varepsilon_{\theta\mu}} f(\varepsilon_{\theta\mu}). \qquad (24.54)$$

The desired energy sum,

$$S_\theta = f(\varepsilon_{\theta\mu}) = n_\theta^0, \qquad (24.55)$$

therefore yields the expected thermal occupation of the sp state θ in the noninteracting ground state.

The same technique can be applied to higher-order diagrams.

$$\Rightarrow \sum_{\gamma\delta} \mathcal{G}_T^{(0)}(\alpha,\gamma;E_n)$$

$$\times (-1)(-1/\beta)^2 \tfrac{1}{2} \sum_{k,m} \sum_{\lambda,\epsilon,\theta} \sum_{\zeta,\xi,\mu} \langle\gamma\lambda|\,V\,|\epsilon\theta\rangle$$

$$\times \mathcal{G}_T^{(0)}(\epsilon,\zeta;E_k)\mathcal{G}_T^{(0)}(\mu,\lambda;E_k+E_m-E_n)$$

$$\times \mathcal{G}_T^{(0)}(\theta,\xi;E_m)\,\langle\zeta\xi|\,V\,|\delta\mu\rangle$$

$$\times \mathcal{G}_T^{(0)}(\delta,\beta;E_n)$$

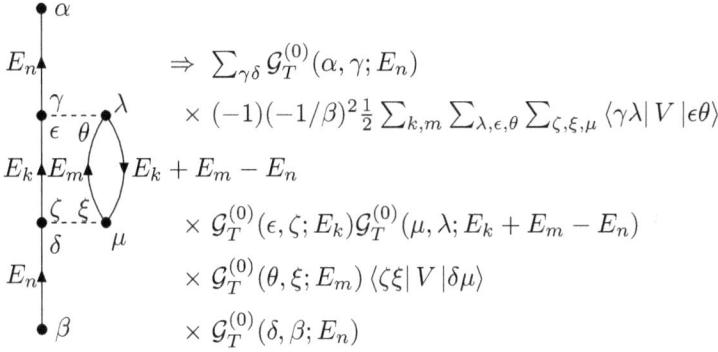

Fig. 24.3   Second-order diagram in the energy formulation.

The second-order self-energy diagram in Fig. 24.3 can be written as

$$\Sigma^{(2)}(\gamma,\delta;E_n) = \frac{1}{2}\sum_{\lambda\epsilon\theta} \langle\gamma\lambda|\,V\,|\epsilon\theta\rangle\,\langle\epsilon\theta|\,V\,|\delta\lambda\rangle\,S_{\lambda\epsilon\theta}, \qquad (24.56)$$

in terms of the double energy sum

$$S_{\lambda\epsilon\theta} = -\frac{1}{\beta^2}\sum_{k,m=-\infty}^{+\infty} \frac{1}{(iE_k-\varepsilon_{\epsilon\mu})}\frac{1}{(iE_m-\varepsilon_{\theta\mu})}\frac{1}{(i(E_k+E_m-E_n)-\varepsilon_{\lambda\mu})}$$

$$= -\frac{1}{\beta^2}\sum_{k,m}\frac{1}{iE_k-\varepsilon_{\epsilon\mu}}G(iE_m). \qquad (24.57)$$

The relevant analytic function is now

$$G(z) = \frac{1}{(z-\varepsilon_{\theta\mu})}\frac{1}{(z+i(E_k-E_n)-\varepsilon_{\lambda\mu})} \qquad (24.58)$$

and has a single pole on the real axis, as well as additional poles in the complex plane. The imaginary part of the latter are at *even* multiples of π/β and cannot coincide with the poles of the Fermi function. One can again consider the complex contour integral $\int_C dz\, f(z)G(z)$ along the same big circle as before, and check that the integrand vanishes sufficiently fast for the residue theorem to be applied in the limit of infinite radius. The result is

$$0 = \sum_m \left(-\frac{1}{\beta}\right)G(iE_m) + \frac{f(\varepsilon_{\theta\mu})-f(\varepsilon_{\lambda\mu})}{\varepsilon_{\theta\mu}-\varepsilon_{\lambda\mu}+i(E_k-E_n)}, \qquad (24.59)$$

where we used the relation $f(\varepsilon_{\lambda\mu} + i(E_n - E_k)) = f(\varepsilon_{\lambda\mu})$. The energy sum in Eq. (24.57) is now simplified, with the aid of Eq. (24.145), to

$$S_{\lambda\epsilon\theta} = -\frac{1}{\beta} \sum_{k=-\infty}^{+\infty} \frac{1}{(iE_k - \varepsilon_{\epsilon\mu})} \frac{f(\varepsilon_{\theta\mu}) - f(\varepsilon_{\lambda\mu})}{(iE_k - iE_n + \varepsilon_{\theta\mu} - \varepsilon_{\lambda\mu})}. \qquad (24.60)$$

The remaining single summation over Matsubara energies can be evaluated with the same procedure, and one finally arrives at

$$
\begin{aligned}
S_{\lambda\epsilon\theta} &= \frac{f(\varepsilon_{\theta\mu}) - f(\varepsilon_{\lambda\mu})}{iE_n + \varepsilon_{\lambda\mu} - \varepsilon_{\theta\mu} - \varepsilon_{\epsilon\mu}} \left(f(\varepsilon_{\epsilon\mu}) - f(iE_n + \varepsilon_{\lambda\mu} - \varepsilon_{\theta\mu}) \right) \\
&= \frac{n_\lambda^0(1 - n_\theta^0)(1 - n_\epsilon^0) + (1 - n_\lambda^0)n_\theta^0 n_\epsilon^0}{iE_n + \varepsilon_{\lambda\mu} - \varepsilon_{\theta\mu} - \varepsilon_{\epsilon\mu}}. \qquad (24.61)
\end{aligned}
$$

24.4   Spectral representations at finite temperature

We clarify the information contained in the sp propagator at finite temperature in Sec. 24.4.1. The relation between the real and imaginary-time propagator is explored in Sec. 24.4.2 through the spectral representation [Landau (1958)] and spectral functions are studied in Sec. 24.4.3.

24.4.1   *Expectation value of operators*

In Sec. 7.4 it was shown that the ground-state expectation value of arbitrary one-body operators can be expressed in terms of the $T = 0$ sp propagator. The temperature propagator $\mathcal{G}_T$ allows likewise the evaluation of the corresponding ensemble averages at finite T. It is sufficient to consider the limit $\tau \to 0^-$ in Eq. (24.6), which selects the removal part of the propagator

$$\hbar \mathcal{G}_T(\alpha, \beta; 0^-) = \frac{1}{Z_G} \mathrm{Tr} \left(e^{-\beta\hat{\Omega}} a_\beta^\dagger a_\alpha \right) = \langle a_\beta^\dagger a_\alpha \rangle. \qquad (24.62)$$

A one-body operator $\hat{O}$ thus has an ensemble average

$$\langle \hat{O} \rangle = \sum_{\alpha,\beta} \langle \alpha | O | \beta \rangle \, \hbar \mathcal{G}_T(\beta, \alpha; 0^-). \qquad (24.63)$$

The same quantity can also be expressed in the imaginary-energy representation. Combining Eqs. (24.51) and (24.62) yields

$$\frac{1}{\beta} \sum_n e^{i\eta E_n} \mathcal{G}_T(\alpha, \beta; E_n) = \langle a_\beta^\dagger a_\alpha \rangle. \qquad (24.64)$$

For the usual case of a Hamiltonian $\hat{H} = \hat{T} + \hat{V}$ containing a one-body part $\hat{T}$ and a two-body interaction $\hat{V}$, it is again possible to construct the ensemble average of $\hat{V}$ solely in terms of the sp propagator (rather than the tp propagator which is needed for a general two-body operator). Taking the derivative with respect to τ of Eq. (24.6), leads to

$$\hbar^2 \frac{\partial \mathcal{G}_T}{\partial \tau}(\alpha, \beta; 0^-) = \frac{1}{Z_G} \text{Tr} \left(e^{-\beta \hat{\Omega}}(-\hat{\Omega} a_\beta^\dagger a_\alpha + a_\beta^\dagger \hat{\Omega} a_\alpha) \right) = \langle a_\beta^\dagger [\hat{\Omega}, a_\alpha] \rangle,$$
(24.65)

in the limit $\tau \to 0^-$. By employing the operator relation [see *e.g.* problem (7) in Ch. 2]

$$\sum_\alpha a_\alpha^\dagger [\hat{\Omega}, a_\alpha] = -(\hat{T} - \mu \hat{N}) - 2\hat{V}$$
(24.66)

one obtains

$$-\hbar^2 \sum_\alpha \frac{\partial \mathcal{G}_T}{\partial \tau}(\alpha, \alpha; 0^-) = \langle \hat{T} - \mu \hat{N} \rangle + 2\langle \hat{V} \rangle.$$
(24.67)

Since the one-body contribution $\langle \hat{T} - \mu \hat{N} \rangle$ is already known through Eq. (24.63), we can calculate the ensemble-averaged interaction energy $\langle \hat{V} \rangle$, and therefore also $\langle \hat{H} \rangle$ and $\langle \hat{\Omega} \rangle$. As a result, the average of $\hat{\Omega}$ becomes

$$\langle \hat{\Omega} \rangle = \frac{1}{2} \sum_{\alpha \beta} \left\{ \langle \alpha | T | \beta \rangle - \delta_{\alpha \beta} \left(\hbar \frac{\partial}{\partial \tau} + \mu \right) \right\} \hbar \mathcal{G}_T(\beta, \alpha; 0^-).$$
(24.68)

In the imaginary-energy representation the τ-derivative in Eq. (24.65) gives rise to

$$\hbar^2 \frac{\partial \mathcal{G}_T}{\partial \tau}(\alpha, \beta; 0^-) = \frac{-1}{\beta} \sum_n iE_n e^{i\eta E_n} \mathcal{G}_T(\beta, \alpha; E_n),$$
(24.69)

where Eq. (24.12) has been used. The corresponding expression for $\langle \hat{\Omega} \rangle$ thus reads

$$\langle \hat{\Omega} \rangle = \frac{1}{2\beta} \sum_{\alpha \beta} \{ \langle \alpha | T | \beta \rangle + \delta_{\alpha \beta} (iE_n - \mu) \} e^{i\eta E_n} \mathcal{G}_T(\beta, \alpha; E_n).$$
(24.70)

We recall that at finite temperature (in contrast to the $T = 0$ case) the grand-canonical potential Ω_G is *not* equal to $\langle \hat{\Omega} \rangle$, since $e^{-\beta \Omega_G} = Z_G = \text{Tr} \, e^{-\beta \hat{\Omega}}$. This implies that Ω_G cannot be generated directly from the propagator $\mathcal{G}_T$. However, it can be determined indirectly by considering a system with variable coupling constant λ,

$$\hat{\Omega}_\lambda = \hat{\Omega}_0 + \lambda\hat{\Omega}_1 \qquad (24.71)$$

varying between $\lambda = 0$, corresponding to the unperturbed situation which is assumed to be known, and $\lambda = 1$ where the full potential $\hat{\Omega}$ is restored. In this case one has

$$Z_G(\lambda) = \text{Tr } e^{-\beta\hat{\Omega}_\lambda} = \sum_n \langle\Psi_n(\lambda)| e^{-\beta\hat{\Omega}_\lambda} |\Psi_n(\lambda)\rangle \qquad (24.72)$$

where the trace can be evaluated with the exact eigenstates $\Psi_n(\lambda)$ of $\hat{\Omega}_\lambda$. The λ-derivative can be performed using the Hellman-Feynman theorem

$$\frac{\partial}{\partial\lambda} Z_G(\lambda) = \sum_n \langle\Psi_n(\lambda)| \left(\frac{\partial}{\partial\lambda} e^{-\beta\hat{\Omega}_\lambda}\right) |\Psi_n(\lambda)\rangle \qquad (24.73)$$

$$= \sum_n \langle\Psi_n(\lambda)| (-\beta\hat{\Omega}_1) |\Psi_n(\lambda)\rangle e^{-\beta\Omega_n(\lambda)} = -\beta Z_G(\lambda)\langle\hat{\Omega}_1\rangle_\lambda,$$

where $\langle.\rangle_\lambda$ is the ensemble average corresponding to $\hat{\Omega}_\lambda$. As a consequence,

$$\frac{\partial}{\partial\lambda}\Omega_G(\lambda) = \langle\hat{\Omega}_1\rangle_\lambda \qquad (24.74)$$

and upon integration

$$\Delta\Omega_G = \Omega_G(1) - \Omega_G(0) = \int_0^1 d\lambda \, \langle\hat{\Omega}_1\rangle_\lambda. \qquad (24.75)$$

In the common situation $\hat{\Omega}_0 = \hat{T} - \mu\hat{N}$ and a two-body perturbation $\hat{\Omega}_1 = \hat{V}$ it is then possible to use Eq. (24.67) to express the integrand in terms of the sp temperature propagator.

24.4.2 *Insertion of exact eigenstates*

The precise content of the temperature propagator is clarified by employing the exact eigenstates $|\Psi_k\rangle$ and eigenvalues Ω_k of the potential $\hat{\Omega}$, to evaluate the trace in Eq. (24.6). For $\tau > 0$ this leads to

$$\mathcal{G}_T(\alpha, \beta; \tau) = -\frac{1}{\hbar Z_G} \text{Tr} \left(e^{-\beta\hat{\Omega}} e^{\hat{\Omega}\tau/\hbar} a_\alpha e^{-\hat{\Omega}\tau/\hbar} a_\beta^\dagger\right)$$

$$= -\frac{1}{\hbar Z_G} \sum_{kl} e^{-\beta\Omega_k} z_{kl\alpha} z_{kl\beta}^* e^{-\tau(\Omega_l - \Omega_k)/\hbar} \qquad (24.76)$$

where the sp transition amplitudes are defined as

$$z_{kl\alpha} = \langle \Psi_k | a_\alpha | \Psi_l \rangle. \tag{24.77}$$

In contrast to the $T = 0$ case, one needs a double sum to completely resolve the propagator: one to evaluate the trace in the ensemble average, and another to insert a complete set of eigenstates of $\hat{\Omega}$ between the removal and addition operator. The $z_{kl\alpha}$ amplitudes are therefore more general, and reflect all possible sp transitions between the eigenstates of $\hat{\Omega}$. Note that the particle number of the eigenstates is not explicitly indicated; the transition amplitudes $z_{kl\alpha}$ automatically select the combinations $N_k = N_l - 1$.

The imaginary energy propagator follows from Eq. (24.14),

$$\mathcal{G}_T(\alpha, \beta; E_n) = -\frac{1}{\hbar Z_G} \sum_{kl} z_{kl\alpha} z^*_{kl\beta} e^{-\beta\Omega_k} \int_0^{\beta\hbar} d\tau\, e^{(\Omega_k - \Omega_l + iE_n)\tau/\hbar}$$

$$= \frac{1}{Z_G} \sum_{kl} z_{kl\alpha} z^*_{kl\beta} \frac{e^{-\beta\Omega_l} + e^{-\beta\Omega_k}}{iE_n + \Omega_k - \Omega_l}. \tag{24.78}$$

This expression is important, since it allows to connect the imaginary-energy propagator with the propagator for real energies that yields information related to excited states. The latter is of course nothing but the FT of the real-time propagator defined in Eq. (24.1),

$$G_T(\alpha, \beta; E) = \int_{-\infty}^{+\infty} d(t - t')\, e^{iE(t-t')/\hbar} G_T(\alpha, \beta; t - t'). \tag{24.79}$$

We first express the real-time propagator by inserting exact eigenstates of $\hat{\Omega}$, in the same way as was done for Eq. (24.76)

$$G_T(\alpha, \beta; t - t') = \frac{1}{i\hbar Z_G} \sum_{kl} z_{kl\alpha} z^*_{kl\beta} e^{i(\Omega_k - \Omega_l)(t-t')/\hbar}$$

$$\times \left\{ \theta(t - t') e^{-\beta\Omega_k} - \theta(t' - t) e^{-\beta\Omega_l} \right\}. \tag{24.80}$$

Substituting this expression into Eq. (24.79) one arrives at the Lehmann representation of the real-energy propagator

$$G_T(\alpha, \beta; E) = \frac{1}{Z_G} \sum_{kl} z_{kl\alpha} z^*_{kl\beta} \left(\frac{e^{-\beta\Omega_k}}{E + \Omega_k - \Omega_l + i\eta} + \frac{e^{-\beta\Omega_l}}{E + \Omega_k - \Omega_l - i\eta} \right). \tag{24.81}$$

24.4.3   *Spectral functions at finite T*

For the following it is useful to introduce the hermitian and antihermitian components of the real-time propagator, defined by

$$\mathcal{H}G_T(\alpha, \beta; E) = \frac{1}{2}\left(G_T(\alpha, \beta; E) + G_T^*(\beta, \alpha; E)\right) \tag{24.82}$$

and

$$\mathcal{A}G_T(\alpha, \beta; E) = \frac{1}{2i}\left(G_T(\alpha, \beta; E) - G_T^*(\beta, \alpha; E)\right), \tag{24.83}$$

which is just the matrix equivalent of the decomposition into real and imaginary parts of a complex number. Employing the Lehmann representation (24.81) and the familiar relation $\frac{1}{x \pm i\eta} = \mathcal{P}\frac{1}{x} \mp i\pi\delta(x)$, these components can be written as

$$\mathcal{H}G_T(\alpha, \beta; E) = \frac{1}{Z_G}\sum_{kl} z_{kl\alpha} z_{kl\beta}^* e^{-\beta\Omega_k} \mathcal{P}\frac{1 + e^{\beta(\Omega_k - \Omega_l)}}{E + \Omega_k - \Omega_l} \tag{24.84}$$

and

$$\begin{aligned}
\mathcal{A}G_T(\alpha, \beta; E) &= \frac{\pi}{Z_G}\sum_{kl} z_{kl\alpha} z_{kl\beta}^* e^{-\beta\Omega_k}\delta(E + \Omega_k - \Omega_l)(e^{\beta(\Omega_k - \Omega_l)} - 1) \\
&= -\frac{\pi}{Z_G}\sum_{kl} z_{kl\alpha} z_{kl\beta}^* e^{-\beta\Omega_k}\delta(E + \Omega_k - \Omega_l)(1 + e^{\beta(\Omega_k - \Omega_l)}) \\
&\quad \times \tanh\left(\frac{\beta E}{2}\right).
\end{aligned} \tag{24.85}$$

From the last equality it is clear that a dispersion relation

$$\mathcal{H}G_T(\alpha, \beta; E) = -\frac{1}{\pi}\mathcal{P}\int_{-\infty}^{+\infty}\frac{dE'}{E - E'}\mathcal{A}G_T(\alpha, \beta; E')\coth\left(\frac{\beta E'}{2}\right) \tag{24.86}$$

holds, linking the hermitian and antihermitian part of the real-time propagator. Equations (24.84)–(24.86) also imply that one can introduce the spectral function matrix

$$S(\alpha, \beta; E) = \frac{1}{Z_G}\sum_{kl} z_{kl\alpha} z_{kl\beta}^* e^{-\beta\Omega_k}\delta(E + \Omega_k - \Omega_l)(1 + e^{-\beta E}) \tag{24.87}$$

as the basic quantity, since it completely determines the real-time propagator through

$$\mathcal{H}G_T(\alpha, \beta; E) = \mathcal{P}\int_{-\infty}^{+\infty}\frac{dE'}{E - E'}S(\alpha, \beta; E') \tag{24.88}$$

and

$$AG_T(\alpha, \beta; E) = -\pi \tanh\left(\frac{\beta E}{2}\right) S(\alpha, \beta; E). \tag{24.89}$$

The spectral function matrix $S(\alpha, \beta; E)$ at finite T is a hermitian matrix with positive eigenvalues. Rewriting it in the form

$$S(\alpha, \beta; E) = \frac{1}{Z_G} \sum_{kl} z_{kl\alpha} z^*_{kl\beta} \delta(E + \Omega_k - \Omega_l)(e^{-\beta\Omega_k} + e^{-\beta\Omega_l}) \tag{24.90}$$

clarifies that its $T \to 0$ limit indeed corresponds to the familiar spectral function matrix at $T = 0$. In this limit the last factor in Eq. (24.90) filters out the ground-state contributions to either the k or l summation. The basic sum rule is fulfilled according to

$$\int_{-\infty}^{+\infty} dE \; S(\alpha, \beta; E) = \frac{1}{Z_G} \sum_{kl} z_{kl\alpha} z^*_{kl\beta}(e^{-\beta\Omega_k} + e^{-\beta\Omega_l})$$

$$= \langle \hat{\rho}_G a_\alpha a^\dagger_\beta + \hat{\rho}_G a^\dagger_\beta a_\alpha \rangle = \delta_{\alpha\beta}. \tag{24.91}$$

Even more satisfactory, the *same* spectral function matrix as defined in Eq. (24.87) or Eq. (24.90) also determines the imaginary-time propagator, as inspection of Eq. (24.78) demonstrates that

$$\mathcal{G}_T(\alpha, \beta; E_n) = \int_{-\infty}^{+\infty} \frac{dE'}{iE_n - E'} S(\alpha, \beta; E'). \tag{24.92}$$

Since approximations to the imaginary-time propagator are readily constructed using the diagrammatic techniques of Sec. 24.3, this would provide a convenient way of calculating the corresponding real-time propagator. The only problem lies in extracting the spectral function, which is a continuous function of energy, from Eq. (24.92) and the value of the imaginary-energy propagator at the infinite but countable set of Matsubara energies. In general, this is a mathematically ill-posed problem. However, the ambiguity is resolved by the fact that the function

$$\mathcal{G}_T(\alpha, \beta; z) = \int_{-\infty}^{+\infty} \frac{dE'}{iz - E'} S(\alpha, \beta; E'), \tag{24.93}$$

is the *only* analytic continuation of $\mathcal{G}_T(\alpha, \beta; E_n)$ which vanishes along any direction as $z \to \infty$ and is analytic everywhere except for the imaginary axis. These requirements are usually fulfilled simply by replacing the Matsubara energy with a general complex argument, in the analytical expression corresponding to a certain approximation for the imaginary-time propagator $\mathcal{G}_T(\alpha, \beta; E_n)$. The analytical continuation thus obtained then coincides automatically with Eq. (24.93), and the underlying spectral function is extracted through

$$S(\alpha, \beta; E) = \frac{1}{2\pi i} \left[\mathcal{G}_T(\alpha, \beta; -iE - \eta) - \mathcal{G}_T(\alpha, \beta; -iE + \eta) \right],$$

$$= \frac{1}{\pi} \mathcal{A}\mathcal{G}_T(\alpha, \beta; -iE - \eta). \tag{24.94}$$

The real-time propagator can then be evaluated through Eqs. (24.88) and (24.89).

As a simple application of Eq. (24.94), the spectral function for a non-interacting system with a temperature propagator given by Eq. (24.25) becomes

$$S^{(0)}(\alpha, \beta; E) = \frac{1}{\pi} \mathcal{A} \sum_i \frac{z_{i\alpha} z_{i\beta}^*}{E - \varepsilon_{i\mu} - i\eta}$$

$$= \sum_i z_{i\alpha} z_{i\beta}^* \delta(E - \varepsilon_{i\mu}). \tag{24.95}$$

The spectral function is also useful when determining the expectation values in Sec. 24.4.1. When the density matrix $\langle a_\beta^\dagger a_\alpha \rangle$ is expressed through Eq. (24.64) in terms of the imaginary-energy propagator, the spectral representation (24.92) can be substituted, and the basic Matsubara energy sum (24.55) leads to

$$\langle a_\beta^\dagger a_\alpha \rangle = \frac{1}{\beta} \sum_n e^{i\eta E_n} \int \frac{dE}{iE_n - E} S(\alpha, \beta; E)$$

$$= \int dE S(\alpha, \beta; E) f(E). \tag{24.96}$$

Likewise one finds, starting from Eq. (24.65),

$$\langle a_\beta^\dagger [\hat{\Omega}, a_\alpha] \rangle = \frac{-1}{\beta} \sum_n iE_n e^{i\eta E_n} \int \frac{dE}{iE_n - E} S(\alpha, \beta; E)$$

$$= \int dE S(\alpha, \beta; E) E f(E), \tag{24.97}$$

allowing the evaluation of the potential energy once the spectral function is known. In order to obtain Eq. (24.97) one should rewrite $iE_n/(iE_n - E) = 1 + E/(iE_n - E)$ and note that the first (constant) term does not survive the Matsubara energy summation since its contribution is of the form

$$\sum_{n=-\infty}^{+\infty} e^{i\eta E_n} = \sum_{n=0}^{+\infty} (e^{(n+\frac{1}{2})z} + e^{-(n+\frac{1}{2})z}) \tag{24.98}$$

with $z = 2i\pi/\beta$. The sum on the right vanishes for any z

$$\sum_{n=0}^{+\infty} (e^{(n+\frac{1}{2})z} + e^{-(n+\frac{1}{2})z}) = \frac{e^{z/2}}{1 - e^z} + \frac{e^{-z/2}}{1 - e^{-z}} = 0, \tag{24.99}$$

as can be seen by evaluating the geometric series.

24.5   Dyson equation

Since the entire diagrammatical perturbation series for the temperature propagator is identical to the one developed for zero temperature, the same analysis applies as in Sec. 8.1. Self-energy diagrams $\Sigma(\gamma, \delta; E_n)$ are obtained by stripping off the external propagators from the diagrams contributing to $\mathcal{G}_T(\alpha, \beta; E_n)$. The same distinction can be made in terms of irreducible self-energy diagrams, and the reducible diagrams which are repetitions of the irreducible ones, connected by a noninteracting propagator. This leads directly to the Dyson equation for the temperature propagator

$$\mathcal{G}_T(\alpha, \beta; E_n) = \mathcal{G}_T^{(0)}(\alpha, \beta; E_n) + \sum_{\gamma\delta} \mathcal{G}_T^{(0)}(\alpha, \gamma; E_n)\Sigma(\gamma, \delta; E_n)\mathcal{G}_T(\delta, \beta; E_n),$$

$$\tag{24.100}$$

where $\Sigma(\gamma, \delta; E_n)$ is the irreducible self-energy.

The analysis of the self-energy that emerges from consideration of the equation of motion of the sp propagator can also be carried out in the same way as in Ch. 9 and will not be given here (see also [Martin and Schwinger (1959)] and [Kadanoff and Baym (1962)]).

In a homogeneous system the usual simplifications (see Sec. 10.4) occur on account of momentum conservation. In a plane-wave basis, the sp propagator and self-energy are diagonal

$$\mathcal{G}_T(\boldsymbol{p}_\alpha, \boldsymbol{p}_\beta; E_n) = \delta_{\boldsymbol{p}_\alpha, \boldsymbol{p}_\beta} \mathcal{G}_T(\boldsymbol{p}_\alpha; E_n), \tag{24.101}$$

where we have suppressed additional quantum numbers. The inverse of the noninteracting propagator in Eq. (24.24) now reads

$$1/\mathcal{G}_T^{(0)}(\boldsymbol{p}; E_n) = iE_n - \varepsilon(\boldsymbol{p}) + \mu \tag{24.102}$$

and the Dyson equation can be solved algebraically as

$$\mathcal{G}_T(\boldsymbol{p}_\alpha; E_n) = \frac{1}{iE_n - \varepsilon(\boldsymbol{p}) + \mu - \Sigma(\boldsymbol{p}; E_n)}. \qquad (24.103)$$

The spectral function can be expressed in terms of the self-energy by applying Eq. (24.94)

$$S(\boldsymbol{p}; E) = \frac{1}{\pi} \text{Im} \frac{1}{E - \varepsilon(\boldsymbol{p}) + \mu - \Sigma(\boldsymbol{p}; -iE - \eta)}. \qquad (24.104)$$

A similar discussion as in Sec. 11.4.2 pertains to the introduction of quasiparticle excitations with energies determined by solving

$$E_Q(\boldsymbol{p}) = \varepsilon(\boldsymbol{p}) - \mu + \text{Re}\,\Sigma(\boldsymbol{p}; -iE_Q(\boldsymbol{p}) - \eta), \qquad (24.105)$$

where the spectral function has its dominant contribution. The width of the quasiparticle excitation is governed by the imaginary part of the self-energy

$$W(\boldsymbol{p}; E) = \frac{1}{\pi} \text{Im}\,\Sigma(\boldsymbol{p}; -iE - \eta). \qquad (24.106)$$

Near the chemical potential and for low temperatures it is small, since Eq. (11.98) implies that the corresponding quantity at $T = 0$ vanishes at the Fermi energy. The width remains finite, however, for nonzero temperatures. For $T \to 0$ and $E \to \mu$ one can show, based on similar phase-space restrictions as in Eq. (11.98), that $W(\boldsymbol{p}; E) \to aE^2 + bT^2$, yielding a nonzero thermal width for the quasiparticle excitations. Examples of spectral functions in an interacting system at finite T are shown in Figs. 24.4 and 24.5 of Sec. 24.7, including a discussion of relevant features.

24.6   Hartree-Fock at finite T

In Sec. 10.1 HF theory was developed as the simplest example of a self-consistent Green's function approach. The extension to finite temperature can be done in the same spirit by considering the self-energy depicted in Fig. 24.2. Evaluating the diagram with the propagator $\mathcal{G}_T$ of the general form in Eq. (24.78), we obtain the HF self-energy

$$\Sigma^{HF}(\gamma, \delta; E_n) = \frac{1}{\beta} \sum_{\epsilon\theta} \langle \gamma\epsilon| V |\delta\theta\rangle \sum_k \mathcal{G}_T(\theta, \epsilon; E_k)e^{i\eta E_k}. \qquad (24.107)$$

Substituting Eq. (24.64) shows that the HF self-energy has a mean-field character

$$\Sigma^{HF}(\gamma, \delta) = \sum_{\epsilon\theta} \langle \gamma\epsilon| V |\delta\theta \rangle \langle a_\epsilon^\dagger a_\theta \rangle, \tag{24.108}$$

representing the average of the tp interaction over the sp density matrix of the ensemble at finite T. As at $T = 0$ the HF self-energy therefore does not depend on the energy.

The HF approximation then requires solving the Dyson equation self-consistently, with the self-energy of Eq. (24.108). Since the latter is independent of the Matsubara energy, it corresponds to an additional static sp potential entering the Dyson equation. The propagator that solves the HF Dyson equation is then by necessity of the noninteracting form (24.25)

$$\mathcal{G}_T^{HF}(\alpha, \beta; E_n) = \sum_i \frac{z_{i\alpha}^{HF} z_{i\beta}^{HF*}}{iE_n - \varepsilon_{i\mu}}, \tag{24.109}$$

where the HF sp orbitals and (absolute) energies are solutions of the eigenvalue problem

$$\sum_\delta \left\{ \langle \gamma| T |\delta \rangle + \Sigma^{HF}(\gamma, \delta) \right\} z_{i\delta}^{HF} = \varepsilon_i^{HF} z_{i\gamma}^{HF} \tag{24.110}$$

containing the self-consistency requirement

$$\Sigma^{HF}(\gamma, \delta) = \sum_{\epsilon\theta} \langle \gamma\epsilon| V |\delta\theta \rangle \left(\sum_i z_{i\theta}^{HF} z_{i\epsilon}^{HF*} f(\varepsilon_i - \mu) \right). \tag{24.111}$$

At finite T the HF potential depends not only on the HF orbitals (as was the case for $T = 0$) but also on the HF energy spectrum, through the thermal occupation factors $f(\varepsilon_{i\mu})$. This has some repercussions for homogeneous systems, where translational invariance implies that the HF sp orbitals are plane waves. As a result, HF theory was rather trivial at $T = 0$ (see Sec. 9.4). The finite-T counterpart of the HF spectrum in Eq. (10.165) is given by

$$\varepsilon^{HF}(\boldsymbol{p}) = \frac{\boldsymbol{p}^2}{2m} + \sum_{\boldsymbol{p}'} \langle \boldsymbol{pp}'| V |\boldsymbol{pp}' \rangle f\left(\varepsilon^{HF}(\boldsymbol{p}') - \mu\right). \tag{24.112}$$

In this case even the determination of the HF spectrum (at fixed chemical potential) represents a self-consistency problem. Figure 24.6 discussed in the next section, contains a typical calculation of the HF momentum distribution in nuclear matter.

24.7   Homogeneous systems with short-range correlations at finite temperature

The nucleon sp propagator in nuclear matter is influenced by the strong short-range correlations induced by realistic nucleon-nucleon interactions. In Sec. 16.3 a general diagrammatic method was discussed that incorporates this physics, which is present whenever the tp interaction has large repulsive components at small interparticle distance. The starting point is the construction of the effective interaction by a resummation of the in-medium ladder diagrams. Then the corresponding vertex function is included in the self-energy. The resulting sp propagator in turn modifies the effective interaction, so that a self-consistency scheme emerges. In this section we analyze the finite-T extension of this method, which was recently implemented by several groups [Bożek (1999); Roth Stoddard (2000); Frick and Müther (2003); Frick (2004); Frick *et al.* (2005); Müther and Dickhoff (2005); Rios (2007)] in order to study properties of nuclear matter at finite temperature including the microscopic calculation of the entropy [Rios *et al.* (2006)]. Compared to the $T = 0$ case, the diagrammatic content remains the same. We will therefore concentrate on deriving the thermal occupation factors accompanying the diagrams, as well as the finite-T versons of tp quantities like the tp propagator or vertex function.

   In the finite temperature theory, higher-order Green's functions are introduced in exactly the same way as the sp propagator in Sec. 24.1, *i.e.* as ensemble averages of time-ordered products of removal and addition operators. A case in point is the two-time tp propagator in the *pphh* channel, studied in Sec. 15.1. The corresponding real-time finite-T propagator is

$$G_{pphh_T}(\alpha, \alpha'; \beta, \beta'; t) = -\frac{i}{\hbar} \left\langle T[a_{\alpha'_\Omega}(t)a_{\alpha_\Omega}(t)a^\dagger_{\beta_\Omega}(0)a^\dagger_{\beta'_\Omega}(0)] \right\rangle. \quad (24.113)$$

Switching to imaginary time, one defines

$$\begin{aligned}
\mathcal{G}_{pphh_T}(\alpha, \alpha'; \beta, \beta'; \tau) &= -\frac{1}{\hbar} \left\langle T_\tau[a_{\alpha'_\Omega}(\tau)a_{\alpha_\Omega}(\tau)a^\dagger_{\beta_\Omega}(0)a^\dagger_{\beta'_\Omega}(0)] \right\rangle \\
&= -\frac{1}{\hbar Z_G} \text{Tr} \left\{ \theta(\tau)e^{(-\beta+\tau/\hbar)\hat{\Omega}}a_{\alpha'}a_\alpha e^{-\tau/\hbar\Omega}a^\dagger_\beta a^\dagger_{\beta'} \right. \\
&\quad \left. + \theta(-\tau)e^{(-\beta-\tau/\hbar)\hat{\Omega}}a^\dagger_\beta a^\dagger_{\beta'}e^{\tau/\hbar\Omega}a_{\alpha'}a_\alpha \right\}. \quad (24.114)
\end{aligned}$$

Following the same arguments as in Sec. 24.1.1, one easily shows that the *pphh* propagator is defined for imaginary-time arguments in the range $-\beta\hbar \leq \tau \leq \beta\hbar$, and that its values for positive and negative arguments are again linked by a quasiperiodic boundary condition, *i.e.* for $-\beta\hbar \leq \tau \leq 0$ one has

$$\mathcal{G}_{pphh_T}(\alpha, \alpha'; \beta, \beta'; \tau) = \mathcal{G}_{pphh_T}(\alpha, \alpha'; \beta, \beta'; \tau + \beta\hbar). \qquad (24.115)$$

Note that this implies periodicity [rather than the antiperiodicity in Eq. (24.11)]. As a consequence, the Fourier series expansion

$$\mathcal{G}_{pphh_T}(\alpha, \alpha'; \beta, \beta'; \tau) = \frac{1}{\hbar\beta} \sum_{n=-\infty}^{+\infty} e^{-i\mathcal{E}_n \tau/\hbar} \mathcal{G}_{pphh_T}(\alpha, \alpha'; \beta, \beta'; \mathcal{E}_n)$$

$$(24.116)$$

now contains the bosonic Matsubara energies

$$\mathcal{E}_n = \frac{2n\pi}{\beta}, \qquad (24.117)$$

that are *even* multiples of $\pi/\hbar$, as could be expected on the basis of the bosonic character of a fermion pair. The Fourier coefficients represent the imaginary-energy propagator

$$\mathcal{G}_{pphh_T}(\alpha, \alpha'; \beta, \beta'; \mathcal{E}_n) = \int_0^{\beta\hbar} d\tau \, e^{i\mathcal{E}_n \tau/\hbar} \mathcal{G}_{pphh_T}(\alpha, \alpha'; \beta, \beta'; \tau). \quad (24.118)$$

The noninteracting *pphh* propagator in imaginary time can be rewritten using Wick's theorem (24.41)

$$\mathcal{G}_{pphh_T}^{(0)}(\alpha, \alpha'; \beta, \beta'; \tau) = -\frac{1}{\hbar} \left\langle T_\tau [a_{\alpha'}(\tau) a_\alpha(\tau) a_\beta^\dagger(0) a_{\beta'}^\dagger(0)] \right\rangle_0 \qquad (24.119)$$

$$= -\hbar \left\{ \mathcal{G}_T^{(0)}(\alpha, \beta; \tau) \mathcal{G}_T^{(0)}(\alpha', \beta'; \tau) - \mathcal{G}_T^{(0)}(\alpha, \beta'; \tau) \mathcal{G}_T^{(0)}(\alpha', \beta; \tau) \right\}$$

as an antisymmetrized product of noninteracting imaginary-time sp propagators. Its imaginary energy representation is determined by the discrete Fourier transform in Eq. (24.118)

$$\mathcal{G}_{pphh_T}^{(0)}(\alpha, \alpha'; \beta, \beta'; \mathcal{E}_n) \qquad (24.120)$$

$$= -\frac{1}{\hbar} \int_0^{\beta\hbar} d\tau \, e^{i\mathcal{E}_n \tau/\hbar} \mathcal{G}_T^{(0)}(\alpha, \beta; \tau) \mathcal{G}_T^{(0)}(\alpha', \beta'; \tau) - (\alpha \leftrightarrow \alpha').$$

Note that the second (exchange) term in Eq. (24.120) arises from the first by interchanging $\alpha \leftrightarrow \alpha'$. Upon substituting the Fourier series (24.14) for the sp propagators and performing the integration over imaginary time, we obtain the discrete convolution

$$\mathcal{G}^{(0)}_{pphh_T}(\alpha, \alpha'; \beta, \beta'; \mathcal{E}_n) \tag{24.121}$$

$$= -\frac{1}{\beta} \sum_{n'} \mathcal{G}^{(0)}_T(\alpha, \beta; E_{n'}) \mathcal{G}^{(0)}_T(\alpha', \beta'; \mathcal{E}_n - E_{n'}) - (\alpha \leftrightarrow \alpha').$$

A key ingredient in the treatment of Sec. 16.3 is the dressed noninteracting *pphh* propagator. The finite-T version can now be readily found by replacing in Eq. (24.121) the noninteracting sp propagators $\mathcal{G}^{(0)}_T$ with the exact (dressed) ones,

$$\mathcal{G}^{f}_{pphh_T}(\alpha, \alpha'; \beta, \beta'; \mathcal{E}_n) \tag{24.122}$$

$$= -\frac{1}{\beta} \sum_{n'} \mathcal{G}_T(\alpha, \beta; E_{n'}) \mathcal{G}_T(\alpha', \beta'; \mathcal{E}_n - E_{n'}) - (\alpha \leftrightarrow \alpha').$$

Substituting the spectral representation (24.92) for $\mathcal{G}_T$, one can recast Eq. (24.122) in terms of the sp spectral function, which is a form more suitable for practical calculations

$$\mathcal{G}^{f}_{pphh_T}(\alpha, \alpha'; \beta, \beta'; \mathcal{E}_n) \tag{24.123}$$

$$= -\frac{1}{\beta} \sum_{n'} \int dE' \frac{S(\alpha, \beta; E')}{iE_{n'} - E'} \int dE'' \frac{S(\alpha', \beta'; E'')}{i\mathcal{E}_n - iE_{n'} - E''} - (\alpha \leftrightarrow \alpha').$$

The sum over Matsubara energies encountered in Eq. (24.123) is of the same type as the one analyzed in Eq. (24.59)

$$\frac{1}{\beta} \sum_{n'} \frac{1}{(iE_{n'} - E')(E_{n'} - i\mathcal{E}_n + E'')} = \frac{f(E') - f(-E'')}{E' + E'' - i\mathcal{E}_n}. \tag{24.124}$$

Since the Fermi function obeys $f(-x) = 1 - f(x)$, the thermal occupation factor appearing in Eq. (24.124) can be expressed alternatively as

$$f(-E'') - f(E') = 1 - f(E') - f(E'')$$
$$= [1 - f(E')][1 - f(E'')] - f(E')f(E''). \tag{24.125}$$

In the final result

$$\mathcal{G}^{f}_{pphh_T}(\alpha, \alpha'; \beta, \beta'; \mathcal{E}_n) = \int dE' \int dE'' \frac{S(\alpha, \beta; E')S(\alpha', \beta'; E'')}{i\mathcal{E}_n - E' - E''}$$
$$\times \{[1 - f(E')][1 - f(E'')] - f(E')f(E'')\} - (\alpha \leftrightarrow \alpha'), \tag{24.126}$$

one can clearly recognize both components (pp and hh propagation) present in $\mathcal{G}^f_{pphh_T}$.

We now have all ingredients to analyze the ladder equation describing scattering of two particles in the medium at finite temperature. For a homogeneous system we replace the general sp basis with a plane-wave basis according to

$$|\alpha\alpha'\rangle \to |\mathbf{k}_\alpha \mathbf{k}_{\alpha'}\rangle = |\mathbf{K}\mathbf{k}\rangle , \qquad (24.127)$$

where $\mathbf{K} = \mathbf{k}_\alpha + \mathbf{k}_{\alpha'}$ represents the conserved center of mass wave vector and $\mathbf{k} = (\mathbf{k}_\alpha - \mathbf{k}_{\alpha'})/2$ is the relative one. The finite-temperature counterpart of Eq. (16.45) then reads

$$\langle \mathbf{k}| \Gamma_{pphh_T}(\mathbf{K},\mathcal{E}_n) |\mathbf{k}'\rangle = \langle \mathbf{k}| V |\mathbf{k}'\rangle \qquad (24.128)$$
$$+ \frac{1}{2} \int \frac{d^3q}{(2\pi)^3} \langle \mathbf{k}| V |\mathbf{q}\rangle \, \mathcal{G}^f_{pphh_T}(\mathbf{K},\mathbf{q};\mathcal{E}_n) \langle \mathbf{q}| \Gamma_{pphh_T}(\mathbf{K},\mathcal{E}_n) |\mathbf{k}'\rangle ,$$

where [compare with Eq. (16.46)]

$$\mathcal{G}^f_{pphh_T}(\mathbf{K},\mathbf{q};\mathcal{E}_n) = \int dE'dE'' \frac{S(\mathbf{K}/2+\mathbf{q};E')S(\mathbf{K}/2-\mathbf{q};E'')}{i\mathcal{E}_n - E' - E''}$$
$$\times \{[1 - f(E')][1 - f(E'')] - f(E')f(E'')\}. \quad (24.129)$$

Since antisymmetrized matrix elements are employed in Eq. (24.128), there is no exchange term required for the noninteracting dressed propagator, like in Eq. (16.45). Note that spin (and isospin) quantum numbers have also been suppressed in Eq. (24.128).

The static part of Γ_{pphh_T} corresponds to the bare interaction and can be isolated according to

$$\langle \mathbf{k}| \Gamma_{pphh_T}(\mathbf{K},\mathcal{E}_n) |\mathbf{k}'\rangle = \langle \mathbf{k}| V |\mathbf{k}'\rangle + \langle \mathbf{k}| \Delta\Gamma_{pphh_T}(\mathbf{K},\mathcal{E}_n) |\mathbf{k}'\rangle , \quad (24.130)$$

while the energy-dependent part can be written in its spectral representation as

$$\langle \mathbf{k}| \Delta\Gamma_{pphh_T}(\mathbf{K},\mathcal{E}_n) |\mathbf{k}'\rangle$$
$$= \int \frac{dE'}{i\mathcal{E}_n - E'} \frac{1}{\pi} \text{Im} \, \langle \mathbf{k}| \Delta\Gamma_{pphh_T}(\mathbf{K},-iE'-\eta) |\mathbf{k}'\rangle . \qquad (24.131)$$

To obtain Eq. (24.131) the same reasoning regarding the uniqueness of the analytic continuation can be followed as the one leading to Eqs. (24.93) and (24.94).

Finally, one must incorporate the effective interaction in the self-energy, in the manner described in Sec. 16.3. The first term in Eq. (24.130) gives rise to the Hartree-Fock like, static term

$$\Sigma_V(\boldsymbol{k}) = \int \frac{d^3k'}{(2\pi)^3} \, \langle \tfrac{1}{2}(\boldsymbol{k}-\boldsymbol{k}')| \, V \, |\tfrac{1}{2}(\boldsymbol{k}-\boldsymbol{k}')\rangle \, n(\boldsymbol{k}), \qquad (24.132)$$

where the finite-T momentum distribution reads

$$n(\boldsymbol{k}) = \frac{1}{\beta} \sum_n \mathcal{G}_T(\boldsymbol{k}; E_n) e^{i\eta E_n}. \qquad (24.133)$$

The dynamic part $\Delta\Gamma_{pphh_T}$ enters the self-energy as

$$\Sigma_{\Delta\Gamma}(\boldsymbol{k}; E_n) = -\frac{1}{\beta} \int \frac{d^3k'}{(2\pi)^3} \sum_{n'} \mathcal{G}_T(\boldsymbol{k}', E_{n'}) \qquad (24.134)$$
$$\times \, \langle \tfrac{1}{2}(\boldsymbol{k}-\boldsymbol{k}')| \, \Delta\Gamma_{pphh_T}(\boldsymbol{K}, E_n + E_{n'}) \, |\tfrac{1}{2}(\boldsymbol{k}-\boldsymbol{k}')\rangle \, .$$

For practical calculations one should replace the sp propagator and the effective interaction by their spectral representations, given respectively by Eq. (24.92) and (24.131). The sum over Matsubara energies is again familiar from Eq. (24.59)

$$\frac{1}{\beta} \sum_{n'} \frac{1}{(iE_{n'} - E')(iE_{n'} + iE_n - E'')} = \frac{f(E') - f(E'' - iE_n)}{E' - E'' + iE_n}$$
$$= \frac{f(E') + b(E'')}{E' - E'' + iE_n}. \qquad (24.135)$$

There now appears a *bosonic* occupation factor $b(x) = 1/(e^{\beta x} - 1)$. Noting that for a fermionic Matsubara energy $e^{i\beta E_n} = -1$, one finds

$$f(E'' - iE_n) = \frac{1}{1 - e^{\beta E''}} = -b(E''). \qquad (24.136)$$

The counterpart of Eq. (16.49) then reads

$$\Sigma_{\Delta\Gamma}(\boldsymbol{k}, E_n) = \int \frac{d^3k'}{(2\pi)^3} \int \frac{dE'dE''}{iE_n + E' - E''} S(\boldsymbol{k}'; E') \qquad (24.137)$$

$$\frac{1}{\pi} \operatorname{Im} \langle \tfrac{1}{2}(\boldsymbol{k}-\boldsymbol{k}')| \, \Delta\Gamma_{pphh_T}(\boldsymbol{K}, -iE'' - \eta) \, |\tfrac{1}{2}(\boldsymbol{k}-\boldsymbol{k}')\rangle \, (f(E') + b(E'')).$$

The integrand in Eq. (24.137) now seemingly has a singularity for $E'' = 0$, where the bosonic $b(E'')$ has a pole. However, the the imaginary part of $\Delta\Gamma_{pphh_T}$ is proportional to $1/b(E'')$, exactly cancelling the singularity and making the integrand well-behaved. This can be seen most easily by considering the imaginary part of the dressed noninteracting propagator in Eq. (24.129)

$$\frac{1}{\pi}\text{Im } \mathcal{G}^f_{pphh_T}(\boldsymbol{K}, \boldsymbol{q}; -iE - \eta) \tag{24.138}$$

$$= \int dE' S(\boldsymbol{K}/2 + \boldsymbol{q}; E') S(\boldsymbol{K}/2 - \boldsymbol{q}; E - E')\{f(E' - E) - f(E')\}.$$

Here we used the alternative version of the thermal occupation factor on the left of Eq. (24.125), which can be further manipulated as

$$f(E' - E) - f(E') = f(E')f(E - E')/b(E). \tag{24.139}$$

This implies that the spectral strength function in Eq. (24.138) is proportional to $1/b(E)$ and vanishes for $E = 0$. The same holds for the effective interaction $\Delta\Gamma_{pphh_T}$ which acquires its imaginary part from $\mathcal{G}^f_{pphh_T}$ through Eq. (24.128).

As an example of the application of this formalism, we now discuss some results obtained by Rios *et al.* for symmetric nuclear matter with the realistic CDBonn potential as sole input. Details on the numerical treatment can be found in [Rios (2007)]. Similar self-consistent results have been obtained in [Bożek (1999); Roth Stoddard (2000); Frick and Müther (2003); Frick (2004)]. In Fig. 24.4 the sp spectral function $S(k; E)$ is shown for three typical momenta, at various nucleon densities and a temperature corresponding to 10 MeV. One recognizes several features already described in Sec. 16.3 for the $T = 0$ case, *e.g.* the concentration of strength in a quasiparticle peak, and the development of the high-energy tail at positive energies, reflecting the short-range correlations present in the bare potential. It should be noted that finite temperature provides an extra width to all spectral distributions, which removes some of the sharp features present at zero temperature. A case in point is the spectral function at $k = k_F$ which at $T = 0$ contains a delta-spike at the Fermi energy, representing the quasiparticle contribution. At finite T the quasiparticle distribution is narrow and located at the chemical potential but still has a finite width. This width decreases with temperature, as can be seen in the temperature dependence of the spectral function shown in Fig. 24.5. Also the zero value of the $T = 0$ spectral function at the Fermi energy for $k \neq k_F$ disappears at finite T: there remains a dip of the spectral function near μ, as seen in the upper and lower panels of Figs. 24.4–24.5. This washing out of sharp features is highly beneficial in practical calculations, as it helps the adequate sampling of the spectral distributions during the iterations to

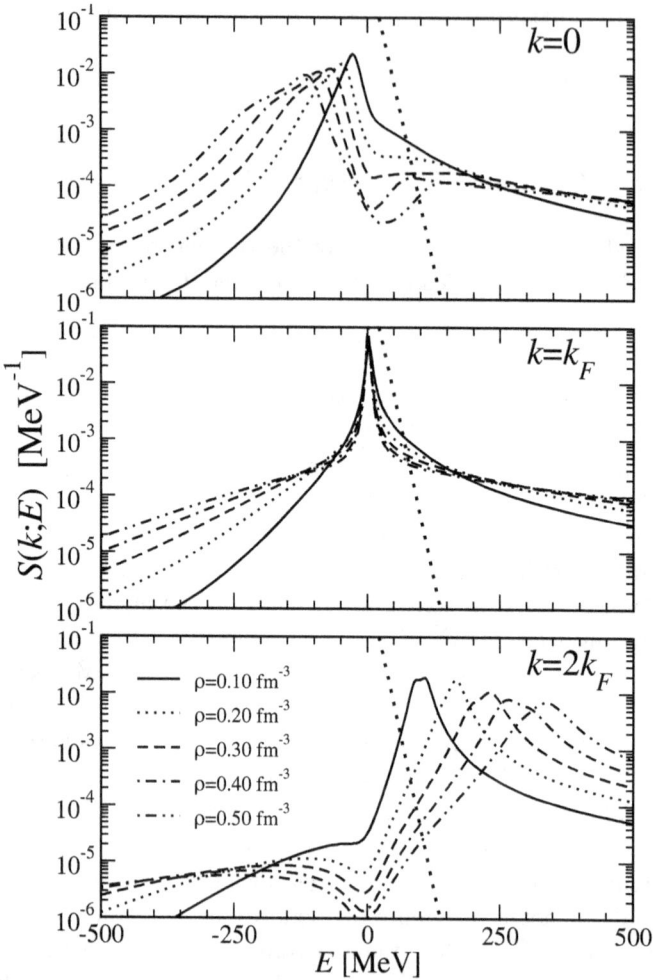

Fig. 24.4   Spectral functions for wave vectors $k = 0, k_F$, and $2k_F$ at a temperature corresponding to 10 MeV for five different densities, identified in the lower panel. The thick dotted line represents the Fermi function at this temperature.

self-consistency. Furthermore, one notes that the temperature dependence of the spectral functions is rather mild, in the sense that the location of the quasiparticle peak hardly changes with temperature, and the temperature effects are confined in a small region around the chemical potential. This makes it reasonable to extrapolate the finite T results to $T = 0$ [Müther and

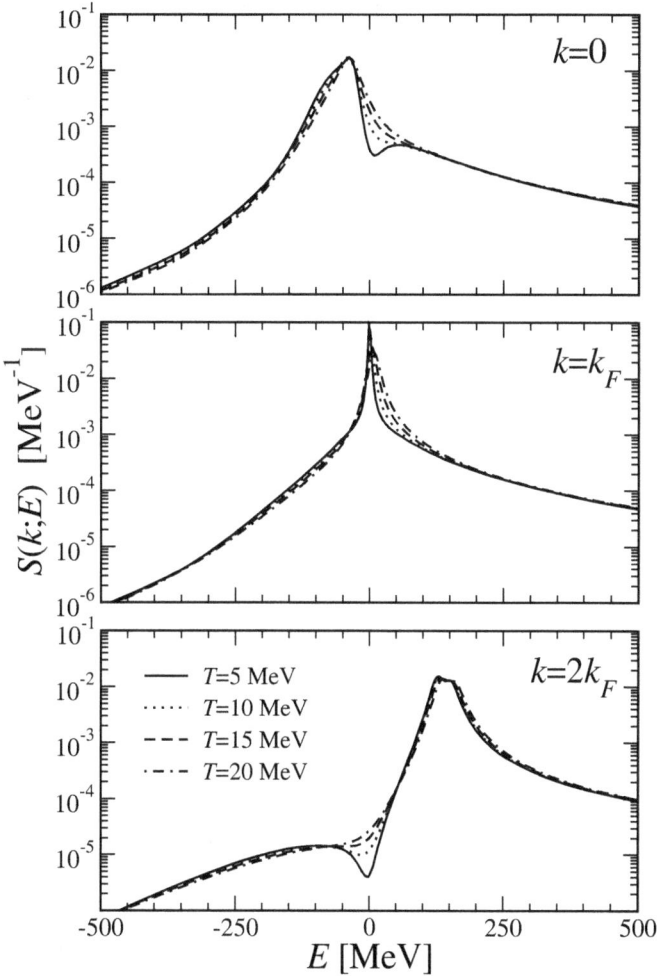

Fig. 24.5 Spectral functions for wave vectors $k = 0, k_F$, and $2k_F$ at a density of 0.16 fm^{-3} for four different temperatures listed in the lower panel.

Dickhoff (2005)] to properly assess the pairing properties with the inclusion of short-range correlations (see Sec. 24.8.2).

The interplay between thermal and correlation effects is nicely illustrated by the momentum distributions depicted in Figs. 24.6–24.7. Note that the momentum distribution $n(k)$ is obtained by integrating the product of the spectral function $S(k; E)$ and the Fermi function $f(E)$. The

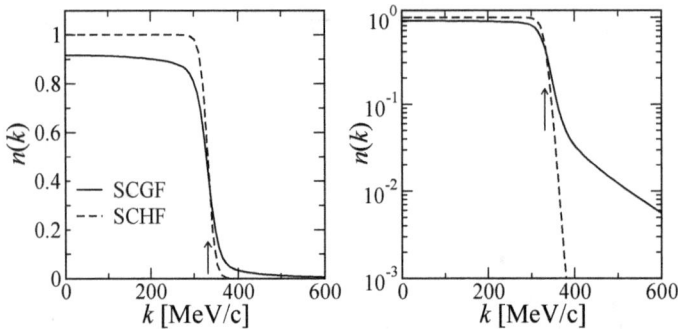

Fig. 24.6   Momentum distribution for a self-consistent ladder calculation (SCGF) compared with a self-consistent HF approximation at a temperature corresponding to 5 MeV
at a density of $\rho = 0.32\,\mathrm{fm}^{-3}$. The arrow denotes the Fermi momentum.

latter is also indicated in Fig. 24.4. In Fig. 24.6 the momentum distribution corresponding to the ladder approximation is compared with the HF
momentum distribution, $1/[1 + \exp(\beta(\varepsilon^{HF}(k) - \mu))]$ (which becomes the
step function $\theta(k - k_F)$ in the limit $T \to 0$). Momenta below (above) k_F are
thermally depleted (occupied) in HF at finite temperature. This depletion
(occupation) is enhanced by the correlation effects in the self-consistent
ladder approach. The discontinuity at k_F of the $T = 0$ correlated momentum distribution is washed out by thermal effects, but there is still a
steep jump near k_F at finite T. The strong temperature dependence is
shown in Fig. 24.7, and is primarily driven by the Fermi function since the
underlying spectral function depends rather weakly on temperature. Note
that at large momenta, the curves at various temperatures converge: The
occupation of these momenta is provided by the short-range repulsion and
tensor components in the interaction, and thermal effects play a minor role
in this region.

The self-consistent Green's function approach also gives access to thermodynamic properties, such as the entropy or free energy per nucleon.
This development and corresponding calculations can be found in [Rios *et
al.* (2006); Rios (2007)]. We refer to this work for further details and only
display results for the free energy and energy per particle in Fig. 24.8 for
the BHF and SCGF approximation at a temperature of 10 MeV as a function of density. As discussed in Sec. 16.1 at $T = 0$, the Fermi energy must
correspond to the minimum of the free energy per particle for a thermodynamically consistent desciption. For fixed density and temperature the

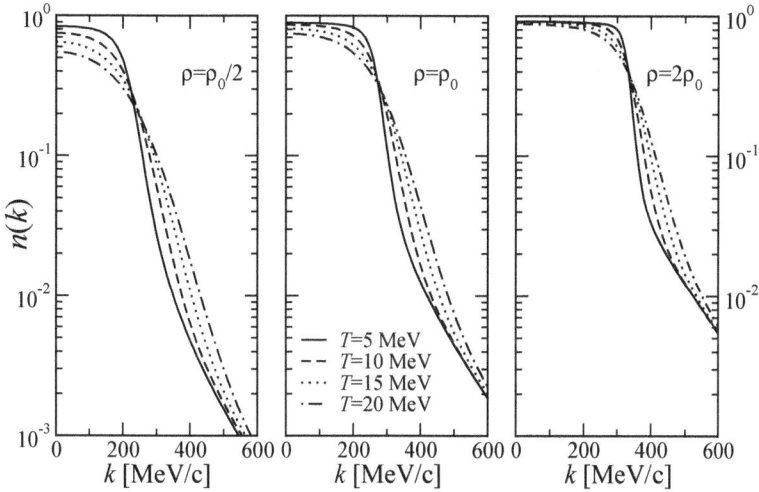

Fig. 24.7 Momentum distributions at three different densities identified in the panels with $\rho_0 = 0.16\,\mathrm{fm}^{-3}$. Results are shown for four different temperatures identified in the middle panel.

microscopic chemical potential $\widetilde{\mu}$ used to evaluate the spectral function, is determined from the normalization condition

$$\rho = \int \frac{d^3 k}{(2\pi)^3} \int_{-\infty}^{\infty} dE\, S(k;E) f(E),\qquad(24.140)$$

suppressing the fermion degeneracy factor, since spin and isospin have been left out of the equations in this section. It should coincide with the macroscopic chemical potential given by the derivative of the free energy with respect to particle number A according to

$$\mu = \frac{\partial F(\rho,T)}{\partial A}.\qquad(24.141)$$

Both versions of the chemical potential are also plotted in Fig. 24.8. The agreement between μ and $\widetilde{\mu}$ is very good for the SCGF approach. As a consequence the Hugenholtz-van Hove theorem is quite well fulfilled and the minimum of F/A and $\widetilde{\mu}$ properly coincide. The BHF approach gives well known inconsistent results exhibiting large differences between the chemical potentials and badly violating the Hugenholtz-van Hove theorem by about 20 MeV.

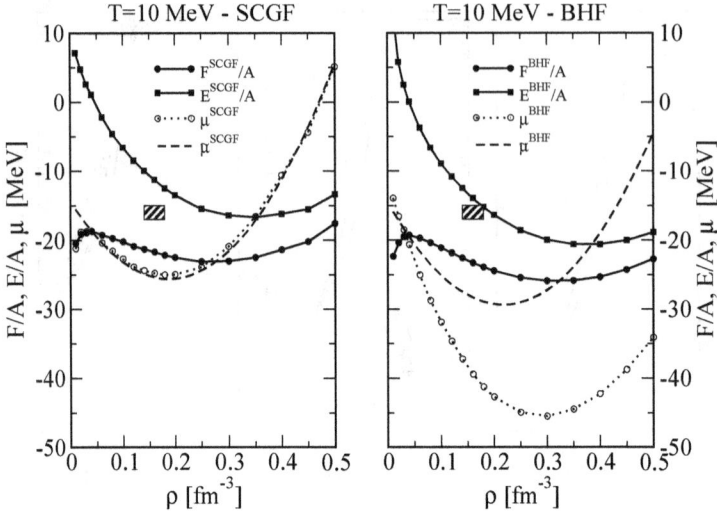

Fig. 24.8   Energy (solid squares) and free energy (solid circles) per particle together with the chemical potentials for SCGF and BHF approximations as a function of density and calculated at a temperature of 10 MeV. The macroscopic chemical potential μ is given by the dashed lines while the microscopic version $\tilde{\mu}$ is characterized by the dotted lines containing open circles. The hatched area represents the empirical saturation point of symmetric nuclear matter.

An extension of finite-temperature SCGF calculations to asymmetric nuclear matter has recently been reported in [Frick *et al.* (2005)]. Of particular interest is the different behavior of the proton and neutron Fermi sea as a function of nucleon asymmetry, discussed in Sec. 23.6. The depletion of these Fermi seas at $k = 0$ is shown in Fig. 24.9 as a function of nucleon asymmetry δ. Clearly protons (neutrons) feel stronger (weaker) correlations with increasing neutron fraction. Its explanation is related to a specific feature of the underlying nucleon-nucleon interaction. In pure neutron matter the main part of the nuclear tensor force does not play a role in contributing to the neutron Fermi sea depletion, since the $^3S_1$-$^3D_1$ component only acts for zero total isospin. Protons are increasingly exposed to this interaction in this limit even though the volume of their Fermi sea is decreasing, leading to the calculated reduction of the occupation $n(0)$ as a function of δ. The associated volume feature of this asymmetry dependence has not yet been identified in finite nuclei in the DOM calculations reported in Sec. 23.6. It could become apparent when nuclei can be studied at larger asymmetry with rare isotope facilities in the future.

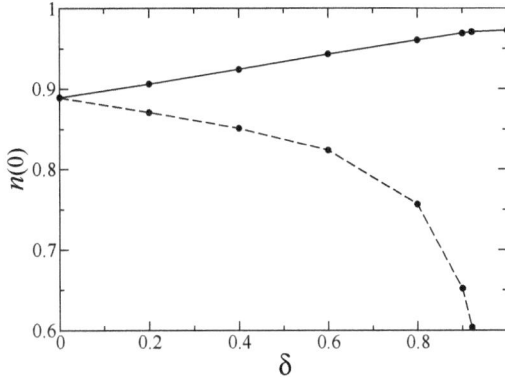

Fig. 24.9 Fermi sea depletion at $k = 0$ for protons (dashed) and neutrons (solid) as a function of nucleon asymmetry at the empirical saturation density of symmetric nuclear matter at a temperature of 5 MeV. The figure is adapted from [Frick *et al.* (2005)].

24.8 Fermion pairing at finite T

The description of fermion systems with pairing correlations at finite temperature follows the same development as the $T = 0$ case. Apart from rule changes, like the ones discussed in Sec. 24.3, there are no additional modifications when the ensemble averages are studied for anomalous propagators and self-energies. A transcription of the material in Ch. 22 therefore leads to the Gorkov equations for the normal and anomalous temperature propagators

$$\mathcal{G}_{11_T}(\boldsymbol{pm}; E_n) = \mathcal{G}_{11_T}^{(0)}(\boldsymbol{pm}; E_n) + \mathcal{G}_{11_T}^{(0)}(\boldsymbol{pm}; E_n)\Sigma_{11}(\boldsymbol{pm}; E_n)\mathcal{G}_{11_T}(\boldsymbol{pm}; E_n)$$
$$+ \mathcal{G}_{11_T}^{(0)}(\boldsymbol{pm}; E_n)\Sigma_{12}(\boldsymbol{pm}; E_n)\mathcal{G}_{21_T}(\boldsymbol{pm}; E_n)$$
$$\mathcal{G}_{21_T}(\boldsymbol{pm}; E_n) = \mathcal{G}_{22_T}^{(0)}(\boldsymbol{pm}; E_n)\Sigma_{21}(\boldsymbol{pm}; E_n)\mathcal{G}_{11_T}(\boldsymbol{pm}; E_n) \qquad (24.142)$$
$$+ \mathcal{G}_{22_T}^{(0)}(\boldsymbol{pm}; E_n)\Sigma_{22}(\boldsymbol{pm}; E_n)\mathcal{G}_{21_T}(\boldsymbol{pm}; E_n).$$

The noninteracting propagators have the form

$$\mathcal{G}_{11_T}^{(0)}(\boldsymbol{pm}; E_n) = \frac{1}{iE_n - \varepsilon_{p\mu}} \qquad (24.143)$$

and

$$\mathcal{G}_{22_T}^{(0)}(\boldsymbol{pm}; E_n) = \frac{1}{iE_n + \varepsilon_{p\mu}}. \qquad (24.144)$$

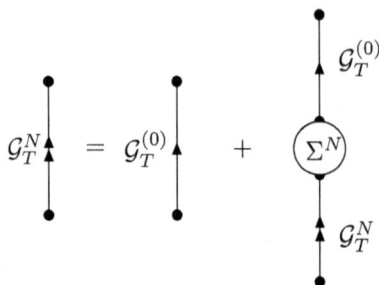

Fig. 24.10 Diagrammatic representation of the sp propagator including only normal self-energy terms in terms of the corresponding self-energy Σ^N and the noninteracting propagator $\mathcal{G}_T^{(0)}$ representing Eq. (24.145).

The solution to Eq. (24.142) is of the same form as Eq. (22.35). We will consider the extension of the BCS gap equation to finite temperature in Sec. 24.8.1 by rearranging the Gorkov equations to first include all normal self-energy insertions [Migdal (1967)]. It is then straightforward to obtain the gap equation at finite temperature. The same formulation is appropriate for systems that exhibit strong medium modifications due to short-range correlations, like the nuclear systems discussed in Sec. 24.8.2.

24.8.1   *Gap equation at finite T*

We start this section by rewriting the Gorkov equations (24.142) for the superfluid propagators by first iterating all contributions of the normal self-energy. Diagrammatically this is accomplished in Fig. 24.10. The line with the two arrows contains all normal self-energy insertions and can represent either $\mathcal{G}_{11_T}^N$ or $\mathcal{G}_{22_T}^N$, where the N superscript denotes that only normal self-energy insertions are included. Schematically we can write this contribution to the normal propagator as

$$\mathcal{G}_T^N(\boldsymbol{p}; E_n) = \mathcal{G}_T^{(0)}(\boldsymbol{p}; E_n) + \mathcal{G}_T^{(0)}(\boldsymbol{p}; E_n)\Sigma^N(\boldsymbol{p}; E_n)\mathcal{G}_T^N(\boldsymbol{p}; E_n) \quad (24.145)$$

for either the 11 or 22 version. The reformulated Gorkov equations then become

$$\mathcal{G}_{11_T}(\boldsymbol{p}m; E_n) = \mathcal{G}_{11_T}^N(\boldsymbol{p}m; E_n) + \mathcal{G}_{11_T}^N(\boldsymbol{p}m; E_n)\Sigma_{12}(\boldsymbol{p}m; E_n)\mathcal{G}_{21_T}(\boldsymbol{p}m; E_n)$$

$$\mathcal{G}_{21_T}(\boldsymbol{p}m; E_n) = \mathcal{G}_{22_T}^N(\boldsymbol{p}m; E_n)\Sigma_{21}(\boldsymbol{p}m; E_n)\mathcal{G}_{11_T}(\boldsymbol{p}m; E_n), \quad (24.146)$$

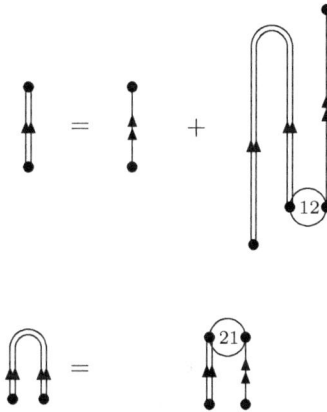

Fig. 24.11 Diagrammatic representation of the propagators $\mathcal{G}_{ij_T}(\boldsymbol{p}m; E_n)$, after the normal self-energy terms have been summed into $\mathcal{G}_{ii_T}^N$. The diagrams illustrate how remaining reducible repetitions of the irreducible self-energies Σ_{12} and Σ_{21} are included.

as graphically illustrated in Fig. 24.11. The equivalence of Eq. (24.146) with the original Gorkov equations (24.142) can be demonstrated diagrammatically order by order. This form is particularly useful when the normal self-energy is responsible for substantial modifications of the noninteracting propagator, as for the nuclear systems discussed in Sec. 24.8.2.

The extension of the BCS gap equation to finite temperature can be obtained from the anomalous propagator in Eq. (24.146) by constructing the lowest-order anomalous self-energy term on both sides of the second equality, or diagrammatically, by closing the corresponding diagrams with the two-body interaction. Employing the form of V used in Eq. (22.10), the generalization of Eq. (22.28) to finite temperature thus becomes

$$\Sigma_{21}(\boldsymbol{p}m) = (-1)\left(-\frac{1}{\beta}\right)\sum_{\boldsymbol{p}'}\frac{W(\boldsymbol{p}-\boldsymbol{p}')}{V}\sum_n e^{i\eta E_n}\mathcal{G}_{21_T}(\boldsymbol{p}'m; E_n) \quad (24.147)$$

$$= \frac{1}{\beta}\sum_{\boldsymbol{p}'}\frac{W(\boldsymbol{p}-\boldsymbol{p}')}{V}\sum_n e^{i\eta E_n}\mathcal{G}_{22_T}^N(\boldsymbol{p}'m; E_n)\Sigma_{21}(\boldsymbol{p}'m)\mathcal{G}_{11_T}(\boldsymbol{p}'m; E_n),$$

noting once more that the gap (Σ_{21}) is energy independent in lowest order. The last equality is generated by substitution of $\mathcal{G}_{21_T}$ from Eq. (24.146). Including only the normal HF self-energy, we may employ

$$\mathcal{G}_{22_T}^N(\boldsymbol{p}m; E_n) \rightarrow \mathcal{G}_{22_T}^{HF}(\boldsymbol{p}m; E_n) = \frac{1}{iE_n + \chi_p}, \quad (24.148)$$

with $\chi_p = \varepsilon_{p\mu} + V_p$. We can anticipate the same form for $\mathcal{G}_{11_T}$ as in Eq. (22.43) so that

$$\mathcal{G}_{11_T}(\boldsymbol{pm}; E_n) \rightarrow \frac{u_p^2}{iE_n - E_p} + \frac{v_p^2}{iE_n + E_p}, \qquad (24.149)$$

with

$$E_p = \sqrt{\chi_p^2 + \Delta_p^2}. \qquad (24.150)$$

Following Eq. (22.37), we write $\Sigma_{21}(\boldsymbol{pm}) = \Delta_p s_m$. The expressions for u_p^2 and v_p^2 are still given by the equivalent of Eq. (22.44). Inserting Eqs. (24.148) and (24.149) in (24.147), we proceed by evaluating all the energy sums, using *e.g.* a decomposition in partial fractions. The final result is the gap equation at finite temperature

$$\Delta_p = -\frac{1}{2} \sum_{p'} \frac{W(\boldsymbol{p} - \boldsymbol{p'})}{V} \frac{\Delta_{p'}}{E_{p'}} \tanh\left(\frac{\beta E_{p'}}{2}\right), \qquad (24.151)$$

which is the generalization of Eq. (22.47) at $T = 0$. We note that Eq. (24.151) reduces to (22.47) in the limit $T \rightarrow 0$.

As in Sec. 22.4 we will study the gap equation for the simplified interaction of Eq. (22.48) that simulates the properties of normal superconductors when c is identified with Ω_D, the Debye energy (see Sec. 22.6.1). Following identical steps that link Eqs. (22.47) and (22.52) and making similar assumptions, we arrive at the corresponding gap equation at finite temperature in the form

$$1 = \frac{\lambda}{2} \int_{-c}^{+c} d\chi \frac{D(\chi)}{\sqrt{\chi^2 + \Delta^2}} \tanh\left(\frac{\beta}{2}\sqrt{\chi^2 + \Delta^2}\right). \qquad (24.152)$$

We will study the limit $T \rightarrow T_c$, where the gap vanishes. As before, we can extract the density of states $D(0)$ at the Fermi energy from the integral. Using the symmetry of the integrand and changing to the integration variable $z = \beta\chi/2$ we find

$$\frac{1}{\lambda D(0)} = \int_0^{\beta c/2} \frac{dz}{z} \tanh z. \qquad (24.153)$$

Integrating by parts yields

$$\frac{1}{\lambda D(0)} = [\ln z \tanh z]_0^{\beta c/2} - \int_0^{\beta c/2} dz \ln z \operatorname{sech}^2 z. \qquad (24.154)$$

For realistic situations the upper limit of the definite integral is large and may be taken ∞. The integral can be looked up in tables. We are then able rewrite Eq. (24.154) in terms of the critical temperature

$$k_B T_c = \frac{2e^\gamma}{\pi} c \, e^{-1/\lambda D(0)} \approx 1.13c \, e^{-1/\lambda D(0)}, \qquad (24.155)$$

where $\gamma \approx 0.5772$ is Euler's constant. Comparison with Eq. (22.54) shows that the ratio of the gap at $T = 0$ and the critical temperature form a material independent ratio given by

$$\frac{\Delta_{T=0}}{k_B T_c} = \pi e^{-\gamma} \approx 1.76, \qquad (24.156)$$

which is in good agreement with experimental data for typical standard superconductors (see [Fetter and Walecka (1971)]).

Other limiting behavior of BCS superconductors includes for $T \ll T_c$

$$\Delta(T) = \Delta_{T=0} - (2\pi \Delta_{T=0} k_B T)^{\frac{1}{2}} \, e^{-\Delta_{T=0}/k_B T} \qquad (24.157)$$

and for $T_c - T \ll T_c$

$$\Delta(T) \approx 3.06 k_B T_c \left(1 - \frac{T}{T_c}\right)^{\frac{1}{2}}. \qquad (24.158)$$

These expressions are derived by studying the low-temperature expansion of the gap equation and its behavior near T_c (see *e.g.* [Abrikosov *et al.* (1975)]). Other important thermodynamic quantities include the electronic specific heat for $T \to 0$

$$\frac{C_s}{V} \approx 2D(0)k_B (2\pi)^{\frac{1}{2}} \left(\frac{\Delta_{T=0}}{k_B T}\right)^{\frac{3}{2}} e^{-\Delta_{T=0}/k_B T}, \qquad (24.159)$$

using the subscript s to identify the superconducting phase. Analytical methods can be employed to generate this result (see *e.g.* [Fetter and Walecka (1971)]). The jump in the electronic specific heat at T_c is also found in this way (with the subscript n identifying the normal phase)

$$\left. \frac{C_s - C_n}{C_n} \right|_{T_c} \approx 1.43, \qquad (24.160)$$

which agrees reasonably with experiment. Complete solutions of the gap equation, including those presented in the next section, must be calculated numerically.

24.8.2   *Gorkov equations with dressed propagators*

The inclusion of short-range correlations in the self-energy for nuclear matter was discussed in Ch. 16. The modifications of the sp propagator at normal density are substantial, leading for example to a quasiparticle strength near the Fermi energy and momentum of about 0.7. The important consequences of the dressing of sp propagation in describing scattering in the medium were explored in Ch. 20. Pairing instabilities encountered in Ch. 15 and standard BCS results for mean-field sp propagators, may thus be modified considerably when such correlations are properly accounted for. The conjecture was investigated in [Müther and Dickhoff (2005)] and we discuss the main features of that analysis and related figures here. The starting point is again the form of the Gorkov equations in Eq. (24.146). We include the full dressing due to the normal self-energy as described in Sec. 24.7. The corresponding spectral representation reads

$$\mathcal{G}_{22_T}^N(\boldsymbol{pm}; E_n) = \int_{-\infty}^{\infty} dE \, \frac{S(p; E)}{iE_n + E}. \tag{24.161}$$

The superfluid propagator yields a corresponding expression given by

$$\mathcal{G}_{11_T}(\boldsymbol{pm}; E_n) = \int_{-\infty}^{\infty} dE' \, \frac{S_s(p; E')}{iE_n - E'}. \tag{24.162}$$

The energy sum in Eq. (24.147) then leads to a generalized denominator in the gap equation defined by

$$-\frac{1}{2E_p} \equiv \frac{1}{\beta} \sum_n e^{i\eta E_n} \int_{-\infty}^{\infty} dE \, \frac{S(p; E)}{iE_n + E} \int_{-\infty}^{\infty} dE' \, \frac{S_s(p; E')}{iE_n - E'}$$

$$= \int_{-\infty}^{\infty} dE \int_{-\infty}^{\infty} dE' \, S(p; E) S_s(p; E') \frac{1 - f(E) - f(E')}{-E - E'}. \tag{24.163}$$

Note that, in spite of the denominator $-E - E'$, the integrand is smooth (see the discussion following Eq. (24.137).

Since for nuclear applications we encounter a strong state dependence, including the possibility of the coupling of different orbital angular momentum, it is necessary to employ the gap equation in a partial wave representation. The formalism developed so far, has focused on pairing with total spin $S = 0$. To allow other pairing quantum numbers, it is necessary to generalize the gap equation of Eq. (24.151) to

$$\Delta_{pmm'} = -\frac{1}{2} \sum_{p'\tilde{m}\tilde{m}'} (pm - pm'|V|p'\tilde{m} - p'\tilde{m}') \frac{\Delta_{p'\tilde{m}\tilde{m}'}}{E_{p'}} \tanh\left(\frac{\beta E_{p'}}{2}\right),$$

(24.164)

where the gap function allows for more general spin structure [Baldo *et al.* (1995)]. Once the total spin of the pairs has been determined, an uncoupled or coupled channel with good total angular momentum and parity must be selected to solve the gap equation. In case of isospin, antisymmetry will then imply the total isospin. For general J the gap function is expanded according to

$$\Delta_{pmm'} = \sum_{\ell m_\ell} (\tfrac{1}{2} \, m \, \tfrac{1}{2} \, m' \, | S \, m + m')$$

(24.165)

$$(S \, m + m' \, \ell \, m_\ell \, | J \, m + m' + m_\ell) \, Y_{\ell m_\ell}(\hat{p}) \Delta_\ell^{JST}(p).$$

If the energy denominator is angle-averaged, as was assumed in Eq. (24.163), the gap function will not depend on the projection of the total angular momentum and the gap equation becomes

$$\Delta_\ell^{JST}(p) = -\sum_{\ell'} \frac{1}{2} \int_0^\infty dp' \, p'^2 \, \langle p\ell|V^{JST}|p'\ell'\rangle \frac{\Delta_{\ell'}^{JST}(p')}{2E_{p'}}.$$

(24.166)

As in earlier chapters the factor $\frac{1}{2}$ is related to the use of antisymmetrized matrix elements of the interaction V. By inserting

$$S(p'; E) = \delta(E - \chi_{p'}),$$

(24.167)

and employing Eq. (22.44) for u_p^2 and v_p^2 to write

$$S_s(p'; E') = \left(\frac{E_{p'} + \chi_{p'}}{2E_{p'}} \delta(E' - E_{p'}) + \frac{E_{p'} - \chi_{p'}}{2E_{p'}} \delta(E' + E_{p'}) \right),$$

(24.168)

one recovers the usual form of the BCS gap equation. If we ignore the difference between the spectral functions S and S_s, we recognize that Eq. (24.166) for the gap function Δ corresponds to the homogeneous scattering equation for the ladder equation at energy $E_{tot} = 0$ and center-of-mass momentum $P = 0$ (see Sec. 24.7). Accordingly, a non-trivial solution of Eq. (24.166) is obtained if, and only if, the ladder equation generates a pole at energy $E_{tot} = 0$, which reflects a bound two-particle state. Fulfillment of this condition generates the pairing solution, thereby demonstrating yet again

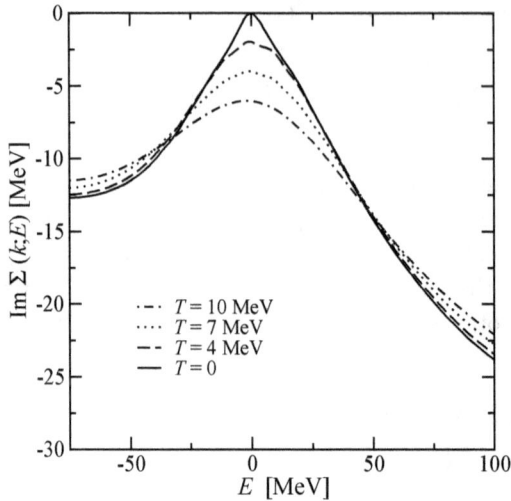

Fig. 24.12   Imaginary part of the self-energy for $k = 225$ MeV/c at saturation density
($\rho_0 = 0.16\text{fm}^{-3}$). Results for temperatures corresponding to 4, 7, and 10 MeV have been
calculated directly. The $T = 0$ plot is generated by extrapolating these results obtained
for the CDBonn interaction.

that the treatment is consistent with the ladder approximation in the non-
superfluid regime discussed in Sec. 24.7. The intimate relation of ladder
diagrams and pairing correlations was explored in [Dickhoff (1988)].

Before discussing pairing, it is helpful to present some illustrative features of the normal self-energy for different realistic nucleon-nucleon interactions. As a typical example we present in Fig. 24.12 the imaginary
part of the self-energy $\Sigma(k; E)$ for nucleons with a fixed momentum k as
a function of the energy variable E. The results have been determined
for symmetric nuclear matter at the empirical saturation density using the
CDBonn [Machleidt (2001)] interaction and are adapted from [Müther and
Dickhoff (2005)]. The energy scale has been constrained to (absolute) energies around the chemical potential μ, since there the self-energy is most
sensitive to the temperature. The numbers for temperatures larger than or
equal to 4 MeV, which are all above T_c (see below), have been obtained
directly from SCGF calculations. They exhibit a rather smooth dependence on the temperature, so that an extrapolation to temperatures below
T_c appears feasible. As an example, the $T = 0$ curve is also included, with
the constraint that the imaginary part of the self-energy vanishes at the
(absolute) energy $\mu = \varepsilon_F$.

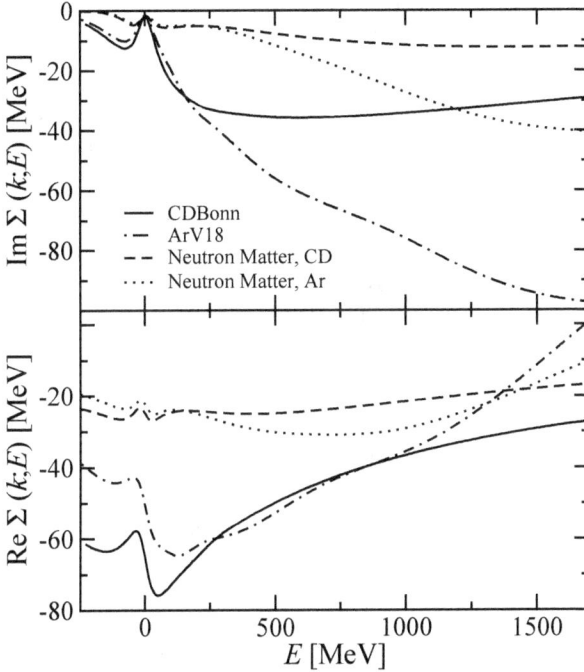

Fig. 24.13 Imaginary (upper) and real (lower panel) part of the self-energy, including the correlated HF term for $k = 225$ MeV/c at saturation density ($\rho_0 = 0.16 \text{fm}^{-3}$). Results for the CDBonn and ArV18 are compared for symmetric nuclear and neutron matter as identified in the upper panel. Also included are results for neutrons with the same momentum in neutron matter at $\rho = 0.08$ fm^{-3}. The temperature for these plots corresponds to 5 MeV.

The imaginary part of the nucleon self-energy is also displayed in the upper panel of Fig. 24.13. The purpose of this figure is to visualize some differences between various models of the nucleon-nucleon interaction and between symmetric nuclear matter and pure neutron matter. A larger interval for the energy variable E is therefore considered. The imaginary parts of the self-energy derived from the CDBonn interaction and the Argonne V18 (ArV18) interaction [Wiringa *et al.* (1995)] are very similar at absolute energies around $E = \mu$. At those energies the ArV18 yields a slightly weaker imaginary part than CDBonn. The differences get larger at positive values of E, where the imaginary part of the self-energy derived from ArV18 reaches a minimum of around -100 MeV at an energy E around 1.7 GeV. The minimum for the CDBonn interaction is only about -35 MeV and occurs at energies E around 0.5 GeV. The ArV18 is thus a stiffer interaction than CDBonn. A further illustration is provided by a HF calculation for nuclear matter at the empirical saturation density that yields a total energy 30 MeV per nucleon for the ArV18, while CDBonn generates 5 MeV per nucleon [Müther and Polls (2000)]. It implies that the generalized HF contribution to the self-energy is more repulsive for ArV18, and a larger part of the attraction is provided by the energy-dependent contribution to the real part of the self-energy. This is immediately obvious, since the energy-dependent contribution to the real part of Σ is connected to the imaginary part by a dispersion relation. The lower panel of Fig. 24.13 for the real part of the self-energy, illustrates this observation: The energy dependence is larger for ArV18 as compared to CDBonn. The weaker attraction of the self-energy derived from CDBonn (for most values of E) reflects the less repulsive contribution of the generalized HF contribution.

Figure 24.13 also displays results for the real and imaginary part of the self-energy for neutrons with the same momentum ($k = 225$ MeV/c) in pure neutron matter. The density of neutron matter is one half of the empirical saturation density of nuclear matter, which implies that these systems have the same Fermi momentum. The imaginary part of the self-energy in neutron matter is weaker than the corresponding one for symmetric nuclear matter, reflecting the dominance of proton-neutron correlations. For both interactions a minimum is obtained around 1.7 GeV. At these high energies the absolute value for the imaginary part of the self-energy is about a factor three larger for ArV18 than for CDBonn. This means that the distribution of sp strength to high energies due to central short-range correlations, is much stronger for the local ArV18 interaction than for the non-local meson-exchange model CDBonn. The results for the lower energies are closer to each other. The differences in the amount of correlations is also reflected in the occupation probability $n(k)$. For symmetric nuclear matter at saturation density, $n(k = 0)$ values of 0.89 and 0.87 are obtained for CDBonn and ArV18, respectively. The corresponding value for neutron matter are $n(k = 0) = 0.968$ and 0.963. Earlier non-self-consistent calculations with older NN interactions tended to yield values of 0.83 for this quantity [Vonderfecht et al. (1991a)] in nuclear matter, with self-consistency raising the number to 0.85 [Roth Stoddard (2000)] in the case of the Reid potential.

For initial orientation we consider the usual BCS approach at $T = 0$ to set the stage. The gap equation (24.166) is then solved in that limit assuming a spectrum of sp energies $\varepsilon(p) = \chi_p + \mu$, which is determined

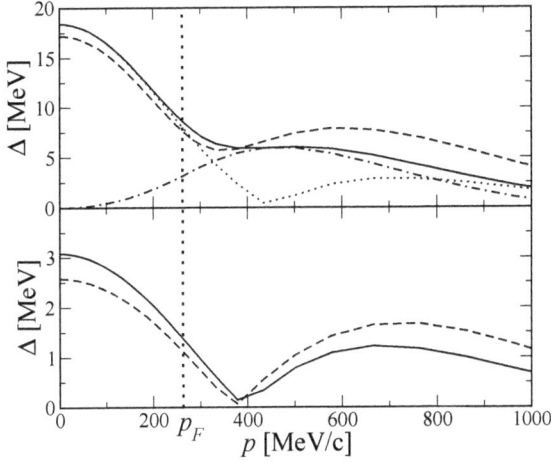

Fig. 24.14   Gap functions $|\Delta_\ell(p)|$ for symmetric nuclear matter ($\rho = 0.16$ fm^{-3}, upper panel) and pure neutron matter ($\rho = 0.08$ fm^{-3}, lower panel) from a solution of the BCS equation (24.166) using the CDBonn and ArV18 interactions at $T = 0$. The vertical dotted line identifies the Fermi-momentum p_F.

from the quasiparticle energies

$$\varepsilon(p) = \frac{p^2}{2m} + \operatorname{Re} \Sigma(p, \varepsilon(p) - \mu)\,, \qquad (24.169)$$

where $\operatorname{Re} \Sigma$ refers to a self-consistent finite-temperature calculation (including ladder diagrams) extrapolated to $T = 0$ according to Fig. 24.12. The extrapolation of the normal self-energy is required since self-consistent ladder equations (including dressing) still yield pairing instabilities below the critical temperature. The spectra of quasiparticle energies are rather similar to the sp spectra used in other work. The corresponding BCS calculations thus involve conventional procedures, as reviewed *e.g.* in [Dean and Hjorth–Jensen (2003)].

The gap functions $|\Delta_\ell(p)|$ are shown in Fig. 24.14. The upper panel shows results for symmetric nuclear matter at saturation density. The partial wave that yields the largest value for Δ and is therefore the relevant one, is the $^3S_1 - {}^3D_1$ channel describing the proton-neutron interaction. The absolute values of the gap functions are displayed for $\ell = 0$ (dotted) and $\ell = 2$ (dot-dashed) (Δ_0 and Δ_2) as well as the total gap function $\Delta = \sqrt{\Delta_0^2 + \Delta_2^2}$ (solid) as a function of the momentum p for the CDBonn interaction [Machleidt (2001)]. In the following we will mainly consider the value of the gap function Δ at the Fermi momentum p_F. We also compare in this figure the results for CDBonn with those for ArV18 [Wiringa *et al.* (1995)] (dashed). For smaller values of p the CDBonn produces larger values for the gap function, while ArV18 leads to larger values for momenta above $p = 400$ MeV/c. It holds true for the $\ell = 0$ and the $\ell = 2$ component and consequently for the total as well. This feature at large values of p is consistent with the notion that ArV18 tends to produce a larger amount of correlations at high momenta and large energies. For lower momenta however, CDBonn generates larger gap functions. Hence, the gap (at the Fermi momentum) from a BCS calculation that uses CDBonn (8.6 MeV) is larger than the value calculated for ArV18 (7.6 MeV), although ArV18 tends to produce more short-range correlations than CDBonn.

The situation is quite similar for neutron-neutron pairing in pure neutron matter, which is displayed in the lower part of Fig. 24.14. Here pairing is dominated by the $^1S_0$ partial wave. At high momenta larger values for the gap function using ArV18 are found, whereas CDBonn yields larger values for $\Delta(p)$ at low momenta. Therefore the value $\Delta(p_F)$ is larger for CDBonn (1.4 MeV) than for ArV18 (1.1 MeV). These values for the neutron-neutron pairing gap are however, much lower than the corresponding values for proton-neutron pairing at the same Fermi momentum. This appears natural, since the proton-neutron interaction is stronger than the neutron-neutron interaction, leading to a bound deuteron and more correlations (see Chs. 16 and 17). On the other hand, one observes proton-proton and neutron-neutron pairing in finite nuclei, while there is hardly any empirical trace of proton-neutron pairing.

The effects of temperature and short-range correlations on the solution of the gap equation will be investigated next. For that purpose we will study the tp propagator of Eq. (24.163) but replace the spectral function of the superfluid phase $S_s(p; E')$ by the normal one $S(p; E')$. This leads to a definition of an average energy denominator $\widetilde{\chi}_p$ of the form

$$\frac{1}{-2\widetilde{\chi}_p} \equiv \int_{-\infty}^{+\infty} dE \int_{-\infty}^{+\infty} dE' \, S(p; E) S(p; E') \frac{1 - f(E) - f(E')}{-E - E'}. \quad (24.170)$$

If we consider this propagator in the limit of the mean-field approximation, *i.e.* $S(p; E) = \delta(E - \chi_p)$ at $T = 0$, it reduces to an energy denominator of the form

$$\frac{1}{-2\widetilde{\chi}_p} \xrightarrow{\text{mf},T=0} \frac{1}{-2|\chi_p|}. \quad (24.171)$$

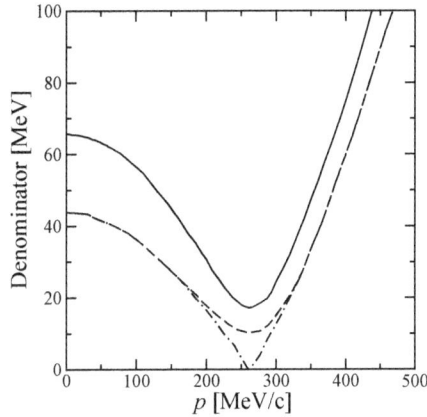

Fig. 24.15   The quantity $\widetilde{\chi}_k$ defined in Eq. (24.170), representing the energy denominator for the propagator of two nucleons in the medium. We display the quasiparticle approximation in the limit $T = 0$ (dot-dashed) and for a finite temperature of $T = 5$ MeV (dashed), and the dressed propagator from fully self-consistent calculations at the same temperature (solid). The figure pertains to symmetric nuclear matter at a density of $\rho = 0.16$ fm^{-3} using the CDBonn interaction.

Consequently, the energy $\widetilde{\chi}_p$ has been defined to exhibit the effects of finite temperature and correlations on the two-particle propagator. Figure 24.15 displays $\widetilde{\chi}_p$, the inverse of this propagator multiplied by -2, and compares it with $|\chi_p|$ using quasiparticle energies (dot-dashed line). The dashed line represents the effects of temperature, *i.e.* the propagator has been calculated using quasiparticle energies and the appropriate Fermi function, while the solid line accounts for finite temperature and correlation effects. The former only yields an enhancement of the effective sp energy $\widetilde{\chi}_p$ for momenta around the Fermi momentum. Adding the effects of correlations in the propagator (solid line), larger values for $\widetilde{\chi}_p$ for all momenta are generated. This reflects the well known feature that finite temperature yields a depletion of the occupation probability of sp states solely for momenta just below the Fermi momentum, while strong short-range correlations provide such a depletion for all momenta of the Fermi sea. In the literature several attempts have been reported to represent the redistribution of the sp strength by a renormalization factor Z_p [Bożek (2002, 2003); Baldo and Grasso (2000); Lombardo *et al.* (2001); Baldo *et al.* (2002)]. The investigation of [Müther and Dickhoff (2005)] demonstrates that there is no simple prescription based on quasiparticle type approximations to

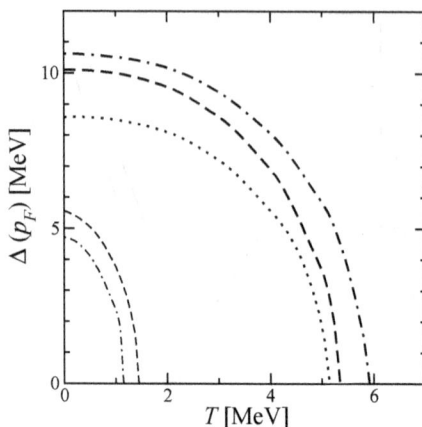

Fig. 24.16   The gap $\Delta(p_F)$ in symmetric nuclear matter as a function of temperature T using the CDBonn interaction. Results are presented for the densities 0.16 (dotted), 0.08 (dot-dashed), and 0.04 fm^{-3} (dashed), with (thin lines) and without taking into account the dressing of the sp propagator due to short-range correlations. The pairing gap disappears at ρ=0.16 fm^{-3}, if dressed propagators are considered.

accomplish this feat and Eq. (24.166) must thus be solved as is.

The gap $\Delta(p_F)$ in symmetric nuclear matter of various densities is presented in Fig. 24.16 as a function of the temperature T. We will first discuss the calculations with the usual BCS approximation in the $^3S_1 - ^3D_1$ partial wave. At the empirical saturation density the CDBonn interaction yields a gap parameter $\Delta(k_F)$ at temperature $T = 0$ of 8.6 MeV (see above), which decreases with increasing temperature until it vanishes at $T = 5.2$ MeV (dotted line). At $\rho = 0.08$ fm^{-3}, which is about half the empirical density, the value of the gap parameter at $T = 0$ is even larger ($\Delta(k_F) = 10.6$ MeV) and the gap calculated within the usual BCS approach disappears only at a temperature of 5.9 MeV (dot-dashed line). This increase of the pairing gap with decreasing density can be related to the momentum-dependence of the pairing gap $\Delta(k)$, as displayed in Fig. 24.14, since the gap function increases with decreasing momentum. Therefore, as the Fermi momentum decreases with density, the value $\Delta(p_F)$ tends to decrease with density. At even lower densities however, the effect is more than compensated by the feature that the phase-space of two-hole configurations decreases with density, so that ultimately the gap parameter will approach the binding energy of the deuteron in the limit of $\rho \to 0$. This explains the decrease of the gap parameter going from $\rho = 0.08$ fm^{-3} to $\rho = 0.04$ fm^{-3} (dashed line).

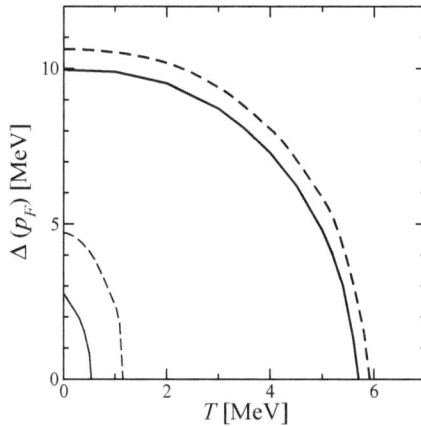

Fig. 24.17 Gap $\Delta(p_F)$ in symmetric nuclear matter at ρ=0.08 fm^{-3} as a function of temperature T using the ArV18 (solid) and the CDBonn interaction (dashed). Dressed results are displayed with thinner lines and exhibit a different T-dependence than typical BCS behavior.

When the effects of short-range correlations are taken into account, the generalized gap equation of Eq. (24.166) does not give a nontrivial solution for symmetrical nuclear matter at $\rho = 0.16$ fm^{-3}. It follows that a proper treatment of correlation effects at normal density yields no proton-neutron pairing as predicted by the usual BCS approach. The effects of short-range correlations tend to decrease with density. As a consequence, non-vanishing gaps for proton-neutron pairing at lower densities are found (see thin lines in Fig. 24.16). This also leads to an increase of the critical temperature and the value of $\Delta(p_F)$ at $T = 0$ going from $\rho = 0.08$ fm^{-3} (thin dot-dashed line) to $\rho = 0.04$ fm^{-3} (thin dashed line). Note that the T-dependence of these gap functions are qualitatively different from the BCS predictions. Differences associated with the various interactions are illustrated in Fig. 24.17. For nuclear matter with a density of $\rho = 0.08$ fm^{-3} the gap parameter is presented as a function of temperature T using the BCS approximation (thick lines) and the generalized gap equation with dressed propagators (thin lines). As already discussed, the ArV18 interaction (solid) yields smaller values for the gap parameter and the critical temperature than the CDBonn interaction (dashed).

Effects of pairing correlations on the spectral function are visualized in Fig. 24.18. As an example we show the spectral function $S(p; E)$ without and $S_s(p; E)$ (multiplied by 2π) with inclusion of pairing correlations for nu-

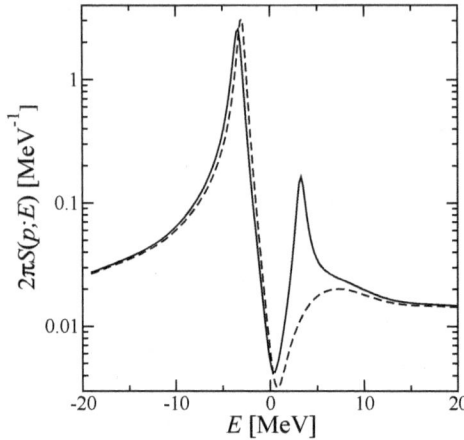

Fig. 24.18   Spectral function for nucleons with momentum $p = 193$ MeV/c with (solid line) and without (dashed line) inclusion of pairing correlations. Results are presented for nuclear matter of $\rho = 0.08$ fm^{-3} at a temperature $T = 0.5$ MeV.

clear matter at $\rho = 0.08$ fm^{-3}. The momentum, $p = 193$ MeV/c, is slightly below the Fermi momentum $p_F = 208$ MeV/c. One observes that pairing enhances the maximum of the spectral distribution at positive values of E substantially and shifts the quasiparticle peak to more negative values of E. Pairing modifies the spectral distribution into the direction which is found in the simple BCS approximation for $S_s(p; E)$ in Eq. (24.168). Note too, that these modifications of the spectral function $S_s(p; E)$ as compared to $S(p; E)$, are limited to a small interval of energies around $E = 0$ (*i.e.* absolute energies around μ) and to momenta close to the Fermi momentum.

Finally, we discuss neutron-neutron pairing in pure neutron matter. We will focus our attention to densities where pairing correlations in the $^1S_0$ partial wave dominate. The gap $\Delta(p_F)$ is displayed in Fig. 24.19 as a function of temperature. Using the BCS approximation with sp energies derived from the quasiparticle energies of self-consistent calculations, a gap at $T = 0$ is found, which, for the range of densities considered, increases with decreasing density. This is in agreement with other calculations, summarized *e.g.* in [Dean and Hjorth Jensen (2003)].

The effects of short-range correlations are weaker in neutron matter than in nuclear matter. Nevertheless, these weaker effects of short-range correlations in neutron matter are sufficient to suppress the formation of a pairing gap in neutron matter at $\rho = 0.08$ fm^{-3}. Such a suppression of

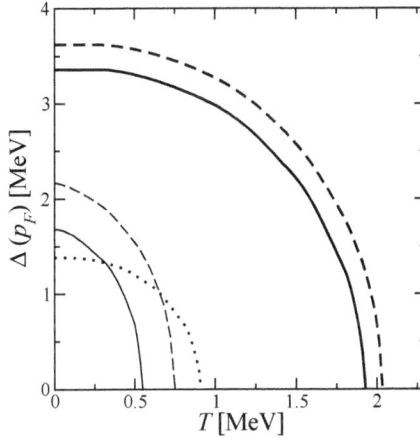

Fig. 24.19 $^1S_0$ gap $\Delta(p_F)$ in neutron matter as a function of temperature T for the CDBonn interaction. Results are presented for the usual BCS approximation at 0.08 (dotted), 0.04 (solid), and 0.02 fm^{-3} (dashed). The generalized gap equation (24.163) only yields a nontrivial solution for 0.04 (thin solid) and 0.02 fm^{-3} (thin dashed).

pairing correlations is observed at smaller densities as well. In this case however, the inclusion of the correlation effects just leads to a reduction of the gap parameter at a given temperature and a reduction of the critical temperature (see Fig. 24.19). Recent work for neutron matter obtained with Monte Carlo and CBF techniques [Fabrocini *et al.* (2005)] indicate that in the density range where the results overlap with those of Fig. 24.19, there is agreement at $T = 0$ with the Monte Carlo numbers.

Summarizing, we conclude that the formation of a pairing gap remains very sensitive to the quasiparticle energies and strength distribution at the Fermi surface and can be suppressed by moderate temperatures. The consequences of short-range correlations are expressed over a larger range of energies and momenta. So while the nonlocal CDBonn interaction is softer with respect to the formation of short-range correlations, it yields larger pairing gaps compared to the local ArV18 model for the NN interaction.

From this sensitivity to different areas in momentum and energy one may conclude that the features of short-range correlations should be rather similar in studies of nuclear matter and finite nuclei. The investigation of pairing phenomena, however, is rather sensitive to the energy spectrum around the Fermi energy, *i.e.* the density of sp states. Shell effects in finite nuclei may therefore lead to quite different results for pairing properties than corresponding studies in infinite matter.

The redistribution of sp strength due to the short-range correlations has a significant effect on the formation of a pairing gap. While the usual BCS approach predicts a gap for proton-neutron pairing in nuclear matter at saturation density as large as 8 MeV, the inclusion of short-range correlations completely suppresses this gap. Correlation effects are weaker at smaller densities, but still lead to a significant quenching of the proton-neutron pairing gap and to a reduction of the critical temperature for the phase transition. Compared to symmetric nuclear matter, correlation effects are weaker in neutron matter. Nevertheless, the inclusion of correlations suppresses the formation of a gap for neutron-neutron pairing at $\rho = 0.08$ fm^{-3} and produces a significant quenching at lower densities.

Present numbers now appear consistent with empirical information on pairing of both isoscalar (proton-neutron) and isovector (proton-proton or neutron-neutron) kind. Note that the calculations do not include polarization contributions to the pairing interaction, which make the gap function complex and energy dependent. A consistent treatment of dressed propagators and such vertex corrections is required for the conservation of symmetries (see [Baym and Kadanoff (1961)] and Ch. 21) but has not been developed to date. In addition, it is quite likely that volume polarization effects in nuclear matter and those in the surface of finite nuclei, will be quite different.

24.9 Exercises

(1) Perform all the steps for bosons, which were discussed in this chapter for the case of fermions, pertaining to the single-particle propagator. In particular, obtain the Lehmann representation and the result for the ground-state energy (and expectation values of one-body operators).
(2) Explore the diagrammatic equivalence of the coupled Dyson equations for the normal and anomalous propagators in the form of Eq. (22.31) and Eq. (24.146).
(3) Perform the energy sums that lead to Eqs. (24.151) and (24.163).
(4) Demonstrate the assertion that Eqs. (24.167) and (24.168), when substituted in Eq. (24.166), lead to the usual form of the BCS gap equation.

Appendix A

Pictures in quantum mechanics

A discussion of the different pictures used in quantum mechanics for the many-particle problem [Fetter and Walecka (1971)], follows the same steps as in one-particle problems [Sakurai (1994)]. As discussed in the Ch. 6, it is possible to use the time-dependent formulation to establish FT insight into the energy properties of the system under study through FT. In addition, it is important to deal with time-dependent interactions when describing the effect of experimental probes, which may transfer energy to the many-particle system.

A.1 Schrödinger picture

In the Schrödinger picture we employ the notation

$$|\Psi_S(t)\rangle = |\Psi(t)\rangle \tag{A.1}$$

to describe the normal time dependence of a state ket. The Schrödinger equation for this many-particle state reads

$$i\hbar \frac{\partial}{\partial t} |\Psi_S(t)\rangle = \hat{H} |\Psi_S(t)\rangle \tag{A.2}$$

with the initial state $|\Psi_S(t_0)\rangle$ at t_0 given. For a time-independent hamiltonian one can obtain the state at t from the one at t_0 in the following way

$$|\Psi_S(t)\rangle = \hat{U}_S(t - t_0) |\Psi_S(t_0)\rangle , \tag{A.3}$$

where

$$\hat{U}_S(t - t_0) = \exp\left\{-\frac{i}{\hbar}\hat{H}(t - t_0)\right\} \tag{A.4}$$

is the time-evolution operator in the Schrödinger picture.

A.2 Interaction picture

One can express the time-independent hamiltonian of the many-body system in terms of

$$\hat{H} = \hat{H}_0 + \hat{H}_1, \tag{A.5}$$

where the problem associated with $\hat{H}_0$ is assumed to be completely solved *e.g.* in terms of the independent-particle model discussed in Ch. 3. In the sp case, the propagator was studied by assuming the one belonging to H_0, $G^{(0)}$ to be known, so that the exact propagator can be expressed in terms of an expansion involving $G^{(0)}$ and the interaction term V. In the present many-particle context, we proceed similarly by assuming knowledge of the time dependence governed by $\hat{H}_0$, so that one can concentrate on the changes introduced by the action of $\hat{H}_1$. At any time t one then defineswe define

$$|\Psi_I(t)\rangle = \exp\left\{\frac{i}{\hbar}\hat{H}_0 t\right\} |\Psi_S(t)\rangle \tag{A.6}$$

the interaction picture state ket in terms of the Schrödinger ket and the noninteracting hamiltonian $\hat{H}_0$. The equation of motion for this ket is given by

$$i\hbar \frac{\partial}{\partial t} |\Psi_I(t)\rangle = -\hat{H}_0 |\Psi_I(t)\rangle + \exp\left\{\frac{i}{\hbar}\hat{H}_0 t\right\} i\hbar \frac{\partial}{\partial t} |\Psi_S(t)\rangle$$

$$= -\hat{H}_0 |\Psi_I(t)\rangle + \exp\left\{\frac{i}{\hbar}\hat{H}_0 t\right\} \left(\hat{H}_0 + \hat{H}_1\right) |\Psi_S(t)\rangle$$

$$= \hat{H}_1(t) |\Psi_I(t)\rangle, \tag{A.7}$$

where

$$\hat{H}_1(t) = \exp\left\{\frac{i}{\hbar}\hat{H}_0 t\right\} \hat{H}_1 \exp\left\{-\frac{i}{\hbar}\hat{H}_0 t\right\}. \tag{A.8}$$

Note that in general $\hat{H}_0$ and $\hat{H}_1$ do not commute. If states are related by Eq. (A.6), then one may obtain a corresponding relation for operators by considering

$$\hat{O}_S |\Psi_S(t)\rangle = |\Psi'_S(t)\rangle \tag{A.9}$$

so that one can write

$$|\Psi'_I(t)\rangle = \exp\left\{\frac{i}{\hbar}\hat{H}_0 t\right\} |\Psi'_S(t)\rangle = \exp\left\{\frac{i}{\hbar}\hat{H}_0 t\right\} \hat{O}_S |\Psi_S(t)\rangle$$

$$= \exp\left\{\frac{i}{\hbar}\hat{H}_0 t\right\} \hat{O}_S \exp\left\{-\frac{i}{\hbar}\hat{H}_0 t\right\} \exp\left\{\frac{i}{\hbar}\hat{H}_0 t\right\} |\Psi_S(t)\rangle$$

$$= \hat{O}_I(t) |\Psi_I(t)\rangle, \tag{A.10}$$

where

$$\hat{O}_I(t) = \exp\left\{\frac{i}{\hbar}\hat{H}_0 t\right\} \hat{O}_S \exp\left\{-\frac{i}{\hbar}\hat{H}_0 t\right\} \tag{A.11}$$

is the operator in the interaction picture that corresponds to $\hat{O}_S$ in the Schrödinger picture. The latter is usually assumed to have no explicit time dependence. This shows that in the interaction picture both state kets and operators have time dependence, while noting that the one for operators is simple. The equation of motion of an operator in the interaction picture may be obtained by considering

$$i\hbar\frac{\partial}{\partial t}\hat{O}_I(t) = \left\{i\hbar\frac{\partial}{\partial t}\exp\left\{\frac{i}{\hbar}\hat{H}_0 t\right\}\right\} \hat{O}_S \exp\left\{-\frac{i}{\hbar}\hat{H}_0 t\right\}$$

$$+ \exp\left\{\frac{i}{\hbar}\hat{H}_0 t\right\} \hat{O}_S \left\{i\hbar\frac{\partial}{\partial t}\exp\left\{-\frac{i}{\hbar}\hat{H}_0 t\right\}\right\}$$

$$= -\hat{H}_0\hat{O}_I(t) + \hat{O}_I(t)\hat{H}_0$$

$$= \left[\hat{O}_I(t), \hat{H}_0\right]. \tag{A.12}$$

An important example involves the addition (removal) operators of particles with sp quantum numbers corresponding to H_0. In this basis

$$\hat{H}_0 = \sum_\lambda \varepsilon_\lambda a_\lambda^\dagger a_\lambda. \tag{A.13}$$

Applying Eq. (A.12), we find

$$i\hbar\frac{\partial}{\partial t}a_{\lambda_I}(t) = \left[a_{\lambda_I}(t), \hat{H}_0\right]$$

$$= \exp\left\{\frac{i}{\hbar}\hat{H}_0 t\right\} \left[a_\lambda, \hat{H}_0\right] \exp\left\{-\frac{i}{\hbar}\hat{H}_0 t\right\}$$

$$= \varepsilon_\lambda a_{\lambda_I}(t), \tag{A.14}$$

with the solution

$$a_{\lambda_I}(t) = e^{-i\varepsilon_\lambda t/\hbar} a_\lambda \qquad (A.15)$$

and correspondingly

$$a^\dagger_{\lambda_I}(t) = e^{i\varepsilon_\lambda t/\hbar} a^\dagger_\lambda. \qquad (A.16)$$

In this basis the two-body interaction in the interaction picture has the form

$$\hat{V}_I(t) = \frac{1}{2} \sum_{\alpha\beta\gamma\delta} (\alpha\beta|V|\gamma\delta)\, a^\dagger_{\alpha_I}(t) a^\dagger_{\beta_I}(t) a_{\delta_I}(t) a_{\gamma_I}(t), \qquad (A.17)$$

and similarly for the auxiliary potential (or external field)

$$\hat{U}_I(t) = \sum_{\alpha\beta} (\alpha|U|\beta)\, a^\dagger_{\alpha_I}(t) a_{\beta_I}(t). \qquad (A.18)$$

These operators have a simple time-dependence, which allows for straight-forward time integrations when a FT is applied. A special role is played by the time-evolution operator in the interaction picture, defined by

$$|\Psi_I(t)\rangle = \hat{\mathcal{U}}(t,t_0)\,|\Psi_I(t_0)\rangle, \qquad (A.19)$$

where the I subscript is suppressed. Clearly

$$\hat{\mathcal{U}}(t_0,t_0) = 1. \qquad (A.20)$$

The explicit construction of this operator is accomplished by considering

$$\begin{aligned}
|\Psi_I(t)\rangle &= \exp\left\{\frac{i}{\hbar}\hat{H}_0 t\right\} |\Psi_S(t)\rangle \\
&= \exp\left\{\frac{i}{\hbar}\hat{H}_0 t\right\} \exp\left\{-\frac{i}{\hbar}\hat{H}(t-t_0)\right\} |\Psi_S(t_0)\rangle \qquad (A.21) \\
&= \exp\left\{\frac{i}{\hbar}\hat{H}_0 t\right\} \exp\left\{-\frac{i}{\hbar}\hat{H}(t-t_0)\right\} \exp\left\{-\frac{i}{\hbar}\hat{H}_0 t_0\right\} |\Psi_I(t_0)\rangle,
\end{aligned}$$

which shows that

$$\hat{\mathcal{U}}(t,t_0) = \exp\left\{\frac{i}{\hbar}\hat{H}_0 t\right\} \exp\left\{-\frac{i}{\hbar}\hat{H}(t-t_0)\right\} \exp\left\{-\frac{i}{\hbar}\hat{H}_0 t_0\right\}. \qquad (A.22)$$

We emphasize again that $\hat{H}$ and $\hat{H}_0$ normally do not commute. Employing this result one can show the following property of the time-evolution operator in the interaction picture

$$\hat{U}^\dagger(t, t_0)\hat{U}(t, t_0) = \hat{U}(t, t_0)\hat{U}^\dagger(t, t_0) = 1, \tag{A.23}$$

demonstrating unitarity

$$\hat{U}^\dagger(t, t_0) = \hat{U}^{-1}(t, t_0). \tag{A.24}$$

We also note that

$$\hat{U}(t_1, t_2)\hat{U}(t_2, t_3) = \hat{U}(t_1, t_3) \tag{A.25}$$

and

$$\hat{U}(t, t_0)\hat{U}(t_0, t) = 1, \tag{A.26}$$

which demonstrates that

$$\hat{U}(t_0, t) = \hat{U}^\dagger(t, t_0). \tag{A.27}$$

For practical applications it is also important to consider an alternative expression for $\hat{U}$. By combining Eqs. (A.7) and (A.19) we find

$$i\hbar\frac{\partial}{\partial t}\hat{U}(t, t_0) = \hat{H}_1(t)\hat{U}(t, t_0). \tag{A.28}$$

Employing the boundary condition in Eq. (A.20), formal integration of this operator equation yields

$$\hat{U}(t, t_0) = 1 - \frac{i}{\hbar}\int_{t_0}^t dt'\ \hat{H}_1(t')\hat{U}(t', t_0). \tag{A.29}$$

Iterating this equation generates an expansion in terms of $\hat{H}_1$

$$\hat{U}(t, t_0) = 1 - \frac{i}{\hbar}\int_{t_0}^t dt'\ \hat{H}_1(t')\left\{1 - \frac{i}{\hbar}\int_{t_0}^{t'} dt''\ \hat{H}_1(t'')\hat{U}(t'', t_0)\right\} \tag{A.30}$$

$$= 1 + \left(\frac{-i}{\hbar}\right)\int_{t_0}^t dt'\ \hat{H}_1(t')$$

$$+ \left(\frac{-i}{\hbar}\right)^2\int_{t_0}^t dt'\int_{t_0}^{t'} dt''\ \hat{H}_1(t')\hat{H}_1(t'') + \ldots$$

$$= \sum_{n=0}^\infty \left(\frac{-i}{\hbar}\right)^n \int_{t_0}^t dt_1\int_{t_0}^{t_1} dt_2 \ldots \int_{t_0}^{t_{n-1}} dt_n \hat{H}_1(t_1)\hat{H}_1(t_2)\ldots\hat{H}_1(t_n).$$

Rewriting the second-order term according to

$$
\hat{\mathcal{U}}_2(t,t_0) = \int_{t_0}^{t} dt' \int_{t_0}^{t'} dt'' \, \hat{H}_1(t')\hat{H}_1(t'')
$$

$$
= \frac{1}{2}\left\{ \int_{t_0}^{t} dt' \int_{t_0}^{t'} dt'' \, \hat{H}_1(t')\hat{H}_1(t'') + \int_{t_0}^{t} dt'' \int_{t''}^{t} dt' \, \hat{H}_1(t')\hat{H}_1(t'') \right\}
$$

$$
= \frac{1}{2}\left\{ \int_{t_0}^{t} dt' \int_{t_0}^{t'} dt'' \, \hat{H}_1(t')\hat{H}_1(t'') + \int_{t_0}^{t} dt' \int_{t'}^{t} dt'' \, \hat{H}_1(t'')\hat{H}_1(t') \right\}
$$

$$
= \frac{1}{2}\left\{ \int_{t_0}^{t} dt' \int_{t_0}^{t} dt'' \left[\theta(t'-t'')\hat{H}_1(t')\hat{H}_1(t'') + \theta(t''-t')\hat{H}_1(t'')\hat{H}_1(t') \right] \right\}
$$

$$
= \frac{1}{2}\int_{t_0}^{t} dt' \int_{t_0}^{t} dt'' \, \mathcal{T}\left[\hat{H}_1(t')\hat{H}_1(t'') \right], \tag{A.31}
$$

introduces the time-ordering operation in the last equality, denoted by $\mathcal{T}$. In this development we employ the option to integrate first over t' and then over t'', followed by interchanging in this term t' and t'', and finally use step functions to extend the integration interval in both terms from t_0 to t. With a little puzzling one can convince oneself that this result can be extended to any order. The complete expansion of $\hat{\mathcal{U}}$ can thus be written as

$$
\hat{\mathcal{U}}(t,t_0) = \sum_{n=0}^{\infty} \left(\frac{-i}{\hbar}\right)^{n} \frac{1}{n!} \int_{t_0}^{t} dt_1 \int_{t_0}^{t} dt_2 ... \int_{t_0}^{t} dt_n \, \mathcal{T}\left[\hat{H}_1(t_1)\hat{H}_1(t_2)...\hat{H}_1(t_n) \right], \tag{A.32}
$$

where the $\mathcal{T}$-operation is extended to order the operator with the latest time farthest to the left, and so on.

A.3 Heisenberg picture

State kets can be made independent of time while assigning time dependence, governed by the full hamiltonian, to operators by employing the Heisenberg picture. We define

$$
|\Psi_H(t)\rangle = \exp\left\{ \frac{i}{\hbar}\hat{H}t \right\} |\Psi_S(t)\rangle, \tag{A.33}
$$

using the full hamiltonian. It is immediately clear then that

$$
i\hbar \frac{\partial}{\partial t} |\Psi_H(t)\rangle = -\hat{H} |\Psi_H(t)\rangle + \hat{H} |\Psi_H(t)\rangle = 0, \tag{A.34}
$$

which confirms that state kets do not depend on time ($|\Psi_H(t)\rangle \equiv |\Psi_H\rangle$). For operators we first consider again

$$\hat{O}_S |\Psi_S(t)\rangle = |\Psi'_S(t)\rangle , \qquad (A.35)$$

so that one can write

$$|\Psi'_H\rangle = \exp\left\{\frac{i}{\hbar}\hat{H}t\right\} |\Psi'_S(t)\rangle \qquad (A.36)$$

$$= \exp\left\{\frac{i}{\hbar}\hat{H}t\right\}\hat{O}_S \exp\left\{-\frac{i}{\hbar}\hat{H}t\right\} \exp\left\{\frac{i}{\hbar}\hat{H}t\right\} |\Psi_S(t)\rangle = \hat{O}_H(t) |\Psi_H\rangle ,$$

where

$$\hat{O}_H(t) = \exp\left\{\frac{i}{\hbar}\hat{H}t\right\}\hat{O}_S \exp\left\{-\frac{i}{\hbar}\hat{H}t\right\} \qquad (A.37)$$

is the operator in the Heisenberg picture that corresponds to $\hat{O}_S$. The equation of motion of an operator in the Heisenberg picture is given by

$$i\hbar\frac{\partial}{\partial t}\hat{O}_H(t) = \left\{i\hbar\frac{\partial}{\partial t}\exp\left\{\frac{i}{\hbar}\hat{H}t\right\}\right\}\hat{O}_S \exp\left\{-\frac{i}{\hbar}\hat{H}t\right\}$$

$$+ \exp\left\{\frac{i}{\hbar}\hat{H}t\right\}\hat{O}_S \left\{i\hbar\frac{\partial}{\partial t}\exp\left\{-\frac{i}{\hbar}\hat{H}t\right\}\right\}$$

$$= -\hat{H}\hat{O}_H(t) + \hat{O}_H(t)\hat{H} = \left[\hat{O}_H(t), \hat{H}\right]$$

$$= \exp\left\{\frac{i}{\hbar}\hat{H}t\right\}\left[\hat{O}_S, \hat{H}\right]\exp\left\{-\frac{i}{\hbar}\hat{H}t\right\}, \qquad (A.38)$$

which shows that if a Schrödinger operator commutes with the hamiltonian, the corresponding Heisenberg operator is a constant of motion. The relation between operators in the interaction and Heisenberg picture, can be obtained by going back to the Schrödinger picture

$$\hat{O}_H(t) = \exp\left\{\frac{i}{\hbar}\hat{H}t\right\}\hat{O}_S \exp\left\{-\frac{i}{\hbar}\hat{H}t\right\}$$

$$= \exp\left\{\frac{i}{\hbar}\hat{H}t\right\}\exp\left\{-\frac{i}{\hbar}\hat{H}_0 t\right\}\hat{O}_I(t) \exp\left\{\frac{i}{\hbar}\hat{H}_0 t\right\}\exp\left\{-\frac{i}{\hbar}\hat{H}t\right\}$$

$$= \hat{\mathcal{U}}(0,t)\hat{O}_I(t)\hat{\mathcal{U}}(t,0). \qquad (A.39)$$

At $t = 0$, state kets in the different pictures coincide

$$|\Psi_H\rangle = |\Psi_S(t=0)\rangle = |\Psi_I(t=0)\rangle \qquad (A.40)$$

as do the operators

$$\hat{O}_S = \hat{O}_H(t=0) = \hat{O}_I(t=0). \tag{A.41}$$

For stationary solutions (energy eigenstates) one has in addition

$$|\Psi_{n_S}(t)\rangle = e^{-iE_n t/\hbar} |\Psi_n\rangle$$
$$= e^{-i\hat{H}t/\hbar} |\Psi_n\rangle, \tag{A.42}$$

which shows that the corresponding states in the Heisenberg picture satisfy the time-independent form of the Schrödinger equation while

$$|\Psi_n\rangle = |\Psi_{n_H}\rangle. \tag{A.43}$$

We also note that

$$|\Psi_H\rangle = |\Psi_I(0)\rangle = \hat{\mathcal{U}}(0, t_0) |\Psi_I(t_0)\rangle, \tag{A.44}$$

which allows one to construct exact eigenstates from interaction picture state kets at an earlier time t_0.

Appendix B

Practical results from angular momentum algebra

This appendix contains a summary of the angular momentum relations that are of relevance for the material in this book. More details can be found *e.g.* in [Brink and Satchler (1994); Edmonds (1996)]. Some prior knowledge of angular momentum algebra (at the level of sp quantum mechanics) is assumed. We will use $\hbar = 1$ and the short-hand notation $[j] = \sqrt{2j+1}$. For an angular momentum operator $\hat{\boldsymbol{j}}$ the simultaneous eigenstates of $\hat{\boldsymbol{j}}^2$ and $\hat{j}_z$ are denoted as $\hat{\boldsymbol{j}}^2 \, |jm\rangle = [j]^2 \, |jm\rangle$ and $\hat{j}_z \, |jm\rangle = m \, |jm\rangle$, with the action of the ladder operators $\hat{j}_\pm = \hat{j}_x \pm i\hat{j}_y$ on the basis vectors given by

$$\hat{j}_\pm \, |jm\rangle = \sqrt{(j \pm m + 1)(j \mp m)} \, |jm \pm 1\rangle. \qquad (B.1)$$

Clebsch–Gordan coefficients

Consider the vector space of the direct product of the eigenstates of two angular momentum operators. The unitary transformation relating the basis of uncoupled (product) states with those coupled to good total angular momentum, is written as

$$|j_1 j_2 JM) = \sum_{m_1 m_2} (j_1 m_1 j_2 m_2 | JM) \, |j_1 m_1; j_2 m_2), \qquad (B.2)$$

in terms of the Clebsch–Gordan (CG) coefficients, which are real numbers. As a consequence, the inverse transformation is

$$|j_1 m_1; j_2 m_2) = \sum_{JM} (j_1 m_1 j_2 m_2 | JM) \, |j_1 j_2 JM). \qquad (B.3)$$

Unitarity is expressed by the orthogonality relations

$$\sum_{m_1 m_2} (j_1 m_1 j_2 m_2 | JM)(j_1 m_1 j_2 m_2 | J'M') = \delta_{J,J'} \delta_{M,M'}, \qquad (B.4)$$

$$\sum_{JM} (j_1 m_1 j_2 m_2 | JM)(j_1 m_1' j_2 m_2' | JM) = \delta_{m_1,m_1'} \delta_{m_2,m_2'}. \quad \text{(B.5)}$$

Note that $(j_1 m_1 j_2 m_2 | JM)$ can be nonzero only if $m_1 + m_2 = M$ and the triangle inequalities $|j_1 - j_2| \le J \le j_1 + j_2$ are fulfilled. Switching the order of the angular momenta, or reversing the sign of all the m's, results in a phase,

$$(-1)^{j_1+j_2-J}(j_1 m_1 j_2 m_2 | JM) = (j_2 m_2 j_1 m_1 | JM) \quad \text{(B.6)}$$
$$= (j_1 - m_1 j_2 - m_2 | J - M). \quad \text{(B.7)}$$

3j-symbols

The CG coefficients are related to the 3j-symbols by

$$(j_1 m_1 j_2 m_2 | j_3 m_3) = (-1)^{j_1-j_2+m_3}[j_3]\begin{pmatrix} j_1 & j_2 & j_3 \\ m_1 & m_2 & -m_3 \end{pmatrix}. \quad \text{(B.8)}$$

The more symmetrical 3j-symbol $\begin{pmatrix} j_1 & j_2 & j_3 \\ m_1 & m_2 & m_3 \end{pmatrix}$ is invariant under permutations of the columns, apart from a phase $(-1)^{j_1+j_2+j_3}$ when the permutation is odd. The same phase arises when the sign of all m_i is reversed

$$\begin{pmatrix} j_1 & j_2 & j_3 \\ m_1 & m_2 & m_3 \end{pmatrix} = (-1)^{j_1+j_2+j_3}\begin{pmatrix} j_1 & j_2 & j_3 \\ -m_1 & -m_2 & -m_3 \end{pmatrix}. \quad \text{(B.9)}$$

Orthogonality is expressed as

$$\sum_{m_1 m_2}\begin{pmatrix} j_1 & j_2 & j_3 \\ m_1 & m_2 & m_3 \end{pmatrix}\begin{pmatrix} j_1 & j_2 & j_3' \\ m_1 & m_2 & m_3' \end{pmatrix} = \frac{\delta_{j_3,j_3'}\delta_{m_3,m_3'}}{[j_3]^2} \quad \text{(B.10)}$$

$$\sum_{j_3 m_3}[j_3]^2\begin{pmatrix} j_1 & j_2 & j_3 \\ m_1 & m_2 & m_3 \end{pmatrix}\begin{pmatrix} j_1 & j_2 & j_3 \\ m_1' & m_2' & m_3 \end{pmatrix} = \delta_{m_1,m_1'}\delta_{m_2,m_2'}. \quad \text{(B.11)}$$

Some special cases are given by

$$\begin{pmatrix} j & j & 0 \\ m & -m & 0 \end{pmatrix} = (jmj-m|00) = \frac{(-1)^{j-m}}{[j]} \quad \text{(B.12)}$$

$$\begin{pmatrix} j & j & 1 \\ m & -m & 0 \end{pmatrix} = \frac{(-1)^{j-m}m}{[j]\sqrt{j(j+1)}} \quad \text{(B.13)}$$

$$\begin{pmatrix} j & j & 2 \\ m & -m & 0 \end{pmatrix} = \frac{(-1)^{j-m}[3m^2 - j(j+1)]}{[j-1][j][j+1]\sqrt{j(j+1)}}. \quad \text{(B.14)}$$

Also note that Eq. (B.9) implies that $\begin{pmatrix} j_1 & j_2 & j_3 \\ 0 & 0 & 0 \end{pmatrix}$ vanishes unless $j_1 + j_2 + j_3$ is even.

6j-symbols

When three angular momenta are involved, the ordering of the couplings (indicated by brackets) becomes important. The unitary transformation relating two different coupling schemes can be expressed in terms of 6j-symbols

$$|(j_1, (j_2 j_3) J_{23}) jm) = \sum_{J_{12}} (-1)^{j_1 + j_2 + j_3 + j} [J_{12}][J_{23}] \qquad \text{(B.15)}$$

$$\times \begin{Bmatrix} j_1 & j_2 & J_{12} \\ j_3 & j & J_{23} \end{Bmatrix} |((j_1, j_2) J_{12}, j_3) jm) .$$

The 6j-symbol $\begin{Bmatrix} j_1 & j_2 & j_3 \\ j_4 & j_5 & j_6 \end{Bmatrix}$ is invariant under permutations of columns, as well as invariant under switching the upper and lower row of two columns, leaving the third column fixed. Orthogonality is expressed as

$$\sum_j [j]^2 \begin{Bmatrix} j_1 & j_2 & j \\ j_3 & j_4 & j' \end{Bmatrix} \begin{Bmatrix} j_1 & j_2 & j \\ j_3 & j_4 & j'' \end{Bmatrix} = \frac{\delta_{j',j''}}{[j']^2}, \qquad \text{(B.16)}$$

and a succession of two changes in the coupling scheme yields the sum rule

$$\sum_j [j]^2 (-1)^{j+j'+j''} \begin{Bmatrix} j_1 & j_2 & j' \\ j_4 & j_3 & j \end{Bmatrix} \begin{Bmatrix} j_1 & j_4 & j'' \\ j_2 & j_3 & j \end{Bmatrix} = \begin{Bmatrix} j_1 & j_2 & j' \\ j_3 & j_4 & j'' \end{Bmatrix}. \qquad \text{(B.17)}$$

Equation (B.15) leads to the extremely useful recoupling formula

$$\sum_m (-1)^{j-m} \begin{pmatrix} j_1 & j_2 & j \\ m_1 & m_2 & m \end{pmatrix} \begin{pmatrix} j_3 & j_4 & j \\ m_3 & m_4 & -m \end{pmatrix} = \qquad \text{(B.18)}$$

$$\sum_{j'm'} [j']^2 (-1)^{j+j_2+j_3+m'} \begin{Bmatrix} j_2 & j_4 & j' \\ j_3 & j_1 & j \end{Bmatrix} \begin{pmatrix} j_1 & j_3 & j' \\ m_1 & m_3 & m' \end{pmatrix} \begin{pmatrix} j_2 & j_4 & j' \\ m_2 & m_4 & -m' \end{pmatrix} .$$

Equivalently, a 6j-symbol can be expressed as a product of four 3j's

$$\begin{Bmatrix} j_1 & j_2 & j_3 \\ j_4 & j_5 & j_6 \end{Bmatrix} = \sum_{\text{all } m_i} (-1)^{j_4 + j_5 + j_6 + m_4 + m_5 + m_6} \qquad \text{(B.19)}$$

$$\times \begin{pmatrix} j_1 & j_2 & j_3 \\ m_1 & m_2 & m_3 \end{pmatrix} \begin{pmatrix} j_1 & j_5 & j_6 \\ m_1 & m_5 & -m_6 \end{pmatrix} \begin{pmatrix} j_4 & j_2 & j_6 \\ -m_4 & m_2 & m_6 \end{pmatrix} \begin{pmatrix} j_4 & j_5 & j_3 \\ m_4 & -m_5 & m_3 \end{pmatrix}$$

A 6j with a zero entry is simply

$$\begin{Bmatrix} j_1 & j_2 & j_3 \\ j_4 & j_5 & 0 \end{Bmatrix} = \frac{(-1)^{j_1+j_2+j_3}}{[j_1][j_2]} \delta_{j_1,j_5} \delta_{j_2,j_4}, \qquad (B.20)$$

provided j_3 obeys the triangle inequalities.

9j-symbols

When four angular momenta are coupled, the unitary transformation relating different coupling schemes contains a 9j-symbol

$$|((j_1, j_3)J_{13}, (j_2 j_4)J_{24})JM) = \sum_{J_{12}J_{34}} [J_{12}][J_{34}][J_{13}][J_{24}] \qquad (B.21)$$

$$\times \begin{Bmatrix} j_1 & j_2 & J_{12} \\ j_3 & j_4 & J_{34} \\ J_{13} & J_{24} & J \end{Bmatrix} |((j_1, j_2)J_{12}, (j_3 j_4)J_{34})JM).$$

The 9j-symbol $\begin{Bmatrix} j_1 & j_2 & j_3 \\ j_4 & j_5 & j_6 \\ j_7 & j_8 & j_9 \end{Bmatrix}$ is invariant under permutations of rows and of columns, apart from a phase $(-1)^{\sum j_i}$ containing all nine elements, when the permutation is odd. Orthogonality is expressed as

$$\sum_{J_{13}J_{24}} [J_{13}]^2 [J_{24}]^2 [J_{12}]^2 [J_{34}]^2 \begin{Bmatrix} j_1 & j_2 & J_{12} \\ j_3 & j_4 & J_{34} \\ J_{13} & J_{24} & J \end{Bmatrix} \begin{Bmatrix} j_1 & j_2 & J'_{12} \\ j_3 & j_4 & J'_{34} \\ J_{13} & J_{24} & J \end{Bmatrix}$$

$$= \delta_{J_{12},J'_{12}} \delta_{J_{34},J'_{34}} \qquad (B.22)$$

and a succession of two changes in the coupling scheme yields the sum rule

$$\sum_{J_{13}J_{24}} [J_{13}]^2 [J_{24}]^2 (-1)^{2j_2+J_{24}+J_{23}-J_{34}} \begin{Bmatrix} j_1 & j_2 & J_{12} \\ j_3 & j_4 & J_{34} \\ J_{13} & J_{24} & J \end{Bmatrix} \begin{Bmatrix} j_1 & j_3 & J_{13} \\ j_4 & j_2 & J_{24} \\ J_{14} & J_{23} & J \end{Bmatrix}$$

$$= \begin{Bmatrix} j_1 & j_2 & J_{12} \\ j_4 & j_3 & J_{34} \\ J_{14} & J_{23} & J \end{Bmatrix} \qquad (B.23)$$

A 9j-symbol with a zero entry can be written as a 6j

$$\begin{Bmatrix} j_1 & j_2 & j \\ j_3 & j_4 & j \\ j' & j' & 0 \end{Bmatrix} = \frac{(-1)^{j_2+j_3+j+j'}}{[j][j']} \begin{Bmatrix} j_1 & j_2 & j \\ j_4 & j_3 & j' \end{Bmatrix}. \qquad (B.24)$$

Spherical tensor operators

A spherical tensor operator of rank j is a set of $2j + 1$ operators A^j_m, $m = -j, .., j$, the members of which transform similarly as the $|jm\rangle$ basis vectors under a rotation. Equivalently, they obey the usual commutation relations with the angular momentum operator,

$$[\hat{j}_z, A^j_m] = mA^j_m; \quad [\hat{j}_\pm, A^j_m] = \sqrt{(j \pm m + 1)(j \mp m)}A^j_{m\pm1}. \quad (B.25)$$

The hermitian conjugate is not a spherical tensor operator, but the set $B^j_m = (-1)^{j+m}(A^j_{-m})^\dagger$ is; *i.e.* an additional phase is required.

The Wigner–Eckart theorem allows to extract the dependence on the m values in a matrix element of a tensor operator

$$(j_1m_1|A^j_m|j_2m_2) = (-1)^{j_1-m_1} \begin{pmatrix} j_1 & j & j_2 \\ -m_1 & m & m_2 \end{pmatrix} (j_1 \| A^j \| j_2). \quad (B.26)$$

The proportionality factor is called the reduced matrix element, and is indicated by double lines in the usual bra-ket notation.

The product of two tensor operators can be coupled to a new spherical tensor operator in the usual way

$$[A^{j_1} \otimes B^{j_2}]^{j_3}_{m_3} = \sum_{m_1m_2} (j_1m_1j_2m_2|j_3m_3)A^{j_1}_{m_1} B^{j_2}_{m_2}. \quad (B.27)$$

For a tensor product of two operators A^{J_a} and B^{J_b} acting in different spaces (*e.g.* operators related to different particles, or a spatial and a spin operator of the same particle), the reduced matrix element can be decomposed as

$$((aj_a, bj_b)j_{ab} \| [A^{J_a} \otimes B^{J_b}]^{J_{ab}} \| (a'j'_a, b'j'_b)j'_{ab}) \quad (B.28)$$

$$= [J_{ab}][j_{ab}][j'_{ab}] \begin{Bmatrix} j_a & j_b & j_{ab} \\ j'_a & j'_b & j'_{ab} \\ J_a & J_b & J_{ab} \end{Bmatrix} (aj_a \| A^{J_a} \| a'j'_a) (bj_b \| B^{J_b} \| b'j'_b).$$

Scalar operators (*e.g.* a rotationally invariant tp interaction) are spherical tensor operators of rank 0. A (pseudo)vector operator $\boldsymbol{A}$ is a spherical tensor operator of rank 1; its spherical components are

$$A^1_0 = A_z; \quad A^1_{\pm1} = \frac{\mp1}{\sqrt{2}}(A_x \pm iA_y). \quad (B.29)$$

Some simple reduced matrix elements are

$$(j \| 1 \| j') = [j]\delta_{j,j'}, \quad (B.30)$$

$$(j \parallel j^1 \parallel j') = [j]\sqrt{j(j+1)}\delta_{j,j'}, \tag{B.31}$$

$$(\tfrac{1}{2} \parallel \sigma^1 \parallel \tfrac{1}{2}) = \sqrt{6}, \tag{B.32}$$

$$(l \parallel Y^L \parallel l') = \frac{(-1)^l}{\sqrt{4\pi}}[l][L][l'] \begin{pmatrix} l & L & l' \\ 0 & 0 & 0 \end{pmatrix}. \tag{B.33}$$

The reduced matrix elements for the transition spin (and isospin) operator in Eq. (14.79) is given by

$$(\tfrac{3}{2} \parallel \mathcal{S}^1 \parallel \tfrac{1}{2}) = 2. \tag{B.34}$$

Bibliography

Abel, W. R., Anderson, A. C. and Wheatley, J. C. (1966) *Phys. Rev. Lett.* **17**, 74.

Abrikosov, A. A., Gorkov, L. P. and Dzyaloshinskii, I. E. (1975) *Methods of Quantum Field Theory in Statistical Physics* (Dover, New York).

Abrikosov, A. A. and Khalatnikov, I. M. (1959) *Rep. Prog. Phys.* **22**, 329.

Alberico, W. M., Ericson, M. and Molinari, A. (1982) *Nucl. Phys.* **A379**, 429.

Alberico, W. M., Molinari, A., De Pace, A., Ericson, M. and Johnson, M. B. (1986) *Phys. Rev. C* **34**, 977.

Allaart, K., Boeker, E., Bonsignori, G., Savoia, M. and Gambhir, Y. K. (1988) *Phys. Rep.* **169**, 209.

Amari, S. (1994) *Ph.D. Thesis, Washington University in St. Louis.*

Amorim. A. and Tjon, J. A. (1992) *Phys. Rev. Lett.* **68**, 772.

Amusia, M. Ya., Cherepkov, N.A., Janev, R.K. and Zivanović, Dj. (1990) *J. Phys. B* **7**, 1435.

Amusia, M. Ya. and Kheifets, A. S. (1985) *J. Phys. B* **18**, L679.

Amusia, M. Ya. (1990) *Atomic Photoeffect* (Plenum Press, New York and London).

Amusia, M. Ya. (1996) *Phys. Rep.* **264**, 7.

Anastasio, M. R., Celenza, L.S., Pong, W.S. and Shakin, C. M. (1983) *Phys. Rep.* **100**, 327.

Anderson, M. H. Ensher, J. R., Matthews, M. R., Wieman, C. E. and Cornell, E. A. (1995) *Science* **269**, 198.

Anderson, P. W. (1963) *Concepts in Solids* (Benjamin, Reading MA).

Anderson, P. W. (1984) *Basic Notions of Condensed Matter Physics* (Benjamin, Menlo Park CA).

Anderson, P. W. (1997) *Theory of High-Temperature Superconductivity in Cuprates* (Princeton University Press, Princeton NJ)

Audi, G., Wapstra, A. H. and Thibault, C. (2003) *Nucl. Phys.* **A729**, 337.

Austern, N. (1965) *Phys. Rev.* **137**, B752.

Aziz, R. A., Nain, V. P. S., Carly, J. C., Taylor, W. J. and McConville, G. T. (1979) *J. Chem. Phys.* **70**, 4330.

Bäckman, S.-O., Chakkalakal, D. A. and Clark, J. W. (1969) *Nucl. Phys.* **A130**, 635.

Balian, R. and Brezin, E. (1969), *Nuovo Cimento* **61B**, 403.

Baldo, M., Cugnon, J., Lejeune, A. and Lombardo, U., (1990) *Nucl. Phys.* **A515**, 409.

Baldo, M. *et al.* (1990) *Phys. Rev. C* **41**, 1748.

Baldo, M., Bombaci I. and Lombardo, U. (1992) *Phys. Lett.* **B283**, 8.

Baldo, M., Lombardo, U. and Schuck, P. (1995) *Phys. Lev. C* **52**, 975.

Baldo, M. (1999) in *Int. Rev. Nucl. Phys.* **8**, 1.

Baldo, M. and Grasso, A. (2000) *Phys. Lett. B* **485**, 115.

Baldo, M., Lombardo, U., Schulze, H.-J. and Zuo Wei (2002) *Phys. Rev C* **66**, 054304.

Baldwin, G. C. and Klaiber, G. S. (1947) *Phys. Rev.* **71**, 3.

Barbieri, C. and Dickhoff, W. H. (2001) *Phys. Rev. C* **63**, 034313; (2002) **65**, 064313.

Barbieri, C. and Dickhoff, W. H. (2003) *Phys. Rev. C* **68**, 014311.

Bardeen, J., Cooper, L. N. and Schrieffer, J. R. (1957) *Phys. Rev.* **108**, 1175.

Barth, von, U. and Holm, B. (1996) *Phys. Rev. B* **54**, 8411.

van Batenburg, M. F. (2001) *Ph.D. Thesis, University of Utrecht.*

Baym, G. (1962) *Phys. Rev.* **127**, 1391.

Baym, G. (2001) *J. Phys. B* **34**, 4541.

Baym, G. and Kadanoff, L. P. (1961) *Phys. Rev.* **124**, 287.

Baym, G. and Pethick, C. (1975) *Annu. Rev. Nucl. Sci.* **25**, 27.

Baym, G. and Pethick, C. (1978) in *The Physics of Liquid and Solid Helium, Part II* Ch. 1, ed. Benneman, K. H. and Ketterson, J. B. (Wiley-Interscience, New York).

Baym, G. and Pethick, C. (1979) *Annu. Rev. Astron. Astr.* **25**, 27.

Becchetti, Jr., F. D. and Greenlees, G. W. (1969) *Phys. Rev.* **182**, 1190.

Bednorz, G. and Müller, K. A. (1986) *Z. Phys.* **B64**, 189.

Beliaev, S. T. (1958) *Sov. Phys. – JETP* **7**, 289.

Bell, J. S. and Squires, E. J. (1959) *Phys. Rev. Lett.* **3**, 96.

Bender, M., Heenen, P-H. and Reinhard P-G. (2003) *Rev. Mod. Phys.* **75**, 121.

Benhar, O. Fabrocini, A. and Fantoni, S. (1989) *Nucl. Phys.* **A505**, 267.

Benhar, O., Fabrocini, A., Fantoni, S. and Sick, I. (1994) *Nucl. Phys.* **A579**, 493.

Berman, B. L. and Fultz, S. C. (1975) *Rev. Mod. Phys.* **47**, 713.

Bethe, H. A. and Bacher, R. F. (1936) *Rev. Mod. Phys.* **8**, 82.

Bethe, H. A. (1956) *Phys. Rev. A* **103**, 1343.

Bishop, R. F., Ghassib, H. B. and Strayer, M. R. (1974) *Phys. Rev. A* **13**, 1570.

Blaizot, J. P. and Ripka, G. (1986) *Quantum Theory of Finite Systems* (MIT Press, Cambridge MA).

Blatt, J. M. and Biedenharn, L. C. (1952) *Phys. Rev.* **86**, 399.

Bloch, C. and Messiah, A. (1962) *Nucl. Phys.* **39**, 95.

Boffi, S., Giusti, C., Pacati, F. D. and Radici, M. (1996) *Electromagnetic Response of Atomic nuclei* (Clarendon, Oxford).

Bogoliubov, N. N. (1947) *J. Phys. USSR* **11**, 23.

Bogoliubov, N. N. (1958) *Sov. Phys. JETP* **7**, 41.

Bohm, D. and Pines, D. (1953) *Phys. Rev.* **92**, 609.

Bohr, A. and Mottelson, B. R. (1998) *Nuclear Structure* (World Scientific, Singapore).

Bohr, A., Mottelson, B. R. and Pines. D. (1958) *Phys. Rev.* **110**, 936.

Born, M. and Oppenheimer, R. (1927) *Ann. Phys. (Leipzig)* **84**, 457.

Borromeo, M., Bonatsos, D., Müther, H. and Polls, A. (1992) *Nucl. Phys.* **A539**, 189.

Bose, S. M. *et al.* (1967) *Phys. Rev.* **155**, 379.

Bòzek, P. (1999) *Phys. Rev. C* **59**, 2619.

Bożek, P. and Czerski, P. (2001) *Eur. Phys. J. A* **11**, 271.

Bòzek, P. (2002) *Phys. Rev. C* **65**, 034327; (2003) *Phys. Lett.* **B551**, 93.

Bradley, C. C. *et al.* (1995) *Phys. Rev. Lett.* **75** , 1687.

Brand, M. G. E., Allaart, K. and Dickhoff, W. H. (1990) *Nucl. Phys.* **A509**, 1.

Brand, M. G. E., Rijsdijk, G. A., Muller, F. A., Allaart, K. and Dickhoff, W. H. (1991) *Nucl. Phys.* **A531**, 253.

Brandow, B. H. (1966) *Phys. Rev.* **152**, 863.

Brandow, B. H. (1971) *Ann. Phys.* **64** , 21.

Brink, D. M. and Satchler, G. R. (1994) *Angular Momentum* (Oxford University Press, 3rd edition).

Brockmann, R. and Machleidt, R. (1990) *Phys. Rev. C* **42**, 1965.

Bromley, D. A., Kuehner, J. A. and Alqvist, E. (1961) *Phys. Rev.* **123**, 878.

Brown, B. A., Hansen, P. G., Sherrill, B. M. and Tostevin, J. A. (2002) *Phys. Rev. C* **65**, 061601.

Brown, G. E. (1969) *Comm. Nucl. Part. Phys.* **3**, 136.

Brown, G. E. (1972) *Many-Body Problems* (North-Holland, Amsterdam).

Brown, G. E. and Jackson, A. D. (1976) *The Nucleon-Nucleon Interaction* (North-Holland, Amsterdam).

Brueckner, K. A. and Levinson, C. A. (1955) *Phys. Rev.* **97**, 1344.

Brueckner, K. A. and Sawada, K. (1957) *Phys. Rev.* **106**, 1117 and 1128.

Brunger, M. J., McCarthy, I. E. and Weigold, E. (1999) *Phys. Rev. A* **59**, 1245.

Burnett, K., Edwards, M. and Clark, C. W. (1999) *Phys. Today* **52**, No. 12, 37.

Butts, D. A. and Rokhsar, D. S. (1997) *Phys. Rev. A* **55**, 4346.

Carey, T. A. *et al.* (1984) *Phys. Rev. Lett.* **53**, 144.

Carlson, J., Pandharipande, V. R. and Wiringa, R. B. (1983) *Nucl. Phys.* **A401**, 59.

Carr Jr., W. J. and Maradudin, A. A. (1964) *Phys. Rev.* **133**, A371.

Castin, Y. and Dum, R. (1998) *Phys. Rev. A* **57**, 3008.

Ceperley, D. M. and Alder, B. J. (1980) *Phys. Rev. Lett.* **45**, 566.

Ceperley, D. M. and Pollock, E. L. (1986) *Phys. Rev. Lett.* **56**, 351.

Ceperley, D. M. (1995) *Rev. Mod. Phys.* **67**, 279.

Charity, R. J., Sobotka, L. G. and Dickhoff, W. H. (2006) *Phys. Rev. Lett.* **97**, 162503.

Charity, R. J., Mueller, J. M., Sobotka, L. G. and Dickhoff, W. H. (2007) *Phys. Rev. C* **76**, 044314.

Chen, J. M., Clark, J. W., Krotscheck, E. and Smith, R. A. (1986) *Nucl. Phys.* **A451**, 509.

Chu, S. (1998) *Rev. Mod. Phys.* **70**, 685.

Có, G., Fabrocini, A. and Fantoni, S. (1994) *Nucl. Phys.* **A568**, 73.

Coester, F., Cohen, S., Day, B. and Vincent, S. M. (1970) *Phys. Rev. C* **1**, 769.

Cohen-Tannoudji, C. (1998) *Rev. Mod. Phys.* **70**, 707.

Cohen-Tannoudji, C., Diu, B. and Laloe, F. (1992) *Quantum Mechanics* (Wiley).

Coon, S. A. *et al.* (1979) *Nucl. Phys.* **A317**, 242.

Cooper, L. N. (1956) *Phys. Rev.* **104**, 1189.

Coplan, M. A., Moore, J. H. and Doering, J. P. (1994) *Rev. Mod. Phys.* **66**, 985.

Cornish, S. L. *et al.* (2000) *Phys. Rev. Lett.* **85**, 1795.

Czerski, P. Dickhoff, W. H., Faessler, A. and Müther, H. (1986) *Phys. Rev. C* **33**, 1753.

Dahlen, N. E. and von Barth, U. (2004) *J. Chem. Phys.* **120**, 6826.

Dalfovo, F., Pitaevskii, L. P. and Stringari, S. (1996) *Phys. Rev. A* **54**, 4213.

Dalfovo, F., Giorgini, S., Pitaevskii, L. P. and Stringari, S. (1999) *Rev. Mod. Phys.* **71**, 463.

Davis, K. B. *et al.* (1995) *Phys. Rev. Lett.* **75**, 3969.

Day, B. D. (1967) *Rev. Mod. Phys.* **39**, 719.

Day, B. D. (1981) *Phys. Rev. C* **24**, 1203.

Day, B. D. and Wiringa, R. B. (1985) *Phys. Rev. C* **32**, 1057.

Dean, D. J. and Hjorth–Jensen, M. (2003) *Rev. Mod. Phys.* **75**, 607.

de Gennes, P. G. (1966) *Superconductivity of Metals and Alloys* (W.A. Benjamin, New York), reissued in *Advanced Book Classics* series in 1999.

DeMarco, B. and Jin, D. S. (1999) *Science* **285**, 1703.

Dewulf, Y., Van Neck, D. and Waroquier, M. (2002) *Phys. Rev. C* **65**, 054316.

Dewulf, Y., Dickhoff, W. H., Stoddard, E. R., Van Neck, D. and Waroquier, M. (2003) *Phys. Rev. Lett.* **90**, 152501.

Dickhoff, W. H., Faessler, A., Meyer-ter-Vehn, J. and Müther, H. (1981) *Phys. Rev. C* **23**, 1154.

Dickhoff, W. H. (1983) *Nucl. Phys.* **A399**, 287.

Dickhoff, W. H. (1988) *Phys. Lett.* **B210**, 15.

Dickhoff, W. H. (1998) *Phys. Rev. C* **58**, 2807.

Dickhoff, W. H., Gearhart, C. C., Roth, E. P., Polls, A. and Ramos, A. (1999) *Phys. Rev. C* **60**, 064319.

Dickhoff, W. H. and Barbieri, C. (2004) *Prog. Part. Nucl. Phys.* **52**, 377.

Dieperink, A. E. L. and de Witt Huberts, P. K. A. (1990) *Ann. Rev. Nucl. Part. Sci.* **40**, 239.

Dirac, P. A. M. (1958) *The Principles of Quantum Mechanics* (Oxford University Press, 4th ed.).

Dolbilkin, S. *et al.* (1965) *Zh. Eksper. Teor. Fiz. Pis'ma (USSR)* **1**, 47.

Doniach, S. and Sondheimer, E. H. (1974) *Green's Functions for Solid State Physicists* (Benjamin, Reading, MA).

Donnelly, R. J., Donnelly, J. A. and Hills, R. N. (1981) *J. Low Temp. Phys.* **44**, 471.

Dreizler, R. M. and Gross, E. K. U. (1990) *Density Functional Theory* (Springer, Berlin).

DuBois, D. F. (1959) *Ann. Phys. (N.Y.)* **7**, 174; **8**, 24.

Dukelsky, J., Esebbag, C. and Schuck, P. (2001) *Phys. Rev. Lett.* **87**, 066403.

Dukelsky, J., Pittel, S. and Sierra G. (2004) *Rev. Mod. Phys.* **76**, 643.

Economou, E. N. (1983) *Green's Functions in Quantum Physics, 2nd edition* (Springer, New York).

Edmonds, A. R. (1996) *Angular Momentum in Quantum Mechanics* (Princeton University press, reissue edition).

Eliashberg, G. M. (1960) *Sov. Phys. JETP* **11**, 696.

Ellis, T. and McClintock, P. V. E. (1985) *Phil. Trans. R. Soc. (London) A* **315**, 259.

Enders, J. *et al.*, *Phys. Rev. C* **67**, 064301.

Ericson, T. and Weise, W. (1988) *Pions and Nuclei* (Clarendon, Oxford).

Fabrocini, A. and Co', G. (2001) *Phys. Rev. C* **63**, 044319.

Fabrocini, A., Fantoni, S., Illarionov, A. Yu. and Schmidt, K. E. (2005) *Phys. Rev. Lett.* **95**, 192501.

Faddeev, L. D. (1961) *Sov. Phys. JETP* **12**, 1014.

Fantoni, S. and Pandharipande, V. R. (1984) *Nucl. Phys.* **A427**, 473.

Fetter, A. L. (1998) *cond-mat*/9811366.

Fetter, A. L. and Walecka, J. D. (1971) *Quantum Theory of Many-Particle Systems* (McGraw-Hill, San Francisco); (2003) Dover Publications.

Feynman, R. P. (1954) *Phys. Rev.* **94**, 267.

Feynman, R. P. (1955) *Progress in Low Temperature Physics*, Vol.1, Ch.2, C. G. Gorter ed., (North Holland, Amsterdam).

Feynman, R. P. and Cohen, M. (1956) *Phys. Rev.* **102**, 1189.

Fock, V. (1930) *Z. Physik* **61**, 126.

Fiedeldey, H. (1966) *Nucl. Phys.* **77**, 149.

Frick, T. and Müther, H. (2003) *Phys. Rev. C* **68**, 034310.

Frick, T. (2004) *Ph.D. Thesis, University of Tübingen*.

Frick, T., Müther, H., Rios, A., Polls, A. and Ramos, A. (2005) *Phys. Rev. C* **71**, 014313.

Fried, D. G. *et al.* (1998) *Phys. Rev. Lett.* **81**, 3811.

Froese Fischer, C., Brage, T. and Jönsson, P. (1997) *Computational Atomic Structure: An MCHF Approach* (IOP Publishing, Bristol).

Frois, B. *et al.* (1977) *Phys. Rev. Lett.* **38**, 152.

Frois, B. and Papanicolas, C. N. (1987) *Ann. Rev. Nucl. Part. Sci.* **37**, 133.

Frölich, H. (1952) *Proc. Roy. Soc.* **A215**, 291.

Frullani, S. and Mougey, J. (1984) *Adv. Nucl. Phys.* **14**, 1.

Fujita, J. and Miyazawa, H. (1957) *Prog. Theor. Phys.* **17**, 360.

Fujita, Y. *et al.* (1988) *Phys. Rev. C* **37**, 45.

Furness, J. B. and McCarthy, I. E. (1973) *J. Phys. B* **6**, L42.

Gade, A. *et al.* (2004) *Phys. Rev. Lett.* **93**, 042501.

Galitskii, V. M. (1958) *Sov. Phys. JETP* **7**, 104.

Galitskii, V. M. and Migdal, A. B. (1958) *Sov. Phys. JETP* **7**, 96.

García-González, P. and Godby, R. W. (2001) *Phys. Rev. B* **63**, 075112.

Gardiner, C. W. (1997) *Phys. Rev. A* **56**, 1414.

Gaudin, M. (1976) *J. Phys. (Paris)* **37**, 1087.

Gavoret, J. and Nozières, P. (1964) *Ann. Phys. (NY)* **28**, 349.

Gell-Mann, M and Low, F. (1951) *Phys. Rev.* **84**, 350.

Gell-Mann, M and Brueckner, K. A. (1957) *Phys. Rev.* **106**, 364.

Georgi, H. (1982) *Lie Algebras in Particle Physics* (Benjamin, Reading MA).

Gersch, H. A., Rodriguez, L. J. and Smith, P. N. (1972) *Phys. Rev. A* **5**, 1547.

Gibbons, P. C., Schnatterly, S. E., Ritsko, J. J. and Fields, J. R. (1976) *Phys. Rev. B* **13**, 2451.

Ginzburg, V. L. and Landau, L. D. (1950) *Zh. Eksp. Theor. Fiz.* **20**, 1064.

Glauber, R. J. and Osland, P. (1979) *Phys. Lett.* **B80**, 401.

Glyde, H. R. and Hernadi, S. I. (1983) *Phys. Rev. B* **28**, 141.

Glyde, H. R., Gibbs, M. R., Stirling, W. G. and Adams, M. A. (1998) *Europhys. Lett.* **43**, 422.

Glyde, H. R. *et al.* (2000a) *Phys. Rev. B* **61**, 1421.

Glyde, H. R., Azuah, R. T. and Stirling, W. G. (2000b) *Phys. Rev. B* **62**, 14337.

Goeppert-Mayer, M. (1949) *Phys. Rev.* **75**, 1969.

Goldhaber, M. and Teller, E. (1948) *Phys. Rev.* **74**, 1046.

Goldstone, J. (1957) *Proc. Roy. Soc. (London)* **A239**, 267.

Golub, G. H. and van Loan, C. F. (1996) *Matrix Computations*, Third Edition, (John Hopkins University Press, Baltimore, Maryland).

Gomes, L. C., Walecka, J. D. and Weisskopf, V. F. (1958) *Ann. Phys.* **3**, 241.

Gorkov, L. P. (1958) *Sov. Phys. JETP* **7**, 505.

Gottfried, K. and Yan, T.-M. (2004) *Quantum Mechanics: Fundamentals* (Springer, 2nd edition).

Grabmayr, P. (1992) *Prog. Part. Nucl. Phys.* **29**, 251.

Gradshteyn, I. S. and Ryzhik, I. M. (1980) *Table of Integrals, Series, and Products* (Academic Press, New York).

Grangé, P. Cugnon, J. and Lejeune, A. (19787) *Nucl. Phys.* **A473**, 365.

Grest, G. S. and Rajagopal, A. K. (1975) *Phys. Rev. B* **12**, 4847.

Greywall, D. S.(1983) *Phys. Rev. B* **27**, 2747.

Greywall, D. S.(1986) *Phys. Rev. B* **33**, 7520.

Griffin, A., Snoke, D. W. and Stringari, S., eds. (1995) *Bose-Einstein Condensation* (Cambridge University Press, Cambridge).

Griffiths, O. J., Hendry, P. C., McClintock, P. V. E. and Nichol, H. A. (2003) *Patterns of Symmetry Breaking*, Vol. 127 of Nato Science Series II: Mathematics, Physics and Chemistry, Ed. H. Arodz *et al.*, (Kluwer, Dordrecht).

Gross, E. K. U., Runge, E. and Heinonen, O. (1991) *Many-Particle Theory* (Adam Hilger, Bristol).

ter Haar, B. and Malfliet, R. (1986) *Phys. Rev. Lett.* **56**, 1237.

Haidenbaur, J. and Plessas, W. (1984) *Phys. Rev. C* **30**, 1822; (1985) **32**, 1424.

Haftel, M. I. and Tabakin, F. (1970) *Nucl. Phys.* **A158**, 1.

Halzen, F. and Martin, A. D. (1984) *Quarks and Leptons* (Wiley, New York).

Handy, N. C., Marron, M. T. and Silverstone, H. J. (1969) *Phys. Rev.* **180**, 45.

Hansen, P. G. and Tostevin, J. A. (2003) *Annu. Rev. Nucl. Part. Sci.* **53**, 219.

Hartree, D. R. (1928) *Proc. Cambridge Phil. Soc.* **24**, 89; 111.

Hau, L. V. *et al.* (1998) *Phys. Rev. A* **58**, R54.

Haxel, O., Jensen, J. H. D. and Suess, H. E. (1949) *Phys. Rev.* **75**, 1766.

Hedin, L. and Lundqvist, S. (1969) *Solid State Phys.* **23**, 1.

Hedin, L., Lundqvist, B. I. and Lundqvist, S. (1967) *Solid State Comm.* **5**, 237.

Heisenberg, W. (1932) *Z. Physik* **77**, 1.

Heisenberg, J. and Blok, H. P. (1983) *Ann. Rev. Nucl. Part. Sci.* **33**, 569.

Hengeveld, W., Dickhoff, W. H. and Allaart, K. (1986) *Nucl. Phys.* **A451**, 269.

Heyde, K. (1990) *The Nuclear Shell Model* (Springer, Heidelberg).

Holm, B. and von Barth, U. (1998) *Phys. Rev. B* **57**, 2108.

Hood, S. T., McCarthy, I. E., Teubner, P. J. O. and Weigold, E. (1973) *Phys. Rev. A* **8**, 2494.

Huang, K. (1987) *Statistical Mechanics, 2nd edition* (Wiley, New York).

Hugenholtz, N. M. and van Hove, L. (1958) *Physica* **24**, 363.

Hugenholtz, N. M. and Pines, D. (1959) *Phys.Rev.* **116**, 489.

Jackson, A. D., Landé, A. and Smith, R. A. (1982) *Phys. Rep.* **86**, 55.

Jackson, A. D., Landé, A., Guitink, R. W. and Smith, R. A. (1985) *Phys. Rev. B 31*, 403.

Jensen, E. and Plummer, E. W. (1985) *Phys. Rev. Lett* **55**, 1918.

Jeukenne, J. P., Lejeune, A. and Mahaux, C. (1976) *Phys. Rep.* **25**, 83.

Jeukenne, J. P., Lejeune, A. and Mahaux, C. (1977) *Phys. Rev. C* **16**, 80.

Jeukenne, J. P. and Mahaux, C. (1983) *Nucl. Phys.* **A394**, 445.

Joachain, C. J. (1975) *Quantum Collision Theory* (North-Holland, Amsterdam).

de Jong, F. and Lenske, H. (1996) *Phys. Rev. C* **54**, 1488.

Kadanoff, L. P. and Baym, G. (1962) *Quantum Statistical Mechanics* (Benjamin, Menlo Park CA), reissued in *Advanced Book Classics* series in 1989.

Kalos, M. H., Lee, M. A., Whitlock, P. A. and Chester, G. V. (1981) *Phys. Rev. B* **24**, 115.

Kamada, K. *et al.* (2001) *Phys. Rev. C* **64**, 044001.

Ketterle, W. (1999) *Phys. Today* **52**, No. 12, 30.

Khalatnikov, I. M. (1965) *An Introduction to the Theory of Superfluidity* (W.A. Benjamin, New York).

Khodel, V. A., Khodel, V. V. and Clark, J. W. (1996) *Nucl. Phys.* **A598**, 390.

Khodel, V. A., Khodel, V. V. and Clark, J. W. (1998) *Phys. Rev. Lett.* **81**, 3828.

Koltun, D. S. (1972) *Phys. Rev. Lett.* **28**, 182.

Koltun, D. S. (1974) *Phys. Rev. C* **9**, 484.

Koltun, D. S. and Eisenberg, J. M. (1988) *Quantum Mechanics of Many Degrees of Freedom* (Wiley, New York).

Koning, A. J. and Delaroche, J. P. (2003) *Nucl. Phys.* **A713**, 231.

Kraeft, W. D., Kremp, D., Ebeling, W. and Röpke, G. (1986) *Quantum Statistics of Charged Particle Systems* (Plenum, New York).

Kramer, G. J. (1990) *Ph. D. Thesis, University of Amsterdam.*

Kramer, G. J. *et al.* (1989) *Phys. Lett.* **B227**, 199.

Kramer, G. J., Blok, H. P. and Lapikás, L. (2001) *Nucl. Phys. A* **679**, 267.

Kromminga, A. J. and Bolsterli, M. (1962) *Phys. Rev.* **128**, 2887.

Kung, C. L., Kuo, T. T. S. and Ratcliff, K. F. (1979) *Phys Rev C* **19**, 1063.

Landau, L. D. (1941) *J. Phys. USSR* **5**, 71; (1947) **11**, 91.

Landau, L. D. (1957) *Sov. Phys. JETP* **3**, 920.

Landau, L. D. (1957) *Sov. Phys. JETP* **5**, 101.

Landau, L. D. (1958) *Sov. Phys. JETP* **7**, 182.

Landau, L. D. (1959) *Sov. Phys. JETP* **8**, 70.

Landau, L. D. and E.M. Lifshitz, E. M. (1977) *Quantum Mechanics, 3rd edition* (Pergamon, Oxford).

Landau, L. D. and E.M. Lifshitz, E. M. (1980) *Statistical Physics, Part 1, 3rd edition* (Pergamon, Oxford).

Lapikás, L. (1993) *Nucl. Phys. A* **553**, 297c.

Lapikás, L. (2005) *private communication.*

Lederer, C. M. and Shirley, V. S. (1978) *Table of Isotopes* (Wiley, New York).

Lee, T. D. and Yang, C. N. (1957) *Phys. Rev.* **105**, 1119.

Lee, T. D., Huang, K. and Yang, C. N. (1957) *Phys. Rev.* **106**, 1135.

Leggett, A. J. (1975) *Rev. Mod. Phys.* **47**, 331.

Leggett, A. J. (2001) *Rev. Mod. Phys.* **73**, 307.

Leggett, A. J. (2004) *Rev. Mod. Phys.* **76**, 999.

Lehmann, H. (1954) *Nuovo Cim.* **11**, 342.

Leuschner, M. *et al.* (1994) *Phys. Rev. C* **49**, 955.

Lewart, D. S., Pandharipande, V. R. and Pieper, S. C. (1988) *Phys. Rev. B* **37**, 4950.

Lewenstein, M. and You, L. (1996) *Phys. Rev. Lett.* **77**, 3489.

Lifshitz, E. M. and Pitaevskii, L. P. (1980) *Statistical Physics, Part 2* (Pergamon, Oxford).

Lindgren, I. and Morrison, J. (1982) *Atomic Many-Body Theory* (Springer, New York).

Lindhard, J. (1954) *Kgl. Danske Videnskab. Selskab, Mat.-fys. Medd.* **28**, 8.

Lohmann, B. and Weigold, E. (1981) *Phys. Lett.* **A86**, 139.

Lombardo, U., Schuck, P. and Zuo, W. (2001) *Phys. Rev C* **64**, 021301(R).

London, F. and London, H. (1935) *Proc. Roy. Soc.* **A149**, 71.

London, F. (1938) *Nature* **141**, 643.

Luttinger, J. M. (1961a) *Phys. Rev.* **119**, 1153.

Luttinger, J. M. (1961b) *Phys. Rev.* **121**, 942.

Ma, Z. Y. and Wambach, J. (1991) *Phys. Lett.* **B256**, 1.

Machleidt, R. (1989) *Adv. Nucl. Phys.* **19**, 191.

Machleidt, R. (2001) *Phys. Rev. C* **63**, 024001.

Macke, W. (1950) *Z. Naturforsch.* **5a**, 192.

Mahan, G. D. (1990) *Many-Particle Physics* (Plenum Press, New York) 2nd edition.

Mahan, G. D. (1994) *Comments Condens. Matter Phys.* **16**, 333.

Mahaux, C., Bortignon, P. F., Broglia, R. A. and Dasso, C. H. (1985) *Phys. Rep.* **120**, 1.

Mahaux, C. and Sartor, R. (1991) *Adv. Nucl. Phys.* **20**, 1.

Manousakis, E., Pandharipande, V. R. and Usmani, Q. N. (1985) *Phys. Rev. B* **31**, 7022; (1991) *Phys. Rev. B* **43**, 13587.

Martin, P. C. and Schwinger, J. (1959) *Phys. Rev.* **115**, 1342.

Martin, W. C., Fuhr, J. R., Kelleher, D. E., Musgrove, A., Podobedova, L., Reader, J., Saloman, E. B., Samsonetti, C. J., Wiese, W. L., Mohr, P. J. and Olsen, K. (2002) *NIST Atomic Spectra Database, Version 2.0* (National Institute of Standards and Technology, Gaithersburg, MD).

Matsubara, T. (1955) *Prog. Theor. Phys.* **14**, 351.

Mattuck, R. D. (1992) *A Guide to Feynman Diagrams in the Many-Body Problem* (Dover, New York).

Mazzanti, F., Polls, A., Boronat, J. and Casulleras, J. (2004) *Phys. Rev. Lett.* **92**, 085301.

McCarthy, I. E., Pascual, R., Storer, P. and Weigold, E. (1989) *Phys. Rev. A* **40**, 3041.

McCarthy, I. E. and Weigold, E. (1991) *Rep. Prog. Phys.* **54**, 789.

McQuarrie, D. A. (1976) *Statistical Mechanics* (Harper & Row, New York).

Meissner, W. and Ochsenfeld, R. (1933) *Naturwiss.* **21**, 787.

Messiah, A. (1999) *Quantum Mechanics* (Dover, New York).

Migdal, A. B. (1957) *Sov. Phys. JETP* **5**, 333.

Migdal, A. B. (1967) *Theory of Finite Fermi Systems and Applications* (Interscience, New York).

Migdal, A. B. (1972) *Sov. Phys. JETP* **34**, 1184.

Migdal, A. B. (1978) *Rev. Mod. Phys.* **50**, 10.

Moniz, E. J. *et al.* (1971) *Phys. Rev. Lett.* **26**, 445.

Moroni, S., Senatore, G. and Fantoni, S. (1997) *Phys. Rev. B* **55**, 1040.

Mougey, J. (1980) *Nucl. Phys.* **A335**, 35.

Müther, H. and Dickhoff, W. H. (1994) *Phys. Rev. C* **49**, R17.

Müther, H. and Dickhoff, W. H. (2005) *Phys. Rev. C* **72**, 054313.

Müther, H., Polls, A. and Dickhoff, W. H. (1995) *Phys. Rev. C* **51**, 3040.

Müther, H. and Polls, A. (2000) *Prog. Part. Nucl. Phys.* **45**, 243.

Nadasen, A. *et al.* (1981) *Phys. Rev. C* **23**, 1023.

Nambu, Y. (1960) *Phys. Rev.* **117**, 648.

Navrátil, P. Vary, J. P. and Barrett, B. R. (2000) *Phys. Rev. C* **62**, 054311.

Negele, J. W. and Orland, H. (1988) *Quantum Many-Particle Systems* (Benjamin, Redwood City, CA).

Negele, J. W. and Yazaki, K. (1981) *Phys. Rev. Lett.* **47**, 71.

Nozières, P. (1997) *Theory of Interacting Fermi Systems* (Addison-Wesley, Reading, Massachusetts).

Nozières, P. and Pines, D. (1990) *The Theory of Quantum Liquids, Vol. II: Superfluid Bose Systems* (Addison-Wesley, Redwood City, CA).

Nozières, P. and Saint James, D. (1982) *J. Physique* **43**, 1133.

Ortiz, G. and Ballone, P. (1994) *Phys. Rev. B* **50**, 1391.

Ortiz, G., Harris, M. and Ballone, P. (1999) *Phys. Rev. Lett.* **82**, 5317.

Osheroff, D. D., Richardson, R. C. and Lee, D. M. (1972) *Phys. Rev. Lett.* **28**, 885.

Osheroff, D. D., Gully, W. J., Richardson, R. C. and Lee, D. M. (1972) *Phys. Rev. Lett.* **29**, 920.

Pais, A. (1986) *Inward Bound* (Oxford University Press, Oxford).

Pan, F., Draayer, J. P. and Ormand, W. E. (1998) *Phys. Lett.* **B422**, 1.

Pandharipande, V. R. and Wiringa, R. B. (1979) *Rev. Mod. Phys.* **51**, 821.

Pandharipande, V. R., Sick, I. and de Witt Huberts, P. K. A. (1997) *Rev. Mod. Phys.* **69**, 981.

Pandya, S. P. (1956) *Phys. Rev.* **103**, 956.

Pauli, W. (1925) *Z. f. Phys.* **31**, 765.

Peirs, K., Van Neck, D. and Waroquier, M. (2002) *J. Chem. Phys.* **117**, 4095.

Penrose, O. and Onsager, L. (1956) *Phys. Rev.* **104**, 576.

Perey, F., and Buck, B. (1962) *Nucl. Phys.* **32**, 353.

Pethick, C. J. and Smith, H. (2003) *Bose-Einstein Condensation in Dilute Gases* (University Press, Cambridge).

Phillips, W. D. (1998) *Rev. Mod. Phys.* **70**, 721.

Piech, K. R. and Levinger, J. S. (1964) *Phys. Rev.* **135**, A332

Pieper, S. C., Wiringa, R. B., and Pandharipande, V. R. (1992) *Phys. Rev. C* **46**, 1741.

Pieper, S. C. and Wiringa, R. B. (2001) *Annu. Rev. Nucl. Part. Sci.* **51**, 53.

Pieper, S. C., Varga, K. and Wiringa, R. B. (2002) *Phys. Rev. C* **66**, 044310.

Pines, D. (1953) *Phys. Rev.* **92**, 626.

Pines, D. (1962) *The Many-Body Problem* (Benjamin, Reading, MA), reissued in *Advanced Book Classics* series in 1997.

Pines, D. (1963) *Elementary Excitations in Solids* (Benjamin, New York, NY).

Pines, D. and Nozières, P. (1966) *The Theory of Quantum Liquids, Vol. I: Normal Fermi Liquids* (Benjamin, New York, NY).

Pippard, A. B. (1953) *Proc. Roy. Soc.* **A216**, 547.

Pitaevskii, L. (1959) *Sov. Phys. JETP* **9**, 830.

Pitaevskii, L. (1960) *Sov. Phys. JETP* **10**, 1267.

Pitaevskii, L. and Stringari, S. (2003) *Bose-Einstein Condensation* (Clarendon, Oxford).

Pollock, E. L. and Ceperley, D. M. (1984) *Phys. Rev. B* **30**, 2555.

Polls, A., Müther, H. and Dickhoff, W. H. (1995) *Nucl. Phys.* **A594**, 117.

Polls, A., Radici, M., Boffi, S., Dickhoff, W. H. and Müther, H. (1997) *Phys. Rev. C* **55**, 810.

Quint, E. N. M. (1988) *Ph.D. Thesis, University of Amsterdam.*

Radici, M., Boffi, S., Pieper, S. C. and Pandharipande, V. R. (1994) *Phys. Rev. C* **50**, 3010.

Raether, H. (1980) *Excitations of Plasmons and Interband Transitions by Electrons* (Springer, Berlin).

Rajamaran, R. and Bethe, H. A. (1967) *Rev. Mod. Phys.* **39**, 745.

Ramos, A., Polls, A. and Dickhoff, W. H. (1989) *Nucl. Phys.* **A503**, 1.

Ramos, A., Dickhoff, W. H. and Polls, A. (1991) *Phys. Rev. C* **43**, 2239.

Rayfield, G. W. (1966) *Phys. Rev. Lett.* **16**, 934.

Reid, R. V. (1968) *Ann. Phys.* **50**, 411.

Redish, E. F. and Villars, F. (1970) *Ann. Phys.* **56**, 355.

Richardson, R. W. (1963) *Phys. Lett.* **B3**, 277; (1968) *J. Math. Phys.* **9**, 1327.

Richardson, R. W. and Sherman, N. (1964) *Nucl. Phys.* **52**, 221.

Rijsdijk, G. A., Allaart, K. and Dickhoff, W. H. (1992) *Nucl. Phys.* **A550**, 159.

Ring, P. and Schuck, P. (1980) *The Nuclear Many-body Problem* (Springer, New York).

Rios, A., Polls, A., Ramos, A. and Müther, H. (2006) *Phys. Rev. C* **74**, 054317.

Rios, A. (2007) *Ph.D. Thesis, University of Barcelona.*

Rohe, D. *et al.* (2004) *Phys. Rev. Lett.* **93**, 182501.

Rombouts, S., Van Neck, D. and Dukelsky, J. (2004) *Phys. Rev. C* **69**, 061303(R).

Roothaan, C. C. J. (1951) *Revs. Mod. Phys.* **23**, 69.

Roth Stoddard, E. (2000) *Ph.D. Thesis, Washington University in St. Louis.*

Ryckebusch, J., Waroquier, M., Heyde, K., Moreau, J. and Ryckbosch, D. (1988) *Nucl. Phys.* **A476**, 237.

Sakurai, J. J. (1967) *Advanced Quantum Mechanics* (Benjamin, Menlo Park, CA).

Sakurai, J. J. (1994) *Modern Quantum Mechanics* (Addison-Wesley, Reading, MA) revised edition.

Samardzic, O., Braidwood, S. W., Weigold, E. and Brunger, M. J. (1993) *Phys. Rev. A* **48**, 4390.

Sawada, K., Brueckner, K. A., Fukada, N. and Brout, R. (1957) *Phys. Rev.* **108**, 507.

Schrieffer, J. R. (1983) *Theory of Superconductivity* (Benjamin, Reading, MA), reissued in *Advanced Book Classics* series in 1999.

Schwandt, P. *et al.* (1982) *Phys. Rev. C* **26**, 55.

Sears, V. F. (1983) *Phys. Rev. B* **28**, 5109.

Sick, I. and de Witt Huberts, P. (1991) *Comm. Nucl. Part. Phys.* **20**, 177.

Singh, B. and Cameron, J. A. (2001) *Nucl. Data Sheets* **92**, 1.

Slater, J. C. (1929) *Phys. Rev.* **34**, 1293.

Slater, J. C. (1930) *Phys. Rev.* **35**, 210.

Sokol, P. in *Bose-Einstein Condensation*, edited by Griffin, A., Snoke, D. W., and Stringari, S. (1995) (Cambridge University Press, Cambridge), p. 51.

Song, H. Q., Baldo, M., Giansiracusa, G. and Lombardo, U. (1998) *Phys. Rev. Lett.* **81**, 1584.

Steiner, P., Höchst, H. and Hüfner, S. in *Photoemission in Solids II*, edited by Ley, L. and Cardona, M. (1979), Topics in Applied Physics Vol. 27 (Springer Verlag, Heidelberg).

Stoks, V. G. J., Klomp, R. A. M., Terheggen, C. P. F. and de Swart, J. J. (1994) *Phys. Rev. C* **49**, 2950.

Streater, R. F. and Wightman, A. S. (2000) *PCT, Spin and Statistics, and All That* (Princeton Univ. Press, Princeton).

Strinati, G. (1988) *Riv. Nuovo Cim.* **11**, 1.

Szabo, A. and Ostlund, N.S. (1989) *Modern Quantum Chemistry: Introduction to Advanced Electronic Structure Theory* (McGraw-Hill, New York).

Takayanagi, K., Shimizu, K. and Arima, A. (1988) *Nucl. Phys.* **A477**, 205.

Taddeucci, T. N. *et al.* (1994) *Phys. Rev. Lett.* **73**, 3516.

Talbot, E. F., Glyde, H. R., Stirling, W. G. and Svensson, E. C. (1988) *Phys. Rev. B* **38**, 11229.

Thouless, D.J. (1961) *Nucl. Phys.* **22**, 78.

Thouless, D.J. and Valatin, J.G. (1961) *Nucl. Phys.* **31**, 211.

Thouless, D. J. (1972) *The Quantum Mechanics of Many-Body Systems* (Academic Press, New York).

Tilley, D. R., Weller, H. R. and Cheves, C. M. (1993) *Nucl. Phys.* **A564**, 1.

Tisza, L. (1938) *Nature* **141**, 913.

Trefz, M., Faessler, A. and Dickhoff, W. H. (1985) *Nucl. Phys.* **A443**, 499.

Usmani, Q. N., Fantoni, S. and Pandharipande, V. R. (1982) *Phys. Rev. B* **26**, 6123.

Valatin, J. G. (1958) *Nuovo Cimento* **7**, 843.

Van Neck, D., Waroquier, M. and Ryckebusch, J. (1991) *Nucl. Phys.* **A530**, 347.

Van Neck, D., Waroquier, M., Dieperink, A. E. L., Pieper, S. C. and Pandharipande, V. R. (1998) *Phys. Rev. C* **57**, 2308.

Van Neck, D., Peirs, K. and Waroquier, M. (2001) *J. Chem. Phys.* **115**, 15.

Van Neck, D., Dewulf, Y. and Waroquier, M. (2001) *Phys. Rev. A* **63**, 062107.

Varner, R. L., Thompson, W. J., McAbee, T. L., Ludwig, E. J. and Clegg, T. B. (1991) *Phys. Rep.* **201**, 57.

Veillard, A. and Clementi, E. (1968) *J. Phys. Chem.* **49**, 2415.

Villars, F. (1967) *Fundamentals in Nuclear Theory* (IAEC, Vienna), Ch. 5.

Vollhardt, D. and Wölfle, P. (1990) *The Superfluid Phases of Helium 3* (Taylor & Francis, London).

Volterra, V. (1959) *Theory of Functionals and of Integral and Integro-differential Equations* (Dover, New York).

Vonderfecht, B. E., Dickhoff, W. H., Polls, A. and Ramos, A. (1991) *Phys. Rev. C* **44**, R1265.

Vonderfecht, B. E., Gearhart, C. C., Dickhoff, W. H., Polls, A. and Ramos, A. (1991) *Phys. Lett.* **B253**, 1.

Vonderfecht, B. E., Dickhoff, W. H., Polls, A. and Ramos, A. (1993) *Nucl. Phys.* **A555**, 1.

Vos, M. and McCarthy, I. E. (1995) *Rev. Mod. Phys.* **67**, 713.

Wagner, G. J. (1986) *AIP Conf. Proc.* **142**, 220.

Wakasa, T. *et al.* (1999) *Phys. Rev. C* **59**, 3177.

Ward, J. C. (1950) *Phys. Rev.* **78**, 182.

Waroquier, M., Ryckebusch, J., Moreau, J., Heyde, K., Blasi, N., van der Werf, S. Y. and Wenes, G. (1987) *Phys. Rep.* **48**, 249.

Weizsäcker, von, C. F. (1935) *Z. Physik* **96**, 431.

Wettig, T. and Jackson, A. D. (1996) *Phys. Rev. B* **53** ,818.

Wheatley, J. C. (1975) *Rev. Mod. Phys.* **47**, 415.

Wieman, C. E. and Cornell, E. A. (1995) *Science* **269**, 198.

Wilks, J. and Betts, D. S. (1987)*An Introduction to Liquid Helium* (Clarendon Press, Oxford).

Wiringa, R. B., Smith, R. A. and Ainsworth, T. L. (1984) *Phys. Rev. C* **29**, 1207.

Wiringa, R. B., Stoks, V. G. J. and Schiavilla, R. (1995) *Phys. Rev. C* **51**, 38.

Wise, J. E. *et al.* (1985) *Phys. Rev. C* **31**, 1699.

de Witt Huberts, P. K. A. (1990) *J. Phys. G: Nucl. Part. Phys.* **16**, 507.

Wong, C. W. and Clement, D. M. (1972) *Nucl. Phys.* **A183**, 210.

Woods, A. D. B. and Cowley, R. A. (1973) *Rep. Prog. Phys.* **36**, 1135.

van der Woude, A. (1987) *Prog. Part. Nucl. Phys.* **18**, 217.

Wu, T. T. (1959) *Phys. Rev.* **115**, 1390.

Wyatt, A. F. G. (1998) *Nature* **391**, 56.

Yang, C. N. (1962) *Rev. Mod. Phys.* **34**, 694.

Yuan, J. (1994) *Ph.D. Thesis, Washington University in St. Louis.*

Yukawa, H. (1935) *Proc. Phys. Math. Soc. Japan* **17**, 48.

Index

www.ingramcontent.com/pod-product-compliance
Lightning Source LLC
Chambersburg PA
CBHW052114230326
41598CB00079B/3668